Birkhäuser

Systems & Control: Foundations & Applications

Serdar Yüksel • Tamer Başar

Stochastic Teams, Games, and Control under Information Constraints

Birkhäuser

Serdar Yüksel
Mathematics & Statistics
Queen's University
Kingston, ON, Canada

Tamer Başar
Coordinated Science Laboratory
University of Illinois Urbana-Champaign
Urbana, IL, USA

ISSN 2324-9749 ISSN 2324-9757 (electronic)
Systems & Control: Foundations & Applications
ISBN 978-3-031-54070-7 ISBN 978-3-031-54071-4 (eBook)
https://doi.org/10.1007/978-3-031-54071-4

Mathematics Subject Classification: 93E20, 93E03, 91A10, 94A15, 93E15

This book is published under the imprint Birkhäuser, www.birkhauser-science.com by the registered company Springer Nature Switzerland AG
The registered company address is: Gewerbestrasse 11, 6330 Cham, Switzerland

To
Adem, Helin, Kate, Sait, and Saibe Yüksel
(S.Y.)
and
Altan, Koray, Eren, Zara, Alya, and Reya
(T.B.)

Preface

The progress in the field of stochastic decision making since the publication of our 2013 book *Stochastic Networked Control Systems: Stabilization and Optimization Under Information Constraints* in the same Series, and expansion and evolution of the topical areas the field encompasses, has given us the impetus to expand on our earlier coverage, to go also beyond stochastic teams to include stochastic games as well.

Given our initial motivation, this book may first be viewed as an addendum to our 2013 book. However, the progress in the field and the encouraging response we have received from the community have led us to make such substantial changes in the coverage (as noted below) that the current book can no longer be viewed as a *2nd edition*. Accordingly, some of the material in our 2013 book have not been included in this book and most of the new material here cannot be found anywhere else in such a complete and unified manner.

Our goal here has been to provide a comprehensive, mathematically rigorous, and accessible treatment of the interaction between information, decision, and control in multi-agent systems. We provide a comprehensive theory and coverage on information structures for stochastic teams as well as stochastic games.

The fields of decentralized stochastic control and networked control have reached a high level of maturity. On decentralization, there has been significant progress on existence, approximations, and optimality analyses, and many outstanding issues have been resolved, even though some still remain open. On networked control, there have been major findings on solution methods for both stabilization and optimization under various information constraints for both linear and non-linear models, and it is perhaps fair to say that the field is slowly converging to a phase where the rate of progress does not have the bursts of initial stages of discovery. Notably, non-linear systems have also found their place toward a broad treatment of information design in control theory and the theory of multi-agent decision making. Importantly, the fields noted above are now important, closely related sub-disciplines of their own, and will likely keep generations of students immersed in the problems involving the interaction of information and control, and decision making in general.

Additionally, we have included in the book a comprehensive treatment of information structures as they arise in stochastic game theory. Game theoretic framework brings in a multitude of additional intricacies with regard to the value and shaping of information, and even the most elementary arguments on the role of information in stochastic teams do not find universal applicability in game theory due to the competitive nature of equilibrium concepts, and fragility of information structure dependent properties, which we cover in detail in the book. We present foundational material regarding the interaction of information and decisions, toward a comprehensive theory of information structures in stochastic teams, games, and (networked) control.

As in the 2013 book, we take information structures both as passively provided (as, for example, in the theory of partially observed Markov decision processes, decentralized stochastic control as well as dynamic game theory) and as actively designed (via encoding and decoding policies when a physical information channel is present, as in communication theoretic problems or signaling game problems). For the latter case, in the context of networked control and signaling games, we may view encoders and decoders also as agents.

Finally, we provide in the book a comprehensive discussion on reinforcement learning theory in the context of the topics presented in the book. Reinforcement learning algorithms use feedback information to improve performance in a control task. The field of reinforcement learning has advanced significantly both in terms of fundamental theoretical contributions and applications. However, it is important to also understand the limitations of learning theory in the presence of model uncertainty and limited information. To this end, we found it appropriate to present several fundamental results toward generating a more complete book. Accordingly, a systematic discussion on reinforcement learning theory under a variety of information structures is presented.

To keep the length of the book within a manageable size, proofs of certain results with particularly tedious derivations have not been included, with relevant references provided as appropriate. Also, we have kept the analysis throughout the book primarily in the discrete-time domain, while we recognize that much research has been done in the continuous-time setup as well. Bibliographic remarks provide some references in this domain.

The book draws and utilizes a diverse set of tools (of both conceptual and analytical nature) from various disciplines, including stochastic control, team theory, game theory, information theory, probability theory, and dynamical systems, and amalgamates them into a unified, coherent, and applicable theory. It could be used as a textbook or as an accompanying text in a graduate-level course on the topic.

Kingston, ON, Canada Serdar Yüksel
Urbana, IL, USA Tamer Başar

Acknowledgements

As in the 2013 book, the work here would not have been possible had we not have the opportunity of interacting with several outstanding scholars and extremely resourceful students.

We would like to thank many individuals whose works, either independently or through collaboration, have shaped the contents of the book. Of particular mention are Fady Alajaji, Gürdal Arslan, Graeme Baker, Diego Cardenas, Giacomo Como, Alexander Condello, Liam Cregg, Philippe Furrer, Nicolas Garcia, Sinan Gezici, Meysam Ghomi, Adina Goldberg, Abhishek Gupta, Ian Hogeboom-Burr, Nuh Aygün Dalkıran Alireza Jalaeian, Andrew Johnston, Ali Devran Kara, Christoph Kawan, Ertan Kazıklı, Jonathan Keeler, Tamás Linder, Aditya Mahajan, Nuno C. Martins, Curtis McDonald, Sean Meyn, Tobias Oechtering, Ayça Özçelikkale, Daniel Quevedo, Naci Saldi, Sina Sanjari, Serkan Sarıtaş, Ryan Simpson, Mikael Skoglund, Nevroz Şen, Sekhar Tatikonda, Richard Wood, Emily Wright, Bora Yongaçoğlu, and Ali Zaidi.

We also had the honor of working with three outstanding scholars who are unfortunately not with us today. Prof. Ari Arapostathis, Prof. Milan Derpich, and Ramiro Zurkowski contributed to several results in the book, both directly and indirectly. It has been a privilege to be able to work with them.

Awni AlTabaa, Liam Cregg, Nicolas Garcia, Meysam Ghomi, Ian Hogeboom-Burr, Ali D. Kara, Ertan Kazıklı, Jonathan Keeler, Curtis McDonald, Lorenzo Miretti, Somnath Pradhan, Naci Saldi, Nevroz Şen Sina Sanjari, Ryan Simpson, Serkan Sarıtaş, Bora Yongaçoğlu, and Omar Mrani-Zentar have read different various sections or versions of the book, and provided substantial technical comments.

The book has also benefitted from technical discussions with many of our colleagues and students, including Tansu Alpcan, Yücel Altuğ, Orhan Arikan, Charalambos D. Charalambous, Debasish Chatterjee, Alexander Condello, Tolga Duman, Atilla Eryilmaz, Adina Goldberg, Orhan Çağri Imer, Akshay Kashyap, Ankur Kulkarni, Marcos M. Vasconcelos, Ather Gattami, Rayadurgam Srikant, Sekhar Tatikonda, Anant Sahai, Demos Teneketzis, and Venu Veeravalli, to whom we also convey our thanks.

Finally, we acknowledge the stimulating, conducive environment of the Coordinated Science Laboratory, University of Illinois at Urbana-Champaign, where the initial seeds of this book project were planted when the first author was a graduate student under the supervision of the second. The bulk of the work was carried out after the first author joined Queen's University, which provided a very stimulating environment in the Department of Mathematics and Statistics for the project to be completed in the company of outstanding scholars and students, notably through the Mathematics and Engineering Program (also known as Apple-Math at Queen's). Bilkent University, where the first author holds an adjunct appointment, also provided a productive venue for collaborations over many years. We thank all these institutions and many of our colleagues there.

Contents

Acronyms and Notations

$\mathrm{trace}(A)$	Trace of a square matrix A
$\det(A)$	Determinant of a square matrix A
1_E	Indicator function for event E
$\mathbb{X}$	A generic vector space
$\langle x, y \rangle$	Inner product between x and y on a Hilbert space
$\mathcal{B}(\mathbb{X})$	The Borel σ-field on $\mathbb{X}$
$\sigma(y)$	σ-field generated by a random variable y
$\mathcal{P}(\mathbb{X})$	Set of probability measures on $\mathcal{B}(\mathbb{X})$
$\mathbb{R}$	Set of real numbers
$\mathbb{R}^n$	Vector space of n-dimensional real vectors
$\mathbb{Z}$	Set of integers
$\mathbb{Z}_+$	Set of non-negative integers
$\mathbb{N}$	Set of positive integers
$\mathcal{Q}$	A space of quantizers
Q	Quantizer or channel depending on context
$\Pi^{comp,i}$	Composite quantization policy for encoder i
Q_t^i	Quantizer used by Agent i at time t
DMi	Decision Maker i
Ai	Agent i
$\underline{\gamma}^i$	Policy of DM i, that is $\{\gamma_t^i, t \geq 0\}$
$\underline{\gamma}$	Ensemble of policies for all decision makers, that is $\{\underline{\gamma}^i\}$
I_t^i or $\mathcal{I}_t^i$	Information variable at Agent i at time t
$\underline{\eta}$	Information structure inducing map $\{\eta^1, \ldots, \eta^N\}$
$E_P^{\underline{\gamma}}\{\cdot\}$	Expectation under policy $\underline{\gamma}$, with initial condition measure P
E_x	Expectation conditioned on an initial condition realization x, or with respect to a random variable x, depending on the context
$Df(x)$ or $J(f(x))$	Jacobian of f at x
$H(\cdot)$	Discrete entropy
$h(\cdot)$	Differential entropy

$I(\cdot;\cdot)$	Mutual information
$D(P_1 \| \| P_2)$	Kullback-Leibler Divergence between P_1 and P_2
$\|x\|$	Euclidean norm of a finite-dimensional real-valued vector x
A' or A^T	transpose of matrix A
$\|S\|$	Cardinality of a set S
$A \setminus B$	Ordered set difference: $\{x : x \in A, x \notin B\}$
$A \triangle B$	$(A \setminus B) \cup (B \setminus A)$
$\ln(x)$ or $\log(x)$	Natural logarithm of positive real x
$\underline{o}$	Zero vector
$\mathcal{T}$	Time/stage index set, $\{1, 2, \ldots, T\}$ or $\{0, 1, \ldots, T-1\}$
$\mathcal{N}$ ($\mathcal{L}$)	Decision maker (DM) index set, $\{1, 2, \ldots, N\}$ ($\{1, 2, \ldots, L\}$)
$u_{[k,s]}$	action (decision) variables from $t = k$ to $t = s$ for $s > k$, $\{u_k, u_{k+1}, \cdots, u_s\}$
$\mathbf{u}$	Collection of actions in $\mathcal{N}$ (or $\mathcal{L}$): $\{u^1, u^2, \ldots, u^N\}$

Chapter 1
Introduction

1.1 Information in Decision and Control Theory, Networked Control, and Stochastic Games

The interaction between information and decision or control is a phenomenon that arises in every decision and control problem. Any performance-driven controller requires information on the unknowns that affect the operation of the underlying system. The quality of the relevant information itself is typically affected by the choice of the control actions in a closed-loop dynamical system. Furthermore, the shaping of measurement signals via coding/signaling can itself be viewed as a control problem where an encoder *control*s the information available at a decoder.

In the theory of state space-based control involving linear systems, the relation between information and control is by now well understood: there exist linear controllers utilizing measurements to generate stabilizing control inputs if the system is both controllable and observable (where these can be relaxed to stabilizability and detectability, respectively, which would also make these conditions necessary for stabilization) [173]. The concept of observability is what qualitatively determines the informational quality and is essentially a property of invertibility (recovering the initial state given measurements), characterized by an (observability) rank condition in both continuous and discrete time. For linear systems, observability is global (as it applies for all initial states) and is universal in the control policies applied, as the control policy does not have any effect on the quality of construction (or recovery) of the state (known as the *no-dual effect* [66] property). However, while the intuition that the controller needs to have sufficient information to be able to arrive at stabilizing policies does carry over to more general non-linear and stochastic setups, simple algebraic characterizations or sharp invertibility conditions that exist for linear systems do not apply to stochastic, non-linear, decentralized and ultimately information-constrained systems. For stochastic non-linear systems, the analysis is more subtle and strict invertibility is often too

S. Yüksel, T. Başar, *Stochastic Teams, Games, and Control under Information Constraints*, Systems & Control: Foundations & Applications,
https://doi.org/10.1007/978-3-031-54071-4_1

restrictive, though stochastic informativeness conditions have been shown to be relevant [176, 422, 520, 522]. Due to the challenges in the analysis which prevent globality as well as control-dependence, more modest and localized definitions are imposed (see e.g. [341, 677, 678]).

Moving now from stabilization to optimization, in the classical theory of partially observed Linear Quadratic Gaussian (LQG) control problems [113, 153, 445], a celebrated result is that the optimal control policy features separation, where the optimal control policy (which turns out to be linear) has the same linear gain matrix that one would obtain for the corresponding fully observed problem but the state would be replaced with the conditional expectation of the state given the information at the controller. This structural result is known as the *separation of estimation and control* (where in more general terms, the separation (of estimation and control) principle is said to hold when an optimal controller exists in a subset of admissible policies where the control depends on the available information only through the conditional expectation of the state given the information), and for this particular case, a more special version of it, known as the *certainty equivalence* principle is at play, where the optimal controller and the estimator can be designed independently. For non-linear systems, however, such structural results do not hold in general, and properties such as neutrality, dual effect, and active probing arise [670, 716, 717]. As noted in [575], *control laws that include active experimentation have been called 'dual' control laws*; active experimentation or probing is performed to reduce the uncertainty regarding a dynamical system. The work [66] establishes conditions under which certainty equivalence is implied by the *no-dual-effect* property, and has also obtained results leading to separation of estimation and control, with refinements reported in [211]. Such properties significantly ease the study of optimal partially observed stochastic control problems; in general, however, these problems require the use of non-linear filtering, which is typically an infinite dimensional non-linear stochastic dynamical system taking values from a space of probability measures [602, 834], and hence are computationally challenging. Nonetheless, despite the underlying challenges, we now know a lot about optimization or stabilization of non-linear stochastic systems under partial information (with increasing information as time evolves) as regards to existence, approximations, robustness, and applicable numerical methods, given an information structure defined by a measurement system. Building on, and generalizing, the results of [122] we can compare and develop a partial ordering of information structures with regard to the performances attained in partially observed optimal stochastic decision and control problems.

A further aspect of informational constraints is with regard to the decentralization aspect of information structures when there are multiple decision makers (also called agents). In statistical decision theory, stochastic control theory, information theory, networked control, game theory, economics, quantum physics and computer science; information structures determine which decision maker knows what information and the informational and correlation properties of agent policies play a critical role in the analysis and design of such systems. Such problems have applications in networked control, information theory, cryptography, economics, and quantum information.

In one class of such systems, different decision makers may have access to only their local or private information even though the probabilistic description may be common knowledge, but the decision makers may have a common goal or objective function to optimize (in which case, the collection of the decision makers is called a *team*). In this case, the analysis becomes challenging as the classical solution methods available and intrinsic concepts defined for the problems discussed earlier (where a decision maker acts repeatedly over time with increasing information) such as the notion of a *state* and *controlled Markovian* structure become far more delicate to define and apply, and ultimately the analytical or numerical methods such as dynamic programming are no longer directly applicable. For example, linear policies may no longer be optimal for decentralized LQG team problems in general, as this routinely accepted belief, which could be characterized as a *folk theorem*, was disproved in a celebrated counterexample [771]. Even basic questions on existence, structure, and approximations for optimality require fundamentally new approaches. These will be covered in detail in this book.

In the setup described so far above, it was assumed that the information acquired by a decision maker is through (passive) measurements defined by stochastic (observation or measurement) kernels from state variables to measurement variables: There is no coding or signaling (into binary streams of data or through modulation into waveforms) or decoding, as the measurements are simply noisy functions of the state (and possibly, control). However, in many emerging applications, measurement kernels are not passive, and are shaped by coding and decoding; the measurements often require binary representations as they pass through digital channels or otherwise are subjected to modulation/demodulation maps. Such coding schemes may entail memory dependence on the distant past, which complicates an optimality analysis as the methods available for classical stochastic control (with either complete state measurements or passive measurements defined by observation kernels as described above) will not be directly applicable: *Networked control* or *information-constrained control* refers to such decentralized control systems in which the decision makers or components are connected through communication channels. Thus, there may be a data link between the sensors (which collect information), the controllers (which make decisions), and the actuators (which execute the controller commands); and the sensors, the controllers and the plant themselves could be geographically separated. With such limitations, many modern control systems become decentralized. Such systems feature multiple decision makers (e.g., sensors, controllers, and encoders) which have access to different and imperfect information, either cooperate with or compete against each other. For such information-constrained control (or networked control) systems, one needs to design encoders and decoders to allow for a controller to be able to have sufficiently informative measurements. The particular criterion of performance, as we will see, crucially affects the conditions imposed on informational content. The presence of active measurement shaping via coding and decoding has necessitated fundamentally new mathematical approaches as well as design paradigms. One should emphasize that the classical theory of information, due to Claude Shannon [664], was designed for non-causal coding settings, which however is not applicable (at least directly) to real-time scenarios with delay constraints and within a control loop. These subtleties will be studied and addressed in detail in the book.

In such networked control problems, one major concern is the characterization of the *minimum* amount of information transfer needed for a satisfactory performance, and particularly for *stability* of the overall system. This information transfer would be between various components of the networked control system. One further criterion for satisfactory control performance is the ability for the controllers to optimally track the plant state, or control the state under a cost criterion, given constraints on the communication, and the construction of dynamic encoding, decoding, and control policies meeting some selected design criteria. Another important problem is the establishment of effective coordination among multiple sensors or multiple controllers/decision makers using minimum possible information exchange. Even in cases when communication resources are not scarce, a strong understanding of the fundamentals (and elucidation of fundamental trade-offs) can be useful in constructing a desirable system architecture, and finally, such an insight can help reduce the computational requirements and complexity.

In many classical as well as emerging applications, there is an additional layer of competition: The agents taking part in the system may have non-aligned goals, or they may have aligned goals but subjective (or inconsistent) probability priors or models on the system. These turn a team problem into a *(non-cooperative) game* problem [87].

In non-cooperative game theory, first one has to go beyond the notion of optimality of a single objective function, and introduce equilibrium solution concepts that take into account strategic interactions among the players. Second, one has to arrive at the realization that the informational dependency of equilibrium values is particularly complicated: First of all, information is no longer, in general, a positive attribute in that having more information does not necessarily imply better performance for the player who receives the additional information. The various solution concepts for equilibria lead to a drastic difference in the nature, or even existence, of solutions. When the information structure is actively shaped by encoding and decoding, many of the results from the team theoretic setting do not carry over. On the other hand, we will see that game theory provides effective and conclusive design methods and solution approaches for various applications involving signaling games, privacy games, reputation games, jamming games, and information theory under subjective/mismatched probabilistic priors at an encoder and a decoder. We will also see that, contrary to the case of nonzero-sum games, zero-sum games allow for a comparison of information structures and for them one can establish stronger regularity properties.

In the book, we also provide a comprehensive discussion on reinforcement learning theory in the context of the general informational themes presented in the book: In many stochastic decision problems, the decision maker(s) may not know the true model (dynamics, information structures or probabilistic description of the present variables) or even the cost functions, and may wish to use past data to obtain an optimal or near optimal solution (that is, via *learning* from past data). In some problems, this may be used as an efficient numerical method to obtain approximately optimal solutions. Yet, in other problems, one may not even know whether the underlying decision problem is a Markov Decision Process (MDP) or a Partially Observed Markov Decision Process (POMDP), or a multi-agent system where other agents are present or not. There are many practical scenarios where one indeed works with data and does not know the structure under which the data is generated and tries to respond to the environment. Reinforcement learning algorithms use past experience and feedback information to improve performance in a control task. In recent years, the field of reinforcement learning has advanced tremendously both in terms of fundamental theoretical contributions and applications. However, one must also recognize the limitations of such algorithms in the presence of model uncertainty and informational constraints. Toward this goal, a systematic study on reinforcement learning theory under a variety of information structures is presented in the book.

This book is a comprehensive undertaking aimed at furthering our understanding of this interaction between information, control, and games in the context of stochastic, decentralized, and networked systems.

The study of the aforementioned classes of problems lies at the intersection of various disciplines of applied mathematics and engineering, namely *control theory* (in view of stability, control, and optimization of dynamical systems and their flows/solutions), *stochastic control* (in view of decision making under uncertainty and under measurability and information structure constraints), *communication, information, and quantization theories* (in view of information exchanges among decision makers), *optimization* (in studying existence, uniqueness, approximability, and geometry of minima), and game theory (as this is the theory behind competitive decision making). Finally, *stochastic (or probability theoretic) analysis* (in view of being the ultimate mathematical tool needed to conduct a study in all of these disciplines in the context of this book) plays an essential role in the understanding, analysis, and synthesis of such systems.

Our aim in this book is to bridge these disciplines in a precise and rigorous manner around the theme of *information structures and their various mathematical properties (such as topological regularity, comparison and approximation properties) in optimal decision making and the optimal design of information structures via coding, decoding, and decision/control.*

The field of *information and control* has had accelerated growth during the past few decades, but many of the problems and challenges that arise were actually obstacles identified already in the 1960s and the 1970s. What makes the situation different today is the accumulation (since that time) of an arsenal of more powerful results and tools from information theory, source coding theory, and the theory of Markov chains and stochastic analysis, directly applicable to the problems at hand. Furthermore, we have a far richer pool of computational (numerical or stochastic) algorithms and resources, making schemes computationally costly or even impossible to implement in the past quite feasible today.

The field had reached some state of maturity in early 2010s and this was our reason for our 2013 book [824] which included much of the existing material at the time. Since then, there has been another very active decade where classical problems have been tackled one after another. In particular, networked control has reached a level of maturity, and many open problems in decentralized stochastic control have been addressed successfully. Furthermore, interaction of game theory and information has been studied extensively leading to new design methods.

We also find it useful and very appropriate to revisit many of the results of the 1970s and the 1980s, and blend them into such a treatise with the current developments, as they have by no means lost their relevance. A similar thinking applies to stochastic dynamic game theory. Notably, we have also realized that many of the discoveries in the 1970s involving game theory and the role information plays seems to have been hidden in the literature and it has been a goal for us to bring some of these findings back to light, with perhaps a more modern interpretation.

The above is the reason why it seems timely to collect and present all this material in such a second book, with a much broader coverage.

Within the framework above, a few essential concepts are visited recurrently in the book:

- A general mathematical theory of information structures in stochastic team theory. The characterization of information structures defined in terms of measurability relations in a given probability space. Comparison and topological properties of information structures for stochastic team optimization problems, and identification of information structures which may lead to a systematic program for generation of optimal policies.
- Probabilistic characterization of dependency properties of decentralized decision policies as a function of information structures and their various relaxations, and implications of these on existence, approximations, and structure of optimal policies in team theory. Systematic methods in arriving at optimal control policies given a decentralized information structure. These involve static reduction methods, various dynamic programming methods, approximation methods, and various convex analysis methods. Arriving at efficient solution methods and stronger structural results (such as symmetry in policies without shared randomness) on optimality or near-optimality, when the number of agents is large or infinite.
- Comparison of information structures and regularity/sensitivity properties of optimal solutions with respect to perturbations in the space of information structures. A recurring theme is the versatility one has in selecting the most appropriate topology to be adopted on control policies for optimal or approximately optimal designs.
- A general mathematical theory of information structures in stochastic game theory, extending the team theoretic counterpart, which includes comparison and topological properties and information-structure dependent refinements of equilibrium concepts, but which also demonstrates fragilities on information structure equivalence properties and on the positive or negative values of information.

- When information is shaped by coding/decoding, optimal design for both stochastic stability of and optimization for networked control systems. Converse results through information theoretic or stochastic and geometric analysis, and constructive algorithms via stochastic drift equations. Stochastic stability may correspond to the existence of an equilibrium distribution, ergodicity of a process, or existence of finite moments. For optimization, a joint analysis of optimal coding and control policies via a variety of approaches, and determination of when separation of estimation, coding, and control can be optimal.
- The operational differences between information theoretic settings (which, in the classical Shannon theoretic sense, requires an infinite ensemble of messages to be transmitted or encoded) and real-time settings in control which do not tolerate delay. Use of information theory in establishing fundamental bounds on information requirements. Optimal information transmission under causality and delay constraints, and jointly optimal channel and controller design for real-time systems, under a variety of information structures. Structural results as well as existence results on optimal policies.
- Signaling games and the lack of robustness in the qualitative nature of solutions between team and game theoretic settings. Effects of cost mismatch and prior mismatch in optimal information design, under both Nash and Stackelberg formulations.
- A comprehensive and systematic treatment of reinforcement learning theory, for both single-agent and multi-agent systems, as well as networked systems, under a variety of information structures, state, action and measurement spaces, and cost criteria.

Intended Audience for the Book

The intended audience for this book includes researchers interested in control theory, information theory, statistics, game theory, and applied mathematics at large.

As indicated earlier, the book adopts a probability theory, information theory, game theory, and stochastic control theory view to decentralized decision making. To comfortably follow the material in the book, the reader should be familiar with linear systems theory (at the first year graduate level), and stochastic processes (again at the first year graduate level). The reader is also expected to have a basic understanding of Markov chains. For those who do not have the requisite background, appropriate references are provided throughout the development in the book, and appendices are included, covering some of this material as well as others.

1.2 Contents of the Book

The book is comprised of 24 chapters, organized into three main themes and five parts, as summarized below. It also has eight appendices, providing background material.

1.2.1 Parts I and II: Mathematical Theory of Information Structures for Stochastic Teams and Games

Part I (Stochastic Teams and a Mathematical Theory of Information Structures), is the longest sub-block, as it lays out the foundations of the rest of the book. It consists of 7 chapters and is primarily concerned with a general mathematical theory of decentralized stochastic control or team theory with respect to information structure dependent properties. The chapters pave the foundations of a general formulation and associated implications on the existence and computation as well as approximation of optimal or person-by-person-optimal solutions, and general results on comparisons and topological properties of information structures. This part can be studied in isolation as an independent resource on stochastic team theory (or decentralized stochastic control). Chapters 2 and 3 present a systematic description and characterization of information structures, as well as solution methods for both static and dynamic teams. Chapter 4 is on a detailed study of decentralized strategic measures, which are probability measures induced on states, measurements, and actions under admissible policies, which also leads to very general existence and approximation, as well as relaxation results. Chapter 5 presents general dynamic programming formulations for stochastic dynamic teams. Chapter 6 focuses on teams with many, and possibly infinitely many, agents. Chapters 7 and 8 study comparison, continuity and existence results on information structures given an optimality criterion, where optimal quantization is also introduced.

Part II (Stochastic Game Theory and Information Structures) consists of 3 chapters, developing a general mathematical theory of information structures in stochastic dynamic game theory in Chaps. 9 and 10, and presenting results on equilibrium solutions, existence, comparison, and regularity (such as continuity) properties in Chap. 11.

1.2.2 Parts III and IV: Information Structure Design Under Data Constraints via Signaling/Coding in Stochastic Teams and Games

Part III (Information Design via Information Theory in Networked Systems and Control), consists of 6 chapters. The first chapter, Chap. 12, presents preliminaries on real-time information theory, and how this deviates from classical Shannon theory. The next two chapters study the problem of stochastic stabilization of networked control systems: Chap. 13 is dedicated to linear models, and the other one, Chap. 14, to non-linear system models.

The remaining 3 chapters focus on optimality criteria; Chap. 15 on optimal zero-delay coding and structural results, Chap. 16 on specialized results tailored for linear models and quadratic criteria, and Chap. 17 on fundamental performance bounds under low distortion criteria.

Part IV (Information and Signaling Games) consists of 3 chapters on coding and decoding, or control, for networked systems under mismatch in either cost criteria or probabilistic descriptions adopted by encoders and decoders, as well as competitive design involving the informational quality of channels connecting encoders and decoders. One of these chapters, Chap. 18, is dedicated to a comprehensive review and further analysis on quadratic signaling games, Chap. 19 is on two special signaling game models involving reputations and privacy games, and Chap. 20 is on zero-sum signaling games.

1.2.3 Part V: Reinforcement Learning Under Full and Partial, Decentralized, and Networked Information

The final Part in the book presents a systematic study of reinforcement learning theory. Chapter 21 studies fully observed MDPs with both finite and continuous spaces. Chapter 22, studies reinforcement learning theory for partially observable models. Chapter 23, for multi-agent and decentralized stochastic control problems, under a variety of information structures with strictly local information or partial global information, introduces several reinforcement learning algorithms and establishes convergence results toward optimality of equilibria. Finally. Chapter 24 studies an application involving networked systems and optimal zero-delay coding.

Part I
Stochastic Teams and a Mathematical Theory of Information Structures

Chapter 2
An Introduction to Stochastic Team Theory and Solutions to Static Teams

In this chapter, we present a general introduction to stochastic team theory, which may also be viewed as a prelude to Chaps. 3–7. Our goal in this chapter is to lay the platform for the book to be accessible to an uninitiated reader, who will then be able to study the more technical chapters to follow. We present numerous results and examples, and introduce various concepts and problems which will all be studied in detail in the chapters to follow.

2.1 Introduction

The goal of this introductory chapter is to present a general mathematical formulation of stochastic teams, first with static and then with dynamic information structures, and to discuss some salient features of these decision problems and associated solution concepts through some simple but illustrative examples.

Stochastic team theory (or decentralized stochastic control) studies problems which consist of a collection of decision makers or agents acting together to optimize a common cost function, but not necessarily sharing all the available information and they may need to use only their local information. In such teams, there exist randomness in the initial states, observations, cost realizations, and/or the evolution of the dynamics. At each time stage, each agent only has partial access to the global information which is defined by the *information structure* (IS) of the problem, to be made precise later in the book.

The chapter discusses both *static* stochastic teams (that is, team decision problems where the information signals received by the decision makers are not affected by actions) and *dynamic* stochastic teams (where the information of at least one decision maker is affected by action). Sections 2.2, 2.2.2, and 2.4 deal with static teams, whereas Sects. 2.3 and 2.3.2 cover dynamic teams. Section 2.2 provides a general formulation for static teams, which is followed by a complete

S. Yüksel, T. Başar, *Stochastic Teams, Games, and Control under Information Constraints*, Systems & Control: Foundations & Applications,
https://doi.org/10.1007/978-3-031-54071-4_2

analysis of a finite stochastic team problem under various information patterns, in Sect. 2.2.2. Section 2.4 provides some general explicit results on existence, uniqueness, and characterization of optimal solutions first for general static teams and then for special classes of teams with Gaussian statistics: those with quadratic and exponentiated quadratic costs. Sections 2.3 and 2.3.2 can be viewed as the counterparts of Sects. 2.2 and 2.2.2 for dynamic teams. First a precise mathematical formulation for dynamic team decision problems is given, in Sect. 2.3, along with various dynamic information structures and appropriate solution concepts, and then an illustrative example of a finite dynamic team is provided in Sect. 2.3.2, within the framework of which some important features of optimal solutions in teams are discussed.

2.2 Static Decision Problems

2.2.1 A General Introduction to Static Decision Problems

Multiple person stochastic decision problems could be formulated with varying degrees of generality, abstraction and rigor, depending on the types of problems to be solved and the intended level of mathematical analysis. As noted, in this chapter our priority is to be accessible and pave the way for the more technical chapters to follow. Common to all possible formulations, however, is the specification of the following basic ingredients which are essential for a mathematical treatment of *decision making under uncertainty*. These are:

1. A set of **decision makers** (synonymously, *agents* or *controllers*), and the sets of alternative **actions** (synonymously, *decisions* or *control actions*) available to them.
2. The **uncertainty** and its probabilistic description.
3. The **information** acquired by each decision maker on the uncertainty and the previous actions.
4. The **payoff** (or *loss*) that accrues to each decision maker as a result of joint actions (over the decision period) and realization of uncertainty.
5. A **solution concept** whereby "best" or "satisfactory" decision rules can be chosen.

Before going into further specification of these entities, let us pause to introduce some terminology and notation which will be needed in the sequel. We will refer to a decision problem as **static** if the information available to each decision maker is independent of the actions of other decision makers (this statement will be made precise later); otherwise, the decision problem is said to be **dynamic**. We will refer to decision makers interchangeably as *agents* or *controllers*, with the ith one denoted **A**i (or **DM**i, since for the static setup we will not be making a distinction between an *agent*, acting multiple times, and a *decision maker*, acting once), where

i takes values in the set $\mathcal{N} := \{1, \ldots, N\}$ where $N \in \mathbb{N}$[1] which is called the *agent (decision maker) set*. The variable under the control of each decision maker will be called the *action* (synonymously, *decision* or *control*) *variable*, and will be denoted by u^i for $\mathbf{A}i$. Each u^i will take values in a given *action set* to be denoted by $\mathbb{U}^i$. Finally, the N-tuple $(u^1, \ldots, u^N)$ will be denoted by $\mathbf{u}$ and the product action space $\mathbb{U}^1 \times \cdots \times \mathbb{U}^N$ by $\mathbb{U}$.

Basic Ingredients of Static Decision Problems
In the static framework we will initially study the class of problems where the action sets, $\mathbb{U}^i$, $i \in \mathcal{N}$, are either (finitely or infinitely) countable, or uncountable but finite dimensional. In the latter case, we take the action set (space) to be isomorphic to the Euclidean[2] space $\mathbb{R}^{m_i}$, for some integer $m_i, i \in \mathcal{N}$ (the spaces thus can be generalized to complete, separable metric spaces; which are also known as Polish metric spaces or their Borel subsets—known as standard Borel spaces—, see Appendix A); furthermore, if there are any constraints imposed on the action variable u^i, we introduce the *action constraint set* S^i, for $\mathbf{A}i$, as a proper subset of $\mathbb{U}^i$.

The uncertainty in the decision problem is captured in the so-called *random state of* **nature**, ξ, which is a random variable (or vector) defined on a given probability space $(\Omega, \mathcal{F}, P_\Omega)$ and taking values in some standard Borel space $(\Xi, \mathcal{B}(\Xi))$. Let P be the probability measure induced by ξ on $(\Xi, \mathcal{B}(\Xi))$, corresponding to P_Ω. To save from notation, the corresponding probability measure will also be denoted by P.

The decision makers do not, in general, have direct access to the true state of nature, but instead observe the value of some other variable, known as the *measurement* (or *information*) *signal*. To define this quantity in precise terms, let us first introduce, for each $i \in \mathcal{N}$, the *information field*, $\mathbf{Y}^i$, for agent $\mathbf{A}i$ as a given subfield of $\mathcal{B}(\Xi)$, generated by a measurable function η^i mapping $(\Xi, \mathcal{B}(\Xi))$ onto $(\mathbb{Y}^i, \mathbf{B}^i)$. This is known as the *information function* for $\mathbf{A}i$, and the N-tuple $\underline{\eta} := (\eta^1, \ldots, \eta^N)$ is called the *information structure* (or *information pattern*) of the decision problem. The information function η^i induces a *partition*, $\mathbf{Y}^i$, of Ξ, and the information (measurement) signal y^i of $\mathbf{A}i$ (which lies in the *measurement set* $\mathbb{Y}^i$) is generated according to η^i, which is functionally written as

$$y^i = \eta^i(\xi) \equiv \tilde{\eta}^i(\omega)\,, \tag{2.1}$$

where the latter relates the measurement signal directly to the original probability space Ω, with elements ω. This is sometimes a more convenient representation to work with, especially if Ω is finite or countable. In that case one can consider Ω and Ξ to be essentially the same set, and thereby view $\mathbf{Y}^i$ also as a partition of Ω, which is a convention we henceforth adopt. In the case of finite probability spaces, we will

[1] The case with $N = \infty$ or finite but very large will be studied in Chap. 6.

[2] Some background material on sets and topological notions can be found in Appendix A.

also adopt the convention, perhaps by a slight abuse of notation and terminology, that the measurement signal y^i can be considered as an element of the partition set $\mathbf{Y}^i$.

The decision makers determine their actions using the measurement signals that they receive, under the *strategies* that they adopt for transforming measurements into actions. The *strategy* (synonymously, *decision rule* (function) or *control law*) of $\mathbf{A}i$ will be denoted by γ^i, and is formally defined as a measurable mapping from $(\Xi, \mathbf{Y}^i)$ into the space $(\mathbb{U}^i, \mathbf{B}_{\mathbb{U}^i})$. This can also be written as a measurable mapping from $(Y^i, \mathbf{B}^i)$ to $(\mathbb{U}^i, \mathbf{B}_{\mathbb{U}^i})$, as we state explicitly below. We denote the set of all such mappings, which also satisfy the additional constraints that may have been imposed on u^i, by Γ^i, to be called the *strategy space* of $\mathbf{A}i$, and note the relationship

$$u^i = \gamma^i(y^i) = \gamma^i(\eta^i[\xi])$$

where the latter relates the action variable to the state of nature, ξ. We will denote the N-tuple $(\gamma^1, \ldots, \gamma^N)$ by $\underline{\gamma}$, and the product strategy space $\Gamma^1 \times \cdots \times \Gamma^N$ by $\mathbf{\Gamma}$. The individual strategy spaces Γ^i may also include the additional structural constraints that may have been imposed on the policies, such as linearity. What may not be allowed in general, however, is for $\mathbf{\Gamma}$ to be *nonrectangular*, that is for the choice out of Γ^i (for some i) to restrict the choice out of $\Gamma^j (j \neq i)$. For example, our formulation (at this point) does not cover "cross-constraints" of the type $f(u^i, u^j) \leq 0, i \neq j$, for some functional f.

Given an $(N{+}1)$-tuple $(\xi, \mathbf{u}) \in \Xi \times \mathbb{U}$, the loss incurred to the decision makers viewed collectively as a *team* will be denoted by $L(\xi, \mathbf{u})$,[3] where the function $L : \Xi \times \mathbb{U} \to \mathbb{R}$ is known as the *loss function* (or the *cost function*) for the team. Its negative, $-L(\xi, \mathbf{u}) =: \mathrm{U}(\xi, \mathbf{u})$, is known as the *payoff function*, which all agents collectively want to "maximize", in a sense to be defined shortly. Implicit here is the assumption that for the team there exists a unique (up to equivalence) utility function which numerically orders different outcomes corresponding to joint actions and realization of the state of nature, in a way consistent with the team's preference ordering among different alternatives.

The loss incurred is generally a random quantity, the randomness appearing through both ξ and $\mathbf{u}$, where the latter depends on ξ through the measurement signals and the strategies adopted by the decision makers. Therefore, one rather works with the expected value of this quantity, which we will be referring to as the *cost function*.[4] Other possible terminologies would be *expected loss function*, *average risk* or *expected cost*, all of which have been used in the literature, which

[3] We will also occasionally be using the letter c to denote the loss or cost function; this should cause no ambiguity when a problem is explicitly defined.

[4] There would be other ways of making the objective function deterministic, such as defining the cost function as the probability of the loss exceeding a given ceiling, or taking it as the supremum of the loss function over $\omega \in \Omega$. We will not be devoting much discussion to such formulations in the book.

we will also use interchangeably. The cost function, $J : \mathbf{\Gamma} \to \mathbb{R}$, is defined on the product strategy space $\mathbf{\Gamma}$ as

$$J(\underline{\gamma}) = \int_{\Xi} L(\xi, \underline{\gamma}(\underline{\eta}[\xi]))P(d\xi) = E[L(\xi, \underline{\gamma}(\underline{\eta}[\xi]))] =: E_{\xi} L(\xi, \underline{\gamma}(\underline{\eta}[\xi])), \tag{2.2}$$

where

$$\underline{\gamma}(\underline{\eta}[\xi]) := (\gamma^1(\eta^1[\xi]), \ldots, \gamma^N(\eta^N[\xi])) \tag{2.3}$$

and E_{ξ} is the operator that takes the expected value of the quantity it precedes, over ξ. To show the explicit dependence of J on also the information structure $\underline{\eta}$, we will sometimes use the notation $J(\underline{\gamma}, \underline{\eta})$, and occasionally use $J(\underline{\gamma}, \underline{\eta}; L, P)$ to also indicate the dependence on the loss function L and the probability distribution P.

The specification of J, along with the product strategy space $\mathbf{\Gamma}$, provides a complete characterization (aside from the solution concept) of a stochastic multiple person decision problem, and is known as the *normal form description*. Note that in this description, the information structure is suppressed and it enters the problem formulation only through the strategy spaces Γ^i, $i \in \mathcal{N}$. The description which lays out explicitly the dependence of the measurement signals on the unknown state of nature is known as the *extensive form description* of the underlying (static) stochastic decision problem. The distinction between these two forms should be more transparent when we introduce dynamic decision problems, later in this chapter.

Notion of Optimality

In the framework laid out above, it would have been possible to endow each decision maker (agent) with a different loss function and also possibly a different subjective probability measure regarding the unknown state of nature. Either of these departures would take us outside the realm of team problems, and necessitate consideration of the more general framework of stochastic (zero-sum or nonzero-sum) games, with associated solution concepts, such as *saddle-point equilibrium* or *Nash equilibrium* [97], to be covered later in the book. In this chapter our interest is in *teams*, with a precise definition being the following:

> A **team** is a collection of individual decision makers who strive for the same goal, using the same (probabilistic) model of the underlying decision process, but not necessarily sharing the same on-line information (such as measurements) on the uncertainty.

For an N-person stochastic team problem, since all agents will be striving towards the same goal, with team preferences quantified in the given loss functional, the most reasonable solution that leads to optimal behavior is the global minimization of the team cost over the product strategy space. Hence, we have the following.

Definition 2.2.1 For a given stochastic team problem with a fixed information structure, $\{J; \Gamma^i, i \in \mathcal{N}\}$, a strategy N-tuple $\underline{\gamma}^* := (\gamma^{1*}, \ldots, \gamma^{N*}) \in \mathbf{\Gamma}$ is an *optimal team decision rule* if

$$J(\underline{\gamma}^*) = \inf_{\underline{\gamma} \in \mathbf{\Gamma}} J(\underline{\gamma}) =: J^*, \tag{2.4}$$

provided that such a strategy exists. The cost level achieved by this strategy, J^*, is the *minimum (or optimal) team cost.*

In the above definition of "optimality in a team", we have taken the information structure as fixed and given *a priori*. Even though the class of systems one typically encounters are primarily of this type, it is possible to consider the information structure of the problem as a variable, alongside the strategies of the agents. In fact, in the theory of organizations (as well as to a large extent in the design of networked systems), the prime goal is to obtain an optimal design for the pair $(\underline{\gamma}; \underline{\eta})$ which is known as the *organizational form* [507]. Of course, to make the problem meaningful, we have to impose some restrictions on $\underline{\eta}$ (such as belonging to some prescribed class of comparable information structures, say Υ), or attach some cost to it. In the absence of such realistic restraints on $\underline{\eta}$, the problem will admit the trivial solution where $\underline{\eta}^*$ (the optimal $\underline{\eta}$) allows the agents to acquire perfect information on the state of nature, ξ, and thereby $\underline{\gamma}^*$ to depend directly on ξ. Under realistic organizational constraints, however, say with $\underline{\eta} \in \Upsilon$, an optimal design $(\underline{\gamma}^*; \underline{\eta}^*) \in \mathbf{\Gamma} \times \Upsilon$ will have the property that there exists no $\underline{\eta} \in \Upsilon$ such that

$$\inf_{\underline{\gamma} \in \mathbf{\Gamma}} J(\underline{\gamma}; \underline{\eta}) < J(\underline{\gamma}^*; \underline{\eta}^*) \tag{2.5}$$

where the cost function J may also include some additional (possibly additive) terms reflecting the costs associated with various $\underline{\eta}$'s. Furthermore, the policy space $\mathbf{\Gamma}$ implicitly depends on the choice out of Υ, so that the product $\mathbf{\Gamma} \times \Upsilon$ is actually not *rectangular*. Note that a natural way of obtaining an optimal organizational form would be to minimize the function $J(\underline{\gamma}^*_{\underline{\eta}}; \underline{\eta})$ over $\underline{\eta} \in \Upsilon$, where $\underline{\gamma}^*_{\underline{\eta}}$ is the team-optimal solution corresponding to the fixed information structure $\underline{\eta}$. We use a subscript on $\underline{\gamma}^*$ here to explicitly point out the fact that the team-optimal solution depends on $\underline{\eta}$ structurally, and in general in a fairly complicated manner, which makes the further optimization of J, as $\underline{\eta}$ varies over Υ, a rather complex problem (not of the standard type), unless the cardinality of Υ is finite. This problem will be studied further in detail in Chap. 8. Additionally, comparison of various information structures will be presented in various contexts in Chap. 7.[5]

One important feature of the team-optimal solution that is worth mentioning at this point (perhaps as a cautionary remark) is that multiple solutions are *not* necessarily *interchangeable*. For a two-person team problem for example, if the pairs of policies (γ^1, γ^2) and (β^1, β^2) are two team-optimal policy pairs, then it

[5] If information constraints involve data-rate limits, as in information and communication theoretic considerations such as measurements over finite capacity information channels, then one needs to jointly study coding, decoding and control policies. This setting will be addressed in Part III of the book.

is not necessarily true that the pair (γ^1, β^2) will also constitute a team-optimal solution. Hence, in case of nonuniqueness of the solution, the agents need to have a common consistent rule as to which one of the possible solutions to adopt, in order to arrive at the optimum. This may require some pre-communication, and pre-commitment to some protocols among the agents.

A weaker solution concept than that of team-optimality introduced in Definition 2.2.1, is that of *person-by-person optimality*, equivalently *Nash equilibrium*, introduced next.

Definition 2.2.2 For a given N-person stochastic team with a fixed information structure, $\{J; \Gamma^i, i \in \mathcal{N}\}$, an N-tuple of strategies $\underline{\gamma}^* := (\gamma^{1^*}, \ldots, \gamma^{N^*})$ constitutes a *Nash equilibrium* (synonymously, a *person-by-person optimal* (pbp optimal) solution) if, for all $\beta \in \Gamma^i$ and all $i \in \mathcal{N}$, the following inequalities hold:

$$J^* := J(\underline{\gamma}^*) \leq J(\underline{\gamma}^{-i*}, \beta), \tag{2.6}$$

where we have adopted the notation

$$(\underline{\gamma}^{-i,*}, \beta) := (\gamma^{1,*}, \ldots, \gamma^{i-1,*}, \beta, \gamma^{i+1,*}, \ldots, \gamma^{N,*}). \tag{2.7}$$

Remark 2.2.1 Nash equilibrium is a weaker solution concept than team-optimality (*cf.* Definition 2.2.1), since satisfaction of the N inequalities (2.6) is clearly necessary but not sufficient for $\underline{\gamma}^*$ to be an optimal team decision rule. But, since every team-optimal solution is necessarily a *pbp* optimal solution, the latter plays an important role in the derivation of the former, as we will see later in the book, with the first demonstration being in Sect. 2.4.

In the next section, we depart from the abstract formulation of the present section and provide an illustrative example which will aid in better understanding of the concepts introduced above.

2.2.2 An Example of a Finite Stochastic Team

A stochastic team problem is said to be *finite* if both the action and the uncertainty sets are finite. In this case (as we have indicated earlier) there is no need to make any distinction between Ω and Ξ, and one may as well work in the original probability space $(\Omega, \mathcal{F}, P_\Omega)$ where the probability measure will be replaced with the probability masses $\{p_j = P_\Omega(\{\omega_j\})\}_{j=1}^{\#}$ where ω_j is an element of Ω with positive probability, and $\# := |\Omega|$, the cardinality of Ω, with those elements of Ω receiving *zero* probability from P_Ω being irrelevant to the decision problem and therefore deleted. By a possible abuse of terminology, we will call the # elements of Ω the states of nature. We note that $\mathcal{F}$ is a collection of all subsets of Ω (hence it has $2^{\#}$ elements), and $\mathbf{Y}^i$ can be taken, without any loss of generality, as a partition of Ω, for each $i \in \mathcal{N}$.

Every two-person finite stochastic static team can be represented by a family of matrices, each matrix (and there will be # of them) corresponding to a different state of nature, ω. The rows of these matrices would correspond to action choices of one agent, say **A**1, the columns would correspond to action choices of the other agent, **A**2, and each entry would be the corresponding loss to the team for that particular ω. This, together with a specification of the class of all possible information signals, $(\mathbf{Y}^1, \mathbf{Y}^2)$, would constitute the *extensive form* description for the team. Such a set-up can also naturally be extended to N-person finite static teams, where now the matrices are replaced by N-dimensional hypercubes.

One approach (and a universally applicable one) towards obtaining the team-optimal solution(s) of such finite static teams is to convert the above *extensive form* into a *normal form* by relating the strategies of the agents directly to the (expected) costs that accrue to the team. As we have indicated earlier, such a formulation would suppress the information signals as well as the role of nature in the decision problem, and it would involve only a single finite, albeit larger dimensional, matrix (or hypercube, if there are more than two agents) whose columns and rows are strategy choices of the agents, and whose lowest entry (or entries) would yield the team-optimal solution. Note that for **A**i the number of alternative strategies (i.e. $|\Gamma^i|$) would be $|\mathbb{U}^i|^{|\mathbf{Y}^i|}$, and hence a derivation based solely on the normal form could easily get intractable if either the number of information signals or the cardinality of the action set for at least one agent is large. It is therefore necessary to look for alternative ways of obtaining the solution, by also exploiting the nature of the information available to the agents. Note that the solution to a finite static stochastic team problem always exists (but it may be nonunique), since it involves optimization over a finite set.

With this prelude, we consider in this section a two-person static stochastic team problem where $\mathbb{U}^1 = \{U(\text{up}), D(\text{down})\}$, $\mathbb{U}^2 = \{L(\text{left}), R(\text{right})\}$, $\Omega = \{\omega_1, \omega_2, \omega_3\}$, $p_1 = p_2 = 0.3$, $p_3 = 0.4$, and the loss matrices are given by

A1 \ A2	L	R
U	1	0
D	3	1

$\omega : \omega_1 \leftrightarrow 0.3$

A1 \ A2	L	R
U	2	3
D	2	1

$\omega_2 \leftrightarrow 0.3$

A1 \ A2	L	R
U	1	2
D	0	2

$\omega_3 \leftrightarrow 0.4$

Under various information structures for the team, we now study the derivation of team-optimal decision rules and some of their properties.

1. Perfect Measurements

Here both agents have access to the true state of nature, and hence $\mathbf{Y}^1 = \mathbf{Y}^2 = \sigma(\{\{\omega_1\}, \{\omega_2\}, \{\omega_3\}\})$, the σ-field generated by the singletons. The cardinality of the strategy spaces (Γ^1 and Γ^2) is $2^3 = 8$ each, and hence the normal form is an

8×8 matrix, requiring a comparison of 64 entries. The normal form of a decision problem could also be called a *pre-commitment model*, since the strategies of the agents tell them what to do under all possible realizations of the information signal, even before the actual state of nature is realized. Observe that

$$J^* := \min_{\gamma^1, \gamma^2} J(\gamma^1, \gamma^2) = \min_{\gamma^1, \gamma^2} E_\omega[L(\omega; \gamma^1(\omega), \gamma^2(\omega))]$$
$$\equiv E_\omega[\min_{u^1, u^2} L(\omega; u^1, u^2)] \quad (2.8)$$

which holds since the agents have perfect measurements on ω and there are finitely many actions. The individual minima of $L(\omega; u^1, u^2)$ are $\min L(\omega_1; u^1, u^2) = L(\omega_1; U, R) = 0$, $\min L(\omega_2; u^1, u^2) = L(\omega_2; D, R) = 1$, and $\min L(\omega_3; u^1, u^2) = L(\omega_3; D, L) = 0$, which lead (from (2.8)) to $J^* = 0.3$ and the unique team-optimal decision rules:

$$\gamma^{1*}(\omega) = \begin{cases} U, & \omega = \omega_1 \\ D, & \text{else} \end{cases} \qquad \gamma^{2*}(\omega) = \begin{cases} L, & \omega = \omega_3 \\ R, & \text{else} \end{cases},$$

which we rewrite symbolically as

$$\underline{\gamma}^* = (UDD, RRL)$$

a convention we adopt (and will henceforth use) for representing strategies in finite spaces.

As a final note we point to the observation that even though the policy pair (UDD, RRL) is unique as a team-optimal solution (which is also, by definition, *pbp* optimal), it is not the unique *pbp* optimal solution. The policy pair (UUD, RLL) is also *pbp* optimal, but it carries the unfavorable cost of 0.6, higher than J^*.

2. Imperfect Identical Measurements
Here we consider the situation where the agents can distinguish only between the pair (ω_1, ω_2) and the singleton ω_3, and hence $\mathbf{Y}^1 = \mathbf{Y}^2 =: \mathbf{Y} = \sigma(\{\{\omega_1, \omega_2\}, \{\omega_3\}\})$. The strategy spaces have 4 elements each, leading to a 4×4 matrix as the normal form. We write out this matrix, for instructional purposes, with the notation $\gamma^i(y^i) = (a, b)$ (with a, b denoting the possible actions of the agents) standing for

$$\gamma^i(y^i) = \begin{cases} a, & y^i = \{\omega_1, \omega_2\} \\ b, & \text{else} \end{cases}$$

A2

		LL	LR	RL	RR
	UU	1.3	1.7	1.3	1.7
A1	UD	0.9	1.7	0.9	1.7
	DU	1.9	2.3	1.0	1.4
	DD	1.5	2.3	0.6*	1.4

The matrix has a unique minimum entry, as indicated, and hence the team problem under the given information pattern admits the unique optimal solution $\underline{\gamma}^* = (DD, RL)$, yielding a cost level of $J^* = 0.6$. Note that this is twice the optimal cost level attained under the perfect state measurements, and we can refer to the difference between the two (informally) as the "value" of the additional measurement which enables the agents to distinguish between the two states ω_1 and ω_2. Note also that in addition to the team-optimal solution given above, the problem admits one other *pbp* optimal solution, which is (UD, LL), with a corresponding (unfavorable) cost level of 0.9.

An alternative derivation for the team-optimal solution, which would involve lower dimensional matrices, follows from a reasoning similar to the one used for the perfect information case. Here the counterpart of (2.8) would be

$$\begin{aligned} J^* := \min_{\gamma^1,\gamma^2} J(\gamma^1, \gamma^2) &= \min_{\gamma^1,\gamma^2} E_\omega[L(\omega; \gamma^1(y), \gamma^2(y))] \\ &= E_y[\min_{u^1,u^2} E_{\omega|y} L(\omega; u^1, u^2)] \end{aligned} \tag{2.9}$$

where we have used the "iterated property" of the conditional expectation: $E_\omega = E_y E_{\omega|y}$ where $E_{\omega|y}$ is the conditional expectation of the random variable it precedes, given that $y \in \mathbf{Y}$ has been observed. Also, since we are operating in finite spaces, expression (2.9) is well-defined and thus we are allowed to interchange the operations of outer expectation (over $y \in \mathbf{Y}$) and minimization (over $\underline{\gamma} \in \Gamma$). Now, the inner minimization in (2.9) involves two matrices, corresponding to two different (and exhaustive) choices for $y : y_1 = \{\omega_1, \omega_2\}$ and $y_2 = \{\omega_3\}$. These matrices, which we may call *conditional cost matrices*, are as follows, with the unique optimal solution indicated in each case:[6]

[6] The entries of the first matrix are obtained from the relationship $E_{\omega|y_1} L(\omega; u^1, u^2) = L(\omega; u^1, u^2) p_{1|1} + L(\omega; u^1, u^2) p_{2|1}$, where $p_{1|1} = P(\omega_1|\{\omega_1, \omega_2\})$ and $p_{2|1} = P(\omega_2|\{\omega_1, \omega_2\})$ are the conditional probabilities, each equal to $1/2$.

A2 (first matrix, $y_1 \leftrightarrow 0.6$)

A1		L	R
	U	1.5	1.5
	D	2.5	1.0*

A2 (second matrix, $y_2 \leftrightarrow 0.4$)

A1		L	R
	U	1	2
	D	0*	2

$y: \quad y_1 \leftrightarrow 0.6 \qquad y_2 \leftrightarrow 0.4$

Since y_1 occurs with probability 0.6 and y_2 with probability 0.4, the (average) optimal team cost is $J^* = (0.6)(1) + (0.4)(0) = 0.6$, attained by the unique pair of decision rules (DD, RL). Note that the first matrix admits one other *pbp* optimal solution (U, L) which, together with the team-optimal solution of the second matrix, leads to a *pbp* optimal solution for the original team, (UD, LL), which is the one found earlier using the 4×4 normal form.

3. No Measurements

When neither agent makes any measurements, $\mathbf{Y}^1$ and $\mathbf{Y}^2$ are trivial σ-fields $\{\emptyset, \Omega\}$, and hence all permissible decision rules are *constant* maps. The normal form is the 2×2 matrix

A2

A1		L	R
	U	1.3*	1.7
	D	1.5	1.4

from which we immediately read: $J^* = 1.3$ and $\underline{\gamma}^* = (U, L)$.

4. Nonidentical Measurements: Perfect for A2 and None for A1

This is the first nonsymmetric information structure that we will be studying. The information sets are $\mathbf{Y}^2 = \sigma(\{\{\omega_1\}, \{\omega_2\}, \{\omega_3\}\})$, $\mathbf{Y}^1 = \sigma(\{\omega_1, \omega_2, \omega_3\})$, leading to *eight* elements for Γ^2 and *two* for Γ^1. The normal form is given by the *two-by-eight* matrix

A2

A1		LLL	LLR	LRL	LRR	RLL	RLR	RRL	RRR
	U	1.3	1.7	1.6	2.0	1.0	1.4	1.3	1.7
	D	1.5	2.3	1.2	2.0	0.9	1.7	0.6*	1.4

and the unique team-optimal solution is, as indicated, $\underline{\gamma}^* = (D, RRL)$, and the value is $J^* = 0.6$. Note that the optimal cost is the same here as in *case* 2, even though the information structures are incomparable. [As compared with *case* 2, here **A**1 has **worse** and **A**2 has **better** information, in the sense that $\mathbf{B}^1_{(4)} \subset \mathbf{B}^1_{(2)}$ and $\mathbf{B}^2_{(4)} \supset \mathbf{B}^2_{(2)}$, where the subscripts on **B** refer to the two different cases, and inclusion is a strict one.[7]]

[7] An equivalent statement would be $\Gamma^1_{(4)} \subset \Gamma^1_{(2)}$ and $\Gamma^2_{(4)} \supset \Gamma^2_{(2)}$. A more formal treatment of comparison of two information structures will be studied in Sect. 7.2.

The question arises now as to whether a procedure similar to those used in *cases 1* and 2 could also be used here to simplify the derivation (that is, to avoid working with a large dimensional matrix, each entry of which has to be computed). Clearly, an identity such as (2.9) cannot be used since the agents do not make identical measurements. However, for each fixed decision rule γ^1 of **A**1 (and there are only *two*) one can obtain a best response (minimizing solution) $T(\gamma^1)$ for **A**2 by using the original matrices, since **A**2 has perfect information:

$$\gamma^1 = U \Rightarrow \gamma^2 = T(U) = (RLL) \Rightarrow J(U, T(U)) = 1.0\ ,$$

$$\gamma^1 = D \Rightarrow \gamma^2 = T(D) = (RRL) \Rightarrow J(D, T(U)) = 0.6\ .$$

The best choice for **A**1 , then, is $\gamma^{1^*} = D$, and the corresponding best response for **A**2 is $\gamma^{2^*} = T(D) = (RRL)$, thus agreeing with what we had obtained earlier.

The above is yet another procedure for obtaining the team-optimal solution in two-person stochastic teams: Fix the policies of one of the agents (preferably the one whose strategy space has fewer elements), obtain the best response of the other agent to each such policy, and compute the corresponding (average) team cost in each case. The lowest such cost is the optimal team cost, and the corresponding policies are the team-optimal decision rules. Such a procedure is always justified because of the following sequence of identities (where we have taken **A**1 as the starting agent):

$$\begin{aligned} J^* &= \min_{\gamma^1}\min_{\gamma^2} E_\omega[L(\omega; \gamma^1(y^1), \gamma^2(y^2))] \\ &\equiv \min_{\gamma^1} E_{y^2}[\min_{u^2} E_{\omega|y^2}[L(\omega; \gamma^1(y^1), u^2)]] \\ &\equiv \min_{\gamma^1} E_\omega[L(\omega; \gamma^1(y^1), T(\gamma^1)(y^2))]. \end{aligned}$$

This would be applicable even if $\mathbf{B}^1$ and $\mathbf{B}^2$ do not satisfy an inclusion relationship (in the particular case above we had $\mathbf{B}^1 \subset \mathbf{B}^2$), but then one has to construct new *conditional cost matrices* ($|\mathbf{Y}^2|$ of them, each of dimension $|\Gamma^1|$-by-$|\mathbb{U}^2|$) in order to obtain the optimal response of **A**2. [See the next case for an information structure of the type where the inclusion does not hold.]

5. Nonidentical Imperfect Measurements

This case will serve to illustrate a point which is sometimes very useful in the derivation of team-optimal solutions. Consider the information structure given by $\mathbf{Y}^1 = \sigma(\{\{\omega_1\}, \{\omega_2, \omega_3\}\})$ and $\mathbf{Y}^2 = \sigma(\{\{\omega_1, \omega_2\}, \{\omega_3\}\})$, where an inclusion property does not hold between $\mathbf{B}^1$ and $\mathbf{B}^2$. This, therefore, does not fall into any of the categories of information structures considered so far in this section (for the specific example). The two methods of derivation here would be:

(i) the direct solution based on the normal form (which is a 4×4 matrix),
(ii) the sequential approach (which involves *two* 4×2 matrices and hence does not offer any savings (and thereby advantage) over the normal form).

These are the two general methods which would be applicable to this class of problems; however, in the present case a simple (but useful) observation yields the solution immediately: The team-optimal decision rules γ^{1*} and γ^{2*} for *case 1* (that is, under perfect measurements) are also well-defined functions on the signal spaces $\mathbf{Y}^1$ and $\mathbf{Y}^2$ above, and hence under the right kind of interpretation, they belong to the strategy spaces Γ^1 and Γ^2 of the present problem. The information structure in *case 1* being *richer* (in fact, the *richest* possible),[8] this observation directly implies that the pair $\{\gamma^{1*} = UD, \gamma^{2*} = RL\}$ is the unique team-optimal solution of the new problem with "coarser" information. Note that the pair (UD, RL) here is indeed the pair (UDD, RRL) of *case 1*, simply rewritten using the adopted convention, on the restricted information space. If we write them out, they both correspond to

$$\gamma^{1*}(y^1) = \begin{cases} U, \ y^1 = \{\omega_1\} \\ \\ D, \ \text{else} \end{cases} \quad ; \quad \gamma^{2*}(y^2) = \begin{cases} R, \ y^2 = \{\omega_1, \omega_2\} \\ \\ L, \ \text{else} \end{cases} .$$

6. "Noisy" Measurements

For reasons which will become clear later, it is useful to distinguish between "imperfect" and "noisy" measurements. The information signals of *cases 2* and *5*, studied above, belong to the former category because they do not bring in additional uncertainty into the problem formulation, other than what exists already in the complete description of the cost matrices. Consider the scenario depicted below, which uses essentially the same team problem as before, but with a different type of information: Agent **A**2 makes no measurements, while **A**1 observes the value of a random variable z, taking two possible (distinct) values, y_1 and y_2. The loss matrices are the same as before, where we now adopt a different symbol, x, to replace ω, with $x_i = \omega_i$. To complete the description of the team problem, we now specify, in the following table, the joint probability mass function (*pmf*) of the pair (x, z), which has to be consistent with the marginal *pmf* of x:

	x_1	x_2	x_3
y_1	0.12	0.21	0.12
y_2	0.18	0.09	0.28

Note that

[8] At this point, this statement should be interpreted as saying "there is no other information structure which provides the agents with more information on the state of nature". The underlying notion will be made precise later.

$$Prob(x_i \mid y_1) = \begin{cases} 4/15, \ i = 1, 3 \\ 7/15, \ i = 2 \end{cases}, \qquad Prob(x_i \mid y_2) = \begin{cases} 18/55, \ i = 1 \\ 9/55, \ \ i = 2 \\ 28/55, \ i = 3 \end{cases}$$

and hence after observing y_1 or y_2 it is not possible for **A**1 to tell, with certainty, the true value of x.

The problem can now be cast in the framework of the general formulation of Sect. 2.2 by constructing an appropriate sample space. Toward this end, let Ω be a set of cardinality 6, with elements ω_{ij} $(i = 1, 2, 3;\ j = 1, 2)$, where ω_{ij} corresponds to the pair (x_i, y_j) and hence $Prob(\omega = \omega_{ij}) = Prob(x = x_i, y = y_j)$. The two possible measurement signals of **A**1 are $y_1^1 = \{\omega_{11}, \omega_{21}, \omega_{31}\}$ and $y_2^1 = \{\omega_{12}, \omega_{22}, \omega_{32}\}$ which together determine the partition $\mathbf{Y}^1$ introduced in Sect. 2.2.[9] We thus have a team problem of the standard type, for which the normal form is

		A1			
		UU	UD	DU	DD
A2	L	1.30∗	1.38	1.42	1.50
	R	1.70	1.70	1.40	1.40

which admits the unique team-optimal solution $\underline{\gamma}^* = (UU, L)$, with a corresponding value of $J^* = 1.30$. An immediate observation here is that this is the same value as that obtained in *case 3*, and hence the additional (noisy) information to **A**1 is of no value to the team. We leave it to the reader to verify that if, instead, agent **A**2 had received this measurement signal, then the team-optimal solution would again be unique and be given by $\underline{\gamma}^* = (D, RL)$, yielding this time a value of $J^* = 1.29$. Hence the same measurement is of some (positive) value to the team, if received by the second agent. As a final scenario, let us consider the information structure under which **both** agents have access to the realization of z (that is they have a complete sharing of information, which makes the problem essentially no different from a single agent stochastic decision problem). In view of the discussion for *case 2*, and especially the relation (2.9), we first form the *conditional cost matrices* corresponding to the two realizations of the measurement signal, y_1 and y_2:

		A2		**A**2	
		L	R	L	R
A1	U	22/15	29/15	64/55∗	83/55
	D	26/15	19/15∗	72/55	83/55
z :		$y_1 \leftrightarrow 0.45$		$y_2 \leftrightarrow 0.55$	

[9] Here, since we have a finite decision problem, we do not distinguish between Ω and Ξ, and hence consider $\mathbf{Y}^1$ as a partition of the sample space Ω.

Then we can readily read, from the above matrices, the unique team-optimal strategy pair: (DU, RL), with a corresponding cost value of 1.21.

It is worth noting that the main feature of this last case, which distinguishes it from the earlier ones, is that the random quantity ω (or, equivalently here, the state of nature ξ) has two identifiable components: the "payoff relevant" part, x, and the information signal, y, with some correlation between them. The role of y is to carry information regarding the true value of x, and it affects the value of the loss function only through the strategy of the agent who receives this information. The advantage of splitting ξ into two components, as above, may not be that obvious at this point, but we will later observe the versatility of such a formulation, especially in the context of infinite decision problems.

2.3 Dynamic Decision Problems

2.3.1 A General Introduction to Dynamic Decision Problems

As mentioned earlier in Sect. 2.2, a decision problem is said to be *dynamic* if the measurements of at least one of the agents involves past actions (of that particular agent or some other agent(s)). In the literature, the connotation "dynamic" is also used to characterize decision problems where an agent acts more than once, even if the measurements do not depend on past actions (the case of *open-loop* information structure). Under mild conditions, to be made precise later in the book, such problems can be converted into static decision problems by essentially working in higher dimensional spaces, but it is generally found convenient to treat them also in the context of truly dynamic problems because of the similarities in the derivation of the optimal solutions. We will have occasions to use both approaches in this book. We describe below an appropriate set-up for the study of truly dynamic decision problems, restricting the exposition to discrete time.

For a truly dynamic problem, we follow the formulation of Sect. 2.2, prior to Eq. (2.1), but now replace the static relationship (2.1) with the dynamic equation

$$y^i = \eta^i(\xi; \mathbf{u}), \quad i \in \mathcal{N} \tag{2.10}$$

where the dependence on $\mathbf{u}$ is assumed to be *strictly causal*, which means that under a given *fixed clock* the information received by each agent can depend only on actions taken in the past. To give this statement a more precise mathematical meaning, let us consider a discrete-time framework where actions are taken at discrete instants of time, $1, 2, \ldots, T$. Let t stand for the generic time variable, and $\mathcal{T}$ denote the (discrete) time set

$$\mathcal{T} := \{1, \ldots, T\}. \tag{2.11}$$

Let u_t^i and y_t^i denote, respectively, the action (decision) variable and the information variable of agent $\mathbf{A}i$ at the time instant $t \in \mathcal{T}$. Furthermore, introduce the notation:

$$\mathbf{u}_t := \{u_t^1, \ldots, u_t^N\}, \quad \mathbf{y}_t := \{y_t^1, \ldots, y_t^N\} \tag{2.12}$$

$$\mathbf{u}_{[t_0,t_1)} \equiv \mathbf{u}_{[t_0,t_1-1]} := \{\mathbf{u}_{t_0}, \mathbf{u}_{t_0+1}, \ldots, \mathbf{u}_{t_1-1}\} \equiv \{u^1_{[t_0,t_1)}, \ldots, u^N_{[t_0,t_1)}\}. \tag{2.13}$$

Then, under the strict causality assumption, (2.10) becomes equivalent to

$$y_t^i = \eta_t^i(\xi; \mathbf{u}_{[1,t)}), \qquad t \in \mathcal{T}, i \in \mathcal{N} \tag{2.14}$$

for some "information functions" η_t^i, $t \in \mathcal{T}$, $i \in \mathcal{N}$. The stochastic variable y_t^i, taking values in $\mathbb{Y}_t^i$, is the *on-line* information available to $\mathbf{A}i$ which he can use in the construction of the decision u_t^i at time t, through an appropriate policy variable $\gamma_t^i : \mathbb{Y}_t^i \to \mathbb{U}_t^i$

$$u_t^i = \gamma_t^i(y_t^i) \equiv \gamma_t^i(\eta_t^i[\xi; \mathbf{u}_{[1,t)}]), \qquad t \in \mathcal{T}, i \in \mathcal{N}. \tag{2.15}$$

A permissible policy γ_t^i is one under which u_t^i becomes a well-defined random variable, defined on the original probability space, and taking values in $S_t^i \subset \mathbb{U}_t^i$, where S_t^i is the *action constraint* set for $\mathbf{A}i$ at time t. Let us denote the set of all such maps by Γ_t^i, which is the *policy space* of $\mathbf{A}i$ at time t. The construction of such a policy space will depend on the problem at hand, and we will see several such constructions throughout the book. At this point let us simply assume that such a construction is given, and rewrite (2.15) in the following compact form:

$$\mathbf{u} = \underline{\gamma}(\underline{\eta}[\xi; \mathbf{u}]), \quad \underline{\gamma} \in \mathbf{\Gamma} := \Gamma^1 \times \ldots \times \Gamma^N \tag{2.16}$$

where Γ^i is the *composite* (over-time) *policy space* of $\mathbf{A}i$:

$$\Gamma^i := (\Gamma_1^i, \Gamma_2^i, \ldots, \Gamma_T^i), \qquad i \in \mathcal{N}.$$

Note that the right-hand side of (2.16) also depends on $\mathbf{u}$, which is what makes dynamic decision problems intrinsically different from the static ones introduced in Sect. 2.2. The equation (2.16) is called a *loop equation*, and a dynamic decision problem is well-defined only if this loop equation has a unique solution for every ξ, that is for some $\underline{\tilde{\gamma}} : \Xi \to \mathbb{U}$,

$$\mathbf{u} = \underline{\tilde{\gamma}}(\xi) \tag{2.17}$$

uniquely solves (2.16). The strict causality condition, or equivalently the structural assumption (2.15), is a condition that guarantees this.

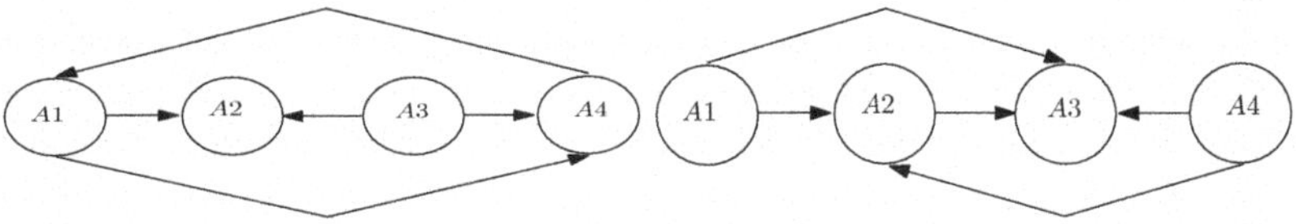

Fig. 2.1 Two scenarios on available action information at stage t

It is possible to relax the strict causality condition and the fixed clock assumption, and replace them by some other conditions under which the loop equations (2.16) still admit a unique solution (see Sect. 3.9). To just provide a flavor of these extensions here, let us note that in (2.15) we can allow u_t^i to depend on u_t^j, $j \neq i$, provided that u_t^j is not allowed to depend on u_t^i either directly or through the actions of other agents. In Fig. 2.1 we have depicted two such scenarios for a four-agent problem with time step t isolated. A pointed arrow indicates that information flows at this stage in the direction of the arrow. The first configuration of Fig. 2.1 does not lead to a well-defined (physically realizable) decision problem because u_t^4 depends on u_t^1 while at the same time u_t^1 depends on u_t^4, hence there is a deadlock. (Clearly closed directed graphs should not be allowed for unique solvability of the loop equations.) The second configuration of Fig. 2.1, on the other hand, depicts an acceptable information exchange[10] since it does not contain any closed direct graph. Hence, the strictly causal relation (2.15) can accommodate such permissible informational relationships among the u_t^i's, and the configuration could be different for each t.

In such a team, if there is a pre-specified order in which the agents act, then such a team is said to be a *sequential team*. However, one can allow these "permissible configurations" to be sample-path dependent (i.e., dependent on ξ), provided that certain measurability conditions are satisfied. This is the situation where the order in which the agents act is determined (partially) by a chance mechanism. Such a dynamic team model is said to be *a non-sequential* team. Non-sequential stochastic team theory will be studied in detail in Sect. 3.9.

Sequential Dynamic Teams in State-Space Form

Now, coming back to (2.14) and (2.15), it is often convenient to introduce an intermediate variable, called the *state variable*, recursively defined by

$$x_{t+1} = f_t(x_t, u_t^1, \ldots, u_t^N; w_t^0), \quad t \in \mathcal{T}, \tag{2.18}$$

[10] Here, and throughout, we are using a finer partition of the time interval in between the two (discrete) time points t and $t+1$, so that agents can select u_t^i, $i \in \mathcal{N}$, in a (partially) sequential order.

where x_1 and $w^0 := \{w_1^0, \ldots, w_T^0\}$ are random exogenous variables with given probability distributions (with the latter being the *system noise*); furthermore, x_{t+1} takes values in a given standard Borel space $\mathbb{X}_{t+1}$. We further introduce the equation

$$y_t^i = g_t^i(x_t, u_{t-1}^1, \ldots, u_{t-1}^N; w_t^i), \qquad i \in \mathcal{N}, \qquad t \in \mathcal{T}, \tag{2.19}$$

where y_t^i is again the measurement of **A**i at stage t, and $\{w_t^i, t \in \mathcal{T}\}$ is known as the measurement noise of **A**i, which has a given probability distribution, possibly different for different agents as to be elucidated below. Note that for $t = 1$, g_t^i would have as argument only x_t and w_t, as controls have not yet been applied. Further note that, the state variable $x_t, t \in \mathcal{T}$, can be eliminated (for $t > 1$) in (2.19) by recursive substitution, so that y_t^i can be expressed solely in terms of the action variables and the primitive random variables:

$$y_t^i = \tilde{g}_t^i(x_1, w_{[1,t-1]}^0, \mathbf{u}_{[1,t-1]}, w_t^i). \tag{2.20}$$

Here, the primitive random variables, or the "states of nature", are

$$\xi := \{x_1, w_{[1,T]}^0, \mathbf{w}_{[1,T]}\}, \tag{2.21}$$

where $\mathbf{w}_{[1,T]}$ is defined similar to $\mathbf{u}_{[1,T]}$, with w replacing u.

The equations above do not completely describe the dynamic decision process, because the construction of the controls and the allowable dependence of the controls on the past measurements and/or actions have not yet been specified. Toward this end, let $\tilde{y}_t^i$ denote some pre-specified subset of the collection $\{\mathbf{y}_{[1,t]}, \mathbf{u}_{[1,t-1]}\}$, possibly a different subcollection for different $i \in \mathcal{N}$. Note that it is possible to find an η_t^i, so that

$$\tilde{y}_t^i = \eta_t^i(\xi; \mathbf{u}_{[1,t-1]}), \quad t \in \mathcal{T}, i \in \mathcal{N},$$

where ξ was introduced earlier by (2.21). The $\tilde{y}_t^i$ defined above can definitely be viewed as a (high-dimensional) vector, and it is precisely the information variable (2.14) where we have not used *tilde* on y simply not to clutter the notation. But the distinction should be clear from context. In static decision problems, there is, of course, no difference between the information and measurement variables, and indeed in Sect. 2.2 we have called y^i as both measurement variable and information variable. In dynamic problems, however, there is a distinction between the two.

The collection of individual information functions $\{\eta_t^i, t \in \mathcal{T}, i \in \mathcal{N}\}$ constitutes the *information structure* of the dynamic decision problem. Perhaps by a slight abuse of notation and terminology, we introduce, for each $t \in \mathcal{T}$ and $i \in \mathcal{N}$, a finite set $\mathcal{I}_t^i$ which specifies which elements of the set of vectors $\{\mathbf{y}_{[1,t]}, \mathbf{u}_{[1,t-1]}\}$ will be used in the construction of the control u_t^i, and we call the collection $\mathcal{I} := \{\mathcal{I}_t^i, i \in \mathcal{N}, t \in \mathcal{T}\}$ again as the *information structure* of the decision problem.

We list below some important information structures which will be used later in the book.

Definition 2.3.1

(1) *Sole prior information* (SPI): **A**i is said to have *SP* information if she makes no measurements and the only information she works with is the prior probability distribution on the random variables. A decision problem has *SP* information if all agents have *SP* information.
(2) *Open-loop* (*OL*) *information*: **A**i is said to have *OL* information if $\mathcal{I}_t^i = \mathcal{I}_1^i$ for all $t \in \mathcal{T}$. A decision problem has *OL information structure* if all agents have OL information (which are not necessarily the same). Note that OL information is different from SPI.
(3) *Complete information sharing* (*CIS*):

$$\mathcal{I}_t^i = \{\mathbf{y}_{[1,t]}, \mathbf{u}_{[1,t-1]}\}, \quad i \in \mathcal{N}, t \in \mathcal{T}.$$

Here there is a complete exchange of present and past measurements as well as past actions.
(4) *Complete measurement sharing* (*CMS*):

$$\mathcal{I}_t^i = \{\mathbf{y}_{[1,t]}\}, \quad i \in \mathcal{N}.$$

Here the past actions are not shared.
(5) n*-step delayed information sharing* (n*DIS*):

$$\mathcal{I}_t^i = \begin{cases} \{y^i_{[t-n+1,t]}, \mathbf{y}_{[1,t-n]}, \mathbf{u}_{[1,t-n]}\}, & t > n \\ \{y^i_{[1,t]}\}, & t \leq n \end{cases} \qquad i \in \mathcal{N}.$$

(6) n*-step delayed measurement sharing* (n*DMS*):

$$\mathcal{I}_t^i = \begin{cases} \{y^i_{[t-n+1,t]}, \mathbf{y}_{[1,t-n]}\}, & t > n \\ \{y^i_{[1,t]}\}, & t \leq n \end{cases} \qquad i \in \mathcal{N}.$$

(7) n*-step delayed control sharing* (n*DCS*):

$$\mathcal{I}_t^i = \begin{cases} \{y^i_{[1,t]}, \mathbf{u}_{[1,t-n]}\}, & t > n \\ \{y^i_{[1,t]}\}, & t \leq n \end{cases} \qquad i \in \mathcal{N}.$$

(8) k*-step periodic information sharing* (k*PIS*):

$$\mathcal{I}_t^i = \begin{cases} \{y^i_{[\lfloor t/k\rfloor k,t]}, \mathbf{y}_{[1,\lfloor t/k\rfloor k]}, \mathbf{u}_{[1,\lfloor t/k\rfloor k]}\}, & t \geq k \\ \{y^i_{[1,t]}\}, & t < k \end{cases} \qquad i \in \mathcal{N}.$$

(9) *Completely decentralized information* (*CDI*):

$$\mathcal{I}_t^i = \{y^i_{[1,t]}\}, \quad i \in \mathcal{N}, t \in \mathcal{T}.$$

Note that this corresponds to *nDMS* with $n = T$.

All the information structures given above are of the *perfect recall* (*PR*) type, in the sense that the agents have full memory of their information in the past. An example of an information structure (*IS*) which is not of the *PR* type is

$$\mathcal{I}_t^i = \{y^i_t\}, \quad i \in \mathcal{N}, t \in \mathcal{T}.$$

Stochastic decision problems whose *IS* s are not of the *PR* type are relatively more difficult to analyze than those with *PR* type *IS*, as we shall see later. Another class of challenging decision problems are those with so-called *nonclassical ISs*. Under such *IS* s an agent sees the action variable of another agent in her information set, or her information is indirectly affected by it, but she does not have access to the measurements/information based on which that action was taken; *nDCS* introduced above may be one such *IS* (when $n \geq 2$), so could *n DMS*, *CDI*, or information structures which are not *PR*. We will say more on such nonclassical *IS*s in the next chapter. Furthermore, we will observe that not all non-classical information structures lead to computational difficulties.

Now, fixing the *IS* of a dynamic decision problem also fixes the strategy (policy) spaces of the agents, as in (2.16). To complete the description as a team problem, we have to specify the cost structure, which we do as follows:

Adopting the description (2.18)–(2.19), also known as the *state-space model*, we associate with the team the loss function[11]

$$L(x_{[1,T+1]}, \mathbf{u}_{[1,T]}) = \sum_{t \in \mathcal{T}} c_t(x_{t+1}, \mathbf{u}_t) \tag{2.22}$$

where each term in the summation is known as the *incremental* (*stagewise*) *loss*. Since $x_{[1,t+1]}$ can be expressed in terms of the primitive random variables and the action variables, replacing u_t^i in (2.22) by $\gamma_t^i(y_t^i)$, $\gamma_t^i \in \Gamma_t^i$, constructed under the given IS, L becomes a function of only ξ (for each fixed $\underline{\gamma} \in \mathbf{\Gamma}$), whose expectation with respect to the subjective probability distribution function of ξ leads as in (2.2) to the cost function:

[11] Here, an alternative form can be $L(x_{[1,T+1]}, \mathbf{u}_{[1,T]}) = \sum_{t \in \mathcal{T}} c_t(x_t, \mathbf{u}_t) + c_{T+1}(x_{T+1})$.

$$J(\underline{\gamma}) = E_{\xi}[L(\xi, \underline{\gamma}(\underline{\eta}[\xi]))]. \tag{2.23}$$

Here L is given by (2.22), with the intermediate variables eliminated by using (2.18) and (2.19).

The function J, along with the product strategy space $\mathbf{\Gamma}$, constitutes the *normal* (*strategic*) form of the dynamic decision problem, and as such is no different (in abstract form) from the normal form introduced in Sect. 2.2 for static multiple person decision problems. Hence, all the solution concepts introduced there, viz. *team-optimality* and *person-by-person optimality* (or *Nash equilibrium*), are equally valid (and relevant) here, which we do not give to avoid repetition. In addition, however, some new features emerge due to the dynamic nature of the information pattern, which use particularly the sequential (extensive form) description of the decision problem. We introduce below two such general features associated with the team-optimal or *pbp* optimal solutions of dynamic team problems.

Definition 2.3.2 Let $D := \{J, \mathbf{\Gamma}, \mathcal{T}\}$ be a dynamic team problem which admits a solution $\underline{\gamma}^* \in \mathbf{\Gamma}$. Let $t > 1$ be an arbitrary point in $\mathcal{T}$, and consider the decision problem $D^{\beta}_{[t,T]}$ which is derived from D by setting $\underline{\gamma}_{[1,t-1]} = \underline{\beta}_{[1,t-1]}$, for an arbitrary $\underline{\beta}_{[1,t-1]} \in \mathbf{\Gamma}_{[1,t-1]}$. Then:

(i) The solution $\underline{\gamma}^* \in \mathbf{\Gamma}$ is *strongly time consistent* (STC) if the subpolicy $\underline{\gamma}^*_{[t,T]}$ constitutes a *solution* to the dynamic team $D^{\beta}_{[t,T]}$, this being so for every $t \in \mathcal{T}, t > 1$, and every permissible $\underline{\beta}_{[1,t-1]} \in \mathbf{\Gamma}_{[1,t-1]}$.
(ii) The solution $\underline{\gamma}^* \in \mathbf{\Gamma}$ is *weakly time consistent* (WTC) if the subpolicy $\underline{\gamma}^*_{[t,T]}$ constitutes a *solution* to the dynamic team $D^{\beta}_{[t,T]}$ when $\underline{\beta}_{[1,t-1]} = \underline{\gamma}^*_{[1,t-1]}$. $\diamond$

Remark 2.3.1 Time-consistency will be re-visited later in the context of game theory. To make this concept easier to interpret, let us consider a partially observed linear Gaussian model with additive control. It is well known that the optimal solution satisfies a separation (of estimation and control) principle, with a Kalman Filter generated conditional mean of the state replacing the state in the linear control policy of the deterministic problem, and further that the Kalman Filter recursions can be run independently of the control policy if the past control actions are available. Accordingly, if the actions are available to the controller at future time stages, one can run the Kalman filter independently of the past policies. On the other hand, if the past actions are not accessible to the controller, the controller can run the Kalman filter only if it knows what the past control policies were. Thus, we conclude that the Kalman filter based optimal controller design is only weakly time consistent if the controller has access to the past measurements; but the optimal controller is strongly time consistent if the actions are also made available. This simple discussion will serve to make the distinction between WTC and STC solutions more explicit for general information structures.

Note that if an equilibrium solution is *STC*, then the past actions do not rein in the present and future actions of the agents under the same solution concept, i.e., the agents have no reason to renege (and deviate from the equilibrium policy or the course of action) even if some inadvertent deviations have taken place in the past. With the WTC solution, however, there is no incentive to renege only if the declared course of action has been followed in the past.

The Intrinsic Model and the Markov Transition Model

Before concluding this section, we should mention that heretofore in our formulation of a dynamic team decision problem, we have allowed an agent to act multiple times, at different time instants, using possibly different information, that is **A**i has $u^i_{[1,T]}$ as her action variable. An alternative (but equivalent) formulation would be to have an agent **A**i be split into T agents, with **A**$i(t)$ (t'th agent in this split, where $t = 1, \ldots, T$) controlling only u^i_t. This would then transform the original N-agent team to an NT-agent team problem, but other than a difference in semantics the two formulations are essentially the same. More details on these different viewpoints to dynamic decision problems can be found in [774, 777, 780] (see Sect. 3.2, where Witsenhausen's *intrinsic model*, as well as other models for dynamic teams, are reviewed).

Another point worth mentioning is that an alternative to the state space model (2.18)–(2.19) exists by properly defining *controlled Markov kernels* for decentralized stochastic control. We will be discussing such formulations in detail in Chap. 5.

2.3.2 An Example of a Finite Dynamic Team

To illustrate some salient aspects of the formulation of dynamic decision problems and to guide the reader on some of the chapters to follow, in the following we consider a finite dynamic team problem with two agents and two stages, and with the agents having the same subjective prior probabilities on the random variables. We will study the derivation of the team-optimal solution under several different *IS* s of the type introduced in the previous section. Since the underlying team is finite (with a finite probability space), a team-optimal solution will exist under all *IS* s.

Now, the description of the stochastic dynamic team follows: At each stage, the control (decision) spaces of the agents have two elements, as in the static team of Sect. 2.2.2.

$$U^1_1 = U^1_2 = \{U(\text{up}), D(\text{down})\}, \quad U^2_1 = U^2_2 = \{L(\text{left},)R(\text{right})\}.$$

The initial state, x_1, is a discrete random variable, taking two values, x_{11} and x_{12}, with respective probabilities 0.4 and 0.6. If $x_1 = x_{1i}$ and $u^1_1 = U$ or D, and $u^2_1 = L$ or R, the loss to the team (i.e., the stagewise loss, $c_1(x_1, u^1_1, u^2_1)$) is given by the "loss matrices"

	A2			A2	
	L	R		L	R
A1 U	1	0	U	1	2
D	3	1	D	0	2
	$x_1 = x_{11}$			$x_1 = x_{12}$	

We denote the first matrix above by LM1 and the second matrix by LM2. The transition to the second stage and the associated cost is now described as follows: Let x_2 take three distinct values, x_{21}, x_{22} and x_{23}, with the rule

$$x_2 = \begin{cases} x_{21}, \text{ if } (u_1^1, u_1^2) = (U, L) \text{ and } x_1 = x_{11} \\ x_{22}, \text{ if } (u_1^1, u_1^2) = (U, L) \text{ and } x_1 = x_{12} \\ x_{23}, \text{ otherwise.} \end{cases}$$

The corresponding cost is determined by x_2, u_2^1, u_2^2, and an independent random variable, in terms of the loss matrices LM1 and LM2 at stage 1, according to the following table.

x_2	loss at stage 2 (c_2)
x_{21}	LM1 $w.p.0.4$, LM2 $w.p.0.6$
x_{22}	LM1 $w.p.0.5$, LM2 $w.p.0.5$
x_{23}	LM1 $w.p.0.3$, LM2 $w.p.0.7$

All random mechanisms are assumed to be independent and the total loss to the team is the arithmetic sum of the stagewise losses, as in (2.22). Note that here we basically have two random variables, x_1 and w_2, say, with the probability distribution of the latter governing the loss structure in the table above, i.e.,

$$\text{Prob}(w_2 = w_{21} \mid x_2) = 1 - \text{Prob}\,(w_2 = w_{22} \mid x_2) = \begin{cases} 0.4, \ x_2 = x_{21} \\ 0.5, \ x_2 = x_{22} \\ 0.3, \ x_2 = x_{23} \end{cases}$$

where w_{21} corresponds to LM1 and w_{22} to LM2. Now we specify the measurements available to the agents: It is assumed that **A**1 knows exactly the value of x_1 at stage 1, and does not make any further measurements (at stage 2). **A**2, on the other hand, makes no measurements at stage 1, but knows precisely the value of w_2 at stage 2; hence

$$y_1^1 = x_1, \quad y_2^1 : \text{void}; \quad y_1^2 : \text{void}, \quad y_2^2 = w_2.$$

For any given permissible policy, the (expected) cost to the team is given by (2.23), and with $\xi := (x_1, w_2)$. The precise form, of course, will depend on the *IS* to be adopted, as delineated below:

1. *Sole prior information* (*SPI*). This will lead to the worst performance for the team, because both agents work under only the given prior information. The policy spaces of the agents are

$$\Gamma^1 = \{Uu, Ud, Du, Dd\}; \quad \Gamma^2 = \{Ll, Lr, Rl, Rr\}$$

where we have used "lower case letters" for the second stage decisions. The only way to solve this team problem is to convert it to *normal form*, which is a 4×4 matrix

A2

		Ll	Lr	Rl	Rr
	Uu	2	2.08	2.2	1.8
A1	Ud	2.38	3.5	3.3	2.5
	Du	2.2	2.2	2.6	2.2
	Dd	3.3	2.9	3.7	3.2

The unique team-optimal solution is (Uu, Rr), with a cost level of $J_{SP} = 1.8$.

2. *Complete measurement sharing* (*CMS*). This is the other extreme case, when the agents know exactly what loss matrix is being optimized at each stage. Since the lowest entries of LM1 and LM2 are both *zero*, the optimal team cost is $J_{CMS} = 0$. The solution would have been the same if, instead, we had the *CIS IS*.
3. *1-step delayed measurement sharing* (*1DMS*). Here the information available to the agents at each stage are

$$\mathcal{I}_2^1 = \mathcal{I}_1^1 = \{x_1\}, \quad \mathcal{I}_1^2 = \emptyset, \quad \mathcal{I}_2^2 = \{x_1, w_2\},$$

and hence $|\Gamma_1^1| = |\Gamma_2^1| = 4$, $|\Gamma_1^2| = 2$, and $|\Gamma_2^2| = 8$. The cardinality of the composite policy spaces are $|\Gamma^1| = 8$,[12] $|\Gamma^2| = 16$, which means that the normal form would be an 8×16 dimensional matrix. One possible approach to the problem would be to compute the entries of this matrix and choose the smallest one as the solution. An alternative approach is a sequential derivation, which makes use of the fact that measurements are shared with a delay of *one* time unit, which is what we discuss below. In Chap. 5 we present systematic methods for arriving at dynamic programming recursions for more general setups of this type.

Suppose that the actions at stage 1 have been taken, and the agents are facing the decision problem (at the second stage) where the value of x_1 is now *common*

[12] This one is not 16 because for each value of x_1, **A**1 has 4 choices, which makes the total 8. Hence, one can replace $\Gamma^1 := \Gamma_1^1 \times \Gamma_2^1$ with a smaller set, without any loss in performance.

knowledge. **A**1 has no other information, and hence his possible actions (policies) are u and d. **A**2, on the other hand, has the additional information, the precise value of w_2, and hence he has *four* possible policies: ll, lr, rl and rr, where lr stands for $u_1^2 = l$, if $w_2 = w_{21}$, and $u_1^2 = r$, otherwise. Now, conditioned on the value of x_1 (which is common knowledge) and the actions taken by the agents at stage 1 (which can also be considered to be common knowledge since we have a (cooperative) team problem and the information based on which these actions were taken is common knowledge), we have the following eight *total cost matrices*

$x_1 = x_{11}$:

$(u_1^1, u_1^2) = (U, L)$

	ll	lr	rl	rr
u	2	2.6	1.6	2.2
d	2.2	3.4	1.4∗	2.6

$(u_1^1, u_1^2) = (U, R)$

	ll	lr	rl	rr
u	1	1.3	0.3∗	0.6
d	2.1	3.3	0.7	1.3

$(u_1^1, u_1^2) = (D, L)$

	ll	lr	rl	rr
u	4	4.3	3.3∗	3.6
d	5.1	6.3	3.7	4.3

$(u_1^1, u_1^2) = (D, R)$

	ll	lr	rl	rr
u	2	2.3	1.3∗	1.6
d	3.1	4.3	1.7	2.3

$x_1 = x_{12}$:

$(u_1^1, u_1^2) = (U, L)$

	ll	lr	rl	rr
u	2	2.5	1.5∗	2
d	2.5	3.5	1.5∗	2.5

$(u_1^1, u_1^2) = (U, R)$

	ll	lr	rl	rr
u	3	3.3	2.3∗	2.6
d	4.1	5.3	2.7	3.3

$(u_1^1, u_1^2) = (D, L)$

	ll	lr	rl	rr
u	1	1.3	0.3∗	0.6
d	2.1	3.3	0.7	1.3

$(u_1^1, u_1^2) = (D, R)$

	ll	lr	rl	rr
u	3	3.3	2.3∗	2.6
d	4.1	5.3	2.7	3.3

In each case, the "starred" entry(ies) denote the minimum entries of the corresponding matrices, which will be carried over to the *first* stage to determine the optimal policies there. Now, at stage 1 **A**1 knows the value of x_1, but **A**2 does not, so that possible policies for **A**1 are UU, UD, DU, DD, while the permissible policies for **A**2 are L and R. Using the optimum entries above, we can construct an equivalent cost matrix at stage 1 through an appropriate averaging process:

A2

		L	R
	UU	1.46	1.5
A1	UD	0.74∗	1.5
	DU	2.22	1.9
	DD	1.5	1.9

This can be viewed as the *optimum cost-to-go matrix* at stage 1, because of the following interpretation that the entries admit. Consider, for example, the entry with the numerical value 2.22: If **A**1 chooses D when $x_1 = x_{11}$ and U when $x_1 = x_{12}$, and **A**2 chooses L, all at stage 1, then whatever choices are made at stage 2 (under the given information) the total (expected) cost can never be lower than 2.22. To arrive at this numerical value, we note that if $x_1 = x_{11}$, $u_1^1 = D$ and $u_1^2 = L$, the cost resulting from an optimum choice of policies at stage 2 would be 3.3 (the lowest entry of the *third* conditional total cost matrix), whereas if $x_1 = x_{12}$, $u_1^1 = U$, $u_1^2 = L$, the optimum total cost would be 1.5 (the lowest entry of the *fifth* matrix). Since the value of x_1 is not available to **A**2 at stage 1, we average these values of x_1, to obtain

$$(0.4)(3.3) + (0.6)(1.5) = 2.22.$$

All other entries of the 4×2 *cost-to-go matrix* can be computed analogously.

Clearly, the minimum cost is $J_{1DMS} = 0.74$, with the unique team-optimal policy being

$$(\gamma_1^{1*}(x_1), \gamma_2^{1*}(x_1)) = \begin{cases} Ud, & x_1 = x_{11} \\ Du, & x_1 = x_{12} \end{cases}$$

$$\gamma_1^{2*} = L, \quad \gamma_2^{2*}(x_1, w_2) = \begin{cases} r, & w_2 = w_{21} \\ l, & w_1 = w_{22} \end{cases}$$

Note that γ_2^{2*} is actually independent of the value of x_1 (which means that even if the measurement y_1^1 had not been shared, the team-optimal solution would still be the same), and that γ_2^{1*} does depend on x_1 (which means that if **A**1 were not allowed to recall the value of x_1 at stage 2, the optimal team cost would have been higher—see the next case).

4. *No sharing, no recall* (*NSR*). Here we have

$$\mathcal{I}_1^1 = \{x_1\}, \mathcal{I}_2^1 = \varphi = \mathcal{I}_1^2, \quad \mathcal{I}_2^2 = \{w_2\}.$$

If we had allowed perfect recall (i.e., $\mathcal{I}_2^1 = \mathcal{I}_1^1$), then the solution would be the one obtained in case 3, as discussed there. However, without perfect recall the solution does not follow from the one in case 3. This information structure is *nonclassical* and hence a simple recursive derivation as in *case 3* is also not possible (more on this will be presented in the following chapters). The only possibility is to construct the normal form for the team, which is characterized by the 8×8 matrix given below.

A2

		Lll	Llr	Lrl	Lrr	Rll	Rlr	Rrl	Rrr
	UUu	2	2.54	1.54	2.08	2.2	2.5	1.5	1.8
	UDu	1.4	1.82	0.82∗	1.24	2.2	2.5	1.5	1.8
	DUu	2.8	3.22	2.22	2.64	2.6	2.9	1.9	2.2
A1	DDu	2.2	2.5	1.5	1.8	2.6	2.9	1.9	2.2
	UUd	2.38	3.46	1.46	2.54	2.86	3.44	1.9	2.5
	UDd	2.14	3.34	0.98	1.82	3.3	4.5	1.9	2.5
	DUd	3.54	4.62	2.38	3.22	3.7	4.9	2.3	2.9
	DDd	3.3	4.5	1.9	2.50	3.7	4.9	2.3	2.9

The unique solution is (UDu, Lrl), with a cost level of $J_{NSR} = 0.82$. Note that UDu stands for

$$\gamma_1^{1*}(x_1) = \begin{cases} U, \; x_1 = x_{11} \\ D, \; x_1 = x_{12} \end{cases}, \quad \gamma_2^{1*} = u,$$

and Lrl denotes the policy

$$\gamma_1^{2*} = L, \quad \gamma_2^{2*}(w_2) \begin{cases} r, \; w_1 = w_{21} \\ l, \; w_1 = w_{22} \end{cases}$$

5. *Open-loop information* (*OLI*). Here the agents use only the measurements they have obtained at stage 1, which is x_1 for **A**1 and no measurement for **A**2. The normal form here is, in fact, a submatrix that can be obtained from the 8×8 matrix of *case 4*. For **A**1 the permissible policies are still the same. For **A**2, however, the permissible ones are Ll, Lr, Rl, Rr, which correspond in *case 4* to Lll, Lrr, Rll, Rrr. Hence, we retain only the *first*, *fourth*, *fifth* and *eighth* columns of the matrix of *case 4*, and the result is the unique team-optimal solution (UDu, Lr), with a cost level of $J_{OL} = 1.24$. We should note in passing that the normal form of *case 1* can also be recovered from the normal form of *case 4*, this time by also eliminating the *second*, *third*, *sixth* and *seventh* rows of the matrix of *case 4*.

Cost Comparisons

Clearly, more information to any one agent in a team will never result in higher optimal team cost. In the latter case we say that the extra information is *useful*. In the context of this specific example, we have the optimum cost comparisons

$$J_{SP} = 1.8 > J_{OL} = 1.24 > J_{NSR} = 0.82 > J_{1DMS} = J_{1DIS} = 0.74$$
$$> J_{CMS} = J_{CIS} = 0$$

The equalities $J_{1DIS} = J_{1DMS}$, $J_{CMS} = J_{CIS}$ hold not only for this specific example, but for general stochastic team problems.

The analysis above also informs us that, due to analytical and computational challenges, one should be methodical in arriving at optimal or near optimal solutions for stochastic dynamic team problems. We will be studying such methods in the following few chapters.

2.4 Team-Optimal Solutions for Static Teams

We present, in this section, a theory for static N-person stochastic teams, by focusing on fundamental issues such as existence, uniqueness and derivation of team-optimal solutions, establishing conditions under which person-by-person (*pbp*) optimal solutions are also team-optimal, and studying the relationships between achievable optimal team costs and information structures in static teams.

Using the terminology and notation introduced in Sect. 2.2, we represent a general team by the $(N + 1)$-tuple $\{J; \Gamma^i, i \in \mathcal{N}\}$, where the cost J is derived from a loss function, using a probability measure P on the states of nature, common to all agents. We consider here the cases where the action spaces $(\mathbb{U}^i, i \in \mathcal{N})$ are either finite or infinite but finite dimensional, and for the latter class we also include the possibility that some hard constraints may be imposed on the decision variables, in which case the action constraint sets $(S^i, i \in \mathcal{N})$ are taken as appropriate closed subsets of the corresponding action spaces.

In Sect. 2.4.1, we consider the class of teams which are either finite or have finite measurement spaces for all agents. For this class, we provide general existence and uniqueness results for team-optimality, and discuss the relationship with *pbp* optimality and the notion of *stationarity* (which is to be defined shortly). In Sect. 2.4.2, we extend this study to teams where the measurement spaces are infinite (but finite dimensional), and develop conditions under which stationarity implies team-optimality. Sections 2.4.3 and 2.4.4 discuss two special, but important, classes of teams: (i) those with quadratic loss functions, first under general probability distributions and then under the Gaussian distribution, and (ii) static teams with exponentiated quadratic loss functions. We also discuss recursive algorithms for the computation of the team-optimal solution in each case.

2.4.1 Teams with Finite Measurement Spaces

We have already identified, in Sect. 2.2.2, one class of team problems for which an optimal solution always exists; namely, *static finite teams, i.e.* teams where both the action and the measurement spaces are finite, or equivalently, the product strategy space $(\boldsymbol{\Gamma})$ is finite. This conclusion would also be valid for dynamic teams where the product strategy space ($\mathbf{\Gamma}$) is finite, since we would be doing a comparison among only a finite number of choices. Two other observations we made in Sect. 2.2.2 were that multiple team-optimal solutions are not necessarily interchangeable (respecting the order), and that a *pbp* optimal solution is not necessarily team-optimal. These two facts are naturally valid not only for finite teams but for infinite teams as well, and further, also for dynamic teams in normal (strategic) form.

If a team problem is not finite, then one has to bring additional structure into the formulation in order to guarantee the existence, as well as the uniqueness, of the solution. Viewing a stochastic team in normal form as one of minimization[13] of a functional, J, over a set, $\mathbf{\Gamma}$, an optimum may fail to exist (in infinite teams) for basically one of two reasons:

(i) The (cost) functional J is unbounded below.
(ii) There exists an infimizing sequence in $\mathbf{\Gamma}$, without any limit in $\mathbf{\Gamma}$.

The former basically says that J^*, defined by the *RHS* of (2.4), is $-\infty$, implying that a sequence can be found in $\mathbf{\Gamma}$ which makes the value of J arbitrarily small (negative). The way to avoid this difficulty is to formulate, from the beginning, a well-defined team problem whose cost is bounded away from $-\infty$. The latter reason, however, cannot be dispensed with that easily since it places some nontrivial restrictions on the topology of the product policy space $\mathbf{\Gamma}$ as well as on the structure of J. In this case J^*, defined by the *RHS* of (2.4), is a finite quantity, but one may only achieve values arbitrarily close to (but larger than) J^*, and never equal to it. This could arise if, for example, the function J has some discontinuities on $\mathbf{\Gamma}$, or $\mathbf{\Gamma}$ is not a *closed* set so that the infimizing sequence may not have a limit in $\mathbf{\Gamma}$. The most general condition that ensures that these two things do not happen is the celebrated *Weierstrass Theorem* (see Appendix A), rephrased below using the team framework, without proof, as more general existence results will be presented and proved in Chap. 4.

Theorem 2.4.1 *The team problem* $\{J; \Gamma^i, i \in \mathcal{N}\}$ *admits a team-optimal solution if the product policy space* $\mathbf{\Gamma}$ *is a compact subset of a metric space, and the cost function* J *is lower semicontinuous (lsc) on* $\mathbf{\Gamma}$ *(with respect to the given metric).*

As one useful application of the above result, consider the class of stochastic team problems which satisfy the following four hypotheses:

$(c.1)$ Each action constraint set S^i ($i \in \mathcal{N}$) is a closed and bounded subset of the action space $\mathbb{U}^i$ ($i \in \mathcal{N}$) which is itself a finite dimensional vector space;
$(c.2)$ $L(\xi; u^1, \ldots, u^N)$ is *almost surely* (*a.s.*) jointly *lsc* in $(u^1, \ldots, u^N) =: \mathbf{u}$, on $\mathbf{U} := \mathbb{U}^1 \times \cdots \times \mathbb{U}^N$;
$(c.3)$ Each measurement set $\mathbb{Y}^i$ ($i \in \mathcal{N}$) is finite, with no element receiving *zero* probability from the probability measure P; or equivalently, for each $i \in \mathcal{N}$, the partition set $\mathbf{Y}^i$ has a finite number of elements, with each element receiving positive probability from P;

[13] For some background material on the optimization of functionals, see Sect. A.5 of Appendix A.

$(c.4)$ $L(\xi; u^1, \dots, u^N)$ is bounded from below, and $E_{\xi|y^i}[L(\xi; u^1, \dots, u^N)]$ is finite for every $y^i \in \mathbb{Y}^i$, $u^j \in \mathbb{U}^j$, $i, j \in \mathcal{N}$.

Then, we have the following theorem (which will be generalized later in Theorem 4.2.4):

Theorem 2.4.2 *For an N-agent static stochastic team problem satisfying $(c.1)$–$(c.4)$ above, there exists at least one team-optimal solution.*

Proof The result follows from Theorem 2.4.1, once we observe that, under the given specifications, the normal form has a *lsc* cost function J on a compact policy space $\boldsymbol{\Gamma}$. We first show the latter, which is equivalent to showing that, for each $i \in \mathcal{N}$, Γ^i is a closed and bounded subset of a finite dimensional space. Towards this end, let $\mathbb{Y}^i$ be generated (without any loss of generality) by the n_i-tuple $\{y_1^i, \dots, y_{n_i}^i\}$, where $n_i := |\mathbb{Y}^i|$ is finite by $(c.3)$. Then, every permissible strategy γ^i for $\mathbf{A}i$ (i.e. every element of Γ^i) can be written as

$$\gamma^i(y^i) = u_j^i, \qquad \text{if} \quad y^i = y_j^i, \quad j = 1, \dots, n_i$$

where each u_j^i lies in S^i. Hence, each strategy can be viewed as an n_i-tuple of vectors $(u_1^i, \dots, u_{n_i}^i)$ belonging to

$$\mathbf{S}^i := \underbrace{S^i \times \cdots \times S^i}_{n_i \text{times}} \subset \underbrace{\mathbb{U}^i \times \cdots \times \mathbb{U}^i}_{n_i \text{times}} =: \mathbb{U}^i$$

which makes Γ^i isomorphic to $\mathbf{S}^i$ which is closed and bounded, and finite dimensional, since it is a finite product of S^i which itself is closed and bounded (by $(c.1)$).

We now show that J is *lsc* on $\mathbf{U} := \mathbb{U}^1 \times \cdots \times \mathbb{U}^N$. To obtain a description of J on $\mathbb{U}$, let us first introduce the notation $\mathbf{y}$ to denote an N-tuple of scalars

$$\mathbf{y} := (y_{t_1}^1, \dots, y_{t_N}^N), \quad t_i \in \{1, \dots, n_i\}, i \in \mathcal{N}$$

where $y_{t_i}^i$ denotes one possible (generic) measurement of $\mathbf{A}i$. By a possible abuse of terminology, we will consider $\mathbf{y}$ as a random quantity, which has $\mathbf{N} := \prod_{i \in \mathcal{N}} n_i$ different realizations. Then, we have the following sequence of equalities:

$$\begin{aligned} J(\underline{\gamma}) &= E_\xi[L(\xi; \gamma^1(\eta^1(\xi)), \dots, \gamma^N(\eta^N(\xi)))] \\ &= E_{\mathbf{y}}[\underbrace{E_{\xi|\mathbf{y}} L(\xi; u_{t_1}^1, \dots, u_{t_N}^N)}_{L\text{av}(\mathbf{y}; u_{t_1}^1, \dots, u_{t_N}^N)}] \qquad (\star) \\ &\equiv \sum_{t; t_i \in \{1, \dots, n_i\}, i \in \mathcal{N}} L\text{av}(\mathbf{y}; u_{t_1}^1, \dots, u_{t_N}^N) Prob(\mathbf{y}) \end{aligned}$$

where the second line follows from the "iterated property" of conditional expectations (see Appendix B) and the adopted convention that $\gamma^i(y^i) = u^i_{t_i}$, when $y^i = y^i_{t_i}$.

Now, under $(c.2)$ and $(c.4)$, $L_{\text{av}}(\mathbf{y}_t; u^1_{t_1}, \ldots, u^N_{t_N})$ is a *lsc* function on $\mathbf{u}$ for each $\mathbf{y}_t$, since it is the integral of a *lsc* function (L) under the conditional measure $Prob(\xi|\mathbf{y})$, as a result of an application of Fatou's lemma (see Theorem B.1.5) by $(c.4)$. In view of this, the last line of $(\star)$ (which is a finite weighted sum of individual *lsc* functions) provides a representation for J on U, which is *lsc*. This then completes the proof of the Theorem. It is worth noting at this point that the result would be true even if condition $(c.2)$ is relaxed somewhat, requiring instead that the function $L_{\text{av}}(\mathbf{y}; \mathbf{u})$ be *lsc* on $\mathbf{u}$, where $L_{\text{av}}(\mathbf{y}; \mathbf{u}) := E_{\xi|\mathbf{y},\mathbf{u}} L(\xi; \mathbf{u})$. ◇

For the general result of Theorem 2.4.2 to be valid, conditions $(c.2)$ and $(c.3)$ cannot be relaxed any further, because the relaxation of $(c.2)$ (with the provision above) would lead to violation of the *lsc* part of Theorem 2.4.1, and the relaxation of $(c.3)$ (meaning that some of the $\mathbb{Y}^i$'s might be infinite sets) leads to policy spaces that are no longer finite dimensional, in which case $(c.1)$ does not imply the compactness of $\mathbf{\Gamma}$. Relaxation of $(c.4)$, on the other hand, would lead to a J that is not necessarily *lsc*. Of course, this condition is automatically satisfied if the underlying probability space is finite. The only condition that can be genuinely relaxed, without affecting the result of the theorem, is $(c.1)$, for which we have the following slightly less restrictive condition:

$(c.1')$ Let $\mathcal{N}_h$ and $\mathcal{N}_s$ be two complementary subsets of $\mathcal{N}$ (*i.e.* $\mathcal{N}_h \cup \mathcal{N}_s = \mathcal{N}$, and $\mathcal{N}_h \cap \mathcal{N}_s = \emptyset$) such that S^i is compact for all $i \in \mathcal{N}_h$, and $S^j \equiv \mathbb{U}^j$ for all $j \in N_s$. Assume that $\sum_{j \in N_s} |u^j| \to \infty$, $L(\xi; u^1, \ldots, u^N) \to \infty$ *a.s.*, for every fixed $u^i \in S^i, i \in N_h$.

Given $(c.4)$, by an application of Fatou's lemma, it can be seen that this condition ensures that u^j, $j \in \mathcal{N}_s$, can be restricted to a (possibly sufficiently large) compact set, thus making the result of Theorem 2.4.2 still valid. Hence we have

Corollary 2.4.1 *An N-agent static stochastic team problem satisfying* $(c.1')$, $(c.2)$, $(c.3)$ *and* $(c.4)$ *admits at least one team-optimal solution.*

Existence results in full generality will be studied in Chap. 4.

Uniqueness

It may be important to determine conditions under which a team-optimal solution is unique, since as we have discussed earlier in Sect. 2.2.2, multiple optima may lead to an inferior outcome if the agents do not have a consistent protocol to resolve the dilemma. Once the existence of an optimum has been established, there would be two ways to verify uniqueness of the solution. One would be to write down a set of necessary conditions to be satisfied by the team-optimal solution and show that these conditions admit at most one solution—as to be discussed later in this subsection. A second way to verify unicity would be to use the normal form for the team, and show strict convexity[14] of J over the product policy space $\mathbf{\Gamma}$, which has

[14] See Appendix A, Sect. A.4, for a definition.

to be a convex set. The following theorem now does precisely that, by relating the (strict) convexity of L to the convexity of J under the hypotheses of Theorem 2.4.2.

Theorem 2.4.3 *In addition to the four hypotheses of Theorem 2.4.2 or of Corollary 2.4.1, let S^i be a convex set for each $i \in \mathcal{N}$, and $L(\xi; \cdot)$ be strictly convex on* $\mathbf{U}$ *a.s.*[15] *Then, the stochastic team problem admits a unique team-optimal solution.*

Proof of Theorem 2.4.3 First note that for $\mathbb{X}$ a finite dimensional vector space and I a finite index set, if f_i, $i \in I$, is a convex (respectively, strictly convex) functional defined on $\mathbb{X}$, then the functional f: $f = \sum_{i \in I} f_i$ is also convex (respectively, strictly convex) on $\mathbb{X}$. Now, the construction given for J, in the proof of Theorem 2.4.2, satisfies the hypotheses of the preceding result with $f_t(\cdot) = L_{\text{av}}(\mathbf{y}_t; \cdot) Prob(\mathbf{y}_t)$, since for each $\mathbf{y}_t$, $L_{\text{av}}(\mathbf{y}_t; \cdot)$ is strictly convex on $\mathbf{u}$ (being the conditional average of an *a.s.* strictly convex functional), and every $u^i_{t_i}$, $t_i \in \{1, \ldots, n_i\}, i \in \mathcal{N}$, appears in at least one of the additive terms in the representation ($\star$) for J. Note, in passing, that $Prob(\mathbf{y}_t)$ may not be positive for every possible N-tuple $(t_1, \ldots, t_N)$, $t_i \in \{1, \ldots, n_i\}, i \in \mathcal{N}$, but for any $j \in \mathbf{N}$, and $t_j \in \{1, \ldots, n_j\}$, $y^j_{t_j}$ will receive positive probability in at least one such sequence, since otherwise this would imply that $Prob(y^j = y^j_{t_j}) = 0$, a contradiction to our initial hypothesis. ◇

The following example serves to illustrate some of the fine points of the results of Theorems 2.4.2 and 2.4.3, and the analyses that led to these results, including the construction ($\star$) used in the proof of Theorem 2.4.2.

Example 2.4.1 Let $N = 2$, $\Xi = \mathbb{U}^1 = \mathbb{U}^2 \equiv \mathbb{R}$, ξ be a (continuous) random variable *uniformly distributed* on the open interval $(0, 2)$, and the loss functional L be given by

$$L(\xi; u^1, u^2) = (u^1)^2 + (u^2)^2 + \xi u^1 u^2 - u^1 - 2u^2.$$

Suppose that **A**1 can tell (through his measurements, and with certainty) whether the realized value of ξ belongs to the open interval $(0, 1)$ or not, and **A**2 can similarly tell whether it belongs to the subinterval $(\frac{1}{2}, \frac{3}{2})$ or not. The question is whether this (static) stochastic team problem admits a team-optimal solution or not, and if it does, whether it is unique and how it can be computed.

Let us first check the conditions of Theorem 2.4.2 and Corollary 2.4.1. Clearly $(c.2)$ and $(c.3)$ are satisfied, where in the latter we choose $\mathbf{y}^1 = \{(0, 1), [1, 2)\}$, $\mathbf{y}^2 = \{(\frac{1}{2}, \frac{3}{2}), (0, \frac{1}{2}] \cup [\frac{3}{2}, 2)\}\}$. We associate the measurement y^i_1 with the first subinterval (in each corresponding partition) and the measurement y^i_2 with the complement (i.e. the second) set. Condition $(c.1)$ is not satisfied, but $(c.1')$ is (with $\mathcal{N}_s = \mathcal{N}$), because, for each $\xi \in (0, 2)$, the *Hessian matrix* of L (see Appendix A,

[15] A random function $L(\xi; \mathbf{u})$ is *a.s. strictly convex* in $\mathbf{u}$ if the set of ξ for which L is not strictly convex in $\mathbf{u}$ is of P_ξ-measure zero.

Sect. A.4),

$$\nabla^2 L(\xi, \mathbf{u}) = \begin{pmatrix} 2 & \xi \\ \xi & 2 \end{pmatrix},$$

is *positive definite* (*p.d.*), implying that $L(\xi, \mathbf{u}) \to \infty$ as $|u^1| + |u^2| \to \infty$, for every fixed $\xi \in (0, 2)$. Note that, even if the open interval $(0, 2)$ is replaced with the closed interval $[0, 2]$ (the distribution still being the same), $(c.1')$ would still be satisfied because, even though $\nabla^2 L(\xi, \mathbf{u})$ is no longer *p.d.* at $\xi = 2$ (in fact, then choosing $u^1 = -u^2$ and letting $u^2 \to \infty$, one can drive L to $-\infty$), the singleton event $\{\xi = 2\}$ receives *zero* probability under the given continuous distribution, and hence the condition holds in the *a.s.* sense. If, however, we had a probability distribution with a *jump* at the point $\xi = 2$, assigning, say, a weight of $\frac{1}{3}$ to that single value, then $(c.1')$ would have been violated.

Now, in the course of the discussion above, we have also established the validity of the two additional hypotheses of Theorem 2.4.3 (the first one trivially, and the second one because of the reason that the Hessian matrix of L is *p.d. a.s.*), from which it follows that the problem indeed admits a unique team-optimal solution. Further discussion on convexity will be presented in Sect. 3.7.

To obtain a characterization of the solution, let us first construct the normal form, following the steps outlined in the proof of Theorem 2.4.2. Noting that the summation in $(\star)$ has *four* terms, with $Prob((y_i^1, y_j^1)) = \frac{1}{4}, i, j = 1, 2$, some algebra leads to the expression

$$J(\underline{\gamma}) = \frac{1}{2}\sum_{i,j=1}^{2} (u_j^i)^2 - \frac{1}{2}\sum_{j=1}^{2} u_j^1 - \sum_{j=1}^{2} u_j^2 + \frac{1}{16}[u_1^1 u_2^2 + 3u_1^1 u_1^2 + 5u_2^1 u_1^2 + 7u_2^1 u_2^2]$$

which is to be minimized with respect to $(u_1^1, u_2^1, u_1^2, u_2^2)$ over $\mathbb{R}^4$. This is a strictly convex functional and is differentiable, which means that the unique solution should satisfy (uniquely) the *stationarity* conditions:

$$\partial J / \partial u_j^i = 0, \quad i, j = 1, 2.$$

These conditions reduce to the set of *four* linear equations:

$$16u_1^1 + 3u_1^2 + u_2^2 = 8, \qquad 16u_2^1 + 5u_1^2 + 7u_2^2 = 8$$

$$3u_1^1 + 5u_2^1 + 16u_1^2 = 16, \qquad u_1^1 + 7u_2^1 + 16u_2^2 = 16$$

which admits the unique solution:

$$u^{1*} := (u_1^{1*}, u_2^{1*}) \approx (0.231, -0.324)$$

$$u^{2*} := (u_1^{2*}, u_2^{2*}) \approx (1.0578, 1.127)$$

with the minimum value being

$$J^* \approx -1.142.$$

It would be instructive to compare this value for the team cost with what would have been achieved if the agents had not made any measurements (i.e. operated in an "open-loop" fashion with no measurements). In such a case, the normal form for the team would be given by the cost functional J_{OL} (where the subscript $_{\mathrm{OL}}$ stands for "open-loop")

$$J_{\mathrm{OL}} = (u^1)^2 + (u^2)^2 - u^1 - 2u^2 + u^1 u^2,$$

since $E[\xi] = 1$. A straightforward minimization of this quadratic (and strictly convex) functional leads to the unique solution

$$u^{1^*} = 0, u^{2^*} = 1 \quad \Rightarrow \quad J^*_{\mathrm{OL}} = -1.$$

Hence, as expected, we observe that the presence of the measurements (which bring the uncertainty in the true value of ξ to intervals of length 1, instead of the original interval of length 2) leads to an improvement. We will make the conclusion later that the former information structure is better than the latter, in a sense to be made precise via Blackwell's theory of comparison of information structures to be generalized to stochastic team problems in Chap. 7.

Another extreme case to consider would be the information structure that provides the agents with the "maximum" information regarding the true value of ξ, which is the measurement signal $y^i = \xi, i = 1, 2$ (i.e. perfect measurement), unless some restrictions are imposed on the information structure. Our results, so far, as embodied in Theorems 2.4.2 and 2.4.3 and Corollary 2.4.1 are not (strictly speaking) applicable to problems of this type, since the measurements belong to infinite sets—this is the topic of the next subsection and future chapters. However, because of the fact that both agents make perfect measurements here, the problem would be easy to analyze, since it is no different from a deterministic (quadratic) optimization problem. As such, this particular (team) problem is not well-defined, since the objective functional is not strictly convex at $\xi = 2$, meaning (in this case) that for $\xi = 2$ the loss functional can be driven to $-\infty$, which in turn implies (in that case) that under the uniform distribution on $(0, 2)$ the cost (average loss) can be made arbitrarily small (negative).[16] The message here is that not only the characterization, but also the existence of a solution in a stochastic team problem could very much depend on the underlying information structure. ◇

We have thus seen, in the preceding example, a constructive procedure for obtaining closed-form solutions to a stochastic static team problem with finite measurement spaces. The question now is whether there are other (alternative) ways of obtaining the solution and also of verifying team-optimality of a given candidate solution without going through the derivation. Such tools would be provided by the *necessary conditions* satisfied by a team-optimal solution, one of which is *person-by-person* (*pbp*) optimality.

Recalling Remark 2.2.1, a *pbp* solution $\underline{\gamma}^* \in \mathbf{\Gamma}$ for a static team problem $(J, \mathbf{\Gamma})$ would be given by

$$\min_{\beta \in \Gamma^i} J(\underline{\gamma}^{-i,*}, \beta) = J(\underline{\gamma}^*), \quad i \in \mathcal{N}, \tag{2.24}$$

which can equivalently be written as

$$\min_{u \in S^i} E_{\xi|y^i} L(\xi; \underline{\gamma}^{-i,*}(\mathbf{y}^{-i}), u) = E_{\xi|y^i} L(\xi; \underline{\gamma}^*(\mathbf{y})), \quad i \in \mathcal{N} \tag{2.25}$$

In other words, we have N separate optimization problems, one for each agent, and in each case the remaining agents' policies frozen at their *pbp* optimal choices. Note that (2.24) is an optimization (of total expectation) in the policy space, whereas (2) is optimization (of conditional expectation) in the action space, for every value of the conditioning variable. If (2.25) admits a unique solution, and if the original problem is known to have a team-optimal solution (as in the case of Theorem 2.4.2 or Corollary 2.4.1), then (2.25) provides an alternative way of obtaining that solution. If, however, (2.25) admits more than one solution, then one would like to determine whether all or some of these are team optimal. Hence, derivation of conditions under which *pbp* optimality implies team optimality are of natural interest. At the outset, one would expect *a.s.* convexity of $L(\xi; \mathbf{u})$ over $\mathbf{u} \in \mathbf{U}$ to play a role here. This is indeed the case, but convexity in itself is not a sufficient condition, as the following example demonstrates:

Consider the purely deterministic loss function $L : \mathbb{R}^2 \to \mathbb{R}$ defined by

$$L(u^1, u^2) = \begin{cases} (u^1)^2 + (1 - u^2)^2, \; u^1 \geq u^2 \\ \\ (u^2)^2 + (1 - u^1)^2, \; u^1 < u^2 \end{cases}$$

---

[16] One can determine the optimal decision rules for each value of $\xi \in (0, 2)$, substitute these unique rules into the given loss functional, and see that its integral over the interval $(0, 2)$ does not exist—which shows that J^* is unbounded under the perfect measurement information structure.

which is strictly convex on $\mathbb{R}^2$. For any fixed $u^2 \in \mathbb{R}$, $\arg\min_{u^1} L(u^1, u^2) = u^2$, and likewise for fixed $u^1 \in \mathbb{R}$, $\arg\min_{u^2} L(u^1, u^2) = u^1$. Hence, there exist infinitely many *pbp* optimal solutions ($u^1 = u^2 = u, u \in \mathbb{R}$), but only one of these, namely ($u^1 = u^2 = \frac{1}{2}$) is team optimal.

The function L above is nondifferentiable at the *pbp* optimal points ($u^1 = u^2$), and in view of this observation one might wonder whether a similar "negative" result can be obtained if the loss function were continuously differentiable. The following Lemma outrules this possibility for deterministic problems, and provides a set of (tight) sufficient conditions for a *pbp* optimal solution to be globally optimal.

Lemma 2.4.1 *Let $L : \mathbb{R}^{m_1} \times \cdots \times \mathbb{R}^{m_N} \to \mathbb{R}$ be a convex (deterministic) loss function, with a pbp optimal solution $\mathbf{u}^\circ := (u^{1^\circ}, \ldots, u^{N^\circ})$. If L is continuously differentiable*[17] *at $\mathbf{u}^\circ$, then $\mathbf{u}^\circ$ is globally (team) optimal.*

Proof From the definition of convexity, we have the inequality

$$L(\alpha \mathbf{v} + (1-\alpha)\mathbf{u}^\circ) \leq \alpha L(\mathbf{v}) + (1-\alpha) L(\mathbf{u}^\circ)$$

for any $\mathbf{v} = (v^1, \ldots, v^N) \in \mathbb{R}^{m_1} \times \cdots \times \mathbb{R}^{m_N}$, and every $\alpha \in [0, 1]$. Rearranging this inequality, we obtain, for $0 < \alpha \leq 1$,

$$\frac{1}{\alpha}[L(\mathbf{u}^\circ + \alpha(\mathbf{v} - \mathbf{u}^\circ)) - L(\mathbf{u}^\circ)] \leq L(\mathbf{v}) - L(\mathbf{u}^\circ),$$

and letting $\alpha \downarrow 0$, we arrive at

$$\sum_{i=1}^{N} \nabla_{u^i} L(\mathbf{u}^\circ)(v^i - u^{oi}) \leq L(\mathbf{v}) - L(\mathbf{u}^\circ), \quad \forall \mathbf{v} \in \mathbb{R}^{m_1} \times \cdots \times \mathbb{R}^{m_N}$$

where the required derivatives exist and the chain rule applies since L is continuously differentiable at the given point. Furthermore, by the *pbp* optimality of $\mathbf{u}^\circ$, all these partial derivatives vanish, leading to

$$L(\mathbf{u}^\circ) \leq L(\mathbf{v}), \quad \forall \mathbf{v} \in \mathbb{R}^{m_1} \times \cdots \times \mathbb{R}^{m_N}$$

which proves global optimality of $\mathbf{u}^\circ$. ◇

This lemma now finds a natural generalization to static stochastic team problems with finite measurement spaces. First, we formally introduce the notion of a "stationary policy".

---

[17] Here, if one is to generalize the space on which L is defined, such as an infinite dimensional space, Fréchet differentiability would be a sufficient condition. In fact, continuous differentiability, and thus continuity of partial derivatives for a finite dimensional function (see Appendix A.4), implies Fréchet differentiability [266]. The key aspect required is that the chain rule in differentiation applies, which is the case for Fréchet differentiable functions. See also [594].

Definition 2.4.1 (Stationary Policy) Given a static stochastic team problem $\{J; \Gamma^i, i \in \mathcal{N}\}$, a policy N-tuple $\underline{\gamma} \in \mathbf{\Gamma}$ is *stationary* if (i) $J(\underline{\gamma})$ is finite, (ii) the N partial derivatives in the following equations are well defined (locally), and (iii) $\underline{\gamma}$ satisfies these equations:

$$\nabla_{u^i}\left(E_{\xi|y^i}[L(\xi; \underline{\gamma}^{-i}(\mathbf{y}^{-i}), u^i)]\right)|_{u^i=\gamma^i(y^i)} = 0, a.s. \quad i \in \mathcal{N}. \tag{2.26}$$

Clearly, the *stationarity condition (2.26)* is a necessary condition for (2.25) if $L(\xi; \mathbf{u})$ is continuously differentiable in each agent's action variable (not necessarily jointly) for every $\xi \in \Xi$, and $S^i, i \in \mathcal{N}$, are open subsets of finite-dimensional vector spaces. It is equivalent to (2.25) if furthermore $S^i, i \in \mathcal{N}$, are convex sets, and $L(\xi; \mathbf{u})$ is convex in $u^i, i \in \mathcal{N}$, for every $\xi \in \Xi$. The following theorem now basically says that if the convexity and continuous differentiability of L is jointly in all the agents' action variables, then a stationary policy is necessarily team-optimal.

Theorem 2.4.4 *For an N-agent static stochastic team problem, let the hypotheses $(c.3)$ and $(c.4)$ be satisfied, S^i be an open convex subset of a finite dimensional vector space, for each $i \in \mathcal{N}$, and $L(\xi, \cdot)$ be convex and continuously differentiable on $\mathbf{S} := S^1 \times \cdots \times S^N$. Under these conditions, if the policy $\underline{\gamma}^\circ$, taking values in $\mathbf{S}$, is stationary, it is team-optimal.*

Proof Using the construction given in the proof of Theorem 2.4.3, J admits a representation on the space U, which is convex and continuously differentiable (this last property follows because by $(c.4)$ the function $L_{av}(\mathbf{y}_t; \cdot)$ is continuously differentiable, and the representation for J is a finite weighted sum of such functions). Then, the result follows by a direct application of Lemma 2.4.1. ◇

Example 2.4.1 Continued.

Returning to the static team of Example 2.4.1, so as to apply Theorem 2.4.4, first by the "monotone convergence theorem" (see Appendix B) the conditional expectation and differentiation in (2.26) can be interchanged, leading to the equivalent stationarity conditions

$$E_{\xi|y^i}[(\partial/\partial u^i)L(\xi; \gamma^j(y^j), u^i)]|_{u^i=\gamma^i(y^i)} = 0, \quad i \neq j, i, j = 1, 2.$$

$\Leftrightarrow$

$$2\gamma^1(y^1) + E[\xi\gamma^2(y^2)|y^1] - 1 = 0$$

$$2\gamma^2(y^2) + E[\xi\gamma^1(y^1)|y^2] - 2 = 0$$

Since y^1 and y^2 each take two different values, this pair of equations is in fact a set of *four* linear equations, identical with the ones encountered earlier. Note that to further simplify the pair of equations above, we can substitute for γ^2 from the second into the first, to arrive at a single equation in terms of γ^1,

$$4\gamma^1(y^1) + 2E[\xi|y^1] - 1 - E[\xi E[\gamma^1(y^1)|y^2]|y^1] = 0$$

which can be solved uniquely for $u_1^1 = \gamma^1(y_1^1)$ and $u_2^1 = \gamma^1(y_2^1)$, to yield $u_1^{1*} \approx 0.231, u_2^{1*} \approx -0.324$. The stationary policies of agent 2 can likewise be obtained. Since the loss function L satisfies all the hypotheses of Theorem 2.4.4, these stationary policies are indeed team-optimal. ◇

2.4.2 Teams on Finite-Dimensional Spaces

We now extend the theory of the previous subsection from finite spaces to a class of uncountable (but finite-dimensional) measurement spaces, with the action spaces again taken as finite dimensional vector spaces. Given such a team problem $\{J; \Gamma^i, i \in \mathcal{N}\}$, at least one of the policy spaces (Γ^i) will be infinite dimensional, which means that condition (c.1) will no longer imply that the policy space $\mathbf{\Gamma}$ is compact. Hence, a counterpart of Theorem 2.4.2 (on the existence of a team solution) will not follow from the given conditions. If $\underline{\gamma}^* \in \mathbf{\Gamma}$ is a team-optimal solution, then it will necessarily satisfy the *pbp* optimality condition (2.25) where now $\mathbb{Y}^i = \mathbb{R}^{r_i}, i \in \mathcal{N}$. If furthermore, the action constraint sets are open, and the function to be minimized in (2.25) is continuously differentiable in the minimizing argument, this being so for all $i \in \mathcal{N}$, then the team solution should satisfy the stationarity conditions (2.26). The question now is whether there exists a counterpart of Theorem 2.4.4, to ensure that every stationary solution is also team-optimal. We first have the following Theorem which provides a set of sufficient conditions for a policy N-tuple to be team-optimal.

Theorem 2.4.5 *Let $\{J; \Gamma^i, i \in \mathcal{N}\}$ be a static stochastic team problem where $\mathbb{U}^i \equiv \mathbb{R}^{m_i}, i \in \mathcal{N}$, the loss function $L(\xi, \mathbf{u})$ is convex and continuously differentiable in $\mathbf{u}$ a.s., and $J(\underline{\gamma})$ is bounded from below on $\mathbf{\Gamma}$. Let $\underline{\gamma}^* \in \mathbf{\Gamma}$ be a policy N-tuple with a finite cost ($J(\underline{\gamma}^*) < \infty$), and suppose that for every $\underline{\gamma} \in \mathbf{\Gamma}$ such that $J(\underline{\gamma}) < \infty$,*

$$\sum_{i \in \mathcal{N}} E\left[\nabla_{u^i} L(\xi; \underline{\gamma}^*(\mathbf{y}))[\gamma^i(y^i) - \gamma^{i*}(y^i)]\right] \geq 0, \tag{2.27}$$

where $E[\cdot]$ denotes the total expectation and the notation $\nabla_{u^i} L(\xi; \underline{\gamma}^*(\mathbf{y}))$ stands for the partial derivatives under policy $\underline{\gamma}^*$. Then, $\underline{\gamma}^*$ is a team-optimal policy, and it is unique if L is strictly convex in $\mathbf{u}$.

Proof First, by the convexity of L, we obtain (as in the proof of Lemma 2.4.1),

$$\frac{1}{\alpha}\left[L(\xi; \underline{\gamma}^*(\mathbf{y}) + \alpha[\underline{\gamma}(\mathbf{y}) - \underline{\gamma}^*(\mathbf{y})]) - L(\xi; \underline{\gamma}^*(\mathbf{y}))\right] \leq L(\xi; \underline{\gamma}(\mathbf{y})) - L(\xi; \underline{\gamma}^*(\mathbf{y})),$$

for all $\alpha \in (0, 1]$. Using the definition of J, this inequality can equivalently be written as (by taking the total expectation):

$$h(\alpha) := \frac{1}{\alpha}[E\{L(\xi; \underline{\gamma}^*(\mathbf{y}) + \alpha[\underline{\gamma}(\mathbf{y}) - \underline{\gamma}^*(\mathbf{y})])\} - J(\underline{\gamma}^*)] \le J(\underline{\gamma}) - J(\underline{\gamma}^*),$$

where $\alpha \in (0, 1]$. Note that both $J(\underline{\gamma})$ and $J(\underline{\gamma}^*)$ are finite, by hypothesis, and the first random variable (i.e., the first loss function) also has a finite expectation for every $\alpha \in (0, 1]$ because of the bound provided by the inequality. Now, due to the convexity of L, its finite integral, $E\{L(\xi; \underline{\gamma}^*(\mathbf{y}) + \alpha[\underline{\gamma}(\mathbf{y}) - \underline{\gamma}^*(\mathbf{y})])\}$ is also convex in α. This leads to the conclusion that (by a property of convex functionals, given in Appendix A, Sect. A.4) $h(\alpha)$ is a monotonically nonincreasing function as $\alpha \downarrow 0$, and furthermore $h(1) \equiv J(\underline{\gamma}) - J(\underline{\gamma}^*)$ is bounded (by hypothesis). It then follows from the *monotone convergence theorem* (see Appendix B) that $\lim_{\alpha \downarrow 0} h(\alpha)$ exists, and the limit and expectation operations can be interchanged. As a consequence of continuous differentiability, this then leads to the inequality

$$\sum_{i=1}^{N} E\{\nabla_{u^i} L(\xi; \underline{\gamma}^*(\mathbf{y}))[\gamma^i(y^i) - \gamma^{i^*}(y^i)]\} \quad \le J(\underline{\gamma}) - J(\underline{\gamma}^*)$$

from which team-optimality of $\underline{\gamma}^*$ follows, since the left-hand-side is nonnegative, by (2.27).

If L were strictly convex in $\mathbf{u}$, a.s., then all the inequalities above would be strict, for $\underline{\gamma} \neq \underline{\gamma}^*$, thus leading to

$$J(\underline{\gamma}^*) < J(\underline{\gamma})$$

which says that $\underline{\gamma}^*$ is the unique team-optimal solution. ◇

Note that the conditions of Theorem 2.4.5 above do not include the stationarity of $\underline{\gamma}^*$, and furthermore inequality (2.27) may not generally be easy to check, since they involve all permissible policies $\underline{\gamma}$ (with finite cost)—generally an uncountable set. It is therefore important to obtain more readily checkable conditions to replace (2.27) and to relate team-optimality to stationarity.

Note that if the following N inequalities hold:

$$E\Big[\nabla_{u^i} L(\xi; \underline{\gamma}^*(\mathbf{y}))[\gamma^i(y^i) - \gamma^{i*}(y^i)] \ge 0, \quad i \in \mathcal{N}, \tag{2.28}$$

then (2.27) would hold.

Then, either one of the two conditions in the Assumption below will accomplish this goal:

Assumption 2.4.1

(c.5) *For all* $\underline{\gamma} \in \Gamma$ *such that* $J(\underline{\gamma}) < \infty$, *the following random variables have well-defined expectations (i.e., are integrable):*

$$\nabla_{u^i} L(\xi; \underline{\gamma}^*(\mathbf{y}))[\gamma^i(y^i) - \gamma^{i*}(y^i)], \quad i \in \mathcal{N}$$

(c.6) Γ^i *is a Hilbert space for each* $i \in \mathcal{N}$, *and* $J(\underline{\gamma}) < \infty$ *for all* $\underline{\gamma} \in \Gamma$. *Furthermore,*

$$E_{\xi|y^i}[\nabla_{u^i} L(\xi; \underline{\gamma}^*(\mathbf{y})] \in \Gamma^i, \quad i \in \mathcal{N}.$$

Of course, (c.6) would be a stronger assumption than (c.5) when Γ^i, $i \in \mathcal{N}$, are taken as Hilbert spaces (by an application of the Cauchy-Schwarz inequality), given the Hilbert space assumption in (c.6). Here we give it as a separate condition because in some problems (such as linear quadratic—as we shall see shortly) (c.6) follows quite readily from the problem formulations.

Theorem 2.4.6 *Let* $\{J; \Gamma^i, i \in \mathcal{N}\}$ *be a static stochastic team problem which satisfies all the hypotheses of Theorem 2.4.5, with the exception of the set of inequalities (2.27). Instead of (2.27), let either* (c.5) *or* (c.6) *be satisfied. Then, if* $\underline{\gamma}^* \in \mathbf{\Gamma}$ *is a stationary policy it is also team optimal. Such a policy is unique if* $L(\xi; \mathbf{u})$ *is strictly convex in* $\mathbf{u}$, *a.s.*

Proof Now, what we need to show is that stationarity of $\underline{\gamma}^*$ implies (under (c.6)) the set of inequalities (2.27). Firstly note that since Γ^i is a vector space, $\gamma^i - \gamma^{i*} \in \Gamma^i$ for every $\gamma^i \in \Gamma^i$, and for every $\beta^i \in \Gamma^i$, there exists a $\gamma^i \in \Gamma^i$ such that $\beta^i = \gamma^i - \gamma^{i*}$. Since $\beta^i \in \Gamma^i \Rightarrow -\beta^i \in \Gamma^i$, the set of inequalities (2.27) become equivalent to

$$E\{\nabla_{u^i} L(\xi; \underline{\gamma}^*(\mathbf{y}))\beta^i(y^i)\} = 0, \quad \forall \beta^i \in \Gamma^i, \quad i \in \mathcal{N}$$

$\Leftrightarrow$

$$E_{y^i}\{E_{\xi|y^i}[\nabla_{u^i} L(\xi; \underline{\gamma}^*(\mathbf{y}))]\beta^i(y^i)\} = 0, \quad \forall \beta^i \in \Gamma^i, \quad i \in \mathcal{N},$$

where the second line follows from the iterated property of conditional expectation, under condition (c.6). Since both product terms above belong to Γ^i which is a Hilbert space, and the equality is required to hold for every element of Γ^i, $i \in \mathcal{N}$, the last line becomes equivalent to

$$E_{\xi|y^i}[\nabla_{u^i} L(\xi; \underline{\gamma}^*(\mathbf{y}))] = 0, \text{ a.s. } y^i, \quad i \in \mathcal{N}.$$

To complete the proof, we now have to show that the stationarity condition (2.26) implies the above, which would be true if we were able to interchange the derivative (which is a limit) and conditional expectation operations. This however is justified (using again the monotone convergence theorem, as in the proof of Theorem 2.4.5), since $J(\gamma)$ is finite for all $\underline{\gamma} \in \mathbf{\Gamma}$ and the conditional expectation above is well-defined (as an element of a Hilbert space). ◇

Theorem 2.4.6 above thus provides an extension of the result of Theorem 2.4.4 from finite to infinite measurement sets. To appreciate some of the fine points of Theorems 2.4.5 and 2.4.6, let us now consider the following example, which was discussed in [429].

Example 2.4.2 Let $N = 2$, $\Xi = \mathbb{U}^1 = \mathbb{U}^2 = \mathbb{R}$, $\xi = x$ be a Gaussian random variable with zero mean and unit variance ($\sim N(0, 1)$), and the loss functional be given by

$$L(x; u^1, u^2) = (u^1 - u^2)^2 e^{x^2} + 2u^1 u^2.$$

Note that L is strictly convex and continuously differentiable in (u^1, u^2) for every value of x. Hence, if the true value of x were known to both agents, the problem would admit a unique team optimal solution: $u^1 = u^2 = 0$, which is also stationary. Since this team-optimal solution does not use the precise value of x, it is certainly optimal also under "no-measurement" information (the other extreme scenario). Note, however, that in this case the only pairs that make $J(\gamma)$ finite, are $u^1 = u^2 = u \in \mathbb{R}$, since

$$E[e^{x^2}] = \frac{1}{\sqrt{2\pi}} \int_{-\infty}^{\infty} e^{+\frac{x^2}{2}} dx = \infty.$$

The set of permissible policies not being an open set, clearly we cannot talk about stationarity in this case. Theorem 2.4.5 (which does not involve stationarity) is applicable here, where inequality (20) is satisfied trivially. Note also that for every $u \in \mathbb{R}$, $u^1 = u^2 = u$ is a *pbp optimal* solution, but only one of these is team optimal.

Now, as a more interesting case consider the measurement scheme:

$$y^1 = x + w^1; \qquad y^2 = x + w^2$$

where w^1 and w^2 are independent random variables uniformly distributed on the interval $[-1, 1]$, which are also independent of x.[18] Clearly, $u^1 = u^2 = 0$ is team-optimal for this case also, but it is not obvious at the outset whether it is stationary or not. Toward this end, let us evaluate (2.26) for $i = 1$ and with $\gamma^2(y^2) = 0$:

$$(\partial/\partial u^1) E_{x, y^2|y^1}[(u^1)^2 e^{\xi^2}] = (\partial/\partial u^1)\{(u^1)^2 E_{x|y^1}[e^{\xi^2}]\} = 2u^1 E_{x|y^1}[e^{\xi^2}]$$

where the last step follows because the conditional probability density of x given y^1 is nonzero only in a finite interval (thus making the conditional expectation finite). By symmetry, it follows that both derivatives in (2.26) vanish at $u^1 = u^2 = 0$, and

[18] Note that here the random state of nature, ξ, is chosen as $(x, w^1, w^2)'$.

hence the team-optimal solution is stationary. It is not difficult to see that in fact this is the only pair of stationary policies. Note that all the hypotheses of Theorem 2.4.6 are satisfied here, under condition $(c.5)$. ◇

In the following, we specialize the above general result to two classes of teams with special cost structures, namely quadratic and exponentiated quadratic loss functions. In both cases the team loss function will be strictly convex and continuously differentiable, so that (2.27) provides a sufficient condition for a policy $\underline{\gamma}^* \in \mathbf{\Gamma}$ to be team-optimal. We will further observe that the conditions of Theorem 2.4.6 are satisfied, so that stationary policies are also team-optimal.

2.4.3 Static Teams with Quadratic Loss and a Hilbert Space Argument to Optimality and Stationarity

Given a probability space $(\Omega, \mathbf{F}, P_\Omega)$, and an associated vector-valued random variable ξ, let $\{J; \Gamma^i, i \in \mathcal{N}\}$ be a static stochastic team problem with the following specifications:

(i) $\mathbb{U}^i \equiv \mathbb{R}^{m_i}, i \in \mathcal{N}$; i.e., the action spaces are unconstrained Euclidean spaces.
(ii) The loss function is a quadratic function of $\mathbf{u}$ for every ξ:

$$L(\xi; \mathbf{u}) = \sum_{i,j \in \mathcal{N}} u^{i'} R_{ij}(\xi) u^j + 2 \sum_{i \in \mathcal{N}} u^{i'} r_i(\xi) + c(\xi) \tag{2.29}$$

where $R_{ij}(\xi)$ is a matrix-valued random variable (with $R_{ij} \equiv R'_{ji}$), $r_i(\xi)$ is a vector-valued random variable, and $c(\xi)$ is a random variable, all generated by measurable mappings on the random state of nature, ξ.
(iii) $L(\xi; \mathbf{u})$ is strictly (and uniformly) convex in $\mathbf{u}$ a.s., i.e., there exists a positive scalar α such that, with $R(\xi)$ defined as a matrix comprised of N blocks, with the ij'th block given by $R_{ij}(\xi)$, the matrix $R(\xi) - \alpha I$ is positive definite a.s., where I is the appropriate dimensional identity matrix.
(iv) $R(\xi)$ is uniformly bounded above, i.e., there exists a positive scalar β such that the matrix $\beta I - R(\xi)$ is positive definite a.s.
(v) $\mathbb{Y}^i \equiv \mathbb{R}^{r_i}, i \in \mathcal{N}$, i.e., the measurement spaces are unconstrained Euclidean spaces.
(vi) $y^i = \eta^i(\xi), i \in \mathcal{N}$, for some appropriate Borel measurable functions $\eta^i, i \in \mathcal{N}$.
(vii) Γ^i is the (Hilbert) space of all Borel measurable mappings of $\gamma^i : \mathbb{R}^{r_i} \to \mathbb{R}^{m_i}$, which have bounded second moments, i.e., $E_{y^i}[\gamma^{i'}(y^i)\gamma^i(y^i)] < \infty$.
(viii) $E_\xi[r_i'(\xi) r_i(\xi)] < \infty, i \in \mathcal{N}$; $E_\xi[c(\xi)] < \infty$.

Definition 2.4.2 A static stochastic team is *quadratic* if it satisfies (i)–(viii) above. It is a *standard quadratic team* if furthermore the matrix R is constant for all ξ (i.e., it is deterministic). If, in addition, ξ is a Gaussian distributed random vector, and $r_i(\xi) = Q_i\xi, \eta^i(\xi) = H^i\xi, i \in \mathcal{N}$, for some deterministic matrices Q_i, H^i, $i \in \mathcal{N}$, the decision problem is a *quadratic-Gaussian team* (more widely known as a linear-quadratic-Gaussian (LQG) team under some further structure on Q_i and H^i).

We now first show that the cost function of a quadratic team given above is bounded and strictly convex on $\mathbf{\Gamma}$.

Proposition 2.4.1 *For a quadratic team,*

(i) $|J(\underline{\gamma})| < \infty$ *for all* $\underline{\gamma} \in \mathbf{\Gamma}$.
(ii) $J(\underline{\gamma})$ *is strictly convex on* $\mathbf{\Gamma}$.

Proof For each $\underline{\gamma} \in \mathbf{\Gamma}$, each component of $u^i = \gamma^i(y^i)$ is a random variable on $(\Omega, \mathcal{F}, P_\Omega)$ with a bounded second moment (i.e., it is a *second-order* random variable), this being true for all $i \in \mathcal{N}$. Now using the fact that the product of any two second-order random variables defined on the same probability space is a well-defined random variable (on the same probability space) with a finite mean value (see Appendix B), it follows that the expected value of the second term of (2.29) is finite. Furthermore, since $R(\xi)$ is uniformly bounded, the expected value of the first term satisfies the bound

$$0 \le E[\sum_{i,j} u^{i'} R_{ij}(\xi)u^j] \equiv E[\mathbf{u}'R(\xi)\mathbf{u}] \le \beta E[\mathbf{u}'\mathbf{u}]$$

where $E\{\mathbf{u}'\mathbf{u}\}$ is finite by the same reasoning as above. Then, it readily follows that $L(\xi; \mathbf{u})$, with $u^i = \gamma^i(y^i)$, $i \in \mathcal{N}$, is a well-defined random variable with a *finite* expectation. Now, since $L(\xi, \mathbf{u})$ is strictly convex in $\mathbf{u}$ for every ξ, we have the strict inequality

$$L(\xi; \tilde{\alpha}\underline{\gamma}(\mathbf{y}) + (1-\tilde{\alpha})\underline{\hat{\gamma}}(\mathbf{y})) < \tilde{\alpha}L(\xi; \underline{\gamma}(\mathbf{y})) + (1-\tilde{\alpha})L(\xi; \underline{\hat{\gamma}}(\mathbf{y}))$$

for all $\tilde{\alpha} \in (0, 1)$, and every $\underline{\gamma}, \underline{\hat{\gamma}} \in \mathbf{\Gamma}$, $\underline{\gamma} \neq \underline{\hat{\gamma}}$. Taking the expected values of both sides, which are finite as shown above, we arrive at

$$J(\tilde{\alpha}\underline{\gamma} + (1-\tilde{\alpha})\underline{\hat{\gamma}}) < \tilde{\alpha}J(\underline{\gamma}) + (1-\tilde{\alpha})J(\underline{\hat{\gamma}})$$

which shows that J is strictly convex. ◇

Now, the stationarity conditions (2.26) associated with the loss functional (2.29) can be evaluated:

$$\left(\nabla_{u^i}\{E_{\xi|y^i}[\sum_{k,j\in\mathcal{N}} u^{k'} R_{kj}(\xi)u^j] + 2u^{i'} E_{\xi|y^i}[r_i(\xi)] + 2E_{\xi|y^i}[\sum_{j\in\mathcal{N},j\neq i} u^{j'} r_j(\xi)] \right.$$
$$\left. + E_{\xi|y^i}[c(\xi)]\}\right)|_{u^i=\gamma^i(y^i)} = 0, \quad i \in \mathcal{N}$$

$$\Leftrightarrow \left(E_{\xi|y^i}[R_{ii}(\xi)]u^i + \sum_{j\in\mathcal{N},j\neq i} E_{\xi|y^i}[R_{ij}(\xi)u^j] + E_{\xi|y^i}[r_i(\xi)]\right)|_{u^i=\gamma^i(y^i)} = 0, i \in \mathcal{N}$$

$$\Leftrightarrow E_{\xi|y^i}[R_{ii}(\xi)]\gamma^i(y^i) + \sum_{j\in\mathcal{N},j\neq i} E_{\xi|y^i}[R_{ij}(\xi)\gamma^j(y^j)] + E_{\xi|y^i}[r_i(\xi)] = 0, i \in \mathcal{N} \tag{2.30}$$

where in going from the first to the second line of the equation we have simply performed vector differentiation with respect to u^i which is outside the conditional expectation, and have also used the fact that $R_{ij} \equiv R'_{ji}$.

Hence, (2.30) constitutes the set of stationarity conditions for the quadratic team. The following theorem now says that the solution is unique and is team-optimal. The approach presented is essentially due to [594].

Theorem 2.4.7 *A quadratic static team (see Definition 2.4.2) admits a unique team-optimal solution $\underline{\gamma}^* \in \mathbf{\Gamma}$, which is also the unique stationary solution satisfying (2.30).*

Proof As one proof approach, one could proceed as follows: Assuming that there exists a stationary solution (i.e., a solution to (2.30)), the uniqueness and team-optimality follow from Theorem 2.4.6, since all its hypotheses are satisfied along with condition (c.6). Hence the proof will be completed if we can show that there exists at least one solution $\underline{\gamma}^* \in \mathbf{\Gamma}$ to (2.30).

On the other hand, in the following, we present a more direct *Projection Theorem* based argument which leads to both the existence of a solution to (2.30), and also the optimality of the solution to (2.30) without invoking Theorem 2.4.6, crucially utilizing the quadratic nature and an associated Hilbert space formulation. Some related material on Hilbert spaces can be found in Sect. A.2).

Let us first note that the quadratic loss function (2.29) can equivalently be written as

$$L(\xi;\mathbf{u}) = \mathbf{u}'R(\xi)\mathbf{u} + 2\mathbf{u}'\mathbf{r}(\xi) + c(\xi)$$
$$\equiv \left(\mathbf{u}+R^{-1}(\xi)\mathbf{r}(\xi)\right)' R(\xi)\left(\mathbf{u}+R^{-1}(\xi)\mathbf{r}(\xi)\right) + c(\xi) - \mathbf{r}(\xi)'R^{-1}(\xi)\mathbf{r}(\xi)$$

where $R(\xi)$ is the matrix whose ij'th block is $R_{ij}(\xi)$, and has an inverse (pointwise) by the a.s. strict convexity of L. Hence, if all agents had perfect access to the precise value of ξ, the minimum value of J would be given by the expected value of the last two terms above. Since this is not the case, the actual minimum value of J will be higher, the difference being due to the error in "approximating the vector $-R^{-1}(\xi)\mathbf{r}(\xi)$ using policies out of $\mathbf{\Gamma}$". An equivalent problem, therefore, is

$$\min_{\underline{\gamma}\in\mathbf{\Gamma}} \tilde{J}(\underline{\gamma}), \quad \tilde{J}(\underline{\gamma}) := E\{\|\underline{\gamma}(\mathbf{y}) + R^{-1}(\xi)\mathbf{r}(\xi)\|^2_{R(\xi)}\} \tag{2.31}$$

and the statement just made (in 'inverted commas') can be given a precise mathematical meaning as follows.

First note that the policy space $\mathbf{\Gamma}$ is the product space $\Gamma^1 \times \cdots \times \Gamma^N$, where each Γ^i is in fact a Hilbert space, with the inner product[19]

$$\langle \alpha^i, \beta^i \rangle_i := E[\alpha^i(y^i)'\beta^i(y^i)] \tag{2.32}$$

(see Appendix A, Sect. A.2, and Appendix B, Sect. B.1). This makes $\mathbf{\Gamma}$ also a Hilbert space, with the inner product

$$\langle \underline{\alpha}, \underline{\beta} \rangle := E\left\{\sum_{i\in\mathcal{N}} \alpha^i(y^i)'\beta^i(y^i)\right\} \equiv E[\underline{\alpha}(\mathbf{y})'\underline{\beta}(\mathbf{y})]. \tag{$\circ$}$$

Note the important restriction that $\underline{\alpha}$ and $\underline{\beta}$ are not allowed to depend on all components of $\mathbf{y}$, because different agents do not have access to the same set of measurements. Now, in order to be able to use (2.31) as a norm compatible with the given inner product, we have to change the ($\circ$) somewhat by weighting it with $R(\xi)$:

$$\langle \underline{\alpha}, \underline{\beta} \rangle = E[\underline{\alpha}(\mathbf{y})'R(\xi)\underline{\beta}(\mathbf{y})]. \tag{$\circ\circ$}$$

This actually changes $\mathbf{\Gamma}$, the space where $\underline{\alpha}$ and $\underline{\beta}$ belong, but because of the given properties of $R(\xi)$, we have an isometry between the two spaces, and therefore can denote the one under the new inner product ($\circ\circ$) also by $\mathbf{\Gamma}$. If every component of $\underline{\gamma}$ were allowed to depend on the entire measurement vector $\mathbf{y}$ (which would be the case if all the agents were to share their measurements), then the set of all permissible $\underline{\gamma}$'s, bounded under the norm induced by ($\circ\circ$) would be a much larger (than $\mathbf{\Gamma}$) space. Let us denote this space by $\mathbf{H}$, and note that it is also a Hilbert space, under the inner product ($\circ\circ$). An important observation now is that $\mathbf{\Gamma}$ is a closed linear subspace of $\mathbf{H}$, *closed* because every convergent sequence in $\mathbf{\Gamma}$ with a limit point will have the limit point in $\mathbf{\Gamma}$. Hence, the team-minimization problem (2.31) is in fact an orthogonal projection problem, one of orthogonally projecting the random vector $\mathbf{x}(\xi) := R^{-1}(\xi)\mathbf{r}(\xi)$ from $\mathbf{H}$ onto $\mathbf{\Gamma}$. The conditions of the

---

[19] This discussion is a prelude to the development in Chap. 5 where we will introduce further topological constructions on team policies toward arriving at existence, near optimality and structural results for optimal solutions under a variety of setups.

orthogonal projection theorem given in Appendix A, Sect. A.2, are satisfied, and therefore there exists a *unique* element of $\mathbf{\Gamma}$ that solves (2.31). Furthermore, by the orthogonality condition this unique element, say $\underline{\gamma}^*$, has the property that

$$\underline{\gamma}^* + \mathbf{x} \perp \gamma, \qquad \forall \gamma \in \mathbf{\Gamma}$$

(see Appendix A, Sect. A.2, for notation and terminology). Using the inner product ($\circ\circ$), this orthogonality relationship can be written as

$$\langle \underline{\gamma}^* + \mathbf{x}, \underline{\gamma} \rangle = E\left\{ [\underline{\gamma}^*(\mathbf{y}) + \mathbf{x}(\xi)]' R(\xi) \underline{\gamma}(\mathbf{y}) \right\} = 0$$

$$\Leftrightarrow$$

$$E\left\{ \sum_{i \in \mathcal{N}} \gamma^i(y^i)' \left[\sum_{j \in \mathcal{N}} R_{ij}(\xi) \gamma^{*j}(y^j) + r_i(\xi) \right] \right\} = 0$$

$$\Leftrightarrow$$

$$E\left\{ \sum_{i \in \mathcal{N}} \gamma^i(y^i)' \left[E_{\xi|y^i}\{R_{ii}(\xi)\} \gamma^{*i}(y^i) + \sum_{j \in \mathcal{N}, j \neq i} E_{\xi|y^i}\left\{ R_{ij}(\xi) \gamma^{*j}(y^j) \right\} + E_{\xi|y^i} r_i(\xi) \right] \right\} = 0$$

where in arriving at the last line we have used the iterative property of conditional expectations. Now, since this equality has to hold for all $\underline{\gamma} \in \mathbf{\Gamma}$, and since $\mathbf{\Gamma}$ is a Hilbert space, it follows that the expression in brackets should vanish for every $i \in \mathcal{N}$[20] which is precisely (2.30). ◇

The proof of the theorem, as presented above, provides us with a new interpretation to the stationarity conditions (2.30). Note that they can be rewritten (compactly) as

$$\mathbb{P} R \underline{\gamma} + \mathbb{P} \mathbf{r} = 0 \tag{2.33}$$

where $\mathbb{P}$ is a linear operator, block diagonal, with the *ii*'th block defined through

$$\mathbb{P}_{ii} \beta^i(\xi) = E_{\xi|y^i} \beta^i(\xi), \quad i \in \mathcal{N}$$

where $\beta^i(\xi)$ is an m_i-dimensional measurable function of ξ, satisfying the boundedness condition $E\{\beta^i(\xi)'\beta^i(\xi)\} < \infty$. As such, the linear operator $\mathbb{P}$ is a *projection operator* defined on a Hilbert space, whose operator norm is one (see Appendix A, Sect. A.2). Note that if the agents had full access to the value of ξ, then the stationarity condition would be

[20] Here we have used the following property of Hilbert spaces: if $\langle \alpha, \beta \rangle = 0$ for all $\beta \in \mathbf{H}$, then $\alpha \equiv 0$.

$$R(\xi)\underline{\gamma}(\xi) + \mathbf{r}(\xi) = 0 \tag{2.34}$$

which of course admits a unique solution since the matrix $R(\xi)$ is invertible for all ξ. Hence, the stationarity equation in the decentralized measurement case is a "projected" version of the one in the centralized full information case, but note, however, that the unique (decentralized) team-optimal solution is <u>*not*</u> a projected version of the centralized one $(-R^{-1}\mathbf{r})$.

The unique team-optimal solution cannot in general be obtained in closed form, and thus one has to turn to some approximation schemes. Viewing (2.30) (or equivalently (2.33) as a fixed-point equation (see Appendix A, Sect. A.6 for details), one approach would be to use successive approximations:

$$\left\{ \begin{array}{ll} \gamma^i_{(k+1)}(y^i) = -[E_{\xi|y^i} R_{ii}(\xi)]^{-1} \left\{ \displaystyle\sum_{j\in\mathcal{N}, j\neq i} E_{\xi|y^i}[R_{ij}(\xi)\gamma^j_{(k)}(y^j)] + E_{\xi|y^i} r_i(\xi) \right\}, \\ \gamma^i_{(0)}(y^i) \quad \equiv 0, \quad i \in \mathcal{N}, \quad k = 0, 1, \ldots \end{array} \right\} \tag{2.35}$$

which is called the *parallel update scheme*, where we have taken the starting points of the iteration as the *zero* function, as an arbitrary choice. This iteration models a dynamic decision process where the agents exchange policy information at every (discrete) point in time, and at the $k+1$'th instant agent i solves the (stochastic) optimization problem

$$\begin{aligned} \min\nolimits_{\gamma^i \in \Gamma^i} & J(\gamma^1_{(k)}, \ldots, \gamma^{i-1}_{(k)}, \gamma^i, \gamma^{i+1}_{(k)}, \ldots, \gamma^N_{(k)}) \\ & = J(\gamma^1_{(k)}, \ldots, \gamma^{i-1}_{(k)}, \gamma^i_{(k+1)}, \gamma^{i+1}_{(k)}, \ldots, \gamma^N_{(k)}), \; i \in \mathcal{N}. \end{aligned}$$

Clearly, if the parallel scheme converges, it will yield the (unique) team-optimal solution in the limit. However, there is generally no guarantee that it will converge, unless some conditions are imposed on the matrix R and the probabilistic structure of the problem. To state two such conditions, let us first write (2.35) in compact form:

$$\underline{\gamma}_{(k+1)} = \mathbf{F}\underline{\gamma}_{(k)} + \hat{\mathbf{r}}, \tag{2.36}$$

where $\mathbf{F}$ is a linear operator mapping $\boldsymbol{\Gamma}$ into itself, and composed of block operators with the diagonal blocks being zero and off-diagonal blocks given by

$$[\mathbf{F}_{ij}\gamma^j](y^i) = -[E_{\xi|y^i} R_{ii}(\xi)]^{-1} E_{\xi|y^i}[R_{ij}(\xi)\gamma^j(y^j)], \quad j \neq i, j \in \mathcal{N}. \tag{2.37}$$

Furthermore, $\hat{\mathbf{r}} \in \boldsymbol{\Gamma}$, with the i'th block vector given by

$$[\hat{\mathbf{r}}(\mathbf{y})]_i = -[E_{\xi|y^i} R_{ii}(\xi)]^{-1} E_{\xi|y^i} r_i(\xi), \quad i \in \mathcal{N}. \tag{2.38}$$

Using the notation introduced in Appendix A, Sect. A.2, let $\ll \mathbf{F} \gg$ denote the *operator norm* of $\mathbf{F}$ and $\rho(\mathbf{F})$ denote its *spectral radius*; furthermore, note the inequality: $\rho(\mathbf{F}) \leq \ \ll \mathbf{F} \gg$. The *Banach* and *Successive Approximation* theorems of Appendix A, Sect. A.6 now readily lead to the following result.

Proposition 2.4.2 *Consider the parallel update scheme (2.35) (equivalently (2.36)) for the general stochastic static team problem.*

(i) The iteration converges for all starting points $\underline{\gamma}_{(0)} \in \mathbf{\Gamma}$ *if, and only if,*

$$\rho(\mathbf{F}) < 1. \tag{2.39}$$

(ii) The iteration converges for all starting points $\underline{\gamma}_{(0)} \in \mathbf{\Gamma}$ *if*

$$\ll \mathbf{F} \gg \ < 1, \tag{2.40}$$

which is therefore a sufficient condition for (2.39). ◇

It is important to note that nonsatisfaction of (2.39) does not necessarily imply that there is no recursive scheme which would compute $\underline{\gamma}^*$; in fact, there may exist nonparallel schemes, or schemes that use relaxation (i.e., higher order memory), that will have better convergence properties. As an example of a nonparallel scheme consider the so-called *sequential scheme* where the agents take their turns, one at a time and in strict order, to re-optimize their policies, i.e., $\gamma^i_{(k+1)}$ is determined through the minimization of

$$\min_{\gamma^i \in \Gamma^i} J(\gamma^1_{(k+1)}, \ldots, \gamma^{i-1}_{(k+1)}, \gamma^i, \gamma^{i+1}_{(k)}, \ldots, \gamma^N_{(k)}).$$

This then leads to the following counterpart of (2.35):

$$\begin{cases} \gamma^i_{(k+1)} = [E_{\xi|y^i} R_{ii}(\xi)]^{-1} \Bigg\{ \sum_{j \in \mathcal{N}, j<i} E_{\xi|y^i}[R_{ij}(\xi)\gamma^j_{(k+1)}(y^j)] \\ \qquad + \sum_{j \in \mathcal{N}, j>i} E_{\xi|y^i}[R_{ij}(\xi)\gamma^j_{(k)}(y^j)] + E_{\xi|y^i} r_i(\xi) \Bigg\}, i \in \mathcal{N}, k = 0, 1 \ldots \\ \gamma^i_{(0)} \quad \equiv 0, \quad i \in \mathcal{N}, i \neq 1. \end{cases} \tag{2.41}$$

Note that this recursion cannot be written in a compact form as in (2.36), and hence a counterpart of Proposition 2.4.2 is not as easy to obtain, unless there is some further structure given on the problem as we will see shortly.

Standard Quadratic Teams

We now study the class of quadratic teams where the matrix $R(\xi)$ is a constant in (ξ), i.e., R is deterministic. The basic equation of stationarity, (2.30), simplifies to

$$\gamma^i(y^i) + \sum_{j\in\mathcal{N}, j\neq i} \tilde{R}_{ij} E_{\xi|y^i}[\gamma^j(y^j)] + E_{\xi|y^i}\tilde{r}_i(\xi) = 0, \quad i \in \mathcal{N} \tag{2.42}$$

where

$$\tilde{R}_{ij} := R_{ii}^{-1} R_{ij}; \quad \tilde{r}_i(\xi) = R_{ii}^{-1} r_i(\xi). \tag{2.43}$$

Clearly, by Theorem 2.4.7, this equation admits a unique solution $\underline{\gamma}^* \in \mathbf{\Gamma}$, whenever the loss function is strictly convex (equivalently, if the constant matrix R is positive definite). The counterpart of the parallel update scheme (2.35) is

$$\gamma^i_{(k+1)}(y^i) = - \sum_{j\in\mathcal{N}, j\neq i} \tilde{R}_{ij} E_{y^j|y^i}[\gamma^j_{(k)}(y^j)] - E_{\xi|y^i}\tilde{r}_i(\xi) \quad i \in \mathcal{N}, k = 0, 1, \ldots \tag{2.44}$$

which we now further study for the case $N = 2$ (i.e., with only two agents). Substituting $\gamma^2_{(k+1)}$ obtained from (2.44) into the same for $i = 1$, we obtain

$$\gamma^1_{(k+2)}(y^1) = \tilde{R}_{12}\tilde{R}_{21} E_{y^2|y^1} E_{y^1|y^2}[\gamma^1_{(k)}(y^1)] + c^1(y^1) \tag{2.45}$$

where

$$c^1(y^1) := -E_{\xi|y^1}[\tilde{r}_1(\xi)] + \tilde{R}_{12} E_{y^2|y^1}[E_{\xi|y^2}\tilde{r}_2(\xi)]. \tag{2.46}$$

Note that if we instead had the sequential update, (2.41), the resulting equation for $i = 1$ would be exactly (2.45) with simply $\gamma^1_{(k)}$ replaced by $\gamma^1_{(k+1)}$. Hence the parallel and sequential update schemes are essentially identical in the case of a two-agent team problem. The following proposition states this result, along with two other useful observations.

Proposition 2.4.3 *For the standard quadratic team with $N = 2$,*

(i) The parallel update schemes (2.44) converges (to a limiting policy pair $\underline{\gamma}^* \in \mathbf{\Gamma}$, which is a team-optimal solution) if, and only if, the single update scheme (2.45) converges.

(ii) If (2.45) converges to a limiting policy $\underline{\gamma}^{1*} \in \Gamma^1$, then $\underline{\gamma}^{1*}$ is the unique team-optimal policy of agent **A**1*, and*

$$\gamma^{2*}(y^2) = -\tilde{R}_{21} E_{y^1|y^2}[\gamma^{1*}(y^1)] - E_{\xi|y^2}\tilde{r}_2(\xi)$$

is the unique team-optimal policy of agent **A**2*.*

(iii) *The parallel and sequential update schemes require the same condition of convergence, which is*

$$\rho(\tilde{R}_{12}\tilde{R}_{21}E_{\xi|y^1}E_{\xi|y^2}) < 1\,,$$

where $\rho(\cdot)$ *is the spectral radius of its linear operator argument mapping* Γ^1 *into itself.*

Proof Parts (i) and (ii) are mere observations and require no proof. Part (iii) follows from the *successive approximation theorem* of Appendix A, Sect. A.6, since in (2.45) $\tilde{R}_{12}\tilde{R}_{21}E_{y^2|y^1}E_{y^1|y^2}$, which can equivalently be written as $\tilde{R}_{12}\tilde{R}_{21}E_{\xi|y^1}E_{\xi|y^2}$ is a linear bounded operator mapping Γ^1 (a Hilbert space) into itself. See also Proposition 2.4.2, and compare (2.45) with (2.33). ◇

The following lemma now paves the way toward showing that the condition of Proposition 2.4.3(iii) is satisfied for all standard quadratic teams.

Lemma 2.4.2 *The loss function (2.29), with* $N = 2$ *and* R_{ij} *constant matrices, is strictly convex if, and only if,* R_{22} *is positive definite and*

$$\rho(\tilde{R}_{12}\tilde{R}_{21}) < 1.$$

Proof Strict convexity of L is equivalent to the positive definiteness of the matrix

$$R := \begin{pmatrix} R_{11} & R_{12} \\ R_{12} & R_{22} \end{pmatrix},$$

which is further equivalent to (by definition)

$$\begin{pmatrix} x \\ y \end{pmatrix}' \begin{pmatrix} R_{11} & R_{12} \\ R_{12}' & R_{22} \end{pmatrix} \begin{pmatrix} x \\ y \end{pmatrix} > 0 \qquad \forall \begin{pmatrix} x \\ y \end{pmatrix} \neq \underline{o}$$

where $\underline{o}$ is the zero vector in $\mathbb{R}^m$, $m := m_1 + m_2$, and x, y have dimensions compatible with the dimensions of the blocks of R. The above can be rewritten as

$$x'R_{11}x + 2x'R_{12}y + y'R_{22}y > 0$$

from which it follows that $R_{11} > 0$, $R_{22} > 0$ are necessary conditions for strict convexity. Now, minimizing this expression with respect to y, we have, by differentiation,

$$y = -R_{22}^{-1}R_{21}x$$

as the unique solution, substitution of which into the original expression leads to

$$\min_y \begin{pmatrix} x \\ y \end{pmatrix}' R \begin{pmatrix} x \\ y \end{pmatrix} = x'(R_{11} - R_{12}R_{22}^{-1}R_{12}')x$$
$$\equiv (R_{11}^{\frac{1}{2}}x)'(I - (R_{11}^{\frac{1}{2}})^{-1}R_{12}R_{22}^{-1}R_{12}'(R_{11}^{\frac{1}{2}})^{-1})(R_{11}^{\frac{1}{2}}x) > 0,$$

for $x \neq 0$, where $R_{11}^{\frac{1}{2}}$ is the unique square root of R_{11}.

The strict inequality holds, for all nonzero x, if, and only if, the matrix

$$I - (R_{11}^{\frac{1}{2}})^{-1}R_{12}R_{22}^{-1}R_{12}'(R_{11}^{\frac{1}{2}})^{-1}$$

is positive definite, which is equivalent to all eigenvalues of the second (nonnegative definite) matrix to be less than one. Hence,

$$\rho((R_{11}^{\frac{1}{2}})^{-1}R_{12}R_{22}^{-1}R_{12}'(R_{11}^{\frac{1}{2}})^{-1}) \equiv \rho(R_{11}^{-1}R_{12}R_{22}^{-1}R_{12}') < 1$$

where we used the fact that for two square matrices A and B, $\rho(AB) = \rho(BA)$. Since $\tilde{R}_{12} = R_{11}^{-1}R_{12}$, $\tilde{R}_{21} = R_{22}^{-1}R_{21} \equiv R_{22}^{-1}R_{12}'$, this completes the proof of the lemma. ◇

This brings us to the following strengthened version of Theorem 2.4.7 for standard quadratic teams with $N = 2$.

Theorem 2.4.8 *For the two-agent standard quadratic team,*

(i) There exists a unique team-optimal solution $\underline{\gamma}^* \in \mathbf{\Gamma}$, which is also the unique solution of (2.42) with $N = 2$.
(ii) Both the parallel and sequential update schemes converge for all starting points in $\mathbf{\Gamma}$.
(iii) Agent **Ai** *'s optimal policy is given by the infinite sum*

$$\gamma^{i*}(y^i) = \sum_{k=0}^{\infty}(\tilde{R}_{ij}\tilde{R}_{ji}E_{y^j|y^i}E_{y^i|y^j})^k c^i(y^i), \quad i, j = 1, 2;\ j \neq i, \qquad (2.47)$$

where c^1 is given by (2.46) and c^2 is defined by the same with 1's and 2's interchanged.

Proof Of course, (i) follows from Theorem 2.4.7, but since (ii) implies (i) in view of Proposition 2.4.3, the independent proof that we will give for (ii) will also provide an alternative proof to this special case of Theorem 2.4.7.

To prove part (ii), it will be sufficient to verify the condition of Proposition 2.4.3(iii). Toward this end, let us first introduce a (Hilbert) space $\hat{\Gamma}^1$ of all m_1-dimensional measurable functions $\widehat{\gamma}(y^1, y^2)$ with bounded second moments: $E_{\mathbf{y}}[|\widehat{\gamma}(y^1, y^2)|^2] < \infty$. Clearly, Γ^1 is a subspace of $\widehat{\Gamma}^1$. Now, the conditional expectation operator $E_{\xi|y^i} =: P^i$ is a projection operator on $\widehat{\Gamma}^1$ (see Appendix B), and hence has operator norm one, for both $i = 1$ and $i = 2$. Since the product $P^1 P^2$ is also a linear bounded operator on $\widehat{\Gamma}^1$, its norm is bounded by,

$$\ll P^1 P^2 \gg \;\leq\; \ll P^1 \gg\ll P^2 \gg= 1.$$

An important observation here is that for any $\widehat{\gamma} \in \widehat{\Gamma}'$, $P^1 P^2 \widehat{\gamma} \in \Gamma^1 \subset \widehat{\Gamma}^1$, and hence we can also view the product operator $P^1 P^2$ as a bounded linear operator mapping Γ^1 into itself. Since $\tilde{R}_{12}\tilde{R}_{21}$ also maps Γ^1 into itself, we have

$$\rho(\tilde{R}_{12}\tilde{R}_{21} P^1 P^2) \leq \rho(\tilde{R}_{12}\tilde{R}_{21})\rho(P^1 P^2)$$

which is the spectral radius inequality on Hilbert spaces (see Appendix A, Sect. A.2). The first product term above is strictly less than *one* by Lemma 2.4.2, and the second term is no greater than *one*, by

$$\rho(P^1 P^2) \leq \;\ll P^1 P^2 \gg\; \leq 1.$$

This completes the proof of (ii), in view of Proposition 2.4.3(iii).

For part (iii) simply note that for $i = 1$, (2.47) is the infinite summation obtained from (2.45) by taking $\gamma^1_{(0)} \equiv 0$, with the limit being a valid (well-defined) element of Γ^1 by part (ii). Clearly the same result holds for $i = 2$. ◇

Remark 2.4.1 The iteration (2.45), or more generally (2.44), is sometimes called the *infinite second guessing* scheme. If the agents had known each other's (optimal) policies, then the iteration would halt after one step. Since this knowledge is not there, they have to estimate (or guess) each other's actions, which would also involve the estimates of each other's estimates, etc., leading in general to an infinite, albeit convergent, sequence.

Even though iteration (2.44) converges for $N = 2$, it does not necessarily converge for $N > 2$. This is mainly due to the fact that strict convexity of L (equivalently, positive definiteness of R) does not imply that[21] $\rho(\tilde{R}) < 1$, unless $N = 2$.

Also one intuitive explanation for this discrepancy is that for $N = 2$ the iteration (2.44) corresponds to the sequence of minimizations

$$J(\gamma^1_{(k+1)}, \gamma^2_{(k)}) = \min_{\gamma^1 \in \Gamma^2} J(\gamma^1, \gamma^2_{(k)}) =: J_{(k+1)}$$

$$J(\gamma^1_{(k+1)}, \gamma^2_{(k+2)}) = \min_{\gamma^2 \in \Gamma^2} J(\gamma^1_{(k+1)}, \gamma^2) =: J_{(k+2)}, \quad k = 0, 1, \ldots$$

[21] Here $\tilde{R}$ is the $(m \times m)$ matrix whose diagonal blocks are zero, and off-diagonal blocks are given by $[\tilde{R}]_{ij} = \tilde{R}_{ij}$, as defined by (2.43).

with the property

$$J_{(1)} \geq J_{(2)} \geq \ldots \geq J_{(k)} \geq J_{(k+1)} \geq \ldots$$

Hence, at every step of the iteration the value of J cannot increase, thus generating a nonincreasing convergent sequence (of costs). This, of course, could also converge to a *pbp*-optimal solution, but we know in this case (as already shown) that a *pbp*-optimal solution (which is also stationary because of the special quadratic structure of the loss function) is also team optimal.

For $N > 2$, however, the iteration (2.44) does not necessarily generate a monotonic cost sequence, nor a subsequence that is monotonic, which is a reason for the failure of (2.44) to converge.

In view of the above, a natural question that arises is whether there exists some other (computational) algorithm that would yield the unique solution of (2.42) (which is known to exist by Theorem 2.4.7). Toward studying this question, let us first rewrite (2.42) as follows, using the compact notation of (2.33):

$$\mathbb{P}R\underline{\gamma} + \mathbb{P}\mathbf{r} = 0. \tag{$*$}$$

Let us add $-\epsilon\underline{\gamma}$ to both sides (where $\epsilon > 0$), and divide throughout by ϵ, to obtain

$$\underline{\gamma} = \mathbb{P}(I - \frac{1}{\epsilon}R)\underline{\gamma} - \frac{1}{\epsilon}\mathbb{P}\mathbf{r} \tag{$**$}$$

where we have used the fact that the projection operator $\mathbb{P}$ and the identity operator (matrix) I commute. Note that ($*$) and ($**$) are in fact identical equations. Now, we associate the following iteration with ($**$):

$$\underline{\gamma}_{(k+1)} = \mathbb{P}(I - \frac{1}{\epsilon}R)\underline{\gamma}_{(k)} - \frac{1}{\epsilon}\mathbb{P}\mathbf{r}, \quad k = 0, 1, \tag{2.48}$$

which, in component form, is, for $i \in \mathcal{N}$, $k = 0, 1, \ldots$,

$$\gamma^i_{(k+1)}(y^i) = (I - \frac{1}{\epsilon}R_{ii})\gamma^i_{(k)}(y^i) - \frac{1}{\epsilon}\sum_{j\in\mathcal{N}, j\neq i} R_{ij}E_{\xi|y^i}\gamma^j(y^j) - \frac{1}{\epsilon}E_{\xi|y^i}r_i(\xi), \tag{2.49}$$

Clearly, if the sequence generated by (2.48) converges to a limit in $\mathbf{\Gamma}$, this also solves ($**$) and equivalently ($*$). Furthermore, (2.42) being a linear iteration, we know from Proposition 2.4.3(i) that the sequence $\{\underline{\gamma}_{(k)}\}$ converges if, and only if

$$\rho(\mathbb{P}(I - \frac{1}{\epsilon}R)) < 1.$$

Since both $\mathbb{P}$ and $(I - \frac{1}{\epsilon}R)$ map $\mathbf{\Gamma}$ into itself, and since $\mathbb{P}$ has operator norm equal to *one*, this inequality will be satisfied if

$$\rho(I - \frac{1}{\epsilon}R) < 1.$$

The matrix R being positive definite, this inequality can always be met by choosing $\epsilon > 0$ sufficiently large. If $\lambda_{\max}(R)$ denotes the maximum eigenvalue of R, choosing $\epsilon > \frac{1}{2}\lambda_{\max}(R)$ will in fact do the job. The specific choice of ϵ within this region may be dictated by other considerations, such as the speed of convergence. Since the smaller the spectral radium $\rho(I - \frac{1}{\epsilon}R)$ is, the "faster" the algorithm will (in general) converge, a reasonable choice of ϵ, with this in mind, is

$$\epsilon = \frac{1}{2}[\lambda_{\max}(R) + \lambda_{\min}(R)]. \tag{2.50}$$

where $\lambda_{\min}(R)$ denotes the minimum eigenvalues of R. These results are now summarized in the following proposition.

Proposition 2.4.4 *For the standard quadratic team with N agents, the parallel update scheme (2.49) converges to the unique team-optimal solution whenever $\epsilon > \frac{1}{2}\lambda_{\max}(R)$. A particular value of ϵ which leads to relatively fast convergence is given by (2.50).*

Proof The result has already been verified prior to the statement of the Proposition. Note that this also provides an alternative proof for Theorem 2.4.7 for the special case of standard quadratic teams. ◇

Remark 2.4.2 The algorithm (2.48) should be viewed (at this point) only as a computational tool, and not carry any significant interpretation in terms of the original team decision problem. There are also other variations of this algorithm which lead to convergence, but further discussion is beyond the scope and goal of this book.

Quadratic-Gaussian Teams

One class of quadratic teams for which the team-optimal solution can be obtained in closed form are those where the random state of nature ξ is a Gaussian random vector. Let us decompose ξ into $N + 1$ block vectors

$$\xi = (x', y^{1'}, y^{2'}, \ldots, y^{N'})' \tag{2.51}$$

of dimensions $r_0, r_1, r_2, \ldots, r_N$, respectively. Being a Gaussian random vector, ξ is completely described in terms of its mean value and covariance matrix, which we specify below:

$$E[\xi] =: \bar{\xi} = (\bar{x}', \bar{y}^{1'}, \ldots, \bar{y}^{N'}) \tag{2.52}$$

$$\text{cov}\,(\xi) =: \Sigma, \text{ with } [\Sigma]_{ij} =: \Sigma_{ij}, \quad i, j = 0, 1, \ldots, N \tag{2.53}$$

$[\Sigma]_{ij}$ denotes the *ij*'th block of the matrix Σ of dimension $r_i \times r_j$, which stands for the cross-variance between the *i*'th and *j*'th block components of ξ. We further assume (in addition to the natural condition $\Sigma \geq 0$) that $\Sigma_{ii} > 0$ for $i \in \mathcal{N}$, which means that the measurement vectors y^i's have nonsingular distributions. To complete the description of the quadratic-Gaussian team, we finally take the linear terms $r_i(\xi)$ in the loss function (2.29) to be linear in x, which makes x the "payoff relevant" part of the state of nature (recall the earlier discussion in Sect. 2.3 on the use of this terminology):

$$r_i(\xi) = D_i x, \quad i \in \mathcal{N} \tag{2.54}$$

where D_i is an $(r_i \times r_0)$ dimensional constant matrix. Note that in Definition 2.4.2 we simply have $Q_i = (D_i, 0, 0)$.

In the characterization of the team-optimal solution for the quadratic-Gaussian team we will need the following important result on the conditional distributions of Gaussian random vectors, which we will have occasion to use also in other chapters in the book. A proof of this result can be found in any standard text on probability theory.

Lemma 2.4.3 *Let z and y be jointly Gaussian distributed random vectors with mean values $\bar{z}$, $\bar{y}$, and covariance*

$$cov\,(z, y) = \begin{pmatrix} \Sigma_{zz} & \Sigma_{zy} \\ \Sigma'_{zy} & \Sigma_{yy} \end{pmatrix} \geq 0, \quad \Sigma_{yy} > 0. \tag{2.55}$$

Then, the conditional distribution of z given y is Gaussian, with mean

$$E[z|y] = \bar{z} + \Sigma_{zy}\Sigma_{yy}^{-1}(y - \bar{y}) \tag{2.56}$$

and covariance

$$cov(z|y) = \Sigma_{zz} - \Sigma_{zy}\Sigma_{yy}^{-1}\Sigma'_{zy} \tag{2.57}$$

The complete solution to the quadratic-Gaussian team is now given in the following.

Theorem 2.4.9 *The quadratic-Gaussian team decision problem as formulated above admits a* unique *team-optimal solution, that is affine in the measurement of each agent:*

$$\gamma^{i*}(y^i) = \Pi^i(y^i - \bar{y}^i) + M^i\bar{x}, \quad i \in \mathcal{N}. \tag{2.58}$$

Here, Π^i is an $(m_i \times r_i)$ matrix ($i \in \mathcal{N}$), uniquely solving the set of linear matrix equations:

$$R_{ii}\Pi^i \Sigma_{ii} + \sum_{j\in\mathcal{N}, j\neq i} R_{ij}\Pi^j \Sigma_{ji} + D_i \Sigma_{0i} = 0, \tag{2.59}$$

and M^i is an $(m_i \times r_0)$ matrix for each $i \in \mathcal{N}$, obtained as the unique solution of

$$\sum_{j\in\mathcal{N}} R_{ij} M^j + D_i = 0, \quad i \in \mathcal{N}. \tag{2.60}$$

Proof Referring back to iteration (2.49), and initializing it with $\gamma^i_{(0)} \equiv 0$, $i \in \mathcal{N}$, it follows from repeated application of Lemma 2.4.3 that $\gamma^i_{(k)}(y^i)$ that is generated by (2.49) is necessarily affine in y^i, for all $k = 1, 2, \ldots$, with the structure given by

$$\gamma^i_{(k)}(y^i) = \Pi^i_{(k)}(y^i - \bar{y}^i) + M^i_{(k)}\bar{x}.$$

By Proposition 2.4.4, this sequence converges, and the limiting solution is necessarily in the form (2.58). Further, by Theorem 2.4.7, this limiting policy should uniquely solve the stationarity equations (2.42). Therefore, all that remains to be done is to substitute (2.58) into (2.42), to arrive at (in view of Lemma 2.4.3):

$$[R_{ii}\Pi^i + \sum_{j\in\mathcal{N}, j\neq i} R_{ij}\Pi^j \Sigma_{ji}\Sigma^1_{ii} + D_i \Sigma_{oi}\Sigma^{-1}_{ii}](y^i - \bar{y}^i) + [\Sigma_{j\in\mathcal{N}} R_{ij} M^j + D_i]\bar{x} \equiv 0,$$

which is an identity for each $i \in \mathcal{N}$. Since $y^i - \bar{y}^i$ and $\bar{x}$ are independent, (2.59) and (2.60) readily follow. Clearly, in view of our reasoning above, the solutions to (2.59) and (2.60) have to be unique. This algebraic result can in fact also be proven directly. For (2.60), it trivially follows because it can be rewritten as

$$RM + D = 0$$

where $M := (M^{1'}, M^{2'}, \ldots, M^{N'})'$; $D := (D^{1'}, D^{2'}, \ldots, D^{N'})'$ and hence the unique solution is

$$M = -R^{-1}D.$$

◇

An important application of the above result is the following static Linear Quadratic Gaussian Problem: Consider a two-controller system evolving in $\mathbb{R}^n$ with the following description: Let x_1 be Gaussian and $x_2 = Ax_1 + B^1u^1_1 + B^2u^2_1 + w_1$

$$y^1_1 = C^1 x_1 + v^1_1,$$

$$y^2_1 = C^2 x_1 + v^2_1,$$

with w, v^1, v^2 zero-mean, i.i.d. disturbances. For $\rho_1, \rho_2 > 0$, let the goal be the minimization of

$$J(\gamma^1, \gamma^2) = E\left[||x_1||_2^2 + \rho_1 ||u_1^1||_2^2 + \rho_2 ||u_1^2||_2^2 + ||x_2||_2^2 \right] \tag{2.61}$$

over the control policies of the form:

$$u_t^i = \mu_t^i(y_1^i), \quad i = 1, 2$$

For such a setting, optimal policies are linear.

A quadratic-Gaussian team is known as a linear-quadratic-Gaussian (LQG) team, if furthermore the measurements have the special structure

$$y^i = H^i x + w^i, \quad i \in \mathcal{N} \tag{2.62}$$

where w^i, $i \in \mathcal{N}$, constitutes an independent sequence of zero-mean Gaussian random vectors, also independent of x. Let us denote the covariance of w^i, known as the measurement noise for agent $\mathbf{A}i$, by $N^i > 0$, $i \in \mathcal{N}$. Note that in this set-up the state of nature is given as

$$\xi = (x', w^{1'}, \ldots, w^{N'})'$$

which is again an $r := \sum_{i=0}^{N} r_i$-dimensional Gaussian random vector.

Now, in view of (2.62), and the independence of the noise sequence, we have

$$\bar{y}^i = H^i \bar{x}, \Sigma_{0i} = \Sigma_{00} H^{i'}, \Sigma_{ij} = H^i \Sigma_{00} H^{j'}, \Sigma_{ii} = H^i \Sigma_{00} H^{i'} + N^i, i \in \mathcal{N}.$$

Clearly, by the positive definiteness of N^i's, Σ_{ii}'s are positive definite, which means that all the hypotheses of Theorem 2.4.9 are satisfied. The following corollary then follows as a special case.

Corollary 2.4.2 *The LQG team decision problem as formulated above admits a unique team-optimal solution given by*

$$\gamma^{i*}(y^i) = \Pi^i (y^i - H^i \bar{x}) + M^i \bar{x}, \quad i \in \mathcal{N}, \tag{2.63}$$

where M^i, $i \in \mathcal{N}$, is the unique solution of (2.60), and Π^i solves uniquely the following version of (2.59):

$$R_{ii} \Pi^i + \left(\sum_{j \in \mathcal{N}, j \neq i} R_{ij} \Pi^j H^j \Sigma_{00} H^{i'} + D_i \Sigma_{00} H^{i'} \right) (H^i \Sigma_{00} H^{i'} + N^i)^{-1} = 0. \tag{2.64}$$

Example 2.4.3 To illustrate the preceding results, consider the two-agent scalar LQG team with loss function

$$L(x, \mathbf{u}) = (u^1 + u^2 + x)^2 + (u^1)^2 + (u^2)^2$$

and measurements

$$y^1 = x + w^1, \quad y^2 = x + w^2$$

under the independent probability measures

$$x \sim N(1, 2), w^1 \sim N(0, 2), w^2 \sim N(0, 1).$$

Direct application of Corollary 2.4.2 leads to the unique team-optimal solution

$$\left.\begin{aligned} \gamma^{1*}(y^1) &= -\tfrac{2}{11}(y^1 - 1) - \tfrac{1}{3} \\ \gamma^{2*}(y^2) &= -\tfrac{3}{11}(y^2 - 1) - \tfrac{1}{3} \end{aligned}\right\}. \tag{$*$}$$

The corresponding minimum team cost can be computed to be

$$J^* := J(\gamma^{1*}, \gamma^{2*}) \simeq 1.42424$$

Note that this is a symmetric team as far as the loss function goes (i.e., $L(u^1, u^2; x) = L(u^2, u^1, x)$), but as far as the measurements go agent **A**1 has "higher" measurement noise than agent **A**2. This is reflected in the team-optimal policies, with the measurement of **A**1 weighted less than the measurement of **A**2 (compare the gain $\frac{2}{11}$ against the gain $\frac{3}{11}$).

If the agents did not have access to any measurements, and thus optimize in the class of constant policies, the unique solution can easily be read off from ($*$) to be

$$\gamma^1_{OL} = \gamma^2_{OL} = -\frac{1}{3}$$

with the corresponding cost being

$$J_{OL} := J(\gamma^1_{OL}, \gamma^2_{OL}) = \frac{7}{3} \cong 2.33.$$

Hence, the decentralized measurements lead to a reduction in the team cost, as compared with the no-measurement (*open-loop*) case. If, on the other hand, the agents shared their measurements, with the team's common measurement now being $\mathbf{y} = (y^1, y^2)'$, the optimum cost should be lower than J^*. To study this specific model, we first note that Theorem 2.4.9 is directly applicable here, with

$$\Sigma_{11} = \Sigma_{12} = \Sigma_{22} = \text{cov}(\mathbf{y}) = \begin{pmatrix} 4 & 2 \\ 2 & 3 \end{pmatrix}; \ \Sigma_{01} = \Sigma_{02} = (2, 2).$$

The unique team-optimal solution can readily be obtained to be

$$\gamma_{sh}^1(\mathbf{y}) = \gamma_{sh}^2(\mathbf{y}) = -\frac{1}{4}[\frac{1}{3}(y^1 - 1) + \frac{2}{3}(y^2 - 1)] - \frac{1}{3}$$

with the corresponding cost being

$$J_{sh} := J(\gamma_{sh}^1, \gamma_{sh}^2) = \frac{4}{3} \cong 1.33.$$

Finally, if both agents had perfect access to the true value of x (the case of perfect measurements), the unique optimal solution would be

$$\gamma_{pr}^1(x) = \gamma_{pr}^2(x) = -\frac{1}{3}x$$

with a cost level of

$$J_{pr} := J(\gamma_{pr}^1, \gamma_{pr}^2) = \frac{1}{3}E[x^2] = 1$$

which is the lowest possible value for J, under any measurement scheme. ◇

2.4.4 Positively Exponentiated Quadratic Loss

Consider again the formulation of the quadratic-Gaussian team in Definition 2.4.2, but with the loss function being a positively exponentiated quadratic function, i.e.,

$$L(\xi; \mathbf{u}) = \theta e^{\frac{\theta}{2}C(\xi;\mathbf{u})}, \quad \theta > 0 \tag{2.65}$$

where C is a strictly convex (in $\mathbf{u}$) function:

$$C(\xi; \mathbf{u}) = \sum_{i,j\in\mathcal{N}} u^{i'} R_{ij} u^j + 2\sum_{i\in\mathcal{N}} u^{i'} D_i x + x' Q x. \tag{2.66}$$

The state of nature, ξ, is a Gaussian vector, as specified earlier by (2.51)–(2.53).

A static team problem with the structure above is known as an *exponential-Gaussian team* or a *linear-exponential-Gaussian team* (LEGT), the latter used especially if the measurements are given in the form (2.62). An exponential (of a quadratic) loss function captures phenomena not obtainable from a quadratic loss function, and is preferred in situations where higher (than second) order moments of the statistical quantities should also be taken into consideration. Such a formulation also has intimate connections with robust control [228].

A team using an exponential quadratic loss function in the construction of policies is called *risk averse* if $\theta > 0$, and *risk preferring* if $\theta < 0$. Here we will discuss only the case $\theta > 0$ because it is only in this case that L is convex in $\mathbf{u}$, which will enable us to apply some of the results of Sect. 2.4.2. The "optimistic" case $\theta < 0$ does not lead to a convex loss function, and hence it is not possible to obtain a general theory to cover this case as well. However, this should not be construed as the $\theta < 0$ case not being well-defined or not interesting. In fact, the stationarity conditions could hold in this case also, but one has to study each problem individually before concluding global (team) optimality.

Returning to the LEGT problem with $\theta > 0$, the first step toward studying its solution would be to obtain a characterization of the stationarity conditions (2.26). To avoid some unnecessary complexity in the analysis to follow, let us take the mean value of ξ to be zero, and furthermore let us restrict ourselves at the outset to linear decision rules (policies) for the agents—the latter will actually create no loss of generality as we shall see later.

Accordingly, let the decision rules be given as

$$\gamma^j(y^j) = A^j y^j, \quad j \in \mathcal{N}. \tag{2.67}$$

Let us fix all but one (say i'th one) as above, and substitute them into (2.66) to obtain

$$C(\xi; u, \{\gamma^j\}_{j\neq i}) = u' R_{ii} u + 2u' T_i' \xi_i + \xi_i' S_i \xi_i =: C_i(\xi_i, u)$$

where u stands for u^i, and T_i, S_i are defined as follows:

$$\xi_i := (x', y^{1'} \dots y^{i-1'} y^{i+1'} \dots y^{N'})'$$

$$T_i' \xi_i = D_i x + \sum_{j\in\mathcal{N}, j\neq i} R_{ij} A^j y^j \tag{2.68}$$

$$\xi_i' S_i \xi_i = x' Q x + \sum_{j,k\in\mathcal{N}; j,k\neq i} y^{j'} A^{j'} R_{jk} A^k y^k + 2 \sum_{j\in\mathcal{N}, j\neq i} y^{j'} A^{j'} D_j x. \tag{2.69}$$

The important point here is that T_i and S_i are constant matrices (not dependent on ξ), but they depend on the policy gain matrices A^j, for all $j \in \mathcal{N}$, except $j = i$.

We now evaluate

$$E_{\xi|y^i} L(\xi; u, \{\gamma^j\}_{j\neq i}) =: J_i(u; y)$$

where, for convenience, we have dropped the superscript from y^i. Using Lemma 2.4.3, the distribution of ξ conditioned on $y^i = y$ is Gaussian, with mean and covariance given by

$$E[\xi\,|\,y] = (\Sigma'_{0i}\Sigma'_{1i}\dots\Sigma'_{Ni})'\Sigma_{ii}^{-1}y =: \widehat{\xi}_i \tag{2.70}$$

$$\text{cov}(\xi_i\,|\,y) = \Sigma^{(i)} - (\Sigma'_{0i}\dots\Sigma'_{Ni})'\Sigma_{ii}^{-1}(\Sigma'_{0i}\dots\Sigma'_{Ni}) =: \widehat{\Sigma}_i \tag{2.71}$$

which we assume to be positive definite. In (2.70) the matrix Σ_{ii} does not appear in $(\dots)$, and likewise in (2.71). Furthermore $\Sigma^{(i)}$ in (2.71) is the covariance of ξ_i, which is the Σ of (2.53) with the $(i+1)$'th row and column block deleted.

Now, aside from a *pdf* normalization constant,

$$J_i(u;\,y) = \int \theta e^{\frac{\theta}{2}C_i(\xi_i;u)}e^{-\frac{1}{2}(\xi_i-\widehat{\xi}_i)'\widehat{\Sigma}_i^{-1}(\xi_i-\widehat{\xi}_i)}d\xi_i$$

where the integration is over the vector ξ_i belonging to an appropriate dimensional Euclidean space. This integral will have a finite value if, and only if, the quadratic term in ξ_i is negative definite, which brings in the condition

$$M_i := \widehat{\Sigma}_i > 0\,. \tag{2.72}$$

Under this condition, the integral can be evaluated (using a property of Gaussian *pdf*'s) to yield (again aside from a positive multiplying constant)

$$J_i(u;\,y) = \theta e^{\tilde{C}_i(\widehat{\xi}_i,u)}$$

where

$$\tilde{C}_i(\widehat{\xi}_i,u) := \frac{\theta}{2}u'R_{ii}u + \frac{1}{2}(\theta T_i u + \widehat{\Sigma}_i^{-1}\widehat{\xi}_i)'M_i^{-1}(\theta T_i u + \widehat{\Sigma}_i^{-1}\widehat{\xi}_i) - \frac{1}{2}\widehat{\xi}_i^{-1}\widehat{\Sigma}_i^{-1}\widehat{\xi}_i.$$

Note that for $\theta > 0$, $\tilde{C}_i$ is strictly convex in u, which implies that $J_i(u;\,y)$ is also strictly convex in u. Minimization of J_i is equivalent to minimization of $\tilde{C}_i$ which, being quadratic, immediately leads to

$$u^i = \gamma^i(y^i) = -(R_{ii} + \theta T_i'M_i^{-1}T_i)^{-1}T_iM_i^{-1}\widehat{\Sigma}_i^{-1}\widehat{\xi}_i =: A^i y^i\,. \tag{2.73}$$

Clearly this solution also satisfies the stationarity condition (2.26), with all other (than i'th) agents' policies fixed as in (2.67).

Note that A^i determined as in (2.73) depends on the fixed gain matrices A^j's, for $j \neq i$; this dependence being through M_i and T_i. Let us denote this relationship by

$$A^i = f^i(A^i,\dots,A^{i-1},A^{i+1},\dots,A^N) \tag{2.74}$$

where f^i is a nonlinear but continuous function, determined uniquely by (2.73). Since agent **A**i was selected arbitrarily in the preceding analysis, a similar function will exist for each agent, so that (2.74) will hold for all $i \in \mathcal{N}$. This readily brings us to the following proposition.

Proposition 2.4.5 *The positively exponentiated Gaussian team problem admits a linear stationary solution if, and only if, there exists a set of matrices A^i, $i \in \mathcal{N}$, mutually satisfying (2.74), and under which (2.72) holds for all $i \in \mathcal{N}$, and $J(\gamma)$ remains finite.*

Proof In view of Definition 2.4.1, the result follows from the analysis that led to the proposition. ◇

Remark 2.4.3 If the random state of nature, ξ, has nonzero mean, as in (2.52), then the policies (2.67) will have to be replaced by the affine structure

$$\gamma^j(y^j) = A^j(y^j - \bar{y}^j) + b^j, \quad i \in \mathcal{N}.$$

Within this structure, one can again proceed through the preceding analysis and arrive at a counterpart of Proposition 2.4.5.

Remark 2.4.4 The boundedness of the cost corresponding to the linear stationary solution can be checked by evaluating the quantity

$$E_{y^i} J_i(A^i y^i; y^i) \tag{$*$}$$

where J_i was defined in the analysis leading to the proposition. We first substitute (2.73) in $C_i(\widehat{\xi}_i, u)$ to obtain

$$\tilde{C}_i(\widehat{\xi}_i, \gamma^i(y^i)) = -\frac{1}{2}\widehat{\xi}_i' N_i \widehat{\xi}_i$$

where

$$N_i := \widehat{\Sigma}_i^{-1} + \widehat{\Sigma}_i^{-1} M_i^{-1} T_i (\frac{1}{\theta} R_{ii} + T_i' M_i^{-1} T_i)^{-1} T_i' M_i^{-1} \widehat{\Sigma}_i^{-1} - \widehat{\Sigma}_i^{-1} M_i^{-1} \widehat{\Sigma}_i^{-1}.$$

Then, we observe that $(*)$ is finite if, and only if, the integral

$$\int \theta e^{-\frac{1}{2}\widehat{\xi}_i' N_i \widehat{\xi}_i} e^{-\frac{1}{2} y' \Sigma_{ii}^{-1} y} dy$$

is finite, where $\widehat{\xi}_i$ is related to y through (2.70). This condition is equivalent to the exponent being negative definite, that is

$$\widehat{\xi}_i' N_i \widehat{\xi}_i + y' \Sigma_{ii}^{-1} y > 0 \qquad \forall y \in \mathbb{R}^{m_i}, y \neq 0. \tag{$**$}$$

Proposition 2.4.5 above leaves a number of questions unanswered. First, we would like to know whether a linear stationary solution, whenever it exists, is team-optimal, and secondly whether there would be other team optimal solutions if the linear stationary solution (*lss*) ceases to exist. Clearly, we would not generally expect the *lss* to exist for all (especially arbitrarily large) values of θ, because of condition (2.72). The theorem below now provides an answer to the first question raised above; the second question is a most difficult one for which no general answer is known as yet.

Theorem 2.4.10 *Let $\underline{\gamma}^* \in \mathbf{\Gamma}$ be the linear stationary solution of Proposition 2.4.5, and let there exist some other linear policy $\underline{\beta} \in \mathbf{\Gamma}$ such that $J(\underline{\beta}) < \infty$. Then, $\underline{\gamma}^*$ is the unique team-optimal solution of the (positively) exponentiated-Gaussian team problem.*

Proof Here we resort to Theorem 2.4.6, which delineates the conditions under which stationarity implies team-optimality. Clearly $L(\xi; \mathbf{u})$ is strictly convex and continuously differentiable (in $\mathbf{u}$), and $J(\underline{\gamma})$ is bounded from below (by zero) for all $\underline{\gamma} \in \mathbf{\Gamma}$. The stationary solution $\underline{\gamma}^*$ has finite cost by hypothesis, and the subset (say, $\hat{\mathbf{\Gamma}}$) of $\mathbf{\Gamma}$ on which J is finite is not a singleton, again by hypothesis. Hence, to apply Theorem 2.4.6, one has to show that condition (c.5) holds for this problem. This is indeed the case, and follows from the fact that the subset $\hat{\mathbf{\Gamma}}$ referred to above is not a singleton. The proof of this result is quite technical, and will not be given here; it can be found in [429]. ◇

Remark 2.4.5 A sufficient condition for the second hypothesis of Theorem 2.4.10 is the following: Choose $\beta \equiv 0$, which is clearly a linear policy. Then, $J(\beta) = E\{\theta e^{\frac{\theta}{2}x'Qx}\}$, which is finite if, and only if, $\Sigma_{00}^{-1} - \theta Q > 0$. Hence, if θ is chosen to be smaller than $1/[\lambda_{\max}(\Sigma_{00})\lambda_{\max}(Q)]$, the second hypothesis is satisfied. Of course, this condition (on θ) can be made less stringent by choosing some other (nonzero) β.

Remark 2.4.6 For the negatively exponentiated Gaussian team problem, Proposition 2.4.5 remains equally valid (now in fact condition (2.72) would be satisfied with a bigger margin on θ), but we do not have the counterpart of Theorem 2.4.10 because of lack of convexity.

Example 2.4.4 To illustrate the main result of Theorem 2.4.10 and to study the restrictions imposed on the parameters of the problem by the various conditions stated there, let us reconsider the static two-agent team problem of Example 2.4.3, with two differences: Now, the loss functional is a positive exponential of the one given there, i.e.,

$$L(x, u) = e^{\theta C(x, \mathbf{u})}$$

$$C(x, \mathbf{u}) = (u^1 + u^2 + x)^2 + (u^1)^2 + (u^2)^2,$$

and the random variable x has zero mean, i.e.,

$$x \sim N(0, 2), \quad w^1 \sim N(0, 2), \quad w^2 \sim N(0, 1).$$

The measurements are still given by

$$y^1 = x + w^1, \quad y^2 = x + w^2.$$

Writing out the stationarity conditions (2.74), we obtain, after some algebra,

$$A^1 = -\frac{1}{2}[1 + A^2 - \theta(A^2)^2]/c_1(\theta, A^2), \quad A^2 = -[1 + A^1 - \theta(A^1)^2]/c_2(\theta, A^1) \tag{$*$}$$

where

$$c_1(\theta, A^2) := 2 - \theta[1 + 2A^2 + 6(A^2)^2] + \theta^2(A^2)^2$$

$$c_2(\theta, A^1) := 3 - \theta[1 + 12(A^1)^2 + 2A^1] + 2\theta^2(A^1)^2.$$

The matrices M_1 and M_2, defined by (2.72) , are given by

$$M_1 = \begin{pmatrix} 2-\theta & -1-\theta A^2 \\ -1-\theta A^2 & 1-2\theta(A^2)^2 \end{pmatrix}, \quad M_2 = \begin{pmatrix} 2-\theta & -\frac{1}{2}-\theta A^1 \\ -\frac{1}{2}-\theta A^1 & \frac{1}{2}-2\theta(A^1)^2 \end{pmatrix},$$

so that condition (2.72) reads

$$0 < \theta < 2, \quad c_1(\theta, A^2) > 0, \quad c_2(\theta, A^1) > 0. \tag{$**$}$$

Trying out two different values of θ, namely $\theta = 1$ and $\theta = \frac{1}{3}$, we find that for the former there is no solution to ($*$) that also satisfies ($**$); for the latter, however, there exists a unique solution ($*$) that also meets ($**$), which is

$$A^{1^*} = -0.236375, \quad A^{2*} = -0.345398 \quad (\theta = \frac{1}{3}).$$

This solution, and the associated value of θ, also satisfy the conditions of Remarks 2.4.4 and 2.4.5, and hence by Theorem 2.4.10 there exists a unique team-optimal solution to the LEGT problem:

$$\gamma^{1*}(y^1) = -0.236375y^1, \quad \gamma^{2*}(y^2) = -0.345398y^2.$$

It is important to note that, as $\theta \to 0$ in ($*$), the nonlinear equations reduce to linear equations:

$$A^1 = -\frac{1}{4}(1 + A^2), \quad A^2 = -\frac{1}{3}(1 + A^1)$$

whose unique solution $(-\frac{2}{11}, -\frac{3}{11})$ is precisely the gain coefficients in the team-optimal solution of Example 2.4.3. Hence in the limit as $\theta \to 0$ we recover the solution of the corresponding LQG team (with loss function $C(x, \mathbf{u})$). This is to be expected because for any positive function C, we have $\lim_{\theta\to 0} \frac{1}{\theta}(e^{\theta C} - 1) = C$.

Remark 2.4.7 A method for solving the set of coupled nonlinear equations $(*)$ of Example 2.4.4 for a particular value of θ, or more generally equations (2.74), is provided by the parallel update scheme

$$A^i_{(k+1)} = f^i(A^1_{(k)}, \ldots, A^{i-1}_{(k)}, A^{i+1}_{(k)}, \ldots, A^N_{(k)}), \quad i \in \mathcal{N}, \quad k = 0, 1, \ldots$$

where the starting point is arbitrary. This set of recursive equations admits exactly the same interpretation as in the LQ team case, and as was the case there, this recursion may not converge (even if the LEG team problem admits a solution) for $N \geq 3$. For $N = 2$, however, the recursion will converge whenever the LEQ team problem admits a solution; this is because at each step this corresponds to an agent's minimization of J by fixing the other agent's policy at its most recently updated value. Since one is basically minimizing a strictly convex functional (in the LEGT problem), the unique minimum, whenever it exists, should be reachable by such a unilaterally cost-minimizing update scheme.

2.5 Concluding Remarks

This chapter has provided a general introduction to stochastic team decision problems and associated solution concepts. Static and dynamic teams have been identified, and in the context of static teams conditions for existence of team-optimal solutions and for person-by-person optimality to imply team-optimality have been obtained. The chapter has also discussed iterative methods for obtaining team-optimal solutions, and illustrated the theory presented with numerical examples. The chapter also serves as an introduction for the material to follow, also by raising or illustrating many of the questions that will be addressed.

2.6 Bibliographic Notes

Team decision theory has its roots in both control theory and economics. Jacob Marschak [506] was perhaps the first to introduce the basic elements of teams, and to provide the first steps toward the development of a systematic *team theory*. Roy Radner [594] followed with a precise mathematical formulation and provided conclusive results to some classes of static teams, establishing connections between person-by-person optimality, stationarity, and team-optimality. Marschak's and Radner's collaborative work culminated in the publication of their influential 1972 book [507]. At the time when such developments were being made, significant progress in the theory of statistical decision theory was also taking place: Bahadur's characterization of information fields and sufficient statistics [53, 54]; Blackwell's sufficient statistics and comparison of experiments results [123]; and Wald's [754], Savage's [642], and Chernoff's [174] contributions to statistical decision theory, among other major developments in probability theory, contributed to the rapid development of team decision theory.

Contributions of Hans Witsenhausen [772, 774, 775, 777, 780] on dynamic teams and characterization of information structures have been crucial in the progress of our understanding of dynamic teams. The reader is referred to Sect. 3.2, where Witsenhausen's *intrinsic model*, as well as other models for dynamic teams, are discussed in detail. Additionally, Sect. 3.9 includes a detailed discussion for non-sequential teams where some important contributions in the literature have been due to Andersland and Teneketzis [26, 27] and Teneketzis [713], in addition to Witsenhausen [772].

Considerations of risk sensitivity motivated researchers to look into team problems with exponentiated loss function, with substantial results in this domain obtained (for teams) by Krainak, Speyer, and Marcus [430]. De Waal and van Schuppen [208] considered extensions to discrete action spaces.

Bagchi and Başar [50] studied stochastic teams in continuous-time as well as non-Gaussian settings for both static and dynamic teams. See [364] for a recent study and an overview of continuous-time LQG teams.

Başar [87] studied team problems and more general nonzero-sum stochastic games when agents do not agree on a common *a priori* probability measure on the primitive random variables, and work under their own subjective views of the environment, with team models in this context necessarily leading to game formulations. This will be discussed further in the book.

The Hilbert space argument presented in Theorem 2.4.7 builds on Radner [594]; this argument was presented in detail earlier in [824, p. 51] and [293].

Further discussion on design of information structures in the context of team theory and economics applications can be found in [35] and [734], among a rich collection of other contributions. A recent application of static tram decision theory to optimal pre-coding design in multi-user communications appears in [542].

In the next chapter, Chap. 3, we will see extensions of the static team theory of this chapter to dynamic teams, where information structures are of paramount importance. We will also consider Witsenhausen's intrinsic model more explicitly in Sect. 3.2. We refer the reader to also Teneketzis [713], in addition to [777], in this context.

In addition to convexity, another class of teams which will lead to desirable computational properties are those that are symmetric or exchangeable. Under such conditions, finite team optimal policies will be symmetric, and in the absence of convexity infinite-agent team policies will be symmetric. These will be studied in detail in Chap. 6

Part of the chapter builds on results from [594] and [429, 430], however with somewhat different proofs for some of the key results. The update schemes considered in Sect. 2.4 are based on [87].

Chapter 3
Information Structures in Decentralized Stochastic Control

3.1 Introduction

In Chap. 2, we introduced stochastic team decision problems with static and dynamic information structures. In the present chapter, and the following two, we will provide a comprehensive mathematical analysis on information structures, which will also lead to answers on existence, analytical and numerical methods, approximations, and structures for optimal solutions to stochastic static and dynamic team problems.

Revisiting Single-Agent Stochastic Control, Challenges in Decentralized Stochastic Control, and a Prelude

To motivate some of the ideas to be presented and employed in the following few chapters, we now re-visit classical, single agent, stochastic control problems, where one has a system of the form

$$x_{t+1} = f(x_t, u_t, w_t), \quad t \in \mathbb{Z}_+, \tag{3.1}$$

where x_t is an $\mathbb{X}$-valued state variable, u_t a $\mathbb{U}$-valued control action variable, w_t a $\mathbb{W}$-valued i.i.d. noise process, and f a measurable function, where $\mathbb{X}, \mathbb{U}, \mathbb{W}$ are appropriate spaces, with all random variables defined on some probability space $(\Omega, \mathcal{F}, P)$. Let the actions be generated using some control policy $\gamma = \{\gamma_t\}$ with

$$u_t = \gamma_t(I_t), \quad t \in \mathbb{Z}_+,$$

where I_t is the information available at t. If

$$I_t = \{x_0, \cdots, x_t; u_0, \cdots, u_{t-1}\},$$

we have a *fully observed* system.

S. Yüksel, T. Başar, *Stochastic Teams, Games, and Control under Information Constraints*, Systems & Control: Foundations & Applications,
https://doi.org/10.1007/978-3-031-54071-4_3

If the agent, or controller, has instead access to only noisy measurements

$$y_t = g(x_t, v_t),$$

with v_t being another i.i.d. noise, and $I_t = \{y_0, \cdots, y_t; u_0, \cdots, u_{t-1}\}$, we then have a *partially observed* system.

We will see that if we view the decision maker acting at time t as **DM**t, with her information I_t (which will be expanding over time), classical stochastic control problems are indeed *classical* in the sense of Witsenhausen, as we discuss in Sect. 3.2.

In the general theory of stochastic control involving one controller acting over time with increasing information, given an optimality criterion (e.g. expected finite horizon cost, discounted cost, average cost, or terminal cost), there are few powerful techniques to establish the existence/computation of optimal policies:

(i) Dynamic programming, Bellman's equations of optimality and measurable selections: Under either weak continuity of the transition probabilities (in both the state and action), or strong continuity (of the transition probability in actions for every state) properties and measurable selection conditions, we have existence results or solutions via explicit computations. The corresponding *measurable selection criteria* have been summarized elaborately in [342, Theorem 3.3.5 and Appendix D] and [343] among many other references, including [347, Theorem 2], [649], and [442]. For infinite-horizon problems, we also have linear programming/convex analytic techniques [342, Chp. 6], [135], applicable under similar conditions.

(ii) Strategic measures approach: For stochastic control problems, *strategic measures* are defined (see [230, 254, 649]) as the set of probability measures induced on the product spaces of the state and action pairs by measurable control policies: Given an initial distribution on the state, and a policy, one can uniquely define a probability measure on the product space. Topological properties, such as measurability and compactness, of sets of strategic measures have been studied in [125, 230, 254, 649].

(iii) For partially observed models, it is known that every such model can be reduced to an equivalent fully observed model [602, 691, 834], where the state is replaced with the conditional probability on the state given the information I_t, with the resulting fully observed MDP with a probability-measure valued state often referred to as a belief-MDP, whose evolution is computed via non-linear filtering. This leads to optimality of *separated* policies, where one first computes the belief and applies an optimal control using the belief variable under a measurable selection condition resulting from weak continuity properties of non-linear filters (see Theorem E.3.2). For linear systems under quadratic loss functions and with Gaussian noise, this approach leads to the use of the celebrated Kalman filter and stronger separation result where the conditional mean is a sufficient statistic for optimal control, which also turns out to be linear [113] (as discussed in Chap. 1). See Appendix E for a further review.

However, all of these approaches crucially require that the information is expanding over time, in the sense that the σ-field generated by the information available to the decision maker at time t is a subset of (or *nested in*) that generated by the information at $t + 1$ for any $t \in \mathbb{Z}_+$: $\sigma(I_t) \subset \sigma(I_{t+1})$. The reason why this is crucial will be made more evident in the chapters to follow, but let us mention briefly at this point that this condition facilitates closedness/compactness conditions of strategic measures and also allows one to use the *iterated expectations* property (see Appendix B) to enable the use of dynamic programming.

In general stochastic team theory or decentralized stochastic control, this nestedness condition may not hold, and hence the standard methods are not typically directly applicable: In the absence of such a nestedness condition, much of the classical and standard analyses on existence/structure/recursions fail. Indeed, as illustrated in Sect. 2.3.1, in decentralized stochastic control a typical model (before being reduced to the intrinsic model) is as follows:

$$x_{t+1} = f(x_t, u_t^1, \ldots, u_t^L, w_t),$$

with the m-th DM having a local control policy

$$u_t^m = \gamma_t^m(I_t^m), t \in \mathbb{Z}_+,$$

where I_t^m denotes the information available to that DM. Decentralized stochastic control theory, accordingly, entails significant challenges that are not directly approachable using the standard machinery of optimal stochastic control.

In this, and the following two chapters, we will present a comprehensive theory on how to approach such problems.

In this chapter, our focus will be on the definition, classification, and general qualitative properties of information structures. In the following two chapters, we will establish decentralized analogues of the above two approaches (i)–(ii) noted above: Chap. 4 on developing a strategic measures approach and Chap. 5 on developing dynamic programming methods for decentralized stochastic control (where the latter will also be seen to be a generalization of (iii) as well). However, we will see that these are significantly more challenging when compared with classical, single-DM, stochastic control theory. Accordingly, new approaches will be introduced, whose utility in arriving at existence, approximations, and structural results on optimal policies will become evident throughout the analysis.

Organization of the Chapter

In this chapter, Sect. 3.2 summarizes a precise and systematic characterization of information structures in view of Witsenhausen's Intrinsic Model, and presents further dynamic team decision models. Section 3.3 introduces non-classical information structures, and highlights the challenges involved in solving stochastic teams which feature such patterns. Section 3.4 presents static reductions of dynamic teams (classified as policy-dependent and policy-independent); the section also discusses stochastic dynamic team problems with nested or partially nested information structures (classical and quasi-classical), presenting also near universal applicability of static reduction methods. Section 3.5 presents a more general intrinsic model involving agents acting possibly multiple times, and a class of agent-wise reductions arrived at via change of measure arguments. Section 3.6 highlights some limitations of static reduction methods in person-by-person-optimality analysis, in that while static reductions always maintain isomorphism between optimal solutions under dynamic teams and their static reductions, this may not hold for person-by-person-optimal solutions or stationary solutions. Section 3.7 is on convexity properties of stochastic team problems and their information structure dependent characterization. Section 3.8 studies expansions of information structures. Finally, Sect. 3.9 studies non-sequential stochastic team theory.

3.2 Witsenhausen's Intrinsic Model and Classification of Information Structures

In this section, we present a characterization of information structures for stochastic team decision problems, building on Hans Witsenhausen's *Intrinsic Model* [777]. In this model (described in discrete time), any action applied at any given time is regarded as applied by an individual decision maker/agent, who acts only once. One advantage of this model, in addition to its generality, is that the definitions regarding information structures can be compactly described.

Suppose that in the decentralized system considered below, there is a pre-defined order in which the decision makers act. Such systems are called *sequential systems*. In the context of a sequential system, the *Intrinsic Model* has three components:[1]

A collection of *measurable spaces* $\{(\Omega, \mathcal{F}), (\mathbb{U}^i, \mathcal{U}^i), (\mathbb{Y}^i, \mathcal{Y}^i), i \in \mathcal{N}\}$, with $\mathcal{N} := \{1, 2, \cdots, N\}$, specifying the system's distinguishable events, and the control and measurement spaces. Here N is the number of control actions taken, and each of these actions is taken by an individual (different) DM (hence, even a DM with perfect recall can be regarded as a separate decision maker every time she acts). The pair $(\Omega, \mathcal{F})$ is a measurable space (on which an underlying probability may be defined). The pair $(\mathbb{U}^i, \mathcal{U}^i)$ denotes the measurable space from which the action, u^i, of **DM**i is selected. The pair $(\mathbb{Y}^i, \mathcal{Y}^i)$ denotes the measurable observation/measurement space for **DM**i.

[1] We note that the intrinsic model of Witsenhausen as defined in [777] provides a set-theoretic characterization of information fields; however, for standard Borel spaces (which includes most of the spaces we will be working with in this book, see Sect. A.2), the model above is equivalent to that of Witsenhausen's. We will utilize this functional representation in our analysis to follow. We also note that Witsenhausen has shown that the intrinsic model can be written as a special case of an equivalent model (also due to Witsenhausen [780]), which has a similar structure as the one above (with functional descriptions on the measurement variables) with only some slight differences.

A *measurement constraint* which establishes the connection between the observation variables and the system's distinguishable events. The $\mathbb{Y}^i$-valued observation variables are given by $y^i = \eta^i(\omega, \mathbf{u}^{[1,i-1]})$, $\mathbf{u}^{[1,i-1]} = \{u^k, k \leq i-1\}$, η^i measurable functions and u^k denotes the action of **DM**k. Hence, the information variable y^i induces a σ-field, $\sigma(\mathcal{I}^i)$ over $\Omega \times \prod_{k=1}^{i-1} \mathbb{U}^k$

A *design constraint* which restricts the set of admissible N-tuple control laws $\underline{\gamma} = \{\gamma^1, \gamma^2, \ldots, \gamma^N\}$, also called *designs* or *policies*, to the set of all measurable control functions, so that $u^i = \gamma^i(y^i)$, with $y^i = \eta^i(\omega, \mathbf{u}^{[1,i-1]})$, and γ^i, η^i measurable functions. Let Γ^i denote the set of all admissible policies for **DM**i and let $\mathbf{\Gamma} = \prod_k \Gamma^k$.

We also introduce a fourth ingredient:

A *probability measure* P defined on $(\Omega, \mathcal{F})$, which describes the measures on the random events in the model.

To complete the description of a stochastic team decision problem, we also need to introduce a cost criterion. Let

$$\underline{\gamma} = \{\gamma^1, \cdots, \gamma^N\}$$

and a cost function be defined as:[2]

$$J(\underline{\gamma}) = E^{\underline{\gamma}}[c(\omega_0, \mathbf{u})] = E[c(\omega_0, \gamma^1(y^1), \cdots, \gamma^N(y^N))], \tag{3.2}$$

for some non-negative measurable loss (or cost) function $c : \Omega_0 \times \prod_k \mathbb{U}^k \to \mathbb{R}_+$. Here, we have the notation $\mathbf{u} = \{u^i, i \in \mathcal{N}\}$, and ω_0 may be viewed as the cost function relevant exogenous variable contained in ω.

Classical, Quasi-Classical (Partially Nested) and Non-classical Information Structures

Earlier, in Sects. 2.2 and 2.3 we had classified information structures as static vs. dynamic. Accordingly, under the intrinsic model, (the information structure of) a team problem is *dynamic* if the information available to at least one DM is affected by the action of at least one other DM. A decentralized team problem is *static*, if the information available at every DM is only affected by exogenous disturbances (Nature) $\omega \in \Omega$; that is, no other DM can affect the information available to any given DM. See Fig. 3.1 for a depiction.

Information structures can also be classified as *classical*, *quasi-classical* or *non-classical*.

[2] Departing from the notation convention of Chap. 2 for the loss function, we use here "c" instead of "L".

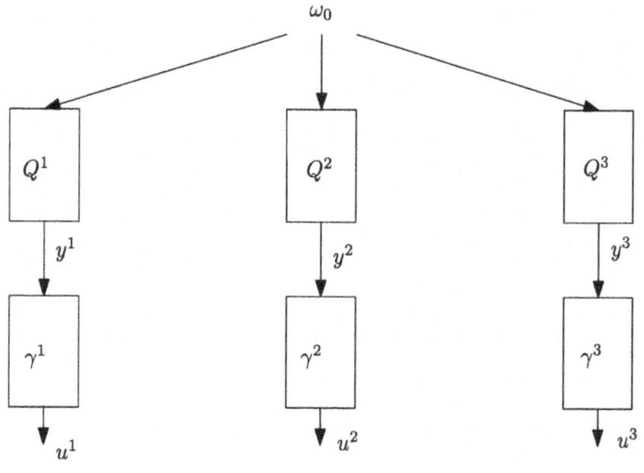

Fig. 3.1 An example of a static information structure. Here, $Q^i(y^i \in \cdot|\omega_0) := P(\eta^i(\omega) \in \cdot|\omega_0)$, $i = 1, 2, 3$

Definition 3.2.1

(i) An Information Structure (IS) is *classical* if y^i contains all of the information available to **DM**k for $k < i$.
(ii) An IS is *quasi-classical* or *partially nested*, if whenever u^k, for some $k < i$, affects y^i through the measurement function η^i, y^i contains y^k (that is $\sigma(y^k) \subset \sigma(y^i)$).
(iii) An IS which is not partially nested is *non-classical*.

As noted, non-classical information structure (*IS*) arises if an agent **DM**i's action affects the information available to another agent **DM**j, who however does not have access to the information available to **DM**i based on which her action was constructed. Another way of stating this is that the information sigma field (over Ω) of agent **DM**j is dependent explicitly on the policy (decision rule, or control law) of **DM**i. The difficulties in developing a comprehensive and broadly applicable theory for such decision problems with non-classical information stems from the fact that actions (controls) exhibit *triple* roles: (i) the control effort of reducing the cost (e.g. as an *actuator*), (ii) improvement of future knowledge of uncertainty, and (iii) *signaling* to agents acting in the future some useful information on relevant random variables, which they do not necessarily acquire—and these roles are generally not aligned.

Stochastic dynamic teams which do not entail non-classical information are generally simpler, since there is no possibility of affecting the information content of measurements or uncertainty at future stages through actions at the present. Such dynamic teams are also known as *neutral*—a term coined by Feldbaum [259]—where (assuming that there is a state space representation, as in (2.18)–(2.19)) the conditional probability distribution of the state vector given past and present measurements, past control actions, and past control laws (or policies) does not depend on the control laws. As captured by Definition 3.2.1, *IS*s which are not of the non-classical type are generally grouped into two categories: *classical* and *quasi-classical*.

Classical ISs include deterministic patterns and centralized information patterns. Referring back to the formulation (2.18)–(2.19) where the state-variable based description was introduced, deterministic patterns arise when the information is not noise-corrupted and may be of the *open-loop* type in which only the initial value of the state is available and no dynamic information is acquired, **or** of the *closed-loop* type where perfect information concerning the current value of the state is also acquired, that is, in (2.19) $y_t^i = x_t$, for all i and t, and there is perfect recall. Centralized patterns arise when all agents exchange their measurements without any delay and also recall the past information. Under the deterministic or centralized stochastic *IS*s, stochastic team problems become equivalent to stochastic control problems and the solution techniques for these have been included in Appendix E; recall the discussion at the beginning of the chapter where we noted that all typical (single-controller) stochastic control problems with fully observed or partially observed models fall into the classical information structure category.

The second type of *IS* which is not non-classical, again as captured by Definition 3.2.1, is of the *quasi-classical* type, also known as *partially nested*. Assuming again the existence of a common clock for all agents, an *IS* is partially nested if an agent's information at a particular stage t can depend on the action of some other agent at some stage $t' < t$ only if she also has access to the information of that agent at stage t'.[3]

As a special case, the static team problems considered in Chap. 2, Sect. 2.4, are partially nested, but not with classical *IS*, unless there is complete sharing of (the static) measurements. The one-step delayed information sharing pattern or the one-step delayed measurement sharing pattern introduced in Sect. 2.3 of Chap. 2 are of the quasi-classical type, and it is also important to note that any performance (for a team) that can be achieved under the former can also be achieved under the latter, and *vice versa*, because they generate the same information sigma fields for each agent, which are further policy invariant (see [92]). In team problems with partially nested information, one talks about *precedence relationships* among agents: an agent $\mathbf{A}i$ is *precedent* to another agent $\mathbf{A}j$ (or $\mathbf{A}i$ *communicates* to $\mathbf{A}j$), if the former agent's actions affect the information of the latter, in which case (to be partially nested) $\mathbf{A}j$ has to have the information based on which the action-generating policy of $\mathbf{A}i$ was constructed.

As remarked on earlier in Sect. 2.3, the order of actions can also be nature-dependent in some settings. As noted, if there is a pre-defined order in which the decision makers act, as above, then we say that a system is *sequential*. Otherwise, the system is non-sequential. Such systems are substantially more difficult to study, since the ambiguities in the order of actions lead to challenges on the interpretation of local information. We will study the non-sequential setting in Sect. 3.9.

[3] Here we can assume without any loss of generality that each agent acts only *once* in the decision process, or equivalently appears only once on the decision tree. If an agent acts more than once, then (as discussed earlier) she can be split into multiple decision makers (DMs), with each DM again acting only once.

3.3 Nonclassical Information Structures and Witsenhausen's Counterexample

In a dynamic stochastic decision problem (or dynamic stochastic team, or stochastic control problem), if the action to be taken at some point in (discrete) time, say k, by an agent **A** i relies on a measurement (information) which is affected by an action taken at some previous point in time, say k' ($k' < k$) by the same agent or by some other agent, and **A** i does not have access to the measurement (information) used in the construction of that previous action, then the underlying decision problem is said to be one with *non-classical information*. Such problems are inherently difficult to solve, and there is no general theory which will aid in the construction of optimal policies that use non-classical information, except for the static reduction method presented earlier, which allows one to view a dynamic team problem as a static one, and structurally consequential (e.g. on existence, structure and regularity results) but computationally of-limited-use methods such as the universal dynamic program to be presented in Chap. 5.

The goal of this section is to discuss the source of these difficulties, convincingly demonstrated by an example introduced by Witsenhausen in a 1968 paper [771], which is known as Witsenhausen's counterexample.

This is called a counterexample, since it counters what may be viewed as a *folk theorem* in optimal estimation, control and filtering: Is it the case that every linear system driven by Gaussian noise under quadratic cost criteria admit solutions that are linear? In the standard theory, involving only classical information structures, the well-known answer to this question is in the affirmative. In 1968, Witsenhausen, however, showed that this is not generally true, through that counter-example.

We discuss here not only that infamous counterexample, but also some variants of it, including ones where the underlying distributions are discrete.

Consider the following two-stage stochastic control or equivalently two-agent dynamic stochastic team problem, where all quantities are scalar (see Fig. 3.2 for flow of information and relationships between different variables).

A random variable, x, with a given distribution is to be transformed into another random variable, $u_0 = \gamma_0(x)$, which is transmitted over a channel, $y = u_0 + w$, with additive noise w of known distribution, the output, y, of which is to be further transformed into another random variable, $u_1 = \gamma_1(y)$. The objective is to choose the transformations γ_0 and γ_1 in such a way that a given performance index, $Q(x, u_0, u_1)$, quadratic in x, u_0, and u_1, is minimized in the average sense. That is, we seek the pair $\gamma^* := (\gamma_0^*, \gamma_1^*)$, if exists, such that

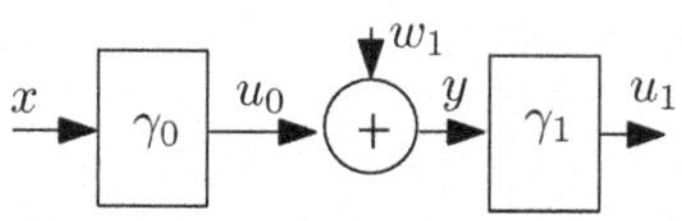

Fig. 3.2 Flow of information in Witsenhausen's counterexample

$$J(\gamma^*) = \inf_{\gamma} J(\gamma) =: J^* \tag{3.3}$$

where

$$J(\gamma) = E\left[Q(x, \gamma_0(x), \gamma_1(y))\right] \tag{3.4}$$

with expectation, $E[\cdot]$, taken over the statistics of x and w, which are assumed to be independent. Furthermore, the minimization is over the space of all Borel-measurable maps, that is, both policies (decision rules) γ_0 and γ_1 are taken to be Borel-measurable maps of the real line onto itself.

This is a stochastic decision problem with *non-classical information*, because the information to be used by the decision rule, γ_1, of the second agent depends on the action, u_0, of the first agent (and thereby on the decision rule of the first agent), but the second agent does not have access to the information of the first agent (that is, x). If we view it as a single-agent problem where the agent acts twice, then it is one where the agent is *memoryless*, that is, she does not remember what she had observed at the earlier stage. As such, these problems belong to the realm of inherently difficult decision problems for which a systematic solution process generally does not exist, one of the main reasons being that due to loss in memory, a standard dynamic programming approach is not possible [774, 777]. We now consider different instances of this class of problems, corresponding to different choices of the performance index Q, and different choices for the distributions for x and w, some of which admit explicit, relatively simpler solutions, while some others do not. Further, in some cases an optimal solution exists (even though not available in closed form) and in other cases only an ϵ-optimal solution exists. Hence, part of the message here is that even though the non-classical nature of the information is generally responsible for the difficulty in obtaining the optimal solution, in some cases the structure of the loss function (performance index) also contributes to the difficulty in solving these problems.

Let the quadratic performance index Q be picked as

$$Q_W(x, u_0, u_1) = k(u_0 - x)^2 + (u_1 - u_0)^2 \,, \tag{3.5}$$

where $k > 0$ is a given parameter. Note that here the first agent wants to stay as close to x as possible, while the second agent wants to stay as close to the action of the first agent, u_0, as possible.

This can also be viewed as a standard discrete-time two-stage stochastic optimal control problem, with state equations (see Fig. 3.3)

$$x_1 = x_0 + v_0 \,, \qquad x_2 = x_1 - v_1 \,, \tag{3.6}$$

measurement equations

$$y_0 = x_0 \,, \qquad y_1 = x_1 + w_1 \,, \tag{3.7}$$

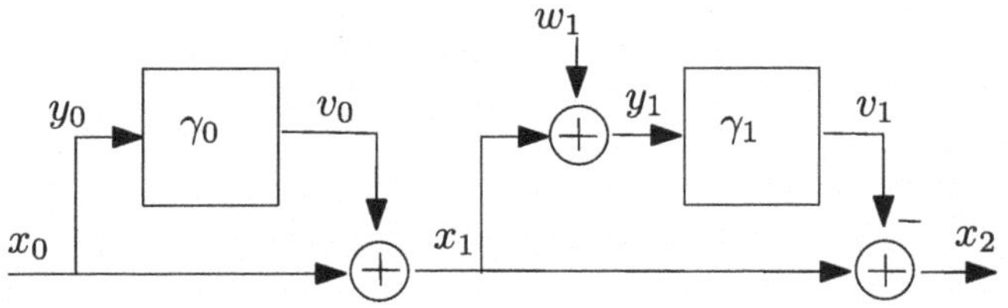

Fig. 3.3 Witsenhausen's counterexample in two-stage state-space form

and memoryless controls

$$v_0 = \gamma_0(y_0)\,, \qquad v_1 = \gamma_1(y_1)\,, \tag{3.8}$$

where μ_0 and μ_1 are the instantaneous measurement output control policies at stages 0 and 1, respectively. This becomes equivalent to the earlier formulation in view of the correspondences

$$u_0 = x_0 + v_0\,, \quad u_1 = v_1\,, \quad x = x_0\,, \quad w = w_1\,, \quad y = y_1\,,$$

if we pick the cost function as

$$\tilde{Q}(x_2, v_0) = (x_2)^2 + k(v_0)^2 \equiv Q_W(x_1 - v_0, x_1, x_1 - x_2)\,.$$

3.3.1 *Witsenhausen's Counterexample with Discrete Distributions*

Now let both x_0 and w_1 admit two-point probability mass functions, each taking values 1 and -1 with equal probability $\frac{1}{2}$. Consider, for each $\epsilon > 0$ sufficiently small, the construction:

$$\gamma_0(x) = x_0 + \epsilon\ \mathrm{sgn}(x) \qquad \gamma_1(y) = \begin{cases} 1+\epsilon, & \text{if } y = 2+\epsilon \text{ or } \epsilon \\ -1-\epsilon, & \text{if } y = -2-\epsilon \text{ or } -\epsilon \end{cases}, \tag{3.9}$$

and note that the corresponding value of J is $k\epsilon^2$. This shows that J, which is a nonnegative quantity, can be made sufficiently close to *zero* by picking ϵ sufficiently small.

Note that if u_1 also had access to the true value of x, then the optimum solution would be $\gamma_0(x) = \gamma_1(x) = x$, resulting in *zero cost*, thus achieving the lowest possible value for J. Now, since u_1 does not have access to the value of x, this being a team problem the true value of x has to be transmitted to u_1 through the action u_0, which is a realization of a policy γ_0; this in turn provides u_1 with the true value of u_0 (even though it is received over a noisy channel) and thus enables the second agent to make the second term of the cost zero. The channel noise being discrete, and u_0 being a continuous variable (taking values on the real line), transmission without any error is possible, albeit at some (arbitrarily small) cost to u_0. It should be clear here that the infimum of J is zero, but it cannot be achieved because in the limit as $\epsilon \to 0$, it is no longer possible to transmit x (and thus u_0, which inevitably depends on it to keep the first term in J small) perfectly.

The phenomenon we have observed above—the true value of a random variable being transmitted by an agent who observes it to another agent who observes only the action of the first agent—is called **signaling** *through control actions*. Stochastic team problems where the information structure is non-classical generally entail *signaling through control actions* (see Sect. 3.8.2 for an explicit example) by at least one agent to other agents who act later, and signaling generally involves a tradeoff or a compromise from an otherwise optimal control action, and this tradeoff cannot generally be formalized in a precise mathematical form, or an explicit characterization cannot always be provided as in this example. This is what contributes to the challenges underlying derivation of team-optimal solutions in dynamic stochastic teams with non-classical information. Furthermore, as we will see in Sect. 3.7.2, team problems with such an information structure are typically non-convex problems.

Hence, the message conveyed by the analysis of the two-agent dynamic team problem of this subsection is that in problems with non-classical information a minimum may not exist, and *signaling* may be essential to achieve a cost arbitrarily close to the infimum. Note that the conclusions would have been the same if instead of (3.5) the loss function was

$$Q'_W(x, u_0, u_1) = k(u_0 - x)^2 + (u_1 - x)^2 , \tag{3.10}$$

which has no cross terms between the two action variables (but IS is still non-classical). Further, if the distributions had been m-point, instead of 2-point, where m is any positive integer, again the conclusion would essentially be the same.

In the next subsection we discuss the original counterexample by Witsenhausen, where the distributions are not discrete, in which case the minimum exists, but the policies that achieve it are not known.

3.3.2 Witsenhausen's Counterexample (in Original Form)

Now, in (3.6)–(3.8) assume that the distributions of the two random variables are Gaussian: x is a Gaussian random variable with mean zero and variance σ_x^2, and w is also a zero-mean Gaussian random variable, with variance σ_w^2, the two again being independent.

Witsenhausen has shown in his 1968 paper [771] that the optimal solution to this problem exists, but there are instances of the problem where the optimum decision rules (μ_0 and μ_1, or equivalently γ_0 and γ_1) are not linear. For the latter, he has shown that for some values of the parameters defining the problem, there exist nonlinear policies which outperform the best linear policies.

A class of such nonlinear policies introduced by Witsenhausen [771], and further improved upon in [62] is

$$u_0 = \gamma_0(x) = \epsilon\,\mathrm{sgn}(x) + \lambda x\,,$$
$$u_1 = \gamma_1(y) = E[\epsilon\,\mathrm{sgn}(x) + \lambda x|y]$$

where ϵ and λ are parameters to be optimized over (in [771] the values are picked as $\lambda = 0$ and $\epsilon = \sigma_x$, and some asymptotics are studied). Clearly, if $\epsilon = 0$, this class of decision rules will be linear, since $E[\lambda x|y]$ will be linear for each λ, however when $\epsilon \neq 0$, the decision rules at both stages will be nonlinear.

To give some indication of how much can be gained by taking $\epsilon \neq 0$, the reader is referred to [62, Table V], where it is shown that one strictly improves upon the linear policies by a non-zero ϵ.

To TB: the costs noted in the 2013 version here from Bansal-Basar'87 were negative as the coefficients defining the quadratic cost are different. I think the message is clear, so there is no need to give particularly detailed numerics.

In particular [62], if $\lambda = 0$, $\epsilon = \sqrt{2/\pi}$ and $k\sigma_x^2 = 1$, as $k \to 0$ the bound on asymptotic performance becomes $(1 - (2/\pi)) = 0.363$. Note that the policy γ_0 above is of the type "linear plus 2-point quantized", which begs the question whether one can improve upon that by using a finer quantizer. For example, one could start with the class of policies

$$u_0 = \gamma_0(x) = \epsilon\,\mathrm{Quant}(x) + \lambda x\,,$$
$$u_1 = \gamma_1(y) = E[\epsilon\,\mathrm{Quant}(x) + \lambda x|y]$$

where *Quant* denotes the quantization operator, and optimize over the parameter values ϵ and λ, as well as over all quantizers.[4] Since the 2-point quantizer is a special case, one could naturally achieve an improvement in performance as a result of this optimization, and in fact one can show [545] that there are instances of the problem where the best performance in the linear class (that is with $\epsilon = 0$, and optimized over λ), could be arbitrarily bad against the best performance in this class; in precise mathematical terms, within the structure above,

$$\sup_{k>0,\sigma>0} \frac{\inf_{\epsilon=0,\lambda} E[Q_W(x,\gamma_0(x),\gamma_1(y))]}{\inf_{\epsilon,\lambda,\mathrm{Quant}} E[Q_W(x,\gamma_0(x),\gamma_1(y))]} = \infty$$

[4] Note that γ_1 above is optimal against the γ_0 picked.

In spite of this improvement, however, the optimal solution to the problem (which exists as mentioned earlier, and to be proved below) is not of the quantized type; on the other hand, as we will see later in Theorem 3.3.1, quantized policies are indeed asymptotically optimal (as the cardinality of the quantizer range increases) and there exist many numerical schemes which apply policies of this (quantized) type.

On Three Proofs of Existence A first proof of existence was given by Witsenhausen himself in [771] using functional analysis and monotonicity properties (via establishing that pointwise limit of an infimizing sequence is optimal). An alternative proof of existence (which is a general method for decentralized stochastic control, and not only specifically tailored to the counterexample) using strategic measures and tightness properties will be presented in Sect. 5.6.1 (via applying Theorem 4.2.4). In the following, for instructional purposes, we present a rather direct proof due to [786].

First note that given any measurable γ_0, the optimal γ_1 is uniquely given by the conditional mean

$$\gamma_1(y) = E[\gamma_0(x)|y],$$

and hence the problem becomes one of minimization with respect to all measurable functions $\gamma_0 = f : \mathbb{R} \to \mathbb{R}$ of the function

$$J(f) := E[k(f(x) - x)^2 + [f(x)]^2 - [E[f(x)|y]]^2] \equiv kE[(z - x)^2] + \text{mmse}(\Phi), \tag{3.11}$$

where Φ is the probability distribution of $z = f(x)$, and mmse(Φ) stands for the minimum mean square error of estimating a random variable z with probability distribution Φ using measurement $y = z + w$, where w is as given before and is independent of z. Note that the second term depends only on Φ, whereas the first term depends on the joint distribution of x and z. Hence, in the process of minimization of (3.11) with respect to f, we could first hold Φ fixed, minimize the first term over the joint distribution of x and z, say $P_{x,z}$, with the marginals fixed at N and Φ respectively, and then minimize the resulting expression with respect to the unknown marginal Q. The outcome of the first minimization (infimization) is

$$k[W_2(N, \Phi)]^2 + \text{mmse}(\Phi), \tag{3.12}$$

where

$$W_2(N, \Phi) := \inf_{\{P_{x,z}: P_x = N, P_z = \Phi\}} \sqrt{E[(z - x)^2]}$$

is the quadratic *Wasserstein* distance[5] between the two probability distributions N and Φ. Hence,

$$\inf_{f} J(f) = \inf_{\Phi}(k[W_2(N, \Phi)]^2 + \text{mmse}(\Phi)) =: \inf_{\Phi} F(\Phi) .$$

For each fixed N, $W_2(N, \Phi)$ is weakly lower semi-continuous in Φ [19] and mmse(Φ) is weakly continuous in Φ [787] (see Theorem 8.7.2 for a related result). Furthermore, if Φ has a bounded second moment $m_2(\Phi)$, with the bound say $\alpha^2 > 0$, then $W_2(N, \Phi)$ is minimized by linearly correlating z with x, more precisely $z = (\alpha/\sigma_x)x$, leading to the value of

$$\min_{\{\Phi : m_2(\Phi) \leq \alpha^2\}} W_2(N, \Phi) = |\alpha - \sigma_x| ,$$

which is an increasing function of α for $\alpha > \sigma_x$. Since mmse(Φ) is non-negative, $|\alpha - \sigma_x|$ is then a lower bound on the cost; consequently as the optimal cost for J is to be finite (e.g. a finite cost is attained under the optimal linear f), for the minimization of $F(\Phi)$, Φ can be restricted to distributions with a uniformly bounded second moment, and thus to a weakly compact set. Then, $\inf_{\Phi} F(\Phi)$ involves minimization of a weakly lower semi-continuous function on a weakly compact set, and hence $F(\Phi)$, and consequently $J(f)$ has a minimum by the Weierstrass theorem.[6] ◇

3.3.3 Numerical Procedures and a Convergence Result for Asymptotic Optimality of Quantized Policies

Even though the existence has been settled, to this date there does not exist an explicit solution to Witsenhausen's counterexample. Nonetheless, the counterexample has attracted significant attention and there have been numerous numerical studies on the problem; an overview is presented at the end of the chapter. Rigorously, near optimality (or asymptotic optimality, as the number of bins grows) of quantized policies for general static reducible stochastic teams will be established in Theorem 4.4.2. Indeed, a slight improvement on the weak lower semi-continuity result noted above due to [786] implies that any weakly dense subset of distributions

[5] Appendix C provides a detailed coverage on some of the notions used here such as the space of probability measures, weak topology on such spaces, and several metrics, including Wasserstein distances.

[6] We further note that the term $W_2(N, \Phi)$ is convex in Φ, but mmse(Φ) is concave in Φ, which makes the function to be minimized, $F(\Phi)$, in general non-convex. A further analysis on non-convexity is given in Sect. 3.7.2 and related results involving optimization on the space of information structures are presented in Chap. 8.

would be near optimal; and Theorem 4.4.2 below will precisely establish this for measures arrived at via quantized policies. We state this result formally in the following.

Theorem 3.3.1 ([625]) *For Witsenhausen's counterexample (in original form), for any* $\epsilon > 0$*, there exists a pair of (quantized) policies* (γ_0, γ_1) *with the ranges of* γ_0 *and* γ_1 *finite, that are* ϵ*-optimal.*

A proof of this result will be presented in Sect. 4.4.2. This result certifies the asymptotic optimality of many numerical and analytical procedures and studies in the literature based on quantization based methods (see [7, 51, 399, 469, 477, 525, 527]).

3.3.4 Generalized Gaussian Test Channel

Within the context of Witsenhausen's counterexample, let us now consider a different choice for the quadratic cost criterion Q:

$$Q_{\mathrm{TC}}(x, u_0, u_1) = k(u_0)^2 + (u_1 - x)^2 , \tag{3.13}$$

where again $k > 0$. Note that here the second agent's objective is to estimate the random variable x in the minimum mean square (MMS) sense, using a measurement that is transmitted over a Gaussian channel where the input to the channel is *shaped* by the first agent who has access to x and has a soft constraint $(kE[(u_0)^2])$ on its action. The version of this problem where the soft constraint is replaced by a hard power constraint, $E[(u_0)^2] \le k$, is known as the *Gaussian Test Channel* (GTC), and in this context γ_0 is the *encoder* and γ_1 the *decoder*, where the latter's optimal choice is clearly the conditional mean of x given y, that is $E[x|y]$. The best encoder for the GTC can be shown to be linear (a scaled version of the source output, x), which in turn leads to a linear optimal decoder. The approach here (as we will discuss further below for a more general, soft-constrained version), which is in fact the only approach known to apply here, is to obtain bounds on the performance using an inequality from information theory involving channel capacity [788] and rate distortion function [111], and then to show that the bound can be achieved using linear policies.[7]

Now, consider the more general version of (3.13):

$$Q_{\mathrm{GTC}}(x, u_0, u_1) = k(u_0)^2 + (u_1 - x)^2 + b_0 u_0 x , \tag{3.14}$$

where b_0 is a parameter. Let

[7] A detailed coverage of information-theoretic notions can be found in Chap. 12. See also Theorem 16.2.2.

$$E[(u_0)^2] =: \alpha \quad \text{and} \quad E[(u_1 - x)^2] =: \beta\,.$$

Then, with J defined as before, by (3.4), and with γ_0 and γ_1 constrained as above, we have the inequalities

$$\begin{aligned}\inf_{\gamma} J(\gamma) &\geq k\alpha + \beta + \inf_{\gamma_0} b_0 E[\gamma_0(x)x] \\ &\geq k\alpha + \beta - |b_0|\sigma_x\sqrt{\alpha}\end{aligned} \tag{3.15}$$

where the second one follows from the Cauchy-Schwarz inequality.

Now, by the data processing theorem [788] (see Lemma 12.3.1), in a linear configuration the mutual information[8] between two random variables closer to each other is no smaller than the mutual information between two random variables farther apart. In our case, this translates to

$$I(x; y) \geq I(x; u_1) \tag{3.16}$$

where $I(\cdot;\cdot)$ stands for mutual information. For each fixed $\alpha > 0$, $I(x; y)$ is bounded from above by the capacity of the channel, $C(\alpha)$, which is known for the Gaussian channel to be [285]

$$C(\alpha) = \frac{1}{2}\log(1 + (\alpha/\sigma_w^2))\,.$$

Further, for each fixed β, the quantity $I(x; u_1)$ is bounded from below by the rate distortion function, $R(\beta)$, for which the expression, when $\beta \leq \sigma_x^2$, is [111]

$$R(\beta) = \frac{1}{2}\log(\sigma_x^2/\beta)\,.$$

In view of (3.16), we have

$$\frac{1}{2}\log(1 + (\alpha/\sigma_w^2)) = C(\alpha) \geq R(\beta) = \frac{1}{2}\log(\sigma_x^2/\beta),$$

leading to the following bound on β: $\beta \geq \sigma_w^2\sigma_x^2/(\alpha + \sigma_w^2)$, which is tight with

$$\gamma_0(x) = -\text{sgn}(b_0)\frac{\sqrt{\alpha}}{\sigma_x}x \tag{3.17}$$

Substitution of this in (3.15) leads to

[8] See Definition 12.3.3 given in Chap. 12.

$$\inf_{\gamma} J(\gamma) \geq k\alpha + \sigma_w^2\sigma_x^2/(\alpha + \sigma_w^2) - |b_0|\sigma_x\sqrt{\alpha} \tag{3.18}$$

Let α^* be the positive value of α that minimizes the bound in (3.18), which exists and is unique. It is a solution of the polynomial equation

$$[2k\sqrt{\alpha} - |b_0|\sigma_x]\,[\alpha + \sigma_x^2]^2 = 2\sigma_w^2\sigma_x^2\sqrt{\alpha}\,. \tag{3.19}$$

Then, when Q is in the structural form (3.14), the solution to (3.4) exists, is linear, and is given by

$$\gamma_0^*(x) = -\mathrm{sgn}(b_0)\frac{\sqrt{\alpha^*}}{\sigma_x}x \tag{3.20}$$

$$\gamma_1^*(y) = E[x|y] = -\frac{\mathrm{sgn}(b_0)\sigma_x\sqrt{\alpha^*}}{\alpha^* + \sigma_w^2}y\,. \tag{3.21}$$

Remark 3.3.1 The main difference between the two problems of Sects. 3.3.2 and 3.3.4 is that Q in the former has a product term between the decision rules of the two agents while in the latter it does not. Hence, it is not only the non-classical nature of the information structure but also the structure of the performance index that determines whether linear policies are optimal in these quadratic dynamic decision problems with Gaussian statistics and non-classical information. Another point to note here is that for a similar formulation, but with discrete distributions, both the generalized test channel model (note that (3.10) is a special case of (3.14)) and the Witsenhausen counterexample system exhibit similar features (not admitting a minimum, but allowing computation of the infimum and construction of explicit policies that achieve a cost arbitrarily close to that value), thus indicating that the nature of the distribution of the primitive random variables also makes a difference in the level of complexity of the solution. More on this in the following sections.

3.4 Static Reduction of Sequential Dynamic Teams: Policy-Dependence Under Partially Nested and Policy-Independence Under Non-classical Information

A powerful method for studying a large class of optimal dynamic team problems is through construction of equivalent problems with static information.

Following Witsenhausen [780] (see [824, Section 3.7]), we say that two information structures are equivalent if: (i) The policy spaces are equivalent/isomorphic in the sense that policies under one information structure are realizable under the other information structure, (ii) the costs achieved under equivalent policies are identical almost surely, and (iii) if there are constraints on the admissible policies, the isomorphism among the policy spaces preserves the constraint conditions.

A large class of sequential team problems admit an equivalent information structure which is static. This is called *static reductions* of dynamic team problems.

3.4.1 Static Reduction I: Classical and Quasi-Classical Information Structures and Their Policy-Dependent Reduction

Under the quasi-classical information, LQG stochastic team problems are tractable by conversion into equivalent static team problems. As an example, consider the following dynamic team with N agents, where each agent acts only once, with $\mathbf{A}k$, $k \in \mathcal{N}$, having the following measurement

$$y^k = C^k \xi + \sum_{i:i \to k} D_{ik} u^i, \tag{3.22}$$

where ξ is an exogenous random variable picked by nature, and $i \to k$ denotes the precedence relation that the action of $\mathbf{A}i$ affects the information of $\mathbf{A}k$ and u^i is the action of $\mathbf{A}i$.

If the information structure is quasi-classical, then

$$\mathcal{I}^k = \{y^k, \{\mathcal{I}^i, i \to k\}\}.$$

That is, $\mathbf{A}k$ has access to the information available to all the signaling agents. Such an IS is equivalent to the IS $\mathcal{I}^k = \{\tilde{y}^k\}$, where $\tilde{y}^k$ is a static measurement given by

$$\tilde{y}^k = \left\{C^k \xi, \{C^i \xi, i \to k\}\right\}. \tag{3.23}$$

Such a conversion can be done provided that the policies adopted by the agents are deterministic, with the equivalence to be interpreted in the sense that any deterministic policy measurable under the original *IS* being measurable also under the new (static) *IS* and vice versa, since the actions are determined by the measurements. The restriction of using only deterministic policies is, however, without any loss of optimality: with policies of all other agents fixed (possibly randomized) no agent can benefit from randomized decisions in such team problems. We will discuss this property further in the following chapter (see, e.g. Theorem 4.1.3).

If the underlying optimization problem is quadratic in all variables, and the random variables are all Gaussian, by such a reduction, the optimization problem can be converted into the class of quadratic Gaussian static team problems considered in Sect. 2.4.3. The team-optimal solution under this new (static) *IS* can then be re-expressed in terms of the original *IS*. We state this formally as a theorem.

Theorem 3.4.1 *Consider a partially nested dynamic team with linear dynamics, zero-mean Gaussian initial state, and zero-mean Gaussian noise under a quadratic cost criterion (also called an LQG team) whose cost function satisfies the hypothesis given in Definition 2.4.2. Such a stochastic dynamic team admits an optimal solution which is linear in the information available to each DM.*

Examples of such an indirect derivation for dynamic teams with quasi-classical information have been given in several papers, such as [350] and [50]. Team-optimal solutions have been constructed under the one-step delayed information sharing pattern [443, 444, 631, 798] as well as under the one-step delayed observation sharing [81] pattern. These do not exhibit *certainty equivalence* or even *separation* as in the case of classical *IS*s, but are recursively computable, involving solutions of coupled linear matrix equations of the type given in Theorem 2.4.9. In view of Definition 2.3.1, we note that while an LQG team with the one-step delayed information sharing pattern, as a quasi-classical information structure, admits an optimal solution which is linear, this does not necessarily hold in general for the n-step delayed information sharing pattern when $n > 1$, unless the dynamics are such that the problem is quasi-classical. We note that generalizations of such results when the time horizon is infinite are partially available under restrictions on linearity, stability, or symmetry [452, 610, 633].

The linearity condition can be relaxed via the following more general condition, which is also due to Ho and Chu [351].

Assumption 3.4.1 *Under a quasi-classical information structure, with*

$$\mathcal{I}^k = \{y^k, \{\mathcal{I}^i, i \to k\}\}.$$

if $y^k = g(\xi, u^{[1,k-1]})$, *then the map* $g(\cdot, u^{[1,k-1]}) : \xi \mapsto y^k$ *is invertible.*

Under this assumption, static reduction is possible.

Policy-Dependence of the Static Reduction In the above, under Assumption 3.4.1, while mapping the policies that are equivalent under the dynamic setup to those that are expressed in terms of exogenous variables in the static-reduced form, we note that the policies' dependence on the exogenous variables explicitly depend on the policies adopted by the preceding DMs. Accordingly, we will refer to the static reduction of partially nested dynamic teams as *policy-dependent static reduction* (as opposed to the *policy-independent* reduction to be presented in the following). Some restrictions and limitations due to such policy-dependence will be studied later in the chapter.

3.4.2 Static Reduction II: Non-classical Information Structures and Their Policy-Independent Reduction

In this sub-section, we introduce another static reduction method, due to Witsenhausen [780], applicable also to non-classical information structures and call it a *policy-independent static reduction*, since this reduction does not depend on the policies adopted.

Consider a (static or dynamic) team setting according to the intrinsic model where each **DM**t measures

$$y^t = g^t(\omega_0, \omega_t, y^1, \ldots, y^{t-1}, u^1, \ldots, u^{t-1}),$$

and the decisions are generated by $u^t = \gamma^t(y^t)$, with $1 \leq t \leq N$. Here, $\omega_0, \omega_1, \cdots, \omega_N$ are primitive (exogenous) variables. We will indeed view, for every $1 \leq n \leq N$, the relation

$$\begin{aligned} &p_n(dy^n|\omega_0, y^1, y^2, \cdots, y^{n-1}, u^1, u^2, \cdots, u^{n-1}) \\ &:= P(dy^n|\omega_0, y^1, y^2, \cdots, y^{n-1}, u^1, u^2, \cdots, u^{n-1}), \end{aligned} \tag{3.24}$$

as a (controlled) stochastic kernel (to be defined later), and through standard stochastic realization results (see e.g., [301, Lemma 1.2], [132, Lemma 3.1], or [44, Lemma F]), we can represent this kernel in a functional form through

$$y^n = g_n(\omega_0, \omega_n, y^1, y^2, \cdots, y^{n-1}, u^1, u^2, \cdots, u^{n-1}), \tag{3.25}$$

for some independent ω_n and measurable g_n.

This team admits an *independent-measurements* reduction provided that the following absolute continuity condition holds: For every $t \in \mathcal{N}$, there exists a reference probability measure Q^t and a function f_t such that for all Borel S:

$$\begin{aligned} &P(y^t \in S|\omega_0, u^1, u^2, \cdots, u^{t-1}, y^1, y^2, \cdots, y^{t-1}) \\ &= \int_S f_t(y^t, \omega_0, u^1, u^2, \cdots, u^{t-1}, y^1, y^2, \cdots, y^{t-1}) Q^t(dy^t), \end{aligned} \tag{3.26}$$

We can then write (since the action of each DM is determined by the measurement variables under a given policy)

$$\begin{aligned} &P(d\omega_0, d\mathbf{y}, d\mathbf{u}) \\ &= P(d\omega_0) \prod_{t=1}^{N} 1_{\{\gamma^t(y^t) \in du^t\}} P(dy^t|\omega_0, u^1, u^2, \cdots, u^{t-1}, y^1, y^2, \cdots, y^{t-1}) \\ &= P(d\omega_0) \end{aligned}$$

$$\prod_{t=1}^{N}\Big(f_t(y^t,\omega_0,u^1,u^2,\cdots,u^{t-1},y^1,y^2,\cdots,y^{t-1})Q^t(dy^t)1_{\{\gamma^t(y^t)\in du^t\}}\Big)$$

$$=\Big(P(d\omega_0)\prod_{m=1}^{N}Q^m(dy^m)1_{\{\gamma^m(y^m)\in du^m\}}\Big)$$

$$\Big(\prod_{t=1}^{N}\Big(f_t(y^t,\omega_0,u^1,u^2,\cdots,u^{t-1},y^1,y^2,\cdots,y^{t-1})\Big)\Big).$$

The cost function $J(\underline{\gamma})$ can then be written as

$$J(\underline{\gamma})$$
$$=\int P(d\omega_0)\prod_{t=1}^{N}(f_t(y^t,\omega_0,u^1,u^2,\cdots,u^{t-1},y^1,y^2,\cdots,y^{t-1})Q^t(dy^t))c(\omega_0,\mathbf{u}) \tag{3.27}$$

with $u^k=\gamma^k(y^k)$ for $1\le k\le N$, and where now the measurement variables can be regarded as independent from each other, and also from ω_0, and by incorporating the $\{f_t\}$ terms into c, we can obtain an equivalent *static team* problem. Hence, the essential step is to appropriately adjust the probability space and the cost function.

Remark 3.4.1 (Change of Measure Formula) Denote the joint probability measure on $(\omega_0,u^1,\ldots,u^N,y^1,\ldots,y^N)$ by P, and the probability measure of ω_0 by $\mathbb{P}^0$. If the preceding absolute continuity condition (3.26) holds, then (under any admissible policy profile $\gamma^1,\cdots,\gamma^N$) there exists a joint reference probability measure $\mathbb{Q}$ on $(\omega_0,u^1,\ldots,u^N,y^1,\ldots,y^N)$ such that the probability measure P is absolutely continuous with respect to $\mathbb{Q}$ ($P\ll\mathbb{Q}$), so that for every Borel set A in $(\Omega_0\times\prod_{i=1}^{N}(\mathbb{U}^i\times\mathbb{Y}^i))$

$$P(A)=\int_A\frac{dP}{d\mathbb{Q}}\mathbb{Q}(d\omega_0,du^1,\ldots,du^N,dy^1,\ldots,dy^N), \tag{3.28}$$

where the reference probability measure

$$\mathbb{Q}(d\omega_0,du^1,\ldots,du^N,dy^1,\ldots,dy^N):=\mathbb{P}^0(d\omega_0)\prod_{i=1}^{N}Q^i(dy^i)1_{\{\gamma^i(y^i)\in du^i\}}, \tag{3.29}$$

leads to a Radon-Nikodym derivative, which is *policy-independent*:

$$\frac{dP}{d\mathbb{Q}}(\omega_0, u^1, \ldots, u^1, y^1 \ldots, y^N) = \prod_{i=1}^{N} f^i(y^i, \omega_0, u^1, \ldots, u^{i-1}, y^1, \ldots, y^{i-1}) \tag{3.30}$$

Indeed, one may slightly relax the condition in (3.26) (which requires the absolute continuity to hold for all $\omega_0, u^1, \cdots, u^{t-1}, y^1, \cdots, y^{t-1}$), to an almost sure existence condition of a derivative under a reference measure, in the sense that (3.30) holds.

The new cost function may now explicitly depend on the measurement values, such that

$$c_s(\omega_0, \mathbf{y}, \mathbf{u}) = c(\omega_0, \mathbf{u}) \prod_{t=1}^{N} f_t(y^t, \omega_0, u^1, u^2, \cdots, u^{t-1}, y^1, y^2, \cdots, y^{t-1}). \tag{3.31}$$

Example 3.4.1 As a common example, consider the additive noise measurement model, for some f, $y^1 = f(x) + w$, with w a noise which admits a probability density function η (not necessarily positive everywhere). Then, for any Borel B,

$$P(y^1 \in B|x) = \int \eta(y^1 - f(x))dy^1 = \int \frac{\eta(y^1 - f(x))}{\kappa(y^1)}\kappa(y^1)dy^1$$

for any κ which is a probability density function that is positive everywhere (such as the standard Gaussian). Thus, the measurement variable can be assumed to be $y^1 \sim \int_{.} \kappa(y)dy$ which is independent of x and has a measure that admits a probability density function κ. Note that here one or both of η and κ can be Gaussian as an important special case.

Example 3.4.2 (Static Reduction of Witsenhausen's Counterexample and Some Implications) Consider now Witsenhausen's counterexample introduced and discussed earlier. Recall that this is a dynamic team problem with non-classical information, where y^1 and w^1 are zero-mean independent Gaussian random variables with unit variance and $u^1 = \gamma^1(y^1)$, $u^2 = \gamma^2(u^1 + w^1)$, and the cost/loss function $c(\omega, u^1, u^2) = k^2(y^1 - u^1)^2 + (u^1 - u^2)^2$ for some $k > 0$: The (policy-independent) static reduction, in this case, proceeds as follows:

$$\begin{aligned}
&\tilde{c}(y^1, y^2, u^1, u^2) \\
&= \int (k(u^1 - y^1)^2 + (u^1 - u^2)^2) Q(dy^1)\gamma^1(du^1|y^1)\gamma^2(du^2|y^2)P(dy^2|u^1) \\
&= \int (k(u^1 - y^1)^2 + (u^1 - u^2)^2) Q(dy^1)\gamma^1(du^1|y^1)\gamma^2(du^2|y^2)\eta(y^2 - u^1)dy^2
\end{aligned}$$

$$= \int \left((k(u^1 - y^1)^2 + (u^1 - u^2)^2)\gamma^1(du^1|y^1)\gamma^2(du^2|y^2)\frac{\eta(y^2 - u^1)}{\eta(y^2)} \right)$$
$$Q(dy^1)\eta(y^2)dy^2$$
$$= \int \left((k(u^1 - y^1)^2 + (u^1 - u^2)^2)\gamma^1(du^1|y^1)\gamma^2(du^2|y^2)\frac{\eta(y^2 - u^1)}{\eta(y^2)} \right)$$
$$Q(dy^1)Q(dy^2) \tag{3.32}$$

where Q denotes a Gaussian measure with zero mean and unit variance and η its density. Such a reduction will be shown to facilitate powerful mathematical analysis later in the book; in particular for this counterexample, this static reduction will be explicitly utilized in Sect. 5.6.1 for an existence result, in Sect. 4.4.2 for an approximate optimality result, and in Sect. 3.7.2 for a non-convexity analysis.

Example 3.4.3 A further example is the generalized Gaussian Test Channel model of Sect. 3.3.4. In this case, the reduction writes as:

$$\int (k(u^1)^2 + (y^1 - u^2)^2)Q(dy^1)\gamma^1(du^1|y^1)\gamma_1(du^2|y^2)P(dy^2|u^1)$$
$$= \int \left((k(u^1)^2 + (y^1 - u^2)^2)Q(dy^1)\gamma^1(du^1|y^1)\gamma_1(du^2|y^2)\frac{\eta(y^2 - u^1)dy^2}{\eta(y^2)} \right)$$
$$\times Q(dy^1)Q(dy^2) \tag{3.33}$$

We note that one can reformulate even a static team to another static one, now with independent measurements, which are also independent from the cost relevant exogenous variable ω_0.[9]

Witsenhausen further considers some special cases, where the DMs may have access to a collection of $\{y^i\}$: In the context of our analysis above, in the **classical** case $I^n = \{y^k, k \leq n\}$ for every **DM**n; in the **independent static** case, $I^n = y^n$ where the measurements are independent; in the **amnesic** case $I^n = y^n$ (e.g., when the team problem is static a priori without reduction, but with possibly dependent measurement variables, which can then be or may not be independent static reducible). A particularly important classification is **static with independent data** where decision makers have access to collections of independent variables.

Remark 3.4.2 The absolute continuity condition (3.26) is in general not restrictive. Indeed, as Witsenhausen notes, a static reduction always holds when the mea-

[9] We will see that this *independent static* reduction will provide remarkable flexibility in facilitating stochastic analysis for existence, approximations, and dynamic programming methods for stochastic dynamic and static teams; this will also prove to be useful in establishing existence of equilibria and comparison of information structures in stochastic game theory

surement variables take values from a countable set, since a reference measure as in Q^i above can always be constructed on the measurement space $\mathbb{Y}^i$ (e.g., $Q^i(z) = \sum_{j\geq 1} 2^{-j} 1_{\{z=m_j\}}$ where $\mathbb{Y}^i = \{m_j, j \in \mathbb{N}\}$) so that the absolute continuity condition (3.26) always holds.

Remark 3.4.3 We will see, in Sect. 8.9 (see also Sect. 3.4.5), that all sequential stochastic team problems are nearly static reducible, where the absolute continuity condition (3.26) is satisfied by a slight perturbation of the measurement channels with negligible impact on the cost.

Remark 3.4.4 It may be instructive to note that the static reduction method considered here is related to the commonly adopted *change of measures* method (for continuous-time classical stochastic control) of applying *Girsanov's measure transformation* [302] to measurements and allowing the decision makers to have independent measurement data under a new probability measure; this has become a standard method in optimal continuous-time stochastic control theory starting perhaps with [107]. We refer the reader to [166] for relations with classical continuous-time stochastic control where the relation with Girsanov's classical measure transformation was recognized. For discrete-time partially observed stochastic control, similar arguments have been introduced in [134] and [136] again in the context of measure transformation. This method has been very useful in existence, approximations, structural and qualitative (such as convexity) properties and near-optimality regarding optimal policies; these will be discussed in detail in the next few chapters. The discussion on the reduction above also appears in [824, Section 3.7].

3.4.3 Further Differences Between Policy-Dependent and Policy-Independent Static Reductions

In the following, we draw attention to some salient differences between the two static-reduction methods introduced above.

(i) For policy-dependent information structures of partially nested teams, in the reduced form, the cost functions are unaltered, which is, however, not the case for the policy-independent reduction. On the other hand, the probability measure on the exogenous variables do not change under the policy-dependent static reduction, but does so in the policy-independent static reduction. Furthermore, for the policy-dependent case the static-reduced policy/map from exogenous variables to actions change depending on the previous policies, but in the policy-independent case this map is unaltered (as the measurement variables are interpreted to be exogenous).

(ii) For policy-dependent information structures, it is essential that policies are deterministic (or there is control sharing consistent with the partial nestedness, that is, whenever measurements are shared, so are the control actions). Policy-independent reduction applies even when the policies are randomized.

(iii) The above observations in (i)–(ii) have implications on convexity properties: if a dynamic team is convex, so is its static reduction for the policy-independent case (including static-measurements with control-sharing reduction). In particular, the policy-dependent case requires that actions be shared. This will be discussed further in Sect. 3.7.

(iv) While for team optimality both static reductions have equivalence properties with the dynamic information setup, for the policy-dependent case, when one considers person-by-person-optimal or stationary policies, significant subtleties emerge: a policy which is person-by-person-optimal in one form may not be so in the other form (unlike the policy-independent case). This will be discussed in Sect. 3.6.

3.4.4 Two Refinements of Policy-Dependent Static Reduction

Static Measurements with Control Sharing Reduction

Consider again the quasi-classical setup where $\mathcal{I}^k = \{y^k, \{\mathcal{I}^i, i \to k\}\}$, so that **A**$k$ has access to the information available to all the signaling agents. We have already observed that such an IS is equivalent to the IS $\mathcal{I}^k = \{\tilde{y}^k\}$, where $\tilde{y}^k$ is a static measurement given by

$$\tilde{y}^k = \Big\{C^k\xi, \{C^i\xi, i \to k\}\Big\}. \tag{3.34}$$

Now, if we require the IS to be

$$\{\tilde{y}^k, \{u^i, i \to k\}\},$$

so that the measurements are static but the actions are also shared, then we will refer to this equivalent model as *Static-Measurements with Control-Sharing Reduction*. Even though this information structure is not static in a strict sense, where only the measurements are so, this reduction will entail a number of useful properties to be studied later in the chapter. Notably, we state the following:

Theorem 3.4.2 ([637]) *Static measurements with control sharing reduction is policy-independent.*

The result stated above follows since $g(\cdot, u^{[1,k-1]})$ in Assumption 3.4.1 is invertible given previous actions $u^{[1,k-1]}$ regardless of the policies of previous DMs.

A Stochastic Relaxation of Partial Nestedness In closing this section, we note that another class of dynamic team problems that can be converted into solvable dynamic optimization problems are those where even though the information structure is non-classical, there is no incentive for signaling because any signaling from say agent **A**i to agent **A**j conveys information to the latter which is "cost irrelevant", that is it does not lead to any improvement in performance. If, for example, given a dynamic stochastic team with information structure $\underline{\eta} = (\eta^1, \ldots, \eta^N)$, possibly non-classical, and the corresponding strategy space $\mathbf{\Gamma} = \Gamma^1 \times \cdots \times \Gamma^N$, and normal form description $(J, \mathbf{\Gamma})$, there is another information structure $\underline{\eta}' = (\eta^{1'}, \ldots, \eta^{N'})$ (quasi-classical or classical), with corresponding composite strategy space $\tilde{\mathbf{\Gamma}} = \tilde{\Gamma}^1 \times \cdots \times \tilde{\Gamma}^N$, where $\mathbf{\Gamma} \supset \tilde{\mathbf{\Gamma}}$, such that

$$\inf_{\underline{\gamma} \in \mathbf{\Gamma}} J(\underline{\gamma}) = \inf_{\underline{\gamma} \in \tilde{\mathbf{\Gamma}}} J(\underline{\gamma}) =: J(\underline{\tilde{\gamma}}^*) , \tag{3.35}$$

then $\underline{\tilde{\gamma}}^*$ provides an optimal-team solution to the original problem even though it utilizes less information. This statement may seem to be a trivial one, but it does play an important role in expanding the class of solvable stochastic dynamic teams by converting some with non-classical information and specific structures into ones with classical information with equality holding in (3.35), which is because the *non-classical* part of the information structure is *performance irrelevant*. This will be discussed further in Sect. 3.8.3.

3.4.5 All Stochastic Dynamic Teams Are Nearly Static (with Independent Measurements) Reducible

Now that we have seen, and will see even further, the benefits of static reduction, the question arises as to whether we can perturb any stochastic dynamic team by adding some additive noise to the measurements to make it static-reducible with arbitrarily small error in the optimal cost: That is, are all dynamic team problems ϵ-away from being static reducible as far as optimal cost is concerned for any $\epsilon > 0$?

We will show, in Sect. 8.9 that this is indeed the case. However, we only announce this result in this chapter, leaving the proof to later, so that the reader is not distracted. The argument that establishes this result is related to garbling and continuity properties of optimal policies on the space of information structures, a topic that will be studied in detail in Chaps. 7 and 8.

3.5 Beyond the One-Shot DM Model: An Agent-Wise Intrinsic Model and Agent-Wise Reductions

Under the intrinsic model for sequential teams introduced by Witsenhausen, as discussed in Sect. 3.2, every DM acts separately and only once. This interpretation may have significant limitations in some applications (especially in dynamic games to be studied later in the book): Depending on the setup, it may be more suitable to consider an agent playing repeatedly over multiple stages as a collection of one-shot DMs acting as a separate individual decision maker/agent.[10] To formalize this framework, where collections of one-shot DMs can be viewed as individual agents, we present an agent-wise intrinsic model in the following.

3.5.1 An Agent-Wise Intrinsic Model with Agents (or Sub-teams of DMs) Acting Multiple Times

We have, as a formal description, the following (see also [637]):

(i) Let $\mathcal{N} := \{1, 2, \ldots, N\}$ denote the set of agents and for each $i \in \mathcal{N}$, introduce a subset TE^i of a set $\mathcal{M} := \{1, 2, \ldots, M\}$ denoting a collection of DMs, **DM**k for $k \in \mathrm{TE}^i$, acting as **A**i (said another way, **A**i encapsulates the collection of DMs indexed by TE^i acting $|\mathrm{TE}^i|$ times, where $|\cdot|$ denotes the cardinality of the set TE^i).

(ii) The observation and action spaces are standard Borel spaces for each **A**i ($i \in \mathcal{N}$), denoted by $\mathbf{Y}^i := \prod_{k\in\mathrm{TE}^i} \mathbb{Y}^i_k$, and $\mathbf{U}^i := \prod_{k\in\mathrm{TE}^i} \mathbb{U}^i_k$, respectively.

(iii) The $\mathbb{Y}^i_k$-valued observation variables are given by $y^i_k = h^i_k(\omega, \{u^p_s\}_{(s,p)\in L^k_i})$, where L^k_i denotes the set of all DMs acting before **DM**k of **A**i (i.e., $(s, p) \in L^k_i$ if **DM**s of **A**p acts before **DM**k of **A**i for all $p \in \mathcal{N}$ and $s \in \mathrm{TE}^p$).

(iv) An admissible policy for each **A**i is denoted by $\boldsymbol{\gamma}^i := \{\gamma^k_i\}_{k\in\mathrm{TE}^i} \in \boldsymbol{\Gamma}^i$ with $u^i_k = \gamma^i_k(y^i_k)$, where the set of admissible policies for each agent is denoted by $\boldsymbol{\Gamma}^i := \prod_{k\in\mathrm{TE}^i} \Gamma^i_k$ for $i \in \mathcal{N}$. An admissible policy tuple for all agents in the game is denoted by $\underline{\boldsymbol{\gamma}} := \{\boldsymbol{\gamma}^1, \cdots, \boldsymbol{\gamma}^N\} \in \boldsymbol{\Gamma}$ where $\boldsymbol{\Gamma} := \prod_{i\in\mathcal{N}} \boldsymbol{\Gamma}^i$.

Building on this formulation, in the following we present further reductions arrived at via change of measure arguments.

3.5.2 Agent-Wise (or Sub-team-wise) Independent-Data, Nested Independent and Partially Nested Independent Reductions

Definition 3.5.1 (Agent-Wise Independent-Data and Nested Reductions) Given a multi-stage stochastic team ($\mathcal{P}^{\mathrm{Multi}}$) with a given information structure, we define the following agent-wise static reductions:

[10] This is especially relevant (and useful) in extension of discrete-time team theory to continuous-time teams.

(i) (Classical Independent-Data Reduction) An independent-data reduction is a change of measure under which the measurements at each agent admit a policy-independent static reduction as in Sect. 3.4.2. Then, these measurements are assembled together for each agent so that at time k, $\mathbf{A}i$ has access to $\{y^i_{[1,k]}, u^i_{[1,k-1]}\}$.

The team problem under this static reduction can be viewed as the one that Witsenhausen has referred to as *a static problem with independent data* [780];

(ii) (Nested, Partially Nested, or Non-Classical Independent-Agent Reduction) A nested reduction is a reduction under which for some reference measure $Q^{\gamma^1,\cdot,\gamma^N}$

$$Q^{\gamma^1,\cdot,\gamma^N}(d\mathbf{u}^{[1,N]}_{[1,T]}, d\mathbf{y}^{[1,N]}_{[1,T]}, d\omega_0) = \prod_{i=1}^{N} Q^{\gamma^i}(du^i_{[1,T]}, dy^i_{[1,T]})P(d\omega_0) \quad (3.36)$$

we have that, with $P^{\gamma^1,\cdot,\gamma^N}$ being the original probability measure induced by the collection of policies $\gamma^1, \cdot, \gamma^N$, $P^{\gamma^1,\cdot,\gamma^N} \ll Q^{\gamma^1,\cdot,\gamma^N}$ and that a new cost function emerges, defined as

$$c(\mathbf{u}^{[1,N]}_{[1,T]}, \mathbf{y}^{[1,N]}_{[1,T]}, \omega_0)\frac{dP}{dQ}(\mathbf{u}^{[1,N]}_{[1,T]}, \mathbf{y}^{[1,N]}_{[1,T]}, \omega_0) \quad (3.37)$$

If for each $\mathbf{A}i$ through $t \in \mathcal{T}$, the information structure is nested, i.e., $\sigma(y^i_t) \subset \sigma(y^i_{t+1})$, we call this an *agent-wise nested independent reduction*. If for each agent, we have only partial nestedness, this is called an *agent-wise partially nested independent reduction*. If some agents are non-classical, then this is called an *agent-wise non-classical independent reduction*.

In particular, under this reduction the *strategic measures* of each agent (or sub-team) are decoupled and in each agent, we have a nested or partially nested information. Note that this generalizes (3.30) to product measures over sub-teams or agents.

(iii) (Non-classical Independent Data Reduction) As a refinement of (ii), a non-classical independent reduction is one so that (i) applies (i.e., there is a change of measure under which the measurements at each agent admits a policy-independent static reduction as in Sect. 3.4.2) for each time stage and each agent, but the agents do not necessarily have access to their earlier information from the previous time stages (e.g., they do not have perfect recall).

3.6 Equivalent Static Reductions Preserve Optimality but May Not Person-by-Person-Optimality or Stationarity

We showed earlier (and we will see further later) that static reduction can be a very useful method for deriving and studying properties of optimal policies in stochastic teams. In this section, we show that static reduction has its limitations when the solution concept is not global optimality but rather only person-by-person-optimality or stationarity. This, in particular, is a concern for policy-dependent reductions.

Recall the definitions of global optimality in Definition 2.2.1 and person-by-person-optimality (Nash equilibrium) in Definition 2.2.2. These apply naturally to both static and dynamic teams. Recall also the definition of stationarity of a policy for a static team in Definition 2.4.1. One can introduce a similar definition for dynamic teams:

Definition 3.6.1 A policy $\underline{\gamma}^* \in \Gamma$ is stationary if, for all $i \in \mathcal{N}$, P-a.s.

$$\nabla_{u^i} E\Bigg[c\bigg(\omega_0, \gamma^{1*}(y^{1*}), \ldots, \gamma^{(i-1)*}(y^{(i-1)*}),$$

$$u^i, \gamma^{(i+1)*}(y^{i+1}_{u^i}), \ldots, \gamma^{N*}(y^N_{u^i})\bigg)\bigg| y^{i*}\Bigg]\Bigg|_{u^i=\gamma^{i*}(y^{i*})} = 0,$$

where for any $j \in \mathcal{N}$,

$$y^{j*} = h^j\Big(\omega_0, \omega_j, \gamma^{1*}(y^{1*}), \ldots, \gamma^{(j-1)*}(y^{(j-1)*}), y^{1*}, \ldots, y^{(j-1)*}\Big),$$

$$y^{j*}_{u^i} = h^j\Big(\omega_0, \omega_j, \gamma^{1*}(y^{1*}),$$

$$\ldots, \gamma^{(i-1)*}(y^{(i-1)*}), u^i, \gamma^{(i+1)*}(y^{(i+1)*}_{u^i}), \ldots, \gamma^{(j-1)*}(y^{(j-1)*}_{u^i}),$$

$$\ldots y^{1*}, \ldots, y^{i*}, y^{(i+1)*}_{u^i}, \ldots, y^{(j-1)*}_{u^i}\Big).$$

We now study the connections between stationary (pbp optimal, globally optimal) policies of dynamic teams and their policy-independent static reductions (see Sect. 3.4.2).

Theorem 3.6.1 ([637]) *Consider a stochastic dynamic team ($\mathcal{P}$) with a policy-independent static reduction.*

(i) A policy $\underline{\gamma}^*$ is pbp optimal (globally optimal) for the dynamic team ($\mathcal{P}$) if and only if $\underline{\gamma}^*$ is pbp optimal (globally optimal) for a policy-independent static reduction of dynamic team ($\mathcal{P}$);
(ii) Let a policy $\underline{\gamma}^*$ satisfy P-almost surely

$$\nabla_{u^i} E^{\gamma^{-i*}}_{\mathbb{Q}}\left[\frac{dP}{d\mathbb{Q}}\bigg| y^i\right]\Bigg|_{u^i=\gamma^{i*}(y^i)} = 0 \qquad \forall i \in \mathcal{N}, \tag{3.38}$$

where $\frac{dP}{d\mathbb{Q}}$ is defined in (3.30). Then, $\underline{\gamma}^*$ is stationary for dynamic team $(\mathcal{P})$ if and only if $\underline{\gamma}^*$ is stationary for a policy-independent static reduction of dynamic team $(\mathcal{P})$.

Proof of Theorem 3.6.1 (i) Since policies do not change under the reduction, the proof of the result for globally and pbp optimal policies follows. (ii) Let us first note the following.

Lemma 3.6.1 ([229, p. 216]) *Consider a probability space $(\tilde{\Omega}, \tilde{\mathcal{F}}, \tilde{\mathbb{P}})$ where $\tilde{\mathbb{P}} \ll \tilde{\mathbb{Q}}$ for some probability measure $\tilde{\mathbb{Q}}$. Let $\mathcal{G} \subset \tilde{\mathcal{F}}$ be a sub-σ-field, and X be an integrable random variable. Then,*

$$E_{\tilde{\mathbb{P}}}[X|\mathcal{G}] = \frac{E_{\tilde{\mathbb{Q}}}[X\frac{d\tilde{\mathbb{P}}}{d\tilde{\mathbb{Q}}}|\mathcal{G}]}{E_{\tilde{\mathbb{Q}}}[\frac{d\tilde{\mathbb{P}}}{d\tilde{\mathbb{Q}}}|\mathcal{G}]} \quad \tilde{\mathbb{P}} - a.s.$$

Now, let $\underline{\gamma}^*$ be a stationary policy for $(\mathcal{P}^D)$. In the following, we show that if $\underline{\gamma}^*$ satisfies (3.38), then it is also stationary under a policy-independent static reduction. Since $\underline{\gamma}^*$ is a stationary policy for $(\mathcal{P}^D)$, using Lemma 3.6.1, we have P-almost surely

$$0 = \nabla_{u^i} E_P^{\gamma^{-i*}}[c(\omega_0, u^1, \ldots, u^N)|y^i]\Big|_{u^i=\gamma^{i*}(y^i)}$$

$$= \nabla_{u^i}\left\{\frac{E_{\mathbb{Q}}^{\gamma^{-i*}}[\tilde{c}(\omega_0, u^1, \ldots, u^N, y^1, \ldots, y^N)|y^i]}{E_{\mathbb{Q}}^{\gamma^{-i*}}[\frac{dP}{d\mathbb{Q}}|y^i]}\right\}\Bigg|_{u^i=\gamma^{i*}(y^i)}. \tag{3.39}$$

Hence,

$$\left\{\frac{\left(\nabla_{u^i} E_{\mathbb{Q}}^{\gamma^{-i*}}[\tilde{c}(\omega_0, u^1, \ldots, u^N, y^1, \ldots, y^N)|y^i]\right) E_{\mathbb{Q}}^{\gamma^{-i*}}[\frac{dP}{d\mathbb{Q}}|y^i]}{\left(E_{\mathbb{Q}}^{\gamma^{-i*}}[\frac{dP}{d\mathbb{Q}}|y^i]\right)^2}\right. \tag{3.40}$$

$$\left. - \frac{E_{\mathbb{Q}}^{\gamma^{-i*}}[\tilde{c}(\omega_0, u^1, \ldots, u^N, y^1, \ldots, y^N)|y^i]\left(\nabla_{u^i} E_{\mathbb{Q}}^{\gamma^{-i*}}[\frac{dP}{d\mathbb{Q}}|y^i]\right)}{\left(E_{\mathbb{Q}}^{\gamma^{-i*}}[\frac{dP}{d\mathbb{Q}}|y^i]\right)^2}\right\}\Bigg|_{u^i=\gamma^{i*}(y^i)}$$

$$= 0. \tag{3.41}$$

Since $\underline{\gamma}^*$ satisfies (3.38), we have that P-almost surely (3.41) is equal to zero, and since $\frac{dP}{d\mathbb{Q}} > 0$ P-almost surely and (3.40) msut be equal to zero P-almost surely, we have P-almost surely

$$\nabla_{u^i} E_{\mathbb{Q}}^{\gamma^{-i*}}[\tilde{c}(\omega_0, u^1, \ldots, u^N, y^1, \ldots, y^N)|y^i]\Big|_{u^i=\gamma^{i*}(y^i)} = 0, \tag{3.42}$$

which implies that $\underline{\gamma}^*$ is a stationary policy for $(\mathcal{P}^D)$ under policy-independent static reductions. For the converse statement, if a policy $\underline{\gamma}^*$ is stationary for $(\mathcal{P}^D)$ under policy-independent static reductions and satisfies (3.38), expressions (3.40) and (3.41) are equal to zero P-almost surely which implies that $\underline{\gamma}^*$ is a stationary policy for $(\mathcal{P}^D)$, and hence, this completes the proof. ◇

We note that (3.38) implies that at a stationary point $\underline{\gamma}^*$, the control has no impact locally on the reduction (on the Radon-Nikodym derivative term $\frac{dP}{d\mathbb{Q}}$ defined in (3.30)). We note that if the team is static, then (3.38) trivially holds since $\frac{dP}{d\mathbb{Q}}$ does not depend on actions.

In the following, we study the connections between stationary (pbp optimal, globally optimal) policies of dynamic teams and their policy-dependent static reductions (studied in Sect. 3.4.1).

Theorem 3.6.2 ([637]) *Consider a stochastic dynamic team $(\mathcal{P}^D)$ with partially nested information structure. Let Assumption 3.4.1 hold. Then, $\underline{\gamma}^{D,*}$ is a globally optimal policy for $(\mathcal{P}^D)$ if and only if $\underline{\gamma}^{S,*}$ is a globally optimal policy for $(\mathcal{P}^S)$ under the policy-dependent static reduction.*

Proof Since the information structure is partially nested and policies are deterministic, under Assumption 3.4.1, there is a bijection from the set of policies Γ^D to the set of policies Γ^S. Therefore, global optimality in one domain implies global optimality in the other domain. ◇

We now provide three examples (see [637]) that serve to demonstrate the subtlety of the connections between stationary (pbp optimal) policies of $(\mathcal{P}^D)$ and $(\mathcal{P}^S)$. These counterexamples show that, in contrast to the case of globally optimal policies, the isomorphism relations between stationary (pbp optimal) policies of $(\mathcal{P}^D)$ and $(\mathcal{P}^S)$ are no longer true, in general (under Assumption 3.4.1).

We first show that corresponding to a policy $\underline{\gamma}^{S,*}$ which is stationary (also pbp optimal) for $(\mathcal{P}^S)$, $\underline{\gamma}^{D,*}$, satisfying the corresponding policy-dependent static reduction, need not be pbp optimal for $(\mathcal{P}^D)$.

Example 3.6.1 Consider a 2-DM stochastic team $(\mathcal{P}^D)$ with $I^1 = \{y_1^D\}$ and $I^2 := \{y_2^D\} = \{y_1^D, \hat{y}_2^D\}$, where $\hat{y}_2^D = \omega_2 + u^1$, and $\omega_2 =: \hat{y}_2^S$ and $y_1^D =: y_1^S = \omega_1$ are primitive random variables. Let the expected cost function be given as

$$E[c(\omega_2, u^1, u^2)] := E[(u^1 - u^2 + \omega_2)^2 - \alpha(u^1)^2], \tag{3.43}$$

for a given $\alpha \in (0, 1)$.

A policy $\underline{\gamma}^{S,*} = (\gamma_1^{S,*}, \gamma_2^{S,*}) = (0, (0, I))$ (where the policy $(0, (0, I))$ denotes $\gamma_1^{S,*} \equiv 0$, $\gamma_{2,1}^{S,*} \equiv 0$, and $\gamma_{2,2}^{S,*}$ is the identity map, I, that is, $u^{1*} = \gamma_1^{S,*}(y_1^S) = 0$ and $u^{2*} = \gamma_2^{S,*}(y_1^S, \hat{y}_2^S) = \hat{y}_2^S$) is pbp optimal for $(\mathcal{P}^S)$.

However, a policy $\underline{\gamma}^{D,*} = (\gamma_1^{D,*}, \gamma_2^{D,*}) = (0, (-\gamma_1^{S*}, I))$ (where the policy $(0, (-\gamma_1^{S*}, I))$ denotes $\gamma_1^{D,*} \equiv 0$, $\gamma_{2,1}^{D,*} = -\gamma_1^{S*}$, and $\gamma_{2,2}^{D,*}$ is the identity map, that is, $u^{1*} = \gamma_1^{D,*}(y_1^D) = 0$ and $u^{2*} = \hat{y}_2^D - \gamma_1^{S,*}(y_1^D)$) is not pbp optimal for $(\mathcal{P}^D)$ since fixing a policy of **DM**2 to $\gamma_2^{D,*}$ such that $u^{2*} = \hat{y}_2^D - \gamma_1^{S,*}(y_1^D)$, the expected cost function will be concave in u^1 ($c(u^1, u^{2*}) = -\alpha(u^1)^2$) and the value will be unbounded from below. We note, however, that $\underline{\gamma}^{D,*}$ is a stationary policy for $(\mathcal{P}^D)$.

In the following example, we show that corresponding to a policy $\underline{\gamma}^{D,*}$ which is stationary (also pbp optimal) for $(\mathcal{P}^D)$, $\underline{\gamma}^{S,*}$ under the corresponding policy-dependent static reduction, need not be pbp optimal for $(\mathcal{P}^S)$.

Example 3.6.2 Consider a 2-DM stochastic team $(\mathcal{P}^D)$ with $I^1 = \{y_1^D\}$ and $I^2 := \{y_2^D\} = \{y_1^D, \hat{y}_2^D\}$, where $\hat{y}_2^D = \omega_2 + u^1$, and $\omega_2 =: \hat{y}_2^S = \omega_1$ and $y_1^D =: y_1^S$ are primitive random variables. Let the expected cost function be given as

$$E[c(\omega_2, u^1, u^2)] := E[\alpha(u^1)^2 + \beta(u^2 - \omega_2)^2 - (u^1 - u^2 + \omega_2)^2], \quad (3.44)$$

for a given $\alpha \in (0, 1)$ and $\beta > 1$.

A policy $\underline{\gamma}^{D,*} = (\gamma_1^{D,*}, \gamma_2^{D,*}) = (0, (0, I))$ (where $\gamma_1^{D,*} \equiv 0$, $\gamma_{2,1}^{D,*} \equiv 0$ and $\gamma_{2,2}^{D,*}$ is the identity map, that is, $u^{1*} = \gamma_1^{D,*}(y_1^D) = 0$ and $u^{2*} = \hat{y}_2^D$) is pbp optimal for $(\mathcal{P}^D)$ since fixing a policy of **DM**2 to $\gamma_2^{D,*}$, the expected cost function will be convex in u^1 ($c(u^1, u^2) = (\alpha + \beta)(u^1)^2$), and fixing a policy of **DM**1 to $\gamma_1^{D,*}$ such that $u^1 = \gamma_1^{D,*}(y_1^D) = 0$, the expected cost function will be convex in u^2 ($c(u^1, u^2) = (\beta - 1)(u^2 - \omega_2)^2$).

However, under the policy-dependent static reduction, a policy $\underline{\gamma}^{S,*} = (\gamma_1^{S,*}, \gamma_2^{S,*}) = (0, (-\gamma_1^{D,*}, I))$, is not pbp optimal for $(\mathcal{P}^S)$ since fixing a policy of **DM**2 to $\gamma_2^{S,*}$ such that $u^2 = \gamma_2^{S,*}(y_1^S, \hat{y}_2^S) = \hat{y}_2^S - \gamma_1^{D,*}(y_1^S)$, the expected cost function will be concave in u^1 ($c(u^1, u^2) = (\alpha - 1)(u^1)^2$).

Next, we provide an example, showing that corresponding to a policy $\underline{\gamma}^{D,*}$ which is stationary (pbp optimal) for $(\mathcal{P}^D)$, a policy $\underline{\gamma}^{S,*}$ under the corresponding policy-dependent static reduction, need not be stationary for $(\mathcal{P}^S)$.

Example 3.6.3 Consider a 2-DM stochastic team $(\mathcal{P}^D)$ with $I^1 = \{y_1^D\}$ and $I^2 = \{y_2^D\} := \{y_1^D, \hat{y}_2^D\}$, where $\hat{y}_2^D = \omega_2 + \sqrt{u^1}$, and ω_2 and $y_1^D = y_1^S := \omega_1$ are primitive random variables. Let the expected cost function be given as

$$E[c(\omega_2, u^1, u^2)] := E[(\sqrt{u^1} - u^2 + \omega_2)^2], \tag{3.45}$$

where $\mathbb{U}^1 = \mathbb{R}_+$.

A policy $\underline{\gamma}^{D,*} = (\gamma_1^{D,*}, \gamma_2^{D,*}) = (0, (0, I))$ (where $\gamma_1^{D,*} \equiv 0$, $\gamma_{2,1}^{D,*} \equiv 0$ and $\gamma_{2,2}^{D,*}$ is the identity map, that is, $u^{1*} = 0$ and $u^{2*} = \hat{y}_2^D$) is stationary for $(\mathcal{P}^D)$.

However, under the policy-dependent static reduction, a policy $\underline{\gamma}^{S,*} = (\gamma_1^{S,*}, \gamma_2^{S,*}) = (0, (\sqrt{\gamma_1^{D*}}, I))$ (where $\gamma_1^{S,*} \equiv 0$, $\gamma_{2,1}^{S,*} = \sqrt{\gamma_1^{D*}}$, and $\gamma_{2,2}^{S,*}$ is the identity map, that is, $u^1 = 0$ and $u^2 = \omega_2 + \sqrt{\gamma_1^{D*}(y_1^S)})$ is not stationary (although it is pbp optimal) for $(\mathcal{P}^S)$. Since fixing the policy of **DM**2 to $\gamma_2^{S,*}$ such that $u^2 = \omega_2$, the derivative of the expected cost function with respect to u^1 is always 1. Hence, the criterion for stationarity does not lead to a solution.

This now brings us to the following theorem.

Theorem 3.6.3 ([637]) *Consider a stochastic dynamic team $(\mathcal{P}^D)$ with partially nested information structure. Let Assumption 3.4.1 hold. Then:*

(i) If $\underline{\gamma}^{D,*}$ is a stationary (pbp optimal) policy for $(\mathcal{P}^D)$, then $\underline{\gamma}^{S,*}$ is not necessarily a stationary (pbp optimal) policy for $(\mathcal{P}^S)$ under the policy-dependent static reduction;

(ii) If $\underline{\gamma}^{S,*}$ is a stationary (pbp optimal) policy for $(\mathcal{P}^S)$, then $\underline{\gamma}^{D,*}$, satisfying the policy-dependent static reduction relation, is not necessarily pbp optimal for $(\mathcal{P}^D)$.

Proof This is a direct consequence of the examples above, where Examples 3.6.2 and 3.6.3 lead to Part(i), and Example 3.6.1 leads to Part(ii). ◇

Sufficient conditions for positive results can also be obtained, in particular if the policies in question are affine; see [637].

3.7 Convexity of Stochastic Team Problems: Information Structure and Cost Dependent Convexity Properties

We have seen earlier (e.g. in Sect. 2.4.2) the benefits of having convex loss/cost functions. In the following we refine such convexity conditions and arrive at weaker conditions for when a team problem may be convex in team policies.

3.7.1 Convexity of Static Team Problems and an Equivalent Representation of Cost Functions

We begin this sub-section with the following definition.

Definition 3.7.1 A (static or dynamic) team problem is convex on $\mathbf{\Gamma}$ if $J(\underline{\gamma}) < \infty$ for all $\underline{\gamma} \in \mathbf{\Gamma}$, and for any $\alpha \in (0, 1)$, $\underline{\gamma_1}, \underline{\gamma_2} \in \mathbf{\Gamma}$,

$$J(\alpha\underline{\gamma_1} + (1-\alpha)\underline{\gamma_2}) \leq \alpha J(\underline{\gamma_1}) + (1-\alpha)J(\underline{\gamma_2})$$

We state the following immediate result without proof, a more general result will be stated in Theorem 3.7.2.

Theorem 3.7.1 *Consider a static team with cost/loss function c and corresponding objective function in normal form J. $J(\underline{\gamma})$ is convex if $c(\omega_0, \mathbf{u})$ is convex in $\mathbf{u}$ for all ω_0, provided that $J(\underline{\gamma}) < \infty$ for all $\underline{\gamma} \in \mathbf{\Gamma}$.*

The condition in Theorem 3.7.1 is not tight, however, since specificity of information structure and measurability could lead to tighter conditions. To substantiate this point, we have the following example.

Example 3.7.1 Let $\Omega = [0, 1]$ and P be the uniform distribution on Ω. Let $N = 2$, and $\mathbb{U}^1 = \mathbb{U}^2 = [1, 2]$. Let

$$c(\omega, u^1, u^2) = 1_{\{\omega\in[0,0.9]\}}\Big((u^1-2)^2+(u^2-2)^2\Big)+1_{\{\omega\in(0.9,1]\}}\Big(\sqrt{1+u^1}+\sqrt{1+u^2}\Big)$$

Now, suppose further that $I^1 = I^2 = \eta^1(\omega) = \eta^2(\omega) = 1_{\{\omega\in[0,0.1)\}}$. It follows that here the team problem is convex, even though $c(\omega, u^1, u^2)$ is not convex on $\{\omega : \omega \in (0.9, 1]\}$, which has a non-zero probability measure. To see this, note that one may view this optimization problem as one of minimizing $J(u_1^1, u_2^1; u_1^2, u_2^2)$, where $u_j^i = \gamma^i(\omega_j)$, with $\omega_1 \equiv \{\omega : \omega \in [0, 0.1)\}$ and $\omega_2 \equiv \{\omega : \omega \in [0.1, 1]\}$. It follows that

$$J(u_1^1, u_2^1; u_1^2, u_2^2) = \sum_{i=1,2} 0.1(u_1^i - 2)^2 + 0.8(u_2^i - 2)^2 + 0.1(\sqrt{u_2^i + 1})$$

The Hessian of J is a diagonal matrix with strictly positive entries, leading to convexity of the problem.

In the following, we will make use of the fact that $u^k \leftrightarrow y^k \leftrightarrow \{\mathbf{y}^{-k}, \omega\}$ form a Markov chain almost surely. Before proceeding further, let us note that the *join* of two σ-fields over some set $\mathbb{X}$ is the coarsest σ-field containing both. The *meet* of two σ-fields is the finest σ-field which is a subset of both. Let $\mathcal{F}^i$ be the σ-field generated by η^i over Ω, and let $\mathcal{F}_c = \bigcap_k \mathcal{F}^k$ be the meet of these fields, which is termed as *common knowledge* by Aumann [45] for finite probabilities spaces. In addition, let $\mathcal{F}_j$ be the *join* of the σ-field, denoted by $\mathcal{F}_j = \sigma(\bigcup_k \mathcal{F}^k)$ or $\bigvee_k \mathcal{F}^k$.

In the following, as earlier, we take the measurement and control action spaces to be standard Borel.

An Equivalent Representation of the Cost Through Iterated Expectations Let us express the expected cost under a given measurable team policy $\underline{\gamma}$ as follows. With the interpretation that $P(u^k \in \cdot|y^k) = 1_{\{u^k=\gamma^k(y^k)\in\cdot\}}$, we obtain from the law of the iterated expectations, with $\mathbf{y} = \{y^1, \cdots, y^N\}$, that

$$E[c(\omega_0, \mathbf{u})] = E\Big[E[c(\omega_0, \mathbf{u})|\mathbf{y}]\Big] \tag{3.46}$$

Under any measurable policy, given $\mathbf{y}$, $\mathbf{u}$ is specified. Thus, with

$$\tilde{c}(y^1, \cdots, y^N, u^1, \cdots, u^N) := E[c(\omega_0, \mathbf{u})|\mathbf{y}],$$

that is $\tilde{c}(y^1, \cdots, y^N, u^1, \cdots, u^N) = \left(\int P(d\omega_0|\mathbf{y})c(\omega_0, u^1, \cdots, u^N)\right)$, the cost function becomes $E[\tilde{c}(y^1, \cdots, y^N, u^1, \cdots, u^N)]$. We will use this representation in the following.

Theorem 3.7.2

(i) If a team problem is convex, then

$$E[c(\omega_0, \mathbf{u})|\mathcal{F}_c]$$

is convex in u almost surely.

(ii) If

$$E[c(\omega_0, \mathbf{u})|\mathcal{F}_j]$$

is convex in u almost surely, then the team problem is convex on the set of team policies that satisfy $J(\underline{\gamma}) < \infty$.

Proof

(i) We will show the contra-positive. Let B be a Borel set such that $P(B) > 0$, $B \in \mathcal{F}_c$, and $E[c(\omega_0, \mathbf{u})|B]$ be non-convex so that there exist $\mathbf{u}$ and $\mathbf{u}'$ and $\lambda \in (0, 1)$ such that

$$E[c(\omega_0, \lambda\mathbf{u} + (1-\lambda)\mathbf{u}')|B] > \lambda E[c(\omega_0, \mathbf{u})|B] + (1-\lambda)E[c(\omega_0, \mathbf{u}')|B]$$

Now, let $\underline{\gamma}$ and $\boldsymbol{\gamma}$ be two team policies so that these only differ on B; and on B $\underline{\gamma} = \mathbf{u}$ and $\boldsymbol{\gamma} = \mathbf{u}'$. Such measurable policies exist, for example by taking $\underline{\gamma}(\omega) = \{0, 0, \cdots, 0\}$ when $\omega \notin B$. These policies are both Borel measurable and admissible given the information structure. Then, $\forall\, \lambda \in (0, 1)$,

$$J(\lambda \underline{\gamma} + (1-\lambda)\boldsymbol{\gamma}) > \lambda J(\underline{\gamma}) + (1-\lambda)J(\boldsymbol{\gamma}),$$

and convexity fails.

(ii) We adopt the equivalent representation (3.46) in this part of the proof. Note that under any measurable policy, the random variable $\tilde{c}(y^1, \cdots, y^N, u^1, \cdots, u^N)$ is measurable on the σ-field generated by $\mathbf{y}$ and thus the join σ-field. The proof then follows from the following. Consider two policies $\underline{\gamma}$ and $\underline{\bar{\gamma}}$ with finite expected costs. It then follows that

$$\begin{aligned}
&J(\lambda \underline{\gamma} + (1-\lambda)\underline{\bar{\gamma}}) \\
&= \int P(d\mathbf{y})\tilde{c}(y^1, \cdots, y^N, \lambda \gamma^1(y^1) + (1-\lambda)\bar{\gamma}^1(y^1), \cdots, \lambda \gamma^N(y^N) \\
&\qquad\qquad + (1-\lambda)\bar{\gamma}^N(y^N)) \\
&\leq \int P(d\mathbf{y})\Big(\lambda \tilde{c}(y^1, \cdots, y^N, \gamma^1(y^1), \cdots, \gamma^N(y^N)) \\
&\qquad\qquad + (1-\lambda)\tilde{c}(y^1, \cdots, y^N, \bar{\gamma}^1(y^1), \cdots, \bar{\gamma}^N(y^N))\Big) \\
&= \lambda J(\underline{\gamma}) + (1-\lambda)J(\underline{\bar{\gamma}})
\end{aligned}$$

◇

We note that Example 3.7.1 satisfies the conditions of Theorem 3.7.2. These conditions will also be used to study Witsenhausen's counterexample further below.

A Generalization of Radner and Krainak et al.'s Theorems

Within the framework above, we can provide a generalization of some of the results presented in Chap. 2, such as Theorem 2.4.5 by utilizing information structure dependent convexity properties. Note that Theorems 2.4.5 and 2.4.6 are not directly applicable to Example 3.7.1. Nonetheless, this result should be viewed in the context of Theorems 2.4.5 and 2.4.6.

Theorem 3.7.3 *Let $\{J; \Gamma^i, i \in \mathcal{N}\}$ be a static stochastic team problem where the cost $E[c(\omega_0, \mathbf{u})|\mathcal{F}_j]$ is convex and continuously differentiable in $\mathbf{u}$ almost surely. Let $\underline{\gamma}^* \in \boldsymbol{\Gamma}$ be a policy N-tuple with a finite cost ($J(\underline{\gamma}^*) < \infty$), and suppose that for every $\underline{\gamma} \in \boldsymbol{\Gamma}$ such that $J(\underline{\gamma}) < \infty$, the following holds:*

$$\sum_{i \in \mathcal{N}} E\{\nabla_{u^i}\tilde{c}(\mathbf{y}, \underline{\gamma}^*(\mathbf{y}))[\gamma^i(y^i) - \gamma^{i*}(y^i)]\} \geq 0, \tag{3.47}$$

where $\tilde{c}(\mathbf{y}, \mathbf{u}) := E[c(\omega_0, \mathbf{u})|\mathcal{F}_j]$. *Then,* $\underline{\gamma}^*$ *is a team-optimal policy, and it is unique if* $\tilde{c}(\mathbf{y}, \mathbf{u})$ *is strictly convex in* $\mathbf{u}$ *almost surely.*

Proof The proof follows by defining the new cost function

$$\tilde{c}(\mathbf{y}, \mathbf{u}) = E[c(\omega_0, \mathbf{u})|\mathcal{F}_j],$$

almost surely. The result then follows as in Theorem 2.4.5. ◇

Theorem 3.7.4 *Let* $\{J; \Gamma^i, i \in \mathcal{N}\}$ *be a static stochastic team problem which satisfies all the hypotheses of Theorem 3.7.3, with the exception of inequality (3.47). Instead of (3.47), let either Assumption 2.4.1(c.5) or Assumption 2.4.1(c.6) be satisfied with* c *replaced with* $\tilde{c}$*. Then, if* $\underline{\gamma}^* \in \boldsymbol{\Gamma}$ *is a stationary policy it is also team optimal. Such a policy is unique if* $E[c(\omega_0, \mathbf{u})|\mathcal{F}_j]$ *is strictly convex in* $\mathbf{u}$*, a.s.*

Proof The proof follows by defining the new cost function $\tilde{c}$ as in the proof of Theorem 3.7.3, and following Theorem 2.4.6. ◇

3.7.2 Convexity of Dynamic Teams

Convexity of the Reduced Model

For a policy-independent static reduction, the policies are equivalent under the original model or the reduction. Such static reduction of a sequential dynamic team problem, if exists, is not unique. However, either all of these are convex or none is. This follows from the fact that the non-uniqueness only arises in the selection of the marginal measure for the independent processes y^i, as discussed in Sect. 3.4.2. For the partially-nested case, control sharing makes static reduction policy-independent by Theorem 3.4.2. The following is then immediate.

Theorem 3.7.5 ([637, 830]) *A stochastic dynamic team problem with a policy-independent static reduction is convex if and only if its static reduction is. This also holds for the policy-dependent reduction (partially-nested) case if control actions are shared whenever measurements are (under static-measurements with control-sharing reduction).*

Non-convexity of the Witsenhausen Counterexample and Its Variants

Consider the static reductions of Witsenhausen's counterexample and the Gaussian Test Channel (3.32)–(3.33). For both (3.32) and (3.33), using the fact that e^{-x^2} is not a convex function, we recognize that this problem is not convex by Theorem 3.7.2(i) (with the common knowledge/information being the trivial σ-algebra consisting of the empty set and its complement), e.g. by considering two deterministic constant policies and their convex combinations.[11]

We note that Witsenhausen states, without proof, in [771] (p. 134) that the counterexample is non-convex in γ^1 for every optimal γ^2 (selected as a best response to γ^1). The discussion above can be viewed as an explicit proof for this result. Note also that for both problems above, linear policies contain pbp optimal policies, but this does not imply global optimality: for the first problem, as presented earlier in the chapter, Witsenhausen had shown the suboptimality of linear policies and for the second problem (3.33), linear policies are indeed optimal (where the proof was based not on variational arguments, but on information theoretic methods). These last observations are not surprising in view of the convexity based analysis in Theorems 2.4.5 and 2.4.6 relating pbp-optimality to global optimality.

3.7.3 *Revisiting Non-classical Information, Witsenhausen's Counterexample and Lack of Convexity Due to Signaling for LQG Teams*

We revisit in this section non-classical information structures discussed earlier, and bring a different perspective to the difficulties underlying decision problems with such structures.

Let us consider dynamic decentralized control systems where multiple controllers with non-shared measurements are present. In such problems, the information structure is generally of non-classical nature, and we have already seen that in such cases there may be an incentive for signaling. Under signaling, the DMs apply their actions to affect the information available at the other decision makers. In this case, the control policies induce stochastic kernels from the exogenous random variable space to the observation space of the signaled DMs.

This is different from the quasi-classical case, where when a signaling DM affects the information of a signaled DM the signaled decision maker already has access to all the information available to the signaling DM. However, for the non-classical case, the problem also features an information transmission aspect, and the signaling DM's objective also includes the design of an optimal measurement channel and thus a communication problem.

[11] Note that here convexity is with respect to the space of deterministic policies. Later on we will consider policies that are proper subsets of probability measures which are equivalent to randomized policies; in this case, the map J is always linear and hence convex in policies. Note that if we take the convex combination of two deterministic policies γ^1 and $\bar{\gamma}^1$ with scaling $\frac{1}{2}, \frac{1}{2}$, the resulting policy would be e.g. $\frac{1}{2}\gamma^1 + \frac{1}{2}\bar{\gamma}^1$. However, if we view them as probability measures, the convex combination will just be $\frac{1}{2}\delta_{\gamma^1(y^1)} + \frac{1}{2}\delta_{\bar{\gamma}^1(y^1)}$.

To make this important issue more explicit, let us consider the following example. Consider a two-controller system evolving in $\mathbb{R}^n$:

$$x_{t+1} = Ax_t + B^1 u_t^1 + B^2 u_t^2 + w_t,$$
$$y_t^1 = C^1 x_t + v_t^1,$$
$$y_t^2 = C^2 x_t + v_t^2,$$

where w, v^1, v^2 are zero-mean, i.i.d. disturbances, and A, B^1, B^2, C^1, C^2 are matrices of appropriate dimensions. For $\rho_1, \rho_2 > 0$, let the objective be the minimization of the cost functional

$$J = E\left[\left(\sum_{t=0}^{T-1} |x_t|^2 + \rho_1 |u_t^1|^2 + \rho_2 |u_t^2|^2\right) + |x_T|^2\right]$$

over control policies of the form:

$$u_t^i = \gamma_t^i(y_{[0,t]}^i, u_{[0,t-1]}^i), \quad i = 1, 2; \quad t = 0, 1, \ldots, T-1.$$

As discussed earlier in Chap. 2, a static LQG team problem (that is, the above with $T = 1$) admits an optimal solution which is linear. The proof for this result has followed the property that the team cost is convex in the joint actions of the DMs, and is continuously differentiable, and as a consequence it suffices to find the unique fixed point. This, in turn, leads (under the Gaussian statistics) to linear optimal strategies for the agents. For a multi-stage problem (say with $T = 2$), however, the cost is in general no-longer *convex* in the action variables of the controllers acting in the first stage $t = 0$ since we have a communications/signaling problem: This is because these actions affect the estimation quality of the other controller in the future stages, if one DM can signal information to the other DM in one stage. We note that this condition is equivalent to $C^1 A^l B^2 \neq 0$ or $C^2 A^l B^1 \neq 0$ with $l + 1$ denoting the delay in signaling with $l = 0$ in the problem considered. If one studies convexity in (normal-form) policies, even under linear policies restriction, such a linear signaling problem in general is not convex (see [571, Section II]). If one views this as a stochastic measurement channel design problem by Corollary 7.3.1 or Theorem 7.3.1, we again lack convexity.

The optimization problem, as such, is not jointly convex in such policies, and finding a fixed point to the stationarity conditions in the optimal policies does not necessarily lead to the conclusion that such policies are optimal. We will revisit this discussion in Sect. 16.3.2.

3.8 Expansion of Information Structures, and Signaling via Control and Stochastic Relaxations of Nestedness

In this section, we further discuss the role of information structures in dynamic stochastic teams, and identify useful refinements that depend on the underlying information structure, probability measure, and the cost structure.

3.8.1 Expansion of Information Structures: A Recipe for Identifying Sufficient Information

We start with a general result on *optimum-performance equivalence* of two stochastic dynamic teams with different information structures.

Proposition 3.8.1 *Let D_1 and D_2 be two stochastic dynamic teams with the same loss function, and differing only in their information structures, $\underline{\eta}_1$ and $\underline{\eta}_2$, respectively, with corresponding composite strategy spaces $\mathbf{\Gamma}_1$ and $\mathbf{\Gamma}_2$, such that $\mathbf{\Gamma}_2 \subseteq \mathbf{\Gamma}_1$. Let D_1 admit a team-optimal solution, denoted by $\underline{\gamma}_1^* \in \mathbf{\Gamma}_1$, with the further property that $\underline{\gamma}_1^* \in \mathbf{\Gamma}_2$. Then $\underline{\gamma}_1^*$ also solves D_2.*

Proof Let J_1 and J_2 be the cost functions of the two teams D_1 and D_2 when expressed in normal form. Note that $\mathbf{\Gamma}_2 \subseteq \mathbf{\Gamma}_1$ says that the strategy space for D_1 is richer than that for D_2. Consider now the following set of equalities and inequalities:

$$J_1(\underline{\gamma}_1^*) = \inf_{\underline{\gamma}_1 \in \mathbf{\Gamma}_1} J_1(\underline{\gamma}_1) \leq \inf_{\underline{\gamma}_1 \in \mathbf{\Gamma}_2} J_1(\underline{\gamma}_1) = \inf_{\underline{\gamma}_2 \in \mathbf{\Gamma}_2} J_2(\underline{\gamma}_2) \leq J_2(\underline{\gamma}_1^*),$$

where the first equality is the definition of $\underline{\gamma}_1^*$; the first inequality follows because infimization over a smaller set cannot lead to a smaller value; the next equality follows because J_1 and J_2 agree on $\mathbf{\Gamma}_2$; and the last inequality follows because as an element of $\mathbf{\Gamma}_2$, $\underline{\gamma}_1^*$ cannot lead to a lower value than the infimum of J_2 on $\mathbf{\Gamma}_2$. ◇

A recipe for utilizing the result above would be:

Given a team problem, say D_2, with *IS* $\underline{\eta}_2$, which is presumably difficult to solve, obtain a *finer IS* $\underline{\eta}_1$, and solve the team problem under this expanded *IS* (assuming that this new team problem is easier to solve). Then, if the team-optimal solution here is adapted to the sigma-field generated by the original coarser *IS*, it solves also the original problem D_2.

To see such a recipe at work, consider the following class of two-agent LQG team problems. The state equation is given by

$$x_{t+1} = Ax_t + B^1 u_t^1 + B^2 u_t^2 + w_t^0, \ t \in \mathcal{T},$$

the measurement of agent **A**1 by

$$y_t^1 = C^1 x_t + w_t^1, \ t \in \mathcal{T},$$

and that of **A2** by

$$y_1^2 = C^2 x_1 + w_1^2,$$

where the initial state x_1, and noise sequences $w^i_{[1,T]}, i = 0, 1$, are independent zero-mean Gaussian random vectors with specified positive-definite covariance matrices, and w_1^2 is also Gaussian zero-mean with a specified positive-definite covariance matrix. The only restriction is that the statistics of w_1^2 are picked, along with the matrices C^1 and C^2, such that y_1^2 is a *garbled* version of y_1^1 with respect to x_1; in other words (in view of Definition 7.2.3) the conditional probability density function of x_1 given (y_1^1, y_1^2) does not depend on y_1^2. One way of satisfying this requirement would be if y_1^2 was a noisy version of y_1^1, obtained for example through

$$y_1^2 = C_3 y_1^1 + v \equiv C_3 C_1 x_1 + C_3 w_1^1 + v =: C_2 x_1 + w_1^2 ,$$

where v is another independent zero-mean Gaussian random vector with positive-definite covariance.

Now given all of the above, let us specify the information structure of the problem as

$$I_t^1 = \{y^1_{[1,t]}, u^2_{[1,T]}\} , \; I_t^2 = \{y_1^2\} , \; t \in \mathcal{T} ,$$

where agent **A**1 keeps her noisy measurement of the current value of the state, and has access to the past, present and future control actions of the other agent, both with perfect recall. **A2**, on the other hand, has open-loop information: she has access to imperfect noisy measurement of the initial value of the state, and uses this in the construction of her policies for all time, which she does in a *pre-commitment mode*, that is she decides on $u^2_{[1,T]} = \gamma^2_{[1,T]}(y_1^2)$ at time $t = 1$, and sticks to it for all t. Note that this is a problem which exhibits non-classical information, because **A**1 observes the actions of **A2** but does not have access to the measurement (y_1^2) used in the construction of these controls, which opens the possibility of signaling, with **A2** signaling the value of y_1^2 to **A**1 through his actions.

An *IS* equivalent to the one above would be the one where the agents' self control actions are also included in the information sets:

$$I_t^1 = \{y^1_{[1,t]}, u^1_{[1,t]}, u^2_{[1,T]}\} , \; I_t^2 = \{y_1^2, u^2_{[1,t]}\} , \; t \in \mathcal{T} ,$$

which is sometimes more convenient to work with. Now, an *IS* which is more informative (finer) than the one above would be the one where **A**1 also has access to y_1^2:

$$\tilde{I}_t^1 = \{y^1_{[1,t]}, y_1^2, u^1_{[1,t]}, u^2_{[1,T]}\} , \; \tilde{I}_t^2 = \{y_1^2, u^2_{[1,t]}\} , \; t \in \mathcal{T} ,$$

This is a partially nested *IS* because **A**1 now knows everything **A**2 knows but not *vice versa*.

To proceed further, let us adopt a loss function for the team:

$$L(x_{[1,T+1]}, u_{[1,T]}) = \sum_{t \in \mathcal{T}} |x_{t+1}|_Q^2 + |u_t|_R^2 ,$$

where $u_t := (u_t^{1'} \ u_t^{2'})'$, $Q \geq 0$ and $R > 0$ are symmetric weighting matrices of appropriate dimensions, and $| \cdot |_Q$ denotes the Euclidean (semi-)norm weighted by Q. The team problem then is minimization of the expected value of L, expressed in terms of the policy variables using the given information. Now, if instead of the original information structure, we take the expanded one $\tilde{I}$, then the problem is readily solvable, and in closed form, as to be shown below, because it is a partially nested LQG team.

We first note that if the weighting matrix R in the loss function is not block diagonal, it can be made one by redefining u_t^1 through a translation that is linear in u_t^2, that is there exists a matrix M of appropriate dimensions, such that with $\tilde{u}_t^1 := u_t^1 + M u_t^2$, $|u_t|_R^2 = |\tilde{u}_t^1|_{R^1}^2 + |u_t^2|_{R^2}^2$, where both R^1 and R^2 are positive-definite matrices. Note that such a transformation is compatible with the original information structure, because **A**1 is allowed to have access to the control actions of **A**2. To keep the notation simple, we will assume henceforth (and clearly without any loss of generality) that the original formulation has the weightings on u^1 and u^2 decoupled, that is that $|u_t|_R^2 = |u_t^1|_{R^1}^2 + |u_t^2|_{R^2}^2$.

Denote the composite strategy spaces corresponding to the original and expanded information structures (that is I and $\tilde{I}$) by $\mathbf{\Gamma}_2$ and $\mathbf{\Gamma}_1$, respectively. Let the corresponding ones for an individual agent be superscripted by the identity of that agent (1 or 2). Recall that a team-optimal solution $\underline{\gamma}^{1*}, \underline{\gamma}^{2*}$ is one that is defined through the relationship (this one under the expanded nIS):

$$J(\underline{\gamma}^{1*}, \underline{\gamma}^{2*}) = \inf_{\underline{\gamma}^1 \in \mathbf{\Gamma}_2^1, \underline{\gamma}^2 \in \mathbf{\Gamma}_2^2} J(\underline{\gamma}^1, \underline{\gamma}^2) \equiv \inf_{\underline{\gamma}^2 \in \mathbf{\Gamma}_2^2} \inf_{\underline{\gamma}^1 \in \mathbf{\Gamma}_2^1} J(\underline{\gamma}^1, \underline{\gamma}^2) , \qquad (\star)$$

where the last expression says that the composite infimization can be carried out sequentially; we can first hold $\underline{\gamma}^2$ fixed as an arbitrary element of $\mathbf{\Gamma}_2^2$, perform minimization with respect to $\underline{\gamma}^1$ over $\mathbf{\Gamma}_2^1$ (let us call this *inner minimization*), and then come back and infimize the resulting expression over $\mathbf{\Gamma}_2^2$. The inner minimization is a discrete-time stochastic control problem with state dynamics

$$x_{t+1} = A x_t + B^1 u_t^1 + B^2 \gamma_t^2 (y_{\bar{1}}^2) + w_t^0 , \ t \in \mathcal{T} ,$$

and loss function

$$L(x_{[1,T+1]}, u_{[1,T]}) = \sum_{t \in \mathcal{T}} |x_{t+1}|_Q^2 + |u_t^1|_{R^1}^2 + |u_t^2|_{R^2}^2 , \qquad (\star\star)$$

where all terms have been defined as before. The controller has access to (y_1^1, y_1^2) at $t = 1$, and to y_t^1 for $t > 1$, and also has access to the realized values of $\gamma^2_{[1,T]}(y_1^2)$ at $t = 1$, and has no restrictions on memory. As such, this is a standard LQG control problem, but with an additional input $B^2\gamma_t^2(y_1^2)$ in the state equation, which however is measurable with respect to the sigma field generated by y_1^2, and hence can be treated as a deterministic quantity (and in fact it **is** given as $B^2u_t^2$) when the expectation of L is conditioned on y_1^2.[12] Note that the third term in the expression for L does not enter into the minimization at this stage.

Now, using standard LQG theory with an additional deterministic driving term in the state equation, given also that $Q \geq 0$ and $R^1 > 0$, the problem of minimization of the conditional expected loss function

$$E[L(x_{[1,T+1]}, \gamma^1_{[1,T]}(\tilde{I}^1_{[1,T]}), \gamma^2_{[1,T]}(y_1^2))|y_1^2]$$

over $\mathbf{\Gamma}_2^1$, for fixed $\gamma^2_{[1,T]}$ admits an optimal solution that is of the form

$$\gamma_t^1(y^1_{[1,t]}, u^1_{[1,t-1]}, y_1^2, u^2_{[1,T]}) = \hat{\gamma}_t^1(\hat{x}_t, u^2_{[1,T]})\,, \qquad (o)$$

where

$$\hat{\gamma}_t^1(\hat{x}_t, u^2_{[1,T]}) = -P_tS_{t+1}A\hat{x}_t - P_t(s_{t+1} + S_{t+1}B^2u_t^2)\,,\ t \in \mathcal{T}\,, \qquad (i)$$

P_t, $t \in \mathcal{T}$, is given by

$$P_t = [R^1 + B^{1'}S_{t+1}B^1]^{-1}B^{1'}\,, \qquad (ii)$$

S_t, $t \in \mathcal{T}$, is an appropriate-dimensional matrix sequence generated by the matrix Riccati difference equation

$$S_t = Q + A'S_{t+1}[I - B^1P_tS_{t+1}]A; \quad S_{T+1} = Q\,, eUnALT \qquad (iii)$$

and s_t, $t \in \mathcal{T}$, is a vector sequence generated by a linear difference equation in retrograde time (where s_t depends linearly on $u^2_{[t+1,T]}$):

$$s_t = A'[I - B^1P_tS_{t+1}]'[s_{t+1} + S_{t+1}B^2u_t^2]; \quad s_{T+1} = 0\,. \qquad (iv)$$

Finally,

$$\hat{x}_t = E[x_t \mid y^1_{[1,t]}, u^1_{[1,t-1]}, y_1^2, u^2_{[1,t-1]}]\,, \qquad (v)$$

[12] For the argument here it is not necessary for the controller to have access to $u^2_{[1,T]}$ since with the expanded *IS* it has access to y_1^2, but shortly we will see that accessibility to $u^2_{[1,T]}$ is crucial to ensure that the optimal solution to the inner minimization is independent of the policy $\underline{\gamma}^2$.

that is, it is the conditional mean of x at time t, given all the past and present measurements and past control actions, and as such is given by the Kalman filter. Note that the solution $\hat{\gamma}_t^1$, $t \in \mathcal{T}$, in (i) does not depend on $\gamma_{[1,T]}^2$ but depends on its realized value $u_{[1,T]}^2$ and also possibly on y_1^2. We next argue that it actually does not depend on the latter. Let $\tilde{x}_t$, $t \in \mathcal{T}$, be generated by

$$\tilde{x}_{t+1} = A\tilde{x}_t + w_t^0 \,, \; \tilde{x}_1 = x_1, \; t \in \mathcal{T} \,,$$

and

$$\tilde{y}_t^1 := C^1\tilde{x}_t + w_t^1 \,, \; t \in \mathcal{T} \,.$$

Then, the conditional probability density function (cpdf) of the random variable x_t given the set of random variables $\{y_{[1,t]}^1, u_{[1,t-1]}^1, y_1^2$, and $u_{[1,t-1]}^2\}$ is the same as that of $\tilde{x}_t$ given $\{\tilde{y}_{[1,t]}^1, y_1^2\}$, that is $p(\tilde{x}_t|\tilde{y}_{[1,t]}^1, y_1^2)$. For $t = 1$, this is the cpdf of x_1 given $\{y_1^1, y_1^2\}$, which is independent of y_1^2 since y_1^2 is a garbled version of y_1^1. For $t > 1$, we have

$$p(\tilde{x}_t|\tilde{y}_{[2,t]}^1, y_1^1, y_1^2) = \int_{x_1} p(\tilde{x}_t|x_1, \tilde{y}_{[2,t]}^1, y_1^1, y_1^2)\, p(x_1|y_1^1, y_1^2)\, dx^1 \,/\, p(\tilde{y}_{[2,t]}^1|y_1^1, y_1^2)\,.$$

The first cpdf above is the cpdf of $w_{[1,t-1]}^0$ which is independent of y_1^2; the second cpdf is independent of y_1^2 because of the garbling condition; and the denominator cpdf is also independent of y_1^2 by the same reasoning as in the first cpdf, because $\tilde{y}_t^1 := C^1\tilde{x}_t + w_t^1$ and w_t^1 is independent of y_1^2. Hence, the Kalman filter does not use the extra measurement y_1^2 when it is a garbled version of y_1^1 with respect to x_1. And consequently the inner minimization problem in ($\star$) would admit the same solution if the information structure for **A**1 did not include y_1^2.

We now focus on the outer minimization problem in ($\star$). First, because of the linear structure of the optimal controller (i), the minimum value of the inner minimization in ($\star$) consists of three general terms: a term which is quadratic in $u_{[1,T]}^2$, a second (bilinear) term which is a product of $u_{[1,T]}^2$ and the primitive random variables, and a third term which depends only on the primitive random variables but not on $u_{[1,T]}^2$. Further, since ($\star\star$) is positive for all $u_{[1,T]}^2$ whenever x_1 has a positive-definite covariance, the minimum value of the inner minimization in ($\star$) is strictly convex in $u_{[1,T]}^2$. Hence, the outer minimization in ($\star$) is a standard static quadratic stochastic optimization problem, which (being strictly convex in the decision variables) admits a unique optimal solution $u_t^2 = \gamma_t^2(y_1^2)$, $t \in \mathcal{T}$, which is linear in y_1^2.

Hence, problem ($\star$) admits a globally optimal solution, $(\underline{\gamma}^{1*} \in \mathbf{\Gamma}_2^1, \underline{\gamma}^{2*} \in \mathbf{\Gamma}_2^2)$, which is linear in the available information to the two agents, and furthermore $\underline{\gamma}^{1*} \in \mathbf{\Gamma}_1^1$ (and naturally also $\underline{\gamma}^{2*} \in \mathbf{\Gamma}_2^1$). In view of Proposition 3.8.1, $(\underline{\gamma}^{1*}, \underline{\gamma}^{2*})$ also solves the original problem with non-classical information.

Before closing the discussion on the solution to the problem, we note that linearity of the state equation and quadratic structure of the loss function did not play any role in the applicability of Proposition 3.8.1, other than that the optimal solution was linear because of the linear-quadratic structure. Hence, provided the basic assumptions of garbling and the specific *IS* structure hold, a similar result can be obtained for nonlinear, non-quadratic problems as well. We remark, however, that both the garbling assumption and the one that allows **A**1 to have full access not only to the past and current values of the actions of **A**2 but also their future values (which entails a pre-commitment mode of operation) are essential for linearity of the optimal solution. The former is needed for the problem with expanded (partially nested) information to have the same minimum as the one with the original non-classical information. The latter is needed because otherwise the recursion (iv) cannot be implemented. If any of these two assumptions do not hold, then there emerges the possibility of **A**2 signaling the value of y_1^2 to **A**1 through her control actions. We discuss below one such class of problems where signaling through control actions becomes instrumental in improving the performance of a team.

3.8.2 Signaling Through Control Actions and Bismut's Method

Consider the following 2-agent stochastic dynamic team problem: u^1 and u^2 are real-valued control actions of agents **A**1 and **A**2, respectively, who have access to information $I^1 = \{u^2\}$ and $I^2 = \{\alpha, x\}$, where α takes two values, 1 and 2, each with equal probability $\frac{1}{2}$, x is a uniform random variable with support set the interval $[-\sqrt{3}, \sqrt{3}]$, α and x are statistically independent, and this is common information to both agents. Hence, $u^1 = \gamma^1(u^2)$ and $u^2 = \gamma^2(\alpha, x)$, where the policies $\gamma^1 \in \Gamma^1$ and $\gamma^2 \in \Gamma^2$ are measurable maps. Note that this is a problem with non-classical information, because **A**1 has access to the action variable of **A**2, but does not have access to the information used by **A**2. Let the loss function for the team be given by

$$L(\alpha, x, u^1, u^2) = \alpha(u^1 + u^2)^2 + (u^2 - x)^2 + (u^1)^2\,, \tag{\#}$$

with the corresponding expected loss (or cost) given by $J(\gamma^1, \gamma^2)$. The goal is to minimize J over $\Gamma^1 \times \Gamma^2$.

Now, if **A**1 instead had the expanded *IS* $\tilde{I}^1 = \{u^2, \alpha, x\}$, then this would be a problem with classical information, and the unique pair of policies would be the ones that minimize (#) for each α and x:

$$\tilde{\gamma}^1(u^2, \alpha, x) = -\frac{1}{1+\alpha}u^2\,, \quad \tilde{\gamma}^2(\alpha, x) = -\frac{1}{2+\alpha}x\,, \tag{\#\#}$$

with the minimum value of L being $\min_{u^1,u^2} L(\alpha, x, u^1, u^2) = \frac{1}{2+\alpha}x^2$, whose expected value, $\frac{7}{24}$, is the minimum of J over the expanded policy spaces.

The question now is whether this expected value can be achieved, exactly or approximately, under the original information structure. Note that **A**1 does not use the additional information on x in (##), but needs the value of α. Hence, if it were possible for **A**1 to decipher the true value of α through her observation of u^2, or put it another way, if **A**2 were able to signal to **A**1 the true value of α through her action u^2 without compromising performance attained under $\tilde{\gamma}^2$ in (##), then the same performance would be achieved under the original *IS*. But a closer look will reveal that since $\tilde{\gamma}^2$ is unique as given, *signaling at zero cost* is not possible. Then, is perfect signaling at some infinitesimal cost is possible? We will argue below that yes, given any $\epsilon > 0$, it is possible to come up with a policy for **A**2 which will lead to a value smaller than $\frac{7}{24} + \epsilon$.

First note that any realized value of $u^2 = \tilde{\gamma}^2(\alpha, x)$ in (##) will be in the interval $[-\frac{\sqrt{3}}{3}, \frac{\sqrt{3}}{3}]$, and signaling the true value of α will require just *one* bit of information. Hence, if **A**2 adopts a strategy of truncating the realized value of $\tilde{\gamma}^2(\alpha, x)$ to n decimal points, and uses an extra $n+1$'th decimal point to convey the true value of α, 1 or 2,[13] then **A**1 will be able to decipher the true value of α from her observation of u^2, and be able to implement the policy $\tilde{\gamma}^1(u^2, \alpha)$. There will be a small deviation from the ultimate performance level of $\frac{7}{24}$, with the *smallness* determined by the value of n. Clearly, given any small $\epsilon > 0$, one can always find a large enough integer n such that the strategy above for **A**2 will result in a cost no larger than $\frac{7}{24} + \epsilon$. But clearly, ϵ cannot be driven to zero, and hence this problem admits only an ϵ-optimal solution, as in the case of the team problem of Sect. 3.3.1, but for a totally different reason.

This type of a result, presented in [119], where part of the action variable in a decimal expansion carries information from one agent to another on an observed random variable through appropriate encoding, resulting in ϵ-optimality, applies to more general problems as long as one agent has perfect access to the action variable of another agent. To put what we have observed above in more precise terms, the ϵ term arises due to the fact that the control policy is to encode information on both the control action and the observation, with as minimum damage as possible to the control action; and this is possible due to the fact that a real number carries infinite amount of information (when information is measured in Shannon information theoretic bits). One way to achieve this is as follows: Since rational numbers are dense in $\mathbb{R}$, for any ϵ', there exists an $n \in \mathbb{Z}$ such that an n-decimal representation which is at most at an ϵ' distance from any real number in a compact set is possible. Therefore, if one is to represent a finite dimensional (say, r-dimensional) control variable in a compact set $\mathbb{U} \subset \mathbb{R}^r$, and a finite dimensional (say, m-dimensional) observation variable in a compact set $\mathbb{Y} \subset \mathbb{R}^m$, these signals can be represented with an arbitrarily small error by a single rational number. Thus, one may embed in this number, the ϵ'-approximate decimal expansions of the numbers to be represented;

---

[13] For example, let a particular realization be $x = 1.7$ and $\alpha = 1$, and n be picked as 7. Then, $\tilde{\gamma}^2(\alpha, x) = -1.7/3$, which would be, using a 7-decimal point truncation, -0.5666666. Then, **A**2's action would be $u^2 = -0.56666661$.

thus, leading to a total of $n(m+r)$ decimal letters, by allocating the most significant nr letters for the control variable. Clearly, the ϵ used in the performance is directly related to the ϵ' used in the truncation argument above.

3.8.3 No-Incentive for Signaling Under Partial Nestedness and Stochastic Partial Nestedness

As we have seen earlier, when the information structure is non-classical, the decision makers may use their actions to communicate with each other. This phenomenon is known as signalling. We have also seen that quasi-classical information structures eliminate the incentive for signaling, since the future decision makers already have access to the information at the signaling decision maker. We now show that such a strict nestedness condition can be further relaxed so that lack of a signaling incentive can be established with a stochastic interpretation and characterization.

Definition 3.8.1 Let $\mathbf{x} = \left[x^1 \; x^2 \; \ldots \; x^L\right]^T$, evolve under the following dynamics

$$x^i_{t+1} = f^i(x^i_t, u^i_t, w^i_t), \qquad y^i_t = g^i(\mathbf{x}_t, v^i_t), \quad i \in \{1, 2, \ldots, L\},$$

$$z_{t+1} = f(z_t, u^1_t, u^2_t, \ldots, u^L_t),$$

for some measurable functions f^i, g^i, f, and with $\{w^i_t\}$ independent state disturbance processes and $\{v^i_t\}$ independent observation noise processes for $i \in \{1, 2, \ldots, L\}$. Here $(x^i_t, w^i_t, v^i_t, i \in \{1, 2, \ldots, L\}, z_0)$ are independent second-order processes. Suppose that each of the L DMs has access to the additional $\{z_t\}$ process:

$$\tilde{y}^i_t = \{y^i_t, z_t\}, \quad i \in \{1, 2, \ldots, L\}$$

If the information available at each controller is such that

$$x^i_t \leftrightarrow (y^i_{[0,t]}, u^i_{[0,t-1]}) \leftrightarrow \{x^j_0, z_0, w^j_{[0,t-1]}, y^j_{[0,t]}, \quad j \neq i\}$$

form Markov chains, so that almost surely,

$$P\Big(dx^i_t | y^i_{[0,t]}, u^i_{[0,t-1]}\Big) = P\Big(dx^i_t | y^i_{[0,t]}, u^i_{[0,t-1]}, \{x^j_0, z_0, w^j_{[0,t-1]}, y^j_{[0,t]}, \quad j \neq i\}\Big)$$

for all t and i, then such an information structure is said to be stochastically decoupled.

Note that this information structure is non-classical since the DMs can signal information to each other (that is, actions affect measurements in the sense discussed earlier), but yet there is no nestedness in the information σ-fields.

Theorem 3.8.1 *Consider a stochastic team problem as above with the objective to be minimized defined as:*

$$E[\sum_{t=0}^{T-1} c_1(x_t^1, u_t^1) + c_2(x_t^2, u_t^2) + \ldots c_L(x_t^L, u_t^L)].$$

If the controllers have stochastically decoupled information structures, the optimization problem can be cast as L decoupled optimization problems with a classical information structure.

Proof Through the law of the iterated expectations

$$E[\sum_{t=0}^{T-1} c_1(x_t^1, u_t^1) + c_2(x_t^2, u_t^2) + \ldots c_L(x_t^L, u_t^L)]$$

$$= E[E[\sum_{t=0}^{T-1} c_1(x_t^1, u_t^1) + c_2(x_t^2, u_t^2) + \ldots c_L(x_t^L, u_t^L)|\tilde{\mathbf{y}}_t, \mathbf{I}_{t-1}]]. \quad (3.48)$$

Here $I_k^i = \{\tilde{y}_s^i, s \leq k\}$ and $\mathbf{I}_k := \{I_k^1, \cdots, I_k^L\}$. We will show that due to the Markov chain condition in Definition 3.8.1, it follows that the above writes as

$$E[\sum_{t=0}^{T-1} \sum_{i=1}^{L} E[c_i(x_t^i, u_t^i)|y_t^i, I_{t-1}^i)]].$$

Since the cost function is also decoupled, it turns out that one can write the dynamic programming recursions as L decoupled optimization problems. Given the new expression, $E[\sum_{t=0}^{T-1} \sum_{i=1}^{L} E[c_i(x_t^i, u_t^i)|y_t^i, I_{t-1}^i]]$, we can write the cost as: $E[\sum_{t=0}^{T-1} \sum_{i=1}^{L} E[c_i(x_t^i, u_t^i)]]$. Thus, the cost can be written as a summation of decoupled independent stochastic optimization problems, involving a *classical* stochastic system. ◇

An example demonstrating the property above is the following.

Example 3.8.1 Consider a dynamical system described by

$$x_{t+1}^1 = a_1 x_t^1 + u_t^1 + w_t^1, x_{t+1}^2 = a_2 x_t^2 + u_t^2 + w_t^2, x_{t+1}^3 = a_3 x_t^3 + u_t^1 + u_t^2 + w_t^3$$

$$y_t^1 = (x_t^1 + v_t^1, x_t^2 + v_t^2 + v_t^{21}, x_t^3 + v_t^{31}),$$

$$y_t^2 = (x_t^1 + v_t^1 + v_t^{12}, x_t^2 + v_t^2, x_t^3 + v_t^{32}),$$

where the variables $x_t^3 + v_t^{31}$ and $x_t^3 + v_t^{32}$ serve as some external random variable z_t. Suppose here that all such external random variables are zero-mean Gaussian, and the goal is the minimization of

$$E\Big[\sum_{t=0}^{T-1}\Big((x_t^1)^2 + (x_t^2)^2 + \rho_1(u_t^1)^2 + \rho_2(u_t^2)^2\Big)\Big],$$

with $\rho_1, \rho_2 > 0$ constants. The control policies are measurable with respect to the sigma-algebra generated by their causal observations and past controls: $I_t^i = \{y_t^i, I_{t-1}^i\}$, with $I_0^i = y_0^i$. This system has a non-classical information structure, as controller 1 affects the observation of controller 2 (the third state x^3 acts as a communication medium between the controllers), but controller 2 cannot recover the information accessible to controller 1. However, the optimal team policy is linear by stochastic partial nestedness and static reduction.

Remark 3.8.1 In the setting discussed in this sub-section, there is an inherent nestedness condition: A signaling DM's private information is not informative given the signaled DM's information even though the signaled DM does not have access to the information at the signaling DM: The nestedness notion here is called *stochastic nestedness* in [806]; the difference between partial nestedness and the setting here is the following. Partial nestedness (or nestedness) is defined independent of the probability measure on the system variables and strictly depends on the information field relationships of the DMs. However, when probabilistic aspects are also considered, a weaker condition identifying the irrelevance of private information at a signaling DM can be established leading to a reduction to a static team. ◇

3.9 Non-sequential Decentralized Stochastic Control: Causality, Solvability and Static Reduction

As mentioned earlier in Sect. 2.3, in some stochastic systems the order of actions of the agents can also be nature-dependent. As discussed heretofore in this chapter, if there is a pre-defined order in which the decision makers act, then the system is *sequential*. Otherwise, the system is non-sequential. As we will see in this section, such non-sequential systems are substantially more difficult to analyze since the ambiguities in the order of actions lead to challenges on the interpretation of local information.

To study such systems, Witsenhausen [772] defines several properties, which will be made more precise in the following: At a high-level, the following will be studied. Witsenhausen defines *causality* to be the property that for each $k \leq N$, the events that a DM acting at time k can distinguish are events that can be induced by the decisions of the first $k-1$ DMs and ω. Furthermore, after $k-1$ DMs have acted, the selection of the kth DM is determined by ω and the actions up to time $k-1$. Nonetheless, the DM acting at time k does not need to know that he/she is in fact that kth DM. Witsenhausen has proved that causality implies *solvability* [772], which ensures that all control action variables in the system are well-defined random variables (as proper measurable functions), and Andersland and Teneketzis [26] introduced *causal-implementability* and *deadlock-freeness* properties which appear to be more general than causality, as they only pertain to the information available to the DMs and do not impose restrictions on the ordering of DMs. All of these properties are defined, and several equivalence results are presented in this section.

Before proceeding with the corresponding definitions, we first introduce a variation on the intrinsic model presented in Sect. 3.2 and some notation that will be used in the definitions.

3.9.1 Intrinsic Model for Non-sequential Stochastic Control

The general intrinsic model that we will introduce will be a minor modification of Witsenhausen's sequential model presented in Sect. 3.2.

The model consists of:

1. The number of DMs in the system, $N \in \mathbb{N}$.
2. A measurable space $(\Omega, \mathcal{F})$ characterizing the underlying event space.
3. $(\mathbb{U}^k, \mathcal{U}^k)$, denoting the standard Borel space from which u^k, the kth control action is selected. The measurable space containing the collective action $\mathbf{u} := (u^1, u^2, \cdots, u^N)$ is denoted by $(\prod_{i=1}^N \mathbb{U}^i, \otimes_{i=1}^N \mathcal{U}^i)$.
4. $(\mathbb{Y}^k, \mathcal{Y}^k)$, denoting the standard Borel space from which y^k, the measurement of **DM**k , takes values.
5. Measurement functions, $y^i = \eta^i(\omega, \mathbf{u})$, which output the measurements of each **DM**i, for $i = 1, 2, \cdots, N$. Let $\mathcal{J}^i$ denote the sigma field induced on $\Omega \times (\prod_{k=1}^N \mathbb{U}^k)$ by the measurement function η^i.
6. A design constraint which restricts the N-tuples of control policies to be measurable given the information structure: $\underline{\gamma} = \{\gamma^1, \gamma^2, \cdots, \gamma^N\}$, with $\gamma^k : (\mathbb{Y}^k, \mathcal{Y}^k) \to (\mathbb{U}^k, \mathcal{U}^k)$, $k = 1, 2, \cdots, N$, so that $u^k = \gamma^k(y^k)$. We let Γ^k denote the set of all such policies for DMk, and $\mathbf{\Gamma} := \prod_{k=1}^N \Gamma^k$ denote the collection of such policies.
7. A probability measure P on $(\Omega, \mathcal{F})$.

We will also find it useful to refine Ω into a collection of signal spaces $\{\Omega_{s_0}, \Omega_0, \cdots, \Omega_N\}$, denoting measurable spaces from which the collection of random variables $\omega_{s_0}, \omega_0, \omega_1, \cdots, \omega_N$ take values. Here, Ω_0-valued ω_0 and Ω_{s_0}-valued ω_{s_0} are the cost and order relevant variables, where ω_0 is the cost-relevant uncertainty, and ω_{s_0} is the uncertainty related to the possibly random ordering of DMs, in the sense that if $c(\omega, u^1, \cdots, u^N)$ is the cost function, $\sigma(\omega_0, u^1, \cdots, u^N)$ is the smallest σ-field over $\Omega \times_{i=1}^N \mathbb{U}^i$ on which c is measurable. A similar discussion applies for ω_{s_0}: Let S_k denote the set of k-DM orderings of $k = 1, 2, \ldots, N$ and let $\psi : \Omega \times (\prod_{i=1}^N \mathbb{U}^i) \to S_N$ denote an ordering function. Then, $\sigma(\omega_{s_0}, u^1, \cdots, u^N)$ is the smallest σ-field over $\Omega \times_{i=1}^N \mathbb{U}^i$ on which ψ is measurable. The variables $\omega_1, \omega_2, \cdots, \omega_N$ represent the noise in the measurements $y^i, i = 1, 2, \cdots, N$. We note that (ω_0, ω_{s_0}) may not be independent with their joint probability distribution given.

Given a cost function $c : \Omega_0 \times \prod_{k=1}^{N} \mathbb{U}^i \to \mathbb{R}_+$, a solution to a stochastic team problem is determination of a policy $\underline{\gamma}$ that achieves $\inf_{\underline{\gamma} \in \Gamma} J(\underline{\gamma})$ with

$$J(\underline{\gamma}) := E^{\underline{\gamma}}[c(\omega_0, \omega_{s_0}, \mathbf{u})].$$

3.9.2 Causality, Causal Implementability and Solvability

We begin by introducing some notation that will be used in the definitions that follow. We denote the ordering of DMs for a particular realized solution path as $s := (s_1, s_2, \ldots, s_N) \in S$, where S denotes the set of all possible N-agent orderings, and $s_k = i$ means that DM^i is the k-th DM to act. Recall that $\mathcal{J}^k$ denotes the sigma-field generated by the information at **DM**k, for $1 \leq k \leq N$. In particular, let $y^i = \eta^i(\omega_0, \omega_i, \mathbf{u})$. Let S_k denote the set of k-DM orderings of $k = 1, 2, \cdots, N$, and $|s|$ denote the range of s, that is the set of indices present in the ordering s, for any $s \in S_{||s||}$, where $||s||$ denotes the cardinality of $|s|$. For any $s \in S_k, k = 1, 2, \cdots, N$, $\mathcal{P}_s$ is defined as the projection of $\Omega \times \prod_{i=1}^{N} \mathbb{U}^i$ onto $\Omega \times (\prod_{i=1}^{k} \mathbb{U}^{s_i})$; i.e. $\mathcal{P}_s(\omega, \mathbf{u}) = (\omega, u^{s_1}, u^{s_2}, \cdots, u^{s_k})$, $\mathcal{P}_{\emptyset}(\omega, \mathbf{u}) = (\omega)$. We let $\mathscr{F}(s)$ denote the (cylindrical) extension of $\mathcal{F} \times (\prod_{i=1}^{k} \mathcal{U}^{s_i})$ to $\Omega \times \prod_{i=1}^{k} \mathbb{U}^i$ for any $s = (s_1, s_2, \cdots, s_k) \in S_k, k = 1, 2, \cdots, N$ (where cylindrical extension is defined as the pre-image of projection). Let $T_j^k : S_k \to S_j$ denote a truncation map that returns the ordering of the first j agents of a k-agent ordering.[14]

We begin with a definition of causality, which will be of particular importance.

Definition 3.9.1 (Causality (C) [772]) An information structure possesses Property C if there exists at least one map $\psi : \Omega \times \prod_{i=1}^{N} \mathbb{U}^i \to S_N$ such that for any $s = (s_1, s_2, \cdots, s_k) \in S_k, k = 1, 2, \cdots, N$,

$$\mathcal{J}^{s_k} \cap [T_k^N \circ \psi]^{-1}(s) \subset \mathscr{F}(T_{k-1}^k(s)) \tag{3.49}$$

Alternatively, for $E \subset \mathcal{J}^{s_k}$ the following holds:

$$E \cap [T_k^N \circ \psi]^{-1}(s) \in \mathscr{F}(|T_{k-1}^k(s)|) \tag{3.50}$$

[14] We note here that in [772], Witsenhausen defines the projection $\mathcal{P}_s$ as being the restriction of $\prod_{i=1}^{N} \mathbb{U}^{s_i}$ to $\prod_{i=1}^{k} \mathbb{U}^{s_i}$, whereas we are using the definition which is adopted in [26, 27]. The reason we use this definition is that in Witsenhausen's treatment in [772], the ordering function ψ is a function from $\prod_{i=1}^{N} \mathbb{U}^i$ to S_N, that is, the ordering function does not take the uncertainty ω as a domain variable. The generalization to include the uncertainty ω as a domain variable is useful as it allows the first DM to be decided by randomness, rather than having a deterministic starting agent as in [772, Lemma 2].

The term $[T_k^N \circ \psi]^{-1}(s)$ denotes the preimage of the mapping $T_k^N \circ \psi : \Omega \times \prod_{i=1}^N \mathbb{U}^i \to S_k$. Therefore, given the ordering $s \in S_k$ of the first k DMs, $[T_k^N \circ \psi]^{-1}(s)$ will be the set of intrinsic outcomes $(\omega, \mathbf{u}) \in \Omega \times \prod_{i=1}^N \mathbb{U}^i$ that maps, through ψ, to an N-DM ordering in which the ordering of the first k agents is s.

An Alternative Characterization of Property C via an Imaginary Sequential Model We can also present an equivalent definition which will be useful later in our analysis [671]: Property C can be represented in terms of the notion of the join of two sigma fields. The difficulty in presenting this definition will mainly be in the notation, which we propose to be altered in order to view the information structure in terms of *imaginary* DMs which are indexed by their time of action rather than their identity. In the following, we view DM^{s_k} not as a specific DM^i, for some realized solution path where $s_k = i$, but instead we consider DM^{s_k} to be the k-th DM to act, regardless of the realized solution path. Therefore, DM^{s_k} is not truly any one physical DM, but rather a collection of DMs, where only one is active as DM^{s_k} for any one realized solution path. In this sense, we also change the actions and measurements of each DM to be represented in terms of the order of the actions, rather than the identity of the DM. The action sets $\mathbb{U}^{s_k}$ and measurement sets $\mathbb{Y}^{s_k}$ are therefore seen as coupled. The information field $\mathcal{J}^{s_k}$, of the k-th DM, is the sigma-field on $\Omega \times \prod_{k=1}^N U^{s_k}$ defined by the following:

$$\mathcal{J}^{s_k} = \sigma\left\{ \bigcup_{i=1}^{N} \{ \mathscr{J}^i \cap [\mathcal{P}_{T_{k-1}^N(s)}]^{-1}(\mathcal{P}_{T_{k-1}^N}(\omega, \mathbf{u})) : \psi(\omega, \mathbf{u})_k = i \} \right\}$$

Now, in this alternative representation we take the join of the sigma field $\mathcal{J}^{s_k}$ with the sigma field generated by the set $[T_k^N \circ \psi]^{-1}(s)$, for all $s = (s_1, s_2, \cdots, s_k) \in S_k, k = 1, 2, \cdots, N$. The sigma field generated by the set $[T_k^N \circ \psi]^{-1}(s)$ refers to the coarsest sigma field over $\Omega \times (\prod_{i=1}^N \mathbb{U}^{s_i})$ for which the mapping $T_k^N \circ \psi$ is measurable. The join of these two sigma fields, denoted by $\bigvee$, is the coarsest sigma field over $\Omega \times \prod_{i=1}^N \mathbb{U}^{s_i}$ containing both. The condition is then that the join of these two fields is a subset of $\mathscr{F}^*(k-1)$, where $\mathscr{F}^*(k-1)$ denotes the cylindrical extension of $\mathcal{F} \times (\prod_{i=1}^{k-1} U^{s_i})$ to $\Omega \times \prod_{i=1}^N U^{s_i}$. Note that the difference between $\mathscr{F}^*(k-1)$ and $\mathscr{F}(T_{k-1}^k(s))$ is that $\mathscr{F}^*(k-1)$ is defined for any given $k \in \{1, 2, \cdots, N\}$, while $\mathscr{F}(T_{k-1}^k(s))$ is only defined if the order $s \in S_k$ is defined as well (or $s \in S_N$ if we apply the map $T_{k-1}^N(s)$). Also, $\mathscr{F}^*(k-1)$ is the cylindrical extension to $\Omega \times \prod_{i=1}^N U^{s_i}$, while $\mathscr{F}(T_{k-1}^k(s))$ is the cylindrical extension to $\Omega \times \prod_{i=1}^N U^i$. The condition is therefore:

$$\mathcal{J}^{s_k} \bigvee \sigma\{[T_k^N \circ \psi]^{-1}(s) : s \in S_k\} \subset \mathscr{F}^*(k-1) \tag{3.51}$$

Note that such an interpretation leads to a sequential model with 'imaginary' agents, as there is always the natural ordering. We state the following lemma.

Lemma 3.9.1 ([671]) *Under any team policy γ, under Property C,*

$$P^{\gamma}(dy^{s_k}|\omega_0, \omega_{s_0}, y^{s_1}, u^{s_1}, \cdots, y^{s_{k-1}}, u^{s_{k-1}})$$
$$= P(dy^{s_k}|\omega_0, \omega_{s_0}, y^{s_1}, u^{s_1}, \cdots, y^{s_{k-1}}, u^{s_{k-1}}) \tag{3.52}$$

That is, the conditional probabilities defining measurement kernels are policy independent.

Proof We have that, for all $1 \leq k \leq N$,

$$P^{\gamma}(dy^{s_k}|\omega_0, \omega_{s_0}, y^{s_1}, u^{s_1}, \cdots, y^{s_{k-1}}, u^{s_{k-1}})$$
$$= \sum_{s \in S_k} P^{\gamma}(dy^{s_k}, s|\omega_0, \omega_{s_0}, y^{s_1}, u^{s_1}, \cdots, y^{s_{k-1}}, u^{s_{k-1}})$$
$$= \sum_{s \in S_k} P^{\gamma}\left(dy^{s_k}\middle| s, \omega_0, \omega_{s_0}, y^{s_1}, u^{s_1}, \cdots, y^{s_{k-1}}, u^{s_{k-1}}\right)$$
$$\times P^{\gamma}(s|\omega_0, \omega_{s_0}, y^{s_1}, u^{s_1}, \cdots, y^{s_{k-1}}, u^{s_{k-1}}))$$
$$= \sum_{s \in S_k} P\left(dy^{s_k}\middle| s, \omega_0, \omega_{s_0}, y^{s_1}, u^{s_1}, \cdots, y^{s_{k-1}}, u^{s_{k-1}}\right) \tag{3.53}$$
$$\times P(s|\omega_0, \omega_{s_0}, y^{s_1}, u^{s_1}, \cdots, y^{s_{k-1}}, u^{s_{k-1}})) \tag{3.54}$$
$$= P(dy^{s_k}|\omega_0, \omega_{s_0}, y^{s_1}, u^{s_1}, \cdots, y^{s_{k-1}}, u^{s_{k-1}}), \tag{3.55}$$

that is, the conditional probability is policy independent. Equations (3.53) and (3.54) critically follow from Property C, via (3.51); as the past control and measurements uniquely determine both the ordering as well as the conditional probability. ◇

This lemma is crucial in allowing for interpreting a non-sequential problem as one with imaginary agents acting in sequence. This result will be utilized later in the section.

Under Property C, it is not necessary for the k-th DM to know which DMs have acted previously, or even that they are the k-th DM, only that the universe (or, *umpire*, if appropriate in this context) knows, and that the DM knows to act at that time. Additionally, since clearly $\mathscr{F}^*(k-1) \subset \mathscr{F}^*(k)$, for each $k = 1, 2, \cdots, N$, we also have that:

$$\sigma\{[T^N_{k-1} \circ \psi]^{-1}(s) : s \in S_{k-1}\} \subset \sigma\{[T^N_k \circ \psi]^{-1}(s) : s \in S_k\},$$

for each $k = 1, 2, \cdots, N$. Therefore, the information fields available to *the universe*, or *umpire* (as is used in [772]) in order to decide the ordering of DMs are nested, i.e., at least as much information can be used to select the index of the k-th DM to act as was used to select the index of the $(k-1)$-th DM.

Due to this condition being an important classification in non-sequential problems, we call this property *random causal sequentiality* (RCS) [671], although implicitly this has already been studied by Witsenhausen in [772, Lemma 1]:

Definition 3.9.2 (Random Causal Sequentiality (RCS)) An information structure possesses property RCS if for any $s = (s_1, s_2, \cdots, s_{||s||}) \in S_{||s||}$, the following holds:

$$[T^N_{||s||} \circ \psi]^{-1} \in \mathscr{F}(|T^{||s||}_{||s||-1}(s)|)$$

Now, in order to be able to have control actions as well defined random variables and define the expected cost function under a given policy in the non-sequential setup, the information structure should possess a property called *solvability-measurability* (SM) [772].

Definition 3.9.3 (Solvability-Measurability (SM) [772]) Property SM holds when, for each $\gamma \in \Gamma$ and $\omega \in \Omega$, there exists one and only one $\mathbf{u} \in \prod_{i=1}^N \mathbb{U}^i$ satisfying the closed-loop equations

$$P_i(\mathbf{u}) = \gamma^i(y^i), i = 1, 2, \cdots, N$$

$$\Rightarrow u^i = \gamma^i(\eta_i(\omega_0, \omega_i, u^1, u^2, \cdots, u^N))$$

Here P_i is the projection (or restriction) of the product $\prod_{k=1}^N \mathbb{U}^k$ to $\mathbb{U}^i$. Furthermore, the Property SM requires that the solution map denoted by $M^\gamma : (\Omega, \mathcal{F}) \to (\prod_{k=1}^N \mathbb{U}^k, \mathcal{B}(\prod_{k=1}^N \mathbb{U}^k))$ is measurable.[15]

Property SM guarantees that the control actions u^i, $i = 1, 2, \cdots, N$, are well defined random variables, given an admissible control policy $\gamma \equiv (\gamma^1, \gamma^2, \cdots, \gamma^N)$. Property SM also ensures that the expectation of the cost function is well defined under a given policy.

A further property in non-sequential stochastic control, called *causal-implementability* (CI), as defined in [26, Definition 2], is necessary and sufficient to ensure that the information available to any DM does not depend on the actions of DMs who act after it [26, Theorem 1]:

Definition 3.9.4 (Causal-Implementability (CI) [26]) An information structure possesses Property CI when there exists at least one map $\psi : \Omega \times \prod_{i=1}^N \mathbb{U}^i \to S_N$ such that for all $k = 1, 2, \cdots, N$, and $(\omega, \mathbf{u}) \in \Omega \times \prod_{i=1}^N \mathbb{U}^i$,

$$\mathscr{I}^{s_k} \cap [\mathcal{P}_{T^N_{k-1}(s)}]^{-1}(\mathcal{P}_{T^N_{k-1}(s)}(\omega, \mathbf{u}))$$

15 We note that this definition of SM is presented differently than in [772], in order to accommodate use of the measurements $y^i, i = 1, 2, \cdots, N$. The closed-loop equations in [772] are given by: $P_i(\mathbf{u}) = \gamma^i(\omega, \mathbf{u}), i = 1, 2, \cdots, N$.

$$\subset \{\emptyset, [\mathcal{P}_{T^N_{k-1}(s)}]^{-1}(\mathcal{P}_{T^N_{k-1}(s)}(\omega, \mathbf{u}))\}$$

where $\mathcal{P}_s$ is defined as the projection of $\Omega \times \prod_{i=1}^N \mathbb{U}^i$ onto $\Omega \times (\prod_{i=1}^k \mathbb{U}^{s_i})$, i.e.,

$$\mathcal{P}_s(\omega, \mathbf{u}) = (\omega, u^{s_1}, u^{s_2}, \cdots, u^{s_k}), \mathcal{P}_\emptyset(\omega, \mathbf{u}) = (\omega).$$

Recall that $T^k_j : S_k \to S_j$ denotes a truncation map that returns the ordering of the first j agents of a k-agent ordering. The term $[\mathcal{P}_{T^N_{k-1}(s)}]^{-1}(\mathcal{P}_{T^N_{k-1}(s)}(\omega, \mathbf{u}))$ will then be a set of the form

$$\{(\omega, \mathbf{u}) : \omega = \omega, u^{s_1} = \hat{u}^{s_1}, u^{s_2} = \hat{u}^{s_2}, \cdots, u^{s_{k-1}} = \hat{u}^{s_{k-1}}\}.$$

The condition therefore states that for any distinguishable set $A \in \mathcal{J}^{s_k}$, the intersection $A \cap [\mathcal{P}_{T^N_{k-1}(s)}]^{-1}(\mathcal{P}_{T^N_{k-1}(s)}(\omega, \mathbf{u}))$ is either equal to $[\mathcal{P}_{T^N_{k-1}(s)}]^{-1}(\mathcal{P}_{T^N_{k-1}(s)}(\omega, \mathbf{u}))$ or $\emptyset$.

Once again an equivalent way to state this definition is in terms of sigma fields. In this alternative approach, we simply state that an information structure possesses Property CI if and only if there exists at least one map $\psi : \Omega \times (\prod_{i=1}^N \mathbb{U}^i) \to S_N$, such that the information field $\mathcal{J}^{s_k}$ of the k-th DM contains only events which are distinguishable by chance, ω, and the actions of previous DMs, $(u^{s_i})_{i<k}$. Mathematically speaking,

$$\mathcal{J}^{s_k} \subset \mathscr{F}^*(k-1) \tag{3.56}$$

Recall that $\mathscr{F}^*(k-1)$ is the cylindrical extension of $\mathcal{F} \times (\prod_{i=1}^{k-1} \mathcal{U}^{s_i})$ to $\Omega \times \prod_{i=1}^N \mathbb{U}^{s_i}$. Following the definition of Property CI, the following property, called *deadlock-freeness* (DF) (also introduced in [26]), is a more easily interpretable and equivalent *high-level* description of Property CI:

Definition 3.9.5 (Deadlock-Freeness (DF) [26]) An information structure possesses Property DF if for each $\gamma \in \Gamma$, and for every $\omega \in \Omega$, there exists an ordering of γ's (i.e., the N control laws), say $\gamma^{s_1(\omega)}, \gamma^{s_2(\omega)}, \cdots, \gamma^{s_N(\omega)}$, such that no control action $u^{s_i(\omega)}$, $i = 1, 2, \cdots, N$, depends on the control actions that follow.

To gain further insight into the definitions above, consider the following two examples from [713].[16]

Example 3.9.1 Let $\Omega = \mathbb{U}^1 = \mathbb{U}^2 = \mathbb{U}^3 = \{0, 1\}$ and

[16] In addition to the Properties SM, DF, C, and CI, [27] defines analogues of these properties which are policy-dependent in the sense that they apply only for a given policy or a subset of policies but not necessarily for all policies.

$$\sigma(\mathcal{I}^1) = \Big\{\emptyset, \Omega \times \mathbb{U}^1 \times \mathbb{U}^2 \times \mathbb{U}^3, \{(\omega, u^1, u^2, u^3) : \omega(1-u^2)u^3 = 1\},$$
$$\{(\omega, u^1, u^2, u^3) : \omega(1-u^2)u^3 = 0\}\Big\},$$
$$\sigma(\mathcal{I}^2) = \Big\{\emptyset, \Omega \times \mathbb{U}^1 \times \mathbb{U}^2 \times \mathbb{U}^3, \{(\omega, u^1, u^2, u^3) : \omega(1-u^3)u^1 = 1\},$$
$$\{(\omega, u^1, u^2, u^3) : \omega(1-u^3)u^1 = 0\}\Big\},$$
$$\sigma(\mathcal{I}^3) = \Big\{\emptyset, \Omega \times \mathbb{U}^1 \times \mathbb{U}^2 \times \mathbb{U}^3, \{(\omega, u^1, u^2, u^3) : \omega(1-u^1)u^2 = 1\},$$
$$\{(\omega, u^1, u^2, u^3) : \omega(1-u^1)u^2 = 0\}\Big\}.$$

This system has a deadlock, since no DM can act as in the first configuration of Fig. 2.1.

Example 3.9.2 Let $\Omega = \mathbb{U}^1 = \mathbb{U}^2 = \{0, 1\}$ and $\sigma(\mathcal{I}^1) = 2^{\Omega \times \mathbb{U}^2}$, $\sigma(\mathcal{I}^2) = 2^{\Omega \times \mathbb{U}^1}$ (where the notation $2^{\mathbb{U}}$ denotes the power set, that is the collection of all subsets of the set $\mathbb{U}$). Consider the following team policy:

$$\gamma^1(\omega, u^2) = 0 \times 1_{\{u^2=0\}} + 1 \times 1_{\{u^2=1\}}$$
$$\gamma^2(\omega, u^1) = 0 \times 1_{\{u^1=0\}} + 1 \times 1_{\{u^1=1\}},$$

where 1_E denotes the indicator function for event E. For this design, consider the realization $\omega = 0$. In this case, $(\omega, u^1, u^2) = (0, 0, 0)$ as well as $(0, 1, 1)$ are acceptable realizations given the policy stated above. A similar setting occurs for $\omega = 1$, since $(1, 0, 0)$ and $(1, 1, 1)$ are acceptable realizations. Hence, for a given cost function c, there does not exist, in general, a well defined (measurable) cost realization variable $c(\omega, u^1, u^2)$ under this policy, and the expectation $E[c(\omega, u^1, u^2)]$ is not well defined given the policy (γ^1, γ^2). This system is not *solvable*.

Property CI, in view of the definitions and discussions above, states that for any possible ordering the information available to each DM at the time it takes its action cannot depend on the action of any DM that has not yet acted. Note that this definition does not impose any restrictions on the information available to select the ordering of DMs, unlike in Property C, which requires that the ordering function has nested information fields. The only difference between Property C and Property CI is indeed that under C, ψ is causal whereas this is not the case under CI. We note that Property SM does not imply Property C. Now, the following theorem captures the relationships among the different properties introduced in this section.

Theorem 3.9.1 *For the case with standard Borel measurement and action spaces, Property CI implies Property C, in the sense that for every non-sequential system with Property CI one can construct an equivalent non-sequential system with Property C. Therefore, we have*

$$C \iff CI \iff DF \Rightarrow SM$$

For the relations $C \Rightarrow CI \iff DF \Rightarrow SM$, see [772, Theorem 1] [26, Theorems 1 and 2]. We note that both of the relations in $C \Rightarrow CI \iff DF$ follow essentially directly from the (alternative and equivalent) σ-field characterizations of these definitions. The relation $C \iff CI$ for standard Borel spaces has recently been reported in [671]. That Property C and CI are near equivalent under countability conditions had been established in [26, Theorem 4].

3.9.3 Static Reduction of Non-sequential Stochastic Dynamic Teams Under Property C

With the above prelude, we can arrive at the following positive result, which builds on Lemma 3.9.1.

Theorem 3.9.2 ([671]) *Suppose that an N-DM non-sequential stochastic control problem possesses Property C for all admissible policies $\gamma \in \Gamma$, and that the set $\mathbb{Y}^i$ is countable for $i = 1, 2, \cdots, N$. Then, the problem admits a policy-independent static reduction.*

The result in Theorem 3.9.2 is stated for the case where $\mathbb{Y}^{s_i}$ is countable. This can be generalized to the standard Borel setup. Let the joint probability measure on $(\omega_0, \omega_{s_0}, y^{s_1}, \ldots, y^{s_N}, u^{s_1}, \ldots, u^{s_N})$ be denoted by P, and the joint probability measure on ω_0 and ω_{s_0} be denoted by $\mathbb{P}^0$. Under an absolute continuity condition, which always holds for the countable measurement space setup (see Remark 3.4.2), given any admissible policy $\gamma = (\gamma^1, \gamma^2, \ldots, \gamma^N)$, there exists a joint reference probability measure $\mathbb{Q}$ on $(\omega_0, \omega_{s_0}, y^{s_1}, \ldots, y^{s_N}, u^{s_1}, \ldots, u^{s_N})$ such that the probability measure P is absolutely continuous with respect to $\mathbb{Q}$. Thus, for every Borel set A in $(\Omega_0, \Omega_{s_0}, \prod_{i=1}^{N}(\mathbb{U}^{s_i} \times \mathbb{Y}^{s_i}))$, we have

$$P(A) = \int_A \frac{dP}{d\mathbb{Q}} \mathbb{Q}(d\omega_0, \omega_{s_0}, dy^{s_1}, \ldots, dy^{s_N}, du^{s_1}, \ldots, du^{s_N}),$$

where

$$\mathbb{Q}(d\omega_0, d\omega_{s_0}, dy^{s_1}, \ldots, dy^{s_N}, du^{s_1}, \ldots, du^{s_N})$$

$$= \mathbb{P}^0(d\omega_0, d\omega_{s_0}) \prod_{i=1}^{N} Q_{s_i}(dy^{s_i}) 1_{\{\gamma^{s_i}(y^{s_i}) \in du^{s_i}\}}, \tag{3.57}$$

leading to a Radon-Nikodym derivative, which is policy-independent:

$$\frac{dP}{d\mathbb{Q}}(\omega_0, \omega_{s_0}, y^{s_1}, \ldots, y^{s_N}, u^{s_1}, \ldots, u^{s_N})$$
$$= \prod_{i=1}^{N} f_{s_i}(y^{s_i}, \omega_0, \omega_{s_0}, y^{s_1}, \ldots, y^{s_{i-1}}, u^{s_1}, \ldots, u^{s_{i-1}}).$$

The new cost function would then be defined as:

$$\mathbf{c}(\omega_{s_0}, \omega_0), y^{[s_1, s_N]}, u^{[s_1, s_N]})$$
$$= f_{s_1}(y^{s_1}, \omega_{s_0}, \omega_0)\Big(\sum_{s_1} P(s_1 | \omega_0, \omega_{s_0})\Big)$$
$$\times f_{s_2}(y^{s_2}, \omega_0, \omega_{s_0}, y^{s_1}, u^{s_1})\Big(\sum_{s_2} P(s_2 | \omega_0, \omega_{s_0}, y^{s_1}, u^{s_1})\Big)$$
$$\times f_{s_N}(y^{s_N}, \omega_{s_0}, \omega_0, y^{s_1}, \cdots, y^{s_{N-1}}, u^{s_1}, \cdots, u^{s_{N-1}})$$
$$\Big(\sum_{s_N} P(s_N | \omega_0, \omega_{s_0}, y^{s_1}, u^{s_1}, \cdots, y^{s_{N-1}}, u^{s_{N-1}})\Big)$$
$$\times \tilde{c}(\omega_0, \omega_{s_0}, s_{[1,N]}, u^{[s_1, s_N]}). \tag{3.58}$$

Theorem 3.9.3 ([671]) *Suppose that an N-DM non-sequential stochastic control problem possesses Property C (for all admissible policies $\gamma \in \Gamma$), and that there exists a reference probability measure $\mathbb{Q}$ on $(\omega_0, \omega_{s_0}, y^{s_1}, \cdots, y^{s_N}, u^{s_1}, \cdots, u^{s_N})$ such that the joint probability measure P on $(\omega_0, \omega_{s_0}, y^{s_1}, \cdots, y^{s_N}, u^{s_1}, \cdots, u^{s_N})$ is absolutely continuous with respect to $\mathbb{Q}$ ($P \ll \mathbb{Q}$) where $\mathbb{Q}$ is given by* (3.57)*. Then the dynamic non-sequential problem admits a policy-independent static reduction.*

Note that here absolute continuity is regarding the measurement kernels of imaginary agents under the Imaginary Sequential Model interpretation. The proof follows from an inductive argument via the chain rule of total probability as in (3.30), where Property C is of critical importance in establishing policy independence of both measurement kernels conditioned on past data and control policies being conditionally independent of past data given local measurements of imaginary agents.

Remark 3.9.1 Reference [671] shows that Property SM is not sufficient for a policy-independent static reduction, but a particular policy-dependent reduction is possible (in which policy appears as a parameter in the reduced cost, leading to issues of impracticality). One further method for decoupling non-sequential problems was proposed in [25], where again policies appear as cost parameters. [671] also shows that a static reduction of partially-nested non-sequential teams satisfying Property C is also possible, where the definition of partial nestedness (studied for the sequential case earlier in the chapter) is adapted to a causal random ordering: In this context, an information structure is quasi-classical or partially nested, if whenever u^{s_k}, for some $k < i$, affects y^{s_i} through the measurement function η^{s_i}, y^{s_i} contains all of the information available to $\mathbf{DM}s_k$. That is $\sigma(y^{s_k}) \subset \sigma(y^{s_i})$, for all $i = 1, 2, \ldots, N$, and for all possible orderings.

3.10 Concluding Remarks

The main theme of this chapter has been information structures and the role they play in the analysis and classification of stochastic dynamic teams. We have witnessed the complexity and challenges involved in obtaining team-optimal solutions for dynamic teams when the information structure is non-classical, which would arise when, for example, the agents do not share information and/or have limited memory. We have also seen that the structures of the cost function and the system dynamics (describing the evolution of the decision process) as well as the overall probabilistic description of the team problem also play important roles in the analysis of team-optimal solutions. We have obtained methods for solving problems with non-classical information structures via static reduction as well as through expanding the information set, obtaining a solution and checking if the solution is realizable according to the original information structure. We studied information-structure and cost dependent convexity and non-convexity properties. We have also presented general static reduction methods which lead to powerful methods for studying dynamic team problems (as will be evident in the following chapters). We finally presented a comprehensive study on non-sequential stochastic team theory.

3.11 Bibliographic Notes

In addition to the extensive references provided in the main body of the chapter, we refer the reader to [6, 30, 81, 92, 349, 444, 494, 499, 561, 631, 631, 736, 774, 798, 806] and [496] for further discussion on and examples of information structures, and solution methods. Partially nested information structures also include the cases where explicit information exchange in a decentralized system among decision makers is faster than information propagation through system dynamics; see [161, 751], and [598]. Related to the notion of partial nestedness, references [60, 751] and [610] have studied sufficiency conditions for tractability and convexity in optimal decentralized control problems. We note that the notion of *quadratic invariance* presented in [610] is equivalent to partial nestedness, and thus the convexity properties are analogous, while in the quadratically invariant setup the analysis is often restricted to linear policies a priori. Computational methods for convex LQ teams via the maximum principle has been reported in [592].

Approximately optimal solutions for weakly coupled teams with non-classical information structures have been considered in [681]. The discussion in the chapter on performance irrelevant signaling builds on and extends [806]. Blackwell's [123] comparison of information structures is a related concept in a single decision maker setting; [495] generalized this view to Witsenhausen's intrinsic model.

Isomorphism properties on optimality and stationary solutions between a closed-loop system and their (policy-independent or policy-dependent) static reductions appear in [637]. Recently, [790] studied optimal teams under open-loop (and, hence, with static information) control for a class of linear convex problems.

Research in continuous-time under decentralized information structures has been relatively recent beyond the perfect-recall setting [240, 369, 603] or the LQG setting. See [166] and [167] for a class of convex problems, and [364, 370] for recent studies on continuous-time stochastic team problems. [591] obtains existence results and develops a discrete-time approximation theory under general information structures, and [370] studies near optimality of decentralized policies for large models.

Section 3.8 builds, in part, on [806]. The material on garbling in the chapter builds, in part, on [507] and [123].

The material on non-sequential stochastic control in Sect. 3.9 builds on [26, 27, 713, 772] and [671].

We will later see in the book that static reductions of dynamic teams provide very effective means to arrive at existence, structural and approximation results. For existence results building on this approach, we refer to [327, 620, 818, 830] and approximations [625].

Chapter 4
Existence, Structure, and Approximations to Optimality: Strategic Measures in Decentralized Stochastic Control

In this chapter, we develop general existence, approximation, and structural results for optimal policies in stochastic dynamic teams by adopting a probabilistic and topological study of information structures.

Probability measures induced on the product spaces of the states, measurements, and actions under admissible policies are called *strategic measures*. Through study of such measures, we will obtain existence, approximation, and structural results for optimal solutions *given an information structure*. We associate various topologies with strategic measures, and study their closedness and compactness properties. We discuss various upper and lower bounding techniques, through relaxations and convex programs, ranging from classically realizable to classically non-realizable (such as quantum and non-signaling) relaxations. We also introduce properly motivated external randomness, and discuss its impact on performance via relaxed control.

Three of the main results of the chapter, stated in Theorem 4.2.4 (for static teams or dynamic teams with an independent-measurements reduction) and Theorems 4.2.5 and 4.2.6 (for sequential teams that do not allow an independent-measurements reduction) appear to be the most general existence results on stochastic team decision theory. However, some slightly weaker versions of these results can also be arrived at through different methods, as laid out in the chapter. We further establish near optimality, in a precise sense, of quantized policies, and finite model approximations for stochastic team decision problems in general standard Borel spaces. Convex relaxations are also presented.

As a general overview the reader can find the hierarchies depicted, and a summary of results presented, in Fig. 4.1.

The results of this chapter have natural follow-ups in Chap. 5, where we introduce various topologies on decision/control strategies, defined independently from information structures, but for which information structures determine whether the topologies entail any utility in arriving at existence, compactness, convexification, approximation or computational results.

S. Yüksel, T. Başar, *Stochastic Teams, Games, and Control under Information Constraints*, Systems & Control: Foundations & Applications,
https://doi.org/10.1007/978-3-031-54071-4_4

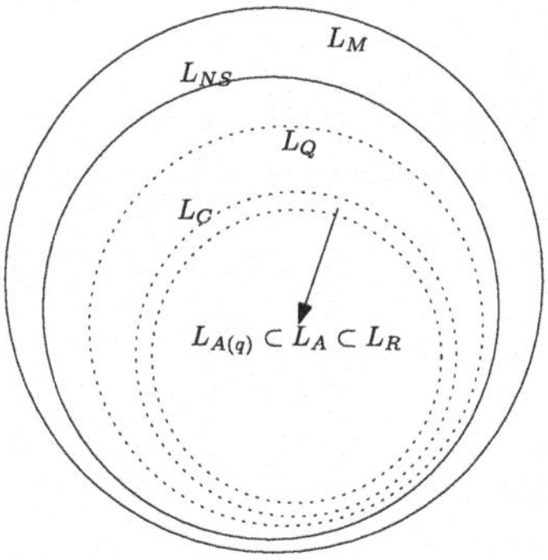

Fig. 4.1 A summary of the results presented in this chapter on decentralized strategic measures. These will be used to arrive at existence, structural, and approximation results for optimal policies in stochastic teams, as well as convex relaxations leading to lower bounds on optimal performance, for a given information structure. (i) $L_{A(q)}(\mu)$, the set of quantized admissible strategic measures, is not closed, nor convex, and is dense in $L_A(\mu)$. (ii) $L_A(\mu)$, the set of strategic measures induced by admissible measurable functions, is not closed, nor convex, but is dense in $L_R(\mu)$ (under independent static reduction). (iii) $L_R(\mu)$, the set of measures generated with local private randomness (i.e. with relaxed policies), is typically not closed or convex, but is closed under independent static reduction (which then leads to a general existence result on optimization problems with lower semi-continuous cost functions). (iv) $L_C(\mu)$, the set of measures with arbitrary common randomness, is convex, but typically not closed but compact when $L_R(\mu)$ is. (v) $L_Q(\mu)$, the set of quantum strategic measures, is convex, but not closed (it is, however, closed under dimension constraints). (vi) $L_{NS}(\mu)$, the set of non-signaling strategic measures, is both closed and convex. (vii) $L_M(\mu)$, the set of measures with local-Markov conditional independence, is both closed and convex

The two approaches presented in this chapter and in Chap. 5, which we may term as *strategic measures approach* vs. *the control topology approach*, lead to complementary results on existence, approximations, and upper and lower bounds and general solution methods in optimal stochastic control.

We start with presentation of decentralized strategic measures and their various properties, in Sect. 4.1. Existence of optimal policies are studied in Sect. 4.2. Convex relaxations on information structures are studied in detail in Sect. 4.3. Near optimality of quantized control policies and approximate models are presented in Sect. 4.4.

4.1 Decentralized Strategic Measures

For classical stochastic control problems, strategic measures are defined (see [230, 585, 650] and [254]) as the set of probability measures induced on the product spaces of the states, measurements, and actions; that is, given an initial state distribution and a policy, one can uniquely define a probability measure on the product space of the states, measurements, and actions. Certain measurability, compactness, and convexity properties of strategic measures for classical stochastic control problems have been studied in [125, 230, 254, 585].

In [830], strategic measures for decentralized stochastic control problems were introduced and many of their properties were established. For decentralized stochastic control problems, considering the set of strategic measures along with compactification and/or convexification of these sets of measures through introduction of private and/or common randomness allows one to associate operationally flexible topologies (such as those leading to a standard Borel space, e.g., weak convergence topology, among others) with the set of strategic measures, as we study in the following subsection.

4.1.1 Measurable Policies as a Subset of Randomized Policies and Strategic Measures

A common method in stochastic control theory is to view a measurable policy as a special case of *relaxed* policies, where relaxation is often achieved through randomization. Such an approach has been ubiquitously adopted in various fields often with different terminologies (e.g., relaxed controls (Young topology) in optimal deterministic control [511, 800], distributional strategies in economics [532, 539], local hidden variables in quantum information theory [71, 181, 761], and optimal quantization [828]). See Appendix F for further discussion on relaxed control topologies.

We recall a relevant representation result [132] given in Appendix F (see also Sect. 5.5.1). Let $\mathbb{X}, \mathbb{M}$ be Borel spaces. Let $\mathcal{P}(\mathbb{X})$ denote the set of probability measures on $\mathbb{X}$. Consider the set of probability measures (F.3). This set is the (Borel measurable) set of the extreme points of the set of probability measures on $\mathbb{X} \times \mathbb{M}$ with a fixed marginal P on $\mathbb{X}$. Furthermore, by (F.4), a stochastic kernel can be viewed as an integral representation over probability measures induced by deterministic policies (i.e., a mixture of deterministic policies). As we will see in Theorem 4.1.1(i), via the Borel isomorphism theorem ([230, Appendix 1] or [223, Chapter 13]), there exists a bijection between $[0, 1]$ and Θ, which will let us arrive at a further operational representation for stochastic kernels.

For a team setup, for any DM k, let

$$\Theta^k := \Big\{ \zeta \in \mathcal{P}(\mathbb{Y}^k \times \mathbb{U}^k) : \zeta = P_k\, Q^{\gamma^k}, \\ Q^{\gamma^k}(\cdot|y^k) = 1_{\{\gamma^k(y^k) \in \cdot\}},\ \gamma^k \in \Gamma^k,\ P_k(\cdot) = P(y^k \in \cdot) \Big\}. \quad (4.1)$$

For a static team, P_k would be fixed; that is, independent of the policies of the preceding DMs. Therefore, in the static case, in view of (F.4), any element $\zeta \in \mathcal{P}(\mathbb{Y}^k \times \mathbb{U}^k)$ with fixed marginal P_k on $\mathbb{Y}^k$ can be expressed as a mixture of Θ^k

$$\zeta(A) = \int_{\Theta^k} \Xi^k(dQ)\, Q(A), \quad A \in \mathcal{B}(\mathbb{Y}^k \times \mathbb{U}^k), \tag{4.2}$$

for some $\Xi \in \mathcal{P}(\Theta^k)$ (where, again, the space Θ^k can be related to $[0, 1]$ as discussed above). In the sequel, we generalize this setup to the set of strategic measures induced by measurable policies and define various relaxed policies that are obtained as a mixture of measurable policies. Indeed, instead of viewing the N-tuple of policies as the joint strategy of DMs, we regard the induced probability distribution on the product space of state, measurements, and actions as the joint strategy, and name it *strategic measure*.

However, we will also recognize that the N-tuple view above is also of both operational and mathematical interest as will be detailed in Sect. 5.5.

4.1.2 Sets of Strategic Measures for Static Teams

Consider a static team problem within the framework of Witsenhausen's intrinsic model. In the following, $B = B^0 \times \prod_{k=1}^N (A^k \times B^k)$ is used to denote the cylindrical Borel sets in $\Omega_0 \times \prod_{k=1}^N (\mathbb{Y}^k \times \mathbb{U}^k)$.

Let $L_A(\mu)$ be the set of strategic measures induced by all admissible measurable policies $\mathbf{\Gamma}$, with $(\omega_0, \mathbf{y}) \sim \mu$; that is, $P \in L_A(\mu) \subset \mathcal{P}\Big(\Omega_0 \times \prod_{k=1}^N (\mathbb{Y}^k \times \mathbb{U}^k)\Big)$ if

$$P(B) = \int_{B^0 \times \prod_{k=1}^N A^k} \mu(d\omega_0, d\mathbf{y}) \prod_{k=1}^N 1_{\{u^k = \gamma^k(y^k) \in B^k\}}, \tag{4.3}$$

for all cylindrical $B \in \mathcal{B}\Big(\Omega_0 \times \prod_{k=1}^N (\mathbb{Y}^k \times \mathbb{U}^k)\Big)$, where $\gamma^k \in \Gamma^k$ for $k = 1, \ldots, N$. Let $L_A(\mu, \underline{\gamma})$ be the strategic measure under a particular strategy $\underline{\gamma} \in \mathbf{\Gamma}$.

The first relaxation is obtained through individual randomization of policies. Specifically, let $L_R(\mu)$ be the set of strategic measures induced by all individually randomized team policies where $(\omega_0, \mathbf{y}) \sim \mu$; that is,

$$L_R(\mu) := \Bigg\{ P \in \mathcal{P}\Big(\Omega_0 \times \prod_{k=1}^N (\mathbb{Y}^k \times \mathbb{U}^k)\Big) :$$
$$P(B) = \int_B \mu(d\omega_0, d\mathbf{y}) \prod_{k=1}^N \Pi^k(du^k | y^k) \Bigg\}, \tag{4.4}$$

where Π^k is taken from the set of all stochastic kernels from $\mathbb{Y}^k$ to $\mathbb{U}^k$ for each $k = 1, \ldots, N$.

Another relaxation, which is stronger than the one above, is obtained by taking a mixture of the elements of $L_A(\mu)$. To this end, define $\Upsilon = [0,1]^N$. We then let

$$L_C(\mu) := \Big\{ P \in \mathcal{P}\Big(\Omega_0 \times \prod_{k=1}^{N} (\mathbb{Y}^k \times \mathbb{U}^k)\Big) :$$
$$P(B) = \int \eta(dz) L_A(\mu, \underline{\gamma}(z))(B),\ \eta \in \mathcal{P}(\Upsilon) \Big\},$$

where $\underline{\gamma}(\cdot)$ takes values from the set of all admissible team policies $\mathbf{\Gamma}$ measurably parametrized by $z \in \Upsilon$ so that the map $\underline{\gamma}(\cdot) : L_A(\mu, \underline{\gamma}(\cdot)) : \Upsilon \to L_A(\mu)$ is Borel measurable, given that $L_A(\mu)$ is a Borel subset of $\mathcal{P}\Big(\Omega_0 \times \prod_{k=1}^{N} (\mathbb{Y}^k \times \mathbb{U}^k)\Big)$ under the weak convergence topology (as we will see in Theorem 4.1.6).

Let $L_{CR,\eta}$ denote the set of strategic measures that are induced by some *fixed* but common independent randomness (with probability measure η) and arbitrary private independent randomness, that is,

$$L_{CR,\eta}(\mu) := \Big\{ P \in \mathcal{P}\Big(\Omega_0 \times \prod_{k=1}^{N} (\mathbb{Y}^k \times \mathbb{U}^k)\Big) :$$
$$P(B) = \int_{B \times \Upsilon} \eta(dz)\mu(d\omega_0, d\mathbf{y}) \prod_k \Pi^k(du^k | y^k, z) \Big\},$$

where Π^k is taken from the set of all stochastic kernels from $\mathbb{Y}^k \times \Upsilon$ to $\mathbb{U}^k$ for each $k = 1, \ldots, N$. Here, the common randomness variable has a fixed measure η.

Let L_{CCR} denote the set of strategic measures that are induced by some arbitrary but common independent randomness and arbitrary private independent randomness, as in $L_C(\mu)$, that is,

$$L_{CCR}(\mu) := \Big\{ P \in \mathcal{P}\Big(\Omega_0 \times \prod_{k=1}^{N} (\mathbb{Y}^k \times \mathbb{U}^k)\Big) :$$
$$P(B) = \int_{B \times \Upsilon} \eta(dz)\mu(d\omega_0, d\mathbf{y}) \prod_k \Pi^k(du^k | y^k, z),\ \eta \in \mathcal{P}(\Upsilon) \Big\}, \tag{4.5}$$

where again Π^k takes place from the set of stochastic kernels from $\mathbb{Y}^k \times \Upsilon$ to $\mathbb{U}^k$ for each $k = 1, \ldots, N$. Here, the common randomness random variable measure η is arbitrary, unlike $L_{CR,\eta}(\mu)$. The following result, essentially from [830], states some structural results regarding the sets of strategic measures introduced above. In particular, it establishes convexity related properties of these sets.

Theorem 4.1.1 *Consider a static team problem. Then, we have the following characterizations.*

(i) $L_R(\mu)$ *has the following representation:*

$$L_R(\mu) = \Big\{ P \in \mathcal{P}\Big(\Omega_0 \times \prod_{k=1}^{N} (\mathbb{Y}^k \times \mathbb{U}^k) \Big) : P(B) = \int U(dz) L_A(\mu, \underline{\gamma}(z))(B),$$

$$U \in \mathcal{P}(\Upsilon), U(dv_1, \cdots, dv_N) = \prod_{k=1}^{N} \eta_k(dv_k), \eta_k \in \mathcal{P}([0,1]) \Big\}; \tag{4.6}$$

that is, $U \in \mathcal{P}(\Upsilon)$ *is constructed by the product of N independent random variables on* [0, 1].

(ii) $L_C(\mu) = L_{CCR}(\mu)$ *and this is a convex set. The set of extreme points of* $L_C(\mu)$ *is* $L_A(\mu)$*. Furthermore,* $L_R(\mu) \subset L_C(\mu)$*.*

(iii) The following equalities hold:

$$\inf_{\underline{\gamma} \in \boldsymbol{\Gamma}} J(\underline{\gamma}) = \inf_{P \in L_A(\mu)} \int P(ds) c(s) = \inf_{P \in L_R(\mu)} \int P(ds) c(s)$$

$$= \inf_{P \in L_C(\mu)} \int P(ds) c(s).$$

In particular, deterministic policies are optimal within the class of randomized policies. In other words, individual as well as common randomness do not lead to improvement in the optimal team cost.

(iv) The sets $L_R(\mu)$ *and* $L_{CR,\eta}(\mu)$ *are not convex. In particular, the presence of independent or (fixed) common randomness do not lead to convexification of the set of strategic measures.*

(v) $L_A(\mu)$, $L_R(\mu)$ *and* $L_C(\mu)$ *are not necessarily weakly closed.*

Proof

(i) Any stochastic kernel from a Polish space into another, $P(dx|y)$, can be realized by some measurable function $x = f(y, v)$ where v is a [0, 1]-valued independent random variable (which can be taken to be uniform without any loss of generality) and f is measurable (see [301, Lemma 1.2], [253, Theorem 1] or [132, Lemma 3.1]). In particular, in the representation (4.2), since Θ^k is a Borel set in a Polish space, by the Borel isomorphism theorem (Appendix 1 in [230] or *Chapter 13* in [223]), there exists a bijection κ between [0, 1] and Θ^k so that (4.2) is replaced with $\mathcal{K}^k(A) = \int p(ds) P^{\kappa(s)}(A)$, for some $p \in \mathcal{P}([0, 1])$ defined by $p(\kappa^{-1}(\cdot)) = \Xi^k(\cdot)$. The result then follows by replacing each of the stochastic kernels with such a representation due to [301].

(ii) For convexity, note that for $\kappa_1 \in L_C(\mu)$, $\kappa_2 \in L_C(\mu)$, and $\alpha \in (0, 1)$, $\alpha\kappa_1 + (1 - \alpha)\kappa_2 \in L_C(\mu)$ since with

$$\kappa_1(B) = \int \eta_1(dz) L_A(\mu, \underline{\gamma}(z)), \quad \eta_1 \in \mathcal{P}(\Upsilon)$$

$$\kappa_2(B) = \int \eta_2(dz) L_A(\mu, \underline{\gamma}(z)), \quad \eta_2 \in \mathcal{P}(\Upsilon),$$

$$(\alpha\kappa_1 + (1-\alpha)\kappa_2)(B) = \int (\alpha\eta_1 + (1-\alpha)\eta_2)(dz) L_A(\mu, \underline{\gamma}(z)), \quad \eta \in \mathcal{P}(\Upsilon)$$

For the extreme sets being in $L_A(\mu)$: Suppose that $\kappa = \int \eta(dz) L_A(\mu, \underline{\gamma}(z))$ is not in $L_A(\mu)$; then it cannot be extreme since for every such κ one can construct κ_1, κ_2 and $\lambda \in (0, 1)$ so that $\kappa = \lambda\kappa_1 + (1-\lambda)\kappa_2$: Partition the support set of κ to two disjoint components $D_1, D_2 \in \mathcal{B}(\Upsilon)$ with $\lambda = \eta(D_1)$ and define $\kappa_1 = \int_{D_1} \frac{\eta(dz)}{\lambda} L_A(\mu, \underline{\gamma}(z))$, $\kappa_2 = \int_{D_2} \frac{\eta(dz)}{1-\lambda} L_A(\mu, \underline{\gamma}(z))$. For the converse, suppose that κ is not extreme: then, there exist α, η_1, η_2 such that $\kappa = \int (\alpha\eta_1 + (1-\alpha)\eta_2)(dz) L_A(\mu, \underline{\gamma}(z))$ and in particular for some set of z values with a positive measure, the measure cannot be represented as: $\delta_z \prod_k 1_{\{\gamma^k(y^k, z) \in \cdot\}}$ for some Dirac δ_z measure: If this were possible, $\eta = \alpha\eta_1 + (1-\alpha)\eta_2$, since this has to be a mixture of two such probability measures.

Then, the inclusion $L_R(\mu) \subset L_C(\mu)$ follows from the representation in (4.6) and the definition in (4.5). The equality $L_C(\mu) = L_{CCR}(\mu)$ follows since both $L_{CCR}(\mu)$ and $L_C(\mu)$ are convex hulls of (or the smallest convex sets containing) $L_R(\mu)$.

(iii) The inequalities

$$\inf_{\underline{\gamma} \in \mathbf{\Gamma}} J(\underline{\gamma}) = \inf_{P \in L_A(\mu)} \int P(ds) c(s)$$

$$\geq \inf_{P \in L_R(\mu)} \int P(ds) c(s) \geq \inf_{P \in L_C(\mu)} \int P(ds) c(s),$$

follow since $L_A(\mu) \subset L_R(\mu) \subset L_C(\mu)$. On the other hand, the integral $\int P(ds) c(s)$ is a linear mapping from the convex set $L_C(\mu)$ to $\mathbb{R}$ (possibly to be extended to include $+\infty$). The extreme points of $L_C(\mu)$ are those that are in $L_A(\mu)$. It then follows from convex analysis (see e.g. [585, Section 2.3]) that

$$\inf_{P \in L_C(\mu)} \int P(ds) c(s) = \inf_{\underline{\gamma} \in \mathbf{\Gamma}} J(\underline{\gamma}).$$

(iv) A convex combination of two measures leads to a new measure which will not necessarily be the product measure of two marginal probability measures: Consider two sets of measures $P^1 P^2$ and $\bar{P}^1 \bar{P}^2$. For $\lambda \in (0, 1)$, a convex combination $\lambda P^1 P^2 + (1-\lambda)\bar{P}^1\bar{P}^2$ cannot be written as the product of their marginals $(\lambda P^1 + (1-\lambda)\bar{P}^1)(\lambda P^2 + 1 - \lambda\bar{P}^2)$.

(v) To establish (v), we present two counterexamples in Sect. 4.2; see Theorem 4.2.2 and Theorem 4.2.3.

◇

With $\mathbb{X}, \mathbb{Y}$ standard Borel spaces, let us recall the following (see Appendix C): we say that a sequence of probability measures (on $\mathbb{X}$) $P_n \to P$ *weakly* if $\int P_n(dx)f(x) \to \int P(dx)f(x)$ for all $f \in C_b(\mathbb{X})$. If this holds for every f measurable and bounded, we say convergence holds *setwise*.

4.1.3 Sets of Strategic Measures for Dynamic Teams in the Absence of Static Reduction

If a stochastic dynamic team admits a static reduction (in particular independent static reduction), then one can define strategic measures by considering an equivalent static problem and characterize the convexity properties of the set of strategic measures, as done in the previous subsection. Our covering in this sub-section excludes such cases, and deals with the more general setting where the dynamic team does not admit a static reduction. Let μ be the distribution of ω_0. Recall that in a dynamic team setting, the distribution of measurements $\mathbf{y}$ is not fixed (and is policy-dependent) as opposed to the static case. We therefore have the following characterization for strategic measures in dynamic sequential teams (consistent with 3.2). Let us introduce, for each $n \in \mathcal{N}$,

$$h_n = \{\omega_0, y^1, u^1, \cdots, y^{n-1}, u^{n-1}, y^n, u^n\},$$

and let $p_n(dy^n|h_{n-1}) := P(dy^n|h_{n-1})$ be the transition kernel characterizing the measurements of DM n under the intrinsic model (see also (3.24)). We note that this may be obtained through the relation:

$$\begin{aligned}
&p_n(y^n \in \cdot|\, \omega_0, y^1, u^1, \cdots, y^{n-1}, u^{n-1}) \\
&:= P\left(\eta^n(\omega, \mathbf{u}^{[1,n-1]}) \in \cdot \,\middle|\, \omega_0, y^1, u^1, \cdots, y^{n-1}, u^{n-1}\right) \\
&= P\left(g^n(\omega_0, \omega_n, u^1, \cdots, u^{n-1}) \in \cdot \,\middle|\, \omega_0, y^1, u^1, \cdots, y^{n-1}, u^{n-1}\right). \qquad (4.7)
\end{aligned}$$

Here ω_n is an independent $[0, 1]$-valued random variable, without any loss. We further note that once a policy is fixed, $p_n(dy^n|h_{n-1})$ represents the conditional distribution of y^n given the past history h_{n-1}. Let $L_A(\mu)$ be the set of strategic measures induced by measurable policies and let $L_R(\mu)$ be the set of strategic measures induced by individually randomized policies for the dynamic team. We have the following characterizations of $L_A(\mu)$ and $L_R(\mu)$ that are quite useful when establishing the closedness of these sets.

Theorem 4.1.2 ([830, Theorem 2.2]) *Consider a stochastic dynamic team problem that does not admit a static reduction. Then, we have the following characterizations:*

(i) *A probability measure* $P \in \mathcal{P}\Big(\Omega_0 \times \prod_{k=1}^{N}(\mathbb{Y}^k \times \mathbb{U}^k)\Big)$ *is a strategic measure induced by a measurable policy (that is in* $L_A(\mu)$*) if and only if, for every* $n = 1, \ldots, N$*, we have (as measurement constraint)*

$$\int P(dh_{n-1}, dy^n)\, g(h_{n-1}, y^n)$$
$$= \int P(dh_{n-1}) \left(\int_{\mathbb{Y}^n} g(h_{n-1}, z)\, p_n(dz|h_{n-1}) \right)$$

and (as design constraint)

$$\int P(dh_n)\, g(h_{n-1}, y^n, u^n)$$
$$= \int P(dh_{n-1}, dy^n) \left(\int_{\mathbb{U}^n} g(h_{n-1}, y^n, a)\, 1_{\{\gamma^n(y^n) \in da\}} \right),$$

for all continuous and bounded functions g *with appropriate arguments, where* $P(d\omega_0) = \mu(d\omega_0)$ *and* $\gamma^n \in \Gamma^n$.

(ii) *A probability measure* $P \in \mathcal{P}\Big(\Omega_0 \times \prod_{k=1}^{N}(\mathbb{Y}^k \times \mathbb{U}^k)\Big)$ *is a strategic measure induced by an individually randomized policy (that is in* $L_R(\mu)$*) if and only if, for every* $n = 1, \ldots, N$*, we have*

$$\int P(dh_{n-1}, dy^n)\, g(h_{n-1}, y^n) = \int P(dh_{n-1}) \left(\int_{\mathbb{Y}^n} g(h_{n-1}, z)\, p_n(dz|h_{n-1}) \right) \tag{4.8}$$

and

$$\int P(dh_n)\, g(h_{n-1}, y^n, u^n)$$
$$= \int P(dh_{n-1}, dy^n) \left(\int_{\mathbb{U}^n} g(h_{n-1}, y^n, a^n)\, \Pi^n(da^n|y^n) \right) \tag{4.9}$$

for all continuous and bounded functions g *with appropriate arguments, where* $P(d\omega_0) = \mu(d\omega_0)$ *and* Π^n *is a stochastic kernel from* $\mathbb{Y}^n$ *to* $\mathbb{U}^n$.

Proof The proof follows from the fact that testing the equalities such as (4.8–4.9) on continuous and bounded functions implies this property for any measurable and bounded function (that is, continuous and bounded functions form a *separating class* (see [116, p. 13] or [241, Theorem 3.4.5]) and thus we recover the measurability properties leading to Witsenhausen's intrinsic model. ◇

Theorem 4.1.3 *Consider a sequential (static or dynamic) team.*

(i) $L_R(\mu)$ *has the following representation so that for any* $P \in L_R(\mu)$*, with* $B = B^0 \times \prod(A^k \times B^k) \in \mathcal{B}(\Omega_0 \times \prod_{k=1}^N (\mathbb{Y}^k \times \mathbb{U}^k))$*,*

$$P(B) = \int U(dz) L_A(\mu, \underline{\gamma}(z))(B)$$

$$U(dv_1, \cdots, dv_N) = \prod_s \eta(dv_s), \eta_s \in \mathcal{P}([0, 1]), \tag{4.10}$$

where $\eta(a, b) = b - a, 0 \le a \le b \le 1$*, that is* η *is the Lebesgue measure on* $[0, 1]$ *and* $\underline{\gamma}(z)$ *is a collection of deterministic policies parametrized by* z*.*

(ii)

$$\inf_{\underline{\gamma} \in \boldsymbol{\Gamma}} J(\underline{\gamma}) = \inf_{P \in L_A(\mu)} \int P(ds)c(s) = \inf_{P \in L_R(\mu)} \int P(ds)c(s)$$

In particular, deterministic policies are optimal within the class of randomized policies.

Proof

(i) Here, we again build on [301, Lemma 1.2]. For any stochastic kernel π^k from $\mathbb{Y}^k$ to $\mathbb{U}^k$, there exists a measurable function $\gamma^k : [0, 1] \times \mathbb{Y}^k \to \mathbb{U}^k$ such that $m\{r : \gamma^k(r, y^k) \in A\} = \Pi^k(u^k \in A | y^k)$, and m is the uniform (Lebesgue measure) on $[0, 1]$. By concatenating N such measures in a product form on $\Upsilon = [0, 1]^N$, we obtain the representation result.

(ii) This follows as in Theorem 4.1.1 (iii), but rather than a convex analytic argument, here we use the integral representation as a consequence of (i); see also Theorem 1-c in [253].

◇

If a sequential team is not classical (and not necessarily strictly *non-classical*), the set of strategic measures is not convex. Note that here the team can be static or dynamic. For the special case of single-agent stochastic control problems with fully available state information at the controller (note that this can be viewed as a single-agent team problem with a classical information structure), convexity of the set of strategic measures has been established in [230] and [584]. We present, in connection with the next theorem, a concise proof for the more general team problems with classical information structures.

Theorem 4.1.4 (Convexity for Classical Information Structures) *If the information structure is classical and is expanded so that when independently randomized policies are allowed,* **DM***i has access to* $\{y^k, u^k, k < i\}$ *and* y^i*, then the set of strategic measures is convex.*[1]

Proof The proof follows from Theorem 4.1.2 and the fact that the condition on the control policies (4.9) is redundant. Nonetheless, to present a direct proof, note that with $N = 1$, convexity is immediate since any stochastic kernel can be realized by a representation of the form $\gamma^1(y^1, r^1)$ for an independent $[0, 1]$ valued r^1. With $N > 1$, consider two strategic measures defined sequentially for k, $1 \le k \le N$:

$$\begin{aligned} P_k^1(d\mathbf{u}, d\mathbf{y}, d\omega_0) = \Pi^1(du^k|y^1, \cdots, y^k, u^1, \cdots, u^{k-1}) \\ \times P(dy^k|y^1, \cdots, y^{k-1}, u^1, \cdots, u^{k-1}, \omega_0) \\ \times P_{k-1}^1(y^1, \cdots, y^{k-1}, u^1, \cdots, u^{k-1}, d\omega_0) \end{aligned} \tag{4.11}$$

$$\begin{aligned} P_k^2(d\mathbf{u}, d\mathbf{y}, d\omega_0) = \Pi^2(du^k|y^1, \cdots, y^k, u^1, \cdots, u^{k-1}) \\ \times P(dy^k|y^1, \cdots, y^{k-1}, u^1, \cdots, u^{k-1}, \omega_0) \\ \times P_{k-1}^2(y^1, \cdots, y^{k-1}, u^1, \cdots, u^{k-1}, d\omega_0) \end{aligned} \tag{4.12}$$

Now, re-writing the above as:

$$P^1(d\mathbf{u}, d\mathbf{y}, d\omega_0) = P^1(y^1, \cdots, y^k, u^1, \cdots, u^k|u^1, y^1)P^1(du^1|y^1)\mu(dy^1, d\omega_0)$$

$$P^2(d\mathbf{u}, d\mathbf{y}, d\omega_0) = P^2(y^1, \cdots, y^k, u^1, \cdots, u^k|u^1, y^1)P^2(du^1|y^1)\mu(dy^1, d\omega_0),$$

we would like to see if $\lambda P^1 + (1 - \lambda)P^2$ can be realized. If we take any convex combination $\lambda P^1(du^1|y^1) + (1 - \lambda)P^2(du^1|y^1)$, then this can be realized without altering any of the policies adopted by the subsequent DMs. ◇

Remark 4.1.1 A celebrated result in game theory, known as Kuhn's theorem [435] (in the context of mixed and behavioral strategies), states that $L_C(\mu)$), which is the convex hull of admissible strategic measures (i.e. those in $L_A(\mu)$), is equivalent to $L_R(\mu)$ when the information structure is classical. We can thus state that this does not hold in the absence of classical-ness, as $L_R(\mu)$ would not be convex (if the information structure is not classical, then convexity fails as discussed above, but the convex hull of admissible policies is, by definition, convex; but the convex hull of $L_R(\mu)$ is $L_C(\mu)$.). See [568, Section 11.4] for further relevant analysis and discussion.

[1] We recall here from Sect. 3.7.2 that the sharing of past control actions when measurements are shared also, was needed for convexity of the expected cost in team policies in normal form (and not necessarily in strategic measures, as studied in the current context).

An implication of the representation results in Theorems 4.1.1 and 4.1.3 is that an optimal policy can be obtained within the class of deterministic team policies.

In [327], Blackwell's Principle of Irrelevant Information Theorem E.2.1 (see [124, 127] and [824, p. 457]) was used to derive such a result in a setting where the randomizations were restricted to be independent:

Theorem 4.1.5 ([327]) *For a sequential dynamic team, any randomized team policy (with independent and private randomization) can be replaced with a policy which is deterministic and performs at least as good as the original (randomized) policy.*

Remark 4.1.2 The different results have their individual merits: (i) Theorem 4.1.5 allows one to replace any sequential team policy with independent randomization with a deterministic one. Note that this result requires *private independent randomness*, that is *common randomness* is not allowed. On the other hand, (ii) Theorems 4.1.1 and 4.1.3, using the powerful theory of convex analysis, establish that even *common independent randomness* does not lead to improvement of performance in stochastic teams or *private randomness* does not for sequential teams; these theorems do not utilize backwards induction, and are thus applicable to teams with countably many DMs (by generalizing the representation result accordingly).

4.1.4 Measurability Properties of Sets of Strategic Measures

As noted earlier, under the weak convergence topology, the set $L_A(\mu)$ introduced by (4.3) is a Borel subset of $\mathcal{P}\left(\Omega_0 \times \prod_k (\mathbb{Y}^k \times \mathbb{U}^k)\right)$. The same is true for $L_R(\mu)$ introduced by (4.4), which is stated in the following theorem. This result will be crucial in the analysis that follows.

Theorem 4.1.6 ([830, Theorem 2.10]) *Consider a sequential (static or dynamic) team.*

(i) The set of strategic measures $L_R(\mu)$ is Borel when viewed as a subset of the space of probability measures on $\Omega_0 \times \prod_{k=1}^N (\mathbb{Y}^k \times \mathbb{U}^k)$ under the topology of weak convergence.
(ii) The set of strategic measures $L_A(\mu)$ is Borel when viewed as a subset of the space of probability measures on $\Omega_0 \times \prod_{k=1}^N (\mathbb{Y}^k \times \mathbb{U}^k)$ under the topology of weak convergence.

Proof In the proof, we will utilize of Theorems C.2.1 and C.2.2, presented in Appendix C.

(i) **Measurability of $L_R(\mu)$**: If $\prod_k \mathbb{Y}^k \times \mathbb{U}^k$ were compact, one could use the characterization in Theorem 4.1.2: For every given g, both sides of (4.8)–(4.9) define measurable functions on $\mathcal{P}(\Omega_0 \times \prod_k \mathbb{Y}^k \times \mathbb{U}^k)$. One can construct a countable collection of *weak-convergence determining* functions on $\Omega_0 \times$

$\prod_k \mathbb{Y}^k \times \mathbb{U}^k$ [241, Theorem 3.4.5]: By Theorems C.2.1 and C.2.2, for each continuous function the set of probability measures which satisfy the equalities in (4.8)–(4.9) is measurable and one only needs a countable number of such equalities, the intersection of the sets of probability measures satisfying all of these has to be Borel. If $\prod_k \mathbb{Y}^k \times \mathbb{U}^k$ is not compact, by the Borel isomorphism theorem, one could construct a bijection between [0, 1] and $\Omega_0 \times \prod_k \mathbb{Y}^k \times \mathbb{U}^k$, and one could follow the proof of Theorem 1 in [230] so that only polynomial functions on [0, 1] are used as test functions on this new space. By the same reasoning as in the compact case, the intersection of the sets of probability measures satisfying countably many equalities of the form (4.8)–(4.9) has to be Borel.

(ii) **Measurability of $L_A(\mu)$**: To establish the measurability of $L_A(\mu)$, we introduce another set, $L^{classical}(\mu)$, as follows. Expand the information structure so that it is now classical such that the information at **DM**k is $\mathcal{I}^k = \{y^k\} \bigcup \cup_{1 \le s \le k-1} \mathcal{I}^s$ for $k > 1$, and write

$$L^{classical}(\mu) := \Bigg\{ P \in \mathcal{P}\Big(\Omega_0 \times \prod_{k=1}^{N} (\mathbb{Y}^k \times \mathbb{U}^k)\Big) : \\ P(B) = \int \mu(d\omega, d\mathbf{y}) \prod_k 1_{\{\gamma^k(\mathcal{I}^k) \in B^k\}} \Bigg\} \tag{4.13}$$

It follows that $L^{classical}(\mu)$ is Borel by Feinberg [254, Theorem 3.2] (which attributes the result to Blackwell [125]) since this information structure may be viewed as a fully observed Markov decision process with $x_k = \mathcal{I}^k$. The measurability of $L_A(\mu)$ follows from the fact that $L_A(\mu) = L_R(\mu) \cap L^{classical}(\mu)$, with both intersected sets being measurable.

◇

Remark 4.1.3 (Implications of Measurability Properties, Witsenhausen's Standard Form, and a Universal Dynamic Program) The Borel measurability properties of $L_R(\mu)$ will be useful in the next chapter (in Sect. 5.4) while arriving at a universal dynamic program by establishing the *universal measurability* and *lower semi-analytic* properties of the value functions.

4.2 Relative Compactness and Closedness of Strategic Measures, and Existence of Optimal Policies

Existence of optimal policies for static and a class of sequential dynamic teams has been studied recently in [327, 620, 830]. More specific setups have been studied in [771, 786, 828] and [824]. Existence of optimal team policies has been established in [167] for a class of continuous-time decentralized stochastic control problems.

For a class of teams which are convex, one can reduce the search space to a smaller parametric class of policies, such as linear policies for quasi-classical linear quadratic Gaussian problems [350, 429, 594].

The following theorem states a general existence result for static teams and for dynamic teams admitting static reduction. Its proof depends on the Weierstrass Extreme Value Theorem.

Theorem 4.2.1 *Consider a static team or the static reduction of a dynamic team with loss function c. Let c be lower semi-continuous in* $\mathbf{u}$ *for every fixed* $\omega_0, \mathbf{y}$ *and* $L_R(\mu)$ *or* $L_C(\mu)$ *be a compact set under weak convergence topology. Then, there exists an optimal team policy. This policy can be chosen deterministic and hence induces a strategic measure in* $L_A(\mu)$.

Remark 4.2.1 Since the cost function c_s in independent static reduction of a dynamic team, as defined in (3.31), also depends on the measurements $\mathbf{y}$, we include $\mathbf{y}$ as an argument in the cost function c in the previous theorem (and drop the subscript).

By Theorem 4.2.1, to prove the existence of optimal team policy, it is sufficient to establish lower semi-continuity of the loss function c and compactness of $L_R(\mu)$ or $L_C(\mu)$ or any other subset of $L_C(\mu)$ that is sufficient for optimality. However, we recall that unless certain conditions are imposed, the conditional independence property dictated by IS is not preserved under weak convergence (indeed this is also true even for setwise convergence which is a stronger notion of convergence). Hence, $L_R(\mu)$ and $L_C(\mu)$ are in general not closed and thus not compact. The following theorem establishes that $L_A(\mu)$ and $L_R(\mu)$ are not closed under the weak convergence topology.

Theorem 4.2.2 ([830, Theorem 2.7]) *Consider a sequence of probability measures* $P_n \in \mathcal{P}(\mathbb{U}^1 \times \mathbb{Y} \times \mathbb{U}^2)$ *so that for all n:*

$$P_n(du^1|y, u^2) = P_n(du^1|y).$$

If $P_n \to P$ *setwise (and thus also weakly), it is not necessarily the case that*

$$P(du^1|y, u^2) = P(du^1|y).$$

That is, conditional independence of u^1 *and* u^2 *given y is not preserved under setwise convergence. In particular,* $L_A(\mu)$ *and* $L_R(\mu)$ *are not (weakly or setwise) closed.*

Before providing a proof for Theorem 4.2.2, let us note that, for each n, we can view P_n as a strategic measure of the following static team problem. In this team problem, Ω is a degenerate space and we have two DMs who are sharing the same measurement y. Therefore, Theorem 4.2.2 states that the set of strategic measures $L_R(\mu)$ is not closed under setwise (and thus weak convergence) topology.

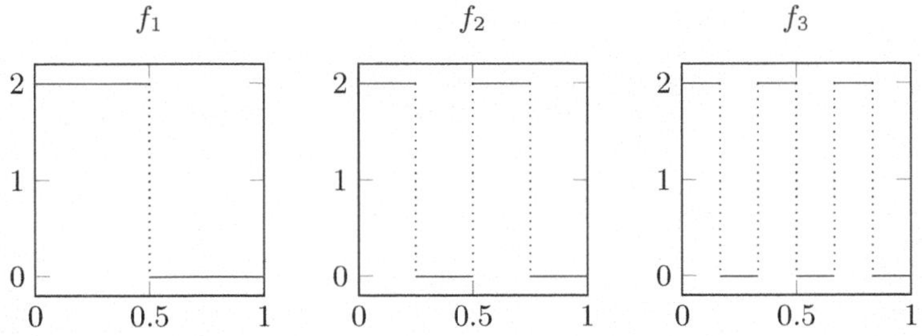

Fig. 4.2 Plots of probability density functions f_1, f_2, and f_3

Proof It suffices to provide a counterexample. We build on an example also to be studied later (in the proof of Theorem 8.3.5, taken from [828]) in the following. Let $\mathbb{Y} = [0, 1]$, $\mathbb{U}^1 = \mathbb{U}^2 = \{0, 1\}$, and $y \sim m$ where m the Lebesgue measure (uniform distribution) on $[0, 1]$. Let

$$L_{nk} = \left[\frac{2k-2}{2n}, \frac{2k-1}{2n}\right), \quad R_{nk} = \left[\frac{2k-1}{2n}, \frac{2k}{2n}\right) \tag{4.14}$$

and define the *square wave* function

$$h_n(t) = \sum_{k=1}^{n} \left(1_{\{t \in L_{nk}\}} - 1_{\{t \in R_{nk}\}}\right). \tag{4.15}$$

Further introduce the probability density functions $f_n(t) = h_n(t) + 1$ (Fig. 4.2).

Let $B_{n,1} = \bigcup_{k=1}^{n} L_{nk}$ and $B_{n,2} = [0, 1] \setminus B_{n,1}$. Define $\{Q_n\}$ as the sequence of 2-cell quantizers given by

$$Q_n(1|y) = 1_{\{y \in B_{n,1}\}}, \quad Q_n(0|y) = 1_{\{y \in B_{n,2}\}}.$$

Let

$$P_n(u^1 = a|y) = P_n(u^2 = a|y) = Q_n(a|y),$$

where $a \in \{0, 1\}$. Define $P \in \mathcal{P}(\mathbb{U}^1 \times \mathbb{Y} \times \mathbb{U}^2)$ as $P(a, A, b) = 1_{\{a=b\}} \frac{1}{2} m(A)$, where $a, b \in \{0, 1\}$ and $A \in \mathcal{B}([0, 1])$.

By the proof of the Riemann-Lebesgue lemma ([768], Thm. 12.21), note that for all $A \in \mathcal{B}([0, 1])$,

$$\lim_{n\to\infty} \int_A Q_n(1|y)m(dy) = \lim_{n\to\infty} \int_A \frac{1}{2} f_n(t)\, m(dt) = \frac{1}{2} m(A),$$

and thus for all $A \in \mathcal{B}([0, 1])$

$$\begin{aligned}
&\lim_{n\to\infty} P_n(u^1 = 1, y \in A, u^2 = 1) \\
&= \lim_{n\to\infty} \int_A P_n(u^1 = 1|y) P_n(u^2 = 1|y) m(dy) \\
&= \lim_{n\to\infty} \int_A P_n(u^1 = 1|y) m(dy) \\
&= \frac{1}{2} m(A) \\
&= P(1, A, 1)
\end{aligned} \tag{4.16}$$

A similar property applies for $(u^1, u^2) = (0, 0)$, $(0, 1)$ and $(1, 0)$ so that

$$\lim_{n\to\infty} P_n(u^1 = a, y \in A, u^2 = b) \to P(a, A, b) = 1_{\{a=b\}} \frac{1}{2} m(A)$$

Thus, $P_n \to P$ setwise. But even though P_n satisfies the conditional independence property that $P_n(u^1 = 1|y, u^2) = Q_n(1|y)$, P does not satisfy the conditional independence property of u^1 and u^2 given y: Under P, u^1 and y are independent but $u^1 = u^2$ and thus $P(u^1 = a|y, u^2 = b) = 1_{\{a=b\}} \neq \frac{1}{2} = P(u^1 = a|y)$. Thus, setwise (and hence weak) convergence does not preserve the conditional independence property. This implies that neither $L_A(\mu)$ nor $L_R(\mu)$ are weakly closed. ◇

We can also show, via a counterexample, that $L_C(\mu)$ is not closed either, with respect to the weak convergence topology. In more concrete terms:

Theorem 4.2.3 *$L_C(\mu)$ is not, in general, weakly or setwise closed.*

Proof Consider the same example as in the proof of Theorem 4.2.2, modified as follows. Let $\mathbb{Y} = [0, 1] \cup (1, 2]$, $\mathbb{U}^1 = \mathbb{U}^2 = \{0, 1, 2, 3\}$. Let x be a $\{0, 1\}$-valued uniformly distributed Bernoulli random variable and let $y \sim 1_{\{x=0\}} v + 1_{\{x=1\}}(v + 1)$ where $v \sim m$ with m being the Lebesgue measure (uniform distribution) on $[0, 1]$. Let L_{nk} and R_{nk} be as given in (4.14) and define the *square wave* function, as before,

$$h_n(t) = \sum_{k=1}^{n} \left(1_{\{t \in L_{nk}\}} - 1_{\{t \in R_{nk}\}}\right).$$

Define also $f_n(t) = h_n(t) + 1$. Let $B_{n,1} = \bigcup_{k=1}^{n} L_{nk}$ and $B_{n,2} = [0, 1] \setminus B_{n,1}$. Define $\{Q_n\}$ as the sequence of 4-cell quantizers given by

$$Q_n(1|y) = 1_{\{y \in B_{n,1}\}}, \quad Q_n(0|y) = 1_{\{y \in B_{n,2}\}}, \qquad y \in [0, 1]$$

and

$$Q_n(3|y) = 1_{\{y \in 1 + B_{n,1}\}}, \quad Q_n(2|y) = 1_{\{y \in 1 + B_{n,2}\}}, \qquad y \in (1, 2].$$

Let

$$P_n(u^1 = a|y) = P_n(u^2 = a|y) = Q_n(a|y),$$

where $a \in \{0, 1, 2, 3\}$. Define $P \in \mathcal{P}(\mathbb{U}^1 \times \mathbb{Y} \times \mathbb{U}^2)$ as

$$P(a, A, b) = \frac{1}{4} 1_{\{a=b, a\in\{0,1\}\}} m(A \cap [0, 1]) + \frac{1}{4} 1_{\{a=b, a\in\{2,3\}\}} m(A \cap (1, 2])$$

where $a, b \in \{0, 1, 2, 3\}$ and $A \in \mathcal{B}([0, 2])$.

We can show that $P_n \to P$ setwise (and thus weakly). In the limit, it is the case that $u^1 = u^2$ almost surely. However, notice that $P \notin L_C(\mu)$: P is not a mixture of conditionally independent random variables given y, with the mixture being independent of y. Here, the mixture representation of P in terms of the two independent $\{0, 1\}$-valued and $\{2, 3\}$-valued random variables is so that the mixing random variable is correlated with y (and provides information on y). ◇

We refer the reader to [14, 340] and [49, 69, 828, Theorem 1.1] for further related results on such intricacies on conditional independence properties. A sufficient condition for compactness of $L_R(\mu)$ under the weak convergence topology was reported in [818, Theorem 5.2].

In the following, we slightly strengthen [818, Theorem 5.2] to allow for lower semi-continuity of the cost function only in the actions.

Before we present a general result on existence, we recall the w-s topology introduced by Schäl [650] (see Appendix C): The w-s topology on the set of probability measures $\mathcal{P}(\mathbb{X} \times \mathbb{U})$ is the coarsest topology under which $\int f(x, u)\nu(dx, du) : \mathcal{P}(\mathbb{X} \times \mathbb{U}) \to \mathbb{R}$ is continuous for every measurable and bounded f which is continuous in u for every x (but unlike weak topology, f does not need to be continuous in x).

Theorem 4.2.4

(i) *Consider a static or a dynamic team that admits an independent static reduction. Let the action sets $\mathbb{U}^i$ be compact. Then, the set $L_R(\mu)$ is compact under weak convergence topology, and is also in the w-s topology (weak in actions, strong in measurements).*

(ii) *Consider a static or a dynamic team that admits an independent static reduction with the loss denoted by c (for either case). Let c be lower semi-continuous in $\mathbf{u}$ for any $\omega_0, \mathbf{y}$. Suppose further that $\mathbb{U}^i$ is $\sigma-$compact (that is, $\mathbb{U}^i = \cup_n K_n$ for a countable collection of increasing compact sets K_n), and (without any loss of generality) the control laws are restricted to those with $E[\varphi^i(u^i)] \leq M$ for some lower semi-continuous $\varphi^i : \mathbb{U}^i \to \mathbb{R}_+$ which satisfies $\lim_{n\to\infty} \inf_{u^i \notin K_n} \varphi^i(u^i) = \infty$. Then, an optimal team policy exists.*

Proof

(i) For each m, consider the strategic measure:

$$P(d\omega_0)\prod_{k=1}^{N}(Q^k\gamma^k)_m(dy^k, du^k).$$

Suppose that $(Q^k\gamma^k)_m(dy^k, du^k)$, converges to $(Q^k\gamma^k)(dy^k, du^k)$ weakly. By Parthasarathy [574, p. 57], the joint product measure $P(d\omega_0)\prod_{k=1}^{N}(Q^k\gamma^k)_m$ (dy^k, du^k) will converge weakly to $P(d\omega_0)\prod_{k=1}^{N}(Q^k\gamma^k)(dy^k, du^k)$; see also [658, Section 5]. Since the marginals on $\prod_k \mathbb{Y}^k$ are fixed, Theorem C.4.1 establishes that the set of all probability measures with a fixed marginal on $\prod_k \mathbb{Y}^k$ is (sequentially) compact under the w-s topology.

(ii) If the function $c_s(\omega_0, \mathbf{y}, \cdot)$ were continuous, this in turn would ensure that the following function (by a truncation and then a limiting argument)

$$\int P(d\omega_0)\prod_{k=1}^{N}(Q^k\gamma^k)(dy^k, du^k)\, c_s(\omega_0, \mathbf{y}, \mathbf{u}),$$

is lower semi-continuous under the w-s topology. Now, by approximating the lower semi-continuous function $c_s(\omega_0, \mathbf{y}, \cdot)$ from below by continuous functions, applying the argument above, and taking the limit, we conclude that the lower semi-continuity also applies when $c_s(\omega_0, \mathbf{y}, \cdot)$ is only lower semi-continuous.

Finally, since the set of admissible strategic measures is sequentially compact under the w-s topology, existence of an optimal team policy follows. The proof for the dynamic case follows analogously by considering its static reduction. ◇ □

Recall that, by Remark 4.1.2, an optimal policy, when it exists, can be assumed to be *deterministic*.

So far, we have presented existence results for static or dynamic teams that admit independent static reduction. In the following, we present existence results for teams that do not admit independent static reduction.

Theorem 4.2.5 ([830, Theorem 2.9]) *Consider a sequential team with a classical information structure with the further property that $\sigma(\omega_0) \subset \sigma(y^1)$ (under every policy, y^1 contains ω_0). Suppose further that $\prod_{k=1}^{N}(\mathbb{Y}^k \times \mathbb{U}^k)$ is compact. If c is lower semi-continuous and each of the kernels p_n (defined in (4.7)) is weakly continuous so that*

$$\int f(y^n)\, p_n(dy^n|\omega_0, y^1, \ldots, y^{n-1}, u^1, \cdots, u^{n-1}) \tag{4.17}$$

is continuous in $\omega_0, y^1, \cdots, y^{n-1}, u^1, \cdots, u^{n-1}$ for every continuous and bounded f, then there exists an optimal team policy which is deterministic.

Proof We first note that when $\prod_k \mathbb{Y}^k \times \mathbb{U}^k$ is compact, the set of all probability measures with a fixed marginal on ω_0 would form a weakly compact set. Therefore, it suffices to ensure the closedness of the set of strategic measures, which leads to compactness of the set. To facilitate such a compactness condition, we first *expand* the information structure so that **DM**k has access to all the previous actions $u^1, \cdots, u^{k-1}$ as well. Later on, we will see that this expansion is actually redundant. With such an expansion, any weakly converging sequence of strategic measures will continue to satisfy (4.9) in the limit due to the fact that there is no conditional independence property in the sequence since all the information is available at **DM**k. On the other hand, to show that for any weakly converging sequence of strategic measures satisfying (4.8) so does the limit, it suffices to show the continuity of

$$\int g(y^n, \omega_0, u^1, \cdots, u^{n-1}) p_n(dy^n | \omega_0, u^1, \cdots, u^{n-1}) \tag{4.18}$$

in $\omega_0, u^1, \cdots, u^{n-1}$. With $f_{\omega_0, u^1, \cdots, u^{n-1}}(y^n) := g(y^n, \omega_0, u^1, \cdots, u^{n-1})$, it follows that as a sequence $(\omega_0, u^1, \cdots, u^{n-1})_k \to (\omega_0, u^1, \cdots, u^{n-1})$, and $y^n_k \to y^n$ the following holds:

$$f_{(\omega_0, u^1, \cdots, u^{n-1})_k}(y^n_k) \to f_{\omega_0, u^1, \cdots, u^{n-1}}(y^n).$$

In other words, $f_{\omega_0, u^1, \cdots, u^{n-1}}(y^n)$ converges *continuously* as it is defined in [658]. Thus, the continuity of (4.17) ensures that (4.18) holds by a generalized convergence theorem given in Theorem C.3.1(i) (see [658, Theorem 3.5]). Thus, the properties (4.8)–(4.9) are preserved under weak convergence. As a result, existence follows from Theorem 4.2.1.

Now, we know that an optimal policy will be deterministic as a consequence of Theorem 4.1.3. Thus, a deterministic policy may not make use of randomization, and therefore **DM**k having access to $\{y^k, y^{k-1}, y^{k-2}, \cdots, \omega_0\}$ is informationally equivalent to him having access to $\{y^k, (y^{k-1}, u^{k-1}), (y^{k-2}, u^{k-2}), \cdots, \omega_0\}$ for an optimal policy. Thus, an optimal team policy exists. ◇

A further existence result along similar lines, though for static teams, is presented next. We note that this theorem generalizes an earlier result in [830, Theorem 3.5] by relaxing the weak continuity condition there on the kernel.

Theorem 4.2.6 ([818, Theorem 5.6]) *Consider a static team with a classical information structure (that is, with an expanding information structure so that $\sigma(y^n) \subset \sigma(y^{n+1})$, $n \geq 1$). Suppose further that $\prod_{k=1}^N (\mathbb{Y}^k \times \mathbb{U}^k)$ is compact. If*

$$\tilde{c}(y^1, \cdots, y^N, u^1, \cdots, u^N) := E[c(\omega_0, \mathbf{u}) | \mathbf{y}, \mathbf{u}]$$

is lower semi-continuous in $\mathbf{u}$ for every $\mathbf{y}$, then there exists an optimal team policy which is deterministic. The same result holds if the information structure is expanded so that both control actions and measurements are available to subsequent decision makers (which makes the information structure dynamic).

Proof Different from Theorem 4.2.5, we eliminate the use of ω_0, and study the properties of the set of strategic measures. Also, different from [830, Theorem 3.5], we relax the assumption of weak continuity. Once again, instead of the weak topology, we will use the w-s topology.

Note that when $\prod_{k=1}^{N} \mathbb{Y}^k \times \mathbb{U}^k$ is compact, the set of all probability measures on $\prod_{k=1}^{N} \mathbb{Y}^k \times \mathbb{U}^k$ forms a weakly compact set. Since the marginals on $\prod_{k=1} \mathbb{Y}^k$ are fixed, Theorem C.4.1 establishes that the set of all probability measures with a fixed marginal on $\prod_{k=1} \mathbb{Y}^k$ is relatively compact under the w-s topology. Therefore, it suffices to ensure the closedness of the set of strategic measures, which leads to sequential compactness of the set under this topology. To facilitate such a compactness condition, we first *expand* the information structure so that **DM**k has access to all the previous actions $u^1, \cdots, u^{k-1}$ as well. Later on, we will see that this expansion is redundant. With this expansion, any w-s converging sequence of strategic measures will continue to satisfy (4.9) in the limit due to the fact that there is no conditional independence property in the sequence since all the information is available at **DM**k. That is, $P_n(du^n|y^n, y_{[0,n-1]}, u_{[0,n-1]})$ satisfies the conditional independence property trivially as all the information is available. On the other hand, for each element in the sequence of conditional probability measures, the conditional probability for the measurements writes as $P(dy^n|y_{[0,n-1]}, u_{[0,n-1]}) = P(dy^n|y^{n-1})$. We now wish to show that this also holds for the w-s limit measure. First we have that for every n, $y^n \leftrightarrow y^{n-1} \leftrightarrow h_{n-1}$ forms a Markov chain. By considering the convergence properties only on continuous and bounded f, as in (4.8), with $P_m \to P$ weakly, we have that

$$\begin{aligned}
&\int P(dy^n|y^{n-1})P_m(dh_{n-1})f(y^n, h_{n-1}) \\
&= \int \Big(P(dy^n|y^{n-1})f(y^n, h_{n-1})\Big)P_m(dh_{n-1}) \\
&\to \int \Big(P(dy^n|y^{n-1})f(y^n, h_{n-1})\Big)P(dh_{n-1}) \\
&= \int P(dy^n|y^{n-1})P(dh_{n-1})f(y^n, h_{n-1})
\end{aligned}$$

Here, $\Big(P(dy^n|y^{n-1})f(y^n, h_{n-1})\Big)$ is not necessarily continuous in y^{n-1}, but it is in h_{n-2} and u^{n-1} by an application of the dominated convergence theorem, and since

$$P_m(dy^1, \ldots, dy^{n-1}, du^1, \ldots, du^{n-1}) \to P(dy^1, \ldots, dy^{n-1}, du^1, \ldots, du^{n-1}),$$

in the w-s sense (setwise in the measurement variable coordinates), convergence holds. Thus, (4.8) is also preserved. Hence, for any w-s converging sequence of strategic measures satisfying (4.8) and (4.9) so does the limit since the team is static and with perfect-recall. By Schäl [649, Theorem 3.7], and the generalization of Portmanteau theorem for the w-s topology, the lower semi-continuity of $\int \mu(d\mathbf{y}, d\mathbf{u})\tilde{c}(\mathbf{y}, \mathbf{u})$ over the set of strategic measures leads to the existence of an optimal strategic measure. As a result, existence follows from steps similar to those in the proof of Theorem 4.2.1. We know that an optimal policy may be taken deterministic without loss (see Remark 4.1.2), therefore **DM**k having access to $\{y^k, y^{k-1}, y^{k-2}, \cdots\}$ is informationally equivalent to him having access to $\{y^k, (y^{k-1}, u^{k-1}), (y^{k-2}, u^{k-2})\}$ for an optimal policy. Thus, an optimal team policy exists. $\diamond$

Remark 4.2.2 The power of this last result may first seem limited. However, some reflection leads to the conclusion that, in the continuous-time theory of stochastic control, a related (but not identical) argument has remarkable consequences. If one makes the measurements independent via a change of measure argument, as in Girsanov's celebrated argument (see [107, 302] and [267]), so that the information structure is first made static, and then makes the information structure classical by considering the actions at time t measurable on the filtration generated by the past noise processes and actions up to time t, the proof of Theorem 4.2.6 can be slightly adapted to show that such a set of measurement-action measures (with a fixed marginal on the measurements) that satisfy conditional independence $u_{[0,t]} \leftrightarrow y_{[0,t]} \leftrightarrow y_s - y_t$ is weakly closed (these are known as wide-sense admissible control policies). Furthermore, the value is continuous in this joint measure on $\{(u, y)_s, s \in [0, T]\}$ and this set of measures is tight. These lead to the compactness-continuity conditions and accordingly an existence result for optimal policies follows. Furthermore, by showing that the set of $\{(u, y)_s, s \geq 0\}$ measures which have quantized support in the measurement variable are dense, one can show also that piece-wise constant control policies are nearly optimal. This allows one to approximate a continuous-time process with a (sampled) discrete-time process. This approach is the essence of the method in [446], though stated somewhat differently.

Remark 4.2.3 (Existence via Agent-Wise Reductions) By merging the results of Theorem 4.2.5 and Theorem 4.2.4, we can also obtain a slight generalization by considering an agent-wise reduction as defined in Definition 3.5.1: If we have an *agent-wise nested independent reduction*, then the set of product measures of individual strategic measures for each agent will be weakly compact and if the cost function defined in (3.37) is lower semi-continuous in actions, then existence follows.

As a recap, we can say that Theorem 4.2.4 (for static teams or dynamic teams with an independent static reduction) and Theorems 4.2.5 and 4.2.6 (for sequential teams that do not allow an independent static reduction) appear to be the most general existence results for sequential team problems considered in this chapter. These results complement one another and cover a very large class of decentralized stochastic control problems.

We next state the following result, which will be useful later in the book, both in the context of comparison of information structures in stochastic teams as well as for the existence of equilibrium solutions in stochastic games and comparison of their information structures.

Theorem 4.2.7 ([356]) *$L_C(\mu)$ is weakly compact whenever $L_R(\mu)$ is.*

Proof Let $\{\varphi_k\}$ be an arbitrary sequence of information structures in $L_C(\mu)$ defined, with $B \in \mathcal{B}\left(\Omega_0 \times \prod_{k=1}^{N}(\mathbb{Y}^k \times \mathbb{U}^k)\right)$ and $\theta \in \{L_A(\mu, \underline{\gamma}), \underline{\gamma} \in \mathbf{\Gamma}\}$, as

$$\varphi_k(B) = \int_B \left(\int \nu_k(d\theta)\theta(dz) \right).$$

Following the relationship between $L_{CCR}(\mu)$ and $L_R(\mu)$, as well as Theorem 4.1.2, we can view $L_C(\mu)$ as the mixture of elements of L_R. The convergence properties of the sequence $\{\varphi_k\}$ can be studied by the convergence of the family of measures

$$\varphi_k(B) = \int_B f_m(z) \left(\int \nu_k(d\theta)\theta(dz) \right),$$

for a fixed countable collection of *weak-convergence determining* functions $f_m : \Omega \times (\prod_{i=1}^{n} \mathbb{X}^i \times \mathbb{Y}^i) \to \mathbb{R}$ [241, Theorem 3.4.5].

Now, using Fubini's Theorem, we can rewrite the above as:

$$\int \nu_k(d\theta) \int f_m(z)\theta(dz),$$

and note that the inner integral is continuous and bounded in θ. Thus we can write this as $\int \nu_k(d\theta)F_m(\theta)$, for the continuous and bounded F_m on $L_R(\mu)$ given with $F_m(\theta) := \int f_m(z)\theta(dz)$. Since $L_R(\mu)$ is weakly compact, the set of probability measures on $\mathcal{B}(L_R(\mu))$ is weakly compact [118]; accordingly, there exists a subsequence of $\{\nu_k\}$, $\{\nu_{k_l}\}$, which converges weakly to some ν in $\mathcal{P}(L_R(\mu))$. Since F_m is continuous and bounded on $L_R(\mu)$, it follows then that $\int \nu_{k_l}(d\theta)F_m(\theta)$ will converge to $\int \nu(d\theta)F_m(\theta)$. This holds for each m, and therefore, we conclude that $\{\varphi_k\}$ has a convergent subsequence in $L_C(\mu)$. Thus, $L_C(\mu)$ is compact. ◇

4.3 Convex Relaxations

In this section, we introduce convex relaxations for static teams as well as dynamic teams that admit independent static reduction by relaxing the assumption of conditional independence among actions. Therefore, in this section, the joint distribution of $\omega_0, \mathbf{y}$ is in product form $\mu(d\omega_0, d\mathbf{y}) = P(d\omega_0) \prod_{k=1}^{N} Q^k(dy^k)$. These relaxations can be classified in increasing order as *quantum-correlated relaxation* and *non-signaling relaxation*. These classes, introduced in [621], are new to the team decision theory literature and these new classes will likely provide novel perspectives and results to team decision theory and decentralized stochastic control.

We saw earlier (see Remark 4.1.2) that individual or common (independent) randomness do not improve optimal team cost, whereas we will see that quantum-correlated and non-signaling relaxations in general improve optimal cost. Moreover, the optimization problem associated with the non-signaling case can be written as a linear program.

The non-signaling type relaxation for team problems was studied in [830] under a different name, building on [22] where it was proved that the set of strategic measures corresponding to the extreme points of non-signaling policies is a strict superset of the set of strategic measures corresponding to measurable policies. This set arises in applications in information theory in the context of converse theorems in multi-terminal source coding (see e.g. Berger-Tung inner-outer bounds [753]). In the quantum information theory literature [151], this correlation structure has also been studied extensively, as we will discuss later in this section.

4.3.1 Non-signaling Relaxation

A joint conditional distribution of actions given observations is *non-signaling* if, for any $i \in \mathcal{N}$, the marginal distribution of action of DM i given its observation does not provide any information about the observations of other agents [71, 510]. Non-signaling has been investigated in quantum mechanics due to its close connection to foundations of quantum mechanics and relativity [588].

Let $L_{NS}(\mu)$ denote the set of non-signaling strategic measures, which are formally defined as follows. P is a non-signaling strategic measure, that is, $P \in L_{NS}(\mu)$, if $P(d\omega_0, d\mathbf{y}) = \mu(d\omega_0, d\mathbf{y})$, and for any subset $\{k_1, \dots, k_M\}$ of $\{1, \dots, N\}$, the actions of the agents in $\{k_1, \dots, k_M\}$ given their measurements are independent of measurements of agents in $\{1, \dots, N\} \setminus \{k_1, \dots, k_M\}$ as well as ω_0; that is, for any $\{k_1, \dots, k_M\} \subset \{1, \dots, N\}$, we have

$$P(du^{k_1}, \dots, du^{k_M} \mid \omega_0, \mathbf{y}) = P(du^{k_1}, \dots, du^{k_M} \mid y^{k_1}, \dots, y^{k_M}). \tag{4.19}$$

We also define a more relaxed version of $L_{NS}(\mu)$, denoted by $L_M(\mu)$ (which we will term as local-Markov correlations), as follows: P is a local-Markov correlations measure, that is, $P \in L_M(\mu)$, if $P(d\omega_0, d\mathbf{y}) = \mu(d\omega_0, d\mathbf{y})$, and for any $k \in \{1, \dots, N\}$, the action u^k of agent k given the measurement y^k is independent of measurements of agents in $\{1, \dots, N\} \setminus \{k\}$ as well as ω_0; that is, we have

$$P(du^k \mid \omega_0, \mathbf{y}) = P(du^k \mid y^k). \tag{4.20}$$

Remark 4.3.1 At first sight, it is tempting to claim that $L_C(\mu)$ is the same as the set of non-signaling strategic measures $L_{NS}(\mu)$ or $L_M(\mu)$. Indeed, in [22] this question was raised, and a counterexample to the claim that $L_C(\mu)$ is equivalent to the set $L_{NS}(\mu)$ was provided, pointing to the fact that the set of extreme points of non-signaling policies is not $L_A(\mu)$, which would imply that non-signaling policies are more general than randomized ones, as $L_A(\mu)$ is the set of extreme points of $L_C(\mu)$ (see Theorem 4.1.1). In the quantum information literature, this result was known some time ago, as we will revisit later in the context of the Clauser-Horne-Shimony-Holt game [180].

It turns out that the non-signaling condition (4.19) can be derived from fewer linear constraints, which will be described below. These constraints indeed enable us to write the optimization problem associated with non-signaling policies as a linear program.

Lemma 4.3.1 ([510, Section II-A]) *A measure P with $P(d\omega_0, d\mathbf{y}) = \mu(d\omega_0, d\mathbf{y})$ is a non-signaling strategic measure if and only if it satisfies the following condition:*

(N) For each $k \in \{1, \dots, N\}$, the marginal distribution of actions excluding u^k is independent of the observation y^k as well as ω_0:

$$P(d\mathbf{u}^{-k}|\omega_0, \mathbf{y}) = P(d\mathbf{u}^{-k}|\mathbf{y}^{-k}). \tag{4.21}$$

Proof Note that each constraint (4.21) is linear in P. The equivalence of (4.19) and (**N**) can then be established as follows. First, it is immediate that (4.19) implies (**N**). Conversely, let P satisfy (**N**). Fix any subset $\{k_1, \dots, k_M\}$ of $\{1, \dots, N\}$. Let $\{l_1, \dots, l_T\} := \{1, \dots, N\} \setminus \{k_1, \dots, k_M\}$. Then, using the condition (**N**), we can prove that

$$P(du^{k_1}, \dots, du^{k_M} \mid \omega_0, \mathbf{y}) = P(du^{k_1}, \dots, du^{k_M} \mid \hat{\omega}_0, \hat{y}^{l_1}, \dots, \hat{y}^{l_T}, y^{k_1}, \dots, y^{k_M}),$$

for all $\omega_0, y^{l_1}, \dots, y^{l_T}$ and $\hat{\omega}_0, \hat{y}^{l_1}, \dots, \hat{y}^{l_T}$, which implies that

$$P(du^{k_1}, \dots, du^{k_M} \mid \omega_0, \mathbf{y}) = P(du^{k_1}, \dots, du^{k_M} \mid y^{k_1}, \dots, y^{k_M}).$$

Indeed, we have

$$\begin{aligned}
&P(du^{k_1}, \dots, du^{k_M} \mid \omega_0, \mathbf{y}) \\
&= \int_{\mathbb{U}^{l_2} \times \dots \times \mathbb{U}^{l_T}} P(d\mathbf{u}^{-l_1} \mid \omega_0, \mathbf{y}) \\
&= \int_{\mathbb{U}^{l_2} \times \dots \times \mathbb{U}^{l_T}} P(d\mathbf{u}^{-l_1} \mid \hat{\omega}_0, \hat{y}^{l_1}, \mathbf{y}^{-l_1}) \text{ (by } (\mathbf{N})) \\
&= P(du^{k_1}, \dots, du^{k_M} \mid \hat{\omega}_0, \hat{y}^{l_1}, \mathbf{y}^{-l_1}) \\
&= \int_{\mathbb{U}^{l_1} \times \mathbb{U}^{l_3} \times \dots \times \mathbb{U}^{l_T}} P(d\mathbf{u}^{-l_2} \mid \hat{\omega}_0, \hat{y}^{l_1}, \mathbf{y}^{-l_1}) \\
&= \int_{\mathbb{U}^{l_1} \times \mathbb{U}^{l_3} \times \dots \times \mathbb{U}^{l_T}} P(d\mathbf{u}^{-l_2} \mid \hat{\omega}_0, \hat{y}^{l_1}, \hat{y}^{l_2}, \mathbf{y}^{-l_1,-l_2}) \text{ (by } (\mathbf{N})) \\
&= P(du^{k_1}, \dots, du^{k_M} \mid \hat{\omega}_0, \hat{y}^{l_1}, \hat{y}^{l_2}, \mathbf{y}^{-l_1,-l_2})
\end{aligned}$$

$$\vdots$$

$$= P(du^{k_1}, \ldots, du^{k_M} \mid \hat{\omega}_0, \hat{y}^{l_1}, \ldots, \hat{y}^{l_T}, y^{k_1}, \ldots, y^{k_M}).$$

This completes the proof. ◇

The following result establishes convexity and a topological property of the sets $L_{NS}(\mu)$ and $L_M(\mu)$.

Theorem 4.3.1 *Both $L_M(\mu)$ and $L_{NS}(\mu)$ are convex and closed under the weak convergence topology.*

Proof Convexity of both sets follows since both $L_M(\mu)$ and $L_{NS}(\mu)$ are intersections of convex sets.

Let $P_n \in L_M(\mu)$ converge to P weakly. Then, for any continuous and bounded g:

$$\int P_n(dy^i, d\mathbf{y}^{-i}, d\omega_0, u^i) g(y^i, \mathbf{y}^{-i}, \omega_0, u^i)$$

$$\to \int P(dy^i, d\mathbf{y}^{-i}, d\omega_0, u^i) g(y^i, \mathbf{y}^{-i}, \omega_0, u^i)$$

Since $P_n \in L_M(\mu)$, we have that

$$\int P_n(dy^i, d\mathbf{y}^{-i}, d\omega_0, u^i) g(y^i, \mathbf{y}^{-i}, \omega_0, u^i)$$

$$= \int \int P_n(d\mathbf{y}^{-i}, d\omega_0 | y^i) P_n(dy^i, du^i) g(y^i, \mathbf{y}^{-i}, \omega_0, u^i).$$

But since the information structure is static, we can write

$$\int P_n(dy^i, d\mathbf{y}^{-i}, d\omega_0, u^i) g(y^i, \mathbf{y}^{-i}, \omega_0, u^i)$$

$$= \int \int \mu(d\mathbf{y}^{-i}, d\omega_0 | y^i) P_n(dy^i, du^i) g(y^i, \mathbf{y}^{-i}, \omega_0, u^i).$$

Now note that,

$$h(y^i, u^i) := \left(\int \mu(d\mathbf{y}^{-i}, d\omega_0 | y^i) g(y^i, \mathbf{y}^{-i}, \omega_0, u^i) \right)$$

is continuous in u^i as g is continuous by the dominated convergence theorem. Since the marginal on y^i is fixed, as noted earlier, [649, Theorem 3.10] (or [58, Theorem 2.5]) establishes that the set of all tight probability measures with a fixed marginal on $\prod_k \mathbb{Y}^k$ is relatively compact under the w-s topology. This implies that P_n converges to P in w-s topology as the sequence $\{P_n\}$ is tight. Hence,

$$\int P_n(dy^i, d\mathbf{y}^{-i}, d\omega_0, u^i) g(y^i, \mathbf{y}^{-i}, \omega_0, u^i) \to \int h(y^i, u^i) P(dy^i, du^i)$$

$$= \int \int \mu(d\mathbf{y}^{-i}, d\omega_0 | y^i) P(dy^i, du^i) g(y^i, \mathbf{y}^{-i}, \omega_0, u^i)$$

and hence $P \in L_M(\mu)$.

By viewing each subset $\{k_1, \ldots, k_m\}$ of $\{1, \ldots, N\}$ as a single DM with a collective action $\mathbf{u}_m := \{u^{k_1}, \ldots, u^{k_m}\}$ and measurement $\mathbf{y}_m := \{y^{k_1}, \ldots, y^{k_m}\}$, applying the same analysis as above shows that each set is closed, and since intersection of closed sets is closed, the closedness of $L_{NS}(\mu)$ follows. As noted above, using the condition (**N**), without any loss, we can only consider the sets of cardinality $N - 1$. ◇

Since the constraints $P(d\mathbf{u}^{-i} | y^i, \mathbf{y}^{-i}, \omega_0) = P(d\mathbf{u}^{-i} | \mathbf{y}^{-i})$, $i \in \mathcal{N}$, in $L_{NS}(\mu)$ for P are linear, the optimization problem associated with non-signaling strategic measures can be written as a linear program on appropriate vector spaces.

4.3.2 *Quantum Correlations and Quantum Team Decision Policies*

An emerging set of correlations, in view of encouraging developments in quantum physics as well as quantum information theory, involves *quantum* correlations.

To introduce quantum-correlated strategic measures, we briefly introduce the mathematical formalism necessary to discuss quantum operations. We refer the reader to books [357, 761] for basics of quantum information and computation. In this section, we assume that all Hilbert spaces are complex and separable.

Quantum physical systems are described by complex Hilbert spaces. For a Hilbert space $\mathcal{H}$ with an inner product $\langle \cdot, \cdot \rangle$, let $\mathcal{L}(\mathcal{H})$ and $\mathcal{D}(\mathcal{H})$ denote the set of Hermitian and positive Hermitian operators with unit trace. A state ρ of a quantum system, living in $\mathcal{H}$, is an element of $\mathcal{D}(\mathcal{H})$; that is, it is a positive Hermitian operator with unit trace. A quantum state ρ is said to be *pure* if it has rank equal to 1. Equivalently, ρ is pure state if there exists a unit vector $u \in \mathcal{H}$ such that

$$\rho = u\, u^*,$$

where u^* is the complex conjugate of the vector u. By spectral decomposition, every quantum state can be written as a mixture of pure states. Hence, the extreme points of $\mathcal{D}(\mathcal{H})$ coincides with the set of pure states.

A measurement on a quantum system is given by a Positive Operator Valued Measure (POVM) [357]; that is, given any measurable outcome space $(\mathbb{X}, \mathcal{X})$, a

POVM is a mapping $M : \mathcal{X} \to \mathcal{L}(\mathcal{H})$ from the σ-algebra $\mathcal{X}$ of $\mathbb{X}$ to the set of Hermitian operators $\mathcal{L}(\mathcal{H})$ on $\mathcal{H}$ such that

(1) $M(\emptyset) = 0,\ \ M(\mathbb{X}) = I$ (i.e., the identity matrix);
(2) $M(B) \geq 0\ \forall B \in \mathcal{X}$;
(3) $M(\bigcup_j B_j) = \sum_j M(B_j)$ for any disjoint collection $\{B_j\}$, where the series is weakly convergent.

Here, weak convergence means that:

$$\lim_{n\to\infty} \left\langle \psi, \sum_{j=1}^{n} M(B_j)\,\psi \right\rangle = \left\langle \psi, \sum_{j=1}^{\infty} M(B_j)\,\psi \right\rangle,$$

for all $\psi \in \mathcal{H}$.

If the measurable outcome space $(\mathbb{X}, \mathcal{X})$ is finite in POVM $M : \mathcal{X} \to \mathcal{L}(\mathcal{H})$, it is sufficient to define POVM $M : \mathbb{X} \to \mathcal{L}(\mathcal{H})$ on $\mathbb{X}$ and extend the definition to $\mathcal{X}$ as follows: for any $A \in \mathcal{X}$, define $M(A) = \sum_{x\in A} M(x)$. For instance, if $\mathbb{X} = \{0, 1\}$, then we only need to define $M(0)$ and $M(1)$. In the finite case, $M : \mathbb{X} \to \mathcal{L}(\mathcal{H})$ should satisfy the following conditions: $M(x) \geq 0\ \forall x \in \mathbb{X}$ and $\sum_x M(x) = \text{Id}$.

When one applies the measurement $M : \mathcal{X} \to \mathcal{L}(\mathcal{H})$ to the quantum system in the state ρ, the probability of obtaining the event $B \in \mathcal{X}$ is given by

$$P(B \mid \rho) = \text{Tr}\big\{M(B)\,\rho\big\}.$$

Note that since $\text{Tr}\{\rho\} = 1$ and $M(B) \geq 0$ for all $B \in \mathcal{X}$, we have $P(B|\rho) \geq 0$ for all $B \in \mathcal{X}$ and $P(\mathbb{X}|\rho) = 1$. Moreover, for any disjoint collection $\{B_j\}$ of events, by (3) we have

$$P(B \mid \rho) = \sum_j P(B_j \mid \rho),$$

where $B := \bigcup_j B_j$. Hence, $P(\cdot|\rho)$ is indeed a probability measure on $\mathbb{X}$. We note that if ρ is a pure state given by a unit vector u, then

$$P(B \mid \rho) = \langle u, M(A)\,u\rangle.$$

In quantum physics, a compound of N quantum systems with the underlying Hilbert spaces $\{\mathcal{H}_i, i = 1, \ldots, N\}$ is represented by the tensor product $\mathcal{H}_1 \otimes \mathcal{H}_2 \otimes \ldots \otimes \mathcal{H}_N$ of the individual Hilbert spaces. Therefore, any state ρ (called the *compound state*) of this compound quantum system is an element of $\mathcal{D}(\mathcal{H}_1 \otimes \mathcal{H}_2 \ldots \otimes \mathcal{H}_N)$. If the state ρ is in product form

$$\rho = \rho_1 \otimes \ldots \otimes \rho_N,$$

then it is called a *product state*. In this case, the individual quantum systems are said to be independent. If ρ is a mixture of product states, it is called *separable*. If the state is not separable, then it is called *entangled*.

Quantum-Correlated Strategic Measures With the above definitions, we can now define quantum-correlated strategic measures. A measure P is a *quantum-correlated* strategic measure if agents have access to a part of a quantum compound state $\rho \in \mathcal{D}(\mathcal{H}_1 \otimes \mathcal{H}_2 \ldots \otimes \mathcal{H}_N)$, where $\{\mathcal{H}_i, i = 1, \ldots, N\}$ is a collection of arbitrary Hilbert spaces, and for each $i = 1, \ldots, N$, DM i makes measurements $M^{i,y^i} : \mathbb{U}^i \to \mathcal{L}(\mathcal{H}_i)$ on the ith part of the state ρ depending on its observation y^i to generate its action u^i as the output of the measurement (recall that $\mathcal{U}^i$ is the Borel σ-algebra on the action space $\mathbb{U}^i$ of DM i); that is, the conditional distribution $P(d\mathbf{u}|\omega_0, \mathbf{y})$ is of the following form:

$$P(d\mathbf{u}|\omega_0, \mathbf{y}) = \mathrm{Tr}\left\{\left(M^{1,y^1}(du^1) \otimes \ldots \otimes M^{N,y^N}(du^N)\right)\rho\right\}.$$

Let $L_Q(\mu)$ denote the set of quantum-correlated strategic measures. We will show that $L_C(\mu)$ is a strict subset of $L_Q(\mu)$. In view of the discussion in the previous paragraph, the strategic measure $P \in L_Q(\mu) \setminus L_C(\mu)$ if the shared state ρ is entangled.

As an example, consider the two-agent team problem with $\mathbb{Y}^1 = \mathbb{Y}^2 = \mathbb{U}^1 = \mathbb{U}^2 = \{0, 1\}$. Suppose that Ω_0 is degenerate. Let $\mathcal{H}_1 = \mathcal{H}_2 = \mathbb{C}^2$. Consider the following states on $\mathbb{C}^2 \times \mathbb{C}^2$:

$$\sigma = \frac{1}{2}E_{1,1} \otimes E_{1,1} + \frac{1}{2}E_{2,2} \otimes E_{2,2}$$

$$\tau = \frac{1}{2}E_{1,1} \otimes E_{1,1} + \frac{1}{2}E_{1,2} \otimes E_{1,2} + \frac{1}{2}E_{2,1} \otimes E_{2,1} + \frac{1}{2}E_{2,2} \otimes E_{2,2},$$

where $E_{a,b}$ is a matrix on $\mathbb{C}^2$ such that $E_{a,b}(i, j) = 1_{\{(i,j)=(a,b)\}}$, for $a, b \in \{1, 2\}$. Note that $E_{1,1}, E_{2,2} \in \mathcal{D}(\mathbb{C}^2)$. Hence, σ is a mixture of product states, and thus it is separable. However, note that $E_{1,2}, E_{2,1} \notin \mathcal{D}(\mathbb{C}^2)$. Therefore, τ is an entangled state (indeed it is maximally entangled state). For any angle θ, let us define

$$\Pi_\theta = \begin{pmatrix} \cos^2(\theta) & \cos(\theta)\sin(\theta) \\ \cos(\theta)\sin(\theta) & \sin^2(\theta) \end{pmatrix} = \begin{pmatrix}\cos(\theta) & \sin(\theta)\end{pmatrix} \begin{pmatrix}\cos(\theta) \\ \sin(\theta)\end{pmatrix}.$$

Using Π_θ, we now define two collections of measurements, $\{M^{1,y^1} : \mathbb{U}^1 \to \mathcal{L}(\mathbb{C}^2); y^1 \in \mathbb{Y}^1\}$, $\{M^{2,y^2} : \mathbb{U}^2 \to \mathcal{L}(\mathbb{C}^2); y^2 \in \mathbb{Y}^2\}$ for DM 1 and DM 2, respectively, as follows:

$$M^{1,0}(0) = \Pi_0, M^{1,0}(1) = \Pi_{\pi/2};\ M^{1,1}(0) = \Pi_{\pi/4}, M^{1,1}(1) = \Pi_{3\pi/4}$$

and

$$M^{2,0}(0) = \Pi_{\pi/8}, M^{2,0}(1) = \Pi_{5\pi/8}; \ M^{2,1}(0) = \Pi_{7\pi/8}, M^{2,1}(1) = \Pi_{3\pi/8}.$$

Namely, if the observation of DM 1 is y^1, then he applies the measurement M^{1,y^1} : $u^1 \mapsto M^{1,y^1}(u^1)$ to the quantum state. Similarly, if the observation of DM 2 is y^2, then he applies the measurement $M^{2,y^2} : u^2 \mapsto M^{2,y^2}(u^2)$ to the quantum state.

Now, if we apply these measurements to the quantum states σ and τ, we obtain two quantum-correlated strategic measures P_σ and P_τ, respectively. The conditional distributions of these quantum-correlated strategic measures on actions given observations can be computed as follows:

$$\begin{aligned}
&P_\sigma(u^1, u^2|y^1, y^2) \\
&= \frac{1}{2}\Big(\mathrm{Tr}\{M^{1,y^1}(u^1)E_{1,1}\}\mathrm{Tr}\{M^{2,y^2}(u^2)E_{1,1}\} + \mathrm{Tr}\{M^{1,y^1}(u^1)E_{2,2}\}\mathrm{Tr}\{M^{2,y^2}(u^2)E_{2,2}\}\Big) \\
&P_\tau(u^1, u^2|y^1, y^2) \\
&= \frac{1}{2}\Big(\mathrm{Tr}\{M^{1,y^1}(u^1)E_{1,1}\}\mathrm{Tr}\{M^{2,y^2}(u^2)E_{1,1}\} + \mathrm{Tr}\{M^{1,y^1}(u^1)E_{1,2}\}\mathrm{Tr}\{M^{2,y^2}(u^2)E_{1,2}\}\Big) \\
&+ \frac{1}{2}\Big(\mathrm{Tr}\{M^{1,y^1}(u^1)E_{2,1}\}\mathrm{Tr}\{M^{2,y^2}(u^2)E_{2,1}\} + \mathrm{Tr}\{M^{1,y^1}(u^1)E_{2,2}\}\mathrm{Tr}\{M^{2,y^2}(u^2)E_{2,2}\}\Big).
\end{aligned}$$

Since $E_{1,1}, E_{2,2} \in \mathcal{D}(\mathbb{C}^2)$, the conditional distribution $P_\sigma(u^1, u^2|y^1, y^2)$ is a mixture of the following independent conditional probability measures

$$P_{1,\sigma}(u^1|y^1)P_{1,\sigma}(u^2|y^2) := \mathrm{Tr}\{M^{1,y^1}(u^1)E_{1,1}\}\mathrm{Tr}\{M^{2,y^2}(u^2)E_{1,1}\}$$

and

$$P_{2,\sigma}(u^1|y^1)P_{2,\sigma}(u^2|y^2) := \mathrm{Tr}\{M^{1,y^1}(u^1)E_{2,2}\}\mathrm{Tr}\{M^{2,y^2}(u^2)E_{2,2}\}.$$

Hence P_σ can be classically realized via common randomness; that is, $P_\sigma \in L_C(\mu)$. However, this is not true for P_τ since $E_{1,2}, E_{2,1} \notin \mathcal{D}(\mathbb{C}^2)$, and thus

$$\mathrm{Tr}\{M^{1,y^1}(u^1)E_{1,2}\}\mathrm{Tr}\{M^{2,y^2}(u^2)E_{1,2}\}$$

and

$$\mathrm{Tr}\{M^{1,y^1}(u^1)E_{2,1}\}\mathrm{Tr}\{M^{2,y^2}(u^2)E_{2,1}\}$$

are not probability measures. Therefore, P_τ cannot be realized classically; that is, $P_\tau \in L_Q(\mu) \setminus L_C(\mu)$.

Here, one can view the following collection as the joint strategy of DMs:

$$\begin{aligned}
&\{\mathcal{H}_i \ (i \in \mathcal{N}), \rho \in \mathcal{D}(\mathcal{H}_1 \otimes \mathcal{H}_2 \ldots \otimes \mathcal{H}_N), \\
&\{M^{i,y^i} : \mathcal{U}^i \to \mathcal{L}(\mathcal{H}_i); y^i \in \mathbb{Y}^i\} \ (i \in \mathcal{N})\Big\}.
\end{aligned}$$

Namely, at the outset, DMs agree on the Hilbert spaces $\mathcal{H}_i$ ($i \in \mathcal{N}$) and share a compound state $\rho \in \mathcal{D}(\mathcal{H}_1 \otimes \mathcal{H}_2 \ldots \otimes \mathcal{H}_N)$ (i.e., an entangled state) to be used in the measurements. Then, they choose decentralized measurements $\{M^{i,y^i} : \mathcal{U}^i \to \mathcal{L}(\mathcal{H}_i); y^i \in \mathbb{Y}^i\}$ ($i \in \mathcal{N}$) without communicating with each other.

Let $L_{Q(d)}(\mu) \subset L_Q(\mu)$ denote the set of quantum-correlated strategic measures where each agent has access to a part of the quantum compound state $\rho \in \mathcal{D}(\mathcal{H}_1 \otimes \mathcal{H}_2 \ldots \otimes \mathcal{H}_N)$ with $\dim(\mathcal{H}_i) \leq d$ for all $i \in \mathcal{N}$.

In the following, we state convexity and some topological properties of the sets $L_Q(\mu)$ and $L_{Q(d)}(\mu)$. We refer the reader to [621] for the following result and for further details.

Theorem 4.3.2

(i) $L_Q(\mu)$ *is a convex set but it is not closed under the weak convergence topology.*
(ii) $L_{Q(d)}(\mu)$ *is not convex if* $L_C(\mu) \not\subset L_{Q(d)}(\mu)$. *In particular,* $L_{Q(1)}(\mu)$ *is not convex and* $L_{Q(2)}(\mu)$ *is not convex for certain models.*
(iii) For each d, $L_{Q(d)}(\mu)$ *is closed under the weak convergence topology if the observation spaces are finite and action spaces are compact.*
(iv) $L_Q(\mu)$ *is a strict super-set of* $\bigcup_d L_{Q(d)}(\mu)$.

The following result is a direct corollary of Theorem 4.3.2-(iii).

Corollary 4.3.1 *Consider a static team decision problem with compact action spaces, and finite measurement spaces (and hence with an independent measurements reduction). Then, the team decision problem* $\inf_{P \in L_{Q(d)}(\mu)} J(P)$ *has an optimal team policy.*

4.3.3 Relations Between Convex Relaxations

Up to this point we have introduced a number of convex strategic measures, we now present a comparison. First, we note that the joint strategies in the set $L_C(\mu)$ indeed correspond to the *local hidden variable* correlations in quantum mechanics. In [232], Einstein, Podolsky, and Rosen have made the point that due to the probabilistic nature of quantum mechanics, the theory cannot be complete and should be supplemented by a local hidden variable to describe the probabilistic nature, where the local hidden variable describes the uncertainties in the experimental setup. In other words, they claimed that $L_C(\mu) = L_Q(\mu)$. However, John Bell, in his influential paper [106], constructed a quantum mechanical system whose statistical behavior cannot be explained via the local hidden variable theory; that is, $L_Q(\mu)$ is a strict super-set of $L_C(\mu)$. This result implies that quantum mechanical systems are intrinsically probabilistic and cannot be explained by the lack of precision in experimental measurements. Experimental verification of Bell's prediction has been established by Alain Aspect in [41]. Additionally, in [588], Popescu and Rohrlich also discussed that correlations (strategic measures in our setting) achieved by quantum mechanical systems can be as large as $L_{NS}(\mu)$ which itself is larger than

$L_Q(\mu)$. However, no experimental verification has been provided for this claim as yet, to our knowledge.

We also note that the lower bounds on the optimal team costs achieved by classical strategies $L_C(\mu)$ are known as *Bell inequalities* in the quantum information theory literature. Bell inequality violations correspond to quantum-correlated strategies that improve the classical optimal team cost [181]. Identifying Bell inequality violations is an active area of research in quantum information theory. One such violation has been established in the CHSH team, to be discussed later. Since the considered sets are convex, by the Hahn-Banach Separating Hyperplanes Theorem [489],[2] identifying Bell inequality violations is equivalent to finding a *separating* function in an inner-product form. Accordingly, identifying Bell inequality violations would be implied by finding a cost function for which the optimal quantum-correlated team cost is strictly smaller than the optimal classical team cost, that is,

$$\inf_{P\in L_Q(\mu)} \int P(ds)\, c(s) < \inf_{P\in L_C(\mu)} \int P(ds)\, c(s).$$

Since both $L_Q(\mu)$ and $L_C(\mu)$ are convex sets, the above inequality states that the hyperplane $\{P : \int P(ds)\, c(s) = J^*\}$ separates the convex sets $L_Q(\mu)$ and $L_C(\mu)$.

Before we establish the relationship between different convex relaxations, we introduce a further information structure, denoted by L_{CJ}, which allows for the common random variable z to be correlated with ω_0 and $\mathbf{y}$; in this case, the information structure is essentially centralized, since the dependence between z and the exogenous variables is arbitrary:

$$\begin{aligned} L_{CJ}(\mu) := \Bigg\{ & P \in \mathcal{P}\Big(\Omega_0 \times \prod_{k=1}^{N} (\mathbb{Y}^k \times \mathbb{U}^k)\Big) : \\ & P(B) = \int_{[0,1]^N \times B^0 \times \prod_{k=1}^{N} A^k} \mu(dz, d\omega_0, d\mathbf{y}) \prod_k 1_{\{u^k = \gamma^k(y^k, z) \in B^k\}}, \\ & \gamma^k(\cdot, z) \in \Gamma^k,\, B \in \mathcal{B}(\Omega_0 \times \prod_k (\mathbb{Y}^k \times \mathbb{U}^k)) \Bigg\}, \end{aligned}$$

where the marginal of $\omega_0, \mathbf{y}$ is fixed as μ.

Theorem 4.3.3 *We have the following relationship between different convex relaxations.*

(i) $L_C(\mu) \subset L_Q(\mu) \subset L_{NS}(\mu) \subset L_M(\mu) \subset L_{CJ}(\mu)$.
(ii) The sets $L_C(\mu), L_Q(\mu), L_{NS}(\mu), L_M(\mu), L_{CJ}(\mu)$ *are convex.*

---

[2] This reasoning will be utilized with a detailed analysis in the proof of Theorem 7.3.4 later in the book.

(iii) The inclusions among the convex sets in (i) are strict.
(iv) There exist problems for which

$$\inf_{P \in L_Q(\mu)} \int P(ds)\, c(s) < \inf_{P \in L_C(\mu)} \int P(ds)\, c(s)$$

$$\inf_{P \in L_{NS}(\mu)} \int P(ds)\, c(s) < \inf_{P \in L_Q(\mu)} \int P(ds)\, c(s)$$

and in particular

$$\inf_{P \in L_{NS}(\mu)} \int P(ds)\, c(s) < \inf_{P \in L_C(\mu)} \int P(ds)\, c(s).$$

One implication of the above is that the convex programs

$$\inf_{P \in L_Q(\mu)} \int P(ds)\, c(s)\ , \quad \inf_{P \in L_{NS}(\mu)} \int P(ds)\, c(s) \text{ or } \inf_{P \in L_M(\mu)} \int P(ds)\, c(s),$$

provide lower bounds to the original problem. In particular, since the optimization problem associated with non-signaling strategic measures can be written as a linear program, it can be solved in polynomial time if the measurement and action spaces are finite.

Proof

(ii) We have already proved the convexity of $L_C(\mu)$ (in Theorem 4.1.1), $L_Q(\mu)$ (in Theorem 4.3.2), $L_{NS}(\mu)$ (in Theorem 4.3.1), and $L_M(\mu)$ (also in Theorem 4.3.1).

(i) Let $P \in L_R(\mu)$; that is,

$$P(d\mathbf{u}|\omega_0, \mathbf{y}) = \prod_{i=1}^{N} \Pi^i(du^i|y^i).$$

Let $\mathcal{H}$ be a Hilbert space with dimension 1. Let $e_w \in \mathcal{H}$ with unit norm. We define

$$\rho = e_w e_w^* \otimes \ldots \otimes e_w e_w^*$$

$$M^{y^i}(A) = \Pi^i(A|y^i)\, e_w e_w^*$$

for all $y^i \in \mathbb{Y}^i$, $A \in \mathcal{B}(\mathbb{U}^i)$, and $i = 1, \ldots, N$. Then, we have

$$P(d\mathbf{u}|\omega_0, \mathbf{y}) = \mathrm{Tr}\Big\{ \Big(M^{y^1}(du^1) \otimes \ldots \otimes M^{y^N}(du^N)\Big)\, \rho \Big\}.$$

This implies that $L_R(\mu) \subset L_Q(\mu)$. Since the convex hull of $L_R(\mu)$ is $L_C(\mu)$ and $L_Q(\mu)$ is convex, we have $L_C(\mu) \subset L_Q(\mu)$.

Let $P \in L_Q(\mu)$; that is, agents have access to a part of a compound quantum state $\rho \in \mathcal{D}(\mathcal{H}_1 \otimes \mathcal{H}_2 \otimes \ldots \otimes \mathcal{H}_N)$, where $\{\mathcal{H}_i, i = 1, \ldots, N\}$ is a collection of arbitrary Hilbert spaces, and, for each $i = 1, \ldots, N$, DM i makes measurements $M^{y^i} : \mathcal{U}^i \to \mathcal{L}(\mathcal{H}_i)$ on the ith part of the state ρ depending on its observations y^i to generate its action as the output of the following measurement:

$$P(d\mathbf{u}|\omega_0, \mathbf{y}) = \mathrm{Tr}\{ \left(M^{y^1}(du^1) \otimes \ldots \otimes M^{y^N}(du^N)\right) \rho\}.$$

We now prove that P satisfies the non-signaling condition. Fix any $k \in \{1, \ldots, N\}$. Then, we have

$$\begin{aligned}
&\int_{\mathbb{U}^k} P(d\mathbf{u} \,|\, \omega_0, y^1, \ldots, y^k, \ldots, y^N) \\
&\quad = \int_{\mathbb{U}^k} \mathrm{Tr}\{ \left(M^{y^1}(du^1) \otimes \ldots \otimes M^{y^k}(du^k) \otimes \ldots \otimes M^{y^N}(du^N)\right) \rho\} \\
&\quad = \mathrm{Tr}\left\{\left(\int_{\mathbb{U}^k} M^{y^1}(du^1) \otimes \ldots \otimes M^{y^k}(du^k) \otimes \ldots \otimes M^{y^N}(du^N)\right)\rho\right\} \\
&\quad = \mathrm{Tr}\left\{\left(M^{y^1}(du^1) \otimes \ldots \otimes \mathrm{Id} \otimes \ldots \otimes M^{y^N}(du^N)\right) \rho\right\} \\
&\quad = \int_{\mathbb{U}^k} \mathrm{Tr}\{ \left(M^{y^1}(du^1) \otimes \ldots \otimes M^{\hat{y}^k}(du^k) \otimes \ldots \otimes M^{y^N}(du^N)\right) \rho\} \\
&\quad = \int_{\mathbb{U}^k} P(d\mathbf{u} \,|\, \omega_0, y^1, \ldots, \hat{y}^k, \ldots, y^N)
\end{aligned}$$

for all $y^1, \ldots, y^{k-1}, y^k, \hat{y}^k, y^{k+1}, \ldots, y^N$ and ω_0. Therefore, P is in $L_{NS}(\mu)$.

(iii) and (iv): We prove (iii) and (iv) by providing a specific example. This example is known as the *Clauser-Horne-Shimony-Holt (CHSH) team problem* [180]. In the CHSH team, we have two agents with binary measurement and action spaces $\{0, 1\}$. The observations are generated independently and uniformly. Hence, there is no state variable in the problem, and thus the problem is automatically static. There is a reward function r which is to be maximized by the team (equivalently $c = -r$ is to be minimized); it is defined by

$$r(y^1, y^2, u^1, u^2) = \begin{cases} 1, & \text{if } u^1 \oplus u^2 = y^1 \cdot y^2 \\ 0, & \text{otherwise.} \end{cases}$$

For this problem, randomized strategic measures $L_C(\mu)$ can achieve the reward level of 0.5 [181]. Quantum-correlated strategic measures can achieve the maximum reward of $\sqrt{2}/2$ [761, Section 6.3.2]. However, non-signaling policies can achieve the maximum reward of 1 using the following policy, which is called the Popescu-Rohrlich (PR) box in the literature [588]:

$$P(u^1, u^2|y^1, y^2) = \begin{cases} 1/2, & \text{if } u^1 \oplus u^2 = y^1 \cdot y^2 \\ 0, & \text{otherwise} \end{cases}$$

It is easy to show that P is non-signaling; that is, u^i is independent of y^j given y^i, for $i, j = 0, 1$ and $i \neq j$. The reward of P is 1, which is the maximum achievable reward by any policy as $0 \leq r \leq 1$. Hence, P is the optimal non-signaling policy. Therefore, for the CHSH team, we have

$$\sup_{P \in L_{NS}(\mu)} \int P(ds)\, r(s) > \sup_{P \in L_Q(\mu)} \int P(ds)\, r(s) > \sup_{P \in L_C(\mu)} \int P(ds)\, r(s)$$

that is, non-signaling policies improve the optimal team performance of quantum-correlated policies, and quantum correlated policies improve the optimal team cost of randomized policies. ◇

Note that to be able to implement the optimal non-signaling policy in the CHSH team, the agents should communicate their observations to a mediator, and then the mediator directs them to apply either the same actions or different actions based on the product of their observations. This kind of communication is, in general, infeasible for team decision problems. Therefore, although allowing non-signaling correlations among actions of agents enables us to formulate the team problem as a linear program (solvable in polynomial time for the finite case), it is in general not realistic to assume that agents can use such policies in real-life applications due to communication constraints dictated by the decentralized information structure. Therefore, non-signaling relaxation can only be used to provide a lower bound to the original team problem.

However, quantum correlated relaxation is indeed an *admissible* extension or relaxation of the classical team problem since it does not require any communication between the agents and a mediator. Hence, obtaining a solution or an approximate solution (as explained in the previous subsection) of a quantum correlated team problem will likely be a significant contribution to team decision theory.

4.4 Near Optimality of Finite Approximations in Decentralized Stochastic Control via Quantization of Measurements and Actions

In this section, we consider finite approximations of static and static-reducible dynamic team problems. Since results of this section will also be applied to static reduction of dynamic teams, as noted in Remark 4.2.1, we assume that the loss/cost function c also takes the measurements $\mathbf{y}$ as an argument (which was not the case in the original problem formulation). Recall that, in the independent static reduction of a dynamic team in (3.31), the reduced cost function c_s is a function of ω_0, $\mathbf{u}$, and $\mathbf{y}$. To obtain finite approximation results, the following assumption is imposed on the components of the model.

Assumption 4.4.1

(a) *The loss function c is continuous in* $(\mathbf{u}, \mathbf{y})$ *for any fixed* ω_0. *In addition, it is bounded on any compact subset of* $\Omega_0 \times \prod_{k=1}^{N}(\mathbb{Y}^k \times \mathbb{U}^k)$.
(b) *For each k,* $\mathbb{U}^k$ *is a closed and convex subset of a completely metrizable locally convex vector space.*
(c) *For each k,* $\mathbb{Y}^k$ *is locally compact.*
(d) *For any subset G of* $\prod_{k=1}^{N}\mathbb{U}^k$, *the function* $w_G(\omega_0, \mathbf{y}) := \sup_{\mathbf{u}\in G} c(\omega_0, \mathbf{y}, \mathbf{u})$ *is integrable with respect to* $\mu(d\omega_0, d\mathbf{y})$, *for any compact subset G of* $\prod_{k=1}^{N}\mathbb{U}^k$ *of the form* $G = \prod_{k=1}^{N} G^k$.
(e) *For any admissible (in the sense of Sect. 3.2)* $\underline{\gamma} \in \mathbf{\Gamma}$ *with* $J(\underline{\gamma}) < \infty$ *and each k, there exists* $u^{k,*} \in \mathbb{U}^k$ *such that* $J(\underline{\gamma}^{-k}, \gamma^k_{u^{k,*}}) < \infty$, *where* $\gamma^k_{u^{k,*}} \equiv u^{k,*}$.

Note that Assumption 4.4.1-(d),(e) hold if the cost function is bounded. Indeed, conditions in Assumption 4.4.1 are quite mild and hold for Witsenhausen's counterexample studied in Sect. 3.3.

In what follows, for any subset G of $\prod_{k=1}^{N}\mathbb{U}^k$, we let

$$\mathbf{\Gamma}_G := \left\{\underline{\gamma} \in \mathbf{\Gamma} : \underline{\gamma}\left(\prod_{k=1}^{N}\mathbb{Y}^k\right) \subset G\right\}$$

and $\mathbf{\Gamma}_{c,G} := \mathbf{\Gamma}_c \cap \mathbf{\Gamma}_G$, where $\mathbf{\Gamma}_c$ denotes the set of continuous strategies. Using these definitions, let us define the following set of strategic measures for any subset G of $\prod_{k=1}^{N}\mathbb{U}^k$:

$$L_A^G(\mu)$$
$$:= \left\{P \in L_A(\mu) : P(B) = \int_{B^0 \times \prod_{k=1}^{N} A^k} \mu(d\omega_0, d\mathbf{y}) \prod_{k=1}^{N} 1_{\{u^k = \gamma^k(y^k) \in B^k\}},\ \underline{\gamma} \in \mathbf{\Gamma}_G\right\}.$$

Let $L_A^{G,c}(\mu)$ denote the set of strategic measures in $L_A^G(\mu)$ induced by continuous policies.

The following result states that there exists a near optimal strategic measure whose support on the product of action spaces, $\prod_{k=1}^{N}\mathbb{U}^k$, is a convex and compact (and thus bounded) subset G of it, and conditional distributions of actions given measurements are induced by continuous policies.

Proposition 4.4.1 *Suppose that Assumption 4.4.1 holds. Then, for any* $\varepsilon > 0$, *there exists a compact subset G of* $\prod_{k=1}^{N}\mathbb{U}^k$ *of the form* $G = \prod_{i=1}^{N} G^i$, *where each* G^i *is convex and compact, such that*

$$\inf_{P \in L_A^{G,c}(\mu)} \int P(ds)\, c(s) < J^* + \varepsilon.$$

Proof Given any strategic measure, using Assumption 4.4.1-(e) and the fact that every measure on a Borel space is tight [574, Theorem 3.2], one can construct a strategic measure in $L_A(\mu)$ whose support on the product of action spaces is convex and compact, and whose cost is $(\varepsilon/2)$-close to the cost of the given strategic measure.

For the new strategic measure, since it has a convex and compact support on the product of action spaces, using Lusin's theorem (see Theorem C.5.1), we can construct a strategic measure induced by continuous policies whose cost function is $\varepsilon/2$-close to the cost of bounded support strategic measure. The proof can be completed by combining these two results. ◇

Since each $\mathbb{Y}^i$ is a locally compact separable metric space, there exists an increasing sequence of compact subsets $\{K_l^i\}$ such that $K_l^i \subset \operatorname{int} K_{l+1}^i$ and $\mathbb{Y}^i = \bigcup_{l=1}^{\infty} K_l^i$ [16, Lemma 2.76], where int D denotes the interior of the set D.

Let d_i denote the metric on $\mathbb{Y}^i$. For each $l \geq 1$, let $\mathbb{Y}_{l,n}^i := \{y_{i,1}, \ldots, y_{i,i_{l,n}}\} \subset K_l^i$ be a $(1/n)$-net in K_l^i. Recall that if $\mathbb{Y}_{l,n}^i$ is a $(1/n)$-net in K_l^i, then for any $y \in K_l^i$ we have $\min_{z \in \mathbb{Y}_{l,n}^i} d_i(y, z) < \frac{1}{n}$. For each l and n, let $q_{l,n}^i : K_l^i \to \mathbb{Y}_{l,n}^i$ be a nearest neighborhood quantizer given by

$$q_{l,n}^i(y) = \arg\min_{z \in \mathbb{Y}_{l,n}^i} d_i(y, z),$$

where ties are broken so that $q_{l,n}^i$ is measurable. If $K_l^i = [-M, M] \subset \mathbb{Y}^i = \mathbb{R}$ for some $M \in \mathbb{R}_+$, the finite set $\mathbb{Y}_{l,n}^i$ can be chosen such that $q_{l,n}^i$ becomes a uniform quantizer. We let $Q_{l,n}^i : \mathbb{Y}^i \to \mathbb{Y}_{l,n}^i$ denote the extension of $q_{l,n}^i$ to $\mathbb{Y}^i$ given by

$$Q_{l,n}^i(y) := \begin{cases} q_{l,n}^i(y), & \text{if } y \in K_l^i, \\ y_{i,0}, & \text{otherwise,} \end{cases}$$

where $y_{i,0} \in \mathbb{Y}^i$ is some auxiliary element.

Define $\Gamma_{l,n}^i = \Gamma^i \circ Q_{l,n}^i \subset \Gamma^i$; that is, $\Gamma_{l,n}^i$ is defined to be the set of all strategies $\tilde{\gamma}^i \in \Gamma^i$ of the form $\tilde{\gamma}^i = \gamma^i \circ Q_{l,n}^i$, where $\gamma^i \in \Gamma^i$. Define also $\mathbf{\Gamma}_{l,n} := \prod_{i=1}^N \Gamma_{l,n}^i \subset \mathbf{\Gamma}$. Note that, for any $i = 1, \ldots, N$, $\Gamma_{l,n}^i$ is the set of policies for DM i which can only use the output levels of the quantizer $Q_{l,n}^i$. In other words, in addition to measurement channel $g^i(dy^i|\omega_0)$ between DM i and the Nature, there is also an analog-to-digital converter (quantizer) between them.

Using these definitions, let us define the following set of strategic measures for any l and n:

$$L_A^{l,n}(\mu)$$

$$:= \left\{ P \in L_A(\mu) : P(B) = \int_{B^0 \times \prod_{k=1}^N A^k} \mu(d\omega_0, d\mathbf{y}) \prod_{k=1}^N 1_{\{u^k = \gamma^k(y^k) \in B^k\}}, \ \underline{\gamma} \in \mathbf{\Gamma}_{l,n} \right\}.$$

The following theorem now states that an optimal (or almost optimal) strategic measure can be approximated with arbitrarily small approximation error for the induced costs by strategic measures in $L_A^{l,n}(\mu)$ for sufficiently large l and n.

Theorem 4.4.1 ([625]) *For any $\varepsilon > 0$, there exist $(l, n(l))$, a compact subset G of $\prod_{k=1}^N \mathbb{U}^k$ of the form $G = \prod_{i=1}^N G^i$, where each G^i is convex and compact, and $P \in L_A^{l,n(l)}(\mu) \bigcap L_A^G(\mu)$ such that*

$$\int P(ds)\, c(s) < J^* + \varepsilon$$

Proof Fix any strategic measure $P \in L_A(\mu)$. Then, by Proposition 4.4.1, there exists a strategic measure $P_G \in L_A^{G,c}(\mu)$ for some compact subset G of $\prod_{k=1}^N \mathbb{U}^k$ of the form $G = \prod_{i=1}^N G^i$, where each G^i is convex and compact, such that

$$\int P_G(ds)\, c(s) \le \int P(ds)\, c(s) + \varepsilon/2.$$

Let $\underline{\gamma} \in \mathbf{\Gamma}_{G,c}$ be the strategy that induces P_G. Then define $\gamma_{l,n}^i = \gamma^i \circ Q_{l,n}^i$ for each $i = 1, \ldots, N$. Note that on any compact subset of $\mathbb{Y}^i$, we have $\gamma_{l,n}^i \to \gamma^i$ as γ^i is continuous. Let $P_{l,n}$ the strategic measure induced by the strategy $(\gamma_{l,n}^1, \ldots, \gamma_{l,n}^N)$.

Note that since $\mu(d\omega_0, d\mathbf{y})$ is tight, for any $\delta > 0$, there exists a compact subset K of $\prod_{k=1}^N \mathbb{Y}^k$ of the form $K = \prod_{i=1}^N K^i$, where each K^i is compact, such that $\mu(\Omega_0 \times K) \ge 1 - \delta$. By choosing a sequence of $\{\delta(l, n)\}$ and the corresponding compact sets $\{K_{l,n}\}$ appropriately, we can prove that

$$\lim_{(l,n)\to\infty} \int P_{l,n}(ds)\, c(s) = \int P_G(ds)\, c(s),$$

as $\gamma_{l,n}^i \to \gamma^i$ on each compact set $K_{l,n}$ and $\mu(\Omega_0 \times K_{l,n}) \ge 1 - \delta(l, n)$. Hence, there exists $(l, n(l))$, such that

$$\int P_{l,n(l)}(ds)\, c(s) \le \int P_G(ds)\, c(s) + \varepsilon/2.$$

This completes the proof. ◇

The above result implies that to compute a near optimal strategic measure for the team problem it is sufficient to consider quantized measurements $\left(Q_{l,n}^1(y^1), \ldots, Q_{l,n}^N(y^N)\right)$ for sufficiently large l and n. Furthermore, this nearly optimal strategic measure can have a compact support of the form $G = \prod_{i=1}^{N} G^i$ on $\prod_{k=1}^{N} \mathbb{U}^k$, where G^i is convex and compact for each $i = 1, \ldots, N$.

For ease of reference, we define

$$L_{A(q)}(\mu) = \bigcup_{(l,n)} L_A^{l,n}(\mu)$$

and introduce $L_{A(c)}(\mu)$ to be the subset of $L_A(\mu)$ defined in (4.3) where the policies are restricted to be in Γ_c.

4.4.1 Finite Measurement Approximate Models

In this subsection, for each (l, n), we define a team model with finite measurement spaces. We prove that, for sufficiently large l and n, an optimal strategic measure of the team model corresponding to (l, n) will provide a strategic measure for the original team model, which is nearly optimal. To this end, fix any (l, n). For the pair (l, n), the corresponding finite measurement team model has the following measurement spaces: $\mathbb{Z}_{l,n}^i := \{y_{i,0}, y_{i,1}, \ldots, y_{i,i_{l,n}}\}$ (i.e., the output levels of $Q_{l,n}^i$), $i \in \mathcal{N}$. The stochastic kernels $g_{l,n}^i(\,\cdot\,|\omega_0)$ from Ω_0 to $\mathbb{Z}_{l,n}^i$ denote the measurement constraints and are given by:

$$g_{l,n}^i(\,\cdot\,|\omega_0) := \sum_{j=0}^{i_{l,n}} g(S_{i,j}^{l,n}|\omega_0)\, \delta_{y_{i,j}}(\,\cdot\,),$$

where $S_{i,j}^{l,n} := \left\{y \in \mathbb{Y}^i : Q_{l,n}^i(y) = y_{i,j}\right\}$. Indeed, $g_{l,n}^i(\,\cdot\,|\omega_0)$ is the push-forward of the measure $g^i(\,\cdot\,|\omega_0)$ with respect to the quantizer $Q_{l,n}^i$.

Let $\Phi_{n,l}^i := \left\{\varphi^i : \mathbb{Z}_{l,n}^i \to \mathbb{U}^i, \varphi^i \text{ measurable}\right\}$ denote the set of measurable policies for DM i and let $\mathbf{\Phi}_{l,n} := \prod_{i=1}^{N} \Phi_{l,n}^i$. The cost function of this team model is $J_{l,n} : \mathbf{\Phi}_{l,n} \to \mathbb{R}_+$, defined by

$$J_{l,n}(\underline{\varphi}) := \int_{\Omega_0 \times \prod_{i=1}^{N} \mathbb{Z}_{l,n}^i} c(\omega_0, \mathbf{y}, \mathbf{u})\, P_{l,n}(d\omega_0, d\mathbf{y}),$$

where $\underline{\varphi} = (\varphi^1, \ldots, \varphi^N)$, $\mathbf{u} = \underline{\varphi}(\mathbf{y})$, and

$$P_{l,n}(d\omega_0, d\mathbf{y}) = P(d\omega_0) \prod_{i=1}^{N} g_{l,n}^i(dy^i|\omega_0) =: \mu_{l,n}(d\omega_0, d\mathbf{y}).$$

For any compact subset G of $\prod_{k=1}^{N} \mathbb{U}^k$, we also define $\mathbf{\Phi}_{l,n}^{G} := \{\underline{\varphi} \in \mathbf{\Phi}_{l,n} : \underline{\varphi}(\prod_{i=1}^{N} \mathbb{Z}_{l,n}^{i}) \subset G\}$.

In order to obtain the desired approximation result, we need to impose the following additional assumption.

Assumption 4.4.2 *For any compact subset G of $\prod_{k=1}^{N} \mathbb{U}^k$ of the form $G = \prod_{i=1}^{N} G^i$, the function w_G is uniformly integrable with respect to the measures $\{\mu_{l,n}\}$; that is,*

$$\lim_{R\to\infty} \sup_{l,n} \int_{\{w_G > R\}} w_G(\omega_0, \mathbf{y})\, d\mu_{l,n} = 0.$$

This assumption is quite mild and is satisfied by the Witsenhausen's counterexample. Let $T_A(\mu_{l,n})$ denote the set of strategic measures induced by measurable policies for this team model. Let $T_A^G(\mu_{l,n})$ be the strategic measures in $T_A(\mu_{l,n})$ whose support on $\prod_{k=1}^{N} \mathbb{U}^k$ is a subset G.

The following theorem is the main result of this subsection. It states that to compute a near optimal strategic measure for the original team problem, it is sufficient to compute an optimal strategic measure for the finite measurement team problem corresponding to sufficiently large l and n.

Theorem 4.4.2 ([625]) *Suppose that Assumptions 4.4.1 and 4.4.2 hold. Then, for any $\varepsilon > 0$, there exists a pair $(l, n(l))$ and a compact subset $G = \prod_{i=1}^{N} G^i$ of $\prod_{k=1}^{N} \mathbb{U}^k$ such that an optimal (or almost optimal) strategic measure $P^{l,n(l)}$ in the set $T_A^G(\mu_{l,n(l)})$ for the $(l, n(l))$ team is ε-optimal for the original team problem when $P^{l,n(l)}$ is extended to $\Omega \times \prod_{k=1}^{N}(\mathbb{Y}^k \times \mathbb{U}^k)$ via quantizers $Q_{l,n(l)}^{i}$; that is,*

$$P_{ex}^{l,n(l)}(\cdot) = \int_{\cdot} \mu(d\omega_0, d\mathbf{y}) \prod_{k=1}^{N} 1_{\{u^k = \gamma^k \circ Q_{l,n(l)}^{k}(y^k) \in \cdot\}}$$

where

$$P^{l,n(l)}(B) = \int_{\cdot} \mu_{l,n(l)}(d\omega_0, d\mathbf{y}) \prod_{k=1}^{N} 1_{\{u^k = \gamma^k(y^k) \in \cdot\}}$$

Proof We first prove the following useful fact. Let $\{\underline{\varphi}_{l,n}\}$ be a sequence of strategies such that $\underline{\varphi}_{l,n} \in \mathbf{\Phi}_{l,n}^{G}$, where $G = \prod_{i=1}^{N} G^i$ and each G^i is convex and compact. For each l and n, define $\underline{\gamma}_{l,n} := \underline{\varphi}_{l,n} \circ Q_{l,n}$, where $Q_{l,n} := (Q_{l,n}^{1}, \ldots, Q_{l,n}^{N})$. Then, we have

$$\lim_{l,n\to\infty} |J_{l,n}(\underline{\varphi}_{l,n}) - J(\underline{\gamma}_{l,n})| = 0.$$

Indeed, let us introduce the following finite measures on $\Omega_0 \times \prod_{k=1}^N \mathbb{Y}^k$:

$$\mu_G(S) := \int_S w_G(\omega, \mathbf{y})\, d\mu,$$

$$\mu_G^{l,n}(S) := \int_S w_G(\omega, \mathbf{y})\, d\mu_{l,n}.$$

Since $\mu_{l,n}$ converges to μ weakly, by Serfozo [658, Theorem 3.5] and Assumption 4.4.2 we have $\mu_G^{l,n} \to \mu_G$ weakly as $l, n \to \infty$. Hence, the sequence $\{\mu_G^{l,n}\}$ is tight. Therefore, there exists a compact subset K of $\Omega_0 \times \prod_{k=1}^N \mathbb{Y}^k$ such that $\mu_G(K^c) < \varepsilon/2$ and $\mu_G^{l,n}(K^c) < \varepsilon/2$ for all l, n. Then, we have

$$\begin{aligned}
&|J_{l,n}(\underline{\varphi}_{l,n}) - J(\underline{\gamma}_{l,n})| \\
&= \left| \int_{\Omega_0 \times \prod_{k=1}^N \mathbb{Y}^k} c(\omega_0, Q_{l,n}(\mathbf{y}), \underline{\gamma}_{l,n}(\mathbf{y}))\, d\mu - \int_{\Omega_0 \times \prod_{k=1}^N \mathbb{Y}^k} c(\omega_0, \mathbf{y}, \underline{\gamma}_{l,n}(\mathbf{y}))\, d\mu \right| \\
&\leq \int_K \left| c(\omega_0, Q_{l,n}(\mathbf{y}), \underline{\gamma}_{l,n}(\mathbf{y})) - c(\omega_0, \mathbf{y}, \underline{\gamma}_{l,n}(\mathbf{y})) \right| d\mu \\
&\qquad + \int_{K^c} w_G(\omega_0, \mathbf{y})\, d\mu + \int_{K^c} w_G(\omega_0, \mathbf{y})\, d\mu_{l,n}.
\end{aligned}$$

The first term in the last expression (to the right of the inequality) goes to zero as $l, n \to \infty$ by the dominated convergence theorem and the fact that c is bounded and continuous in $\mathbf{y}$. The second term is less than ε by tightness. Since ε is arbitrary, this completes the proof of the fact. Note that this fact implies that for any $P^{l,n} \in T_A^G(\mu_{l,n})$, we have

$$\lim_{l,n\to\infty} \left| \int P^{l,n}(ds)\, c(s) - \int P_{\text{ex}}^{l,n}(ds)\, c(s) \right| = 0.$$

Note also that using this fact, it follows that

$$\liminf_{l,n\to\infty} \inf_{P \in T_A^G(\mu_{l,n})} \int P(ds)\, c(s) \geq J^*.$$

Since the extension of the strategic measures $T_A^G(\mu_{l,n})$ to $\Omega_0 \times \prod_{k=1}^N (\mathbb{Y}^k \times \mathbb{U}^k)$ forms the set of strategic measures $L_A^{l,n}(\mu) \bigcap L_A^G(\mu)$, by Theorem 4.4.1, we can complete the proof. Indeed, let $P_{\text{ex}}^{l,n(l)} \in L_A^{l,n}(\mu) \bigcap L_A^G(\mu)$ be the ε optimal policy for the original problem. Then we have

$$J^* + \varepsilon \geq \limsup_{l,n\to\infty} \int P_{\text{ex}}^{l,n(l)}(ds)\, c(s)$$

$$= \limsup_{l,n\to\infty} \int P^{l,n(l)}(ds)\, c(s)$$

$$\geq \limsup_{l,n\to\infty} \inf_{P\in T_A^G(\mu_{l,n})} \int P(ds)\, c(s).$$

Hence

$$J^* + \varepsilon \geq \limsup_{l,n\to\infty} \inf_{P\in T_A^G(\mu_{l,n})} \int P(ds)\, c(s) \geq \liminf_{l,n\to\infty} \inf_{P\in T_A^G(\mu_{l,n})} \int P(ds)\, c(s) \geq J^*.$$

This completes the proof. ◇

We interpret this result as saying that the space of information structures obtained by quantizing the original information structure is approximately optimal: denseness under weak convergence implies denseness of strategic measures as far as optimality is concerned.

Remark 4.4.1 Results proved in this subsection on approximation of static team problems can be directly applied to static reductions of dynamic teams if static reductions satisfy Assumption 4.4.1 and Assumption 4.4.2.

We caution the reader that the optimal expected cost is not continuous under weak convergence of information structures in general, as will be noted in Sect. 8.3. For the approximation result, however, since the approximation of the information structure through quantization only approaches the optimal solution from above in a particular way, the result is applicable.

4.4.2 Specific Results for Witsenhausen's Problem

The approximation results above hold for the counterexample of Witsenhausen studied in Sect. 3.3.2. For Witsenhausen's counterexample, the analysis follows as a corollary of the results presented above, by first showing that the static reduction of the counterexample fits into the setup presented earlier (which we commented on during the presentation of the previous subsection following Assumptions 4.4.1 and 4.4.2).

For any $l \in \mathbb{R}_+$, let $L := [-l, l]$ and let $q_{l,n}$ denote the uniform quantizer on L having n output levels; that is,

$$q_{l,n} : L \to \{y_1, \ldots, y_n\} \subset L$$

where $y_j = -l + (j - \frac{1}{2})\Delta$, $j = 1, \ldots, n$, and

$$q^{-1}(y_j) = \left[y_j - \frac{\Delta}{2}, y_j + \frac{\Delta}{2}\right),$$

where $\Delta = \frac{2l}{n}$. Let us extend $q_{l,n}$ to $\mathbb{R}$ by mapping $\mathbb{R} \setminus L$ to $y_0 = 0$. For each (l, n), let $\mathsf{Z}_{l,n} := \{y_0, y_1, \ldots, y_n\}$ (i.e., output levels of the extended $q_{l,n}$) and define the probability measure $\mathbb{P}_{l,n}$ on $\mathsf{Z}_{l,n}$ as

$$\mathbb{P}_{l,n}(y_i) = \kappa(q_{l,n}^{-1}(y_i))$$

where κ is the standard Gaussian measure $\mathcal{N}(0, 1)$. Moreover, let $\Pi_{l,n}^i := \{\pi^i : \mathsf{Z}_{l,n} \to \mathbb{U}^i, \pi^i \text{measurable}\}$ and define an approximate finite model cost as

$$J_{l,n}(\gamma_0, \gamma_1) := \sum_{j,i=0}^{n} \tilde{c}(y_i, y_j, \gamma_0(y_i), \gamma_1(y_j))\mathbb{P}_{l,n}(y_i)\mathbb{P}_{l,n}(y_j)$$

where $\tilde{c}$ is the static reduction of Witsenhausen's counterexample presented in (3.32).

We then have the following refinement of Theorem 3.3.1.

Theorem 4.4.3 *For any $\varepsilon > 0$, there exist $(l, n(l))$ and $m \in \mathbb{R}_+$ such that an optimal policy (γ_0^*, γ_1^*) in the set $\Pi_{l,n(l)}^{1,M} \times \Pi_{l,n(l)}^{2,M}$ for the cost $J_{l,n(l)}$ is ε-optimal for Witsenhausen's counterexample when (γ_0^*, γ_1^*) is extended to $\mathbb{Y}^1 \times \mathbb{Y}^2$ via $\gamma_i = \gamma_i^* \circ q_{l,n(l)}$, $i = 1, 2$, where $M := [-m, m]$ and $\Pi_{l,n(l)}^{i,M} := \{\pi^i \in \Pi_{l,n(l)}^i : \pi^i(Z_{l,n}) \subset M\}$.*

4.5 Concluding Remarks

In this chapter, we have revisited information structures within a probability theoretic and topological context. We have provided precise definitions of information structures, placed various topologies on them, and discussed closedness and compactness properties. We have presented existence and approximation results for optimal decision/control policies. We have discussed various upper and lower bounding techniques, through relaxations and convex programs, ranging from classically realizable and classically non-realizable (such as quantum and non-signaling) relaxations. Figure 4.1 has a depiction of a summary of some of the results on strategic measures.

We showed that externally provided randomness should be properly motivated, and when one introduces relaxed control; which is a common solution technique in discrete-time and continuous-time classical stochastic control, this should be done cautiously since such randomness may indeed lead to unacceptable performance.

As noted earlier, Theorem 4.2.4 (for static teams or dynamic teams with an independent-measurements reduction) and Theorems 4.2.5 and 4.2.6 (for sequential teams that do not allow an independent-measurements reduction) appear to be the most general existence results for the problems considered in this chapter. However, some slightly weaker versions of these results can still be arrived at through different methods, as laid out in the chapter and as will be presented in the next one.

We considered in this chapter only sequential decentralized stochastic control. As noted earlier, if there is a pre-defined order in which the DMs act, then we say that a system is *sequential*; otherwise, the system is non-sequential. The study of non-sequential decentralized stochastic control systems in the context of what we studied throughout this chapter remains a challenging but promising open research avenue, though all results that apply for static teams naturally hold also for those causal non-sequential systems that satisfy static reducibility conditions presented in Sect. 3.9.

4.6 Bibliographic Notes

With regard to the strategic measures approach presented here, we note that for classical stochastic control problems such measures were defined (see [230, 585, 650] and [254]) as the set of probability measures induced on the product (sequence) spaces of the state and action pairs by measurable control policies: Given an initial distribution on the state, and a policy, one can uniquely define a probability measure on the product space. Certain measurability, compactness and convexity properties of strategic measures for single decision maker problems were studied in [125, 230, 254, 585].

In the economics and game theory literatures, information structures have been studied extensively. Within the broad framework of stochastic games, stochastic team problems are also termed as *identical interest games*. In these literatures, $L_C(\mu)$ appears in the analysis of Aumann's correlated equilibrium [46]. Common and independent randomness discussions appear in the analysis of comparison of information structures [470]. For further discussions, including a multi-stage generalization known as communication equilibria, see [270]. For a detailed treatment of these issues, we refer the reader to [532].

A sufficient condition for compactness of $L_R(\mu)$ under the weak convergence topology was reported in [327]. The results in [327] also apply to static reduction of sequential dynamic teams, and a class of teams with non-compact action spaces which however have moment-type cost functions leading to a tightness condition on the set of strategic measures with a finite cost. In particular, the existence result applies to Witsenhausen's counterexample studied in Sect. 3.3. We note that Theorem 4.2.4 [818] is more general with a more direct proof. We thus note, for completeness, that existence of optimal policies for static and a class of sequential dynamic teams had been studied in [327, 771, 786, 818, 828, 830]. Conditions for optimality have been established in [167] for a class of continuous-time decentralized stochastic control problems. We also note, for (classical) partially observed models, existence results in [121, 267] with a discrete-time review in [818]. For a class of teams which are convex, one can reduce the search space to a smaller parametric class of policies, such as linear policies for quasi-classical linear quadratic Gaussian problems [350, 429, 594].

For a further reading on approximation of Witsenhausen's counterexample, we refer the reader to [51, 62, 305, 469, 477, 525, 527]. A comprehensive review on quantized approximations is available in [628]. We refer the reader to [304, 306] in which approximations of a class of static team problems are considered. On numerical methods for Witsenhausen's counterexample, in addition to the references provided in Sect. 3.3.3, we note [7] for a local optimality analysis. We also note that there exist information theoretic methods and multi-dimensional relaxations. Keeping in mind the subtle but consequential differences between one-shot, finite-dimensional and infinite-dimensional coding schemes to be presented in Sect. 12.4.1, relaxations of the counterexample have been presented in [178, 323, 463, 565]. Further convex relaxation studies include [438]. A study on (lack of) dual effect properties, in the sense of invariance on the information regarding noise variables with respect control policies under general information structures for both sequential and non-sequential systems has been investigated in [74].

As a further remark on quantum correlations and models for multi-agent systems, recall that an important property for non-classical information structures is that the information fields at the decision makers are not nested. With the recognition that the information fields generated by local measurements lead to subtle conditions on solvability and causality, an alternative probabilistic model, based on quantum mechanics, for describing such problems has been proposed in [68] and [67]. At least for causal systems (i.e., with Property C), we had seen in the previous chapter that a classical probabilistic description is sufficient to study non-sequential causal systems.

Decentralized detection, involving multiple distributed sensors, is an important application of stochastic team theory [740], also studied in the context of large numbers of decision makers [722].

Some of this chapter builds on [625, 830] and [621].

Chapter 5
Control Policy Topologies and Dynamic Programming Methods in Decentralized Stochastic Control

5.1 Introduction

In this chapter, which is companion to the previous Chap. 4, towards our theme on arriving at optimal solutions and their various properties, we adopt a *control topologies* approach. To this end, we introduce several topologies on decision/control strategies, for which information structures determine whether a particular topology is useful in arriving at analytical and computational methods as well as existence, compactness or approximation results. These topologies include those generated by a product metric of individual controllers viewed as subsets of appropriate probability measures, relaxed control policies and wide-sense admissible control policies, weak* topology on randomized policies, policies satisfying conditional independence properties leading to a universal dynamic program, and topologies that satisfy a closed-ness property under weak convergence of marginal probability measures.

In the chapter, we highlight and emphasize the point that one has freedom in deciding which particular topology to choose to associate with control policies. Presence of such a flexibility leads to versatility on analytical methods, approximations, and optimality.

Accordingly, in this chapter, rather than studying the probability measures induced on the actions, measurements and state processes as in the previous chapter, we consider individual decision maker policies separately. This approach can be named and viewed as an N-tuple approach or a control topology approach as the goal will be to consider the product space involving individual control policies. Dynamic programming methods will be seen as a special instance of this approach. Furthermore, as we will see in the following chapter, this approach will also be instrumental in arriving at optimality and structural results for stochastic teams with a large number of (and even infinitely many) decision makers.

S. Yüksel, T. Başar, *Stochastic Teams, Games, and Control under Information Constraints*, Systems & Control: Foundations & Applications,
https://doi.org/10.1007/978-3-031-54071-4_5

We note here that this approach was already instrumental in the proof of Theorem 2.4.7 where the team policy was given a product Hilbert space topology, as we will also briefly recall later in the chapter.

Our analysis will start with three approaches to dynamic programming; the first one, in Sect. 5.2, requires some increasing common information among the agents. The other two, presented in Sects. 5.3 and 5.4, are applicable to any sequential dynamic program. In Sect. 5.5 we study static teams and various ways to place control topologies. Section 5.6 views the classical information structure as an instance of the theory presented (for general information structures) in the book to compare the findings with classical results in single-agent stochastic control theory.

5.2 Dynamic Programming I: Common Information as Information State

As we mentioned in Sect. 3.1, in the theory of classical stochastic control, a standard approach is Dynamic Programming via backward induction, building on Bellman's optimality principle. As reviewed in Appendix E, this approach requires one to introduce a state space, an action space, a controlled transition kernel, a cost function, and an optimality criterion.

In a dynamic team problem, if all decision makers (DMs) have access to the same relevant information, then essentially we have a centralized decision problem, where all DMs can be lumped into a single DM, and can be viewed as such, and a classical stochastic control problem can be formulated. If, however, the DMs do not share all the same relevant information, then we may end up with a computationally intractable program to generate an optimal solution. A possible approach toward establishing a tractable program is through the construction of a controlled Markov chain where the controlled Markov state may now live in a larger state space (for example, the space of probability measures) and the actions are elements of function spaces. This controlled Markov construction may lead to computational feasibility in generating optimal policies.

The above entails a *dynamic programming approach*, which has been adopted extensively in the literature since the 1970s (see for example, [6, 43, 177, 567, 797, 806], and generalized and formalized in [558, 560]) through the use of a team-policy which uses common information to generate partial functions for each DM to generate their actions using local information. Thus, in the dynamic programming approach, a separation of team decision policies in the form of a two-tier architecture, *a higher-level controller* and a *lower-level controller*, can be established with the use of information common to all DMs. In the following, we present the ingredients of such an approach, as formalized in [560] and termed *the common information approach*:

1. Elimination of irrelevant information at the DMs: In this step, irrelevant local information at the DMs, say at **DM**k, is identified as follows. By letting the

policies of other DMs to be arbitrary, the policy of **DM**k can be optimized as a best-response function, and data at **DM**k irrelevant to this optimization can be removed. As an application, Theorems 15.3.1 and 15.3.3, to be presented later in the book, capture this step.
2. Construction of a coordinated system: This step identifies the common information and local/private information at the DMs, after Step 1 above has been carried out for all DMs. A *fictitious coordinator* (*higher-level controller*) uses the common information to generate team policies, which in turn dictates the (*lower-level*) DMs what to do with their local information. The construction of quantization policies in Sect. 12.2.2 (see also (15.3)) demonstrate this step.
3. Formulation of the cost as a Markov Decision Process (MDP), in view of the coordinator's optimal control problem. The conditional probability of the constructed *state* given the previous *action* and *state* has to be policy-independent, as in (E.2) in the Appendix, and the equivalent cost function is to be a measurable function of the *state* and *action*. Theorems 15.3.2, 15.3.4 and 15.6.1 capture this step.
4. Solution of the MDP leads to the structural results for the coordinator to generate optimal team policies, which in turn dictates the DMs what actions to take given their local information realizations. Theorem 15.6.1 captures this step.
5. Establishment of the equivalence between the solution obtained and the original problem, and translation of the optimal policies. Any coordination strategy can be realized in the original system. Note that, even though there is no real coordinator, such a coordination can be realized implicitly, due to the presence of common information.

In addition to the examples to be presented in Chap. 15, we will provide here a further explicit setting with such a recipe at work, in the context of the *k-stage periodic belief sharing pattern* in the next section. In particular, Lemma 5.2.1 and Lemma 5.2.2 will highlight this approach.

Example: k-Stage Periodic Belief Sharing Pattern
Here, we will use the term *belief* for a probability measure-valued random variable. This terminology has been used particularly in the artificial intelligence and computer science communities, which we adopt here.

Consider a joint process $\{x_t, y_t, t \in \mathbb{Z}_+\}$, where we assume for simplicity, but without much loss of conceptual generality, that the spaces where x_t, y_t take values in are finite dimensional real-valued or countable. They are generated by

$$x_{t+1} = f(x_t, u_t^1, \ldots, u_t^L, w_t),$$

$$y_t^i = g(x_t, v_t^i),$$

where x_t is the state, $u_t^i \in \mathbb{U}^i$ is the control action, and $(w_t, v_t^i, 1 \leq i \leq L)$ are second order, zero-mean, mutually independent, i.i.d. noise processes. We also assume, for simplicity here, that the state noise w_t either has a probability mass function, or a probability measure with a density function. To minimize the notational clutter, $P(x)$ will denote the probability mass function for discrete-valued spaces or probability density function for continuous spaces.

Suppose that there is a common information vector $\mathcal{I}_t^c$ at some time t, which is available to all DMs. At times $ks - 1$, with $k > 0$ fixed, and $s \in \mathbb{Z}_+$, the DMs share all their information: $\mathcal{I}_{ks-1}^c = \{\mathbf{y}_{[0,ks-1]}, \mathbf{u}_{[0,ks-1]}\}$ and further $\mathcal{I}_0^c = \{P(x_0)\}$, that is at time 0 the DMs have the same *a priori* belief on the initial state. Hence, at time t, **DM**i has access to $\{y^i_{[ks,t]}, \mathcal{I}^c_{ks-1}\}$.

Until the next *common* observation instant $t = k(s+1) - 1$ we can regard the individual decision functions specific to **DM**i to be $\{u_t^i = \gamma_s^i(y^i_{[ks,t]}, \mathcal{I}^c_{ks-1})\}$; we let $\boldsymbol{\gamma}_s$ denote the ensemble of such decision functions for all DMs and let $\underline{\gamma}$ denote the team policy.

It then suffices to generate $\boldsymbol{\gamma}_s$ for all $s \geq 0$, as the decision outputs conditioned on $y^i_{[ks,t]}$, under $\gamma_s^i(y^i_{[ks,t]}, \mathcal{I}^c_{ks-1})$, can then be generated. In such a case, we can define $\boldsymbol{\gamma}_s(., \mathcal{I}^c_{ks-1})$ to be the joint team decision rule mapping $\mathcal{I}^c_{ks-1}$ into the set of action vectors: $\{\gamma_s^i(y^i_{[ks,t]}, \mathcal{I}^c_{ks-1}), i \in \mathcal{L} = \{1, 2 \ldots, L\}, t \in \{ks, ks+1, \ldots, k(s+1)-1\}\}$.

Let $[0, T-1]$ be the decision horizon, where T is divisible by k. Let the objective of the DMs be the joint minimization of

$$E^{\gamma^1,\gamma^2,\ldots,\gamma^L}[\sum_{t=0}^{T-1} c(x_t, u_t^1, u_t^2, \ldots, u_t^L)] =: E_{x_0}^{\underline{\gamma}} \sum_{t=0}^{T-1} c(x_t, \mathbf{u}_t) =: J(\underline{\gamma}),$$

over all policies $\gamma^1, \gamma^2, \ldots, \gamma^L$. The cost function $J(\underline{\gamma})$ can be expressed as: $E^{\underline{\gamma}}[\sum_{s=0}^{\frac{T}{k}-1} \bar{c}(\boldsymbol{\gamma}_s(., \mathcal{I}^c_{ks-1}), \bar{x}_s)]$ with $\bar{x}_s := x_{ks}$, and

$$\bar{c}(\boldsymbol{\gamma}_s(., \mathcal{I}^c_{ks-1}), \bar{x}_s) = E_{\bar{x}_s}^{\underline{\gamma}}[\sum_{t=ks}^{k(s+1)-1} c(x_t, \mathbf{u}_t)]$$

Lemma 5.2.1 ([806]) *Consider the system setup above. Let $\mathcal{I}_t^c$ be a common information vector supplied to the DMs regularly every k time stages, so that they have common memory with a control policy generated as described above. Then, $\{\bar{x}_s := x_{ks}, \boldsymbol{\gamma}_s(\cdot, \mathcal{I}^c_{ks-1}), s \geq 0\}$ forms a controlled Markov chain.*

Proof Let $\boldsymbol{\gamma}_{[0,s]} = \{\boldsymbol{\gamma}_m(., \mathcal{I}^c_{km-1}), 0 \leq m \leq s\}$. It follows that

$$\begin{aligned}
&P(\bar{x}_{s+1}|\boldsymbol{\gamma}_{[0,s]}, \bar{x}_{[0,s]}) \\
&= \sum_{(\mathbf{x},\mathbf{y})_{[ks,k(s+1)-1]}} P(\bar{x}_{s+1}, x_{[ks,k(s+1)-1]}, \mathbf{y}_{[ks,k(s+1)-1]}|\boldsymbol{\gamma}_{[0,s]}, \bar{x}_{[0,s]}) \\
&= \sum_{(\mathbf{x},\mathbf{y})_{[ks,k(s+1)-1]}} \Bigg(\prod_{m=ks}^{k(s+1)-1} P(x_{m+1}|x_m, \boldsymbol{\gamma}_s(\mathbf{y}_{[ks,m]}, \mathcal{I}^c_{ks-1}), \bar{x}_s, \boldsymbol{\gamma}_{[0,s-1]}, \bar{x}_{[0,s-1]})
\end{aligned}$$

$$P(y_m^1|x_m)P(y_m^2|x_m)\dots P(y_m^L|x_m)\Big)$$

$$= \sum_{(\mathbf{x},\mathbf{y})_{[ks,k(s+1)-1]}} \Big(\prod_{m=ks}^{k(s+1)-1} P(x_{m+1}|x_m, \boldsymbol{\gamma}_s(\mathbf{y}_{[ks,m]}, \mathcal{I}^c_{ks-1}))$$

$$P(y_m^1|x_m)P(y_m^2|x_m)\dots P(y_m^L|x_m)\Big)$$

$$= P(\bar{x}_{s+1}|\boldsymbol{\gamma}_s(., \mathcal{I}^c_{ks-1}), \bar{x}_s).$$

◇

In view of Lemma 5.2.1, we have the following separation result.

Lemma 5.2.2 ([806]) *Let $\mathcal{I}^c_t$ be a common information vector supplied to the DMs regularly every k time steps. There is no loss in performance if $\mathcal{I}^c_{ks-1}$ is replaced by $P(\bar{x}_s|\mathcal{I}^c_{ks-1})$.*

Proof As we observed earlier, the cost can be written as a function of additive costs:

$$J_{x_0}(\underline{\gamma}) = E^{\underline{\gamma}}_{x_0}[\sum_{s=0}^{\frac{T}{k}-1} \bar{c}(\boldsymbol{\gamma}_s, \bar{x}_s)],$$

with

$$\bar{c}(\boldsymbol{\gamma}_s, \bar{x}_s) = E^{\underline{\gamma}}_{\bar{x}_s}[\sum_{t=ks}^{k(s+1)-1} c(x_t, \mathbf{u}_t)].$$

For the minimization of a stage-additive cost in Partially Observed Markov Chains, it suffices to transform the state to an equivalent state of conditional distributions as discussed in Appendix E. Hence $P(\bar{x}_s|\mathcal{I}^c_{ks-1})$ acts as a sufficient statistic. ◇

An implication then is that one can apply dynamic programming with a fixed state space (that is, with a non-increasing size in time). The above motivates us to introduce the following pattern.

Definition 5.2.1 (*k*-Stage Periodic Belief Sharing Pattern [806]) An information pattern in which the DMs share their posterior beliefs to reach a joint belief about the system state is called a *belief sharing information pattern*. If belief sharing occurs periodically every k-stages ($k > 1$), the DMs also share the control actions they applied in the last $k - 1$ stages, together with intermediate belief information. In this case, the information pattern is called the *k-stage periodic belief sharing information pattern*.

We note that such a dynamic programming approach is also applicable to several of the information structures presented in Definition 2.3.1, such as the *n-step delayed information sharing*. We note, however, some subtleties beyond the one-step delayed setting [558, 736, 774, 797] where a natural conjecture [774] to separation does not apply when $n \geq 2$ [736].

For $k > 1$, it should be noted that, the exchange of the control actions is essential. When belief-sharing occurs at every stage, then controls can be generated by each of the DMs, and hence the control actions need not be shared.

The above, and related dynamic programming methods, however require that there is some history sharing in the past. In the following two sections, we present dynamic programming methods applicable to any sequential stochastic control problem.

5.3 Dynamic Programming II: Witsenhausen's Standard Form for Sequential Finite Teams

Witsenhausen [775] introduced a general approach establishing that *any* sequential team optimization problem may admit a formulation appropriate for a dynamic programming analysis. This approach has since been called Witsenhausen's Standard Form.

Witsenhausen's standard form proceeds as follows, within the general (intrinsic model of Sect. 3.2) framework we have introduced and adopted.

(i) State: $x_1 = \omega$, and for $2 \leq t \leq N$, $x_t = \{\omega, u^1, \ldots, u^{t-1}\}$.

(i') Extended State $\pi_t \in \mathcal{P}(\Omega \times \prod_{i=1}^{t-1} \mathbb{U}^i)$: For every Borel $B \in \Omega \times \prod_{i=1}^{t-1} \mathbb{U}^i$,

$$\pi_t(B) = E_{\pi_t}[1_{\{(\omega,u^1,\ldots,u^{t-1})\in B\}}].$$

(ii) Control Action: γ^t is $\sigma(y^t)$-measurable and $\mathbb{U}^t$-valued, for $1 \leq t \leq N$.

(iii) (Total) Cost: $E[c(\omega, u^1, \ldots, u^N)]$.

(iv) Transition Dynamics/Kernel are defined with

$$x_t = g_t(x_{t-1}; \gamma^{t-1}(x_{t-1})).$$

As defined above, π_t is $\mathcal{P}(\Omega \times \prod_{i=1}^{t-1} \mathbb{U}^i)$-valued. Let us endow this space of probability measures with the total variation norm. If one views the set of such probability measures as a subset of the set of all signed countably additive finite measures, one could consider a duality pairing under the weak*-topology [227, Section 4.6] (which would correspond to the classical weak convergence topology in probability theory (see Appendix C) if the state and the measurement spaces were compact [489]). However, even in the absence of such compactness conditions, with Ψ_t being the set of bounded continuous functionals on $\Omega \times \prod_{i=1}^{t} \mathbb{U}^i$, $\psi_{t-1} \in \Psi_{t-1}$ defines a linear and continuous (bounded) map on the linear space of signed measures, where $\mathcal{P}(\Omega \times \prod_{i=1}^{t-1} \mathbb{U}^i)$ is viewed as a subset, through the relation

$$\langle \psi_{t-1}, \pi_t \rangle := \int \pi_t(d\omega, du^1, \ldots, du^{t-1}) \psi_{t-1}(\omega, u^1, \ldots, u^{t-1}).$$

Here, ψ_t can be viewed as a co-state variable defined by the relation

$$\langle \psi_t, [T_t(\gamma^t)]\pi_t \rangle = \langle [T_t^*(\gamma^t)]\psi_t, \pi_t \rangle \tag{5.1}$$

with $T_t(\gamma^t) : \mathcal{P}(\Omega \times \prod_{i=1}^{t-1} \mathbb{U}^i) \to \mathcal{P}(\Omega \times \prod_{i=1}^{t} \mathbb{U}^i)$ defined by

$$\pi_{t+1}(E) = [T_t(\gamma^t)](\pi_t)(E) = \pi_t\Big(\{x_t : g_t(x_t; \gamma^t) \in E\} \Big)$$

and $T_t^* : \Psi_t \to \Psi_{t-1}$ is adjoint to T_t defined through (5.1):

$$[T_t^*(\gamma^t)]\psi_t = \psi_{t-1}, \tag{5.2}$$

with the terminal condition

$$\psi_N = c(\omega, u^1, \cdots, u^N), \tag{5.3}$$

and the (expected) cost induced by a team policy being

$$\langle \psi_N, [T_N(\gamma^N)]\pi_N \rangle.$$

Given the above, Witsenhausen states the following (necessity condition):

Theorem 5.3.1 ([775, Thm. 1]) *If a team policy $\underline{\gamma}^* = \{\gamma^{1,*}, \cdots, \gamma^{N,*}\}$ is optimal, then with π_t^* being the associated state variables, we have that for all $1 \le t \le N$,*

$$J(\underline{\gamma}^*) = \min_{\gamma^t} \langle \psi_t^*, [T_t(\gamma^t)]\pi_t^* \rangle \tag{5.4}$$

where ψ_t^* is obtained inductively through (5.2) with γ^t taken to be $\gamma^{t,*}$ and terminal condition given by (5.3).

Proof Let us write

$$\pi_t = [T_{t-1}(\gamma^{t-1})] \circ [T_{t-2}(\gamma^{t-2})] \circ \ldots [\circ T_1(\gamma^1)]\pi_0$$

Note that π_t does not depend on $\gamma^t, \gamma^{t+1}, \ldots, \gamma^N$. Introducing

$$\psi_t = [T_{t+1}^*(\gamma^{t+1})] \circ [T_{t+2}^*(\gamma^{t+2})] \circ \ldots \circ [T_N^*(\gamma^N)]\psi_N,$$

we note that ψ_t does not depend on $\gamma^1, \ldots, \gamma^t$. Thus, fixing all $\gamma^i, i \neq t$ to be from the set $\{\gamma^{1,*}, \cdots, \gamma^{N,*}\} \setminus \{\gamma^{t,*}\}$, the optimal γ^i has to be minimizing (5.4). ◇

Dynamic Programming Using Witsenhausen's Standard Form We note that while Witsenhausen [775, Thm. 1] provides only a necessity condition of optimality, one could also obtain a sufficiency condition through backward induction (by not fixing the control policies prior to any time t to be optimal *a priori*). In particular, we offer the following reasoning: Through backwards induction as in Bellman's principle of optimality, first for $t = N$, for any γ^t, and admissible state π_t

$$\langle \psi_t^*, [T_t(\gamma^{t,*})]\pi_t \rangle = \langle [T_t^*(\gamma^{t,*})]\psi_t^*, \pi_t \rangle \leq \langle \psi_t^*, [T_t(\gamma^t)]\pi_t \rangle,$$

and with $\psi_{t-1}^* = [T_t^*(\gamma^{t,*})]\psi_t$, and dynamics $\pi_t = [T_{t-1}(\gamma^{t-1})]\pi_{t-1}$, we have that for $1 \leq t \leq N-1$, if the following holds,

$$\langle \psi_{t-1}^*, [T_{t-1}(\gamma^{t-1,*})]\pi_{t-1} \rangle \leq \langle \psi_{t-1}^*, [T_{t-1}(\gamma^{t-1})]\pi_{t-1} \rangle,$$

then we have an optimal policy. By induction, the optimal cost would be equal to

$$\langle \psi_1^*, [T_1(\gamma^{1,*})]\pi_1 \rangle.$$

Well-posedness of a dynamic program or existence results were not established in Witsenhausen's Standard Form [775] for standard Borel models. Indeed, Witsenhausen notes that *...a deeper understanding of these problems, including an existence theory for nonclassical patterns, will require more insight into some delicate questions in analysis....* This motivates the construction in the next section.

5.4 Dynamic Programming III: Universal Dynamic Program for all Sequential Teams in General Spaces

In this section, one of our goals is to address some of the issues raised and left open by Witsenhausen as described in the previous section by presenting a different construction also utilizing the mathematical analysis presented in earlier chapters. In the following, we first introduce a controlled Markov model based on the strategic measures formulation studied in the previous chapter, which will be invoked further below. This model allows for the recursions to be written explicitly and will lead to desirable measurability, and later, existence, properties. In doing this, we will follow [818].

As in Chap. 3, we split ω into ω_0 and the rest (recall that ω represents the entire exogenous uncertainty acting on the system, including the initial state, and the system and measurement noises): ω_0 is the smallest (in dimension) random variable whose sigma-field generated over Ω, together with those generated by $u^1, \ldots, u^N$ on their respective action spaces, c is measurable on.

5.4.1 *Controlled Markov Model Formulation for the Dynamic Program*

We provide below the formulation of a controlled Markov model, which will allow for execution of the universal dynamic program.

(i) State: $x_t = \{\omega_0, u^1, \cdots, u^{t-1}, y^1, \cdots, y^t\}$, $1 \le t \le N$.

(i') Extended State: $\pi_t \in \mathcal{P}(\Omega_0 \times \prod_{i=1}^{t} \mathbb{Y}^i \times \prod_{i=1}^{t-1} \mathbb{U}^i)$ where, for Borel $B \in \Omega_0 \times \prod_{i=1}^{t} \mathbb{Y}^i \times \prod_{i=1}^{t-1} \mathbb{U}^i$,

$$\pi_t(B) := E_{\pi_t}[1_{\{(\omega_0, y^1, \cdots, y^t; u^1, \cdots, u^{t-1}) \in B\}}].$$

Thus, $\pi_t \in \mathcal{P}(\Omega_0 \times \prod_{i=1}^{t} \mathbb{Y}^i \times \prod_{i=1}^{t-1} \mathbb{U}^i)$ where the space of probability measures is endowed with the weak convergence topology.

(ii) Control Action: Given π_t, the *action* $\hat{\gamma}^t$ is a probability measure in $\mathcal{P}(\Omega_0 \times \prod_{k=1}^{t} \mathbb{Y}^k \times \prod_{k=1}^{t} \mathbb{U}^k)$ that satisfies the conditional independence relation:

$$u^t \leftrightarrow y^t \leftrightarrow x_t = (\omega_0, y^1, \cdots, y^t; u^1, \cdots, u^{t-1}) \tag{5.5}$$

(that is, for every Borel $B \in \mathbb{U}^i$, almost surely under $\hat{\gamma}^t$, the following holds:

$$P(u^t \in B | y^t, (\omega_0, y^1, \cdots, y^t; u^1, \cdots, u^{t-1})) = P(u^t \in B | y^t)$$

with the restriction

$$x_t \sim \pi_t.$$

Denote by $\Gamma^t(\pi_t)$ the set of all such probability measures. Any $\hat{\gamma}^t \in \Gamma^t(\pi_t)$ defines, for almost every realization y^t, a conditional probability measure on $\mathbb{U}^t$. When the notation does not lead to any confusion or ambiguity, we will denote the action at time t by $\gamma^t(du^t|y^t)$, which is understood to be consistent with $\hat{\gamma}^t$.

(ii') Control Action for Static (or Static Reduced) Teams with Independent Measurements: Given π_t, the *action* $\hat{\gamma}^t$ is a probability measure on $\mathbb{Y}^t \times \mathbb{U}^t$ with a fixed marginal $P(dy^t)$ on $\mathbb{Y}^t$, that is $\pi_t^{\mathbb{Y}^t}(dy^t) = P(dy^t)$. Denote by $\Gamma^t(\pi_t^{\mathbb{Y}^t})$ the set of all such probability measures. As above, when the notation does not lead to any confusion or ambiguity, we will denote the action at time t by $\gamma^t(du^t|y^t)$, which is understood to be consistent with $\hat{\gamma}^t$.

(iii) (Total) Cost: $E[c(\omega_0, u^1, \ldots, u^N)] =: \langle c, \hat{\gamma}^N(\pi_N) \rangle$, with:

$$\left(\hat{\gamma}^N(\pi_N)\right)(B) := E_{\hat{\gamma}^N(\pi_N)}[1_{\{(\omega_0, y^1, \cdots, y^N; u^1, \cdots, u^N) \in B\}}],$$

for $1 \le t \le N$.

(iv) Transition Kernel: $T_t(\gamma^t) : \mathcal{P}(\Omega_0 \times \prod_{i=1}^{t} \mathbb{Y}^i \times \prod_{i=1}^{t-1} \mathbb{U}^i) \to \mathcal{P}(\Omega_0 \times \prod_{i=1}^{t+1} \mathbb{Y}^i \times \prod_{i=1}^{t} \mathbb{U}^i)$, so that

$$\pi_{t+1} = T_t(\gamma^t)(\pi_t),$$

where

$$T_t(\gamma^t)(\pi_t)(A_1, A_2, A_3)$$
$$= \int_{x_t \in A_1} \int_{u^t \in A_2} \int_{y^{t+1} \in A_3} P(dy^{t+1}|\omega_0, y^1, \cdots, y^t; u^1, \cdots, u^t)\gamma^t(du^t|y^t)\pi_t(dx_t),$$

for Borel A_1, A_2, A_3. We recall again that $P(dy^{t+1}|\omega_0, y^1, \cdots, y^t; u^1, \cdots, u^t)$ is as given in (4.7), and the control actions γ^t are as defined above.

We may refer to policies satisfying (5.5) *Locally-Conditionally-Independent Policies*. The following is a useful supporting result.

Theorem 5.4.1 *For each given π_t, $\Gamma^t(\pi_t)$ is a weakly compact set. Thus, the action space is compact.*

Proof This can be viewed as a special instance of Theorem 4.3.1(i): π_t is fixed as we alter the conditional measure

$$P(u^t \in B|y^t, (\omega_0, y^1, \cdots, y^t; u^1, \cdots, u^{t-1})) = P(u^t \in B|y^t)$$

◇

Remark 5.4.1 With regard to items (ii) and (ii') in the controlled Markov model formulation, we again note that defining relaxed/randomized policies in decentralized stochastic control (as a way to facilitate the optimality analysis as in Chap. 4) is a subtle exercise. One should not overly relax the probabilistic dependence and conditional independence properties among the decision makers too much: an illuminating example is what is known as the Clauser-Horne-Shimony-Holt game [180] presented in the previous chapter (see also [22] and Remark 4.3.1) where a related convex class of information structures were introduced for which the optimal cost under the relaxed policies is strictly lower than the optimal cost for the original problem.

Some further differences between the above formulation and Witsenhausen's standard form are as follows.

(a) In Witsenhausen's standard form, there is a deterministic update equation $x_t = g_t(x_{t-1}; \gamma^{t-1}(x_{t-1}))$, whereas in the formulation here we do not have such a measurable deterministic map, since ω is not part of the state. Furthermore, γ^t is allowed to be randomized.
(b) Although it is not necessary for the analysis (we will apply a standard Bellman optimality argument with the aforementioned construction, unlike

Witsenhausen's approach), to be consistent with Witsenhausen's formulation, we can also define $\psi_t \in \Psi_t$ (as a bounded measurable functional from $\Omega_0 \times \prod_{k=1}^{t} \mathbb{Y}^k \times \prod_{k=1}^{t} \mathbb{U}^k$ to $\mathbb{R}$) as a co-state variable. First, we can view the space of probability measures as a subset of the locally convex space [613] of finite signed measures with the following notion of convergence: we say that $\nu_n \to \nu$ if $\int f(x)\nu_n(dx) \to \int f(x)\nu(dx)$ for every continuous and bounded function f. With the transition kernel given as (5.6), we can define the recursions and the adjoint variable ψ_t iteratively, with

$$\langle \psi_t, [T_t(\gamma^t)]\pi_t \rangle = \langle [T_t^*(\gamma^t)]\psi_t, \pi_t \rangle \tag{5.6}$$

where $T_t^*(\gamma^t) : \Psi_t \to \Psi_{t-1}$ is adjoint to T_t defined through (5.6) (so that $[T_t^*(\gamma^t)]\psi_t = E^{\gamma^t}[\psi_t | x_t])$ -a more explicit representation is presented further below using the properties of conditional expectation- and

$$\psi_N(\omega_0, y^1, \ldots, y^N, u^1, \ldots, u^N) := c(\omega_0, u^1, \ldots, u^N). \tag{5.7}$$

(c) Unlike Witsenhausen's standard form which endows the set of probability measures with total variation, here one endows the set with the weak convergence topology. This way a standard Borel space is obtained.

In the formulation introduced, an explicit representation would be the following.

Example 5.4.1 (Explicit Representation as a Terminal Time Problem) Given $\gamma^t, \cdots, \gamma^N$, we have

$$\begin{aligned}
&\langle \psi_{t-1}, [T(\gamma^{t-1})]\pi_{t-1} \rangle \\
&= \int \pi_{t-1}(d\omega_0, dy^1, \ldots, y^{t-1}, du^1, \ldots, du^{t-2}) \\
&\quad \times E\Bigg[c\Big(\omega_0, u^1, \ldots, u^{t-2}, \gamma^{t-1}(y^{t-1}), \gamma^t(Y^t), \cdots, \gamma^N(Y^N)\Big) \\
&\qquad\qquad \Bigg| \omega_0, u^1, \cdots, u^{t-2}, y^1, \cdots, y^{t-1}\Bigg]
\end{aligned}$$

Here, $\{Y^i\}$ variables are displayed with capital letters to highlight that these are random given ω_0 (and $u^1, \cdots, u^{t-2}, y^1, \cdots, y^{t-1}$), unlike what a corresponding expression would look like under Witsenhausen's formulation when they would be deterministic given ω.

The following result is immediate by construction, through a standard Bellman optimality argument (recall the analysis at the end of the previous section).

Theorem 5.4.2 *Consider a team policy $\underline{\gamma}^* = \{\gamma^{1,*}, \cdots, \gamma^{N,*}\}$ that satisfies, for $1 \leq t \leq N$, and every admissible π_t,*

$$J_t(\pi_t) = \inf_{\gamma^t \in \Gamma^t(\pi_t)} \langle \psi_t^*, [T_t(\gamma^t)]\pi_t \rangle \tag{5.8}$$

where ψ_t^* is obtained inductively through (5.6) with γ^t taken to be $\gamma^{t,*}$ and final condition (5.7). Then, $\underline{\gamma}^*$ is optimal.

5.4.2 Well-Posedness of the Dynamic Program (5.8)

To establish well-posedness of the dynamic programming recursions we will revisit the concepts of universal measurability and analytic functions.

It suffices for a function to be universally measurable (and not necessarily Borel measurable) for its integration with respect to some probability measure to be well-defined: a function f is μ-measurable if there exists a Borel measurable function g which agrees with f μ-a.e. A function that is μ-measurable for every probability measure μ is called universally measurable. A set B is universally measurable if the function $1_{\{x \in B\}}$ is universally measurable.

A measurable image of a Borel set is called an *analytic* set [230, Appendix 2]. We note that this is evidently equivalent to the seemingly more restrictive condition of being a *continuous* image of some Borel set. The image of a Borel set under a measurable function, and hence an analytic set, is universally measurable.

The integral of a universally measurable function is well-defined and is equal to the integral of a Borel measurable function which is μ-almost equal to that function (with μ being the measure for integration). While applying dynamic programming, we often seek to establish the existence of measurable functions through the operation:

$$J_t(x_t) = \inf_{u \in \mathbb{U}(x_t)} \left(c(x_t, u) + \int J_{t+1}(x_{t+1}) Q(dx_{t+1}|x_t, u) \right)$$

However, here we need a stronger condition than universal measurability for the recursions to be well-defined: a function f is called lower semi-analytic if $\{x : f(x) < c\}$ is analytic for each scalar $c \in \mathbb{R}$. Lower semi-analytic functions are universally measurable [230].

Theorem 5.4.3 ([230]) *Let $i : \mathbb{S} \to \mathbb{X}$ and $f : \mathbb{S} \to \mathbb{R}$ be Borel measurable functions. Then,*

$$v(x) = \inf_{\{z: i(z)=x\}} f(z)$$

is lower semi-analytic.

Note that [230, p. 85]

$$i^{-1}(\{x : v(x) < c\}) = \{z : f(z) < c\}$$

The set $\{z : f(z) < c\}$ is Borel, and thus if i is also Borel, it follows that v is lower semi-analytic. An application of this result leads to the following.

Theorem 5.4.4 *Consider* $G = \{(x, u) : u \in \mathbb{U}(x) \subset \mathbb{U}\}$ *a measurable subset of* $\mathbb{X} \times \mathbb{U}$. *Let* $v : \mathbb{X} \times \mathbb{U} \to \mathbb{R}$ *be Borel measurable. Then, the map*

$$\kappa(x) := \inf_{(x,u)\in G} v(x, u),$$

is lower semi-analytic (and thus universally measurable).

The proof follows from the fact that $i^{-1}(\{x : \kappa(x) < c\}) = \{(x, z) : v(x, z) < c\}$, where i is the projection of Gn onto $\mathbb{X}$, which is a continuous operation; the image may not be measurable but as a measurable mapping of a Borel set, it is analytic and thus the function v is lower semi-analytic and thus universally measurable. The next result is due to [114, Theorems 7.47 and 7.48].

Theorem 5.4.5 *Let* E_1, E_2 *be Borel sets, and let for every* $x \in E_1$, $E_2(x)$ *be a subset of* E_2 *such that the set* $\{(x, u) : x \in E_1, u \in E_2(x)\}$ *is a Borel set in* $E_1 \times E_2$.

(i) Let $g : E_1 \times E_2 \to \mathbb{R}$ *be lower semi-analytic. Then,*

$$h(e_1) = \inf_{e_2\in E_2(e_1)} g(e_1, e_2)$$

is lower semi-analytic.

(ii) Let $g : E_1 \times E_2 \to \mathbb{R}$ *be lower semi-analytic. Let* $Q(de_2|e_1)$ *be a stochastic kernel. Then,*

$$f(e_1) := \int g(e_1, e_2)Q(de_2|e_1)$$

is lower semi-analytic.

We now return to the question of the well-posedness of the dynamic programming formulation introduced in the previous subsection, together with an ϵ-optimality result.

Theorem 5.4.6

(a) The dynamic programming equations (5.8) introduced in Theorem 5.4.2 are well defined.

(b) For every $\epsilon > 0$, *an* ϵ*-optimal policy exists in* $\Gamma^t(\pi_t)$ *(Markov policies according to the controlled Markov model introduced).*

Proof We first prove (a).

Step (i) We utilize the formulation introduced in Sect. 5.4 with the standard Borel action set $\hat{\gamma}^t$ and apply backwards induction. Let

$$\pi_N(\cdots) = P(d\omega_0, dy^1, \cdots, dy^N, du^1, \cdots, du^{N-1}).$$

Recall that at time N, we minimize over $\hat{\gamma}^N \in \Gamma^N(\pi_N)$ which is the set of all probability measures which satisfies

$$u^N \leftrightarrow y^N \leftrightarrow (\omega_0, y^1, \cdots, y^{N-1}; u^1, \cdots, u^{N-1})$$

We observe that the graph

$$\{\pi_t, \Gamma^t(\pi_t)\},$$

is measurable. Note that $\{(\pi_N, \Gamma^N(\pi_N))\}$ is in fact $L_R(\mu)$, which is a Borel set by Theorem 4.1.6, and for each $1 \leq t \leq N$, we will have the restriction of this measurable set to $(\Omega_0, \prod_{k=1}^t \mathbb{Y}^k, \prod_{m=1}^{t-1} \mathbb{U}^m)$.
Theorem 5.4.5(i) thus ensures that

$$J_N^*(\pi_N) = \inf_{\gamma^N \in \Gamma^N(\pi_N)} \langle \psi_N^*, T_N(\gamma^N)\pi_N \rangle$$

is a lower semi-analytic function.

Step (ii) For $\inf_{\gamma^{N-1}} \langle \psi_{N-1}^*, T_{N-1}(\gamma^{N-1})\pi_{N-1} \rangle$, with

$$\psi_{N-1}^*(x_{N-1}) = E\Big[c(\omega, u^1, \ldots, \gamma^{t-1}(y^{t-1}), \gamma^{N-1}(y^{N-1}), \gamma^{N,*}(Y^N)) \Big| \omega_0, y^1, \ldots, y^N, u^1, \ldots, u^{N-1}\Big] \quad (5.9)$$

we also obtain a lower semi-analytic function. This follows because of the inductive nature of dynamic programming: Theorem 5.4.5(ii) ensures that $E[J(\pi_N)|\pi_{N-1}, \gamma^{N-1}]$ is lower semi-analytic. Note that $P(d\pi_N|\pi_{N-1}, \gamma^{N-1})$ is a well defined stochastic kernel. In particular, through the sequence of iterated expectations, we write

$$\inf_{\gamma^1} E^{\gamma^1}\Big[\inf_{\gamma^2} E^{\gamma^2}\Big[\cdots \inf_{\gamma^N} E^{\gamma^N}\Big[c(\omega_0, u^1, \cdots, u^N)|\pi_N\Big] \cdots \Big|\pi_2\Big]\Big|\pi_1\Big] \quad (5.10)$$

At each iteration, we apply first Theorem 5.4.5 (i) and then (ii) to ensure lower semi-analyticity. This implies that the equations introduced in Theorem 5.4.2 are well defined.

(b) Since the initial probability measure π_0 is given, the dynamic program (5.8) admits an ϵ-optimal solution [666]. □

Remark 5.4.2 The infima in (5.10) can be replaced to be over deterministic γ^n instead of $\hat{\gamma}^n$; see Remark 4.1.2

5.4.3 *Existence of Optimal Policies Through the Dynamic Programming Formulation*

Given the dynamic programming formulation, the discussion in the previous subsection leads to the question of whether the dynamic programming approach can lead to new existence and/or regularity properties on optimal policies.

A Counterexample on Non-existence Under Weak Continuity of Kernels and its Analysis

In the following, we revisit Sect. 3.3.1 and the example presented there on non-existence of an optimal policy. This example has the same cost function (3.5) and information structure as in Witsenhausen's original counterexample, but with only discrete distributions. The linear dynamics and the flow of dynamics and information were presented in Figs. 3.3 and 3.2.

Recall that in the discrete distribution case, $x_0 = x$ and w_1 were $\{-1, 1\}$-valued independent uniformly distributed random variables.

For this setup, via the policies given in (3.9), the cost could be made arbitrarily close to *zero* by picking ϵ sufficiently small. Thus, the infimum of J was 0, but this cannot be achieved because in the limit as $\epsilon \to 0$, it was no longer possible to transmit x. There does not exist an optimal policy!

In view of this, a formal negative result is thus the following, the proof of which is the counterexample presented above.

Theorem 5.4.7 *Consider a sequential dynamic team viewed under the controlled Markov (universal dynamic programming) formulation of Sect. 5.4.1. Suppose that*

$$E[\int f(z)\pi_{n+1}(dz)|\pi_n, \hat{\gamma}^n]$$

is (weakly) continuous in $(\pi_n, \hat{\gamma}^n)$ for every continuous and bounded f. Suppose that all of the measurement and action spaces and Ω_0 are compact and the cost function c is continuous. Even for such a team, with only two decision makers, an optimal team policy does not necessarily exist.

We now go through the arguments to see why continuity/compactness conditions may not hold under such weak continuity conditions as presented in Theorem 5.4.7.

Let $\pi_N(\cdots) = P(d\omega_0, dy^1, \cdots, dy^N, du^1, \cdots, du^{N-1})$. Let $\hat{\gamma}^N$ be the joint measure defined by the control action γ^N. Note that the action space depends on the state. In particular, through an application of Theorem 5.4.5(i), by writing $e_1 = \pi_N$ and $e_2 = \gamma^N(\pi_N)$, given that E_2 is Borel,

$$h(e_1) = \inf_{e_2 \in E_2} g(e_1, e_2)$$

will be lower semi-analytic. With some analysis one can show that the action space is compact.

Through the compactness condition, we know that for every π_N there exists an optimal γ^N. A question then is whether γ^N can be taken to be measurable in π_N. This holds because: $\Gamma^N(\pi_N)$ is compact and its graph is measurable, as a consequence of the following result.

Lemma 5.4.1 ([347, Theorem 2], [442, 649]) *Let $\mathbb{X}, \mathbb{U}$ be Polish spaces and $\Gamma = (x, \psi(x))$ where $\psi(x) \subset \mathbb{U}$ is such that, $\psi(x)$ is compact for each $x \in \mathbb{X}$ and Γ is a Borel measurable set in $\mathbb{X} \times \mathbb{U}$. Let $c(x, u)$ be a continuous function on $\psi(x)$ for every x. Then, there exists a Borel measurable function $f : \mathbb{X} \to \mathbb{U}$ such that*

$$c(x, f(x)) = \min_{u \in \psi(x)} c(x, u)$$

By this result, it follows that $\hat{\gamma}^N$ can be measurably selected given π_N. Thus, an optimal policy exists for this time stage. Furthermore, $J_N(\pi_N)$ is not only lower semi-analytic, but is actually Borel measurable (this also follows by realizing that the final stage decision needs to be an optimal estimator given the specifics of the problem).

However, $J_N(\pi_N)$ is not in general continuous and it is this condition that fails in the counterexample.

A sufficient condition for continuity is the following upper semi-continuity condition for the set-valued map Γ^N, mapping each realization π_N to the set $\Gamma^N(\pi_N)$. For some related results see [342, Appendix D, Proposition D.5]).

Lemma 5.4.2 $\inf_{\hat{\gamma}^N \in \Gamma^N(\pi_N)} J_N(\pi_N, \hat{\gamma}^N)$ *is continuous in π_N if (i) $\Gamma^N(\pi_N)$ is an upper semi-continuous set-valued map so that if $\pi_{N,n} \to \pi_N$ and $\hat{\gamma}_n^{N,*} \to \hat{\gamma}^{N,*}$ with $\hat{\gamma}_n^{N,*} \in \Gamma^N(\pi_{N,n})$ then $\hat{\gamma}^{N,*} \in \Gamma^N(\pi_N)$ and (ii) $\cup_{\pi_{N,n} \to \pi_N} \Gamma^N(\pi_{N,n})$ is pre-compact (that is, tight) for every sequence $\pi_{N,n} \to \pi_N$.*

Remark 5.4.3 We note that condition (ii) in the statement of Lemma 5.4.2 above is sometimes imposed in the definition of upper semi-continuity of set-valued of maps. Nonetheless, for completeness, we have made this explicit as a further condition.

For a class of static teams, this condition is applicable as we study more explicitly in the following.

An Existence Result for Static Teams or Static-Reducible Dynamic Teams (with Independent Measurements)

In view of the discussion above, a positive result is the following. This result presents a complementary approach to that given in the proof of Theorem 4.2.4 (though with a slightly weaker result in comparison).

Note that for such static teams under the model introduced in Sect. 5.4.1 we view the control action space $\Gamma^t(\pi_t^{\mathbb{Y}^t})$ at time t as the set of probability measures on $\mathbb{Y}^t \times \mathbb{U}^t$ with a fixed marginal $P(dy^t)$ on $\mathbb{Y}^t$.

Theorem 5.4.8 *Consider a static team or a dynamic team that admits a static reduced form with independent measurements reduced cost c_s (see (3.31)). Suppose that all of the action spaces are compact and c_s is continuous and bounded. Then, an optimal team policy exists.*

Proof

Step (i) As presented in Sect. 5.4.1, we view the control action space as the set of all joint measures on $\mathbb{Y}^i \times \mathbb{U}^i$ with a fixed marginal on y^i. Since the team is static (with independent measurements), this decouples the action spaces from the actions of the previous DMs.

Step (ii) With the new formulation, we show that the optimal cost function $J_n(\pi_n)$ is continuous for every $1 \leq n \leq N$. This follows through an inductive argument as follows. For the last stage N, this follows from the continuity of the cost function in $J_N(\pi_N, \gamma^N)$, as we establish below.

The key step in the proof is to show that we have the desired continuity under the new state and action space construction. In particular, let $\pi_{N-1,m} \to \pi_{N-1}$ and $(Q\gamma)_m^N \to (Q\gamma)^N$ (where convergence is in the weak sense). With an abuse of notation, suppose that x_{N-1} also contains u^{N-1} for the analysis below. Then, for every continuous and bounded f, with

$$\int \pi_{N-1,m}(dx_{N-1})Q(dy^N)\gamma_m^N(du^N|y^N)f(x_{N-1}, y^N, u^N)$$

$$= \int \pi_{N-1,m}(dx_{N-1})h_{N-1}^m(x_{N-1})$$

with

$$h_{N-1}^m(x) := \int \Big(Q(dy^N)\gamma_m^N(du^N|y^N)f(x, y^N, u^N)\Big),$$

and

$$h_{N-1}(x) := \int \Big(Q(dy^N)\gamma^N(du^N|y^N)f(x, y^N, u^N)\Big),$$

we show that $h^m_{N-1}(x_m)$ is a continuously converging function (as it is defined in [658]) in the sense that for every $x_m \to x$,

$$h^m_{N-1}(x_m) \to h_{N-1}(x)$$

This follows from the following. First write

$$h^m_{N-1}(x_m) - h_{N-1}(x) = \Big(h^m_{N-1}(x_m) - h_{N-1}(x_m)\Big) + \Big(h_{N-1}(x_m) - h_{N-1}(x)\Big)$$

The second term converges to zero by the dominated convergence theorem. The first term converges to zero since

$$\Big(h^m_{N-1}(x_m) - h_{N-1}(x_m)\Big)$$
$$= \int \Big(Q(dy^N)\gamma^N_m(du^N|y^N) - Q(dy^N)\gamma^N(du^N|y^N)\Big) f(x_m, y^N, u^N)$$

Here, $(Q\gamma)^N_m \to (Q\gamma^N)$ weakly and $f(x_m, y^N, u^N) \to f(x, y^N, u^N)$ continuously that is, as a mapping from $\mathbb{Y}^N \times \mathbb{U}^N$ to $\mathbb{R}$, $f_m(y^N, u^N) := f(x_m, y^N, u^N) \to f(y^N, u^N) := f(x, y^N, u^N)$ as $m \to \infty$. Thus, the convergence holds by a generalized dominated convergence theorem given in Theorem C.3.1(i) (see [658, Theorem 3.5] or [454, Theorem 3.5]). Thus, continuity of $J_N(\pi_N, \gamma^N)$ in (π_N, γ^N) holds.

Step (iii) Now that we have established the continuity of $J_N(\pi_N, \gamma^N)$, it follows that by the compactness of Γ^N (which is independent of the *state* of the system),

$$J_N(\pi_N) := \inf_{\gamma^N \in \Gamma^N} J_N(\pi_N, \gamma^N)$$

is continuous in π_N. Furthermore, an optimal selection exists for γ^N through the measurable selection condition given in Lemma 5.4.1.

Step (iv) In fact, what we have in Step (iii) also implies that

$$J_{N-1}(\pi_{N-1}, \gamma_{N-1}) = E[J_N(\pi_N)|\pi_{N-1}, \gamma^{N-1}],$$

is continuous in $(\pi_{N-1}, \gamma^{N-1})$. This follows because π_N is determined by $(\pi_{N-1}, \gamma^{N-1})$, and thus what needs to be shown is that with $\pi_N = G(\pi_{N-1}, \gamma^{N-1})$,

$$\int \Big(G(\pi_{N-1}, \gamma^{N-1})\Big)(dx_N) f(x_N)$$

is continuous in π_{N-1}, γ^{N-1}. Note that

$$\pi_N(dx_{N-1}, du^{N-1}, dy^N) = \pi_{N-1}(d\bar{x}_{N-1})(Q\gamma)^{N-1}(dy^{N-1}, du^{N-1})Q^N(dy^N),$$

where we use the notation $\bar{x}_{N-1} = \{\omega_0, y^1, \ldots, y^{N-2}, u^1, \ldots, u^{N-2}\}$ as $x_{N-1} \setminus y^{N-1}$. Through a similar argument as in **Step (ii)**, continuity can be established.

Step (v) By induction, $J_1(\pi_0)$ can be established. Furthermore, we obtain that the optimal cost is continuous in π_0.

□

We formally state the implication in **Step (v)** as a theorem.

Theorem 5.4.9 *Under the hypotheses of Theorem 5.4.8, the optimal team cost is continuous in π_0, the prior measure on (ω_0, y^1).*

5.5 Control Topology via a Cartesian Product of Individual Agent Policies

In this section, we revisit the approach presented in Sect. 2.2 where we viewed the team policy simply as an N-tuple of individual policies, though here with more general topologies considered for individual policies.

5.5.1 On Classical Deterministic and Stochastic Control and Utility of Young Measures

It is instructive to first mention the significant consequences of the use of various topologies that are already well-known in classical control theory (when there is a single controller who has access to the state variable). In deterministic nonlinear, geometric, and continuous-time control, structural properties such as stabilizability, controllability, and reachability are drastically impacted by the restrictions on the classes of allowable controls (e.g., continuous, Lipschitz, finitely differentiable, or smooth control functions in the state or time, the latter when control is open-loop [146, 371, 615, 678]) and naturally the control topology induced is dictated by the class of admissible controls.

For optimal control, to allow for continuity/compactness arguments, imposing *a priori* compactness over spaces of measurable functions often creates an artificial restriction, e.g., via the Arzela-Ascoli theorem, which necessarily entails restrictive conditions on equi-continuity properties of the considered policies.

In deterministic optimal control theory, relaxed controls [759, 800] allow for the mathematical analysis on continuity-compactness to be applied with no artificial restrictions on the classes of control policies considered. A particularly consequential approach is via the study of topologies on *Young measures* defined by randomized/relaxed controls [526, 800],[162, Section 2.1],[759, p. 254],[511] where one views the topology on control policies to be identified with the weak convergence topology of a measure defined on a product space with a fixed marginal at an input/state space (typically the Lebesgue measure in optimal deterministic control). We refer the reader to Appendix F for further discussions on the Young topology.

Let us consider an open-loop controller, where the control is only a function of the time variable. We let $\nu(dt, du)$ be a measure on $[0, T] \times \mathbb{U}$ where the first marginal $\lambda(dt)$ is the normalized Lebesge measure on the time interval $[0, T]$, and let $\nu(du|t) = 1_{\{\gamma(t) \in du\}}$ be the conditional measure induced by deterministic open loop control. So, any deterministic open-loop control is embedded via:

$$\nu(dt, du) = \lambda(dt)\, 1_{\{\gamma(t) \in du\}}.$$

If we allow for randomized policies, we obtain the set $\mathcal{P}_\lambda([0, T] \times \mathbb{U})$ of all probability measures with fixed marginal on $[0, T]$. This set is weakly closed, whose extreme points are those induced by deterministic policies (as was also discussed in the context of (F.3)). Thus, any deterministic optimal control problem, which can be written in an integral form and has a lower semi-continuous cost function in actions, will have an optimal solution, which will then be deterministic as these form the extreme points of randomized controls. It can also be shown that such policies are dense in the space of randomized policies, in addition to forming the extreme points in the set of randomized policies (see e.g., [105] [451], and [539, 19, Theorem 3], but also many texts in optimal stochastic control where denseness of deterministic controls have been established inside the set of relaxed controls [130]). We refer the reader to [511] for further discussions on this topic. The following example builds on these, with somewhat different arguments.

Let $\mathbb{X} = \mathbb{R}, \mathbb{U} = [0, 1]$, and $f : \mathbb{X} \times \mathbb{U} \to [0, 1]$ and $c : \mathbb{X} \times \mathbb{U} \to [0, 1]$ be measurable functions continuous in the control action variable. Consider the following optimal control problem:

$$\inf_{\substack{\gamma_t:\mathbb{X}\to\mathbb{U} \\ u_t=\gamma_t(x_t)}} \int_0^1 c(x_t, u_t)\, \lambda(dt), \quad \text{subject to} \quad \frac{dx}{dt} = f(x_t, u_t) \tag{5.11}$$

The natural space to consider is the set of all control functions which depends on the current state, where the only restriction is measurability. However, allowing for measurability only does not facilitate continuity/compactness arguments since, as noted above, imposing compactness on a space of functions is an unnecessarily restrictive condition. Accordingly, one often cites appropriate but tedious measurable selection theorems building on optimality equations through dynamic programming.

On the other hand, given any state-feedback control and given a fixed initial state x_0, there exists a corresponding open-loop control which leads to the same value for (5.11) as the state-feedback control. Hence, such deterministic control problems, as in (5.11) may as well be formulated over the space of open-loop controls, as far as existence is concerned.

Accordingly, we consider open-loop controls and those which are relaxed. With $T = 1$, let $\mathcal{P}_\lambda([0,T]\times\mathbb{U})$ be the set of relaxed open-loop policies (Young measures). We place the weak convergence topology on $([0,T]\times\mathbb{U})$, and call this the *Young topology* on relaxed controls.

Now consider the space $C([0,1];\mathbb{X}) \times \mathcal{P}_\lambda([0,T]\times\mathbb{U})$, where $C([0,1];\mathbb{X})$ is the space of continuous functions from $[0,1]$ to $\mathbb{X}$. We endow this space with the product topology, with the first component being under the supremum norm and the second under Prohorov metric (or any weak convergence inducing metric) (i.e., the Young topology).

Note that the cost (5.11) is now continuous on $C([0,1];\mathbb{X}) \times \mathcal{P}_\lambda([0,T]\times\mathbb{U})$. Note also that since f is uniformly bounded, we have that the set $\mathcal{A}$ of all admissible sample paths of the state $x : [0,1] \to \mathbb{X}$ is equicontinuous, and thus, by the Arzela-Ascoli theorem, $\mathcal{A}$ is relatively compact in $C([0,1];\mathbb{X})$. Accordingly, our space of interest $\mathcal{A} \times \mathcal{P}_\lambda([0,T]\times\mathbb{U})$ is a relatively compact subset of $C([0,1];\mathbb{X}) \times \mathcal{P}_\lambda([0,T]\times\mathbb{U})$. Define now

$$H = \{(x,m) \in C([0,1];\mathbb{X}) \times \mathcal{P}_\lambda([0,T]\times\mathbb{U}) :$$
$$x_t - \int_0^t f(x_s,u)\, m_s(du)\, \lambda(ds) = 0\Big\},$$

where $m_s(du) = m(du|s)$. Now note that this set is closed under the topology defined on $C([0,1];\mathbb{X}) \times \mathcal{P}_\lambda([0,T]\times\mathbb{U})$ and a relatively compact subset of $C([0,1];\mathbb{X}) \times \mathcal{P}_\lambda([0,T]\times\mathbb{U})$. Hence, H is compact. The problem then is to find an optimal $(x,m) \in H$ which attains the solution of (5.11), reformulated as:

$$\inf_{(x,m)\in H} \int_0^1 c(x_t,u)\, m(dt,du)$$

This is continuous in (x,m) by an application of the generalized weak convergence theorem under continuous convergence (see Theorem C.3.1 or [658, Theorem 3.5] or [454, Theorem 3.5]). Therefore, there exists an optimal solution to the problem.

We also note that, as we will see in detail in Example 8.5.1 (building on [828, p. 878]), the set of deterministic controls is not weakly closed (or setwise closed) under Young's relaxation. In fact, there are control problems where an optimal solution exists among relaxed controls but not in deterministic controls [451].

In both continuous-time and discrete-time stochastic optimal control, generalizing what we covered above, by placing an appropriate topology on control policies, can lead to general existence results, and also approximation results. In continuous-time optimal stochastic control, it has been shown that under mild conditions, induced costs are continuous in control policies under a related topology introduced by Borkar [131] (see Appendix F), which will also be studied further in Sect. 5.5.3, under several cost criteria. A significant practical implication is that one can discretize both time and space to obtain near-optimal policies [590].

5.5.2 Trading-Off Continuity with Compactness: A Product Space Approach on Individual Control Policies

To make the ideas in this section more crisp, consider the set of individually randomized strategic measures $L_R(\mu)$ studied in Sect. 4.1.2. In this set, a policy of a DM k is a stochastic kernel $\Pi^k(du^k|y^k)$ from $\mathbb{Y}^k$ to $\mathbb{U}^k$. Instead of studying continuity and compactness properties regarding strategic measures, we can focus on individual control policies and measurements, and their Cartesian products. Let us introduce, as in Sect. 4.1.1,

$$\Theta^i = \{v \in \mathcal{P}(\mathbb{Y}^i \times \mathbb{U}^i) : v(dy^i \times \mathbb{U}^i) = \xi^i(dy^i)\} \tag{5.12}$$

to be the distributional strategy for DM i, which is the joint probability measure with a fixed marginal ξ^i on the measurement variable y^i. Note that if the problem is an independent static reduction of a dynamic team, then one can choose ξ^i as the reference measure Q^i in the static reduction.

By our earlier analysis, we know that Θ^i (or a subset of Θ^i that is sufficient for optimality) is compact under the weak topology, or the w-s topology if $\mathbb{U}^i$ is compact or if this subset satisfies some moment condition. Hence, since the marginals on measurements are fixed, we can view the weak topology or w-s topology on Θ^i as a topology on the set of policies Γ^i. In particular, any sequence of policies:

$$\underline{\gamma}_n := (\gamma_n^1, \cdots, \gamma_n^N)$$

will have a converging subsequence, $\underline{\gamma}_{n_k}$, to a limit $\underline{\gamma}$, and therefore, closedness (in particular compactness) is not an issue under this topology. In other words, the IS is always preserved under this topology, which is not the case in strategic measure approach as it has been shown in Theorem 4.2.2.

Note that if the measurements are independent, then strong existence results can be obtained using this approach. This is precisely the same condition obtained via the strategic measures approach studied in Theorem 4.2.4.

On the other hand, what is not clear is whether we have, in general,

$$\lim_{n\to\infty} J(\underline{\gamma}_n) = J(\underline{\gamma})$$

under this topology (this is always true when we place a topology on the set of strategic measures). In particular, if we have a general $\mu(d\omega_0, d\mathbf{y})$, it is not clear if we have this *joint continuity* condition. The following example will demonstrate this subtlety through a negative implication. Therefore, there is a trade-off between continuity and compactness in these approaches:

(i) The strategic measure approach provides continuity but may not preserve IS (and thus compactness);

(ii) Conversely, a product topology on sets of individual policies leads to a compact policy space but entails loss of continuity.

Example 5.5.1 Consider the counterexample presented in Theorem 4.2.2. By considering each individual DM's policy separately, we have that each will converge individually; that is,

$$P_n(dy^1, du^1) \to P(dy^1, du^1), \;\; P_n(dy^2, du^2) \to P(dy^2, du^2).$$

Now, consider a cost function $c(\omega_0, u^1, u^2) = (u^1 - u^2)^2$. Note that, for each n, under P_n, actions u^1 and u^2 are the same (i.e., $u^1 = u^2$) given $y^1 = y^2 = y$, and thus $J(\underline{\gamma}_n) = \int P_n(ds)\, c(s) = 0$ for every n. However, in the limit where u^i and $y = y^i$ $(i = 1, 2)$ are independent under P, we have $J(\underline{\gamma}_n) = \int P(ds)\, c(s) = 1/2$. This is a consequence of viewing individual policies separately without their joint convergence properties. Thus, we do not even have lower semi-continuity. Hence, while we have established compactness, we have lost lower semi-continuity by placing the topologies directly on control policies and not on the strategic measures.

5.5.3 A Weak-Topology on Randomized Team Decision/Control Policies*

In this sub-section, inspired by the topology on policies for classical stochastic control problems introduced by Borkar [131] (see also [34, Section 2.4]), we study a topology on the set of randomized policies for static teams or dynamic teams that admit static reduction studied in [620] (see Appendix F). By an abuse of notation, we denote the set of randomized policies for DM i by Γ^i, which can be written as $\Gamma^i = \{\gamma^i : \gamma^i$ is a measurable function from $\mathbb{Y}^i$ to $\mathcal{P}(\mathbb{U}^i)\}$, where $\mathcal{P}(\mathbb{U}^i)$ is endowed with the Borel σ-algebra generated by weak convergence topology. We will use the latter characterization of Γ^i when introducing the topology. For a metric space $\mathbb{E}$, let $C_0(\mathbb{E})$ be the Banach space of all continuous real-valued functions on $\mathbb{E}$ vanishing at infinity, endowed with the norm

$$\|g\|_\infty = \sup_{e \in \mathbb{E}} |g(e)|.$$

Recall that $\mathcal{M}(\mathbb{E})$ and $\mathcal{P}(\mathbb{E})$ denote the set of all finite signed measures and probability measures on $\mathbb{E}$. Let $\mathbb{E}_1$ and $\mathbb{E}_2$ be two metric spaces. For any $\mu \in \mathcal{M}(\mathbb{E}_1 \times \mathbb{E}2)$, let $\mathrm{Proj}_{\mathbb{E}_1}(\mu)(\cdot) = \mu(\cdot \times \mathbb{E}_2)$ be the marginal of μ on $\mathbb{E}_1$. For a Banach space $\mathbb{K}$, let $\mathbb{K}^*$ denote its topological dual. For any i, we now introduce a topology on Γ^i. To this end, we impose the following assumption.

Assumption 5.5.1

(a) Ω_0, $\mathbb{Y}^i$, and $\mathbb{U}^i$ $(i = 1, \ldots, N)$ are locally compact.

(b) *For all i, the measurement channel has the characterization $g^i(dy_i|\omega_0) = g^i(y_i, \omega_0)\, \mu^i(dy^i)$ for some probability measure μ^i on $\mathbb{Y}^i$; that is, $g^i(dy_i|\omega_0)$ is absolutely continuous with respect to $\mu^i(dy^i)$ for all ω_0.*

By the Riesz representation theorem [269, Theorem 7.17], the topological dual of $(C_0(\mathbb{U}^i), \|\cdot\|_\infty)$ is $(\mathcal{M}(\mathbb{U}^i), \|\cdot\|_{TV})$, where $\|\cdot\|_{TV}$ is the total variation norm on $\mathcal{M}(\mathbb{U}^i)$. We now first define the set of w^*-measurable functions from $\mathbb{Y}^i$ to $\mathcal{M}(\mathbb{U}^i)$. Subsequently, we will prove that the set of randomized policies Γ^i for DM i is a bounded subset of this set. A function $\gamma : \mathbb{Y}^i \to \mathcal{M}(\mathbb{U}_i)$ is called w^*-measurable if the mapping

$$\mathbb{Y}^i \ni y \mapsto \int d\gamma(y)(du)\, g(u) \in \mathbb{R}$$

is measurable for all $g \in C_0(\mathbb{U}^i)$ [165, p. 18]. Let $\mathcal{L}\big(\mu^i, \mathcal{M}(\mathbb{U}^i)\big)$ denote the set of all such functions. With this definition, we now define the following set

$$\mathcal{L}_\infty\big(\mu^i, \mathcal{M}(\mathbb{U}^i)\big) := \Big\{\gamma \in \mathcal{L}\big(\mu^i, \mathcal{M}(\mathbb{U}^i)\big) : \|\gamma\|_\infty := \operatorname*{ess\,sup}_{y \in \mathbb{Y}^i} \|\gamma(y)\|_{TV} < \infty\Big\},$$

where ess sup is taken with respect to the measure μ^i. Note that μ_i is the reference probability measure in Assumption 5.5.1-(b) for the measurement channel g^i.

Let $L_1\big(\mu^i, C_0(\mathbb{U}^i)\big)$ denote the set of all (Bochner-)integrable [216] functions from $\mathbb{Y}^i$ to $C_0(\mathbb{U}^i)$, endowed with the norm

$$\|f\|_1 := \int_{\mathbb{Y}^i} \|f(y)\|_\infty\, \mu^i(dy^i).$$

By using the fact that $C_0(\mathbb{U}^i)^* = \mathcal{M}(\mathbb{U}^i)$, one can prove that the topological dual of $\big(L_1\big(\mu^i, C_0(\mathbb{U}^i)\big), \|\cdot\|_1\big)$ can be identified by $\big(\mathcal{L}_\infty\big(\mu^i, \mathcal{M}(\mathbb{U}^i)\big), \|\cdot\|_\infty\big)$ [165, Theorem 1.5.5, p. 27]; that is,

$$L_1\big(\mu^i, C_0(\mathbb{U}^i)\big)^* = \mathcal{L}_\infty\big(\mu^i, \mathcal{M}(\mathbb{U}^i)\big).$$

Using this identification, we can equip $\mathcal{L}_\infty\big(\mu^i, \mathcal{M}(\mathbb{U}^i)\big)$ with the w^*-topology induced by $L_1\big(\mu^i, C_0(\mathbb{U}^i)\big)$. We write $\gamma_n \rightharpoonup^* \gamma$, if γ_n converges to γ in $\mathcal{L}_\infty\big(\mu^i, \mathcal{M}(\mathbb{U}^i)\big)$ with respect to the w^*-topology. This topology is equivalent to Borkar's topology introduced in [131] (see [34, Section 2.4]), which was introduced for randomized Markov policies to study continuous-time stochastic control problems. Note that we can identify the set of randomized policies, Γ^i, as a bounded subset of $\mathcal{L}_\infty\big(\mu^i, \mathcal{M}(\mathbb{U}^i)\big)$:

$$\Gamma^i = \Big\{\gamma \in \mathcal{L}_\infty\big(\mu^i, \mathcal{M}(\mathbb{U}^i)\big) : \gamma(y) \in \mathcal{P}(\mathbb{U}^i)\ \mu^i - \text{a.e.}\Big\}.$$

Hence, we can equip Γ^i with the relative w^*-topology inherited by the w^*-topology on $\mathcal{L}_\infty(\mu^i, \mathcal{M}(\mathbb{U}^i))$.

Now, we derive some properties of this topology on Γ^i. Firstly, since $\mathcal{P}(\mathbb{U}^i)$ is bounded in $\mathcal{M}(\mathbb{U}^i)$, by the Banach-Alaoglu Theorem [269, Theorem 5.18], Γ^i is relatively compact with respect to the w^*-topology. Since $C_0(\mathbb{U}^i)$ is separable, by [344, Lemma 1.3.2], Γ^i is also relatively sequentially compact. However, Γ^i is not closed with respect to the w^*-topology. Indeed, let $\mathbb{Y}^i = \mathbb{U}^i = \mathbb{R}$. Define $\gamma_n(y^i)(\cdot) := \delta_n(\cdot)$ and $\gamma(y^i)(\cdot) := 0(\cdot)$, where δ_n denotes Dirac-delta measure at point n and $0(\cdot)$ denotes the degenerate (or trivial) measure on $\mathbb{U}^i$; that is, $0(D) = 0$ for all $D \in \mathcal{B}(\mathbb{R})$. Let $g \in L_1(\mu^i, C_0(\mathbb{U}^i))$. Then we have

$$\begin{aligned}
\lim_{n\to\infty} \int_{\mathbb{Y}^i} \left(\int_{\mathbb{U}^i} \gamma_n(y)(du)\, g(y)(u) \right) \mu^i(dy) &= \lim_{n\to\infty} \int_{\mathbb{Y}^i} g(y)(n)\, \mu^i(dy) \\
&= \int_{\mathbb{Y}^i} \lim_{n\to\infty} g(y)(n)\, \mu^i(dy) \text{ (as } \|g(y)\|_\infty \text{ is } \mu^i\text{-integrable)} \\
&= 0 \text{ (as } g(y) \in C_0(\mathbb{U}^i)).
\end{aligned}$$

Hence, $\gamma_n \rightharpoonup^* \gamma$. But, $\gamma \notin \Gamma^i$, and thus Γ^i is not closed. Hence, as opposed to the topology introduced in Sect. 5.5 on randomized policies, here the set of policies is not closed under the w^*-topology; that is, the information structure constraints may not be preserved. However, as we will see in the sequel, under this topology, the cost J is lower semi-continuous, which is in general not the case for the topology introduced in Sect. 5.5.2.

However, by imposing additional assumptions on the components of the problem, we may first ensure closedness of the set of policies or subset of it that is sufficient for optimality under this topology using lower semi-continuity of the cost. Then, we can prove the existence of an optimal policy. This result will be complementary to the existence results presented in Sect. 4.2.

In the remainder of this analysis, Γ^i is equipped with this topology. In addition, $\mathbf{\Gamma}$ has the product topology induced by these w^*-topologies; that is, $\underline{\gamma}_n$ converges to $\underline{\gamma}$ in $\mathbf{\Gamma}$ with respect to the product topology if and only if $\gamma_n^i \rightharpoonup^* \gamma^i$ for all $i = 1, \ldots, N$. In this case, we write $\underline{\gamma}_n \rightharpoonup^* \underline{\gamma}$. Note that $\mathbf{\Gamma}$ is sequentially relatively compact under this topology.

To this end, for any $L > 0$, we define

$$\mathbf{\Gamma}_L := \left\{ \underline{\gamma} \in \mathbf{\Gamma} : J(\underline{\gamma}) < J^* + L \right\}$$

and

$$S_L := \left\{ \lambda \in \mathcal{P}\left(\Omega_0 \times \prod_{i=1}^N (\mathbb{Y}^i \times \mathbb{U}^i) \right) : \right.$$

$$\lambda(d\omega_0, d\mathbf{y}, d\mathbf{u}) = \mu(d\omega_0) \prod_{i=1}^{N} \gamma^i(du^i|y^i)\mu^i(dy^i), \underline{\gamma} \in \mathbf{\Gamma}_L \Big\}.$$

For each $i = 1, \ldots, N$, we define $S_L^i := \mathrm{Proj}_{\mathbb{Y}^i \times \mathbb{U}^i}(S_L)$. In order to prove that $\mathbf{\Gamma}_L$ is closed with respect to the w^*-topology, we impose the assumption below.

Assumption 5.5.2 *For some $L > 0$, S_L^i is tight for $i = 1, \ldots, N$.*

We provide in the following theorem several conditions that imply Assumption 5.5.2.

Theorem 5.5.1 ([620, Theorem 4]) *Suppose that either of the following two conditions hold:*

(i) *$\mathbb{U}^i$ is compact for all i.*
(ii) *If $\mathbb{U}^i$ is not compact,*

(a) *The cost function c satisfies the following condition: for each j, for any $M \geq 0$, and for any compact $K \subset \Omega_0 \times \prod_{k=1}^{N} \mathbb{Y}^k \times \prod_{k=1}^{j-1} \mathbb{U}^k$, there exists a compact $L \subset \mathbb{U}^j$ such that*

$$\inf_{K \times L^c \times \prod_{k=j+1}^{N} \mathbb{U}^k} c(\omega_0, \mathbf{y}, \mathbf{u}) \geq M.$$

and

(b) *For all j, $g^j > 0$ and g^j is lower semi-continuous.*

Then, Assumption 5.5.2 holds.

The following theorem establishes the existence of an optimal team decision rule using the w^*-topology.

Theorem 5.5.2 ([620, Theorem 6]) *Suppose that Assumptions 5.5.1 and 5.5.2 hold. Moreover, c is lower semi-continuous and measurement channels $g^i(dy^i|\omega_0)$ are continuous with respect to the total variation distance. Then, there exists $\underline{\gamma}^* \in \mathbf{\Gamma}_L$ which is optimal.*

Proof When the cost function c is a compactly supported continuous function, by the Stone-Weierstrass Theorem [453, Lemma 6.1], c can be uniformly approximated by functions of the form

$$\sum_{j=1}^{k} r_j \prod_{i=1}^{N} f_{j,i} g_{j,i}, \tag{5.13}$$

where $r_j \in C_0(\Omega_0)$, $f_{j,i} \in C_0(\mathbb{Y}^i)$, and $g_{j,i} \in C_0(\mathbb{U}^i)$ for each $j = 1, \ldots, k$, and $i = 1, \ldots, N$, are compactly supported. The cost J is continuous with respect to the w^*-topology if c is of the form in (5.13) [620, Theorem 5]. Since any compactly supported function can be uniformly approximated by such functions, J is also continuous if c is compactly supported. Note that if c is lower semi-continuous, then c can be approximated pointwise from below by compactly supported functions (see the proof of [344, Proposition 1.4.18]). Hence, if c is lower semi-continuous, then J is also lower semi-continuous with respect to the w^*-topology by monotone convergence theorem. Now, we prove that $\mathbf{\Gamma}_L$ is closed under the w^*-topology, which will complete the proof since $\mathbf{\Gamma}_L$ is relatively compact. To this end, let

$$\underline{\gamma}_n \rightharpoonup^* \underline{\gamma},$$

for some $\underline{\gamma} \in \prod_{i=1}^N \mathcal{L}_\infty\big(\mu^i, \mathcal{M}(\mathbb{U}^i)\big)$ (recall that $\mathbf{\Gamma}$ is not closed with respect to w^*-topology). If $\gamma^i \in \mathcal{P}(\mathbb{U}^i)$ μ^i-a.e. for all i, then $\underline{\gamma}$ is in $\mathbf{\Gamma}_L$, and thus $\mathbf{\Gamma}_L$ is closed.

Fix any i. Note that the sequence $\{\gamma_n^i \otimes \mu^i\}$ is tight as it is a subset of S_L^i. Thus, there exists a further subsequence, denoted for simplicity by $\{\gamma_l^i \otimes \mu^i\}$, that converges weakly to some $\lambda \in \mathcal{P}(\mathbb{Y}^i \times \mathbb{U}^i)$. This implies that $\gamma^i \otimes \mu^i = \lambda$, and thus $\gamma^i \otimes \mu^i(\mathbb{Y}^i \times \mathbb{U}^i) = 1$. Hence, $\gamma^i \in \mathcal{P}(\mathbb{U}^i)$ μ^i-a.e. and $\mathbf{\Gamma}_L$ is closed. This completes the proof. □

Remark 5.5.1 As a related argument, we refer the reader to [436, 437] where weak topology is used in a similar context. In these works, [437, Lemma 4.9] and [436, Lemma A.1] can be exploited to prove the existence of optimal policies. Indeed, if one can prove that actions are square integrable under any strategy, the set of actions is weakly relatively compact, and the cost function is lower semi-continuous, then the existence of optimal policies can be deduced.

Remark 5.5.2 As noted earlier, three theorems stated in the previous chapter, Theorem 4.2.4 (for static teams or dynamic teams with an independent-measurements reduction) and Theorems 4.2.5 and 4.2.6 (for sequential teams that do not allow an independent-measurements reduction) are the most general existence results and thus generalize the analysis presented in this subsection as well (in particular, Theorem 4.2.4).

5.5.4 Hilbert Space Construction for Static LQG Teams

It is instructive to recall that for the proof of Theorem 2.4.7 involving static LQG teams, the team policy was given a product Hilbert space topology: Γ is the product space $\Gamma^1 \times \cdots \times \Gamma^N$, where each Γ^i was a Hilbert space, with the inner product

$$\langle \alpha^i, \beta^i \rangle_i := E[\alpha^i(y^i)'\beta^i(y^i)]$$

This makes Γ also a Hilbert space, with the inner product

$$\langle \underline{\alpha}, \underline{\beta} \rangle := E\left\{\sum_{i \in \mathcal{N}} \alpha^i(y^i)'\beta^i(y^i)\right\} \equiv E[\underline{\alpha}(\mathbf{y})'\underline{\beta}(\mathbf{y})].$$

Via this topology on the controls, the optimality of an affine solution was established there via the Projection theorem.

We emphasize that this construction has interesting properties, in view of the discussion presented so far, in that it satisfies simultaneously both *closedness* (this can be verified e.g. via [335, Prop 2.9(a,b)]) and *continuity*, and thus offers a consequential distinction with the constructions presented in Sect. 5.5.2. Recall that the issue in the product metric approach studied in Sect. 5.5.2 was that the cost was not continuous, which is not the case in the Hilbert space formulation (since the cost criterion here is formulated via the inner product which defines the criterion of convergence).

Remark 5.5.3 (Extended Weak Convergence, Topology of Information, and Adapted Topologies)
A versatile topology, which has evidently been used in a variety of contexts in stochastic analysis, is roughly given by the following convergence notion. Consider a stochastic process converging to another one in the following sense: all finite dimensional marginals converge weakly and the conditional kernels on the future random variables given the past converge weakly as well when conditional kernels are viewed as measure-valued stochastic processes.

This notion has been applied to different problems. Aldous has termed it *extended weak convergence* [14] and Hellwig has named it *the information topology* [340]; these have recently been shown to be equivalent in discrete-time [49, Theorem 1.1]. In addition to the conditional independence preservation [340] [325], applications in robust stochastic control also follow from this discussion, with details being very context-specific, as in [48, 104, 389, 454].

In relation to our context, such a convergence notion again requires strong continuity conditions under all admissible policies. In the same spirit of the discussion in Sect. 5.5.2, even though continuity holds under convergence with this notion, compactness will require more restrictive conditions. With this approach, the recent work [325] (see also [69]) has established existence results for a setup where either the measurements are countable or there is a common information among DMs which is countable space-valued and with the private information satisfying an absolute continuity condition. As noted earlier in Chap. 4, static reduction applies to such settings and the results presented in that chapter (notably Theorem 4.2.4) apply leading to more relaxed results. We recall also that the use of w-s topology in Theorem 4.2.4 relaxes the requirements of continuity.

5.6 Some Applications and Revisiting Existence Results for Classical (Single-DM) Stochastic Control

Some of the results from this chapter as well as Chap. 4 can be utilized to establish existence results for some important special instances and classical results. These also serve as validations of the general theory presented here when tested under classical stochastic control problems.

5.6.1 Witsenhausen's Counterexample with Gaussian Variables

Consider Witsenhausen's counterexample as depicted in Figs. 3.3 and 3.2: This is a dynamic non-classical team problem with y^1 and w^1 zero-mean independent Gaussian random variables with unit variance and $u^1 = \gamma^1(y^1)$, $u^2 = \gamma^2(u^1 + w^1)$ and the cost function $c(\omega, u^1, u^2) = k(y^1 - u^1)^2 + (u^1 - u^2)^2$ for some $k > 0$. Witsenhausen's counterexample can be expressed, through a change of measure argument (also due to Witsenhausen) as follows. Recall that the static reduction for this problem was given in (3.32), where Q denotes a Gaussian measure with zero mean and unit variance and η its density. Since the optimal policy for **DM**2 is $\gamma^2(y^2) = E[u^1|y^1]$ and $E[(E[u^1|y^1])^2] \leq E[(u^1)^2]$, it is evident with a two-stage analysis (see [327, p. 1701]) that without any loss we can restrict the policies to be so that $E[(u^i)^2] \leq M$ for some finite M, for $i = 1, 2$; this ensures a weak compactness condition on both $\hat{\gamma}^1$ and $\hat{\gamma}^2$. Since the reduced cost $\left((k(u^1 - y^1)^2 + (u^1 - u^2)^2)\frac{\eta(y^2 - u^1)}{\eta(y^2)}\right)$ is continuous in the actions, Theorem 4.2.4 applies.

5.6.2 Existence for Partially Observable Markov Decision Processes (POMDPs)

Consider a partially observable stochastic control problem with the following dynamics:

$$x_{t+1} = f(x_t, u_t, w_t), \qquad y_t = g(x_t, v_t).$$

Here, x_t is the $\mathbb{X}$-valued state, u_t is the $\mathbb{U}$-valued control, and y_t is the $\mathbb{Y}$-valued measurement variable, where we will assume that these three spaces are finite dimensional real vector spaces. Furthermore, (w_t, v_t) are i.i.d. noise processes and $\{w_t\}$ is independent of $\{v_t\}$. The controller only has causal access to $\{y_t\}$: A deterministic admissible control policy γ is a sequence of functions $\{\gamma_t\}$ so that $u_t = \gamma_t(y_{[0,t]}; u_{[0,t-1]})$. The goal is to minimize

$$E_{x_0}^{\gamma}[\sum_{t=0}^{T-1} c(x_t, u_t)],$$

for some continuous and bounded $c : \mathbb{X} \times \mathbb{U} \to \mathbb{R}_+$.

Such a problem can be viewed as a decentralized stochastic control problem with increasing information, that is, one with a *classical* information structure.

As discussed in Sect. E.3, any POMDP can be reduced to a (completely observable) MDP whose states are the posterior state distributions or *beliefs* of the observer. A standard approach for solving such problems then is to reduce the partially observable model to a fully observable one (also called the belief-MDP) by defining $\pi_t(A) := P(x_t \in A | y_{[0,t]}, u_{[0,t-1]})$, $A \in \mathcal{B}(\mathbb{X})$, and observing that (π_t, u_t) is a controlled Markov chain where π_t is $\mathcal{P}(\mathbb{X})$-valued with $\mathcal{P}(\mathbb{X})$ being the space of probability measures on $\mathbb{X}$ under the weak convergence topology. Through such a reduction, existence results can be established by obtaining conditions which would ensure that the controlled Markovian kernel for the belief-MDP is weakly continuous, that is if $\int F(\pi_{t+1}) P(d\pi_{t+1} | \pi_t = \pi, u_t = u)$ is jointly continuous (weakly) in π and u for every continuous and bounded function F on $\mathcal{P}(\mathbb{X})$. Theorem E.3.2 presents sufficient conditions for weak continuity (also know as the weak Feller property) of the belief-MDP.

In the context of the model presented, if $f(\cdot, \cdot, w)$ is continuous and g has the form: $y_t = g(x_t) + v_t$, with g continuous and w_t admitting a continuous density function η, an existence result can be established building on the measurable selection criteria under weak continuity (see Sect. E.2.4), provided that $\mathbb{U}$ is compact.

On the other hand, through Theorem 4.2.6, such an existence result can also be established by obtaining a static reduction under the aforementioned conditions. Indeed, through (3.31), with η denoting the density of v_n, we have $P(y_n \in B | x_n) = \int_B \eta(y - g(x_n)) dy$. With η and g continuous and bounded, taking $y^n := y_n$, by writing $x_{n+1} = f(x_n, u_n, w_n) = f(f(x_{n-1}, u_{n-1}, w_{n-1}), u_n, w_n)$, and iterating inductively to obtain

$$x_{n+1} = h_n(x_0, \mathbf{u}_{[0,n-1]}, \mathbf{w}_{[0,n-1]}),$$

for some h_n which is continuous in $\mathbf{u}_{[0,n-1]}$ for every fixed $x_0, \mathbf{w}_{[0,n-1]}$, one obtains a reduced cost (3.31) that is a continuous function in the control actions. Theorem 4.2.6 then implies the existence of an optimal control policy. This reasoning is also applicable when the measurements are not additive in the noise but with $P(y_n \in B | x_n = x) = \int_B m(y, x) \eta(dy)$ for some m continuous in x and η a reference measure.

5.6.3 *Revisiting Fully Observable Markov Decision Processes with the Construction Presented in the Chapter*

Consider a fully observed Markov decision process where the goal is to minimize

$$E_{x_0}^{\gamma}[\sum_{t=0}^{T-1} c(x_t, u_t)],$$

for some continuous and bounded $c : \mathbb{X} \times \mathbb{U} \to \mathbb{R}_+$. Suppose that the controller has access to $x_{[0,t]}, u_{[0,t-1]}$ at time t. This system can always be viewed as a sequential team problem with a classical information structure. Under the assumption that the transition kernel according to the usual formulation, that is $P(dx_1|x_0 = x, u_0 = u)$ is weakly continuous (in the sense discussed in the previous application above), it follows that the transition kernel according to the formulation introduced in Sect. 5.4 is also weakly continuous by an application of [658, Theorem 3.5]. It follows that the condition in Lemma 5.4.2 holds when $\mathbb{U}$ is compact, and hence the existence of an optimal policy follows. A similar analysis is applicable when one considers the case where $P(dx_1|x_0 = x, u_0 = u)$ is strongly continuous in u for every fixed state x and the bounded cost function is continuous only in u (this is another typical setup where measurable selection conditions hold [342, Chapter 3]).

5.7 Concluding Remarks

We have presented in this chapter various topologies on decision/control strategies defined independent of information structures, but for which information structures determine whether the topologies have utility in arriving at existence, compactness, convexification or approximation results. We have applied these topological constructions to obtain various dynamic programming formulations and existence results as well as to be able to study teams with large numbers of DMs (as we will do in the next chapter).

5.8 Bibliographic Notes

As we discussed in Sect. 3.1, in the context of single DM stochastic control, there are a few main techniques utilized to establish the existence and structure of optimal policies for finite-horizon and infinite-horizon problems. One such approach is based on *dynamic programming* and the corresponding *measurable selection criteria*, which has been covered elaborately in [342, Theorem 3.3.5 and Appendix D] and [343] among many other references. Another method is based on the properties of *strategic measures* [230, 254, 585, 650]. These two techniques provide comprehensive sufficient conditions for existence results. There are also techniques based on empirical measures, which are particularly useful for infinite-horizon problems, for both the average cost as well as discounted cost criteria, leading to linear programming/convex analytical conditions (e.g., [342, Chp. 6][135]). However, all of these approaches in general crucially require that the information is non-contracting over time, in the sense that the σ-field generated by the information available at the decision maker at time n is a subset of that generated by the information at $n + 1$, and so on. In decentralized stochastic control, this nestedness condition does not occur in general, and hence the classical methods are not typically applicable.

That dynamic programming can be a useful tool for a class of dynamic teams has been known since the beginning of the 1970s. A possible approach toward establishing a tractable dynamic program is through the construction of a controlled Markov chain where the controlled Markov state may now live in a larger state space and the actions are elements in possibly function spaces. Such a *dynamic programming approach* has been adopted extensively in the literature (see for example, [6, 43, 177, 452, 736, 774, 797, 806] and significantly generalized and termed as the *common information approach* in [558] and [560]) through the use of a team-policy which uses common information to generate partial functions for each DM to generate its actions using local information. This construction requires a common knowledge among decision makers, which is a unifying assumption in the aforementioned contributions in the literature. Witsenhausen [775] and [818] developed general dynamic programming algorithms, which are conceptually instructive and mathematically (though not necessarily computationally) consequential (in particular on existence and recursive analysis, as we saw earlier in the chapter). Another related method, refining the standard form of [775] utilizing common knowledge among agents, is presented in [557].

It is also worth mentioning some related efforts in the stochastic control literature dealing with systems under partial information, e.g. [369, 603]. These provide a dynamic programming formulation for a class of stochastic control problems with *perfect recall* leading to a generalized form for a Bellman (martingale) optimality principle for a special class of systems, see [369, Definition 4.1, Theorem 5.7] (see also [240] for a controlled-Markovian and perfect-recall setup). The assumption that the information structure is classical (due to the perfect-recall property, see Sect. 5.3) in these papers allows for explicit optimality equations for each sample path. Also along the dynamic programming based approaches, related contributions for a class of decentralized differential stochastic control systems (in continuous-time) are [166] and [167] which have established the Bellman (martingale) and dynamic programming equations to arrive at optimality conditions. Notably, these results also establish conditions for the global optimality of person-by-person-optimal policies for a class of convex problems.

Some of the analyses provided in this chapter appear in [818] (see also [497]) and [621].

Chapter 6
Many-Agent Convex and Non-convex Exchangeable (Mean-Field) Teams and Optimality of Symmetric Policies

In this chapter, we focus on teams with a high, and even infinite, population of DMs. We will be building on the results presented until now to arrive at optimality and structural results for such stochastic teams. We will see that often working with a high population of DMs leads to a tractable mathematical analysis for optimality and near-optimality analysis.

The results in this chapter will also make us appreciate the versatility of placing appropriate team decision and control topologies on policy spaces.

A major result of the chapter is the optimality of symmetric or exchangeable team policies for teams with infinitely many agents, and near optimality of such policies for large, but finite, teams. We will investigate the convex and non-convex setups separately, where the latter will require a more tedious analysis which is applicable to far more general models, whereas the former will give us stronger structural results for finite teams (in particular, on the optimality of symmetric policies for teams with finitely many DMs). In the infinite agent limit, optimality of symmetric policies holds both for convex exchangeable and non-convex exchangeable team problems.

We will build on the decentralized model studied in Chap. 3. Section 6.1 motivates the problem and presents the setup. Section 6.2 studies the convex static setup. Section 6.3 is on the convex dynamic case. Section 6.4 studies large teams which are not necessarily convex, but exchangeable; this covers a much larger class of models. For each of the setups, we will obtain structural results on optimal policies, existence results and approximation results. In particular, we will establish the optimality of symmetric policies in the limit of infinitely many agents and the existence of such optimal policies.

S. Yüksel, T. Başar, *Stochastic Teams, Games, and Control under Information Constraints*, Systems & Control: Foundations & Applications,
https://doi.org/10.1007/978-3-031-54071-4_6

6.1 Introduction

In the following, we motivate the general setup to be considered, in the static context (later to be generalized to the dynamic setup). Consider the following team decision problem with N DMs. We will adopt the intrinsic model presented in Sect. 3.2.

Problem $(\mathcal{P}_N)$ Let $\mathcal{N} = \{1, \ldots, N\}$, $\underline{\gamma}_N = (\gamma^1, \cdots, \gamma^N)$, and $\Gamma_N = \prod_{i=1}^N \Gamma^i$. Define an expected cost function under $\underline{\gamma}_N$ as

$$J_N(\underline{\gamma}_N) = E^{\underline{\gamma}_N}[c(\omega_0, \mathbf{u})] \tag{6.1}$$

Consider, in particular, a more specific form of (6.1) as follows: Let $\underline{\gamma}_N = \{\gamma^1, \cdots, \gamma^N\}$, and $\Gamma_N = \prod_i^N \Gamma^i$, the space of admissible policies for the team with N-DMs. Consider the following special structure of the additive cost, where the coupling between the actions is through an aggregate/average term:

$$J_N(\underline{\gamma}_N) = \frac{1}{N} E^{\underline{\gamma}_N}\Bigg[\sum_{i=1}^{N} c(\omega_0, u^i, \frac{1}{N}\sum_{p=1}^{N} u^p)\Bigg]. \tag{6.2}$$

When the number of DMs is infinite, the following would be a natural extension of (6.2).

Problem $(\mathcal{P}_\infty)$ Consider a stochastic team with countably infinite number of decision makers, that is, $\mathcal{N} = \mathbb{N}$. Let $\Gamma = \prod_{i \in \mathbb{N}} \Gamma^i$ and $\underline{\gamma} = (\gamma^1, \gamma^2, \ldots)$. Let $c : \Omega_0 \times \mathbb{U} \times \mathbb{U} \to \mathbb{R}_+$. Define the expected cost under $\underline{\gamma}$ as

$$J(\underline{\gamma}) = \limsup_{N \to \infty} \frac{1}{N} E^{\underline{\gamma}}\Bigg[\sum_{i=1}^{N} c\bigg(\omega_0, u^i, \frac{1}{N}\sum_{p=1}^{N} u^p\bigg)\Bigg]. \tag{6.3}$$

One of our goals here is to establish conditions for a team policy to be optimal, and also connect the optimal cost and policies for $(\mathcal{P}_\infty)$ and $(\mathcal{P}_N)$. We will add further structure to the problems, such as symmetry, exchangeability or convexity properties depending on the setup. Note that, as mentioned earlier, the cost function reveals a dependency among several DMs in terms of their actions via their average; such a cost dependency is known as a *mean-field* dependency.

We note that the cost function given in (6.3) can actually be made more general by considering dependency on the empirical measure of the control actions; in this case, it is only required that the cost be weakly continuous in the empirical measure. That is, in (6.3), c would now take the form

$$c(\omega_0, u^i, \mu^N),$$

with

$$\mu^N(B) = \frac{1}{N}\sum_{p=1}^{N} \delta_{u^p}(B)$$

for Borel $B \subset \mathbb{U}$. The analysis to be presented in the chapter is directly applicable to this more general case as well.

We will investigate the relations between the sequence of solutions to (6.2) and the solution to (6.3). We will study static and dynamic settings separately. In the dynamic case, we will also study the setting where the flow of the system may also have an evolution of the *mean-field* type.

6.2 Convex Static Teams with Large Numbers of DMs

The following absolute continuity condition will be useful when an independent static reduction is utilized (as in Chap. 3).

Assumption 6.2.1 *For any* **DM**i*, there exists a probability measure* Q^i *on* $\mathbb{Y}^i$ *and a function* f^i *such that for all Borel sets* $S = S^1 \times \cdots \times S^N$*, we have*

$$P((y^1, \ldots, y^N) \in S\big|\omega_0) = \prod_{i=1}^{N} \int_{S^i} f^i(y^i, \omega_0, y^1, \ldots, y^{i-1}) Q^i(dy^i). \tag{6.4}$$

As noted earlier in Chap. 3, Assumption 6.2.1 allows us to reduce the original static team problem to another static problem where the observation of each DM is independent of observations of other DMs and also independent of ω_0. Hence, under Assumption 6.2.1, we can focus on each **DM**i separately and identify Γ^i via the set of probability measures, as in (4.1):

$$\Theta^i := \left\{ P \in \mathcal{P}(\mathbb{U}^i \times \mathbb{Y}^i) \middle| P(du^i, dy^i) = 1_{\{\gamma^i(y^i) \in du^i\}} Q^i(dy^i), \gamma^i \in \Gamma^i \right\} \tag{6.5}$$

As noted earlier in Chap. 4, the above set is the set of extreme points of the set of probability measures on $(\mathbb{U}^i \times \mathbb{Y}^i)$ with fixed marginals Q^i on $\mathbb{Y}^i$ and it inherits Borel measurability and topological properties of that Borel measurable set. We define convergence on policies as $\gamma^i_n \to \gamma^i$ if $1_{\{\gamma^i_n(y^i) \in du^i\}} Q^i(dy^i) \to 1_{\{\gamma^i(y^i) \in du^i\}} Q^i(dy^i)$ (in the weak convergence topology) as $n \to \infty$ for each DM. We will also allow for randomized (relaxed) policies. Accordingly, each individual control policy $\gamma^i \in \Gamma^i$ is an element of the set of probability measures $\mathcal{P}(\mathbb{U}^i \times \mathbb{Y}^i)$ with a fixed marginal, Q^i, on $\mathbb{Y}^i$.

Sufficient Conditions of Optimality

In the following, we present sufficient conditions of team optimality for $(\mathcal{P}_\infty)$. Toward that end we will be following Sect. 2.4.2 and note that the results to be presented here are generalizations of the theorems presented there to the setting of static teams with countably infinite number of decision makers.

We first have the following assumption, under which we state and prove two main results.

Assumption 6.2.2 *Let*

A1 The function $c(\omega_0, u^i, \frac{1}{N}\sum_{p=1}^{N} u^p)$ is an $\mathbb{R}_+$-valued jointly convex function of second and third arguments and differentiable in u^i with continuous partial derivatives, for every $\omega_0 \in \Omega_0$.

A2 For some $\underline{\gamma}^* \in \Gamma$,

$$\lim_{N\to\infty} \frac{1}{N}\sum_{i=1}^{N} E^{\underline{\gamma}^*}\left[c(\omega_0, u^i, \frac{1}{N}\sum_{p=1}^{N} u^p)\right] < \infty. \tag{6.6}$$

We note that the cost function being differentiable in u^i means that it is totally differentiable in u^i, i.e., $\frac{d}{du^i}c(\omega_0, u^i, \frac{1}{N}\sum_{p=1}^{N} u^p) = \frac{\partial}{\partial u^i}c(\omega_0, u^i, \mu_N) + \frac{1}{N}\frac{\partial}{\partial \mu_N}c(\omega_0, u^i, \mu_N)$ where we write $\mu_N = \frac{1}{N}\sum_{p=1}^{N} u^p$.

Theorem 6.2.1 ([632]) *Assume that (A1) holds and (A2) holds for $\underline{\gamma}^* \in \Gamma$. If for all $\underline{\gamma} \in \Gamma$ with $J(\underline{\gamma}) < \infty$,*

$$\mathit{lim\ sup}_{N\to\infty} \frac{1}{N} E\left[\sum_{i=1}^{N}\sum_{k=1}^{N} \frac{d}{du^k}c(\omega_0, \gamma^{i*}, \mu_N^*)(\gamma^k - \gamma^{k*})\right] \geq 0, \tag{6.7}$$

where $\mu_N^* = \frac{1}{N}\sum_{p=1}^{N} \gamma^{p*}(y^p)$, then $\underline{\gamma}^*$ is a globally optimal team policy for $(\mathcal{P}_\infty)$.

The proof is in the Appendix. In some applications, (6.7) (as in Theorem 2.4.5) can be difficult to check since it must be satisfied for all $\underline{\gamma} \in \Gamma$ with $J(\underline{\gamma}) < \infty$. In the next sub-section, we address this issue by introducing a constructive approach for static teams with countably infinite number of DMs as a limit of a sequence of team optimal policies of the corresponding static teams with a finite number of DMs. We first present a useful general result.

Theorem 6.2.2 *Let $\underline{\gamma}_N^* \in \Gamma_N$ be an optimal policy for $(\mathcal{P}_N)$. If there exists $\underline{\gamma}^* \in \Gamma$, with $J(\underline{\gamma}^*) < \infty$, satisfying*

$$\mathit{lim\ sup}_{N\to\infty} J_N(\underline{\gamma}_N^*) \geq J(\underline{\gamma}^*), \tag{6.8}$$

then $\underline{\gamma}^*$ is a globally team optimal policy for $(\mathcal{P}_\infty)$.

Proof We have

$$J(\underline{\gamma}^*) \leq \limsup_{N\to\infty} \frac{1}{N} \sum_{i=1}^{N} E^{\underline{\gamma}_N^*} \Big(c(\omega_0, u^i, \mu_N) \Big) \tag{6.9}$$

$$= \limsup_{N\to\infty} \inf_{\underline{\gamma}_N \in \Gamma_N} \frac{1}{N} \sum_{i=1}^{N} E^{\underline{\gamma}_N} \Big(c(\omega_0, u^i, \mu_N) \Big) \tag{6.10}$$

$$= \limsup_{N\to\infty} \inf_{\underline{\gamma} \in \Gamma} \frac{1}{N} \sum_{i=1}^{N} E^{\underline{\gamma}} \Big(c(\omega_0, u^i, \mu_N) \Big) \tag{6.11}$$

$$\leq \inf_{\underline{\gamma} \in \Gamma} \limsup_{N\to\infty} \frac{1}{N} \sum_{i=1}^{N} E^{\underline{\gamma}} \Big(c(\omega_0, u^i, \mu_N) \Big) \tag{6.12}$$

$$= \inf_{\underline{\gamma} \in \Gamma} J(\underline{\gamma}),$$

where $\mu_N := \frac{1}{N} \sum_{p=1}^{N} u^p$ and (6.9) follows from (6.8), and (6.10) is true since γ_N^* is a team optimal policy for $(\mathcal{P}_N)$. Furthermore, (6.11) follows from the fact that $[\underline{\gamma}|_N : \underline{\gamma} \in \Gamma] = \Gamma_N$, where $\underline{\gamma}|_N$ is $\underline{\gamma}$ restricted to its first N components. ◇

Remark 6.2.1 Under (A2), one can replace (6.8) with

$$\limsup_{N\to\infty} \frac{1}{N} \sum_{i=1}^{N} \left[E^{\underline{\gamma}_N^*} \Big(c(\omega_0, u^i, \mu_N) \Big) - E^{\underline{\gamma}^*} \Big(c(\omega_0, u^i, \mu_N) \Big) \right] \geq 0. \tag{6.13}$$

6.2.1 Asymptotically Optimal Policies as a Limit of Finite Team Optimal Policies

In the following, we present a sufficient condition for (6.8). In conjunction with this, we also have a constructive method to obtain optimal policies using asymptotic analysis.

Theorem 6.2.3 ([632]) *Assume that*

(a) For every N, there exist $\underline{\gamma}_N^* \in \Gamma_N$ for $(\mathcal{P}_N)$,

(b) For some $B \in \mathcal{F}$ of P measure 1, for all $\omega \in B$ and for every fixed $y^i(\omega)$, $\gamma_N^{i*}(y^i)$ converges to $\gamma_\infty^{i*}(y^i)$ uniformly in $i = 1, 2, \ldots, N$, i.e.,

$$\lim_{N\to\infty} \sup_{1\leq i\leq N} |\gamma_N^{i*}(y^i) - \gamma_\infty^{i*}(y^i)| = 0 \; P - a.s.,$$

(c) There exists a P-integrable function $g(\omega_0, \underline{y})$ such that, for every N,

$$\frac{1}{N}\sum_{i=1}^{N} c\left(\omega_0, \gamma_{\infty}^{i*}(v^i), \frac{1}{N}\sum_{p=1}^{N} \gamma_{\infty}^{p*}(v^p)\right) \leq g(\omega_0, \underline{v}),$$

where $\underline{v} = (v^1, v^2, \ldots)$.

Then $\underline{\gamma}^*$*, a team optimal policy for* $(\mathcal{P}_\infty)$*, is a pointwise limit of* $\underline{\gamma}_N^*$*, an optimal policy for* $(\mathcal{P}_N)$*, i.e.,* $\gamma^{i*}(v^i) = \lim_{N\to\infty} \gamma_N^{i*}(v^i) = \gamma_{\infty}^{i*}(v^i)$ *P-almost surely.*

In the following sub-section, we show that for symmetric teams sufficient conditions of optimality can be satisfied quite effortlessly.

6.2.2 Symmetric Teams

In the following, we present sufficient conditions for team optimality in symmetric and mean-field teams. The concept of symmetry has been studied in a variety of contexts; see e.g., [202, 555].

Definition 6.2.1 (Exchangeable Teams) An N-DM team is *exchangeable* if the value of the expected cost function (see (6.1)) is invariant under every permutation of policies.

Definition 6.2.2 (Symmetrically Optimal Teams) A team is *symmetrically optimal,* if for every given policy, there exists an identically symmetric policy (i.e., each DM has the same policy) which performs at least as good as the given policy.

In the following, we characterize the symmetry of the general setup for $(\mathcal{P}_N)$. First, we recall the definition of an *exchangeable* finite set of random variables.

Definition 6.2.3 Random variables $x^1, x^2, \ldots, x^N$ are *exchangeable* if any permutation, σ, of the set of indices $\{1, \ldots, N\}$ does not change the joint probability measures of the random variables, i.e., $P(dx^{\sigma(1)}, dx^{\sigma(2)}, \ldots, dx^{\sigma(N)}) = P(dx^1, dx^2, \ldots, dx^N)$.

Lemma 6.2.1 ([632]) *For a fixed* N*, consider an* N*-DM team defined as* $(\mathcal{P}_N)$ *(see* (6.1)*) and let the cost function be a convex function of* $\underline{u}_N$ *P-almost surely. Assume that the cost function is exchangeable P-almost surely with respect to the actions, i.e., for any permutation of indices,* σ*, P-almost surely* $c(\omega_0, u^1, \ldots, u^N) = c(\omega_0, u^{\sigma(1)}, \ldots, u^{\sigma(N)})$*. If* $\mathbb{U}$ *is convex, and observations of DMs are exchangeable conditioned on* ω_0*, then the team is symmetrically optimal.*

Proof Any permutation of policies does not change the value of $J_N(\underline{\gamma}_N)$ since

$$J_N(\underline{\gamma}_N^{\sigma}) = \int c(\omega_0, u^1, \ldots, u^N) P_N(dv^1, \ldots, dv^N|\omega_0)$$

$$
\begin{aligned}
&\times 1_{\{(\gamma^{\sigma(1)}(v^1),\ldots,\gamma^{\sigma(N)}(v^N))\}}(du^1,\ldots,du^N)P(d\omega_0)\\
&= \int c(\omega_0, u^{\sigma(1)},\ldots,u^{\sigma(N)})\\
&\quad \times 1_{\{(\gamma^{\sigma(1)}(v^{\sigma(1)}),\ldots,\gamma^{\sigma(N)}(v^{\sigma(N)}))\}}(du^{\sigma(1)},\ldots,du^{\sigma(N)})\\
&\quad \times P_N(dv^{\sigma(1)},\ldots,dv^{\sigma(N)}|\omega_0)P(d\omega_0) \qquad (6.14)\\
&= \int c(\omega_0, u^1,\ldots,u^N)1_{\{(\gamma^1(v^1),\ldots,\gamma^N(v^N))\}}(du^1,\ldots,du^N)\\
&\quad \times P_N(dv^1,\ldots,dv^N|\omega_0)P(d\omega_0) = J_N(\underline{\gamma}_N),
\end{aligned}
$$

where (6.14) follows from the assumption that the cost function is exchangeable with respect to the actions, and the hypothesis that observations of DMs are P-almost surely exchangeable conditioned on the random variable ω_0. Let $\underline{\gamma}_N^* = (\gamma^{1*}, \gamma^{2*}, \ldots, \gamma^{N*})$ be a given team policy for $(\mathcal{P}_N)$ (see (6.1)). Consider $\underline{\tilde{\gamma}}_N$ as a convex combination of all possible permutations of policies by averaging them, $\sigma \in \Sigma$, where Σ is the set of all possible permutations. Since $\mathbb{U}$ is convex, $\underline{\tilde{\gamma}}_N$ is a legitimate control policy. Following from convexity of the cost function P-almost surely, we have for $\alpha_\sigma = \frac{1}{|\Sigma|}$ (where $|\Sigma|$ denotes the cardinality of Σ),

$$
\begin{aligned}
J_N(\underline{\tilde{\gamma}}_N) &:= J_N\Big(\sum_{\sigma\in\Sigma}\alpha_\sigma \underline{\gamma}_N^{*,\sigma}\Big) \leq \sum_{\sigma\in\Sigma}\alpha_\sigma J_N(\underline{\gamma}_N^{*,\sigma})\\
&= \sum_{\sigma\in\Sigma}\alpha_\sigma J_N(\underline{\gamma}_N^{*}) \;= J_N(\underline{\gamma}_N^{*}),
\end{aligned}
$$

where the inequality follows from convexity of the cost function P-almost surely for every fixed realization of observations, since we have

$$
\begin{aligned}
&E\left[c\left(\omega_0, \sum_{\sigma\in\Sigma}\alpha_\sigma(\underline{\gamma}_N^{*,\sigma})^1(y^1),\ldots,\sum_{\sigma\in\Sigma}\alpha_\sigma(\underline{\gamma}_N^{*,\sigma})^N(y^N)\right)\right]\\
&\leq E\left[\sum_{\sigma\in\Sigma}\alpha_\sigma c\left(\omega_0, (\underline{\gamma}_N^{*,\sigma})^1(y^1),\ldots,(\underline{\gamma}_N^{*,\sigma})^N(y^N)\right)\right]\\
&= \sum_{\sigma\in\Sigma}\alpha_\sigma E\left[c\left(\omega_0, (\underline{\gamma}_N^{*,\sigma})^1(y^1),\ldots,(\underline{\gamma}_N^{*,\sigma})^N(y^N)\right)\right],
\end{aligned}
$$

where $(\underline{\gamma}_N^{*,\sigma})^j$ denotes the j-the component of $\underline{\gamma}_N^{*,\sigma}$, and the inequality above follows from Jensen's inequality since the cost function is convex P-almost surely. Hence, the team is symmetrically optimal. ◇

In the following, we present another characterization of symmetrically optimal teams; this appears to be a standard result; however, a proof is included for completeness.

Lemma 6.2.2 *For a fixed N, consider an N-DM team defined as $(\mathcal{P}_N)$ (see* (6.1)*) and let the cost function be convex in $\underline{u}_N$ P-almost surely. Assume that the action set for each DM is convex. If the expected cost function (see* (6.1)*) is exchangeable with respect to the policies, then the team is symmetrically optimal.*

Proof Let $\underline{\gamma}_N^* = (\gamma^{1*}, \gamma^{2*}, \ldots, \gamma^{N*})$ be a given team policy for $(\mathcal{P}_N)$ (see (6.1)). According to the definition of exchangeable teams, any permutation of policies, say $\hat{\underline{\gamma}}_N^* = (\gamma^{i_1*}, \gamma^{i_2*}, \ldots, \gamma^{i_N*})$, does not change the value of the expected cost function, and hence achieve the same expected cost as the one induced by $\underline{\gamma}_N^*$. Consider $\tilde{\underline{\gamma}}_N$ as a uniform randomization among all possible permutations of optimal policies. Since $\mathbb{U}$ is convex, $\tilde{\underline{\gamma}}_N$ is a legitimate control policy. By convexity of the cost function, through Jensen's inequality, and the fact that any permutation of optimal policies preserves the value of the cost function, we have $J_N(\tilde{\underline{\gamma}}_N) \leq J_N(\underline{\gamma}_N^*)$. Since $\tilde{\underline{\gamma}}_N$ is also identically symmetric, the proof is completed. ◇

Theorem 6.2.4 ([632]) *Consider an N-DM team defined as $(\mathcal{P}_N)$. Let the action sets be convex and the cost function be convex in the second and third arguments P-almost surely. If observations are exchangeable conditioned on ω_0, then the team is symmetrically optimal.*

Proof The cost function defined in $(\mathcal{P}_N)$ is exchangeable in actions, and hence under convexity of the action sets and the cost function, and following from the hypothesis that observations are exchangeable conditioned on ω_0, the proof is completed using Lemma 6.2.1. ◇

Theorem 6.2.4 will now be utilized in our analysis to follow.

6.2.3 *Optimal Solutions for Mean-Field Teams as Limits of Optimal Policies for Finite Symmetric Teams*

In the following, we present results for symmetrically optimal static teams in the high population regime, and particularly when the number of DMs is countably infinite. First, we focus on the case when the observations of DMs are identical and independent, and then we deal with the case of dependent observations under additional assumptions.

The first result in this vein is the next theorem which is obtained under the assumption that observations are independent and identically distributed. The proof uses a measure concentration argument to establish convergence.

Theorem 6.2.5 ([632]) *Consider a team problem defined as* $(\mathcal{P}_\infty)$ *(see* (6.3)*) with convex cost function in the second and third arguments* P*-almost surely. Let the action sets be compact and convex for each DM, and* y^i*s be i.i.d. random variables. If there exists a sequence of optimal policies for* $(\mathcal{P}_N)$*,* $\{\gamma_N^*\}_N$*, which converges (for every DM due to the symmetry) pointwise to* γ_∞^* *as* $N \to \infty$*, then* γ_∞^* *(which is identically symmetric) is an optimal policy for* $(\mathcal{P}_\infty)$*.*

A proof is provided in the Appendix.

In the following, we relax the hypothesis that observations of decision makers are independent. A proof is given in the appendix.

Proposition 6.2.1 *Consider a team problem defined as* $(\mathcal{P}_\infty)$ *with cost function being jointly convex in the second and third arguments* P*-almost surely. Let the action sets be compact and convex for each DM, and* $y^i = h(x, z^i)$*, where* z^i*s are i.i.d. random variables. If there exists a sequence of optimal policies for* $(\mathcal{P}_N)$*,* $\{\gamma_N^*\}_N$*, which converges pointwise to policy* γ_∞^* *as* $N \to \infty$*, then* γ_∞^* *(which is identically symmetric) is an optimal policy for* $(\mathcal{P}_\infty)$*.*

A further relaxation is the following, where we no longer assume that the action sets are compact; this is particularly important for LQG models as we will see later. We omit the proof, but just note that it builds on a tightness argument.

Theorem 6.2.6 ([632]) *Consider a team defined as* $(\mathcal{P}_\infty)$ *with the cost function jointly convex in the second and third arguments* P*-almost surely. Let the action sets be convex for each DM. Let* y^i*s be i.i.d. random variables. If there exists a sequence of optimal policies for* $(\mathcal{P}_N)$*,* $\{\gamma_N^*\}_N$*, which converges pointwise to a policy* γ_∞^* *as* $N \to \infty$*, and for some* $\delta > 0$*,* $\sup_{N \geq 1} E(|\gamma_N^*(y^1)|^{1+\delta}) < \infty$*, then* γ_∞^* *(which is identically symmetric) is an optimal policy for* $(\mathcal{P}_\infty)$*.*

6.2.4 An Existence Theorem on Globally Optimal Policies for Mean-Field Team Problems

An implication of the set of results above is an existence result on globally optimal policies for mean-field problems. Theorem 6.2.5 had stated that if a pointwise limit as $N \to \infty$ of a sequence of optimal policies for $(\mathcal{P}_N)$ exists, then this limit is a globally optimal policy for the mean-field team $(\mathcal{P}_\infty)$.

The following result builds on this insight. The proof is not included here, but it builds on showing that there exists a converging subsequence of optimal policies whose limit constitutes an optimal policy for the mean-field limit. We note, however, that the limit (and thus the optimal) policy *is not necessarily deterministic*.

Theorem 6.2.7 ([632]) *Consider* $(\mathcal{P}_\infty)$ *(see* (6.3)*) with the cost function being jointly convex in the second and third arguments* P*-almost surely. Let the action sets be convex for each DM. Assume further that, without any loss, the optimal* *control laws are restricted to those with $E(\varphi_i(u^i)) \leq K$ for some finite K, where $\varphi_i : \mathbb{U}^i \to \mathbb{R}_+$ is lower semi-continuous. If y^i's are i.i.d. random variables, then there exists an optimal policy for $(\mathcal{P}_\infty)$.*

6.2.5 Examples

In this sub-section, we consider two special classes of mean-field teams, namely LQG and LQ static teams with coupling between states of nature, and then we consider LQG symmetric static teams with coupling between control actions.

Example 6.2.1 (Static Quadratic Gaussian Teams with Coupling Between States of Nature) Consider the following mean-field static team where **DMi** has the observation:

$$y^i = x^i + z^i, \tag{6.15}$$

where $\{z^i\}_{i\in\mathbb{N}}$ and $\{x^i\}_{i\in\mathbb{N}}$ are i.i.d. zero mean Gaussian random variables with positive variances. Let $\{z^i\}_{i\in\mathbb{N}}$ be independent of $\{x^i\}_{i\in\mathbb{N}}$. The expected cost function is defined as

$$J(\underline{\gamma}) = \limsup_{N\to\infty} \frac{1}{N} E^{\underline{\gamma}}\left[\sum_{i=1}^{N} R(u^i)^2 + Q(u^i \dot{-} x^i - \mu_N)^2\right], \tag{6.16}$$

where $\mu_N := \frac{1}{N}\sum_{k=1}^{N} x^k$, and $R > 0$ and $Q > 0$ are scalars.

Proposition 6.2.2 *For LQG static mean-field teams as formulated above, under the measurement scheme* (6.15), *a globally optimal policy for $(\mathcal{P}_\infty)$, $\gamma_\infty^{i*}(y^i)$, is achieved as the limit $N \to \infty$ of $\gamma_N^{i*}(y^i)$, the sequence of optimal solutions for $(\mathcal{P}_N)$.*

Proof Sketch We invoke Theorem 6.2.3 to prove the theorem. The stationary policy is obtained as

$$\gamma_N^{i*} = \frac{Q}{R+Q}(1+\frac{1}{N})E(x^i|y^i),$$

where the result follows from the assumption that x^is are independent of z^i's and x^k's, $k \neq i$ for every i, and the assumption that the random variables are zero-mean. Stationary policies are team optimal for $(\mathcal{P}_N)$ in this formulation (following our analysis in Sect. 2.4.2). We have $\gamma_\infty^{i*}(y^i) = \frac{Q}{R+Q}E(x^i|y^i)$. Since y^is are zero mean Gaussian random variables, we have $E(x^i|y^i) = \Sigma_{x^i y^i}\Sigma_{y^i y^i}^{-1} y^i := Ky^i$, where Σ_{XY} is defined as a covariance of two random variables X and Y. We have P-a.s.,

$$\sup_{1\le i\le N} |\gamma_N^{i*}(y^i) - \gamma_\infty^{i*}(y^i)| = \frac{Q}{R+Q} \sup_{1\le i\le N} |\frac{1}{N} E(x^i|y^i)|$$

$$= \frac{KQ}{R+Q} \sup_{1\le i\le N} |\frac{1}{N} y^i| \xrightarrow[N\to\infty]{} 0, \tag{6.17}$$

where (6.17) follows from

$$\lim_{N\to\infty} \sup_{1\le i\le N} \frac{1}{N^2}(y^i)^2 \le \lim_{N\to\infty} \frac{1}{N^2}\sum_{i=1}^N (y^i)^2 = 0 \; P-a.s,$$

where the first inequality is true since $(y^i)^2$s are non-negative, and equality follows from the strong law of large numbers (SLLN) since y^is are i.i.d. and have a finite variance, hence, (b) holds. A slight analysis then completes the proof via Theorem 6.2.3. ◇

Example 6.2.2 (Static Non-Gaussian Teams) Let the observation scheme be (6.15), where $\{z^i\}_{i\in\mathbb{N}}$ and $\{x^i\}_{i\in\mathbb{N}}$ are i.i.d. zero mean random variables with finite variances. Let $\{z^i\}_{i\in\mathbb{N}}$ be independent of $\{x^i\}_{i\in\mathbb{N}}$. The expected cost function is defined as (6.16). Let R, Q be positive scalars.

Proposition 6.2.3 *For LQ static teams as formulated above, under the measurement scheme* (6.15), $\gamma_\infty^{k*}(y^k) = (R+Q)^{-1}QE(x^k|y^k)$ *is globally optimal for* ($\mathcal{P}_\infty$) *and is obtained as the limit of* $\gamma_N^{k*}(v^k)$ *as* $N\to\infty$.

The proof follows from Theorem 6.2.1 and Theorem 6.2.3.

Example 6.2.3 (LQG Symmetric Teams with Coupling Between Control Actions) Let

$$\underline{v}^i = H^i\underline{x} + \underline{z}^i, \tag{6.18}$$

where $\{\underline{z}^i\}_{i\in\mathbb{N}}$ is independent zero mean Gaussian random vectors also independent of $\underline{x}$, with covariance $\Sigma_{jj} = N^0 > 0$. Define $\omega = (\underline{x}, \underline{z}^1, \underline{z}^2, \ldots)$, and $\omega_0 := \underline{x}$ where $\underline{x}$ is a Gaussian random vector with covariance $E(\underline{x}\underline{x}^T) = \Sigma_{00}$. The expected cost function is given as:

$$J(\underline{\gamma}) = \limsup_{N\to\infty} \frac{1}{N} E^{\underline{\gamma}}\Big[\sum_{i=1}^N (u^i)^T R u^i - 2\sum_{i=1}^N (u^i)^T D$$

$$\times (\underline{x} + \frac{1}{N}\sum_{k=1}^N u^k) + (\underline{x} + \frac{1}{N}\sum_{k=1}^N u^k)^T Q(\underline{x} + \frac{1}{N}\sum_{k=1}^N u^k)\Big], \tag{6.19}$$

where R is an appropriate dimensional positive definite matrix, and D and Q are appropriate dimensional positive semi-definite matrices, and $R > 2D$. In the following, we follow steps in Sect. 2.4.2.

Lemma 6.2.3 *Consider an N-DM LQG team formulated above, under the measurement scheme* (6.18), *the global optimal policy for $(\mathcal{P}_N)$ is linear, i.e., $\gamma_N^{k*}(y^k) = \pi_N^k y^k$. Here, π_N^k, $k = 1, \ldots, N$, is a matrix with appropriate dimensions, and which is obtained by solving the following parallel update scheme,*

$$\pi_{N,(i)}^k = -L_N\left[S^k + \frac{1}{N}\sum_{p=1,p\neq k}^{N} \pi_{N,(i)}^p H^p S^k\right], \quad k = 1, \ldots, N, \tag{6.20}$$

where $L_N := (R+\frac{Q}{N^2}-\frac{2D}{N})^{-1}(\frac{Q}{N}-D)$, $S^k := \Sigma_{00}(H^k)^T(H^k\Sigma_{00}(H^k)^T + \Sigma_{kk})^{-1}$ and the initial points of the iterations are considered as zero functions.

The following builds on Proposition 6.2.1 and Theorem 6.2.6.

Proposition 6.2.4 *Consider $(\mathcal{P}_\infty)$ with the expected cost* (6.19). *Under the following measurement scheme*

$$\underline{v}^i = H\underline{x} + \underline{z}^i, \tag{6.21}$$

where $\underline{z}^i$s are i.i.d. Gaussian random vectors, $\gamma_\infty^{i*}(v^i) = \pi_\infty^* v^i$ is an optimal policy for $(\mathcal{P}_\infty)$ and is the pointwise limit of $\gamma_N^{i*}(v^i) = \pi_N^* v^i$, an optimal policy for $(\mathcal{P}_N)$.

6.3 Convex Dynamic Teams with Large Numbers of DMs

We now consider the dynamic team setup, where we will impose the condition that the optimization problem is convex in policies (see Sect. 3.7). Subsequently, in Sect. 6.4, we will be relaxing this condition and study non-convex dynamic teams.

We first briefly introduce the setup for stochastic (symmetric) dynamic teams with finite as well as infinitely many DMs. The set $\mathcal{N}$ denotes the collection of agents (which can be a finite set $\{1, 2, \ldots, N\}$ or a countably infinite set $\mathbb{N}$). Consider a collection of DMs, as earlier, viewed as a single **A**i ($i \in \mathcal{N}$) acting at different time instances ($t \in \mathcal{T}$), where the set $\mathcal{T}$ denotes the number of times each DM can act (the set $\mathcal{T}$ can be a finite set $\{1, 2, \ldots, T\}$ or a countably infinite set $\mathbb{N}$, depending on whether the horizon is finite or infinite). Define state dynamics and observations for each **DM**i at time t as

$$x_{t+1}^i = f_t(x_t^i, u_t^i, w_t^i), \tag{6.22}$$

$$y_t^i = h_t(x_{[0,t]}^i, u_{[0,t-1]}^i, v_{[0,t]}^i), \tag{6.23}$$

where the functions f_t and h_t are measurable. The information structure of **DM**i at time t is $I_t^i = \{y_t^i\}$ (or $I_t^i = \{y_k^i\}_{k\in K_t}$ for a subset $K_t \subseteq \{0, \dots, t\}$), and $\zeta_t^i := (w_t^i, v_t^i)$ (with $\zeta_0^i := (x_0^i, w_0^i, v_0^i)$) denotes uncertainty corresponding to dynamics and observations at time t for **DM**i which are exogenous random vectors.

In view of symmetry in the above setup (since functions f_t and h_t are identical across the DMs), we assume that spaces where state, action, observation, and disturbance variables take values from are symmetric (identical) across players (also across times for simplicity of our analysis) and are Borel subsets of appropriate dimensional Euclidean spaces $\mathbb{X} \subseteq \mathbb{R}^m$, $\mathbb{U} \subseteq \mathbb{R}^{m'}$, $\mathbb{Y} \subseteq \mathbb{R}^{m''}$, $\mathbb{W}$, and $\mathbb{V}$, where m, m', and m'' are positive integers. Now, a dynamic stochastic team problem with finite (N) number of DMs (also referred to as a pre-limit mean-field stochastic team problem) is defined as follows:

Problem ($[\mathcal{P}_T^{N,\mathbf{MF}}$) An N-DM finite-horizon stochastic team problem is described with an expected cost function under policy $\underline{\gamma}_T^{1:N}$ as

$$J_T^N(\underline{\gamma}_T^N) = \frac{1}{N}\sum_{t=0}^{T-1}\sum_{i=1}^{N} E^{\underline{\gamma}_T^{1:N}}\left[c\left(\omega_0, x_t^i, u_t^i, \frac{1}{N}\sum_{p=1}^{N} u_t^p, \frac{1}{N}\sum_{p=1}^{N} x_t^p\right)\right], \tag{6.24}$$

for some Borel measurable cost function $c : \Omega_0 \times \mathbb{X} \times \mathbb{U} \times \mathbb{U} \times \mathbb{X} \to \mathbb{R}_+$, where ω_0 is an exogenous random vector and $\underline{\gamma}_T^{1:N} = \gamma_{[0,T-1]}^{1:N}$.

We now introduce a stochastic team problem with a countably infinite number of DMs.

Problem ($\mathcal{P}_T^{\infty,\mathbf{MF}}$) A mean-field team problem is described with an expected cost function under policy $\underline{\gamma}_T$ as

$$J_T^{\infty}(\underline{\gamma}_T) = \limsup_{N\to\infty} \frac{1}{N}\sum_{t=0}^{T-1}\sum_{i=1}^{N} E^{\underline{\gamma}_T}\left[c\left(\omega_0, x_t^i, u_t^i, \frac{1}{N}\sum_{p=1}^{N} u_t^p, \frac{1}{N}\sum_{p=1}^{N} x_t^p\right)\right], \tag{6.25}$$

for some Borel measurable cost function $c : \Omega_0 \times \mathbb{X} \times \mathbb{U} \times \mathbb{U} \times \mathbb{X} \to \mathbb{R}_+$, where $\underline{\gamma}_T^i = \gamma_{[0,T-1]}^i$ for $i \in \mathbb{N}$ and $\underline{\gamma}_T = \{\underline{\gamma}_T^i\}_{i\in\mathbb{N}}$.

In this section, we define exchangeable teams and symmetric information structures (as a refinement of those in the preceding section), and show that for convex exchangeable dynamic teams with finite horizons, optimal policies exhibit a symmetry structure. For any number of DMs, this symmetry structure is more relaxed when compared with the previous section, which focused on problems under a static information structure, and is applicable to dynamic teams which may not admit a static reduction, as long as convexity in policies holds.

6.3.1 Finite Horizon Convex Dynamic Team Problems with a Symmetric Information Structure

We first characterize symmetry in dynamic team problems. Following Sect. 3.5.1, by considering a collection of DMs as a single Agent ($i = 1, \ldots, N$) acting at different time instances ($t = 0, \ldots, T-1$), we can define the team problem with (NT)-DMs as a team with N-Agents:

(i) Let the observation and action spaces be standard Borel spaces and be identical for each Agent $\mathbf{A}i$, ($i = 1, \ldots, N$), with $\mathbb{Y}_i := \mathbf{Y} = \prod_{t=0}^{T-1} \mathbb{Y}^t$, $\mathbb{U}_i := \mathbf{U} = \prod_{t=0}^{T-1} \mathbb{U}^t$, respectively. The sets of all admissible policies are denoted by $\Gamma = \prod_{i=1}^{N} \Gamma_i = \prod_{i=1}^{N} \prod_{t=0}^{T-1} \Gamma^t$.

(ii) For $i = 1, \ldots, N$, $y_t^i := h_t^i(x_0^{1:N}, \zeta_{[0,t]}^{1:N}, u_{[0,t-1]}^{1:N})$ represents the observation of $\mathbf{A}i$ at time t (h_t^i's are Borel measurable functions).

(iii) Let $(\underline{\zeta}^{1:N}) := (\underline{\zeta}^1, \ldots, \underline{\zeta}^N)$, where $\underline{\zeta}^i := (x_0^i, \zeta_{[0,T-1]}^i)$ denotes all the uncertainty associated with $\mathbf{A}i$ including his initial states. We assume that $(\underline{\zeta}^i)$ takes values in Ω_ζ (where at each time instance t, it takes values in Ω_{ζ_t}). Let μ denote the law of $\underline{\zeta}^{1:N}$.

(iv) Let the expected cost function be defined as $J_N(\underline{\gamma}^{1:N}) = E^{\underline{\gamma}^{1:N}}[c(\underline{\zeta}^{1:N}, \underline{u}^{1:N})]$, for some Borel measurable cost function $c : \prod_{i=1}^{N}(\Omega_\zeta \times \mathbb{U}_i) \to \mathbb{R}_+$, where $\underline{\gamma}^{1:N} = (\underline{\gamma}^1, \underline{\gamma}^2, \ldots, \underline{\gamma}^N)$ and $\underline{\gamma}^i = \gamma_{[0,T-1]}^i$ for $i = 1, \ldots, N$.

Now, we define the symmetric information structure (note that symmetric information structures can be classical, partially nested, or non-classical).

Definition 6.3.1 (Symmetric Information Structure) Let the information of $\mathbf{A}i$ acting at time t be described as $I_t^i := \{y_t^i\}$. The information structure of a sequential N-DM team problem is *symmetric* if $y_t^i = h_t(x_0^i, x_0^{-i}, \zeta_{[0,t]}^i, \zeta_{[0,t]}^{-i}, u_{[0,t-1]}^i, u_{[0,t-1]}^{-i})$ for some h_t which is identical for all $i = 1, \ldots, N$ (note that the arguments of the function depend on i).

We note that the above definition can be generalized to be applicable to teams with countably infinite DMs and infinite horizon problems.

The symmetric information structure can also be interpreted and defined as a graph, relating $\mathbf{A}$s and their information through directed edges. Consider $G(V, \mu)$ as a directed graph with $V = \{1, \ldots, NT\}$ nodes and where $\mu \subset V \times V$ determines the directed edges between the nodes; this represents the dependency notation in the information of nodes, i.e., (i, j) denotes a directed edge from i to j, $i \to j$, meaning that u^i affects y^j through the relation $y^i = h^i(\omega, \underline{u}^{[1,i-1]})$ defined in the intrinsic model. We denote by $\downarrow j$ the set of nodes i such that $i \to j$ (ancestors), and $\downarrow\downarrow j := \{\downarrow j\} \cup \{j\}$. Similarly, we can define descendants by $\uparrow j$. We can again define a collection of DMs as a single agent $\mathbf{A}i$, ($i = 1, \ldots, N$), acting at different time instances ($t = 0, \ldots, T-1$) on a graph with a symmetric information structure (an example is in Fig. 6.1). Assume that

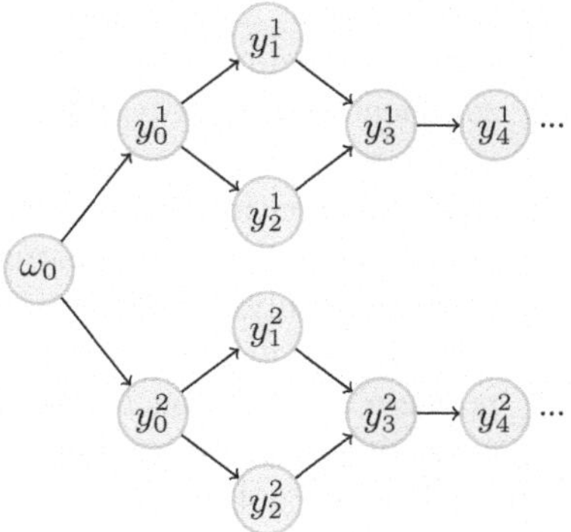

Fig. 6.1 An example of the graph structure of a symmetric dynamic team

(i) there exists a node $\{i\}$ (root node), ω_0. Each sub-graph represents a single agent acting at time instances $t = 0, \dots, T-1$, and there exists a finite number of sub-graphs $G_p(\hat{V}, \hat{\mu})$ such that $\cup_{p=1}^N G_p \cup \{i\} = G$, where G_p's are isomorphic (see e.g., [767]) for all $p = 1, \dots, N$, i.e., for every node with directed edges in each sub-graph there exists a unique node with identical directed edges in the corresponding sub-graphs, where $\hat{V} = \{0, \dots, T-1\}$, and G_p^k refers to a node k in G_p for all $p = 1, \dots, N$ and $k = 0, \dots, T-1$,

(ii) sharing of the information is symmetric across sub-graphs, i.e., for $p, s = 1, \dots, N$, and $k, j = 0, \dots, T-1$, and for every edge from a node G_p^k to a node G_s^j, there exists an edge from a node G_p^k to nodes G_{-p}^j, where G_{-p}^j denotes $(G_1^j, \dots, G_{p-1}^j, G_{p+1}^j, \dots, G_N^j)$, and also there exist edges from nodes G_{-p}^k to a node G_p^j.

Assumption 6.3.1 *For any permutation σ of the set $\{1, \dots, N\}$, we have for all ω_0,*

$$c(\omega_0, (\underline{\zeta}^\sigma)^{1:N}, (\underline{u}^\sigma)^{1:N}) = c(\omega_0, \underline{\zeta}^{1:N}, \underline{u}^{1:N}), \tag{6.26}$$

where $(\underline{\zeta}^\sigma)^{1:N} := (\underline{\zeta}^{\sigma(1)}, \dots, \underline{\zeta}^{\sigma(N)})$ and $(\underline{u}^\sigma)^{1:N} := (\underline{u}^{\sigma(1)}, \dots, \underline{u}^{\sigma(N)})$.

Next, we again recall some definitions and results from Sect. 3.7 on convexity of static and dynamic team problems, tailored to the analysis in this chapter.

Definition 6.3.2 An N-Agent team problem (static or dynamic) is convex in policies if for any two team policies $\underline{\gamma}_T^{1:N}$ and $\underline{\tilde{\gamma}}_T^{1:N}$ in the set $\{\underline{\gamma}_T^{1:N} \in \Gamma : J(\underline{\gamma}_T^{1:N}) < \infty\}$, and for any $\alpha \in (0,1)$, we have

$$J_T(\alpha \underline{\gamma}_T^{1:N} + (1-\alpha)\underline{\tilde{\gamma}}_T^{1:N}) \le \alpha J_T(\underline{\gamma}_T^{1:N}) + (1-\alpha) J_T(\underline{\tilde{\gamma}}_T^{1:N}).$$

Consider a sequential team problem, and assume that action sets are convex, and $J(\underline{\gamma}) < \infty$ for all $\underline{\gamma} \in \Gamma$ (or alternatively, restrict the action sets to those leading to finite cost). Then,

(i) [by Theorems 3.7.1 and 3.7.2] for static team problems, convexity of the cost function in actions, that is P-almost sure convexity in $\underline{u}^{1:N}$ of $c(\underline{\zeta}^{1:N}, \underline{u}^{1:N})$, is sufficient for convexity of the team problem in policies,
(ii) [by Theorem 3.7.5] for dynamic team problems with a policy-independent static reduction, convexity of the team problem in policies is equivalent to convexity of its static reduction. For partially nested dynamic teams with a static reduction (more generally, for stochastically partially nested team problems [830, Section 3.3]) if the statically reduced cost function is convex in actions, then the team problem is convex on Γ (see Sect. 3.7 for a cautionary remark on information expansion by control action sharing so that the reduction is policy-independent via Theorem 3.4.2).

6.3.2 *Optimality of Symmetric Policies for Convex Dynamic Teams with a Symmetric Information Structure*

In the following, we define notions of exchangeable and symmetrically optimal teams analogous to those presented for the static setup.

Definition 6.3.3 (Exchangeable Teams) An N-Agent team is *exchangeable* if the value of the expected cost function is invariant under every permutation of policies of DMs, i.e., $J_T(\underline{\gamma}_T^1, \underline{\gamma}_T^2, \ldots, \underline{\gamma}_T^N) = J_T((\underline{\gamma}_T^\sigma)^1, \ldots, (\underline{\gamma}_T^\sigma)^N)$.

Definition 6.3.4 (Symmetrically Optimal Teams) A team is *symmetrically optimal*, if for each given policy $\underline{\gamma}_T = (\underline{\gamma}_T^1, \ldots, \underline{\gamma}_T^N)$, there exists an identically symmetric policy (i.e., each DM has the same policy, $\underline{\tilde{\gamma}}_T = (\underline{\tilde{\gamma}}_T^1, \ldots, \underline{\tilde{\gamma}}_T^N)$, and $\underline{\tilde{\gamma}}_T^i = \underline{\tilde{\gamma}}_T^j$ for all $i, j = 1, \ldots, N$,) which performs at least as good as the given policy.

Remark 6.3.1 The concepts of exchangeable and symmetrically optimal dynamic teams here involve agents and not individual DMs acting at each time stage: in particular, the value of the expected cost function may not be invariant under exchanging γ_t^i with γ_k^j for $k \neq t, k, t = 0, \ldots, T-1$, and for $i, j = 1, \ldots, N$.

In the following, we provide a characterization for exchangeable and symmetrically optimal dynamic teams.

Theorem 6.3.1 *Consider a dynamic team problem with a symmetric information structure under Assumption 6.3.1. If*

(a) the action set of each DM is convex,
(b) $(\underline{\zeta}^1, \ldots, \underline{\zeta}^N)$ are exchangeable, and
(c) for all policies $\gamma \in \Gamma$, and for all $A = A^1 \times \cdots \times A^N$ where $A^i \in \mathcal{Y}^i$,

$$\prod_{t=0}^{T-1} P\Big(y_t^{1:N} \in A \Big| x_0^{1:N}, \zeta_{[0,t-1]}^{1:N}, y_{[0,t-1]}^{1:N}, \gamma_0^1(y_0^1), \ldots, \gamma_{t-1}^1(y_{t-1}^1), \ldots, \gamma_{t-1}^N(y_{t-1}^N)\Big) \tag{6.27}$$

$$= \prod_{t=0}^{T-1}\prod_{i=1}^{N} P\Big(y_t^i \in A^i \Big| x_0^i, \zeta_{[0,t-1]}^i, y_{\downarrow t}^{\downarrow\downarrow i}, \gamma_{\downarrow t}^{\downarrow\downarrow i}(y_{\downarrow t}^{\downarrow\downarrow i})\Big),$$

where $\downarrow\downarrow (1 : N)$ *denotes* $(\downarrow\downarrow 1, \ldots, \downarrow\downarrow N)$ *and* $y_{\downarrow t}^{\downarrow\downarrow i}$ *corresponds to the observations of* $\mathbf{A}j$*'s* $j = 1, \ldots, N$ *(including* $\mathbf{A}i$ *itself), at time instances* $p = 0, \ldots, t-1$ *where the action of* $\mathbf{A}j$*'s at time* p *affects the observation of* $\mathbf{A}i$ *at time* t *(*$\gamma_{\downarrow t}^{\downarrow\downarrow(1:N)}(y_{\downarrow t}^{\downarrow\downarrow(1:N)})$*) can be defined similarly), then the team problem is exchangeable. Furthermore, if the team problem is convex in policies, then it is symmetrically optimal.*

Proof We first show that for any permutation $\sigma \in S$,

$$J_T((\underline{\gamma}_T^\sigma)^1, \ldots, (\underline{\gamma}_T^\sigma)^N) = J_T(\underline{\gamma}_T^1, \ldots, \underline{\gamma}_T^N),$$

i.e., the team is exchangeable. We have,

$$J_T\left((\underline{\gamma}_T^\sigma)^1, \ldots, (\underline{\gamma}_T^\sigma)^N\right)$$

$$= \int c\left(\underline{\zeta}^{1:N}, (\underline{\gamma}_T^\sigma)^1(\underline{y}^1), \ldots, (\underline{\gamma}_T^\sigma)^N(\underline{y}^N)\right) \mu(dx_0^{1:N}, d\zeta_{[0,T-1]}^{1:N})$$

$$\times \prod_{t=0}^{T-1}\prod_{i=1}^{N} P\left(dy_t^i \Big| x_0^i, \zeta_{[0,t-1]}^i, y_{\downarrow t}^{\downarrow\downarrow i}, (\gamma^\sigma)_{\downarrow t}^{\downarrow\downarrow i}(y_{\downarrow t}^{\downarrow\downarrow i})\right) \tag{6.28}$$

$$= \int c\left((\underline{\zeta}^\sigma)^{1:N}, (\underline{\gamma}_T^\sigma)^1((\underline{y}^\sigma)^1), \ldots, (\underline{\gamma}_T^\sigma)^N((\underline{y}^\sigma)^N)\right) \mu(d(x_0^\sigma)^{1:N}, d(\zeta_{[0,T-1]}^\sigma)^{1:N})$$

$$\times \prod_{t=0}^{T-1}\prod_{i=1}^{N} P\left(d(y^\sigma)_t^i \Big| (x_0^\sigma)^i, (\zeta_{[0,t-1]}^\sigma)^i, (y^\sigma)_{\downarrow t}^{\downarrow\downarrow i}, (\gamma^\sigma)_{\downarrow t}^{\downarrow\downarrow i}((y^\sigma)_{\downarrow t}^{\downarrow\downarrow i})\right) \tag{6.29}$$

$$= \int c\left(\underline{\zeta}^{1:N}, \underline{\gamma}_T^1(\underline{y}^1), \ldots, \underline{\gamma}_T^N(\underline{y}^N)\right) \mu(dx_0^{1:N}, d\zeta_{[0,T-1]}^{1:N})$$

$$\times \prod_{t=0}^{T-1}\prod_{i=1}^{N} P\left(dy_t^i \Big| x_0^i, \zeta_{[0,t-1]}^i, y_{\downarrow t}^{\downarrow\downarrow i}, \gamma_{\downarrow t}^{\downarrow\downarrow i}(y_{\downarrow t}^{\downarrow\downarrow i})\right) \tag{6.30}$$

$$= J_T(\underline{\gamma}_T^1, \ldots, \underline{\gamma}_T^N),$$

where (6.28) follows from condition (c). Equality (6.29) follows from exchanging $\underline{y}^i$ and $\underline{\zeta}^i$ with $(\underline{y}^\sigma)^i$ and $(\underline{\zeta}^\sigma)^i$ by relabeling them. Let $\underline{\gamma}_T^* = (\underline{\gamma}_T^{1*}, \ldots, \underline{\gamma}_T^{N*})$ be a given policy. Consider $\underline{\tilde{\gamma}}_T$ as a convex combination of all possible permutations of policies by averaging them. Since action spaces are convex by condition (a), $\underline{\tilde{\gamma}}_T$ is a control policy. Following from convexity of the cost function in policies, we have

$$J_T(\underline{\tilde{\gamma}}_T) := J_T\Big(\sum_{\sigma \in S} \frac{1}{|S|} \underline{\gamma}_T^{*,\sigma}\Big) \leq \sum_{\sigma \in S} \frac{1}{|S|} J_T(\underline{\gamma}_T^{*,\sigma}) = J_T(\underline{\gamma}_T^*),$$

where $|S|$ denotes the cardinality of the set S and the inequality above follows from the hypothesis that the team problem is convex on Γ, and the last equality follows from exchangeability of the team problem. This implies that the team is symmetrically optimal, and completes the proof. ◇

Lemma 6.3.1 *Consider a dynamic team problem with a symmetric partially nested information structure (see Definition 6.3.1) which admits a static reduction. Under Assumption 6.3.1, and Assumptions (a), (b), (c) of Theorem 6.3.1, if the cost function is jointly convex on $\underline{u}^1, \ldots, \underline{u}^N$ P-almost surely, then the team is symmetrically optimal.*

We note that here by symmetry we mean symmetry across all DMs.

Proof The proof follows from Theorem 6.3.1 since the team is convex on Γ. ◇

It follows that if a static reduction of an exchangeable, symmetrically optimal, dynamic team exists, then it is exchangeable and symmetrically optimal.

6.3.3 Convex Mean-Field Teams with a Symmetric Information Structure

In the following, we establish global optimality results for convex mean-field dynamic teams with a symmetric information structure. Consider the dynamics given in (6.22) and (6.23) and problems defined as $\mathcal{P}_T^{N,\mathrm{MF}}$ and $\mathcal{P}_T^{\infty,\mathrm{MF}}$.

Toward solving the two problems, and establishing connections between their solutions, we first introduce a number of assumptions, some paralleling those introduced earlier in the context of static mean-field teams.

Assumption 6.3.2 *Functions f_t and h_t are continuous in the states and actions, and f_t's are bounded.*

Assumption 6.3.3 *The cost function in* (6.42), $c : \Omega_0 \times \mathbb{X} \times \mathbb{U} \times \mathbb{U} \times \mathbb{X} \to \mathbb{R}_+$, *is jointly continuous in the second, third, fourth, and fifth arguments, where $\mathbb{X}$, $\mathbb{U}$ denote, respectively, the state space and action space of the agents at each time instant for all ω_0.*

Assumption 6.3.4 *We have the following.*

(i) $(x_0^1, x_0^2, \ldots)$ *are i.i.d. random vectors conditioned on* ω_0,

(ii) for $t = 0, \ldots, T-1$, $\{w_t^i\}_{i\in\mathbb{N}}$ *are i.i.d. random vectors, and for* $i \in \mathbb{N}$, $\{w_t^i\}_{t=0}^{T-1}$ *are mutually independent, and independent of* ω_0 *and* $(x_0^1, x_0^2, \ldots)$. *For* $t = 0, \ldots, T-1$, $\{v_t^i\}_{i\in\mathbb{N}}$ *are i.i.d. random vectors, and for* $i \in \mathbb{N}$, $\{v_t^i\}_{t=0}^{T-1}$ *are mutually independent, and independent of* ω_0, $(x_0^1, x_0^2, \ldots)$, *and* w_t^i*'s for* $i \in \mathbb{N}$ *and* $t = 0, \ldots, T-1$.

Assumption 6.3.5 *Conditioned on* ω_0, $(x_0^1, x_0^2, \ldots)$ *are exchangeable random vectors.*

We first have the following result.

Lemma 6.3.2 ([633]) *Consider a team problem defined as* $(\mathcal{P}_T^{N,MF})$ *with a symmetric information structure and assume that it is convex in policies. Let the action space of each DM be compact and convex. Under Assumption 6.3.2, Assumption 6.3.3, and Assumption 6.3.4, the team is symmetrically optimal.*

Proof The proof follows from Theorem 6.3.1. ◇

The next result establishes a connection between the solutions of the two team problems.

Theorem 6.3.2 ([633]) *Consider the dynamic team problem introduced in* $(\mathcal{P}_T^{\infty,MF})$ *with* $(\mathcal{P}_T^{N,MF})$ *having a symmetric information structure for every N. Assume that for every N the latter is convex in policies. Let the action space of each agent be compact and convex. Let either of the following two conditions hold:*

(i) Assumptions 6.3.3, 6.3.2, and 6.3.4.

(ii) Assumptions 6.3.2, 6.3.3, 6.3.4(ii), and 6.3.5.

Then, if there exists a sequence of optimal policies for $(\mathcal{P}_T^{N,MF})$, $\{\underline{\gamma}_T^{*,N}\}_N$, *which converges (for every agent, due to the symmetry) pointwise to* $\underline{\gamma}_T^{*,\infty}$ *as* $N \to \infty$, *then* $\underline{\gamma}_T^{*,\infty}$ *(which is an identically symmetric policy) is an optimal policy for* $(\mathcal{P}_T^{\infty,MF})$.

The proof follows from steps similar to those used in the static setup (e.g. Theorem 6.2.5), and is thus omitted.

An implication of Theorem 6.3.2 is an existence result on globally optimal policies for dynamic mean-field team problems. In particular, we will establish in the following the existence of a converging subsequence, in an appropriate sense, for a sequence of optimal policies for finite teams with an increasing number of agents. What we have (in Theorem 6.3.3 to follow) is pointwise convergence; also by Theorem 6.3.2, if a sequence of optimal policies for $(\mathcal{P}_T^{N,MF})$, $\{\underline{\gamma}_T^{*,N}\}_N$, converges pointwise, a globally optimal policy exists.

To this end, we allow agents to apply randomized policies. For each agent (**A***i* for $i \in \mathbb{N}$), a probability measure $P \in \mathcal{P}(\Omega_0 \times \mathbb{X} \times \prod_{t=0}^{T-1}(\mathbb{W} \times \mathbb{V}) \times \prod_{t=0}^{T-1}(\mathbb{U} \times \mathbb{Y}))$ is a probability measure induced by a randomized policy if and only if for every $t = 0, \ldots, T-1$,

$$\int g(\omega_0, x_0^i, \zeta_{[0,t-1]}^i, y_{[0,t]}^i, u_{[0,t]}^i) P(d\omega_0, dx_0^i, d\zeta_{[0,t-1]}^i, dy_{[0,t]}^i, du_{[0,t]}^i)$$

$$= \int g(\omega_0, x_0^i, \zeta_{[0,t-1]}^i, y_{[0,t]}^i, u_{[0,t]}^i) \mu^i(dx_0^i, d\zeta_{[0,t-1]}^i|\omega_0) P(d\omega_0) \tag{6.31}$$

$$\times \prod_{k=0}^{t} \Pi_k^i(du_k^i|y_k^i) p_k^i(dy_k^i|\omega_0, x_0^i, \zeta_{0:k-1}^i, y_{0:k-1}^i, u_{0:k-1}^i),$$

for some stochastic kernel Π_k^i on $\mathbb{U}$ given $\mathbb{Y}$, for all continuous and bounded g, where p_k^i is the transition kernel characterizing the observations of $\mathbf{A}i$ at time t,

$$p_k^i\left(y_k^i \in \cdot \middle| \omega_0, x_0^i, \zeta_{0:k-1}^i, y_{0:k-1}^i, u_{0:k-1}^i\right)$$

$$:= P\left(h_k(x_{0:k}^i, u_{\downarrow k}^i, v_{0:k}^i) \in \cdot \middle| \omega_0, x_0^i, \zeta_{0:k-1}^i, y_{0:k-1}^i, u_{0:k-1}^i\right),$$

and μ^i is a fixed probability measure on initial states and disturbances of $\mathbf{A}i$ conditioned on ω_0.

For the next result, we introduce an absolute continuity assumption on observations of DMs.

Assumption 6.3.6 *For every* $\mathbf{A}i$ *and* $t = 0, \ldots, T-1$, *there exists a function* $\psi_t^i : \mathbb{Y} \times \Omega_0 \times \mathbb{X} \times \prod_{k=0}^{t-1}(\mathbb{W} \times \mathbb{V}) \times \prod_{k=0}^{t-1}(\mathbb{Y} \times \mathbb{U}) \to \mathbb{R}_+$ *continuous in actions and a probability measure* ν_t^i *on* $\mathbb{Y}$ *such that for all Borel sets* $A = A^1 \times \cdots \times A^N$,

$$P(y_t^{1:N} \in A|\omega_0, x_0^{1:N}, \zeta_{[0,t-1]}^{1:N}, y_{[0,t-1]}^{1:N}, u_{[0,t-1]}^{1:N})$$

$$= \prod_{i=1}^{N} \int_{A^i} \psi_t^i(y_t^i, \omega_0, x_0^i, \zeta_{[0,t-1]}^i, y_{[0,t-1]}^i, u_{[0,t-1]}^i) \nu_t^i(dy_t^i).$$

This assumption allows us to obtain an independent measurements reduction. For example, if v_t^i for all $i \in \mathbb{N}$ and $t = 0, \ldots, T-1$ are i.i.d. with a probability measure admitting a density function so that the observation of each $\mathbf{A}i$ at time t is $y_t^i = \tilde{h}_t(x_t^i, u_{\downarrow t}^i) + v_t^i$, where $\tilde{h}_t$ is continuous, then Assumption 6.3.6 holds. We now have the following theorem.

Theorem 6.3.3 ([633]) *Consider a dynamic mean-field team defined in* ($\mathcal{P}_T^{\infty,MF}$) *with* ($\mathcal{P}_T^{N,MF}$) *having a symmetric information structure for every* N. *Assume that for every* N, *the team problem is convex in policies, and the action sets are convex. Assume further that, without any loss, the optimal policies are restricted to those with* $E(\varphi_i(u^i)) \leq K$ *for some finite* K, *where* $\varphi_i : \mathbb{U} \to \mathbb{R}_+$ *is lower semi-continuous (moment condition). Under Assumption 6.3.2, Assumption 6.3.3, and Assumption 6.3.4, if either*

(i) Assumption 6.3.6 holds (with no further assumptions on the information structure of each **Ai** *for $i \in \mathbb{N}$ through time $t = 0, \ldots, T-1$), or*

(ii) for each **Ai** *for $i \in \mathbb{N}$ through time $t = 0, \ldots, T-1$, there exists a static reduction with the classical information structure (i.e., under the reduction, the information structure is expanding such that $\sigma(y_t^i) \subset \sigma(y_{t+1}^i)$),*

then, there exists an optimal policy for ($\mathcal{P}_T^{\infty,MF}$).

Remark 6.3.2 Since the space of policies that are deterministic (where Π_k^i in (6.31) are indicator functions) is not closed under weak convergence (see Example 8.5.1 to be presented later), we allow for randomization in the policies and therefore the optimal limit policy is not necessarily deterministic according to the above result; however, it is identical for each DM.

Proof Sketch We use individually randomized policies and show that for every sequence of N-DM optimal policies, there exists a subsequence which converges to an optimal independently randomized policy for the mean-field limit under an appropriate topology defined by the product topology where each coordinate is endowed with the weak convergence topology. First, one can show that for each N-Agent team problem, optimal policies are deterministic and symmetric, and one can consider independently randomized policies induced by such policies $\{P_N\}_N$ (where $P_N \in \mathcal{P}(\mathbf{Y} \times \mathbf{U})$ for each DM satisfying (6.31)), as the sequence to be studied. We also define the sequence of empirical measures induced by these policies, Q_N.

Then, one shows that for every sequence of policies satisfying a moment condition, there exists a subsequence such that both policies for each DM, $\{P_n\}_n$, and empirical measures induced by these policies, $\{Q_n\}_n$, (where $n \in$ ▯ is the index set of a convergent subsequence), converge weakly to a limit P-almost surely; that is, for a set of P-measure 1, for every bounded function g which is continuous in actions and observations, $P(\lim_{n\to\infty} \left|\int g dQ_n - \int g dQ\right| \Big| \omega_0) = 0$. To this end, one can show that for each agent, $\{P_n\}_n$ is tight, and then show that the empirical measures induced by these policies converge weakly to a limit P-almost surely.

One can then show that the set of policies for each DM is closed under the weak convergence topology; hence, the limit policy satisfies the required measurability/conditional independence conditions (that is, the limit policy satisfies (6.31)). Finally, one can use the lower semi-continuity argument to show that the expected cost function under the induced limit policy is less than or equal to the expected cost achieved under the sequence of N-DM optimal policies. ◇

Remark 6.3.3 For the existence result, to show that the set of policies induced by independent randomized policies for each DM through time $t = 0, \ldots, T-1$ (see (6.31)) is closed under the weak convergence topology, one utilizes the arguments in the proof of Theorem 4.2.4. We note that the extension of the existence result of Theorem 4.2.4 to the setup here is not obvious since the conclusion here in Theorem 6.3.3 cannot be made rigorously without considering the technical steps involving infinite dimensions and limit arguments.

6.4 Non-convex Exchangeable Teams and Mean-Field Optimality of Randomized Symmetric Policies with No Common Randomness

In this section, we address non-convex mean-field teams. One of the main difficulties in studying such non-convex team problems lies in showing that globally optimal policies (for mean-field team problems) are symmetric (identical for each DM). This difficulty stems from the observation that, in general, globally optimal policies are not symmetric for non-convex pre-limit N-DM team problems. This is in contrast to the convex mean-field teams where symmetry can be established for both pre-limit N-DM and mean-field team problems as we established earlier.

To arrive at existence and structural results on optimal policies, we first introduce a topology on control policies, which involves various relaxations given the decentralized information structure. For a general setup of stochastic team problems with N DMs, under exchangeability of observations of DMs and the cost function, we show that without loss of global optimality, the search for optimal policies can be restricted to those that are N-exchangeable. Then, by extending N-exchangeable policies to infinitely exchangeable ones, establishing a convergence argument for the induced costs, and using a de Finetti type theorem, we establish the existence of an optimal decentralized policy for static and dynamic teams with countably infinite number of DMs, which turns out to be symmetric (i.e., identical) and randomized. In particular, unlike previous sections, convexity of the cost in policies is not assumed. Finally, we show near optimality of symmetric independently randomized policies for finite N-DM team problems and thus establish approximation results for N-DM weakly coupled stochastic teams.

6.4.1 Exchangeable Static Teams with Large Numbers of DMs

To provide the essence of the ideas, we will primarily consider N-DM static team problems, but a sketch of the arguments for the dynamic case will also be presented (which will only entail a similar level of mild technical generalization as we have seen in the convex case considered in the previous two sections).

As we consider exchangeable team problems, we let action and observation spaces be identical through DMs, $\mathbb{U}^i = \mathbb{U} \subseteq \mathbb{R}^n$ and $\mathbb{Y}^i = \mathbb{Y} \subseteq \mathbb{R}^m$ for all $i \in \mathcal{N}$, where n and m are positive integers.

Our focus is on a class of exchangeable team problems satisfying an exchangeability assumption on the cost function.

Assumption 6.4.1 *The cost function is exchangeable with respect to actions for all ω_0, i.e., for any permutation σ of $\{1, \ldots, N\}$, $c(\omega_0, u^1, \ldots, u^N) = c(\omega_0, u^{\sigma(1)}, \ldots, u^{\sigma(N)})$ for all ω_0.*

We have the following expected cost criterion for the N-DM problem $\mathcal{P}_N$:

$$E^{\underline{\gamma}_N}\left[\frac{1}{N}\sum_{i=1}^{N} c\left(\omega_0, u^i, \frac{1}{N}\sum_{p=1}^{N} u^p\right)\right]. \tag{6.32}$$

Clearly, $\frac{1}{N}\sum_{i=1}^{N} c(\omega_0, u^i, \frac{1}{N}\sum_{p=1}^{N} u^p)$ satisfies Assumption 6.4.1. As noted earlier, the cost function c in (6.32), in fact, can be made more general by considering any cost function $c : \Omega_0 \times \mathbb{U} \times \tilde{\mathbb{U}} \to \mathbb{R}_+$, depending on the empirical measure of the control actions under mild continuity conditions, that is,

$$c(\omega_0, u^i, \Xi(\frac{1}{N}\sum_{p=1}^{N} \delta_{u^p})), \tag{6.33}$$

where $\Xi : \mathcal{P}(\mathbb{U}) \to \tilde{\mathbb{U}}$ is weakly continuous, $\delta_{(\cdot)}$ is Delta-Dirac measure, and $\tilde{\mathbb{U}}$ is a subset of appropriate dimensional Euclidean space. However, for clarity in presentation, we will follow (6.32). We will use the same notation for the cost function c as in (6.1).

Now, we introduce a stochastic team problem with countably infinite number of DMs, $\mathcal{P}_\infty$: Consider a stochastic team with countably infinite number of DMs, that is, $\mathcal{N} = \mathbb{N}$. Let $\Gamma := \prod_{i\in\mathbb{N}} \Gamma^i$ and $\underline{\gamma} := (\gamma^1, \gamma^2, \ldots)$. Let the expected cost under $\underline{\gamma}$ be given as

$$J(\underline{\gamma}) = \limsup_{N\to\infty} E^{\underline{\gamma}}\left[\frac{1}{N}\sum_{i=1}^{N} c\left(\omega_0, u^i, \frac{1}{N}\sum_{p=1}^{N} u^p\right)\right], \tag{6.34}$$

for some Borel measurable cost function $c : \Omega_0 \times \mathbb{U} \times \mathbb{U} \to \mathbb{R}_+$.

Later on, we will allow DMs to apply randomized policies and provide a reformulation of the team problems with randomized policies. Our first goal here is to establish the existence of a symmetric (identical) randomized globally optimal policy for static mean-field team problems ($\mathcal{P}_\infty$). To this end, we first establish N-exchangeability of randomized optimal policies for ($\mathcal{P}_N$) and symmetry for optimal randomized policies of ($\mathcal{P}_\infty$). Then, in Theorem 6.4.2, using symmetry, we establish an existence result for ($\mathcal{P}_\infty$).

Control Topologies and a De Finetti Theorem on the Policy Space

Assumption 6.2.1 allows us to reduce the problem as a static team problem where now the observation of each DM is independent of observations of other DMs and also independent of ω_0 (since under the measure transformation (6.4), a probability measure on the observation of each DM is Q^i, which is independent of observations of other DMs and ω_0). Hence, under Assumption 6.2.1, we can focus on each **DMi** separately via the set of probability measures as in (6.5). Accordingly, as earlier, we define convergence on policies as $\gamma_n^i \to \gamma^i$ iff $1_{\{\gamma_n^i(y^i)\in du^i\}} Q^i(dy^i) \to 1_{\{\gamma^i(y^i)\in du^i\}} Q^i(dy^i)$ (in the weak convergence topology) as $n \to \infty$ for each DM. We will also allow for randomized (relaxed) policies.

Now that we have a standard Borel space formulation for Γ^i for each $i \in \mathcal{N}$, we can define the following set of Borel probability measures on the set of admissible policies Γ_N as $L^N := \mathcal{P}(\Gamma_N)$, where Borel σ-field $\mathcal{B}(\Gamma^i)$ is induced by the topology defined above. Define the set of randomized policies induced by common randomness as:

$$L_{\mathrm{CO}}^N := \Bigg\{ P_\pi \in L^N \Bigg| \text{for all } A_i \in \mathcal{B}(\Gamma^i) :$$

$$P_\pi(\gamma^1 \in A_1, \dots, \gamma^N \in A_N) = \int_{z\in[0,1]} \prod_{i=1}^N P_\pi^i(\gamma^i \in A_i | z)\eta(dz),\ \eta \in \mathcal{P}([0,1]) \Bigg\},$$

where η is the distribution of common, but independent (from intrinsic exogenous system variables), randomness. Note that conditioned on z, policies are independent. It can be shown that the sets of randomized policies L_{CO}^N and L^N are identical, and hence, the set of randomized policies L^N corresponds to randomized policies induced by individual and common randomness (and can be seen as an alternative representation of Eq. (4.5), tailored for the analysis of this chapter via a De Finetti theorem to be presented). Since individual and common randomness do not improve the optimal expected cost (by Theorem 4.1.1), the relaxation of the problem to sets of randomized policies L^N is a legitimate relaxation for team problems with N-DMs.

We refine the concept of *exchangeability* (from Definition 6.2.3) for an infinite collection of random variables and then policies, as follows:

Definition 6.4.1 Random variables $x^1, x^2, \dots, x^N$ defined on a common probability space are *N-exchangeable* if for any permutation σ of the set $\{1, \dots, N\}$,

$$P\left(x^{\sigma(1)} \in A^1, x^{\sigma(2)} \in A^2, \dots, x^{\sigma(N)} \in A^N\right)$$
$$= P\left(x^1 \in A^1, x^2 \in A^2, \dots, x^N \in A^N\right)$$

for any measurable $\{A^1, \dots, A^N\}$, and $(x^1, x^2, \dots)$ is *infinitely-exchangeable* if it is *N-exchangeable* for all $N \in \mathbb{N}$.

Now, we define the set of exchangeable probability measures on policies as:

$$L_{\mathrm{EX}}^N := \Bigg\{ P_\pi \in L^N \Bigg| \text{for } A_i \in \mathcal{B}(\Gamma^i) \text{ and for all } \sigma \in S_N :$$

$$P_\pi(\gamma^1 \in A_1, \ldots, \gamma^N \in A_N) = P_\pi(\gamma^{\sigma(1)} \in A_1, \ldots, \gamma^{\sigma(N)} \in A_N)\bigg\},$$

where S_N is the set of permutations of $\{1, \ldots, N\}$. We note that L^N_{EX} is a convex subset of L^N.

We also define the set $L^N_{\text{CO,SYM}}$ as the set of identical probability measures on policies induced by a common randomness:

$$L^N_{\text{CO,SYM}} := \bigg\{P_\pi \in L^N \bigg| \text{for } A_i \in \mathcal{B}(\Gamma^i) :$$

$$P_\pi(\gamma^1 \in A_1, \ldots, \gamma^N \in A_N) = \int_{z \in [0,1]} \prod_{i=1}^{N} P_\pi(\gamma^i \in A_i | z)\eta(dz),\ \eta \in \mathcal{P}([0, 1])\bigg\},$$

where we drop the index i on P_π to capture the property that the independent randomization is identical through DMs. Also, define the set of probability measures on policies with only private independent randomness as:

$$L^N_{\text{PR}} := \bigg\{P_\pi \in L^N \bigg| \text{for } A_i \in \mathcal{B}(\Gamma^i) : P_\pi(\gamma^1 \in A_1, \ldots, \gamma^N \in A_N)$$

$$= \prod_{i=1}^{N} P^i_\pi(\gamma^i \in A_i)\bigg\}.$$

Finally, define the set of probability measures on policies with identical and independent randomness:

$$L^N_{\text{PR,SYM}} := \bigg\{P_\pi \in L^N \bigg| \text{for } A_i \in \mathcal{B}(\Gamma^i) : P_\pi(\gamma^1 \in A_1, \ldots, \gamma^N \in A_N)$$

$$= \prod_{i=1}^{N} P_\pi(\gamma^i \in A_i)\bigg\}.$$

For a team with countably infinite number of DMs, define sets of probability measures L, L_{EX}, L_{CO}, $L_{\text{CO,SYM}}$, L_{PR}, $L_{\text{PR,SYM}}$ similarly using the Ionescu-Tulcea extension theorem by iteratively adding new coordinates to the probability measure. We define the set of probability measures L on the infinite product Borel spaces $\Gamma = \prod_{i \in \mathbb{N}} \Gamma^i$ as:

$$L := \mathcal{P}(\Gamma). \tag{6.35}$$

Next, we define the set of infinitely exchangeable probability measures on policies as:

$$L_{\mathrm{EX}} := \Bigg\{ P_\pi \in L \,\Bigg|\, \text{for } A_i \in \mathcal{B}(\Gamma^i) \text{ and for all } N \in \mathbb{N}, \text{ and for all } \sigma \in S_N : \\ P_\pi(\gamma^1 \in A_1, \ldots, \gamma^N \in A_N) = P_\pi(\gamma^{\sigma(1)} \in A_1, \ldots, \gamma^{\sigma(N)} \in A_N) \Bigg\},$$

and also define

$$L_{\mathrm{CO}} := \Bigg\{ P_\pi \in L \,\Bigg|\, \text{for } A_i \in \mathcal{B}(\Gamma^i) : \\ P_\pi(\gamma^1 \in A_1, \gamma^2 \in A_2, \ldots) = \int_{z\in[0,1]} \prod_{i\in\mathbb{N}} P^i_\pi(\gamma^i \in A_i|z)\eta(dz), \ \eta \in \mathcal{P}([0,1]) \Bigg\}.$$

Note that L_{CO} is a convex subset of L and its extreme points are in the set of probability measures on policies with private independent randomness:

$$L_{\mathrm{PR}} := \Bigg\{ P_\pi \in L \,\Bigg|\, \text{for } A_i \in \mathcal{B}(\Gamma^i) : P_\pi(\gamma^1 \in A_1, \gamma^2 \in A_2, \ldots) \\ = \prod_{i\in\mathbb{N}} P^i_\pi(\gamma^i \in A_i) \Bigg\}.$$

Also, we define

$$L_{\mathrm{CO,SYM}} := \Bigg\{ P_\pi \in L \,\Bigg|\, \text{for } A_i \in \mathcal{B}(\Gamma^i) : \\ P_\pi(\gamma^1 \in A_1, \gamma^2 \in A_2, \ldots) = \int_{z\in[0,1]} \prod_{i\in\mathbb{N}} P_\pi(\gamma^i \in A_i|z)\eta(dz), \ \eta \in \mathcal{P}([0,1]) \Bigg\},$$

and further

$$L_{\mathrm{PR,SYM}} := \Bigg\{ P_\pi \in L \,\Bigg|\, \text{for } A_i \in \mathcal{B}(\Gamma^i) : P_\pi(\gamma^1 \in A_1, \gamma^2 \in A_2, \ldots) \\ = \prod_{i\in\mathbb{N}} P_\pi(\gamma^i \in A_i) \Bigg\}.$$

In view of the introduced topology and sets of Borel probability measures on policies (sets of randomized polices), we now establish a connection between L_{EX} and $L_{\mathrm{CO,SYM}}$ using the classical de Finetti theorem, which will note that infinitely-exchangeable randomized policies are mixtures of conditionally i.i.d. randomized policies.

Theorem 6.4.1 (A De Finetti Theorem for Individual Policy Process Sequence [636]) *Any infinitely-exchangeable randomized policy $P_\pi \in L_{\mathrm{EX}}$ is in the set of* *randomized policies* $L_{CO,SYM}$ *(*$P_\pi \in L_{CO,SYM}$*), i.e., for any* $P_\pi \in L_{EX}$*, there exists a* $[0, 1]$*-valued random variable* Z *such that for any* $A_i \in \mathcal{B}(\Gamma^i)$

$$P_\pi(\gamma^1 \in A_1, \gamma^2 \in A_2, \dots) = \int_{z \in [0,1]} \prod_{i \in \mathbb{N}} \tilde{P}_\pi(\gamma^i \in A_i | z) \eta(dz), \quad \eta \in \mathcal{P}([0, 1]), \tag{6.36}$$

where for every fixed z*,* $\tilde{P}_\pi(\cdot|z) \in \mathcal{P}(\Gamma^i)$*.*

Proof In view of the introduced weak convergence topology on Γ^i, we have that Γ^i is a closed subset of the Borel space $\mathcal{P}(\mathbb{U} \times \mathbb{Y})$, and hence, Γ^i is Borel for $i \in \mathbb{N}$. The proof then follows from [384, Theorem 1.1] since $\Gamma = \prod_{i=1}^{\infty} \Gamma^i$ is Borel. We note that the de Finetti representation in [384, Theorem 1.1] is of the form $P_\pi(\gamma^1 \in A_1, \gamma^2 \in A_2, \dots) = \int_{\mathcal{P}(\Gamma^i)} \prod_{i=1}^{\infty} m(A^i)\hat{\eta}(dm)$ for $\hat{\eta} \in \mathcal{P}(\mathcal{P}(\Gamma^i))$ which can be written as in (6.36). The reason for this is that $\mathcal{P}(\Gamma^i)$ is an (uncountable) standard Borel space [114, Corollary 7.25.1], and hence, by the Borel-isomorphism [114, Proposition 7.16], it is Borel isomorphic to $[0, 1]$. ◇

Structural Results for the Pre-limit and the Limit Models as $N \to \infty$

The following is a useful structural result for N-DM team problems. It shows that N-exchangeable policies are optimal under mild conditions. In contrast to Theorem 6.2.4, the absence of convexity here **does not allow for the optimality of symmetric/identical and deterministic** policies for each N. As $N \to \infty$, however, we will see further below that the asymptotic results will coincide.

Lemma 6.4.1 *For a fixed* N*, consider an* N*-DM static team. Assume that* $\bar{L}^N$ *is an arbitrary convex subset of* L^N*. Under Assumption 6.2.1 and that the cost function is exchangeable with respect to actions for all* ω_0*, if observations of DMs are exchangeable conditioned on* ω_0*, then*

$$\inf_{P_\pi \in \bar{L}^N} \int P_\pi(d\underline{\gamma})\mu^N(d\omega_0, d\underline{y})c^N(\underline{\gamma}, \underline{y}, \omega_0)$$
$$= \inf_{P_\pi \in \bar{L}^N \cap L^N_{EX}} \int P_\pi(d\underline{\gamma})\mu^N(d\omega_0, d\underline{y})c^N(\underline{\gamma}, \underline{y}, \omega_0). \tag{6.37}$$

Assumption 6.4.2 *The cost function* $c : \Omega_0 \times \mathbb{R}^n \times \mathbb{R}^n \to \mathbb{R}_+$ *is jointly continuous in its second and third arguments for all* ω_0*.*

Under mild conditions, we can show that the optimal expected costs induced by L^N_{EX} and L_{EX} become closer as N goes to infinity. Hence, by Lemma 6.4.1, under symmetry, this allows us to show that without loss of global optimality, optimal policies of static mean-field teams with countably infinite number of DMs can be considered to be an infinitely exchangeable type. This key result, stated next, builds on [215, Theorem 13] and [15, Proposition 7.20] (see also [383] for more general results).

Lemma 6.4.2 ([636]) *Suppose that Assumption 6.2.1 and Assumption 6.4.2 hold. Assume further that* $\mathbb{U}$ *is compact and the cost function is bounded. If observations of DMs are i.i.d. random vectors conditioned on* ω_0*, then*

$$\limsup_{N\to\infty} \inf_{P_\pi^N \in L_{EX}^N} \int P_\pi^N(d\underline{\gamma})\mu^N(d\omega_0, d\underline{y})c^N(\underline{\gamma}, \underline{y}, \omega_0)$$

$$= \limsup_{N\to\infty} \inf_{P_\pi \in L_{EX}} \int P_{\pi,N}(d\underline{\gamma})\mu^N(d\omega_0, d\underline{y})c^N(\underline{\gamma}, \underline{y}, \omega_0), \tag{6.38}$$

where $P_{\pi,N}$ *is the marginal of the* $P_\pi \in L_{EX}$ *to the first* N *components and* μ^N *is the marginal of the fixed probability measure on* $(\omega_0, y^1, y^2, \dots)$ *to the first* $N+1$ *components.*

We now establish the existence of a randomized optimal policy for $(\mathcal{P}_\infty)$, which is symmetric.

Theorem 6.4.2 ([636]) *Consider a static team problem* $(\mathcal{P}_\infty)$ *where Assumption 6.2.1 and Assumption 6.4.2 hold. Assume further that* $\mathbb{U}$ *is compact. If observations of DMs are i.i.d. random vectors conditioned on* ω_0*, then there exists a randomized optimal policy* P_π^* *for* $(\mathcal{P}_\infty)$ *which is in* $L_{PR,SYM}$*, that is*

$$\min_{P_\pi \in L_{PR,SYM}} \limsup_{N\to\infty} \int P_{\pi,N}(d\underline{\gamma})\mu^N(d\omega_0, d\underline{y})c^N(\underline{\gamma}, \underline{y}, \omega_0)$$

$$= \inf_{P_\pi \in L_{PR}} \limsup_{N\to\infty} \int P_{\pi,N}(d\underline{\gamma})\mu^N(d\omega_0, d\underline{y})c^N(\underline{\gamma}, \underline{y}, \omega_0).$$

Following Lemmas 6.4.1, 6.4.2, and Theorem 6.4.2, thanks to Theorem 6.4.1, we can show that without losing global optimality, optimal policies for mean-field teams can be considered to be symmetric and privately randomized ($L_{\text{PR,SYM}}$).

We emphasize that a *strategic measures* approach would not be feasible for arriving at this solution since exchangeability in the actions is not sufficient to ensure that the *dominating* random variable (in the De Finetti representation) is independent of the intrinsic randomness in the system.

By Theorem 6.2.5, Proposition 6.2.1 and Theorem 6.2.7, we recall again that if one also has convexity in the cost as well as action sets $\mathbb{U}^i$, then one can also establish that for every finite N, the optimal policies are symmetric and deterministic, but in the infinite limit, randomization may be required.

We now present an example where Theorem 6.4.2 can be applied but the existence result for the convex setup cannot be applied because the assumption that $\mathbb{U}$ is convex is violated.

Example 6.4.1 Consider a team problem with the following expected cost function

$$J(\underline{\gamma}) = \limsup_{N \to \infty} E^{\underline{\gamma}}\Big[\Big((\frac{1}{N}\sum_{i=1}^{N} u^i) - \frac{1}{2}\Big)^2\Big],$$

where σ-field is $\sigma(y^i) = \{\emptyset, \Omega\}$ (this corresponds to a team setup where DMs have no measurements, hence measurable functions (policies) are constant functions), and we consider $u^i \in \{0, 1\}$ for each DM. Clearly, an optimal policy that achieves zero (the lowest possible value) is the one where a matching partition (such as even numbers vs. odd numbers) among DMs picking $u^i = 0$ and $u^i = 1$. One can see that there is an optimal policy in $L_{\mathrm{PR,SYM}}$ since each DM can choose independently an action zero or one with probability half and this achieves the expected cost of zero; however, there is no identically deterministic policy that achieves zero expected cost. We note also that the problem is not a convex problem as the action sets are not convex.

Convexity is a restrictive condition which may not hold in many applications, notably in dynamic teams with non-classical information, as will be briefly discussed next.

6.4.2 *Dynamic Exchangeable Teams*

The analysis above also applies to dynamic exchangeable teams. One particular advantage of such an analysis over what was covered by Theorem 6.3.3 is that when a stochastic team is coupled through the dynamics, due to the resulting non-classical information structure, the optimization problem will generally be non-convex (see Sect. 3.7). Therefore, the results in Theorem 6.4.2 for exchangeable teams and its dynamic counterpart to be presented below, in the absence of a convexity requirement, is particularly important for optimal stochastic dynamic teams under general information structures.

Now, Theorem 6.4.2 applies identically for this setup by replacing DMs with agents acting over multiple time stages (via re-defining action and measurement spaces), paralleling the approach presented for convex dynamic teams which generalized the results for convex static teams to the dynamic case. The generalization is only technical given the static setup analysis. We provide a sketch of the ideas and refer the reader to [636] for a detailed analysis.

We will also establish the existence of a symmetric (identical) randomized globally optimal policy for mean-field dynamic team problems where DMs are weakly coupled through the average of states and actions in dynamics and/or the cost function. Again, we consider exchangeable teams, and hence, we let action, observation, and state spaces, respectively, be identical through DMs $i \in \mathcal{N}$, and for simplicity, also through time $t = 0, \ldots, T-1$, $\mathbb{U}_t^i = \mathbb{U} \subseteq \mathbb{R}^n$, $\mathbb{Y}_t^i = \mathbb{Y} \subseteq \mathbb{R}^{n'}$, $\mathbb{X}_t^i = \mathbb{X} \subseteq \mathbb{R}^{n''}$ for all $i \in \mathcal{N}$ and $t = 0, \ldots, T-1$, where n, n' and n'' are positive integers. Introduce the state dynamics and observation dynamics of DMs as follows:

$$x_{t+1}^i = f_t\left(x_t^i, u_t^i, \frac{1}{N}\sum_{p=1}^{N} x_t^p, \frac{1}{N}\sum_{p=1}^{N} u_t^p, w_t^i\right), \tag{6.39}$$

$$y_t^i = h_t\left(x_{[0,t]}^i, u_{[0,t-1]}^i, v_{[0,t]}^i\right), \tag{6.40}$$

where functions f_t and h_t are measurable functions and v_t^i and w_t^i are random vectors representing uncertainties in the state dynamics and the observations.

As noted earlier, the state dynamics (6.39) here also can be made more general by considering any measurable function f_t (for $t = 0, \ldots, T-1$), depending on the empirical measures of the states and control actions under mild continuity conditions, that is,

$$f_t\left(x_t^i, u_t^i, \Xi^x(\frac{1}{N}\sum_{p=1}^{N} \delta_{x_t^p}), \Xi^u(\frac{1}{N}\sum_{p=1}^{N} \delta_{u_t^p}), w_t^i\right), \tag{6.41}$$

where $\Xi^u : \mathcal{P}(\mathbb{U}) \to \tilde{\mathbb{U}}$ and $\Xi^x : \mathcal{P}(\mathbb{X}) \to \tilde{\mathbb{X}}$ are weakly continuous, and $\tilde{\mathbb{U}}$ and $\tilde{\mathbb{X}}$ are subsets of appropriate dimensional Euclidean spaces. However, for clarity in the presentation, we will focus on (6.39), noting that the generalization will be immediate. A similar generalization is possible for the cost function (6.42) below, as was discussed for the static setup.

Let the admissible policies $(\gamma_{[0,T-1]}^i)_{i\in\mathcal{N}}$ (with $\gamma_{[0,T-1]}^i := (\gamma_0^i, \ldots, \gamma_{T-1}^i)$) be measurable control functions so that $u_t^i = \gamma_t^i(y_t^i)$ for all $i \in \mathcal{N}$ and $t = 0, \ldots, T-1$.

Problem $(\mathcal{P}_T^N)$ Consider N-DM mean-field dynamic teams with the expected cost function under $\underline{\gamma}^{1:N}$ given as

$$J_T^N(\underline{\gamma}^{1:N}) = E^{\underline{\gamma}^{1:N}}\left[\frac{1}{N}\sum_{t=0}^{T-1}\sum_{i=1}^{N} c\left(\omega_0, x_t^i, u_t^i, \frac{1}{N}\sum_{p=1}^{N} u_t^p, \frac{1}{N}\sum_{p=1}^{N} x_t^p\right)\right], \tag{6.42}$$

where $\underline{\gamma}^{1:N} := (\gamma_{[0,T-1]}^1, \ldots, \gamma_{[0,T-1]}^N)$, $\gamma_{[0,T-1]}^i := (\gamma_0^i, \ldots, \gamma_{T-1}^i)$, and $\omega_0 : (\Omega, \mathcal{F}, \mathbb{P}) \to (\Omega_0, \mathcal{F}_0)$ is a cost-related random variable, where Ω_0 is a Borel space with its Borel σ-field $\mathcal{F}_0$.

Problem $(\mathcal{P}_T^\infty)$ Consider mean-field dynamic teams with the expected cost function under $\underline{\gamma}$ given as

$$J_T^\infty(\underline{\gamma}) = \limsup_{N\to\infty} J_T^N(\underline{\gamma}^{1:N}), \tag{6.43}$$

where $\underline{\gamma} := (\gamma_{[0,T-1]}^1, \gamma_{[0,T-1]}^2, \ldots)$ and $\underline{\gamma}^{1:N} := (\gamma_{[0,T-1]}^1, \ldots, \gamma_{[0,T-1]}^N)$.

We can define globally optimal policies for $(\mathcal{P}_T^N)$ and $(\mathcal{P}_T^\infty)$. Again, we allow DMs to apply randomized policies and provide a description of the problems within randomized policies. One can establish the existence of a symmetric (identical through DMs) randomized globally optimal policy for $(\mathcal{P}_T^\infty)$. Similar to the static case, one first establishes N-exchangeablity of randomized optimal policies for $(\mathcal{P}_T^N)$ and symmetry for optimal randomized policies of $(\mathcal{P}_T^\infty)$. Then, using symmetry, we can establish the existence of a solution to $(\mathcal{P}_T^\infty)$. The solution technique for dynamic problems is similar to the static one, which requires more technical arguments and additional assumptions.

Control Topologies

We again allow randomization in policies, but first we recall two reduction conditions (independent and nested reduction) studied in Sect. 3.5.2 that enable us to define sets of Borel probability measures on randomized policies for dynamic teams with different information structures by considering a policy of a single agent **A**i ($i = 1, \ldots, N$) acting at different time instances ($t = 0, \ldots, T-1$). We adopt these definitions to the model studied here.

Assumption 6.4.3 *One of the following conditions holds:*

(i) (Independent reduction): for every $N \in \mathbb{N} \cup \{\infty\}$ *and for* $i = 1, \ldots, N$ *and* $t = 0, \ldots, T-1$, *there exists a probability measure* τ_t^i *on* $\mathbb{Y}^t$ *and a function* $\psi_t^i : \mathbb{Y}^t \times \Omega_0 \times \prod_{p=1}^N (\prod_{k=0}^{t-1} \Omega_{\zeta_k} \times \prod_{k=0}^{t-1} (\mathbb{U}^k \times \mathbb{Y}^k)) \to \mathbb{R}_+$ *such that for all Borel sets* A^i *on* $\mathbb{Y}^t$ *(with* $A = A^1 \times \cdots \times A^N$*)*

$$\begin{aligned} &\nu_t^N(A|\omega_0, x_0^{1:N}, \zeta_{[0,t-1]}^{1:N}, y_{[0,t-1]}^{1:N}, u_{[0,t-1]}^{1:N}) \\ &= \prod_{i=1}^N \int_{A^i} \psi_t^i(y_t^i, \omega_0, x_0^{1:N}, \zeta_{[0,t-1]}^{1:N}, y_{[0,t-1]}^{1:N}, u_{[0,t-1]}^{1:N}) \tau_t^i(dy_t^i). \end{aligned}$$

(ii) (Nested reduction): for every $N \in \mathbb{N} \cup \{\infty\}$ *and for* $i = 1, \ldots, N$ *and* $t = 0, \ldots, T-1$, *there exists a probability measure* $\bar{\eta}_t^i$ *and a function* φ_t^i *such that for all Borel sets* A^i *on* $\mathbb{Y}^t$ *(with* $A = A^1 \times \cdots \times A^N$*)*

$$\begin{aligned} &\nu_t^N(A|\omega_0, x_0^{1:N}, \zeta_{[0,t-1]}^{1:N}, y_{[0,t-1]}^{1:N}, u_{[0,t-1]}^{1:N}) \\ &= \prod_{i=1}^N \int_{A^i} \varphi_t^i(y_t^i, \omega_0, x_0^{-i}, \zeta_{[0,t-1]}^{-i}, y_{[0,t-1]}^{-i}, u_{[0,t-1]}^{-i}) \bar{\eta}_t^i \\ &(dy_t^i | x_0^i, \zeta_{[0,t-1]}^i, y_{[0,t-1]}^i, u_{[0,t-1]}^i), \end{aligned}$$

and for each **DM**i *through time* ($t = 0, \ldots, T-1$)*, there exists a static reduction with the classical information structure (i.e., under the reduction,* *the information structure of each DM through time is expanding such that* $\sigma(y_t^i) \subset \sigma(y_{t+1}^i)$ *for* $t = 0, \ldots, T-1$*) with* $y_t^i \sim \eta_t^i$.

We note again that Assumption 6.4.3(i) allows us to obtain an independent measurements reduction both through DMs and through time. Assumption 6.4.3(ii) holds if an independent static reduction exists through DMs and there exists a nested static reduction for each DM through time, i.e., under the reduction, the information is expanding for each DM through time. In view of the above reduction conditions, we introduce a suitable topology for randomized policies. Under Assumption 6.4.3(i), we define convergence on policies as (via Young topologies, see Appendix F):

$$\underline{\gamma}_n^i \xrightarrow{n\to\infty} \underline{\gamma}^i \quad \text{if} \quad \gamma_{t,n}^i(du_t^i|y_t^i)\tau_t^i(dy_t^i) \xrightarrow[\text{weakly}]{n\to\infty} \gamma_t^i(du_t^i|y_t^i)\tau_t^i(dy_t^i) \quad t{=}0,\ldots,T-1.$$

Under Assumption 6.4.3(ii), we define convergence on policies as:

$$\underline{\gamma}_n^i \xrightarrow{n\to\infty} \underline{\gamma}^i \quad \text{if} \quad \gamma_{t,n}^i(du_t^i|y_{[0,t]}^i)\eta_t^i(dy_{[0,t]}^i) \xrightarrow[\text{weakly}]{n\to\infty} \gamma_t^i(du_t^i|y_{[0,t]}^i)\eta_t^i(dy_{[0,t]}^i)$$

Hence, under Assumption 6.4.3, we define all the sets of randomized policies defined in Sect. 6.4.1 for the dynamic teams by considering $\underline{\gamma}^i$.

Structural Results for the Pre-limit and the Limit Models as $N \to \infty$

Let the information of DM^i acting at time t be described as $I_t^i := \{y_t^i\}$. Recall that the information structure of a sequential N-Agent team problem is *symmetric* if

$$y_t^i = h_t(x_0^i, x_0^{-i}, \zeta_{[0,t]}^i, \zeta_{[0,t]}^{-i}, u_{[0,t-1]}^i, u_{[0,t-1]}^{-i}),$$

for some h_t which is identical for all $i = 1, \ldots, N$ (note that the arguments of the function depend on i) and $b^{-i} = (b^1, \ldots, b^{i-1}, b^{i+1}, \ldots, b^N)$ for $b = x_0, \zeta_{[0,t]}, u_{[0,t-1]}$.

To this end, we focus on a more general setup of team problems with randomized policies $P_\pi \in L^N$ as

$$J_N^\pi(\underline{\gamma}^{1:N}) := \int P_\pi(d\underline{\gamma})\mu^N(d\omega_0, d\underline{\zeta})c^N(\underline{\zeta}, \underline{\gamma}, \underline{y}, \omega_0)\nu^N(d\underline{y}|\underline{\zeta}, \underline{\gamma}, \omega_0)$$

$$:= \int \Big(\int c(\omega_0, \underline{\zeta}^{1:N}, \underline{u}^1, \ldots, \underline{u}^N) \prod_{i=1}^N \underline{\gamma}^i(d\underline{u}^i|\underline{y}^i) \Big)$$

$$P_\pi(d\underline{\gamma}^1, \ldots, d\underline{\gamma}^N)\mu^N(d\omega_0, d\underline{\zeta}^{1:N})$$

$$\times \prod_{t=0}^{T-1} \nu_t^N \left(dy_t^{1:N} \middle| \omega_0, x_0^{1:N}, \zeta_{[0,t-1]}^{1:N}, y_{[0,t-1]}^{1:N}, u_{[0,t-1]}^{1:N} \right), \tag{6.44}$$

where $c^N(\underline{\zeta}, \underline{\gamma}, \underline{y}, \omega_0) := \int c(\omega_0, \underline{\zeta}^{1:N}, \underline{u}^1, \ldots, \underline{u}^N) \prod_{i=1}^N \underline{\gamma}^i(d\underline{u}^i|\underline{y}^i)$ and the following assumption holds.

Assumption 6.4.4

(a) $(\underline{\zeta}^1, \ldots, \underline{\zeta}^N)$ *are exchangeable conditioned on* ω_0*;*

(b) for all $t = 0, \ldots, T-1$*, and all Borel sets* A^i *on* $\mathbb{Y}^t$ *(with* $A = A^1 \times \cdots \times A^N$*)*

$$\nu_t^N \left(A \middle| \omega_0, x_0^{1:N}, \zeta_{[0,t-1]}^{1:N}, y_{[0,t-1]}^{1:N}, u_{[0,t-1]}^{1:N} \right)$$

$$= \prod_{i=1}^N \nu_t^i \left(A^i \middle| \omega_0, x_0^i, \zeta_{[0,t-1]}^i, y_{[0,t-1]}^{1:N}, u_{[0,t-1]}^{1:N} \right),$$

where ν_t^i *is a stochastic kernel of the observation of* **DM***i at time* t*,* y_t^i*, induced by* h_t*.*

Lemma 6.4.3 *Consider a dynamic team problem with a symmetric information structure. Let Assumptions 6.3.1 and 6.4.4 hold. Assume that* $\bar{L}^N$ *is an arbitrary convex subset of* L^N*. Then,*

$$\inf_{P_\pi \in \bar{L}^N} \int P_\pi(d\underline{\gamma}) \mu^N(d\omega_0, d\underline{\zeta}) c^N(\underline{\zeta}, \underline{\gamma}, \underline{y}, \omega_0) \nu^N(d\underline{y}|\underline{\zeta}, \underline{\gamma}, \omega_0)$$

$$= \inf_{P_\pi \in \bar{L}^N \cap L_{EX}^N} \int P_\pi(d\underline{\gamma}) \mu^N(d\omega_0, d\underline{\zeta}) c^N(\underline{\zeta}, \underline{\gamma}, \underline{y}, \omega_0) \nu^N(d\underline{y}|\underline{\zeta}, \underline{\gamma}, \omega_0).$$

In the following, we establish the existence of a globally optimal randomized policy for dynamic mean-field teams with infinitely many DMs. Define state dynamics and observations as (6.39) and (6.40). The IS of **DM**i at time t is $I_t^i = \{y_t^i\}$, and $\zeta_t^i := (w_t^i, v_t^i)$ (with $\zeta_0^i := (x_0^i, w_0^i, v_0^i)$) denotes the uncertainty corresponding to dynamics and observations at time t for **DM**i which are exogenous random vectors with standard Borel spaces. First, we re-state the infinite DM mean-field team problem and its pre-limit within randomized policies.

Problem $(\mathcal{P}_T^N)$ Consider an N-DM dynamic team with the expected cost under a randomized policy $P_\pi^N \in L^N$ given as:

$$\int P_\pi^N(d\underline{\gamma}) \mu^N(d\omega_0, d\underline{\zeta}) c^N(\underline{\zeta}, \underline{\gamma}, \underline{y}, \omega_0) \nu^N(d\underline{y}|\underline{\zeta}, \underline{\gamma}, \omega_0)$$

$$:= \int \Bigg(\int \frac{1}{N} \sum_{t=0}^{T-1} \sum_{i=1}^N c\left(\omega_0, x_t^i, u_t^i, \frac{1}{N} \sum_{p=1}^N u_t^p(y_t^p), \frac{1}{N} \sum_{p=1}^N x_t^p \right) \prod_{k=1}^N \underline{\gamma}^k(d\underline{u}^k|\underline{y}^k) \Bigg)$$

$$\times P_{\pi}^{N}(d\underline{\gamma}^{1}, \ldots, d\underline{\gamma}^{N})\mu^{N}(d\omega_0, d\underline{\zeta}^{1:N})$$

$$\prod_{t=0}^{T-1} \nu_t^N\left(dy_t^{1:N} \middle| \omega_0, x_0^{1:N}, \zeta_{[0,t-1]}^{1:N}, y_{[0,t-1]}^{1:N}, u_{[0,t-1]}^{1:N}\right), \tag{6.45}$$

where

$$c^N(\underline{\zeta}, \underline{\gamma}, \underline{y}, \omega_0) := \int \frac{1}{N}\sum_{t=0}^{T-1}\sum_{i=1}^{N} c\left(\omega_0, x_t^i, u_t^i, \frac{1}{N}\sum_{p=1}^{N} u_t^p(y_t^p), \frac{1}{N}\sum_{p=1}^{N} x_t^p\right)$$
$$\prod_{k=1}^{N} \underline{\gamma}^k(d\underline{u}^k|\underline{y}^k).$$

Problem ($\mathcal{P}_T^{\infty}$) Consider an infinite-DM static team with the following expected cost under a randomized policy $P_\pi \in L$ given as:

$$\limsup_{N\to\infty} \int P_{\pi,N}(d\underline{\gamma})\mu^N(d\omega_0, d\underline{\zeta})c^N(\underline{\zeta}, \underline{\gamma}, \underline{y}, \omega_0)\nu^N(d\underline{y}|\underline{\zeta}, \underline{\gamma}, \omega_0) \tag{6.46}$$

where $P_{\pi,N}$ is the restriction of $P_\pi \in L$ to its first N components and μ^N is the marginal of the fixed probability measure on $(\omega_0, \underline{\zeta}^1, \underline{\zeta}^2, \ldots)$ to the first $N+1$ components.

Assumption 6.4.5 *Assume that Assumption 6.4.3 holds with functions ψ_t^i and φ_t^i of the following forms for every $i \in \mathcal{N}$ and $t = 0, \ldots, T-1$:*

$$\psi_t^i\left(y_t^i, \omega_0, x_0^i, \zeta_{[0,t-1]}^i, y_{[0,t-1]}^i, u_{[0,t-1]}^i, \frac{1}{N}\sum_{p=1}^{N} u_{[0,t-1]}^p, \frac{1}{N}\sum_{p=1}^{N} x_{[0,t]}^p\right),$$

$$\varphi_t^i\left(y_t^i, \omega_0, \frac{1}{N}\sum_{p=1}^{N} u_{[0,t-1]}^p, \frac{1}{N}\sum_{p=1}^{N} x_{[0,t]}^p\right),$$

where ψ_t^i is continuous in the last three arguments (actions and the empirical mean of actions and states) and φ_t^i is continuous in the last two arguments (the empirical mean of actions and states).

Before presenting the main result for dynamic mean-field teams, we present sufficient conditions under which the expected cost function induced by randomized optimal policies in L_{EX}^N and L_{EX} are equal as N goes to infinity, and hence, following from Lemma 6.4.3, under symmetry, this shows that without loss of global optimality, optimal policies of dynamic mean-field teams can be considered to be of infinitely-exchangeable type.

Lemma 6.4.4 *Consider the team problem ($\mathcal{P}_T^N$) where Assumptions 6.4.5, 6.3.2, and 6.3.4 hold. Assume further that the cost function is bounded. Then,*

$$\limsup_{N\to\infty} \inf_{P_\pi^N \in L_{EX}^N} \int P_\pi^N(d\underline{\gamma})\mu^N(d\omega_0, d\underline{\zeta})c^N(\underline{\zeta}, \underline{\gamma}, \underline{y}, \omega_0)\nu^N(d\underline{y}|\underline{\zeta}, \underline{\gamma}, \omega_0)$$

$$= \limsup_{N\to\infty} \inf_{P_\pi \in L_{EX}} \int P_{\pi,N}(d\underline{\gamma})\mu^N(d\omega_0, d\underline{\zeta})c^N(\underline{\zeta}, \underline{\gamma}, \underline{y}, \omega_0)\nu^N(d\underline{y}|\underline{\zeta}, \underline{\gamma}, \omega_0), \tag{6.47}$$

where $P_{\pi,N}$ is the restriction of $P_\pi \in L_{EX}$ to its first N components, and μ^N is the marginal of the fixed probability measure on $(\omega_0, \underline{\zeta}^1, \underline{\zeta}^2, \ldots)$ to the first $N+1$ components.

In the following, we establish an existence and structural result for an optimal randomized policy for ($\mathcal{P}_T^\infty$).

Theorem 6.4.3 *Consider a mean-field team problem ($\mathcal{P}_T^\infty$) with ($\mathcal{P}_T^N$) having a symmetric information structure for every N. Let Assumptions 6.4.5, 6.3.2, and 6.3.4 hold. Then, there exists an optimal randomized policy P_π^* for (P_T^∞) which is in $L_{PR,SYM}$,*

$$\inf_{P_\pi \in L_{PR,SYM}} \limsup_{N\to\infty} \int P_{\pi,N}(d\underline{\gamma})\mu^N(d\omega_0, d\underline{\zeta})c^N(\underline{\zeta}, \underline{\gamma}, \underline{y}, \omega_0)\nu^N(d\underline{y}|\underline{\zeta}, \underline{\gamma}, \omega_0)$$

$$:= \limsup_{N\to\infty} \int P_{\pi,N}^*(d\underline{\gamma})\mu^N(d\omega_0, d\underline{\zeta})c^N(\underline{\zeta}, \underline{\gamma}, \underline{y}, \omega_0)\nu^N(d\underline{y}|\underline{\zeta}, \underline{\gamma}, \omega_0)$$

$$= \inf_{P_\pi \in L} \limsup_{N\to\infty} \int P_{\pi,N}(d\underline{\gamma})\mu^N(d\omega_0, d\underline{\zeta})c^N(\underline{\zeta}, \underline{\gamma}, \underline{y}, \omega_0)\nu^N(d\underline{y}|\underline{\zeta}, \underline{\gamma}, \omega_0).$$

6.4.3 *Approximate Optimality of Symmetrically Independent Policies for Exchangeable N-DM Stochastic Teams*

Furthermore, it can also be shown that independently randomized symmetric policies are near optimal for large N: We show that for large N, symmetric policies are nearly optimal and the restriction of the optimal infinite solution to the finite team problem is nearly optimal for large N. To present ideas more effectively, we first introduce the following set of probability measures on policies as:

$$L_{\mathrm{D}}^N := \left\{ P_\pi \in L^N \middle| \text{for all } A_i \in \mathcal{B}(\Gamma^i) : P_\pi(\gamma^1 \in A_1, \ldots, \gamma^N \in A_N) \right.$$

$$= \prod_{i=1}^{N} 1_{\{\tilde{\gamma}^i \in A_i\}}, \text{ for } \tilde{\gamma}^i \in \Gamma^i \Big\},$$

where the above set corresponds to Dirac-delta measures in L_{PR}^N.

Theorem 6.4.4 ([636]) *Consider a static team problem $(\mathcal{P}_N)$ where Assumption 6.4.2 holds and observations $(y^i)_{i\in\mathcal{N}}$ are i.i.d. conditioned on ω_0. Assume further that the cost function is bounded and action sets are compact. Then,*

(i)

$$\inf_{P_\pi^N \in L_{PR,SYM}^N} \int P_\pi^N(d\underline{\gamma})\mu^N(d\omega_0, d\underline{y})c^N(\underline{\gamma}, \underline{y}, \omega_0)$$
$$\leq \inf_{P_\pi^N \in L_{CO}^N} \int P_\pi^N(d\underline{\gamma})\mu^N(d\omega_0, d\underline{y})c^N(\underline{\gamma}, \underline{y}, \omega_0) + \epsilon_N, \tag{6.48}$$

and

$$\inf_{P_\pi^N \in L_{PR,SYM}^N} \int P_\pi^N(d\underline{\gamma})\mu^N(d\omega_0, d\underline{y})c^N(\underline{\gamma}, \underline{y}, \omega_0)$$
$$\leq \inf_{P_\pi^N \in L_D^N} \int P_\pi^N(d\underline{\gamma})\mu^N(d\omega_0, d\underline{y})c^N(\underline{\gamma}, \underline{y}, \omega_0) + \epsilon_N, \tag{6.49}$$

where $\epsilon_N \to 0$ as N goes to infinity.

(ii) If $P_\pi^* \in L_{PR,SYM}$ is a randomized optimal policy of $(\mathcal{P}_\infty)$, then there exist $\bar{\epsilon}_N \geq 0$ where $\bar{\epsilon}_N \to 0$ as N goes to infinity, and

$$\int P_{\pi,N}^*(d\underline{\gamma})\mu^N(d\omega_0, d\underline{y})c^N(\underline{\gamma}, \underline{y}, \omega_0)$$
$$\leq \inf_{P_\pi^N \in L_D^N} \int P_\pi^N(d\underline{\gamma})\mu^N(d\omega_0, d\underline{y})c^N(\underline{\gamma}, \underline{y}, \omega_0) + \epsilon_N + \bar{\epsilon}_N, \tag{6.50}$$

where $P_{\pi,N}^*$ is the restriction of P_π^* to the first N components.

The main idea for establishing Part(i) is to use Lemmas 6.4.1 and 6.4.2 to provide an approximation of the optimal expected cost by restricting the search for randomized policies to those that are restrictions of randomized policies in L_{EX} to the N first components. We note that since the set of policies L_{D}^N is not a convex subset of the set of randomized policies L^N, (6.48) does not immediately imply (6.49) using Lemma 6.4.1 but the result can be established using an extreme point argument, and since policies in L_{D}^N are optimal among all randomized policies L_{PR}^N for N-DM teams. Part(ii) follows from Part(i) and Theorem 6.4.2, using the fact that a randomized optimal policy $P_\pi^* \in L_{\mathrm{PR,SYM}}$ of $(\mathcal{P}_\infty)$ provides an approximation for the optimal expected cost when the search for randomized optimal policy for N-DM teams is restricted to those in $L^N_{\mathrm{PR,SYM}}$.

Similarly, we can address the same problem and present approximation results of optimal policies for symmetric dynamic N-DM team problems: If P^*_π is a (randomized) symmetric optimal policy for $(\mathcal{P}^\infty_T)$, then there exist $\epsilon_N \geq 0$, with $\epsilon_N \to 0$ as $N \to \infty$, such that $P^*_\pi|_N$ is ϵ_N-optimal for $(\mathcal{P}^N_T)$ where $P^*_\pi|_N$ is the restriction of P^*_π to the first N DMs. This builds on the symmetry results and analysis for $(\mathcal{P}^\infty_T)$. We state the following.

Theorem 6.4.5 ([636]) *Consider a dynamic team problem $(\mathcal{P}^N_T)$. Let Assumptions 6.4.5, 6.3.2, and 6.3.4 hold. If the cost function is bounded, then*

(i)

$$\begin{aligned}
&\inf_{P^N_\pi \in L^N_{PR,\,SYM}} \int P^N_\pi(d\underline{\gamma})\mu^N(d\omega_0, d\underline{\zeta})c^N(\underline{\zeta}, \underline{\gamma}, \underline{y}, \omega_0)\nu^N(d\underline{y}|\underline{\zeta}, \underline{\gamma}, \omega_0) \\
&\leq \inf_{P^N_\pi \in L^N_{CO}} \int P^N_\pi(d\underline{\gamma})\mu^N(d\omega_0, d\underline{\zeta})c^N(\underline{\zeta}, \underline{\gamma}, \underline{y}, \omega_0)\nu^N(d\underline{y}|\underline{\zeta}, \underline{\gamma}, \omega_0) + \epsilon_N,
\end{aligned} \tag{6.51}$$

and

$$\begin{aligned}
&\inf_{P^N_\pi \in L^N_{PR,\,SYM}} \int P^N_\pi(d\underline{\gamma})\mu^N(d\omega_0, d\underline{\zeta})c^N(\underline{\zeta}, \underline{\gamma}, \underline{y}, \omega_0)\nu^N(d\underline{y}|\underline{\zeta}, \underline{\gamma}, \omega_0) \\
&\leq \inf_{P^N_\pi \in L^N_{D}} \int P^N_\pi(d\underline{\gamma})\mu^N(d\omega_0, d\underline{\zeta})c^N(\underline{\zeta}, \underline{\gamma}, \underline{y}, \omega_0)\nu^N(d\underline{y}|\underline{\zeta}, \underline{\gamma}, \omega_0) + \epsilon_N,
\end{aligned} \tag{6.52}$$

where $\epsilon_N \to 0$ as N goes to infinity.

(ii) If $P^*_\pi \in L_{PR,SYM}$ is a randomized optimal policy of $(\mathcal{P}^\infty_T)$, then there exist $\bar{\epsilon}_N \geq 0$ where $\bar{\epsilon}_N \to 0$ as N goes to infinity and

$$\begin{aligned}
&\int P^*_{\pi,N}(d\underline{\gamma})\mu^N(d\omega_0, d\underline{\zeta})c^N(\underline{\zeta}, \underline{\gamma}, \underline{y}, \omega_0)\nu^N(d\underline{y}|\underline{\zeta}, \underline{\gamma}, \omega_0) \\
&\leq \inf_{P^N_\pi \in L^N_{D}} \int P^N_\pi(d\underline{\gamma})\mu^N(d\omega_0, d\underline{\zeta})c^N(\underline{\zeta}, \underline{\gamma}, \underline{y}, \omega_0)\nu^N(d\underline{y}|\underline{\zeta}, \underline{\gamma}, \omega_0) + \epsilon_N + \bar{\epsilon}_N,
\end{aligned}$$

where $P^*_{\pi,N}$ is the restriction of P^*_π to the first N components.

6.5 Appendix: Proofs

6.5.1 Proof of Theorem 6.2.1

Under (A1), the required derivatives in (6.7) in the direction of u^i exist. Now, we use the convexity property to justify interchanging the expectation and the differentiation similar to the analysis in Sect. 2.4, and then use (6.6) and (6.7) to establish global optimality of $\underline{\gamma}^*$ for $(\mathcal{P}_\infty)$. Under (A1), we have for every $\alpha \in (0, 1]$,

$$\sum_{i=1}^{N} c(\omega_0, \gamma^{i*} + \alpha\delta^i, \mu_N^* + \frac{\alpha}{N}\sum_{p=1}^{N}\delta^p) - c(\omega_0, \gamma^{i*}, \mu_N^*)$$

$$\leq \alpha \sum_{i=1}^{N}\Big(c(\omega_0, \gamma^i, \mu_N) - c(\omega_0, \gamma^{i*}, \mu_N^*)\Big),$$

where $\mu_N = \frac{1}{N}\sum_{p=1}^{N}\gamma^p(y^p)$ and $\delta^i = \gamma^i - \gamma^{i*}$. Let

$$h_N^{\omega_0}(\alpha) := \frac{1}{\alpha}\Bigg[\frac{1}{N}\sum_{i=1}^{N} c(\omega_0, \gamma^{i*} + \alpha\delta^i, \mu_N^* + \frac{\alpha}{N}\sum_{p=1}^{N}\delta^p) - c(\omega_0, \gamma^{i*}, \mu_N^*)\Bigg].$$

Then, $h_N^{\omega_0}(\alpha)$ is a monotone non-increasing function as $\alpha \to 0$ in $\alpha \in [0, 1]$ and bounded from above by $h_N^{\omega_0}(1)$. Thus, $h'_{+,N}(\omega_0, 0) := \lim_{\alpha\to 0} h_N^{\omega_0}(\alpha)$ exists. Since $h_N^{\omega_0}(\alpha)$ is a monotonic non-increasing function as $\alpha \to 0$ in $\alpha \in [0, 1]$ and bounded above by $h_N^{\omega_0}(1)$, we can use the monotone convergence theorem to interchange the limit and the expectations

$$\lim_{\alpha\to 0} E(h_N^{\omega_0}(\alpha)) = E(\lim_{\alpha\to 0} h_N^{\omega_0}(\alpha)) = E(h'_{+,N}(\omega_0, 0)). \tag{6.53}$$

We have $E(h'_{+,N}(\omega_0, 0)) = \frac{1}{N}E(\sum_{i=1}^{N}\sum_{k=1}^{N} c_{u^k}(\omega_0, \gamma^{i*}, \mu_N^*)\delta^k)$. Define

$$F_{\underline{\gamma}_N}^N(\alpha) := \frac{1}{N}E\Bigg(\sum_{i=1}^{N} c(\omega_0, \gamma^{i*} + \alpha\delta^i, \mu_N^* + \frac{\alpha}{N}\sum_{p=1}^{N}\delta^p)\Bigg).$$

Note that $F_{\underline{\gamma}_N}^N(\alpha)$ exists for $\alpha \in [0, 1]$, since $E(h_N^{\omega_0}(\alpha)) \leq E(h_N^{\omega_0}(1)) < \infty$, and $E(\frac{1}{N}\sum_{i=1}^{N} c(\omega_0, \gamma^{i*}, \mu_N^*)) < \infty$. Therefore, one can write $F'^N_{+\underline{\gamma}_N}(0) = \lim_{\alpha\to 0} E(h^{\omega_0}(\alpha))$, and

$$F'^{N}_{\underline{\gamma}^{+}_{N}}(0) = \frac{1}{N} E\left(\sum_{i=1}^{N} \sum_{k=1}^{N} c_{u^k}(\omega_0, \gamma^{i*}, \mu_N^*)(\gamma^k - \gamma^{k*}) \right).$$

Thus, we can write

$$J(\underline{\gamma}) - J(\underline{\gamma}^*) = \limsup_{N\to\infty} F^N_{\underline{\gamma}_N}(1) - \limsup_{N\to\infty} F^N_{\underline{\gamma}_N}(0) \tag{6.54}$$

$$= \limsup_{N\to\infty} F^N_{\underline{\gamma}_N}(1) - \liminf_{N\to\infty} F^N_{\underline{\gamma}_N}(0) \tag{6.55}$$

$$\geq \limsup_{N\to\infty} \frac{F^N_{\underline{\gamma}_N}(1) - F^N_{\underline{\gamma}_N}(0)}{1} \tag{6.56}$$

$$\geq \limsup_{N\to\infty} F'^{N}_{\underline{\gamma}^{+}_{N}}(0) \geq 0, \tag{6.57}$$

where (6.55) follows from (A2) and (6.6), and $-\liminf_{N\to\infty} a_N = \limsup_{N\to\infty} -a_N$, $\limsup_{N\to\infty} a_N + \limsup_{N\to\infty} b_N \geq \limsup_{N\to\infty}(a_N + b_N)$ imply (6.56), and (6.57) holds since $F^N_{\underline{\gamma}_N}(.)$ is a convex function using [223, Corollary 6.3.3], and since $a_N \geq b_N$ then $\limsup_{N\to\infty} a_N \geq \limsup_{N\to\infty} b_N$. Finally, the last inequality follows from (6.7); hence, $J(\underline{\gamma}) - J(\underline{\gamma}^*) \geq 0$, and the proof is completed. ◇

6.5.2 *Proof of Theorem 6.2.5*

Action spaces and the cost function are convex, and following from the hypothesis that y^i's are i.i.d. random variables (hence they are exchangeable conditioned on ω_0) and the result of Theorem 6.2.4, one can consider a sequence of N-DM teams which are symmetrically optimal that defines $(\mathcal{P}_N)$ (see (6.2)) and whose limit is identified with $(\mathcal{P}_\infty)$. Define empirical measures on actions and observation of $Q_N(B) := \frac{1}{N}\sum_{i=1}^{N} \delta_{\zeta^i_N}(B)$, and $\tilde{Q}_N(B) := \frac{1}{N}\sum_{i=1}^{N} \delta_{\zeta^i_\infty}(B)$, where $B \in \mathcal{Z} := \mathbb{U} \times \mathbb{V}$, $\zeta^i_N := (\gamma^*_N(y^i), y^i)$, $\zeta^i_\infty := (\gamma^*_\infty(y^i), y^i)$, and $\delta_Y(\cdot)$ is the Dirac measure for any random variable Y. In the following, we first show that $\tilde{Q}_N$ converges weakly to $Q = Law(\zeta^i_\infty)$ P-almost surely, and then we show that (6.8) holds, and we invoke Theorem 6.2.2.

Step 1 For every $g \in C_b(\mathcal{Z})$, where we denote $C_b(X)$ as the space of continuous and bounded functions in X, we have

$$\lim_{N\to\infty} P\left(\left| \int g dQ_N - \int g d\tilde{Q}_N \right| \geq \epsilon \right)$$

$$\leq \epsilon^{-1} \lim_{N\to\infty} \frac{1}{N} \sum_{i=1}^{N} E\left[\left|g(\gamma_N^*(y^i), y^i) - g(\gamma_\infty^*(y^i), y^i)\right|\right] \tag{6.58}$$

$$= \epsilon^{-1} \lim_{N\to\infty} E\left[\left|g(\gamma_N^*(y^i), y^i) - g(\gamma_\infty^*(y^i), y^i)\right|\right] \tag{6.59}$$

$$= \epsilon^{-1} E\left[\lim_{N\to\infty} \left|g(\gamma_N^*(y^i), y^i) - g(\gamma_\infty^*(y^i), y^i)\right|\right] = 0, \tag{6.60}$$

where (6.58) follows from Markov's inequality, the triangle inequality and the definition of the empirical measure, and (6.59) follows from the hypothesis that y^is are identical random variables. Since g is bounded and continuous, the dominated convergence theorem implies (6.60). Hence, for every subsequence there exists a subsubsequence such that $|\int g dQ_{N_{k_l}} - \int g d\tilde{Q}_{N_{k_l}}|$ converges to zero P-almost surely as $l \to \infty$. On the other hand, since y^is are i.i.d. random variables, the strong law of large numbers implies that $\tilde{Q}_N$ converges weakly to Q P-almost surely, that is $|\int g d\tilde{Q}_N - \int g dQ|$ converges to zero P-almost surely for every $g \in C_b(\mathcal{Z})$. Hence, through choosing a suitable subsequence, Q_{N_k} converges P-almost sure weakly to Q since for every continuous and bounded function g, we have P-a.s.,

$$\lim_{N\to\infty} \left|\int g dQ_N - \int g dQ\right|$$
$$\leq \lim_{N\to\infty} \left(\left|\int g dQ_N - \int g d\tilde{Q}_N\right| + \left|\int g d\tilde{Q}_N - \int g dQ\right|\right) = 0. \tag{6.61}$$

Step 2 Following from [658, Theorem 3.5] or [454, Theorem 3.1] using the fact that the cost function is non-negative and continuous, we have

$$\limsup_{N\to\infty} \frac{1}{N} \sum_{i=1}^{N} E\left[c\left(\omega_0, \gamma_N^*(y^i), \frac{1}{N}\sum_{i=1}^{N} \gamma_N^*(y^i)\right)\right]$$
$$\geq \liminf_{N\to\infty} E\left[E\left[\int_{\mathcal{Z}} c\left(\omega_0, u, \int_{\mathbb{U}} u Q_N(du \times \mathbb{V})\right) Q_N(du, dv)\middle|\omega_0\right]\right]$$
$$\geq E\left[E\left[\liminf_{N\to\infty} \int_{\mathcal{Z}} c\left(\omega_0, u, \int_{\mathbb{U}} u Q_N(du \times \mathbb{V})\right) Q_N(du, dv)\middle|\omega_0\right]\right]$$
$$\geq E\left[E\left[\int_{\mathcal{Z}} c\left(\omega_0, u, \int_{\mathbb{U}} u Q(du \times \mathbb{V})\right) Q(du, dv)\middle|\omega_0\right]\right], \tag{6.62}$$

where the first inequality follows from the definition of Q_N and replacing limsup by liminf. The second inequality follows from Fatou's lemma. In the following, we justify (6.62). Since Q_N converges weakly to Q P-almost surely, using continuous mapping theorem [116, page 20], we have that $Q_N(du \times \mathbb{V})$ converges weakly to $Q(du \times \mathbb{V})$ P-almost surely, and hence, the compactness of $\mathbb{U}$ implies

$\int_{\mathbb{U}} uQ_N(du \times \mathbb{V}) \to \int_{\mathbb{U}} uQ(du \times \mathbb{V})$ P-almost surely, and continuity of the cost function P-almost surely implies that $c(\omega_0, u, \int_{\mathbb{U}} uQ_N(du \times \mathbb{V}))$ converges to $c\left(\omega_0, u, \int_{\mathbb{U}} uQ(du \times \mathbb{V})\right)$ P-almost surely. Define a non-negative bounded sequence $G_N^M := \min\{M, c\left(\omega_0, u, \int_{\mathbb{U}} uQ_N(du \times \mathbb{V})\right)\}$, where $G_M^N \uparrow G^N := c\left(\omega_0, u, \int_{\mathbb{U}} uQ_N(du \times \mathbb{V})\right)$ as $M \to \infty$; then we have P-almost surely

$$\begin{aligned}
&\liminf_{N\to\infty} \int_{\mathbb{Z}} c\left(\omega_0, u, \int_{\mathbb{U}} uQ_N(du \times \mathbb{V})\right) Q_N(du, dv) \\
&= \lim_{M\to\infty} \liminf_{N\to\infty} \int_{\mathbb{Z}} c\left(\omega_0, u, \int_{\mathbb{U}} uQ_N(du \times \mathbb{V})\right) Q_N(du, dv) \\
&\geq \lim_{M\to\infty} \liminf_{N\to\infty} \int_{\mathbb{Z}} G_N^M Q_N(du, dv) \\
&= \lim_{M\to\infty} \int_{\mathbb{Z}} G^M Q(du, dv) \\
&= \int_{\mathbb{Z}} c\left(\omega_0, u, \int_{\mathbb{U}} uQ(du \times \mathbb{V})\right) Q(du, dv),
\end{aligned}$$

where the first inequality follows from the definition of G_N^M and the second equality is true using [658, Theorem 3.5] since G_N^M is bounded (hence it is uniformly Q_N-integrable) and continuously converges to G^M, and the monotone convergence theorem implies the last equality. Hence, (6.62) holds, which implies (6.8), and thus the proof is completed using Theorem 6.2.2. ◇

6.5.3 *Proof of Proposition 6.2.1*

Since z^i's are i.i.d. random variables, and observations, $y^i = h(x, z^i)$, have identical distributions (but are not independent), similar to the proof of Theorem 6.2.5, using symmetry, one can show that (6.60) holds. In the following, we show that (6.61) and (6.62) hold.

$$\lim_{N\to\infty} P\left(\left|\int g d\tilde{Q}_N - \int g dQ\right| \geq \epsilon\right)$$

$$\leq \frac{1}{\epsilon^2} \lim_{N\to\infty} E\left[\left|\frac{1}{N}\sum_{i=1}^{N} g(\gamma_\infty^*(y^i), y^i) - E(g(\gamma_\infty^*(y^1), y^1))\right|^2\right] \tag{6.63}$$

$$= \lim_{N\to\infty} (\epsilon)^{-2} E\left[E\left[\left|\frac{1}{N}\sum_{i=1}^{N} L(\gamma_\infty^*(v^i), v^i)\right|^2 \middle| x\right]\right] \tag{6.64}$$

$$= 0, \tag{6.65}$$

where $L(\gamma^*_\infty(y^i), y^i) := g(\gamma^*_\infty(y^i), y^i) - E(g(\gamma^*_\infty(y^1), y^1)|x)$, and (6.63) follows from Chebyshev's inequality, and (6.64) follows from the law of iterated expectations. The structure $y^i = h(x, z^i)$ implies conditional independence of y^i's given x, and hence, using the law of large numbers and since $g \in C_b(\mathcal{Z})$, we have (6.65), and this implies that $\tilde{Q}_{N_k}$ converges weakly to $Law(\zeta^i_\infty|x)$ P-almost surely as $k \to \infty$. Hence, through choosing a suitable subsequence, $Q_{N_{k_l}}$ converges P-almost sure weakly to $Q = Law(\zeta^i_\infty|x)$ as $l \to \infty$ and the rest of the proof to justify (6.62) is the same as that of Theorem 6.2.5. ◇

6.6 Concluding Remarks

We have studied, in this chapter, teams with large numbers of (including infinitely many) DMs. For convex and non-convex but exchangeable settings, we have obtained structural results on optimal policies, existence results, and approximation results. For each of the settings, we established the existence of optimal policies for the mean-field limit and the structure of optimal policies, which turn out to have a symmetry structure across the DMs/agents. For the convex setup, however, we have also shown symmetry of optimal policies for finite N-DM problems. In the absence of convexity for the exchangeable setting, we have established near optimality of symmetric policies for large, but finite, N-DM/agent problems.

6.7 Bibliographic Notes

Mean-field games (see e.g., [361, 362, 456]) can be viewed as limit models of symmetric non-zero-sum non-cooperative finite player games with a mean-field interaction. We note that in team problems, person-by-person optimality (Nash equilibrium when viewed as games) does not in general imply global optimality both for N-DM teams and teams with countably infinite number of DMs. As we have mentioned, for static teams, a sufficient condition is the convexity of the cost function, subject to minor regularity conditions [429]. However, mean-field teams under decentralized information structures generally correspond to dynamic team problems with non-classical information structures, hence, mean-field team problems may be non-convex even under the convexity of the cost function due to non-classical information structures. Hence, person-by-person optimality is generally inconclusive for global optimality. The existence of Nash equilibria (and hence, person-by-person-optimal solutions for team problems) has been established for mean-field games in [70, 159, 447, 456, 480], among others.

Some relevant studies on the existence and convergence of equilibria from the mean-field games literature are the following: In [157], for one-shot mean-field games, under regularity assumptions on the cost function, it has been shown that mixed Nash strategies of N-player symmetric games converge through a subsequence to a limit (which is a weak-solution of the mean-field limit). In [265], through a concentration of measures argument, it has been shown that a subsequence of symmetric local approximate Nash equilibria for N player games converges to a solution for the mean-field game under the assumption that the normalized occupational measures converges weakly to a deterministic measure. Furthermore, using a similar method in [448], assumptions on equilibrium policies of large population mean-field symmetric stochastic differential games have been presented to allow for convergence of asymmetric approximate Nash equilibria to a weak solution of the mean-field game [448, Theorem 2.6] in the presence of common randomness. Using martingale methods and relaxed controls (see also [159, 265, 447, 448]), an existence result and a limit theory have been established for controlled McKean-Vlasov dynamics [449]. We note that in [159, 447–449], it has been assumed that each player has full access to the information available to all players, i.e., the controls are functions of all initial states, driving Wiener processes of all players, and common randomness (also called *open-loop* controllers in the mean-field games' literature). We note that in the setup of this book, under these strategies, the information structure corresponds to static information. The equilibria with respect to *closed-loop* (in the team problem setting, with respect to dynamic information structure) is different since the deviating player can still influence the information of other players and hence it can influence the average of states or actions substantially, as we have seen earlier.

We also note additional related works by Lacker [449, 450] where either convergence of open-loop controllers, or convergence of Nash equilibria induced by closed-loop controllers (where controls are measurable path-dependent functions of states, $u_t^i = \varphi(t, x_{[0,t]})$, where $x_{[0,t]} = (x^1_{[0,t]}, \ldots, x^N_{[0,t]})$ and φ is a measurable function) or Markovian controllers ($u_t^i = \varphi(t, x_t)$, where $x_t = (x_t^1, \ldots, x_t^N)$) have been considered. In [450], the information structure is classical (a centralized problem) since players have access to all the information available).

In Lemma 6.3.2 and Theorem 6.3.2, we considered a non-classical information structure for teams defined as $(\mathcal{P}_T^{\infty,\mathrm{MF}})$ with a convex expected cost in policies. For teams defined as $(\mathcal{P}_T^{\infty,\mathrm{MF}})$ with a symmetric partially nested information structure which admits static reduction, the above result holds, and similar to the proof of Theorem 6.3.2, it can be proven under the assumption that the cost functions is convex in actions (since convexity of the cost function in actions is a sufficient condition for convexity of the expected cost function in policies for this class of problems).

Related to mean-field games, in the economic theory literature, in [509, 653], anonymous games have been studied and existence and characterization of structural properties of Cournot-Nash equilibria have been established. This Cournot-Nash equilibrium concept corresponds to a mean-field equilibrium for a (one-shot) static problem (see also [376] for sequential anonymous games). However, such an equilibrium does not necessarily imply global optimality in the context of team problems since, as discussed earlier, person-by-person optimality does not in general imply global optimality in the absence of convexity. Social optima for mean-field linear quadratic Gaussian control problems under both centralized and restricted decentralized information structures have been considered in [31, 360, 363, 758]. We also note a result in [804] where two large teams compete in a mean-field competition game. We refer readers to [154, 158] for a literature review and a detailed summary of some recent results on mean-field games and social optima problems. We also refer the reader to [711] for risk-sensitive mean-field problems

Parts of our results in this chapter correspond to Linear Quadratic Gaussian (LQG) teams. In [350], it has been shown that for teams with a finite number of DMs, dynamic teams with a partially nested information structure can be reduced to a static one ([350, 780]) where Radner's theorem concludes global optimality of linear policies for LQG team problems, as we have seen earlier. However, for average cost infinite horizon, partially nested, LQG dynamic team problems so far there has been no universal result establishing that a globally optimal policy is linear, time-invariant, and stabilizing, and this has been often imposed a priori: In [610], the problem of designing a linear, time-invariant, stabilizing, state feedback optimal policy for decentralized $\mathcal{H}_2$-optimization problems, which satisfy the quadratically invariance property, has been addressed by reparametrizing the problem as a convex problem (via Youla parameterization). In [609], it has been shown that for sequential team problems involving linear systems, quadratic invariance and the partially nested property are equivalent. For a class of partially ordered (POSET) systems, state space techniques have been utilized to obtain optimal, linear, time-invariant, state feedback controllers for $\mathcal{H}_2$-optimization problems with sparsity constraints [661]. A similar result has been established in [695] where linearity and time invariance have been imposed apriori. In [473], $\mathcal{H}_2$-optimization output feedback problems with two-DMs have been considered and optimality results have been established when the optimal policies are restricted to linear, time invariant, stabilizing policies. However, the results in [473, 610, 661, 695] are inconclusive regarding global optimality. A further contribution here is to consider average cost infinite horizon dynamic team problems without restricting the set of policies to those that are linear, time-invariant, and stabilizing unlike the results in [473, 610, 661, 695]. We note again that the optimality of linear policies for infinite horizon LQG problems is open in its generality and we provided here positive results for a sub-class of such problems.

Many of the results of this chapter have appeared in [632, 633, 636], and in part [497, 818]) and [621].

Chapter 7
Comparison of Information Structures and Their Blackwell Ordering

7.1 Introduction

In this chapter and the following one, we will continue with our coverage of information structures but from a different perspective than what was considered in Chaps. 4 and 5 (where properties *given a fixed information structure* were studied). In these two chapters, we will view information structures as design variables and study both qualitative and quantitative properties of information structures as they relate to optimal stochastic team decision problems.

Measurement Channels as Information Structures and Their Comparison and Regularity Properties

To motivate the discussion further, let us consider a dynamic decision problem in the single-agent context. Consider a dynamical system in discrete-time (as single-agent, time-invariant counterparts of (2.18) and (2.19))

$$x_{t+1} = f(x_t, u_t, w_t),$$
$$y_t = g(x_t, v_t), t \in \mathbb{Z}_+,$$

for some measurable functions f, g, with $\{w_t\}$ being i.i.d. system noise process and $\{v_t\}$ an i.i.d. measurement disturbance process, which are independent of x_0 and each other. Here, we assume the state, measurement, and action spaces to be Borel subsets of finite dimensional Euclidean spaces.

In the above, we can view g as inducing a measurement channel Q, which is a stochastic kernel or a regular conditional probability measure from $\mathbb{X}$ to $\mathbb{Y}$ in the sense that $Q(\cdot\,|x)$ is a probability measure on the (Borel) σ-algebra $\mathcal{B}(\mathbb{Y})$ on $\mathbb{Y}$ for every $x \in \mathbb{X}$, and $Q(A|\cdot) : \mathbb{X} \to [0, 1]$ is a Borel measurable function for every $A \in \mathcal{B}(\mathbb{Y})$.

S. Yüksel, T. Başar, *Stochastic Teams, Games, and Control under Information Constraints*, Systems & Control: Foundations & Applications,
https://doi.org/10.1007/978-3-031-54071-4_7

As before, an *admissible policy* is a sequence of control functions $\{\gamma_t, t \in \mathbb{Z}_+\}$ such that γ_t is measurable with respect to the σ-algebra generated by the information variables

$$I_t = \{y_{[0,t]}, u_{[0,t-1]}\}, \quad t \in \mathbb{N}, \qquad I_0 = \{y_0\}.$$

where

$$u_t = \gamma_t(I_t), \quad t \in \mathbb{Z}_+ \tag{7.1}$$

are the $\mathbb{U}$-valued control actions. Let $\boldsymbol{\Gamma}$ be the space of all admissible policies.

With the above setup, let the objective be one of minimization of the cost

$$J(P, Q, \underline{\gamma}) = E_P^{Q,\underline{\gamma}}\left[\sum_{t=0}^{T-1} c(x_t, u_t)\right], \tag{7.2}$$

over the set of all admissible policies $\underline{\gamma}$, where $c : \mathbb{X} \times \mathbb{U} \to \mathbb{R}$ is a Borel measurable stagewise cost function and $E_P^{Q,\underline{\gamma}}$ denotes the expectation with initial state probability measure given by P, under policy $\underline{\gamma}$ and given channel Q.

See Fig. 7.1 for an illustration with $T = 1$.

For $P \in \mathcal{P}(\mathbb{X})$ and $Q \in \mathcal{Q}$ we let PQ denote the joint distribution induced on $(\mathbb{X} \times \mathbb{Y}, \mathcal{B}(\mathbb{X} \times \mathbb{Y}))$ by channel Q with input distribution P:

$$PQ(A) = \int_A Q(dy|x)P(dx), \quad A \in \mathcal{B}(\mathbb{X} \times \mathbb{Y}).$$

We adopt the convention that given a probability measure μ, the notation $z \sim \mu$ means that z is a random variable with distribution μ.

Comparison of Measurement Channels (Stochastic Kernels) The question we ask is when one can compare two measurement channels Q^1, Q^2 such that

$$\inf_{\underline{\gamma} \in \boldsymbol{\Gamma}} J(P, Q^1, \underline{\gamma}) \leq \inf_{\underline{\gamma} \in \boldsymbol{\Gamma}} J(P, Q^2, \underline{\gamma}),$$

for all measurable and bounded cost functions c in (7.2).

We will study this question in this chapter and also generalize it to the multi-agent setup.

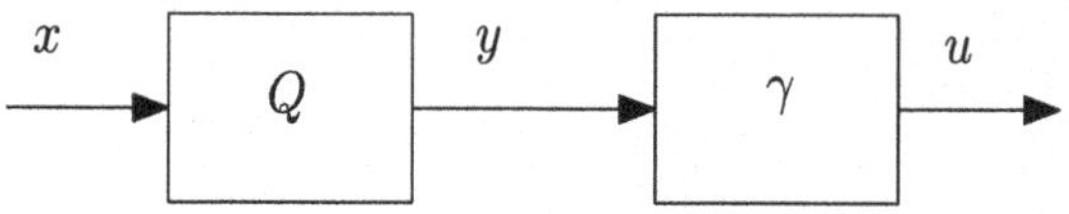

Fig. 7.1 Measurement channel as an information structure

We also present two further questions below, which will be studied in the next chapter.

Continuity on the Space of Measurement Channels (Stochastic Kernels) Suppose that $\{Q_n, n \in \mathbb{N}\}$ is a sequence of communication channels converging in some sense to a channel Q. Then a second question we ask is when does

$$Q_n \to Q$$

imply

$$\inf_{\underline{\gamma} \in \mathbf{\Gamma}} J(P, Q_n, \underline{\gamma}) \to \inf_{\underline{\gamma} \in \mathbf{\Gamma}} J(P, Q, \underline{\gamma})?$$

Existence of Optimal Measurement Channels and Quantizers Let $\mathcal{Q}$ be a set of communication channels. A second question we ask is when do there exist minimizing and maximizing channels for the optimization problems

$$\inf_{Q \in \mathcal{Q}} \inf_{\underline{\gamma}} E_P^{Q,\underline{\gamma}} \left[\sum_{t=0}^{T-1} c(x_t, u_t) \right]$$

and

$$\sup_{Q \in \mathcal{Q}} \inf_{\underline{\gamma}} E_P^{Q,\underline{\gamma}} \left[\sum_{t=0}^{T-1} c(x_t, u_t) \right].$$

If solutions to these problems exist, are they unique?

These questions will also be generalized to multi-agent contexts both in this chapter and in Chap. 8.

In Sect. 7.2 we motivate the notion of the value of information structures in stochastic decision problems. In particular, we will view the information structure as a *design variable*. We will investigate values of measurement channels and develop a general partial order (comparison) of information structures in Sect. 7.3. Notably, in this chapter, we will develop a Blackwell ordering of information structures, which also generalizes Blackwell's analysis from finite space setups to standard Borel spaces. For the forward direction to Blackwell ordering, we present three alternative proof methods to show that garbling leads to inferior performance. We then present a converse theorem on the comparison of information structures in Sect. 7.3.2 building on a convex analytical geometric argument. The multi-agent setup is studied in Sect. 7.4, where the comparison results are generalized via the notion of correlated garbling, to be introduced.

These results will establish that comparison of information structures is equivalent to a (possibly correlated) garbling property between them.

7.2 Value of Information and Preliminary Results on Comparison of Information Structures

In this section, some general discussion on the value of information structures will be presented, with explicit examples and some preliminary results. Throughout this section, we adopt the general framework of Sect. 2.2, unless otherwise stated, where we also take the information structure to be static.

The results in this section will then be generalized (also with converse statements) in the subsequent analysis covered in Sects. 7.3 and 7.4.

If, for a stochastic team decision problem, we allow the flexibility of having more than one IS for the team, a natural question that arises is "*when is one information structure better than another*?" To shed some light on this question, let us first introduce the notation $R(\underline{\gamma}; \underline{\eta})$ (as in Sect. 2.2) as the cost function of the team (to replace "J" which we had used all along), to recognize the dependence of the team performance (or cost) also on the information structure $\underline{\eta}$. This function R may also include some additional terms quantifying costs due to the acquisition of information, i.e., costs directly associated with various $\underline{\eta}$'s, as in

$$R(\underline{\gamma}; \underline{\eta}) := \tilde{J}(\underline{\gamma}; \underline{\eta}) + \kappa(\underline{\eta}) \equiv E_\xi[L(\xi, \underline{\gamma}(\underline{\eta}[\xi]))] + \kappa(\underline{\eta}) \tag{7.3}$$

where $\tilde{J}$ is the team (expected) cost under $\underline{\gamma}$ and $\underline{\eta}$, without any cost on information, and $\kappa(\underline{\eta}) \geq 0$ is the cost due to the choice of the particular information structure $\underline{\eta}$. We note that as $\underline{\eta}$ changes, the policy space of the team, $\mathbf{\Gamma}$, may also change, and hence to make this dependence explicit, we will write the policy space as $\mathbf{\Gamma}(\underline{\eta})$.

Clearly, being a scalar valued function, (7.3) provides a strict ordering of all permissible information structures for a given team problem (with a fixed loss function L and a fixed cost of information κ) according to the best that can be achieved. This leads us to the following first comparison (strict ordering) of *IS*s for a given team problem.

Definition 7.2.1 For a stochastic team problem with a fixed cost structure as given by (7.3) but a variable information structure (*IS*), an *IS* $\underline{\eta}$ is *better* (or *more valuable*) than another *IS* $\underline{\eta}'$ if

$$\inf_{\underline{\gamma} \in \mathbf{\Gamma}(\underline{\eta})} R(\underline{\gamma}; \underline{\eta}) < \inf_{\underline{\gamma} \in \mathbf{\Gamma}(\underline{\eta}')} R(\underline{\gamma}; \underline{\eta}').$$

Two *IS*s $\underline{\eta}$ and $\underline{\eta}'$ are of *equal value* if the inequality above is an equality. ◇

The strict ordering provided by *Definition 7.2.1* is dependent on the underlying cost structure (i.e., L and c) and it would be desirable to obtain an ordering that is valid for all reasonable loss functions (e.g., for all strictly convex loss functions, all continuous and bounded costs functions or all measurable and bounded cost functions). The following definition which provides only a partial ordering, serves this purpose:[1]

Definition 7.2.2 Let $\mathcal{L}$ be a given class of loss functions for a given stochastic team problem with fixed information cost and variable *IS*. Then, an *IS* $\underline{\eta}$ is *uniformly better* (*uniformly more valuable, more informative*) than another *IS* $\underline{\eta}'$ (with respect to the class $\mathcal{L}$) if, under the cost structure (7.3),

$$\inf_{\underline{\gamma}\in\mathbf{\Gamma}(\underline{\eta})} R(\underline{\gamma};\underline{\eta}) \leq \inf_{\underline{\gamma}\in\mathbf{\Gamma}(\underline{\eta}')} R(\underline{\gamma};\underline{\eta}')$$

for all $L \in \mathcal{L}$, with strict inequality for at least one $L \in \mathcal{L}$. A somewhat weaker notion which allows the possibility for the inequality above to be an equality for all $L \in \mathcal{L}$ is "$\underline{\eta}$ is *uniformly at least as valuable as (least as informative as)* $\underline{\eta}'$" (with respect to the class $\mathcal{L}$). ◇

Now, to obtain a sufficient condition for an *IS* to be *uniformly at least as valuable as* another *IS* , let us recall (from Sect. 2.2, Eq. (24.2)) that given an *IS* $\underline{\eta} = (\eta^1, \ldots, \eta^N)$, there is an associated set of measurements $\mathbf{y} = (y^1, \ldots, y^N)$, which can also be viewed as random variables (or vectors) defined directly on the original probability space $(\Omega, \mathbf{F}, P)$. Let $\sigma(y^i)$ be the smallest sigma field (on Ω) with respect to which y^i is measurable. Clearly, $\sigma(y^i) \subseteq \mathbf{F}$. To indicate the dependence of this sigma-field on η^i, we will use a subscript on σ: $\sigma_\eta(y^i)$. We now state a relatively immediate result.

Proposition 7.2.1 *Suppose we are given a static stochastic N-agent team problem with two possible ISs $\underline{\eta}$ and $\underline{\eta}'$. Let $\mathcal{L}$ be a class of loss functions with the property that for $L \in \mathcal{L}$, the random variable $L(\xi, \underline{\gamma}(\underline{\eta}[\xi]))$ has a well-defined (possibly infinite) expected value for every $\underline{\gamma} \in \mathbf{\Gamma}(\underline{\eta})$ and every $\underline{\gamma} \in \mathbf{\Gamma}(\underline{\eta}')$. Then, if $\kappa(\eta) = \kappa(\eta')$, and*

$$\sigma_\eta(y^i) \supseteq \sigma_{\eta'}(y^{i'}) \quad \textit{for all } i \in \mathcal{N}, \textit{then}$$

IS $\underline{\eta}$ is uniformly at least as valuable as $\underline{\eta}'$.

Proof Suppose that the result is not true. Then, there exists a loss function L such that

$$\inf_{\underline{\gamma}\in\mathbf{\Gamma}(\underline{\eta})} \tilde{J}(\underline{\gamma};\underline{\eta}) > \inf_{\underline{\gamma}\in\mathbf{\Gamma}(\underline{\eta}')} \tilde{J}(\underline{\gamma};\underline{\eta}'). \tag{$*$}$$

[1] If the objective is not minimization of risk, but stabilization of a dynamical system (naturally, with T taken the set of all positive integers), then obtaining a total order is possible under certain criteria, as information theoretic notions may be applicable.More on this will be said later in the book in Chap. 13.

But under the given sigma-field inclusion, $\Gamma^i(\eta^i) \supseteq \Gamma^i(\eta^{i'})$, $i \in \mathcal{N}$, which implies that $\mathbf{\Gamma}(\underline{\eta}) \supseteq \mathbf{\Gamma}(\underline{\eta}')$. The infimum over a larger set cannot lead to a higher value (note that the function $\tilde{J}$ does not depend on $\underline{\eta}$ or $\underline{\eta}'$ *explicitly*—but *only* through $\underline{\gamma}$), and hence the inequality ($*$) leads to a contradiction. ◇

The sigma-field inclusion condition of *Proposition 7.2.1* is also known as the *fineness condition* (i.e., the sigma-field under one *IS* being finer (strictly speaking, *not coarser*) than under another *IS*), and clearly it is necessary that $\kappa(\eta) = \kappa(\eta')$ for the result to hold in this generality. It is possible to produce simple examples which show that the result could fail if some non-constant cost is attached to the *IS* (see [824, p. 80]).

As noted earlier while introducing Witsenhausen's intrinsic model, the sigma-field inclusion condition of *Proposition 7.2.1* is in fact equivalent to the presence of N-functions $h^i : Y^i \to Y^{i'}$ such that

$$y^{i'} = h^i(y^i) = h^i(\eta^i[\xi]) \equiv \eta^{i'}(\xi), \quad i \in \mathcal{N}$$

The result of *Proposition 7.2.1*, although useful, is typically too restrictive and is not always directly applicable to situations where our intuition tells us that one particular *IS* should be *uniformly better* than another. As a case in point, consider the standard formulation of a stochastic team with measurement vectors $y^1, \ldots, y^N$ (defined on the common probability space $(\Omega, \mathbf{F}, \mathbf{P})$), generating an information structure $\underline{\eta}$. Suppose that the measurement of one of the agents, say **A**1, is corrupted by additive noise (independent of $y^1, \ldots, y^N$, as well as of the "payoff-relevant" part of the random state of nature, x), thus leading to the new information structure

$$\eta' : \quad y^{1'} = y^1 + w^{1'}, \quad y^{i'} = y^i, \quad i = 2, \ldots, N.$$

Here we say that $y^{1'}$ is a *garbled* version of the measurement (or information) y^1 of **A**1, and as a consequence $\underline{\eta}'$ is a garbled version of the *IS* $\underline{\eta}$. Now, since $w^{1'}$ is independent noise, we would expect $\underline{\eta}'$ to be less valuable (less informative) to the team (and particularly, $y^{1'}$ to be less informative to **A**1) than $\underline{\eta}$, and hence the team should not be able to achieve better performance under $\underline{\eta}'$ than under $\underline{\eta}$. To prove this, however, *Proposition 7.2.1* cannot be directly used, because we do not necessarily have an inclusion relationship between $\sigma(y^1)$ and $\sigma(y^{1'})$. However, the Proposition can still be used (indirectly) to arrive at a more general result. To this end, let us first make precise what it means for a random vector to be a garbled version of another random vector.

Definition 7.2.3 Let x, y, y' be three random variables defined on a common probability space $(\Omega, \mathbf{F}, P)$. Let $P_{y'|y,x}$ be the conditional probability distribution of y' given (y, x). Then, y' is a *garbled version* of y *with respect to* x, if the following Markov chain condition

$$x \leftrightarrow y \leftrightarrow y'$$

applies (equivalently $P_{y'|y,x} = P_{y'|y}$).

It is useful first to consider the one-agent stochastic optimization problem where the roles of the three random vectors x, y, y' used in *Definition 7.2.3* are more transparent:

Proposition 7.2.2 *Consider a one-agent stochastic optimization problem with loss function $L(x, u)$, two separate ISs η and η' corresponding to static measurements y and y', respectively, and no cost of information. Let $\xi := (x, y, y')$ be a random vector defined on the probability space $(\Omega, \mathbf{F}, P)$, and let $\Gamma(\eta)$ and $\Gamma(\eta')$ be two policy spaces for the agent, corresponding to the ISs η and η', which are consistent with any constraint that might have been imposed on the action variable $u \in U$. Further assume that for every $\gamma \in \Gamma(\eta)$ and $\gamma' \in \Gamma(\eta')$, the loss functions $L(x, \gamma(y))$ and $L(x, \gamma'(y'))$ are integrable (i.e., have well-defined expected values). Then, if y' is a garbled version of y with respect to x, η is uniformly at least as valuable as η'.*

We will present alternative proofs later in Theorems 7.3.2 and 7.3.3 in the context of more general setups; nonetheless, the following analysis is instructive as the proof follows an argument based on Blackwell's *Irrelevant Information Theorem* (see Theorem E.2.1).

Proof Let $y'' := (y, y')$, with corresponding *IS* denoted by η''. Let $\Gamma(\eta'')$ be the class of all permissible policies for the agent under the *IS* η''. Clearly, $\Gamma(\eta) \subseteq \Gamma(\eta'')$ and $\Gamma(\eta') \subseteq \Gamma(\eta'')$, and furthermore $\sigma(y'')$ includes both $\sigma(y)$ and $\sigma(y')$. Hence, by *Proposition 7.2.1*, η'' is uniformly at least as valuable as both η' and η. Particularly

$$\inf_{\gamma'' \in \Gamma(\eta'')} E_\xi L(x, \gamma''(y, y')) \le \inf_{\gamma' \in \Gamma(\eta')} E_\xi L(x, \gamma'(y')) .$$

Now, given any $\gamma'' \in \Gamma(\eta'')$, note the following set of equalities:

$$\begin{aligned} E_\xi L(x, \gamma''(y, y')) &= E_{y,y'} E_{x|y,y'} L(x, \gamma''(y, y')) = E_{y,y'} E_{x|y} L(x, \gamma''(y, y')) \\ &=: E_{y,y'} \tilde{L}(y, \gamma''(y, y')) , \end{aligned}$$

where the *first* equality follows from the Bayes rule: $E_{x,y,y'} \equiv E_{y,y'} \cdot E_{x|y,y'}$, the *second* equality follows from condition (2) of *Definition 7.2.3* under the "garbling" hypothesis of the *Proposition*, and the *third* equality defines $\tilde{L}$ (as a measurable function of (y, y')) through its inner conditional expectation (of x given y). Likewise, for any $\gamma \in \Gamma(\eta)$, $E_{x|y} L(x, \gamma(y)) = \tilde{L}(y, \gamma(y))$. Now, concentrating on $\tilde{L}(y, \gamma(y))$ and $\tilde{L}(y, \gamma''(y, y'))$, it follows from Blackwell's *Irrelevant Information*

Theorem (see Theorem E.2.1) that for any given $\gamma'' \in \Gamma(\eta'')$, there exists a $\gamma \in \Gamma(\eta)$ such that

$$E_y \tilde{L}(y, \gamma(y)) \le E_{y,y'} \tilde{L}(y, \gamma''(y, y')),$$

and hence that

$$\inf_{\gamma \in \Gamma(\eta)} E_{x,y} L(x, \gamma(y)) \le \inf_{\gamma \in \Gamma(\eta'')} E_\xi L(x, \gamma(y, y')),$$

and since $\Gamma(\eta) \subseteq \Gamma(\eta'')$, we have to have an equality here. Therefore,

$$\inf_{\gamma \in \Gamma(\eta)} E_{x,y} L(x, \gamma(y)) = \inf_{\gamma'' \in \Gamma(\eta'')} E_\xi L(x, \gamma''(y, y')) \le \inf_{\gamma' \in \Gamma(\eta')} E_\xi L(x, \gamma'(y')),$$

which establishes the desired result. ◇

Note the role x plays in the result of *Proposition 7.2.2*: it is the *payoff-relevant* portion of the random state of nature. If we have a genuine team problem, with $N(\ge 2)$ agents, then this result would not immediately extend because the payoff-relevant portion of the random state of Nature will be a different random vector to different agents; specifically, with **A***i* isolated, and under a fixed *IS* $\underline{\eta}$, the vector $\xi^i := (x, y^1, \ldots, y^{i-1}, y^{i+1}, \ldots, y^N)$ will be the payoff-relevant portion of ξ from the point of view of **A***i*. On the positive side, this says that in an N-agent static team problem, if *only one* agent's information changes, so that it is a garbled version of the old one, then the new (garbled) *IS* cannot lead to improved performance (which follows directly from *Proposition 7.2.2*). An iterative use of this feature leads to the following more general result where the measurements of all or a subset of the agents are allowed to change.

Proposition 7.2.3 *Consider an* N*-agent stochastic team problem, under the hypothesis of Proposition 7.2.2, naturally extended to N agents, and with* $\underline{\eta} := \{y^1, \ldots, y^N\}$ *and* $\underline{\eta}' := \{y^{1'}, \ldots, y^{N'}\}$*, where* $y^{i'}$ *is a garbled version of* y^i *with respect to*

$$\xi_i' := (x, y^{1'}, y^{2'}, \ldots, y^{i+1'}, \ldots, y^{N'}), \quad i \in \mathcal{N}.$$

Then, $\underline{\eta}$ *is uniformly at least as valuable as* $\underline{\eta}'$.[2]

We will present a more general result in Theorem 7.4.1 where *correlated garblings* (see Definition 7.4.1) are also allowed and a converse result is also obtained; nonetheless, a concise proof, via a backward induction argument, is presented next.

[2] Here, we allow for the possibility that y^i and $y^{i'}$ are identical for some $i \in \mathcal{N}$.

Proof For the sake of simplicity, let us assume that all the minima in the following development exist. If a minimum does not exist, then it is possible to consider an infimizing sequence, and the main result again goes through. Consider the following sequence of equalities and inequalities, where we let $\underline{\gamma}' \in \mathbf{\Gamma}(\underline{\eta}')$ denote a team-optimal solution under the *IS* $\underline{\eta}'$:

$$
\begin{aligned}
\min_{\underline{\gamma}\in\mathbf{\Gamma}(\underline{\eta}')} E_\xi L(x, \underline{\gamma}(\mathbf{y}')) &= \min_{\beta\in\Gamma^N(\eta^{N'})} E_\xi L(x, \{\gamma^{j'}(y^{j'})\}_{j=1}^{N-1}, \beta(y^{N'})) \\
&\geq \min_{\beta\in\Gamma^N(\eta^N)} E_\xi L(x, \{\gamma^{j'}(y^{j'})\}_{j=1}^{N-1}, \beta(y^N)) \\
&\equiv E_\xi L(x, \{\gamma^{j'}(y^{j'})\}_{j=1}^{N-1}, \gamma^{N*}(y^N)) \\
&\geq \min_{\beta\in\Gamma^{N-1'}(\eta^{N-1'})} E_\xi L(x, \{\gamma^{j'}(y^{j'})\}_{j=1}^{N-2}, \beta(y^{N-1'})), \gamma^{N*}(y^N)) \\
&\geq \min_{\beta\in\Gamma^{N-1}(\eta^{N-1})} E_\xi L(x, \{\gamma^{j'}(y^{j'})\}_{j=1}^{N-2}, \beta(y^{N-1})), \gamma^{N*}(y^N)) \\
&\equiv E_\xi L(x, \{\gamma^{j'}(y^{j'})\}_{j=1}^{N-2}, \gamma^{N-1*}(y^{N-1}), \gamma^{N*}(y^N)) \\
&\geq \ldots\ldots \geq \min_{\beta\in\Gamma^1(\eta^1)} E_\xi L(x, \beta(y^1), \{\gamma^{i*}(y^i)\}_{i=2}^N) \\
&\geq \min_{\underline{\gamma}\in\mathbf{\Gamma}(\underline{\eta})} E_\xi L(x, \underline{\gamma}(\mathbf{y})).
\end{aligned} \tag{7.4}
$$

In the above, the first equality is a property of sequential minimization, the next inequality follows from *Proposition 7.2.2* by identifying x there with $\xi'_N := (x, y^{1'}, \ldots, y^{N-1'})$, y with y^N, and y' with $y^{N'}$, and by making use of the given hypothesis that $y^{N'}$ is a garbled version of y^N with respect to ξ'_N. The next identity defines $\gamma^{N*} \in \Gamma^N(\eta^N)$, and the following inequality says that there may be some other element of $\Gamma^{N-1}(\eta^{N-1'})$ (other than $\gamma^{N-1'}$) that provides a lower value for the given cost. The next inequality again follows from *Proposition 7.2.2* since $y^{N-1'}$ is a garbled version of y^{N-1} with respect to ξ'_{N-1}, and the identity again defines $\gamma^{N-1*} \in \Gamma^{N-1}(\eta^{N-1})$. An induction argument then shows that the minimum of $J(\underline{\gamma})$ over $\mathbf{\Gamma}(\underline{\eta})$ cannot be higher than the minimum over $\mathbf{\Gamma}(\underline{\eta}')$, thus completing the proof. ◇

Remark 7.2.1 An important special case under which the "garbling" condition of *Proposition 7.2.3* is satisfied occurs when $\underline{\eta}$ and $\underline{\eta}'$ are related by

$$y^{i'} = y^i + w^{i'}, \qquad i \in \mathcal{N}$$

where $w^{i'}, i \in \mathcal{N}$, is a sequence of random vectors that are independent of the $N+1$ tuple $(x, y^1, \ldots, y^N)$. Then, clearly for every $i \in \mathcal{N}$,

$$P_{y^{i'}|y^i,\xi_i'} = P^{w^{i'}}_{y^{i'}-y^i}$$

where $P^{w^{i'}}$ is the probability distribution function of $w^{i'}$, which shows that $y^{i'}$ is a garbled version of y^i with respect to ξ_i'. ◇

In view of *Remark 7.2.1* above, we have the following useful result, which we state as a corollary to *Proposition 7.2.3.*

Corollary 7.2.1 *Consider the LQG team problem of Sect. 2.4.3, under the measurement scheme (2.62), where the measurement noises $w^i \sim N(0, N^i)$, $N^i > 0, i \in \mathcal{N}$, are independent. Let the optimum team cost be given by $J^*(N^1, \ldots, N^N)$ as a function of the noise covariances. Then, J^* is a nondecreasing function of the N^i's, under the matrix partial ordering.*[3] ◇

7.3 Blackwell Ordering and Comparison of Information Structures

The following result has important consequences in decentralized stochastic control problems as will be elaborated on later. We will consider the setup in Sect. 7.1.

Theorem 7.3.1 ([826]) *Consider the single-stage version of the problem introduced in Sect. 7.1 (that is, with $T = 1$), and let the integral $\int c(x, \gamma(y))PQ(dx, dy)$ exist for all $\gamma \in \mathbf{\Gamma}$ and $Q \in \mathcal{Q}$. Then, the function*

$$J(P, Q) = \inf_{\gamma\in\mathbf{\Gamma}} E_P^{Q,\gamma}[c(x, u)]$$

is concave in Q.

Proof For $\alpha \in [0, 1]$ and $Q', Q'' \in \mathcal{Q}$, let $Q = \alpha Q' + (1 - \alpha)Q'' \in \mathcal{Q}$, i.e.,

$$Q(A|x) = \alpha Q'(A|x) + (1 - \alpha)Q''(A|x)$$

for all $A \in \mathcal{B}(\mathbb{Y})$ and $x \in \mathbb{X}$. Noting that $PQ = \alpha PQ' + (1 - \alpha)PQ''$, we have

$$\begin{aligned} J(P, Q) &= J(P, \alpha Q' + (1 - \alpha)Q'') \\ &= \inf_{\gamma\in\mathbf{\Gamma}} E_P^{Q,\gamma}[c(x, u)] \\ &= \inf_{\gamma\in\mathbf{\Gamma}} \int c(x, \gamma(y))PQ(dx, dy) \end{aligned}$$

[3] In other words, $N^{i'} - N^i \geq 0$ (nonnegative definite), $i \in \mathcal{N}$, implies $J^*(N^{1'}, \ldots, N^{N'}) \geq J^*(N^1, \ldots, N^N)$. In words, the more "noisy" the measurements are, the higher is the team cost.

$$
\begin{aligned}
&= \inf_{\gamma \in \mathbf{\Gamma}} \left(\alpha \int c(x, \gamma(y)) P Q'(dx, dy) \right. \\
&\qquad \left. +(1-\alpha) \int c(x, \gamma(y)) P Q''(dx, dy) \right) \\
&\geq \inf_{\gamma \in \mathbf{\Gamma}} \left(\alpha \int c(x, \gamma(y)) P Q'(dx, dy) \right) \\
&\qquad + \inf_{\gamma \in \mathbf{\Gamma}} \left((1-\alpha) \int c(x, \gamma(y)) P Q''(dx, dy) \right) \\
&= \alpha J(P, Q') + (1-\alpha) J(P, Q'') \qquad (7.5)
\end{aligned}
$$

proving that $J(P, Q)$ is concave in Q. ◇

The following result is a folk theorem in statistical decision theory whose proof is similar to that of Theorem 7.3.1.

Proposition 7.3.1 ([826]) *The function*

$$V(P) := \inf_{u \in \mathbb{U}} \int c(x, u) P(dx), \qquad (7.6)$$

is concave in P, under the assumption that c is measurable and bounded.

7.3.1 Blackwell Ordering: Forward

We now use the preceding observations to revisit a classical result in statistical decision theory and comparison of experiments, due to David Blackwell [123], who considered a finite $\mathbb{X}$. In a single DM setting, we refer to the probability space induced on $\mathbb{X} \times \mathbb{Y}$ as an information structure.

Definition 7.3.1 *An information structure induced by some channel Q_2 is garbled (or stochastically degraded) with respect to another one, Q_1, if there exists a channel Q' on $\mathbb{Y} \times \mathbb{Y}$ such that*

$$Q_2(B|x) = \int_{\mathbb{Y}} Q'(B|y) Q_1(dy|x), \;\; B \in \mathcal{B}(\mathbb{Y}), \;\; P \text{ a.s. } x \in \mathbb{X}.$$

We also define the following notion:

Definition 7.3.2 *An information structure μ is more informative than another information structure Φ if*

$$\inf_{\gamma \in \Gamma} E_P^{\Phi,\gamma}[c(x, u)] \geq \inf_{\gamma \in \Gamma} E_P^{\mu,\gamma}[c(x, u)]$$

for all single-stage single-agent decision problems $(c(x, u), \mathbb{U})$.

We emphasize that in Definition 7.3.2, $\mathbb{U}$ is also a design variable for the decision problem. For instance, if $\mathbb{U}$ were a singleton, then the comparison of information structures would be meaningless. With this in mind, we state Blackwell's classical result in the following.

In view of Proposition 7.3.1, we have the following. The proof will build on a concavity argument.

Theorem 7.3.2 *[Blackwell [123]] If Q_2 is stochastically degraded with respect to Q_1, then the information structure induced by channel Q_1 is more informative with respect to the one induced by channel Q_2 in the sense that*

$$\inf_{\gamma} E_P^{Q_2,\gamma}[c(x,u)] \geq \inf_{\gamma} E_P^{Q_1,\gamma}[c(x,u)],$$

for all measurable and bounded cost functions c.

Proof Let $(x, y^1) \sim PQ_1$, y^2 be such that $\Pr(y^2 \in B | x = x, y^1 = y) = Q'(B|y)$ for all $B \in \mathcal{B}(\mathbb{Y})$, $y^1 \in \mathbb{Y}$, and $x \in \mathbb{X}$. Then x, y^1, and y^2 form a Markov chain in this order and therefore $P(dy^2|y^1, x) = P(dy^2|y^1)$ and $P(x|dy^2, y^1) = P(x|y^1)$ a.s.. Thus we have, with V being as given in (7.6),

$$\begin{aligned}
J(P, Q_2) &= \int V\big(P(\cdot|y^2)\big)P(dy^2) \\
&= \int V\Big(\int P(\cdot|y^1)P(dy^1|y^2)\Big)P(dy^2) \\
&\geq \int \Big(\int P(dy^1|y^2)V\big(P(\cdot|y^1)\big)\Big)P(dy^2) \\
&= \int V\big(P(\cdot|y^1)\big)\Big(\int P(dy^1|y^2)P(dy^2)\Big) \\
&= \int V\big(P(\cdot|y^1)\big)P(dy^1) = J(P, Q_1).
\end{aligned}$$

where in arriving at the inequality, we used Proposition 7.3.1 and Jensen's inequality. ◇

The comparison argument applies also for multi-stage settings with $T > 1$, that is the original problem introduced in Sect. 7.1. The proof will now build on yet another argument, based on *simulation*.

Theorem 7.3.3 *For the multi-stage problem (7.2), if Q_2 is stochastically degraded with respect to Q_1, then the information structure induced by channel Q_1 is more informative with respect to the one induced by channel Q_2 in the sense that for all measurable and bounded cost functions c in (7.2),*

$$J(P, Q_1) \leq J(P, Q_2)$$

Proof We will follow an argument based on simulation via randomization. As in the proof of Theorem 7.3.2, let for $t \geq 0$, $(x_t, y_t^1) \sim PQ_1$, y_t^2 be such that $\Pr(y_t^2 \in B | x = x, y_t^1 = y) = Q'(B|y)$ for all $B \in \mathcal{B}(\mathbb{Y})$, $y_t^1 \in \mathbb{Y}$, and $x_t \in \mathbb{X}$. Then x_t, y_t^1, and y_t^2 form a Markov chain in this order. Now, the cost value achieved under any given policy $\underline{\gamma}^2$ which is measurable under the channel Q_2, can also be achieved by a randomized decision policy of the DM under channel Q_1. That is, given a policy $\underline{\gamma}^2 = \{\gamma_t^2(y_{[0,t]}^2), 0 \leq t - 1\}$, the DM can generate a randomized decision policy $\underline{\gamma}^1 = \{\gamma_t^1(y_{[0,t]}^1), 0 \leq t \leq T - 1\}$ such that for all $B \in \mathcal{B}(\mathbb{U})$,

$$P(\gamma_t^1(y_{[0,t]}^1) \in B) = \int_{\tilde{y}_t^2} 1_{\{\gamma^2(\tilde{y}_{[0,t]}^2) \in B\}} P(d\tilde{y}_{[0,t]}^2 | y_{[0,t]}^1),$$

where $\tilde{y}_{[0,t]}^2$ has the same probability measure as $y_{[0,t]}^2$, generated recursively for $t \geq 0$. Hence, any policy under Q_2 can be simulated by a DM under Q_1 through a randomized policy. Through a dynamic programming argument and by Theorem E.2.1 (Blackwell's Irrelevant Information Theorem), for every randomized policy there exists a deterministic policy under Q_1 which is at least as good as the randomized policy. Hence, the result follows. ◇

Remark 7.3.1 Blackwell's informativeness provides a partial order in the space of measurement channels; that is, not every pair of two channels can be compared. We will later see that if the goal is not the minimization of a cost function, but that of stochastic stabilization of an open-loop unstable linear system over a communication channel in an appropriate sense, then one can obtain a total order on the space of channels, with the additional flexibility of allowing encoding and decoding operations. ◇

7.3.2 Blackwell Ordering: Converse

Establishing the converse to Blackwell ordering (i.e., less informativeness implies garbling) is significantly more challenging. For the case with general spaces, related results are attributed to [129], and [160], [690], which relate an ordering of information structures in terms of dilatations and their relation with comparisons under concave functions defined on conditional probability measures. A very concise yet informative review can be found in [460, p. 130–131] and a more comprehensive review is in [721].

When $\mathbb{X}$ is finite, Blackwell showed that the above condition also has a converse theorem if P has positive measure on each element of $\mathbb{X}$: For an information structure to be more informative, weak stochastic degradedness is a necessary condition.

In the following, we study the converse in a general setup involving standard Borel spaces. We note that the setup differs slightly from that of Blackwell in

[123], contributing to the fact that this is a *partial* converse to Blackwell's result. In Blackwell's original setup with finite $\mathbb{X}$, information structures could be compared over different priors on $\mathbb{X}$ as the comparison would apply uniformly to all such prior measures that satisfy a positivity condition on each of the finitely many outcomes. In the setup here, since the space is possibly uncountable, we consider a fixed prior measure on $\mathbb{X}$.

Theorem 7.3.4 (Converse to Blackwell Ordering [353]) *Consider the single-stage single-agent decision problem, where the measurement is absolutely continuous in the sense of (3.26); i.e., there exists a function f and a reference probability measure $\bar{Q}$ such that for all Borel S:*

$$P(y \in S|x) = \int_S f(y,x)\bar{Q}(dy).$$

If $\mathbb{Y}$ is compact and an information structure μ is more informative than another information structure ν over all single-agent decision problems with compact standard Borel action spaces and bounded cost functions $c : \mathbb{X} \times \mathbb{U} \to \mathbb{R}$ that are continuous in u for every x, then ν must be a garbling of μ in the sense of Definition 7.3.1.

The proof is technical and is given in the appendix. The argument builds on showing that the set of garblings forms a compact convex set, and via the Hahn-Banach Separation Theorem defined on an appropriate locally convex space of measures, if the garbling condition does not hold, then one can construct a decision problem under which the order of inequality does not hold.

7.4 Multi-Agent Setting and Correlated Garblings

We now consider a multi-agent (team) setting. Consider the setting presented in Chap. 4. Let us first define a notion of garbling with potentially common randomness for a team, generalizing the notion from Definition 7.3.1:

Definition 7.4.1 An information structure ν is a correlated garbling of an information structure μ with potentially common randomness if there exists $\eta \in \mathcal{P}(\Upsilon)$, where $\Upsilon = [0,1]^N$, such that:

$$\nu(B) = \int_{B\times\Upsilon} \eta(dz)\mu(d\omega_0, d\bar{y}) \prod_k \Pi^k(d\tilde{y}^k|y^k, z), \quad \forall B \in \mathcal{B}(\Omega \times \prod_{i=1}^N \mathbb{Y}^i)$$

for some stochastic kernels Π^k from $(\mathbb{Y}^k \times \Upsilon)$ to $\mathbb{Y}^k$, for $k = 1, \ldots, n$.

Note that here η is an arbitrary probability measure, as in the definition of $L_C(\mu)$. Assume $L_C(\mu)$ is closed (see Theorem 4.2.7), a sufficient condition being the independent static reducibility via Theorem 4.2.4 under Assumption 7.4.1 stated below.

Let $\mathcal{E}$ denote the set of all team problems with compact action spaces and cost functions that are continuous in player actions for every ω_0.

For the following result, we assume that the measurements of the players are either independent or by a change of measure argument, under absolute continuity, have been turned into a static information structure with independent measurements:

Assumption 7.4.1 *We have that*

$$\mu(d\omega_0, dy^1, \cdots, dy^N) \ll \mu_0(d\omega_0) \prod_{i=1}^{N} \bar{Q}^i(dy^i) \tag{7.7}$$

for some reference measures μ_0 and $\bar{Q}^i$ so that the Radon-Nikodym derivative

$$\frac{d(\mu(d\omega_0, dy^1, \cdots, dy^N))}{d(\mu_0(d\omega_0) \prod_{i=1}^{N} \bar{Q}^i(dy^i))}(\omega_0, y^1, \cdots, y^N) =: f(\omega_0, y^1, \cdots, y^N)$$

exists.

We now state the main theorem of this section. The proof is given in the appendix and follows closely that of Theorem 7.3.4.

Theorem 7.4.1 *[356] For team problems under Assumption 7.4.1, an information structure μ is more informative than an information structure ν if only if ν is a garbling of μ for each DM, with potentially common randomness among the garblings (see Definition 7.4.1).*

7.5 Appendix: Proofs

7.5.1 Proof of Theorem 7.3.4

Under the conditions of the theorem, an optimal policy (which is also deterministic) exists for every information structure, a result that will be established in Theorem 8.3.1.

Step (1) Let ζ be the fixed probability distribution on $\mathbb{X}$ for any given team decision problem. Take information structures $\mu, \nu \in \{\mathcal{P}(\mathbb{X} \times \mathbb{Y}) : P_{\mathbb{X}} = \zeta\}$, where μ is more informative than ν in Blackwell's sense (i.e. $J(\zeta, \mu, \gamma^*) \geq J(\zeta, \nu, \gamma^*)$ over all decision problems with bounded cost functions c that are continuous in u).

Take the space $K\mu$, a subset of $\mathcal{P}(\mathbb{X} \times \mathbb{Y})$, to be the space of all possible garblings of μ, where the garblings are from $\mathbb{Y}$ to $\mathbb{Y}$.

Step (2) We now establish the weak compactness of the space of all garbled information structures. First, observe that the set of all induced garblings on the product space (involving all of K) inducing probability measures of the form

$$P_K(dx, dy, d\tilde{y}) := \mu(dx, dy)K(d\tilde{y}|y)$$

leads to a weakly pre-compact space in the space of probability measures on $\mathbb{X} \times \mathbb{Y} \times \mathbb{Y}$. If closedness can also be established, this would lead to a weakly compact space. This follows from the proof of Theorem 4.2.6 or that of Theorem 4.3.1: since the marginals on $\mathbb{X} \times \mathbb{Y}$ are fixed, any limit of a weakly converging sequence will also satisfy the property that the limit is a garbling of the original information structure. For completeness, we present the following: With $P_K(dx, dy, d\tilde{y}) = K(d\tilde{y}|y)\mu(dx, dy)$, consider a weakly converging sequence $P_{K_n}(dx, dy, d\tilde{y})$. We will show that the weak limit also admits such a garbled structure. Let $P_{K_n}(dx, dy, d\tilde{y})$ converge weakly to $P(dx, dy, d\tilde{y})$. Then, for every continuous and bounded g,

$$\int g(x, y, \tilde{y}) P_K(dx, dy, d\tilde{y}) = \int \left(\int g(x, y, \tilde{y})\mu(dx|dy) \right) P_{K_n}(dy, d\tilde{y}).$$

Since the marginal on y is fixed, even though the function $\int g(x, y, \tilde{y})\mu(dx|dy)$ is only measurable and bounded in y and is continuous in $\tilde{y}$, w-s convergence is equivalent to weak convergence of $P_{K_n}(dy, d\tilde{y})$ (by Theorem C.4.1) and as a result we have that

$$\int \left(\int g(x, y, \tilde{y})\mu(dx|dy) \right) P_{K_n}(dy, d\tilde{y})$$
$$\to \int \left(\int g(x, y, \tilde{y})\mu(dx|dy) \right) P(dy, d\tilde{y}).$$

As a result, P decomposes as $P(dx, dy, d\tilde{y}) = \mu(dx, dy)\tilde{K}(d\tilde{y}|y)$ for some $\tilde{K}$. This establishes the weak compactness of the garbled information structure in the product space $\mathbb{X} \times \mathbb{Y} \times \mathbb{Y}$.

Now, take the projection of this space onto the measures on the first and the third coordinates; as a continuous image of a weakly compact set, this map will also be compact and leads to the space $K\mu$.

Finally, $K\mu$ is convex, since the space of stochastic kernels is convex. As a result, the space $K\mu$ of all possible garblings of μ is a convex and compact subset of $\mathcal{P}(\mathbb{X} \times \mathbb{Y})$ under the weak convergence topology.

Now, assume that there does not exist a stochastic kernel $\kappa \in K$ such that:

$$\nu = \kappa\mu,$$

that is, we assume that ν is not a garbling of μ and then proceed with a proof by contradiction. Then, $K\mu \cap \nu = \emptyset$. That is, $\nu \notin K\mu$.

Step (3) We now use the Hahn-Banach Separation Theorem for Locally Convex Spaces by treating the space of probability measures $\mathcal{P}(\mathbb{X}\times\mathbb{Y})$ as a locally convex space of measures (see [613, Theorem 3.4]). As such, since the spaces $K\mu$ and $\{\nu\}$ are subsets of this space and are convex, closed and compact, in addition to being disjoint, we can separate them using a continuous linear map from $\mathcal{P}(\mathbb{X}\times\mathbb{Y})$ to $\mathbb{R}$.

To apply [613, Theorem 3.4], we require local convexity of $\mathcal{P}(\mathbb{X} \times \mathbb{Y})$, and accordingly we define the locally convex space of probability measures with the following notion of convergence: We say that $\nu_n \to \nu$ if $\int f(x, y)\nu_n(dx, dy) \to \int f(x, y)\nu(dx, dy)$ for every measurable and bounded function which is continuous in y for every x. We note that our measures must still have a fixed marginal ζ on $\mathbb{X}$.

Since continuous and bounded functions *separate* probability measures (in the sense that, if the integrations of two measures with respect to continuous functions are equal, the measures must be equal), it follows from [613, Theorem 3.10] that we can represent every continuous linear map on $\mathcal{P}(\mathbb{X} \times \mathbb{Y})$ using the form $\int f(x, y)\nu(dx, dy)$ for some measurable and bounded function $f(x, y)$ continuous in y for every x. It also follows from [613, Theorem 3.10] that, given this notion of convergence, $\mathcal{P}(\mathbb{X} \times \mathbb{Y})$ is a locally convex space.

Therefore, we have the following statement from combining [613, Theorem 3.4] and [613, Theorem 3.10]: there exists a measurable and bounded function (continuous in y) $f : \mathbb{X} \times \mathbb{Y} \to \mathbb{R}$ and constants $D_1, D_2 \in \mathbb{R}$ where $D_1 < D_2$ such that:

$$\langle \nu, f\rangle \leq D_1, \; \langle \kappa\mu, f\rangle \geq D_2, \;\; \forall \kappa \in K,$$

where we have used the following notation:

$$\langle \nu, f\rangle = \int_{\mathbb{X}\times\mathbb{Y}} f(x, y)\nu(dx, dy).$$

This leads to the following inequality: $\langle \nu, f\rangle < \langle \kappa\mu, f\rangle, \; \forall\kappa \in K$.

Step (4) Now consider the class of decision problems with bounded cost functions, continuous in the actions, with compact $\mathbb{Y}$, $\mathbb{U}$, where $\mathbb{U} = \mathbb{Y}$. This is clearly a subset of all decision problems considered so far in the proof. Now let $f(x, y)$ be the separating function found above. Consider a problem in this particular subclass where $f(x, y)$ is the cost function (which is valid since $\mathbb{U} = \mathbb{Y}$ and $f(x, y)$ is bounded continuous in y). We note that $\langle \nu, f\rangle$ returns the expected value with cost function $f(x, y)$ under information structure ν when the agent adopts the identity policy $\gamma^{id}(y) = y$. We note the following:

$$\int_{\mathbb{X}\times\mathbb{Y}} f(x, y)\nu(dx, dy) < \int_{\mathbb{X}\times\mathbb{Y}} f(x, y)\kappa\mu(dx, dy), \;\; \forall\kappa \in K,$$

and hence,

$$\int_{\mathbb{X}\times\mathbb{Y}} f(x,y)\nu(dx,dy) < \inf_{\kappa\in K}\int_{\mathbb{X}\times\mathbb{Y}} f(x,y)\kappa\mu(dx,dy)$$

$$= \inf_{\kappa\in K}\int_{\mathbb{X}\times\mathbb{Y}} f(x,y')\int_{\mathbb{Y}}\kappa(dy'|y)\mu(dx,y)$$

$$= \inf_{\kappa\in K}\int_{\mathbb{X}\times\mathbb{Y}} f(x,\kappa(\cdot|y))\mu(dx,dy),$$

where $f(x,\kappa(\cdot|y)) := \int_{\mathbb{Y}} f(x,y')\kappa(dy'|y)$. Recalling that $\kappa(\cdot|y)$ has a functional representation $\gamma(y) = g(y,\omega)$ for some independent noise variable ω, and since K is the space of all stochastic kernels from $\mathbb{Y}$ to $\mathbb{Y}$, we can observe that this gives us:

$$\inf_{\kappa\in K}\int_{\mathbb{X}\times\mathbb{Y}} f(x,\kappa(\cdot|y))\mu(dx,dy) = \inf_{\gamma\in\Gamma}\int_{\mathbb{X}\times\mathbb{Y}} f(x,\gamma(y))\mu(dx,dy).$$

Since we allow for randomized policies, this minimization is equivalent to finding the optimal policy $\gamma^* \in \Gamma$ for the cost function $f(x,y)$ under information structure μ. And thus, we have

$$\int_{\mathbb{X}\times\mathbb{Y}} f(x,y)\nu(dx,dy) = J(\zeta,\nu,\gamma^{id}) < J(\zeta,\mu,\gamma^*)$$

$$= \inf_{\kappa\in K}\int_{\mathbb{X}\times\mathbb{Y}} f(x,\kappa(y))\mu(dx,dy).$$

Since we have identified a problem where, when applying its optimal policy, μ performs worse than ν does under some policy, we have contradicted the fact that μ is better than ν. Therefore, there must exist a $\kappa \in K$ such that $\nu = \kappa\mu$, and thus ν is a garbling of μ. ◇

7.5.2 Proof of Theorem 7.4.1

Part 1 First we establish the forward direction. Let μ be an information structure, and let ν be another information structure such that the measurements under ν are given by $y^i_\nu = h^i(y^i_\mu, Z)$ for $i = 1,\ldots,n$ and some independent randomness Z, that may be common between players.

We observe that any outcome achievable under ν is achievable under μ, since outcomes are determined by policies, which are measurable functions from measurements to actions, and the measurements under ν can be simulated using the measurements from μ and the independent randomness Z, and thus the resulting expected outcomes can also be achieved under μ.

Part 2 We now study the converse direction. Let $\mu, \nu \in \mathcal{P}(\Omega_0 \times \mathbb{Y}^1 \times \cdots \times \mathbb{Y}^N)$, and assume that μ is more informative than ν. Since this means the expected cost is lower under μ than under ν for all team problems, in particular it is true for teams where $\mathbb{U}^i = \mathbb{Y}^i$. We will fix $\mathbb{U}^i = \mathbb{Y}^i$ and evaluate this class of teams for the remainder of the proof.

Let $K_Z\mu$ denote the space of all garblings of μ with common randomness allowed. I.e.

$$K_Z\mu = \{\nu \in \mathcal{P}(\Omega_0 \times \mathbb{Y}^1 \times \cdots \times \mathbb{Y}^N) : y^i_\nu = h^i(y^i_\mu, Z)\}$$

for some independent randomness Z that may be common between DMs.

We assume that $\nu \in K_Z\mu$ and proceed with a proof by contradiction.

Note that since $L_C(\mu)$ is closed under Assumption 7.4.1, via Theorem 4.2.7. Furthermore, this set is convex and with $\mathbb{Y}^i = \mathbb{U}^i$ here, we have that $K_Z\mu$ will be closed and convex. This follows from noticing that $L_C(\mu)$ is the set of all measures on $(\Omega \times \prod_{i=1}^N (\mathbb{Y}^i \times \mathbb{U}^i))$ with fixed marginal μ on $(\Omega \times \prod_{i=1}^N \mathbb{Y}^i)$, where the measure on the action spaces is created through applying all possible policies (which we can view as garblings) with common randomness; since each player's action space and measurement space are assumed to be the same, this means that $K_Z\mu$ is the projection of $L_C(\mu)$ onto the state and action spaces, and thus $K_Z\mu$ is also closed and convex.

As in the proof of Theorem 7.3.4, we use the Hahn-Banach Separation Theorem for Locally Convex Spaces. To apply this, we require local convexity of $\mathcal{P}(\Omega_0 \times \prod_{k=1}^N \mathbb{Y}^k)$, and thus we define the locally convex space of probability measures with the following notion of convergence: We say that $P_n \to P$ if

$$\int f(\omega_0, y^1, \ldots, y^N) P_n(d\omega_0, dy^1, \ldots, dy^N) \\ \to \int f(\omega_0, y^1, \ldots, y^N) P(d\omega_0, dy^1, \ldots, dy^N),$$

for every measurable and bounded function which is continuous in $(y^1, \ldots, y^N)$ for every ω_0. We note that our measures must still have a fixed marginal ζ on Ω_0.

Since continuous and bounded functions *separate* probability measures (in the sense that, if the integrations of two measures with respect to continuous functions are equal, the measures must be equal), we can represent every continuous linear map on our space using the form, following [613, Theorem 3.10]:

$$F = \int f(\omega_0, y^1, \ldots, y^N)\nu(d\omega_0, dy^1, \ldots, dy^N),$$

for some measurable and bounded function $f(\omega_0, y^1, \ldots, y^N)$ continuous for every ω_0. It also follows that, given this notion of convergence, $\mathcal{P}(\Omega_0 \times \prod_{k=1}^N \mathbb{Y}^k)$ is a locally convex space.

Thus, since ν is compact and convex and $K_Z\mu$ is closed and convex, following a combination of [613, Theorem 3.4] and [613, Theorem 3.10], we have that there exists a continuous and bounded function $f : \Omega_0 \times \mathbb{Y}^1 \times \cdots \times \mathbb{Y}^N \to \mathbb{R}$ such that there exist constants $D_2 > D_1$ with

$$\langle \nu, f \rangle < D_1, \langle \kappa_Z\mu, f \rangle > D_2 \qquad \forall \kappa_Z\mu \in K_Z\mu.$$

That is to say,

$$\int f(\omega_0, y^1, \ldots, y^N)\nu(d\omega_0, dy^1, \ldots, dy^N)$$
$$< \int f(\omega_0, y^1, \ldots, y^N)\kappa_Z\mu(d\omega_0, dy^1, \ldots, dy^N).$$

Now, since $\mathbb{Y}^i = \mathbb{U}^i$ we can view f as the cost function for some problem, where $\langle \nu, f \rangle$ gives the expected cost for the team with information structure ν when each DM plays the identity policy.

We note that because the inequality holds for all $\kappa_Z\mu \in K_Z\mu$, in particular it holds for the minimum over all such measures, and thus, we have the following:

$$\begin{aligned}\langle \tilde{\gamma}\nu, f \rangle &< \inf_{\kappa_Z\mu \in K_Z\mu} \langle \kappa_Z\mu, f \rangle \\ &= \inf_{\kappa_Z\mu \in K_Z\mu} \int f(\omega_0, y^1, \ldots, y^N)\kappa_Z\mu(d\omega_0, dy^1, \ldots, dy^N)\end{aligned}$$

But, this is equivalent to finding the optimal strategic measure in $L_C(\mu)$.

Thus, we have that, for the cost function f, the performance under the identity policy under information structure ν is strictly better than the performance under the optimal policy for information structure μ. This contradicts the fact that μ is more informative than ν. Therefore, we must have that $\nu \in K_Z\mu$. □

7.6 Concluding Remarks

In this chapter, we have studied comparison of information structures for both single-agent stochastic decision and multi-agent stochastic team decision problems with general spaces. We have obtained characterizations of their Blackwell ordering, with both forward and converse directions.

7.7 Bibliographic Notes

Comparison of information structures (or statistical experiments/observation channels) were characterized by Blackwell in his seminal paper [122]. For finite stochastic decision problems, Blackwell established necessary and sufficient conditions for comparing information structures: an information structure is said to be "better" than another one if, for any cost function involving a hidden state variable and the player's decision, the expected optimal cost cannot increase when switching from the former information structure to the latter, where for selecting an action the decision maker has access to only the output of a possibly noisy observation channel mapping the hidden state to the observation variable. Blackwell's characterization was that an information structure is better than another one if and only if the latter is a garbled version of the former one. Since then, there has been significant interest in the comparison, value and regularity of information structures in various research communities and into more general models [460, 690, 721]. For finite games, [579] generalized Blackwell's result to zero-sum games with a complete characterization. This result was generalized to the standard Borel zero-sum game setup in [353], where a partial converse of Blackwell's information structure ordering and relations with a well-known result due to Strassen [690] were also established. Le Cam generalized Blackwell's criteria through what is now known as the *Le Cam deficiency*, which allows for the proximity of any two information structures to be determined [459, 460, 595].

Cartier, Fell, and Meyer relate an ordering of information structures in terms of dilatations (where the hidden variable x does not appear explicitly in the analysis) in [160]. A concise yet informative review is in [460, p. 130–131]. A detailed discussion on comparisons of information structures along the same approach is present in the comprehensive book [721].

Regarding the concavity properties of the optimization problems in measurement channels developed in this chapter and in [828], relevant results in the setting of quantization problems have been studied in [329]; see [35] for a discussion in the context of team theory.

The results in Sect. 7.4 generalize [470] and [824, Chapter 4]. We also add the following remark on the converse to Blackwell: Strassen, in [690, Theorem 2], has presented a related result that is often invoked when comparison of experiments is studied in infinite dimensional probability spaces. For completeness, we present a discussion in the following.

Let Ω be a convex compact metrizable subset of a locally convex topological vector space. For Borel probability measures μ and ν write $\mu \prec \nu$ if and only if for all $y \in \mathcal{S} = \{\text{all continuous concave functionals on } \Omega\}$, $\int y \, d\mu \geq \int y \, d\nu$.

Theorem 7.7.1 *[690, Theorem 2] $\mu \prec \nu$ if and only if there is a dilatation P such that $\nu = P\mu$, where a dilatation P is a Markov kernel from Ω to Ω such that for all continuous affine functions z on Ω, $zP = z$.*

The condition $zP = z$ means that for any continuous affine function z on Ω: $\int_{\Omega} z(r)P(\omega, dr) = z(\omega), \quad \forall \omega \in \Omega$.

Theorem 7.7.1 does not lead to a converse to Blackwell's theorem in the generality presented in Theorem 7.3.4: Let Ω be the space of probability measures on $\mathbb{X}$. Let μ be an information structure that is more informative than another information structure ν in Blackwell's sense. Let us restrict ourselves to decision problems where $\mathbb{U}$ is compact. Let Q_{μ} and Q_{ν} be the measurement channels for the DM under information structures μ and ν, respectively. By definition, we have for all measurable and bounded cost functions c continuous in the actions:

$$\inf_{\gamma \in \Gamma} \int \zeta(dx)Q_{\mu}(dy|x)c(x, \gamma(y)) \leq \inf_{\eta \in \Gamma} \int \zeta(dx)Q_{\nu}(dy|x)c(x, \eta(y))$$

Let $P^{\mu}(dy)Q(dx|y)$ be the alternative disintegration of the information structure μ following Bayes' rule. Likewise, perform the same disintegration for ν. Then we can rewrite the above equation as (due to the measurable selection conditions as in the proof of Theorem 3.1 in [828]):

$$\int P^{\mu}(dy)(\inf_{u \in \mathbb{U}} \int Q(dx|y)c(x, u)) \leq \int P^{\nu}(dy)(\inf_{u \in \mathbb{U}} \int Q(dx|y)c(x, u)), \tag{7.8}$$

Define $\Pi^{\mu}(A) := \int_{\mathbb{Y}} P^{\mu}(dy)1_{Q^{\mu}(\cdot|A)}$. We note that Π^{μ} is a probability measure on Ω. Define Π^{ν} similarly. Then (7.8) becomes

$$\int \Pi^{\mu}(d\pi)(\inf_{u \in \mathbb{U}} \int \pi(dx)c(x, u)) \leq \int \Pi^{\nu}(d\pi)(\inf_{u \in \mathbb{U}} \int \pi(dx)c(x, u)),$$

with the interpretation that $\pi(dx) = Q(dx|y)$. Let $W^*(\pi) = \inf_{u \in \mathbb{U}} \int \pi(dx)c(x, u)$. Then we can rewrite this once again as:

$$\int \Pi^{\mu}(d\pi)W^*(\pi) \leq \int \Pi^{\nu}(d\pi)W^*(\pi).$$

Since Π^{μ} and Π^{ν} are probability distributions on Ω, and W^* is a function over Ω, we will have $\nu \prec \mu$ in Strassen's sense if the above inequality holds for all continuous and concave functions over Ω.

We can show that W^* is continuous and concave in π provided that additionally c is continuous *both* in x and u: Let $\pi_n \to \pi$ weakly. Let u_n^* be optimal for π_n. Then:

$$|\int c(x,u_n^*)\pi_n(dx) - \int c(x,u^*)\pi(dx)|$$
$$\leq \max(\int c(x,u_n^*)(\pi_n(dx) - \pi(dx)), \int c(x,u^*)\pi_n(dx) - \pi(dx)).$$

We note that $\int c(x, u_n^*)(\pi_n(dx) - \pi(dx))$ goes to 0 following [658, Theorem 3.5] or [454, Theorem 3.5]. The second term converges to zero by the weak convergence of π_n to π. We emphasize the requirement that c is continuous in both x and u, whereas in Theorem 7.3.4 only continuity in u was required (one can construct a simple counterexample, even when $\mathbb{U}$ is a singleton to show that continuity in x is necessary for this argument to hold). Concavity of W^* in the conditional measure $\pi(dx)$ follows from Proposition 7.3.1. Now, if one can show that by using all bounded continuous cost functions c and compact action spaces $\mathbb{U}$, the space of all continuous and concave functions on Ω is spanned by the space of all W^* functions, then a converse can be attained through Strassen's result. We note here that every concave and upper semi-continuous W can be written as an infimum of a family of affine functions (Fenchel-Moreau Theorem [606]) and an analysis can be pursued towards this direction at least for the case where c can be assumed to be continuous in both variables and the condition on W is to be relaxed in Strassen's theorem. However, due to the conditions of upper semi-continuity of W^* and the joint continuity of c noted earlier in both the state and actions, the applicability of Strassen's theorem to the setting considered in the chapter does not seem to hold in the generality reported in the chapter.

In the chapter, Theorem 7.3.2 follows from [123]. Some of the results presented in this chapter are based on [828], [825]. Converse to Blackwell for general spaces appears in [353].

Chapter 8
Continuity, Existence and Topological Properties of Optimization on the Space of Information Structures

8.1 Introduction

In this chapter, we will continue with our discussion of information structures (from the previous chapter) but from a different perspective: studying their regularity (such as continuity) and topological properties. In particular, we study topological and structural properties of optimal cost on spaces of information structures. We investigate values of measurement channels and the problem of channel optimization for a class of stochastic control problems. In addition to establishing existence and structural results for optimal policies, the topological properties (such as continuity on space of information structures) developed in this chapter will also be useful in assessing robustness of designs when a probabilistic description of a model is not fully accurate. We will introduce quantizers as a special class of measurement channels, and obtain several useful properties and existence results for optimal quantizers (quantizers will be studied further in the book in Part III). We will also show that all stochastic dynamic teams are nearly static reducible (with policy-independence). We will also investigate the effects of the prior measure on static and dynamic team decision problems.

The chapter is organized as follows. Following this introduction, Sect. 8.2 deals with topological characterization of information structures. Continuity of optimal solutions in channels for single-stage optimization problems is considered in Sect. 8.3, and existence analysis is provided in Sect. 8.4. Quantizers, viewed as a subclass of measurement channels, are studied in detail in view of continuity and existence properties in Sect. 8.5. The multi-stage case is studied in Sect. 8.6. Continuity when the prior probability is also viewed as a variable is studied in Sect. 8.7. The multi-agent setting is studied in Sect. 8.8. Finally, as a culmination of some of the efforts in this chapter and the previous one, we show that all stochastic dynamic teams are nearly static reducible (with policy-independence) in Sect. 8.9.

S. Yüksel, T. Başar, *Stochastic Teams, Games, and Control under Information Constraints*, Systems & Control: Foundations & Applications,
https://doi.org/10.1007/978-3-031-54071-4_8

8.2 Topological Characterization of Measurement Channels

In this section, we provide topologies for measurement channels. Such constructions will be used extensively in view of existence of optimal channels, as well as optimal quantizers to be considered both in this chapter and later in the book.

Consider the setting of Sect. 7.1 and Fig. 7.1. In the following, we have $\mathbb{X} = \mathbb{R}^n$ and $\mathbb{Y} = \mathbb{R}^m$, and $\mathcal{Q}$ denotes the set of all measurement channels (stochastic kernels) with input space $\mathbb{X}$ and output space $\mathbb{Y}$. A review on convergence notions involving sequences of probability measures can be found in Appendix C. In the following, we essentially adopt the Young topology framework studied in Sect. 5.5.1, see Appendix F.

Definition 8.2.1 (Convergence of Channels)

(i) A sequence of channels $\{Q_n\}$ converges to a channel Q *weakly at input* P if $PQ_n \to PQ$ weakly.
(ii) A sequence of channels $\{Q_n\}$ converges to a channel Q *setwise at input* P if $PQ_n \to PQ$ setwise, i.e., if $PQ_n(A) \to PQ(A)$ for all Borel sets $A \subset \mathbb{X}\times\mathbb{Y}$.
(iii) A sequence of channels $\{Q_n\}$ converges to a channel Q in *total variation at input* P if $PQ_n \to PQ$ in total variation, i.e., if $\|PQ_n - PQ\|_{TV} \to 0$.

If we introduce the equivalence relation $Q \equiv Q'$ if and only if $PQ = PQ'$, $Q, Q' \in \mathcal{Q}$, then the convergence notions in Definition 8.2.1 only induce the corresponding topologies on the resulting equivalence classes in $\mathcal{Q}$, instead of the entire $\mathcal{Q}$. Since in most of the development the input distribution P is fixed, there should be no confusion when we talk about the induced topologies (resp. metrics) on $\mathcal{Q}$.

The preceding definition was explicitly dependent on the input distribution P. See Theorem F.3.1 for the relation between the convergences under different input measures. The next lemma provides a set of sufficient conditions which may be easier to verify, independent of the input distribution. The proof is provided in the appendix.

Lemma 8.2.1

(i) If $\{Q_n(\cdot\,|x)\}$ converges to $Q(\cdot\,|x)$ weakly for P-a.e. x, then $PQ_n \to PQ$ weakly.
(ii) If $\{Q_n(\cdot\,|x)\}$ converges to $Q(\cdot\,|x)$ setwise for P-a.e. x, then $PQ_n \to PQ$ setwise.
(iii) If $\{Q_n(\cdot\,|x)\}$ converges to $Q(\cdot\,|x)$ in total variation for P-a.e. x, then $PQ_n \to PQ$ in total variation.

The conditions in Lemma 8.2.1 are almost universal in the choice of input probability measures; that is, the convergence characterizations will be independent of the input distributions if each of the conditions is replaced with convergence of $\{Q_n(\cdot\,|x)\}$ to $Q(\cdot\,|x)$ for all $x \in \mathbb{X}$. This is particularly useful when the input distribution is unknown, or when it may change. The latter can occur in multi-stage stochastic control problems.

In the following, we provide a few examples [828], where we follow the setting of Sect. 7.1.

1. Consider the case where the measurement channel has the form $y_t = Cx_t + v_t$, where $\{v_t\}$ is an i.i.d. noise (measurement disturbance) process. Suppose that $v_t \sim f_{\theta_0}$ for some $\theta_0 \in \Theta$, where $\Theta \subset \mathbb{R}^d$ is a parameter set and $\{f_\theta : \theta \in \Theta\}$ is a parametric family of densities such that $f_{\theta_n}(v) \to f_{\theta_0}(v)$ for all $v \in \mathbb{R}^n$ and any sequence of parameters θ_n such that $\theta_n \to \theta_0$. Then by Scheffé's theorem [133] f_{θ_n} converges to f_{θ_0} in the L_1 sense, and consequently, the sequence of corresponding additive channels $Q_n(\cdot|x)$ converges to the channel $Q(\cdot|x)$ (corresponding to f_θ) in total variation for all x.
2. Consider again the measurement channel $y_t = Cx_t + v_t$, but assume this time that we only know that v_t has a density (that is, its measure is absolutely continuous with respect to the Lebesgue measure) p (which is unknown to us). However, suppose that we are provided with a sequence of independent realizations for the noise process. Thus, with these independent observations $v_1, \ldots, v_n$ from the noise process, we can use any of the consistent nonparametric methods, e.g., [212], to obtain an estimate p_n which converges (with probability one) to p in the L_1 sense as $n \to \infty$. The corresponding sequence of estimated channels $Q_n(\cdot|x)$ converges to the true channel $Q(\cdot|x)$ in total variation for all x with probability one.
3. Now suppose that the observation channel Q is such that $Q(\cdot|x)$ admits a conditional density $f(y|x)$ for all $x \in \mathbb{R}^n$. Given observations $(x_1, y_1), \ldots, (x_n, y_n)$ drawn independently from the distribution PQ, there exists a sequence of nonparametric conditional density estimates $f_n(y|x)$ such that, as $n \to \infty$,

$$\int \left(\int |f_n(y|x) - f(y|x)|\, dy \right) P(dx) \to 0,$$

with probability one [328]. This immediately implies that the channels Q_n corresponding to these estimates converge to Q in total variation at input P.
4. Finally, assume again the additive measurement model $y_t = Cx_t + v_t$, where now we do not have any information about the distribution μ of v_t. In this case there are no methods to consistently estimate μ in total variation from independent samples $v_1, \ldots, v_n$ [213]. The same negative conclusion applies for setwise convergence as well. However, the empirical distribution μ_n of the samples converges weakly to μ with probability one (via characterizing weak convergence by studying integrals of a countable family of functions, see Appendix C; and the strong law of large numbers). The corresponding estimated observation channels $Q_n(\cdot|x)$ converge weakly to the true channel $Q(\cdot|x)$ for all x with probability one.

8.3 Single Stage: Continuity of the Optimal Cost in Channels

In this section, we study continuity properties under total variation, setwise convergence, and weak convergence, for the single-stage case. Thus, we investigate the continuity of the functional

$$J(P, Q) = \inf_{\underline{\gamma}} E_P^{Q,\underline{\gamma}}\big[c(x_0, u_0)\big]$$
$$= \inf_{\gamma \in \Gamma} \int_{\mathbb{X}\times\mathbb{Y}} c(x, \gamma(y))Q(dy|x)P(dx)$$

in the channel $Q \in \mathcal{Q}$, where Γ is the collection of all Borel measurable functions mapping $\mathbb{Y}$ into $\mathbb{U}$. Note that by our previous notation, $\underline{\gamma} = \gamma$ is an admissible first-stage control policy. As before, $\mathcal{Q}$ denotes the set of all channels with input space $\mathbb{X}$ and output space $\mathbb{Y}$.

Our results in this section as well as subsequent sections in this chapter will utilize one or more of the assumptions on the cost function c and the (Borel) set $\mathbb{U} \subset \mathbb{R}^k$:

Assumption 8.3.1 *Consider the following:*

(A1) *The function* $c : \mathbb{X} \times \mathbb{U} \to \mathbb{R}$ *is bounded and continuous on* $\mathbb{X} \times \mathbb{U}$.
(A2) *The function* $c : \mathbb{X} \times \mathbb{U} \to \mathbb{R}$ *is non-negative, measurable, and bounded.*
(A3) *The function* $c : \mathbb{X} \times \mathbb{U} \to \mathbb{R}$ *is non-negative, measurable, bounded, and continuous on* $\mathbb{U}$ *for every* $x \in \mathbb{X}$.
(A4) $\mathbb{U}$ *is a compact set.*
(A5) $\mathbb{U}$ *is a convex set.*

Before proceeding further, we look for conditions under which an optimal control policy exists; i.e, when the infimum in $\inf_\gamma E_P^{Q,\gamma}[c(x, u)]$ is a minimum. The following result is proved in the appendix of the chapter.

Theorem 8.3.1 ([828]) *Suppose that Assumptions A3 and A4 hold. Then, there exists an optimal control policy for any channel* $Q \in \mathcal{Q}$.

8.3.1 Continuity Properties Under Weak Convergence

Theorem 8.3.2 $J(P, Q)$ *may not be sequentially continuous under weak convergence of channels even for continuous loss functions* c *and compact* $\mathbb{X}$, $\mathbb{Y}$, *and* $\mathbb{U}$.

Proof The following example demonstrates that $J(P, Q)$ may not be sequentially continuous under weak convergence of channels even for continuous loss functions c and compact $\mathbb{X}$, $\mathbb{Y}$, and $\mathbb{U}$.

Let $\mathbb{X} = \mathbb{Y} = \mathbb{U} = [0, 1]$. Suppose that the cost is given as $c(x, u) = (x - u)^2$ and assume that P is a discrete distribution with two atoms:

$$P = \frac{1}{2}\delta_0 + \frac{1}{2}\delta_1,$$

where δ_0 is the delta measure at point 0, that is, $\delta_0(A) = 1_{\{0 \in A\}}$ for every Borel set A, where 1_E denotes the indicator function of event E. Similarly, δ_1 is the delta measure at point 1. Let $\{Q_n\}$ be a sequence of channels given by

$$Q_n(\cdot\,|x) = \begin{cases} \delta_{\frac{1}{n}} & \text{if} \quad x \geq \frac{1}{n}, \\ \delta_0 & \text{if} \quad x < \frac{1}{n}. \end{cases} \tag{8.1}$$

In this case, the optimal control policy, which is unique up to changes in points of measure zero, is

$$\gamma_n(y) = 1_{\{y \geq \frac{1}{n}\}}, \qquad n \in \mathbb{N},$$

leading to a cost of 0. We observe that the limit of the sequence $\{Q_n(\cdot\,|x)\}$ is given by

$$Q(\cdot\,|x) = \delta_0 \quad \text{for all } x \in \mathbb{R}. \tag{8.2}$$

It can now be shown [828] that $Q_n \to Q$ weakly at input P. However, the limit of the sequence of channels cannot distinguish between the inputs, since the channel output always equals 0. Thus, even though

$$J(P, Q_n) = 0, \quad \text{for all } n \geq 1,$$

the cost of $Q = \lim_n Q_n$ is

$$J(P, Q) = \frac{1}{4}$$

since, letting $(X, Y) \sim PQ$, we have $\gamma(y) = E[X|Y = y] = 1/2$ for all y. ◇

Upper semi-continuity, however, can be established.

Theorem 8.3.3 ([828]) *Suppose that assumptions A1 and A5 hold. If $\{Q_n\}$ is a sequence of channels converging weakly at input P to a channel Q, then*

$$\limsup_{n\to\infty} J(P, Q_n) \leq J(P, Q),$$

that is, $J(P, Q)$ is upper semi-continuous on $\mathcal{Q}$ under weak convergence.

Proof Let μ be an arbitrary probability measure on $(\mathbb{X} \times \mathbb{Y}, \mathcal{B}(\mathbb{X} \times \mathbb{Y}))$, and let $\mu_{\mathbb{Y}}$ be its second marginal, i.e., $\mu_{\mathbb{Y}}(A) = \mu(\mathbb{X} \times A)$ for $A \in \mathcal{B}(\mathbb{Y})$. Let $g \in \Gamma$ be arbitrary. By Lusin's theorem [612, Thm. 2.24] (see Theorem C.5.1), there exists a continuous function $f : \mathbb{Y} \to \mathbb{U}$ such that

$$\mu_{\mathbb{Y}}\{y : f(y) \neq g(y)\} < \epsilon.$$

Letting $B = \{y : f(y) \neq g(y)\}$ we obtain

$$\int \left|c(x, g(y)) - c(x, f(y))\right| \mu(dx, dy) = \int_{\mathbb{X} \times B} \left|c(x, g(y)) - c(x, f(y))\right| \mu(dx, dy)$$
$$< \epsilon \cdot c^*,$$

where $c^* = \sup_{x,u} c(x, u) < \infty$ by assumption *A1*, so that

$$\int c(x, f(y))\mu(dx, dy) < \int c(x, g(y))\mu(dx, dy) + c^*\epsilon. \tag{8.3}$$

Let $\mathcal{C}$ be the set of continuous functions from $\mathbb{Y}$ into $\mathbb{U}$. Define

$$j(\mu, \mathcal{C}) = \inf_{\gamma \in \mathcal{C}} \int c(x, \gamma(y))\mu(dx, dy), \qquad j(\mu, \Gamma) = \inf_{\gamma \in \Gamma} \int c(x, \gamma(y))\mu(dx, dy)$$

and note that $j(\mu, \mathcal{C}) \geq j(\mu, \Gamma)$ since $\mathcal{C} \subset \Gamma$. By (8.3), $j(\mu, \mathcal{C})$ is upper bounded by the right-hand-side of (8.3). Since g in (8.3) was arbitrary, we obtain $j(\mu, \mathcal{C}) \leq j(\mu, \Gamma) + c^*\epsilon$, which in turn implies $j(\mu, \mathcal{C}) \leq j(\mu, \Gamma)$ since $\epsilon > 0$ was arbitrary. Hence $j(\mu, \mathcal{C}) = j(\mu, \Gamma)$.

Applying the above first to PQ_n and then to PQ, we obtain

$$\begin{aligned}
&\lim \sup_{n \to \infty} \inf_{\gamma \in \Gamma} \int c(x, \gamma(y)) PQ_n(dx, dy) \\
&= \lim \sup_{n \to \infty} \inf_{f \in \mathcal{C}} \int c(x, f(y)) PQ_n(dx, dy) \\
&\leq \inf_{f \in \mathcal{C}} \lim \sup_{n \to \infty} \int c(x, f(y)) PQ_n(dx, dy) \\
&= \inf_{f \in \mathcal{C}} \int c(x, f(y)) PQ(dx, dy) \\
&= \inf_{\gamma \in \Gamma} \int c(x, \gamma(y)) PQ(dx, dy)
\end{aligned}$$

where the next to last equality holds since PQ_n converges weakly to PQ. ◇

Weak Continuity Along De-Garbling Sequences Recall the notion of garbling in Definition 7.3.1. Let us define Q_n to be a de-garbling sequence if Q_n is a garbling of Q_{n+1}. An example would be $y = x + \frac{1}{n}v$, where v is an independent Gaussian random variable. As n increases, the channel is de-garbled (see e.g. Corollary 7.2.1). Thus, by Blackwell's comparison theorem, we know that $J(P, Q_n) \geq J(P, Q_{n+1})$.

Theorem 8.3.4 ([352]) *Suppose that Assumptions A1 and A5 hold. Let $Q_n \to Q$ weakly at input P be such that Q_n is a de-garbling sequence so that Q_n is a garbling of Q_m for $m > n$, as well as Q. Then, we have that J is weakly continuous along this sequence, i.e.,*

$$\lim_{n\to\infty} J(P, Q_n) = J(P, Q)$$

Proof By Theorem 8.3.3, we have upper semi-continuity under weak convergence. On the other hand, we have that $J(P, Q_n) \geq J(P, Q)$ by the de-garbling property that Q_n is a garbling of Q, and therefore

$$\lim\inf_{n\to\infty} J(P, Q_n) = \lim_{n\to\infty} J(P, Q_n) \geq J(P, Q),$$

and lower semi-continuity also holds, resulting in continuity. ◇

Remark 8.3.1 This result's essence will turn out to be very consequential later in Sect. 8.9: Notice that, for real valued state and measurement spaces spaces, a measurement channel defined with $y = f(x)$ is not absolutely continuous with respect to a reference probability measure, in the sense of (3.26); however, for every $m \in \mathbb{N}$, $y = x + \frac{1}{m}v$ with v a Gaussian independent random variable is absolutely continuous with respect to a Gaussian reference measure. Thus the measure transformation technique is applicable if one allows for an arbitrarily small error while studying optimal stochastic control problems of this type (by slightly adding noise and studying this problem, whose optimal expected cost is arbitrarily close to the solution of the original problem).

8.3.2 Continuity Properties Under Setwise Convergence

We state the following for setwise convergence.

Theorem 8.3.5 ([828])

(i) Under assumption A2, the optimal cost

$$J(P, Q) := \inf_{\gamma\in\Gamma} E_P^{Q,\gamma}[c(x, u)]$$

is sequentially upper semi-continuous on the set of communication channels $\mathcal{Q}$ under setwise convergence.

(ii) $J(P, Q)$ may not be sequentially continuous under setwise convergence of channels even for continuous loss functions and compact $\mathbb{X}$, $\mathbb{Y}$, and $\mathbb{U}$.

Proof

(i) Let $\{Q_n\}$ converge setwise to Q at input P. Then

$$\begin{aligned}
&\lim_{n\to\infty}\sup\inf_{\gamma\in\Gamma}\int c(x,\gamma(y))PQ_n(dx,dy)\\
&\le \inf_{\gamma\in\Gamma}\lim\sup_{n\to\infty}\int c(x,\gamma(y))PQ_n(dx,dy)\\
&= \inf_{\gamma\in\Gamma}\int c(x,\gamma(y))PQ(dx,dy),
\end{aligned}$$

where the equality holds since c is bounded.

(ii) The following counter-example demonstrates that $J(P,Q)$ may not be sequentially continuous also under setwise convergence of channels even for continuous cost functions and compact $\mathbb{X}$, $\mathbb{Y}$, and $\mathbb{U}$.

Let $\mathbb{X}=\mathbb{Y}=\mathbb{U}=[0,1]$. Assume that X has the two-point distribution

$$P=\frac{1}{2}\delta_0+\frac{1}{2}\delta_1.$$

Let $Q(\cdot\,|x)=U([0,1])$ for all x, so that if $(x,y)\sim PQ$; then y is independent of x and has the uniform distribution on $[0,1]$. Let $c(x,u)=(x-u)^2$.

By independence, $E[x|y]=E[x]=1/2$, and thus,

$$\begin{aligned}
J(P,Q)&=\min_{\gamma\in\Gamma}E[(x-\gamma(y))^2]=E[(x-E[x|y])^2]\\
&=\frac{1}{2}\left(1-\frac{1}{2}\right)^2+\frac{1}{2}\left(0-\frac{1}{2}\right)^2=\frac{1}{4}.
\end{aligned}$$

For $n\in\mathbb{N}$ and $k=1,\dots,n$, consider the intervals

$$L_{nk}=\left[\frac{2k-2}{2n},\frac{2k-1}{2n}\right),\qquad R_{nk}=\left[\frac{2k-1}{2n},\frac{2k}{2n}\right) \tag{8.4}$$

and the "square wave" function

$$h_n(t)=\sum_{k=1}^{n}\left(1_{\{t\in L_{nk}\}}-1_{\{t\in R_{nk}\}}\right).$$

Since $\int_0^1 h_n(t)\,dt=0$ and $|h_n(t)|\le 1$, the function $f_n(t)=\big(1+h_n(t)\big)1_{\{t\in[0,1]\}}$ is a probability density function. Furthermore, the standard proof of the Riemann-Lebesgue lemma (for example [768], Thm. 12.21) implies that

$$\lim_{n\to\infty}\int_0^1 h_n(t)g(t)\,dt=0\quad\text{for all } g\in L_1([0,1],\mathbb{R})$$

and therefore

$$\lim_{n\to\infty}\int_0^1 f_n(t)g(t)\,dt = \int_0^1 g(t)\,dt \quad \text{for all } g \in L_1([0,1],\mathbb{R}). \tag{8.5}$$

In particular, we obtain that the sequence of probability measures induced by the sequence $\{f_n\}$ converges setwise to $U([0,1])$.

Now, for every n, introduce a channel as follows

$$Q_n(\cdot\,|x) = \begin{cases} U([0,1]), & x = 0 \\ \sim f_n, & x = 1. \end{cases}$$

Then $Q_n(\cdot|x) \to Q$ setwise for $x = 0$ and $x = 1$, and thus $PQ_n \to PU([0,1])$ setwise. However, letting $(x, y_n) \sim PQ_n$, a simple calculation shows that the optimal policy for PQ_n is

$$\gamma_n(y) = E[x|y_n = y] = \begin{cases} 0, & y \in \bigcup_{k=1}^n R_{nk} \\ \frac{2}{3}, & y \in \bigcup_{k=1}^n L_{nk} \end{cases}$$

and therefore for every $n \in \mathbb{N}$,

$$\begin{aligned} J(P, Q_n) &= \min_{\gamma\in\Gamma} E[(x - \gamma(y_n))^2] \\ &= \frac{1}{2}\int_0^1 (0 - \gamma_n(y))^2\,dy + \frac{1}{2}\int_0^1 (1 - \gamma_n(y))^2 f_n(y)\,dy = \frac{1}{6}. \end{aligned}$$

Thus, the optimal cost value is not continuous under setwise convergence. ◇

8.3.3 *Continuity Under Total Variation*

We have continuity under the stronger notion of total variation:

Theorem 8.3.6 ([828]) *Under assumption A2, the optimal cost $J(P, Q)$ is continuous on the set of communication channels $\mathcal{Q}$ under the topology of total variation.*

Proof Assume that $Q_n \to Q$ in total variation at input P. Let $\epsilon > 0$ and pick the ϵ-optimal policies γ_n and γ under channels Q_n and Q, respectively. That is, letting $\hat{J}(Q', \gamma') = E_P^{Q',\gamma'}[c(x,u)]$ for any $\gamma' \in \Gamma$ and $Q' \in \mathcal{Q}$, we have $\hat{J}(Q_n, \gamma_n) < J(P, Q_n) + \epsilon$ and $\hat{J}(Q, \gamma) < J(P, Q) + \epsilon$.

Considering first the case, $J(P, Q_n) < J(P, Q)$, we have

$$\begin{aligned} J(P, Q) - J(P, Q_n) &< J(P, Q) - \hat{J}(Q_n, \gamma_n) + \epsilon \\ &< \hat{J}(Q, \gamma_n) - \hat{J}(Q_n, \gamma_n) + \epsilon. \end{aligned}$$

By a symmetric argument it follows that

$$|J(P,Q) - J(P,Q_n)| \leq \max\left(\hat{J}(Q,\gamma_n) - \hat{J}(Q_n,\gamma_n), \hat{J}(Q_n,\gamma) - \hat{J}(Q,\gamma)\right) + \epsilon \tag{8.6}$$

Now, since c is bounded, it follows that for any $\gamma' \in \Gamma$,

$$\begin{aligned} &|\hat{J}(Q_n,\gamma') - \hat{J}(Q,\gamma')| \\ &= \left|\int c(x,\gamma'(y))PQ_n(dx,dy) - \int c(x,\gamma'(y))PQ(dx,dy)\right| \\ &\leq \|c\|_\infty \|PQ_n - PQ\|_{TV}. \end{aligned}$$

Together with (8.6), this implies that $|J(P,Q_n) - J(P,Q)| \leq \|c\|_\infty \|PQ_n - PQ\|_{TV} + \epsilon$. Since $\epsilon > 0$ was arbitrary, we obtain $|J(P,Q_n) - J(P,Q)| \leq \|c\|_\infty \|PQ_n - PQ\|_{TV}$, and therefore $J(P,Q_n) \to J(P,Q)$ as claimed. ◇

Remark 8.3.2 We can relax the boundedness condition for c to a *uniform integrability* condition. Let $Q_n \to Q$ and let for every n, γ_n be optimal for Q_n, γ optimal for Q. If uniformly for $\{Q_n,\gamma_n\}$ and Q,γ, we have that for every $\epsilon > 0$, there exists an $L < \infty$ such that

$$\left|E_{\nu_0}^{Q^n,\gamma^n}[c(x,u)] - E_{\nu_0}^{Q^n,\gamma^n}[c(x,u)1_{(c(x,u)\leq L)}]\right| \leq \epsilon,$$

with c non-negative and measurable, then continuity holds. See the proof of Theorem 16.1.2 for further details on such an analysis.

Thus, total variation, although a strong metric, is useful in establishing continuity.

8.3.4 Further Sufficient Conditions for Continuity Under Weak Convergence

We have seen in Theorem 8.3.4 that weak convergence of information structures may imply continuity under restrictive conditions. We have also seen in Theorem 8.3.6 that total variation is a metric which provides general continuity properties. A careful analysis of the proof of Theorem 8.3.6 should reveal that we essentially need a uniform convergence property for setwise convergence to also be sufficient for continuity. That is, we need to have

$$\lim_{n\to\infty} \sup_{\gamma\in\mathcal{F}_s} \left|\int \left(\int Q(dy|x)c(x,\gamma(y)) - \int Q_n(dy|x)c(x,\gamma(y))\right)P(dx)\right| = 0,$$

for a class of admissible policies $\mathcal{F}_s$ to be able to have continuity under setwise convergence. Thus, one important question of practical interest, is the following: What type of stochastic control problems, cost functions, and allowable policies lead to solutions which admit such a uniform convergence principle under setwise convergence? Some partial answers to this question can be found in [720].

Likewise, a parallel discussion applies to weak convergence under the assumption that for every Q_n and for Q, corresponding optimal policies γ_n and γ are continuous and are assumed to be from a restricted class of policies $\mathcal{F}_w$. One needs to have

$$\int_{\mathbb{X}\times\mathbb{Y}} c(x, \gamma_n(y)) Q_n(dy|x) P(dx) \to \int_{\mathbb{X}\times\mathbb{Y}} c(x, \gamma(y)) Q(dy|x) P(dx)$$

A sufficient condition for this is the following form of uniform weak convergence:

$$\lim_{n\to\infty} \sup_{\gamma\in\mathcal{F}_w} \left| \int_{\mathbb{X}\times\mathbb{Y}} c(x, \gamma(y)) Q_n(dy|x) P(dx) - \int_{\mathbb{X}\times\mathbb{Y}} c(x, \gamma(y)) Q(dy|x) P(dx) \right| = 0.$$

Two application areas of the above set of results in stochastic control would be learning and identification:

When one does not know the system dynamics, such as the observation channel, one typically attempts to learn the channel via test inputs or empirical observations. Let $\{(x_i, y_i),\ i \in \mathbb{N}\}$ be an $\mathbb{X} \times \mathbb{Y}$-valued i.i.d sequence generated according to some distribution μ. Defining for every measurable $B \subset \mathbb{X} \times \mathbb{Y}$, and $n \in \mathbb{N}$, the empirical occupation measures

$$\mu_n(B) = \frac{1}{n} \sum_{i=1}^{n} 1_{\{(x_i, y_i)\in B\}},$$

one has $\mu_n(B) \to \mu(B)$ almost surely (a.s.) by the strong law of large numbers. However, it is generally not true that $\mu_n \to \mu$ setwise a.s. (e.g., μ_n never converges to μ setwise when either x_i or y_i has a nonatomic distribution), in which case μ_n cannot converge to μ in total variation. A number of related examples were considered in Sect. 8.2.

Again by the strong law of large numbers, for any μ-integrable function f on $\mathbb{X} \times \mathbb{Y}$, one has, almost surely,

$$\lim_{n\to\infty} \int f(x, y) \mu_n(dx, dy) = \int f(x, y) \mu(dx, dy)$$

In particular, $\mu_n \to \mu$ weakly with probability one [223].

In the learning theoretic context, the convergence of optimal costs under μ_n to the cost optimal for μ is called the *consistency of empirical risk minimization* (see [735] for an overview). In particular, if the cost function and the allowable control policies $\mathcal{F}_l$ are such that

$$\lim_{n\to\infty} \sup_{\gamma\in\mathcal{F}_l} \left| \int c(x,\gamma(y))\mu_n(dx,dy) - \int c(x,\gamma(y))\mu(dx,dy) \right| = 0,$$

then we arrive at consistency.

A class of measurable functions $\mathcal{E}$ is called a *Glivenko-Cantelli class* [224], if the integrals with respect to the empirical measures converge almost surely to the integrals with respect to the true measure uniformly over $\mathcal{E}$. Thus, if

$$\mathcal{G} = \{\gamma : c(x,\gamma(y)) \in \mathcal{E}\},$$

where $\mathcal{E}$ is a class of Glivenko-Cantelli family of functions, then we could establish consistency. One example of a Glivenko-Cantelli family of real functions on $\mathbb{R}^N$ is the family $\{f : \|f\|_{BL} \leq M\}$ for some $0 < M < \infty$, where $\|\cdot\|_{BL}$ denotes the bounded Lipschitz norm [224].

Thus, if we place an appropriate restriction on the class of control policies and cost functions, we can have continuity under weak convergence of information structures as well as consistency in learning, and robustness to errors in the probabilistic description of information structures. We again point out that in Sect. 8.9, we will show that when perturbations in information channels are made via carefully designed garbling sequences, then continuity under even weak convergence of information structures can be attained under mild conditions.

8.4 Single Stage: Compactness Properties of Spaces of Channels

In this section, again for the single-stage problem, we study characterizations of compactness (or sequential compactness), which will be useful in obtaining existence results, and which may also be useful in obtaining approximation results.

In the following, we will obtain conditions for sequential compactness of channels under setwise convergence and total variation convergence. Note that sequential compactness under total variation convergence implies setwise sequential compactness, which in turn implies weak sequential compactness.

We first discuss setwise convergence. A set of probability measures $\mathcal{M}$ on some measurable space is said to be *setwise sequentially precompact* if every sequence in $\mathcal{M}$ has a subsequence converging setwise to a probability measure (not necessarily in $\mathcal{M}$). For two finite measures ν and μ defined on the same measurable space, we write $\nu \leq \mu$ if $\nu(A) \leq \mu(A)$ for all measurable sets A.

We have the following condition for setwise (pre)compactness (see also pages 305–306 in [227])

Lemma 8.4.1 ([128, Thm. 4.7.25]) *Let μ be a finite measure on a measurable space $(\mathbb{T}, \mathcal{A})$. Assume that a set of probability measures $\Psi \subset \mathcal{P}(\mathbb{T})$ satisfies*

$$P \leq \mu, \quad \textit{for all } P \in \Psi.$$

Then Ψ is setwise precompact.

As before, $PQ \in \mathcal{P}(\mathbb{X} \times \mathbb{Y})$ denotes the joint probability measure induced by input P and channel Q, where $\mathbb{X} = \mathbb{R}^n$ and $\mathbb{Y} = \mathbb{R}^m$. A simple consequence of the preceding majorization criterion is the following.

Lemma 8.4.2 ([828]) *Let ν be a finite measure on $\mathcal{B}(\mathbb{X} \times \mathbb{Y})$ and let P be a probability measure on $\mathcal{B}(\mathbb{X})$. Suppose that $\mathcal{Q}$ is a set of channels such that*

$$PQ \leq \nu, \quad \textit{for all } Q \in \mathcal{Q}.$$

Then $\mathcal{Q}$ is setwise sequentially precompact at input P in the sense that any sequence in $\mathcal{Q}$ has a subsequence $\{Q_n\}$ such that $Q_n \to Q$ setwise at input P for some channel Q.

Proof By Lemma 8.4.1, the set of joint measures $\mathcal{M} = \{PQ : Q \in \mathcal{Q}\}$ is setwise sequentially precompact, that is, any sequence in $\mathcal{M}$ has a subsequence $\{PQ_n\}$ converging to some $\hat{P}$ setwise. Furthermore, since the first marginal of PQ_n is P for all n, the first marginal of $\hat{P}$ is also P (since $PQ_n(A \times \mathbb{Y}) \to \hat{P}(A \times \mathbb{Y})$ for all $A \in \mathcal{B}(\mathbb{X})$). Now let Q be a regular conditional probability measure satisfying $\hat{P} = PQ$, and the result follows for the subsequential convergence of Q_n. ◇

For a probability density function p on $\mathbb{R}^N$, we let P_p denote the induced probability measure: $P_p(A) = \int_A p(x)\,dx$, $A \in \mathcal{B}(\mathbb{R}^N)$. The next lemma provides a sufficient condition for precompactness.

Lemma 8.4.3 ([828]) *Let μ be a finite Borel measure on $\mathbb{R}^N$ and let $\mathcal{F}$ be an equicontinuous and uniformly bounded family of probability density functions. Define $\Psi \subset \mathcal{P}(\mathbb{R}^N)$ by*

$$\Psi = \{P_p : P_p \leq \mu,\ p \in \mathcal{F}\}.$$

Then Ψ is precompact under total variation.

Proof See Sect. 8.10.3. ◇

The next result is an analogue of Lemma 8.4.2 and has an essentially identical proof.

Lemma 8.4.4 ([828]) *Let $\mathcal{Q}$ be a set of channels such that $\{PQ : Q \in \mathcal{Q}\}$ is a precompact set of probability measures under total variation. Then $\mathcal{Q}$ is precompact under total variation at input P.*

The following theorem, when combined with the preceding results, leads to sufficient conditions for the existence of best and worst channels when the given family of channels $\mathcal{Q}$ is closed under the appropriate convergence notion.

Theorem 8.4.1 ([828])

(i) *There exists a worst channel in $\mathcal{Q}$, that is, a solution to the maximization problem*

$$\sup_{Q\in\mathcal{Q}} J(P,Q) = \sup_{Q\in\mathcal{Q}} \inf_{\gamma} E_P^{Q,\gamma}[c(x,u)]$$

when the set $\mathcal{Q}$ is weakly sequentially compact and assumptions A1, A4, and A5 hold.

(ii) *There exists a worst channel in $\mathcal{Q}$ when the set $\mathcal{Q}$ is setwise sequentially compact and assumption A2 holds.*

(iii) *There exist best and worst channels in $\mathcal{Q}$, that is, solutions to the minimization problem $\inf_{Q\in\mathcal{Q}} J(P,Q)$ and the maximization problem $\sup_{Q\in\mathcal{Q}} J(P,Q)$ when the set $\mathcal{Q}$ is compact under total variation and assumption A2 holds.*

Proof Under the stated conditions, we have sequential upper semi-continuity or continuity (Theorems 8.3.3, 8.3.5, and 8.3.6) under the corresponding topologies. By sequential compactness, the existence of the cost maximizing (worst) channel follows when $J(P,Q)$ is upper-semicontinuous, while the existence of the cost minimizing (best) channel follows when $J(P,Q)$ is lower-continuous in Q. ⋄

Remark 8.4.1 The existence of worst channels is a useful result for the robust control or game-theoretic approach to optimization problems. If the problem is formulated as a game where the uncertainty in the set is regarded as a maximizer and the controller is the minimizer, one could search for a max-min solution, which we have just proven to exist. We note that, in information theory, problems of similar nature have been considered in the context of mutual information and jamming games [99, 682]; this subject will be studied in detail in Chap. 20. ⋄

8.5 Quantizers as a Class of Channels

In this section, we consider the problem of convergence and optimization of quantizers. Quantization will be a recurrent subject throughout the rest of the book.

We begin with the definition of a quantizer.

Definition 8.5.1 An M-cell *vector quantizer*, Q, is a (Borel) measurable mapping from a subset of $\mathbb{X} = \mathbb{R}^n$ to the finite set $\{1, 2, \ldots, M\}$, characterized by a measurable partition $\{B_1, B_2, \ldots, B_M\}$ such that $B_i = \{x : Q(x) = i\}$ for $i = 1, \ldots, M$. The B_i's are called the cells (or bins) of Q.

Remark 8.5.1

(i) For later convenience we allow for the possibility that some of the cells of the quantizer are empty.
(ii) Traditionally, in source coding theory, a quantizer is a mapping $Q : \mathbb{R}^n \to \mathbb{R}^n$ with a finite range. Thus q is defined by a partition and a reconstruction value in $\mathbb{R}^n$ for each cell in the partition. That is, for given cells $\{B_1, \ldots, B_M\}$ and reconstruction values $\{q^1, \ldots, q^M\} \subset \mathbb{R}^n$, we have $Q(x) = q^i$ if and only if $x \in B_i$. In the definition above, we do not include the reconstruction values.

A quantizer Q with cells $\{B_1, \ldots, B_M\}$ can also be characterized as a stochastic kernel Q from $\mathbb{X}$ to $\{1, \ldots, M\}$ defined by

$$Q(i|x) = 1_{\{x \in B_i\}}, \quad i = 1, \ldots, M,$$

so that $Q(x) = \sum_{i=1}^{M} q^i Q(i|x)$. We denote by $\mathcal{Q}_D(M)$ the space of all M-cell quantizers represented in the channel form. In addition, we let $\mathcal{Q}(M)$ denote the set of stochastic kernels from $\mathbb{X}$ to $\{1, \ldots, M\}$, i.e., $Q \in \mathcal{Q}(M)$ if and only if $Q(\cdot|x)$ is a probability distribution on $\{1, \ldots, M\}$ for all $x \in \mathbb{X}$, and $Q(i|\cdot)$ is Borel measurable for all $i = 1, \ldots, M$. Note that $\mathcal{Q}_D(M) \subset \mathcal{Q}(M)$, and by our definition $\mathcal{Q}_D(M-1) \subset \mathcal{Q}_D(M)$ for all $M \geq 2$. We note that elements of $\mathcal{Q}(M)$ are sometimes referred to as random quantizers.

Lemma 8.5.1 ([828]) *The set of quantizers $\mathcal{Q}_D(M)$ is setwise sequentially precompact at any input P.*

Proof The proof follows from Lemma 8.4.2 and the interpretation above viewing a quantizer as a channel. In particular, a majorizing finite measure ν is obtained by defining $\nu = P \times \lambda$, where λ is the counting measure on $\{1, \ldots, M\}$ (note that $\nu(\mathbb{R}^n \times \{1, \ldots, M\}) = M$). Then for any measurable $B \subset \mathbb{R}^n$ and $i = 1, \ldots, M$, we have $\nu(B \times \{i\}) = P(B)\lambda(\{i\}) = P(B)$ and thus

$$PQ(B \times \{i\}) = P(B \cap B_i) \leq P(B) = \nu(B \times \{i\}).$$

Since any measurable $D \subset \mathbb{X} \times \{1, \ldots, M\}$ can be written as the disjoint union of the sets $D_i \times \{i\}$, $i = 1, \ldots, M$, with $D_i = \{x \in \mathcal{X} : (x, i) \in D\}$, the above implies $PQ(D) \leq \nu(D)$. ◇

The following lemma provides a useful result.

Lemma 8.5.2 ([828]) *A sequence $\{Q_n\}$ in $\mathcal{Q}(M)$ converges to a Q in $\mathcal{Q}(M)$ setwise at input P if and only if*

$$\int_A Q_n(i|x)P(dx) \to \int_A Q(i|x)P(dx) \quad \textit{for all } A \in \mathcal{B}(\mathbb{X}) \textit{ and } i = 1, \ldots, M.$$

Proof The lemma follows by noticing that for any $Q \in \mathcal{Q}(M)$ and measurable $D \subset \mathbb{X} \times \{1, \ldots, M\}$,

$$PQ(D) = \int_D Q(dy|x)P(dx) = \sum_{i=1}^{M} \int_{D_i} Q(i|x)P(dx)$$

where $D_i = \{x \in \mathcal{X} : (x, i) \in D\}$. ◇

Theorem 8.5.1 *The space of quantizers $\mathcal{Q}_D(M)$ is not closed under setwise convergence.*

As proof of this theorem, the following example shows that the space of quantizers $\mathcal{Q}_D(M)$ is not closed under setwise convergence:

Example 8.5.1 Let $\mathbb{X} = [0, 1]$ and P the uniform distribution on $[0, 1]$. Recall the definition $L_{nk} = \left[\frac{2k-2}{2n}, \frac{2k-1}{2n}\right)$ in (4.14) and let $B_{n,1} = \bigcup_{k=1}^{n} L_{nk}$ and $B_{n,2} = [0, 1] \setminus B_{n,1}$. Define $\{Q_n\}$ as the sequence of 2-cell quantizers given by

$$Q_n(1|x) = 1_{\{x \in B_{n,1}\}}, \quad Q_n(2|x) = 1_{\{x \in B_{n,2}\}}.$$

This then implies that for all $A \in \mathcal{B}([0, 1])$,

$$\lim_{n\to\infty} \int_A Q_n(dy|x)P(dx) = \lim_{n\to\infty} \int_0^1 \frac{1}{2} f_n(t)\, dt = \frac{1}{2} P(A),$$

and thus, by Lemma 8.5.2, Q_n converges setwise to Q given by $Q(1|x) = Q(2|x) = \frac{1}{2}$ for all $x \in [0, 1]$. However, Q is not a (deterministic) quantizer.

Definition 8.5.2 The class of *finitely randomized quantizers* $\mathcal{Q}_{FR}(M)$ is the convex hull of $\mathcal{Q}_D(M)$, i.e., $Q \in \mathcal{Q}_{FR}(M)$ if and only if there exist $k \in \mathbb{N}$, $Q_1, \ldots, Q_k \in \mathcal{Q}_D(M)$, and $\alpha_1, \ldots, \alpha_k \in [0, 1]$ with $\sum_{i=1}^{k} \alpha_i = 1$, such that

$$Q(i|x) = \sum_{j=1}^{k} \alpha_j Q_j(i|x), \quad \text{for all } i = 1, \ldots, M \text{ and } x \in \mathbb{X}.$$

The next result says that $\mathcal{Q}(M)$ is the (sequential) closure of the convex hull of $\mathcal{Q}_D(M)$.

Theorem 8.5.2 ([828]) *For any $Q \in \mathcal{Q}(M)$ there exists a sequence $\{\hat{Q}_n\}$ of finitely randomized quantizers in $\mathcal{Q}_{FR}(M)$ which converges to Q setwise at any input P.*

The preceding theorem has important implications in that it tells us that the space of deterministic quantizers is a "basis" for the space of communication channels between $\mathbb{X}$ and $\{1, \ldots, M\}$ in an appropriate sense. For the case when both spaces are finite, we can obtain a result reminiscent of the Birkhoff-von Neumann Theorem [749], the proof of which immediately follows from the Krein-Milman Theorem [75].

Theorem 8.5.3 ([828]) *Let $\mathbb{X}, \mathbb{Y}$ be finite spaces and let $\mathcal{Q}$ be the space of stochastic kernels (matrices) from $\mathbb{X}$ to $\mathbb{Y}$. Then every $Q \in \mathcal{Q}$ can be expressed as a convex combination of (deterministic) quantizers from $\mathbb{X}$ to $\mathbb{Y}$.*

Remark 8.5.2 (Extreme Point Property of Quantizers) Related to Theorem 8.5.3, it is worth stating a further representation result due to Borkar [132] (see also [141]). This was already discussed in Sect. 4.1.1 in further detail: Consider the set of probability measures

$$\Theta := \{\zeta \in P(\mathbb{R}^n \times \mathbb{M}) : \zeta = PQ, Q \in \mathcal{Q}\}, \tag{8.7}$$

on $\mathbb{R}^n \times \mathbb{M}$ having fixed input marginal P, equipped with weak topology. This set is the (Borel measurable) set of the extreme points on the set of probability measures on $\mathbb{R}^n \times \mathbb{M}$ with a fixed input marginal P. Hence, the set of all stochastic kernels from $\mathbb{R}^n$ to $\mathbb{M}$ with fixed input marginal measure P on $\mathbb{R}^n$ is such that any element $\mathcal{K}$ in this space can be expressed in the form

$$\mathcal{K}(A) = \int \xi(dQ) PQ(A), \quad A \in \mathcal{B}(\mathbb{R}^N \times \mathbb{M})$$

for some $\xi \in \mathcal{P}(\Theta)$. ◇

In the following we again consider Euclidean spaces and show that an optimal channel can be replaced with an optimal quantizer without any loss in performance.

Proposition 8.5.1 ([828]) *For any $Q \in \mathcal{Q}(M)$, there exists a $Q' \in \mathcal{Q}_D(M)$ with $J(P, Q') \leq J(P, Q)$. If there exists an optimal channel in $\mathcal{Q}(M)$, then there is a quantizer in $\mathcal{Q}_D(M)$ that is optimal.*

Proof We need to prove only the first statement. For a policy $\gamma : \{1, \ldots, M\} \to \mathbb{U} = \mathbb{X}$ (with finite cost) define for all i,

$$\bar{B}_i = \{x : c(x, \gamma(i)) \leq c(x, \gamma(j)), \quad j = 1, \ldots, M\}.$$

Letting $B_1 = \bar{B}_1$ and $B_i = \bar{B}_i \setminus \bigcup_{j=1}^{i-1} B_j$, $i = 2, \ldots, M$, we obtain a partition $\{B_1, \ldots, B_M\}$ and a corresponding quantizer $Q' \in \mathcal{Q}_D(M)$. It is easy to see that $E_P^{Q',\gamma}[c(x, u)] \leq E_P^{Q,\gamma}[c(x, u)]$ for any $Q \in \mathcal{Q}(M)$. ◇

The following shows that setwise convergence of quantizers implies convergence under total variation.

Theorem 8.5.4 ([828]) *Let $\{Q_n\}$ be a sequence of quantizers in $\mathcal{Q}_D(M)$ which converges to a quantizer $Q \in \mathcal{Q}_D(M)$ setwise at P. Then, the convergence is also under total variation at P.*

Proof See Sect. 8.10.4. ◇

We have seen in the preceding set of results that, without any further restriction, the convergence of quantizers may not lead to desirable continuity properties: We observed that, for example, the space of quantizers is not closed.

To alleviate this aspect, in the following we consider quantizers with convex codecells and an input distribution that is absolutely continuous with respect to the Lebesgue measure on $\mathbb{R}^n$ [330]. We note that such quantizers are commonly used in practice, and for a large class of cost functions they are known to contain optimal policies among all possible quantizers. For cost functions of the form $c(x, u) = \|x - u\|^2$ for $x, u \in \mathbb{R}^n$, the cells of optimal quantizers (if they exist) will be convex by Lloyd-Max conditions of optimality [487]; see [330] for further results on convexity of bins for entropy constrained quantization problems. We will study such quantizers later in the book while establishing the structure and existence of optimal dynamic quantizers (in Chaps. 15 and 16).

Now, suppose that $Q \in \mathcal{Q}_D(M)$ with cells $B_1, \ldots, B_M$, each of which is a convex subset of $\mathbb{R}^n$. By the separating hyperplane theorem [490], there exist pairs of complementary closed half spaces $\{(H_{i,j}, H_{j,i}) : 1 \le i, j \le M, i \ne j\}$ such that for all $i = 1, \ldots, M$,

$$B_i \subset \bigcap_{j \ne i} H_{i,j}.$$

Each $\bar{B}_i := \bigcap_{j \ne i} H_{i,j}$ is a closed convex polytope and by the absolute continuity of P one has $P(\bar{B}_i \setminus B_i) = 0$ for all $i = 1, \ldots, M$. One can thus obtain a (P–a.s) representation of Q by the $M(M-1)/2$ hyperplanes $h_{i,j} = H_{i,j} \cap H_{j,i}$.

Let $\mathcal{Q}_c(M)$ denote the collection of M-cell quantizers with convex cells. Since one can represent such a hyperplane h by a vector $(a_1, \ldots, a_m, b) \in \mathbb{R}^{n+1}$ with $\sum_k |a_k|^2 = 1$ such that $h = \{x \in \mathbb{R}^n : \sum_i a_i x_i = b\}$, it is possible to obtain a parametrization over $\mathbb{R}^{(M-1)(n+1)}$ of all such quantizers in $\mathcal{Q}_c$.

Consider a sequence $\{Q_n\}$ in $\mathcal{Q}_c(M)$. It can be shown (see the proof of Thm. 1 in [330]) that using the above appropriate parametrization of the separating hyperplanes, a subsequence Q_{n_k} can be chosen which converges to a $Q \in \mathcal{Q}_c(M)$ in the sense that $P(B_i^{n_k} \triangle B_i) \to 0$ for all $i = 1, \ldots, M$, where the $B_i^{n_k}$ and the B_i are the cells of Q_{n_k} and Q, respectively. In view of the proof of Theorem 8.5.4, we obtain the following.

Theorem 8.5.5 ([828]) *The set $\mathcal{Q}_c(M)$ is compact under total variation at any input measure P that is absolutely continuous with respect to the Lebesgue measure on $\mathbb{R}^n$.*

We can now state an existence result for optimal quantization.

Theorem 8.5.6 ([828]) *Let P be absolutely continuous. Then, there exists a quantizer Q with M cells that minimizes $J(P, Q) = \inf_\gamma E_P^{Q,\gamma}[c(x, u)]$ under assumption A2, with Q restricted to $\mathcal{Q}_c(M)$.*

Proof The existence result follows from Theorems 8.4.1 and 8.5.5. ◇

The above result and the topological construction will be useful and used extensively when we search for optimal dynamic quantizers later in the book.

Remark 8.5.3 In the standard quantization literature, finding an optimal quantizer entails finding optimal codecells and corresponding reconstruction points. Our formulation here does not require the existence of optimal reconstruction points (i.e., existence of an optimal policy γ). ◇

Remark 8.5.4 (Further Existence Results on Optimal Quantizers) It is worth noting that for the existence of an optimal quantizer and reconstruction values for

$$\inf_Q \inf_\gamma E_P^{Q,\gamma}[c(x, u)],$$

the condition that the source admits a density can be relaxed, provided that the cost function is lower-semicontinuous in u. We refer the reader to [2] for cost functions of the form $c(x, u) = c(|x-u|)$, and [482] for more general costs under the assumption that $\mathbb{U}$ is compact and there exists $\bar{u}$ such that $\int P(dx)c(x, \bar{u}) < \infty$. ◇

Remark 8.5.5 (Near Optimality of Quantized Approximations, Successive Refinement and a Total Variation Convergence Argument) When one views quantizers as a collection of point masses, if the source has a density, then the total variation distance between the source and its quantized approximation always remains as 2. On the other hand, as the quantization gets finer, one could make the approximation essentially be in total variation: Let $y = Q_m(x)$, where Q_m is a uniform quantization of $[0, 1]$ into 2^m bins, $\{\mathcal{B}_k, k = 1, \cdots, 2^m\}$. In this case, the quantizers are garblings of one another and the limit is the fully informative channel: $Q(dy|x) = \delta_x(dy)$. In this case, we only have weak convergence of the kernels.

On the other hand, building on [396, 434] we can add a further perturbation to Q_m so that for each bin $\tilde{Q}(d\tilde{y}|y) = \text{Unif}(\mathcal{B}_k)$ if $y \in \mathcal{B}_k$, so that the composite kernel from x to the garbled channel output $\tilde{y}$ has a density across the bin corresponding to each of the 2^m bins. In this case, under a mild continuity condition, the measures on $\tilde{y}$ and x will then converge to one another in total variation as m increases. Furthermore, since $\tilde{Q}$ is a garbling of Q, the performance under Q is no less than the performance under $\tilde{Q}$. Therefore, we conclude that for optimal decision making, total variation convergence, and hence continuity, holds. We state this formally in the following.

Theorem 8.5.7 *Let $Q_m(dy) = \sum_{k=1}^{2m-1}(k + \frac{1}{2})2^{-m} 1_{\{k2^{-m} \leq y < (k+1)2^{-m}\}}$ be a sequence of de-garbling channels. Suppose that $Q(dy|x) = f(x, y)\lambda(dy)$ for* *some reference measure* λ *and density function* $f(x, y)$ *where* $f(x, \cdot)$ *is continuous. Let* QQ_m *be a composite garbled channel defined by*

$$QQ_m(dy|x) = \int Q_m(dy'|y)Q(dy|x)$$

Then, under assumption A2, we have $\lim_{m\to\infty} J(P, QQ_m) = J(P, Q)$.

Proof It follows that, by adding a further channel, $\tilde{Q}_m(dy'|\tilde{y})$ which distributes y' uniformly over $[k2^{-m}, (k+1)2^{-m})$, we have that

$$QQ_m\tilde{Q}_m(dy|x) = \int \tilde{Q}_m(dy'|\tilde{y})Q_m(dy'|y)Q(dy|x) \to Q(dy|x)$$

in total variation, and hence by Lemma 8.2.1, we have convergence in total variation. Therefore, we have that $\lim_{m\to\infty} J(P, QQ_m\tilde{Q}_m) = J(P, Q)$. On the other hand, we have, by garbling,

$$J(P, Q) \le J(P, QQ_m) \le J(P, QQ_m\tilde{Q}_m)$$

which then establishes the result. $\diamond$

8.6 The Multi-Stage Case

We now consider the general stochastic control problem with T stages (as in (7.2) in Sect. 7.1). It should be noted that the effect of a control policy applied at any given time-stage presents itself in two ways, in the cost incurred at the given time-stage and the effect on the process distribution (and hence, the estimation error at the controller regarding the true state of the system) at future time-stages. This is known as the dual effect of control [66].

The next theorem establishes the continuity of the optimal cost in the measurement channel under some regularity conditions. Note that the existence of best and worst channels follows under an appropriate compactness condition as in Theorem 8.4.1 (iii). We first need the following definition.

Definition 8.6.1 A sequence of channels $\{Q_n\}$ converges to a channel Q *uniformly* in total variation if

$$\lim_{n\to\infty} \sup_{x\in\mathbb{X}} \left\|Q_n(\cdot\,|x) - Q(\cdot\,|x)\right\|_{TV} = 0.$$

Note that in the special but important case of noise-additive measurement channels, uniform convergence in total variation is equivalent to the weaker condition that $Q_n(\cdot\,|x) \to Q(\cdot\,|x)$ in total variation for each x. When the additive noise is absolutely continuous with respect to the Lebesgue measure, uniform convergence in total variation is equivalent to requiring that the noise density corresponding to Q_n converges in the L_1 norm to the density corresponding to Q. For example, if the noise density is estimated from n independent observations using any of the L_1 consistent density estimates described in e.g. [212], then the resulting Q_n will converge (with probability one) uniformly in total variation.

Theorem 8.6.1 ([828]) *Consider the cost function (7.2) with arbitrary $T \in \mathbb{N}$. Suppose that Assumption A2 holds. Then, the optimization problem is continuous in the observation channel in the sense that if $\{Q_n\}$ is a sequence of channels converging to Q uniformly in total variation, then*

$$\lim_{n\to\infty} J(P, Q_n) = J(P, Q).$$

Proof Let $\epsilon > 0$ and pick ϵ-optimal policies $\underline{\gamma}^n = \{\gamma_0^n, \gamma_1^n, \dots, \gamma_{T-1}^n\}$ and $\underline{\gamma} = \{\gamma_0, \gamma_1, \dots, \gamma_{T-1}\}$ for channels Q_n and Q, respectively. That is, using the notation in (7.2), we have $J(P, Q_n, \underline{\gamma}^n) < J(P, Q_n) + \epsilon$ and $J(P, Q, \underline{\gamma}) < J(P, Q) + \epsilon$. The argument used to obtain (8.6) gives

$$\begin{aligned}&|J(P, Q) - J(P, Q_n)|\\ &\quad < \max\Big(J(P, Q, \underline{\gamma}^n) - J(P, Q_n, \underline{\gamma}^n), J(P, Q_n, \underline{\gamma}) - J(P, Q, \underline{\gamma})\Big) + \epsilon.\end{aligned}$$

We will next show that both terms in the maximum above converge to zero. First we consider the term

$$J(P, Q^n, \underline{\gamma}^n) - J(P, Q, \underline{\gamma}^n) = \sum_{t=0}^{T-1} E_P^{Q^n, \underline{\gamma}^n}[c(x_t, u_t)] - E_P^{Q, \underline{\gamma}^n}[c(x_t, u_t)]. \tag{8.8}$$

Under policy $\underline{\gamma}^n = \{\gamma_0^n, \gamma_1^n, \dots, \gamma_{T-1}^n\}$, we have $u_t = \gamma_t^n(y_{[0,t]}, u_{[0,t-1]})$. We suppress in the notation the dependence of u_t on $\gamma_0^n, \dots, \gamma_{t-1}^n$ and write $u_t = \gamma_t^n(y_{[0,t]})$.

For $t = 0, \dots, T-1$, and $k = 0, \dots, t$, define $\zeta_{k,t}^n : \mathbb{X}^k \times \mathbb{Y}^k \to \mathbb{R}$ by setting

$$\zeta_{t,t}^n(x_{[0,t]}, y_{[0,t]}) := c(x_t, \gamma_t^n(y_{[0,t]}))$$

and defining recursively for $k = t-1, \dots, 0$,

$$\begin{aligned}&\zeta_{k,t}^n(x_{[0,k]}, y_{[0,k]})\\ &:= \int P(dx_{k+1}|x_k, \gamma_k^n(y_{[0,k]}))Q_n(dy_{k+1}|x_{k+1})\zeta_{k+1,t}^n(x_{[0,k+1]}, y_{[0,k+1]}).\end{aligned}$$

Note that $\|\zeta_{t,t}^n\|_\infty \le \|c\|_\infty$ and thus $\|\zeta_{k,t}^n\|_\infty \le \|c\|_\infty$ for all $k = t-1, \dots, 0$.

Fix $0 \le k \le t$ and consider a system such that the observation channel is Q at stages $0, \dots, k-1$ and Q_n at stages $k, k+1, \dots, t$. Let μ_k^n denote the distribution of the resulting process segment $(x_{[0,k]}, y_{[0,k]})$ under policy $\underline{\gamma}^n$ (by definition $\mu_0^n = PQ_n$). Also under policy $\underline{\gamma}^n$, let ν_k^n denote the distribution of $(x_{[0,k]}, y_{[0,k]})$ if the observation channel is Q for all the stages $0, \ldots, t$. Then, we have

$$E_P^{Q^n,\underline{\gamma}^n}[c(x_t, u_t)] = \int \mu_0^n(dx_0, dy_0)\zeta_{0,t}^n(x_0, y_0)$$

and

$$E_P^{Q,\underline{\gamma}^n}[c(x_t, u_t)] = \int \nu_t^n(dx_{[0,t]}, dy_{[0,t]})\zeta_{t,t}^n(x_{[0,t]}, y_{[0,t]}).$$

Note that by construction, for all $k = 1, \ldots, t$,

$$\int \mu_k^n(dx_{[0,k]}, dy_{[0,k]})\zeta_{k,t}^n(x_{[0,k]}, y_{[0,k]})$$
$$= \int \nu_{k-1}^n(dx_{[0,k-1]}, dy_{[0,k-1]})\zeta_{k-1,t}^n(x_{[0,k-1]}, y_{[0,k-1]}).$$

Thus each term in the sum on the right hand side of (8.8) can be expressed as a telescopic sum, which in turn can be bounded term-by-term, as follows:

$$\begin{aligned}
&\left|E_P^{Q^n,\underline{\gamma}^n}[c(x_t, u_t)] - E_P^{Q,\underline{\gamma}^n}[c(x_t, u_t)]\right| \\
&= \left|\sum_{k=0}^{t} \int \mu_k^n(dx_{[0,k]}, dy_{[0,k]})\zeta_{k,t}^n(x_{[0,k]}, y_{[0,k]}) \right.\\
&\qquad \left. - \int \nu_k^n(dx_{[0,k]}, dy_{[0,k]})\zeta_{k,t}^n(x_{[0,k]}, y_{[0,k]})\right| \\
&\leq \sum_{k=1}^{t} \|\mu_k^n - \nu_k^n\|_{TV}\|\zeta_{k,t}^n\|_\infty \leq \|c\|_\infty \sum_{k=1}^{t} \|\mu_k^n - \nu_k^n\|_{TV}.
\end{aligned} \tag{8.9}$$

For any Borel set $B \subset \mathbb{X}^{k+1} \times \mathbb{Y}^{k+1}$, define $B(x_{[0,k]}, y_{[0,k-1]}) = \{y_k \in \mathbb{Y} : (x_{[0,k]}, y_{[0,k]}) \in B\}$, so that

$$\begin{aligned}
|\mu_k^n(B) - \nu_k^n(B)| &= \left|\int \nu_{k-1}^n(dx_{[0,k-1]}, dy_{[0,k-1]}) \int P(dx_k|x_{k-1}, \gamma_{k-1}^n(y_{[0,k-1]})) \right.\\
&\qquad \left. \Big(Q_n(B(x_{[0,k]}, y_{[0,k-1]})|x_k) - Q(B(x_{[0,k]}, y_{[0,k-1]})|x_k)\Big)\right| \\
&\leq \sup_{x_k \in \mathbb{X}} \|Q_n(\cdot\,|x_k) - Q(\cdot\,|x_k)\|_{TV}.
\end{aligned}$$

The preceding bound and the uniform convergence of $\{Q_n\}$ imply $\lim_n \|\mu_k^n - \nu_k^n\|_{TV} = 0$ for all k. Combining this with (8.9) and (8.8) yields

$$J(P, Q^n, \underline{\gamma}^n) - J(P, Q, \underline{\gamma}^n) \to 0.$$

Replacing $\underline{\gamma}^n$ with $\underline{\gamma}$ we can use an identical argument to show that $J(P, Q^n, \underline{\gamma}) \to J(P, Q, \underline{\gamma})$. Since $\epsilon > 0$ in (8.8) was arbitrary, the proof is complete. ◇

We obtained the continuity of the optimal cost on the space of channels equipped with a more stringent notion for convergence in total variation. This result and its proof indicate that further technical complications arise in multi-stage problems. Likewise, upper semi-continuity under weak convergence and setwise convergence require more stringent uniformity assumptions.

On the other hand, the concavity property applies directly also to the multi-stage case. That is, $J(P, Q)$ is concave in the space of channels; the proof of this result follows that of Theorem 7.3.1.

Remark 8.6.1 One related approach regarding the multi-stage case is to consider adaptive observation channels. For example, one may aim to design optimal adaptive quantizers for a control problem. In this case, Markov Decision Process tools can be used for obtaining existence conditions for optimal channels and quantizers. This approach will be adopted in Chap. 15 and structural as well as existence results will be presented for optimal policies. ◇

8.7 Continuity and Robustness to the Prior Probability Model

In the chapter so far, we have considered the setting where the prior P was fixed and the channels were part of the design. One could also consider the case when the prior is not specified *a priori*, and one can even consider the case where the assumed prior for the optimal control design is an incorrect one, raising the question of robustness to incorrect priors and probabilistic models. The next three subsections address these variations around the nominal model.

8.7.1 Single-Stage Setup

We first consider the single-stage setup. The nominal prior measure will be denoted by $P \in \mathcal{P}(\mathbb{X})$, the measurement channel will be $Q \in \mathcal{Q}$ and the control action will be from the admissible control policies, Γ. Here, Γ is just the set of functions from $\mathbb{Y}$ to $\mathbb{U}$ that are measurable with respect to the σ-algebra generated by $I_0 = \{Y_0\}$. The optimal single stage cost will be defined as,

$$J^*(P, Q) = \inf_{\gamma \in \Gamma} \int_{\mathbb{X} \times \mathbb{Y}} c(x, \gamma(y)) Q(dy|x) P(dx).$$

By Theorem 8.3.1, an optimal γ exists for any given prior.

Continuity Properties Under Weak Convergence

The first result below shows that $J^*(P, Q)$ is not necessarily continuous in prior measures under weak convergence. We will then present an upper semi-continuity result and some sufficiency continuity conditions for weak convergence in prior measures.

Theorem 8.7.1 ([388]) *Given a channel $Q \in \mathcal{Q}$, $J^*(P, Q)$ is not necessarily continuous in P under weak convergence. This holds even when $\mathbb{X}$, $\mathbb{Y}$, and $\mathbb{U}$ are compact and when c is bounded and continuous on $\mathbb{X} \times \mathbb{U}$.*

The proof follows from a counterexample [388]. The theorem below says that if we place some continuity restrictions on the measurement channel, then the continuity of optimal single stage cost function under weak convergence of prior measures can be guaranteed. We first need an assumption on continuity of Q.

Assumption 8.7.1 *Q is continuous in total variation in the sense that as $x_n \to x$, $\|Q(dy|x_n) - Q(dy|x)\|_{TV} \to 0$.*

Theorem 8.7.2 ([388]) *Let a cost function $c : \mathbb{X} \times \mathbb{U} \to \mathbb{R}$ be given and let Assumption 8.7.1 hold. Further, let $c(x, u)$ be bounded and continuous on $\mathbb{X} \times \mathbb{U}$, and $\mathbb{U}$ be compact. If $P_n \to P$ weakly, then $J^*(P_n, Q) \to J^*(P, Q)$.*

The proof of this result is presented in the appendix. We note here that a directly related result due to Wu and Verdú [787] establishes continuity of the MMSE error (that is with $c(x, u) = \|x - u\|^2$) under weak convergence when the channel is additive, the additive noise has a finite variance and it admits a continuous and bounded density function (this leads to a channel which satisfies Assumption 8.7.1); recall that this result was utilized in the analysis of (3.11). In this case, the regularity conditions of optimal γ (which satisfies smoothness properties) for a given prior allows one to eliminate the boundedness condition on c. These results should also be interpreted in view of the analysis in Sect. 8.3.4.

Continuity Properties Under Total Variation Convergence

The proof of the following result builds on arguments similar to those used in the proof of Theorem 8.3.6.

Theorem 8.7.3 *Let $c(x, u)$ be bounded, continuous on $\mathbb{U}$ for every x, and $\mathbb{U}$ be compact. Then, the optimal cost $J^*(P, Q)$ is continuous on the set of input distributions, $\mathcal{P}(\mathbb{X})$, under the topology of total variation. In other words, if $\|P_n - P\|_{TV} \to 0$, then $|J^*(P_n, Q) - J^*(P, Q)| \to 0$.*

8.7.2 Robustness to Incorrect Priors and Mismatch Bounds for Single Stage Problems

Consider the single stage stochastic control problem with cost function c and initial distribution P, which we denote by $\Xi = (c, P)$. Consider a variation of this problem, with the only difference being that now the initial distribution is $\widetilde{P}$. The DM DM computes an optimal policy, $\widetilde{\gamma}^*$, for the problem $\widetilde{\Xi} := (c, \widetilde{P})$ and applies it to Ξ. The question we address is whether it is possible to approximate the loss in performance, that is, whether we can bound $J(P, Q, \widetilde{\gamma}^*) - J^*(P, Q)$? Such a situation naturally arises when the initial distribution, P, is uncertain, and the DM has a prior belief, $\widetilde{P}$, which is perhaps based on an incorrect initial model.

Proposition 8.7.1 ([388]) *Assume that $c : \mathbb{X} \times \mathbb{U} \to \mathbb{R}$ is nonnegative, measurable, and bounded. Let $\widetilde{\gamma}^*$ be an optimal (or ε-optimal) control policy for the single stage stochastic control problem $\widetilde{\Xi} = (c, \widetilde{P}, Q)$, where c is a cost function, $\widetilde{P}$ is an initial distribution, and Q is a measurement channel. Let P be another probability distribution on $\mathbb{X}$. Then, $|J(P, Q, \widetilde{\gamma}) - J^*(P, Q)| \leq 2\|c\|_\infty \|P - \widetilde{P}\|_{TV}$.*

Proposition 8.7.2 ([388]) *Let an initial distribution $P \in \mathcal{P}(\mathbb{X})$, and a cost function $c : \mathbb{X} \times \mathbb{U} \to \mathbb{R}$ be given. Assume that $P_n \to P$ weakly and let $\gamma^*_{P_n}$ be an optimal (or ε-optimal) policy for the control problem $\Xi_n = (c, P_n)$. Then, it is not necessarily true that $J(P, Q, \gamma^*_{P_n}) \to J^*(P, Q)$ as $n \to \infty$. This result holds even if c is bounded and continuous in both x and u.*

We next present a positive result for weak convergence.

Theorem 8.7.4 ([388]) *Suppose that $c(x, u)$ is bounded, $\mathbb{U}$ is compact and Assumption 8.7.1 holds. Then, as $P_n \to P$ weakly, $\lim_{n\to\infty} |J(P, Q, \gamma^*_{P_n}) - J^*(P, Q)| \to 0$, that is the system is robust to errors in the priors under weak convergence.*

8.7.3 Continuity and Robustness to Incorrect Priors and Mismatch Bounds for Multi-Stage Problems

For multi-stage problems, the dynamics of the model and the informativeness of measurements should be considered while studying the effects of changes in the priors. We consider the following cost criteria:

1. Infinite-horizon discounted cost

$$J_\beta(\mu, \gamma) = E^{\mu,\gamma}\left[\sum_{i=0}^{\infty} \beta^i c(x_i, u_i)\right] \quad \beta \in [0, 1)$$

2. Infinite-horizon average cost

$$J_\infty(\mu,\gamma) = \lim\sup_{T\to\infty} \frac{1}{T} E^{\mu,\gamma}\left[\sum_{i=0}^{T-1} c(x_i,u_i)\right]$$

For a given prior μ, consider the optimal costs:

$$J_\beta^*(\mu) = \inf_{\gamma\in\Gamma} J_\beta(\mu,\gamma), \qquad J_\infty^*(\mu) = \inf_{\gamma\in\Gamma} J_\infty(\mu,\gamma)$$

An optimal control policy γ^μ for a given prior μ is the control policy which achieves the infimum of the expected cost over all admissible control policies,

$$J_\beta(\mu,\gamma^\mu) = \inf_{\gamma\in\Gamma} J_\beta(\mu,\gamma) = J_\beta^*(\mu)$$

Continuity Under conditions given in Theorem E.3.2 in the Appendix, we have that the optimal cost is continuous in the priors.

Robustness In accordance with the earlier discussion on robustness, consider again the situation where the controller falsely thinks the prior of a system is ν, when in reality it is μ. Then the controller will implement the policy γ^ν which is optimal with respect to ν but incurs an expected cost of $J_\beta(\mu,\gamma^\nu)$. If the controller had utilized the correctly designed policy, the cost would have been $J_\beta^*(\mu)$. In accordance with the earlier robustness analysis, the criteria of interest are the differences

$$J_\beta(\mu,\gamma^\nu) - J_\beta^*(\mu) \tag{8.10}$$

$$J_\infty(\mu,\gamma^\nu) - J_\infty^*(\mu) \tag{8.11}$$

for the discounted cost problem and the average cost problem, respectively.

Theorem 8.7.5 ([388, Theorem 3.2][522, Theorem 3.8]) *Assume the cost function c is bounded, non-negative, and measurable. Let γ^ν be the optimal control policy designed with respect to a prior ν. For both $A=\beta$ or $A=\infty$, we have*

$$J_A(\mu,\gamma^\nu) - J_A^*(\mu) \le 2\frac{\|c\|_\infty}{1-\beta}\|\mu-\nu\|_{TV} \tag{8.12}$$

For the multi-stage setting, unlike the single-state setting, access to data can help the controller overrule an incorrect initial prior; this property is known as filter stability. [522] has introduced the following stochastic non-linear observability definition, under which filter stability applies.

Definition 8.7.1 (Stochastic Observability for Non-Linear Systems [522]) A POMDP is called one step observable (universal in admissible control policies) if for every $f \in C_b(\mathcal{X})$ and every $\epsilon > 0$ there exists a measurable and bounded function g such that

$$\|f(\cdot) - \int_{\mathcal{Y}} g(y)Q(dy|\cdot)\|_{\infty} < \epsilon \tag{8.13}$$

Under such an observability condition, and its more relaxed variations, [522] (see also [521]) has shown that the effect of an incorrect initial prior in fact may be completely absent for the average cost problem. Furthermore, robustness bounds are also applicable under weak convergence, provided that the channel is continuous under total variation (Assumption 8.7.1) [388, Theorem 3.3].

8.8 Multi-Agent Setting

In this section, we consider the multi-agent case. For a succinct presentation, we will consider changes in both the priors and the channels, and view the entire measure on the intrinsic variables as the information structure.

Consider $N \in \mathbb{Z}_{\geq 1}$ agents in a static team problem. Let $\mathbb{X}$ denote a standard Borel state space, with $x \sim \zeta$ an $\mathbb{X}$-valued random variable, which is the state of nature where ζ is the prior distribution on x, and which is common knowledge to all agents. We denote by $\mathbb{Y}^i$ agent i's standard Borel measurement space, for $i \in \{1, \ldots, N\}$. For each agent, we define their measurement y^i, which is a $\mathbb{Y}^i$-valued random variable, as $y^i = g^i(x, v^i)$ for some noise variable v^i (and which, without any loss of generality, can be taken to be $[0, 1]$-valued). By stochastic realization arguments, the above formulation is equivalent to viewing y^i as being generated by a measurement channel Q^i, which is a Markov transition kernel from $\mathbb{X}$ to $\mathbb{Y}^i$.

In this section, we will refer to the joint probability measure on $\mathbb{X} \times \mathbb{Y}^1 \times \cdots \times \mathbb{Y}^N$ by $\mu(dx, dy^1, \ldots, dy^N)$, and define this as the *information structure*.

For $\zeta \in \mathcal{P}(\mathbb{X})$ and kernel Q, we let ζQ denote the joint distribution induced on $(\mathbb{X} \times \mathbb{Y}, \mathcal{B}(\mathbb{X} \times \mathbb{Y}))$ by channel Q with input distribution ζ:

$$\zeta Q(A) = \int_A Q(dy|x)\zeta(dx), \quad A \in \mathcal{B}(\mathbb{X} \times \mathbb{Y}).$$

Each agent also has a standard Borel action space $\mathbb{U}^i$, and the team cost function $c : \mathbb{X} \times \mathbb{U}^1 \times \cdots \times \mathbb{U}^N \to \mathbb{R}$. The agent's goal is to minimize the expected cost functional:

$$J^i(\mu, \gamma^1, \ldots, \gamma^N) := E^{\mu, \bar{\gamma}}[c^i(x, \gamma^1(y^1), \ldots, \gamma^N(y^N))]$$

where the agents select their policies from the set of all admissible *policies* $\Gamma^i := \{\gamma : \mathbb{Y}^i \to \mathbb{U}^i\}$, which are measurable functions from an agent's measurement space to its action space. We refer to $u^i = \gamma^i(y^i)$ as the action of the agent, and γ^i as its policy.

Assumption 8.8.1

(A1) The cost function c is measurable and bounded.
(A2) The cost function c is continuous and bounded.
(A3) The information structure is absolutely continuous with respect to a product measure:

$$P(dy^1, dy^2, \cdots, dy^N, dx) \ll \bar{Q}^1(dy^1)\bar{Q}^2(dy^2)\cdots \bar{Q}^N(dy^N)\zeta(dx),$$

for reference probability measures $\bar{Q}^i$, $i = 1, 2, \cdots$. That is, e.g., for $N = 2$, there exists an integrable f which satisfies for every Borel A, B, C,

$$P(y^1 \in B, y^2 \in C, x \in A) = \int_{A,B,C} f(x, y^1, y^2)\zeta(dx)\bar{Q}^1(dy^1)\bar{Q}^2(dy^2).$$

(A4) The action space of each agent is compact.
(A5) The bounded measurable cost function is continuous in agents' actions for every state of nature x.
(A6) The individual agent channels Q^i are fixed and independent (i.e., given x, all the measurement variables y^i are conditionally independent). Furthermore, each Q^i is continuous in total variation in the sense of Assumption 8.7.1, i.e., as $x_m \to x$, then $\|Q^i(\cdot|x) - Q^i(\cdot|x_m)\|_{TV} \to 0$.
(A7) The action spaces for the agents are convex subsets of $\mathbb{R}^k$, for some k.

The induced cost under a collection of policies $\gamma^1, \cdots, \gamma^N$ for the team problem is given by

$$J(c, \mu, \bar{\gamma}) =: \int c(x, \gamma^1(y^1), \ldots, \gamma^N(y^N))\mu(dx, dy^1, \ldots, dy^N),$$

the infimum of which over all admissible policies is denoted by $J^*(c, \mu)$, which will be referred to also as the value of the information given structure. General existence results for team-optimal policies were established in Chaps. 4 and 5.

In this section, we address the following question: if $\mu_m \to \mu$ under weak, setwise, or total variation convergence, what can be said about the regularity of $J^*(c, \mu_m)$?

Theorem 8.8.1 ([354]) *Consider an N-DM team and let Assumptions 8.8.1-A3, A4 and A5 hold. Then, the optimal expected cost, $J^*(c, \mu)$, is continuous under total variation convergence of information structures.*

Proof An optimal solution exists for any given information structure under the assumptions made (see Theorem 4.2.4 and Sect. 4.2). Let μ and ν be two information structures, and let $\gamma_\mu^{i,*}$ and $\gamma_\nu^{i,*}$ denote agent i's individual strategy as part of the team-optimal strategies $\bar{\gamma}_\mu^*$, $\bar{\gamma}_\nu^*$ under each respective information structure. Let the cost function be bounded in absolute value by M.

Without loss of generality, assume that $J^*(c, \mu) - J^*(c, \nu) \geq 0$. Then,

$$J^*(c, \mu) - J^*(c, \nu) = \int c(x, \gamma_\mu^{1,*}(y^1), \dots, \gamma_\mu^{N,*}(y^N))\mu(dx, dy^1, \dots, dy^N)$$
$$- \int c(x, \gamma_\nu^{1,*}(y^1), \dots, \gamma_\nu^{N,*}(y^N))\nu(dx, dy^1, \dots, dy^N)$$
$$\leq \int c(x, \gamma_\nu^{1,*}(y^1), \dots, \gamma_\nu^{N,*}(y^N))\mu(dx, dy^1, \dots, dy^N)$$
$$- \int c(x, \gamma_\nu^{1,*}(y^1), \dots, \gamma_\nu^{N,*}(y^N))\nu(dx, dy^1, \dots, dy^N) \leq M\|\mu - \nu\|_{TV},$$

where the first inequality holds by perturbing the strategy under μ to be $\bar{\gamma}_\nu^*$, rather than the team-optimal policy $\bar{\gamma}_\mu^*$, and the second inequality holds by the definition of total variation distance. The result then follows. ◇

Theorem 8.8.2 ([354]) *Under Assumptions 8.8.1-A2, A3, A4 and A6, the optimal expected cost, $J^*(c, \mu)$, is continuous under weak convergence of the priors.*

Proof Let ζ_m be a sequence of priors on $\mathbb{X}$ converging weakly to ζ. Without loss of generality, assume that $J^*(c, Q^1 \dots Q^N \zeta) - J^*(c, Q^1 \dots Q^N \zeta_m) \geq 0$. Following the same steps as in the proof of Theorem 8.8.1, we get

$$J^*(c, Q^1 \dots Q^N \zeta) - J^*(c, Q^1 \dots Q^N \zeta_m)$$
$$\leq \int (\zeta - \zeta_m)(dx) \int c(x, \gamma_m^{1,*}(y^1), \dots, \gamma_m^{N,*}(y^N))Q^1(dy^1|x) \dots Q^N(dy^N|x).$$

An equicontinuity argument (as in the discussion following (8.24)) completes the proof. ◇

Theorem 8.8.3 ([354]) *Let Assumptions 8.8.1-A3, A4, A5 and A7 hold. Let μ_m be a sequence of information structures converging weakly to an information structure μ, and let the prior be fixed. Then, the optimal expected cost function $J^*(c, \mu)$ is upper semicontinuous in μ under weak convergence. If the prior is not fixed, and if we impose continuity in x also (under Assumption 8.8.1-A2), the optimal expected cost function is upper semicontinuous.*

Proof Let $\bar{\Gamma} = \Gamma^1 \times \cdots \times \Gamma^N$. Let $\mu_{\mathbb{Y}^j}$ be the marginal of μ on its $(j+1)$th component. Let γ^j be an arbitrary policy for **DM**j. The action spaces for DMs are convex by Assumption A7. Let $M = \|c\|_\infty$. Then, by Lusin's theorem (see Theorem C.5.1), for any $\epsilon > 0$, there exists a continuous function $f^j : \mathbb{Y}^j \to \mathbb{U}^j$ such that $\mu_{\mathbb{Y}^j}(\{y^j : f^j(y^j) \neq \gamma^j(y^j)\}) < \epsilon$.

Letting B^j denote the set $B^j = \{y^j : f^j(y^j) \neq \gamma^j(y^j)\}$, and proceeding with this for $j = 1, \dots, N$, we have:

$$\int c(x, f^1(y^1), \ldots, f^N(y^N))\mu(dx, dy^1, \ldots, dy^N)$$
$$< \int c(x, \gamma^1(y^1), \ldots, \gamma^N(y^N))\mu(dx, dy^1, \ldots, dy^N) + NM\epsilon. \qquad (8.14)$$

Denote by Γ_C^j the policy space for **DM**j in which the agent's policy is continuous from $\mathbb{Y}^j$ to $\mathbb{U}^j$. We note that the value for a team problem will be the same regardless of whether the DMs use policies out of Γ or Γ_C: That the value must be lower when Γ is used is a consequence of $\Gamma_C \subset \Gamma$. That the value must be lower under Γ_C is a consequence of (8.14), since the γ^j policies, as well as ϵ, were arbitrary. Thus, in view of the fact that μ_m converges weakly to μ:

$$\lim \sup_{m\to\infty} \inf_{\bar{\gamma}\in\bar{\Gamma}} \int c(x, \gamma^1(y^1), \ldots, \gamma^N(y^N))\mu_m(dx, dy^1, \ldots, dy^N)$$
$$= \lim \sup_{m\to\infty} \inf_{\bar{\gamma}_C\in\bar{\Gamma}_C} \int c(x, \gamma_C^1(y^1), \ldots, \gamma_C^N(y^N))\mu_m(dx, dy^1, \ldots, dy^N)$$
$$\leq \inf_{\bar{\gamma}_C\in\bar{\Gamma}_C} \lim \sup_{m\to\infty} \int c(x, \gamma_C^1(y^1), \ldots, \gamma_C^N(y^N))\mu_m(dx, dy^1, \ldots, dy^N)$$
$$= \inf_{\bar{\gamma}_C\in\bar{\Gamma}_C} \int c(x, \gamma_C^1(y^1), \ldots, \gamma_C^N(y^N))\mu(dx, dy^1, \ldots, dy^N)$$
$$= \inf_{\bar{\gamma}\in\bar{\Gamma}} \int c(x, \gamma^1(y^1), \ldots, \gamma^N(y^N))\mu(dx, dy^1, \ldots, dy^N).$$

In the inequality above we do not need continuity of c in x since weak convergence of a product measure with a fixed marginal is equivalent to w-s convergence of the joint measure, by Theorem C.4.1.

If the prior is not fixed, then we impose continuity in x also under Assumption 8.8.1-A2. ◇

Theorem 8.8.4 ([354]) *Let Assumption 8.8.1-A3, A4 and A5 hold. Let μ_m be a sequence of information structures converging setwise to an information structure μ. Then the value function $J^*(c, \mu)$ is upper semicontinuous in μ under setwise convergence.*

Proof The proof closely follows that of Theorem 8.8.3, but Lusin's theorem is not invoked in this case. By interchanging the limit superior and the infimum, and then applying the fact that $\{\mu_m\} \to \mu$ setwise, we get:

$$\lim \sup_{m\to\infty} \inf_{\bar{\gamma}\in\bar{\Gamma}} \int c(x, \gamma^1(y^1), \ldots, \gamma^N(y^N))\mu_m(dx, dy^1, \ldots, dy^N)$$
$$\leq \inf_{\bar{\gamma}\in\bar{\Gamma}} \lim \sup_{m\to\infty} \int c(x, \gamma^1(y^1), \ldots, \gamma^N(y^N))\mu_m(dx, dy^1, \ldots, dy^N)$$
$$= \inf_{\bar{\gamma}\in\bar{\Gamma}} \int c(x, \gamma^1(y^1), \ldots, \gamma^N(y^N))\mu(dx, dy^1, \ldots, dy^N).$$

◇

8.9 All Stochastic Dynamic Teams Are Nearly Static Reducible via Additive Perturbation of Information Structures

In this section, we address the following question: Is it possible to perturb any discrete-time sequential (single-agent or multi-agent) stochastic control problem by adding some arbitrarily small additive (e.g. Gaussian or otherwise) noise to the measurements to make the measurement kernels absolutely continuous with respect to a fixed reference measure, so that a change-of-measure (also known as a static reduction) can be made with arbitrarily small error in the optimal cost? That is, are all sequential stochastic decision problems ϵ-away from being static reducible as far as optimal cost is concerned, for any $\epsilon > 0$? We will show that this is indeed the case when the cost function is bounded and continuous in agents' actions and the action spaces are convex.

We note that such a result is not direct and in fact surprising given the general lack of continuity presented in Theorem 8.3.2 in the space of information structures under weak convergence (which is indeed the notion considered in our positive result when we will, under weak convergence, approach a Dirac delta measure by a sequence of Gaussian measures with decreasing variances) and this negative result was also shown in Theorem 8.3.5 to hold even under setwise perturbations of measurement channels. To obtain our positive result in this section, we will consider a cautious construction where Lusin's theorem will be utilized twice, first to approximate functions and second to approximate stochastic kernels with their continuous and weakly continuous versions, respectively.

We will follow the setup presented in Sect. 3.2. Recall that for a dynamic team problem the probability measure on the measurements are policy-dependent, as opposed to the static case. Recall also (4.7) which defines the measurement kernel.

For the purpose of this section, we will assume $\mathbb{Y}^i \subset \mathbb{R}^{k_i}$, for some $k_i \in \mathbb{Z}_{\geq 1}$, for every $i \in \mathcal{N}$, and we will require admissible policies be defined from $\mathbb{R}^{k_i}$ to $\mathbb{U}^i$. For our analysis, we will allow for an agent's measurements through the agent's channel p_i to be garbled by an additive Gaussian noise channel τ_m^i, which provides the DM with measurement $z^i = y^i + q_m^i$, where $q_m^i \in \mathbb{R}^{k_i}$ is a vector of independent identically distributed $\mathbb{R}$-valued Gaussian noise with mean 0 and variance $1/m$. This is the measurement that the DM then uses when selecting its action $u^i = \gamma^i(z^i)$ and becomes the measurement that can affect future DMs' measurements through their channels $p_{i+1}, \ldots, p_N$. We denote the DM channels under the garbled setup by $p_1\tau_m^1, \ldots, p_N\tau_m^N$. We denote the team-optimal cost (i.e. the expected cost achieved under optimal selection of DM policies) under this setup with $J^*(P, p_1\tau_m^1, \ldots, p_N\tau_m^N)$, and the team-optimal cost under the original setup with $J^*(P, p_1, \ldots, p_N)$.

We now introduce the following assumptions (where actually only A1 and A2 will be needed for our main result):

Assumption 8.9.1 *Consider the following:*

(A1) The cost function c is continuous and bounded.
(A2) Each **DM***i's action space* $\mathbb{U}^i$ *is a convex subset of* $\mathbb{R}^{n_i}$*, for some* $n_i \in \mathbb{Z}_{\geq 1}$.
(A3) Each **DM***i's measurement channel given in (4.7) is weakly continuous in the following sense: if* $(z_m^1, u_m^1, \ldots, z_m^{i-1}, u_m^{i-1}) \to (z^1, u^1, \ldots, z^{i-1}, u^{i-1})$*, then*

$$p_i(y^i \in \cdot|\omega_0, z_m^1, u_m^1, \ldots, z_m^{i-1}, u_m^{i-1}) \to p_i(y^i \in \cdot|\omega_0, z^1, u^1, \ldots, z^{i-1}, u^{i-1})$$

weakly for all $\omega_0 \in \Omega_0$.

We will show that under Assumption 8.9.1(A1) and (A2)

$$\lim_{m\to\infty} J^*(P, p_1\tau_m^1, \ldots, p_N\tau_m^N) = J^*(P, p_1, \ldots, p_N).$$

From this, it follows that any sequential dynamic team problem is ϵ-away in optimal cost from being static reducible.

First, an auxiliary result will be stated and proved under the additional assumption Assumption 8.9.1(A3), in Theorem 8.9.1. Then, this result will be relaxed to include measurement channels that do not necessarily satisfy Assumption 8.9.1(A3) in Theorem 8.9.2, which is the main result of this section.

Our approach is related to perturbing information structures under the weak convergence topology and garbling of information structures. A minor variation of the proof of Theorem 7.3.3 leads to the following:

Lemma 8.9.1 *For sequential dynamic team problems, if each agent's information under a set of measurement channels* $(p_1, \ldots, p_N)$ *is a garbling of their information under another set of measurement channels* $(\tilde{p}_1, \ldots, \tilde{p}_N)$ *then the team-optimal expected cost cannot be strictly lower under* $(p_1, \ldots, p_N)$ *than under* $(\tilde{p}_1, \ldots, \tilde{p}_N)$.

By garbling, as we studied earlier in Chap. 7, we mean that each agent's measurement under $(p_1, \ldots, p_N)$ is such that $y^i_{p_i} = g^i(y^i_{\tilde{p}_i}, v_i)$ for some measurable g^i and independent noise v_i, for all $i \in \mathcal{N}$, where $y^i_{\tilde{p}_i}$ is the measurement under the original channel $\tilde{p}_i$.

Next, we present a supporting result whose proof will be utilized for the main result of this section, Theorem 8.9.2, to be presented immediately after.

Theorem 8.9.1 *[355] Assume that Assumption 8.9.1(A1, A2, and A3) hold. For any* $\epsilon > 0$*, any sequential dynamic team problem is* ϵ*-away in optimal expected cost from being static reducible.*

Theorem 8.9.2 *[355] Assume that Assumption 8.9.1(A1 and A2) hold. For any* $\epsilon > 0$*, any sequential dynamic team problem is* ϵ*-away in optimal expected cost from being static reducible.*

We note that the proofs of the theorems above are also constructive: the solution and the expected cost obtained for the perturbed system are realizable (under a randomized policy) for the original model. Detailed proofs for both theorems are

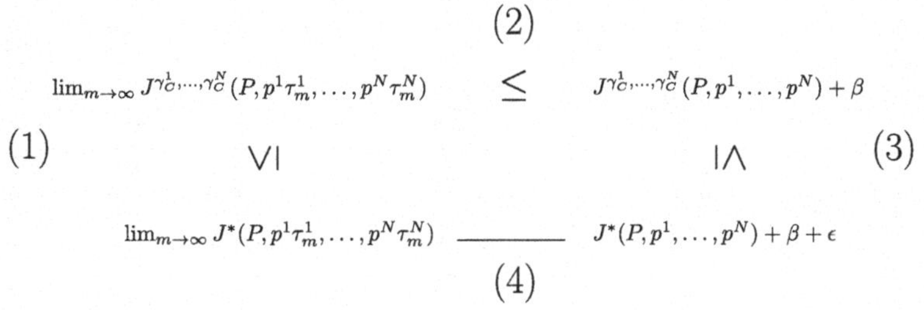

Fig. 8.1 Arguments of the Proof of Theorem 8.9.2

presented in the Appendix. For the convenience of the reader, we provide below a proof program for Theorem 8.9.2 in the following.

Proof Program The proof will first use Lusin's theorem to construct continuous policies $\gamma_C^1, \ldots, \gamma_C^N$ for the DMs that approximate their equilibrium policies on a set of $1 - \epsilon_i$, for any $\epsilon_i > 0$. Then, it will be shown using a generalized dominated convergence theorem that the garbled information structure converges weakly to the ungarbled information structure. These steps will then be combined to prove the desired statement for Theorem 8.9.1. The relaxation in Theorem 8.9.2 will use Lusin's theorem an additional time to show that each DM's channel can be approximated by a channel that satisfies Assumption 8.9.1(A3) on a set of measure $1 - \beta_i$, for any $\beta_i > 0$.

Figure 8.1 depicts the general proof program. Here $J^{\bar{\gamma}}(\cdot)$ denotes the expected cost when the policies are given specified by $\bar{\gamma}$. (1) holds because a continuous collection $\gamma_C^1, \ldots, \gamma_C^N$ cannot achieve a cost better than the optimal cost under channels $p_1\tau_m^1, \ldots, p_N\tau_m^N$, for any m. (2) holds by applying Step 2, where $\beta > 0$ is the (arbitrarily small) error term caused by approximating the channels with weakly continuous kernels. (3) holds by Step 1, because $\gamma_C^1, \ldots, \gamma_C^N$ approximate the δ-optimal policies $\gamma^{1,*}, \ldots, \gamma^{N,*}$, except on a set of small measure, resulting in an arbitrarily small error term $\epsilon > 0$. After being defined, $\gamma_C^1, \ldots, \gamma_C^N$ are fixed throughout the proof, and in Step 2, the application of Lusin's theorem to approximate the DM channels uses these fixed continuous policies, as they affect the measures on the DM action spaces. The terms in (4) are related by Lemma 8.9.1, with the relationship being $J^*(P, p_1\tau_m^1, \ldots, p_N\tau_m^N) \geq J^*(P, p_1, \ldots, p_N)$.

8.10 Appendix: Proofs

8.10.1 Proof of Lemma 8.2.1

(i) Since $c(x, \cdot)$ is continuous and bounded on $\mathbb{Y}$ for all x, we have

$$\lim_{n\to\infty}\int_{\mathbb{X}\times\mathbb{Y}} c(x,y)PQ_n(dx,dy) = \lim_{n\to\infty}\int_{\mathbb{X}}\left(\int_{\mathbb{Y}} c(x,y)Q_n(dy|x)\right)P(dx)$$
$$= \int_{\mathbb{X}}\left(\int_{\mathbb{Y}} c(x,y)Q(dy|x)\right)P(dx)$$
$$= \int_{\mathbb{X}\times\mathbb{Y}} c(x,y)PQ(dx,dy)$$

where first we used Fubini's Theorem, and then the dominated convergence theorem (see Appendix A) and the fact that $\int_{\mathbb{X}} c(x,y)Q_n(dy|x)$ is bounded and converges to $\int_{\mathbb{X}} c(x,y)Q(dy|x)$ for P-a.e. x.

(ii) Let $A \in \mathcal{B}(\mathbb{X}\times\mathbb{Y})$ and for x, let $A_x = \{y : (x,y) \in A\}$. As in the previous proof,

$$\lim_{n\to\infty} PQ_n(A) = \lim_{n\to\infty}\int_{\mathbb{X}} Q_n(A_x|x)P(dx)$$
$$= \int_{\mathbb{X}} Q(A_x|x)P(dx) = PQ(A)$$

by the dominated convergence theorem, since $\lim_{n\to\infty} Q_n(A_x|x) = Q(A_x|x)$ for P-a.e. x.

(iii) We have

$$\sup_{A\in\mathcal{B}(\mathbb{X}\times\mathbb{Y})} |PQ_n(A) - PQ(A)|$$
$$= \sup_{A\in\mathcal{B}(\mathbb{X}\times\mathbb{Y})}\left|\int_{\mathbb{X}} Q_n(A_x|x)P(dx) - \int_{\mathbb{X}} Q(A_x|x)P(dx)\right|$$
$$\leq \sup_{A\in\mathcal{B}(\mathbb{X}\times\mathbb{Y})}\int_{\mathbb{X}} \left|Q_n(A_x|x) - Q(A_x|x)\right|P(dx)$$
$$\leq \int_{\mathbb{X}} \sup_{B\in\mathcal{B}(\mathbb{Y})} \left|Q_n(B|x) - Q(B|x)\right|P(dx).$$

Since $\sup_{B\in\mathcal{B}(\mathbb{Y})} \left|Q_n(B|x) - Q(B|x)\right| \to 0$ for P-a.e. x, an application of the dominated convergence theorem completes the proof. ◇

8.10.2 *Proof of Theorem 8.3.1*

We have

$$J(P,Q) = \inf_{\gamma\in\Gamma}\int_{\mathbb{X}\times\mathbb{Y}} c(x,\gamma(y))Q(dy|y)P(dx).$$

Let $(x, y) \sim PQ$ and let $P(\cdot\,|y)$ be the (regular) conditional distribution of x given y. If $(PQ)_{\mathbb{Y}}$ denotes the distribution of y, then

$$J(P, Q) = \inf_{\gamma \in \Gamma} \int_{\mathbb{Y}} \int_{\mathbb{X}} c(x, \gamma(y)) P(dx|y)(PQ)_{\mathbb{Y}}(dy)$$
$$= \int_{\mathbb{Y}} \left(\inf_{u \in \mathbb{U}} \int_{\mathbb{X}} c(x, u) P(dx|y) \right) (PQ)_{\mathbb{Y}}(dy),$$

where the validity of the second equality is explained below.

By assumption *A3*, c is bounded and $c(x, u_n) \to c(x, u)$ if $u_n \to u$ for all x; thus by the dominated convergence theorem

$$\int_{\mathbb{X}} c(x, u_n) P(dx|y) \to \int_{\mathbb{X}} c(x, u) P(dx|y)$$

proving that $g(u, y) = \int_{\mathbb{X}} c(x, u) P(dx|y)$ is continuous in u for each y. Since $\mathbb{U}$ is compact, there exists $\gamma^*(y) \in \mathbb{U}$ such that $g(\gamma^*(y), y) = \inf_{u \in \mathbb{U}} g(u, y)$. A standard argument shows that $\gamma^* : \mathbb{Y} \to \mathbb{U}$ can be taken to be measurable (see, e.g., Appendix D of [342]) and we have

$$J(P, Q) = \int_{\mathbb{X} \times \mathbb{Y}} c(x, \gamma^*(y)) Q(dy|x) P(dx).$$

◇

8.10.3 Proof of Lemma 8.4.3

By Lemma 8.4.1, Ψ is setwise sequentially precompact, and thus any sequence in Ψ has a subsequence $\{P_n\}$ such that $P_n \to P$ setwise for some $P \in \mathcal{P}(\mathbb{R}^N)$. P is clearly absolutely continuous with respect to the Lebesgue measure on $\mathbb{R}^N$, and so it admits a density p.

Let p_n be the density of P_n. It suffices to show that

$$\lim_{n \to \infty} \|p_n - p\|_1 = 0 \tag{8.15}$$

since $\|p_n - p\|_{TV} = 2\|p_n - p\|_1 = 2\int |p_n(x) - p(x)|\, dx$.

Pick a sequence of compact sets $K_j \subset \mathbb{R}^N$ such that $K_j \subset K_{j+1}$ for all $j \in \mathbb{N}$, and $\bigcup_j K_j = \mathbb{R}^N$. Since the collection of densities $\{p_n\}$ is uniformly bounded and equicontinuous, it is precompact in the supremum norm on each K_j by the Arzelà-Ascoli theorem [223]. Thus there exist subsequences $\{p_{n_k^j}\}$ such that

$$\lim_{k \to \infty} \sup_{x \in K_j} |p_{n_k^j}(x) - p^j(x)| = 0$$

for some continuous $p^j : K_j \to [0, \infty)$.

Since the K_j are nested, one can choose $\{p_{n_k^{j+1}}\}$ to be a subsequence of $\{p_{n_k^j}\}$ for all $j \in \mathbb{N}$. Then p^{j+1} coincides with p^j on K_j and we can define $\hat{p}$ on $\mathbb{R}^N$ by setting $\hat{p}(x) = p^j(x)$, $x \in K_j$. We can now use Cantor's diagonal method to pick an increasing sequence of integers $\{m_i\}$ which is a subsequence of each $\{n_k^j\}$, and thus

$$\lim_{i\to\infty} p_{m_i}(x) = \hat{p}(x), \quad \text{for all } x \in \mathbb{R}^N. \tag{8.16}$$

Note that by construction the convergence is uniform on each K_j (and $\hat{p}$ is continuous). By uniform convergence $P_{p_{m_i}}(A) \to P_{\hat{p}}(A)$ for all Borel subsets A of K_j. The setwise convergence of P_n to P_p implies $P_{p_{m_i}}(A) \to P_p(A)$ for all Borel sets, and thus we must have $p = \hat{p}$ almost everywhere. This and (8.16) imply via Scheffé's theorem [117] that

$$\|p_{m_j} - p\|_1 \to 0$$

which completes the proof. ◇

8.10.4 Proof of Theorem 8.5.4

Let $B_1^n, \ldots, B_M^n$ be the cells of Q_n. Since $Q_n \to Q$ setwise at input P, we have $PQ_n(B \times \{i\}) \to PQ(B \times \{i\})$ for any $B \in \mathcal{B}(\mathbb{X})$. Since $PQ_n(B \times \{i\}) = \int_B 1_{\{x\in B_i^n\}} P(dx)$, we obtain

$$P(B \cap B_i^n) \to P(B \cap B_i), \quad \text{for all } i = 1, \ldots, M.$$

If $B_1, \ldots, B_M$ are the cells of Q, the above implies $P(B_j \cap B_i^n) \to P(B_j \cap B_i)$ for all $i, j \in \{1, \ldots, M\}$. Since both $\{B_i^n\}$ and $\{B_n\}$ are partitions of $\mathbb{X}$, we obtain

$$P(B_i^n \triangle B_i) \to 0 \quad \text{for all } i = 1, \ldots, M,$$

where $B_i^n \triangle B = (B_i^n \setminus B) \cup (B \setminus B_i^n)$. Then we have

$$\begin{aligned}
&\|PQ_n - PQ\|_{TV} \\
&= \sup_{f:\|f\|_\infty \le 1} \left| \sum_{i=1}^M \left(\int_{\mathbb{X}} f(x,i) Q_n(i|x) P(dx) - \int_{\mathbb{X}} f(x,i) Q(i|x) P(dx) \right) \right| \\
&= \sup_{f:\|f\|_\infty \le 1} \left| \sum_{i=1}^M \int_{\mathbb{X}} f(x,i) \left(1_{\{x\in B_i^n\}} - 1_{\{x\in B_i\}} \right) P(dx) \right|
\end{aligned}$$

$$\leq \sup_{f:\|f\|_\infty\leq 1} \sum_{i=1}^{M} \int_{B_i^n \triangle B_i} |f(x,i)|P(dx)$$

$$\leq \sum_{i=1}^{M} P(B_i^n \triangle B_i) \to 0 \tag{8.17}$$

and convergence in total variation follows. ◇

8.10.5 *Proof of Theorem 8.7.2*

Proof We first show that under the stated assumptions, the following holds.

$$\lim_{k\to\infty} \sup_{\gamma} \left| \int Q(dy|x_k)c(x_k, \gamma(y)) - \int Q(dy|x)c(x, \gamma(y)) \right| = 0 \tag{8.18}$$

for $x_k \to x$. To see this, write

$$\lim_{k\to\infty} \sup_{\gamma} \left| \int Q(dy|x_k)c(x_k, \gamma(y)) - \int Q(dy|x)c(x, \gamma(y)) \right|$$

$$\leq \lim_{k\to\infty} \sup_{\gamma} \left| \int Q(dy|x_k)c(x_k, \gamma(y)) - \int Q(dy|x)c(x_k, \gamma(y)) \right| \tag{8.19}$$

$$+ \lim_{k\to\infty} \sup_{\gamma} \left| \int Q(dy|x)c(x_k, \gamma(y)) - \int Q(dy|x)c(x, \gamma(y)) \right| \tag{8.20}$$

The term in (8.19) converges to 0 by Assumption (8.7.1), since we have

$$\lim_{k\to\infty} \sup_{\gamma} \left| \int Q(dy|x_k)c(x_k, \gamma(y)) - \int Q(dy|x)c(x_k, \gamma(y)) \right|$$
$$\leq \lim_{k\to\infty} \|Q(.|x_k) - Q(.|x)\|_{TV}$$
$$= 0$$

By assumption the action space $\mathbb{U}$ is compact. Therefore, $c(x_k, u) \to c(x, u)$ uniformly on $\mathbb{U}$. Thus, (8.20) also goes to 0.

Let the optimal policies be given by γ_n^* and γ^* for initial distributions P_n and P respectively. We now consider two cases:
Firstly, if $J^*(P_n, Q) < J^*(P, Q)$ then

$$J^*(P, Q) - J^*(P_n, Q) \leq J(P, Q, \gamma_n^*) - J(P_n, Q, \gamma_n^*). \tag{8.21}$$

Secondly, if $J^*(P, Q) < J^*(P_n, Q)$ then

$$J^*(P_n, Q) - J^*(P, Q) \leq J(P_n, Q, \gamma^*) - J(P, Q, \gamma^*). \tag{8.22}$$

Inequalities (8.21) and (8.22) are combined to give

$$\begin{aligned} &|J^*(P, Q) - J^*(P_n, Q)| \\ &\quad \leq \max(J(P, Q, \gamma_n^*) - J(P_n, Q, \gamma_n^*), J(P_n, Q, \gamma^*) - J(P, Q, \gamma^*)). \end{aligned} \tag{8.23}$$

We now note the following:

$$\begin{aligned} &|J^*(P_n, Q) - J^*(P, Q)| \\ &\quad \leq \max\left(\left|\int (P_n - P)(dx) \int Q(dy|x)c(x, \gamma_n^*(y))\right|,\right. \\ &\qquad \left.\left|\int (P_n - P)(dx) \int Q(dy|x)c(x, \gamma^*(y))\right|\right), \end{aligned} \tag{8.24}$$

For continuity, we need (8.24) to tend to 0 as $n \to \infty$.

Let F be a family of functions from normed linear spaces $\mathbb{S}$ to $\mathbb{T}$. The family F is said to be *equicontinuous at a point* $x_0 \in S$ if, for every $\epsilon > 0$, there exists a $\delta > 0$ such that $|f(x) - f(x_0)| \leq \epsilon$ for all $f \in F$ and for every x such that $|x - x_0| \leq \delta$. The family F is said to be *equicontinuous* if it is equicontinuous at each $x \in S$. A consequence of weak convergence is that if f_n is a uniformly bounded and equicontinuous family, then $P_n \to P$ weakly implies that $\int (P_n - P)(dx) f_n(x) \to 0$ [223, Corollary 11.3.4]. Condition (8.18) ensures that the sequence of functions $\int Q(dy|x_k)c(x_k, \gamma_n(y))$ is equicontinuous as it gives us a uniform continuity over the family of all admissible policies. ◇

8.10.6 Proof of Theorem 8.9.1

Step 1 Let $p_i(y^i \in \cdot|\omega_0, y^1, u^1, \ldots, y^{i-1}, u^{i-1})$ denote each respective DM's measurement channel for $i \in \mathcal{N}$. Let $\tau_m^1, \ldots, \tau_m^N$ be independent additive Gaussian noise channels with variance $1/m$ and mean 0 for each DM. We will use y^i to denote the measurement by **DM***i* through channel p_i, and z^i to denote the corrupted measurement of y^i through additive Gaussian noise channel τ_m^i.

Let $\bar{\gamma}^* = (\gamma^{1,*}, \ldots, \gamma^{N,*})$ be a δ-team-optimal policy (where $\delta > 0$ is arbitrary) given cost function c *when the Gaussian noise channels are not used*. I.e. $\bar{\gamma}^*$ is a policy that achieves a cost of $J^*(P, p_1, \ldots, p_N) + \delta$, or lower. Let μ denote the joint probability measure on $\Omega_0 \times \mathbb{Y}^1 \times \ldots \mathbb{Y}^N$ when the DMs use channels $p_1, \ldots, p_n$, and ν_m denote the measure on $(\Omega_0 \times \mathbb{Y}^1 \times \mathbb{R}^{k_1} \times \ldots \mathbb{Y}^N \times \mathbb{R}^{k_N})$ when the DMs use channels $p_1\tau_m^1, \ldots, p_N\tau_m^N$.

We will now apply Lusin's theorem (Theorem C.5.1) for each DM to allow for the restriction of each DM's policy $\gamma^{i,*}$ to be continuous on a subset of $\mathbb{Y}^i$ of arbitrarily large marginal measure under policy $\bar{\gamma}^*$.

There is now the question of whether changing the policy affects the marginal on the measurements. This will be handled via a total probability argument: For each DM, we will fix all prior DMs' policies as their personal policy from the δ-team-optimal strategy $\bar{\gamma}^*$, while all future DMs' policies can be left as arbitrary measurable functions. We require all prior DMs' policies to be fixed in order to fix the measure on **DM**i's measurement space $\mathbb{Y}^i$, since **DM**i's measurement may depend on previous DMs' actions (and thus policies). Future DMs' policies are irrelevant to the application of Lusin's theorem, as the theorem guarantees that the marginal on the subset of $\mathbb{Y}^i$ where $\gamma^{i,*}$ is not continuous is small, which is unaffected by future DMs' measures on their measurement spaces.

For the first DM, by Lusin's theorem, there exists a set A^1 such that $\gamma^{1,*}$ is continuous on $(\mathbb{Y}^1 \setminus A^1)$ and

$$\int_{\Omega_0 \times A^1 \times \mathbb{Y}^2 \times \cdots \times \mathbb{Y}^N} P(d\omega_0) p_1(dy^1|\omega_0) p_2(dy^2|\omega_0, y^1, f^1(y^1)) \ldots$$
$$p_N(dy^N|\omega_0, y^1, f^1(y^1), \ldots, y^{N-1}, f^{N-1}(y^{N-1})) < \epsilon_1,$$

for any measurable $f^1, \ldots, f^{N-1}$.

For the second DM, for $\epsilon_2 > 0$, given fixed policy $\gamma^{1,*}$ for DM 1, there exists a set A^2 such that $\gamma^{2,*}$ is continuous on $(\mathbb{Y}^2 \setminus A^2)$ and

$$\int_{\Omega_0 \times \mathbb{Y}^1 \times A^2 \times \mathbb{Y}^3 \times \cdots \times \mathbb{Y}^N} P(d\omega_0) p_1(dy^1|\omega_0) p_2(dy^2|\omega_0, y^1, \gamma^{1,*}(y^1)) \ldots$$
$$p_N(dy^N|\omega_0, y^1, \gamma^{1,*}(y^1), y^2, f^2(y^2), \ldots, y^{N-1}, f^{N-1}(y^{N-1})) < \epsilon_2,$$

for any measurable $f^2, \ldots, f^{N-1}$.

Continuing in this way, for the Nth DM, for $\epsilon_N > 0$, given fixed $\gamma^{1,*}, \ldots, \gamma^{N-1,*}$, there exists a set A^N such that $\gamma^{N,*}$ is continuous on $(\mathbb{Y}^N \setminus A^N)$ and

$$\int_{\Omega_0 \times \mathbb{Y}^1 \times \cdots \times \mathbb{Y}^{N-1} \times A^N} P(d\omega_0) p_1(dy^1|\omega_0) p_2(dy^2|\omega_0, y^1, \gamma^{1,*}(y^1)) \ldots$$
$$p_N(dy^N|\omega_0, y^1, \gamma^{1,*}(y^1), \ldots, y^{N-1}, \gamma^{N-1,*}(y^{N-1})) < \epsilon_N.$$

Furthermore, following Tietze's extension theorem, Theorem C.5.2, it follows that there exist continuous policies $\gamma_C^1, \ldots, \gamma_C^N$, such that $\gamma_C^i(x) = \gamma^{i,*}(x), \ \forall x \in (\mathbb{Y}^i \setminus A^i)$ for each $i \in \{1, \cdots, N\}$.

Step 2 Define the following terms: $h_i := (\omega_0, y^1, \gamma^{1,*}(y^1), \ldots, y^{i-1}, \gamma^{i-1,*}(y^{i-1}))$, where $h_1 = (\omega_0)$. $r_i := (\omega_0, z^1, \gamma^{1,*}(z^1), \ldots, z^{i-1}, \gamma^{i-1,*}(z^{i-1}))$, where $r_1 = (\omega_0)$. We now study the following limit, for any fixed y^1, ω_0:

$$\lim_{m\to\infty} \int \tau_m^1(dz^1|y^1)p_2(dy^2|r_2)\tau_m^2(dz^2|y^2)\ldots$$
$$p_N(dy^N|r_N)\tau_m^N(dz^N|y^N)c(\omega_0, \gamma_C^1(z^1), \ldots, \gamma_C^N(z^N))$$

Letting $G_m^N(y^N, z^{N-1}, \ldots, z^1) := \int \tau_m^N(dz^N|y^N)c(\omega_0, \gamma_C^1(z^1), \ldots, \gamma_C^N(z^N))$ (recalling that ω_0 is fixed), we can observe that G_m^N is a sequence of real-valued and bounded functions, and that if $(y_m^N, z_m^{N-1}, \ldots, z_m^1) \to (y^N, z^{N-1}, \ldots, z^1)$, then

$$G_m^N(y_m^N, z_m^{N-1}, \ldots, z_m^1) \to c(\omega_0, \gamma_C^1(z^1), \ldots, \gamma_C^{N-1}(z^{N-1}), \gamma_C^N(y^N)),$$

i.e. G_m^N continuously converges (as it is defined in [658]) to

$$G^N(y^N, z^{N-1}, \ldots, z^1) := c(\omega_0, \gamma_C^1(z^1), \ldots, \gamma_C^{N-1}(z^{N-1}), \gamma_C^N(y^N)).$$

The important aspect is that the measurement z^N is now replaced by y^N in the cost function. This follows from the fact that $\tau_m^N(\cdot|y^N)$ converges weakly to the Dirac delta measure at y^N (see e.g. [366, Example 12.21] for a related argument) and that the DMs' policies are continuous and the cost function is continuous and bounded. Continuing, we have:

$$\lim_{m\to\infty} \int \tau_m^1(dz^1|y^1)p_2(dy^2|r_2)\ldots$$
$$p_N(dy^N|r_N)\tau_m^N(dz^N|y^N)c(\omega_0, \gamma_C^1(z^1), \ldots, \gamma_C^N(z^N))$$
$$= \lim_{m\to\infty} \int \tau_m^1(dz^1|y^1)p_2(dy^2|r_2)\ldots\tau_m^{N-1}(dz^{N-1}|y^{N-1})$$
$$\times \int p_N(dy^N|r_N)G_m^N(y^N, z^{N-1}, \ldots, z^1).$$

Let $H_m^N(z^{N-1}, \ldots, z^1) := \int p_N(dy^N|r_N)G_m^N(y^N, z^{N-1}, \ldots, z^1)$. We can see that, since G_m^N is a sequence of real-valued and bounded functions, H_m^N is also such a sequence. Furthermore, if $(z_m^{N-1}, \ldots, z_m^1) \to (z^{N-1}, \ldots, z^1)$, then $H_m^N(z_m^{N-1}, \ldots, z_m^1)$ continuously converges to $H^N(z^{N-1}, \ldots, z^1) := \int p_N(dy^N|r_N)G^N(y^N, z^{N-1}, \ldots, z^1)$ by the weak continuity assumption on the channel p_N and the continuity of c (and thus G_m^N).

Continuing in this way, for decreasing i, we iteratively define G_m^i for $i \in \{1, \ldots, N-1\}$ as

$$G_m^i(y^i, z^{i-1}, \ldots, z^1) := \int \tau_m^i(dz^i|y^i)H_m^{i-1}(z^{i-1}, \ldots, z^1)$$

and H_m^i for $i \in \{1, \dots, N-1\}$ as

$$H_m^i(z^{i-1}, \dots, z^1) := \int p_i(dy^i|r_i) G_m^i(y_m^i, z_m^i, \dots, z_m^1).$$

We note that each of these functions are real-valued and bounded, and will continuously converge. In particular,

$$G_m^i(y^i, z^{i-1}, \dots, z^1)$$
$$\to \int \left(\prod_{j=i}^{N-1} p_j(dy^j|h_j) \right) c(\omega_0, \gamma_C^1(z^1), \dots, \gamma_C^{i-1}(z^{i-1}), \gamma_C^i(y^i), \dots, \gamma_C^N(y^N)),$$

and each H_m^i will converge to $\int p_i(dy^i|r_i) G^i(y^i, z^{i-1}, \dots, z^1)$.

Thus, eventually we arrive at:

$$\lim_{m\to\infty} \int \tau_m^1(dz^1|y^1) p_2(dy^2|r_2) \dots$$
$$p_N(dy^N|r_N) \tau_m^N(dz^N|y^N) c(\omega_0, \gamma_C^1(z^1), \dots, \gamma_C^N(z^N))$$
$$= \lim_{m\to\infty} \int \tau_m^1(dz^1|y^1) H_m^2(z^1),$$

where H_m^2 is a continuously converging sequence of continuous and bounded real-valued functions.

Thus, since $\tau_m^1(\cdot|y^1)$ converges weakly to the Dirac delta measure at y^1, applying the generalized dominated convergence result in Theorem C.3.1, we have that:

$$\lim_{m\to\infty} \int \tau_m^1(dz^1|y^1) H_m^2(z^1)$$
$$= H^2(y^1)$$
$$= \int p_2(dy^2|h_2) p_3(dy^3|h_3) \dots p_N(dy^N|h_N) c(x, \gamma_C^1(y^1), \dots, \gamma_C^N(y^N)).$$

Step 3 In this step, we complete the proof. First note that

$$\lim_{m\to\infty} E_{\tau_m^1,\dots,\tau_m^N}^{P,p_1,\dots,p_N}[c(\omega_0, \gamma_C^1, \dots, \gamma_C^N)]$$
$$= \lim_{m\to\infty} \int c(\omega_0, \gamma_C^1(z^1), \dots, \gamma_C^N(z^N))$$
$$\times P(d\omega_0) p_1(dy^1|r_1) \tau_m^1(dz^1|y^1) \dots p_N(dy^N|r_N) \tau_m^N(dz^N|y^N)$$

$$= \lim_{m\to\infty} \int P(d\omega_0)p_1(dy^1|\omega_0) \int c(\omega_0, \gamma_C^1(z^1), \ldots, \gamma_C^N(z^N))\tau_m^1(dz^1|y^1)$$
$$\times\, p_2(dy^2|r_2)\tau_m^2(dz^2|y^2)\ldots p_N(dy^N|r_N)\tau_m^N(dz^N|y^N)$$
$$= \int c(\omega_0, \gamma_C^1(y^1), \ldots, \gamma_C^N(y^N))P(d\omega_0)p_1(dy^1|h_1)p_2(dy^2|h_2)\ldots p_N(dy^N|h_N),$$

where the final equality holds by the convergence result from Step 2. Furthermore, we have that:

$$\mu(\Omega_0 \times (\mathbb{Y}^1 \setminus A_1) \times \cdots \times (\mathbb{Y}^N \setminus A_N)) \geq 1 - \sum_{i=1}^{N} \mu_{\mathbb{Y}^i}(A_i) = 1 - \sum_{i=1}^{N} \epsilon_i.$$

Letting $M := \|c\|_\infty \in \mathbb{R}_+$, we have that:

$$\int_{\Omega_0\times\mathbb{Y}^1\times\cdots\times\mathbb{Y}^N} c(\omega_0, \gamma_C^1(y^1), \ldots, \gamma_C^N(y^N))P(d\omega_0)p_1(dy^1|h_1)p_2(dy^2|h_2)\ldots$$
$$p_N(dy^N|h_N)$$
$$\leq \Bigg(\int_{\Omega_0\times(\mathbb{Y}^1\setminus A_1)\times\cdots\times(\mathbb{Y}^N\setminus A_N)} c(\omega_0, \gamma^{1,*}(y^1), \ldots, \gamma^{n,*}(y^N))$$
$$\times\, P(d\omega_0)p_1(dy^1|h_1)p_2(dy^2|h_2)\ldots p_N(dy^N|h_N)\Bigg) + M\sum_{i=1}^{N}\epsilon_i$$
$$\leq \Bigg(\int_{\Omega_0\times\mathbb{Y}^1\times\cdots\times\mathbb{Y}^N} c(\omega_0, \gamma^{1,*}(y^1), \ldots, \gamma^{n,*}(y^N))$$
$$\times\, P(d\omega_0)p_1(dy^1|h_1)p_2(dy^2|h_2)\ldots p_N(dy^N|h_N)\Bigg)$$
$$+ 2M\sum_{i=1}^{N}\epsilon_i$$
$$\leq J^*(P, p_1, \ldots, p_N) + 2M\sum_{i=1}^{N}\epsilon_i + \delta.$$

The above holds since on $(\mathbb{Y}^i \setminus A_i)$, γ_C^i is equivalent to $\gamma^{i,*}$.
Since each $\epsilon_i > 0$ and δ are arbitrary, we have that:

$$\lim_{m\to\infty} E_{\tau_m^1,\ldots,\tau_m^N}^{P,p_1,\ldots,p_N}[c(\omega_0, \gamma_C^1, \ldots, \gamma_C^N)] \leq J^*(P, p_1, \ldots, p_N).$$

Since we know that, for any m, $J^*(P, p_1\tau_m^1, \ldots, p_N\tau_m^N) \le E^{P, p_1\tau_m^1, \ldots, p_N\tau_m^N}$ $[c(\omega_0, \gamma_C^1, \ldots, \gamma_C^N)]$ by definition (since $(\gamma_C^1, \ldots, \gamma_C^N)$ are not necessarily optimal policies for these measurement channels), it follows that:

$\lim_{m\to\infty} J^*(P, p_1\tau_m^1, \ldots, p_N\tau_m^N) \le J^*(P, p_1, \ldots, p_N)$.

Applying Lemma 8.9.1, we know that for every $m \in \mathbb{Z}$,

$$J^*(P, p_1\tau_m^1, \ldots, p_N\tau_m^N) \ge J^*(P, p_1, \ldots, p_N).$$

Thus, we get $\lim_{m\to\infty} J^*(P, p_1\tau_m^1, \ldots, p_N\tau_m^N) = J^*(P, p_1, \ldots, p_N)$. This completes the proof. ◇

8.10.7 *Proof of Theorem 8.9.2*

Step 1 Exactly as in the proof of Theorem 8.9.1, we use Lusin's theorem to construct sets $(\mathbb{Y}^1 \setminus A^1), \ldots, (\mathbb{Y}^N \setminus A^N)$ on which $\gamma^{1,*}, \ldots, \gamma^{N,*}$ are respectively continuous, and we use Tietze's extension theorem to approximate $\bar{\gamma}^*$ with $\gamma_C^1, \ldots, \gamma_C^N$. We will fix this continuous collection of functions in the following.

Step 2 We now apply Lusin's theorem again, however this time we apply it to each DM's channel to approximate each channel with a weakly continuous kernel.

Note that a stochastic kernel (as a regular conditional probability) is defined with the following property: for every $(\omega_0, y^1, u^1, \ldots, y^{i-1}, u^{i-1})$, $p^i(y^i \in \cdot|\omega_0, y^1, u^1, \ldots, y^{i-1}, u^{i-1})$ is $\mathcal{P}(\mathbb{Y}^i)$-valued and for every Borel set $B \subset \mathbb{Y}^i$, $p^i(y^i \in B|\cdot) : \omega_0, y^1, u^1, \ldots, y^{i-1}, u^{i-1} \to \mathbb{R}$ is Borel-measurable.

Following [114, Proposition 7.26], the above property is equivalent to the the kernel $p^i(y^i \in \cdot|\omega_0, y^1, u^1, \ldots, y^{i-1}, u^{i-1})$ being a Borel-measurable map from $\Omega_0 \times \mathbb{Y}^1 \times \mathbb{U}^1 \times \cdots \times \mathbb{Y}^{i-1} \times \mathbb{U}^{i-1}$ to $\mathcal{P}(\mathbb{Y}^i)$ (which is endowed with the weak convergence topology).

Because $\mathbb{Y}^i$ is standard Borel, $\mathcal{P}(\mathbb{Y}^i)$ is a separable metric space, and can be viewed as the space of probability measures $\mathcal{P}(\mathbb{Y}^i)$ as a convex subset of a locally convex space [613, Chapter 3] of signed measures defined on $\mathcal{B}(\mathbb{Y}^i)$, where we define the locally convex space of signed measures with the following notion of convergence: We say that $\nu_n \to \nu$ if $\int f(y^i)\nu_n(dy^i) \to \int f(y^i)\nu(dy^i)$ for every continuous and bounded function $f : \mathbb{Y}^i \to \mathbb{R}$.

Thus, Theorems C.5.1 and C.5.2 apply with the continuous extension being probability measure-valued (by Theorem C.5.2).

Therefore for any $\beta_i > 0$, for fixed $\gamma_C^1, \ldots, \gamma_C^{i-1}$, and $p_1, \ldots, p_{i-1}$, there exists a channel p_{iC} that satisfies Assumption 8.9.1(A3), in the sense that if $(\omega_m, y_m^1, u_m^1, \ldots, y_m^{i-1}, u_m^{i-1}) \to (\omega_0, y^1, u^1, \ldots, y^{i-1}, u^{i-1})$, then

$$p_{iC}(y^i \in \cdot\,|\omega_m, y_m^1, u_m^1, \ldots, y_m^{i-1}, u_m^{i-1}) \to p_{iC}(y^i \in \cdot\,|\omega_0, y^1, u^1, \ldots, y^{i-1}, u^{i-1}),$$

weakly. We emphasize here that the DM policies are fixed as the continuous policies developed in Step 1; these policies, in conjunction with fixing the previous DMs' channels, determine the joint probability distribution on $\Omega_0 \times \mathbb{Y}^1 \times \mathbb{U}^1 \times \cdots \times \mathbb{Y}^{i-1} \times \mathbb{U}^{i-1}$. Furthermore,

$$p_{iC}(\cdot|\omega_0, y^1, u^1, \ldots, y^{i-1}, u^{i-1}) = p_i(\cdot|\omega_0, y^1, u^1, \ldots, y^{i-1}, u^{i-1}),$$

on a set of measure $1 - \beta_i$ on $\Omega_0 \times \mathbb{Y}^1 \times \mathbb{U}^1 \times \cdots \times \mathbb{Y}^{i-1} \times \mathbb{U}^{i-1}$. Let D^i denote this set of measure $1 - \beta_i$, and use D^i_j to denote the projection of the set D^i on its jth component.

Let $E := (\cap_{j=1}^N D_1^j \times \cap_{j=2}^N D_2^j \times \mathbb{R}^{k_1} \times \cap_{j=2}^N D_3^j \times \cdots \cap_{j=N-1}^N D_{2N-1}^j \times D_{2N}^N \times \mathbb{R}^{k_N} \times D_{2N+1}^N)$. I.e., $E \subset (\Omega_0 \times \mathbb{Y}^1 \times \mathbb{R}^{k_1} \times \mathbb{U}^1 \times \ldots \mathbb{Y}^N \times \mathbb{R}^{k_N} \times \mathbb{U}^N)$, where $p_i = p_{iC}$ on E, for every $i \in \mathcal{N}$, and E is constructed using the intersections of the spaces of arbitrarily large measure on which the channels are weakly continuous. Let E^c denote the complement of E in $(\Omega_0 \times \mathbb{Y}^1 \times \mathbb{R}^{k_1} \times \mathbb{U}^1 \times \ldots \mathbb{Y}^N \times \mathbb{R}^{k_N} \times \mathbb{U}^N)$. Let $M := \|c\|_\infty \in \mathbb{R}_+$. Now, applying the fact that the marginal measure on each D^i is known, we note that, given fixed $\gamma_C^1, \ldots, \gamma_C^N$, for any $m \in \mathbb{Z}_{\geq 1}$:

$$\begin{aligned}
&\int_{E^c} c(\omega_0, u^1, \ldots, u^N) P(d\omega_0) p_1(dy^1|r_1) \tau_m^1(dz^1|y^1) \gamma_C^1(du^1|z^1) \ldots \\
&\qquad\qquad p_N(dy^N|r_N) \tau_m^N(dz^N|y^N) \gamma_C^N(du^N|z^N) \\
&\leq M \int_{E^c} P(d\omega_0) p_1(dy^1|r_1) \tau_m^1(dz^1|y^1) \gamma_C^1(du^1|z^1) \ldots \\
&\qquad\qquad p_N(dy^N|r_N) \tau_m^N(dz^N|y^N) \gamma_C^N(du^N|z^N) \\
&\leq M \sum_{i=1}^N \beta_i.
\end{aligned}$$

Step 3 Assume that $p_1, \ldots, p_N$ satisfy Assumption 8.9.1(A3); this step follows identically to Step 2 from Theorem 8.9.1. Once again, the conclusion is that, for any ω_0 and y^1:

$$\begin{aligned}
&\lim_{m \to \infty} \int \tau_m^1(dz^1|y^1) p_2(dy^2|r_2) \tau_m^2(dz^2|y^2) \ldots \\
&\qquad p_N(dy^N|r_N) \tau_m^N(dz^N|y^N) c(\omega_0, \gamma_C^1(z^1), \ldots, \gamma_C^N(z^N)) \\
&= \int p_2(dy^2|h_2) p_3(dy^3|h_3) \ldots p_N(dy^N|h_N) c(x, \gamma_C^1(y^1), \ldots, \gamma_C^N(y^N)).
\end{aligned}$$

Step 4 Now, we can verify the main result. We have that:

$$\lim_{m\to\infty} E^{P,p_1\tau_m^1,\ldots,p_N\tau_m^N}[c(\omega_0,\gamma_C^1,\ldots,\gamma_C^N)]$$

$$= \lim_{m\to\infty}\int c(\omega_0,\gamma_C^1(z^1),\ldots,\gamma_C^N(z^N))$$

$$\times P(d\omega_0)p_1(dy^1|r_1)\tau_m^1(dz^1|y^1)\ldots p_N(dy^N|r_N)\tau_m^N(dz^N|y^N)$$

$$= \lim_{m\to\infty}\int c(\omega_0,u^1,\ldots,u^N)$$

$$\times P(d\omega_0)p_1(dy^1|r_1)\tau_m^1(dz^1|y^1)\gamma_C^1(du^1|z^1)\ldots$$

$$p_N(dy^N|r_N)\tau_m^N(dz^N|y^N)\gamma_C^N(du^N|z^N)$$

$$\leq \left(\lim_{m\to\infty}\int_E c(\omega_0,u^1,\ldots,u^N)P(d\omega_0)p_1(dy^1|r_1)\tau_m^1(dz^1|y^1)\gamma_C^1(du^1|z^1)\ldots\right.$$

$$\left. p_N(dy^N|r_N)\tau_m^N(dz^N|y^N)\gamma_C^N(du^N|z^N)\right) + M\sum_{i=1}^N \beta_i$$

$$= \left(\int_E c(\omega_0,u^1,\ldots,u^N)P(d\omega_0)p_1(dy^1|h_1)\gamma_C^1(du^1|y^1)p_2(dy^2|h_2)\ldots\right.$$

$$\left. p_N(dy^N|h_N)\gamma_C^N(du^N|y^N)\right) + M\sum_{i=1}^N \beta_i$$

$$\leq \left(\int c(\omega_0,u^1,\ldots,u^N)P(d\omega_0)p_1(dy^1|h_1)\gamma_C^1(du^1|y^1)p_2(dy^2|h_2)\ldots\right.$$

$$\left. p_N(dy^N|h_N)\gamma_C^N(du^N|y^N)\right) + 2M\sum_{i=1}^N \beta_i$$

$$= \int c(\omega_0,\gamma_C^1(y^1),\ldots,\gamma_C^N(y^N))P(d\omega_0)p_1(dy^1|h_1)p_2(dy^2|h_2)\ldots p_N(dy^N|h_N)$$

$$+ 2M\sum_{i=1}^N \beta_i$$

$$\leq \int c(\omega_0,\gamma^{1,*}(y^1),\ldots,\gamma^{N,*}(y^N))P(d\omega_0)p_1(dy^1|h_1)p_2(dy^2|h_2)\ldots$$

$$p_N(dy^N|h_N)$$

$$+ 2M\sum_{i=1}^N \epsilon_i + 2M\sum_{i=1}^N \beta_i$$

$$= J^*(P,p_1,\ldots,p_N) + \delta + 2M(\sum_{i=1}^N \epsilon_i + \sum_{i=1}^N \beta_i).$$

The first steps follow using Step 2, via the fact that, if $(z^1, u^1, y^2, z^2, u^2, \ldots, y^N, z^N, u^N) \in E$, then, given $\gamma_C^1, \ldots, \gamma_C^N$, each $p_i = p_{iC}$ on E, and each p_{iC} satisfies Assumption 8.9.1(A3). Furthermore, the measure of the complement of E is uniformly small for every m. We also use a similar argument as in the proof of Theorem 8.9.1 to swap the continuous policies with $\bar{\gamma}^*$ for the final inequality.

Since each $\epsilon_i > 0$, $\beta_i > 0$, and δ are arbitrary, and $\gamma_C^1, \ldots, \gamma_C^N$ are not necessarily optimal policies under $p_1\tau_m^1, \ldots, p_N\tau_m^N$, we have that:

$$\lim_{m\to\infty} J^*(P, p_1\tau_m^1, \ldots, p_N\tau_m^N) \leq J^*(P, p_1, \ldots, p_N).$$

Applying Lemma 8.9.1, we know that for every $m \in \mathbb{Z}$,

$$J^*(P, p_1\tau_m^1, \ldots, p_N\tau_m^N) \geq J^*(P, p_1, \ldots, p_N).$$

Thus we get $\lim_{m\to\infty} J^*(P, p_1\tau_m^1, \ldots, p_N\tau_m^N) = J^*(P, p_1, \ldots, p_N)$. This completes the proof. ◇

8.11 Concluding Remarks

This chapter has studied structural and topological properties of stochastic team problems in the space of measurement channels, information structures, and quantizers, first for single-agent (single-stage and multi-stage (i.e. stochastic control)) problems, and then multi-agent ones. Continuity, compactness and existence results have been established both for measurement channels as well as quantizers (viewed as a suitable subset of measurement channels).

We have restricted the coverage to Euclidean state spaces, but this is not essential as many (but not all) of the results in this chapter can be extended to the case where $\mathbb{X}$, $\mathbb{Y}$, and $\mathbb{U}$ are Polish spaces. In particular, all the results in Sect. 8.3 carry through without any change, except Theorem 8.3.3. The results of Sect. 8.4 hold for this more general setup (however, in Lemma 8.4.3 we need the additional condition that the space be σ-compact). Likewise, most of the results in Sect. 8.5 on quantization hold more generally (in fact, Theorem 8.5.2 holds for an arbitrary measurable space), but two of the main results, Theorems 8.5.5 and 8.5.6, do require the assumption that $\mathbb{X}$ is a finite-dimensional Euclidean space.

For multi-agent problems (that is, stochastic teams), we have presented continuity properties of value functions with respect to information structures. One of the results is that the value function is continuous under total variation convergence of information structures. The change in expected value when switching between two information structures is bounded above by the product of the total variation distance and the L^∞ norm of the cost function. The value function is upper semicontinuous under setwise convergence for measurable and bounded cost functions, and upper semicontinuous under weak convergence for bounded and continuous cost functions when the agents' action spaces are convex.

As a further positive, and consequential, result, we have shown that all sequential team problems are nearly static reducible. This paves the way for significant versatility in numerical and analytical methods for the study of stochastic dynamic teams.

Generalizations of some of these results to stochastic games will be presented in Chap. 11.

8.12 Bibliographic Notes

On the theory of optimal quantization, [1], [317] study the effects of uncertainties in the input distribution and consider robustness in the quantizer design. References [481] and [586] study the consistency of optimal quantizers based on empirical data for an unknown source. Another study, [723], deals with topological properties and the existence of optimal quantizers in the context of decentralized detection. In [549], some discussions on the topology of information channels can be found. In [787], continuity and other functional properties of minimum mean square estimation problems under Gaussian channels have been studied.

References [168, 600, 676] and [727] considered similar settings where the uncertain probability measure is considered for the product space in multi-stage control. Authors of these papers have considered both optimal control and estimation, and the related problem of optimal control design when the channel is unknown. In particular, [676] has studied the existence of optimal continuous estimation policies and worst-case channels under a relative entropy constraint characterizing the uncertainty in the system. In [600], the total variation norm is considered as a measure of uncertainty, and an inf-sup policy has been determined (thus, the setup considered is that of a min-max problem for the generation of optimal control policies). Similarly, there are connections with robust detection, such as those studied in [365] and [587], when the source distribution to be detected belongs to some set.

Some of the results presented in this chapter are based on [828], [825] and [354], as mentioned at the points where relevant results have been presented.

Part II
Stochastic Game Theory and Information Structures

Chapter 9
An Introduction to Stochastic Game Theory

9.1 Introduction

An introduction to stochastic game theory with static information, as included in this chapter, follows the basic lines of the first two sections of Chap. 2, which were devoted to *stochastic teams*, with however some rather subtle differences in the setup and solution concepts. Accordingly, in the following, we introduce the game setup, paralleling the flow of Chap. 2 for teams, intentionally allowing for redundancies or repetitions, so that the coverage is self-contained while also bringing forth connections to the team formulation.

We first provide in the next section a general formulation for static decision/game problems with appropriate solution concepts. Then, in Sect. 9.3, we discuss illustrative examples, which are appropriate counterparts of the ones in Sect. 2.2.2 for games. Specifically, we extend the example model considered in Sect. 2.2.2 along two directions, the first one being a game obtained by probabilistic multi-modeling, and the second one a zero-sum game displaying direct conflicts of interest, which also leads to a nonzero-sum game when discrepancies exist (among the two players) regarding the probabilities attached to different states of nature.

9.2 A General Framework for Static Stochastic Games

As noted in Sect. 2.2 for the basic setup for static decision problems, we will refer to a decision problem as **static** if the information available to each decision maker (DM) is independent of the actions of other DMs; otherwise, the decision problem is said to be **dynamic**. We will refer to DMs interchangeably as *agents* or *players*, or *controllers*, with the ith one denoted by $\mathbf{A}i$, where i takes values in the set $\mathcal{N} := \{1, \ldots, N\}$ which is called the *agent* (*player*) *set*.

S. Yüksel, T. Başar, *Stochastic Teams, Games, and Control under Information Constraints*, Systems & Control: Foundations & Applications,
https://doi.org/10.1007/978-3-031-54071-4_9

As before, the connotation "agents" or "decision makers" is used primarily to refer to DMs in a **team** decision problem (*i.e.*, typically when both the payoff (or cost) and the probabilistic description of uncertainty are common to all DMs), while the word "players" is often reserved to refer to DMs in intrinsic **game** situations, even though we will also occasionally use "agents" or "decision makers" in a game setting. The variable under the control of each DM is called *action* (synonymously, *decision* or *control variable*), which will be denoted by u^i for **A***i*. Each u^i will take values in a given *action set* to be denoted by $\mathbb{U}^i$. Finally, the N-tuple $(u^1, \ldots, u^N)$ will be denoted by $\mathbf{u}$ and the product action space $\mathbb{U}^1 \times \cdots \times \mathbb{U}^N$ by $\mathbb{U}$.

We assume that all random variables are defined on a probability space $(\Omega, \mathbf{F}, P)$. Different from the basic formulation of teams, here we also allow for situations when the DMs do not *see the world* in exactly the same way, resulting in different (subjective) probability measures for different DMs. To accommodate this feature, we will sometimes use a superscript i on P, to denote the subjective view by **A***i*. We also let $\mathbf{P}$ denote the class of all possible probability measures (or distribution functions) that could be adopted by individual DMs. Note however that, even though the subjective probability measures might be different, there is an underlying true probability space, regardless of whether it is consistent with the subjective perceptions or not. This aspect will be discussed further in the next chapter.

Let **A***i* have access to a $\mathbb{Y}^i$-valued measurement $y^i = \eta^i(\xi(\omega))$, where for some standard Borel space Ξ, the Ξ-valued ξ is the (collection of all) exogenous random variable relevant to the system, also called the state of nature. As in Sect. 2.2, the players determine their actions using the measurement signals that they receive, under the *strategies* that they adopt for transforming measurements into actions. The *strategy* (synonymously, *decision rule* (function) or *control law*) of **A***i* will be denoted by γ^i, and is formally defined as a measurable mapping from the measurable space $(\mathbb{Y}^i, \mathcal{B}(\mathbb{Y}^i))$ to $(\mathbb{U}^i, \mathcal{B}(\mathbb{U}^i))$, where $\mathcal{B}(\cdot)$ denotes the Borel sigma-field over $(\cdot)$. We denote the set of all such mappings, which also satisfy additional constraints that may have been imposed on u^i, by Γ^i, to be called the *strategy space* of **A***i*, and note the relationship

$$u^i = \gamma^i(y^i) = \gamma^i(\eta^i(\xi)) \tag{9.1}$$

where the latter relates the action variable to the state of nature, ξ. We will denote the N-tuple $(\gamma^1, \ldots, \gamma^N)$ by $\underline{\gamma}$, and the product strategy space $\Gamma^1 \times \cdots \times \Gamma^N$ by $\mathbf{\Gamma}$. The individual strategy spaces Γ^i may also include the additional structural constraints that may have been imposed on the policies, such as linearity.

Given an $(N{+}1)$-tuple $(\xi, \mathbf{u}) \in \Xi \times \mathbb{U}$, the loss incurred to **A***i* will be denoted by $c^i(\xi, \mathbf{u})$, where the function $c^i : \Xi \times \mathbf{U} \to \mathbf{R}$ is known as the *loss function* of **A***i*. Its negative, $-c^i(\xi, \mathbf{u}) =: \mho^i(\xi, \mathbf{u})$, is known as the *payoff function* of **A***i*, which he strives to "maximize", in the sense to be defined shortly. Implicit here is the assumption that each DM can formulate a unique (up to equivalence) utility function, which numerically orders different outcomes corresponding to joint actions and realization of the state of nature, in a way consistent with his preference ordering among different alternatives.

The loss incurred by each DM is generally a random quantity, the randomness appearing through both ξ and $\mathbf{u}$, where the latter depends on ξ through the measurement signals and the strategies adopted by the DMs. Therefore, one rather works with the expected value of this quantity, which we will be referring to as the *cost function*. Other possible terminologies would be *expected loss function*, *average risk* or *expected cost*, all of which have been used in the literature, and we have also used in this book. The cost function for $\mathbf{A}i$, $J^i : \mathbf{\Gamma} \to \mathbf{R}$, is defined on the product strategy space $\mathbf{\Gamma}$:

$$J^i(\underline{\gamma}) = \int_{\Xi} c^i(\xi, \underline{\gamma}(\underline{\eta}(\xi)))dP^i(\xi) =: E_{\xi}c^i(\xi, \underline{\gamma}(\underline{\eta}[\xi])), \tag{9.2}$$

where

$$\underline{\gamma}(\underline{\eta}(\xi)) := (\gamma^1(\eta^1(\xi)), \ldots, \gamma^N(\eta^N(\xi)))$$

and E_{ξ} is the operator that takes the expected value of the quantity it precedes, over ξ. To also show the explicit dependence of J^i on the information structure $\underline{\eta}$, we will use the notation

$$J^i(\underline{\gamma}, \underline{\eta}),$$

and also occasionally use $J^i(\underline{\gamma}, \underline{\eta}; c^i, P^i)$ to also indicate the dependence on the loss function c^i and the subjective probability distribution (for $\mathbf{A}i$) P^i.

The specification of the N-tuple $\{J^i, i \in \mathcal{N}\}$, along with the product strategy space $\mathbf{\Gamma}$, provides a complete characterization (aside from the solution concept) of a stochastic multiple person decision problem, and is known as the *normal form description*. Note that in this description, the information structure is suppressed and it enters the problem formulation only through the strategy spaces Γ^i, $i \in \mathcal{N}$. The description which lays out explicitly the dependence of the measurement signals on the unknown state of nature is known as the *extensive form description* of the underlying (static) stochastic decision (game) problem, which is typically more appropriate for finite space models.

Different Classifications and Solution Concepts

In the general framework laid out above, as discussed earlier in the book, particularly in Chap. 2, a stochastic multiple person decision problem is a **team** (synonymously, *team decision problem*) if all DMs have the same cost function, J; that is, $J^i \equiv J$ for all $i \in \mathcal{N}$. This entails both the individual loss functions c^i and the subjective probabilities P^i to be identical for all agents.

If either (some of) the c^i's or P^i's (or both) are different, then the decision problem is no longer a team, since this leads to at least one of the J^i's to be different from the rest. In this case the decision problem is called an *N-person nonzero-sum stochastic game* or simply a **stochastic game**. When $N=2$, a stochastic game is called a **zero-sum** *stochastic game* if either $J^1 \equiv -J^2$, or this relation can be made to hold by multiplying J^1 by a positive constant and/or adding to it a constant. In the latter case the terminology "constant-sum" is also used.

For an N-person stochastic team problem, since all agents will be striving toward the same goal, with team preferences quantified in the given common cost functional, the most reasonable solution that leads to optimal behavior is the global minimization of the team cost over the product strategy space. As noted in Definition 2.2.1, this is called the team-optimal solution. A further concept, that of person-by-person-optimality, was considered in Definition 2.2.2.

One important feature of the team-optimal solution that is worth noting at this point (perhaps as a cautionary remark, and in contrast to the solution concept for zero-sum games, as we will see shortly) is that multiple solutions are *not* necessarily *interchangeable*. For a two-person team problem for example, if the pairs of policies (γ^1, γ^2) and (β^1, β^2) are two team-optimal policy pairs, then it is not necessarily true that the pair (γ^1, β^2) will also constitute a team-optimal solution. Hence, in case of nonuniqueness of the solution, the agents have to have a common consistent rule as to which one of the possible solutions to adopt, in order to arrive at the optimum. This may require some precommunication, and precommitment to some protocols among the agents.

Moving on to defining equilibrium solutions for stochastic games, we first note that while in a team problem the agents indulge themselves in full cooperation toward achieving global optimal performance (though still under the informational constraints imposed by the given information structure), in zero-sum games the players have absolutely no incentive to cooperate, since one player's gain is necessarily a loss to the other player. Hence, for a zero-sum stochastic game in normal form, with an objective function J to be minimized by **A**1 and maximized by **A**2, the most reasonable equilibrium solution would be the **saddle point** of J on $\boldsymbol{\Gamma}$.

Definition 9.2.1 *For a given two-person zero-sum stochastic game with a fixed information structure, $\{J; \Gamma^1, \Gamma^2\}$, a strategy pair $\underline{\gamma}^* := (\gamma^{1^*}, \gamma^{2^*}) \in \boldsymbol{\Gamma}$ is in saddle-point equilibrium (synonymously, a saddle-point solution) if, for all $\gamma^1 \in \Gamma^1$ and $\gamma^2 \in \Gamma^2$,*

$$J(\gamma^{1^*}, \gamma^2) \leq J(\gamma^{1^*}, \gamma^{2^*}) =: J^* \leq J(\gamma^1, \gamma^{2^*}). \tag{9.3}$$

The value J^* is the saddle-point value (or simply the value) of the game.

Existence of saddle-point equilibria in zero-sum games will be established later in Theorem 11.2.2.

An important feature of the saddle-point solution, which will be seen to be quite consequential, is the following property, also referred to as the "rectangular" property of multiple saddle points.

Theorem 9.2.1 (Ordered Interchangeability [773, Fact 7]) *Multiple saddle points are* ***orderly interchangeable****, that is if (η^1, η^2) and (β^1, β^2) are two saddle-point strategy pairs, then the two pairs (η^1, β^2) and (β^1, η^2) also constitute saddle points.*

Proof By (9.3), for all admissible strategy pairs (γ^1, γ^2) and (ζ^1, ζ^2), the following hold

$$J(\eta^1, \gamma^2) \le J(\eta^1, \eta^2) \le J(\gamma^1, \eta^2), \tag{9.4}$$

$$J(\beta^1, \zeta^2) \le J(\beta^1, \beta^2) \le J(\zeta^1, \beta^2). \tag{9.5}$$

Then, by letting $\zeta^2 = \eta^2$ and $\zeta^1 = \eta^1$ in (9.5) and $\gamma^2 = \beta^2$ in (9.4), and pulling (9.4) to the right-hand side of (9.5) using the common term $J(\eta^1, \beta^2)$, we have for all admissible γ^1:

$$J(\beta^1, \eta^2) \le J(\beta^1, \beta^2) \le J(\eta^1, \beta^2) \le J(\eta^1, \eta^2) \le J(\gamma^1, \eta^2), \tag{9.6}$$

from which it follows that the strategy β^1 minimizes J with maximizer's strategy set at η^2. Using similar (reciprocal) arguments, we have, for each admissible ζ^2,

$$J(\beta^1, \zeta^2) \le J(\beta^1, \beta^2) \le J(\eta^1, \beta^2) \le J(\eta^1, \eta^2) \le J(\beta^1, \eta^2), \tag{9.7}$$

which says that η^2 maximizes J, with minimizer's strategy fixed at β^1. Thus, (β^1, η^2) constitutes a saddle point. A similar argument establishes saddle-point property of (η^1, β^2). The last statement of the theorem follows by letting $\gamma^1 = \beta^1$ in (9.6) and $\zeta^2 = \eta^2$ in (9.7). $\diamond$

Existence of a saddle point implies that the two quantities below (called, respectively, the *upper value* and the *lower value* of the game) are equal:

$$\bar{J} = \inf_{\gamma^1 \in \Gamma^1} \sup_{\gamma^2 \in \Gamma^2} J(\gamma^1, \gamma^2)$$

$$\underline{J} = \sup_{\gamma^2 \in \Gamma^2} \inf_{\gamma^1 \in \Gamma^1} J(\gamma^1, \gamma^2).$$

If the upper and lower values are not equal (in which case $\bar{J} > \underline{J}$), then one introduces *security strategies* for the two players, defined as follows.

Definition 9.2.2 *For a two-person zero-sum stochastic game, $\gamma^{1*} \in \Gamma^1$ is a minimax strategy for* $A1$, *if*

$$\sup_{\gamma^2 \in \Gamma^2} J(\gamma^{1^*}, \gamma^2) = \bar{J}.$$

Likewise, $\gamma^{2^*} \in \Gamma^2$ is a maximin strategy for **A**2*, if*

$$\inf_{\gamma^1 \in \Gamma^1} J(\gamma^1, \gamma^{2^*}) = \underline{J}.$$

Remark 9.2.1 Whenever they exist, the minimax and maximin strategies become saddle-point strategies if the upper and lower values are equal, which also shows that, whenever it exists, the saddle-point value, J^*, has to be unique, as also stated in Theorem 9.2.1. Precise conditions under which such an existence result holds are given later in Theorem 11.2.2. Typically, existence conditions require the presence of mixed (or randomized) policies or strategies so that the actions are probability measure valued. As we saw earlier in Chap. 4, any stochastic kernel from a Polish space to another one, $P(dx|y)$, can be realized by some measurable function $x = f(y, v)$ where v is a [0, 1]-valued independent random variable. Accordingly, a randomized or mixed strategy can also be viewed as a behavioral strategy (which makes explicit use of the randomness in v). See the earlier Remark 4.1.1 for some related discussion.

For a general N-person (nonzero-sum) game, and when there is no possibility of cooperation among the players (such as formation of coalitions), a reasonable equilibrium solution concept is the one provided by *Nash (1951)*:

Definition 9.2.3 *For a given N-person stochastic nonzero-sum game with a fixed information structure, $\{J^i; \Gamma^i, i \in \mathcal{N}\}$, an N-tuple $\underline{\gamma}^* := (\gamma^{1^*}, \ldots, \gamma^{N^*})$ of strategies is in Nash equilibrium (synonymously, a Nash solution), if for all $\beta \in \Gamma^i$ and all $i \in \mathcal{N}$, the following inequalities hold:*

$$J^{i^*} := J^i(\underline{\gamma}^*) \leq J^i(\underline{\gamma}^{-i^*}, \beta), \tag{9.8}$$

where we have adopted the notation

$$(\underline{\gamma}^{-i^*}, \beta) := (\gamma^{1^*}, \ldots, \gamma^{i-1^*}, \beta, \gamma^{i+1^*}, \ldots, \gamma^{N^*}). \tag{9.9}$$

The N-tuple of values $\{J^{i^*}, i \in \mathcal{N}\}$ is a noncooperative (Nash) equilibrium outcome of the stochastic game.

Remark 9.2.2 For $N = 2$ and with $J^1 \equiv -J^2 =: J$, every Nash equilibrium solution is also a saddle-point solution of the associated zero-sum game. For general N and with $J^1 \equiv \cdots \equiv J^N =: J$, Nash equilibrium is a weaker solution concept than team-optimality (*cf. Definition 2.2.1*), since in a team problem Nash equilibrium dictates that, for all $\beta \in \Gamma^i$ and $i \in \mathcal{N}$,

$$J(\underline{\gamma}^*) \leq J(\underline{\gamma}^{-i^*}, \beta). \tag{9.10}$$

Satisfaction of these N inequalities is clearly necessary but not sufficient for $\underline{\gamma}^*$ to be an optimal team decision rule. As we saw in Definition 2.2.2, decision rules for team problems, which are in Nash equilibrium, are called *person-by-person (pbp) optimal.*

Remark 9.2.3 As in the case of the saddle-point equilibrium concept, Nash equilibrium *in pure strategies* may not exist, even in finite games. In view of this, one extends the definition of Nash equilibrium to mixed or behavioral strategies. See Chap. 4 for an extensive discussion on such randomized strategies.

Multiple Nash equilibria do not generally enjoy the appealing "ordered interchangeability" property (Theorem 9.2.1) of saddle-point equilibria, and furthermore the cost pairs corresponding to different Nash equilibria are in general different and even non-comparable. It is therefore important to trim down the number of Nash equilibria a game admits, based on some rational basis. One such (partial) ordering is the one provided by the notion of *dominance*:

Definition 9.2.4 *A strategy N-tuple $\underline{\gamma}$ dominates (or is better than) another strategy N-tuple $\underline{\beta}$ if $J^i(\underline{\gamma}) \le J^i(\underline{\beta})$, for all $i \in \mathcal{N}$, with strict inequality for at least one i. A Nash equilibrium policy $\underline{\gamma}^*$ is admissible if it is not dominated by any other Nash equilibrium policy. A Nash equilibrium is called dominant (or pay-off dominant) if it dominates every other equilibrium.*

We conclude the presentation of this section by introducing a second noncooperative solution concept for two-person nonzero-sum games, corresponding to the case when the roles of the players are not symmetric. This formulation (originated by *von Stackelberg*) models situations where one player, called the *leader*, can announce his strategy ahead of time (and is to remain committed to it) and impose it on the other player, the *follower*, who reacts rationally by optimizing his cost functional. For a precise mathematical description, let $T : \Gamma^1 \to \Gamma^2$ be any map satisfying the relation

$$J^2(\gamma^1, T(\gamma^1)) := \min_{\gamma^2 \in \Gamma^2} J^2(\gamma^1, \gamma^2) \tag{9.11}$$

which is called a *rational reaction function* for the follower. Then we have:

Definition 9.2.5 *For a two-person nonzero-sum game of the type introduced in this section, let there exist a unique mapping $T : \Gamma^1 \to \Gamma^2$ satisfying (9.11). Then, $\gamma^{1^*} \in \Gamma^1$ is a Stackelberg decision rule for* **A**1*, the leader, and $\gamma^{2^*} = T(\gamma^{1^*})$ is a corresponding optimal response for* **A**2*, the follower, where*

$$J^{1^*} := J^1(\gamma^{1^*}, \gamma^{2^*}) = \min_{\beta \in \Gamma^1} J^1(\beta, T(\beta)). \tag{9.12}$$

The pair $(\gamma^{1^*}, \gamma^{2^*})$ is also called the Stackelberg solution of the game with **A**1 *as the leader, and the quantity J^{1^*} is the leader's Stackelberg cost.*

Remark 9.2.4 When the rational reaction function T is multi-valued (*i.e.* if the solution to the minimization problem in (9.11) is not unique for every $\gamma^1 \in \Gamma^1$), then we have what is known as a "rational reaction set" for the follower, on which the leader's best *security strategy* [1] (in the Stackelberg sense) as well as his best *cooperative strategy* can be defined. Furthermore, the notion of "*mixed-strategy* Stackelberg equilibria" is also a relevant concept here, as in the cases of saddle-point and Nash equilibria. In fact, here it gains more relevance because it is possible to formulate examples, in both finite and infinite games, where, by using a mixed strategy, the leader can achieve a *better* cost than the J^{1*} given above (by (9.12). For further elaboration on this feature of the Stackelberg solution we refer the reader to *Başar and Olsder* [97]; see also Remark 18.1.1 in the context of Stackelberg signaling games.

Remark 9.2.5 It may be insightful to note that for the special case of two-person teams (that is when $J^1 \equiv J^2$) the notion of a Stackelberg solution coincides with that of team-optimality, even if the mapping T in (9.12) is not unique. This, in fact, provides one method for the derivation of team-optimal solutions, which involves the solution of two single-agent optimization problems, one of which is "parameterized" by the policies of the other. Note that the solution would be independent of the order in which these two optimizations are performed. A similar equivalence exists between the Stackelberg and saddle-point solutions, whenever $J^1 \equiv -J^2$, where again any of the two players could optimize given the rational reaction function of the other player, and this works for either player.

In the following section, we provide illustrative examples which will aid in better understanding of some of the concepts introduced above. Several features that we will observe in the context of these examples will also help motivate the studies in the chapters to follow.

9.3 An Illustrative Example of a Finite Stochastic Game: Variations on the Team Theoretic Examples

We now consider two variations on the basic example of Sect. 2.2.2, both of which will lead to game formulations and thereby provide a setting for a discussion of the equilibrium solution concepts introduced in Sect. 9.2.

[1] The best security strategy safeguards the leader against the uncertainty brought about by the non-uniqueness of the rational reaction function. The best cooperative strategy, on the other hand, uses this uncertainty to the leader's advantage as though the follower were cooperating with him on his indifference curve; for further discussion and results on this extended Stackelberg solution, see *Başar and Olsder* [97].

9.3.1 Multiple Subjective Probabilities

As a variation on the basic finite stochastic team problem of Sect. 2.2.2, let us consider the situation where the agents do not have the same probabilistic outlook on the world, and while they agree on the numerical values for the entries of the loss matrices, corresponding to the three possible states of nature, they do not assign the same probabilities to the three realizations. Without any loss of generality for the discussion to follow, let us take the probability vector $\mathbf{p}^1$, for **A**1, to be the same as the one given in Sect. 2.2.2, whereas for **A**2 let us consider a different vector of probabilities, $\mathbf{p}^2$, and assume that this is common information to both. Since $\mathbf{p}^1 \neq \mathbf{p}^2$, the costs (expected losses) that accrue to the agents for any given pair of strategies will be different, and hence the problem can no longer be treated as a team, except in some special cases as discussed below.

Nash Game
Under a general information structure, the underlying stochastic decision problem will be a nonzero-sum game with cost functions J^1, J^2:

$$J^i(\gamma_1, \gamma_2) := E^i_\omega L(\omega; \gamma^1(y^1), \gamma^2(y^2))$$

where the superscript on E designates the expectation under the subjective probability distribution (for ω) adopted by **A**i, $i = 1, 2$. If we adopt the *Nash equilibrium* solution concept, then via (9.8), using the "iterated property" of the conditional expectation, we have

$$J^{1^*} = E^1_{y^1}\{\min_{u^1} E^1_{\omega|y^1} L(\omega; u^1, \gamma^{2^*}(y^2))\}$$

$$J^{2^*} = E^2_{y^2}\{\min_{u^2} E^2_{\omega|y^2} L(\omega; \gamma^{1^*}(y^1), u^2)\}$$

from which it follows that

$$\gamma^{1^*}(y^1) \equiv \arg\min_{u^1} E^1_{\omega|y^1} L(\omega; u^1, \gamma^{2^*}(y^2)) \tag{9.13}$$

$$\gamma^{2^*}(y^2) \equiv \arg\min_{u^2} E^2_{\omega|y^2} L(\omega; \gamma^{1^*}(y^1), u^2). \tag{9.14}$$

This now motivates the introduction of the following condition on the underlying conditional probability measures:

Condition 9.3.1 *The linear (conditional expectation) operators $E^1_{\omega|y^2}$ and $E^2_{\omega|y^2}$ are identical, for all $y^2 \in \mathbf{Y}^2$. In words, the conditional probability distribution of ω given y^2 is the same under both subjective probability measures (in this case the pmf's $\mathbf{p}^1$ and $\mathbf{p}^2$), this being so for every $y^2 \in \mathbf{Y}^2$.*

For future reference, we also state the counterpart of this condition for **A**1's information set:

Condition 9.3.2 *The linear (conditional expectation) operators* $E^1_{\omega|y^1}$ *and* $E^2_{\omega|y^1}$ *are identical, for all* $y^1 \in \mathbf{Y}^1$. *In words, the conditional probability distribution of* ω *given* y^1 *is the same under both subjective probability measures (in this case the pmf's* $\mathbf{p}^1$ *and* $\mathbf{p}^2$*), this being so for every* $y^1 \in \mathbf{Y}^1$.

An immediate consequence of the pair of Nash inequalities, given by (9.13)–(9.14), is that whenever Condition 9.3.1 holds, every team solution obtained under $\mathbf{p}^1$ will satisfy (9.13) and (9.14), and hence constitute a Nash equilibrium (NE) solution to the game. Analogously, if Condition 9.3.2 holds, then every team solution obtained under $\mathbf{p}^2$ will constitute a NE solution to the game. The game may admit other Nash equilibria, however, since the Nash solution corresponds to the weaker concept of *person-by-person (pbp) optimality*. Hence, whenever this equivalence condition between the two conditional distributions under the two measures is satisfied, then the "Nash game" derived from a team problem through subjective probabilities can essentially be considered as a team problem and be solved as such, including the derivation of all its *pbp* optimal solutions. A typical example of a situation where the Condition 9.3.1 holds, regardless of the underlying measures, is when **A**2 has perfect information on ω (such as *cases 1* and *4* of Sect. 2.2.2). As a somewhat more elaborate example we can consider *case 5* of Sect. 2.2.2, with $\mathbf{p}^2 = (0.45, 0.45, 0.1)$, for which Condition 9.3.1 is again satisfied and therefore the team-optimal solution given there is a nondominated (admissible) NE solution. We emphasize here the word "a" for the reason that it is quite possible for the game to admit other admissible NE solutions (different from the team-optimal solutions) in spite of the fact that the team-optimal solutions are (by definition) the only nondominated *pbp* optimal solutions of the original team. The following example demonstrates this feature of the solution:

Example 9.3.6 Consider the basic problem of Sect. 2.2.2 as the nominal team, with the *pmf*'s given as $\mathbf{p}^1 = (0.3, 0.3, 0.4)$, $\mathbf{p}^2 = (0.45, 0.45, 0.1)$, and under the information pattern $\mathbf{Y}^1 = \sigma(\{\omega_1, \omega_2\}, \omega_3)$, $\mathbf{Y}^2 = \sigma(\{\omega_1, \omega_2, \omega_3\})$. Note that the *pmf*'s together with the given information pattern satisfy Condition 9.3.2, and hence the set of all deterministic (pure-strategy) Nash equilibria for the stochastic game should coincide with the set of all *pbp* optimal solutions of the team problem formulated under $\mathbf{p}^2$ and using the same information structure as given. To see this explicitly, we now write down the cost matrices for the two agents, using the convention adopted earlier for finite strategies:

J^1 (A1 rows, A2 columns):

	L	R
UU	1.3	1.7
UD	0.9	1.7
DU	1.9	1.4
DD	1.5	1.4

J^2 (A1 rows, A2 columns):

	L	R
UU	1.45	1.55
UD	1.35	1.55
DU	2.35	1.1
DD	2.25	1.1

The normal form of the nominal team problem is the matrix indicated by J^2, which admits two team-optimal solutions: (DU, R) and (DD, R), leading to the optimal cost level of 1.1. The problem also admits a *pbp* optimal solution in deterministic (pure) strategies: (UD, L), with a cost level of 1.35. It is interesting to note that there is also a mixed-strategy *pbp* optimal solution:

$$\big(UD(\frac{23}{27}), DD(\frac{4}{27})), (L(\frac{1}{3}), R(\frac{2}{3})\big),$$

with an (inferior) average cost of $\frac{4.45}{3} \approx 1.483$. Clearly, the three (pure-strategy) *pbp* optimal solutions, namely (DU,R), (DD,R), and (UD,L), also provide NE solutions for the stochastic game with normal form (J^1, J^2), leading to the cost pairs (1.4, 1.1), (1.4, 1.1), and (0.9, 1.35), respectively, and these are the only pure-strategy Nash equilibria of the stochastic game. We see that the second one and the third equilibria are "noncomparable"; neither dominated, and both admissible. ◇

Now, if the stated condition on the equivalence between the two operators $E^1_{\omega|y^2}$ and $E^2_{\omega|y^2}$ (or the operators $E^1_{\omega|y^1}$ and $E^2_{\omega|y^1}$, if **A2**'s problem is the nominal team, as in the previous setting) does not hold, then neither a *pbp* optimal solution of the nominal team is necessarily a NE solution for the game, nor a Nash solution of the game is necessarily a *pbp* optimal solution for the team. Then, one has to work with the normal form of the resulting NE game, and obtain its equilibrium solution. It may also happen that the problem may not admit any (pure-strategy) equilibrium, in which case one has to resort to mixed-strategy equilibria, which always exists in finite games (see Sect. 10.5). The following example illustrates this point.

Example 9.3.7 Replace the basic team problem of Sect. 2.2.2 with the simpler team where ω now takes two values, ω_1 and ω_2, with loss matrices and the subjective probabilities for the two agents given as follows:

A2 (left table), A1 rows:

	L	R
U	1.5	1.0
D	3.0	1.0

A2 (right table):

	L	R
U	3.0	2.0
D	0.0	1.0

$\mathbf{p^1}$: $\omega_1 \leftrightarrow 0.7$ $\omega_2 \leftrightarrow 0.3$

$\mathbf{p^2}$: $\omega_1 \leftrightarrow 0.3$ $\omega_2 \leftrightarrow 0.7$

Let us take the information structure to be "no measurement" for both agents. Then, the normal form for the resulting nonzero-sum game is

A2, A1 rows (J^1):

	L	R
U	1.95	1.30
D	2.10	1.00

A2 (J^2):

	L	R
U	2.55	1.70
D	0.90	1.00

It follows by inspection that the game does not admit any pure-strategy NE. Using a standard procedure for 2×2 bi-matrix games (see[97, Chapter 3]), we obtain the unique mixed-strategy NE to be $\left((\frac{2}{19}, \frac{17}{19}); (\frac{2}{3}, \frac{1}{3})\right)$, where the first pair of numbers is the pair of probabilities with which **A**1 chooses U and D (and in that order), and the second pair likewise corresponds to the probabilities with which L and R are chosen by **A**2. The (average) equilibrium cost pair in mixed strategies is $\left(J_{av}^{1\,*} \approx 1.40;\ J_{av}^{2\,*} \approx 1.07\right)$. $\diamond$

Stackelberg Game

The Nash equilibrium is not the only possible noncooperative solution concept that can be applied to the nonzero-sum game obtained as a result of multiple subjective probabilities. In an asymmetric mode of play, that is if one player (say **A**1) is dominant in the decision process, the Stackelberg solution would be more relevant. Toward studying this equilibrium concept for the problem under consideration, we first note that Condition 9.3.1 also plays a role here since, under it, (9.11) can be written as

$$
\begin{aligned}
J^2(\gamma^1, T(\gamma^1)) &= E^2_{y^2}\{\min_{u^2} E^2_{\omega|y^2} L(\omega; \gamma^1(y^1), u^2)\} \\
&\equiv E^2_{y^2}\{\min_{u^2} E^1_{\omega|y^2} L(\omega; \gamma^1(y^1), u^2)\}
\end{aligned}
$$

implying that the follower's reaction function (or reaction set, in the case when the mapping T above is nonunique) is the same as in the team problem formulated under **A**1's subjective probability. Hence, also in view of *Remark 9.2.5*, under Condition 9.3.1 the set of all Stackelberg solutions with **A**1 as the leader is equivalent to the set of all team-optimal solutions obtained when both agents adopt **A**1's subjective probability. Note that this is a stronger result than the one obtained for the Nash game, and furthermore there is no need to go outside the class of pure strategies for the leader since no mixed strategy pair can dominate the team-optimal solution.

Clearly, if both Condition 9.3.1 and Condition 9.3.2 hold, then the Stackelberg solution with either player as leader will be identical, and will be the same as the team-optimal solution with either player's subjective probability. This, indeed, turns out to be the case in Example 9.3.6, where the only Stackelberg solution (with either player as the leader) is (UD, L)—the unique team-optimal solution. For Example 9.3.7, however, neither Condition 9.3.1 nor Condition 9.3.2 holds, and the individual Stackelberg solutions are:

A1 leader: $(U, R) \to (1.3, 1.7)$
A2 leader: $(D, R) \to (1, 1)$.

Note the interesting feature that **A**1 does better (in fact, the best) as follower than as leader. Both players would in fact prefer the leadership of **A**2, which is in a sense a stable situation in the choice of leadership. However, if we have the scenario where **A**1's leadership is enforced, the question comes up whether as leader he can do better than his performance using pure strategies, by implementing some mixed strategy, perhaps getting closer to what he can achieve under **A**2's leadership, which is in fact (in this game) the very best he can achieve (1 being the lowest entry of his cost matrix). Now note that since Condition 9.3.1 does not hold here, we have a genuine game problem, and hence in view of *Remark 9.2.4* there is the possibility of existence of a mixed strategy for **A**1 as leader that leads to a better performance than his pure-strategy Stackelberg solution. This is indeed the case here, by choosing the mixed strategy $(\epsilon + \frac{2}{19}, -\epsilon + \frac{17}{19})$, where the first entry is the probability with which the option U is selected and ϵ is a sufficiently small positive quantity, **A**1 (as leader) can asymptotically approach the average cost level of $\frac{19.6}{19}$, which is clearly more favorable to him as compared with his pure-strategy performance of 1.3, and very close to the best he can hope to achieve (which is 1, as we mentioned above). We leave it to the reader to show that **A**1 can come arbitrarily close to the value $\frac{19.6}{19}$ but can never achieve it, which is often the case with mixed-strategy Stackelberg solutions.

9.3.2 Directly Conflicting Goals

We have seen in the previous subsection how discrepancies in the subjective probabilities adopted by agents in a team could lead to intrinsic game situations, even though the loss functions are identical. Another way of deviation from a team framework would be if the agents perceive the entries in the loss matrices differently, and/or the goals set for them differently. Such a departure from the team framework would naturally also lead to a game formulation, where again the Nash and Stackelberg solution concepts would be applicable. Here, we will consider an extreme case of such a departure where, in the context of the basic example of Sect. 2.2.2, we will assume that the goals are directly conflicting, in the sense that while **A**1 is still a minimizer, **A2** now attempts to maximize the same quantity. The loss matrices are the same as in Sect. 2.2.2, and the probability weights are common to both players, as given there. The resulting decision problem is now a zero-sum game, for which we study below the saddle-point solution under the six types of information patterns considered for the team problem of Sect. 2.2.2. Hence, the normal forms for the game are the same ones obtained earlier, with the only difference being that we are now interested in the saddle-point instead of the team-optimal solution.

The complete set of solutions for the six zero-sum games corresponding to the three information patterns introduced in Sect. 2.2.2 are now presented below, in the same order as they appear there.

1. *Perfect measurements.* Since **A2** is maximizing, the saddle-point relation is

$$E_\omega\{\min_{u^1} L(\omega; u^1, \gamma^{2^*}(\omega))\} = E_\omega\{\max_{u^2} L(\omega; \gamma^{1^*}(\omega), u^2)\} \equiv J(\gamma^{1^*}, \gamma^{2^*}) =: J^*.$$

   This relation ensures that we may proceed by first obtaining the saddle point of each matrix (corresponding to each state of nature), and then composing the saddle point of the original game from the individual solutions. Following this procedure, we obtain two solutions: (UDD, LLR), (UDU, LLR), with a *value* of 1.7. [Note that this value is **cost** to **A**1 and **payoff** to **A2**.]
2. *Imperfect identical measurements.* In this case, we have

$$E_y\{\min_{u^1} E_{\omega|y} L(\omega; u^1, \gamma^{2^*}(y))\} = E_y\{\max_{u^2} E_{\omega|y} L(\omega; \gamma^{1^*}(y), u^2)\}$$
$$\equiv J(\gamma^{1^*}, \gamma^{2^*}) =: J^*,$$

   and the procedure used just following (16) applies here, with the only difference being that team-optimality is now replaced by saddle-point equilibrium. We again have two saddle-point solutions: (UD, LR), (UU, LR), with a value of 1.7. Comparing the *values* of the two games corresponding to *cases* 1 & 2, we arrive at the conclusion that the extra information provided by *case 1* (to both players) over what is available in *case 2* is "of no value" to either player.
3. *No measurements.* Using the normal form obtained in Sect. 2.2.2, we observe that there is no saddle point in pure strategies. The mixed saddle-point solution is $\big((0.2, 0.8); (0.6, 0.4)\big)$, with an average value of $J^*_{\text{av}} = 1.46$. Note that this value is lower than that of *cases* 1 and 2, which means that the extra information (provided that it is shared by both players) helps **A2** and is harmful to **A**1.

4. *Perfect information for* **A**2*and none for* **A**1. The unique saddle-point solution of the given normal form is (U, LRR), with a value of $J^* = 2$. The interpretation here is that the extra information received by **A**2 (over the 'no information') helps him.[2]
5. *Nonidentical measurements.* In this case recall that $Y^1 = \sigma(\{\omega_1\}, \{\omega_2, \omega_3\})$ and $Y^2 = \sigma(\{\omega_1, \omega_2\}, \{\omega_3\})$. We now observe that one of the two saddle-point solutions obtained in *case 1* (the first one) is compatible with this information pattern, which implies (by going through an argument similar to that in Sect. 2.2.2) that a saddle-point solution here is precisely the first one of *case 1*, simply rewritten for the given information pattern: (UD, LR), $J^* = 1.7$. Furthermore, the problem cannot admit any other saddle-point solution because of the ordered interchangeability property (*i.e. Theorem 9.2.1*).
6. *Noisy measurements.* Using the normal form obtained in Sect. 2.2.2, we observe that the game does not admit any saddle point in pure strategies. The unique mixed-strategy saddle-point solution is: $\left((\frac{1}{21}, \frac{20}{21}); (\frac{5}{7}, \frac{2}{7})\right)$, $J^*_{\text{av}} = \frac{99}{70} \approx 1.4143$. Even if the measurement is "noisy", we would expect **A**1 to do no worse than in the "no-measurement" case. This is indeed the case here; he actually does slightly better (*i.e.* he experiences a slight decrease in the average cost).

In the zero-sum game, as considered above, if the probability weights attached to different states of nature are not the same for the two players, then in general the problem has to be treated as a nonzero-sum game where again the Nash or Stackelberg solution concepts become relevant. However, there do exist nontrivial cases where, in spite of the discrepancies in the subjective probabilities by the two players, the saddle-point solution would still be applicable. Here, Condition 9.3.1 (or Condition 9.3.2) still plays an important role, and under it every NE solution of the game derived from a nominal zero-sum game under different subjective probabilities (for the two players) is a saddle-point solution of the zero-sum game that uses only **A**1's probability weightings. To see this, we first write down the counterpart of (2.19) when the nominal problem is a zero-sum game:

$$\gamma^{1^*}(y^1) \equiv \arg\min_{u^1} E^1_{\omega|y^1}[L(\omega; u^1, \gamma^{2^*}(y^2))]$$

$$\gamma^{2^*}(y^2) \equiv \arg\max_{u^2} E^2_{\omega|y^2}[L(\omega; \gamma^{1^*}(y^1), u^2)]$$

[2] This property will formally be established in Sect. 11.2: In a zero-sum game one can obtain a Blackwell order on information structures, a corollary of which is that more information does not hurt the player receiving it while the other's information is unaltered. This may seem immediate first; however one should be cautious that, unlike in team theory, here equilibrium does not allow an argument of the type *one can choose to ignore the additional information and thus no information cannot hurt.* The discussions in the next two chapters will present further insight and mathematical reasoning.

Note that these are both necessary and sufficient for a pair $(\gamma^{1^*}, \gamma^{2^*})$ to be in Nash equilibrium for the game $\{J^1, J^2\}$, where J^i is obtained by taking the expected value of L under **A**i's probability measure. But, under Condition 9.3.1, the superscript ^{2} in (2.23b) can be changed to ^{1}, thus yielding precisely the relations for a saddle point of a zero-sum game defined under **A**1's probability measure. Naturally, if Condition 9.3.2 holds, then the same would be true, with the zero-sum game defined now under **A**2's probability weightings.

If the Nash equilibrium solution above is replaced by the Stackelberg solution concept, then precisely the same kind of features follow even if the rational reaction function T in (9.12) is not unique. We leave it to the reader to verify this property of the solution.

If the above conditions are not satisfed, then the zero-sum game with different probability weightings for the players has to be treated as a genuine nonzero-sum game, in which case the Nash and Stackelberg solutions will generally not be identical. As an illustration of this point, consider Example 9.3.7 but with **A**2 now acting as a maximizer. Under the given subjective probabilities, and with the "no measurement" information structure, the normal form is the nonzero-sum game $\{J^1, -J^2\}$ where both players are now minimizers. This new game admits two (pure-strategy) Nash equilibria: (U, L) and (D, R), with cost pairs of $(1.95, -2.55)$ and $(1, -1)$, respectively, both of which are admissible. The first of these is also the Stackelberg solution computed under **A**2's leadership, and the second one is the Stackelberg solution that uses **A**1's leadership.

9.4 Concluding Remarks

In this chapter, we have presented a general introduction to stochastic games with static information, viewed as a game theoretic counterpart of static stochastic teams covered in Chap. 2.

9.5 Bibliographic Notes

There are many excellent texts on game theory, including [97, 281, 336, 345, 475, 532, 568, 570], that focus on different aspects of the theory. Game theory is a very broad area of applied mathematics, which by now spans various disciplines and academic journals. Accordingly, the foci of different societies and disciplines, and corresponding texts, may often be slightly or even drastically different.

Chapter 10
Information Structures in Stochastic Dynamic Games and Informational Properties of Equilibria

This chapter deals with information structures and various information structure dependent properties of equilibrium solutions in stochastic dynamic games.

Study of information structures in stochastic dynamic games leads to many technical and conceptual obstacles that need to be cautiously addressed. Indeed, there exist very few general studies on information structures in game theory, paralleling those for stochastic teams. Many of the standard texts on game theory study dynamic games with perfect-recall (classical) information structures, and many studies consider the extensive tree form representations, where subtleties in informational ambiguities cannot be depicted effectively.

Lack of a general treatment of information structures in stochastic dynamic games is not without reason: information structures for general stochastic games entail far more challenges and exhibit many nuances that are not present in the theory of stochastic teams, a field with formidable challenges in itself as demonstrated comprehensively so far in the book.

This chapter does not resolve all such challenges for games, but serves to bring the relatively complete discussion in the context of sequential team theory studied earlier in the book to the theory of games, and paves the path for a comprehensive theory of information structures in stochastic dynamic games.

In Sect. 10.1, information structures for stochastic games are introduced. Solution concepts for stochastic dynamic games are presented in Sect. 10.2. Some information structure dependent properties of equilibria are presented in Sect. 10.3, whose Sect. 10.3.4 studies isomorphism properties of equilibria under equivalent information structures (such as via static reduction). The LQG setting is studied in detail in Sect. 10.4. Existence of equilibria is studied in Sect. 10.5, and the problem of subjective priors in dynamic games is considered in Sect. 10.6.

S. Yüksel, T. Başar, *Stochastic Teams, Games, and Control under Information Constraints*, Systems & Control: Foundations & Applications,
https://doi.org/10.1007/978-3-031-54071-4_10

10.1 Information Structures in Stochastic Game Theory

10.1.1 A Mathematical Characterization of Information Structures for Stochastic Dynamic Games

In the following, we introduce a characterization of information structures, building, on but different from, Witsenhausen's intrinsic model (Sect. 3.2), tailored for stochastic dynamic games, resulting in subtle, but crucial, differences.

An Intrinsic Model for Sequential Games (Generalizing Witsenhausen's One-Shot-DM Formulation)

Suppose that in the decentralized system considered, involving N players, there is a pre-defined order in which the players act. Such systems will again be called *sequential*, as was also the case in Sect. 3.2. There will be some salient differences, however, with Witsenhausen's characterization, which for example requires the presence of a single probability space model and belief system, unlike what we present below. Accordingly, the game theoretic characterization we will adopt has the following ingredients.

(i) A collection of *measurable spaces* $\{(\Omega, \mathcal{F}), (\mathbb{U}^i, \mathcal{U}^i), (\mathbb{Y}^i, \mathcal{Y}^i), i \in \mathcal{N}\}$, with $\mathcal{N} := \{1, 2, \cdots, N\}$, specifying the system's distinguishable events, and the control and measurement spaces. Here $N = |\mathcal{N}|$ is the number of control actions taken, with each of these actions taken by a different individual player (when the fact that a player acts only once needs to be emphasized, we will use *DM* instead of *Player* to make this explicit). The pair $(\Omega, \mathcal{F})$ is a measurable space (on which a probability measure may be defined). The pair $(\mathbb{U}^i, \mathcal{U}^i)$ denotes the measurable space from which the action, u^i, of **DM**i is selected. The pair $(\mathbb{Y}^i, \mathcal{Y}^i)$ denotes the measurable observation/measurement space for **DM**i.

(ii) A *measurement constraint* which establishes the connection between the observation variables and the system's distinguishable events. The $\mathbb{Y}^i$-valued observation variables are given by $y^i = \eta^i(\omega, \mathbf{u}^{[1,i-1]})$, where $\mathbf{u}^{[1,i-1]} := \{u^k, k \leq i-1\}$, η^i are measurable functions, and u^i denotes the action of **DM**i, $i \in \mathcal{N}$. Hence, the information variable y^i induces a σ-field, $\sigma(\mathcal{I}^i)$, over $\Omega \times \prod_{k=1}^{i-1} \mathbb{U}^k$.

(iii) A *design constraint*, which restricts the set of admissible N-tuple control laws $\underline{\gamma} = \{\gamma^1, \gamma^2, \ldots, \gamma^N\}$, also called *designs* or *policies*, to the set of all (admissible) measurable control functions, so that $u^i = \gamma^i(y^i)$. We will occasionally also allow for *randomized* policies so that $u^i = \gamma^i(y^i, r^i)$, where r^i is a randomization variable. Let Γ^i denote the set of all such admissible or randomized policies for **DM**i and let $\Gamma := \prod_k \Gamma^k$. Note that in the randomized case, we also allow the y^i variables to be accompanied by individual randomization variables r^i, taken to be [0, 1]-valued without any loss of generality.

(iv) A *probability measure* P defined on $(\Omega, \mathcal{F})$, which describes the measures on the random events in the model.

(v) A collection of loss or cost functions $c^i : \Omega_0 \times \mathbb{U} \to \mathbb{R}$, where $\Omega_0 \subset \Omega$.

As noted earlier, we will also allow for the players to have subjective priors. Hence, as a generalization of (iv), we have:

(vi) A collection of subjective probability measures $\{P^i\}$ on $(\Omega, \mathcal{F})$ which describes the subjective beliefs on each Player i on the random events in the model. This subjective randomness may involve the cost functions of the other players in the system as well as the actual exogenous random variables.

Under this model, (the information structure of) a sequential game problem is *dynamic* if the information available to at least one player is affected by the action of at least one other player. A decentralized problem is *static*, if the information available at every player is only affected by exogenous disturbances (Nature) $\omega \in \Omega$; that is, no other player can affect the information at any given player.

As in team theory, information structures in games can also be classified as *classical*, *quasi-classical* or *nonclassical*: An Information Structure (IS) $\{y^i, 1 \leq i \leq N\}$ is *classical* if y^i contains all of the information available to **DM**k for $k < i$. An IS is *quasi-classical* or *partially nested*, if whenever u^k, for some $k < i$, affects y^i through the measurement function η^i, y^i contains y^k (that is $\sigma(y^k) \subset \sigma(y^i)$). An IS which is not partially nested is *nonclassical*.

Figure 3.1 of the team setting would also be relevant for static stochastic games.

10.1.2 An Agent-Wise Intrinsic Model for Games with Players Acting Multiple Times

Under the intrinsic model presented above, every DM acts separately and only once. As also noted in Sect. 3.5, this interpretation has significant limitations: Depending on the setup, it may be more suitable to consider an agent playing repeatedly over multiple stages as a collection of one-shot DMs acting as a team (so that collections of teams may be viewed as taking part in a game). To formalize this for sequential games where collections of one-shot DMs cooperate among themselves as a team, to play sequentially against other collections of DMs (other agents), we introduce N-player games, where each agent is a collection of (one-shot) DMs (similar to the setting of Sect. 3.5). Hence, we have, as a formal description, the following:

(i) Let $\mathcal{N} := \{1, 2, \ldots, N\}$ denote the set of players and for each $i \in \mathcal{N}$, introduce a subset (a team) TE^i of a set $\mathcal{M} := \{1, 2, \ldots, M\}$ denoting a collection of DMs, **DM**k for $k \in \mathrm{TE}^i$, acting as player i (or Agent **A**i) (said another way, **A**i encapsulates the collection of DMs indexed by TE^i acting $|\mathrm{TE}^i|$ times, where

$|\cdot|$ denotes the cardinality of the set TE^i). The observation and action spaces are standard Borel spaces for each $\mathbf{A}i$ ($i \in \mathcal{N}$), denoted by $\mathbf{Y}^i := \prod_{k\in \mathrm{TE}^i} \mathbb{Y}^i_k$, and $\mathbf{U}^i := \prod_{k\in \mathrm{TE}^i} \mathbb{U}^i_k$, respectively.

(ii) The $\mathbb{Y}^i_k$-valued observation variables are given by $y^i_k = h^i_k(\omega, \{u^p_s\}_{(s,p)\in L^k_i})$, where L^k_i denotes the set of all DMs acting before $\mathbf{DM}k$ of $\mathbf{A}i$ (i.e., $(s, p) \in L^k_i$ if $\mathbf{DM}s$ of $\mathbf{A}p$ act before $\mathbf{DM}k$ of $\mathbf{A}i$ for all $p \in \mathcal{N}$ and $s \in \mathrm{TE}^p$).

(iii) An admissible policy for each $\mathbf{A}i$ is denoted by $\boldsymbol{\gamma}^i := \{\gamma^k_i\}_{k\in\mathrm{TE}^i} \in \boldsymbol{\Gamma}^i$ with $u^i_k = \gamma^i_k(y^i_k)$, where the set of admissible policies for each agent is denoted by $\boldsymbol{\Gamma}^i := \prod_{k\in\mathrm{TE}^i} \Gamma^i_k$ for $i \in \mathcal{N}$. An admissible policy tuple for all players in the game is denoted by $\underline{\boldsymbol{\gamma}} := \{\boldsymbol{\gamma}^1, \cdots, \boldsymbol{\gamma}^N\} \in \boldsymbol{\Gamma}$ where $\boldsymbol{\Gamma} := \prod_{i\in\mathcal{N}} \boldsymbol{\Gamma}^i$. When needed, admissible policies are generalized to randomized policies by appending independent [0, 1]-valued random variables to each y^i_k.

In addition to the above, we consider the direct generalizations of (iv)–(v)–(vi) of the previous sub-section to the agent-wise intrinsic model introduced here, so that an underlying probability measure is considered (as in (iv)), a collection of loss functions (one for each agent) is present (as in (v)), and a collection of subjective probability measures (one for each agent) is to be considered (as in (vi)).

Remark 10.1.1 In a dynamic game, where a player may act repeatedly at numerous time stages (with different and possibly increasing information as time progresses), the above can be viewed to be such that all player actions at different times can be considered to be taken by different players, but who share the same subjective probability (though only the prior probability, which may be revised by the increasing information), and the same cost function of interest. That is, $\mathbf{A}i$ corresponding to the controller/player acting repeatedly may share the subjective cost and probability measures. In this case, the setup can be viewed as one where teams of players would be playing against other teams of players. If the information is expanding for this *aggregate* player, we will have a nested information structure for such players. If the information is not expanding for this subset of players (that is, if a player acting repeatedly does not have perfect recall), then this would amount to a situation where teams of players are playing against teams of players with the set of team policies not necessarily being convex in view of Remark 4.1.1, in the absence of relaxations. This will be discussed further, in particular in Sect. 11.6.

10.2 Solution Concepts for Dynamic Games

Solution concepts for stochastic dynamic games entail richness and complexity (or flexibility) which lead to further subtleties, as we present in this section. This complexity is perhaps the primary reason why studying information structures in stochastic dynamic games is particularly challenging.

We refer the reader to the texts [97, 281, 532, 568] for more detailed analysis on some of the following, though in different setups and contexts. In view of the information structure modeled above, we present the following concise description of solution concepts.

The general framework we have provided for stochastic dynamic games in the previous section would not be complete without the introduction of solution concepts so as to pin down the underlying structure according to which "equilibrium" policies of the players can be selected. We have already introduced such game-theoretic equilibrium solution concepts in Chap. 9 in the context of static games, particularly Nash and Stackelberg equilibria for non-zero-sum games, and saddle-point equilibria for zero-sum games. These apply equally well to dynamic games provided that they are brought to strategic (normal) form, but the richness of information structures in the latter lead to additional solution concepts, or refinements of standard ones, that exploit the sequential form introduced in the previous section.

Accordingly, we work in this section with the model introduced in the previous section, and particularly in Sect. 10.1.2. Toward that end, we first introduce the cost function for Player i, given by

$$J^i(\underline{\gamma}) = E^{i,\underline{\gamma}}[c^i(\omega_0, \mathbf{u})] = \int P^i(d\omega)c^i(\omega, \mathbf{u}), \qquad i \in \mathcal{N},$$

where $u^i = \gamma^i(\cdot)$, and to avoid notational complexity, we embed all the uncertainty in ω. As discussed in Sects. 10.1.1 and 10.1.2, the cost or the probabilistic model may be subjective (that is, player-dependent). In the following, we also take, as introduced earlier, Γ^i to be the policy space of Player i over the entire game, compatible with the information structure for that player as well as her action sets.

10.2.1 Nash Equilibrium

The definition of Nash equilibria, for a game in strategic form, is as presented earlier in Definition 9.2.3. For an N-player game, a family of policies $\{\gamma^{*,1}, \gamma^{*,2}, \cdots, \gamma^{N,*}\}$ is said to constitute a **Nash equilibrium** if, for all $i \in \mathcal{N}$,

$$J^i(\gamma^{*,i}, \gamma^{*,-i}) \leq J^i(\gamma^i, \gamma^{*,-i}) \quad \forall \gamma^i \in \Gamma^i. \tag{10.1}$$

10.2.2 Stackelberg Equilibrium

The definition of Stackelberg equilibria is as presented earlier in Definition 9.2.5. We again naturally have two players, designated as a leader and a follower, where the leader announces her strategy and the follower responds to the announced strategy. The leader is not allowed to backtrack from this commitment, unlike this possibility in the case of Nash equilibrium.

The pair of policies $\{\gamma^{*,1}, \gamma^{*,2}\}$ is said to constitute a **Stackelberg equilibrium** if

$$J^1(\gamma^{*,1}, \gamma^{*,2}(\gamma^{*,1})) \leq J^1(\gamma^1, \gamma^{*,2}(\gamma^1)) \quad \forall \gamma^1 \in \Gamma^1, \tag{10.2}$$

where $\gamma^{*,2}(\gamma^1)$ satisfies[1]

$$J^2(\gamma^1, \gamma^{*,2}(\gamma^1)) \leq J^2(\gamma^1, \gamma^2(\gamma^1)) \quad \forall \gamma^2 \in \Gamma^2. \tag{10.3}$$

As indicated earlier, sequential forms of dynamic games allow for further refinements on both solution concepts defined above for games in strategic form. We discuss in the remaining subsections of this section some of these refinements.[2]

10.2.3 Perfect Bayesian Equilibrium and Backward Induction

The Perfect Bayesian Equilibrium concept requires that the players are sequentially rational at every instant of time (of decision in a sequential game) in the sense that the expected future costs/pay-offs given the local information state are optimized in an expectation (Bayesian) sense for all players, conditioned on (almost every, under the induced probability/strategic measure) history/information. This leads to the notion of *continuation costs/pay-offs*, which results when, given the past, the future is considered optimally by every player.

As such, the Perfect Bayesian Equilibrium requires sequential rationality, and in some sense, there is again *commitment*: future players know and trust that the earlier players have acted according to their optimal policies and they respond optimally themselves. Accordingly, earlier players are committed to their policies and the future ones respond optimally. In some situations, the past policies do not matter, in which case the solution will turn out to be strongly-time-consistent. More on this will be presented later in the chapter.

This approach, in particular, leads to a backward induction or dynamic programming reasoning for the computation of equilibrium policies, an approach which has been applied commonly for such setups.

---

[1] We note that here we do not imply that there is a unique follower reaction determined by (10.3), typically such a $\gamma^{*,2}(\gamma^1)$ will be non-unique. In this context, the discussion and refinements presented in Remark 9.2.4 apply to the setup here as well.

[2] An instructive setup, also highly relevant to the theme of the book, where the distinction between a Nash and a Stackelberg equilibrium solution concepts is mathematically and operationally significant is in the context of signaling games. This subtle difference of commitment distinguishes Nash equilibria (as in the classical setup studied in [189]) from Stackelberg equilibria in signaling games, as discussed in detail in Chap. 18 (where the Stackelberg setup is also called Bayesian Persuasion).

For clarity in presentation, let us first consider the setup of Sect. 10.1.1. Considering $\gamma^{[1,N-1]}$, γ^N is optimized and then, inductively proceeding with given $\gamma^{[1,N-2]}$, $\gamma^{[N-1,N]}$ are optimized (with γ^N already determined) in terms of the best-response maps for each Player. We note that this backwards induction is in terms of policies, and requires a careful construction of controlled Markov dynamics. In fact, the Universal Dynamic Programming algorithm presented in Sect. 5.4 is applicable as a dynamic program for general sequential games under the perfect Bayesian equilibrium concept (for a related analysis, see [163]).

The discussion above can also be generalized to the setup presented in Sect. 10.1.2 where players can act over multiple time stages.

In this context, the closely related notion of **subgame perfection** in games with complete and common information corresponds to a solution concept where each DM responds optimally for all past realizations (almost surely). Often, in stochastic games with incomplete information, one cannot achieve this in such a strict sense (as the past realizations of states, actions and measurements are not available to a player at the time of control policy execution) and the performance criterion in an expectation sense conditioned on the given (local) information is the appropriate one. Accordingly, Perfect Bayesian Equilibrium is a more appropriate concept where the future given the local information state is optimized in an expectation sense. A further restriction is known as *sequential equilibrium*, to be discussed further below.

The discussion above also highlights the intricacies with respect to time-consistency of solutions: If the information structure does not have perfect recall (and classical information) then the future decision makers have to run a Bayesian estimate with the knowledge of announced policies and not the executed policies. Thus, if the executed policies are made public, we can claim strong-time-consistency, but if they are not made public upon execution, we cannot always claim strong-time-consistency of perfect Bayesian policies. See Remark 2.3.1 for an illuminating example from classical single-controller stochastic control theory.

We now consider the following example.

Example 10.2.1 The concept of non-credible threat demonstrates rather effectively the differences between the Nash equilibrium concept and its perfect Bayesian or subgame perfect equilibria refinements. Consider the following *Entry Game*. A potential entrant, **A**1, decides on whether to enter a market or not. The actions of **A**1 are visible to **A**2. The market is controlled by a monopolist, **A**2. If the entrant enters the market, the monopolist can start a price war (i.e., fight) or share the market. The table below presents some sample utility (presented as negative of the cost) values. It follows that there are two Nash equilibria: (Enter / Share) and (Out/Fight). On the other hand, only the first equilibrium is the one that is subgame perfect (and thus arrived via backwards induction): if **A**1 enters, **A**2's rational response would be to share, and not fight. The fact that **A**1 acts before **A**2, and **A**2 observes this action gives advantage to **A**1. The subgame perfect equilibrium here is also Stackelberg with **A**1 as leader. Since **A**2 observes the action of **A**1, Stackelberg with **A**1 as leader makes sense, which gives him the highest possible payoff. If, on the other hand, **A**2 is leader (regardless of whether he observes **A**1's action or not, in this case), then he could announce Fight and commit to it (before **A**1 acts), in which case he ends up with the highest payoff, 2, and **A**1 ends up with the lower payoff of 0. And this corresponds to the other NE. A conclusion is that backwards induction is not applicable for all Nash equilibria, but it applies for subgame perfect equilibria. This discussion also implies that, in the language of information structures studied, even partially nested information structures may admit Nash equilibria that are not subgame perfect and not every equilibrium can be arrived at via dynamic programming.

A2 (columns), **A1** (rows)

	Fight	Share
Enter	-1	1
Out	0	0

$-J^1$

A2

	Fight	Share
Enter	-1	1
Out	2	2

$-J^2$

◇

10.2.4 Markov Perfect Equilibria

Markov perfect equilibrium is one where there is a controlled Markovian state, and each player responds optimally given the current state of the Markovian play. Here, the policies are restricted to those that are Markov, that is, they use the state as a sufficient statistic at each time stage. One should be cautious in considering this definition: the **state of a Markovian model** can be rather design-dependent with significant level of mathematical flexibility, as we have seen earlier in the book (in Chap. 5).

10.2.5 Time-Consistency Revisited

Recall Definition 2.3.2, revised here only slightly in view of the game theoretic formulation. First consider the one-shot DM setup considered with Sect. 10.1.1.

Definition 10.2.1 The collection of policies $\underline{\gamma}^* = \{\gamma^{1,*}, \cdots, \gamma^{N,*}\}$ is weakly time consistent (WTC) if $\gamma^{k,*}$ is a best response to $\{\beta^1, \cdots, \beta^{k-1}\}$ when $\{\beta^1, \cdots, \beta^{k-1}\} = \{\gamma^{1,*}, \cdots, \gamma^{k-1,*}\}$, for every $k \in \mathcal{N} \setminus \{1\}$. The solution $\underline{\gamma}^* = \{\gamma^{1,*}, \cdots, \gamma^{N,*}\}$ is strongly time consistent (STC) if $\gamma^{k,*}$ is a best response to every admissible $\{\beta^1, \cdots, \beta^{k-1}\}$.

Now, we generalize the definition above and consider the multi-stage setup studied in Sect. 10.1.2 where players act over multiple time stages.

Definition 10.2.2 Let there be a multi-stage dynamic game, played over the time horizon $\mathcal{T}$, which admits an equilibrium solution $\underline{\gamma}^* \in \boldsymbol{\Gamma}$. Let $t > 1$ be an arbitrary point in $\mathcal{T}$, and consider the decision problem $D^{\beta}_{[t,T]}$ which is derived from D by setting $\underline{\gamma}_{[1,t-1]} = \underline{\beta}_{[1,t-1]}$, for an arbitrary admissible $\underline{\beta}_{[1,t-1]} \in \boldsymbol{\Gamma}_{[1,t-1]}$. Then:

(i) The solution $\underline{\gamma}^* \in \boldsymbol{\Gamma}$ is *strongly time consistent* (STC) if the subpolicy $\underline{\gamma}^*_{[t,T]}$ constitutes an *equilibrium* to the dynamic game $D^{\beta}_{[t,T]}$, this being so for every $t \in \mathcal{T}, t > 1$, and every permissible $\underline{\beta}_{[1,t-1]} \in \boldsymbol{\Gamma}_{[1,t-1]}$.
(ii) The solution $\underline{\gamma}^* \in \boldsymbol{\Gamma}$ is *weakly time consistent* (WTC) if the subpolicy $\underline{\gamma}^*_{[t,T]}$ constitutes an *equilibrium* to the dynamic game $D^{\beta}_{[t,T]}$ when $\underline{\beta}_{[1,t-1]} = \underline{\gamma}^*_{[1,t-1]}$.

◇

As in the team theoretic setup, if an equilibrium solution is *STC*, then the past policies do not rein in the present and future policies of the players under the same solution concept, i.e., the players have no reason to renege (and deviate from the equilibrium policy or the course of action) even if some inadvertent deviations have taken place in the past. With the WTC solution, however, there is no incentive to renege only if the declared course of action has been followed in the past.

The above discussion also highlights the distinction between partially nested information structures and the partially nested information structures with control sharing. In general, we will have strong time consistency in the latter case but not in the former one. See Remark 2.3.1 for a discussion on the information structure dependence properties of WTC and STC solutions. STC solutions often require the presence of the Markovian state information (under *some* controlled Markov formulation of the problem), a special case of which is the perfect recall condition. We refer the reader to [88] for a detailed discussion on time-consistency in dynamic games.

Remark 10.2.1 (A Discussion on Relations Between Different Solution Refinements) It is evident from the above that the different refinements of Nash equilibria are related through inclusion or equivalence relations under certain conditions. For example, every perfect Bayesian equilibrium has to be WTC for stochastic dynamic games with the information structures in Sect. 10.1.1. This may also hold for the STC case, under stronger conditions (if the Bayesian beliefs are also performed on events with measure zero under the supposed equilibrium policies and backwards-induction optimality is considered and, as noted earlier, the executed policies are announced). A converse, from WTC to perfect Bayesianness, may hold when a controlled Markov optimality argument is feasible via the arguments presented above and in Chap. 5, tailored to the game theoretic setup.

In Sect. 10.1.2, the discussion was more nuanced. Not every WTC policy is subgame-perfect or perfect Bayesian: A Stackelberg solution with commitment to be studied below is such an example; such a policy cannot be arrived at by backward induction yet it is consistent by design of commitment (in the policy space for the leader with no backwards induction with regard to the follower's policies). See [400] and [261] for related discussions. Thus, we can state that perfect Bayesianness is a strictly stronger refinement than the WTC property and not every WTC solution can be arrived at via dynamic programming or backward induction.

As noted earlier, there is a close relationship between Perfect Bayesianness and subgame perfection, which often has been studied under the assumption of perfect recall and finite game setups in the literature. Nonetheless, we will not present a comprehensive discussion on comparison of these concepts, because even the notion of Markovianness can admit a rather broad interpretation, as we have seen earlier in the book in Chap. 5 and a complete theory is not yet available, perhaps in part because a game theoretic counterpart of the Universal Dynamic Programming studied in Sect. 5.4 is yet to be established in the most general form possible.

10.2.6 Global vs. Stage-Wise Stackelberg Equilibria for Multi-Stage Games

Consider a multi-stage game where Player 1 and Player 2 each act N times:

$$x_{k+1} = f(x_k, u_k^1, u_k^2, w_k)$$

with w_k an i.i.d. noise. Let policies of the players be $\gamma^i = \{\gamma_k^i, k = 0, .., N-1\}, i = 1, 2$. Each γ_k^i maps her local information, denoted with I_k^i, to some space of actions so that $u_k^i = \gamma_k^i(I_k^i)$.

Suppose that the Players wish to minimize, for some $T > 1$:

$$J^i(x, \gamma^1, \gamma^2) = E^{\gamma^1,\gamma^2}[\sum_{k=0}^{T-1} c^i(x_k, u_k^1, u_k^2)], \qquad i = 1, 2. \tag{10.4}$$

In view of the definitions involving Stackelberg and perfect Bayesian equilibria, we present two further refinements in the following. Let us make the setup more concrete.

Definition 10.2.3 (Stackelberg with Leader Commitment (Global Solutions)) Player 1, as leader, announces the *entire policy profile* γ^1. Player 2, as follower, responds to this entire profile (possibly via backwards induction); Player 1 optimizes his policy profile accordingly, resulting in equilibria (provided one exists).

The Stackelberg definition above implies that Player 1 announces the entire collection of policies (over the time horizon) and Player 2 responds as a follower. Thus, Player 2 would respond to a fixed committed collection of Player 1 policies for future time stages. This definition would not necessarily lead to a dynamic programming formulation. Furthermore, the solution would not necessarily be strongly time consistent or even weakly time consistent. As a special case, if the state information x_t is available, and Player 1 is restricted to use only Markov commitment policies; then, from Player 2's perspective, he is faced with a classical MDP and an optimal solution can be obtained with again memoryless (Markov) policies. However, from Player 1's perspective, the evaluation of an equilibrium policy will be highly complicated due to the dependence of Player 2's policies on Player 1's future policies even if the structure of Player 2's policy is Markovian.

This gets even further complicated if Player 1's policy is arbitrary: In this case, Player 2's best response does not have to be Markovian (state-based) even if the full state were available. While such an equilibrium is computationally challenging to find, as our discussion below reveals, this equilibrium may significantly benefit the leader. See [97, 108] for further discussions and applications. In the following, we present an illustrative example.

Example 10.2.2 ([97, Example 7.1]) Consider a two-stage game:

$$x_2 = x_1 - u^2, \qquad x_3 = x_2 - u^1$$

with cost functionals

$$c^1(x_{[1,3]}, u^1, u^2) = (x_3)^2 + 2(u^1)^2 + \beta(u^2)^2, \qquad \beta \geq 0$$

$$c^2(x_{[1,3]}, u^1, u^2) = (x_3)^2 + (u^2)^2$$

Player 1, the leader, acts at time stage 2 and has access to x_1, x_2 prior to selecting u^2. The follower, Player 2, acts at stage 1 and has access to x_1. In this case, it can be seen that obtaining a solution through variational methods is rather difficult as it involves coupled non-linear optimality equations: given any announced policy of the leader (Player 1), Player 2 will respond, but Player 1's policy explicitly uses the action of Player 2 as an input which itself depends on the policy of Player 1 and this dependence *a priori* may not have any regularity properties (except measurability).

Rather than solving this problem via optimization or variational methods (which are not applicable due to lack of apriori smoothness), [97] develops an approach of first obtaining a lower bound, by imposing that Player 2 also minimize c^1, instead of c^2 (which then serves as a lower bound for the cost of Player 1 under any policy: note that this defines a team problem with the common cost being c^1) and then showing that Player 1 can enforce this global minimum by selecting a particular control policy (via threatening the follower): A resulting commitment Stackelberg solution pair then is

$$\gamma^{1,*}(x_1, x_2) = 1_{\{x_2 = \frac{3\beta}{3\beta+2}x_1\}} \frac{1}{3} x_2 + 1_{\{x_2 \neq \frac{3\beta}{3\beta+2}x_1\}}(-k x_2)$$

$$\gamma^{2,*}(x_1) = \frac{2}{3\beta + 2} x_1.$$

where $k > 0$, otherwise arbitrary. Notably, $\gamma^{1,*}$ is not a policy that uses only x_2! The form of the policy shows that Player 1 threatens Player 2 to act as he wishes so that $x_2 = \frac{3\beta}{3\beta+2} x_1$ is satisfied. Thus, this policy is not of Markov type, it entails memory, and this policy cannot be arrived at via backwards induction under the Perfect Bayesian Stackelberg Equilibrium concept. It turns out that the induced policy can also be realized by the linear policy:

$$\tilde{\gamma}^{1,*}(x_1, x_2) = \frac{1}{3} x_2 + \frac{2\beta - 3}{3\beta}(x_2 - \frac{3\beta}{3\beta + 2} x_1)$$

with the same $\gamma^{2,*}$ as above. Furthermore, one can also obtain non-linear Stackelberg solutions which lead to the same equilibrium value for the leader. ◇

While in Example 10.2.2 above, the leader managed to achieve his global optimum in an equilibrium, the solution does not always have to be the global minimum; by which we mean the corresponding team problem with the team cost being that of the leader. Thus, not all admissible values can be enforced by the leader via threat. Such a scenario is given in [97, Example 7.2] where the follower acts twice.

Consider now Example 10.2.1 as a two-stage Stackelberg game where Player 2 is the leader announcing his policy (but Player 1 acts first, prior to Player 2 in the 2nd stage). If we view Player 2 as committed to play Fight (which would not be feasible if we had subgame perfection as discussed in the example), he receives a better reward. Commitment pays off for the leader as a non-credible threat! Recall that this policy, as noted in the Example above, is not perfect Bayesian -this is why the threat is said to be *non-credible*-. This policy is not weakly or strongly time-consistent: If Player 1 instead enters, Player 2's response should be *share*; so *fight always* is not weakly time consistent.

In many applications, however, the leader may not have dominance over the entire horizon, but only stage-wise, leading to an implementation of the leader-follower decision-making process stage-wise, via backwards induction under the perfect Bayesian solution concept. Following [83], we will refer to such a solution concept *Stagewise Stackelberg Equilibrium*. We note that [97] also refers to such an equilibrium as *Feedback Stackelberg*. Feedback Stackelberg equilibria for continuous-time differential games were studied in [93].

Definition 10.2.4 (Stagewise Stackelberg Equilibrium) A Stagewise Stackelberg Equilibrium (or Feedback Stackelberg Equilibrium) is an equilibrium arrived at via backward induction. Let the time horizon be $T - 1$. In this case, Player 1 (as the leader) at the terminal time stage announces a policy, γ^1_{T-1}, which may depend on the entire past (in anticipation of Player 2's response). Player 2 (as the follower) at the terminal time stage responds with an optimal policy (as anticipated by Player 1).

A value for the stage equilibrium $T-1$ is then computed; and the same recursions are applied for earlier time stages leading to a (possibly generalized) dynamic program, with a static Stackelberg game solved at each step of the iteration.

In case of a controlled Markov model where the state information is available, if Player 1 announces γ^1_{T-1}, which may depend on the entire past, Player 2 would respond with an optimal policy which would be Markovian. Player 1's response at time $T-1$ to this anticipated response of Player 2 will also be Markovian if full state is available. In the case of full state information, then, the equilibrium value at $T-1$ will only depend on x_{T-1}. This facilitates a tractable backward induction. In this case, the solution will be strongly time consistent. If full state is not available, one can again define a Stagewise Stackelberg Equilibrium (which may perhaps also be called a *Perfect Bayesian Stackelberg equilibrium*) with the same order of reasoning, but the solution will necessarily lead to a complicated backward induction and the solution may only be weakly time consistent.

It is also evident that every Stagewise Stackelberg Equilibrium induces a policy for Player 1. If Player 1 commits to this policy, the best response of Player 2 will be unaltered under a Stackelberg Equilibrium with Leader Commitment.

The observations above lead to the following theorem.

Theorem 10.2.1

(i) *A Stagewise Stackelberg (Feedback Stackelberg) Equilibrium is also admissible under Stackelberg Equilibrium with Leader Commitment. This inclusion may be strict in that the set of Stackelberg Equilibria with Leader's Commitment (Global Stackelberg) may include equilibria that are not Perfect Bayesian Stackelberg.*

(ii) *Under perfect recall and full state information, a Perfect Bayesian Stackelberg Equilibrium can be generated via backward induction. This solution is strongly time consistent. In the general case, the solution will be weakly time consistent.*

(iii) *Not every Stackelberg Equilibrium with Leader's Commitment is strongly or weakly time consistent. Not every such solution can be obtained via backward induction.*

10.2.7 Saddle-Point (Nash) Equilibria via Markov Policies Under State Information for Zero-Sum Games

Consider again (10.4) and assume the zero-sum setup so that $c^1 = -c^2$ and that the information pattern allows for perfect state access such that $\sigma(x_t) \subset \sigma(I^i_t)$ for both players.

In this case, observe that if one player adopts a policy which is Markovian (that only depends on the current state and time), the best response of the remaining player will also have the same structure as that player will be solving a classical stochastic control problem as studied in Theorem E.2.2. Therefore, if an equilibrium exists among Markov policies then this would be a saddle point leading to the unique value of the game under any other equilibria (possibly under non-Markovian policies). As will also be revisited in Theorem 10.3.1, this follows from the ordered interchangeability property (Theorem 9.2.1).

10.2.8 Some Additional Equilibrium Refinements

Sequential Equilibrium

Sequential equilibrium is a stricter version of perfect Bayesian equilibrium, in the sense that previous moves by players may not have been executed according to the actual equilibrium (PBE) policy. Thus, a player may need to make a decision at a future stage at an information state which would normally not be reachable had the prior actions been consistent with the equilibrium policies. Accordingly, the updates on events that should not have occurred must also be considered; however, we caution the reader that there exist subtle distinctions between this concept and strong time consistency presented earlier with regard to how the zero-measure events are to be handled (that is, the conditional probabilities and resulting equilibria on measure zero events). There are further refinements and conditions presented in the literature, see e.g. [568, Chapter 12] for a critical and comprehensive review. A notable concept is that of *Trembling Hand Perfect Equilibrium* which requires robustness of the game to genuine (non-strategic) perturbations in the execution of equilibrium policies.

(Bayesian) Subjective Equilibrium

Introduced by Kalai and Lehrer [382], an equilibrium in a stochastic dynamic game is *subjective* if each player responds to the environment in a Bayesian fashion. A player's response must be consistent with her belief of the other players, via the *closed-loop* effects of the response of the other players to her policy. We note that perfect Bayesianness may not necessarily hold under this equilibrium concept.

Mean-Field Equilibrium

For a class of stochastic games where the players are coupled through mean-field terms in either the cost or the dynamics (or both), the notion of mean-field games [361, 362, 456]) can be viewed as limit models of symmetric non-zero-sum non-cooperative N-player games with a mean field interaction as $N \to \infty$. The mean-field approach designs policies for both cases of games with infinitely many players, as well as games with large numbers of players, where the equilibrium policies for the former are typically shown to be ϵ-equilibrium for the latter. One significance of mean-field equilibrium when applied to large but finite population stochastic dynamic games is that it leads to an approximate Nash equilibrium solution when an exact Nash equilibrium cannot be obtained because of the complex strategic interactions among players when the information structure is decentralized across players. This is analogous to our findings in near-optimality analysis presented for large teams in Chap. 6.

(Non-Bayesian) Subjective Equilibrium Under Beliefs of Models or Exogenous Uncertainty

In many applications in which agents face complex environments, they respond to their environments under imperfect or incorrect modeling assumptions. A subjective equilibrium in which a player believes that the closed-loop effects of her actions do not exist, generalizing the notion of *price-taking equilibrium* [322], is called *subjective equilibrium under beliefs of exogenous uncertainty* [38]. For large numbers of decision makers, under mean-field dynamics and cost-coupling, it can be shown that such an equilibrium is indeed a Nash equilibrium. A further related subjective equilibrium concept, where best-responses under an assumed structural model are consistent with the perceived model, has been presented in [794]. A further related equilibrium concept will be studied, in the context of reinforcement learning, in Chap. 23.

10.3 Information Structure Dependent Properties of Equilibria in Game Theory

In the following we present some information-structure dependent properties of equilibria. We emphasize a number of game theoretic attributes on information structures which lead to richness, as well as complexity and challenges, in the study of information structures in stochastic games.

10.3.1 Value of Additional Information Can Be Positive or Negative in Non-Zero Sum Games

Relationships between ISs and non-cooperative (particularly, Nash) equilibria in non-zero-sum games are quite complex, with one of the paramount characteristics on that front being the following: More information to a given player does not necessarily result in better performance for that player, as to be demonstrated through Examples 10.3.1 and 10.3.2 in the following. In general games, information can have both positive and negative value to a player since additional information can lead to a perturbation which is not necessarily monotone due to the presence of competitive and strategic interactions among players, unlike the case of stochastic teams.

Example 10.3.1 ([352, 356] An Example with Static Information Structure) Let $\mathbb{X} = [0, 1]$, with prior distribution ζ defined by the continuous uniform distribution. Let $\mathbb{Y}^1 = \mathbb{Y}^2 = [0, 1]$.

We define the action spaces of the respective players as $\mathbb{U}^1 = \mathbb{U}^2 = [0, 1]$, and the cost functions as:

$$c^1(x, u^1, u^2) = \begin{cases} (x - u^1)^2 + 2, & u^2 = x \\ 2(x - u^1)^2 + (u^1 - u^2)^2, & u^2 \neq x \end{cases}$$

$$c^2(x, u^1, u^2) = \begin{cases} (x - u^2)^2 + 2, & u^1 = x \\ 2(x - u^2)^2 + (u^1 - u^2)^2, & u^1 \neq x \end{cases}$$

Under measurement channels where neither player has any information regarding x, (which could be achieved for instance by measurement channels returning $y^1 = y^2 = 0, \forall x \in \mathbb{X}$), there exists a unique Nash equilibrium solution given by $u^1 = u^2 = \frac{1}{2}$. Uniqueness can be seen as follows: the event that $u^1 = x$ or $u^2 = x$ has zero measure and therefore the complementary conditions are active for defining the cost in the above. Note though that in this case the cost is essentially a team cost since minimization over u^1 of

$$2(x - u^1)^2 + (u^1 - u^2)^2$$

will lead to the same solution as the minimization of

$$2(x - u^1)^2 + (u^1 - u^2)^2 + 2(x - u^2)^2.$$

The same applies for c^2, and the players solve essentially the same problem. Therefore, we have a standard static quadratic team problem which admits a unique optimal solution (see Theorem 2.4.7) with $u^1 = \frac{1}{2} = u^2$. This leads to a cost of $\frac{3}{2}$.

However, when the player measurement channels are such that both players have perfect information regarding x (e.g. through $y^1 = y^2 = x$), we first see that another equilibrium solution arises: $u^1 = u^2 = x$. One can verify that this is a Nash equilibrium by noting that if Player 2 holds their strategy constant as $u^2 = x$, then Player 1 will play the game with cost function $|x - u^1| + 2$, which is minimized by playing $u^1 = x$. The same holds true for Player 2 in the reverse case where Player 1's strategy is fixed. We now make the point that this is the unique such deterministic Nash equilibrium:

The Nash equilibrium under no information, $u^1 = u^2 = \frac{1}{2}$, is no longer an equilibrium under perfect information. We can see this by noting that if Player 2 holds her strategy constant as $u^2 = 1/2$, then Player 1 will almost surely be tasked with minimizing $2(x-u^1)^2+(u^1-\frac{1}{2})^2$. However, now that Player 1 knows x, and because the first term is weighted more heavily, the cost is minimized by Player 1 adopting $u^1=\frac{2}{3}x+\frac{1}{3}u^2$.

If neither of the equalities in $u^1=x=u^2$ were to hold, we would again reduce to a team problem and in this case, we would have

$$\gamma^1(x)=\frac{2}{3}x+\frac{1}{3}\gamma^2(x), \qquad \gamma^2(x)=\frac{2}{3}x+\frac{1}{3}\gamma^1(x)$$

leading to, by uniqueness (see Theorem 2.4.7), $\gamma^1(x)=x=\gamma^2(x)$, which however would take us back to the inefficient equilibrium given above. Thus, the unique pure strategy Nash equilibrium for the full-information problem is at $u^1=u^2=x$. This results in an expected cost of 2 for each player.

This example has demonstrated that in a static game it is possible for more information to hurt both players.

Example 10.3.2 ([102] An Example with Dynamic Information Structure) Consider a card drawn at random from a deck, where its color can be either red or black, each with probability 1/2. Player 1 first declares his guess of the color, and then, after hearing what Player 1 guessed, Player 2 submits her guess for the color. If players guess the same color, the payout is for \$2 each, whereas if one player guesses correctly, that player receives a payout of \$6 and the other player receives \$0.

In the case where both players are uninformed about the color of the card, the expected payout is \$3 each, as Player 1's optimal strategy is arbitrary, and Player 2's optimal strategy is to guess the opposite color of what Player 1 guessed.

In the case where both players are informed of the color of the card prior to declaring their guess, the equilibrium for the game occurs when both players guess the true color of the card. In this case, the expected payout becomes \$2 for each player.

Thus, for general non-zero sum games, more information can hurt some or even all of the players in a system; see also [3, 90, 312, 348, 387]. As a further example, the reader is referred to [94] on the comparison of information structures for team-like (LQG) and zero-sum like (quadratic duopoly) games. We recall again that for team problems the value of information is never negative, and for zero-sum game problems we will see that more information does not hurt the player receiving it, as will be discussed further later in the chapter.

10.3.2 Salient Information-Structure Dependent Properties

1. The team theoretic or stochastic control argument of a player *choosing to ignore the additional information* or *expansion of information structures as a recipe for*

optimization certification (as in Sect. 3.8) do not apply in games, even in zero-sum games.

2. Equivalence of information structures, which was utilized crucially in our study of optimal team theory, no longer holds in general (in particular, for policy-dependent static reductions of information structures): As we have seen in Chap. 3, if a team optimal solution exists, then its reduction is also team-optimal. This equivalence no longer holds for non-zero-sum or zero-sum games (for equilibrium solutions) in general: Dynamic games and their static reductions lead to subtle differences and non-equivalence results. If information is expanded, e.g., by control-sharing, then there may be infinitely many, essentially non-unique, equilibria (in that, the induced costs are different) equilibria and there may not be one that is the most desirable for every player. Even existence properties are not preserved by information expansion, which was a property taken for granted in team theory.
3. Not every Nash equilibrium can be arrived at via backward induction / dynamic programming, even when the information structure is classical (that is, with perfect-recall). Accordingly, refinements to solution concepts are to be introduced.
4. We will see in Chap. 11 that, for general non-zero sum games, equilibrium values are not continuous in information structures even under total variation, unlike the setting in team theory studied in Chap. 8.

10.3.3 Zero-Sum Games Admit Several Regularity Properties

Zero-sum games, however, feature several positive attributes regarding information structure-dependent properties of equilibrium solutions:

1. Under mild conditions, every zero-sum game has a value (see, in particular, Theorems 11.2.1 and 11.2.2).
2. Optimality is greatly facilitated by the versatile ordered interchangeability property (see Theorem 9.2.1). This eliminates the equilibrium selection problem that even team problems suffer from.
3. A complete theory of ordering of ISs will be presented in Theorem 11.3.2, generalizing Blackwell's theorem covered in Theorem 7.3.4. In particular, an additional information cannot hurt a player, though this argument is quite subtle in the game case: once again the *choose to ignore the additional information and thus no information cannot hurt* argument is no longer applicable,[3] even though by Corollary 11.3.1 the implication that more information cannot hurt turns out to be correct via a much more delicate (geometric) argument.

[3] Of course, unless the presence of this additional information is not known by the other player.

4. Dynamic games and their static reductions, under existence conditions for both, will still maintain certain equivalence conditions. However, existence properties may not be isomorphic; that is, a solution may exist in a dynamic game but not in its static-reduced counterpart.
5. We will see in Chap. 11 that, as in team problems considered in Chap. 8, equilibrium values for zero-sum games are continuous in ISs under total variation and share similar (though, slightly refined) semi-continuity properties under weak/setwise convergences of ISs.

In view of the above, we can somehow interpret zero-sum games as an intermediate category of games between teams and genuine non-zero sum games as far as their informational properties are concerned.

10.3.4 Lack of Equilibrium Equivalence Under Policy-Dependent vs. Policy-Independent Static Reductions

In Sect. 3.4, we studied static reduction of dynamic teams, and established policy-dependent and policy-independent reductions. We have seen in multiple contexts the benefit of pursuing such a reduction. However, as seen in Sect. 3.6, equilibrium properties of a given team policy in a stochastic team may not be preserved under static reductions and in particular Nash equilibrium or stationarity properties are not isomorphic even when ISs are equivalent. The examples presented in Sect. 3.6 apply to game problems as well (by viewing person-by-person-optimality as Nash equilibrium). Furthermore, sufficient conditions establishing isomorphism relations can be obtained [182, 637]

In particular, we note that for the partially-nested setting (with policy-dependent static reduction) we have the following differences.

1. *Equilibrium solutions are not isomorphic under information expansions or static reductions.* Under quasi-classical patterns, unlike team problems, information expansions drastically change the nature of solutions in games: quasi-classicalness (or partial nestedness) does not typically allow for expansion of ISs even under deterministic policies without altering the nature of the game.
2. *Equilibrium solutions may have to be randomized unlike team optimality and hence static reduction may be inapplicable in the absence of control-sharing.* One other crucial difference with team theory is that since by our analysis earlier optimal policies are, without any loss, deterministic, static reductions are well-posed under quasi-classical information patterns. On the other hand, for games randomization is generally essential even for existence of equilibria, and accordingly, policy-dependent static reductions would be ill-posed for such problems.

On the other hand, the *policy-independent static reduction* considered earlier in view of the absolute continuity condition (3.26) will maintain its versatility in the game theoretic setup also.

1. Equilibria are isomorphic under (policy-independent) static reduction.
2. Stationary solutions are isomorphic under (3.38) adapted to players' individual criteria (see Theorem 3.6.1(ii)).
3. We will arrive at very general existence results for saddle point equilibria in zero-sum games in Theorem 11.2.2.
4. When we consider *partially nested information structures with control sharing (whenever measurements are shared)*, the results apply as in the policy-independent case (following Theorem 3.4.2).

Nonetheless, existence of a saddle point with dynamic IS does not necessarily imply the existence of a saddle-point in the corresponding static reduction, as exemplified below:

Example 10.3.3 ([637]) Consider a zero-sum game, with $u^1 = \gamma^1(y^1)$ and $u^2 = \gamma^2(y^1, y^2)$ and $y^2 = u^1 + v^2$ where y^1, v^2 are i.i.d. random variables with finite second moments. Let Player 2 be the minimizer and Player 1 the maximizer for the expected cost criterion

$$E[(u^1 + v^2 - u^2)^2 - \alpha(u^1)^2],$$

with $\alpha \in (0, 1)$. Note that the IS is quasi-classical. Now note that with the given information, Player 2 can make the first term zero, leaving Player 1 to maximize the second term, which would be zero. It is not difficult to see then that the game admits a unique saddle-point solution under the given dynamic information, which is $(\gamma^1 \equiv 0, \gamma^2(y^1, y^2) = y^2)$.

The static reduction is: $u^1 = \gamma^1(y^1)$ and $u^2 = \gamma^2(y^1, v^2)$. In this case, since Player 2 does not have u^1 in his information set, he cannot prevent Player 1 from driving the cost (which is now strictly convex for all γ^2) to infinity. Hence, there is no saddle-point equilibrium for the static reduction.

We next provide another example of a game along a similar theme, this time a deterministic nonzero-sum game, which admits Nash equilibrium with dynamic information, but not with static information:

Example 10.3.4 Consider a deterministic quadratic non-zero sum game, where Players 1 and 2 wish to, respectively, minimize the following cost functions, where b is a nonzero real number, and u^i, controlled by Player i, $i \in \{1, 2\}$, takes values on the real line:

$$c^1(u^1, u^2) = (u^1 - u^2 - b)^2,$$

$$c^2(u^1, u^2) = (u^1 - u^2)^2.$$

First viewing this as a static game, it is straightforward to see that the unique best response maps for Player 1 and Player 2 are, respectively, $u^1 = u^2 + b$ and $u^2 = u^1$, which however have no point of intersection (that is, no fixed point, since $b \neq 0$), and hence there exists no Nash equilibrium solution.

On the other hand, if we instead have dynamic information, where Player 2 has access to the action of Player 1, then Player 2's optimal (c^2-minimizing) policy, which is unique and independent of what Player 1 does, is $\gamma^2(u^1) = u^1$. This makes c^1 independent of u^1, and hence any real number for Player1 is trivially an optimal choice for him. Hence, dynamic information of the type given leads to a NE, non-unique in policies, but with a unique pair of NE costs ($J^1 = b^2, J^2 = 0$).[4] If we reverse the roles of the players in the dynamic information, and have Player 1 have access to Player 2's action, then it is straightforward to see that the unique NE policy for Player 1 is $\gamma^1(u^2) = u^2 + b$, with this time any real number for Player 2 being a NE policy for him. Hence, the NE policies are again non-unique, but the pair of NE cost values are unique, namely $J^1 = 0$, $J^2 = b^2$.

In particular, if the information structure is classical, and if we expand the information structure for future players, the way we did in Chap. 3, even the existence of solutions is not an invariant.

10.3.5 Deterministic Games and Information Structures

For deterministic problems, such subtleties had been observed earlier in the literature [78, 79, 630, 776], where the analysis is often more direct. Note that the static reduction of a deterministic dynamic game would be a game under open-loop deterministic policies.

1. As in the stochastic case, if a deterministic team optimal solution exists then its reduction is also team-optimal (where the reduction is an open-loop policy). This is no longer true for non-zero-sum or zero-sum games.
2. Open-loop and closed-loop solutions share some stronger isomorphism properties in that, whenever an equilibrium is realizable in open-loop for both zero-sum and non-zero-sum games; the same strategies also constitute closed-loop policies (but not every closed-loop realization is an equilibrium). See [97, Prop. 6.6] and [776, Proposition 5],[630], [773]. The reason is that if one player adopts a closed-loop policy while all the other players are applying open-loop policies, the game is unaltered except by actions of the said player. But these actions have to be optimal, so the open-loop solution will still be optimal [97, 630]. This applies for all the players, leading to the conclusion.
3. However, not every closed-loop equilibrium is realizable in open-loop. There are even cases where no equilibrium exists in open-loop but one does exist in closed-

[4] These are strongly time-consistent NE. There is possibility (in general) of other NE, but in this case it turns out that there aren't any others.

loop as we saw above. A policy can be represented in different informational forms (open-loop / closed loop / using redundancy etc.) and how it is represented, by way of the reactions of the other players, can drastically affect existence properties [78].

4. Comparison of ISs is significantly more direct, when compared with the stochastic case studied in the next chapter: For deterministic zero-sum games: if an equilibrium exists and if more information is provided to a given player, then the equilibrium values are unaltered [776, Prop. 2].
5. For non-zero sum games, equilibria may even be essentially non-unique, and different equilibria may lead to different equilibrium cost values. For zero-sum games, however, informational non-uniqueness leads to essentially unique equilibria (as noted earlier).
6. Linear quadratic games, even zero-sum games, may admit solutions which are non-linear [76].

Building on insights from [79] and [773], we present now relationships between saddle-point equilibria (SPE) of deterministic zero-sum dynamic games (ZSDGs) under different ISs. Consider the class of ZSDGs with system dynamics

$$x_{n+1} = f(x_n, u_n^1, u_n^2)$$

and cost functional

$$J(\gamma^1, \gamma^2) = c_N(x_N) + \sum_{k=0}^{N-1} c_k(x_k, u_k^1, u_k^2).$$

Three information structures, and thus three types of policies, are of interest:

(i) Open-loop policies ($\underline{\gamma}^o$): Each player has access only to x_0 for all time stages.
(ii) Closed-loop no memory ($\underline{\gamma}^m$): Each player has access to only the current value of the state at all time stages (Markov feedback policies)
(iii) Closed-loop ($\underline{\gamma}^c$): Each player has access to the entire past and current state information at each point in time, that is : $\{x_{[0,k]}\}$ at time k.

The following theorem now captures relationships between saddle-point solutions under these three different information structures. In the statement of the theorem, we will define an equivalence relation between two policies to be with respect to their open-loop representations; for example, $\gamma^{1,m} \equiv \gamma^{1,o}$ means that the open-loop representation of $\gamma^{1,m}$ is $\gamma^{1,o}$ and they lead to the same strategic measure or solution trajectory.

Theorem 10.3.1 ([79])

(a) If the deterministic zero-sum dynamic game formulated above admits a unique pure (deterministic) saddle point (Markov feedback) solution $\{\gamma^{1,m}, \gamma^{2,m}\}$ and if $\{\gamma^{1,o}, \gamma^{2,o}\}$ is any open-loop saddle-point solution, then

(i) $\gamma^{1,m} \equiv \gamma^{1,o}$.
(ii) $\gamma^{2,m} \equiv \gamma^{2,o}$.
(iii) $\{\gamma^{1,o}, \gamma^{2,o}\}$ *is the unique open-loop saddle point solution.*

(b) Let $\{\gamma^{1,m}, \gamma^{2,m}\}$ *be, as in (a) above, the unique Markov feedback saddle-point solution of the zero-sum dynamic game. Furthermore, let* $\{\gamma^{1,c}, \gamma^{2,c}\}$ *be any saddle-point solution in closed-loop policies. Then,*

(i) $\gamma^{1,m} \equiv \gamma^{1,c}$ *(i.e., their open-loop representations are the same).*
(ii) $\gamma^{2,m} \equiv \gamma^{2,c}$.

(c) Any open-loop saddle-point equilibrium also constitutes a saddle-point equilibrium within the class of closed-loop policies.

At the heart of a proof for the result above is the uniqueness property of equilibria and then the ordered interchangeability property (Theorem 9.2.1): the best response to an open-loop policy or a closed-loop policy of an opponent, can be expressed as an open-loop (static reduced) policy. For c) see e.g. [97, Prop. 6.6], [776, Proposition 5], [78, 630, 773].

The above of course also holds for the team-optimal setting: for deterministic team problems every closed-loop representation of an optimal open-loop solution is also optimal.

10.4 Linear Quadratic Games and Dependence of Equilibrium Solutions on Information Structures

In this section, we focus on linear-quadratic (LQ) games (in some cases with Gaussian distribution–so-called linear-quadratic-Gaussian (LQG) games), and study existence and uniqueness of Nash equilibrium with partial and decentralized information, for both the static and dynamic games, as well as when players have subjective probabilities.

10.4.1 Static LQG Games: Nash Equilibria

In this subsection, we consider a general class of N-player static stochastic games where each player has a cost function quadratic in the actions of the players and the state of Nature, x, which is a second-order random vector with a given distribution. Player i has access to a measurement, y^i, of x, corrupted with noise, v^i, which is again a second-order random vector with a given distribution, for $i \in \mathcal{N} := \{1, 2, \ldots, N\}$. The random vector x is assumed to be independent of the noise sequence v^i, $i \in \mathcal{N}$, and these random vectors are of arbitrary (but fixed) dimensions. Specifically, we have

$$y^i = \eta^i(x, v^i),$$

for some measurable map $\eta^i, i \in \mathcal{N}$. Now, with $u^i = \gamma^i(y^i),\ i \in \mathcal{N}$, with arbitrary (but fixed) dimensions, and with admissible policies $\gamma^i \in \Gamma^i,\ i \in \mathcal{N}$ defined so that the resulting actions are second-order random vectors (this actually makes Γ^i a Banach space), the cost function for Player j, $j \in \mathcal{N}$, is given by

$$J^j(\underline{\gamma}) = E[(u^j)^T C_j x + \frac{1}{2}(u^j)^T D_{jj} u^j + \sum_{i \neq j}(u^j)^T D_{ji} u^i],$$

where C_j, D_{jj}, D_{ji} are matrices of appropriate dimensions, with $i, j \in \mathcal{N}, i \neq j$, and $D_{jj} > 0$. Toward the goal of obtaining Nash equilibrium, note that this is the most general quadratic cost function for Player j in terms of the $(N+1)$-tuple $(x, u^1, \ldots, u^N)$, since adding any other terms involving x and $u^i, i \neq j$ will not affect the best response function of Player j (that is these additional terms are irrelevant to the minimization problem faced by Player j, with fixed $\gamma^i, i \neq j$. Now, given that J^j is strictly convex in $\gamma^j \in \Gamma^j$ (since $D_{jj} > 0$), straightforward minimization of the quadratic cost function $J^j(\gamma^j, \gamma^{-j})$ over $\gamma^j \in \Gamma^j$ for each fixed $(N-1)$-tuple $\gamma^{-j} \in \Gamma^{-j}$, yields the unique best response function for Player j, as captured by the lemma below.

Lemma 10.4.1 *In the static stochastic quadratic game formulated above, and toward construction of its Nash equilibria, the best response function for Player j is unique in Γ^j, and given by, for all $j \in \mathcal{N}$,*

$$\gamma^{j,*} = -D_{jj}^{-1} C_j E[x|y^j] - D_{jj}^{-1} \sum_{j \neq i} D_{ji} E[\gamma^i(y^i)|y^j] \tag{10.5}$$

Hence, establishing the existence of a Nash equilibrium amounts to establishing existence of a solution to (10.5), and the Nash equilibrium is unique if and only if the solution to the coupled set of Eq. (10.5) is unique. Note that since conditional expectation is a linear operator, (10.5) defines, for each $j \in \mathcal{N}$, an affine mapping from Γ^{-j} to Γ^j, and if we concatenate (10.5) over $j \in \mathcal{N}$, then we have an affine mapping from Γ into itself, and hence we would be looking for a fixed point of that affine map (and uniqueness of fixed point implies, and is implied by, uniqueness of Nash equilibrium).

To proceed further with such an analysis, let us focus on the case of a 2-player game ($N = 2$). In this case, we can actually bring the fixed-point analysis down to space Γ^1 (or equivalently Γ^2), instead of doing it on the product space $\Gamma = \Gamma^1 \times \Gamma^2$. Toward that end, we can substitute γ^1 from (10.5) into the same for $N = 1$, or equivalently swap the roles of the players, leading to

$$\gamma^i(y^i) = k_i(y^i) + K_i \gamma^i(y^i), \tag{10.6}$$

where

$$k_i(y^i) = -D_{ii}^{-1}C_iE[x|y^i] + D_{ii}^{-1}D_{ij}D_{jj}^{-1}C_2E[E[x|y^j]|y^i], i = 1, 2; j \neq i,$$

and the linear operator K_i is given by

$$K_i(\cdot) = D_{ii}^{-1}D_{ij}D_{jj}^{-1}D_{ji}E[E[\cdot|y^j]|y^i], \quad i = 1, 2; j \neq i$$

Theorem 10.4.1 ([80]) *For $i = 1, 2$, let C_i denote the set of all real matrices similar to $D_{ii}^{-1}D_{ij}D_{jj}^{-1}D_{ji}$. If there exists at least one matrix $\Lambda \in C_1$ or $\Lambda \in C_2$ with the property*

$$\Lambda^T\Lambda < I,$$

then the fixed-point Eq. (10.6) (or equivalently (10.5) with $N = 2$) admit a unique solution, and consequently the two-player stochastic game admits a unique Nash equilibrium.

The proof builds on establishing the best-response maps as a contraction mapping in a Banach space of square-integrable admissible control policies. Crucial in this analysis is the fact that conditional expectation itself is a non-expansive map. See Proposition 2.4.3 and the subsequent analysis for a related analysis.

In general, it is not possible to obtain an explicit analytical solution to (10.6), even though as we have mentioned earlier, and as in Sect. 2.4.3, one could apply iterative numerical methods with guarantees of convergence to the fixed point. Such a Banach iteration argument applied to (10.6) when all random vectors are jointly Gaussian (that is, the underlying game is an LQG one) leads to the conclusion that the fixed point is linear/affine in the measurements (simply start the iteration with $0 \in \Gamma^i$, and observe that every generated term in the iteration will be linear/affine, since conditional expectations involving Gaussian random vectors are linear (if zero mean) or affine (if means are non-zero). Toward a precise statement of that special case, let us pick the measurement vectors as:

$$y^i = H_ix + v^i, \qquad x \sim \mathcal{N}(\bar{x}, Q), v^i \sim \mathcal{N}(0, R_i), \quad R_i > 0, \quad i = 1, 2,$$

where x, v^1, v^2 are taken to be independent. Then, we have the following result for the LQG static game with two players, as a special case of Theorem 10.4.1.

Theorem 10.4.2 ([80, Theorem 2]) *Consider the two-player LQG game as formulated above, and assume that the condition in Theorem 10.4.1 holds. Then, the unique Nash equilibrium policy (among all admissible policies) for Player i is given by*

$$\gamma^{i,*}(y^i) = A_i\bar{x} + B_i(\hat{x}^i - \bar{x}),$$

where

$$A_i = -\left(I - D_{ii}^{-1} D_{ij} D_{jj}^{-1} D_{ji}\right)^{-1} D_{ii}^{-1}\left(C_i - D_{ij} D_{jj}^{-1} C_j\right), \quad i \neq j, i = 1, 2$$

$$\hat{x}^i = E[x|y^i] = \bar{x} + QH_i^T(H_i QH_i^T + R_i)^{-1}(y^i - H_i\bar{x}), \qquad i = 1, 2$$

and B_1 is the unique solution of the (Lyapunov-type) matrix equation

$$B_1 + PB_1L = M,$$

where

$$\begin{aligned}
P &= -D_{11}^{-1} D_{12} D_{22}^{-1} D_{21},\\
L &= QH_1^T(H_1QH_1^T + R_1)^{-1}H_1QH_2^T(H_2QH_2^T + R_2)^{-1}H_2\\
M &= -D_{11}^{-1}C_1 + D_{11}^{-1}D_{12}D_{22}^{-1}C_2QH_2^T(H_2QH_2^T + R_2)^{-1}H_2
\end{aligned}$$

and

$$B_2 = -D_{22}^{-1}D_{21}B_1QH_1^T(H_1QH_1^T + R_1)^{-1}H_1 - D_{22}^{-1}C_2$$

We have mentioned prior to the statement of the theorem that one way of justifying that the Nash equilibrium policies are affine (and therefore are of the form as in the theorem) is to start the Banach iteration (whose convergence is guaranteed by the Banach contraction mapping theorem under the condition of Theorem 10.4.1) with the *zero* policy, and see that every term in the iteration is then affine in the measurement of the player who conducts the iteration.

One additional powerful attribute of the contraction argument presented is that *even if the players start the iteration with non-linear policies*, the generated sequence is guaranteed to converge to the (same) unique fixed point, which is of the linear/affine type as above.

For $N > 2$, in principle, one can still arrive at optimal solutions but the direct substitution is more complicated. We recall here that analogous derivations involving the contraction properties of best-response maps in the team theoretic context were presented in Sect. 2.4 (notably Proposition 2.4.4) and the analysis here follows essentially the same steps.

10.4.2 LQG Multi-Stage Non-zero-Sum Games: Information Expansion May Lead to Non-unique Nash Equilibrium Costs

In this subsection, we move on to multi-stage (dynamic) LQG nonzero-sum games, and study properties under two different information structures which are inconsequential in stochastic dynamic teams, but make a significant difference in the structures of NE policies in games.

Specifically, we consider the one-step delayed observation sharing pattern as well as the one-step delayed sharing pattern (see Definition 2.3.1), where both are quasi-classical information structures and the latter is an expanded version of the former, including also control actions. We will show in the following that unlike the case of dynamic teams, for games informational expansion leads to significant complexity in the nature of NE solutions.

Unlike global optimality in teams, Nash equilibria under quasi-classical information structures are not equivalent to those under their static reduction. Equilibria is essentially non-unique, if there is redundancy of information. In particular, under the one-step delayed observation sharing pattern, we will establish that NE solution may be unique, but under the one-step delayed sharing pattern there will be uncountably many *essentially non-unique* solutions (by essentially non-unique, we mean that different NE solutions lead to different payoff/cost values), and hence the non-uniqueness is consequential, and not just a different way of realizing the same policy or cost, as was made evident in Example 10.2.2).

Let us now consider the classes of multi-stage N-Player LQG non-zero sum games with one-step-delay sharing and one-step-delay observation sharing ISs, as discussed in [80]. The state dynamics are described by

$$x_{t+1} = A_t x_t + \sum_{i=1}^{N} B_t^i u_t^i + w_t, \tag{10.7}$$

where u_t^i is an appropriate-dimensional action variable of Player i at time t, A_t and B_t^i are appropriate dimensional matrices and $\{w_t\}_t$ is a sequence of zero-mean mutually independent Gaussian random vectors also independent of a zero-mean Gaussian random vector x_0, the initial state. Observations (measurements) of each player over time are defined by

$$y_t^i = H_t^i x_t + v_t^i, \quad i \in \mathcal{N} := \{1, \dots, N\}, \tag{10.8}$$

where H_t^i is an appropriate dimensional matrix, and $\{v_t^i\}_t$ for each $i \in \mathcal{N}$ is a sequence of zero-mean mutually independent Gaussian random vectors, also independent of $\{w_t\}_t$ and x_0. Covariances of all random vectors are assumed to be positive definite. Let $\mathbf{y}_{[0,t]} := \{y^1_{[0,t]}, \dots, y^N_{[0,t]}\}$ and $\mathbf{u}_{[0,t]} := \{u^1_{[0,t]}, \dots, u^N_{[0,t]}\}$. For a specified IS ($I_t^i$ for Player i at time t, to be introduced shortly), let γ^i denote an admissible policy for Player i, so that $u_t^i = \gamma_t^i(I_t^i)$. Then, the expected quadratic cost function for Player i is defined as

$$J^i(\underline{\gamma}) := E\left[\sum_{t=0}^{T-1}\left(x_t' Q_t^i x_t + \sum_{j=1}^{N} (u_t^j)' R_t^{j,i} u_t^j\right) + x_T' Q_T^i x_T\right],$$

where Q_t^i, R_t^{ji} are nonnegative definite weighting matrices, with $R_t^{ii} > 0$.

Now, the partially nested information structures of concern here are:

One-step-delay observation sharing: $I_t^{i,\mathrm{DOS}} := \{\mathbf{y}_{[0,t-1]}, y_t^i\}$;
One-step-delay sharing: $I_t^{i,\mathrm{DS}} := \{\mathbf{y}_{[0,t-1]}, \mathbf{u}_{[0,t-1]}, y_t^i\}$;
Closed-loop local measurement feedback: $I_t^{i,\mathrm{LMF}} := \{\mathbf{y}^i_{[0,t-1]}\}$

Theorem 10.4.3 ([80, 81]) *Consider the class of LQG dynamic NZSGs introduced above.*

(i) [81, Theorem 5] If the IS is $I_t^{i,DOS}$ for all $t \in \mathcal{T}$ and $i \in \mathcal{N}$, then, under some sufficient (contraction) conditions involving the matrices that appear in the state dynamics and the cost functions of players, there exists a unique Nash equilibrium, which turns out to be linear (affine if the random vectors have nonzero-means) and in the form

$$\gamma_t^{i*}(I_t^{i,DOS}) = \Lambda_t^{i,i} y_t^i + \sum_{j=1}^{N}\sum_{k=0}^{t-1} \Lambda_t^{i,j} y_k^j,$$

where $\Lambda_t^{i,j}$ is an appropriate dimensional matrix for each $t \in \{0, \ldots, T-1\}$ and $i, j \in \mathcal{N}$.

(ii) [80] If the IS is $I_t^{i,DS}$, then Nash equilibrium policies are essentially non-unique.

The proof builds, similar to the analysis in Sect. 10.4.4, on establishing the best-response maps as a contraction mapping in a Banach space of properly defined square-integrable policies but doing this in an intricate way recursively backwards, where sufficient conditions for the contraction have been introduced in [80, (4.26a) and (4.26b)] for two-player games. As before, crucial in this analysis is the fact that conditional expectation itself is a non-expansive map, but this property is used at each stage of the game, and temporal decomposition of the dynamic game is not possible, which makes the verification more complicated. Item (ii) of Theorem 10.4.3 can be demonstrated by the following example:

Example 10.4.1 ([80]) Following the general formulation and notational convention of this subsection, consider a scalar 2-stage stochastic game with 2 players:

$$x_1 = x_0 + u_0^1 + u_0^2, \qquad x_0 \sim \mathcal{N}(0, 1)$$

$$x_2 = x_1 + u_1^1$$

$$J^1(\gamma^1, \gamma^2) = E^{\gamma^1,\gamma^2}[(x_2)^2 + (u_1^1)^2 + (u_0^1)^2]$$

$$J^2(\gamma^1, \gamma^2) = E^{\gamma^1,\gamma^2}[(x_2)^2 + (u_0^2)^2]$$

$$y_1^1 = x_1 + v_1^1, \qquad y_0^1 = x_0 + v_0^1, \qquad y_0^2 = x_0 + v_0^2$$

with $v_j^i \sim \mathcal{N}(0,1)$, independent for $i = 1, 2$; $j = 0, 1$, and independent of x_0. If Player 1 has access to y_0^1 at stage 0, and in addition to y_1^1, y_0^2 at stage 1, and Player 2 (who acts only at stage 0) has access to y_0^2, then the game features one-step delay observation sharing pattern (Player 1 having access to Player 2's measurement with one-step delay), and Theorem 10.4.3 directly applies. Working out the details (which can be found in [80, Example 1], we have the unique Nash equilibrium:

$$\gamma_0^{1,*}(y_0^1) = \frac{-39}{269} y_0^1, \qquad \gamma_0^{2,*}(y_0^2) \qquad = \frac{-35}{269} y_0^2$$

$$\gamma_1^{1,*}(y_0^1, y_0^2, y_1^1) = \frac{-19}{269} y_0^1 - \frac{41}{269}\frac{1}{2} y_0^2 - \frac{1}{8} y_1^1.$$

We note that it is possible to show that the policies above are indeed in NE by verifying directly the pair of inequalities

$$J^1(\gamma^{1,*}, \gamma^{2,*}) \le J^1(\gamma^1, \gamma^{2,*}), \quad J^2(\gamma^{1,*}, \gamma^{2,*}) \le J^1(\gamma^{1,*}, \gamma^2), \quad \forall(\gamma^1, \gamma^2)$$

Now, let us allow Player 1 also to have access to the action of Player 2, u_0^2, which then makes the IS one-step delay sharing, because of also the inclusion of action information. In this case, one can show that the following family of policies, parametrized by p, constitute a NE for each $p > \frac{-8}{23}$:

$$\gamma_0^{1,*}(y_0^1) = -\frac{39 + 8p}{269 + 88p} y_0^1, \qquad \gamma_0^{2,*}(y_0^2) = -\frac{35 + 40p}{269 + 88p} y_0^2$$

$$\gamma_1^{1,*}(y_0^1, y_0^2, y_1^1, u_0^2) = -\frac{19 + 8p}{269 + 88p} y_0^1 - \frac{41 - 8p}{269 + 88p}\frac{1}{2} y_0^2 - \frac{1}{8} y_1^1$$
$$+ p[u_0^2 + \frac{35 + 40p}{269 + 88p} y_0^2]$$

Again the NE property of these policies for each p in the range specified can be verified directly on the NE inequalities, but derivation of these policies is quite intricate, with details in [40]. We should also note that the above is not the only set of NE policies under the one-step delayed sharing pattern; one can also construct nonlinear policies using a similar process. We finally note that if $p = 0$ in the family above, we recover the unique NE solution under the observation sharing pattern as a special case. ◇

The above highlights distinctions and subtleties in dynamic games when compared with dynamic team problems. Nonetheless, linearity can still constitute Nash equilibria even under closed-loop local measurement feedback (with a non-classical information structure): $I_t^{i,\text{LMF}} := \{\mathbf{y}^i_{[0,t-1]}\}$. From Witsenhausen's counterexample, however, we know that in a team theoretic setup linear policies would not be optimal, though a Nash equilibrium is a weaker concept for stochastic teams, as emphasized earlier in the book. Uniqueness, in general does not hold, which can also be seen by considering Witsenhausen's counterexample as an instance of such a problem (where a team optimal non-linear policy also constitutes a Nash equilibrium).

Example 10.4.2 Consider the following 2-player LQG nonzero-sum dynamic game:

$$\begin{aligned} x_{t+1} &= Ax_t + Bu_t^1 + Bu_t^2 + w_t \\ y_t^i &= C^i x_t + v_t^i, \qquad i = 1, 2 \end{aligned}$$

where all the exogenous variables are zero-mean Gaussian. Let the cost functions of the players be:

$$J^i(\underline{\gamma}) = E^{\underline{\gamma}}[\sum_{k=0}^{N-1} x_k^T Q^i x_k + (u_k^i)^T R^i u_k^i]$$

where $Q^i \geq 0$, $R^i > 0$.

We now consider the closed-loop local measurement feedback $I_t^{i,\text{LMF}} := \{\mathbf{y}^i_{[0,t-1]}\}$, introduced earlier.

Proposition 10.4.1 *Under closed-loop local measurement feedback, linear policies may constitute Nash equilibria (without a priori restriction to linear policies).*

Proof Take $i = 2$ and let, for a sequence of matrices $\{\mathcal{H}_t^i, \quad t \geq 0\}$, $\mathbf{A}i$ apply a linear policy via

$$u_t^i = \mathcal{H}_t^i(y^i_{[0,t]}).$$

Then, from the perspective of $\mathbf{A}1$, we have an LQG problem with dynamics

$$x_{t+1} = Ax_t + Bu_t^1 + B\left(\mathcal{H}_t^2(y^2_{[0,t]})\right) + w_t$$

In this case, opening up

$$\begin{aligned} y_t^i &= C^i\left(A^t x_0 + \sum_{k=0}^{t-1} A^{t-k-1}(Bu_k^1 + Bu_k^2 + w_k)\right) \\ &= C^i\left(A^t x_0 + \sum_{k=0}^{t-1} A^{t-k-1}\left(Bu_k^1 + B(\mathcal{H}_k^i(y^i_{[0,k]})) + w_k\right)\right) \end{aligned} \tag{10.9}$$

and inductively expressing y_t^i in terms of the previous measurements, we arrive at a quadratic linear Gaussian optimization problem in $\{u_k^1, k \geq 0\}$ with *classical information structure* (see Chap. 3), whose solution will be linear/affine in the measurements (via backwards induction). This can be expressed as some $\gamma_t^1(I_t^1) = \mathcal{H}_t^1(y_{[0,t]}^1), \quad t \geq 0$, plus a possibly constant term for each time stage. Here, $\mathcal{H}_t^1$ will explicitly depend on $\mathcal{H}_t^2$. What is important is that γ^1 is linear/affine.

Alternatively if **A**1 fixes a policy of the form

$$\gamma_t^1(I_t^1) = \mathcal{H}_t^1(y_{[0,t]}^1), \quad t \geq 0,$$

then, the best response of **A**2 will have a linear/affine policy of the form

$$\mathcal{H}_t^2(y_{[0,t]}^2)$$

plus a possibly constant term. If there is a fixed point, then these will form Nash equilibria. ◇

Showing whether a fixed point for the example above exists and actually computing it entail a complicated analysis, since unlike the one-step delayed pattern one cannot in general obtain a state/filter equation based representation of optimal policies (unlike the partially nested information pattern with control sharing, e.g. under $I_t^{i,\mathrm{DS}}$). In particular, equations to solve for the linear gain terms will actually be non-linear in these gains.

Remark 10.4.1 (Linear Policies May Entail Optimality in LQG Games with Non-classical Information Structures) In Sect. 20.1.2, we will revisit the counterexample of Witsenhausen, introduced and studied extensively in Sect. 3.3, but now under a game theoretic formulation [89] as a zero-sum game with a slightly different quadratic criterion given by (20.1), but consistent with the original information structure. We will show that unlike the counterexample, the (saddle-point) equilibrium solution is attained by (unique) linear policies for both players.

10.4.3 Linear Quadratic Games with Subjective Priors

As mentioned earlier, non-zero sum games can arise even when the nominal decision problem is a team, but the subjective probabilities of individual agents/players are different. To go a bit deeper into this, consider a nominal team problem with team cost:

$$\begin{aligned} c(x, u^1, u^2) = (u^1)^T R_1(u^1) + (u^2)^T R_1(u^2) + 2(u^1)^T R_{12} u^2 \\ +2(u^1)^T S_1 x + 2(u^1)^T S_2 x \end{aligned}$$

where $R_1 > 0$, $R_2 > 0$ and $y^i = \eta^i(x, v^i)$ for some independent noise v^i. Suppose that P^i is the prior of Player i on the random vectors (x, v^1, v^2) and the policy space of Player i is the collection of policies measurable with respect to $\sigma(y^i)$. Suppose further that these priors are mutually absolutely continuous and the Radon-Nikodym derivatives $\frac{dP^i}{dP^j}$ are uniformly bounded, for $i, j = 1, 2$.

Now, for every policy γ^2, the best response of Player 1 to $u^2 = \gamma^2(y^2)$, will be given by the minimization of

$$E^1[(u^1)^T R_1(u^1) + (u^2)^T R_1(u^2) + 2(u^1)^T R_{12}u^2 + 2(u^1)^T S_1 x + 2(u^1)^T S_2 x|y^1]$$

If we do the same with the players' roles switched, we will end up with the following fixed point equation, for $i, j = 1, 2, i \neq j$, for equilibrium policies $\gamma^{*,i}$, $i = 1, 2$:

$$\begin{aligned}\gamma^{*,i}(y^i) &= -R_i^{-1}E^i\Big[R_{12}\gamma^{*,j}(y^j) + S_i x|y^i\Big] \\ &= -R_i^{-1}\Big(R_{12}E^i[\gamma^{*,j}(y^j)|y^i] - R_i^{-1}S_i E^i[x|y^i]\Big) \qquad (10.10)\end{aligned}$$

To study whether such a fixed point exists, one can follow an iterative process similar to what was considered in (2.41)–(2.45), leading to results similar to those presented in Proposition 2.4.3 as well as in Theorem 10.4.2. In the Gaussian case, policies obtained will be linear/affine. As earlier, under some contraction conditions, there will be a unique solution to the fixed-point equation even in the class of non-linear policies (using the non-expansiveness of the conditional expectation operator), which then shows that for the Gaussian case the unique equilibrium solutions are affine. Notice that the Radon-Nikodym condition is to ensure that the cost functions are integrable under both priors; for the Gaussian case this holds by direct verification without an additional uniformity condition. See [87] for further discussions on this problem, which also includes generalizations for Stackelberg games with inconsistent priors. We note here the striking result that even for inconsistent Gaussian priors, a Stackelberg version of the quadratic cost problem may admit non-linear solutions [87, Section V]. In Chap. 18 we will arrive at a similar result in the context of quadratic Gaussian signaling games with subjective priors.

We note that for an iterative process, the players *do not* need to know the subjective prior of their opponents. This discussion further justifies and emphasizes the importance of iterative best-response dynamics.

10.4.4 Stackelberg Equilibria for Static and Dynamic LQG Games and Optimality of Affine Policies

We conclude the section on LQG games with an analysis of Stackelberg solutions. Here, we follow [85].

Consider two agents, **A**1 and **A**2, with the following objective functions:

$$
\begin{aligned}
c^1(x, u^1, u^2) &= \frac{1}{2}(u^1)^T D_{11} u^1 + (u^1)^T D_{12} u^2 + \frac{1}{2}(u^2)^T D_{13} u^2 \\
&\quad + (u^1)^T C_{11} x + (u^2)^T C_{12} x + (u^1)^T k_{11} + (u^2)^T k_{12} \\
c^2(x, u^1, u^2) &= \frac{1}{2}(u^2)^T D_{22} u^2 + (u^2)^T D_{21} u^1 + \frac{1}{2}(u^1)^T D_{23} u^1 \\
&\quad + (u^1)^T C_{21} x + (u^2)^T C_{22} x + (u^1)^T k_{21} + (u^2)^T k_{22}
\end{aligned}
$$

where x is n-dimensional zero-mean Gaussian, u^1 and u^2 are multi-dimensional real valued decision variables, and the matrices are of appropriate dimensions. We first consider a static information structure with, for agent $\mathbf{A}i$,

$$
y^i = H_i x + v^i
$$

where v^i is zero-mean Gaussian and y^i is a real, possibly vector valued, random variable.

In this case, **A**1 (the leader) announces her policy and **A**2 responds to this announced policy.

Recall from our analysis in Chap. 2 that if the cost criteria are aligned under mild positive definiteness conditions on the cost matrices, by Radner's theorem, the optimal team policy would be unique. A Stackelberg solution would be a team optimal solution under such an alignment. It would be interesting to see whether the optimality of linear policies carries over to the mis-aligned quadratic criteria.

For the static setting, since there is no information transmission or signaling aspects to be considered, **A**2's best response to the leader's announced policy will be unique, and given by:

$$
\gamma^2(y^2) = -D_{22}^{-1} C_{22} E[x|y^2] - D_{22}^{-1} k_{22} - D_{22}^{-1} D_{21} E[\gamma^1(y^1)|y^2]
$$

Given this, **A**1 needs to optimize her expected cost. Toward that end, one has to substitute the above best response into c^1, and optimize its expected value over all policies for **A**1. This is a functional optimization problem where γ^1 also enters under a conditional expectation; that is, via $E[\gamma^1(y^1)|y^2]$. In [85], via perturbing the policy γ^1, a necessary condition is obtained, which leads to a unique optimal solution under a contraction condition, which is linear in y^1 (or affine in case of non-zero means), which then makes the follower's corresponding policy from the preceding expression also linear (or affine) (in y^2), given that $E[y^i|y^j]$ is linear (or affine) in y^j, for $j, i = 1, 2, j \neq i$, since they are jointly Gaussian ([85, Equation (10)]).

For the multi-stage case, under a one-step-delay information sharing pattern and where u^2 gets realized prior to u^1, via the non-standard approach of defining a team policy under the cost of the leader, as seen in Sect. 10.2.6, optimality of linear policies can be attained.

Finally, we comment on the non-classical setup. Since a team optimal solution in Witsenhausen's counterexample has to be a Stackelberg solution, Witsenhausen's counterexample shows that the Stackelberg solutions for LQG game problems under non-classical information structures are not necessarily linear.

10.5 Existence of Equilibria: Implications of the Geometric Properties of Information Structures and Strategic Measures Approach

The topological constructions of policies presented in Chaps. 4 and 5 will directly lead to existence results for Nash equilibria as well. Notably, the compactness, convexity and locally convex space formulation of sets of probability measures under either weak or weak-setwise topologies will be consequential. The geometric analysis presented in Chap. 4 finds applications also in game theory (as in the concept of correlated equilibria [46], arrived at via the presence of common randomness).

On the other hand, in stochastic team theory our analysis on existence was based on Weierstrass theory: lower semi-continuity and compactness properties under appropriate topologies. In game theory, however, the primary tool is fixed-point theory, which often makes use of the control-topology approach presented in Chap. 5 (where each player is provided with a particular policy/strategy space, as opposed to an analysis based on the strategic measure on the general play). Also, as we had discussed earlier, Nash equilibria correspond to person-by-person-equilibria in teams, and therefore unless one can establish uniqueness of such equilibria, global optimality in teams cannot be arrived at via such a fixed point analysis. Indeed, for general games, there often exist multiple equilibria.

The following result essentially appears in [59]; its proof as given here, however, builds upon ideas presented earlier in the book. We assume that the policy spaces are randomized, and adopt the notation of Sect. 3.4.2. We also note that, as we have seen in Sect. 3.4.2, Nash equilibria are isomorphic under (policy-independent) static reduction, justifying our approach adopted in the proof.

Theorem 10.5.1 *Consider a static game (or policy-independent static reducible dynamic games with independent measurements reduction) so that* $y^i \sim Q^i, i \in \mathcal{N}$, *with subjective costs* $\{J^i(\gamma^i, \gamma^{-i}), \quad i \in \mathcal{N}\}$ *given by*

$$J^i(\underline{\gamma}) = \int \prod_u Q^i(dy^i)\gamma^i(du^i|y^i)c^i(\omega_0, \mathbf{y}, u^i, u^{-i}),$$

where $c^i(\omega_0, \mathbf{y}, \cdot)$ *is bounded and continuous in the actions* $\mathbf{u}$, *and for each* i, $\mathbb{U}^i$ *is a compact subset of a Polish space. Then, the game admits a Nash equilibrium.*

Proof The proof is an application of the following. ◇

Theorem 10.5.2 (Kakutani-Fan-Glicksberg [16, Corollary 17.55]) *Let* K *be a compact non-empty convex subset of a locally convex Hausdorff space and let the set-valued map* $\varphi : K \to 2^K$ *have convex values and have a closed graph* $G = \{(x, \varphi(x)), x \in K\}$ *(i.e., if* $x_n \to x$, *and if* $\varphi(x_n) \to z$, *then* $(x, z) \in G$*). Then,* φ *admits a fixed point.*

In the application of the above to Theorem 10.5.1, we take φ as a collection of set-valued best-response maps from $\prod_i \Gamma^i$ into itself:

$$\varphi(\gamma^1, \cdots, \gamma^N) = \{\mathrm{BR}(\gamma^{-1}), \cdots, \mathrm{BR}(\gamma^{-N})\},$$

where

$$\mathrm{BR}(\gamma^{-i}) = \{\arg\min_{\kappa^i \in \Gamma^i} J^i(\kappa^i, \gamma^{-i})\}$$

To ensure that the best response maps admit solutions, we build on our topological analysis in Chap. 4 which leads to a convex compact policy space. We identify, almost surely, every admissible policy γ^i with a probability measure on the product space, so that $\bar{Q}^i\gamma^i$ is a probability measure on $\mathbb{Y}^i \times \mathbb{U}^i$ with fixed marginal $Q^i(dy^i)$ on $\mathbb{Y}^i$. Let Γ^i denote the space of all such measures since every $Q^i\gamma^i$ can be identified with an element in Γ^i almost surely. We endow the policy Γ^i with the w-s topology, weak in the action and setwise in the measurement y^i. The policy space $Q^i\gamma^i$ is convex and (weakly and w-s) compact.

As in the proof of Theorem 7.3.4, we will utilize the locally convex topology on $\mathcal{P}(\mathbb{Y}^i \times \mathbb{U}^i)$, and so we define the locally convex space of probability measures with the following notion of convergence: We say that $\nu_n \to \nu$ if $\int f(y^i, u^i)\nu_n(dy^i, du^i) \to \int f(y^i, u^i)\nu(dy^i, du^i)$ for every measurable and bounded function which is continuous in u^i for every y^i. We note that our measures must still have fixed marginals on $\mathbb{Y}^i$. Thus, the probability measure set here is to be viewed as a compact convex subset of the set of signed measures defined as a locally convex space with the notion above. Recall again the w-s topology (see Sect. C.4) on the set of probability measures $\mathcal{P}(\mathbb{Y}^i \times \mathbb{U}^i)$; this is the coarsest topology under which $\int f(y^i, u^i)\nu(dx, du) : \mathcal{P}(\mathbb{Y}^i \times \mathbb{U}^i) \to \mathbb{R}$ is continuous for every measurable and bounded f which is continuous in u^i for every y^i. Now, since the exogenous variables are fixed (by Theorem C.4.1), weak convergence in this setting is equivalent to w-s convergence, and continuity in the exogenous variable is not needed.

Due to linearity, we have that the best-response maps are convex valued.

The cost J^i is weakly continuous on Γ^i for every fixed γ^{-i}. Note that since Γ^i is compact, $\min_{\gamma^i \in \Gamma^i} J^i(\gamma^i, \gamma^{-i})$ is continuous in γ^{-i}. Then, consider a sequence of actions $(\bar{Q}^{-i}\gamma^{-i})_m(dy^{-i}, du^{-i})$ which converges to $(Q^{-i}\gamma^{-i})(dy^{-i}, du^{-i})$ weakly. The marginals on $\omega_0, \mathbf{y}$ are fixed. We have continuity of J^i in $Q^i\gamma^i$ in the w-s topology and, by the equivalence in this setting, the weak topology. Then, $\min_{\gamma^i \in \Gamma^i} J(\gamma^i, \gamma_m^{-i}) \to \min_{\gamma^i \in \Gamma^i} J(\gamma^i, \gamma^{-i})$. Now, if γ_m^i is a best response to γ_m^{-i} and if $\gamma_{m_k}^i$ is a weakly converging subsequence with weak limit γ^i, we have that γ_i is a best response to γ^{-i}. In particular if $\underline{\gamma}_{m_k} \to \underline{\gamma}$, then $J^i(\underline{\gamma}_m) \to J^i(\underline{\gamma})$. The above then shows that the map φ is upper semi-continuous (and that the graph of the best response map is closed; if $(Q^{-i}\gamma^{-i})_m$ converges to a limit $(Q^{-i}\gamma^{-i})$ and if $Q^i\gamma^i$ is a best response sequence, some subsequence of which converges to a limit, then the limit is a best response to $(Q^{-i}\gamma^{-i})$). We refer the reader to also Lemma 5.4.2 and [342, Appendix D, Proposition D.5]).

Thus, by Theorem 10.5.2, an equilibrium exists. ◇

We note that Theorem 10.5.2 generalizes some of the more frequently cited theorems known as the Kakutani's Fixed Point Theorem, Brouwer's Fixed Point Theorem, Schauder's Fixed Point Theorem, and Tychonoff's Fixed Point Theorem.

A few remarks are now in order.

Remark 10.5.1 In the context of zero-sum games, we will see a more refined analysis in Theorem 11.2.1 (which is an adaptation of the above) and a strict refinement in Theorem 11.2.2 (which utilizes the zero-sum structure explicitly for a complementary result without an absolute continuity, or independent measurements, condition on the information structure).

Remark 10.5.2 (Teams Playing Against Teams and the Problem-Dependent Role of Common Randomness) In case a team of players plays against another team of players, existence may not hold due to lack of convexity in the set of *team* policies taking part in the game when convexity of the policy space is violated: Indeed, Anantharam and Borkar [22] present an example where such a situation arises. Related to this, conditions on convexity and correlations (including relaxations and quantum correlations) were presented in Chap. 4. In fact, existence results on equilibria can directly be obtained when one has nested reduction (see Assumption 6.4.3) for dynamic games as one can define a topology on team policies which satisfies both convexity and compactness under mild conditions; see Theorem 4.1.1(ii) for convexity and Theorem 4.2.7 for compactness of the set of strategic measures $L_C(\mu)$. Accordingly, existence of equilibria can be established for problems involving a team of agents playing against another team of agents by directly generalizing Theorem 10.5.1. In the context of zero-sum games, such a result will be presented in Theorem 11.6.1 utilizing compactness and convexity of policies within a team (obtained via common randomness), and once again without an absolute continuity condition among different team marginal measures (via Theorem 4.2.7), but only among agents within a team (unlike what would be needed for a general nonzero-sum game). The conditions for optimal policies become remarkably less tedious, though the mathematical analysis still is, when teams with infinitely many agents play with one another in an exchangeable game, in that there is no longer common randomness needed for the existence of equilibrium policies; this will be studied in detail in Sect. 11.6.

Remark 10.5.3 (Correlated Equilibrium) The Nash equilibrium concept when each player (though not teams of DMs) is provided with common randomness leads to the concept of *Correlated Equilibrium* [46]. We saw earlier that correlated equilibrium is related to $L_C(\mu)$ defined in (4.6). As noted in Remark 10.1.1, when multiple actions are considered, individual deviations at each stage may not determine equilibria. With this intuition in mind, it is not difficult to conclude that the set of correlated equilibria is larger than the set of all Nash equilibria. For further discussions, including a multi-stage generalization known as communication equilibria, see [270]. For a detailed treatment of these issues, we refer the reader to [532, p. 131].

10.6 Subjective Priors in Dynamic Games Revisited: Common Knowledge, Agreement, Robustness and Learning

As noted earlier in Sect. 9.3.1 and above in Sect. 10.4.3, if the priors held by different DMs are subjective (different), even if the cost functions are identical, we would typically end up with a genuine game problem. To make this more explicit, let Q denote a measurement channel, and define the following cost:

$$E^{\gamma}_{\pi}[c(x,u)] = \int \pi(dx)Q(dy|x)c(x,\gamma(y)).$$

Now, given a fixed policy γ, the costs $E^{\gamma}_{\pi}[c(x,u)]$ and $E^{\gamma}_{\mu}[c(x,u)]$ lead to different values since the prior measures π and μ are different [87]. Therefore $E^{\gamma}_{\cdot}[c(x,u)]$ defines a (measurable) function on the space of priors (under weak convergence), leading to possibly different values for different priors given a fixed policy γ. In particular, with multiple DMs present, a cost function to be minimized will not be a common cost when viewed as a function of team policies.

On the other hand, when a game is a multi-stage one, the players may update their priors given the increasing amount of data available to them. For some setups different players may know that their priors are different, but for others, this may not be the case. It is an important consideration to study how these priors are updated with increasing information. In this section, we will investigate such properties of subjective priors in a stochastic dynamic game and establish connections with subjective equilibria, classical Bayesian theory and merging theory, non-linear filtering, and robust stochastic control.

10.6.1 Common Knowledge

We now discuss an important notion introduced by Aumann [45]. Let $(\Omega, \mathcal{F}, P)$ be a probability space, that is Ω a sample space, $\mathcal{F}$ a $\sigma-$field of subsets of Ω, and P a probability measure, where we first take Ω to be a finite space. Let $E \in \mathcal{F}$ denote an event and further let two information variables measurable on this probability space be available to two decision makers DM 1 and DM 2, such that $\mathcal{I}^1$ is available to DM 1 and $\mathcal{I}^2$ is available to DM 2.

Let $\mathcal{F}^i = \sigma(\mathcal{I}^i)$ denote the sigma-field generated by $\mathcal{I}^i$, $i = 1, 2$ (hence $\mathcal{F}^i$ is generated by a partition of Ω). Let us say that DM i knows $E \in \mathcal{F}$ at ω (and denote this by $\{\omega \in E\} \in \mathcal{F}^i$) if there exists a set $B^i \in \mathcal{F}^i$ such that $\omega \in B^i \subset E$. Assume that the partitions of each DM (induced by the local information variables) are known to both DMs. Aumann introduced the notion of an event E being *common knowledge* between DM 1 and DM 2 at $\omega \in E$ as follows: Whenever $\omega \in E$ happens, DM 1 knows E, DM 2 knows E, DM 1 knows that DM 2 knows E, DM 2 knows that DM 1 knows E, and so on. An event E is common knowledge if it is common knowledge for all $\omega \in E$.

Aumann made the following characterization of common knowledge based on information fields: If $E \in \mathcal{F}^1 \cap \mathcal{F}^2$, then it is common knowledge. This is not only sufficient, but also necessary: Suppose that E is common knowledge and some $\omega \in E$ takes place. As stated earlier, the event that DM 1 knows E at ω means that there exists some set $B \in \mathcal{F}^1$ such that $\omega \in B \subset E$. Now, if E is common knowledge, it must be that $\{\omega : \{\omega \in E\} \in \mathcal{F}^1\} \in \mathcal{F}^2$, that is DM 2 knows that DM 1 knows that E took place, for all $\omega \in E$. For this to happen for every $\omega \in E$, it must be that there exists a finite index set T_2 such that $E = \cup_{t \in T_2} E_t^2$, where $E_t^2 \in \mathcal{F}^2, t \in T_2$ and for all $t \in T_2$, $E_t^2 \subset E$ (for, otherwise, there would exist some $\omega \in E$ such that DM 2 could not know if DM 1 knows whether E happened or not). A parallel argument applies for DM 1 for another finite index set. Hence, $E \in \mathcal{F}^1 \cap \mathcal{F}^2$. Thus, an event $E \in \mathcal{F}$ is common knowledge *only if* $E \in \sigma(\mathcal{I}^1) \cap \sigma(\mathcal{I}^2)$. The *if* direction, on the other hand, follows from the observation that $\{\{\omega \in E\} \in \mathcal{F}^1\} \in \mathcal{F}^2$ (since $\{\omega : \{\omega \in E\} \in \mathcal{F}^1\} = E$ and $E \in \mathcal{F}^2$) and $\{\omega : \omega \in \{\{\omega \in E\} \in \mathcal{F}^1\} \in \mathcal{F}^2\} \in \mathcal{F}^1$ and so on, for every such finite iteration.

Hence, an event $E \in \mathcal{F}$ is common knowledge if and only if $E \in \sigma(\mathcal{I}^1) \cap \sigma(\mathcal{I}^2)$. We note that this relation does not depend on the particular probability measure P.

This equivalence between the definition of common knowledge and the sigma-field characterization can be extended to cases where Ω is an uncountable space. However, one technical issue is that sets (of one knows that the other knows) of the form $\{\{\omega \in E\} \in \sigma(\mathcal{I}^1)\}$ may not be an event in $\mathcal{F}$, since this may involve uncountable unions. By incorporating the probability measure P, [563] has provided an approach by defining an equivalence class among the elements in $\mathcal{F}$ by excluding null events, allowing a generalized definition of Aumann to apply to uncountable settings and [144] has considered *common knowledge with probability 1*, through a completion of sigma-fields by adding null events to the information sets of DMs (note that if one partitions Ω into measurable sets with positive probability, the partition is a countable one).

Hereafter, we will adopt the definition of common knowledge based on the sigma-field characterization. That is, an event $E \in \mathcal{F}$ is common knowledge if $E \in \sigma(\mathcal{I}^1) \cap \sigma(\mathcal{I}^2)$. In most cases, however, common knowledge with probability 1 is sufficient since null events do not affect the expected costs.

A celebrated result (*agreement theorem*) of Aumann in this context is that if decision makers have a common prior measure, and if their posterior probability measures on some event (that is, their conditional expectation on the indicator function for an event) is common knowledge, then these posteriors (their conditional expectations) should be equal. We will shortly see a generalization of this result in the context of asymptotic agreement.

10.6.2 Asymptotic Agreement and Common Knowledge with Common Priors but Different Posteriors

If the posteriors of two DMs are not common knowledge, and yet the DMs have a common prior, then a common posterior can be obtained through communication. In control and economics literatures, agreement on such a common knowledge has typically been arrived at through iterative exchange of decisions.

Let X be an integrable random variable (so that $E[|X|] < \infty$), such as 1_A, with A denoting an event. Suppose that two decision makers DM 1 and DM 2 have access to some local random variables (such as measurements on X) defined on a common probability space and correlated with X, and exchange their conditional probability measures over time. Suppose further that:

the information σ-fields at each DM is increasing: $\mathcal{F}^i_t \subset \mathcal{F}^i_{t+1}$

for all $n \in \mathbb{N}$, there exists $m > n$ such that $\mathcal{F}^i_m$ contains information on $E[X|\mathcal{F}^j_n]$, $i, j = 1, 2$. That is, the DMs exchange their estimates (but not their raw data) infinitely often.

Note that by the smoothing property of conditional expectation,

$$E[E[X|\mathcal{F}^i_{n+1}|\mathcal{F}^i_n]] = E[X|\mathcal{F}^i_n],$$

and that by Jensen's inequality

$$\sup_n E[|E[X|\mathcal{F}^i_n]|] \leq E[E[|X||\mathcal{F}^i_n]] = E[|X|] < \infty.$$

Hence, $(E[X|\mathcal{F}^i_n], \mathcal{F}^i_n)$ is a martingale sequence and the submartingale convergence theorem (Theorem D.2.2) leads to the conclusion that $\lim_{n\to\infty} E[X|\mathcal{F}^i_n]$ exists almost surely for $i = 1, 2$. Let $\mathcal{F}^i_\infty$ denote the smallest sigma-algebra containing $\mathcal{F}^i_n$ for all $n \in \mathbb{N}$. Hence, the almost sure limit is $E[X|\mathcal{F}^i_\infty]$. Furthermore, $E[X|\mathcal{F}^i_n]$

is a uniformly integrable martingale sequence since X is integrable (see Theorem 3.3.2 in [133]) and as a consequence one also has that

$$E[|E[X|\mathcal{F}_n^i] - E[X|\mathcal{F}_\infty^i]|] \to 0.$$

Since $E[X|\mathcal{F}_n^j]$ is $\mathcal{F}_m^i$ measurable, it is also $\mathcal{F}_\infty^i$ measurable and likewise for all n. Hence $E[X|\mathcal{F}_\infty^j]$ is $\mathcal{F}_\infty^i$ measurable, and $E[X|\mathcal{F}_\infty^i]$ is measurable on both $\mathcal{F}_\infty^1$ and $\mathcal{F}_\infty^2$. Thus, it is measurable on $\cap_{i=1,2}\mathcal{F}_\infty^i$. Hence the expectations are asymptotically equal and are common knowledge.

A similar reasoning applies for a larger number of DMs as well. Such a reasoning has been considered by Borkar and Varaiya in [139] (see also Geanakoplos and Polemarchakis [296], Tsitsiklis and Athans [725], Li and Başar [476] and Teneketzis and Varaiya [715], among other contributions and approaches in the literature).

10.6.3 Inconsistent Priors, Merging with Increasing Information, and Bayesian and Non-Bayesian Subjective Equilibria

Consider again a two DM setting. If the priors are different, *the fact that they are different* may or may not be common knowledge. The DMs may run some agreement protocol to exchange information. In such a case, the following possibilities emerge:

(i) Each DM may run an agreement protocol assuming that the other DM has the same prior as himself [715] or have a probabilistic belief on the prior of the other DM [164]. In these cases, the DMs may realize that their probability models are inconsistent, and yet reach an agreement or may not realize that the models are inconsistent but reach an agreement or reach different conclusions. In other scenarios the DMs may *agree to disagree*, that is they may realize that their models are different, and a communication exchange protocol will not provide any further information that would change the agreement status, or lack of it.

(ii) [Merging of Opinions and Bayesian Learning] The agents may run Bayesian estimators and asymptotically achieve common knowledge in the more restricted sense of agreement for events in the future (which Blackwell and Dubins [126] refer to as *merging*) which may lead to a subjective equilibrium.

This argument builds on a martingale formulation as the Radon-Nikodym derivative between the assumed strategic measure and the true one is an integrable random variable (under the true measure) defining a martingale as the common information grows. More precisely: Let P^μ and P^ν be two probability measures on $(\mathbb{X}^{\mathbb{Z}_+}, \mathcal{B}(\mathbb{X}^{\mathbb{Z}_+})$, where the product space is endowed with the product topology. Let $P^\nu \ll P^\mu$. Recall that this implies that $P^\mu(B) = 0 \implies P^\mu(B) = 0$ and that

$\exists f : \mathbb{X}^{\mathbb{Z}_+} \to \mathbb{R}_+$ so that for all Borel $D \in \mathcal{B}(\mathbb{X}^{\mathbb{Z}_+})$

$$P^{\nu}(D) = \int_D f(x_{[0,\infty)}) P^{\mu}(dx_{[0,\infty)}),$$

i.e., f is the Radon-Nikodym derivative: $\frac{dP^{\nu}}{dP^{\mu}}(x_{[0,\infty)}) =: f(x_{[0,\infty)})$. Blackwell and Dubins [126] refer to the following result as *merging*:

Theorem 10.6.1 ([126]) *P^{μ} almost surely (and therefore P^{ν} almost surely)*

$$\lim_{n\to\infty} \|P^{\nu}(dx_{[n+1,\infty)}|X_{[0,n]}) - P^{\mu}(dx_{[n+1,\infty)}|X_{[0,n]})\|_{TV} = 0$$

For a related information theoretic angle via relative entropy based arguments, see Barron [73].

When the DMs have inconsistent priors, they can update their probability measures for events taking place as a function of future random variables with increasing information, and in some settings the learned information can overrule the priors leading to a form of consistency. Building on Blackwell and Dubins's theorem above [126], Kalai and Lehrer [380] have shown that a sufficient condition for agreement on such conditional probabilities in the case of strictly common (knowledge) observations is that the DMs' prior beliefs are absolutely continuous with respect to each other (that is, each of the DMs assigns a positive prior probability to an event in the future play path on which the other DM also has a positive prior probability) (see also [382]).

However, in such game theoretic models with decisions and control impacting the flow, such an absolute continuity condition has been shown to be too restrictive for many applications [550] [271]: The reason is that when one responds to a subjective model, an optimal response restricts the true play path, or strategic measure, as an optimal path is to be pursued; and this best response will typically define a strategic measure with respect to which the beliefs of the other agent is not absolutely continuous. In addition, such an approach entails impractical Bayesian updates.

(iii) Even an identical information can make DMs with subjective priors further apart from each other if the underlying system evolves [521, Example 3.4] after the measurement of information. This comment may perhaps serve as a caution to over-reliance on data in (centralized or decentralized) decision making and control, and highlights the crucial dependence on priors and their over-rullability by increasing information (which may be viewed as an *observability* property). Even for single DM settings, robustness to incorrect priors typically require such observability properties (see Definition 8.7.1 and [388, 522, 731].

(iv) [Non-Bayesian Subjective Equilibria via Structural Model Assumptions] Different from the (Bayesian) subjective equilibria of [382], a setting where subjectivity enters is when players cannot detect the falseness of their modeling assumptions unless they actively probe the environment: the environment variables generated at a subjective equilibrium may be distributionally consistent with their beliefs when viewed in isolation as far as the marginal probability measures on the environment variables are concerned, hence the agents would not have an incentive to deviate to an alternative strategy. At such an equilibrium, called *subjective equilibrium under beliefs of exogenous uncertainty* [38], each player chooses its decisions under the (incorrect) belief that a stochastic environment process driving the system is exogenous whereas in actuality this process is a solution of closed-loop dynamics affected by each individual player. Players observe past realizations of the environment variables and their local information. At equilibrium, if players are given the full distribution of the stochastic environment process as if it were an exogenous process, they would have no incentive to unilaterally deviate from their strategies. Existence and further properties of such a concept has been studied in [38], where it is also shown that in the infinite population limit this solution concept converges to a mean-field equilibrium.

There may be other structural assumptions or mechanisms via which the agents may arrive at subjective equilibria. For example, if all agents assume, somewhat incorrectly that, they take part in a controlled Markov model where their measurements are the Markovian state variables, then the standard Q-learning algorithm indeed converges under mild conditions and through iterative play [391, 672], one can show that for a large class of games, the players indeed converge to an equilibria which may be defined to be subjective, in view of the discussion above [794, 796]. This will be studied in Chap. 23.

(v) For one-shot problems where merging does not occur and players are oblivious to each other's subjectivity (provided that mutual absolute continuity applies between the priors), in the presence of policy announcements at each iteration, iterative update algorithms can still lead to equilibria, as we discussed in Sect. 10.4.3.

10.7 Concluding Remarks

In this chapter, we have presented a systematic analysis toward a mathematical theory of information structures in stochastic dynamic games where our development has paralleled the corresponding mathematical theory for stochastic team theory. Our analysis has highlighted fragilities and informational subtleties that game theory problems intrinsically possess. However, many connections and similarities were also established between the team theoretic and game theoretic settings.

10.8 Bibliographic Notes

As noted at the beginning of the chapter, there does not exist a complete treatment of the impact of different information structures on equilibria in stochastic games.

With a few exceptions, the standard texts on game theory study dynamic games with perfect-recall (classical) information structures, and many studies consider the extensive tree form representations, where subtleties in informational ambiguities cannot be depicted effectively. Some of the most general studies are [97, 281, 336, 532, 568] which however restrict the class of games considered in one way or another.

Accordingly, a comprehensive theory comparable to that for team theory does not currently exist in the game theory literature. Nonetheless, we would like to emphasize that the analysis in [568, Section 11.1] presents an abstraction touching on some of the concepts presented by the model introduced in this chapter (as well as the non-sequential model as presented in Sect. 3.9 with regard to causality and solvability properties) and even on the equivalence properties between information structures without formalizing some of the concepts, while explicitly noting on [568, p. 204] that *the literature on games with imperfect recall is very small.*

While writing this chapter, we have reflected on how the 1970s was a very active period on study of information structures in dynamic teams and games (see e.g. [77–81, 630, 773, 776]). It would be desirable to see that the evidently dissipated interest in many of such classical studies may be rejuvenated given our detailed coverage here, interpreted within the general framework of the book toward a comprehensive theory of information structures we have aimed to present in the book.

The recent works [378] and [346] introduce a variation of Witsenhausen's information structures in a game theoretic context and presents detailed overviews on this subject including the general tree model presented in [17].

For subtleties involving various Stackelberg solution concepts in connection with ISs, we refer the reader to [93, 97, 98, 108, 572].

On the existence of equilibria and uniqueness properties and various results on repeated games, we refer the reader to [336] for an accessible, yet very comprehensive, coverage.

On the negative value of information, see, in addition to the examples given in the text [102, 352, 356], [3, 90, 312, 348, 387]. There exists a very rich and fertile theory on 'information economics'. This is often with regard to strategic interactions among DMs and can be viewed as an information structure design, where the agents also act as encoders. Some aspects of this will be discussed later in the context of signaling games in Chap. 18.

See [295] for a general discussion on common knowledge and agreement, and [547] for a comprehensive analysis on agreement and learning, both in the Bayesian and non-Bayesian contexts. We refer the reader to Rubinstein's Electronic Email game [611], for an interesting example which demonstrates that common knowledge is not robust to noise.

We have already seen earlier in the book that static reductions provide very effective means to arrive at existence, structural and approximation results. For existence results building on this approach, please see [327, 620, 818, 830] and approximations [625]. For game problems, for an application on existence, see [353]. [637] has highlighted the fragility of informational equivalence and reductions in stochastic game problems and person-by-person-optimal solutions in stochastic team and game problems.

For some team and game problems with sufficient history sharing, dynamic programming methods have also been adopted in the literature (see for example, [43], [797], [177], [6], [806], and these have been unified and generalized in [326, 558–560]) through the use of a team-policy which uses common information to generate partial functions for each DM to generate their actions using local information. Thus, in the dynamic programming approach, a separation of team decision policies in the form of a two-tier architecture, *a higher-level controller* and a *lower-level controller*, can be established with the use of common knowledge. On the other hand, it has also been shown that for any sequential team problem with arbitrary information structures, a dynamic programming formulation can be formulated, see [775] for countable spaces and [818] for general spaces which also lead to very general existence results. While for team problems these lead to global optimality; for game problems, these only lead to subgame perfect equilibria. Recent studies involving information structures in dynamic teams and games, in view of optimality and equilibrium conditions under various solution concepts include [705, 710, 738], where the analysis in [710] can be viewed as a weakly time consistent Markov perfect strategy under decentralization. Recent advances on dynamic games have been collected in [100] and [101].

Several of the results presented in this chapter are based on [80, 81]. The result of Theorem 10.5.1 appears in [59, Theorem 3.1] though the proof here is tailored towards the material presented earlier in the book.

Chapter 11
Information Dependent Properties of Equilibria: Existence, Comparison, Continuity and Team-against-Team Games

11.1 Introduction

In this chapter, we study four information structure dependent aspects of stochastic games: (i) on refined conditions on existence of saddle-point equilibria, (ii) on comparison and Blackwell ordering of information structures, (iii) on continuity properties of equilibrium values in information structures, and (iv) games where teams consisting of multiple (and even infinitely many) agents play against under agents.

Consider the class of single-stage static games with $N \in \mathbb{Z}_{\geq 1}$ players. We will follow the same structure considered in Sect. 8.8. Let x be an $\mathbb{X}$-valued random variable with probability measure ζ. Let $\mathbb{Y}^i$ be Player i's measurement space, for $i \in \{1, \ldots, N\}$. For Player i, we define the measurement y^i with $y^i = g^i(x, v^i)$ for some noise variable v^i. We denote the joint probability measure on $\mathbb{X} \times \mathbb{Y}^1 \times \cdots \times \mathbb{Y}^N$ by $\mu(dx, dy^1, \ldots, dy^N)$, and define this as the *information structure*.

Each player also has an action space $\mathbb{U}^i$, and a cost function $c^i(x, u^1, \ldots, u^N)$: $\mathbb{X} \times \mathbb{U}^1 \times \cdots \times \mathbb{U}^N \to \mathbb{R}$. The player's goal is to minimize her expected cost functional for a given game G with an underlying information structure μ:

$$J^i(G, \mu, \gamma^1, \ldots, \gamma^N) := E^{\mu,\bar{\gamma}}[c^i(x, \gamma^1(y^1), \ldots, \gamma^N(y^N))]$$

where the players select their policies from the set of all admissible *policies* $\Gamma^i := \{\gamma : \mathbb{Y}^i \to \mathbb{U}^i\}$, which are measurable functions from a player's measurement space to her action space. We refer to $u^i = \gamma^i(y^i)$ as the action of the player, and γ^i as her policy.

In this context, let us recall that various properties relevant to this context were discussed earlier in the book. Specifically:

S. Yüksel, T. Başar, *Stochastic Teams, Games, and Control under Information Constraints*, Systems & Control: Foundations & Applications,
https://doi.org/10.1007/978-3-031-54071-4_11

Existence We had obtained existence results on optimal solutions for team problems (in Chap. 4) and equilibria for general stochastic games (in Theorem 10.5.1) earlier in the book. In this chapter, we will obtain refinements of Theorem 10.5.1 for zero-sum games.

Comparison of Information Structures We have seen in Chap. 7 that in statistical decision theory in single decision-maker problems, an information structure is now worse than another one if for any cost function involving a hidden state variable and an action variable which is restricted to be conditionally independent from the state given some measurement, the solution value under the former is not worse than that under the latter. We have also seen that more information never hurts a decision maker and have obtained a partial Blackwell ordering. A similar conclusion was made for team problems.

In the following, we will see that for stochastic games such an ordering is not in general possible since additional information can lead to changes in equilibria with positive or negative incremental values to a player. However, for zero-sum games a partial order characterization is possible, as we will show in this chapter.

Continuity Properties of Equilibrium Values in Information Structures In the single DM setting, there is naturally a single information channel. For a fixed prior, in Chap. 8, we had shown that the value function is continuous under total variation convergence of information structures for single-player decision problems, (which can also be viewed as a special subset of zero-sum games where one player's action is irrelevant to the cost function). Further, counterexamples were presented which also revealed that continuity is not necessarily preserved under weak or setwise convergence, but the value function is upper semicontinuous under both. The case where priors also change was also studied, where conditions for optimality under both total variation and weak convergence were established: total variation convergence of the priors always leads to convergence, but continuity under the weak convergence of priors requires a total variation continuity on the information channels; see Sect. 8.7. We also had addressed stochastic team problems and had obtained continuity results (both positive and negative) in Sect. 8.8.

In this chapter, building on the results summarized above from several earlier chapters in the book, we will also establish continuity and regularity properties of equilibrium values for stochastic games in the spaces of information structures (ISs). In our analysis, we will also allow the prior probability measures to be varying. In this context, (i) we will show that the value function for a zero-sum game is continuous in total variation convergence of the IS if the game has a bounded measurable cost function. In addition, when the prior is fixed, the value function is either upper or lower semicontinuous in setwise convergence of an IS, if the sequence of ISs is either a minimizer-garbling or a maximizer-garbling sequence. The same results hold for weak convergence of ISs, when the assumptions on the cost functions are such that they are continuous and bounded, and the player action spaces are convex. If the channels are fixed and continuous in total variation, continuity under weak convergence of the prior holds. (ii) The same results hold for weak convergence of the information structures, when the assumptions on the cost function are such that it is continuous and bounded, and the player action spaces are convex. If the channels are fixed and continuous in total variation, continuity under weak convergence of the prior holds. (iii) A counterexample reveals that players in general non-zero-sum games may not have value functions that are continuous in total variation convergence of ISs, even for bounded cost functions. (iv) We obtain generalizations of these results when teams of agents play against each other in a zero-sum setting, where existence of equilibria will be established and comparisons of and continuity in ISs will be addressed.

Team-Against-Team Games As an application of many of the results presented in the book on both stochastic games and teams, we will consider setups where teams of agents take part in a game against other teams consisting of multiple agents. We will consider zero-sum, non-zero-sum, and exchangeable setups separately, as well as the case with teams consisting of infinitely many agents, and discuss the role of common randomness.

Existence of saddle-point equilibria for zero-sum stochastic games is covered in Sect. 11.2. Section 11.3 studies the comparison of ISs for zero-sum games in standard Borel spaces. Section 11.4 establishes continuity properties, noted above. Section 11.5 shows that for non-zero sum stochastic games, such regularity properties do not apply even under the most stringent topologies (such as total variation convergence) and that value of information is not necessarily positive for such games (unlike in zero-sum games, or in stochastic teams). Finally, Sect. 11.6 studies zero-sum, non-zero-sum, and exchangeable games involving teams of agents playing against one another.

11.2 Stochastic Zero-Sum Games and Existence of Saddle-Point Equilibria

In the following, we present conditions on existence of saddle-point equilibria in zero-sum stochastic games. These conditions will complement and refine the ones on general non-zero sum games studied in Theorem 10.5.1. Now, consider a two-agent setting with measurements for the agents given as follows.

$$y^i = g^i(x, v^i), \quad i = 1, 2,$$

where the noise variables v^1 and v^2 are independent. Suppose that g^i induces a channel Q^i for $i = 1, 2$ as described earlier and DM i has only access to y^i. Let $\underline{\gamma} = \{\gamma^1, \gamma^2\}$ denote the measurable policies of the agents.

Given fixed $\mathbb{X}$, $\mathbb{Y}^1$, $\mathbb{Y}^2$, and ζ such that $x \sim \zeta$, a game $G = (c, \mathbb{U}^1, \mathbb{U}^2)$ is a triple of a measurable and bounded cost function $c : \mathbb{X} \times \mathbb{U}^1 \times \mathbb{U}^2 \to \mathbb{R}$ and the action spaces of the players, $\mathbb{U}^1$, $\mathbb{U}^2$.

We will state some assumptions on ISs, but we note that not all of these will be needed for all the results to follow, due to the zero-sum nature of the game (unlike the general game setup presented in Theorem 10.5.1).

Assumption 11.2.1 *The information structure is absolutely continuous with respect to a product measure:*

$$P(dy^1, dy^2, dx) \ll \bar{Q}^1(dy^1)\bar{Q}^2(dy^2)\zeta(dx),$$

for some reference probability measures $\bar{Q}^i$, $i = 1, 2$. *That is, there exist probability measures* $\bar{Q}^1$, $\bar{Q}^2$, *and an integrable* f *which satisfies for every Borel* A, B, C

$$P(y^1 \in B, y^2 \in C, x \in A) = \int_{A,B,C} f(x, y^1, y^2)\zeta(dx)\bar{Q}^1(dy^1)\bar{Q}^2(dy^2).$$

Assumption 11.2.2 *The following conditional independence (or Markov) condition holds:*

$$P(dy^1, dy^2, dx) = Q^1(dy^1|x)Q^2(dy^2|x)\zeta(dx).$$

where the measurements of agents are absolutely continuous so that for each $i = 1, 2$, *there exists a non-negative function* f^i *and a reference probability measure* $\bar{Q}^i$ *such that for all Borel* S:

$$Q^i(y^i \in S|x) = \int_S f^i(y^i, x)\bar{Q}^i(dy^i).$$

We note that Assumption 11.2.2 implies Assumption 11.2.1, but this assumption often allows for a simpler interpretation of some of the analysis to be presented. The main general results will be presented in the absence of either of the assumptions above.

Let the joint measure $P(dy^1, dy^2, dx)$ define the *information structure (IS)* for the game and let us denote this with μ. For a zero-sum game with the conditional independence assumption in Assumption 11.2.2, an IS μ consists of private ISs μ^1 and μ^2 defined with Q^i, $i = 1, 2$. Define μ^i as the joint probability measure induced on $\mathcal{P}(\mathbb{X} \times \mathbb{Y}^i)$ by the measurement channel Q^i with input distribution $\zeta(dx)$. For our analysis, we will allow policies to be randomized with independent randomness. Which is to say, the set of all admissible measurable policies Γ^i will be the set of all measurable functions γ^i, where $u^i = \gamma^i(y^i, \omega^i)$ for some independent noise variable ω^i. Admissible randomized policies are stochastic kernels from $\mathbb{Y}^i$ to $\mathbb{U}^i$. Under conditional independence, let us define the following cost functional for a single-stage setup:

$$J(\zeta, \mu^1, \gamma^1, \mu^2, \gamma^2) = E_\zeta^{Q^1, Q^2, \underline{\gamma}}[c(x, u^1, u^2)]$$
$$= \int_{\mathbb{X}\times\mathbb{Y}} c(x, \gamma^1(y^1), \gamma^2(y^2)) Q^1(dy^1|x) Q^2(dy^2|x)\zeta(dx).$$

Suppose that DM 1 (the minimizer) wishes to minimize the cost, and DM 2 (the maximizer) wishes to maximize it. Let γ^1 and γ^2 be defined as earlier for each DM.

Definition 11.2.1 Given an information structure μ, $(\gamma^{1,*}, \gamma^{2,*})$ is a saddle-point equilibrium (SPE) for the zero-sum game if

$$\inf_{\gamma^1\in\Gamma^1} J(\zeta, \mu^1, \gamma^1, \mu^2, \gamma^{2,*}) = J(\zeta, \mu^1, \gamma^{1,*}, \mu^2, \gamma^{2,*})$$
$$= \sup_{\gamma^2\in\Gamma^2} J(\zeta, \mu^1, \gamma^{1,*}, \mu^2, \gamma^2).$$

Let $V_G^\mu(\gamma^1, \gamma^2)$ be the expected value of the cost function c for the maximizer, for some game G, given information structure μ and strategies (γ^1, γ^2) for the minimizer and maximizer, respectively:

$$V_G^\mu(\gamma^1, \gamma^2) = \int c(x, \gamma^1(y^1), \gamma^2(y^2)) Q^1(dy^1|x) Q^2(dy^2|x)\zeta(dx).$$

Let $V^*(G, \mu)$ be $V_G^\mu(\gamma^1, \gamma^2)$ where (γ^1, γ^2) are chosen to be the saddle-point equilibrium strategies for the players.

11.2.1 Existence of Saddle-Point Equilibria (SPE)

In the finite space case, by Theorem 10.5.1 equilibrium solutions always exist [748] (see also e.g. [97, Theorem 4.4]), but this does not hold true in general [548]. Theorem 11.2.1 below gives sufficient conditions for equilibrium solutions to exist for games with incomplete information.

Theorem 11.2.1 (Existence of SPE [353, Theorem 3.1] [59, Theorem 3.1]) *For a zero-sum game of the type introduced earlier in this chapter, assume that Assumption 11.2.1 holds. Further, let the following hold:*

(i) The action spaces of players, $\mathbb{U}^1$, $\mathbb{U}^2$, are compact.
(ii) The cost function c is bounded and continuous in players' actions, for every state of nature x.

Then, a saddle-point equilibrium exists under possibly randomized policies, and thus there exists a value of the zero-sum game.

The proof is given in the appendix, which entails an approach to topologize and convexify/compactify the policy spaces.

For completeness, we also present the following theorem, which is a mild relaxation of the theorem above, removing the requirements for the *independent-measurements* reduction and in particular the absolute continuity which facilitates this reduction. A proof is given in the appendix.

Theorem 11.2.2 ([500]) *(also [59, Theorem 3.4] [353, Theorem 3.2])[Existence of Equilibria with a Further Relaxation] For a zero-sum game of the type introduced earlier in this chapter, assume the following hold.*

(i) The action spaces of players, $\mathbb{U}^1, \mathbb{U}^2$*, are compact.*
(ii) The cost function c *is bounded and continuous in players' actions, for every state of nature* x*.*

Then a saddle-point equilibrium exists under possibly randomized policies, and thus there exists a value of the zero-sum game.

11.3 Comparison of Information Structures in Zero-Sum Games

We have the following definitions.

Definition 11.3.1 For fixed $\mathbb{X}, \mathbb{Y}^1, \mathbb{Y}^2$ and ζ such that $x \sim \zeta$, we say that an information structure μ is *better (or no worse) for the maximizer* than information structure ν (written as $\nu \lesssim \mu$) over all games in a given class of games $\mathbb{G}$ if and only if for all games G in $\mathbb{G}$:

$$V^*(G, \mu) \geq V^*(G, \nu).$$

We denote by $\kappa^i \mu$ the information structure in which player i's information from μ is *garbled* by a stochastic kernel κ^i. We let -i denote the other player in the game. Explicitly, this means the information structure becomes:

$$(\kappa^i \mu)(B, dy^{-i}, dx) = \int_{\mathbb{Y}^i} \kappa^i(B|y^i)\mu(dy^i, dy^{-i}, dx), \; B \in \mathcal{B}(\mathbb{Y}^i). \tag{11.1}$$

We use K^i to denote the space of all such stochastic kernels κ^i for player i. We first present a result involving finite spaces. This will be generalized to standard Borel spaces later on.

Theorem 11.3.1 ([578]) *Let* $\mathbb{X}, \mathbb{Y}^1, \mathbb{Y}^2$ *be finite spaces. For any two ISs* μ *and* ν*,* μ *is better for the maximizer than* ν *over all games with finite action spaces* $\mathbb{U}^1, \mathbb{U}^2$ *if and only if there exist kernels* $\kappa^i \in K^i, i = 1, 2,$ *such that*

$$\kappa^1 \nu = \kappa^2 \mu,$$

In particular, under Assumption 11.2.2, we have the more explicit characterization with

$$\kappa^1 Q^1_\nu = Q^1_\mu \quad \text{and} \quad Q^2_\nu = \kappa^2 Q^2_\mu,$$

where Q^i_μ and Q^i_ν are the measurement channels for player i under information structures μ and ν, respectively.

In the following we will present a standard Borel generalization of this result along with a proof. We will utilize Theorem 7.3.4.

For fixed $\mathbb{X}$ with $x \sim \zeta$, and fixed $\mathbb{Y}^1, \mathbb{Y}^2$, let $\tilde{\mathbb{G}}_\zeta(\mathbb{X}, \mathbb{Y}^1, \mathbb{Y}^2)$ be the class of games for which the players have compact action spaces and the cost function is bounded and continuous in players' actions for every state x.

Lemma 11.3.1 ([353]) *Given fixed $\mathbb{X}$, ζ, $\mathbb{Y}^1$, and $\mathbb{Y}^2$, for any IS μ and any kernels $\kappa^i \in K^i$, we have*

$$\kappa^2\mu \lesssim \mu \quad \textit{and} \quad \mu \lesssim \kappa^1\mu,$$

over all games in $\tilde{\mathbb{G}}_\zeta(\mathbb{X}, \mathbb{Y}^1, \mathbb{Y}^2)$.

A proof is given in the appendix. We also develop the following converse result, whose proof is also in the appendix.

Lemma 11.3.2 ([353]) *Assume that $\mathbb{X}, \zeta, \mathbb{Y}^1, \mathbb{Y}^2$ are given and $\mathbb{Y}^1, \mathbb{Y}^2$ are compact. Further, let ISs ν and μ be given. If $\nu \lesssim \mu$ over all games in $\tilde{\mathbb{G}}_\zeta(\mathbb{X}, \mathbb{Y}^1, \mathbb{Y}^2)$, then there exist kernels $\kappa^i \in K^i$ such that:*

$$\kappa^1\nu = \kappa^2\mu.$$

Furthermore, under Assumption 11.2.2, we have the more explicit characterization with

$$\kappa^1 Q^1_\nu = Q^1_\mu \quad \text{and} \quad Q^2_\nu = \kappa^2 Q^2_\mu,$$

where Q^i_μ and Q^i_ν are the measurement channels for player i under information structures μ and ν, respectively.

The above then leads to the following theorem.

Theorem 11.3.2 ([353]) *Assume that $\mathbb{X}, \zeta, \mathbb{Y}^1, \mathbb{Y}^2$ are given and $\mathbb{Y}^1, \mathbb{Y}^2$ are compact. Further, let ISs ν and μ be given. Then, μ is better for the maximizer than ν ($\nu \lesssim \mu$) over all games in $\tilde{\mathbb{G}}_\zeta(\mathbb{X}, \mathbb{Y}^1, \mathbb{Y}^2)$ if and only if there exist kernels $\kappa^i \in K^i$ such that:*

$$\kappa^1\nu = \kappa^2\mu.$$

Proof The *if* direction follows directly from Lemma 11.3.1:

$$\nu \lesssim \kappa^1 \nu = \kappa^2 \mu \lesssim \mu$$

The *only if* direction is given in Lemma 11.3.2. ◇

Corollary 11.3.1 *Consider the setup of Theorem 11.3.2. The value of additional information to a player is never negative for her in any zero-sum game in* $\tilde{\mathbb{G}}_\zeta(\mathbb{X}, \mathbb{Y}^1, \mathbb{Y}^2)$.

Proof If μ is an information structure which is more informative for the maximizer than another information structure ν, we know there exists a kernel κ^2 such that $\nu = \kappa^2\mu$, and thus know that $V^*(G, \mu) \geq V^*(G, \nu)$, and κ^2 is a well-defined map since we can map any additional information to a fixed number. A parallel argument applies for the minimizer. ◇

We note that the value of information is not always positive to a player, since there exist situations (such as when the action set is a singleton) when there would be no change in performance despite additional information.

Remark 11.3.1 The comparison result brings with it the following intuitive interpretation: An IS μ is better (i.e., no worse) for the maximizer than another IS ν if and only if one of the following holds:

1. ν^2 is a non-identity garbling of the maximizer's channel from μ, and the minimizer's channel is the same under both.
2. μ^1 is a non-identity garbling of the minimizer's channel from ν, and the maximizer's channel is the same under both.
3. ν^2 is a non-identity garbling of the maximizer's channel from μ, and μ^1 is a non-identity garbling of the minimizer's channel from ν.
4. The information structures are identical.

In plain terms, this has the following interpretation: In zero-sum games, improving or hurting both players' information structures will never give a general benefit to either player over all games. The only time a player will not do worse under a new information structure is if it only makes his channel better, only makes his opponent's channel worse, makes his channel better and his opponent's channel worse, or is identical to the previous information structure (and the player is guaranteed to not do worse if any of these conditions holds).

11.4 Continuity Properties of Equilibrium Values in the Space of Information Structures

Consider a zero-sum game where again Player 1's goal is to minimize the expected cost, while Player 2's goal is to maximize it.

Given fixed $\mathbb{X}$, $\mathbb{Y}^1$, $\mathbb{Y}^2$, and ζ such that $x \sim \zeta$, a game $g = (c, \mathbb{U}^1, \mathbb{U}^2)$ is a triple of a measurable cost function $c : \mathbb{X} \times \mathbb{U}^1 \times \mathbb{U}^2 \to \mathbb{R}$ and standard Borel action spaces for each player $\mathbb{U}^1$, $\mathbb{U}^2$.

Let $J^*(g, \mu)$ denote $J(g, \mu, \gamma^1, \gamma^2)$ when (γ^1, γ^2) are chosen as saddle-point equilibrium strategies for the players.

First, we show that continuity in total variation implies continuity in expected value for players in zero-sum games.

Theorem 11.4.1 ([354])

(i) *For a fixed zero-sum game which satisfies Assumption 8.8.1-A3, A4 and A5, the value function is continuous in information structures under total variation convergence of information structures, in the sense that if $\|\mu_n - \mu\|_{TV} \to 0$, then $|J^*(g, \mu_n) - J^*(g, \mu)| \to 0$ for all μ_n satisfying Assumption 8.8.1-A3.*

(ii) *The value function is not necessarily continuous under weak or setwise convergence of information structures even under Assumption 8.8.1-A2 and A4 and a fixed prior.*

We now present our results for weak and setwise convergences of information structures in zero-sum games. Recall the definition of garbling given in (11.1) and consider the following.

Definition 11.4.1 For a fixed prior ζ, a sequence of information structures $\{\mu\}_{k=1}^{\infty}$ is a *maximizer-garbling* (or a minimizer-degarbling) sequence if for every $j \in \{1, \ldots, \infty\}$ there exist kernels κ_j^1 and κ_j^2 such that $\mu_{j+1}\kappa_j^2 = \kappa_j^1\mu_j$.

Definition 11.4.2 For a fixed prior ζ, a sequence of information structures $\{\mu\}_{k=1}^{\infty}$ is a *minimizer-garbling* (or a maximizer-degarbling) sequence if for every $j \in \{1, \ldots, \infty\}$ there exist kernels κ_j^1 and κ_j^2 such that $\kappa_j^1\mu_{j+1} = \mu_j\kappa_j^2$.

We note that, following the results in Theorems 11.3.1 and 11.3.2, the value function for any zero-sum game will be monotonically decreasing for a maximizer-garbling sequence and monotonically increasing for a minimizer-garbling sequence.

Next, we present examples of maximizer-garbling sequences.

Example 11.4.1 Consider a two-player zero-sum game with state space $\mathbb{X} = [0, 1]$ endowed with the uniform distribution as ζ. Let $\mathbb{Y}^1 = \mathbb{Y}^2 = \mathbb{R}$.

For $m \in \mathbb{Z}_{\geq 0}$, define μ_m by the following independent measurement channels for the players: Player 1 receives measurement y^1, which is drawn randomly from a Gaussian distribution with mean x and variance $1 + \frac{1}{m}$; Player 2 receives measurement y^2 which is drawn randomly from a Gaussian distribution with mean x and variance $1 - \frac{1}{m+1}$. Since Gaussian distributions with higher variances are garblings of Gaussian distributions with lower variances, this sequence is a maximizer-garbling sequence, since for any $i \in \mathbb{Z}_{\geq 0}$, the maximizer's channel under μ_{i+1} is a garbling of his channel under μ_i, while the minimizer's channel under μ_i is a garbling of her channel under μ_{i+1}. Furthermore, this sequence converges weakly to the information structure in which both players' channels have variance 1. One can also construct such an example where we have $y^1 = x +$

v_m^1, and if the variance of v_m^1 decreases to zero, we will have weak convergence to a point distribution.

Another example involves noiseless but quantized measurement channels: $y^1 = Q_m^1(x)$, where Q_m^1 is a uniform quantization of $[0, 1]$ into 2^m bins. In this case, the quantizers are garblings of one another and the limit is the fully informative channel: $Q(dy|x) = \delta_x(dy)$.

Theorem 11.4.2 ([354]) *Let Assumption 8.8.1-A2, A3, A4 and A7 hold and the prior be fixed. Let μ_m be a sequence of information structures converging weakly to information structure μ. If the sequence is a maximizer-garbling, then the value function is lower semicontinuous. If the sequence is a minimizer-garbling, then the value function is upper semicontinuous.*

Theorem 11.4.3 ([354]) *Consider a fixed prior and let Assumption 8.8.1-A3, A4 and A5 hold. Let μ_m be a sequence of information structures converging setwise to information structure μ. If the sequence is a maximizer-garbling, then the value function is lower semicontinuous (in the sense that if $\mu_n \to \mu$ weakly and μ_n, μ satisfy Assumption 8.8.1-A3, then $\liminf_{n\to\infty} J^*(g, \mu_n) \geq J^*(g, \mu)$). If the sequence is a minimizer-garbling, then the value function is upper semicontinuous.*

We can also show that when the player channels are fixed and independent and the prior varies, the value function is continuous under weak convergence of the prior. This generalizes a single-player result from [388].

Theorem 11.4.4 ([354]) *Under Assumption 8.8.1-A2, A4 and A6, the value function is continuous under weak convergence of the priors.*

11.5 General Games: Lack of Continuity and Lack of Positive Value of Information

We recall that continuity and comparison results on information structures for stochastic teams were studied in Chaps. 8 and 7, respectively. In the previous section, we studied such properties in the context of zero-sum games. In this section, we consider general non-zero sum stochastic games.

While we have shown that zero-sum games and team problems both exhibit continuity under total variation convergence of information structures, here we present a counterexample that reveals that this is not true for general non-zero-sum games.

Example 11.5.1 Consider a two-player game with state space $\mathbb{X} = [-1, 1]$ endowed with the continuous uniform distribution as ζ. Let $\mathbb{Y}^1 = \{1, 2, 3\}$ and $\mathbb{Y}^2 = \{0\}$, and $\mathbb{U}^1 = \mathbb{U}^2 = [-1, 1]$. For $m \in \mathbb{Z}_{\geq 1}$, define information structure μ_m by channels Q_m^1 and Q_m^2 for Players 1 and 2, respectively, where Q_m^1 is a quantizer with three bins:

$$y_m^1 = \begin{cases} 1, & x \in [-1, -\frac{1}{2} - \frac{1}{8m}) \\ 2, & x \in [-\frac{1}{2} - \frac{1}{8m}, \frac{1}{2} + \frac{1}{4m}] \\ 3, & x \in (\frac{1}{2} + \frac{1}{4m}, 1] \end{cases}$$

and Q_m^2 returns $y^2 = 0$ for all $x \in \mathbb{X}$.

Following Theorem 8.5.4, since our information structure is defined by quantizers that converge setwise at input ζ, $\{\mu_m\} \to \mu$ in total variation, where μ is defined by the quantizer for Player 1, which sorts x into the bins $[-1, -1/2)$, $[-1/2, 1/2]$, and $(1/2, 1]$, and player 2 has the same channel that always returns $y^2 = 0$.

Now consider the following cost functions for the two players:

$$c^1(x, u^1, u^2) = (x - u^1)^2 - (u^2)^2, \qquad c^2(x, u^1, u^2) = \begin{cases} (u^2)^2, & u^1 = 0 \\ (u^2 - 1)^2, & u^1 \neq 0 \end{cases}$$

For each m, Player 1's optimal strategy is to minimize $(x - u^1)^2$, and thus his optimal policy is to play $u^1 = E_{\mu_n}[x|y^1]$. Due to the asymmetry of Player 1's information channel, $u^1 \neq 0$ for all m. Thus, Player 2's optimal strategy is to play $u^2 = 1$, and her expected cost is 0.

However, under μ, Player 1 will play 0 with probability 1/2. Thus, Player 2's optimal strategy is to play $u^2 = 1/2$, and her expected cost is 1/4. Thus, the value function for Player 2 is not continuous under total variation convergence of the information structure.

We can thus conclude that general non-zero-sum games do not necessarily exhibit continuity in information structures under total variation convergence (and therefore also under weak or setwise convergence of information structures) for games with measurable and bounded cost functions.

11.6 Team Against Team Games: Zero-Sum, General and Exchangeable Games, and Role of Common Randomness

We now consider an important special setup, which may also be viewed as an application of many of the results presented in the book on both stochastic games and teams, where two (or more) teams participate in a game. An intrinsic model for settings of this type was presented in Sect. 10.1.2.

Consider a stochastic game with a finite number of DMs forming two teams who play against one another. The team, Team$i$ or PL^i, $i = 1, 2$ consists of DMs indexed by $\mathcal{N}_i = \{1, \ldots, N_i\}$, with policies denoted by $\underline{\gamma}_N^i := (\gamma_1^i, \cdots, \gamma_{N_i}^i)$ and policy spaces $\Gamma_N^i := \prod_{k=1}^{N_i} \Gamma_k^i$.

Let the expected cost function for each team, PL^i, under a policy profile $\underline{\gamma}_N^{1:2} := (\underline{\gamma}_N^1, \underline{\gamma}_N^2)$ be given by

$$J_N^i(\underline{\gamma}_N^{1:2}) = \mathbb{E}^{\underline{\gamma}_N^{1:2}}\left[c^i(\omega_0, \underline{u}_N^{1:2})\right], \tag{11.2}$$

for some measurable cost function c^i, $i = 1, 2$.

As noted earlier, lack of convexity in the set of team policies can lead to games without a saddle point or Nash equilibrium. For problems of this type, recall that the existence of Nash equilibrium under general conditions follows the analysis in Remark 10.5.2, provided convexity conditions hold leading to applicability of the Kakutani-Fan-Glicksberg fixed point theorem (Theorem 10.5.2). In the following, we will specialize to two distinct setups.

We first discuss the zero-sum setting, in the next subsection. This will be followed by the non-zero sum setting, where we will also take the sizes of each team to be arbitrarily large, and even unbounded. A utility of the latter case will be that common randomness will no longer be needed for equilibrium behavior.

11.6.1 Zero-Sum Games Between Teams of Finitely Many Agents: Existence of Equilibria and Comparison and Regularity in Information

As before, consider the setup where Team 1 (the minimizing team) consists of N_1 DMs and Team 2 (the maximizing team) consists of N_2 DMs. The two teams compete in a single-stage zero-sum game (with static information structure). Here we use a subscript i to denote the team to which a DM belongs.

The teams in the game have the (completely mis-aligned) cost function $c : \Omega_0 \times \mathbb{U}_1^1 \times \cdots \times \mathbb{U}_{N_1}^1 \times \mathbb{U}_1^2 \times \cdots \times \mathbb{U}_{N_2}^2 \to \mathbb{R}$: Team 1's goal is to minimize the expected cost and Team 2's goal is to maximize it.

The information structure μ for this game is a joint probability measure on $\Omega_0 \times \mathbb{Y}_1^1 \times \cdots \times \mathbb{Y}_{N_1}^1 \times \mathbb{Y}_1^2 \times \cdots \times \mathbb{Y}_{N_2}^2$. We use μ_i to denote the marginal of μ on $\Omega_0 \times \mathbb{Y}_1^i \times \cdots \times \mathbb{Y}_{N_i}^i$ for $i = 1, 2$ (i.e. μ_i captures the information available to team i).

We use the notation V and V^*, rather than J and J^*, to denote the value and equilibrium value of a game, respectively.

Definition 11.6.1 Given an information structure μ, let $\mathbf{\Gamma}_i$ denote the set of team policies of Team i, with elements $\bar{\gamma}_i$, $i = 1, 2$. We say that $(\bar{\gamma}^{*,1}, \bar{\gamma}^{*,2})$ is an equilibrium for the team-against-team game if

$$\inf_{\bar{\gamma}_1 \in \mathbf{\Gamma}^1} V(\mu, \bar{\gamma}^1, \bar{\gamma}^{*,2}) = V(\mu, \bar{\gamma}^{*,1}, \bar{\gamma}^{*,2}) = \sup_{\bar{\gamma}^2 \in \mathbf{\Gamma}^2} V(\mu, \bar{\gamma}^{*,1}, \bar{\gamma}^2).$$

Existence of Equilibria in Team-Against-Team Zero-Sum Games

We have earlier seen in Theorems 11.2.1 and 11.2.2, two different sets of results on existence of saddle-point equilibria; the former on existence of saddle-point equilibria in two-player zero-sum games under an absolute continuity condition, and the latter not requiring this. In the following, we present a generalization (to team games) of the relaxed version where absolute continuity among the different team-players does not necessarily apply, but this only applies within each team (to ensure the convexity of the team policies of a given team). Of course the case where absolute continuity applies across all players would be a special case.

For $i = 1, 2$, for a fixed information structure μ, let μ_i be the marginal of μ on $\Omega_0 \times \prod_{j=1}^{N_i} \mathbb{Y}^i_j$. Here, μ_i represents the portion of the information structure μ that is relevant for Team i's action selection (as they do not have access to the opposing team's measurements). We then define $L_C(\mu_i)$ following the usual definition for a single team (see Sect. 4.1.2 or Sect. 7.4), allowing for actions to be selected using independent (but possibly shared between players) randomness.

We thus consider the following assumption.

Assumption 11.6.1 *Equation (7.7) holds for both μ_1 and μ_2. That is,*

$$\mu_i(d\omega_0, dy^i_1, \cdots, dy^i_{N_i}) \ll \mu^i_0(d\omega_0) \prod_{i=1}^{n} \bar{Q}_i^{N_i}(dy^i_{N_i})$$

for some reference measures μ^i_0 and $\bar{Q}_i^m$, $m = 1, \cdots, N_i$.

As noted in the context of (7.7), this condition is important in establishing that $L_R(\mu_i)$ is compact under the weak convergence topology. It then follows that $L_C(\mu_i)$ is closed under Assumption 7.4.1, via Theorem 4.2.7.

Note again that here no assumptions have been made regarding independence of the teams' measurement channels (only within a team an absolute continuity condition is needed), as no restrictions were made on the structure of μ. However, it is assumed that any randomness used in action selection is independent between the two teams.

Theorem 11.6.1 ([356]) *For a given team-against-team zero-sum game, assume that the following hold:*

(i) Team policies are selected such that the strategic measures for Teams 1 and 2 belong to $L_C(\mu_1)$, $L_C(\mu_2)$, respectively.
(ii) $L_C(\mu_1)$ and $L_C(\mu_2)$ are compact under weak convergence (a sufficient condition being Assumption 11.6.1 and compactness of action sets).
(iii) The cost function c is bounded and continuous in players' actions for any $\omega_0 \in \Omega_0$.

Then, an equilibrium exists.

A proof sketch is provided in the Appendix.

Comparison of Information Structures in Team-Against-Team Zero-Sum Games

Now, we can prove an extension of Blackwell's comparison of information structures similar to that in Theorems 11.3.1 and 11.3.2. This also generalizes Theorem 7.4.1 with a similar proof program.

Let $\tilde{\mathcal{G}}$ denote the set of all team-against-team zero-sum games for which the cost function is continuous and bounded in players' actions for every state.

Theorem 11.6.2 ([356]) *We are given Ω_0, P, and compact measurement and action spaces, and information structures ν and μ for which Assumption 11.6.1 holds. Recall Definition 7.4.1 on correlated garbling of information. Then, μ is more informative for the maximizing team than ν over all games in $\tilde{\mathcal{G}}$ if and only if the following hold:*

(i) The measurement of each player in Team 1 under ν is given by $\tilde{y}_i^1(\nu) = h^i(y_i^1(\mu), Z_1)$ for $i = 1, \dots, n$, and some independent randomness Z_1 (with possibly dependent coordinate components). Denote this information structure by $\kappa_1\mu$.

(ii) The measurement of each player in Team 2 under μ is given by $\tilde{y}_i^2(\mu) = h^i(y_i^2(\nu), Z_2)$ for $i = 1, \dots, n$, and some independent randomness Z_2 (with possibly dependent coordinate components). Denote this information structure by $\kappa_2\nu$.

Equivalently, μ is more informative for the maximizing team than ν if and only if $\kappa_1\mu = \kappa_2\nu$ for some κ_1 and κ_2.

The proof closely follows that of Theorem 11.3.2, since via compactness and convexity of team policies, one directly follows the analysis in Theorem 11.3.2.

This theorem provides a general result on the comparison of information structures for both teams and zero-sum games. When only cost functions where one team's actions are irrelevant to the cost are considered, the result reduces to that of team problems; when each team is restricted to having only one player, the result reduces to that of a standard zero-sum game. When both of these restrictions are applied, the result reduces to the original single-player decision problem studied by Blackwell. Each of these special cases was studied earlier in the book.

Continuity of Value in Information Structures in Team-Against-Team Zero-Sum Games

We now present a continuity result on the value of the game under the total variation convergence of information structures for zero-sum games involving teams of agents, generalizing Theorem 11.4.1. A proof sketch is given in the Appendix.

Theorem 11.6.3 ([356]) *Suppose that the hypotheses of Theorem 11.6.1 hold so that an equilibrium exists. Then, the value function $J^*(c, \mu)$ is continuous under* *total variation convergence of information structures provided that perturbations in information satisfy the conditions of existence.*

Remark 11.6.1 The results of this section can also be applied to multi-stage problems provided that some absolute continuity conditions hold so that appropriate static or nested reductions are applicable (see e.g. Assumption 6.4.3 as well as Sect. 3.5.2) so that convexity and compactness properties can be established, as we have seen earlier in the book. We also recall that policy-independent static reduction preserves Nash, and thus also, saddle-point, equilibria. See also Remark 10.5.2.

11.6.2 Non-Zero-Sum Games Between Teams of Finitely Many Agents: Existence of Equilibrium

The analysis on existence presented in Theorem 11.6.1 above applies almost identically to the non-zero-sum case. Here, the teams in the game have cost functions $c^i : \Omega_0 \times \mathbb{U}_1^1 \times \cdots \times \mathbb{U}_{N_1}^1 \times \mathbb{U}_1^2 \times \cdots \times \mathbb{U}_{N_2}^2 \to \mathbb{R}$, $i = 1, 2$. As earlier, the information structure μ for this game is a joint probability measure on $\Omega_0 \times \mathbb{Y}_1^1 \times \cdots \times \mathbb{Y}_{N_1}^1 \times \mathbb{Y}_1^2 \times \cdots \times \mathbb{Y}_{N_2}^2$. We use μ_i to denote the marginal of μ on $\Omega_0 \times \mathbb{Y}_1^i \times \cdots \times \mathbb{Y}_{N_i}^i$ for $i = 1, 2$ (i.e. μ_i captures the information available to team i).

Definition 11.6.2 Given an information structure μ, let $\mathbf{\Gamma}_i$ denote the set of team policies of Team i, with elements $\bar{\gamma}_i$, $i = 1, 2$. We say that $(\bar{\gamma}^{*,1}, \bar{\gamma}^{*,2})$ is a Nash equilibrium for the team-against-team game if

$$\inf_{\bar{\gamma}_i \in \mathbf{\Gamma}^i} J(\mu, \bar{\gamma}^i, \bar{\gamma}^{*,-i}) = J(\mu, \bar{\gamma}^{*,1}, \bar{\gamma}^{*,2}).$$

We also impose the following condition (noting that this was not required for the zero-sum game setup).

Assumption 11.6.2

$$\begin{aligned} &\mu(d\omega_0; dy_1^1, \ldots, dy_2^{N_1}; dy_2^1, \ldots, dy_2^{N_2}) \\ &\ll \zeta(d\omega_0)\mu_1(dy_1^1, \ldots, dy_2^{N_1})\mu_2(dy_2^1, \ldots, dy_2^{N_2}). \end{aligned}$$

Theorem 11.6.4 *For a given team-against-team non-zero-sum game, assume that the following conditions hold:*

(i) Team policies are selected such that the strategic measures for Teams 1 and 2 belong to $L_C(\mu_1)$, $L_C(\mu_2)$, respectively.
(ii) $L_C(\mu_1)$ and $L_C(\mu_2)$ are compact under weak convergence (a sufficient condition being Assumption 11.6.1 and compactness of action sets).
(iii) Assumption 11.6.2 holds.

(iv) *The cost functions c^i, $i = 1, 2$, are bounded and continuous in players' actions for any $\omega_0 \in \Omega_0$.*

Then, a Nash equilibrium exists.

The proof follows along the arguments noted in the proof of Theorem 10.5.1 and Remark 10.5.2.

11.6.3 Nash Equilibria for Exchangeable Team Against Team Games, the Mean-Field Limit and Absence of Common Randomness

We now study stochastic mean-field games among a finite number of teams, each with large finite as well as infinite number of decision makers. We will crucially utilize the results presented in Sect. 6.4. A practical implication of the analysis here will be that common randomness is not needed in the infinite population limit.

We first study the static setup. Consider a stochastic game with a finite number of DMs, i.e., for each PL^i, $\mathcal{N}_i = \{1, \ldots, N_i\}$, $\underline{\gamma}^i_N := (\gamma^i_1, \cdots, \gamma^i_{N_i})$ and $\Gamma^i_N := \prod_{k=1}^{N_i} \Gamma^i_k$. Let the expected cost function for each PL^i under a policy profile $\underline{\gamma}^{1:2}_N := (\underline{\gamma}^1_N, \underline{\gamma}^2_N)$ be given by

$$J^i_N(\underline{\gamma}^{1:2}_N) = \mathbb{E}^{\underline{\gamma}^{1:2}_N}\left[c^i(\omega_0, \underline{u}^{1:M}_N)\right] \tag{11.3}$$

for some Borel measurable cost function $c^i : \Omega_0 \times \prod_{j=1}^{2} \prod_{k=1}^{N_j} \mathbb{U}^j \to \mathbb{R}_+$. $\mathbb{E}^{\underline{\gamma}^{1:2}_N}$ denote the expectation with respect to $\mathbb{P}$ by taking actions induced by $\underline{\gamma}^{1:2}_N$. We define ω_0 as the Ω_0-valued, cost function relevant, exogenous random variable, taking values from a Borel space Ω_0 with its Borel σ-field $\mathcal{F}_0$. In the above, we used the notation $k : j$ to denote $\{k, \ldots, j\}$, and $\underline{u}^i_N := u^i_{1:N_i}$.

For general exchangeable static games, the cost functions of the players satisfy the following exchangeability condition.

Assumption 11.6.3 *The function c^i is (separately) exchangeable with respect to actions of DMs for all ω_0 and $i \in \mathcal{M}$, i.e., for any permutations σ and τ of $\{1, \ldots, N_1\}$ and $\{1, \ldots, N_2\}$, respectively,*

$$c^i\left(\omega_0, u^1_{1:N_1}, u^2_{1:N_2}\right) = c^i\left(\omega_0, u^1_{\sigma(1):\sigma(N_1)}, u^2_{\tau(1):\tau(N_2)}\right) \ \forall \omega_0,$$

where $u^1_{\kappa(1):\kappa(N_1)} = u^1_{\kappa(1)}, \ldots, u^1_{\kappa(N_1)}$ for $\kappa = \sigma$ or τ.

Let $\mathcal{P}(\cdot)$ denote the space of probability measures, and $\delta_{\{\cdot\}}$ denote the Dirac delta. For our main results when the number of DMs is driven to infinity, we consider a special case of the preceding game, where the costs satisfy Assumption 11.6.3 and are given by

$$c^i\left(\omega_0, u^1_{1:N_1}, u^2_{1:N_2}\right)$$
$$=\frac{1}{N_i}\sum_{k=1}^{N_i}\tilde{c}^i_k\left(\omega_0, u^i_k, \Xi^1(\frac{1}{N_1}\sum_{p=1}^{N_1}\delta_{u^1_p}), \Xi^2(\frac{1}{N_2}\sum_{p=1}^{N_2}\delta_{u^2_p})\right), \tag{11.4}$$

for some weakly continuous functions $\Xi^i : \mathcal{P}(\mathbb{U}^i) \to \tilde{\mathbb{U}}^i$ with $\tilde{\mathbb{U}}^i$, which is a subset of appropriate dimensional Euclidean space, and for some Borel measurable cost functions $\tilde{c}^i : \Omega_0 \times \mathbb{U}^i \times \prod_{j=1}^2 \tilde{\mathbb{U}}^j \to \mathbb{R}_+$ that satisfies

$$\frac{1}{N_i}\sum_{k=1}^{N_i}\tilde{c}^i_k\left(\omega_0, u^i_k, \Xi^1(\frac{1}{N_1}\sum_{p=1}^{N_1}\delta_{u^1_p}), \Xi^2(\frac{1}{N_2}\sum_{p=1}^{N_2}\delta_{u^2_p})\right)$$
$$=\frac{1}{N_i}\sum_{k=1}^{N_i}\tilde{c}^i_{\beta(k)}\left(\omega_0, u^i_k, \Xi^1(\frac{1}{N_1}\sum_{p=1}^{N_1}\delta_{u^1_p}), \Xi^2(\frac{1}{N_2}\sum_{p=1}^{N_2}\delta_{u^2_p})\right), \tag{11.5}$$

for every permutation β. When the number of DMs drives to infinity, we consider the special case with $c^i_k = \tilde{c}^i$ for some functions $\tilde{c}^i$. We note that weak continuity of Ξ^i is not required for the exchangeability condition, but it is required for the main theorems to be presented.

Mean-Field Static Game $\mathcal{P}_\infty$

Consider a stochastic game with countably infinite number of DMs, i.e., let $\mathcal{N}_i = \mathbb{N}$, $\Gamma^i := \prod_{k\in\mathbb{N}}\Gamma^i_k$ and $\underline{\gamma}^i := (\gamma^i_1, \gamma^i_2, \ldots)$. Let the expected cost of PL^i under a policy profile $\underline{\gamma}^{1:2}$ be given by

$$J^i_\infty(\underline{\gamma}^{1:2}) = \limsup_{N_1,N_2\to\infty}$$
$$\mathbb{E}^{\underline{\gamma}^{1:2}}\left[\frac{1}{N_i}\sum_{k=1}^{N_i}c^i\left(\omega_0, u^i_k, \Xi^1(\frac{1}{N^1}\sum_{p=1}^{N_1}\delta_{u^1_p}), \Xi^2(\frac{1}{N_2}\sum_{p=1}^{N_2}\delta_{u^2_p})\right)\right],$$

for some weakly continuous functions $\Xi^i : \mathcal{P}(\mathbb{U}^i) \to \tilde{\mathbb{U}}^i$ and Borel measurable cost functions $c^i : \Omega_0 \times \mathbb{U}^i \times \prod_{j=1}^2 \tilde{\mathbb{U}}^j \to \mathbb{R}_+$.

Assumption 11.6.4 *For every $N_1, N_2 \in \mathbb{N}\cup\{\infty\}$, let $\tilde{\mu}^N$ be the conditional distribution of observations $\underline{y}^{1:2}$ given ω_0. There exists a probability measure Q^i_k on $\mathbb{Y}^i$ and a bounded function $f^i_k : \mathbb{Y}^i \times \Omega_0 \times \prod_{j=1,\neq k}^{N_i}\mathbb{Y}^i \times \prod_{j=1}^{N_{-i}}\mathbb{Y}^{-i} \to \mathbb{R}_+$ for all $i \in \mathcal{M}$ and $k \in \mathcal{N}_i$ such that for all Borel sets B^i_k in $\mathbb{Y}^i$ (with $B := \prod_{i=1}^2 B^i_1 \times \cdots \times B^i_{N_i}$)*

$$\tilde{\mu}^N(B|\omega_0) = \prod_{i=1}^{2}\prod_{k=1}^{N_i}\int_{B_k^i} f_k^i\left(y_k^i, \omega_0, y_{-k}^i, \underline{y}^{-i}\right) Q_k^i(dy_k^i), \tag{11.6}$$

where $y_{-k}^i = \{y_{1:N_i}^i\}\backslash\{y_k^i\}$.

Under this assumption, as before, via a change of measures argument, the observations of the DMs are independent and also independent of ω_0. Consider also the following assumptions.

Assumption 11.6.5

(i) $(y_k^1)_{k\in\mathcal{N}_1}$ *and* $(y_k^2)_{k\in\mathcal{N}_2}$ *are independent, conditioned on* ω_0.
(ii) For all $i \in \mathcal{M}$, $(y_k^i)_{k\in\mathcal{N}_i}$ *have an identical distribution, conditioned on* ω_0.

Assumption 11.6.6 *For* $i \in \mathcal{M}$,

(i) $\mathbb{U}^i$ *is compact.*
(ii) $c^i(\omega_0, \cdot, \cdot, \cdot)$ *in* (11.4), *is continuous for all* ω_0.

Under Assumption 11.6.5, we can rewrite (11.6) as follows:

$$\tilde{\mu}^N(B|\omega_0) = \prod_{i=1}^{2}\prod_{k=1}^{N_i}\int_{B_k^i} \hat{f}^i\left(y_k^i, \omega_0\right) Q^i(dy_k^i), \tag{11.7}$$

where $\hat{f}^i : \mathbb{Y}^i \times \Omega_0 \to \mathbb{R}_+$ and Q^i are identical for all DMs within PL^i for $i = 1, 2$.

Next, the set of exchangeable randomized policies $L_{\mathsf{EX},N}^i \subseteq L_N^i$ for each PL^i is given by

$$\begin{aligned} L_{\mathsf{ex},N}^i := \Bigg\{ P_\pi \in L_N^i \Bigg| &\forall A_k \in \mathcal{B}(\Gamma_k^i) \text{ and } \sigma \in S_{N_i} : \\ &P_\pi(\gamma_1^i \in A_1, \ldots, \gamma_{N_i}^i \in A_{N_i}) \\ &= P_\pi(\gamma_{\sigma(1)}^i \in A_1, \ldots, \gamma_{\sigma(N_i)}^i \in A_{N_i}) \Bigg\}, \end{aligned} \tag{11.8}$$

where S_{N_i} is the set of all permutations of $\{1, \ldots, N_i\}$. $L_{\mathsf{EX},N}^i$ is a convex subset of L_N^i. Let $L_{\mathsf{CO,\,SYM},N}^i$ be the set of all symmetric (identical) randomized policies induced by common and individual randomness:

$$\begin{aligned} L_{\mathsf{CO,\,SYM},N}^i := \Bigg\{ P_\pi \in L_N^i \Bigg| \forall A_k \in \mathcal{B}(\Gamma_k^i) : P_\pi(\gamma_1^i \in A_1, \ldots, \gamma_{N_i}^i \in A_{N_i}) \\ = \int_{z^i\in[0,1]} \prod_{k=1}^{N_i} \tilde{P}_\pi(\gamma_k^i \in A_k|z)\eta^i(dz^i), \quad \eta^i \in \mathcal{P}([0,1]) \Bigg\}, \end{aligned}$$

where for all $k \in \mathcal{N}_i$, conditioned on z^i, $\tilde{P}_\pi \in \mathcal{P}(\Gamma^i_k)$ corresponds to an independent randomized policy for each DM^i_k ($k \in \mathcal{N}_i$), which is symmetric among PL^i's DMs. Note that $L^i_{\text{CO, SYM},N} \subset L^i_{\text{EX},N}$.

Randomized Policies

When the number of DMs for each PL is driven to infinity, we define the corresponding sets L^i, $L^i_{\text{CO CO}}$, L^i_{EX}, and $L^i_{\text{CO, SYM}}$ using the Ionescu-Tulcea extension theorem. The set of randomized policies L on the infinite product Borel spaces $\Gamma^i := \prod_{k\in\mathbb{N}} \Gamma^i_k$ is defined as $L^i := \mathcal{P}(\Gamma^i)$. Let

$$L^i_{\text{CO}} := \left\{P_\pi \in L^i \middle| \forall A_k \in \mathcal{B}(\Gamma^i_k) : P_\pi(\gamma^i_1 \in A_1, \gamma^i_2 \in A_2, \ldots) \right.$$
$$\left. = \int_{z^i\in[0,1]} \prod_{k\in\mathbb{N}} P_{\pi,k}(\gamma^i_k \in A_k|z^i)\eta^i(dz^i), \quad \eta^i \in \mathcal{P}([0,1])\right\}.$$

Let the set of all infinitely exchangeable randomized policies L^i_{EX} be given by

$$L^i_{\text{ex}} := \left\{P_\pi \in L^i \middle| \forall A_k \in \mathcal{B}(\Gamma^i_k), N_i \in \mathbb{N}, \text{ and } \sigma \in S_{N_i} : \right.$$
$$P_\pi(\gamma^i_1 \in A_1, \ldots, \gamma^i_{N_i} \in A_{N_i})$$
$$\left. = P_\pi(\gamma^i_{\sigma(1)} \in A_1, \ldots, \gamma^i_{\sigma(N_i)} \in A_{N_i})\right\},$$

and

$$L^i_{\text{CO, SYM}} := \left\{P_\pi \in L^i \middle| \forall A_k \in \mathcal{B}(\Gamma^i_k) : P_\pi(\gamma^i_1 \in A_1, \gamma^i_2 \in A_2, \ldots) \right.$$
$$\left. = \int_{z^i\in[0,1]} \prod_{k\in\mathbb{N}} \tilde{P}_\pi(\gamma^i_k \in A_k|z^i)\eta^i(dz^i), \quad \eta^i \in \mathcal{P}([0,1])\right\}.$$

We also define the following set of randomized policies with only private independent randomness:

$$L^i_{\text{PR}} := \left\{P_\pi \in L^i \middle| \forall A_k \in \mathcal{B}(\Gamma^i_k) : \text{for } P_{\pi,k} \in \mathcal{P}(\Gamma^i_k) \right.$$
$$\left. P_\pi(\gamma^i_1 \in A_1, \gamma^i_2 \in A_2, \ldots) = \prod_{k\in\mathbb{N}} P_{\pi,k}(\gamma^i_k \in A_k)\right\},$$

and the symmetric ones by

$$L^i_{\mathsf{PR,\ SYM}} := \left\{ P_\pi \in L^i \middle| \forall A_k \in \mathcal{B}(\Gamma^i_k) : \text{for } \tilde{P}_\pi \in \mathcal{P}(\Gamma^i_k) \right.$$

$$\left. P_\pi(\gamma^i_1 \in A_1, \gamma^i_2 \in A_2, \ldots) = \prod_{k \in \mathbb{N}} \tilde{P}_\pi(\gamma^i_k \in A_k) \right\}.$$

We note that L^i_{PR} and $L^i_{\mathsf{PR,\ SYM}}$ are not convex sets; however, L^i_{PR} ($L^i_{\mathsf{PR,\ SYM}}$) contains the set of extreme points of the convex set L^i_{CO} ($L^i_{\mathsf{CO,\ SYM}}$).

In the following, we first present two lemmas on convexity, compactness, and relationships between the preceding set of exchangeable randomized policies. The following two lemmas will lead to Theorems 11.6.5 and 11.6.6.

Lemma 11.6.1 *[635] Consider $L^i_{\mathsf{EX},N}$, $L^i_{\mathsf{CO,\ SYM},N}$, L^i_{EX}, and $L^i_{\mathsf{CO,\ SYM}}$. Then, the following hold:*

(i) $L^i_{\mathsf{CO,\ SYM}} = L^i_{\mathsf{EX}}$.
(ii) $L^i_{\mathsf{CO,\ SYM},N} \subseteq L^i_{\mathsf{EX},N}$*; however, in general* $L^i_{\mathsf{CO,\ SYM},N} \neq L^i_{\mathsf{EX},N}$.

Lemma 11.6.2 *Let $\mathbb{U}^i$ be compact. Then, $L^i_{\mathsf{EX},N}$ and L^i_{EX} are convex and compact.*

Proof We only provide a proof sketch. The complete proof can be found in [635, Section 6.1]. Since $\mathbb{U}^i$ is compact, we can show that $L^i_{\mathsf{EX},N}$ is tight. It remains to show that $L^i_{\mathsf{EX},N}$ is closed. Suppose that a sequence $\{P^{i,n}_\pi\}_n$ converges as $n \to \infty$ to $P^{i,\infty}_\pi$. Also, suppose that a permutation of $\{P^{i,n}_\pi\}_n$, converges as $n \to \infty$ to $P^{\sigma,i,\infty}_\pi$. Since $\{P^{i,n}_\pi\}_n$ are exchangeable, their limits are equal, i.e., $P^{\sigma,i,\infty}_\pi = P^{i,\infty}_\pi$ for any permutation. This implies that $L^i_{\mathsf{EX},N}$ is closed and completes the proof. ◇

Existence of an Exchangeable Nash Equilibrium (NE) for $\mathcal{P}_N$

The expected costs $J^i_{\pi,N}(P^1_\pi, P^2_\pi)$ for $\mathcal{P}_N$ under $(P^1_\pi, P^2_\pi) \in L^1_N \times L^2_N$ are given by

$$\int P^1_\pi(d\underline{\gamma}^1) P^2_\pi(d\underline{\gamma}^2) \mu^N(d\omega_0, d\underline{y}) c^i_N(\underline{\gamma}, \underline{y}, \omega_0) \tag{11.9}$$

where μ^N is the joint probability measure of $\omega_0, dy^1_{1:N_1}, dy^2_{1:N_2}$ and

$$c^i_N(\underline{\gamma}, \underline{y}, \omega_0) := \int c^i(\omega_0, u^1_{1:N_i}, u^2_{1:N_i}) \prod_{i=1}^{2} \prod_{k=1}^{N_i} \gamma^i_k(du^i_k | y^i_k).$$

Next, we introduce the following assumptions on DMs' action spaces, costs, and observations.

Assumption 11.6.7 *For all $i = 1, 2$,*

(i) $\mathbb{U}^i$ is compact.
(ii) $c^i(\omega_0, \cdot, \dots, \cdot)$ is continuous and (uniformly) bounded for all ω_0.
(iii) $(y_k^i)_{k\in\mathcal{N}_i}$ are N_i-exchangeable, conditioned on ω_0.
(iv) $(y_k^1)_{k\in\mathcal{N}_1}$ and $(y_k^2)_{k\in\mathcal{N}_2}$ are mutually independent, conditioned on ω_0.

We note that since i.i.d. random variables are (infinitely) exchangeable, Assumption 11.6.5 implies Assumptions 11.6.7 (iii) and (iv). The following theorem establishes existence of an exchangeable NE of $\mathcal{P}_N$.

Theorem 11.6.5 ([635]) *Consider the game $\mathcal{P}_N$ with a given IS. Let Assumptions 11.6.3, 11.6.4 and 11.6.7 hold. Then, there exists an exchangeable NE (P_π^{1*}, P_π^{2*}) for $\mathcal{P}_N$ among all policies in $L_N^1 \times L_N^2$, i.e., $(P_\pi^{1*}, P_\pi^{2*}) \in L_{\mathsf{EX},N}^1 \times L_{\mathsf{EX},N}^2$.*

Proof We provide a sketch. For a complete proof, see [635, Section 6.2]. The proof proceeds in two steps. In step 1, we use Theorem 10.5.2, to show that there exists a NE (P_π^{1*}, P_π^{2*}) among all policies in $L_{\mathsf{EX},N}^1 \times L_{\mathsf{EX},N}^2$. Then, in step 2, we show that (P_π^{1*}, P_π^{2*}) constitutes a NE among all randomized policies in $L_N^1 \times L_N^2$.

Step 1. We define $\Phi : \prod_{i=1}^2 L_{\mathsf{EX},N}^i \to 2^{\prod_{i=1}^2 L_{\mathsf{EX},N}^i}$ as the best response correspondence $\Phi(P_\pi^1, P_\pi^2) := \prod_{i=1}^2 \mathrm{BR}(P_\pi^{-i})$, where

$$\mathrm{BR}(P_\pi^{-i}) := \left\{ P_\pi^i \in L_{\mathsf{ex},N}^i \,\middle|\, \forall \hat{P}_\pi^i \in L_{\mathsf{ex},N}^i \right.$$
$$\left. J_\pi^i(P_\pi^i, P_\pi^{-i}) \le J_\pi^i(\hat{P}_\pi^i, P_\pi^{-i}) \right\} \tag{11.10}$$

for $i = 1, 2$. By Lemma 11.6.2, the set $L_{\mathsf{ex},N}^i$ is non-empty, convex, and compact. We show that Φ admits a fixed point, which will be denoted by (P_π^{1*}, P_π^{2*}). To this end, we first show that joint continuity of the induced costs under joint convergences of individual team policies, as a consequence of Assumptions 11.6.7 and 11.6.4, leads to upper semi-continuity of best response maps. Then, we invoke Kakutani-Fan-Glicksberg fixed point theorem to establish the existence of a NE among all exchangeable policy profiles.

Step 2. Then, we use the results in Sect. 6.4.1 by utilizing exchangeability of the costs and observations in Assumption 11.6.7, and linearity of expected costs in randomized policy profiles to show that for any $P_\pi^{-i*} \in L_{\mathsf{EX},N}^{-i}$,

$$\inf_{\hat{P}_\pi^i \in L_N^i} J_\pi^i(\hat{P}_\pi^i, P_\pi^{-i*}) = \inf_{\hat{P}_\pi^i \in L_{\mathsf{EX},N}^i} J_\pi^i(\hat{P}_\pi^i, P_\pi^{-i*}). \tag{11.11}$$

This implies that (P_π^{1*}, P_π^{2*}) constitutes a NE for the $\mathcal{P}_N$, and completes the proof.

◇

Existence of a Symmetric NE for $\mathcal{P}_\infty$

The expected costs $J^i_{\pi,N}$ for $\mathcal{P}_N$ with the mean-field interaction under a randomized policy profile $(P^1_\pi, P^2_\pi) \in L^1_N \times L^2_N$ can be defined in a way similar to (11.9), with

$$c^i_N(\underline{\gamma}, \underline{y}, \omega_0) = \int \prod_{i=1}^{2} \prod_{k=1}^{N_i} \gamma^i_k(du^i_k | y^i_k)$$

$$\times \frac{1}{N_i} \sum_{k=1}^{N_i} c^i \left(\omega_0, u^i_k, \Xi^1(\frac{1}{N_1} \sum_{p=1}^{N_1} \delta_{u^1_p}), \Xi^2(\frac{1}{N_2} \sum_{p=1}^{N_2} \delta_{u^2_p}) \right). \tag{11.12}$$

We note that the above cost is a special case of $\mathcal{P}_N$ introduced in (11.9), since it has a special structure satisfying the conditions of Assumption 11.6.3.

Next, the expected costs for $\mathcal{P}_\infty$ under a randomized policy profile $(P^1_\pi, P^2_\pi) \in L^1 \times L^2$ are given by

$$J^i_{\pi,\infty}(P^1_\pi, P^2_\pi) := \limsup_{N_1, N_2 \to \infty}$$

$$\int P^1_{\pi,N_1}(d\underline{\gamma}^1) P^2_{\pi,N_2}(d\underline{\gamma}^2) \mu^N(d\omega_0, d\underline{y}) c^i_N(\underline{\gamma}, \underline{y}, \omega_0), \tag{11.13}$$

where P^i_{π,N_i} is the marginal of $P^i_\pi \in L^i$ to the first N_i components, and μ^N is the distribution of $(\omega_0, y^1_{1:N_1}, y^2_{1:N_2})$.

The following theorem establishes existence of a NE for $\mathcal{P}_\infty$ that is symmetric and independent.

Theorem 11.6.6 ([635]) *Consider the game $\mathcal{P}_\infty$ with a given IS. Under Assumptions 11.6.4, 11.6.5, and 11.6.6, there exists an independently randomized symmetric NE profile (P_π^{1*}, P_π^{2*}), for $\mathcal{P}_\infty$ among all policies in $L^1 \times L^2$, i.e., $(P_\pi^{1*}, P_\pi^{2*}) \in L^1_{\mathsf{PR,\,SYM}} \times L^2_{\mathsf{PR,\,SYM}}$.*

We now provide some insight into the proof of Theorem 11.6.6. One uses Theorem 11.6.5 to guarantee the existence of an (infinitely)-exchangeable NE for $\mathcal{P}_N$ among policies in this class. Then, by compactness of L^i_{EX}, we extract a converging sequence, which by the de Finetti representation theorem, can be considered to be in $L^i_{\mathsf{CO,\,SYM}}$. Then, we use the fact that observations are conditionally i.i.d., and continuity of the costs in Assumptions 11.6.5 and 11.6.6, to show that the limit policy constitutes a NE for $\mathcal{P}_\infty$ among (infinitely)-exchangeable. Finally, as for Theorem 6.4.2, we use the de Finetti theorem and an argument used in [634, Lemma 2] to show that we can restrict the search over infinitely exchangeable policies in finding best response policies for $\mathcal{P}_\infty$ for a fixed infinitely exchangeable policy of the other players.

We note that in contrast to Theorem 11.6.5 where we established the existence of exchangeable randomized policies, Theorem 11.6.6 establishes the existence of a symmetric randomized NE with only independent randomness. The reason is that the set of randomized policies with only independent randomness in not a convex subset of L^i_N (as we have seen in Chap. 4), and hence, the Kakutani-Fan-Glicksberg fixed-point theorem (Theorem 10.5.2) does not apply. However in the limit as the number of DMs drives to infinity, by the optimality results, we can consider a single DM as a representative DM for each team with countably infinite number of DMs, who all apply the same policy as the representative DM. This allows us to use the Kakutani-Fan-Glicksberg fixed point theorem (Theorem 10.5.2) on the corresponding set of randomized policies with independent randomization (for a single DM), which is a convex and compact set. Once one shows a joint continuity result, Theorem 10.5.2 leads to equilibrium existence with the desired structure.

Approximations of a Symmetric NE for $\mathcal{P}_N$

By Theorem 11.6.5, there exists an exchangeable NE for $\mathcal{P}_N$ with exchangeable costs. This NE is not necessarily symmetric, independently randomized, and varies by the number of DMs in the game. On the other hand, Theorem 11.6.6 established the existence of a NE for $\mathcal{P}_\infty$ that is symmetric and independently randomized. In view of this, we address the following question: *Does there exist a scalable (with respect to the number of DMs) approximate NE for $\mathcal{P}_N$with the mean-field interaction that is symmetric and independently randomized?* We answer this question in the affirmative by showing that a symmetric independently randomized NE for $\mathcal{P}_\infty$ constitutes an approximate NE for $\mathcal{P}_N$.

Theorem 11.6.7 ([635]) *Consider the games $\mathcal{P}_N$ and $\mathcal{P}_\infty$ with a given IS. Let Assumptions 11.6.4, 11.6.5, and 11.6.6 hold. Then, an independently randomized symmetric NE for $\mathcal{P}_\infty$ constitutes an $\epsilon_{N_{1:2}}$-NE for $\mathcal{P}_N$ among all policies in $L^1_N \times L^2_N$, where $\epsilon_{N_{1:2}} \to 0$ as $N_1, N_2 \to \infty$.*

The discussion above also applies to the dynamic case under appropriate change of measure arguments, similar to those presented in Sect. 6.4.2, by generalizing the static results to this setting (see [635]).

11.7 Appendix: Proofs

11.7.1 Proof of Theorem 11.2.1

Step (1): By Assumption 11.2.1, we can apply an *independent-measurements reduction* following the analysis in Sect. 3.4, see Fig. 11.1.
As noted earlier, the main benefit of this approach is to define a compact/convex policy space for the players. To complete this reformulation, we note that the following holds for some function f and reference probability measures $\bar{Q}^i$:

$$P(dx, dy^1, dy^2, du^1, du^2)$$
$$= \zeta(dx) f(x, y^1, y^2) \bar{Q}^1(dy^1) 1_{\{\gamma^1(y^1) \in du^1\}} \bar{Q}^2(dy^2) 1_{\{\gamma^2(y^2) \in du^2)\}}.$$

where $1_{\{.\}}$ is the indicator function. Thus, the value function for the game can be written as:

$$V_G^{\mu}(\gamma^1, \gamma^2) = \int f(x, y^1, y^2) c(x, u^1, u^2) \bar{Q}^1(dy^1) \bar{Q}^2(dy^2) \zeta(dx).$$

We then create a new cost function $c(x, u^1, u^2, y^1, y^2) = c(x, u^1, u^2) f(x, y^1, y^2)$.
Let $x \in \mathbb{X}$ be the random state of nature. Let γ^1, γ^2 be the policies for the players, and u^1, u^2 be the resulting actions chosen by the players. Here, we allow for policies γ^i where u^i is chosen in a random way, i.e. $u^i = \gamma^i(y^i, \omega^i)$, where ω^i is some $[0, 1]$-valued independent random variable.

Step (2): Let c be the reformulated cost function of this game, under the new product probability measure. We have

$$V_G^{\mu}(\gamma^1, \gamma^2) = \int c(x, u^1, u^2, y^1, y^2)(\bar{Q}^1\gamma^1)(dy^1, du^1)(\bar{Q}^2\gamma^2)(dy^2, du^2)\zeta(dx).$$

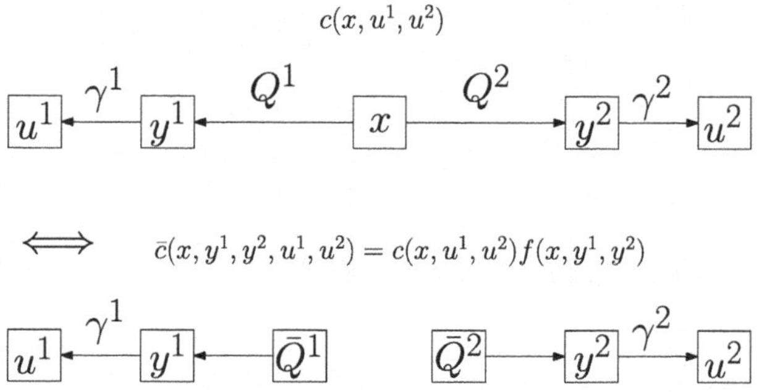

Fig. 11.1 Reformulation of two information structures (with respect to an independent measurements reduction)

Here, $(\bar{Q}^1\gamma^1)(dy^1, du^1)$ and $(\bar{Q}^2\gamma^2)(dy^2, du^2)$ are the probability measures induced on the measurement and the action variables. By independence due to the reduction, we can consider the expected cost as a function of the reduced-form policies: $V_G^{\mu}(\gamma^1, \gamma^2) = V_G^{\mu}(\bar{Q}^1\gamma^1, \bar{Q}^2\gamma^2)$. Now, without loss of generality, we fix $\bar{Q}^1\gamma^1$, allowing us to express the above equation in the following form:

$$V_G^{\mu}(\bar{Q}^1\gamma^1, \bar{Q}^2\gamma^2)$$
$$= \int (\bar{Q}^2\gamma^2)(dy^2, du^2) \int c(x, u^1, u^2, y^1, y^2)(\bar{Q}^1\gamma^1)(dy^1, du^1)\zeta(dx).$$

Let $\int c(x, u^1, u^2, y^1, y^2)(\bar{Q}^1\gamma^1)(dy^1, du^1)\zeta(dx)$ be defined as $\bar{c}(u^2, y^2)$.
Now that we have an independent-measurements reduction, we will identify, almost surely, every admissible policy with a probability measure on the product space: we adopt the view that, given a game (G, μ), $\bar{Q}^i\gamma^i$ is a probability measure on $\mathbb{Y}^i \times \mathbb{U}^i$ with fixed marginal $\bar{Q}^i(dy^i)$ on $\mathbb{Y}^i$. Let Γ^i denote the space of all such measures since every $\bar{Q}^i\gamma^i$ can be identified with an element of Γ^i almost surely. The pairing of an information structure and a policy induces a probability measure P on the five-tuple: $(\mathbb{X}, \mathbb{Y}^1, \mathbb{Y}^2, \mathbb{U}^1, \mathbb{U}^2)$, with

$$P(dx, dy^1, dy^2, du^1, du^2) = \gamma^1(du^1|y^1)\gamma^2(du^2|y^2)Q^1(dy^1|x)Q^2(dy^2|x)\zeta(dx).$$

This construction allows us to obtain a proper topology to work with for spaces of policies with desirable convexity and compactness properties.
Recall again the w-s topology (see Appendix C) on the set of probability measures $\mathcal{P}(\mathbb{X} \times \mathbb{U})$; this is the coarsest topology under which $\int f(x, u)\nu(dx, du) : \mathcal{P}(\mathbb{X} \times \mathbb{U}) \to \mathbb{R}$ is continuous for every measurable and bounded f which is continuous in u for every x (but unlike weak topology, f does not need to be continuous in x). Now, since the exogenous variables are fixed, weak convergence in this setting is equivalent to w-s convergence, and continuity in the exogenous variable is not needed here. Consider a sequence of actions $(\bar{Q}^2\gamma^2)_m(dy^2, du^2)$ which converges to $(\bar{Q}^2\gamma^2)(dy^2, du^2)$ weakly. We have that $\bar{c}(u^2, y^2)$ is continuous in u^2. Since μ is fixed, the marginals on $\mathbb{Y}^2$ are fixed. Therefore, by Theorem C.4.1, we can use the w-s topology on the set of probability measures $\mathcal{P}(\mathbb{Y}^2 \times \mathbb{U}^2)$. And thus we have continuity of $V_G^{\mu}(\bar{Q}^1\gamma^1, \bar{Q}^2\gamma^2)$ in $\bar{Q}^2\gamma^2$ in the w-s topology and, by the equivalence in this setting, the weak topology.
This also holds for continuity in $\bar{Q}^1\gamma^1$ in the reverse case where we fix $\bar{Q}^2\gamma^2$. Therefore, in general, we have that $V_G^{\mu}(\cdot, \cdot)$ is continuous in $(\bar{Q}^i\gamma^i)$ when $(\bar{Q}^{-i}\gamma^{-i})$ is fixed.

Step (3): Let $\Gamma = \{\lambda \in \mathcal{P}(\mathbb{Y} \times \mathbb{U}) : \lambda_{\mathbb{Y}} = \bar{Q}\}$ be our reduced policy space, where $\bar{Q}$ is the fixed marginal of the measure λ on $\mathbb{Y}$. Following Sect. 4.1.1, the space of all $\bar{Q}^i\gamma^i$ (which we denote by Γ^i) is compact under weak convergence (and is convex).

Step (4): We observe that $V_G^{\mu}(\bar{Q}^1\gamma^1, \bar{Q}^2\gamma^2)$ is linear and hence is both concave and convex in each entry. For completeness, we establish this linearity result. Take $\theta \in (0, 1)$. Then, without loss of generality, we fix $\bar{Q}^1\gamma^1$ and obtain the following:

$$\begin{aligned}
&V_G^{\mu}(\bar{Q}^1\gamma^1, \theta\bar{Q}^2\gamma^2 + (1-\theta)\tilde{Q}^2\tilde{\gamma}^2) \\
&= \int (\theta\bar{Q}^2\gamma^2 + (1-\theta)\tilde{Q}^2\tilde{\gamma}^2)(dy^2, du^2) \int_{y^2,u^2} \bar{c}(x, u^2, y^2) \\
&= \int (\theta\bar{Q}^2\gamma^2)(dy^2, du^2) \int_{y^2,u^2} \bar{c}(x, u^2, y^2) \\
&\quad + \int (1-\theta)(\tilde{Q}^2\tilde{\gamma}^2)(dy^2, du^2) \int_{y^2,u^2} \bar{c}(x, u^2, y^2) \\
&= \theta V_G^{\mu}(\bar{Q}^1\gamma^1, \bar{Q}^2\gamma^2) + (1-\theta) V_G^{\mu}(\bar{Q}^1\gamma^1, \tilde{Q}^2\tilde{\gamma}^2).
\end{aligned}$$

Under the weak topology, the space of probability measures is a metric space, and thus the spaces Γ^i are Hausdorff spaces. Since $V_G^{\mu}(\bar{Q}^1\gamma^1, \bar{Q}^2\gamma^2)$ is continuous, and convex/concave in the compact Hausdorff spaces Γ^i, we have the following equality [245, Theorem 1]:

$$\min_{Q^1\gamma^1} \max_{Q^2\gamma^2} V_G^{\mu}(Q^1\gamma^1, Q^2\gamma^2) = \max_{Q^2\gamma^2} \min_{Q^1\gamma^1} V_G^{\mu}(Q^1\gamma^1, Q^2\gamma^2).$$

This establishes a (saddle-point) equilibrium for the game. ◇

11.7.2 Proof of Theorem 11.2.2

Here, we let $\mu_{\mathbb{Y}^i}$ denote the marginal of the information structure μ on $\mathbb{Y}^i$. We will combine our policies with these marginals to form the product measures $\mu_{\mathbb{Y}^1}(dy^1)\gamma^1(du^1|y^1)$ and $\mu_{\mathbb{Y}^2}(dy^2)\gamma^2(du^2|y^2)$ on the players' measurement and action spaces. We will denote these measures by $(\mu_{\mathbb{Y}^1}\gamma^1)$ and $(\mu_{\mathbb{Y}^2}\gamma^2)$

Similar to the previous proof, without loss of generality, we fix Player 2's strategy γ^2. Then, we have:

$$\begin{aligned}
V_G^{\mu}(\gamma^1, \gamma^2) = &\int \mu_{\mathbb{Y}^1}(dy^1)\gamma^1(du^1|y^1) \\
&\left(\int_{\mathbb{X}\times\mathbb{Y}^2\times\mathbb{U}^2} \mu(dx, dy^2|dy^1)\gamma^2(du^2|y^2)c(x, u^1, u^2)\right)
\end{aligned}$$

Let

$$\int_{\mathbb{X}\times\mathbb{Y}^2\times\mathbb{U}^2} \mu(dx, dy^2|y^1)\gamma^2(du^2|y^2)c(x, u^1, u^2) =: \bar{c}(u^1, y^1).$$

Note that, by assumption, $\bar{c}(u^1, y^1)$ is bounded, and is continuous in u^1. Furthermore, it is also evident that $\bar{c}(u^1, y^1)$ is measurable in y^1. By the same arguments as in the proof of the preceeding theorem, via the machinery of the w-s topology, we can show that $\bar{c}(u^1, y^1)$ is continuous in $(\mu_{\mathbb{Y}^1}\gamma^1)$ under w-s convergence, and thus also under weak convergence. This also holds for continuity in $(\mu_{\mathbb{Y}^2}\gamma^2)$ in the reverse case where we fix γ^1. Note also that the convexity and compactness of the strategic measure here would follow as a consequence of Theorem 4.3.1 which shows that conditional independence properties are preserved when only one player perturbs the strategic measure (unlike the case where multiple players make simultaneous perturbations in Theorems 4.2.2 or 4.2.3).

The space of all $\mu^i_{\mathbb{Y}^i}\gamma^i$ is compact under weak convergence, and we see that $V^{\mu}_{G}(\mu_{\mathbb{Y}^1}\gamma^1, \mu_{\mathbb{Y}^2}\gamma^2)$ is linear, and hence is both concave and convex in each entry.

The existence of a (saddle-point) equilibrium for the game then follows by Fan [245, Theorem 1]. ◇

11.7.3 *Proof of Lemma 11.3.1*

Let us consider the first relation.

Take an arbitrary zero-sum game $G \in \tilde{\mathbb{G}}_{\zeta}(\mathbb{X}, \mathbb{Y}^1, \mathbb{Y}^2)$ with cost function c and action spaces $\mathbb{U}^1$ and $\mathbb{U}^2$. Let (γ^1, γ^2) be the Bayesian Nash equilibrium policies for the players under the information structure $\kappa^2\mu$, and (η^1, η^2) be the Bayesian Nash equilibrium policies under the information structure μ. By our assumption on $\tilde{\mathbb{G}}_{\zeta}(\mathbb{X}, \mathbb{Y}^1, \mathbb{Y}^2)$, these policies exist [Theorem 11.2.1]. Let Q^i be the measurement channel for player i under information structure μ.

The expected value of the cost for the maximizer under the first information structure is:

$$V^{\kappa^2\mu}_{G}(\gamma^1, \gamma^2) = \int_{\mathbb{X}\times\mathbb{Y}^1\times\mathbb{Y}^2} c(x, \gamma^1(y^1), \gamma^2(y^2))\kappa^2\mu(dx, dy^1, dy^2).$$

By definition, the equilibrium solution (γ^1, γ^2) for G under $\kappa^2\mu$ is given by the solution to the min-max problem:

$$\min_{\theta^1\in\Gamma^1}\max_{\theta^2\in\Gamma^2} V^{\kappa^2\mu}_{G}(\theta^1, \theta^2).$$

Therefore, since γ^1 is the minimizing policy under $\kappa^2\mu$, by perturbing the minimizer's policy γ^1 to be the policy $\eta^1 \in \Gamma^1$, we have the following inequality (i.e. we make the minimizer no longer play her optimal policy):

$$V^{\kappa^2\mu}_{G}(\gamma^1, \gamma^2) \le V^{\kappa^2\mu}_{G}(\eta^1, \gamma^2).$$

We now compare the two quantities $V_G^{\kappa^2\mu}(\eta^1, \gamma^2)$ and $V_G^{\mu}(\eta^1, \eta^2)$. To do so, fix η^1 across both terms and consider a cost function $\tilde{c}(x, \theta^2(y^2)) : \mathbb{X} \times \mathbb{U}^2 \to \mathbb{R}$ such that $\tilde{c}(x, \theta^2(y^2)) = c(x, \eta^1(y^1), \theta^2(y^2))\ \forall\, \theta^2 \in \Gamma^2$. That is, by holding the minimizer's strategy fixed as η^1, we reduce c to $\tilde{c}$ such that we now have a cost function that only reflects dependence on the maximizer's policy when the minimizer's policy is held at η^1. Such a function $\tilde{c}$ clearly exists, as the value of $\eta^1(y^1)$ is only dependent on x (potentially in some stochastic way, in that it depends on $Q^1(y|x)$), when η^1 (and μ^1) are fixed, and so can be absorbed into the dependency of $\tilde{c}$ on x.

We now focus on the single-player decision problem for the maximizer given by cost function $\tilde{c}$ and information structures $(\kappa^2\mu)^2$ and μ^2 (which we use to denote the maximizer's private information structures present in $\kappa^2\mu$ and μ, respectively, i.e. the marginals on $(\mathbb{X} \times \mathbb{Y}^2)$). This is a single-player decision problem and as such can be treated using the forward direction to Blackwell's ordering of information structures [123, Theorem 2], which holds in this infinite-dimensional case [824]. Since $\tilde{c}$ and c are equivalent over all strategies in Γ^2, we know that γ^2 and η^2 are still optimal policies for the maximizer to play under the respective information structures for this game. Thus, since $(\kappa^2\mu)^2$ is a garbling of μ^2 by channel κ^2, and since $\tilde{c}(x, \eta^2(y^2)) = c(x, \eta^1(\mu^1(x)), \eta^2(y^2))$, we have that:

$$V_G^{\kappa^2\mu}(\eta^1, \gamma^2) = \int_{\mathbb{X}\times\mathbb{Y}^2} \tilde{c}(x, \gamma^2(y^2))\kappa^2\mu(dx, dy^2)$$

$$\leq \int_{\mathbb{X}\times\mathbb{Y}^2} \tilde{c}(x, \eta^2(y^2))\mu(dx, dy^2) = V_G^{\mu}(\eta^1, \eta^2).$$

Putting this all together, we have $V_G^{\kappa^2\mu}(\gamma^1, \gamma^2) \leq V_G^{\kappa^2\mu}(\eta^1, \gamma^2) \leq V_G^{\mu}(\eta^1, \eta^2)$. Since this is true for any arbitrary game $G \in \tilde{\mathbb{G}}_\zeta(\mathbb{X}, \mathbb{Y}^1, \mathbb{Y}^2)$, we have that $\kappa^2\mu \lesssim \mu$.

A nearly identical argument can be applied to show that $\mu \lesssim \kappa^1\mu$. ◇

11.7.4 Proof of Lemma 11.3.2

Let (γ^1, γ^2) be an equilibrium solution under ν, and let (η^1, η^2) be an equilibrium solution under μ. Let Q_ν^i and Q_μ^i be the measurement channel for player i under the information structures ν and μ, respectively. As in Lemma 11.3.1, these equilibria exist and are solutions of the standard min-max problem.

Therefore, we have the following inequality:

$$V_G^{\nu}(\gamma^1, \eta^2) \leq \max_{\theta^2 \in \Gamma^2} V_G^{\nu}(\gamma^1, \theta^2) = V_G^{\nu}(\gamma^1, \gamma^2).$$

Likewise, we can determine the following:

$$V_G^{\mu}(\eta^1, \eta^2) = \min_{\alpha^1 \in \Gamma^1} V_G^{\nu}(\alpha^1, \eta^2) \leq V_G^{\mu}(\gamma^1, \eta^2).$$

In addition, by assumption that $\nu \lesssim \mu$ we have that:

$$V_G^{\nu}(\gamma^1, \gamma^2) \leq V_G^{\mu}(\eta^1, \eta^2).$$

From above, one observes that $V_G^{\nu}(\gamma^1, \eta^2) \leq V_G^{\mu}(\eta^1, \eta^2)$. In the same manner as in Lemma 11.3.1, we hold η^2 fixed across both terms and develop a reduced single-player cost function $\tilde{c}$. Once again, we use ν^i and μ^i to denote the private (i.e. marginal) information structure for Player i under ν and μ, respectively. We then have a single-player decision problem where we note that γ^1 and η^1 are still the optimal policies for the minimizer for each respective information structure:

$$\begin{aligned} J(\zeta, \nu^1, \gamma^1) &= \int_{\mathbb{X} \times \mathbb{Y}^1} \tilde{c}(x, \gamma^1(y^1))\nu(dx, dy^1) \\ &\leq \int_{\mathbb{X} \times \mathbb{Y}^1} \tilde{c}(x, \eta^1(y^1))\mu(dx, dy^1) = J(\zeta, \mu^1, \eta^1). \end{aligned}$$

Since the inequality $V_G^{\nu}(\gamma^1, \eta^2) \leq V_G^{\mu}(\eta^1, \eta^2)$ holds true for every arbitrary zero-sum game $G \in \tilde{\mathbb{G}}_{\zeta}(\mathbb{X}, \mathbb{Y}^1, \mathbb{Y}^2)$, it holds for every game in the subclass $\hat{\mathbb{G}}$, defined here to be all games in $\tilde{\mathbb{G}}_{\zeta}(\mathbb{X}, \mathbb{Y}^1, \mathbb{Y}^2)$ where the action space of the maximizer is $\mathbb{U}^2 = \{0\}$. Moreover, we see that for any arbitrary bounded single-player cost function that is continuous in the player's action, $\bar{c}(x, u^1) : \mathbb{X} \times \mathbb{U} \to \mathbb{R}$, there exists a two-player cost function $\hat{c}(x, u^1, u^2)$ corresponding to some game in $\hat{\mathbb{G}}$ such that $\hat{c}(x, u^1, u^2) = \bar{c}(x, u^1) \; \forall \, u^1 \in \mathbb{U}^1$ (following naturally from the fact that $u^2 = 0$ for these games). One such construction of $\hat{c}$ would be $\hat{c}(x, u^1, u^2) = \bar{c}(x, u^1) + u^2$; when played in $\hat{\mathbb{G}}$, $\hat{c}(x, u^1, u^2) = \bar{c}(x, u^1) \forall u^1 \in \Gamma^1$. We note that since $\bar{c}(x, u^1)$ is continuous in u^1 for all x and is bounded, it is valid for a game in $\hat{\mathbb{G}} \subset \tilde{\mathbb{G}}_{\zeta}(\mathbb{X}, \mathbb{Y}^1, \mathbb{Y}^2)$ to use $\hat{c}(x, u^1, u^2)$.

Therefore, $\bar{c}$ is a valid single-player reduction of $\hat{c}$ when the maximizer's strategy is held fixed. Since this process can be done for any single-player cost function $\hat{c}$, we see that for given $\mathbb{U}^i$ and $\mathbb{X}$, the reduction of all measurable and bounded two-player cost functions that are continuous in players' actions with one player playing a fixed strategy, is surjective on the entire space of bounded single-player cost functions that are continuous in the player's action. Therefore, the inequality above will hold over all single-player cost functions that are bounded and continuous in action, since it holds over all arbitrary two-player games $G \in \tilde{\mathbb{G}}_{\zeta}(\mathbb{X}, \mathbb{Y}^1, \mathbb{Y}^2)$.

Lastly, we note that since c is framed such that a higher quantity is better for the maximizer, the minimizer wants to minimize the value of $\tilde{c}$. Therefore, from the minimizer's perspective, the inequality above indicates that she can never perform worse under ν than under μ over all single-player problems with bounded cost functions that are continuous in the player's action. Thus, by the converse direction to Blackwell's ordering of information structures [123, Theorem 6] which, by the assumptions of the Lemma and the restrictions on the class $\tilde{\mathbb{G}}_\zeta(\mathbb{X}, \mathbb{Y}^1, \mathbb{Y}^2)$ (namely compactness of $\mathbb{Y}^i$ and $\mathbb{U}^i$), holds in this infinite setup due to Theorem 7.3.4, we have that μ^1 must be a garbling of ν^1.

In a similar manner, in view of the inequality $V_G^\nu(\gamma^1, \gamma^2) \le V_G^\mu(\gamma^1, \eta^2)$, we conclude that ν^2 must be a garbling of μ^2.

Therefore, we have that the minimizer's channel in ν is garbled in μ and the maximizer's channel in μ is garbled in ν. Combining these two properties leads to the desired equality for some $\kappa^i \in K^i$: $\kappa^1\nu = \kappa^2\mu$. ◇

11.7.5 Proof of Theorem 11.4.1

(i) By Theorem 11.2.1, Assumptions 8.8.1-A3, A4 and A5 guarantee that an equilibrium exists under μ and ν.

Note that if μ_n satisfies Assumption 8.8.1-A3 and $\mu_n \to \mu$ in total variation, then μ also has to satisfy Assumption 8.8.1-A3 (to see this, let $\mu_n \ll \kappa$ for each n implying that $\kappa(B) = 0 \implies \mu_n(B) = 0$. If there were to exist B with $\mu(B) > 0$, but with $\kappa(B) = 0$, then this would imply that $\mu_n(B) = 0$ for all n but not for the limit measure $\mu(B)$. This would violate total variation convergence).

Assume that μ and ν are such that $\|\mu - \nu\|_{TV} \le \epsilon$ for some $\epsilon > 0$. Without loss of generality, assume that $J^*(g, \mu) - J^*(g, \nu) \ge 0$ for some game g (a symmetric argument can be applied in the case where $J^*(g, \mu) - J^*(g, \nu) \le 0$). Then, we have the following:

$$
\begin{aligned}
&J^*(g, \mu) - J^*(g, \nu) \\
&= \int c(x, \gamma_\mu^{1,*}(y^1), \gamma_\mu^{2,*}(y^2))\mu(dx, dy^1, dy^2) \\
&\quad - \int c(x, \gamma_\nu^{1,*}(y^1), \gamma_\nu^{2,*}(y^2))\nu(dx, dy^1, dy^2) \\
&\le \int c(x, \gamma_\nu^{1,*}(y^1), \gamma_\mu^{2,*}(y^2))\mu(dx, dy^1, dy^2) \\
&\quad - \int c(x, \gamma_\nu^{1,*}(y^1), \gamma_\mu^{2,*}(y^2))\nu(dx, dy^1, dy^2).
\end{aligned}
$$

The inequality comes from perturbing the minimizer's equilibrium strategy in the first term, making the expected cost larger, and perturbing the maximizer's equilibrium strategy in the second term, making the expected cost smaller. Then, viewing $\gamma_\nu^{1,*}$ and $\gamma_\mu^{2,*}$ as fixed, we can view $c(\cdot, \gamma_\nu^{1,*}(\cdot), \gamma_\mu^{2,*}(\cdot))$ as a measurable and bounded function from $\mathbb{X} \times \mathbb{Y}^1 \times \mathbb{Y}^2 \to \mathbb{R}$. Let $M \in \mathbb{R}_{\ge 0}$ be such that $\|c\|_\infty \le M$. We have:

$$\int c(x, \gamma_\nu^{1,*}(y^1), \gamma_\mu^{2,*}(y^2))\mu(dx, dy^1, dy^2) \\ - \int c(x, \gamma_\nu^{1,*}(y^1), \gamma_\mu^{2,*}(y^2))\nu(dx, dy^1, dy^2) \\ \leq M\|\mu - \nu\|_{TV} \leq M\epsilon.$$

Thus, if $\|\mu_n - \mu\|_{TV} \to 0$, then $|J^*(g, \mu) - J^*(g, \nu)| \to 0$.

(ii) For stochastic control problems (which can be viewed as single-player games), Theorem 8.3.5 (building on [828, Sections 3.1.1 and 3.2.1]) established that value functions are not necessarily continuous under weak or setwise convergence, via a counterexample. This counterexample directly extends to the zero-sum game case: one can extend such a single-player game to a two-player zero-sum game where the action of one player has no impact on the cost: e.g. $|\mathbb{U}^2| = 1$ so that Player 2 has no freedom to select a control action and the information structure of Player 1 changes. ◇

11.7.6 Proof of Theorem 11.4.2

Consider a minimizer-garbling sequence. By A2, our cost function is bounded; let $M \in \mathbb{R}_{\geq 0}$ such that $\|c\|_\infty = M$. Let $\mu_{\mathbb{Y}^j}$ be the marginal of μ on its $(j+1)$th component. Let γ^j be an arbitrary policy for player j. We note that every Polish space is second countable (i.e. has a countable basis), since it is separable and metrizable. Thus, every standard Borel space is second countable. Then, by Lusin's theorem, using the fact that $\mathbb{U}^j$ is convex by assumption, we have that for any $\epsilon > 0$, there exists a continuous function $f^j : \mathbb{Y}^j \to \mathbb{U}^j$ such that

$$\mu_{\mathbb{Y}^j}(\{y^j : f^j(y^j) \neq \gamma^j(y^j)\}) < \epsilon. \tag{11.14}$$

Letting $B^2 := \{y^2 : f^2(y^2) \neq \gamma^2(y^2)\}$, we have:

$$\int_{\mathbb{X}\times\mathbb{Y}^1\times\mathbb{Y}^2} c(x, \gamma^1(y^1), \gamma^2(y^2))\mu(dx, dy^1, dy^2) - M\epsilon \\ < \int_{\mathbb{X}\times\mathbb{Y}^1\times\mathbb{Y}^2} c(x, \gamma^1(y^1), f^2(y^2))\mu(dx, dy^1, dy^2) \\ < \int_{\mathbb{X}\times\mathbb{Y}^1\times\mathbb{Y}^2} c(x, \gamma^1(y^1), \gamma^2(y^2))\mu(dx, dy^1, dy^2) + M\epsilon \tag{11.15}$$

This is true since the cost function is bounded by $[-M, M]$ and thus the expected cost on $\mathbb{X} \times B^i \times \mathbb{Y}^{-i}$ is bounded by $[-M\epsilon, M\epsilon]$. Denote by Γ_C^j the policy space for player j in which the player's policy is continuous from $\mathbb{Y}^j$ to $\mathbb{U}^j$. We note that if Player 1's strategy is fixed, Player 2 receives the same cost if supremizing over policies in Γ^2 or Γ^2_C: that the value must be lower when Γ^2 is used is a consequence of $\Gamma^2_C \subset \Gamma^2$. That the value must be less than or equal to that under Γ^2 when Γ^2_C is used is a consequence of (11.15), since the policies, as well as ϵ, were arbitrary. The same reasoning applies in the reverse case. We now observe that, for any $m \in \mathbb{Z}_{>0}$:

$$\inf_{\gamma^1 \in \Gamma^1_C} \sup_{\gamma^2 \in \Gamma^2} \int c(x, \gamma^1(y^1), \gamma^2(y^2)) \mu_m(dx, dy^1, dy^2)$$

$$= \inf_{\gamma^1 \in \Gamma^1} \sup_{\gamma^2 \in \Gamma^2} \int c(x, \gamma^1(y^1), \gamma^2(y^2)) \mu_m(dx, dy^1, dy^2). \tag{11.16}$$

To prove this, let $\gamma^{1,*}$ and $\gamma^{2,*}$ be the equilibrium policies under μ_m. Then, following the discussion on Lusin's theorem, for any $\epsilon > 0$, there exists a continuous policy $\hat{\gamma}^1 \in \Gamma^1_C$, which is equal to $\gamma^{1,*}$ except on a set B^1 such that $\mu_{m,\mathbb{Y}^1}(B^1) = \epsilon$. If Player 1 plays this strategy, then for any strategy that Player 2 selects in Γ^2, the value for the game will increase by at most $2M\epsilon$. This is because, if Player 2's arbitrary new strategy $\hat{\gamma}^2$ results in a change in performance on $A := \{\mathbb{X} \times B^1 \times \mathbb{Y}^2\}$, the maximum possible difference in expected cost on this set is $2M\epsilon$. If Player 2's new strategy results in better performance on $C := \{\mathbb{X} \times (\mathbb{Y}^1 \setminus B^1) \times \mathbb{Y}^2\}$, then the difference between playing $\hat{\gamma}^2$ and $\gamma^{2,*}$ is at most $2M\epsilon$, because $\hat{\gamma}^1$ is identical to $\gamma^{1,*}$ on this set, and thus a larger difference would necessarily contradict the fact that $\gamma^{2,*}$ is Player 2's best response to $\gamma^{1,*}$ under μ (since Player 2 could lose by at most $2M\epsilon$ on set A while employing this strategy against $\gamma^{1,*}$ instead of playing $\gamma^{2,*}$, and thus gaining by more than $2M\epsilon$ on C, Player 2 would guarantee better expected performance playing $\hat{\gamma}^2$ against $\gamma^{1,*}$ over playing $\gamma^{2,*}$, violating the Nash-equilibrium condition). Thus, since Player 1 has a continuous strategy that guarantees a worst possible outcome of a $2M\epsilon$ increase in expected cost, we know that:

$$\inf_{\gamma^1 \in \Gamma^1_C} \sup_{\gamma^2 \in \Gamma^2} \int c(x, \gamma^1(y^1), \gamma^2(y^2)) \mu_m(dx, dy^1, dy^2)$$

$$\leq \inf_{\gamma^1 \in \Gamma^1} \sup_{\gamma^2 \in \Gamma^2} \int c(x, \gamma^1(y^1), \gamma^2(y^2)) \mu_m(dx, dy^1, dy^2) - 2M\epsilon$$

We also know that since $\Gamma^1_C \subset \Gamma^1$, we have:

$$\inf_{\gamma^1 \in \Gamma^1_C} \sup_{\gamma^2 \in \Gamma^2} \int c(x, \gamma^1(y^1), \gamma^2(y^2)) \mu_m(dx, dy^1, dy^2)$$

$$\geq \inf_{\gamma^1 \in \Gamma^1} \sup_{\gamma^2 \in \Gamma^2} \int c(x, \gamma^1(y^1), \gamma^2(y^2)) \mu_m(dx, dy^1, dy^2).$$

Since ϵ was arbitrary, (11.16) follows. Applying a similar argument again, it follows that:

$$\inf_{\gamma^1\in\Gamma^1_C}\sup_{\gamma^2\in\Gamma^2_C}\int c(x,\gamma^1(y^1),\gamma^2(y^2))\mu_m(dx,dy^1,dy^2)$$
$$=\inf_{\gamma^1\in\Gamma^1}\sup_{\gamma^2\in\Gamma^2}\int c(x,\gamma^1(y^1),\gamma^2(y^2))\mu_m(dx,dy^1,dy^2). \tag{11.17}$$

Now take an arbitrarily large $M\in\mathbb{Z}_{\geq 0}$. Then, we have:

$$\inf_{\gamma^1\in\Gamma^1}\sup_{\gamma^2\in\Gamma^2}\int c(x,\gamma^1(y^1),\gamma^2(y^2))\mu_M(dx,dy^1,dy^2)$$
$$=\inf_{\gamma^1\in\Gamma^1_C}\sup_{\gamma^2\in\Gamma^2_C}\int c(x,\gamma^1(y^1),\gamma^2(y^2))\mu_M(dx,dy^1,dy^2)$$
$$\leq\inf_{\gamma^1\in\Gamma^1_C}\sup_{\gamma^2\in\Gamma^2_C}\limsup_{M\to\infty}\int c(x,\gamma^1(y^1),\gamma^2(y^2))\mu_M(dx,dy^1,dy^2) \tag{11.18}$$
$$=\inf_{\gamma^1\in\Gamma^1_C}\sup_{\gamma^2\in\Gamma^2_C}\int c(x,\gamma^1(y^1),\gamma^2(y^2))\mu(dx,dy^1,dy^2)$$
$$=\inf_{\gamma^1\in\Gamma^1_C}\sup_{\gamma^2\in\Gamma^2}\int c(x,\gamma^1(y^1),\gamma^2(y^2))\mu(dx,dy^1,dy^2).$$

The first and final equalities follow from (11.17). The inequality (11.18) follows from the fact that the value function is monotonically increasing in M: this is because the sequence of information structures is a minimizer-garbling sequence. Therefore, the value function will be monotonically increasing in M by the forward direction to Blackwell's ordering of information structures (Theorem 7.3.2).[1] The second equality follows from the weak convergence of $\{\mu\}_m$ to μ. Since the above holds for all $M\in\mathbb{Z}_{\geq 0}$, it will also hold in the limit of M, and thus:

$$\limsup_{m\to\infty}\inf_{\gamma^1\in\Gamma^1}\sup_{\gamma^2\in\Gamma^2}\int c(x,\gamma^1(y^1),\gamma^2(y^2))\mu_m(dx,dy^1,dy^2)$$
$$\leq\inf_{\gamma^1\in\Gamma^1}\sup_{\gamma^2\in\Gamma^2}\int c(x,\gamma^1(y^1),\gamma^2(y^2))\mu(dx,dy^1,dy^2).$$

[1] It is achieving this inequality that requires the sequence of information structures to be a minimizer-garbling sequence, rather than an arbitrary weakly converging sequence of information structures.

Thus, the value function is upper semicontinuous. A similar proof applies for maximizer-garbling sequences. ◇

11.7.7 Proof of Theorem 11.4.3

Consider a maximizer-garbling sequence. We have that:

$$\liminf_{m\to\infty} \inf_{\gamma^1\in\Gamma^1} \sup_{\gamma^2\in\Gamma^2} \int c(x,\gamma^1(y^1),\gamma^2(y^2))\mu_m(dx,dy^1,dy^2)$$

$$\geq \sup_{\gamma^2\in\Gamma^2} \liminf_{m\to\infty} \inf_{\gamma^1\in\Gamma^1} \int c(x,\gamma^1(y^1),\gamma^2(y^2))\mu_m(dx,dy^1,dy^2).$$

The inequality above follows from interchanging the limit inferior and the supremum. Following a reasoning similar to that in the proof of Theorem 11.4.2, we have the following for any fixed policy $\gamma^2 \in \Gamma^2$:

$$\liminf_{m\to\infty} \inf_{\gamma^1\in\Gamma^1} \int c(x,\gamma^1(y^1),\gamma^2(y^2))\mu_m(dx,dy^1,dy^2)$$

$$\geq \inf_{\gamma^1\in\Gamma^1} \int c(x,\gamma^1(y^1),\gamma^2(y^2))\mu(dx,dy^1,dy^2).$$

This gives us

$$\liminf_{m\to\infty} \inf_{\gamma^1\in\Gamma^1} \sup_{\gamma^2\in\Gamma^2} \int c(x,\gamma^1(y^1),\gamma^2(y^2))\mu_m(dx,dy^1,dy^2) \tag{11.19}$$

$$\geq \sup_{\gamma^2\in\Gamma^2} \inf_{\gamma^1\in\Gamma^1} \int c(x,\gamma^1(y^1),\gamma^2(y^2))\mu(dx,dy^1,dy^2).$$

Thus, the value function is lower semicontinuous. A similar proof leads to the result for minimizer-garbling sequences. ◇

11.7.8 Proof of Theorem 11.4.4

Let ζ_m be a sequence of priors on $\mathbb{X}$ converging weakly to ζ. Following the same reasoning as in the proof of Theorem 11.4.1, if for some j, $J^*(g, Q^1\zeta Q^2) - J^*(g, Q^1\zeta_j Q^2) \geq 0$, then we have

$$J^*(g, Q^1\zeta Q^2) - J^*(g, Q^1\zeta_j Q^2)$$

$$\leq \int c(x, \gamma_j^{1,*}(y^1), \gamma^{2,*}(y^2)) Q^1(dy^1|x) Q^2(dy^2|x) \zeta(dx)$$
$$- \int c(x, \gamma_j^{1,*}(y^1), \gamma^{2,*}(y^2)) Q^1(dy^1|x) Q^2(dy^2|x) \zeta_j(dx)$$
$$= \int (\zeta - \zeta_j)(dx) \int c(x, \gamma_j^{1,*}(y^1), \gamma^{2,*}(y^2)) Q^1(dy^1|x) Q^2(dy^2|x).$$

Similarly, if $J^*(g, Q^1 \zeta Q^2) - J^*(g, Q^1 \zeta_j Q^2) \leq 0$, we get

$$J^*(g, Q^1 \zeta_j Q^2) - J^*(g, Q^1 \zeta Q^2)$$
$$\leq \int (\zeta_j - \zeta)(dx) \int c(x, \gamma^{1,*}(y^1), \gamma_j^{2,*}(y^2)) Q^1(dy^1|x) Q^2(dy^2|x).$$

Thus, in general we have

$$|J^*(g, Q^1 \zeta Q^2) - J^*(g, Q^1 \zeta_m Q^2)| \leq$$
$$\max\{|\int (\zeta - \zeta_m)(dx) \int c(x, \gamma_m^{1,*}(y^1), \gamma^{2,*}(y^2)) Q^1(dy^1|x) Q^2(dy^2|x)|,$$
$$|\int (\zeta - \zeta_m)(dx) \int c(x, \gamma^{1,*}(y^1), \gamma_m^{2,*}(y^2)) Q^1(dy^1|x) Q^2(dy^2|x)|\}. \tag{11.20}$$

Following the proof of Theorem 8.7.2, as the channels are continuous in total variation,

$$\int c(x, \gamma_m^{1,*}(y^1), \gamma^{2,*}(y^2)) Q^1(dy^1|x) Q^2(dy^2|x),$$

and

$$\int c(x, \gamma^{1,*} y^1), \gamma_m^{2,*}(y^2)) Q^1(dy^1|x) Q^2(dy^2|x)$$

are equicontinuous families of functions. Using this fact and that ζ_m converges weakly to ζ, by Dudley [223, Corollary 11.3.4] it follows that both terms in (11.20) converge to zero, and thus $|J^*(g, Q^1 \zeta Q^2) - J^*(g, Q^1 \zeta Q^2)| \to 0$. ◇

11.7.9 Proof of Theorem 11.6.1

In what follows, we assume that a team policy $\bar{\gamma}^i = (\gamma_1^i, \ldots, \gamma_{N_i}^i)$ is such that $u_k^i = \gamma_k^i(y_k^i, z^i)$, for some independent randomness $z_i \in [0, 1]^{N_i}$ for each team, for

$k = 1, \ldots, N_i, i = 1, 2$. That is, policies are selected such that the induced strategic measures are in $L_C(\mu_1)$ and $L_C(\mu_2)$. Note that the measure on $(\Omega \times \prod_{j=1}^{N_1} \mathbb{Y}_j^1 \times \prod_{k=1}^{N_2} \mathbb{Y}_k^2)$ is fixed for both teams by the information structure μ, and thus $L_C(\mu_i)$ has a fixed marginal on the state and measurement spaces.

Let $L_C(\mu_i, \bar{\gamma}_i)$ be team i's strategic measure induced by team policy $\bar{\gamma}_i$, for $i = 1, 2$. Denote by S_i the projection of the space $L_C(\mu_i)$ onto $\prod_{k=1}^{N_i}(\mathbb{Y}_k^i \times \mathbb{U}_k^i)$ (i.e. its second to final components), eliminating the state space from the measure and simplifying notation for convenience. Since the measure on the exogenous variables is fixed by the information structure μ, S_i will inherit convexity (as conditional independence will be preserved under convex combinations given that the opponent team policy is fixed) and weak compactness from $L_C(\mu_i)$ following from the analysis in the proof of Theorem 11.2.2; see Theorem 4.3.1 which shows that conditional independence is preserved when only one player perturbs the strategic measure (unlike the case where multiple players make simultaneous perturbations in Theorem 4.2.3).

Now, without loss of generality, we will fix Team 2's policy as $\bar{\gamma}_2^*$ (which is an arbitrary policy that induces a strategic measure s_2^* in $L_C(\mu_2)$). We then have:

$$V_G^{\mu}(s_1, s_2^*) = \int s_1(dy_1^1, \ldots, dy_{N_1}^1, du_1^1, \ldots, du_{N_1}^1) \ldots \tag{11.21}$$

$$\left(\int \mu^{\bar{\gamma}_2^*}(d\omega_0, dy_1^2, \ldots, dy_{N_2}^2, du_1^2, \ldots, du_{N_2}^2 | y_1^1, \ldots, y_{N_1}^1)\right.$$

$$\left. c(\omega_0, u_1^1, \ldots, u_{N_1}^1, u_1^2, \ldots, u_{N_1}^2)\right), \tag{11.22}$$

where $\mu^{\bar{\gamma}_2^*}$ denotes the strategic measure induced by Team 2's policy $\bar{\gamma}_2^*$. Denote the bracketed term as $\bar{c}(y_1^1, \ldots, y_{N_1}^1, u_1^1, \ldots, u_{N_1}^1)$. We can see that, by assumption, $\bar{c}$ is continuous in $u_1^1, \ldots, u_{N_1}^1$, and furthermore it is measurable in $y_1^1, \ldots, y_{N_1}^1$.

Thus, following as in the proof of Theorem 11.2.2, we establish compactness of individual team policies and continuity in the policies.

Following compactness and convexity of S_1 and S_2, by the continuity (and linearity in each entry) of the game in the team policies, by Fan's minimax theorem, [245, Theorem 1], we have that a saddle-point equilibrium exists. ◇

11.7.10 Proof of Theorem 11.6.3

Following the proof of Theorem 11.4.1 and viewing the common randomness variable as an exogenous integration term, the result follows from similar steps. We note again that when selecting a team policy $\bar{\gamma}$ under an information structure, we view the selection of the policy as the selection of both the independent randomness

$z \in [0,1]^N$ and the maps $\gamma^i : \mathbb{Y}^i \times [0,1]^N \to \mathbb{U}^i$ for each player. Since z is independent of the other random variables, we will, with a slight abuse of notation, allow the dependence of z to be absorbed into the γ^i terms, and thus denote them as maps from $\mathbb{Y}^i$ to $\mathbb{U}^i$.

Theorem 11.6.1 shows that an equilibrium exists under μ and ν. Assume that μ and ν are such that $\|\mu - \nu\|_{TV} \leq \epsilon$ for some $\epsilon > 0$. Without loss of generality, assume that $J^*(g,\mu) - J^*(g,\nu) \geq 0$ for some game g (a symmetric argument can be applied to the case where $J^*(g,\mu) - J^*(g,\nu) \leq 0$). Then, we have the following:

$$
\begin{aligned}
& J^*(g,\mu) - J^*(g,\nu) \\
&= \int c(x, \bar{\gamma}^1_\mu, \bar{\gamma}^2_\mu)\mu(dx; dy^1_1, \cdots, dy^1_{N_1}; dy^2_1, \cdots, dy^2_{N_2}) \\
&\qquad - \int c(x, \bar{\gamma}^1_\nu, \bar{\gamma}^2_\nu)\nu(dx; dy^1_1, \cdots, dy^1_{N_1}; dy^1_2, \cdots, dy^2_{N_2}) \\
&\leq \int c(x, \bar{\gamma}^1_\nu(y^1), \bar{\gamma}^2_\mu(y^2))\mu(dx; dy^1_1, \cdots, dy^1_{N_1}; dy^2_1, \cdots, dy^2_{N_2}) \\
&\qquad - \int c(x, \bar{\gamma}^1_\nu(y^1), \bar{\gamma}^2_\mu(y^2))\nu(dx; dy^1_1, \cdots, dy^1_{N_1}; dy^2_1, \cdots, dy^2_{N_2}).
\end{aligned}
$$

The inequality comes from perturbing the minimizer team's equilibrium strategy in the first term, making the expected cost larger, and perturbing the maximizer team's equilibrium strategy in the second term, making the expected cost smaller. Since the functions to be integrated are identical, by considering the symmetric argument for the case with $J^*(g,\mu) - J^*(g,\nu) \leq 0$, we have that as $\|\mu_n - \mu\|_{TV} \to 0$, $|J^*(g,\mu_n) - J^*(g,\mu)| \to 0$. ◇

11.7.11 Proof of Theorem 11.6.6

We provide a sketch of the proof of Theorem 11.6.6. For a complete proof, see [635, Section 6.3]. The proof proceeds in two steps. In step 1, we show that there exists a policy profile $(P^{1*}_{\pi,\infty}, P^{2*}_{\pi,\infty})$ that constitutes a NE for $\mathcal{P}_\infty$, among all randomized policies in $L^1_{\mathsf{PR,\,SYM}} \times L^2_{\mathsf{PR,\,SYM}}$. In step 2, we show that the policy profile $(P^{1*}_{\pi,\infty}, P^{2*}_{\pi,\infty})$ constitutes a NE for $\mathcal{P}_\infty$, among all randomized policies in $L^1 \times L^2$.

S1. We first restrict the search in finding a NE for $\mathcal{P}_\infty$ to randomized policies belonging to $L^1_{\mathsf{PR,\,SYM}} \times L^2_{\mathsf{PR,\,SYM}}$. Since observations are conditionally i.i.d. and the costs are continuous (by Assumptions 11.6.5 and 11.6.6) by the strong law of large numbers, we can rewrite the expected costs $J^i_{\pi,\infty}(P^1_\pi, P^2_\pi)$ for any (P^1_π, P^2_π) in $L^1_{\mathsf{PR,\,SYM}} \times L^2_{\mathsf{PR,\,SYM}}$ as

$$
\begin{aligned}
&J^i_{\pi,\infty}(P^1_\pi, P^2_\pi) \\
&:= \bar{J}^i_{\pi,\infty}(P^i_{\pi,\mathsf{R}}, P^{-i}_{\pi,\mathsf{R}}, \Lambda^i, \Lambda^{-i}) \qquad (11.23)\\
&:= \int c^i\left(\omega_0, u^i_{\mathsf{R}}, \Xi^1\left(\Lambda^1(du^1_{\mathsf{R}})\right), \Xi^2\left(\Lambda^2(du^2_{\mathsf{R}})\right)\right) \Lambda^i(du^i_{\mathsf{R}}) \\
&\times \mathbb{P}^0(d\omega_0) \prod_{s=1}^{2} \hat{f}^s\left(y^s_{\mathsf{R}}, \omega_0\right) Q^s(dy^s_{\mathsf{R}}) P^s_{\pi,\mathsf{R}}(d\gamma^s_{\mathsf{R}}) \gamma^s_{\mathsf{R}}(du^s_{\mathsf{R}}|y^s_{\mathsf{R}}),
\end{aligned}
$$

where the sub-index R denotes the representative DM for each PL^i. In the above,

$$
P^i_\pi(\underline{\gamma}^i \in \cdot) = \prod_{k \in \mathbb{N}} P^i_{\pi,\mathsf{R}}(\gamma^i_k \in \cdot), \tag{11.24}
$$

for some $P^i_{\pi,\mathsf{R}}(\gamma^i_k \in \cdot)$ belonging to the set $\mathcal{P}(\Gamma^i_{\mathsf{R}})$, and

$$
\Lambda^i(\cdot) = \mathcal{L}(u^i_{\mathsf{R}}). \tag{11.25}
$$

We note that under Assumptions 11.6.4 and 11.6.5, observations of DMs within players are i.i.d. and also independent of ω_0 via using a change of measure argument in (11.7). In view of the above costs, a policy profile (P^1_π, P^2_π) of the forms (11.24) constitutes a NE for $\mathcal{P}_\infty$ if $P^i_{\pi,\mathsf{R}} \in \mathrm{BR}(P^{-i}_{\pi,\mathsf{R}}, \Lambda^1, \Lambda^2)$ for $i = 1, 2$ with

$$
\begin{aligned}
\mathrm{BR}(P^{-i}_{\pi,\mathsf{R}}, \Lambda^1, \Lambda^2) := \Bigg\{ P^i_{\pi,\mathsf{R}} \in \mathcal{P}(\Gamma^i_{\mathsf{R}}) \Bigg| \forall \hat{P}^i_{\pi,\mathsf{R}} \in \mathcal{P}(\Gamma^i_{\mathsf{R}}) \\
\bar{J}^i_{\pi,\infty}(P^i_{\pi,\mathsf{R}}, P^{-i}_{\pi,R}, \Lambda^i, \Lambda^{-i}) \leq \bar{J}^i_{\pi,\infty}(\hat{P}^i_{\pi,\mathsf{R}}, P^{-i}_{\pi,\mathsf{R}}, \Lambda^i, \Lambda^{-i}) \Bigg\},
\end{aligned}
$$

and the consistency condition (11.25) holds. We note that the above best response map is different from that defined in (11.10), this is because, in the above, each representative DM policy of a player is a best response to the policy of the representative DM of the other player and the mean-field terms of both players. Define the best response correspondence $\Phi : \prod_{i=1}^2 \mathcal{P}(\Gamma^i_{\mathsf{R}}) \to 2^{\prod_{i=1}^2 \mathcal{P}(\Gamma^i_{\mathsf{R}})}$ as $\Phi = \Psi \circ \Xi$ with $\Xi(P^1_{\pi,\mathsf{R}}, P^2_{\pi,\mathsf{R}}) := (\Lambda^1, \Lambda^2)$ satisfying the consistency condition (11.25), and $\Psi(\Lambda^1, \Lambda^2) = \prod_{i=1}^2 \mathrm{BR}(P^{-i}_{\pi,\mathsf{R}}, \Lambda^1, \Lambda^2)$. We have $\mathcal{P}(\Gamma^i_{\mathsf{R}})$ is non-empty, compact and convex. Then, one shows that Φ admits a fixed point, which implies that there exists a symmetric NE for $\mathcal{P}_\infty$ among policies belonging to $L^1_{\mathsf{PR,\ SYM}} \times L^2_{\mathsf{PR,\ SYM}}$.

S2. It suffices to show that for fixed $P^{-i*}_{\pi,\infty}$ and Λ^{-i}_∞, the following equality holds:

$$
\inf_{P^i_\pi \in L^i_{\mathsf{PR,\ SYM}}} J^i_{\pi,\infty}(P^i_\pi, P^{-i*}_{\pi,\infty}) = \inf_{P^i_\pi \in L^i} J^i_{\pi,\infty}(P^i_\pi, P^{-i*}_{\pi,\infty}).
$$

By fixing $P_{\pi,\infty}^{-i*}$ and Λ_{∞}^{-i}, PL^i faces a mean-field team problem. Then, one uses the results in Sect. 6.4.1.

◇

11.7.12 Proof of Theorem 11.6.7

We provide a sketch. For a complete proof, see [635, Section 6.4]. By Theorem 11.6.6, there exists a symmetric NE for $\mathcal{P}_\infty$. Let denote this NE by $(P_\pi^{1*}, P_\pi^{2*}) \in L^1_{\text{PR, SYM}} \times L^2_{\text{PR, SYM}}$. Similar to in the proof of Theorem 11.6.6, using the strong law of large numbers, we get

$$J^i_{\pi,\infty}(P_\pi^{1*}, P_\pi^{2*}) = \limsup_{N_i \to \infty} \inf_{P^i_{\pi,N} \in L^i_{\text{pr, sym}}|_{N_i}} J^i_{\pi,N}(P^i_\pi, P_\pi^{-i*})$$

for $i = 1, 2$. On the other hand, by the strong law of large numbers since $(P_\pi^{1*}, P_\pi^{2*}) \in L^1_{\text{PR, SYM}} \times L^2_{\text{PR, SYM}}$, we also get

$$J^i_{\pi,\infty}(P_\pi^{1*}, P_\pi^{2*}) = \lim_{N_i \to \infty} J^i_{\pi,N}(P_\pi^{1*}, P_\pi^{2*}), \quad \forall i = 1, 2. \tag{11.26}$$

Hence, there exists $\hat{\epsilon}_{N_{1:2}} > 0$, converging to zero as $N_1, N_2 \to \infty$ such that for $i = 1, 2$

$$\left| J^i_{\pi,N}(P_\pi^{i*}, P_\pi^{-i*}) - \inf_{P^i_{\pi,N} \in L^i_{\text{pr, sym}}|_{N_i}} J^i_{\pi,N}(P^i_\pi, P_\pi^{-i*}) \right| \le \hat{\epsilon}_{N_{1:2}}.$$

This implies that (P_π^{1*}, P_π^{2*}) constitutes an $\hat{\epsilon}_{N_{1:2}}$-NE for $\mathcal{P}_N$ among symmetric policies $L^1_{\text{PR, SYM}}|_{N_1} \times L^2_{\text{PR, SYM}}|_{N_2}$. Again using the strong law of large numbers, we get for $i = 1, 2$

$$\begin{aligned} &\limsup_{N_i \to \infty} \inf_{P^i_{\pi,N} \in L^i_N} J^i_{\pi,N}(P^i_\pi, P_\pi^{-i*}) \\ &= \limsup_{N_i \to \infty} \inf_{P^i_{\pi,N} \in L^i_{\text{PR, SYM}}|_{N_i}} J^i_{\pi,N}(P^i_\pi, P_\pi^{-i*}). \end{aligned} \tag{11.27}$$

Hence, there exists $\epsilon_{N_{1:2}} > 0$, converging to zero as $N_1, N_2 \to \infty$ such that for $i = 1, 2$

$$\left| J^i_{\pi,N}(P_\pi^{i*}, P_\pi^{-i*}) - \inf_{P^i_{\pi,N} \in L^i_N} J^i_{\pi,N}(P^i_\pi, P_\pi^{-i*}) \right| \le \epsilon_{N_{1:2}}. \tag{11.28}$$

This implies that $(P_{\pi}^{1*}, P_{\pi}^{2*})$ constitutes an $\epsilon_{N_{1:2}}$-NE for $\mathcal{P}_N$ among all policies in $L_N^1 \times L_N^2$, and completes the proof. ◇

11.8 Concluding Remarks

For zero-sum games, we have presented a generalization of Blackwell's ordering for games with standard Borel spaces, and also obtained various continuity properties of values of a game on the spaces of information structures.

Combined with the analysis in Chap. 8, we have now presented continuity properties of value functions and equilibrium solutions in zero-sum, team, and general game problems with respect to information structures. We have shown that the value functions of both zero-sum games and team problems are continuous under total variation convergence of information structures. In both cases, the change in expected value when switching between two information structures is bounded above by the product of the total variation distance and the L^{∞} norm of the cost function. For team problems, it was shown earlier that the value function is upper semicontinuous under setwise convergence for measurable and bounded cost functions, and upper semicontinuous under weak convergence for bounded and continuous cost functions when the players' action spaces are convex.

We have shown in this chapter that, for zero-sum games, the value function is upper semicontinuous for minimizer-garbling sequences of information structures, and lower semicontinuous for sequences of maximizer-garbling information structures under weak (if the cost function is continuous and bounded and the action spaces are convex) or setwise (if the cost function is measurable and bounded) convergence.

A counterexample revealed that players in general non-zero-sum games may not have value functions that are continuous under total variation convergence of information structures, even when the cost functions are bounded.

The results have also been generalized to games where teams of decision makers play against each other in the context of zero-sum games and non-zero-sum games. Once compactness and convexity properties were established, we observed that the generalization was natural. The critical role of common randomness has also been studied and it has been shown that this role is absent in infinite exchangeable games.

11.9 Bibliographic Notes

Some comprehensive texts studying informational aspects of stochastic game theory are [97, 281, 532, 568] which also study various informational regularity and existence properties of equilibrium solutions for stochastic games as well as continuity and comparison results for zero-sum games [532, Chapter 3].

The existence of a value for zero-sum games with incomplete information has been studied rather extensively. Milgrom and Weber present an existence result for more general games in [539, Theorem 1]. The comprehensive treatment in [532, Proposition III.4.2.] imposes continuity in all the variables (unlike what was presented in the chapter). Furthermore, [532, Proposition III.4.2.] builds on a topology construction on policies which is different from what we presented here; regarding the construction in [532] we would like to caution that in the absence of absolute continuity conditions on the information structure, this construction may lead to a lack of closedness on the sets of admissible policies (or *strategic measures*) as the counterexample [830, Theorem 2.7] reveals. For a more detailed discussion, we refer the reader to Sect. 5.5.2 (see also [621, Sect. 7.2]). We also refer the reader to [500] which presents relaxed conditions, where continuity in the exogenous variables are not imposed and the absolute continuity condition is relaxed. Finally, [59, Theorem 3.4] presents a generalization where action sets are information dependent. The essence of these results were presented, independently, in [353, Theorem 3.2]; this result was stated and proven in this chapter as Theorem 11.2.2.

Bassan et al. [102] further provided sufficient conditions for games to have the 'positive value of information property', where providing additional information to some or all players results in greater or equal payoffs for all players. Gossner and Mertens highlighted zero-sum games as a particularly interesting class to study in the context of ordering information structures in games and did preliminary work on this ordering [312]. For comparison of information structures in zero-sum games with finite measurement and action spaces, Pęski provided necessary and sufficient conditions, and thus a complete characterization [578]. Prior to Pęski's results, De Meyer, Lehrer, and Rosenberg had shown that the value of information is positive in zero-sum games, albeit with a slightly different setup than Pęski, where their payoff depended on an individual 'type' for each player rather than a common state of nature; their results were applicable for infinite action spaces and finite type spaces [206]. Furthermore, Lehrer and Shmaya studied a 'malevolent nature' zero-sum game played between nature and a player in a finite setting, and characterized a partial ordering of information structures for these games [205]. We also note the following references on topological and continuity properties of information structures in single-agent and multi-agent team problems [818, 828, 830]. A recent comprehensive study on the value and topological properties of information structures in zero-sum games, which also generalizes [578] to the countably infinite probability space setup, is [297].

The analysis in [233] shows uniform continuity of the value function for zero-sum games under convergence of player information fields using a notion of convergence of sigma fields developed by Boylan in [142], under Lipschitz continuity assumptions on the cost function; whereas [234] establishes continuity of the value function in the prior distribution on the state space for zero-sum games using the total variation metric. A somewhat different notion of ϵ-equilibrium (a uniform ϵ equilibrium concept where ϵ-proximity, with ϵ being uniform for each conditional expected cost/reward given the private information realization, regardless of how unlikely the realizations are) has been studied in [379] and

[607] for finite games, where it has been shown that total variation continuity in information does not hold under this concept.

For zero-sum games involving teams of agents, as emphasized explicitly in [22], lack of convexity in the set of team policies can lead to games which do not have a saddle-point. We allowed in this chapter each team to select a team policy in which players may share common randomness, where the randomness was independent between the two teams. Allowing this, rather than only independently randomized policies for each team, ensured that the team policy spaces are convex, which was crucial in demonstrating the existence of an equilibrium.

Some of the results in this chapter appear in [353, 354, 356, 635].

Part III
Information Design via Information Theory in Networked Systems and Control

Chapter 12
Coding for Control and Connections with Information Theory

Starting with this chapter, in Parts III and IV, we will present results on information structure constraints determined by the presence of data-rate constrained channels; in particular, the measurement channels will have embedded in them a coder and a decoder. Accordingly, different from the setup in Parts I and II, in the treatment in this chapter the input to the channel can be shaped.

This additional flexibility brings out fundamental challenges in optimality and equilibrium analysis, which motivates the detailed analyses to follow.

12.1 Introduction

In this chapter, we study quantizers and encoders, and decoders (to be viewed as part of a measurement system). The notion of a quantizer was introduced formally in Sect. 8.5. We will discuss further properties of quantizers, and their performance, and bring a perspective where we view quantizers as decision variables. The chapter is also concerned with the derivation of fundamental bounds in connection with stabilizability of a linear system over a communication channel. The ideas and results presented here will be used throughout the rest of the book.

The chapter introduces the notion of real-time coding and defines the selection of a quantizer function as a decision problem in Sect. 12.2. In Sect. 12.3, a review of basic operational definitions in information theory is presented. Section 12.4.1 highlights the subtle differences between the performances of optimal coding for a single random variable and the limit performance of a sequence of optimal codes for blocks of random variables as the block length becomes unbounded (a common view adopted in Shannon's formulation of information theory). Performance bounds of quantizers for causal and non-causal coding of unstable processes are studied in Sect. 12.4.2. Fundamental lower bounds for stabilization are presented in Sect. 12.5.

S. Yüksel, T. Başar, *Stochastic Teams, Games, and Control under Information Constraints*, Systems & Control: Foundations & Applications,
https://doi.org/10.1007/978-3-031-54071-4_12

12.2 Quantization and Real-Time Coding

12.2.1 Real-Time Coding

In real-time applications such as remote control of time-sensitive processes, causality in encoding and decoding is a natural limitation. As discussed earlier, there is a natural causal ordering of events in a controlled process, consisting of measurement, estimation, and actuation. All these events need to take place in real time and not with significant delay. In the following, we provide a characterization of information structures in such systems.

Essential in communications problems is the embedding of information into a finite set, possibly with loss of information. This is done through quantization.

Before proceeding further, recall from Definition 8.5.1 that a quantizer Q is a (Borel measurable) function from a topological space $\mathbb{X}$ to a finite index set $\mathcal{M} := \{1, 2, \ldots, M\}$. We define the bins or cells in such a quantizer as the sets

$$\mathcal{B}_i = \{x \in \mathbb{X} : Q(x) = i\}, \quad i \in \mathcal{M}.$$

Thus, a quantizer partitions its domain set. Occasionally, quantization bins are represented by *reconstruction values*. Traditionally, in source coding theory, a quantizer is also characterized by a collection of reconstruction values in addition to a set of partitions. According to our model, this corresponds to assigning a sequence of vectors $\{q^i \in \mathbb{A}\}$ (for some set $\mathbb{A}$ which is typically $\mathbb{X}$ itself) such that:

$$Q(x) = \sum_{i \in \mathcal{M}} q^i 1_{\{x \in \mathcal{B}_i\}}.$$

Thus, one could regard the above to be a composition of a quantizer and a decoding function $\mathcal{D} : \mathcal{M} \to \mathbb{A}$ defined by:

$$\mathcal{D}(k) = \sum_{i \in \mathcal{M}} q^i 1_{\{i=k\}}.$$

When reconstruction values are specified *a priori*, we will explicitly include them in the definition of the quantizer. An example of a quantizer, which will be re-visited later in the book, is the following : A *(modified) uniform quantizer* $Q_K^\Delta : \mathbb{R} \to \mathbb{R}$ with step size Δ and $K + 1$ (with K even) number of bins satisfies the following for $k = 1, 2 \ldots, K$ (see Fig. 12.1):

$$Q_K^\Delta(x) = \begin{cases} (k - \frac{1}{2}(K+1))\Delta, & \text{if } x \in [(k - 1 - \frac{1}{2}K)\Delta, (k - \frac{1}{2}K)\Delta) \\ (\frac{1}{2}(K-1))\Delta, & \text{if } x = \frac{1}{2}K\Delta \\ 0, & \text{if } x \notin [-\frac{1}{2}K\Delta, \frac{1}{2}K\Delta]. \end{cases} \tag{12.1}$$

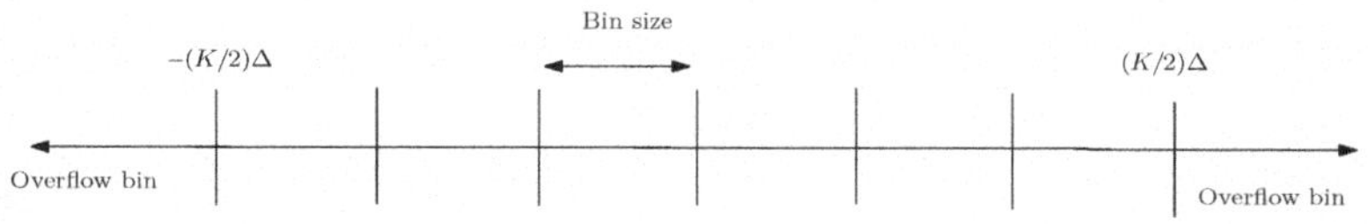

Fig. 12.1 A modified uniform quantizer. There is a single overflow bin

where we have $\mathcal{M} = \{1, 2, \ldots, K + 1\}$. The quantizer-decoder mapping thus described corresponds to a uniform quantizer with bin size Δ. The interval $[-K/2, K/2]$ is termed the *granular region* of the quantizer, and $\mathbb{R} \setminus [-K/2, K/2]$ is named the *overflow region* of the quantizer (see Fig. 12.1). We will refer to this quantizer as a *(modified) uniform quantizer*, since the overflow region is assigned a single bin.

Typically, it is assumed that a quantizer is followed by an encoder. Let $\mathbb{M}$ be a finite set, typically taken as $\{0, 1\}$, that is, the binary alphabet. We refer to an encoder $\mathcal{E}$ as a mapping from $\mathcal{M}$ to $\mathbb{M}^*$, where $\mathbb{M}^*$ denotes the set of all finite-length sequences with elements in $\mathbb{M}$. The sequences in $\mathbb{M}^*$ are called *codewords*. If the lengths of the codewords corresponding to elements in $\mathcal{M}$ under an encoder $\mathcal{E}$ are all equal, the encoder is said to be a *fixed-rate* encoder. If the lengths are different, the encoder is said to be *variable-rate*. Hence, we define the *fixed-rate* rate of a quantizer or an encoder by the (base-2) logarithm of the number of cells, which is $\lceil \log_2(|\mathcal{M}|) \rceil$ for the quantizer described in (12.1).

Unless explicitly stated, we will always assume that a quantizer is followed by an encoder.

12.2.2 Information Structures for Real-Time Encoders and Controllers: Policies, Actions and Measurability

This subsection considers a typical optimal causal encoding/quantization setup in a networked control system. For simplicity of the setup, we consider only two encoders, for a decentralized system, and use this system to introduce the causality and measurability constraints in quantizer design.

We begin by providing a description of the system model. Consider a partially observed Markov process, defined on a probability space, again, $(\Omega, \mathcal{F}, P)$, and described by the following discrete-time equations for $t \geq 0$:

$$x_{t+1} = f(x_t, u_t, w_t), \tag{12.2}$$

$$y_t^i = g^i(x_t, v_t^i), \tag{12.3}$$

for (Borel) measurable functions $f, g^i, i = 1, 2$, with $\{x_0, w_t, v_t^i, i = 1, 2\}$ random variables, which are mutually independent across time and space and whose distribution functions are available at the decision makers. We further have $x_t \in \mathbb{X}$,

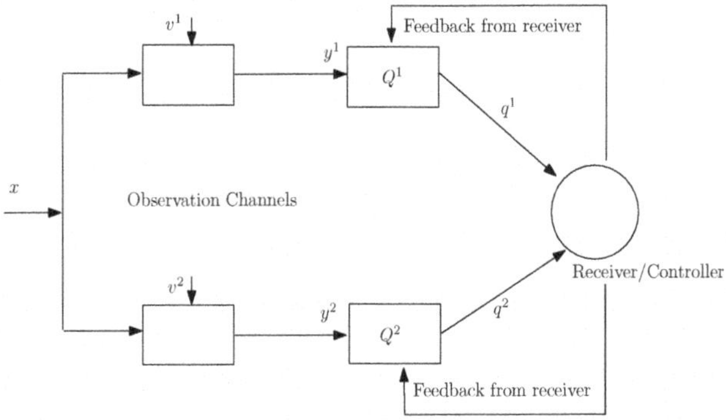

Fig. 12.2 Partially observed source under a decentralized structure

and $y_t^i \in \mathbb{Y}^i$, $u_t \in \mathbb{U}$, where $\mathbb{X}$, $\mathbb{Y}^i$, $\mathbb{U}$ are complete, separable, metric spaces (Polish spaces), and thus, include countable spaces or $\mathbb{R}^n$, $n \in \mathbb{N}$.

Consider a scenario where an encoder, Encoder i, is located at one end of a measurement channel characterized by (12.3), this being for $i = 1, 2$. The encoders transmit their information to a receiver (see Fig. 12.2), over a discrete noiseless channel with finite capacity, and hence, they have to quantize their information.

We let $\Pi^{comp,i}$ denote a *composite quantization policy* for Encoder i, defined as a sequence of functions $\{Q_t^{comp,i}, t \geq 0\}$ which are causal such that the quantization output at time t, q_t^i, under $\Pi^{comp,i}$ is generated by a function of its local information, that is, a mapping measurable with respect to the sigma-algebra generated by

$$I_t^i = \{y_{[0,t]}^i, q_{[0,t-1]}^i, z_{[0,t-1]}^i\}, \quad t \geq 1, \qquad I_0^i = \{y_0^i\},$$

to $\mathcal{M}_t^i$, where

$$\mathcal{M}_t^i := \{1, 2, \ldots, |\mathcal{M}_t^i|\},$$

for $0 \leq t \leq T-1$ and $i = 1, 2$. Here z^i denotes some additional side information available, such as feedback from the receiver.

Let $\mathbb{I}_t^i$ denote the space in which the realization I_t^i takes values in, such that $I_t^i \in \mathbb{I}_t^i$, $t \geq 0$. Hence,

$$Q_t^{comp,i} : \mathbb{I}_t^i \to \mathcal{M}_t^i.$$

Alternatively, and equivalently, we can express the policy $\Pi^{comp,i}$ as a composition of a *Quantization Policy* $\underline{\gamma}^i$ and a *Quantizer*. A quantization policy of encoder i, $\underline{\gamma}^i$, is a sequence of functions $\{\gamma_t^i\}$, such that for each $t \geq 0$, γ_t^i is a mapping from the information space $\mathbb{I}_t^i$ to a space of quantizers $\mathbb{Q}_t^i$ to be clarified further below. A quantizer is subsequently used to generate the quantizer output. That is for every t and i, $\gamma_t^i(I_t^i) \in \mathbb{Q}_t^i$ and for every realization $I_t^i \in \mathbb{I}_t^i$,

$$Q_t^{comp,i}(I_t^i) = (\gamma_t^i(I_t^i))(I_t^i), \tag{12.4}$$

mapping the information space to $\mathcal{M}_t^i$ in its most general form. Even though there may appear to be duplicated information in (12.4) (since first a map is used to pick a quantizer, and then the quantizer maps the available information to outputs), it is possible to eliminate any existing informational redundancy: A quantizer action will be generated based on the common information at the encoders and the receiver, and the quantizer will map the relevant private information at the encoder to the quantization output. Such a separation is without any loss in the space of all composite quantization policies. To establish this explicitly, let the information at the receiver at time $t \geq 0$ be $I_t^r = \{q^1_{[0,t]}, q^2_{[0,t]}\}$. Let the common information, under feedback information, at the encoders and the receiver be I_t^c. Thus, we can express any measurable composite quantization policy as:

$$Q_t^{comp,i}(I_t^i) = (\gamma_t^i(I_t^c))(I_t^i \setminus I_t^c),$$

mapping the information space to $\mathcal{M}_t^i$.

We note that any composite quantization policy $Q_t^{comp,i}$ can be expressed in the form above; that is, there is no loss in the space of possible such policies, since for any $Q_t^{comp,i}$, one could define

$$\gamma_t^i(I_t^c)(\cdot) := Q_t^{comp,i}(I_t^c, \cdot).$$

Thus, we let DM^i have policy $\underline{\gamma}^i = \{\gamma_t^i\}$ and under this policy generate quantizer actions $\{Q_t^i, t \geq 0\}$, $Q_t^i \in \mathbb{Q}_t^i$ (Q_t^i is the quantizer used at time t). Under action Q_t^i, the encoder generates q_t^i, as the *quantization output* at time t.

The receiver (or the controller), upon receiving the information from the encoders, generates its decision at time t, also causally: An admissible causal receiver policy is a sequence of measurable functions $\underline{\gamma}^0 = \{\gamma_t^0\}$ such that

$$\gamma_t^0 : \prod_{s=0}^{t} \left(\mathcal{M}_s^1 \times \mathcal{M}_s^2\right) \to \mathbb{U}, \qquad t \geq 0$$

where $\mathbb{U}$ denotes the decision set for the receiver.

In the following, we let the bold letters denote the ensemble of letters in the sense that $\mathbf{Q}_t = \{Q_t^1, Q_t^2\}$ and $\mathbf{q}_{[0,t]} = \{q^1_{[0,t]}, q^2_{[0,t]}\}$. Then, $u_t = \gamma_t^0(\mathbf{q}_{[0,t]})$ for $t \geq 0$.

Case With Noisy Channels

A more general setup that we will consider in the book is the setting which involves noisy channels. We consider first memoryless noisy channels (in the following definitions, we assume feedback is not present; minor adjustments can be made to capture the case with feedback).

Definition 12.2.1 A Discrete Memoryless Channel (DMC) is characterized by a discrete input alphabet $\mathcal{M}$, a discrete output alphabet $\mathcal{M}'$, and a conditional probability mass function $P(q'|q)$, from $\mathcal{M} \times \mathcal{M}'$ to $\mathbb{R}$ which satisfies the following. Let $q_{[0,n]} \in \mathcal{M}^{n+1}$ be a sequence of input symbols, let $q'_{[0,n]} \in \mathcal{M}'^{n+1}$ be a sequence of output symbols, where $q_k \in \mathcal{M}$ and $q'_k \in \mathcal{M}'$ for all k and let P^{n+1}_{DMC} denote the joint mass function on the $n+1$-tuple input and output spaces. It follows that $P^{n+1}_{DMC}(q'_{[0,n]}|q_{[0,n]}) = \prod_{k=0}^{n} P_{DMC}(q'_k|q_k), \forall q_{[0,n]} \in \mathcal{M}^{n+1}, q'_{[0,n]} \in \mathcal{M}'^{n+1}$, where q_k, q'_k denote the kth component of the vectors $q_{[0,n]}, q'_{[0,n]}$, respectively.

Definition 12.2.2 A real Continuous Memoryless Channel (CMC) is characterized by a continuous (Borel) input alphabet $\mathcal{M} \subset \mathbb{R}^{s_1}$, and a continuous (Borel) output alphabet $\mathcal{M}' \subset \mathbb{R}^{s_2}$, (where $s_1, s_2 \in \mathbb{N}$) and a regular conditional probability measure $p(A|m)$, from $\mathcal{M} \times \mathcal{B}(\mathcal{M}')$ to $\mathbb{R}$, where $\mathcal{B}(\mathcal{M}')$ is the Borel σ-algebra over $\mathcal{M}'$. Let $m_{[0,n]} \in \mathcal{M}^{n+1}$ be a sequence of input symbols, $\{m_0, m_1, \ldots, m_n\}$, and let A_0^n be a sequence of Borel subsets in $\mathcal{B}(\mathcal{M}')$, $\{A_0, A_1, \ldots, A_n\}$, where $m_k \in \mathcal{M}$ and $A_k \in \mathcal{B}(\mathcal{M}')$ for all k. Let p^{n+1}_{CMC} be the probability measure on $\mathcal{B}(\mathcal{M}^{n+1} \times \mathcal{M}'^{n+1})$. A CMC from $\mathcal{M}^{n+1}$ to $\mathcal{M}'^{n+1}$ satisfies the following for all $0 \le k \le n$: $p^{n+1}_{CMC}(A_k|m_{[0,k]}, A_{[0,k-1]}) = p(A_k|m_k)$, for almost all $m_{[0,k]} \in \mathcal{M}^{k+1}$, where m_k denotes the kth component of the vector $m_{[0,n]}$. ◇

Channels can also have memory. We state the following for both discrete and continuous-alphabet channels.

Definition 12.2.3 A discrete channel (continuous channel) with memory is characterized by a sequence of discrete (continuous) input alphabets $\mathcal{M}^{n+1}$, discrete (continuous) output alphabets $\mathcal{M}^{n+1}$, and a sequence of regular conditional probability measures $P_n(dq'_{[0,n]}|q_{[0,n]})$, from $\mathcal{M}^{n+1}$ to $\mathcal{M}'^{n+1}$.

Typically, discrete channels admit finite alphabets. Unless stated otherwise, while considering discrete channels, we will assume channels with finite alphabets.

As before, consider a scenario where an encoder, Encoder i, is located at one end of a measurement channel characterized by (12.3), this being so for $i = 1, 2$. The encoders transmit their information to a receiver (see Fig. 12.3), over noisy channels.

In this context, we will also let $\Pi^{comp,i}$ denote a *composite encoding policy* for Encoder i, defined as a sequence of functions $\{Q_t^{comp,i}, t \ge 0\}$ which are causal such that the output at time t, q_t^i, under $\Pi^{comp,i}$ is generated by a function of its local information, that is, a mapping measurable with respect to the sigma-algebra generated by

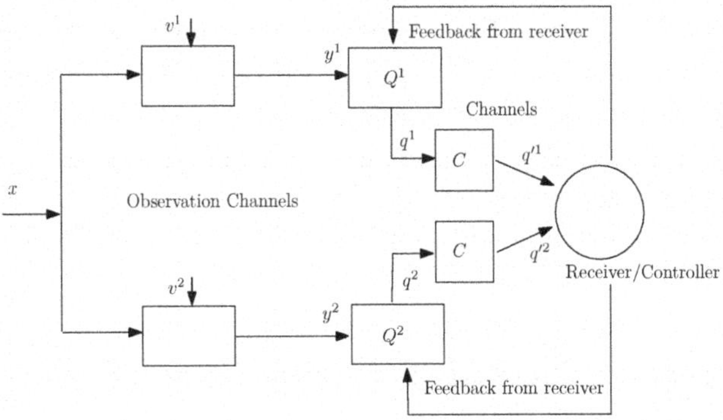

Fig. 12.3 Partially observed source under a decentralized structure with noisy channels

$$I_t^i = \{y^i_{[0,t]}, q^i_{[0,t-1]}, z^i_{[0,t-1]}\}, \quad t \geq 1, \qquad I_0^i = \{y_0^i\},$$

to $\mathcal{M}_t^i$, where $\mathcal{M}_t^i := \{1, 2, \ldots, |\mathcal{M}_t^i|\}$, for $0 \leq t \leq T-1$ and $i = 1, 2$. Here z^i denotes some additional side information available, such as feedback from the receiver.

As before, let the information at the receiver at time $t \geq 0$ be $I_t^r = \{q'^1_{[0,t]}, q'^2_{[0,t]}\}$. Let the common information, under feedback information, at the encoders and the receiver be I_t^c. Thus, we can express any measurable composite encoding policy as:

$$Q_t^{comp,i}(I_t^i) = (\gamma_t^i(I_t^c))(I_t^i \setminus I_t^c),$$

mapping the information space to $\mathcal{M}_t^i$. Thus, we let DM i have policy $\underline{\gamma}^i$ and under this policy generate quantizer actions $\{Q_t^i, t \geq 0\}$, $Q_t^i \in \mathbb{Q}_t^i$ (Q_t^i is the quantizer used at time t). Under action Q_t^i, the encoder generates q_t^i, as the *quantization output* at time t. The channel, maps q_t to q_t' in a stochastic fashion as described in Definitions 12.2.1–12.2.3.

The receiver (or the controller), upon receiving the information from the channel, generates its decision at time t, also causally: An admissible causal receiver policy is a sequence of measurable functions $\underline{\gamma}^0 = \{\gamma_t^0\}$ such that $u_t = \gamma_t^0(\mathbf{q}'_{[0,t]})$ where $u_t \in \mathbb{U}$.

Design Objective

With the above formulation, one typical objective functional for the decision makers would be the following for some $T \in \mathbb{N}$:

$$\inf_{\Pi^{comp}, \underline{\gamma}^0} E_{\nu_0}^{\Pi^{comp}, \underline{\gamma}^0}[\sum_{t=0}^{T-1} c(x_t, u_t)],$$

over all policies Π^{comp}, $\underline{\gamma}^0$ with initial distribution ν_0 on x_0. Here $c(x_t, u_t)$, is a non-negative function and $u_t = \gamma_t^0(\mathbf{q}'_{[0,t]})$ for $t \geq 0$. As an example, this setup includes, for the case when $\mathbb{X} = \mathbb{R}^n$, the quadratic cost function: $c(x, u) = |x - u|^2$.

Another objective would be to ensure that the dynamical system $\{x_t\}$ is stochastically stable in some appropriate sense.

Both of these considerations will be made precise later in the book.

12.3 Information Theoretic Preliminaries and Performance of Quantizers

In this section, we collect relevant notions and results from information theory.

12.3.1 Information Theoretic Preliminaries

We now present a number of fundamental information theoretic concepts. In this subsection, to make the following definitions easier to interpret, we will denote random variables with capital letters.

Definition 12.3.1 Let X be an $\mathbb{X}-$valued random variable, where $\mathbb{X}$ is a countable set. The *entropy* of X is defined as

$$H(X) = -\sum_{z \in \mathbb{X}} P(z) \log_2(P(z)),$$

where P is the probability mass function (pmf) of the random variable x. If Y is another $\mathbb{X}-$valued random variable, the *conditional entropy of X given $Y = y_0$* is defined by

$$H(X \mid Y = y_0) = -\sum_{z \in \mathbb{X}} P(z \mid y = y_0) \log_2(P(z \mid y = y_0)),$$

where $P(\cdot \mid y = y_0)$ is the conditional pmf of X given $Y = y_0$. The *conditional entropy of X given Y* is defined by the average of such conditional entropies such that

$$H(X \mid Y) = \sum_{y_0 \in \mathbb{X}} P(y = y_0) H(X \mid Y = y_0).$$

Definition 12.3.2 Let X be an $\mathbb{X}-$valued random variable, where $\mathbb{X} = \mathbb{R}^n$, and the probability measure induced by X is absolutely continuous with respect to the Lebesgue measure. The *differential entropy* of X is defined by

$$h(X) = -\int_{\mathbb{X}} p(x) \log_2(p(x))dx\,,$$

where $p(\cdot)$ is the probability density function (pdf) of X. If Y is another random variable with density (or mass function) p_y, the *conditional differential entropy of X given $Y = y_0$* is defined by

$$h(X \mid Y = y_0) = -\int_{y_0} p(z \mid y = y_0) \log_2(p(z \mid y = y_0))dz\,,$$

where $p(\cdot \mid y = y_0)$ is the conditional pdf of X given $Y = y_0$. The *conditional differential entropy of X given Y* is defined by the average of such conditional entropies such that

$$h(X \mid Y) = \int_{y_0 \in \mathbb{X}} p_y(y_0) h(X \mid Y = y_0) dy_0.$$

An important property of entropy (and differential entropy) is the chain rule: $H(X, Y) = H(X) + H(Y|X)$, where $H(X, Y)$ is the entropy for the vector random variable (X, Y). If X and Y are independent, $H(Y|X) = H(Y)$. A further important result is that conditioning on a random variable cannot increase entropy: $H(X|Y) \leq H(X)$; conditioning on a particular realization, however, may increase the entropy.

Definition 12.3.3 The *Mutual Information* between a discrete (continuous) random variable X, and another discrete (continuous) random variable Y, defined on a common probability space, is defined as

$$I(X; Y) = H(X) - H(X|Y)\,,$$

where $H(X)$ is the entropy of X (differential entropy if X is a continuous random variable), and $H(X|Y)$ is the conditional entropy of X given Y (conditional differential entropy if X is a continuous random variable).

For more general settings, that is, for cases including when the random variables are continuous, discrete or a mixture of the two or for settings where the random variables take values in Polish spaces, mutual information is defined as

$$I(X; Y) := \sup_{Q_1, Q_2} I(Q_1(X); Q_2(Y)),$$

where Q_1 and Q_2 are quantizers with finitely many bins (see Chapter 5 in [314]). An important relevant result is the following. Let X be a random variable and Q be a quantizer applied to X. Then, $H(Q(X)) = I(X; Q(X)) = h(X) - h(X|Q(X))$.

Definition 12.3.4 (Relative Entropy) The *Kullback-Leibler Divergence* or *Relative Entropy* between two probability measures P_1 and P_2 such that P_1 is absolutely continuous with respect to P_2 is defined as:

$$D(P_1||P_2) = \int \log_2(f(x))P_1(dx),$$

where $f(x) = \frac{dP_1}{dP_2}(x)$ is the Radon-Nikodym derivative of P_1 with respect to P_2, that is for all events A, $P_1(A) = \int_A f(x)P_2(dx)$.

Mutual information between two random variables X (with probability measure P_1) and Y (with probability measure P_2) is also given by $D(P_1P_2||P_1 \times P_2)$, that is the relative entropy between the joint measure and a measure which is the product of the measures of X and Y, denoted here by $P_1 \times P_2$.

An important information-theoretic property is the data-processing inequality.

Lemma 12.3.1 *Let X, Y, Z be three random variables which form a Markov chain in the specified order, that is $X \leftrightarrow Y \leftrightarrow Z$. Then, $I(X;Y) \geq I(X;Z)$.*

An immediate consequence of Lemma 12.3.1 is that given two random variables X, Y, for any measurable function pairs f, g, the following holds: $I(f(X); g(Y)) \leq I(X;Y)$.

An important property of differential entropy is the entropy-power inequality, given as follows.

Lemma 12.3.2

(a) Let X, Y be independent random variables and let $Z = X + Y$. Then,

$$2^{2h(Z)} \geq 2^{2h(X)} + 2^{2h(Y)},$$

where equality holds if X, Y are Gaussian.

(b) Let X, Y be conditionally independent given U and let $Z = X + Y$. Then,

$$2^{2h(Z|U)} \geq 2^{2h(X|U)} + 2^{2h(Y|U)},$$

where $h(\cdot|U)$ denotes the conditional differential entropy.

Further properties relating entropy and moments are also available, see e.g. [540, Lemma 1] and [491].

12.3.2 Fixed or Variable Rates of a Quantizer/Encoder

As discussed earlier, the *fixed-rate* rate of a quantizer or encoder is defined by the (base-2) logarithm of the number of cells, which is $\lceil \log_2(|\mathcal{M}|) \rceil$ for the quantizer described in (12.1). There is also a rate definition in terms of the expected number of bits to be used; this is called the *variable-rate* rate of a quantizer.

Let X be an $\mathbb{X}-$valued random variable, with $\mathbb{X} finite$, and let P denote the probability measure induced by X. Suppose that we wish to represent this variable in terms of binary numbers and an encoder maps $\mathbb{X}$ to a space of binary sequences. The variable-rate rate captures the average number of bits needed to represent the encoder outputs: That is, let $l(x)$ denote the length of a binary string (codeword) which is used to represent the symbol $x \in \mathbb{X}$. Then, the variable-rate rate of this encoder is given by

$$E[l(X)] = \sum_{x \in \mathbb{X}} P(x) l(x).$$

An important result in source-coding theory is the following.

Proposition 12.3.1 *[188] Let X be an $\mathbb{X}$−valued random variable where $\mathbb{X}$ is a finite set. The variable-rate rate of any encoder which leads to a noiseless representation of the random variable satisfies*

$$H(X) \leq E[l(X)].$$

Furthermore, there exists an encoder with variable-rate rate:

$$E[l(X)] \leq H(X) + 1,$$

which leads to a noiseless representation of the source X.

Proposition 12.3.1 thus represents a lower bound on the average-rate of quantization outputs when a source is quantized. When a source is quantized, the quantized output can be coded with a variable-rate encoder and the entropy of the quantization output serves as a lower bound on the average length of the codewords. Note that, if an encoder simultaneously encodes a sequence of n i.i.d. random variables $x_{[0,n-1]}$, then, with increasing values of n, entropy becomes an arbitrarily close measure for the minimum average variable-rate information rate needed for noiseless representation since for every n there exists a code with average-rate satisfying

$$H(X_0) = \frac{1}{n} H(X_{[0,n-1]}) \leq \frac{1}{n} E[l(X_{[0,n-1]})] \leq H(X_0) + \frac{1}{n}.$$

12.3.3 Rate-Distortion Theory

Shannon's rate-distortion function is defined operationally as follows: Given a componentwise $\mathbb{X}$−valued stochastic process $\{X_t, t \in \mathbb{Z}_+\}$, a *rate-distortion* pair (R, D) is achievable if given the source process $\{X_t\}$, and a distortion function $\rho : \mathbb{X} \times \mathbb{X} \to \mathbb{R}_+$, there exist sequences of

1. Quantizers (Encoders): $\mathcal{E}_n : \mathbb{X}^n \to \mathcal{M}(n)$ with $|\mathcal{M}(n)| \leq 2^{Rn}$,
2. Decoders: $\mathcal{D}_n : \mathcal{M}(n) \to \mathbb{X}^n$ such that $\hat{X}_t = \mathcal{D}_n(q_{[0,n-1]})$, $0 \leq t \leq n-1$ with

$$D_n := \frac{1}{n} E[\sum_{t=0}^{n-1} \rho(X_t, \hat{X}_t)],$$

and $\lim_{n\to\infty} D_n \leq D$.

The quantity

$$\inf\{R : (R, D) \quad \text{is achievable}\}$$

is the (operational) *Rate-distortion* function of the source at the distortion level D. The quantity $\inf\{D : (R, D) \quad \text{is achievable}\}$ is the (operational) *Distortion-rate function* at rate R.

12.3.4 Channel Coding and Shannon Capacity

Given a channel, a rate R is achievable (transmittable) if there exists a sequence (R, ϵ_n), with $\epsilon_n \to 0$ as $n \to \infty$, such that for every n, there exist (Fig. 12.4):

1. A set of messages $\mathcal{M}(n) := \{1, 2, 3 \ldots, M(n)\}$ such that $|\mathcal{M}(n)| = M(n) \geq 2^{Rn}$,
2. A channel coder

$$\mathcal{E}_n : \mathcal{M}(n) \to \mathcal{M}^n \tag{12.5}$$

and a decoder: $\mathcal{D}_n : \mathcal{M}'^n \to \mathcal{M}(n)$, with average error probability

$$P_e := \frac{1}{|\mathcal{M}(n)|} \sum_{c \in \mathcal{M}(n)} P(\mathcal{D}_n(q'_{[0,n-1]}) \neq c | c \quad \text{is transmitted}) \leq \epsilon_n.$$

Given a channel, the supremum rate R that can be achieved is called the (operational) *Shannon capacity* of the channel. We note that for DMCs the above characterization holds for the worst-case error probability criterion also, in addition to the average error criterion (see [188, Theorem 7.7.1].

We emphasize that the above are *operational definitions* for rate-distortion and capacity. When the source and the channels belong to further special classes (for example those which are ergodic [744], [741]), then the operational definitions stated above also admit what is known as *single-letter mathematical characterizations*.

If the source sequence is independent and identically distributed, the *rate distortion function*, $R(D)$, of the random variable X (whose i.i.d. realizations form the sequence) is the minimum (or infimum) value of the mutual information over the class of stochastic kernels from the input to the output, subject to the constraint

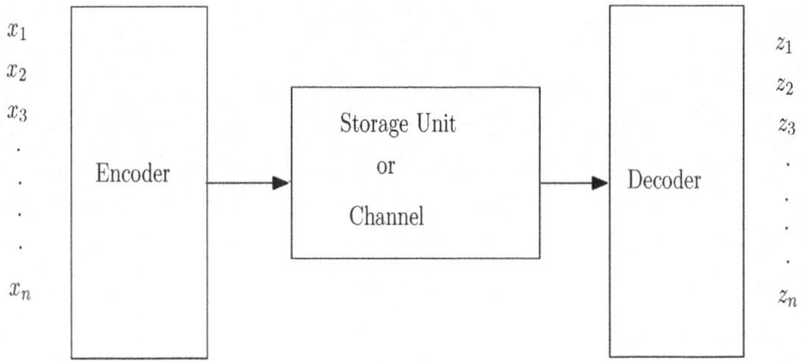

Fig. 12.4 Shannon's setup assumes block coders which are delay insensitive

that distortion is not higher than $D \in \mathbb{R}_+$:

$$R(D) = \inf_{P(d\hat{x}|x):E[\rho(X,\hat{X})] \leq D} I(X;\hat{X}), \tag{12.6}$$

where the infimum is over the set of all stochastic kernels $P(d\hat{x}|x)$ from $\mathbb{X}$ to $\mathbb{X}$. We also note that the inverse kernel $P(dx|\hat{x})$ (from the reconstruction space to the source alphabet space) is known as the *backward test channel* (and thus, this exhibits a duality relationship between the channel coding problem and the source coding problem).

A memoryless Gaussian random source constitutes an important example of the above. For a Gaussian source with variance σ_x^2, the rate-distortion function is given by $R(D) = (1/2)\log_2(\sigma_x^2/D)$ for $D \leq \sigma_x^2$.

Likewise, suppose that we have a discrete memoryless channel. The capacity of such a memoryless channel can be expressed as

$$C = \sup_{P(q)} I(q;q'), \tag{12.7}$$

where the supremum is over the set of all admissible channel input distributions $P(q)$.

Two important types of channels are the binary symmetric channel and the binary erasure channel.

A binary symmetric channel is defined by the transition probabilities: $P(q' = 0|q = 1) = P(q' = 1|q = 0) = \epsilon$. The capacity of such a channel is obtained when the input probability is such that $P(q = 0) = P(q = 1) = 1/2$, leading to the capacity value of $1 - H_b(\epsilon)$, where the binary entropy function H_b is defined as, for $\epsilon \in [0, 1]$,

$$H_b(\epsilon) = -\epsilon \log_2(\epsilon) - (1 - \epsilon)\log_2(1 - \epsilon).$$

A binary erasure channel is defined by the transition probabilities $P(q' = 0|q = 0) = P(q' = 1|q = 1) = 1 - \epsilon$ for some $\epsilon \in [0, 1]$, and the existence of an erasure symbol e such that $P(q' = e|q = 0) = P(q' = e|q = 1) = \epsilon$. The capacity of such a channel is also obtained when the input probability is such that $P(q = 0) = P(q = 1) = 1/2$ leading to the capacity value of $(1 - \epsilon)$.

Such channels and their generalizations will be revisited later. Gaussian sources and channels are further important examples. In Chap. 16, these are studied in further detail.

Channels with Feedback

Communication systems may operate with feedback, that is, encoders can use the previous channel outputs to select channel inputs in a causal fashion. In this case, the encoder sequences in (12.5) are to be replaced by

$$\mathcal{E}_n = \left\{ \mathcal{E}_{n,k} : \mathcal{M}(n) \times \mathcal{M}'^k \to \mathcal{M}_k \right\},$$

for $0 \leq k \leq n-1$ such that $q_k = \mathcal{E}_{n,k}(c, q'_{[0,k-1]})$. Such feedback does not increase the capacity of memoryless channels, but does in general increase the capacity of channels with memory [188]. Furthermore, feedback may allow for the use of practical coding schemes with significantly less complexity and typically improves the *reliability of channels*, which is a measure of the rate of improvement in error probability as a function of the number of channel uses.

12.4 Two Salient Aspects where *Information Theory in Control* Differs from *Classical Shannon Theory*

12.4.1 Infinite-Dimensional Coding vs. Finite-Dimensional Coding

The practical or operational applicability of the information-theoretic notions (such as capacity and the rate-distortion function) generally requires processing of sequences of random variables with unbounded block-length, whereas in real-time control applications (of main concern in this book) excess delay in processing cannot be tolerated. The distinction between the two is quite subtle and we aim to make this more transparent in the following.

Toward this end, we will emphasize the differences between the settings of Shannon's rate-distortion function and the problem of *distortion-constrained entropy minimization* considered in [329] and [332], among other references.

Consider a real valued random variable x with probability measure P. Suppose that we wish to quantize this source subject to a constraint on the rate of communication. This constraint can involve either a fixed-rate or a variable-rate measure of information rate.

Let the quantizer have bins $\{\mathcal{B}_i\}$ and let the reconstruction values be $\{\gamma(Q(x))\}$, where γ denotes a decoder function. To economize notation, we will denote the reconstruction value by $\{Q(x)\}$.

Let $J(Q) = E[\rho(x, Q(x))] = \int \rho(x, Q(x))P(dx)$, where $\rho : \mathbb{R} \times \mathbb{R} \to \mathbb{R}_+$ is a distortion function. The distortion under quantizer Q is given by:

$$J(Q) = \sum_{i=1}^{|\mathcal{M}|} \int_{\mathcal{B}_i} \rho(x, q^i)P(dx)$$

with $q^i = Q(x)$ for $x \in \mathcal{B}_i$.

We define the distortion-constrained fixed-rate function (or *operational rate-distortion function* [320]) as

$$R_{FQ}(D) = \inf\{\lceil \log_2(|Q(\mathbb{R})|)\rceil : J(Q) \leq D\},$$

where $|Q(\mathbb{R})|$ denotes the cardinality of the quantizer index set.

The entropy under quantizer Q is (which, as discussed earlier, is a measure of the minimum average rate needed for lossless representation of quantizer outputs):

$$H(Q) := -\sum_i P(\mathcal{B}_i) \log_2(P(\mathcal{B}_i))$$

We define the distortion-constrained entropy function as:

$$R_Q(D) = \inf_{Q \in \mathcal{Q}} \{H(Q) : J(Q) \leq D\},$$

where $\mathcal{Q}$ is the space of all quantizers Q, possibly with countably infinite number of bins. For such a setting, the number of quantizer bins in an optimal quantizer can be finite or infinite [331]. It is evident that $R_{FQ}(D) \geq R_Q(D)$, since fixed-rate coding is a special case of variable-rate coding.

Note that in the expression of $R(D)$, see (12.6), the infimization is over the space of stochastic kernels, whereas in the quantization framework, it is over only the space of quantizers. Furthermore, when $\hat{x}$ admits a discrete probability measure and $P(\hat{x}|x) = 1_{\{\hat{x}=Q(x)\}}$, $I(x; \hat{x})$ reduces to $H(\hat{x})$. Thus, the essential difference is the space of optimization. It is a direct consequence of the larger space of optimization in the rate-distortion formulation that $R_Q(D) \geq R(D)$ where the inequality is strict in many of the applications in information theory. Furthermore, even though the rate-distortion function is a convex function of distortion, the distortion-constrained entropy function is not a convex function of distortion (see [329], [332]; where in the former study [329] an analysis for a uniform source has been explicitly considered). Furthermore, the set of quantizers is not convex. Note also that we have already discussed some further properties of quantizers in Sect. 8.5.

As discussed in Sect. 12.3.3, from a practical or an operational viewpoint, in the rate-distortion theoretic analysis, an increasing sequence of blocks of random variables with unbounded length is considered, and compression is performed based on these sequences. The rate-distortion theoretic performance values are obtained as the limit of the performances attained for these increasing blocks of variables. The rate-distortion function leads to an asymptotically tight lower value, since through a solution to (12.6), a test-channel is constructed and the induced *backward test channel* can be used to construct an asymptotically efficient codebook, utilizing the properties of laws of large numbers: Only for quantization of vector valued processes with ergodic behavior, in the limit of large dimensions, does one observe an equivalence between the operational rate-distortion function, distortion-constrained entropy function, and the rate-distortion function [320].

For a quantization problem, however, a single realization of a random variable is observed, and the output is generated only for the single realization.

We note that, in view of the above, even when an independent sequence is encoded, block encoding may perform strictly better than memoryless quantization. We will, however, see later in Chap. 15 that for causal (zero-delay) coding of independent sources, scalar (memoryless) coding is as good as any causal block encoder of arbitrary length.

Another important difference between finite-length and infinite-length problems is on joint source-channel coding. In the limit of large block lengths, one can design an optimal quantizer/encoder/decoder scheme (for the minimization of some average distortion criterion subject to a rate-constraint) by first compressing an ergodic source using a rate-distortion achieving code (up to an arbitrarily small error), and applying a capacity achieving channel code to map the compression outputs to channel inputs, when the channel is also ergodic. This is known as *source-channel separation*. However, such a setup is not generally optimal for finite blocklength settings and the quantization and channel encoding operations need to be analyzed jointly for optimal performance. We will revisit this topic later in Chap. 16.

We end the section by noting that, although typical information-theoretic approaches require long blocks, important special cases provide optimal performance even under delay-limited settings: One popular example is the problem of transmitting a Gaussian source over a Gaussian channel, the so called *Gaussian Test Channel (GTC) Problem*, introduced and discussed in Chap. 3. As stated there, the no-memory property of the GTC is due to the fact that the source and channel pairs are matched in the sense that capacity and rate-distortion achieving pairs are identical, leading to optimality (known as the *matching property*) . We will discuss this topic also further in Chap. 16.

12.4.2 Coding for Stable vs. Unstable Sources

In the next few chapters we will study information requirements, and coding and control algorithms for stabilization of open-loop unstable linear systems controlled over communication channels. Toward obtaining such results, in the following, we provide a partial review of the contributions in the information theory literature for such sources.

There have been important contributions in the information theory literature on non-causal coding of non-stationary/unstable sources: Consider the following Gaussian auto-regressive (AR) process:

$$x_t = -\sum_{k=1}^{m} a_k x_{t-k} + w_t,$$

where $\{w_t\}$ is i.i.d. zero-mean, Gaussian random sequence with variance $E[w_t^2] = \sigma^2$. If the roots of the polynomial $H(z) = 1 + \sum_{k=1}^{m} a_k z^{-k}$ are all in the interior of the unit circle, then the process is asymptotically stationary and its rate distortion function (with the distortion being the expected, normalized Euclidean error) is given parametrically by the following [313], obtained by considering the asymptotic distribution of the eigenvalues of the correlation matrix:

$$D_\theta = \frac{1}{2\pi}\int_{-\pi}^{\pi} \min(\theta, \frac{1}{g(w)})dw,$$

$$R(D_\theta) = \frac{1}{2\pi}\int_{-\pi}^{\pi} \max(1/2(\log \frac{1}{\theta g(w)}), 0)dw,$$

with $g(w) = \frac{1}{\sigma^2}|1 + \sum_{k=1}^{m} a_k e^{-ikw}|^2$. If at least one root is on or outside the unit circle, the analysis is more involved as the asymptotic eigenvalue distribution contains unbounded components. It has been shown in [313], [339], and [318] that using the properties of the eigenvalues as well as Jensen's formula for integrations along the unit circle, that $R(D_\theta)$ above should be replaced by:

$$R(D_\theta) = \frac{1}{2\pi}\int_{-\pi}^{\pi} \max\left(\frac{1}{2}\log(\frac{1}{\theta g(w)}), 0\right)dw + \sum_{k=1}^{m} \frac{1}{2}\max(0, \log(|\rho_k|^2) \tag{12.8}$$

where $\{\rho_k\}$ are the roots of the polynomial H.

Thus, an important finding in the above literature is that the logarithms of the unstable poles in such linear systems appear in the rate-distortion formulations, an issue which has also been observed in the networked control literature, which we will discuss further below. It is important to emphasize that the underlying coding schemes are non-causal, that is the encoder has access to the entire ensemble before the encoding begins, or the coding is a sliding-block/sliding-window scheme with a finite degree of non-causality.

In contrast with information theory, due to sensitivity to delay, the control theory literature has mainly considered causal/zero-delay coding for unstable (or non-stationary) sources, in the context of networked control systems. For specific settings, in [782] and Baillieul [55], and for more general contexts, in [707], [706]), and [552], lower bounds needed for stabilization over communication channels under various assumptions on the system noise and channels have been obtained; such results are often referred to as *data-rate theorems*. Such theorems state that for stabilizability under information constraints, in the mean-square sense, an average rate per time stage needed for stabilizability has to be at least the sum of the logarithms of the unstable poles/eigenvalues in the system, that is:

$$R \geq \sum_{k=1}^{m} \frac{1}{2} \max\left(0, \log(|\rho_k|^2)\right)$$

This result could be compared with (12.8). In particular, together with the information theoretic bound (12.8), these suggest that the rate requirement is not due to causality, but due to the intrinsic evolution of the (differential) entropy in an unstable system. A more general result will be presented in the next section, in Theorem 12.5.1. Further extensions of these results will be provided in the following two chapters.

More precisely, in the next section, we will provide a derivation of the result above. In Chap. 13, we will show, using stochastic stabilization arguments, that the rate bound is tight for a large class of channels, and investigate further criteria such as stationarity, ergodicity and existence of finite moments, under a variety of system models. A non-linear system generalization will be presented in Chap. 14.

12.4.3 Adaptive vs. Non-adaptive Coding, Uniform Distribution Approximation and Design Considerations

Related to the discussion in this section, we note that for a large class of open-loop unstable sources one also needs to enforce adaptive encoders (see Theorem 13.2.1), whereas for the design of coders for stable processes such an adaptation is not necessary for stabilization. This distinction rules out, for example, the approximation of quantization errors for linear systems as additive noise models (where the quantization error is modeled as orthogonal or independent of the source); which is a common design tool for communications engineering (justified e.g. via dithering as reviewed in Sect. 16.3) but not suitable for control theoretic applications. In the next chapters, we present stochastic stabilization via adaptive coding and we establish their rate optimality.

Adaptation, however, requires the encoder to have access to what the decoder knows; in the absence of channel feedback or indirect noiseless acknowledgement feedback from the control actions, adaptation will not be feasible. In this case, when faced with unbounded noise, one needs to consider the entire space space to be encoded [823]. When the channel is a continuous-channel, this can be done with no difficulty (e.g. when the channel is Gaussian or erasure channel modeled as a packet-loss channel, as in [652] and [367]).

We present a detailed discussion on quantization methods, and their information theoretic relaxations in Sect. 16.3.

12.5 A Fundamental Bound on Information Rates for Real-Time Stabilization over Noiseless Channels

As a special case of the model considered earlier in (12.2)–(12.3), we consider here a multi-dimensional linear system connected over a noiseless channel:

$$x_{t+1} = Ax_t + Bu_t + w_t, \quad y_t = x_t, \tag{12.9}$$

where $x_t \in \mathbb{R}^n$ is the state at time t, u_t is the control input, and $\{w_t\}$ is a sequence of i.i.d. $\mathbb{R}^n$-valued zero-mean second-order random variables. Here A is the system matrix with at least one eigenvalue greater than 1 in magnitude, that is, the system is open-loop unstable. We assume that (A, B) is a stabilizable pair.

Let the quantizer, as described earlier, map its information to a finite set $\mathcal{M}_t$ for $t \geq 0$. We define a quantizer, a quantization policy and a controller policy accordingly (see Fig. 12.5).

Theorem 12.5.1 *For the problem formulated above, with x_0 a second-order random variable, let $R_{avg}(T) = \frac{1}{T}\sum_{t=0}^{T-1} R_t$, be the per-stage average quantization rate at time $T-1$, where R_t is the quantization rate at time t (thus, encoding is possibly variable-rate). For this setup, under any stabilizing control and causal quantization policy, for either*

$$lim\ sup_{t\to\infty} \frac{1}{t} \log(E[|x_t - E[x_t|q_{[0,t]}]|^2]) \leq 0,$$

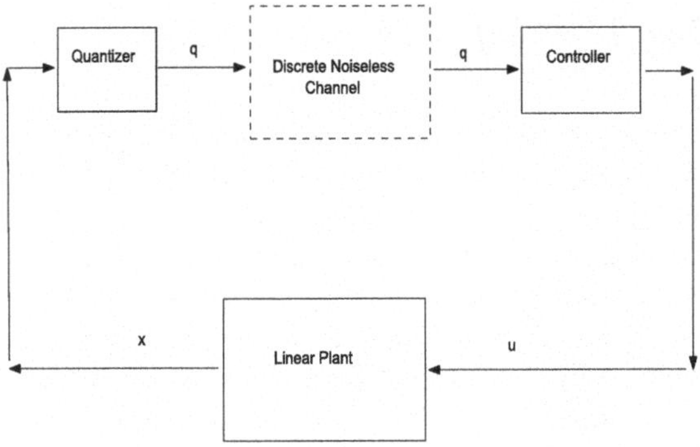

Fig. 12.5 Control over a finite-rate noiseless channel with quantized observations at the controller

or

$$\lim_{t\to\infty}\sup \frac{1}{t}\log(E[|x_t|^2]) \leq 0,$$

to hold, the per-stage average quantization rate must satisfy the inequality

$$\lim_{T\to\infty}\inf R_{avg}(T) \geq \sum_{|\lambda_i|>1} \log_2(|\lambda_i|),$$

where $\{\lambda_i, 1 \leq i \leq n\}$ *are the eigenvalues of* A*, with each repeated eigenvalue (if any) taken as a distinct element in the sequence.*

Proof See section "Proof of Theorem 12.5.1" in Appendix. ◇

A further general result is the following.

Theorem 12.5.2 *With* x_0 *a second-order random variable, let* $\lim_{T\to\infty} R_{avg}(T) = C$*. For the controlled state process* $\{x_t\}$ *to be asymptotically mean stationary (AMS) (see Definition D.3.5 in Appendix D), it must be that* $C \geq \sum_{|\lambda_i|>1} \log_2(|\lambda_i|)$.

The proof of this theorem is almost identical to that of Theorem 14.2.3, and is therefore omitted.

Remark 12.5.1 A refinement of the theorems above will be discussed in the context of stabilization over a general class of noisy communication channels in Chap. 13 (see e.g. Theorem 13.6.2 and to more general system architectures). These will be generalized to non-linear systems in Chap. 14 (see Theorems 14.2.3, 14.3.2, and 14.3.3 for the AMS criterion; and under an asymptotic ergodicity criterion, see Theorems 14.2.5, 14.4.1, 14.4.2 and 14.4.3). Furthermore, sufficiency results and constructive stabilizing coding and control policies will also be presented in the following chapters.

12.6 Appendix: Proofs

12.6.1 Proof of Theorem 12.5.1

The proof follows from [811] and [822]. The matrix A can always be *block-diagonalized* with two blocks, where the first block has only stable eigenvalues, and the second one unstable eigenvalues. For the stable modes, one need not use the channel, and hence for the remaining discussion and analysis we can assume, without any loss of generality, that A has only unstable eigenvalues.

Let the expected square of the Euclidean norm of the random vector x_t be denoted by D_t, and the covariance matrix of the same be denoted by C_t. Thus, $D_t = \text{trace}(C_t)$. We also note that among random vectors with a fixed covariance matrix, the differential entropy is maximized by a jointly Gaussian distribution, which in turn has a finite entropy [188]. Since entropy serves as a lower bound on the average rate (see Proposition 12.3.1), the average per-stage quantization rate satisfies the following for $T \in \mathbb{N}$:

$$
\begin{aligned}
T R_{avg}(T) &\geq H(q_{[0,T-1]}) = \sum_{t=1}^{T-1} H(q_t | q_{[0,t-1]}) + H(q_0) \\
&\geq \sum_{t=1}^{T-1} \Big(H(q_t | q_{[0,t-1]}) - H(q_t | x_t, q_{[0,t-1]}) \Big) + H(q_0) \\
&= \sum_{t=1}^{T-1} I(x_t; q_t | q_{[0,t-1]}) + H(q_0) \\
&= \sum_{t=1}^{T-1} \Big(h(x_t | q_{[0,t-1]}) - h(x_t | q_{[0,t]}) \Big) + H(q_0) \\
&= \sum_{t=1}^{T-1} \Big(h(Ax_{t-1} + w_{t-1} + Bu_{t-1} | q_{[0,t-1]}) - h(x_t | q_{[0,t]}) \Big) + H(q_0) \\
&= \sum_{t=1}^{T-1} \Big(h(Ax_{t-1} + w_{t-1} | q_{[0,t-1]}) - h(x_t | q_{[0,t]}) \Big) + H(q_0) \\
&\geq \sum_{t=1}^{T-1} \Big(h(Ax_{t-1} + w_{t-1} | q_{[0,t-1]}, w_{t-1}) - h(x_t | q_{[0,t]}) \Big) + H(q_0) \\
&= \sum_{t=1}^{T-1} \Big(h(Ax_{t-1} | q_{[0,t-1]}, w_{t-1}) - h(x_t | q_{[0,t]}) \Big) + H(q_0) \\
&= \sum_{t=1}^{T-1} \Big(h(Ax_{t-1} | q_{[0,t-1]}) - h(x_t | q_{[0,t]}) \Big) + H(q_0) \qquad (12.10) \\
&= \sum_{t=1}^{T-1} \Big(\log_2(|\det(A)|) + h(x_{t-1} | q_{[0,t-1]}) - h(x_t | q_{[0,t]}) \Big) + H(q_0) \\
&= \Big(\sum_{t=1}^{T-1} \log_2(|\det(A)|) \Big) + h(x_0 | q_0) - h(x_{T-1} | q_{[0,T-1]}) + H(q_0) \\
&\geq \Big(\sum_{t=1}^{T-1} \log_2(|\det(A)|) \Big) + h(x_0 | q_0) - h(x_{T-1}) + H(q_0) \qquad (12.11)
\end{aligned}
$$

$$\geq \left(\sum_{t=1}^{T-1} \log_2(|\det(A)|)\right) + h(x_0|q_0) - \frac{1}{2}\log\left((2\pi e)^n \det(C_{T-1})\right) + H(q_0)$$

$$\geq \left(\sum_{t=1}^{T-1} \log_2(|\det(A)|)\right) + h(x_0|q_0) - \frac{1}{2}\log\left((2\pi e)^n (\frac{1}{n} D_{T-1})^n\right) + H(q_0) \tag{12.12}$$

The first inequality follows since discrete entropy is always non-negative, and the second and third inequalities follow from the fact that conditioning does not increase entropy (see [188]). The fourth inequality follows from the property that the Gaussian measure maximizes the entropy for a given covariance matrix. Equation (12.10) follows from the assumption that $\{w_t\}$ is an i.i.d. process, and (12.11) uses the fact that conditioning does not increase entropy (this last step could be skipped if the goal was to only show the subexponential growth of the estimation error in the sense that $\limsup_{t\to\infty} \frac{1}{t}\log(E[|x_t - E[x_t|q_{[0,t]}]|^2]) \leq 0$). In (12.12) we use $\det(C_{T-1}) \leq (\frac{1}{n}\mathrm{trace}(C_{T-1}))^n$ (which is a consequence of *the inequality of arithmetic and geometric means* applied to eigenvalues). Sinc $\limsup_{t\to\infty} \frac{1}{t}\log(D_t) \leq 0$, and $h(x_0) < \infty$ we obtain

$$\lim \inf_{T\to\infty} R_{avg}(T) \geq \sum_{|\lambda_i|>1} \log_2(|\lambda_i|).$$

◇

12.7 Concluding Remarks

In this chapter, information structures for encoders and controllers have been introduced, with the setup presented in this chapter to be used throughout the book. In the following chapters of *Part III* of the book, stochastic stabilization under such measurability and information rate constraints will be covered. In *Part IV* of the book, optimization of coding and control policies will be studied in detail, under such information constraints.

In the chapter, the connections between real-time coding and information theoretic approaches have been investigated and subtle differences have been highlighted. Further, properties of quantizers, and their performance have been studied. The chapter has also been concerned with the derivation of fundamental bounds in connection with stabilizability of a linear system over a communication channel.

12.8 Bibliographic Notes

A comprehensive tutorial on quantization and source coding theory can be found in [320].

In the information theory literature, stochastic stability results were established mostly for stationary sources, which are already in some appropriate sense stable sources. In this literature, the stability of the estimation errors as well as the encoder state processes have been studied. These systems mainly involve causal and non-causal coding (block coding, as well as sliding-block coding) of stationary sources [421], [309], [419], and asymptotically mean stationary sources [268]. Real-time settings such as sigma-delta quantization schemes have also been considered in the literature; see for example [783] among others. We refer the reader to [318] regarding rate-distortion results for such non-stationary processes and on the methods used in [313] and [339]. Berger [109] obtained the rate-distortion function for Wiener processes, and in addition, developed a two-part coding scheme, which was later generalized in [618] and [619] to unstable Markov processes driven by bounded noise. The scheme in [109] exploits the independent increment property of Wiener processes. References [809], [808] obtained ergodicity results for open loop unstable systems controlled over noisy or noiseless channels.

Uniform scalar quantizers, in addition to their simplicity, are also approximately optimal in the limit of high-rate distortion [303] with respect to the mean-square distortion measure [822]. [303] has shown that the uniform quantizer followed by an entropy coder is at most 0.255 bits worse than any optimal quantizer, if one exists, in the context of entropy constrained causal quantization, a situation which is also applicable in linear control systems.

The fundamental lower bound in Theorem 12.5.1 is a generalization of the results established in [552] and [707], building on [811] and [822]. Part of this chapter follows from [810] and [822].

Chapter 13
Stochastic Stabilization of Linear Systems Under Information Constraints

This chapter and the companion Chap. 14 address the problem of stabilization of linear and non-linear stochastic control systems under information constraints, where the constraints manifest themselves via the presence of communication channels, with finite-capacity, connecting the plant to be controlled with the controllers. A general setup is presented in Fig. 14.1.

13.1 Introduction

In this chapter, we present necessary and sufficient conditions for stochastic stabilization of networked control systems over finite-capacity channels (Fig. 13.1).

The system to be considered in this chapter will be linear. In this context, as before, we will refer by a *Coding Policy* Π, a sequence of functions $\{\gamma_t^e, t \geq 0\}$ which are causal such that the channel input at time t, $q_t \in \mathcal{M}$, under Π^{comp}, is generated by a function of its local information, that is,

$$q_t = \gamma_t^e(\mathcal{I}_t^e),$$

where $\mathcal{I}_t^e = \{x_{[0,t]}, q'_{[0,t-1]}\}$ and $q_t \in \mathcal{M}$, the channel input alphabet given by $\mathcal{M} := \{1, 2, \ldots, M\}$, for $0 \leq t \leq T-1$.

The channel maps q_t to q'_t in a stochastic fashion so that $P(q'_t|q_t, q_{[0,t-1]}, q'_{[0,t-1]})$ is a conditional probability measure on $\mathcal{M}'$ for all $t \in \mathbb{Z}_+$. If this expression is equal to $P(q'_t|q_t)$, the channel is said to be a memoryless channel, that is, the past variables do not affect the channel output q'_t given the current channel input q_t.

The receiver/controller, upon receiving the information from the channel, generates its decision at time t, also causally: An admissible causal controller policy is a sequence of functions $\gamma = \{\gamma_t\}$ such that

S. Yüksel, T. Başar, *Stochastic Teams, Games, and Control under Information Constraints*, Systems & Control: Foundations & Applications,
https://doi.org/10.1007/978-3-031-54071-4_13

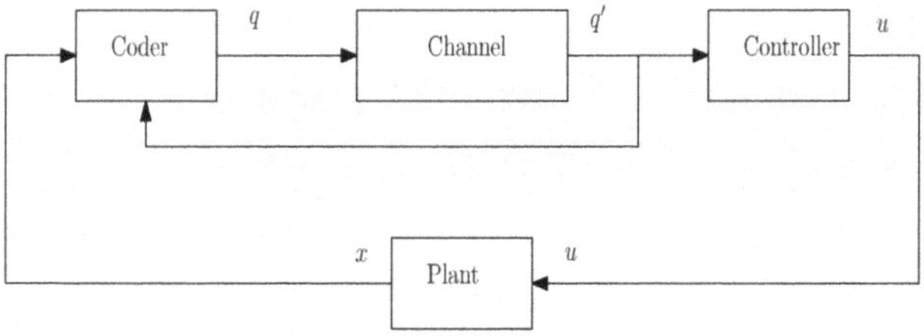

Fig. 13.1 Control of a system over a noisy channel

$$\gamma_t : \mathcal{M}'^{t+1} \to \mathbb{R}^m, \qquad t \geq 0,$$

so that $u_t = \gamma_t(q'_{[0,t]})$.

We will consider linear systems within both fully-observed and partially observed settings (see Fig. 12.3 for a model more general than the one depicted above), and obtain conditions for stochastic stabilizability. In such systems, there is an information rate constraint on the quantizers. We use the random-time state-dependent drift criteria introduced in Appendix G to obtain stabilizing schemes. The rate requirements under such an approach meet the fundamental lower bound given in Theorem 12.5.1.

We then consider stochastic stabilization of linear systems over various types of noisy channels: these are erasure channels, general discrete channels (memoryless as well as with memory) with noiseless feedback, and a class of discrete and continuous channels without any feedback (Gaussian channels will be considered in detail in Chap. 16). We will present strong forms of stability and ergodicity, which will come at the expense of further complexities in the design and analysis when compared with noiseless channels.

In particular, the chapter identifies conditions on the channels for which there exist coding and control policies such that a controlled linear system (driven possibly by unbounded noise) with state x_t is stochastically stable in one or more of the following senses:

The closed-loop state $\{x_t\}$ is asymptotically stationary or asymptotically mean stationary (AMS) (see Definition D.3.5, and Appendix D).

The state $\{x_t\}$ and the coding and control parameters lead to an ergodic process (and even a positive Harris recurrent Markov chain).

$\lim_{T\to\infty} \frac{1}{T} \sum_{t=0}^{T-1} |x_t|^2$ exists and is finite almost surely (this will be referred to as *quadratic stability*). This can be generalized to moment stability so that for $m \in \mathbb{N}$, $\lim_{T\to\infty} \frac{1}{T} E[\sum_{t=0}^{T-1} |x_t|^m] < \infty$.

For the above criteria:

(i) We will obtain a tight converse theorem on stabilizability in the AMS sense over a large class of noisy channels. As a main result, we will see that the Shannon capacity provides the appropriate measure on whether a system can be stabilized or not under arbitrary admissible coding and control policies under this criterion.

(ii) We establish constructive proofs for various channels; where via martingale and Markov chain methods, tightness of Shannon capacity is established for a large class of channels. For the noiseless, finite-rate, channel setting, we present a comprehensive analysis involving also partially observed models, multi-dimensional models, and multi-sensor models. Following noiseless channels, our stabilization analysis involves the erasure channel, which is perhaps the simplest non-trivial noisy channel. This channel is also important in many engineering applications, since it is used as a common model, such as in control over communication networks arising in many industrial systems. Our treatment of the erasure channel allows us to obtain stronger results regarding stochastic stability of linear systems when compared with general noisy channels.

We will see that the Shannon capacity of the erasure channel is an almost sufficient condition for asymptotic stationarity as well as for the positive Harris recurrence of the state and the quantizer process. For quadratic (finite second moment) stabilization, however, the conditions on the channel are more stringent. Similar conclusions will be arrived at in the context of more general discrete memoryless channels (DMCs) with feedback in that the Shannon capacity presents a tight boundary between necessity and sufficiency when the goal is the AMS property or the presence of an invariant probability measure for the closed loop process.

(iii) We will arrive at lower bounds via information theoretic methods and stochastic analysis. We note that, in Chap. 14, while generalizing the results in this chapter to non-linear systems, we will also present an alternative lower bounding method via stochastic geometric growth analysis, which is also applicable to linear systems. However, since the information theoretic method is slightly more general in the linear setup, we will only present this approach in this chapter.

(iv) For sufficiency, we will present adaptive coding and establish the optimality via a random-time state-dependent stochastic drift analysis of Appendix G. We will also show, however, that in the absence of feedback the design problem is particularly more challenging (unless the channel is a continuous one, such as a Gaussian channel (which is studied in further detail in Chap. 16)) since the entire state space is to be encoded (the absence of which leads to transience) via a possibly time-invariant, non-adaptive design. We also present a review of existing design methods in the literature, both in this chapter as well as in Sect. 16.3.

The treatment in this chapter starts with Sect. 13.2 which lays out the setting. Section 13.2.1 establishes stochastic stability for a scalar setting, Sect. 13.2.2 considers multi-dimensional systems, and Sect. 13.2.3 investigates partially observed settings. Section 13.2.4 studies the multi-sensor (multi-terminal) setting.

Then, we move on to noisy channels: we study stabilization over channels with noiseless feedback in Sect. 13.3, where we obtain converse theorems on stabilizability over a noisy channel. We then consider stabilization over erasure channels in Sect. 13.4, and later more general DMCs in Sect. 13.5, and establish ergodicity and quadratic stability results using the theory developed in Appendix G. Channels with memory and multi-dimensional systems are investigated in Sect. 13.6. Section 13.7 investigates the case with noisy reverse channels without any feedback. An appendix, constituting Sect. 13.8, includes proofs of some of the main results.

13.2 The Case of Discrete Noiseless Channels

In this section, we consider stabilization of linear noisy systems described by the dynamics

$$x_{t+1} = Ax_t + Bu_t + w_t, \tag{13.1}$$

where $x_t \in \mathbb{R}^n$, (A, B) is controllable, and the noise process $\{w_t\}$ is a zero-mean i.i.d. sequence of random vectors, with stabilization to be carried out over noiseless channels with finite capacity. We assume also that x_0 is a second-order random variable.

The conditions we will impose on the zero-mean i.i.d. noise sequence will prove to be important in the goals to be achieved. The noise sequence is required to satisfy one of the following assumptions:

Assumption 13.2.1 *Each w_t admits a probability measure ν which is absolutely continuous with respect to the Lebesgue measure λ on $\mathbb{R}^n$ (that is, the measure admits a density), and for every $D \in \mathcal{B}(\mathbb{R}^n)$ with positive Lebesgue measure, $\nu(D) > 0$. Furthermore, $E[|w_t|^2] < \infty$.*

The second assumption is slightly stronger.

Assumption 13.2.2 *Each w_t has a probability measure ν which admits a density, and for every $D \in \mathcal{B}(\mathbb{R}^n)$ with positive Lebesgue measure, $\nu(D) > 0$. Furthermore, $E[|w_t|^{2+\epsilon}] < \infty$ for some $\epsilon > 0$.*

Finally, we may also take the noise to be Gaussian to arrive at further results (given that a Gaussian random variable has all of its moments finite).

Assumption 13.2.3 *$\{w_t\}$ is a sequence of Gaussian random variables.*

Under Assumption 13.2.1, we will establish stationarity and ergodicity. Under Assumption 13.2.2 (or the stronger Assumption 13.2.3), we will establish the existence of finite moments.

Before discussing the more general case in Sect. 13.2.2, to acquire some insight we first discuss the scalar version described by the following equation:

$$x_{t+1} = ax_t + bu_t + w_t. \tag{13.2}$$

As depicted in Fig. 12.5 and generally described in Sect. 12.2.2, our goal is to stabilize such a system by designing quantization and controller policies, where the quantizers at a given time map the information to a finite set finite set $\mathcal{M}$.

13.2.1 Stochastic Stability Analysis for a Scalar System

A uniform quantizer: $Q_K^{\Delta} : \mathbb{R} \to \mathbb{R}$ with step size Δ and $K+1$ (with K even) number of bins satisfies the following for $k = 1, 2 \ldots, K$ (see Fig. 12.1):

$$Q_K^{\Delta}(x) = \begin{cases} (k - \frac{1}{2}(K+1))\Delta, & \text{if } x \in [(k-1-\frac{1}{2}K)\Delta, (k-\frac{1}{2}K)\Delta) \\ (\frac{1}{2}(K-1))\Delta, & \text{if } x = \frac{1}{2}K\Delta \\ 0, & \text{if } x \notin [-\frac{1}{2}K\Delta, \frac{1}{2}K\Delta]. \end{cases} \tag{13.3}$$

We refer to the interval $[-(K/2)\Delta, (K/2)\Delta]$ the granular region of the quantizer, and the complement of this set as the overflow region on the quantizer.

Adaptive Quantizers and a Zooming Scheme

A general class of quantization policies involves *adaptive quantizers*: Following the terminology of Chap. 5, such a quantizer policy picks quantizers whose selection involves memory. Use of such quantizers is motivated by the following result whose proof can be found in the appendix.

Theorem 13.2.1 *Consider (13.2), with* $|a| > 1$. *A fixed quantizer leads to a transient Markov chain in the sense that* $P_x(\tau_S < \infty) < 1$ *where for some* $s > 0$, $S = (-\infty, s)$ *is an open set containing the origin,* $x \notin S$ *and* $\tau_S := \inf(t > 0 : x_t \in S)$.

Let $\mathbb{S}$ be a set of states for a quantizer state S. Let $F : \mathbb{S} \times \mathbb{R} \to \mathbb{S}$ be a state update-function. An adaptive quantizer may have the following state update equations: $S_{t+1} = F(Q_t(x_t), S_t)$. Here, Q_t is the quantizer applied at time t, x_t is the input to the quantizer Q_t, and S_t is the *state* of the quantizer. Such a quantizer is implementable since the updates can be performed at both the encoder and the decoder.

A particular class of adaptive quantizers has been introduced by Goodman and Gersho [309]. One type has the following form (see (13.3) and Fig. 12.1) with Q_K^{Δ} being a uniform quantizer with $K+1$ bins and bin-size Δ, and $\hat{Q}$ determining the updates in the bin-size of the uniform quantizer as a function of the source and the current bin size:

$$q_t = Q_K^{\Delta_t}(x_t), \qquad \Delta_{t+1} = \Delta_t \hat{Q}(q_t, \Delta_t) \tag{13.4}$$

Here Δ_t characterizes the uniform quantizer, as it is the bin size of the quantizer at time t. In the model that we will consider, $\hat{Q}$ (which we will replace with the term $\bar{Q}$) will take the ratio between x and Δ, and Δ as its input.

In the analysis, we will consider such adaptive quantizers, whose realizations will be from a class of uniform quantizers, considered earlier in (12.1). Thus, in this setting, we modify the description of a traditional uniform quantizer by assigning the same value when the state is in the overflow region of the quantizer.

The quantizer system is connected over a noiseless channel with a finite capacity to an estimator (controller) as depicted in Fig. 12.5. The controller has only access to the information it has received through the channel. The controller in the model estimates the state and then applies its control action. As such, the problem reduces to a state estimation problem since (13.2) is controllable. Hence, stability of the estimation error is equivalent to stability of the state itself.

Stochastic Stability Analysis

In the following, we will consider a discrete noiseless channel with capacity R (hence with $2^R = |\mathcal{M}| \in \mathbb{Z}_+$).

We will establish the following theorem in this context.

Theorem 13.2.2 *[373, 808]*

(i) Consider an adaptive quantizer applied to the linear control system described by (13.2), under Assumption 13.2.1. If the noiseless channel has capacity,

$$R = \log_2(\lceil |a| + \epsilon \rceil + 1),$$

for some $\epsilon > 0$, then there exists an adaptive quantization policy and a control policy which makes (x_t, Δ_t) stochastically stable in the sense of positive Harris recurrence (in particular, with a unique invariant probability measure).

(ii) If we sample the process (13.2) with a period of T and apply the above result, with a^T replacing a, then we have

$$R_T := \frac{\log_2(\lceil |a^T| + \epsilon \rceil + 1)}{T},$$

and as $\lim_{T\to\infty} R_T = \log_2(|a|)$, we have that $\log_2(|a|)$ is a near tight bound.

(iii) If in addition, we have that Assumption 13.2.2 holds, the above quantization and control scheme with the given rate makes x_t quadratically stable, and, in particular, x_t has a finite second moment under its invariant measure.

(iv) Under Assumption 13.2.3, x_t has all its moments finite under its invariant measure.

In the following, we present the essential arguments leading to Theorem 13.2.2. Note that (ii) follows from (i) via a direct argument [373].

Theorem 13.2.3 ([808]) *Consider an adaptive quantizer applied to the linear control system described by (13.2), under Assumption 13.2.1. If the noiseless channel has capacity,*

$$R = \log_2(\lceil |a| + \epsilon \rceil + 1),$$

for some $\epsilon > 0$, there exists an adaptive quantization policy such that there exists a compact set $S \subset \mathbb{R}^2$ so that with $\tau_S = \min(t > 0 : x_t \in S)$:

$$\sup_{(x,\Delta)\in S} E[\tau_S | x_0 = x] < \infty,$$

which makes S a recurrent set. Furthermore

$$P_{(x,\Delta)}(\tau_S < \infty) = 1$$

for all (x, Δ) pairs visited by the chain.

For a constructive proof, one considers adaptive quantizers for which the quantizer bins are updated as follows. The quantizer bin sizes will get enlarged until the state process hits the granular region of the quantizer, that is, the joint process hits the set $\{(x_t, \Delta_t) : |h_t| \leq 1\}$, where

$$h_t = x_t / (\frac{K}{2} \Delta_t) \tag{13.5}$$

In this case, the quantizer will be said to be in the *perfect-zoom* (or *in-view*) phase. Due to the effect of the system noise, occasionally the state will be in the overflow region of the quantizer, leading to an *under-zoom* phase.

Now, with $K = \lceil |a| + \epsilon \rceil$, $R = \log_2(K + 1)$, let us introduce $R' = \log_2(K)$. We will consider the following update rules. For $t \geq 0$ and with $\Delta_0 > L$ for some $L \in \mathbb{R}_+$, and $\hat{x}_0 \in \mathbb{R}$, consider:

$$u_t = -\frac{a}{b}\hat{x}_t, \quad \hat{x}_t = Q_K^{\Delta_t}(x_t), \quad \Delta_{t+1} = \Delta_t \bar{Q}(|\frac{x_t}{\Delta_t 2^{R'-1}}|, \Delta_t) \tag{13.6}$$

If we use $\delta, \epsilon, \alpha > 0$ with $\alpha < 1$ and $L > 0$ such that,

$$\begin{aligned} \bar{Q}(x, \Delta) &= |a| + \delta \quad \text{if} \quad |x| > 1 \\ \bar{Q}(x, \Delta) &= \alpha \quad \text{if} \quad 0 \leq |x| \leq 1, \Delta \geq L \\ \bar{Q}(x, \Delta) &= 1 \quad \text{if} \quad 0 \leq |x| \leq 1, \Delta < L, \end{aligned} \tag{13.7}$$

we will show that a recurrent set exists. We note that the above imply that $\Delta_t \geq L\alpha =: L'$ for all $t \geq 0$.

The stability for such a scheme can be established using random-time stochastic drift conditions. This is because the quantizer helps reduce the uncertainty on the system state only when the state is in the *granular* region of the quantizer. The times when the state is in this region are random stopping times (defined on the filtration generated by the state and quantizer processes).

Theorem 13.2.4 ([808]) *Under the setup of Theorem 13.2.3, for the adaptive quantizer in (13.6), if the quantizer bin sizes are such that their (base−2) logarithms are integer multiples of some scalar s, and* $\log_2(\bar{Q}(\cdot,\cdot))$ *take values in integer multiples of s where the integers taken are relatively prime (that is they share no common divisors except for 1), then the process* $\{(x_t, \Delta_t)\}$ *is a positive (Harris) recurrent Markov chain (and has a unique invariant distribution).*

The relative prime condition, which is a mild technical condition, allows the state space for the quantizer bins to be countable, leading to irreducibility and consequently positive Harris recurrence. The proof is given in the appendix.

The following results are on moment stability, whose proofs can be found in the appendix. Note that more restrictive assumptions (Assumptions 13.2.2 and 13.2.3) are imposed on the tail distribution of the i.i.d. noise process $\{w_t\}$.

Theorem 13.2.5 ([808]) *Under the setups of Theorems 13.2.3, 13.2.4, and Assumption 13.2.3, it follows that* $\lim_{t\to\infty} E[x_t^2] < \infty$, *and this limit is independent of the initial state of the system. Furthermore,* $\lim_{t\to\infty} E[\Delta_t^2] < \infty$.

This result also holds under Assumption 13.2.2 with essentially the same arguments as in the proof of Theorem 13.2.5.

Theorem 13.2.6 ([373]) *Under the setups of Theorems 13.2.3, 13.2.4, and Assumption 13.2.2, it follows that* $\lim_{t\to\infty} E[x_t^2] < \infty$, *and this limit is independent of the initial state of the system. Furthermore,* $\lim_{t\to\infty} E[\Delta_t^2] < \infty$.

We also have the following.

Theorem 13.2.7 *Let the assumptions of Theorem 13.2.4 and Assumption 13.2.3 hold. In addition, assume that*

$$|a|^m \left(\frac{1}{(2^R - 1)^m}\right) < 1,$$

for some $m \in \mathbb{N}$. *Then, with the adaptive quantization policy (13.7) and initial condition* (x_0, Δ_0), *we have*

$$\lim_{t\to\infty} \mathbb{E}[|x_t|^m] = \mathbb{E}_\pi[|x_0|^m] < \infty .$$

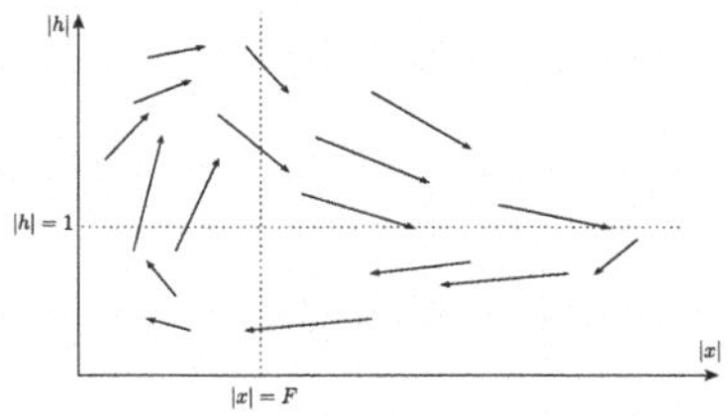

Fig. 13.2 Drift in the Markov Process. When under-zoomed, the error increases on average and the quantizer *zooms out*; when perfectly-zoomed (i.e., in-view), the error decreases and the quantizer *zooms in*

Application of the Theory of Random-Time State-Dependent Stochastic Drift

We now make explicit the connection with the general theory for random-time stochastic drift introduced in Appendix G. The Markov chain considered is (x_t, Δ_t) (see Lemma 13.8.1).

Figure 13.2 provides some intuition on the construction of stopping times and the Lyapunov functions, where $h_t = x_t/(2^{R'-1}\Delta_t)$ was introduced in (13.5). The arrows shown in the figure denote the mean one-step increments of (x_t, h_t). That is, the arrow ν with base at (x, h) is defined by

$$\nu = \mathbb{E}[(x_{t+1}, h_{t+1}) - (x_t, h_t) \mid (x_t, h_t) = (x, h)]$$

With $F > 0$ fixed, and with $F' = F2^{-(R'-1)}$, two sets are used to define the small set in the drift criteria: $C_x = \{x : |x| \leq F\}$ for $F > 0$, and $C_\Delta = \{\Delta : \Delta \leq F'\}$. Let $C_h = \{h : |h| \leq 1\}$, and assume that $F > 0$ is chosen sufficiently large so that $(x_t, \Delta_t) \in C_x \times C_\Delta$ whenever $(x_t, h_t) \in C_x \times C_h$. When x_t is outside C_x and h_t outside C_h (the *under-zoomed phase* of the quantizer), there is a drift of h_t towards C_h. When the process x_t reaches C_h (the *perfectly-zoomed (or in-view) phase* of the quantizer), then the process drifts towards C_x.

In the model considered, the controller can receive meaningful information regarding the state of the system when the source lies in the granular region of the quantizer, that is, $x_t \in [-\frac{1}{2}K\Delta_t, \frac{1}{2}K\Delta_t]$.

The times at which these events occur form an increasing sequence of *stopping times*. We will apply the drift criteria presented in Sect. G.1.3 for these random stopping times. In particular, we define the sequence of stopping times as:

$$\tau_0 = 0, \quad \tau_{z+1} = \inf\{k > \tau_z : |h_k| \leq 1\}, \quad z \in \mathbb{N}. \tag{13.8}$$

These are the times when information reaches the controller regarding the value of the state when the state is in the granular region of the quantizer.

Using this construction of stopping times, we will find appropriate Lyapunov functions (such as $V(x, \Delta) = \log(\Delta^2)$ and $V(x, \Delta) = \Delta^2$), and drift functions

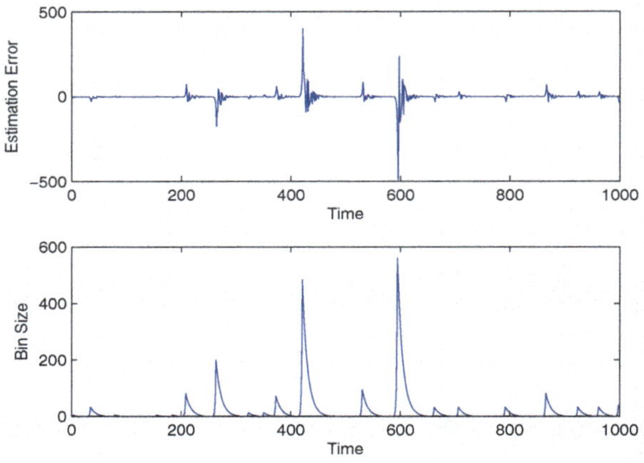

Fig. 13.3 Sample path for a stochastically stable quantizer. The variables picked are as follows: $L' = 1, \epsilon = 0.3, \delta = 0.25, \eta = 0.02$

which will lead to stationarity as well as finite moments for the Markov process $\{(x_t, h_t)\}$ through the satisfaction of special instances of Theorem G.1.4. See the appendix for details.

Simulation For a simulation study, we consider a linear system with the following dynamics:

$$x_{t+1} = 2.2x_t + u_t + w_t,$$

where $\{w_t\}$ are i.i.d., zero-mean Gaussian with variance 1. We use the zooming quantizer with rate $\log_2(4) = 2$, since 4 is the smallest integer larger than or equal to $\lceil 2.2 \rceil + 1$. We have taken $L' = 1$. Figure 13.3 corroborates the stochastic stability result, by showing the under-zoomed and perfectly-zoomed phases, with the peaks in the plots showing the under-zoom phases.

13.2.2 *The Multi-Dimensional Case*

The scheme introduced in the previous section is also applicable to the multi-dimensional setup. Let us revisit (13.1), where $x_t \in \mathbb{R}^n$ is the state at time t, $u_t \in \mathbb{R}^m$ is the control action, and $\{w_t\}$ is a sequence of i.i.d. $\mathbb{R}^n$-valued zero-mean random variables satisfying Assumption 13.2.2. Here A is the system matrix with at least one eigenvalue greater than 1 in magnitude, that is, the system is open-loop unstable. We also assume at this point that the eigenvalues are real.

Without any loss of generality, we assume A to be in Jordan form. Because of this, we allow w_t to have correlated components, that is the correlation matrix $E[w_t w_t^T]$ is not necessarily diagonal. We may also assume that (A, B) is controllable, and for ease in presentation we assume that B is invertible mainly to make the stopping time analysis easier to pursue (if B is not invertible, the sampled system can be made to have an invertible control matrix, with a periodic scheme of period at most n).

Stabilizability for the diagonalizable case immediately follows from the discussion for scalar systems, since the analysis for the scalar case is applicable to each of the subsystems along the eigenvectors. The possibly correlated noise components will lead to the recurrence analysis discussed earlier. For such a setup, the stopping times can be arranged to be identical for each mode, for the case when the quantizer captures all the state components. Once this is satisfied, the drift conditions will be obtained. The non-diagonalizable Jordan case, however, is more involved. The approach for the scalar system still applies, but it needs to be appropriately generalized.

We now make this discussion more precise. One can consider two approaches for stabilization of a multi-dimensional system. We first discuss one approach which we will refer to as the *block-coding approach*.

Consider the following system

$$\begin{bmatrix} x^1_{t+1} \\ x^2_{t+1} \end{bmatrix} = \begin{bmatrix} \lambda & 1 \\ 0 & \lambda \end{bmatrix} \begin{bmatrix} x^1_t \\ x^2_t \end{bmatrix} + B \begin{bmatrix} u^1_t \\ u^2_t \end{bmatrix} + \begin{bmatrix} w^1_t \\ w^2_t \end{bmatrix}. \tag{13.9}$$

The approach entails quantizing the components in the system according to the adaptive quantization rule provided earlier, that is, we modify the scheme in (13.6) as follows: For $i = 1, 2$, let $R' = R'_i = \log_2(2^{R_i} - 1) = \log_2(K_i)$ (that is, the same rate is used for quantizing the components with the same eigenvalue). For $t \geq 0$ and with $\Delta^1_0, \Delta^2_0 \in \mathbb{R}$, consider:

$$u_t = -B^{-1} A \hat{x}_t,$$

$$\begin{bmatrix} \hat{x}^1_t \\ \hat{x}^2_t \end{bmatrix} = \begin{bmatrix} Q^{\Delta^1_t}_{K_1}(x^1_t) \\ Q^{\Delta^2_t}_{K_2}(x^2_t) \end{bmatrix}, \tag{13.10}$$

$$\Delta^1_{t+1} = \Delta^1_t \bar{Q}(|h^1_t|, |h^2_t|, \Delta^1_t), \qquad \Delta^2_{t+1} = \Delta^2_t \bar{Q}(|h^1_t|, |h^2_t|, \Delta^2_t), \tag{13.11}$$

with, for $i = 1, 2$, $\delta^i, \epsilon^i, \eta^i > 0$, $\eta^i < \epsilon^i$ and $L^i > 0$ such that

$$\bar{Q}(x, y, \Delta) = |\lambda| + \delta^i \quad \text{if} \quad |x| > 1, \quad \text{or} \quad |y| > 1$$

$$\bar{Q}(x, y, \Delta) = \frac{|\lambda|}{2^{R'_i} - \eta^i} \quad \text{if} \quad 0 \leq |x| \leq 1, |y| \leq 1, \Delta^i > L^i$$

$$\bar{Q}(x, y, \Delta) = 1 \quad \text{if} \quad 0 \leq |x| \leq 1, |y| \leq 1, \Delta^i \leq L^i$$

Note that, the above imply that $\Delta^i_t \geq L^i \frac{|\lambda|}{2^{R'_i} - \eta^i} =: L'^i$. We also assume that for some sufficiently large η_Δ, $\Delta^1_0 = \eta_\Delta \Delta^2_0$, which leads to the result that $\Delta^1_t = \eta_\Delta \Delta^2_t$ for all $t \geq 0$. See Fig. 13.4 for a depiction of the quantizer used at a particular time.

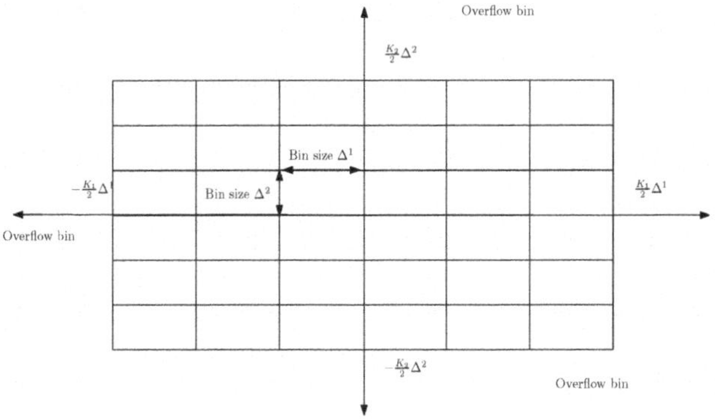

Fig. 13.4 A uniform vector quantizer. There is a single overflow bin

Instead of (13.8), the sequence of stopping times is now defined as follows:

$$\tau_0 = 0, \quad \tau_{z+1} = \inf\{k > \tau_z : |h_k^i| \leq 1, i \in \{1, 2, \ldots, n\}\}, \quad z \in \mathbb{Z}_+,$$

where $h_t^i = \frac{x_t^i}{\Delta_t^i 2^{R_i'-1}}$. Here Δ^i is the bin size of the quantizer in the direction of the eigenvector x^i, with rate R_i'.

With this approach, the drift criterion applies almost identically as it does for the scalar case. The extensions of Theorems 13.2.3, 13.2.4 and 13.2.5 are then immediate.

We thus have the following result. The proof for a single Jordan block with real eigenvalues is in the appendix. The extension to multiple Jordan blocks as well as systems having complex eigenvalues in the real Jordan canonical form follows from similar arguments by coupling the modes having complex conjugate pairs in a single vector quantizer through the establishment of a geometric measure bounding (majorizing) the probability measure of a subsequent stopping time.

Theorem 13.2.8 ([373]) *Consider the multi-dimensional system (13.1). If the average rate satisfies*

$$R > \sum_{|\lambda_i|>1} \log_2(|\lambda_i|),$$

there exists a stabilizing scheme leading to a Markov chain with a bounded second moment in the sense that lim sup$_{t\to\infty} E[|x_t|_2^2] < \infty$.

Proof See Sect. 13.8.4. ◇

Remark 13.2.1 In an alternative scheme, one could consider an approach that we call *sequential stabilization of the scalar components*: We could adopt a lower to upper sequential approach, considering stabilized modes at particular stopping times as noise with a finite moment. Using an inductive argument, one can first start with the lowest mode (in the matrix diagonal) of the system, and stabilize that mode so that there is a finite invariant second moment of the state. We can then view this mode as a second-order disturbance for the one upper mode and obtain conditions for stabilizability for this mode, by employing the same approach, since the lower mode is a positive Harris recurrent Markov process. We note that the random process for the upper mode might not have Markov dynamics for its marginal, but the joint system consisting of all modes and quantizer parameters will be Markov. Furthermore, the lower modes, viewed as noise, will no longer be independent from the state process. For example, the effective disturbance affecting the stochastic evolution of a repeated mode is no longer Gaussian. However appropriate moment bounds (such as Jensen's and Hölder's inequality) can be applied to carry out the analysis. Such a sequential stabilization approach will be adopted for noise-free multi-sensor systems in Sect. 13.2.4, and for Gaussian channels in Chap. 16. ◇

13.2.3 The Partially Observed Case

The results in the previous sections have direct counterparts in the partially observed case, where along with (13.1), we have the measurement equation

$$y_t = Cx_t + v_t. \tag{13.12}$$

Here, (A, C) is observable and the noise terms $\{w_t\}, \{v_t\}$ are i.i.d, mutually independent, and Gaussian. The encoder can run an observer, which is given by a Kalman filter. In Appendix E.4 we discuss the reduction to a fully observed model in further detail.

Suppose that $E[w_t w_t^T] = W > 0$ and $E[v_t v_t^T] = V > 0$. We let

$$m_t := E[x_t | y_{[0,t-1]}, u_{[0,t-1]}]$$

$$\Sigma_{t|t-1} := E[(x_t - m_t)(x_t - m_t)^T | y_{[0,t-1]}, u_{[0,t-1]}]$$

Introduce $\tilde{m}_t = E[x_t | y_{[0,t]}, u_{[0,t-1]}]$. In Appendix E.4, we show that

$$\tilde{m}_t = m_t + \Sigma_{t|t-1} C^T (C\Sigma_{t|t-1} C^T + V)^{-1} (y_t - Cm_t),$$

or

$$\begin{aligned}\tilde{m}_t &= A\tilde{m}_{t-1} \\ &\quad + \Sigma_{t|t-1} C^T (C\Sigma_{t|t-1} C^T + V)^{-1} (CA(x_{t-1} - \tilde{m}_{t-1}) + v_t) \\ \Sigma_{t+1|t} & \\ &= A\Sigma_{t|t-1} A^T + W - (A\Sigma_{t|t-1} C^T)(C\Sigma_{t|t-1} C^T + V)^{-1} (C\Sigma_{t|t-1} A^T)\end{aligned} \tag{13.13}$$

with $\Sigma_{0|-1} = E[x_0 x_0']$. The zero-mean variable $x_t - \tilde{m}_t$ is orthogonal to $\mathcal{I}_t = \{y_{[0,t]}, u_{[0,t-1]}\}$, in the sense that the error is independent of the information available at the controller: the information available at the encoder with regard to the state is Gaussian for all time stages, consequently independence and orthogonality of $x_t - \tilde{m}_t$ and $\mathcal{I}_t$ are equivalent.

Then, $\tilde{m}_t$ is a fully observed system driven by an independent (but not identical) Gaussian noise process. Asymptotically, however, the independent noise process converges (in total variation) to a stationary distribution since $\Sigma_{t|t-1}$ converges to a constant matrix under the setup considered. The stability analysis for (13.1) can then be applied for (13.13) given the uniform bounds on the noise variances, in view of the convergence argument. Hence, the state process hits a compact set infinitely often, and the state moments are uniformly bounded. As a result, we have the following.

Theorem 13.2.9 ([373]) *Consider the partially observed multi-dimensional system (13.1) and (13.12). If the average rate satisfies*

$$R > \sum_{|\lambda_i|>1} \log_2(|\lambda_i|),$$

there exists a stabilizing scheme leading to a Markov chain with a bounded second moment in the sense that lim $\sup_{t\to\infty} E[|x_t|^2] < \infty$.

Remark 13.2.2 The discussions here and in Sect. 13.2.2 apply to multi-sensor settings using a quantizer construction similar to that in Sect. 13.2.2. In this case, the state-space is partitioned based on the observable modes at different sensors and coordination among the quantizers for each sensor is established with limited information exchange. In Sect. 13.2.4, this setup is discussed. ◇

13.2.4 Multi-Sensor and Multi-Controller Case: Decentralized Stabilization

For completeness, we present here a brief discussion for the decentralized setting. We refer the reader to [824, Chapter 9] for a detailed analysis on multi-sensor and multi-controller setups.

A fundamental result in control theory is that a controllable and observable single station (or centralized) linear time-invariant (LTI) system can be stabilized through an observer (equivalently, dynamic output) feedback. These results do not generalize directly to decentralized systems with multiple sensors or controllers. Decentralization presents further intricacies in large part because of the issues related to signaling in view of information transmission through control actions.

Consider the class of multi-station n-dimensional discrete-time LTI systems, described by

$$x_{t+1} = Ax_t + \sum_{j=1}^{L} B^j u_t^j ,$$

$$y_t^i = C^i x_t , \quad 1 \le i \le L, \ t = 0, 1, \ldots \tag{13.14}$$

where it is assumed that the joint system is stabilizable and detectable, that is $(A, [B^1|B^2|\ldots|B^L])$ is stabilizable and $(A, [(C^1)'|(C^2)'|\ldots|(C^L)']')$ is detectable, but the individual pairs (A, B^i) may not be stabilizable or (A, C^i) may not be detectable, for $1 \le i \le L$. Here, $x_t \in \mathbb{R}^n$ is the state of the system, $u_t^i \in \mathbb{R}^{m_i}$ is the control applied by station i, and $y_t^i \in \mathbb{R}^{p_i}$ is the observation available at station i, all at time t. Without any loss of generality, we assume the system matrix A to be in Jordan form. The initial state x_0 is unknown, but is known to be generated according to some probability distribution which is supported on a compact set $\mathcal{X}_0 \subset \mathbb{R}^n$.

The information available to station i at time t is

$$I_t^i = \{y_{[0,t]}^i, u_{[0,t-1]}^i\}, \tag{13.15}$$

where, as before, $u_{[0,t-1]}^i$ denotes $\{u_0^i, u_1^i, \ldots, u_{t-1}^i\}$ and $y_{[0,t]}^i = \{y_0^i, y_1^i, \ldots, y_t^i\}$. Hence, for each time t, each station has access to only its measurement and also has full memory on its past measurements and actions. If γ_t^i denotes the strategy (policy, control law) of station i at time t, we have $u_t^i = \gamma_t^i(I_t^i)$. We assume that the entire system dynamics is common information to all agents (stations).

Definition 13.2.1 The decentralized system described in (13.14) is stabilizable under the decentralized information structure if there exists a set of policies $\underline{\gamma}$ such that, with $\{x_t^{\underline{\gamma}}, \ t \ge 0\}$ denoting the state trajectory under $\underline{\gamma}$, $|x_t^{\underline{\gamma}}| \to 0$ as $t \to \infty$ almost surely.

Let us introduce the notation

$$\mathcal{K}^i := \begin{bmatrix} B^i \ AB^i \ \cdots \ A^{n-1}B^i \end{bmatrix}$$

$$\mathcal{O}^i := \begin{bmatrix} (C^i)' \ (C^iA)' \cdots (C^iA^{n-1})' \end{bmatrix}'$$

and let the controllable and unobservable subspaces at station i be denoted by K^i and N^i, respectively, where K^i is the range space of $\mathcal{K}^i$ and N^i is the null-space of $\mathcal{O}^i$. We will, by a slight abuse of notation, refer to the subspace orthogonal to N^i as the observable subspace at the ith station and will denote it by O^i.

We now introduce some basic ingredients and notation used throughout the chapter. Let $\mathbb{U} \subset \mathbb{R}^n$, $\mathbb{V} \subset \mathbb{R}^n$ be Euclidean subspaces. We adopt the following notation:

$$\mathbb{U} \cup \mathbb{V} = \{x : x = \alpha u + \beta v, u \in \mathbb{U}, v \in \mathbb{V}, \alpha, \beta \in \mathbb{R}\}$$

$$\mathbb{U} \cap \mathbb{V} = \{x : x \in \mathbb{U}, x \in \mathbb{V}\}$$

$$\mathbb{U} - \mathbb{V} = \{u : u \in \mathbb{U}, u'v = 0, \forall v \in \mathbb{V}\},$$

With the above definitions, for a vector space $S \subset \mathbb{R}^n$, we have $S^C = \mathbb{R}^n - S$, as the orthogonal complement of S. We denote by $P_{\mathbb{U}}(x) : \mathbb{R}^n \to \mathbb{U}$, the orthogonal projection of a vector x onto the subspace $\mathbb{U} \subset \mathbb{R}^n$. We denote by

$$[v_1, v_2, \ldots, v_m] := \{\sum_{i=1}^{m} \alpha_i v_i, \alpha_i \in \mathbb{R}\},$$

the space spanned by the vectors $\{v_1, v_2, \ldots, v_m\}$.

For two sets Ψ and Ξ, $\Psi \backslash \Xi = \{\eta : \eta \in \Psi, \eta \notin \Xi\}$, is the standard set difference.

By *modes* of a linear system, we refer to the subspaces (eigenspaces) which are invariant in the absence of control; as such, when all the eigenvalues of the system matrix A are distinct, the eigenvectors uniquely identify the modes of the system. In case the geometric multiplicity of an eigenvalue is less than its algebraic multiplicity, generalized eigenvectors (generalized modes) span the eigenspace corresponding to a particular eigenvalue.

We next introduce some relevant graph-theoretic notions: A *directed graph* $\mathcal{G}$ consists of a set of vertices, $\mathcal{V}$, and a set of directed edges, $(a, b) \in \mathcal{E}$, such that $a, b \in \mathcal{V}$. A path in $\mathcal{G}$ of length d consists of a sequence of d directed edges such that each edge is connected. A graph in which there exists a path from any node to any other node is a strongly connected graph. We define the minimum distance between two sets of nodes $S_1, S_2 \subset \mathcal{G}$ as $d(S_1, S_2) = \sum_{i \in S_1} \min\{d(i, j), j \in S_2\}$, where $d(i, j)$ denotes the minimum number of paths between node i and j (if such a finite number exists), with the trivial case being $d(i, i) = 0$ for all nodes.

In particular, through lifting (that is, by viewing the system as a higher-dimensional linear system in a sampled setting), one can apply linear time-invariant policies. In the following, we state a theorem on the universality of linear-time varying controllers for decentralized stabilization. To facilitate the argument, we introduce the notion of a quotient system [425]. Toward this end, we first discuss the notion of connectivity. For two stations i and j, if $K^i \not\subset N^j$, then there exist control signals generated at station i which are observed at station j. We denote this by $i \to j$. This is equivalent to the condition that $C^j A^l B^i \neq 0$, for some $0 \leq l \leq n-1$, as we state in the following:

Lemma 13.2.1 *$i \to j$ if and only if $C^j (A)^l B^i \neq 0$, for at least one l, $1 \leq l \leq n$.*

The above shows that station i can send information to station j through control actions $\{u_t^i\}$. One may construct a directed communication graph given the above relationship, by considering also the possibility that two stations may be connected through other stations. In the following, we will, by an abuse of notation, use $i \to j$ to indicate that there is a path from station i to station j, possibly through other stations.

If every station is connected to every other station in the sense above, the system is said to be *strongly connected*. If the network of stations is not strongly connected, it can be uniquely represented as a disjoint union of strongly connected subsystems.

As a consequence of signaling, each subsystem can be effectively regarded as a single centralized subsystem. Let N be the number of strongly connected subsystems in the system. Let us order the strongly connected subsystems by an order of precedence, that is subsystems i and j satisfy $i < j$ only if station i cannot signal to station j. This relation provides the direction of information flow among the subsystems. By concatenating the system as a composition of the modes controlled by strongly connected subsystems and regarding each subsystem as a single block, and using Kalman's canonical decomposition (see [173]) the system matrix can then be transformed to a block-upper triangular form as follows (see [186] or [770]):

$$A = \begin{bmatrix} \tilde{A}^1 & * & * \cdots & * \\ 0 & \tilde{A}^2 & * \cdots & * \\ \vdots & \vdots & \vdots \cdots & \vdots \\ 0 & 0 & 0 \cdots & \tilde{A}^N \end{bmatrix}, \quad B = \begin{bmatrix} \tilde{B}^1 & * & * \cdots & * \\ 0 & \tilde{B}^2 & * \cdots & * \\ \vdots & \vdots & \vdots \cdots & \vdots \\ 0 & 0 & 0 \cdots & \tilde{B}^N \end{bmatrix},$$

$$C = \begin{bmatrix} \tilde{C}^1 & * & * \cdots & * \\ 0 & \tilde{C}^2 & * \cdots & * \\ \vdots & \vdots & \vdots \cdots & \vdots \\ 0 & 0 & 0 \cdots & \tilde{C}^N \end{bmatrix} \tag{13.16}$$

where each of the subsystems $(\tilde{A}^k, \tilde{B}^k, \tilde{C}^k)$ is strongly connected and $*$ denotes some possibly non-zero matrix. Now, the stabilizability of each of these subsystems, in a sequential manner, leads to the result that such a system is stabilizable if and only if all of the individual strongly connected subsystems are stabilizable (by effectively centralized policies in each strongly connected component).

Theorem 13.2.10 *[308] The decentralized system described in (13.14) is stabilizable under the decentralized information structure (13.15) if and only if each of the strongly connected subsystems $(\tilde{A}^k, \tilde{B}^k, \tilde{C}^k)$ are stabilizable-detectable.*

Furthermore, linear time-varying policies do not bring in any loss, as is captured in the following result.

Theorem 13.2.11 *[308] The decentralized system described in (13.14) is stabilizable under the decentralized information structure (13.15) if and only if it is stabilizable by periodic linear time-varying controllers.*

Remark 13.2.3 Even though the proof of Theorem 13.2.10 in [308] does not use information theoretic ideas, this result has an information theoretic flavor. In particular, the theorem suggests that the controllers which can stabilize a given mode should either have local information about that mode, or be provided information about that mode through signaling by those stations which can signal to them. Such a signaling approach also suggests a constructive linear time-varying control structure for Theorem 13.2.11. ◇

Now, suppose that the controllers are connected to the plant over a discrete-noiseless channel. In this case, the control signals u^i are coded and decoded over discrete noiseless channels with finite capacity. Hence, the applied control and transmitted messages follow a coding (i.e., binary representation) and a decoding process. We assume here fixed-rate encoding, that is, the rate is defined as the (base-2) logarithm of the number of symbols to be transmitted: The coding process of the controller at station i is a mapping measurable with respect to the sigma-algebra generated by I_t^i to $\mathcal{M}_t^i = \{1, 2, \ldots, |\mathcal{M}_t^i|\}$, which is the quantizer codebook at station i at time t, and $|\cdot|$ denotes the cardinality function. Hence, at each time t, station i sends $\log_2(|\mathcal{M}_t^i|)$ bits over the channel to the plant.

From an operational viewpoint, we assume that there is a decoder at the actuator who receives the corresponding controller commands and applies the actions. This can be regarded as an actuator located in the plant who receives the commands from the controller. Hence, the channel is between the controller and the actuator. One may view the system in Fig. 13.5.

We now consider the following problem: Let $\mathcal{R}$ denote the set of average rates on L sensor and controller channels which lead to stabilization, that is,

$$\mathcal{R} = \Bigg\{(R^i, i \in 1, 2, \ldots, L) : \\ \exists \{u^1_{[0,\infty)}, u^2_{[0,\infty)}, \ldots, u^L_{[0,\infty)}\}, \lim_{T\to\infty} ||x_T||_\infty = 0\Bigg\}$$

where $R^i = \limsup_{T\to\infty} \frac{1}{T}\sum_{t=0}^{T-1} \log_2(|\mathcal{M}_t^i|)$. Then, what is the minimum *average total rate* $\mathbf{R} := \min_{\mathcal{R}}\{\sum_{i=1}^{L} R^i\}$, such that decentralized stabilization is possible?

We state the following [824, Chapter 9].

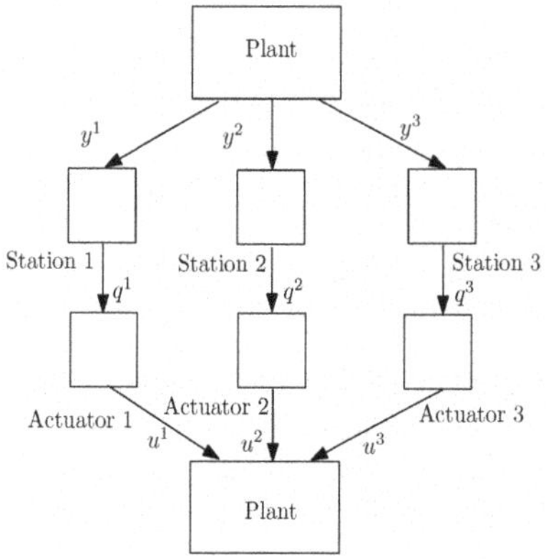

Fig. 13.5 Multi-station (multi-controller) system structure

Proposition 13.2.1 *Consider (13.14), where the system is jointly controllable and jointly observable. Suppose that the system matrix A is in Jordan form, where each Jordan block admits distinct eigenvalues. Then, for each Jordan block, there exists at least one controller which can observe the entire eigenspace, and there exists at least one controller which can control the eigenspace.*

Assumption 13.2.4 *The system matrix A is in Jordan form, where each Jordan block admits distinct eigenvalues.*

Theorem 13.2.12 *Suppose that Assumption 13.2.4 holds. (i) Let A be such that the eigenvalues are real. Then, a tight lower bound on the total rate required,* **R**, *between the controllers and the plant for stabilizability is given by*

$$\left\{ \sum_{|\lambda_i|>1} (\eta_{M_i}) \Big(\log_2(|\lambda_i|) \Big) \right\}, \tag{13.17}$$

where

$$\eta_{M_i} = \min_{l,m \in \{1,2,\ldots,L\}} \Big\{ D^*(l,m) : l \to m, [x^i] \subset O^i \cup O^m, \\ [x^i] \subset K^m, D^*(l,m) = d(l,m) + 1. \Big\} \tag{13.18}$$

(ii) There exist stabilizing coding and control policies whose sum rate is arbitrarily close to the lower bound in (13.17). Hence this bound is asymptotically achievable.

We note that the first result of the theorem above admits a max-flow min-cut interpretation over a temporal graph [290]. One can approach the problem as information transfer over a network, where the rate of information flow across any cut is less than the mutual information between the inputs on either side of the cut conditioned on the inputs on the other side of the cut.

Multi-Sensor Structure with a Centralized Controller

We next discuss the case where there is a central controller which receives information from multiple sensors. Consider the following LTI system (see Fig. 13.6)

$$\begin{aligned} x_{t+1} &= Ax_t + Bu_t\,, \ t \geq 0, \\ y_t^i &= C^i x_t\,, \ t \geq 0\,, \ i = 1, \ldots, L\,, \end{aligned} \tag{13.19}$$

where (A, B) is stabilizable, $(A, [(C^1)' \ldots (C^L)']')$ is detectable, and the initial state x_0 is a random vector with a known continuous distribution over a compact support. The information received by the sensors is quantized and sent to the controller, which has access to only the information sent by the sensors. Suppose

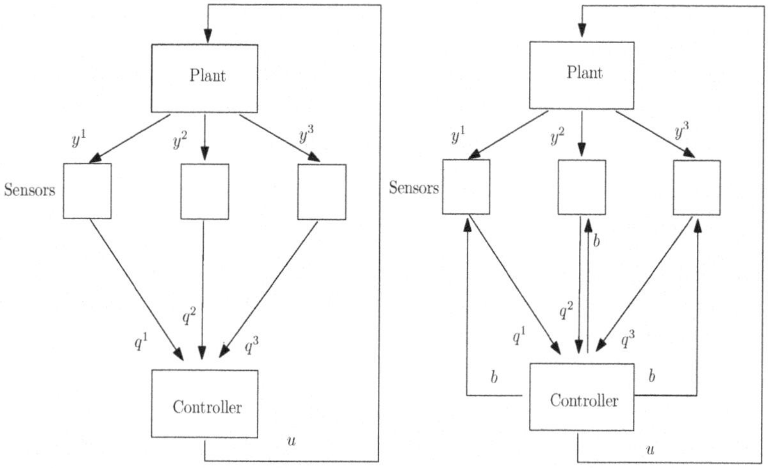

Fig. 13.6 Multi-sensor system architecture: (left) noise-free system case and (right) the case with unbounded noise (where b represents a one-bit coordination symbol for simultaneous zooming in and out)

also that the controller is connected to the plant through a noiseless infinite capacity channel.

The system can be represented, using Kalman's canonical decomposition, in an upper-triangular form (see [373]) so that the system has the following form, where $*$ may or may not be zero and $(\tilde{A}^k, (\tilde{C}^k)', 1 \leq k \leq L)$ are detectable pairs.

$$A = \begin{bmatrix} \tilde{A}^1 & * & * \cdots & * \\ 0 & \tilde{A}^2 & * \cdots & * \\ \vdots & \vdots & \vdots \cdots & \vdots \\ 0 & 0 & 0 \cdots & \tilde{A}^L \end{bmatrix}, \qquad C = \begin{bmatrix} \tilde{C}^1 & * & * \cdots & * \\ 0 & \tilde{C}^2 & * \cdots & * \\ \vdots & \vdots & \vdots \cdots & \vdots \\ 0 & 0 & 0 \cdots & \tilde{C}^L \end{bmatrix} \tag{13.20}$$

The controller can stabilize the components sequentially by receiving information from the sensors, as before, by moving up in the matrix. The following result then essentially follows from Theorem 13.2.12, through a sequential stabilization argument.

Theorem 13.2.13 *For the system considered in (13.19), for asymptotic stability the average total rate for every stabilizing coding and control policy satisfies* $R \geq \sum_{|\lambda_i|>1} \log_2(|\lambda_i|)$*, and this bound is tight in the sense that there exists a stabilizing coding and control policy who average total rate satisfies* $R > \sum_{|\lambda_i|>1} \log_2(|\lambda_i|)$.

One further aspect of the difference between the multi-sensor case and the multi-controller case is that for the former there is no issue of signaling. This allows one to obtain tight results when there is both system and observation noise, as we discuss in the following.

Multi-Sensor Systems Driven by Unbounded Noise

Another setting we consider is a multi-sensor system of the form

$$\begin{aligned} x_{t+1} &= Ax_t + Bu_t + w_t \,, \; t \geq 0, \\ y_t^i &= C^i x_t + v_t^i \,, \; t \geq 0 \,, \; i = 1, \ldots, L \,, \end{aligned} \tag{13.21}$$

where (A, B) is stabilizable, $(A, [(C^1)' \ldots (C^L)']')$ is detectable, $\{v_t^i, w_t\}$ are i.i.d. noise processes with $E[|v_t^i|^{2+\epsilon}] < \infty$, $E[|w_t|^{2+\epsilon}] < \infty$, and Assumption 13.2.5, to be introduced next, is applicable.

Assumption 13.2.5 *Suppose that A is in Jordan form. For every Jordan block in A, there exists a sensor which observes the entire block.*

We note that here the Jordan blocks do not necessarily have distinct eigenvalues unlike Assumption 13.2.4.

Through the random-time state-dependent drift analysis of Appendix G, and obtaining an upper-diagonal structure as in (13.20), we can extend Theorem 13.2.8 to the multi-sensor case. However, there are two aspects which need to be addressed: Coordination among the sensors for the stopping time analysis and the coupling between the *observable modes* of the sensors. In view of Theorem 13.2.8 and the construction in Sect. 13.2.2, we can let all the sensor quantizers zoom-out and zoom-in simultaneously by allowing a *zoom-out symbol* to be exchanged (see Fig. 13.7). Assumption 13.2.5 facilitates the separation of observable modes which partitions the (unstable portion of the) state space $\mathbb{R}^n$.

To facilitate the random-time drift analysis, with T denoting a sampling period in the system, a sequence of stopping times for zooming-in events can be defined as

$$\tau_0 = 0, \quad \tau_{z+1} = \inf\{k2nT > \tau_z : |h^i_{k2nT}| \leq 1, i \in \{1, 2, \ldots, n\}\}, \quad z \in \mathbb{Z}_+,$$

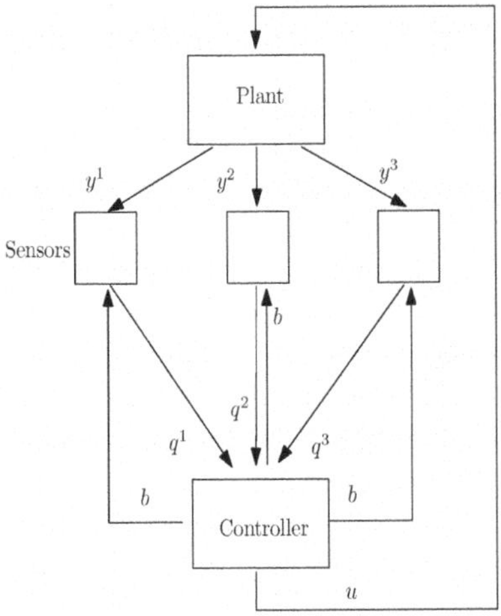

Fig. 13.7 Multisensor system structure with unbounded noise. Here b represents the one-bit coordination symbol for simultaneous zooming in and out

where $h_t^i = \frac{\bar{x}_t^i}{\Delta_t^i 2^{R_i'-1}}$. Here Δ^i is the bin size of the quantizer in the direction of the eigenvector x^i, with rate used R_i' for a corresponding eigenvalue, and $\bar{x}_t^i$ is an estimate of the state component x_t^i, computed by a corresponding sensor.

Under update rules for the bin sizes as in (13.10), we can establish a finite geometric measure which dominates the stopping time distribution. In view of these, by taking T as a free parameter, we can establish the following.

Theorem 13.2.14 ([373]) *Consider the system in (13.21). To achieve asymptotic stability in the sense that*

$$lim\ sup_{t\to\infty} E[|x_t|_2^2] < \infty,$$

it suffices to have the average total rate satisfy $R > \sum_{|\lambda_i|>1} \log_2(|\lambda_i|)$.

The results can be extended to cases where Assumption 13.2.5 does not hold, for which case upper and lower bounds on the information rates can be provided to facilitate the drift arguments. In this case, the structure of the system in (13.20) plays an important role in the analysis. In particular, if the eigenvalues of the matrices $\tilde{A}^1, \cdots, \tilde{A}^L$ are ordered in decreasing magnitude, then Theorem 13.2.14 holds without Assumption 13.2.5. The reader is referred to [373] for details.

13.3 Stabilization of Linear Systems Over Noisy Channels: Fundamental Bounds

In this section, we discuss the noisy channel setting. We note that the Gaussian channel setting will be considered separately in its full generality in Chap. 16.

13.3.1 Control and Communication Model

We first consider a scalar LTI discrete-time system, with the multi-dimensional case relegated to Sect. 13.6. Here, the scalar system is described by

$$x_{t+1} = ax_t + bu_t + w_t, \qquad t \in \mathbb{Z}_+ \tag{13.22}$$

where x_t is the state at time t, u_t is the control input, the initial state x_0 is a zero-mean second order random variable, and $\{w_t\}$ is a sequence of zero-mean i.i.d. Gaussian random variables, also independent of x_0. We assume that the system is open-loop unstable and controllable, that is $|a| \geq 1$ and $b \neq 0$.

This system is connected over a noisy channel to a controller, as shown in Fig. 13.8. The controller has access to the information it has received through the

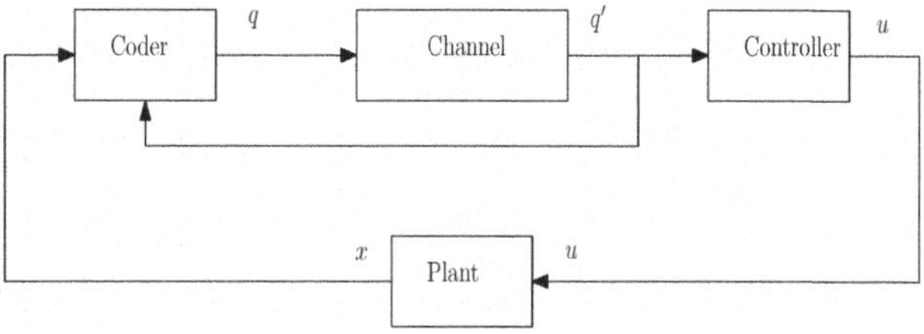

Fig. 13.8 Control of a system over a noisy channel

channel. A source coder maps the source symbols, state values, to corresponding channel inputs. The quantizer outputs are transmitted through a channel, after passing through a channel encoder. We recall here Definitions 12.2.1 and 12.2.3.

The receiver has access to noisy versions of the quantizer/coder outputs for each time, which we denote by $q'_t \in \mathcal{M}'$. The quantizer and the channel encoder policies are causal so that the channel input at time t, q_t, is generated using the information vector I^e_t available at the encoder:

$$I^e_t = \{I^e_{t-1}, x_t, q_{t-1}, q'_{t-1}\}, \quad t \in \mathbb{N}, \qquad I^e_0 = \{x_0\}.$$

That is, the encoder has access to noiseless feedback from the channel output. The control policy at time t, also causal, is measurable with respect to the sigma-algebra generated by I^c_t:

$$I^c_t = \{I^c_{t-1}, q'_t\}, \quad t \in \mathbb{N}, \qquad I^c_0 = \emptyset,$$

and is a mapping to $\mathbb{R}$. We assume that the source model, channel model, the probabilistic description of the noise variables and the initial state distribution ν_0 are available to both the quantizer/channel encoder and the controller.

We will call such coding and control policies *admissible*.

We start our analysis with a converse theorem. This will be discussed first in the context of discrete memoryless channels.

13.3.2 Converse Theorem on Stochastic Stability Over a Discrete Memoryless Channel (DMC)

We have the following converse theorem, which generalizes Theorem 12.5.1. This result is further generalized to a class of channels with memory in Theorem 13.6.2.

Theorem 13.3.1 ([811]) *Suppose that a linear plant given as in (13.22) controlled over a DMC, under some admissible coding and controller policy, satisfies the condition*

$$lim\ inf_{T \to \infty} \frac{1}{T} h(x_T) \leq 0, \tag{13.23}$$

where h denotes the entropy function. Then, the channel capacity C must satisfy

$$C \geq \log_2(|a|).$$

Proof This is a special case of Theorem 13.6.2 which deals with the multi-dimensional case. ◇

Remark 13.3.1 The condition (13.23) is a weak one. For example a stochastic process whose second moment grows subexponentially in time, $\lim \inf_{T \to \infty} \frac{\log(E[x_T^2])}{T} \leq 0$, satisfies this condition. ◇

We now present a supporting result, building on Matveev [513].

Lemma 13.3.1 *[513] Suppose that a linear plant given as in (13.22) is controlled over a DMC. If*

$$C < \log_2(|a|),$$

then

$$lim\ sup_{T \to \infty} P(|x_T| \leq b(T)) \leq \frac{C}{\log_2(|a|)},$$

for all $b(T) > 0$ *such that* $\lim_{T \to \infty} \frac{1}{T} \log_2(b(T)) = 0$.

Proof See the proof of Proposition 13.6.1 in Sect. 13.8.6, which considers the more general setting of channels with memory. ◇

With this lemma at hand, we can prove the following.

Theorem 13.3.2 ([813]) *Suppose that a linear plant given as in (13.22) is controlled over a DMC. If, under some causal encoding and controller policy, the state process is AMS, then the channel capacity C must satisfy*

$$C \geq \log_2(|a|).$$

Proof See the proof of Theorem 13.6.3 in Sect. 13.8.7 ◇

In the following sections, we will see that the condition $C \geq \log_2(|a|)$ in Theorems 13.3.1 and 13.3.2 is almost sufficient as well for stability in the AMS sense. Toward this goal, we first discuss the erasure channel with feedback, and then consider more general DMCs, followed by a class of channels with memory. We will also investigate quadratic stability.

13.4 Stochastic Stabilization Over Erasure Channels with Feedback

For the linear system (13.22), we first consider a particular and important class of DMCs, the erasure channel, in the following.

The details of the setup considered are specified as follows: The channel source consists of state values from $\mathbb{R}$. The source output is, as before, quantized. We use the same set of quantizers as in the previous chapter, defined by (12.1), repeated below:

$$Q_K^{\Delta}(x) = \begin{cases} (k - \frac{1}{2}(K+1))\Delta, & \text{if } x \in [(k-1-\frac{1}{2}K)\Delta, (k-\frac{1}{2}K)\Delta) \\ (\frac{1}{2}(K-1))\Delta, & \text{if } x = \frac{1}{2}K\Delta \\ 0, & \text{if } x \notin [-\frac{1}{2}K\Delta, \frac{1}{2}K\Delta]. \end{cases} \tag{13.24}$$

The quantizer outputs are transmitted through a memoryless erasure channel, after being subjected to a bijective mapping, which is performed by the channel encoder. The channel encoder maps the quantizer output symbols to corresponding channel inputs $q \in \mathcal{M} := \{1, 2 \ldots, K+1\}$. A channel encoder at time t, denoted by $\mathcal{E}_t$, maps the quantizer outputs to $\mathcal{M}$ such that $\mathcal{E}_t(Q_t(x_t)) = q_t \in \mathcal{M}$.

The controller/decoder has access to noisy versions of the encoder outputs for each time, which we denote by $\{q'\} \in \mathcal{M} \cup \{e\}$, with e denoting the erasure symbol, generated according to a probability distribution for every fixed $q \in \mathcal{M}$. The channel transition probabilities are given by:

$$P(q' = i | q = i) = p, \qquad P(q' = e | q = i) = 1 - p, \qquad i \in \mathcal{M}.$$

At each time $t \in \mathbb{Z}_+$, the controller/decoder applies a mapping $\mathcal{D}_t : \mathcal{M} \cup \{e\} \to \mathbb{R}$, given by

$$\mathcal{D}_t(q_t') = \mathcal{E}_t^{-1}(q_t') \times 1_{\{q_t' \neq e\}} + 0 \times 1_{\{q_t' = e\}}.$$

Let $\{\Upsilon_t\}$ denote a binary sequence of i.i.d. random variables, representing the erasure process in the channel, where the event $\Upsilon_t = 1$ indicates that the signal is transmitted with no error through the channel at time t. Let $p = \mathbb{E}[\Upsilon_t]$ denote the probability of success in transmission.

The following key assumptions are imposed throughout this section: Given $K \geq 2$ introduced in the definition of the quantizer, define the *rate variables*

$$R := \log_2(K+1) \qquad R' = \log_2(K) \tag{13.25}$$

We fix positive scalars δ, α satisfying

$$|a|2^{-R'} < \alpha < 1, \tag{13.26}$$

and

$$\alpha(|a| + \delta)^{p^{-1}-1} < 1. \tag{13.27}$$

Similar to (13.6) and (13.7), we consider the following update rules. For $t \in \mathbb{Z}_+$ and with $\Delta_0 \in \mathbb{R}$ selected arbitrarily, consider:

$$\begin{aligned} u_t &= -\frac{a}{b}\hat{x}_t, \\ \hat{x}_t &= \mathcal{D}_t(q'_t) = \Upsilon_t Q_K^{\Delta_t}(x_t), \\ \Delta_{t+1} &= \Delta_t \bar{Q}(\Delta_t, |\frac{x_t}{\Delta_t 2^{R'-1}}|, \Upsilon_t). \end{aligned} \tag{13.28}$$

Here, $\bar{Q} : \mathbb{R} \times \mathbb{R} \times \{0, 1\} \to \mathbb{R}$ is defined below, where $L > 0$ is a constant.

$$\begin{aligned} &\bar{Q}(\Delta, h, p) = |a| + \delta \quad \text{if} \quad |h| > 1, \quad \text{or} \quad p = 0 \\ &\bar{Q}(\Delta, h, p) = \alpha \quad \text{if} \quad 0 \leq |h| \leq 1, p = 1, \Delta > L \\ &\bar{Q}(\Delta, h, p) = 1 \quad \text{if} \quad 0 \leq |h| \leq 1, p = 1, \Delta \leq L \end{aligned}$$

The update equations above imply that

$$\Delta_t \geq L\alpha =: L'. \tag{13.29}$$

Without any loss of generality, we assume that $L' \geq 1$.

We note that given the channel output $q'_t \neq e$, the controller can simultaneously deduce the realization of Υ_t and the event $\{|h_t| > 1\}$, where $h_t := \frac{x_t}{\Delta_t 2^{R'-1}}$. This is due to the fact that if the channel output is not the erasure symbol, the controller knows that the signal is received with no error. If $q'_t = e$, however, then the controller applies 0 as its control input and enlarges the bin size of the quantizer.

Stochastic Stability and Positive Harris Recurrence

Lemma 13.4.1 *Under (13.28), the process (x_t, Δ_t) is a Markov chain.*

Proof The system's state evolution can be expressed as

$$x_{t+1} = ax_t - a\hat{x}_t + w_t,$$

where $\hat{x}_t = \Upsilon_t Q_K^{\Delta_t}(x_t)$. It follows that the process (x_t, Δ_t) evolves as a nonlinear state space model:

$$x_{t+1} = a(x_t - \Upsilon_t Q_K^{\Delta_t}(x_t)) + w_t$$
$$\Delta_{t+1} = \Delta_t \bar{Q}(\Delta_t, |\frac{x_t}{2^{R'-1}\Delta_t}|, \Upsilon_t). \tag{13.30}$$

in which (w_t, Υ_t) is i.i.d.. Thus, the pair (x_t, Δ_t) forms a Markov chain. ◇

The following result is established in the appendix to the chapter, based on the stochastic stability results of Appendix G.

Proposition 13.4.1 *[829] If* (13.25) *holds, then there exists a compact set $A \times B \subset \mathbb{R}^2$ satisfying the finite-mean return property*

$$\sup_{(x,\Delta)\in A\times B} \mathbb{E}_{x,\Delta}[\tau_{A\times B}] < \infty,$$

and the recurrence condition $P_{(x,\Delta)}(\tau_{A\times B} < \infty) = 1$ for any admissible (x, Δ).

A result on the existence and uniqueness of an invariant probability measure is the following. It basically establishes irreducibility and aperiodicity, which leads to positive Harris recurrence, by Proposition 13.4.1.

Theorem 13.4.1 ([829]) *For an adaptive quantizer satisfying* (13.25)*, suppose that the quantizer bin sizes are such that their base-2 logarithms are integer multiples of some scalar s, and $\log_2(\bar{Q}(\cdot))$ takes values in integer multiples of s. Then the process (x_t, Δ_t) forms a positive Harris recurrent Markov chain. If the integers taken are relatively prime (that is they share no common divisors except 1), then the invariant probability measure is independent of the value of the integer multiplying s.*

We note that the (Shannon) capacity of such an erasure channel is given by $\log_2(K+1)p$ [188]. From (13.25)–(13.27), the following is obtained.

Theorem 13.4.2 *If $\log_2(K)p > \log_2(|a|)$, then α, δ exist such that the conditions of Theorem 13.4.1 are satisfied.*

Remark 13.4.1 Thus, the Shannon capacity of the erasure channel is an almost sufficient condition for the positive Harris recurrence of the state and the quantizer process. We will see that under the weaker notions of ergodicity or asymptotic mean stationarity, this result applies to a large class of memoryless channels and a class of channels with memory and feedback (see Theorem 13.5.3): There is a direct relationship between asymptotic mean stationarity and the Shannon capacity of the channel used in the system. As we shall see, however, for quadratic or finite moment stability, Shannon capacity is typically not sufficient. ◇

Quadratic Stability

Under slightly stronger conditions, we obtain a finite second moment:

Theorem 13.4.3 ([829]) *Suppose that the assumptions of Theorem 13.4.1 hold, and in addition the following bound holds,*

$$a^2\Big(1 - p + \frac{p}{(2^R - 1)^2}\Big) < 1. \tag{13.31}$$

Then, for each initial condition (x_0, Δ_0),

$$\lim_{t\to\infty} \mathbb{E}[x_t^2] = \mathbb{E}_\pi[x_0^2] < \infty .$$

By the positive Harris recurrence property, the above holds also in a sample-path sense almost surely, leading to quadratic stability.

The above theorem also has a converse in the sense that the rate condition in (13.31) is tight (up to the transmission of an additional symbol).

Theorem 13.4.4 ([541]) *A necessary condition for mean square stability is:*

$$a^2\Big(1 - p + \frac{p}{2^{2R}}\Big) < 1.$$

Proof By the property that the Gaussian measure maximizes the entropy over the set of all random variables with a fixed variance, we have

$$E[x_t^2 | q'_{[0,t-1]}] \geq \frac{1}{2\pi e} 2^{2h(a^t x_0 | q'_{[0,t-1]})}. \tag{13.32}$$

Conditioned on the event L_k of k successful transmissions at $t \geq k$ transmissions, there are kR bits at the controller by time t. Since $kR \geq I(x_0; q'_{[0,t-1]}, L_k) = h(x_0) - h(x_0|q'_{[0,t-1]}, L_k)$, it follows that, $2^{2h(a^t x_0|q'_{[0,t-1]},L_k)} \geq a^{2t} 2^{2h(x_0)}/2^{2kR}$. Conditional entropy is an average of these realized conditional entropies (see Definition 12.3.2). A necessary condition for the boundedness of the second moment is then $\liminf_{t\to\infty} E[\frac{a^{2t}}{2^{2R_t}}] < 1$, where R_t is the total number of bits successfully received by time t. By considering combinatorially the events of erasures up until time t, we obtain

$$a^{2t} \sum_{k=0}^{t} \binom{t}{k} p^k (1-p)^{t-k} \frac{1}{2^{2Rk}} = (a^2(\frac{p}{2^{2R}} + 1 - p))^{2t} < 1,$$

as a necessary condition. ◇

Thus, the sufficiency condition in Theorem 13.4.3 almost meets this bound except for the additional symbol transmitted for the under-zoom events. We note that the average rates can be made arbitrarily close to the rate required in (13.4.4) by sampling the control system with larger periods. Such a relaxation of the sampling period, however, would lead to a process which is not Markov, but an AMS process which is quadratically stable.

We now consider the mth moment case. In this case, the proof is almost identical to that of Theorem 13.4.3 and is thus omitted.

Theorem 13.4.5 ([829]) *Consider the scalar system in (13.22). Let* $m \in \mathbb{N}$, *suppose that the assumptions of Theorem 13.4.1 hold, and in addition we have the inequality,*

$$|a|^m \left(1 - p + \frac{p}{(2^R - 1)^m}\right) < 1.$$

Then, with the adaptive quantization policy considered, and given the initial condition (x_0, Δ_0),

$$\lim_{t \to \infty} \mathbb{E}[|x_t|^m] = \mathbb{E}_\pi[|x_0|^m] < \infty .$$

13.4.1 Connections with Random-Time Drift Criteria

As it was done in Sect. 13.2.1, we point out the connections between the results above and the random-time drift criteria.

The process (x_t, Δ_t) will once again form a Markov chain. Now, in the model considered, the controller can receive meaningful information regarding the state of the system when two events occur concurrently: the channel carries information with no error, and the source lies in the granular region of the quantizer; that is, $x_t \in [-\frac{1}{2}K\Delta_t, \frac{1}{2}K\Delta_t)$ and $\Upsilon_t = 1$. The times at which both of these events occur form an increasing sequence of random stopping times, defined as

$$\tau_0 = 0, \quad \tau_{z+1} = \inf\{k > \tau_z : |h_k| \le 1, \Upsilon_k = 1\}, \quad z \in \mathbb{N}. \tag{13.33}$$

In the proofs of the stability theorems, we will apply Theorem G.1.4 for these stopping times. These are the times when information reaches the controller regarding the value of the state when the state is in the granular region of the quantizer. As in Fig. 13.2, we introduce $C_x = \{x : |x| \le F\}, \quad C_h = \{h : |h| \le 1\}$, for some sufficiently large F value, which will serve as a small set as well as a recurrent set.

In view of Theorem G.1.4, first without an irreducibility assumption, we will establish recurrence of the set $C_x \times C_h$ by defining a Lyapunov function of the form $V(x_t, \Delta_t) = \frac{1}{2}\log_2(\Delta^2) + B_0$ for some $B_0 > 0$. Later we will establish the irreducibility of the Markov chain by imposing a countability condition. The details are given in the appendix.

13.4.2 Simulation

Consider the linear system (13.22) with $a = 2.5$, $b = 1$, and $\{w_t\}$ a sequence of i.i.d., zero-mean Gaussian variables with $E[w_t^2] = 1$. The erasure channel has error probability $1 - p = 0.1$. For stability with a finite second moment, we employ a quantizer with rate

$$\log_2(\lceil \sqrt{\frac{p}{\frac{1}{a^2} - (1-p)}} \rceil + 1) = \log_2(5)$$

bits, or a uniform quantizer with 5 bins. Here, we have taken $L' = 1$. Figure 13.9 illustrates the stochastic stability results presented in Theorems 13.4.1 and 13.4.3. The plots show the under-zoomed and perfectly-zoomed phases, with the peaks in the plots showing the under-zoom phases. For the plot with 5 levels, the system is positive Harris recurrent, since the update equations are such that $\alpha = 0.629$, $\delta = 0.025$, and $\log_2(\bar{Q}(\cdot)) \in \{-0.6744, 0, 1.363\}$. These values satisfy the irreducibility condition since $-0.6744 = -(1/2)1.363$, and the communication conditions are satisfied. Furthermore,

$$\alpha(|a| + \delta)^{\frac{1}{p}-1} < 0.698 < 1.$$

One notable aspect is that increasing the bit rate by only two bits in Fig. 13.9 leads to a much more desirable sample path. Moreover, by increasing the rate, the severity of rare events is reduced.

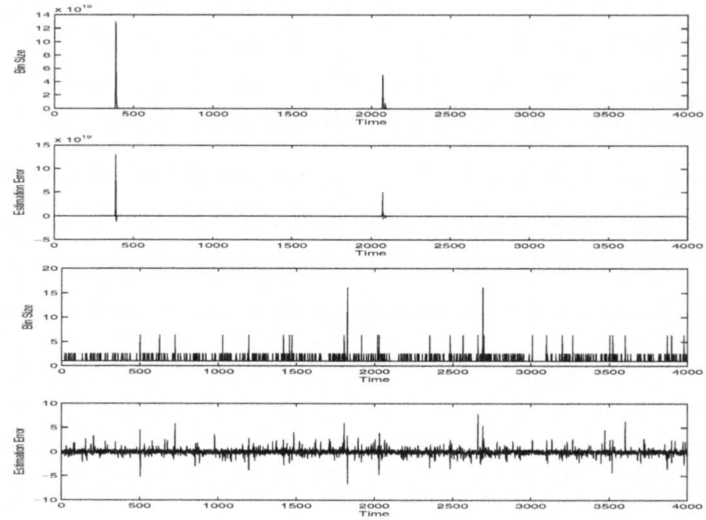

Fig. 13.9 Sample paths for a stochastically stable system with a 5-bin quantizer and a 17-bin quantizer

13.5 Stochastic Stabilization Over General DMCs with Feedback

In this section, we consider more general DMCs. The construction will be more tedious, but the essence of the analysis for the erasure channel applies to this setting as well. Proofs of several of the results in this section can be found in [824, Chapter 8].

We first note that the condition $C \geq \log_2(|a|)$ in Theorem 13.3.1 is almost sufficient for strong forms of stability.

Theorem 13.5.1 ([811]) *For the existence of a compact coordinate recurrent set (see Definition D.3.4), the following is sufficient: The channel capacity C satisfies: $C > \log_2(|a|)$.*

We now consider the following update algorithm which is used in the proof of Theorem 13.5.1. Let n be a given block length. Consider a class of uniform quantizers, defined by two parameters, with bin size $\Delta > 0$, and an even number $K(n) \geq 2$ (see Fig. 12.1). Define the uniform quantizer as follows: For $k = 1, 2 \ldots, K(n)$,

$$Q^{\Delta}_{K(n)}(x) = \begin{cases} (k - \frac{1}{2}(K(n) + 1))\Delta, & \text{if } x \in [(k-1-\frac{1}{2}K(n))\Delta, (k-\frac{1}{2}K(n))\Delta) \\ (\frac{1}{2}(K(n) - 1))\Delta, & \text{if } x = \frac{1}{2}K(n)\Delta \\ \mathcal{Z}, & \text{if } x \notin [-\frac{1}{2}K(n)\Delta, \frac{1}{2}K(n)\Delta]. \end{cases}$$

where $\mathcal{Z}$ is the overflow symbol in the quantizer. Let $\{x : Q^{\Delta}_{K(n)}(x) \neq \mathcal{Z}\}$ be the *granular region* of the quantizer.

At every sampling instant $t = kn, k = 0, 1, 2, \ldots,$ the source coder $\mathcal{E}^s_t$ quantizes output symbols in $\mathbb{R} \cup \{\mathcal{Z}\}$ to a set $\mathcal{M}(n) = \{1, 2, \ldots, K(n)+1\}$. A channel encoder $\mathcal{E}^c_t$ maps the elements in $\mathcal{M}(n)$ to corresponding channel inputs $q_{[kn,(k+1)n-1]} \in \mathcal{M}^n$. For each time $t = kn - 1, k = 1, 2, 3, \ldots,$ the channel decoder applies a mapping $\mathcal{D}_{tn} : \mathcal{M}'^n \to \mathcal{M}(n)$, such that

$$c'_{(k+1)n-1} = \mathcal{D}_{kn}(q'_{[kn,(k+1)n-1]}).$$

Finally, the controller runs an estimator:

$$\hat{x}_{kn} = (\mathcal{E}^s_{kn})^{-1}(c'_{(k+1)n-1}) \times 1_{\{c'_{(k+1)n-1} \neq \mathcal{Z}\}} + 0 \times 1_{\{c'_{(k+1)n-1} = \mathcal{Z}\}}.$$

Hence, when the decoder output is the overflow symbol, the estimation output is 0.

As in the previous two chapters, at time kn the bin size Δ_{kn} is taken to be a function of the previous state $\Delta_{(k-1)n}$ and the past n channel outputs. Further, the encoder has access to the previous channel outputs, thus making such a quantizer implementable at both the encoder and the decoder.

With $K(n) > \lceil |a|^n \rceil$, $R = \log_2(K(n) + 1)$, let us introduce $R'(n) = \log_2(K(n))$ and let

$$R'(n) > n \log_2(\frac{|a|}{\alpha}),$$

for some $\alpha, 0 < \alpha < 1$ and $\delta > 0$. When clear from the context, we will drop the index n in $R'(n)$. We will consider the following update rules in the controller actions and the quantizers. For $t \geq 0$ and with $\Delta_0 > L$ for some $L \in \mathbb{R}_+$, and $\hat{x}_0 \in \mathbb{R}$, consider, for $t = kn$, $k \in \mathbb{N}$

$$\begin{aligned} u_t &= -1_{\{t=(k+1)n-1\}} \frac{a^n}{b} \hat{x}_{kn}, \\ \Delta_{(k+1)n} &= \Delta_{kn} \bar{Q}(\Delta_{kn}, c'_{(k+1)n-1}), \end{aligned} \tag{13.34}$$

where c' denotes the decoder output variable. If we use $\delta > 0$ and $L > 0$ such that,

$$\begin{aligned} \bar{Q}(\Delta, c') &= (|a| + \delta)^n & \text{if} \quad & c' = \mathcal{Z}, \\ \bar{Q}(\Delta, c') &= \alpha^n & \text{if} \quad & c' \neq \mathcal{Z}, \Delta \geq L, \\ \bar{Q}(\Delta, c') &= 1 & \text{if} \quad & c' \neq \mathcal{Z}, \Delta < L, \end{aligned} \tag{13.35}$$

we can show that a recurrent set exists. Note that the above implies that $\Delta_t \geq L\alpha^n =: L'$ for all $t \geq 0$.

Thus, we have three main events: When the decoder output is the overflow symbol, the quantizer is zoomed out (with a coefficient of $(|a| + \delta)^n$). When the decoder output is not the overflow symbol $\mathcal{Z}$, the quantizer is zoomed in (with a coefficient of α^n) if the current bin size is greater than or equal to L, and otherwise the bin size does not change.

In the following, we make the quantizer bin size process space countable and as a result establish the irreducibility of the sampled process (x_{tn}, Δ_{tn}).

Theorem 13.5.2 *For an adaptive quantizer satisfying the conditions of Theorem 13.5.1, suppose that the quantizer bin sizes are such that their logarithms are integer multiples of some scalar s, and $\log_2(\bar{Q}(\cdot))$ takes values in integer multiples of s. Suppose the integers taken are relatively prime (that is they share no common divisors except for 1). Then the sampled process (x_{tn}, Δ_{tn}) forms a positive Harris recurrent Markov chain at sampling times on the space of admissible quantizer bins and state values.*

Theorem 13.5.3 *Under the conditions of Theorems 13.5.1 and 13.5.2, the process $\{x_t, \Delta_t\}$ is n-stationary, n-ergodic and hence AMS. That is, if the channel capacity C satisfies: $C > \log_2(|a|)$, there exists a coding and control policy such that the process $\{x_t, \Delta_t\}$ is n-stationary, n-ergodic and AMS.*

Proof The proof follows from the observation that a positive Harris recurrent Markov chain is recurrent and stationary. It uses the property that if a sampled process is a positive Harris recurrent Markov chain, and if the intersampling time is fixed, with a time-homogenous update in the inter-sampling times, then the process is mixing, $n-$ergodic and $n-$stationary. ◇

Quadratic Stability and Finite Second Moment

We now discuss quadratic and finite moment stability (see Definition D.3.8). For a given coding scheme with block-length n, a message set $\mathcal{M}(n) = \{1, 2, \ldots, K(n)+1\}$, and a decoding function $\gamma : \mathcal{M}'^n \to \{1, 2, \ldots, K(n)+1\}$, we have three types of errors:

Type I-A: Error from a granular symbol to another granular symbol. A bound for such an error is:

$$P^e_{g|g}(n) := \max_{c \in \mathcal{M}(n) \setminus \mathcal{Z}} P(\gamma(q'_{[0,n-1]}) \neq c, \gamma(q'_{[0,n-1]}) \neq \mathcal{Z} | c \quad \text{is transmitted})$$

Type I-B: Error from a granular symbol to $\mathcal{Z}$:

$$P^e_{g|g}(n) := \max_{c \in \mathcal{M}(n) \setminus \mathcal{Z}} P(\gamma(q'_{[0,n-1]}) = \mathcal{Z} | c \quad \text{is transmitted})$$

Type II: Error from $\mathcal{Z}$ to a granular symbol:

$$P^e_{g|\mathcal{Z}}(n) := P(\gamma(q'_{[0,n-1]}) \neq \mathcal{Z} | \mathcal{Z} \quad \text{is transmitted})$$

Type II error will be shown to be crucial in the analysis of the error exponent. Types I-A and I-B will play an important role in establishing the drift properties. The following theorem captures the main results for establishing quadratic stability.

Theorem 13.5.4 ([811]) *A sufficient condition for quadratic stability (for the joint (x_t, Δ_t) process) over a DMC is that:*

$$\lim_{n \to \infty} (\frac{1}{n} \log(P^e_{\mathcal{Z}|g}(n)) + 2\log(|a| + \delta) < 0,$$

$$\lim_{n \to \infty} (\kappa \frac{1}{n} \log(P^e_{g|\mathcal{Z}}(n)) + 2\log(|a| + \delta) < 0,$$

$$\lim_{n \to \infty} (\kappa \frac{1}{n} \log(P^e_{g|g}(n)) + 2\log(|a| + \delta) + 2\kappa \log(\alpha) < 0,$$

$$R'(n) > n \log_2(|a|/\alpha)$$

and

$$\kappa < \frac{1}{\log_{\frac{|a|+\delta}{|a|}}(\frac{|a|+\delta}{\alpha})}.$$

Let

$$\bar{P}_e(n) := \max_{c \in \mathcal{M}(n)} P(\gamma(q'_{[0,n-1]}) \neq c|c \quad \text{is transmitted}).$$

When the block-length is clear from context, we drop the explicit dependence on n. We have the following corollary to Theorem 13.5.4.

Corollary 13.5.1 *A sufficient condition for quadratic stability (for the joint* (x_t, Δ_t) *process) over a DMC is:*

$$\lim_{n \to \infty} (\kappa \frac{1}{n} \log(\bar{P}_e(n)) + 2 \log(|a| + \delta) < 0,$$

with rate $R'(n) > n \log_2(\frac{|a|}{\alpha})$.

Remark 13.5.1 For a DMC with block length n, Shannon's random coding [284] leads to:

$$P_e(n) \leq e^{-nE(R)+o(n)},$$

uniformly for all codewords $c \in \{1, 2, \ldots, \mathcal{M}(n)\}$ with c' being the decoder output (thus, the random exponent also applies uniformly over the set). Here $\frac{o(n)}{n} \to 0$ as $n \to \infty$ and $E(R) > 0$ for $0 < R < C$. Thus, under the above conditions, the exponent under random coding should satisfy $E(R) > \frac{2\log_2(|a|+\delta)}{\kappa}$. ◇

Remark 13.5.2 (Converse Results for Quadratic Stability) For quadratic stability over erasure channels, we observed in Theorem 13.4.4 that a converse theorem exists, and the proposed scheme achieves the converse result. For general DMCs, however, a tight converse result on quadratic stabilizability is not yet available. One primary reason is that the error exponents of fixed length block codes with noiseless feedback for general DMCs are not currently known. We note here that the channel reliability or error exponent of DMCs is typically improved with feedback, unlike the capacity of DMCs. Some partial results have been reported in [218] (in particular, the sphere packing upper bound is tight for a class of symmetric channels for rates above a critical rate even with feedback), see *Chapter 10* of [194] among others. Related references addressing partial results include [530] and [531] which consider lower bounds on estimation error moments for transmission of a single variable over a noisy channel (in the context of this chapter, this single variable may correspond to the initial state x_0). A further related notion for quadratic stability is the notion of *any-time capacity* introduced in [617] and [619], which is discussed further in Sect. 13.10. ◇

Zero-Error Transmission for $\mathcal{Z}$
An important practical setup would be the case when $\mathcal{Z}$ is transmitted with no error and is not confused with messages from the granular region. The following captures this.

Assumption 13.5.1 $P^e_{\mathcal{Z}|g}(n) = P^e_{g|\mathcal{Z}}(n) = 0$ *for* $n \geq n_0$ *for some* $n_0 \in \mathbb{N}$.

Theorem 13.5.5 *Under Assumption 13.5.1, a sufficient condition for quadratic stability is:*

$$\lim_{n\to\infty} (\bar{P}_e(n))(|a| + \delta)^{2n} < 1,$$

with rate $R'(n) > n\log_2(\frac{|a|}{\alpha})$ *and* $\kappa > 1/2$.

It is worth emphasizing that, the reliability of sending the symbol $\mathcal{Z}$ for the under-zoom phase allows a relaxation in the overall channel reliability requirements.

13.6 Channels with Memory and Multi-Dimensional Sources

Definition 13.6.1 Channels are said to be of **Class A** type, if

they satisfy the following Markov chain condition:

$$q'_t \leftrightarrow q_t, q_{[0,t-1]}, q'_{[0,t-1]} \leftrightarrow \{x_0, w_t, t \geq 0\},$$

for all $t \geq 0$, and
their capacity with feedback is given by:

$$C = \lim_{T\to\infty} \max_{\{P(q_t|q_{[0,t-1]}, q'_{[0,t-1]}),\quad 0\leq t\leq T-1\}} \frac{1}{T} I(q_{[0,T-1]} \to q'_{[0,T-1]}),$$

where the directed mutual information is defined by

$$I(q_{[0,T-1]} \to q'_{[0,T-1]}) = \sum_{t=1}^{T-1} I(q_{[0,t]}; q'_t | q'_{[0,t-1]}) + I(q_0; q'_0).$$

DMCs naturally belong to this class. For DMCs, feedback does not increase the capacity [188]. Such a class also includes finite state stationary Markov channels which are indecomposable [576], and non-Markov channels which satisfy certain symmetry properties [657]. Further examples can be found in [708] and [200].

Theorem 13.6.1 ([811]) *Suppose that a linear plant given by (13.22) is controlled over a **Class A** type noisy channel with feedback. If the channel capacity (with feedback) is less than* $\log_2(|a|)$*, then*

(i) the following condition

$$lim\ inf_{T\to\infty}\frac{1}{T}h(x_T) \le 0,$$

cannot be satisfied under any policy, and

(ii) the state process cannot be AMS under any policy. ◇

Proof This is a special case of Theorems 13.6.2 and 13.6.3, and hence its proof is postponed until later. ◇

Remark 13.6.1 The result above is *negative*, but one can also obtain a positive one: If the channel capacity is greater than $\log_2(|a|)$ and there is a positive error exponent (uniform over all transmitted messages, as in Theorem 14 of [576]), then there exists a coding scheme leading to an AMS state process provided that the channel restarts itself with the transmission of every new block (either independently or as a Markov process). ◇

Remark 13.6.2 If the channel is not information stable, then information spectrum methods lead to pessimistic realizations of capacity (known as the lim inf *in probability* of the normalized information density, see [708, 744]). We do not consider such channels in this book, although the approach here is generalizable to some cases when the channel state is Markov and the worst case initial input state is considered as in [576]. ◇

Higher-Order Plants

The result for the scalar problem has a natural counterpart in the multi-dimensional setting. Consider the linear system described by

$$x_{t+1} = Ax_t + Bu_t + Gw_t, \tag{13.36}$$

where $x_t \in \mathbb{R}^N$ is the state at time t, $u_t \in \mathbb{R}^m$ is the control input, and $\{w_t\}$ is a sequence of zero-mean i.i.d. $\mathbb{R}^d$-valued Gaussian random vectors. Here A is the square system matrix with at least one eigenvalue greater than or equal to 1 in magnitude, that is, the system is open-loop unstable. Furthermore, (A, B) and (A, G) are controllable pairs.

In the following we assume that all eigenvalues $\{\lambda_i, 1 \le i \le N\}$ of A are unstable, that is have magnitudes greater than or equal to 1. There is no loss here since if some eigenvalues are stable, by a similarity transformation, the unstable modes can be decoupled from the stable ones and one can instead consider a lower dimensional system; stable modes are already recurrent.

Consider a multi-dimensional linear system as in (13.36) with all eigenvalues unstable, that is $|\lambda_i| \ge 1$ for $i = 1, \ldots, N$. We have the following results:

Theorem 13.6.2 *For such a system controlled over a* **Class A** *type noisy channel with feedback, if the channel capacity (with feedback) satisfies* $C < \sum_i \log_2(|\lambda_i|)$*, there does not exist a stabilizing (coding and control) policy leading to* $\text{lim inf}_{T\to\infty}\frac{1}{T}h(x_T) \leq 0$.

Proof See Sect. 13.8.5. ◇

Lemma 13.6.1 *For such a system controlled over a* **Class A** *type noisy channel with feedback, if* $C < \log_2(|A|)$*, then,*

$$\text{lim sup}_{T\to\infty} P(|x_T| \leq b(T)) \leq \frac{C}{\log_2(|A|)} > 0,$$

for all $b(T) > 0$ *such that* $\lim_{T\to\infty}\frac{1}{T}\log_2(b(T)) = 0$.

Proof See Sect. 13.8.6 ◇

Given this lemma, we state the following.

Theorem 13.6.3 *Consider such a system controlled over a* **Class A** *type noisy channel with feedback. If there exists some encoding and controller policy so that the state process is AMS, then the channel capacity (with feedback) C must satisfy* $C \geq \log_2(|A|)$.

Proof See Sect. 13.8.7 ◇

For sufficiency, we will assume that A is a diagonalizable matrix (a sufficient condition for which is that its eigenvalues are distinct real).

Theorem 13.6.4 ([811]) *Consider a multi-dimensional system with a diagonalizable matrix A. If the Shannon capacity of the DMC used in the controlled system satisfies* $C > \sum_{|\lambda_i|>1}\log_2(|\lambda_i|)$*, there exists a stabilizing scheme in the AMS sense.*

Regarding channels with memory, the discussions in Remark 13.6.1 also apply for this setting.

Remark 13.6.3 Theorem 13.6.4 can be extended to the case where the matrix A is not diagonalizable, in the same spirit as in Theorem 13.2.8, by constructing stopping times in view of the coupling between modes sharing a common eigenvalue. ◇

13.7 Stabilization with Noisy Forward and Feedback/Reverse Channels: When Adaptation Is Infeasible

Generalizing the coverage so far in the chapter, we can also consider stabilization of open-loop unstable continuous-time linear stochastic systems when communications between the plant and the controller (forward communication), and the

Fig. 13.10 Control over noisy forward and reverse channels

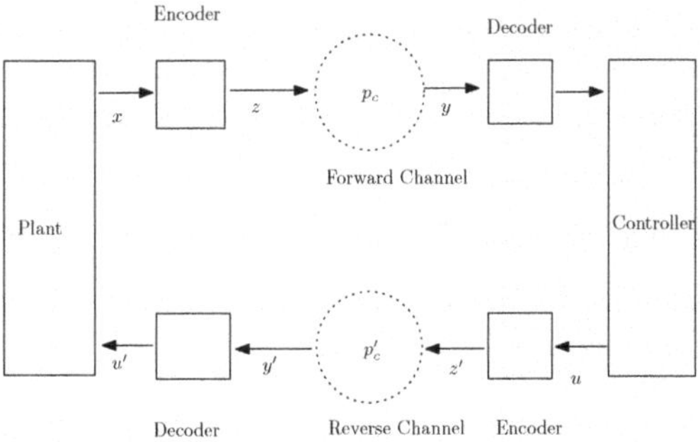

controller and the plant (reverse communication) are conducted over noisy channels which are either discrete memoryless or continuous memoryless (see Fig. 13.10).

In such setups, due to the absence of noise-free feedback (either directly from the channel or from the control), adaptation in coding/decoding may no longer be feasible (as the encoder and the decoder would not be able to synchronize their updates on the adaptive quantizers), and the entire state space is to be encoded this time in a time-invariant fashion. For such a time-invariant map, however; in a realistic setup, codewords corresponding to rare events would need to take longer to be transmitted over a channel, and thus a state-dependent drift is to be considered for stochastic stabilization. For such a setup, a new coding scheme, analyzed via a version of the state-dependent stochastic drift considered in Appendix G, is presented in [823] (see also [824, Chapter 8]).

At this point, we also note that there exist several designs which use time-invariant maps encoding the entire state space. These often follow uniform or logarithmic quantizers (possibly via randomization, through dithering), followed by an entropy coder. See Sect. 16.3 for a general discussion.

13.8 Appendix: Proofs

13.8.1 Proof of Theorem 13.2.1

We provide a proof for the case $a > 1$; a similar line of argument applies for $a < -1$. Suppose that $|u_t| \leq U$ for some finite U. Let a Lyapunov function be picked as $V(x) = \gamma^{-x}$, defined for positive x and with $\gamma > 1$. Define a process $\tilde{x}_{t+1} = \bar{a}\tilde{x}_t + w_t$, where w_t is almost surely the same noise process acting on the original system and $a > \bar{a} > 1$ with $x_0 = \tilde{x}_0$. It follows that $x_t \geq \tilde{x}_t$ for $t \geq 1$ almost surely if the initial condition is greater than $U/(a - \bar{a})$. Let us pick such an initial condition. Observe that for all $\tilde{x} > U/(a - \bar{a})$, $E[V(\tilde{x}_{t+1})|\tilde{x}_t = \tilde{x}] \leq V(\tilde{x})$, since $E[\gamma^{-(\bar{a}\tilde{x}+w_t)}] = \gamma^{-\bar{a}\tilde{x}}E[\gamma^{-w}]$. For all $\tilde{x} \in \{\tilde{x} : \gamma^{(\bar{a}-1)\tilde{x}} > E[\gamma^{-w}]\} =: S^C$ (where $S^C = \mathbb{R} \setminus S$) the Lyapunov condition (G.6) in Theorem G.1.8 holds. Since the function V is strictly decreasing, $P_{\tilde{x}}(\tau_S < \infty) < 1$ for $\tilde{x} \in S^C$. This implies that $P_x(\tau_S < \infty) < 1$ for the original chain. ◇

13.8.2 Proof of Theorem 13.2.3

Toward the proof, we will first state a supporting result.

Lemma 13.8.1 *Let $\mathcal{B}(\mathbb{R} \times \mathbb{R}_+)$ denote the Borel σ–field on $\mathbb{R} \times \mathbb{R}_+$. Then,*

$$P\Big((x_t, \Delta_t) \in (C \times D) | (x_{t-1}, \Delta_{t-1}), ..., (x_0, \Delta_0)\Big)$$

$$= P\Big((x_t, \Delta_t) \in (C \times D) | (x_{t-1}, \Delta_{t-1})\Big),$$

$\forall (C \times D) \in \mathcal{B}(\mathbb{R} \times \mathbb{R}_+)$, i.e. (x_t, Δ_t) is a Markov chain.

Let us define $h_t := \frac{x_t}{\Delta_t 2^{R'-1}}$. Consider the following sets: $C_x = \{x : |x| \leq E\}$, $C_h = \{h : |h| \leq 1\}$, with $E_1 = 2^{R'-1}L'$. Further, let another set be $C'_x = \{x : |x| \leq F\}$, with a sufficiently large F value, to be derived below. We will study the expected number of time stages between visits of $\{(x_t, h_t)\}$ to $C'_x \times C_h$.

We first show that the sequence $\{h_t, t \geq 0\}$ visits C_h infinitely often with probability 1 and the expected length of the excursion is uniformly bounded over all possible values of $(x, h) \in C'_x \times C_h$. Let $V(h_t) = h_t^2$ serve as a Lyapunov function. Define a sequence of stopping times for the perfect-zoom case with (where the initial state is perfectly zoomed)

$$\tau_0 = 0, \quad \tau_{z+1} = \inf\{k > \tau_z : |h_k| \leq 1\}, \quad z \in \mathbb{Z}_+$$

We have that, if $|h_t| > 1$ (under-zoomed) $E[h_{t+1}^2 | h_t, x_t] \leq \frac{(a^2 + \frac{E[w_1^2]}{|x_t|^2})}{(a+\delta)^2}(h_t)^2$. Since when $|h_t| > 1$, we have that $|x_t| > 2^{R'-1}L\alpha$, it follows that

$$E[h_{t+1}^2 | h_t, x_t] \leq (\frac{a^2 + \frac{E[w_1^2]}{E_1^2}}{(a+\delta)^2})(h_t)^2.$$

If $|h_t| \leq 1$, then

$$E[h_{t+1}^2] \leq \frac{a^2 \frac{(\Delta_t)^2}{4} + E[w_1^2]}{(\Delta_t 2^{R'-1})^2}(\frac{1}{\alpha})^2 \leq \frac{a^2 \frac{L'^2}{4} + E[w_1^2]}{(L' 2^{R'-1})^2}(\frac{1}{\alpha})^2 =: K_1,$$

where $L' = L\frac{|a|}{|a|+\epsilon-\eta}$ (this is a lower bound on Δ_t). Hence, it follows that

$$E[h_{t+1}^2 - h_t^2|h_t, x_t] \leq -\rho h_t^2 + K_1 1_{\{|h_t|\leq 1\}}, \tag{13.37}$$

where 1_U is the indicator function for event U with $\rho = 1 - \frac{(a^2+\frac{E[w_1^2]}{E^2})}{(a+\delta)^2}$. Since for $A, B > 0$, $A^2 + B^2 \leq (A+B)^2$ it follows that letting L' such that $\sqrt{\frac{E[w_t^2]}{L'2^{R'-1}}} < \delta$ will ensure $\rho > 0$. Now, let us define $K_1' := K_1 + 1$, $M_0 := V(h_0)$, and for $t \geq 1$,

$$M_t := V(h_t) - \sum_{i=0}^{t-1}(-\rho + K_1' 1_{\{h_i \in C_h\}})$$

Define a stopping time: $\tau^N = \min(N, \min\{i > 0 : V(h_i) + \rho t \geq N\}, \min\{i > 0 : V(h_i) \leq 1\})$. Since, $E[M_{t+1}|(x_s, h_s), s \leq t] \leq M_t, \quad \forall t \geq 0$, it follows that, $\{M_t\}$ is a supermartingale sequence. The stopping time τ^N is bounded and the supermartingale sequence is integrable for $t \leq \tau^N$. Hence, we have, by the Martingale Optional Sampling Theorem (see Appendix D.2): $E[M_{(\tau^N)}] \leq E[M_0]$. Thus, we obtain $E[\sum_{i=0}^{\tau^N-1} \rho] \leq V(h_0) + K_1' E[\sum_{i=0}^{\tau^N-1} 1_{\{h_i \in C_h\}}]$, leading to $\rho E[\tau^N - 1 + 1] \leq V(h_0) + K_1'$, and by the Monotone Convergence Theorem,

$$\rho \lim_{N\to\infty} E[\tau^N] = \rho E[\tau] \leq V(h_0) + K_1' \leq 1 + K_1',$$

$$E[\tau_{z+1} - \tau_z] \leq (1 + K_1')/\rho, \tag{13.38}$$

uniformly for $h_{\tau_z} \in C_h$. The above also indicates that if the initial condition is not zoomed, in finite time the process will reach a perfectly zoomed state almost surely. By the strong Markov property (x_{τ_z}, h_{τ_z}) is also a Markov chain as $\{\tau_z < n\} \in \mathcal{F}_n$, the filtration generated by the quantizer state and the quantizer output at time n. The probability that $\tau_{z+1} \neq \tau_z + 1$, is upper bounded by:

$$\begin{aligned}
&P\Big(\{|a|\Delta_{\tau_z}/2 + w_{\tau_z} \geq (2^{R'-1})\Delta_{\tau_z}\frac{1}{\alpha}\}\\
&\qquad\qquad \bigcup\{-|a|\Delta_{\tau_z}/2 + w_{\tau_z} \leq -(2^{R'-1})\Delta_{\tau_z}\frac{1}{\alpha}\}\Big)\\
&= P\Big((w_{\tau_z})^2 > (\frac{\Delta_{\tau_z}}{2}(\frac{2^{R'}}{\alpha} - |a|))^2\Big)\\
&=: P_e(\Delta_{\tau_z})
\end{aligned} \tag{13.39}$$

If $\tau_{z+1} \neq \tau_z + 1$, this means that the error is increasing on the average and the system is once-again under-zoomed at time $t = \tau_z + 1$: $x_{\tau_z+1} = a x_{\tau_z} + w_{\tau_z}$ with

$\Delta_{\tau_z+1} = \frac{|a|}{|a|+\epsilon-\eta}\Delta_{\tau_z}$ (when $\Delta_{\tau_z} \geq L$). With some positive probability, the quantizer will still be in the perfect-zoom phase: $\tau_{z+1} = \tau_z + 1$. In case perfect-zoom is lost, there is a uniform bound on when the zoom is expected to be recovered. It follows that, conditioned on increment in the error, until the next stopping time, the process will increase exponentially, and hence

$$x_{\tau_{z+1}} = a^{\tau_{z+1}-\tau_z}(x_{\tau_z} + \sum_{t=0}^{\tau_{z+1}-\tau_z-1} a^{-t-1} w_{t+\tau_z}).$$

We now show that there exist $\psi > 0$, $|G| < \infty$ such that

$$E[\log(\Delta^2_{\tau_{z+1}})|\Delta_{\tau_z}, h_{\tau_z}] \leq \log(\Delta^2_{\tau_z}) - \psi + G1_{\{|\Delta_{\tau_z}| \leq F\}} \tag{13.40}$$

Now, it follows that,

$$E[\log(\Delta^2_{\tau_{z+1}})|\Delta_{\tau_z}, h_{\tau_z}] \leq \log(\Delta^2_{\tau_z}) - \psi + G1_{\{|\Delta_{\tau_z}| \leq F'\}},$$

with

$$F' = \frac{\sqrt{E[w_1^2]}\sqrt{(2(1+K_1')/\rho)\log(|a|+\delta)}}{K_2\sqrt{-\psi - 2\log(\alpha)}},$$

and $G = 2\log(F') + 2((1+K_1')/\rho)\log(|a|+\delta) + \psi$ for some $\psi > 0$. Hence, we have obtained another drift condition for the sampled Markov chain.

Together with (13.38), this leads to, by Theorem G.1.6 (which follows from Theorem G.1.4 and Corollary G.1.1, except the irreducibility condition), the following result: The newly constructed process Δ_{τ_z} hits the set $\{\Delta_t : |\Delta_t| \leq F'\}$ infinitely often with finite expected return time. This is equivalent to x_{τ_z} hitting the set $C'_x = \{x : |x| \leq F := 2^{R'-1}F'\}$ with finite expected return time.

In the above, we assumed that the initial state is perfectly zoomed. If the initial condition is not in a perfect-zoom phase, with probability 1, in finite time the state process will move to this phase by (13.37) and the subsequent discussion. ◇

13.8.3 Proof of Theorem 13.2.5

Toward the proof, we state a number of supporting results:

Lemma 13.8.2 *Let $z = 0$, $\tau_z = 0$ and $\tau_1 = \tau_{z+1} - \tau_z$. It follows that, for $k \in \mathbb{Z}_+$,*

$$P(\tau_1 > k|x_0, \Delta_0) \leq Ce^{-((\zeta^{k-1}N-1/2)\Delta_0)^2/(2\sigma'^2)},$$

uniformly for $|h_0| \leq 1$ with

$$\sigma'^2 = \frac{E[w_1^2]}{1-|a|^{-2}}, \quad \zeta = \frac{|a|+\delta}{|a|}, \quad N = (2^{R'-1}\frac{\alpha}{|a|}),$$

$$C = \sigma' \frac{2}{\sqrt{2\pi}(2N-1)\Delta_0/2}.$$

Now, since a decaying exponential decays faster than a polynomial, and the bound above is decreasing in Δ_0, there exists an $M < \infty$ such that, with $r > (|a|+\delta)^2$, and for all $\Delta_0 \geq L'$,

$$P(\tau_1 = k) < Ce^{-((\zeta^{k-2}N-1/2)\Delta_0)^2/(2\sigma'^2)} \leq Mr^{-k}, \quad \forall k \in \mathbb{Z}_+ \tag{13.41}$$

Lemma 13.8.3 *If for some $\gamma > 0$, $b < \infty$, the following holds*

$$\gamma E[\sum_{k=0}^{\tau_1-1} \Delta_k^2 | x_0, \Delta_0] \leq \Delta_0^2 - E[\Delta_{\tau_1}^2 | x_0, \Delta_0] + b 1_{\{(\Delta_0, h_0) \in (C_x' \times C_h)\}},$$

then $\lim_{t\to\infty} E[\Delta_t^2 | x_0, \Delta_0] < \infty$.

The proof of this result follows from Theorem G.1.4. Let us now note that (with a simple bounding argument in the last inequality)

$$E[\sum_{t=0}^{\tau_1-1} \Delta_t^2 | x_0, \Delta_0] = \sum_{l=1}^{\infty} P(\tau_1 = l) \sum_{k=0}^{l-1} E[\Delta_k^2 | \tau_1 = l, x_0, \Delta_0]$$

$$\leq \Delta_0^2 \sum_{l=1}^{\infty} P(\tau_1 = l) \sum_{k=0}^{l-1} (|a|+\delta)^{2k}$$

$$\leq \Delta_0^2 \left(\sum_{l=1}^{\infty} M(r^{-l}) \frac{(|a|+\delta)^{2l} - 1}{(|a|+\delta)^2 - 1} \right)$$

$$\leq M\Delta_0^2 \left(\frac{1}{1 - r^{-1}(|a|+\delta)^2} - \frac{1}{1-r^{-1}} \right) \frac{1}{(|a|+\delta)^2 - 1}.$$

Toward obtaining a bound on $E[\Delta_{\tau_1}^2 | x_0, \Delta_0]$,

$$E[\Delta_{\tau_1}^2 | x_0, \Delta_0]$$
$$= P(\tau_1 = 1) E[\Delta_{\tau_1}^2 | \tau_1 = 1, x_0, \Delta_0] + P(\tau_1 > 1) E[\Delta_{\tau_1}^2 | \tau_1 > 1, x_0, \Delta_0]$$
$$\leq P(\tau_1 = 1) \Delta_0^2 (\frac{|a|}{|a| + \epsilon - \eta})^2 + P(\tau_1 > 1) \sum_{k=2}^{\infty} P(\tau_1 = k) E[\Delta_k^2 | \tau_1 = k, x_0, \Delta_0]$$

$$\leq P(\tau_1 = 1)\Delta_0^2(\frac{|a|}{|a|+\epsilon-\eta})^2 + P(\tau_1 > 1)\sum_{k=2}^{\infty} Mr^{-k}(|a|+\delta)^{2k}\Delta_0^2$$

$$\leq P(\tau_1 = 1)\Delta_0^2(\frac{|a|}{|a|+\epsilon-\eta})^2$$

$$+P(\tau_1 > 1)M((|a|+\delta)^2 r^{-1})^2\frac{\Delta_0^2}{1-r^{-1}(|a|+\delta)^2}$$

Thus, we require, for some $\gamma > 0$, and sufficiently large Δ_0

$$\left(P(\tau_1 = 1)\Delta_0^2\alpha^2 + P(\tau_1 > 1)M((|a|+\delta)^2 r^{-1})^2\frac{\Delta_0^2}{1-r^{-1}(|a|+\delta)^2}\right)$$

$$< -\gamma M\Delta_0^2\left((\frac{1}{1-r^{-1}(|a|+\delta)^2} - \frac{1}{1-r^{-1}})\frac{1}{(|a|+\delta)^2-1}\right) + \Delta_0^2$$

Since $\lim_{\Delta_0\to\infty} P(\tau_1 > 1) = 0$, $P_e(\Delta_{\tau_z}) \leq E[d_{\tau_z}^2]/(K_2\Delta_{\tau_z})^2$, for some sufficiently small γ, say:

$$\gamma < \frac{(1-\alpha^2)}{\left(M(\frac{1}{1-r^{-1}(|a|+\delta)^2} - \frac{1}{1-r^{-1}})\frac{1}{(|a|+\delta)^2-1}\right)},$$

the desired stability result follows for $\{\Delta_t\}$, that is $\lim_{t\to\infty} E[\Delta_t^2|x_0, \Delta_0] < \infty$. We now have the final result:

Lemma 13.8.4 *If* $R > \log_2(\lceil|a|+\epsilon\rceil + 1)$, *then* $\lim_{t\to\infty} E[x_t^2|x_0, \Delta_0] < \infty$.

Thus, with $R > \log_2(\lceil|a|+\epsilon\rceil + 1)$, stability with a finite second moment is achieved. Finally, the limit is independent of the initial distribution since the chain is irreducible, by Theorem 13.2.4. This completes the proof of the theorem. ◇

13.8.4 Proof of Theorem 13.2.8

We assume, without any loss of generality, that all the eigenvalues are unstable. Let P denote P_{x_0,Δ_0}. We first note that, by an application of the union bound,

$$P(\tau_1 > k) \leq P(\cup_{k=1}^{n}(|h_k^i| > 1)|\text{zoom until} \quad k))$$

$$\leq \sum_{k=1}^{n} P(|h_k^i| > 1|\text{zoom until} \quad k).$$

Consider a two-dimensional Jordan block example considered in (13.9). In the following, since $|x_0^1 - \hat{x}_0^1| < \Delta_0^1/2$, we will, by an abuse of notation, let $|x_0^1| \leq \Delta_0^1/2$ and likewise, $|x_0^2| \leq \Delta_0^2/2$. Let $\delta^1 = \delta^2 = \delta$. For $P(|h_k^1| > 1)$, we have the following:

$$
\begin{aligned}
&P(|h_k^1| > 1) \\
&\leq P\Big(|(\sum_{i=0}^{k-1} \lambda^{-i-1} w_i^1 + \sum_{i=0}^{k-1}\sum_{j=0}^{i-1} \lambda^{-j-2} w_j^2)| \\
&\quad > \Delta_0^1/2\Big((\frac{|\lambda| + \delta}{|\lambda|})^{k-1} \frac{\alpha}{|\lambda|} 2^{R_1'} - (1 + k|\lambda|^{-1}\eta_\Delta)\Big)\Big)
\end{aligned} \tag{13.42}
$$

Under Assumption 13.2.2, the bound in (13.42) can be used to obtain:

$$
\begin{aligned}
&P\Big(|(\sum_{i=0}^{k-1} \lambda^{-i-1} w_i^1 + \sum_{i=0}^{k-1}\sum_{j=0}^{i-1} \lambda^{-j-2} w_j^2)|^{2+\epsilon} \\
&\quad > \Big(\Delta_0^1/2(\frac{|\lambda| + \delta}{|\lambda|})^{k-1} \frac{\alpha}{|\lambda|} 2^{R_1'} - (1 + k|\lambda|^{-1}\eta_\Delta)\Big)^{2+\epsilon}\Big).
\end{aligned} \tag{13.43}
$$

Now, it can be shown that $|(\sum_{i=0}^{k-1} \lambda^{-i-1} w_i^1 + \sum_{i=0}^{k-1}\sum_{j=0}^{i-1} \lambda^{-j-2} w_j^2)|^{2+\epsilon} \leq C_B k^{2(2+\epsilon)}$ for some $C_B < \infty$. By an application of Markov's inequality and with the condition $R_1' > \log_2(\frac{|a|}{\alpha})$ the drift criteria and the analysis in Sect. 13.8.3 follows, and the finite moment conditions are satisfied.

A parallel argument applies for dimensions higher than 2, leading to the desired conclusion. For a complete proof, see [373]. ◇

13.8.5 *Proof of Theorem 13.6.2*

For channels of the type **Class A** (which includes the discrete memoryless channels (DMC) as a special case), the capacity is given by:

$$
C = \lim_{T\to\infty} \max_{\{P(q_i | q_{[0,t-1]}, q'_{[0,t-1]})\}} \frac{1}{T} I(q_{[0,T-1]} \to q'_{[0,T-1]}).
$$

where

$$
I(q_{[0,T-1]} \to q'_{[0,T-1]}) = \sum_{t=1}^{T-1} I(q_{[0,t]}; q'_t | q'_{[0,t-1]})
$$

Let us define $R_T = \max_{\{P(q_t|q_{[0,t-1]},q'_{[0,t-1]}),0\le t\le T-1\}} \frac{1}{T}\sum_{t=0}^{T-1} I(q'_t;q_{[0,t]}|q'_{[0,t-1]})$. Observe that for $t > 0$:

$$\begin{aligned}
I(q'_t; q_{[0,t]}|q'_{[0,t-1]}) &= H(q'_t|q'_{[0,t-1]}) - H(q'_t|q_{[0,t]}, q'_{[0,t-1]}) \\
&= H(q'_t|q'_{[0,t-1]}) - H(q'_t|q_{[0,t]}, x_t, q'_{[0,t-1]}) \quad (13.44) \\
&\ge H(q'_t|q'_{[0,t-1]}) - H(q'_t|x_t, q'_{[0,t-1]}) \\
&= I(x_t; q'_t|q'_{[0,t-1]}). \quad (13.45)
\end{aligned}$$

Here, (13.44) follows from the assumption that the channel is of **Class A** type (and the control actions are determined by channel outputs). It follows that since for two sequences $\{a_n\}$, $\{b_n\}$ with $a_n \ge b_n$, we have $\limsup_n a_n \ge \limsup_n b_n$, and R_T is assumed to have a limit:

$$\lim_{T\to\infty} R_T$$

$$\ge \limsup_{T\to\infty} \frac{1}{T}\left(\sum_{t=1}^{T-1} I(x_t; q'_t|q'_{[0,t-1]})) + I(x_0; q'_0)\right)$$

$$= \limsup_{T\to\infty} \frac{1}{T}\left(\sum_{t=1}^{T-1} \left(h(x_t|q'_{[0,t-1]}) - h(x_t|q'_{[0,t]})\right) + I(x_0; q'_0)\right)$$

$$= \limsup_{T\to\infty} \frac{1}{T}\left(\sum_{t=1}^{T-1} \left(h(Ax_{t-1} + Gw_{t-1} + Bu_{t-1}|q'_{[0,t-1]}) - h(x_t|q'_{[0,t]})\right) + I(x_0; q'_0)\right)$$

$$= \limsup_{T\to\infty} \frac{1}{T}\left(\sum_{t=1}^{T-1} \left(h(Ax_{t-1}+Gw_{t-1}|q'_{[0,t-1]}) - h(x_t|q'_{[0,t]})\right) + I(x_0; q'_0)\right) \quad (13.46)$$

$$\ge \limsup_{T\to\infty} \frac{1}{T}\left(\sum_{t=1}^{T-1} \left(h(Ax_{t-1} + Gw_{t-1}|q'_{[0,t-1]}, w_{t-1}) - h(x_t|q'_{[0,t]})\right) + I(x_0; q'_0)\right) \quad (13.47)$$

$$= \limsup_{T\to\infty} \frac{1}{T}\left(\sum_{t=1}^{T-1} \left(h(Ax_{t-1}|q'_{[0,t-1]}, w_{t-1}) - h(x_t|q'_{[0,t]})\right) + I(x_0; q'_0)\right)$$

$$= \limsup_{T\to\infty}\frac{1}{T}\left(\sum_{t=1}^{T-1}\left(h(Ax_{t-1}|q'_{[0,t-1]}) - h(x_t|q'_{[0,t]})\right) + I(x_0;q'_0)\right) \quad (13.48)$$

$$= \limsup_{T\to\infty}\frac{1}{T}\left(\sum_{t=1}^{T-1}\left(\log_2(|A|)+h(x_{t-1}|q'_{[0,t-1]})-h(x_t|q'_{[0,t]})\right)+I(x_0;q'_0)\right)$$

$$= \limsup_{T\to\infty}\frac{1}{T}\left(\left(\sum_{t=1}^{T-1}\log_2(|A|)\right)+h(x_0|q'_0)-h(x_{T-1}|q'_{[0,T-1]})+I(x_0;q'_0)\right)$$

$$= \log_2(|A|)-\liminf_{T\to\infty}\left(\frac{1}{T}h(x_{T-1}|q'_{[0,T-1]})\right) \quad (13.49)$$

$$\geq \log_2(|A|) - \liminf_{T\to\infty}\left(\frac{1}{T}h(x_{T-1})\right) \geq \log_2(|A|). \quad (13.50)$$

Equality (13.46) follows from the fact that the control action is a function of the past channel outputs, (13.47) follows from the fact that conditioning does not increase entropy, (13.48) follows from the observation that $\{w_t\}$ is an independent process. (13.50) follows from conditioning. The other (intermediate) lines follow from the properties of mutual information. By the hypothesis, $\liminf_{t\to\infty}\frac{1}{t}h(x_t) \leq 0$, it must be that $\lim_{T\to\infty} R_T \geq \log_2(|A|)$. Thus, the capacity also needs to satisfy this bound. ◇

13.8.6 Proof of Lemma 13.6.1

Before stating the proof we note that a further, more general, result is presented in Lemma 14.2.1, in the context of non-linear systems. The proof follows from that of Theorem 13.6.2, with the following differences (where we build on [513]).

Define the event for $K > 0$ so that $P(|x_0| < K) > 0$, as

$$\mathcal{S}_\eta^K = \{\omega : |x_0| \leq K, w = \eta, i.e., w_k = \eta_k, \eta_k \in \mathbb{R}^p, k \geq 0\},$$

such that the noise realizations are fixed and deterministic. Here, $|\cdot|$ denotes the l_2 norm. In the following, we will drop the subscript and superscripts and let $P_{\mathcal{S}}$ or $P(\cdot|\mathcal{S})$ denote the conditional probabilities given the event $\mathcal{S}_\eta^K$. We recall here that $w_t, t \geq 0$, and x_0 are assumed to be independent.

Define $\bar{x}_t = x_t - \sum_{k=0}^{t-1} A^{t-k-1}Bu_t$ to be the control-free state such that $\bar{x}_0 = x_0$ and $\bar{x}_{t+1} = A\bar{x}_t + Gw_t$. By Definition 13.6.1, we note that the capacity expression satisfies

$$C = \lim_{T\to\infty} \max_{\{P(q_t|q_{[0,t-1]},q'_{[0,t-1]}),\quad 0\leq t\leq T-1\}} \frac{1}{T}I(q_{[0,T-1]} \to q'_{[0,T-1]})$$

$$= \lim_{T\to\infty} \max_{\{P(q_t|q_{[0,t-1]},q'_{[0,t-1]}),\quad 0\le t\le T-1\}} \frac{1}{T} I(q_{[0,T-1]} \to q'_{[0,T-1]}|\mathcal{S}),$$

where the conditional directed information is given by

$$I(q_{[0,T-1]} \to q'_{[0,T-1]}|\mathcal{S}) = \sum_{t=1}^{T-1} I(q_{[0,t]}; q'_t|q'_{[0,t-1]}, \mathcal{S}) + I(q_0; q'_0|\mathcal{S}).$$

This result is a consequence of the equivalence of the expressions under a capacity achieving channel input policy and the Markov chain condition in Definition 13.6.1. Now, instead of (13.44), the following can be established

$$\begin{aligned} I(q'_t; q_{[0,t]}|q'_{[0,t-1]}, \mathcal{S}) &= H(q'_t|q'_{[0,t-1]}, \mathcal{S}) - H(q'_t|q_{[0,t]}, q'_{[0,t-1]}, \mathcal{S}) \\ &= H(q'_t|q'_{[0,t-1]}, \mathcal{S}) - H(q'_t|q_{[0,t]}, \bar{x}_T, q'_{[0,t-1]}, \mathcal{S}) \qquad (13.51) \\ &\ge H(q'_t|q'_{[0,t-1]}, \mathcal{S}) - H(q'_t|\bar{x}_T, q'_{[0,t-1]}, \mathcal{S}) \\ &= I(\bar{x}_T; q'_t|q'_{[0,t-1]}, \mathcal{S}). \end{aligned}$$

Here, (13.51) holds by conditioning and by the fact that the system variables do not affect the channel as a consequence of Definition 13.6.1. We also note that here the events are conditioned on the event realization $\mathcal{S}$. Thus, $TR_T \ge \sum_{t=0}^{T-1} I(\bar{x}_T; q'_t|q'_{[0,t-1]}, \mathcal{S}) = I(\bar{x}_T; q'_{[0,T-1]}|\mathcal{S})$. Let us write

$$I(\bar{x}_T; q'_{[0,T-1]}|\mathcal{S}) = h_{\mathcal{S}}(\bar{x}_T) - h_{\mathcal{S}}(\bar{x}_T|q'_{[0,T-1]}),$$

where the notation $h_{\mathcal{S}}(\cdot)$ denotes the conditioning on the event realization $\mathcal{S}$. The following then holds:

$$h_{\mathcal{S}}(\bar{x}_T|q'_{[0,T-1]}) \ge h_{\mathcal{S}}(\bar{x}_T) - TR_T. \qquad (13.52)$$

Furthermore,

$$\begin{aligned} h_{\mathcal{S}}(\bar{x}_T) &= h_{\mathcal{S}}(A^T(x_0 + \sum_{k=0}^{T-1} A^{-k-1}Gw_k)) \\ &= \log_2(|A|)T + h_{\mathcal{S}}((x_0 + \sum_{k=0}^{T-1} A^{-k-1}Gw_k)) \qquad (13.53) \end{aligned}$$

Observe that since u_t is a function of $q'_{[0,t]}$ for all $t \geq 0$,

$$h_{\mathcal{S}}(\bar{x}_T|q'_{[0,T-1]}) = h_{\mathcal{S}}(x_T|q'_{[0,T-1]}) \leq h_{\mathcal{S}}(x_T, \mathcal{Y}|q'_{[0,T-1]}),$$

where $\mathcal{Y}$ is a binary random variable which is 1 if $|x_T| \leq b(T)$ and 0 otherwise. Let

$$P(\mathcal{Y} = 1|\mathcal{S}) = P(|x_T| \leq b(T)|\mathcal{S}) =: p_T^{\mathcal{S}}.$$

Then,

$$\begin{aligned} h_{\mathcal{S}}(x_T, \mathcal{Y}|q'_{[0,T-1]}) &= h_{\mathcal{S}}(x_T|q'_{[0,T-1]}, \mathcal{Y}) + h_{\mathcal{S}}(\mathcal{Y}|q'_{[0,T-1]}) \\ &\leq h_{\mathcal{S}}(x_T|q'_{[0,T-1]}, \mathcal{Y}) + 1, \end{aligned}$$

since $\mathcal{Y}$ is binary. We have that

$$\begin{aligned} h_{\mathcal{S}}(x_T|q'_{[0,T-1]}, \mathcal{Y}) \leq\; & p_T^{\mathcal{S}} \frac{n}{2} \log_2(2\pi e b^2(T)) \\ & +(1 - p_T^{\mathcal{S}}) h_{\mathcal{S}}\left(x_T \middle| q'_{[0,T-1]}, |x_T| \geq b(T)\right) \end{aligned}$$

and

$$\begin{aligned} & h_{\mathcal{S}}\left(x_T \middle| q'_{[0,T-1]}, |x_T| > b(T)\right) \\ & = h_{\mathcal{S}}\left(A^T(x_0 + \sum_{k=0}^{T-1} A^{-k-1}(w_k + u_k)) \middle| q'_{[0,T-1]}, |x_T| > b(T)\right) \\ & = h_{\mathcal{S}}\left(A^T(x_0) \middle| q'_{[0,T-1]}, |x_T| > b(T)\right) \qquad (13.54) \\ & \leq \log_2(|A|)T + \frac{n}{2} \log_2(2\pi e K^2). \qquad (13.55) \end{aligned}$$

Here, (13.54) follows since $\mathcal{S}$ determines the noise realizations and the channel outputs determine the control action realizations given a control policy; (13.55) follows from the fact that the Gaussian measure maximizes the differential entropy among random variables with a fixed covariance matrix, which in this case is further bounded by the effects of the event $\mathcal{S}$ as $|x_0|$ is bounded by K. We have then, by (13.52),(13.53), and (13.55), and by adjusting T and $\mathcal{S}$ accordingly,

$$\begin{aligned} \liminf_{T\to\infty} \frac{1}{T}\Bigg(& 1 + (1 - p_T^{\mathcal{S}})(\log_2(|A|)T + \frac{n}{2} \log_2(2\pi e K^2)) \qquad (13.56) \\ & + p_T^{\mathcal{S}} \frac{n}{2} \log_2(2\pi e b^2(T))\Bigg) \geq \log_2(|A|) - C, \end{aligned}$$

It then follows that, $\limsup_{T\to\infty} P(|x_T| \leq b(T)|\mathcal{S}) \leq \frac{C}{\log_2(|A|)}$, for all $b(T)$ such that $\log_2(b(T))/T = 0$. Thus,

$$\begin{aligned}
&\limsup_{T\to\infty} P(|x_T| \leq b(T)) \\
&\leq \limsup_{T\to\infty} P(|x_T| \leq b(T), |x_0| \leq K) + \limsup_{T\to\infty} P(|x_T| \leq b(T), |x_0| \geq K) \\
&\leq \limsup_{T\to\infty} P(|x_T| \leq b(T), |x_0| \leq K) + P(|x_0| \geq K) \\
&= \limsup_{T\to\infty} \int P(d\eta) P_{\mathcal{S}^K_\eta}(|x_T| \leq b(T)) + P(|x_0| \geq K) \\
&\leq \int P(d\eta) \limsup_{T\to\infty} P_{\mathcal{S}^K_\eta}(|x_T| \leq b(T)) + P(|x_0| \geq K) \qquad (13.57) \\
&\leq \int P(d\eta) \frac{C}{\log_2(|A|)} + P(|x_0| \geq K) \\
&= \frac{C}{\log_2(|A|)} + P(|x_0| \geq K)
\end{aligned}$$

where we use Fatou's lemma in (13.57) and the fact that (13.56) holds for every restriction of the noise realizations η and K values. Since an individual probability measure is tight, $\lim_{K\to\infty} P(|x_0| \geq K) = 0$, the right hand side can be made arbitrarily close to $\frac{C}{\log_2(|A|)}$ and the result follows. ◇

13.8.7 Proof of Theorem 13.6.3

If the process is AMS (see Sect. D.3.1), then there exists a stationary measure $\bar{P}$ such that

$$\lim_{T\to\infty} \frac{1}{T} \sum_{k=1}^{T} P(T^{-k} D) = \bar{P}(D), \qquad (13.58)$$

for all (cylinder) events D. Let for $b_B \in \mathbb{R}_+$, $B \in \mathcal{B}(\mathbb{R}^N)$ be given by $B = \{x : |x| \leq b_B\}$ and $X_n(z) = z_n$ be the coordinate function (see Sect. D.3.1) where $z = \{z_0, z_1, z_2, \cdots\}$.

If, by Lemma 13.3.1

$$\limsup_{T\to\infty} P(|x_T| \leq b_B) \leq \frac{C}{\log_2(|A|))} < 1, \qquad (13.59)$$

holds for all $b_B \in \mathbb{R}_+$, then $\bar{P}_n(B) < \frac{C}{\log_2(|A|)}$ for all compact B, where $\bar{P}_n$ is the marginal probability on the nth coordinate defined as

$$\bar{P}_n(B) = \bar{P}(x : |X_n(x)| \leq b_B).$$

But then $\bar{P}_n$, as an individual probability measure, must be tight [116], therefore, for every $\delta > 0$ there exists $b_B < \infty$ such that $\bar{P}_n(B) \geq 1 - \delta$. But, by (13.58), this would imply that $\limsup_{t\to\infty} P(T^{-t}B) = \limsup_{t\to\infty} P(|x_t| \in B) \geq 1 - \delta$, leading to a contradiction with (13.59) for $\delta < 1 - \frac{C}{\log_2(|A|)}$. Hence, the AMS property cannot be achieved. ◇

13.8.8 Proof of Proposition 13.4.1

As in (13.33), we introduce a sequence of stopping times as

$$\tau_0 = 0, \quad \tau_{z+1} = \inf\{k > \tau_z : |h_k| \leq 1, p_k = 1\}, \quad z \in \mathbb{N}$$

By the strong Markov property and the nature of the stopping times, (x_{τ_z}, h_{τ_z}) is also a Markov chain. In the following, we show that there exist $b_0 > 0$, $b_1 < \infty$ such that

$$\mathbb{E}[\log(\Delta^2_{\tau_{z+1}})|\Delta_{\tau_z}, h_{\tau_z}] \leq \log(\Delta^2_{\tau_z}) - b_0 + b_1 1_{\{|\Delta_{\tau_z}| \leq F\}} \tag{13.60}$$

We first bound the probability $P(\tau_{z+1} - \tau_z \geq k|\Delta_{\tau_z}, h_{\tau_z})$ from above and below in the following two lemmas.

Lemma 13.8.5 *[829, Lemma 5.2] The discrete probability measure $P(\tau_{z+1} - \tau_z = k \mid x_{\tau_z}, \Delta_{\tau_z})$ has the upper bound*

$$P(\tau_{z+1} - \tau_z \geq k|x_{\tau_z}, \Delta_{\tau_z}) \leq (1-p)^{k-1} + G_k(\Delta_{\tau_z}),$$

where $G_k(\Delta_{\tau_z}) \to 0$ as $\Delta_{\tau_z} \to \infty$ uniformly in x_{τ_z}. ◇

Lemma 13.8.6 *The discrete probability measure $P(\tau_{z+1} - \tau_z = k \mid x_{\tau_z}, \Delta_{\tau_z})$ has the lower bound*

$$P(\tau_{z+1} - \tau_z \geq k|x_{\tau_z}, \Delta_{\tau_z}) \geq (1-p)^{k-1},$$

for all realizations of $x_{\tau_z}, \Delta_{\tau_z}$.

We can now invoke Theorem G.1.6. With the candidate Lyapunov function $V_0(x_t, \Delta_t) = \log(\Delta_t^2) + B_0$, for $\Delta_{\tau_z} > L$, we have that

$$\mathbb{E}[V_0(x_{\tau_{z+1}}, \Delta_{\tau_{z+1}}) \mid x_{\tau_z}, \Delta_{\tau_z}] = B_0 + P(\tau_{z+1} - \tau_z = 1)\Big(2\log(\alpha) + \log(\Delta^2_{\tau_z})\Big)$$

$$+ \sum_{k=2}^{\infty} \log(\Delta^2_{\tau_z+k}) P(\tau_{z+1} - \tau_z = k \mid x_{\tau_z}, \Delta_{\tau_z})$$

Thus, the drift satisfies:

$$\begin{aligned}
&\mathbb{E}[V_0(x_{\tau_{z+1}}, \Delta_{\tau_{z+1}}) \mid x_{\tau_z}, \Delta_{\tau_z}] - V_0(x_{\tau_z}, \Delta_{\tau_z}) \\
&= \sum_{k=1}^{\infty} 2\log((|a|+\delta)^{(k-1)}\alpha) P(\tau_{z+1} - \tau_z = k \mid x_{\tau_z}, \Delta_{\tau_z}) \\
&= 2\log(\alpha) + 2\sum_{k=1}^{\infty}(k-1)\log(|a|+\delta) P(\tau_{z+1} - \tau_z = k \mid x_{\tau_z}, \Delta_{\tau_z}). \quad (13.61)
\end{aligned}$$

By the summability of $\sum_{k=1}^{\infty} G_k(\Delta_{\tau_z})$ [829], and the dominated convergence theorem,

$$\begin{aligned}
&\lim_{\Delta_{\tau_z}\to\infty} \sum_{k=1}^{\infty}(k-1)((1-p)^{k-1} + G_k(\Delta_{\tau_z}) - (1-p)^k) \\
&\quad = \sum_{k=1}^{\infty} \lim_{\Delta_{\tau_z}\to\infty}(k-1)((1-p)^{k-1} + G_k(\Delta_{\tau_z}) - (1-p)^k) \\
&\quad = \sum_{k=1}^{\infty} p(1-p)^{k-1}(k-1) = p^{-1} - 1
\end{aligned}$$

Provided (13.27) holds, it follows that for some $b_0 > 0$,

$$\begin{aligned}
&\lim_{\Delta_{\tau_z}\to\infty} \left\{\mathbb{E}[V_0(x_{\tau_{z+1}}, \Delta_{\tau_{z+1}}) \mid x_{\tau_z}, \Delta_{\tau_z}] - V_0(x_{\tau_z}, \Delta_{\tau_z})\right\} \\
&= 2 \lim_{\Delta_{\tau_z}\to\infty} \left\{\log(\alpha) + \sum_{k=1}^{\infty}(k-1)\log(|a|+\delta) P(\tau_{z+1} - \tau_z = k \mid x_{\tau_z}, \Delta_{\tau_z})\right\} \\
&\le -b_0\,. \quad (13.62)
\end{aligned}$$

For Δ_{τ_z} in a compact set and lower bounded by L' defined by (13.29), $\mathbb{E}[\log(\Delta^2_{\tau_{z+1}}) \mid x_{\tau_z}, \Delta_{\tau_z}]$ is uniformly bounded. This follows from the representation of the drift given in (13.61). Finally, since,

$$G_k(\Delta_{\tau_0}) \le (1-p)^{k-1}\Gamma_m \Delta_{\tau_0}^{-2m} \frac{1}{1-((1-p)\xi^{2m})^{-1}},$$

it follows that $\sum_{k=1}^{\infty} G_k(\Delta_{\tau_0})k < \infty$ and as a result

$$\sup_{x_{\tau_z}, \Delta_{\tau_z}} \mathbb{E}_{x_{\tau_z}, \Delta_{\tau_z}}[\tau_{z+1} - \tau_z] < \infty. \quad (13.63)$$

Consequently, under the bound (13.27), there exist $b_0 > 0$, $b_1 < \infty$, $F' > 0$ such that,

$$\mathbb{E}[V_0(x_{\tau_{z+1}}, \Delta_{\tau_{z+1}})|x_{\tau_z}, \Delta_{\tau_z}] \leq V_0(x_{\tau_z}, \Delta_{\tau_z}) - b_0 + b_1 1_{\{|\Delta_{\tau_z}| \leq F'\}}. \quad (13.64)$$

With (13.63) and (13.64), Corollary G.1.1 leads to positive Harris recurrence.

Together with Lemma G.1.6 and Lemma 13.8.5, (13.63) and (13.64) imply the existence of a set $C_x \times C_\Delta$ satisfying the property that

$$\sup_{(x,\Delta) \in C_x \times C_\Delta} \mathbb{E}_{x,\Delta}[\tau_{C_x \times C_\Delta}] < \infty.$$

To verify the property $P_{(x,\Delta)}(\tau_{C_x \times C_h} < \infty) = 1$ for any admissible (x, Δ), the argument follows, as before, from the construction of

$$\Theta_k(\Delta, x) := P(\tau_1 \geq k \mid x, \Delta),$$

where

$$\tau_1 = \inf(k > 0 : |x_k| \leq 2^{R'-1}\Delta_k, x_0 = x, \Delta_0 = \Delta)$$

and observing that $\Theta_k(\Delta, x)$ is majorized by a geometric measure with similar steps. Once a state which is perfectly zoomed, that is which satisfies $|x_t| \leq 2^{R'-1}\Delta_t$, is visited, the stopping time analysis can be used to verify that from any initial condition the recurrent set is visited in finite time with probability 1. ◇

13.9 Concluding Remarks

In this chapter, we have considered constructions of quantizers leading to stochastic stability for an open-loop unstable system driven by noise. In particular, we have shown that such quantizers are rate-efficient, in addition to being simple and fixed-rate. One important result in this context is the observation that there exists a finite limit for the sequence of second moments $\{E[|x_t|^2], \quad t \in \mathbb{Z}_+\}$. The chapter has also considered stochastic stabilization over noisy channels of linear systems driven by unbounded noise, and established tight conditions for asymptotic mean stationarity and ergodicity, as well as sufficient conditions for stability in the sense of having a finite average second moment for the state process. One message was that Shannon capacity and the stochastic stabilizability (in the sense of the AMS property or ergodicity) are inherently related. This relationship leads to stronger results in the context of discrete noiseless channels and erasure channels: For these channels, we have established not only asymptotic mean stationarity and n-ergodicity, but also stationarity and ergodicity. We have also presented conditions for quadratic and finite second moment stability. We note again the assumption that the system noise does not need to be Gaussian. For the second moment stability, a sufficiently light tail which would provide a geometric bound on the stopping times is sufficient. For the AMS property, this is not needed. For a noiseless DMC, a finite second moment for the system noise is sufficient for the existence of an invariant probability measure. To establish irreducibility, we require, however, that the noise admit a density which is positive everywhere. A detailed discussion involving noisy channels has also been presented, summarizing and further extending some of the findings in [824, Chapters 7-9].

13.10 Bibliographic Notes

We refer the reader to [824, Chapters 7-9] for a detailed literature review. In the following, we provide a brief discussion. References [55, 552, 707, 739, 782] are among the first publications which obtained minimum lower bounds needed for stabilization over communication channels, under various sets of assumptions on the system noise and channels, also known as *data-rate theorems*. Such theorems, typically, state that for stabilizability of linear systems under information constraints the minimum rate needed for stabilizability has to be at least the sum of the logarithms of the unstable poles/eigenvalues in the system; that is:

$$\sum_{k=1}^{m} \frac{1}{2} \max(0, \log(|\lambda_k|^2)). \tag{13.65}$$

Such results were then generalized in a variety of directions as extensively reviewed in [29, 273, 508, 517, 554, 824] as well as the relatively recent studies [417, 541, 597, 616, 799].

Part of this chapter is based on [373, 808–811, 821, 823], and [513]. The results on erasure channels are primarily based on [829], and [805].

Chapter 14
Stabilization of Non-Linear Systems: Information Theoretic vs. Geometric Analysis

In this chapter, we consider non-linear models controlled under information constraints. More specifically, this chapter is concerned with necessary and sufficient conditions on information channels in a networked non-linear stochastic control system for which there exist coding and control policies such that the controlled system is stochastically stable in one or more of the following senses: (i) The state $\{x_t\}$ and the coding and control parameters lead to a stochastically stable process (in senses to be made precise), and (ii) $\{x_t\}$ is asymptotically stationary, or asymptotically mean stationary (AMS), and (iii) $\{x_t\}$ is (asymptotically) ergodic.

We will pursue a similar proof program as in the linear case presented in Chap. 13 in that we will arrive at lower bounds and upper bounds.

For the lower bounds, however, we will present two alternative approaches: an information theoretic approach and a dynamical systems/geometric approach. We will see that the bounds obtained through these two approaches present complementary conditions, which are consistent and which collapse to identical results for the linear setup, recovering the lower bounds presented in Chap. 13. For upper bounds, as in the previous chapter, we will again consider stochastic drift analysis. Different from the linear case, however, in this case we will not be able to establish the equivalence of upper and lower bounds, but we will be able to generalize the arguments to include a much broader class of systems for which the linear case can be viewed as a special case.

The two approaches utilized will reveal different strengths and open problems, for future research. Our analysis based on the information theoretic approach will allow us to consider very general channels and sources, but will lead to more restrictive bounds. On the other hand, the geometric approach will let us to obtain stronger bounds for more specific sources and channels.

We will also see that the AMS criterion leads to an analysis which is more tedious when compared with the more stringent, but mathematically convenient, criterion of asymptotic ergodicity.

S. Yüksel, T. Başar, *Stochastic Teams, Games, and Control under Information Constraints*, Systems & Control: Foundations & Applications,
https://doi.org/10.1007/978-3-031-54071-4_14

Section 14.1 introduces the models considered. Section 14.2 is on the information theoretic approach towards stochastic stabilization. The stochastic geometric growth method is studied in Sect. 14.3, where the ergodic setup is presented in Sect. 14.4. A constructive achievability analysis is given in Sect. 14.5. A comparison between the information theoretic and geometric approaches is provided in Sect. 14.6.

14.1 Introduction

Consider an N-dimensional controlled non-linear system described by the discrete-time equations

$$x_{t+1} = f(x_t, u_t, w_t), \tag{14.1}$$

for a (Borel measurable) function f, with $\{w_t\}$ being an independent and identically distributed (i.i.d) system noise process with $w_t \sim \nu$.

This system is connected over a noisy channel with a finite capacity to a controller, as shown in Fig. 14.1. The controller has access to the information it has received through the channel. A source coder maps the source symbols, state values, to corresponding channel inputs. The channel inputs are transmitted through a channel; we assume that the channel is a finite alphabet channel with input alphabet $\mathcal{M}$ and output alphabet $\mathcal{M}'$.

As before, we refer by a *Coding Policy* Π, a sequence of functions $\{\gamma_t^e, t \geq 0\}$ which are causal such that the channel input at time t, $q_t \in \mathcal{M}$, under Π^{comp}, is generated by a function of its local information, that is,

$$q_t = \gamma_t^e(\mathcal{I}_t^e),$$

where $\mathcal{I}_t^e = \{x_{[0,t]}, q'_{[0,t-1]}\}$ and $q_t \in \mathcal{M}$, the channel input alphabet given by $\mathcal{M} := \{1, 2, \ldots, M\}$, for $0 \leq t \leq T - 1$.

The channel maps q_t to q'_t in a stochastic fashion so that $P(q'_t | q_t, q_{[0,t-1]}, q'_{[0,t-1]})$ is a conditional probability measure on $\mathcal{M}'$ for all $t \in \mathbb{Z}_+$. If this expression is equal to $P(q'_t | q_t)$, the channel is said to be a memoryless channel, that is, the past variables

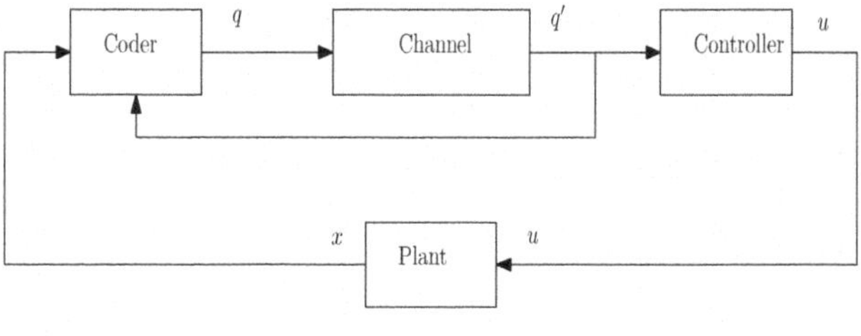

Fig. 14.1 Control of a system over a noisy channel

do not affect the channel output q'_t given the current channel input q_t. Even though in this chapter we will consider discrete alphabet channels, the analysis is also applicable to a large class of continuous alphabet channels (through an appropriate quantized approximation of the channel; see e.g. [236]).

The receiver/controller, upon receiving the information from the channel, generates its decision at time t, also causally: An admissible causal controller policy is a sequence of functions $\gamma = \{\gamma_t\}$ such that

$$\gamma_t : \mathcal{M}'^{t+1} \to \mathbb{R}^m, \qquad t \geq 0,$$

so that $u_t = \gamma_t(q'_{[0,t]})$.

14.2 Information Theoretic Approach to Arriving at Lower Bounds and Fundamental Limits

14.2.1 Sublinear Entropy Growth and a Generalization of Bode's Integral Formula for Non-linear Systems

Here, instead of a general $\mathbb{R}^N$-valued non-linear state model

$$x_{n+1} = f(x_n, u_n, w_n), \tag{14.2}$$

we will consider non-linear systems in either one of the following three forms:

$$x_{n+1} = f(x_n, w_n) + Bu_n, \tag{14.3}$$

$$x_{n+1} = f(x_n) + Bu_n + w_n, \tag{14.4}$$

$$x_{n+1} = f(x_n, u_n) + w_n. \tag{14.5}$$

We will also have an occasion to study non-linear systems of a fourth form:

$$x_{n+1} = f(x_n, w_n) + B(x_n)u_n. \tag{14.6}$$

In all of the models above, x_n is the $\mathbb{R}^N$-valued state, w_n is the $\mathbb{R}^N$-valued noise variable, u_n is $\mathbb{R}^s$ valued, and w_n is assumed to be an i.i.d. process with $w_n \sim \nu$.

We assume throughout that f is measurable and continuously differentiable in the state variable. For a possibly non-linear differentiable function $f : \mathbb{R}^N \to \mathbb{R}^m$, the Jacobian matrix of f is an $n \times m$ matrix function consisting of partial derivatives of f such that

$$J(f(x))(i, j) = \frac{\partial (f(x))_i}{\partial x_j}, \quad 1 \leq i \leq m, 1 \leq j \leq n.$$

We note that another common notation for the Jacobian is $Df(x)$.

We have the following two assumptions.

Assumption 14.2.1 *In the models considered above* $f(\cdot, w) : \mathbb{R}^N \to \mathbb{R}^N$ *is invertible for every realization of* w.

In the second assumption below, $|J(f)|$ will denote the absolute value of the determinant of the Jacobian. Furthermore, with $f_w(x) = f(x, w)$, we define $J(f(x, w)) := J(f_w(x))$.

Assumption 14.2.2 *There exist* $M_1 \in \mathbb{R}$ *and* $L_1 \in \mathbb{R}$ *so that for all* x, w

$$L_1 \leq \log_2(|J(f(x, w))|) \leq M_1$$

The following is our first result; it provides conditions for sublinear entropy growth (in time) which implies quadratic stability. The result will also be used in the next section and its proof leads to a generalization of Bode's Integral Formula as discussed further below. Let $\pi_t(B) = P(x_t \in B)$ for all Borel B.

Theorem 14.2.1 *[816] Consider the networked control problem over a Class A channel. Let (i) f have the form in (14.3), (ii) Assumptions 14.2.1 and 14.2.2 hold, and (iii)* x_0 *have finite differential entropy. Then,*

(a) If there is an admissible coding and control policy such that

$$lim\ inf_{t\to\infty} h(x_t)/t \leq 0,$$

it must be that

$$C \geq lim\ inf_{T\to\infty} \frac{1}{T} \sum_{t=0}^{T-1} \int \pi_t(dx) \Big(\int \nu(dw) \log_2(|J(f(x, w))|) \Big) \tag{14.7}$$

(b) If there is an admissible coding and control policy such that

$$lim\ sup_{t\to\infty} h(x_t)/t \leq 0,$$

it must be true that

$$C \geq lim\ sup_{T\to\infty} \frac{1}{T} \sum_{t=0}^{T-1} \int \pi_t(dx) \Big(\int \nu(dw) \log_2(|J(f(x, w))|) \Big) \tag{14.8}$$

(c) In either of the cases (a) and (b), if $L := \inf_{x,w} \log_2 |J(f(x, w))|$, *then* $C \geq L$.

Remark 14.2.1 The condition $\limsup_{t\to\infty} h(x_t)/t \leq 0$ is a weak one. For example a stochastic process whose second moment grows subexponentially so that $\limsup_{T\to\infty} \frac{\log(E[x_T^2])}{T} \leq 0$, satisfies this condition. Hence, *quadratic stability* implies this condition. ◇

Remark 14.2.2 In the theorem, we would have arrived at the same results if we had replaced $\limsup_{t\to\infty} h(x_t)/t \leq 0$ with $\limsup_{t\to\infty} \frac{1}{t} h(x_t|q'_{[0,t-1]}) \leq 0$. This condition would be more relevant for state estimation problems, where the goal is not necessarily to make the state stable, but to make the estimation error stable (where u_t would be the state estimate and $x_t - u_t$ would be the estimation error). Since $h(x_t|q'_{[0,t-1]}) \leq h(x_t)$, it is evident that the condition $h(x_t)/t \leq 0$ implies that $h(x_t|q'_{[0,t-1]})/t \leq 0$. ◇

Remark 14.2.3 We note that if the system had been of the model (14.6), the expression involving $J(f(x,w))$ would explicitly depend on the control policy, which would in turn depend possibly on the entire past channel outputs, making the expression computationally more involved. ◇

14.2.2 A Generalization of Bode's Integral Formula for Non-linear Systems

The proof of Theorem 14.2.1 reveals an interesting connection with and generalization of Bode's Integral Formula (and what is known as the *waterbed effect*) [538] [42, Theorem 11.1] to non-linear systems, which we state formally in the following. The result also suggests that an appropriate generalization for non-linear systems is through an information theoretic approach that recovers Bode's original result for the linear case as we discuss further below.

Theorem 14.2.2 *[816] Let (i) f have the form in (14.3), (ii) Assumption 14.2.1 hold, and (iii) x_0 have finite differential entropy. If there is an admissible coding and control policy with* $\limsup_{t\to\infty} h(x_t)/t \leq 0$, *then*

$$\begin{aligned} &\limsup_{T\to\infty} \frac{1}{T} I(q_{[0,T-1]} \to q'_{[0,T-1]}) \\ &\geq \limsup_{T\to\infty} \frac{1}{T} \sum_{t=0}^{T-1} \int \pi_t(dx) \Big(\int \nu(dw) \log_2(|J(f(x,w))|) \Big) \end{aligned} \tag{14.9}$$

Proof This follows directly from Eqs. (14.41),(14.43) and (14.50). □

Remark 14.2.4 (Reduction to Bode's Integral Formula for Linear Systems and Gaussian Noise) If the system considered is linear with all open-loop eigenvalues unstable, the channel is an additive noise channel so that $q'_t = q_t + v_t$ for some stationary Gaussian noise, and time-invariant control policies are considered leading to a stable system, then with the more common notation of $y_t = q'_t$, the right hand side of (14.9) would be the sum of the unstable eigenvalues of the linear system matrix. For a stationary Gaussian process [see [188], page 274] the entropy rate can be written as

$$\frac{1}{2}\log(2\pi e) + \int_{-1/2}^{1/2} \frac{1}{2}\log(S(f))df$$

with S denoting the spectral density of the process. Now, (14.41) becomes

$$\begin{aligned} I(q'_t; q_{[0,t]}|q'_{[0,t-1]}) &= h(q'_t|q'_{[0,t-1]}) - h(q'_t|q_{[0,t]}, q'_{[0,t-1]}) \\ &= h(q'_t|q'_{[0,t-1]}) - h(v_t|v_{[0,t-1]}), \end{aligned}$$

and thus the left hand side of (14.9) reduces to the difference between the entropy rate of the process q'_t (that is, $\lim_{t\to\infty} h(q'_t|q'_{[0,t-1]})$) and that of the stationary noise process v_t (that is, $\lim_{t\to\infty} h(v_t|v_{[0,t-1]})$). Then, the left hand side of (14.9) equals

$$\int_{-1/2}^{1/2} \frac{1}{2}\log(\frac{S_y(f)}{S_v(f)})df,$$

which then is equal to the integral of the log-sensitivity function (corresponding to the transfer function from the measurement disturbance process v_t to the output process q'_t). This leads to the celebrated *Bode's Integral Formula*. In the context of linear systems, earlier extensions of this formula have been studied in [237] with an information theoretic interpretation under the restriction to linear policies (see e.g. Theorem 4.6 in [237]), in [508], under more general possibly non-linear stabilizing control policies which lead to a stationary process, and in [528] and [840] for a class of non-linear noise-free systems. For a further modern treatment, see [246]. ◇

14.2.3 Asymptotic Mean Stationarity and Ergodicity

Consider now the system (14.4), under some admissible policy, controlled over a channel.

Assumption 14.2.3 *Assume that*

$$\begin{aligned} M &:= \sup_{x\in\mathbb{R}^N} \log_2|J(f(x))| < \infty, \\ L &:= \inf_{x\in\mathbb{R}^N} \log_2|J(f(x))| > -\infty. \end{aligned}$$

Lemma 14.2.1 ([816]) *Consider the system (14.4) controlled over a Class A type noisy channel with feedback with $h(x_0) < \infty$ under Assumption 14.2.3. If $C < L$, under any admissible policy,*

$$lim\ sup_{T\to\infty} P(|x_T| \leq b(T)) \leq \frac{M - (L - C)}{M},$$

for all $b(T) > 0$ *such that* $\lim_{T\to\infty} \frac{1}{T} \log_2(b(T)) = 0$.

The proof of Lemma 14.2.1 is in Sect. 14.7.2 of the Appendix. An implication of this result follows; this generalizes Theorem 13.6.3. The proof is in the Appendix.

Theorem 14.2.3 ([816]) *Consider the system (14.4) controlled over a Class A type noisy channel with feedback, and let Assumption 14.2.1 hold. If, under some causal encoding and controller policy, the state process is AMS, then the channel capacity* C *must satisfy* $C \geq L$.

14.2.4 Stationarity and Positive Harris Recurrence Under Structured (Stationary) Policies

In many applications, one uses a state-space formulation for coding and control policies. In the following, we will consider stationary update rules which have the form

$$\begin{aligned} q_t &= \gamma^e(x_t, m_t) \\ u_t &= \gamma^d(m_t, q'_t), \\ m_t &= \eta(m_{t-1}, q'_{t-1}), \end{aligned} \tag{14.10}$$

for functions γ^e, γ^d, and η. Here, m is an $\mathbb{S}$-valued memory or *quantizer state* variable. A large class of adaptive encoding policies have this form. This includes, delta modulation, differential pulse coded modulation (DPCM), adaptive differential pulse coded modulation (ADPCM), Goodman-Gersho type adaptive quantizers (see e.g. [419, 421]), as well as the coding schemes used for stabilization of networked control systems under fixed-rate codes [808]. Even further, jointly optimal source and channel codes for zero-delay coding schemes under infinite horizon optimization criteria also have the form above (where $\mathbb{S}$ is a space of probability measures [484]). We now present a necessary structural result on the encoders.

A Necessary Structural Result on the Encoders

Let m_t, the state of the encoder, take values in $\mathbb{S}$. Consider (14.4). A stabilizing time-invariant encoder/decoder/controller policy given (14.10), in general, cannot have $|\mathbb{S}| < \infty$.

Theorem 14.2.4 *Consider (14.4) with scalar x_t, and w_t with a probability measure ν such that it has a density positive everywhere and $E_\nu[\gamma^{-w}] < \infty$ for some $\gamma > 1$. Suppose that there exists $K > 0$ so that $\inf_{x>K} \frac{df}{dx}(x) > 1$ and $\frac{df}{dx}(x)$ is bounded. Then, a finite cardinality for $\mathbb{S}$, under (14.10), leads to a transient system in the sense that $P_x(\tau_S < \infty) < 1$ where for some $s > 0$, $S = (-\infty, s)$ is an open set containing the origin, $x > s$ and $\tau_S := \inf(t > 0 : x_t \in S)$. A similar result applies for the condition $\sup_{x<-K} \frac{df}{dx}(x) < -1$, with $S = (s, \infty)$ for some $s < 0$ and $x < s$.*

Proof Let $\inf_{x>K} \frac{df}{dx}(x) > \bar{a} > 1$. It follows from $f(x) = f(K) + \int_K^x \frac{df}{dx}(s)ds$ that for some $M < \infty$, $f(x) \geq M + \bar{a}x$ for $x > K$. Since both q'_t and m_t can take finitely many values, there exists U such that $|u_t| \leq U$ for all t. Let, with $\gamma > 1$, a Lyapunov function be picked as $V(x) = \gamma^{-x}$, defined for positive x. Now, it follows that for sufficiently large x: $E[V(x_{t+1})|x_t = x] \leq V(x)$, since $E[\gamma^{-(f(x)+u_t+w_t)}] = E[\gamma^{-f(x)}\gamma^{-u_t}\gamma^{-w_t}] \leq E[\gamma^{-(M-U)}\gamma^{-(\bar{a}x+w_t)}] = \gamma^{-(M-U)}\gamma^{-\bar{a}x}E[\gamma^{-w}]$, for all $x \in \{x : \gamma^{(\bar{a}-1)x} > E[\gamma^{-w}]\gamma^{-M+U)}\}$. Due to the additive noise process, the source can escape any bounded interval with a non-zero probability. As a result, by Theorem G.1.8, transience follows. □

Transience prohibits the existence of a stationary probability measure. The discussion above is parallel to Theorem 7.3.1 in [824] for linear systems. Related to the discussion above, for linear systems, the unboundedness of second moments in Proposition 5.1 in [552] and the transience of such a controlled state process was established in Theorem 4.2 in [823]. We also note that [596] studied conditions for stabilization when the control actions are uniformly bounded, the controlled multi-dimensional system is marginally stable and is driven by noise with unbounded support.

Stationarity and Ergodicity

Now, instead of asymptotic mean stationarity, we will consider the more stringent condition of (asymptotic) stationarity of the controlled source process. For ease in presentation we will assume that m_t takes values in a countable set, even though the extension to more general spaces is possible, without substantial changes in the derivation of the results.

Lemma 14.2.2 *If the channel is memoryless, the process (x_t, m_t) is a Markov chain.*

Proof For any $t \in \mathbb{N}$,

$$P(dx_t, m_t|x_s, m_s, s \leq t-1)$$
$$= \sum P(dx_t, m_t, q'_{t-1}|x_s, m_s, s \leq t-1)$$

$$= \sum P(dx_t|x_{t-1}, \gamma^d(m_{t-1}, q'_{t-1}))P(q'_{t-1}|\gamma^e(x_{t-1}, m_{t-1}))P(m_t|q'_{t-1}, m_{t-1})$$

$$= \sum P(dx_t, m_t, q'_{t-1}|x_{t-1}, m_{t-1}) = P(dx_t, m_t|x_{t-1}, m_{t-1}) \tag{14.11}$$

where we use (14.10). □

In the following, we assume that the channel is memoryless. For the Markov chain (x_t, m_t), let $\pi_t(B) = E[1_{\{(x_t \in B\}}]$ for all Borel B, that is, π_t is the marginal occupation probability for the state process x_t.

Theorem 14.2.5 ([816]) *Suppose that the encoding, control and the memory update laws are given by (14.10). Let (i) f have the form (14.3), (ii) Assumptions 14.2.1 and 14.2.2 hold, and (iii) $h(x_0) < \infty$. For the positive Harris recurrence of the process x_t, m_t (which implies the existence of a unique invariant measure π (and thus ergodicity)), it must be that*

$$C \geq \int \pi(dx)\Big(\int \nu(dw) \log_2(|J(f(x, w))|)\Big), \tag{14.12}$$

provided that lim sup$_{t\to\infty}\frac{1}{t}h(x_t) \leq 0$.

Remark 14.2.5 If one considers a more general control-affine model such as of the form (14.6) with $x_{t+1} = f(x_t, w_t) + B(u_t)x_t$, the condition would read as:

$$C \geq \int \pi(dx, m, q')\Big(\int \nu(dw) \log_2\Big(\Big|J\Big(f(x, w) + B(\gamma^d(m, q'))x\Big)\Big|\Big)\Big),$$

where π is invariant for the (enlarged) Markov chain (x_t, m_t, q'_t). ◇

14.3 Stochastic and Geometric Volume Growth Approach

14.3.1 Some Intuition via a Visit to the Deterministic Setup

For deterministic nonlinear systems, *invariance entropy*, a concept introduced by Colonius and Kawan [184], measures the smallest average data rate of a noiseless channel above which a compact subset Q of the state space can be made invariant by a controller receiving its state information through this channel. The essence of the idea behind this concept is as follows: If the controller has n bits of information available, it can distinguish between at most 2^n different states, and hence generate at most 2^n different control inputs.

To explain the concept of *invariance entropy* [184] (or an evidently equivalent notion of topological feedback entropy [553]), let $\mathbb{X}$ be a topological space, $\mathbb{U}$ a non-empty set, and

$$x_{n+1} = f(x_n, u_n) \tag{14.13}$$

be a discrete-time control system on $\mathbb{X}$, where $f : \mathbb{X} \times \mathbb{U} \to \mathbb{X}$ is a map with the property that $f_u := f(\cdot, u) : \mathbb{X} \to \mathbb{X}$ is continuous for each $u \in \mathbb{U}$. We merge all solutions of (14.13) into one map $\varphi : \mathbb{Z}_+ \times \mathbb{X} \times \mathbb{U}^{\mathbb{Z}_+} \to \mathbb{X}$,

$$\varphi(n, x, \underline{u}) := f_{u_{n-1}} \circ \cdots \circ f_{u_0}(x),$$

where $\mathbb{U}^{\mathbb{Z}_+}$ is the set of all sequences in $\mathbb{U}$ and $\underline{u} = (u_0, u_1, \ldots)$.

Let $Q \subset \mathbb{X}$ be a compact set with non-empty interior satisfying the following *strong invariance condition* (essentially, this is assumption SI in [553, p. 1586]): For every $x \in Q$ there is $u_x \in \mathbb{U}$ with $f(x, u_x) \in \text{int}Q$.

For a number $\tau \in \mathbb{Z}_+$, a set $\mathcal{S} \subset \mathbb{U}^\tau$ is called (τ, Q)-spanning if for every $x \in Q$ there is $\underline{u} \in \mathcal{S}$ such that $\varphi(j, x, \underline{u}) \in \text{int}Q$ for $j = 1, \ldots, \tau$. The minimal cardinality of such a set is denoted by $r_{\text{inv}}(\tau, Q)$ and the *invariance entropy* of Q is defined by

$$h_{\text{inv}}(Q) := \lim_{\tau \to \infty} \frac{1}{\tau} \log r_{\text{inv}}(\tau, Q),$$

where subadditivity guarantees the existence of the limit. We note also that in [243] it was proved that $h_{\text{inv}}(Q) = h_{\text{fb}}(Q)$, where $h_{\text{fb}}(Q)$ denotes topological feedback entropy [553].

Theorem 14.3.1 ([553] (see also [243])) *Consider system (14.13). Suppose that a sensor measures its states and is connected to a controller via a noiseless digital channel which carries one discrete-valued symbol per sampling interval, selected from a coding alphabet S_k of time-varying size. If the transmission data rate*

$$R = lim\ inf_{k \to \infty} \frac{1}{k} \sum_{j=0}^{k-1} \log |S_j|$$

of the channel satisfies $R > h_{\text{inv}}(Q)$, then a coder-controller pair exists, which renders Q invariant. If $R < h_{\text{inv}}(Q)$, then no such coder-controller pair exists.

An important result in the deterministic formulation is with regard to stabilization to a point: Under the assumptions that (i) f has the form in (14.13) with $\mathbb{X} = \mathbb{R}^N$, $U = \mathbb{R}^m$ and has continuous partial derivatives, (ii) there exists an equilibrium pair (x^*, u^*), i.e., $x^* = f(x^*, u^*)$, (iii) a local strong invariability condition is satisfied which relates the size of an invariant set and the size of a control action set in the sense that for any $\epsilon > 0$, there exist $\rho > 0$ so that for all $\epsilon' \in (0, \rho]$, the set $\{x : |x - x^*| \leq \epsilon'\}$ is strongly invariant with the control action set $U = \{u : |u - u^*| \leq \epsilon\}$, and (iv) the pair (A, B) is controllable, where A, B are the Jacobians of f with respect to state and control at x^*, u^*, [553] has shown that for convergence to the equilibrium an average rate $R > \sum_{|\lambda_i|>1} \log(|\lambda_i|)$ is sufficient, where $\{\lambda_i\}$ are the eigenvalues of the Jacobian of f_{u^*} at x^*.

Consequently, the number of control inputs needed to achieve the control objective (on a finite time interval) is a measure for the necessary information. The definition of invariance entropy thus reads

$$h_{\mathrm{inv}}(Q) := \lim_{\tau \to \infty} \frac{1}{\tau} \log r_{\mathrm{inv}}(\tau, Q),$$

where $r_{\mathrm{inv}}(\tau, Q)$ is the minimal number of control inputs needed to achieve invariance of Q on the time interval $[0, \tau]$ for arbitrary initial states in Q. It is relatively immediate to observe that the growth rate of $r_{\mathrm{inv}}(\tau, Q)$ is directly related to the rate of volume expansion for subsets of Q under the evolution of the system. Indeed, the faster the volume is expanded, the more coding regions, and hence the more different control inputs, would be necessary to keep the whole volume inside Q. Since, for every reasonable stabilization objective, it is necessary to keep certain volumes bounded (or even shrink them to zero), the same ideas as used in the definition of invariance entropy should work universally for stabilization over discrete channels. This intuition was rigorously verified in a number of publications, including [184, 199, 243, 404].

Stochastic Setup

We next demonstrate that an approach similar to the one discussed above is also applicable to stochastic systems, stochastic channels, and to stochastic stability, and this can also be utilized to recover the deterministic theory as a special case.

As before, we will have two criteria: Asymptotic mean stationarity (AMS) and ergodicity. We will see that one can arrive at complementary conditions.

As an auxiliary quantity to derive lower bounds on the necessary channel capacity for generating an AMS state process, we discuss a new concept of stabilization entropy, introduced in [408] (and an ergodic theoretic generalization in Sect. 14.4 introduced in [287]), inspired by both invariance entropy and measure-theoretic entropy of dynamical systems, in particular by a characterization of the latter due to [401] and a generalization thereof developed in [599]. Roughly speaking, stabilization entropy looks at the exponential growth rate of the number of length-n control sequences necessary to keep the state inside some set for a certain fraction of the number n of times with a certain positive probability. The corresponding set, the frequency of times, and the probability are parameters that can be adjusted, and the relation to channel capacity can only be established for certain choices of these parameters.

If $X^{\mathbb{Z}_+}$ is the set of all sequences in some set X, we write $\bar{x} = (x_t)_{t \in \mathbb{Z}_+}$ for elements of $X^{\mathbb{Z}_+}$ and $\theta : X^{\mathbb{Z}_+} \to X^{\mathbb{Z}_+}$ for the left shift operator, i.e.,

$$(\theta \bar{x})_t = x_{t+1} \quad \text{for all } t \in \mathbb{Z}_+,\ \bar{x} \in X^{\mathbb{Z}_+}.$$

Moreover, we write $\bar{x}_{[0,t]} = (x_0, x_1, \ldots, x_t)$ for $t \in \mathbb{Z}_+$ and $\mathcal{B}(X)$ for the Borel σ-field of a topological space X.

We assume that all measurable spaces are standard Borel and all random variables associated with a given control system are modeled on a common (standard Borel) probability space $(\Omega, \mathcal{F}, P)$. The standard Borel space assumption leads to useful universal measurability properties which are utilized in this chapter (see the earlier discussion in Sect. 5.4.2): A measurable image of a Borel set is called an *analytic* set [230, App. 2]. We note here that this is evidently equivalent to the seemingly more restrictive condition of being a *continuous* image of a Borel set. The following property will be utilized in our analysis: The image of a Borel set under a measurable map, and hence an analytic set, is universally measurable [230].

Throughout the chapter, we consider a stochastic control system

$$x_{t+1} = f(x_t, u_t, w_t), \quad t = 0, 1, 2, \ldots \tag{14.14}$$

This defines a measurable map $f : \mathbb{R}^N \times U \times W \to \mathbb{R}^N$, where $\mathbb{R}^N$ is endowed with the Borel σ-field $\mathcal{B}(\mathbb{R}^N)$, $(U, \mathcal{F}_U)$ is a measurable space and $(W, \mathcal{F}_W, \nu)$ a probability space. The noise is modeled by an i.i.d. sequence $(w_t)_{t \in \mathbb{Z}_+}$ of random variables on $(W, \mathcal{F}_W)$ with associated probability measure ν. The initial state x_0 is modeled by another random variable with probability measure π_0 on $(\mathbb{R}^N, \mathcal{B}(\mathbb{R}^N))$ and is assumed to be independent of $(w_t)_{t \in \mathbb{Z}_+}$.

We write $\varphi(t, x_0, \bar{u}, \bar{w})$, $t \in \mathbb{Z}_+$, for the unique trajectory with initial value $x_0 \in \mathbb{R}^N$ associated with the noise realization $\bar{w} \in W^{\mathbb{Z}_+}$ and the control sequence $\bar{u} \in U^{\mathbb{Z}_+}$.

We assume that an encoder, knowing the states $x_0, x_1, \ldots, x_t$ at time $t \in \mathbb{Z}_+$, transmits at time $t \in \mathbb{Z}_+$ a symbol q_t through a noiseless discrete channel to a decoder/controller. We assume that the decoder receives the signals without delay. The finite coding alphabet is denoted by $\mathcal{M}$ and the capacity of the channel is

$$C = \log \#\mathcal{M}.$$

Thus, at time t, the controller has the symbol string $q_{[0,t]} = (q_0, q_1, \ldots, q_t) \in \mathcal{M}^{t+1}$ available to generate the control input u_t. Any coding and control policy of this form is called a *causal coding and control policy*. A more general setup including a noisy channel will be introduced and studied in Sect. 14.3.4.

The control objective considered serves to make the state process $(x_t)_{t \in \mathbb{Z}_+}$ asymptotically mean stationary (AMS) (See Definition D.3.5).

14.3.2 A New Concept: Stabilization Entropy

Definition 14.3.1 [408] For any Borel set $B \subset \mathbb{R}^N$, $T \in \mathbb{Z}_{>0}$ and $\rho, r \in (0, 1)$, a set $S \subset U^T$ is called (T, B, ρ, r)-*spanning* if there exists a set $\tilde{\Omega} \in \mathcal{F}$ with $P(\tilde{\Omega}) \geq 1 - \rho$ so that for every $\omega \in \tilde{\Omega}$ there is $\bar{u} \in S$ with

$$\frac{1}{T}\#\{t \in [0; T-1] : \varphi(t, x_0(\omega), \bar{u}, \bar{w}(\omega)) \in B\} \geq 1 - r. \tag{14.15}$$

We write $s_B(T, \rho, r)$ to denote the smallest cardinality of a (T, B, ρ, r)-spanning set (where $s_B(T, \rho, r) = \infty$ if no finite (T, B, ρ, r)-spanning set exists) and define the (B, ρ, r)-*stabilization entropy of system* (14.14) *by*

$$h_B(\rho, r) := \limsup_{T\to\infty} \frac{1}{T} \log s_B(T, \rho, r).$$

Some remarks about this definition are now in order:

Remark 14.3.1

(i) As noted earlier, in (14.15), $\varphi(\cdot, x_0(\omega), \bar{u}, \bar{w}(\omega))$ denotes the solution of the recursion $x_{t+1} = f(x_t, u_t, w_t)$ with $x_0 = x_0(\omega)$, with u_t being the t-th element in the sequence $\bar{u}$ and $w_t = w_t(\omega)$, $\bar{w}(\omega) = (w_t(\omega))_{t\in\mathbb{Z}_+}$. In particular, the control sequences $\bar{u}$ in the above definition are not generated by a coding and control policy. Indeed, $h_B(\rho, r)$ is an intrinsic quantity of the open-loop system.

(ii) The existence and finiteness of (T, B, ρ, r)-spanning sets are not immediately clear from the definition. However, as we will see below, in relevant cases this is guaranteed. In general, we always have $0 \leq h_B(\rho, r) \leq \infty$.

(iii) There are some immediate monotonicity properties of the function $h_B(\cdot, \cdot)$. Namely, if r or ρ become smaller, $h_B(\rho, r)$ increases. This in particular implies the existence of corresponding limits as $r \to 0$ and $\rho \to 0$ (which may be infinite).

We now present a key lemma which relates the channel capacity necessary for stabilization to the stabilization entropy. In particular, it shows that finite (T, B, ρ, r)-spanning sets exist for appropriate choices of B, ρ, r, provided that the AMS property can be achieved.

Lemma 14.3.1 ([408]) *Assume that the AMS property is achieved via a causal coding and control policy over a noiseless channel of capacity C. Then, for every Borel set $B \subset \mathbb{R}^N$ with $0 < Q(B) < 1$, and all sufficiently small $\varepsilon > 0$, we have*

$$C \geq h_B\left(\frac{1+\frac{\varepsilon}{2}}{1+\varepsilon}, (1+\varepsilon)Q(B^c)\right).$$

If $Q(B) = 1$, then for all $r \in (0, 1)$ and $\varepsilon > 0$ sufficiently small, we have

$$C \geq h_B\left(\frac{1+\frac{\varepsilon}{2}}{1+\varepsilon}, (1+\varepsilon)r\right).$$

Lemma 14.3.1, while sounding technical, has significant consequences, since it allows for the application of volume-growth arguments that have been used in the literature for deterministic settings.

14.3.3 Volume-Expanding Systems

Here, we assume throughout that the measure π_0 of the random variable x_0 is absolutely continuous with respect to the Lebesgue measure m on $\mathbb{R}^N$ and that the associated density is essentially bounded.

Consider a system of the form

$$x_{t+1} = f(x_t) + u_t + w_t \tag{14.16}$$

with $U = W = \mathbb{R}^N$ and an injective C^1-map $f : \mathbb{R}^N \to \mathbb{R}^N$ satisfying (with $J(f(x))$ denoting the Jacobian of f at x)

$$|\det J(f(x))| \geq 1 \quad \text{for all } x \in \mathbb{R}^N. \tag{14.17}$$

Theorem 14.3.2 ([408]) *Consider system* (14.16) *satisfying* (14.17) *and* $\pi_0 \ll_b m$. *Assume that the AMS property is achieved with an associated AMS measure Q via a causal coding and control policy over a noiseless channel of capacity C. Then for all Borel sets $B \subset \mathbb{R}^N$ with $0 < m(B) < \infty$, we have*

$$C \geq Q(B) \log \inf_{x \in B} |\det J(f(x))|. \tag{14.18}$$

Remark 14.3.2 The preceding theorem recovers, as a special case, Theorem 14.2.3, which shows that $C \geq \inf_{x \in \mathbb{R}^N} \log |\det J(f(x))|$. However, the result there is more general with regard to the allowed class of channels.

Remark 14.3.3 In the inequality (14.18), we see a trade-off between the Q-measure of the set B and the infimal volume growth on B. If some characteristics of the measure Q are known, one can try to optimize the lower bound by a careful choice of B. We also note that

$$\int Q(\mathrm{d}x) \log |\det J(f(x))| \geq Q(B) \inf_{x \in B} \log |\det J(f(x))|$$

holds for all Borel sets B, where the left-hand side is the expected volume expansion with respect to the AMS measure Q. Hence, it is tempting to conjecture that also the integral above is a lower bound on the capacity. Under the stronger criterion of asymptotic ergodicity, we will show further below that this is indeed the case.

The next corollary shows that imposing further properties on the AMS measure Q can lead to more concrete bounds.

Corollary 14.3.1 *Consider system* (14.16) *satisfying* (14.17) *and* $\pi_0 \ll_b m$. *Assume that the AMS property is achieved via a noiseless channel of capacity C and the measure Q satisfies for some* $M, p > 0$ *the moment constraint*

$$\int Q(\mathrm{d}x)|x|^p \leq M.$$

Then the channel capacity satisfies

$$C \geq \sup_{\kappa^p \geq M} \left(1 - \frac{M}{\kappa^p}\right) \min_{|x| \leq \kappa} \log|\det J(f(x))|. \tag{14.19}$$

Proof Consider the set $B := \overline{B_\kappa(0)}$ for a fixed $\kappa > 0$. By Markov's inequality, the moment constraint implies $Q(B) \geq 1 - \frac{M}{\kappa^p}$. Hence, Theorem 14.3.2 implies the assertion. ◇

We provide below two examples.

Example 14.3.1 For a linear system with $f(x) = Ax$, $A \in \mathbb{R}^{N \times N}$ satisfying $|\det A| \geq 1$, our result implies the well-known relation (cf. [811, 824])

$$C \geq \log|\det A| = \sum_\lambda n_\lambda \log|\lambda|$$

with summation over all eigenvalues λ of A with associated multiplicities n_λ. By a simple decoupling argument this can be refined to show that $C \geq \sum_\lambda \max\{0, n_\lambda \log|\lambda|\}$ ◇

The next example shows that for nonlinear systems the supremum in (14.19) is not necessarily attained as $\kappa \to \infty$, i.e., the lower bound (14.18) indeed expresses a trade-off between the measure of B and the minimal volume expansion on B.

Example 14.3.2 Consider a map $f : \mathbb{R} \to \mathbb{R}$ with derivative

$$f'(x) = \begin{cases} 2 & \text{if } |x| \leq 1 \\ 2^{\frac{1}{\sqrt{|x|}}} & \text{if } |x| > 1 \end{cases}$$

and note that $|f'(x)| = f'(x) > 1$ for all $x \in \mathbb{R}$. Since f' is symmetric and monotonically decreasing on $[0, \infty)$, we obtain

$$\min_{|x| \leq \kappa} \log|f'(x)| = \log|f'(\kappa)| \quad \text{for all } \kappa > 0.$$

Corollary 14.3.1, applied with $M = p = 1$ thus yields the capacity bound

$$C \geq \sup_{\kappa \geq 1} \left(1 - \frac{1}{\kappa}\right) \frac{1}{\sqrt{\kappa}}.$$

A straightforward analysis shows that this supremum is attained as a maximum at $\kappa = 3$, and hence $C \geq 2/(3\sqrt{3})$. ◇

14.3.4 The Noisy Channel Case

For discrete noiseless channels, the key idea combining the volume-growth based approaches for deterministic models with the stochastic system setup was the observation that the number of control sequences is bounded from above by the total number of received messages. This approach clearly does not directly apply to the noisy channel setup, for there can be an arbitrarily large number of possibly distinct received channel outputs, but these may not carry reliable information. In the following, we develop a new method to address this for a discrete memoryless channel (DMC).

Figure 14.1 shows the control loop, using a DMC with feedback for data transmission from the encoder to the controller. The channel has a finite input alphabet $\mathcal{M}$ and a finite output alphabet $\mathcal{M}'$. The channel input q_t at time t is generated by a function γ_t^e so that $q_t = \gamma_t^e(x_{[0,t]}, q'_{[0,t-1]})$. The channel maps q_t to q'_t in a stochastic fashion so that $P(q'_t \in \cdot | q_t, q_{[0,t-1]}, q'_{[0,t-1]}) = P(q'_t \in \cdot | q_t)$ is a conditional probability measure on $\mathcal{M}'$ for all $t \in \mathbb{Z}_+$, for every realization $q_t, q_{[0,t-1]}, q'_{[0,t-1]}$. The controller, upon receiving the information from the channel, generates its decision at time t, also causally: $u_t = \gamma_t^c(q'_{[0,t]})$.

Consider a DMC with channel capacity C. The following property, known as the *strong converse*, holds, see [415] and [194, Problem 10.17]: For any $R > C$, under any coding policy:

$$\lim_{T\to\infty} p_e(T) = 1, \tag{14.20}$$

where $p_e(T)$ is the average probability of error among 2^{RT} equally likely messages after the channel is used T times under coding and decoding policies admissible according to the standard information-theoretic formulation of communication with noiseless feedback, cf. [662].

Now we consider a scalar system of the form

$$x_{t+1} = f(x_t) + u_t + w_t \tag{14.21}$$

with a C^1-function $f : \mathbb{R} \to \mathbb{R}$ satisfying

$$|f'(x)| \geq 1 \quad \text{for all } x \in \mathbb{R}. \tag{14.22}$$

Our main result reads as follows.

Theorem 14.3.3 ([408]) *Consider system* (14.21) *satisfying* (14.22). *Assume that* $\pi_0 \ll m$ *with* p *denoting the density with respect to* m, *that* $K := \text{supp}(\pi_0)$ *is a compact interval, and*

$$p_{\min} := \operatorname*{ess\,inf}_{x \in K} p(x) > 0, \qquad p_{\max} := \operatorname*{ess\,sup}_{x \in K} p(x) < \infty.$$

Then, if the AMS property is achieved via a causal coding and control strategy over a DMC of capacity C, *we have*

$$C \geq \inf_{x \in \mathbb{R}} \log |f'(x)|.$$

At this point it may be instructive to explain the proof approach which builds on the construction of an auxiliary coding problem that relates the number -per time stage- of distinct control actions (in a similar spirit that was the basis of the definition of stabilization entropy) to an information transmission problem and in turn to an analysis on channel capacity with feedback; by considering the fact that the number of informative messages per time stage to be transmitted with regard to the initial state cannot be less than the desired bound. The coding problem is related to a channel coding theorem via optimal transport inequalities.

We next state the following variation where the initial measure may have non-compact support.

Theorem 14.3.4 ([408]) *Consider system* (14.21) *satisfying* (14.22). *Assume that* $\pi_0 \ll m$ *with* p *denoting the density with respect to* m, *and that for every* $\epsilon > 0$, *there exists a compact interval* K_ϵ *such that,* $\pi_0(K_\epsilon) \geq 1 - \epsilon$, *and with*

$$p_{\min}^{K} := \operatorname*{ess\,inf}_{x \in K} p(x) > 0, \qquad p_{\max}^{K} := \operatorname*{ess\,sup}_{x \in K} p(x) < \infty,$$

the following condition holds:

$$\lim_{\epsilon \to 0} \frac{\int_{\mathbb{R} \setminus K_\epsilon} p(x) \mathrm{d}x}{p_{\min}^{K_\epsilon}} = 0. \tag{14.23}$$

Then, if the AMS property is achieved via a causal coding and control strategy over a DMC of capacity C, *we have*

$$C \geq \inf_{x \in \mathbb{R}} \log |f'(x)|.$$

The proof of this result parallels that of Theorem 14.3.3, with some differences, however, which we point out in the proof in the Appendix.

Remark 14.3.4 A sufficient condition for (14.23) is that p is differentiable, positive everywhere, and monotone decreasing in either direction as $|x|$ increases for sufficiently large values of $|x|$, and $\lim_{|x|\to\infty} p'(x)/p(x) = \infty$. This follows from an application of L'Hospital's theorem to the expression

$$\lim_{x\to\infty} \frac{\int_{|s|>x} p(s)\mathrm{d}s}{\min(p(x), p(-x))}.$$

Probability densities which decay faster than an exponential (such as the Gaussian) satisfy this condition. An exponential density (if one-sided, the denominator will just be $p(x)$) keeps this ratio a constant as $|x|$ increases and densities with a heavier tail than an exponential do not satisfy this condition.

14.4 Case with Ergodicity via the Stochastic Geometric Method

In this section, we consider the criterion that the closed-loop process is AMS Ergodic (see Definition D.3.6).

Consider the system

$$x_{t+1} = f(x_t, w_t) + u_t \tag{14.24}$$

where x_t and u_t are $\mathbb{R}^N$-valued for some $N \in \mathbb{Z}_+$ and w_t takes values in a standard probability space $\mathbb{W}$. For a fixed $w \in \mathbb{W}$, let us denote the map $x \mapsto f(x, w)$ by f_w. Suppose also that the following holds:

(A1) The map $f : \mathbb{R}^N \times \mathbb{W} \to \mathbb{R}^N$ is Borel measurable.
(A2) The noise process $(w_t)_{t\in\mathbb{Z}_+}$ is i.i.d. By abuse of notation, ν denotes both the law of any individual w_t, as well as the process measure.
(A3) The map $f_w : \mathbb{R}^N \to \mathbb{R}^N$ is C^1 and injective for any $w \in \mathbb{W}$.
(A4) The initial state x_0 is random and independent of the noise process. We write π_0 for the associated probability measure.
(A5) The measure π_0 is absolutely continuous with respect to the N-dimensional Lebesgue measure m, and its density (which exists by the Radon-Nikodym theorem) is bounded.
(A6) There is a constant $c > 0$ with $|\det J(f_w(x))| > c$ for all $x \in \mathbb{R}^N$ and $w \in \mathbb{W}$.

We write $(\Omega, \mathcal{F}, P)$ for the probability space on which both x_0 and w_t are defined.

Theorem 14.4.1 ([287]) *Consider system* (14.34) *satisfying assumptions (A1)–(A6). Suppose the system is controlled over a discrete noiseless channel with capacity C and a coding and control policy achieves that the state process is AMS ergodic with asymptotic mean Q. Then, the capacity must satisfy*

$$\int\int \log |\det J(f_w(x))|\, Q(dx)\nu(dw) \leq C.$$

Our second main theorem relaxes the condition of the channel being noiseless. On the other hand, the class of nonlinear systems considered is more restrictive.

Theorem 14.4.2 *Consider the scalar system*

$$x_{t+1} = f(x_t, w_t) + u_t$$

satisfying assumptions (A1)–(A5). Additionally, suppose that the following holds:

1. $|f_w'(x)| \geq 1$ *for every* $x \in \mathbb{R}$.
2. *The support of* π_0 *is a compact interval* $K \subseteq \mathbb{R}$.
3. *The essential infimum and supremum of the density of* π_0*, denoted by* $\rho_{\min}$ *and* $\rho_{\max}$*, respectively, satisfy* $0 < \rho_{\min} \leq \rho_{\max} < \infty$.

Suppose that the system is controlled over a discrete memoryless channel with feedback of capacity C and a causal coding and control policy results in the state process being AMS ergodic with asymptotic mean Q. Then, the channel capacity must satisfy

$$\int\int \log |f_w'(x)| Q(dx)\nu(dw) \leq C. \tag{14.25}$$

14.4.1 Proof Ideas Under the Ergodicity Criterion

We now provide some intuition for the proofs of the main theorems the first of which is based on a generalization of the notion of stabilization entropy introduced in Definition 14.3.1. The technical proofs are provided in the appendix.

Consider system (14.34) given (an open-loop) control sequence $u := (u_t)_{t\in\mathbb{Z}_+}$, a noise realization $w := (w_t)_{t\in\mathbb{Z}_+}$, and an initial state $x_0 \in \mathbb{R}^N$; note that the trajectory $x := (x_t)_{t\in\mathbb{Z}_+} \in (\mathbb{R}^N)^{\mathbb{Z}_+}$ of the state is uniquely determined. Let us denote this trajectory by $\varphi(\cdot, x_0, u, w)$ so that for any $t \in \mathbb{Z}_+$,

$$x_t = \varphi(t, x_0, u, w).$$

We want to find a subset of control sequences that allow to render certain subsets of the state space invariant in a probabilistic sense. This leads to the next definitions of spanning sets and stabilization entropy for finite collections of subsets of $\mathbb{R}^N$ and $\mathbb{W}$, respectively, which generalize Definition 14.3.1, where a single set was considered.

Definition 14.4.1 Let $B \in \mathcal{B}(\mathbb{R}^N)$ and $D \in \mathcal{B}(\mathbb{W})$ be finite disjoint unions of Borel sets $B_1, \ldots, B_n$ and $D_1, \ldots, D_m$, respectively. Let also R denote a collection of numbers $r_{k,l} \in [0, 1]$ for $k \in \{1, \ldots, n\}$ and $l \in \{1, \ldots, m\}$ satisfying

$$1 - r := \sum_{k=1}^{n} \sum_{l=1}^{m} (1 - r_{k,l}) \in [0, 1].$$

Fix $T \in \mathbb{Z}_+$ and $\rho \in (0, 1)$. A set of control sequences $S \subseteq (\mathbb{R}^N)^T$ is called (T, B, D, ρ, R)-spanning if there exists $\tilde{\Omega} \in \mathcal{F}$ such that the following conditions hold:

$P(\tilde{\Omega}) \geq 1 - \rho$.
For each $\omega \in \tilde{\Omega}$, there exists a control sequence $u \in S$ such that

$$\frac{1}{T}|\{t \in [0; T-1] : \\ (\varphi(t, x_0(\omega), u, w(\omega)), w_t(\omega)) \in B_k \times D_l\}| \geq 1 - r_{k,l}$$

holds for all k and l.

Note that we have abused the notation here by calling a set (T, B, D, ρ, R)-spanning instead of $(T, (B_k)_{k=1}^n, (D_l)_{l=1}^m, \rho, R)$-spanning. When doing so, there is the underlying assumption that the partitions of B and D are fixed.

In the above definition, the fact that all random variables are modeled on a common probability space ensures that given ω, the initial state and the noise sequence of length T are deterministic. Intuitively speaking, a subset of control sequences of length T is (T, B, D, ρ, R)-spanning if the probability that, for all k, l, we can maintain the state variable in B_k and the noise variable in D_l for at least $1 - r_{k,l}$ percent of the time, is at least $1 - \rho$. We want to use the size of spanning sets to quantify the difficulty of a control task, which leads to the next definition.

Definition 14.4.2 (Multi-Set Stabilization Entropy) For the system (14.34), define the (B, D, ρ, R)-stabilization entropy by

$$h(B, D, \rho, R) := \limsup_{T\to\infty} \frac{1}{T} \log s(T, B, D, \rho, R),$$

where $s(T, B, D, \rho, R)$ denotes the smallest cardinality of a (T, B, D, ρ, R)-spanning set. We define this quantity to be ∞ if no or no finite spanning set exists.

It is immediate that finite (T, B, D, ρ, R)-spanning sets need not exist. As we will see however, they do exist in desired scenarios. The following lemma is instrumental.

Lemma 14.4.1 *Consider system (14.34) with the assumptions of Theorem 14.4.1 (i.e., a coding and control policy exists over a noiseless channel of capacity* $C = \log|\mathcal{M}|$ *which makes the state process AMS ergodic). Let*

$B := \bigsqcup_{k=1}^{n} B_k \in \mathcal{B}(\mathbb{R}^N)$ *and* $D := \bigsqcup_{l=1}^{m} D_l \in \mathcal{B}(\mathbb{W})$ *be finite disjoint unions of Borel sets,*
$\rho \in (0, 1)$ *be arbitrary.*

Next, define the sequence of numbers

$$R_\epsilon := (r_{k,l})_{1\le k\le n, 1\le l\le m},$$

where

$$r_{k,l} := \begin{cases} (1+\epsilon)(1 - Q(B_k)\nu(D_l)) & \text{if } Q(B_k)\nu(D_l) \in (0,1) \\ 1 & \text{if } Q(B_k)\nu(D_l) = 0 \\ \epsilon & \text{if } Q(B_k)\nu(D_l) = 1 \end{cases}$$

and observe that for $\epsilon > 0$ *small enough, the following conditions are satisfied:*

(i) $1 - r := \sum_{k=1}^{n}\sum_{l=1}^{m}(1 - r_{k,l}) \in [0, 1]$.
(ii) $1 - (1+\epsilon)(1 - Q(B_k)\nu(D_l)) \in (0, 1)$ *for all* k, l *with* $Q(B_k)\nu(D_l) \in (0, 1)$.

Thus, for such a small ϵ*, the generalized stabilization entropy* $h(B, D, \rho, R_\epsilon)$ *is well-defined. (Of course,* r *and the* $r_{k,l}$*'s are* ϵ*-dependent, but we drop this from the notation.) Then, for all sufficiently small* $\epsilon > 0$*, the capacity must satisfy*

$$h(B, D, \rho, R_\epsilon) \le C. \tag{14.26}$$

Proof The case $Q(B_k)\nu(D_l) \in \{0, 1\}$ for some k, l turns out to be trivial, and hence no proof is needed. The proof of the remaining cases is based on the observation that the joint process of state and noise satisfies the same sample path limit property as an ergodic process; in particular

$$P\Big(\Big\{\omega \in \Omega : \lim_{T\to\infty} \frac{1}{T}\sum_{t=0}^{T-1} \mathbb{1}_{B_k}(x_t(\omega))\mathbb{1}_{D_l}(w_t(\omega)) = Q(B_k)\nu(D_l),\ \forall k, l\Big\}\Big) = 1.$$

This allows us to find numbers $I_0, J_0 \in \mathbb{Z}_+$ such that

$$\frac{1}{I_0} \le \epsilon(1 - Q(B_k)\nu(D_l)),\ \forall k, l$$

and for all $T \ge J_0$, the set

$$\tilde{\Omega}_T := \Big\{\omega \in \Omega : \Big|\frac{1}{N}\sum_{t=0}^{N-1} \mathbb{1}_{B_k}(x_t(\omega))\mathbb{1}_{D_l}(w_t(\omega)) \\ - Q(B_k)\nu(D_l)\Big| < \frac{1}{i}, \ \forall k, l \text{ whenever } N \geq T\Big\}$$

satisfies $P(\tilde{\Omega}_T) \geq 1 - \epsilon$. For an arbitrary $T \geq J_0$, we define the set of control sequences

$$S_T := \{u_{[0;T-1]}(\omega) : \omega \in \tilde{\Omega}_T\}.$$

It can be shown that S_T is a $(T, B, D, \rho, R_\epsilon)$-spanning with $|S_T| \leq |\mathcal{M}|^T$, which implies the desired estimate. □

14.4.2 A Utility of Geometric Analysis: Refined Bounds via Coordinate Splitting

We earlier established that under mild technical assumptions, the information rate required for asymptotic ergodicity must not be smaller than the logarithm of the determinant of the system linearization, averaged over the noise and ergodic state measures. Here, we will establish that for systems with a stable component, it suffices to consider only the unstable dimensions, providing a refinement on the general channel capacity bound for a large class of systems.

Let us motivate the discussion with two examples. Consider

$$x_{t+1} = f(x_t, w_t) + Bu_t \tag{14.27}$$

Suppose that system (14.27) is controlled over a noiseless channel with capacity C, and made to be asymptotically ergodic with AMS mean Q. Then under slightly different technical assumptions, Theorems 14.4.1, 14.4.2, and 14.2.5 establish that

$$\int\int \log_2 |\det D_x f(x, w)| Q(dx)\nu(dw) \leq C \tag{14.28}$$

where ν is the distribution of the i.i.d. noise. This bound, however, is in general not tight, as the following two examples illustrate.

Example 14.4.1 Consider the two dimensional linear system given by

$$\begin{bmatrix} x_{t+1} \\ y_{t+1} \end{bmatrix} = \begin{bmatrix} 2 & 0 \\ 0 & 1/2 \end{bmatrix} \begin{bmatrix} x_t \\ y_t \end{bmatrix} + w_t + u_t, \tag{14.29}$$

where u_t and w_t take values in $\mathbb{R}^2$, and the noise is i.i.d with zero mean. The LHS of (14.28) is easily seen to equal zero for this system, thus providing a vacuous bound on channel capacity. We know (e.g. from the previous chapter) that a tight bound on linear systems is the log-sum of the unstable eigenvalues. Note that by replacing $1/2$ in the above matrix with any number no smaller than one, the bound in (14.28) recovers the tight linear bound. As we will see, the refinement of the channel capacity bound in this section will recover the tight bound in the general linear systems case (thus, also with stable eigenvalues).

Consider now a second, more interesting, example:

Example 14.4.2 Consider the system $(x_t, y_t)_{t\in\mathbb{Z}_+}$ in $\mathbb{R}^2$ evolving with scalar-valued i.i.d. noise according to

$$\begin{bmatrix} x_{t+1} \\ y_{t+1} \end{bmatrix} = \begin{bmatrix} (x_t^3 + x_t)(1 + y_t^2) \\ \frac{1}{2} y_t + w_t \end{bmatrix} + u_t \tag{14.30}$$

with x_0 and y_0 independent and admitting bounded densities. We note that the y-component of the above system is stochastically stable. Moreover, the presence of the y-term in the dynamics of the x-component cannot be modeled as noise, as the i.i.d. assumptions required for data-rate theorems of non-linear systems to hold is not satisfied.

Suppose the above system is made asymptotically ergodic via a coding and control policy with AMS mean Q. We compute

$$Df_w(x, y) = \begin{bmatrix} (3x^2 + 1)(1 + y^2) & (x^3 + x)2y \\ 0 & 1/2 \end{bmatrix} \tag{14.31}$$

and apply (14.28) to obtain

$$\int \log_2 |\frac{(1 + y^2)}{2}(3x^2 + 1)| dQ)(x, y) \leq C. \tag{14.32}$$

Note that there is a factor of $1/2$ coming from the stable second component in the integrand. It seems sensible that the bound should hold without this factor, as the coding and control policy need not be concerned with the stochastically stable component. Indeed, the result in to be presented shortly establishes that the above bound holds when removing the factor of $1/2$ and is therefore a strict refinement for certain systems.

Consider a subset $p \subseteq \{1, .., N\}$ of indices listed in increasing order as $p_1 < p_2 < \cdots < p_{|p|}$. Let $z_1 < \cdots < z_{N-|p|}$ denote the elements in $\{1, .., N\} \setminus p$. We define the permutation $\psi_p : \mathbb{R}^N \to \mathbb{R}^N$ by

$$\psi_p(x_1, ..., x_N)_i = \begin{cases} x_{p_i} & i \leq |p| \\ x_{z_{i-|p|}} & i > |p| \end{cases}$$

for $i \in \{1, .., N\}$. Also, let $\pi_p : \mathbb{R}^N \to \mathbb{R}^{|p|}$ denote the natural projection of coordinates $p_1, .., p_{|p|}$. For a map $f : \mathbb{R}^N \to \mathbb{R}^N$, a set p as above, and a fixed vector $(y_1, .., y_{N-|p|})$ we define the map $f^p(\cdot, y_1, .., y_{N-|p|}) : \mathbb{R}^{|p|} \to \mathbb{R}^{|p|}$ by

$$f^p(x, y_1, .., y_{N-|p|}) := \pi_p(f(\psi_p^{-1}(x, y_1, .., y_{N-|p|}))) \tag{14.33}$$

where $x \in \mathbb{R}^{|p|}$. As an example, consider $N = 4$, $p = \{2, 4\}$, a fixed vector (y_1, y_2), and a function $f : \mathbb{R}^4 \to \mathbb{R}^4$ written as $f = (f_1, f_2, f_3, f_4)$ for maps $f_i : \mathbb{R}^4 \to R$. Then

$$f^p(x_1, x_2, y_1, y_2) = (f_2(y_1, x_1, y_2, x_2), f_4(y_1, x_1, y_2, x_2)).$$

This notation allows us to precisely state our main result. Consider the system

$$x_{t+1} = f(x_t, w_t) + Bu_t \tag{14.34}$$

where x_t is $\mathbb{R}^N$-valued for some $N \in \mathbb{Z}_+$, $B \in \mathbb{R}^{N \times N'}$, u_t is $\mathbb{R}^{N'}$-valued, and w_t takes values in a standard probability space $\mathbb{W}$. For a fixed $w \in \mathbb{W}$, let us denote the map $x \mapsto f(x, w)$ by f_w. Suppose that the following holds:

Assumption 14.4.1

(i) *The state evolution map f is Borel measurable.*
(ii) *The noise process $(w_t)_{t \in \mathbb{Z}_+}$ is i.i.d. By abuse of notation, ν denotes both the i.i.d. measure on $\mathcal{B}(\mathbb{W})$ and the noise process measure on $\mathcal{B}(\mathbb{W}^{\mathbb{Z}_+})$.*
(iii) *The map $f_w(\cdot) : \mathbb{R}^N \to \mathbb{R}^N$ is C^1 and injective for any $w \in \mathbb{W}$.*
(iv) *The initial state $x_0 \in \mathbb{R}^N$ is random and independent of the noise process, and its law π_0 admits a bounded density.*
(v) *The set* $\Gamma = \Big\{ p \subseteq \{1, .., N\} : \exists c_p > 0 \mathit{with} |\det Df_w^p(x_{p_1}, .., x_{p_{|p|}}, x_{z_1}, .., x_{z_{N-|p|}})| > c_p, \forall x \in \mathbb{R}^N, w \in \mathbb{W} \Big\}$ *is non-empty.*
(vi) *Letting $\{t_1, .., t_s\} \in \Gamma$ be arbitrary and writing the random initial state as $x_0 = (x_0^1, .., x_0^N)$, there exists a set S consisting of possible realizations of the initial state components not indexed by Γ such that the event that these non-indexed initial states take a realization in S has non-zero probability, and the law of $(x_0^{t_1}, .., x_0^{t_s})$ admits a bounded density when conditioned on the event that the non-indexed initial states take on a realization $\hat{x}$, for any $\hat{x} \in S$.*

Theorem 14.4.3 ([288]) *Consider system* (14.34) *satisfying Assumption 14.4.1(i)–(vi), controlled over a noiseless channel with finite alphabet $\mathcal{M}$ and capacity $C := \log_2 |\mathcal{M}|$. If there exists a coding and control policy which renders the state process*

$(x_t)_{t\in\mathbb{Z}_+}$ *AMS ergodic (asymptotically ergodic) with asymptotic mean Q, then we must have that*

$$\max_{p\in\Gamma} \int\int \log |detDf_w^p(x_{p_1}, ..., x_{p_{|p|}}, x_{z_1}, ..., x_{z_{N-|p|}})| Q(dx_1, ..., dx_N)\nu(dw) \leq C, \tag{14.35}$$

where the Jacobian above is the $|p| \times |p|$ matrix of partial derivatives of $f_w^p(\cdot, x_{z_1}, ..., x_{z_{N-|p|}})$ evaluated at $(x_{p_1}, ..., x_{p_{|p|}}) \in \mathbb{R}^{|p|}$.

Here we provide only a sketch of the proof of the theorem, since it follows essentially the steps of the proof of Theorem 14.4.1, with some differences that entail splitting of coordinates, which we point out in the sketch of the proof. But, before doing this, we make below two useful remarks.

Remark 14.4.1 By taking $p = \{1, .., M\}$ (if $\{1, .., M\} \in \Gamma$), we recover the bound (14.28). For a large class of systems, however, it is clear that Theorem 14.4.3 is a strict refinement, as can be seen by noting that in Example 14.4.2, taking $p = \{1\}$ recovers the sharper bound

$$\int \log_2 |(1 + y^2)(3x^2 + 1)| Q(dx, dy) \leq C. \tag{14.36}$$

It is clear that for linear systems, the new bound recovers the tight bound such as the one in Example 14.4.1.

Remark 14.4.2 The technical assumption (vi) is satisfied if the initial state has independent components each admitting a bounded density. Suppose now that for a given system, assumptions (i)–(v) of Theorem 14.4.3 are satisfied, but (vi) only holds for certain subsets of Γ. Then, the theorem will still hold; however, the max in (14.35) should be taken only over subsets of Γ for which assumption (vi) holds.

Proof Sketch We first fix an integer $m \leq M$ and view the map f_w as a function of two vectors, i.e. we decompose the state into a pair (x, y) where $x \in \mathbb{R}^m$ and $y \in \mathbb{R}^{N-m}$. Consider the control system in Theorem 14.4.3 and note that for a fixed sequence of controls $u := (u_t)_{t\in\mathbb{Z}_+}$, a fixed sequence of noise terms $w := (w_t)_{t\in\mathbb{Z}_+}$, and a fixed initial state $x_0 \in \mathbb{R}^N$, the state process $(x_t)_{t\in\mathbb{Z}_+}$ is deterministic. Let us introduce the notation $\varphi(t, x_0, u, w) := x_t$ for every $t \in \mathbb{Z}_+$. Letting π_m and π_{N-m} denote the natural projection of $\mathbb{R}^N$ onto the first m and last $N - m$ coordinates, respectively, we further define $\varphi^m(t, x_0, u, w) := \pi_m(x_t)$ and $\varphi^{N-m}(t, x_0, u, w) := \pi_{N-m}(x_t)$ so that

$$\varphi(t, x_0, u, w) = (\varphi^m(t, x_0, u, w), \varphi^{N-m}(t, x_0, u, w)) \text{ for every } t \in \mathbb{Z}_+.$$

Proof then follows steps similar to that of Theorem 14.4.1, except that in this case we condition on the initial state realization of the elements in the last $N - m$ coordinates and consider the stochastic flow only in the first m coordinates, which is ergodic by hypothesis. We then consider all possible partitions. For details, see [288]. □

14.5 Achievability: Discrete Noiseless Channels and a Stationary and Ergodic Construction

In this section, we provide achievability results and a stabilizing coding/control policy. As discussed earlier, the study of non-linear systems have typically considered noise-free controlled systems; e.g. [56, 479], and [207]. As also noted earlier, for noise-free systems, it typically suffices to only consider a sufficiently small *invariant* neighborhood of an equilibrium point to obtain stabilizability conditions, which is not necessarily the case when the system is driven by an additive noise process.

Theorem 14.5.1 *Consider a non-linear system of the form (14.5), where $\{w_t\}$ is a sequence of zero-mean Gaussian random vectors and there exists a control function $\kappa(z)$ such that $|f(x,\kappa(z))|_\infty \leq |a||x-z|_\infty$ for all $x, z \in \mathbb{R}^N$, with $\kappa(0) = 0$. For the stationarity and ergodicity of $\{x_t\}$ (and thus with a unique invariant probability measure), it suffices that $C > N\log_2(|a|) + 1$.*

Before providing a proof of this result, we make a useful remark.

Remark 14.5.1 It may be possible in general to reduce the rate requirements by the use of variable-rate encoding schemes; for example, if there exists a compact region outside of which the constant a can be upper bounded by a smaller number, a region-dependent quantization rate can be applied which can reduce the average data rate required for system stability. In this chapter, since there is an explicit channel, our focus has been on fixed-rate coding schemes. ◇

Proof The proof follows essentially from the approach developed in [808] and [373] with extension to non-linear analysis. Consider the case with $N = 2$. Let $\Delta > 0$ denote the bin size for a uniform quantizer and let for each coordinate $x^i \in \mathbb{R}, i = 1, 2$;

$$Q_K^\Delta(x^i) = \begin{cases} (k - \frac{1}{2}(K+1))\Delta, & \text{if } x^i \in [(k-1-\frac{1}{2}K)\Delta, (k-\frac{1}{2}K)\Delta) \\ (\frac{1}{2}(K-1))\Delta, & \text{if } x^i = \frac{1}{2}K\Delta \\ 0, & \text{if } x^i \notin [-\frac{1}{2}K\Delta, \frac{1}{2}K\Delta]. \end{cases} \tag{14.37}$$

and define $\mathbf{Q}^\Delta(x) = (Q_K^\Delta(x^1), Q_K^\Delta(x^2))$ if $Q_K^\Delta(x^i) \neq 0$ for $i = 1, 2$ and $\mathbf{Q}^\Delta(x) = 0$ if $Q_K^\Delta(x^i) = 0$ for some i. Thus, the number of symbols in the image of $\mathbf{Q}^\Delta$ is $K^2 + 1$ (and not $(K+1)^2$). The quantizer outputs are transmitted through a memoryless erasure channel, after being subjected to a bijective mapping, which is performed by the channel encoder. The channel encoder maps the quantizer output symbols to corresponding channel inputs $q \in \mathcal{M} := \{1, 2 \ldots, K^2 + 1\}$. A channel encoder at time t, denoted here by $\mathcal{E}_t$, maps the quantizer outputs to $\mathcal{M}$ such that $\mathcal{E}_t(Q_t(x_t)) = q_t \in \mathcal{M}$. For $i = 1, 2$, let $R' = \log_2(K)$. For $t \geq 0$ and with $\Delta_0^1 = \Delta_0^2 \in \mathbb{R}$, define $h_t^i = \frac{x_t^i}{\Delta_t^i 2^{R'-1}}$, and with

$$\hat{x}_t = \begin{bmatrix} \hat{x}_t^1 \\ \hat{x}_t^2 \end{bmatrix},$$

consider:

$$u_t = -\kappa(\hat{x}_t),$$

$$\begin{bmatrix} \hat{x}_t^1 \\ \hat{x}_t^2 \end{bmatrix} = \begin{bmatrix} Q_{K_1}^{\Delta_t^1}(x_t^1) \\ Q_{K_2}^{\Delta_t^2}(x_t^2) \end{bmatrix} 1_{\{\max_i |h_i| \leq 1\}} + \begin{bmatrix} 0 \\ 0 \end{bmatrix} 1_{\{\max_i |h_i| > 1\}}, \tag{14.38}$$

$$\Delta_{t+1}^1 = \Delta_t^1 \bar{Q}(|h_t^1|, |h_t^2|, \Delta_t^1, \Delta_t^2), \qquad \Delta_{t+1}^2 = \Delta_t^2 \bar{Q}(|h_t^1|, |h_t^2|, \Delta_t^1, \Delta_t^2), \tag{14.39}$$

with, for $i = 1, 2$, $\delta > 0$ $\alpha \in (0, 1)$, $L > 0$ such that

$$\begin{aligned}
&\bar{Q}(x, y, \Delta^1, \Delta^2) = |a| + \delta \quad \text{if} \quad |x| > 1, \quad \text{or} \quad |y| > 1 \\
&\bar{Q}(x, y, \Delta^1, \Delta^2) = \alpha \quad \text{if} \quad |x| \leq 1, |y| \leq 1; \quad \Delta^1 > L, \Delta^2 > L \\
&\bar{Q}(x, y, \Delta^1, \Delta^2) = 1 \quad \text{if} \quad |x| \leq 1, |y| \leq 1; \quad \Delta^1 \leq L \quad \text{or} \quad \Delta^2 \leq L
\end{aligned}$$

Note that, the above imply $\Delta_t^i \geq \alpha L$. To make the state space for the bin size process countable as in [824] [808], we take that $\log_2(\bar{Q}(\cdot))$ take values in integer multiples of s where the integers taken are relatively prime (that is they share no common divisors except for 1); see Lemma 7.6.2 in [824]. □

We state the following without proof.

Lemma 14.5.1 *The process* (x_t, Δ_t) *is a Markov chain.*

We define a sequence of stopping times as follows:

$$\tau_0 = 0, \quad \tau_{z+1} = \inf\{k > \tau_z : |h_k^i| \leq 1, i \in \{1, 2\}\}, \quad z \in \mathbb{Z}_+.$$

By the strong Markov property and the nature of the stopping times, (x_{τ_z}, h_{τ_z}) is also Markov. In the following, we show that there exist $b_0 > 0$ and $b_1 < \infty$ such that

$$\mathbb{E}[\log(\Delta_{\tau_{z+1}}^2)|\Delta_{\tau_z}, h_{\tau_z}] \leq \log(\Delta_{\tau_z}^2) - b_0 + b_1 1_{\{|\Delta_{\tau_z}| \leq F\}} \tag{14.40}$$

We first bound, in the following lemma, the probability $P(\tau_{z+1} - \tau_z \geq k|\Delta_{\tau_z}, h_{\tau_z})$ from above.

Lemma 14.5.2 *The discrete probability measure $P(\tau_{z+1} - \tau_z = k \mid x_{\tau_z}, \Delta_{\tau_z})$ has the upper bound*

$$P(\tau_{z+1} - \tau_z \geq k | x_{\tau_z}, \Delta_{\tau_z}) \leq M(\Delta_{\tau_z}) r^{-k},$$

for some $r > 1$ and $\lim_{\Delta \to \infty} M(\Delta) = 0$.

We will now invoke Theorem G.1.4 and Corollary G.1.1 in the Appendix to the book. With the candidate Lyapunov function $V_0(x_t, \Delta_t) = \log(\Delta_t^2)$, for $\Delta_{\tau_z} > L$, we have

$$\begin{aligned}
\mathbb{E}[V_0(x_{\tau_{z+1}}, \Delta_{\tau_{z+1}}) \mid x_{\tau_z}, \Delta_{\tau_z}] &= P(\tau_{z+1} - \tau_z = 1)\Big(2\log(\alpha) + \log(\Delta_{\tau_z}^2)\Big) \\
&\quad + \sum_{k=2}^{\infty} \log(\Delta_{\tau_z + k}^2) P(\tau_{z+1} - \tau_z = k \mid x_{\tau_z}, \Delta_{\tau_z}) \\
= P(\tau_{z+1} - \tau_z = 1)\Big(2\log(\alpha) + \log(\Delta_{\tau_z}^2)\Big) & \\
&\quad + \sum_{k=2}^{\infty} 2(\log_2(\alpha) + (k-1)(|a| + \delta) M(\Delta) r^{-k}
\end{aligned}$$

Now, $\lim_{\Delta_0 \to \infty} P(\tau_{z+1} - \tau_z = 1 | \Delta_0, x_0) = 1$ uniformly in $|x_0| \leq 2^{R'-1}\Delta_0$. As a result, the drift condition of Theorem G.1.1 holds. We need to ensure, however, the small/petite set property of compact sets to establish positive Harris recurrence. A compact set for this chain is petite due to the countability of the values that Δ_t takes and the *uniform countable additivity* property of the Markov chain due to the presence of the additive Gaussian noise, as in (D.2) and the continuity of f in x. This argument applies for N-dimensional systems as well with $N > 2$. This completes the proof of Theorem 14.5.1. □

Remark 14.5.2 The approach adopted in the proof of Theorem 14.5.1 applies for more general channels (such as erasure channels or discrete memoryless channels) subject to more tedious error bounds, where we could follow the arguments presented in the previous chapter. ◇

14.6 Discussion: Comparison of the Information Theoretic and Stochastic Growth (Dynamical Systems) Approaches

We present in this section a comparison between the three sections of the chapter: Sect. 14.2 on the information theoretic approach, Sect. 14.3 on the stochastic geometric approach for the AMS criterion, and Sect. 14.4 on the stochastic geometric approach for the ergodicity criterion.

We have two sets of comparisons.

For the AMS criterion, Theorem 14.2.3 was arrived at via the information theoretic method. Theorem 14.3.2 (and the noisy channel generalization Theorem 14.3.3) was arrived at via the geometric method. Theorem 14.2.3 allowed for more general channels, but Theorem 14.3.2 is more relaxed.

For the ergodicity criterion, the information theoretic method culminated in Theorem 14.2.5. Theorems 14.4.1 and 14.4.2 established ergodicity via the stochastic geometric method. In particular, Theorem 14.4.3 allowed for splitting the dynamical system into those that are stable and unstable via a novel approach building on a stochastic geometric argument.

We note that obtaining the sharper bound established in Theorem 14.4.3 using only information theoretic methods does not appear to be possible, with the main impediment being the fact that when splitting the state and conditioning on a past state realization (or some other sufficient information), the unstable and stable state components may not be independent random variables, which was required for the information theoretic methods to apply in this context. The geometric analysis proved to be versatile in allowing us to enhance the results.

In summary, below is some of the benefits of the stochastic geometric approach:

(i) Refined stochastic stability results applicable to a more general class of system models (Theorem 14.3.2) and more refined stability criteria such as the AMS property in combination with moment conditions (see Corollary 14.3.1);
(ii) A direct derivation, building on volume growth arguments, applicable to a plethora of criteria;
(iii) More refined bounds for a large class of systems through trading-off growth rates with the measures of sets under the coordinate projection of a stationary measure (see Theorem 14.3.2);
(iv) The unification of the theory developed for deterministic systems controlled over noise-free communication channels with their stochastic counterparts, involving both stochastic nonlinear dynamical systems and noisy communication channels (see Theorem 14.3.3);
(v) For the ergodicity criterion, when we compare Theorems 14.4.1 and 14.4.2 with Theorem 14.2.5; Theorem 14.4.1 is more general in the sense that it applies to arbitrary causal coding and control policies, not just Markov ones. Moreover, it does not require the assumption of sublinear growth of the differential entropy of the state process. Theorem 14.2.5 assumes that the state process is positive Harris recurrent, which implies unique ergodicity, while Theorem 14.4.1 only assumes ergodicity of the AMS measure. On the other hand, compared with Theorems 14.4.2 and 14.2.5 considers a more general class of channels (involving memory) as well as systems taking values in higher dimensions. Theorem 14.4.3 allowed for splitting the dynamical system into those that are stable and unstable via a novel approach. This is a consequential result in that the entropy growth of a system may be

negative but this may be due to a presence of a stable mode compensating the growth in the direction of an unstable direction. This was already observed for the linear systems case by the sum of the unstable eigenvalues analysis in Theorem 12.5.2. A non-linear generalization of such a splitting turned out to be challenging via information theoretic methods, but the stochastic geometric method allowed for such a generalization.

(vi) On the other hand, the information theoretic method allows for the well-developed machinery of Shannon theory. A stochastic dynamical system generalization of some of the concepts, such as conditional entropy, requires rather indirect arguments. As we have seen in the use of optimal transport in the proof of Theorem 14.3.3, many of the ideas need to be established from scratch.

14.7 Appendix: Proofs

14.7.1 Proof of Theorem 14.2.1

Recall that for channels of the type Class A (which includes discrete memoryless channels (DMC) as a special case), the capacity is given by:

$$C = \lim_{T\to\infty} \max_{\{P(q_t|q_{[0,t-1]},q'_{[0,t-1]})\}} \frac{1}{T} I(q_{[0,T-1]} \to q'_{[0,T-1]})$$

where

$$I(q_{[0,T-1]} \to q'_{[0,T-1]}) = \sum_{t=1}^{T-1} I(q_{[0,t]}; q'_t|q'_{[0,t-1]}) + I(x_0; q'_0).$$

Let us define $R_T = \max_{\{P(q_t|q_{[0,t-1]},q'_{[0,t-1]}),0\le t\le T-1\}} \frac{1}{T}\sum_{t=0}^{T-1} I(q'_t; q_{[0,t]}|q'_{[0,t-1]})$. Observe that for $t > 0$:

$$\begin{aligned} I(q'_t; q_{[0,t]}|q'_{[0,t-1]}) &= H(q'_t|q'_{[0,t-1]}) - H(q'_t|q_{[0,t]}, q'_{[0,t-1]}) \\ &= H(q'_t|q'_{[0,t-1]}) - H(q'_t|q_{[0,t]}, x_t, q'_{[0,t-1]}) \quad (14.41) \\ &\ge H(q'_t|q'_{[0,t-1]}) - H(q'_t|x_t, q'_{[0,t-1]}) \\ &= I(x_t; q'_t|q'_{[0,t-1]}). \quad (14.42) \end{aligned}$$

Here, (14.41) follows from the assumption that the channel is of Class A type.

(a) Consider the following

$$\lim_{T\to\infty} R_T \geq \limsup_{T\to\infty} \frac{1}{T}\left(\sum_{t=1}^{T-1} I(x_t; q'_t | q'_{[0,t-1]})) + I(x_0; q'_0)\right) \tag{14.43}$$

$$= \limsup_{T\to\infty} \frac{1}{T}\left(\sum_{t=1}^{T-1}\left(h(x_t|q'_{[0,t-1]}) - h(x_t|q'_{[0,t]})\right) + I(x_0; q'_0)\right)$$

$$= \limsup_{T\to\infty} \frac{1}{T}\sum_{t=1}^{T-1}\left(h(x_t|q'_{[0,t-1]}) - h(x_t|q'_{[0,t]})\right)$$

$$= \limsup_{T\to\infty} \frac{1}{T}\sum_{t=1}^{T-1}\left(h(f(x_{t-1}, w_{t-1}) + Bu_{t-1}|q'_{[0,t-1]}) - h(x_t|q'_{[0,t]})\right)$$

$$= \limsup_{T\to\infty} \frac{1}{T}\sum_{t=1}^{T-1}\left(h(f(x_{t-1}, w_{t-1})|q'_{[0,t-1]}) - h(x_t|q'_{[0,t]})\right)$$

$$= \limsup_{T\to\infty} \frac{1}{T}\sum_{t=1}^{T-1}\left(\sum_{\zeta_{[0,t-1]}} h(f(x_{t-1}, w_{t-1})|q'_{[0,t-1]} = \zeta_{[0,t-1]})\right.$$

$$\left. P(q'_{[0,t-1]} = \zeta_{[0,t-1]}) - h(x_t|q'_{[0,t]})\right) \tag{14.44}$$

$$\geq \limsup_{T\to\infty} \frac{1}{T}\sum_{t=1}^{T-1}\left(\sum_{\zeta_{[0,t-1]}} h(f(x_{t-1}, w_{t-1})|q'_{[0,t-1]} = \zeta_{[0,t-1]}, w_{t-1})\right.$$

$$\left. P(q'_{[0,t-1]} = \zeta_{[0,t-1]})\right) - h(x_t|q'_{[0,t]}) \tag{14.45}$$

$$= \limsup_{T\to\infty} \frac{1}{T}\sum_{t=1}^{T-1}\left(\left(\sum_{\zeta_{[0,t-1]}}\right.\right.$$

$$\int h(f(x_{t-1}, w)|q'_{[0,t-1]} = \zeta_{[0,t-1]}, w_{t-1} = w)$$

$$\left.\left. \nu(dw) \times P(q'_{[0,t-1]} = \zeta_{[0,t-1]})\right) - h(x_t|q'_{[0,t]})\right) \tag{14.46}$$

$$= \limsup_{T\to\infty} \frac{1}{T}\sum_{t=1}^{T-1}\left(\sum_{\zeta_{[0,t-1]}} P(q'_{[0,t-1]} = \zeta_{[0,t-1]})\right.$$

$$\times\left(\int \nu(dw)\left(\int P(dx_{t-1}|q'_{[0,t-1]} = \zeta_{[0,t-1]}, w_{t-1} = w)\right.\right.$$

$$\log_2(|J(f(x_{t-1}, w))|)$$

$$+h(x_{t-1}|q'_{[0,t-1]} = \zeta_{[0,t-1]}, w_{t-1} = w)\Big)\Big) - h(x_t|q'_{[0,t]})\Big) \tag{14.47}$$

$$= \limsup_{T\to\infty} \frac{1}{T} \sum_{t=1}^{T-1} \Big(\sum_{\zeta_{[0,t-1]}} P(q'_{[0,t-1]} = \zeta_{[0,t-1]})$$

$$\times\Big(\int \nu(dw) \Big(\int P(dx_{t-1}|q'_{[0,t-1]} = \zeta_{[0,t-1]})$$

$$\log_2(|J(f(x_{t-1}, w))|) + h(x_{t-1}|q'_{[0,t-1]} = \zeta_{[0,t-1]})\Big)\Big) - h(x_t|q'_{[0,t]})\Big) \tag{14.48}$$

$$= \limsup_{T\to\infty} \frac{1}{T} \sum_{t=1}^{T-1} \Big(\Big(\sum_{\zeta_{[0,t-1]}} P(q'_{[0,t-1]}$$

$$= \zeta_{[0,t-1]}) \int P(dx_{t-1}|q'_{[0,t-1]} = \zeta_{[0,t-1]})$$

$$\times \int \nu(dw) \log_2(|J(f(x_{t-1}, w))|)\Big) + h(x_{t-1}|q'_{[0,t-1]}) - h(x_t|q'_{[0,t]})\Big) \tag{14.49}$$

$$= \limsup_{T\to\infty} \frac{1}{T} \Big(\sum_{t=0}^{T-1} \int \pi_t(dx) \Big(\int \nu(dw) \log_2(|J(f(x, w))|)\Big)$$

$$-h(x_{T-1}|q'_{[0,T-1]})\Big)$$

$$\geq V - \liminf_{T\to\infty} \frac{1}{T} h(x_{T-1}|q'_{[0,T-1]}) \tag{14.50}$$

Here,

$$V := \liminf_{T\to\infty} \frac{1}{T} \sum_{t=0}^{T-1} \int \pi_t(dx) \Big(\int \nu(dw) \log_2(|J(f(x, w))|)\Big) \tag{14.51}$$

Equations (14.44) and (14.46) follow from the definition of conditional entropy, (14.45) follows from conditioning on the random variable w_{t-1}. Equations (14.47)–(14.48) follow from the fact that $x_{t-1} \leftrightarrow q'_{[0,t-1]} \leftrightarrow w_{t-1}$ is a Markov chain and the following property. For every realization $q'_{[0,t-1]} = \zeta_{[0,t-1]}$,

$$h(f(x_{t-1}, w_{t-1})|q'_{[0,t-1]} = \zeta_{[0,t-1]}, w_{t-1} = w)$$
$$= h(f^w(x_{t-1})|q'_{[0,t-1]} = \zeta_{[0,t-1]}, w_{t-1} = w)$$

$$= \int P(dx_{t-1}|q'_{[0,t-1]} = \zeta_{[0,t-1]}, w_{t-1} = w) \log_2(|J(f_w(x_{t-1}))|)$$

$$+h(x_{t-1}|q'_{[0,t-1]} = \zeta_{[0,t-1]}, w_{t-1} = w) \tag{14.52}$$

$$= \int P(dx_{t-1}|q'_{[0,t-1]} = \zeta_{[0,t-1]}) \log_2(|J(f(x_{t-1}, w))|)$$

$$+h(x_{t-1}|q'_{[0,t-1]} = \zeta_{[0,t-1]}),$$

where $f_w(x) := f(x, w)$ is an invertible function for every w, and as a result (14.52) follows from the entropy formula for invertible functions of a random variable (see, e.g., p. 167 of [683] and Lemma 4 in [840]) and the last line follows from the condition $x_{t-1} \leftrightarrow q'_{[0,t-1]} \leftrightarrow w_{t-1}$. Equation (14.49) follows from Fubini's theorem by Assumption 14.2.2.

By the hypothesis, $\liminf_{t\to\infty}\frac{1}{t}h(x_t) \leq 0$, it must be that $\lim_{T\to\infty} R_T \geq V$. Thus, the capacity also needs to satisfy this bound.

In the above derivation, (14.50) follows from the fact that for two sequences a_n, b_n:

$$\limsup_{n\to\infty}(a_n + b_n) \geq \limsup_{n\to\infty} a_n + \liminf_{n\to\infty} b_n. \tag{14.53}$$

(b) If $\limsup_{t\to\infty} h(x_t|q'_{[0,t-1]})/t \leq 0$, (14.50) can be applied through (14.53) with V defined as

$$V := \limsup_{n\to\infty}\frac{1}{T}E\Bigg[\Bigg(\sum_{t=1}^{T-1}\int P(dx_t|q'_{[0,t-1]})$$

$$\Bigg(\int \nu(dw)\log_2(|J(f(x_t, w))|)\Bigg)\Bigg)\Bigg]$$

and in (14.50), $\liminf_{T\to\infty}\frac{1}{T}h(x_{T-1}|q'_{[0,T-1]})$ being replaced with lim sup of the same expression. □

14.7.2 Proof of Lemma 14.2.1

Define the event for $K > 0$ so that $P(|x_0| < K) > 0$ as

$$\mathcal{S}_\eta^K = \{\omega : |x_0| \leq K, w = \eta, i.e., w_k = \eta_k, \eta_k \in \mathbb{R}^p, k \geq 0\},$$

such that the noise realizations are fixed and deterministic. In the following, we will drop the subscript and superscripts, and let $P_{\mathcal{S}}$ or $P(\cdot|\mathcal{S})$ denote the conditional probabilities given the event $\mathcal{S}_{\eta}^{K}$.We recall here that $w_t, t \geq 0$ and x_0 are assumed to be independent.

By Definition 13.6.1, first note that the capacity expression satisfies

$$C = \lim_{T\to\infty} \max_{\{P(q_t|q_{[0,t-1]},q'_{[0,t-1]}),\quad 0\le t\le T-1\}} \frac{1}{T} I(q_{[0,T-1]} \to q'_{[0,T-1]})$$

$$= \lim_{T\to\infty} \max_{\{P(q_t|q_{[0,t-1]},q'_{[0,t-1]}),\quad 0\le t\le T-1\}} \frac{1}{T} I(q_{[0,T-1]} \to q'_{[0,T-1]}|\mathcal{S}), \quad (14.54)$$

where the conditional directed information is given by

$$I(q_{[0,T-1]} \to q'_{[0,T-1]}|\mathcal{S}) = \sum_{t=1}^{T-1} I(q_{[0,t]}; q'_t|q'_{[0,t-1]}, \mathcal{S}) + I(q_0; q'_0|\mathcal{S}).$$

Here, (14.54) is a result of the following: Consider any encoder policy

$$P^* = \{P^*(q_0), P^*(q_1|q_0, q'_0), \cdots, P^*(q_t|q_{[0,t-1]}, q'_{[0,t-1]}), \cdots\}.$$

For any $T \in \mathbb{N}$, note that almost surely the following holds:

$$\begin{aligned}
&P(q'_t|q'_{[0,t-1]}, \mathcal{S}) \\
&= \sum_{q_{[0,t]}} P(q'_t, q_{[0,t]}|q'_{[0,t-1]}, \mathcal{S}) \\
&= \sum_{q_{[0,t]}} P(q'_t|q_{[0,t]}, q'_{[0,t-1]}, \mathcal{S}) P(q_{[0,t]}|q'_{[0,t-1]}, \mathcal{S}) \\
&= \sum_{q_{[0,t]}} P(q'_t|q_{[0,t]}, q'_{[0,t-1]}) P(q_{[0,t]}|q'_{[0,t-1]}, \mathcal{S}) \qquad (14.55)\\
&= \sum_{q_{[0,t]}} P(q'_t|q_{[0,t]}, q'_{[0,t-1]}) P^*(q_t|q'_{[0,t-1]}, q_{[0,t-1]}, \mathcal{S}) P(q_{[0,t-1]}|q'_{[0,t-1]}, \mathcal{S}) \\
&= \sum_{q_{[0,t]}} P(q'_t|q_{[0,t]}, q'_{[0,t-1]}) P^*(q_t|q'_{[0,t-1]}, q_{[0,t-1]}) P(q_{[0,t-1]}|q'_{[0,t-1]}, \mathcal{S}) (14.56)\\
&= \sum_{q_{[0,t]}} P(q'_t|q_{[0,t]}, q'_{[0,t-1]}) P^*(q_t|q'_{[0,t-1]}, q_{[0,t-1]}) P(q_{[0,t-1]}|q'_{[0,t-1]}) \qquad (14.57)\\
&= P(q'_t|q'_{[0,t-1]}) \qquad (14.58)
\end{aligned}$$

where (14.55) follows from Definition 13.6.1, (14.56) from the structure of a coding policy, and (14.57) from the following inductive argument. Note that $P(q_0, q'_0|\mathcal{S}) = P(q_0, q'_0)$. If $P(q_{[0,t-1]}, q'_{[0,t-1]}|\mathcal{S}) = P(q_{[0,t-1]}, q'_{[0,t-1]})$, it follows that

$$
\begin{aligned}
&P(q_{[0,t]}, q'_{[0,t]}|\mathcal{S}) \\
&= P(q'_t|q_{[0,t]}, q'_{[0,t-1]}, \mathcal{S})P^*(q_t|q'_{[0,t-1]}, q_{[0,t-1]}, \mathcal{S})P(q_{[0,t-1]}, q'_{[0,t-1]}|\mathcal{S}) \\
&= P(q'_t|q_{[0,t]}, q'_{[0,t-1]})P^*(q_t|q'_{[0,t-1]}, q_{[0,t-1]})P(q_{[0,t-1]}, q'_{[0,t-1]}|\mathcal{S}) \\
&= P(q'_t|q_{[0,t]}, q'_{[0,t-1]})P^*(q_t|q'_{[0,t-1]}, q_{[0,t-1]})P(q_{[0,t-1]}, q'_{[0,t-1]}) \\
&= P(q_{[0,t]}, q'_{[0,t]}) \qquad (14.59)
\end{aligned}
$$

As a result, (14.57) simplifies to (14.58) by eliminating the conditioning on $\mathcal{S}$ and (14.54) holds.

We now use a similar argument as in (14.50), but need to modify the steps due to the conditioning on $\mathcal{S}$:

$$
\begin{aligned}
&\lim_{T\to\infty} R_T \\
&\geq \lim\sup_{T\to\infty}\frac{1}{T}\left(\sum_{t=1}^{T-1} I(x_t; q'_t|q'_{[0,t-1]})|\mathcal{S}) + I(x_0; q'_0|\mathcal{S})\right) \\
&= \lim\sup_{T\to\infty}\frac{1}{T}\left(\sum_{t=1}^{T-1}\left(h_{\mathcal{S}}(x_t|q'_{[0,t-1]}) - h_{\mathcal{S}}(x_t|q'_{[0,t]})\right) + I(x_0; q'_0|\mathcal{S})\right) \\
&= \lim\sup_{T\to\infty}\frac{1}{T}\left(\sum_{t=1}^{T-1}\left(h_{\mathcal{S}}(f(x_{t-1}) + Bu_{t-1} + w_{t-1}|q'_{[0,t-1]}) - h_{\mathcal{S}}(x_t|q'_{[0,t]})\right)\right. \\
&\qquad \left. + I(x_0; q'_0|\mathcal{S})\right) \\
&= \lim\sup_{T\to\infty}\frac{1}{T}\left(\sum_{t=1}^{T-1}\left(h_{\mathcal{S}}(f(x_{t-1}) + Bu_{t-1}|q'_{[0,t-1]}) - h_{\mathcal{S}}(x_t|q'_{[0,t]})\right)\right. \\
&\qquad \left. + I(x_0; q'_0|\mathcal{S})\right) \\
&= \lim\sup_{T\to\infty}\frac{1}{T}\left(\sum_{t=1}^{T-1}\left(h_{\mathcal{S}}(f(x_{t-1})|q'_{[0,t-1]}) - h_{\mathcal{S}}(x_t|q'_{[0,t]})\right) + I(x_0; q'_0|\mathcal{S})\right) \\
&= \lim\sup_{T\to\infty}\frac{1}{T}\left(\sum_{t=1}^{T-1} E[\left(P_{\mathcal{S}}(dx_{t-1}|q'_{[0,t-1]}) \log_2(|J(f(x_{t-1}))|)]\right.\right. \\
&\qquad \left.\left. + h_{\mathcal{S}}(x_{t-1}|q'_{[0,t-1]}) - h_{\mathcal{S}}(x_t|q'_{[0,t]})\right) + I(x_0; q'_0|\mathcal{S})\right) \qquad (14.60) \\
&\geq V_{\mathcal{S}} - \lim\inf_{T\to\infty}\left(\frac{1}{T} h_{\mathcal{S}}(x_{T-1}|q'_{[0,T-1]})\right) \qquad (14.61)
\end{aligned}
$$

$$\geq V_{\mathcal{S}} - \liminf_{T \to \infty} \left(\frac{1}{T} h_{\mathcal{S}}(x_{T-1}) \right) \geq V_{\mathcal{S}}. \tag{14.62}$$

Here,

$$V_{\mathcal{S}} = \liminf_{T \to \infty} E\left[\frac{1}{T} \left(\sum_{t=1}^{T-1} P_{\mathcal{S}}(dx_{t-1} | q'_{[0,t-1]}) \log_2(|J(f(x_{t-1}))|) \right) \right], \tag{14.63}$$

and (14.60) follows from the fact that

$$h_{\mathcal{S}}(f(x_{t-1}) | q'_{[0,t-1]}) = E\left[\int P_{\mathcal{S}}(dx_{t-1} | q'_{[0,t-1]}) \log_2(|J(f(x_{t-1}))|) \right]$$
$$+ h_{\mathcal{S}}(x_{t-1} | q'_{[0,t-1]})$$

Thus, with $V_{\mathcal{S}} \geq L$, it follows that

$$\liminf_{T \to \infty} \left(\frac{1}{T} h_{\mathcal{S}}(x_{T-1} | q'_{[0,T-1]}) \right) \geq L - C \tag{14.64}$$

We now seek to obtain a bound on $h_{\mathcal{S}}(x_{T-1} | q'_{[0,T-1]})$. As in [513], note that

$$h_{\mathcal{S}}(x_T | q'_{[0,T]}) \leq h_{\mathcal{S}}(x_T, \mathcal{Y} | q'_{[0,T]}),$$

where $\mathcal{Y}$ is a binary random variable which is 1 if $|x_T| \leq b(T)$ and 0 otherwise. Let

$$P_{\mathcal{S}}(\mathcal{Y} = 1) = P_{\mathcal{S}}(|x_T| \leq b(T)) =: p_T^{\mathcal{S}}.$$

Then,

$$h_{\mathcal{S}}(x_T, \mathcal{Y} | q'_{[0,T]}) = h_{\mathcal{S}}(x_T | q'_{[0,T]}, \mathcal{Y}) + H_{\mathcal{S}}(\mathcal{Y} | q'_{[0,T]})$$
$$\leq h_{\mathcal{S}}(x_T | q'_{[0,T]}, \mathcal{Y}) + 1,$$

since $\mathcal{Y}$ is binary. We have that

$$h_{\mathcal{S}}(x_T | q'_{[0,T]}, \mathcal{Y}) \leq p_T^{\mathcal{S}} \frac{n}{2} \log_2(2\pi e b^2(T)) \tag{14.65}$$
$$+ (1 - p_T^{\mathcal{S}}) h_{\mathcal{S}}\left(x_T \middle| q'_{[0,T]}, |x_T| \geq b(T) \right)$$

and

$$
\begin{aligned}
&h_{\mathcal{S}}\left(x_T \middle| q'_{[0,T]}, |x_T| > b(T)\right) \\
&= h_{\mathcal{S}}\left(f(x_{T-1}) + Bu_{T-1} + w_{T-1} \middle| q'_{[0,T]}, |x_T| > b(T)\right) \\
&\leq h_{\mathcal{S}}\left(f(x_{T-1}) + Bu_{T-1} + w_{T-1} \middle| q'_{[0,T-1]}, |x_T| > b(T)\right) \qquad (14.66) \\
&= h_{\mathcal{S}}\left(f(x_{T-1}) + w_{T-1} \middle| q'_{[0,T]-1}, |x_T| > b(T)\right) \\
&= h_{\mathcal{S}}\left(f(x_{T-1}) \middle| q'_{[0,T]-1}, |x_T| > b(T)\right) \qquad (14.67) \\
&= E\left[\int P_{\mathcal{S}}\left(dx_{T-1} \middle| q'_{[0,T-1]}, |x_T| > b(T)\right) \log_2\left(\left|J(f(x_{T-1}))\right|\right)\right] \\
&\qquad + h_{\mathcal{S}}\left(x_{T-1} \middle| q'_{[0,T-1]}, |x_T| > b(T)\right) \qquad (14.68) \\
&\leq E\left[\int P_{\mathcal{S}}\left(dx_{T-1} \middle| q'_{[0,T-1]}, |x_T| > b(T)\right) \log_2\left(\left|J(f(x_{T-1}))\right|\right)\right] \\
&\qquad + h_{\mathcal{S}}\left(x_{T-1} \middle| q'_{[0,T-2]}, |x_T| > b(T)\right) \\
&\leq M + h_{\mathcal{S}}\left(x_{T-1} \middle| q'_{[0,T-2]}, |x_T| > b(T)\right) \\
&\quad \vdots \\
&\leq MT + h_{\mathcal{S}}\left(x_0 \middle| |x_T| > b(T)\right).
\end{aligned}
\tag{14.69}
$$

Here (14.67) follows due to the fact that $\mathcal{S}$ determines the noise realizations. We note that non-linearity of f adds further technical challenges when compared with the linear systems setup.[1] Here, M is the supremum of $\log_2(|J(f(x))|)$. In the above derivation in (14.68), we use the fact that f is invertible. In the last inequality, we use the fact that the entropy of a random variable with a fixed covariance is upper

[1] Two technical intricacies here are as follows: For differential entropy (unlike discrete entropy) the relationship $h(x+y) \leq h(x) + h(y)$ does not in general hold for random variables x, y; this is why first a conditioning on $\mathcal{S}$ is taken in the proof. Furthermore, we cannot obtain an upper bound by taking out the conditioning on the event $|x_T| > b(T)$, since conditioning on a single event may decrease or increase entropy; note that conditioning on a random variable, however, does not increase the entropy.

bounded by the entropy of a Gaussian random variable with the same covariance, and that $|x_0|$ conditioned on $\mathcal{S}$ is upper bounded by K^2.

Thus, by (14.64–14.65) and (14.69) we have

$$\liminf_{T\to\infty}\frac{1}{T}\Bigg(1+(1-p_T^{\mathcal{S}})(MT+h_{\mathcal{S}}(x_0\Big||x_T|>b(T)))$$
$$+p_T^{\mathcal{S}}\frac{n}{2}\log_2(2\pi e b^2(T))\Bigg)\geq L-C, \tag{14.70}$$

Since $h_{\mathcal{S}}\Big(x_0\Big||x_T|>b(T)\Big)\leq (n/2)\log_2(2\pi e K^2)$, it follows that for all K and η

$$\limsup_{T\to\infty}P_{\mathcal{S}_\eta^K}(|x_T|\leq b(T))\leq\frac{M-(L-C)}{M},$$

for all $b(T)$ such that $\lim_{T\to\infty}\log_2(b(T))/T=0$. But now

$$\begin{aligned}
&\limsup_{T\to\infty}P(|x_T|\leq b(T))\\
&\leq \limsup_{T\to\infty}P(|x_T|\leq b(T),|x_0|\leq K)\\
&+\limsup_{T\to\infty}P(|x_T|\leq b(T),|x_0|\geq K)\\
&\leq \limsup_{T\to\infty}P(|x_T|\leq b(T),|x_0|\leq K)+P(|x_0|\geq K)\\
&=\limsup_{T\to\infty}\int P(d\eta)P_{\mathcal{S}_\eta^K}(|x_T|\leq b(T))+P(|x_0|\geq K)\\
&\leq\int P(d\eta)\limsup_{T\to\infty}P_{\mathcal{S}_\eta^K}(|x_T|\leq b(T))+P(|x_0|\geq K) \qquad (14.71)\\
&\leq\int P(d\eta)\frac{M-(L-C)}{M}+P(|x_0|\geq K)\\
&=\frac{M-(L-C)}{M}+P(|x_0|\geq K)
\end{aligned}$$

where we use Fatou's lemma in (14.71) and the fact that (14.70) holds for every restriction of the noise realizations η and K values. Since an individual probability measure is tight, $\lim_{K\to\infty}P(|x_0|\geq K)=0$, the right hand side can be made arbitrarily close to $\frac{M-(L-C)}{M}$ and the result follows. □

14.7.3 Proof of Theorem 14.2.5

Proof First note that

$$\begin{aligned} C &\geq I(q_t, q_t') = H(q_t') - H(q_t'|q_t) \\ &\geq H(q_t'|m_t) - H(q_t'|q_t) = H(q_t'|m_t) - H(q_t'|q_t, x_t, m_t) \\ &\geq H(q_t'|m_t) - H(q_t'|x_t, m_t) = I(q_t'; x_t|m_t) \end{aligned}$$

Hence,

$$C \geq \liminf_{T\to\infty} \frac{1}{T} \sum_{t=0}^{T-1} I(q_t'; x_t|m_t)$$

$$= \liminf_{T\to\infty} \frac{1}{T} \sum_{t=0}^{T-1} \Big(h(x_t|m_t) - h(x_t|m_t, q_t') \Big)$$

$$= \liminf_{T\to\infty} \Bigg(\frac{1}{T} \sum_{t=1}^{T-1} \bigg(h\Big(f(x_{t-1}, w_{t-1}) + Bu_{t-1}|m_t \Big) - h(x_t|m_t, q_t') \bigg)$$

$$+ I(q_0'; x_0|m_0) \Bigg)$$

$$= \liminf_{T\to\infty} \frac{1}{T} \sum_{t=1}^{T-1} h\Big(f(x_{t-1}, w_{t-1}) + Bu_{t-1})|m_t \Big) - h(x_t|m_t, q_t')$$

$$\geq \liminf_{T\to\infty} \frac{1}{T} \sum_{t=1}^{T-1} h\Big(f(x_{t-1}, w_{t-1}) + Bu_{t-1}|m_t, m_{t-1}, q_{t-1}' \Big) - h(x_t|m_t, q_t')$$

$$= \liminf_{T\to\infty} \frac{1}{T} \sum_{t=1}^{T-1} h\Big(f(x_{t-1}, w_{t-1}) + Bu_{t-1}|m_{t-1}, q_{t-1}' \Big) - h(x_t|m_t, q_t')$$

$$\geq \liminf_{T\to\infty} \frac{1}{T} \sum_{t=1}^{T-1} h\Big(f(x_{t-1}, w_{t-1}) + Bu_{t-1}|w_{t-1}, m_{t-1}, q_{t-1}' \Big) - h(x_t|m_t, q_t')$$

$$= \liminf_{T\to\infty} \frac{1}{T} \sum_{t=1}^{T-1} h\Big(f(x_{t-1}, w_{t-1})|w_{t-1}, m_{t-1}, q_{t-1}' \Big) - h(x_t|m_t, q_t')$$

$$= \liminf_{T\to\infty} \frac{1}{T} \sum_{t=1}^{T-1} \bigg(\int \nu(dw) \Big(\sum P(m_{t-1} = m, q_{t-1}' = q')$$

$$\times \int P(x_{t-1} \in dx|m_{t-1} = m, q_{t-1}' = q', w_{t-1} = w) \log_2(|J(f(x, w_{t-1}))|)$$

$$+ h(x_{t-1}|m_{t-1} = m, q_{t-1}' = q', w_{t-1} = w) \Big) - h(x_t|m_t, q_t') \bigg)$$

$$= \liminf_{T\to\infty}\frac{1}{T}\sum_{t=1}^{T-1}\Big(\int \nu(dw)\Big(\sum P(m_{t-1}=m, q'_{t-1}=q')$$

$$\times \int P(x_{t-1}\in dx|m_{t-1}=m, q'_{t-1}=q')\log_2(|J(f(x,w_{t-1}))|)$$

$$+h(x_{t-1}|m_{t-1}=m, q'_{t-1}=q')\Big) - h(x_t|m_t, q'_t)\Big) \tag{14.72}$$

$$= \liminf_{T\to\infty}\frac{1}{T}\sum_{t=1}^{T-1}\Big(\int \nu(dw_{t-1})\Big(\int \pi_{t-1}(dx)\log_2(|J(f(x,w_{t-1}))|)$$

$$+h(x_{t-1}|m_{t-1}, q'_{t-1})\Big) - h(x_t|m_t, q'_t)\Big) \tag{14.73}$$

$$= \liminf_{T\to\infty}\frac{1}{T}\sum_{t=1}^{T-1}\int \pi_{t-1}(dx)\int \nu(dw)\log_2(|J(f(x,w))|)$$

$$-h(x_{T-1}|m_{T-1}, q'_{T-1}) \tag{14.74}$$

$$\geq \liminf_{T\to\infty}\frac{1}{T}\Big(\sum_{t=1}^{T-1}\int \pi_{t-1}(dx)\Big(\int \nu(dw)\log_2(|J(f(x,w))|)\Big) - h(x_{T-1})\Big)$$

$$\geq \liminf_{T\to\infty}\frac{1}{T}\sum_{t=1}^{T-1}\int \pi_{t-1}(dx)\Big(\int \nu(dw)\log_2(|J(f(x,w))|)\Big)$$

$$-\limsup_{T\to\infty}\frac{1}{T}h(x_{T-1})$$

$$\geq \liminf_{T\to\infty}\frac{1}{T}\sum_{t=1}^{T-1}\int \pi_{t-1}(dx)\Big(\int \nu(dw)\log_2(|J(f(x,w))|)\Big) \tag{14.75}$$

$$= \liminf_{T\to\infty}\int \pi_0(dx)E_x\Big[\frac{1}{T}\sum_{t=0}^{T-1}\int \pi_{t-1}(dx)$$

$$\Big(\int \nu(dw)\log_2(|J(f(x,w))|)\Big)\Big] \tag{14.76}$$

$$\geq \int \pi_0(dx)\liminf_{T\to\infty}E_x\Big[\frac{1}{T}\sum_{t=0}^{T-1}\int \pi_{t-1}(dx)$$

$$\Big(\int \nu(dw)\log_2(|J(f(x,w))|)\Big)\Big] \tag{14.77}$$

$$= \int \pi_0(dx)\Big(\int \pi(dz)\Big(\int \nu(dw)\log_2(|J(f(z,w))|)\Big)\Big) \tag{14.78}$$

$$= \int \pi(dx)\Big(\int \nu(dw)\log_2(|J(f(x,w))|)\Big) \tag{14.79}$$

In the second and third inequalities (beyond the first one, which is explained prior to the equations above), we use the fact that conditioning on a random variable does not increase the entropy and the update laws (14.10). The equality (14.73) holds since for every w, the map $f(.,w)$ is invertible (here $J(f(x,w_t))$ is the Jacobian for the realized value of w_t) and that w_t is an independent noise process using the laws of total probability. Here (14.72) follows due to the independence of w_t, (14.74) follows from Fubini's Theorem since $\log_2(|J(f(x,w))|)$ is bounded, (14.77) follows from Fatou's lemma given the assumption that $\log_2(|J(f(x,w))|)$ is bounded from below, and (14.78) follows from positive Harris recurrence (see Theorem 4.3.1 in [344]). □

14.7.4 Proof of Lemma 14.3.1

We fix a causal coding and control policy which achieves the AMS property over a noiseless channel of capacity C. For a given set $B \in \mathcal{B}(\mathbb{R}^N)$ with $0 < Q(B) < 1$, the AMS property implies

$$\lim_{T\to\infty}\frac{1}{T}\sum_{t=0}^{T-1} P(x_t \in B^c) = 1 - Q(B) =: r. \tag{14.80}$$

Since $P(x_t \in B^c) = E[\mathbb{1}_{B^c}(x_t)]$, this can also be written as

$$\lim_{T\to\infty} E\Big[\frac{1}{T}\sum_{t=0}^{T-1}\mathbb{1}_{B^c}(x_t)\Big] = r.$$

We pick $\varepsilon \in (0, (1-r)/r)$ and choose $T_0 > 0$ so that

$$E\Big[\frac{1}{T}\sum_{t=0}^{T-1}\mathbb{1}_{B^c}(x_t)\Big] \le \Big(1+\frac{\varepsilon}{2}\Big)r, \quad \forall T \ge T_0.$$

By Markov's inequality, this implies that for every $T \ge T_0$, the event

$$\tilde{\Omega}_T := \Big\{\omega \in \Omega : \frac{1}{T}\sum_{t=0}^{T-1}\mathbb{1}_{B^c}(x_t(\omega)) \le (1+\varepsilon)r\Big\}$$

occurs with probability $P(\tilde{\Omega}_T) \ge \varepsilon/(2(1+\varepsilon))$, since

$$P\Big(\frac{1}{T}\sum_{t=0}^{T-1}\mathbb{1}_{B^c}(x_t) > (1+\varepsilon)r\Big) \le \frac{E[\frac{1}{T}\sum_{t=0}^{T-1}\mathbb{1}_{B^c}(x_t)]}{(1+\varepsilon)r}$$

$$\le \frac{1+\frac{\varepsilon}{2}}{1+\varepsilon} = 1 - \frac{\varepsilon}{2(1+\varepsilon)}.$$

Note that for every $\omega \in \tilde{\Omega}_T$, the number of t's in $[0; T-1]$ satisfying $x_t(\omega) \in B^c$ is $\le (1+\varepsilon)rT$. Now for every $T \ge T_0$ consider the set

$$S_T := \{\bar{u}_{[0,T-1]}(\omega) \in U^T : \omega \in \tilde{\Omega}_T\}$$

of control sequences generated in the time interval $[0; T-1]$ provided that $\omega \in \tilde{\Omega}_T$ and $x_0 = x_0(\omega)$, $\bar{w} = \bar{w}(\omega)$. Since the maximum number of different messages that can be transmitted in the time interval $[0; T-1]$ is $(\#\mathcal{M})^T$, we have

$$\#S_T \le (\#\mathcal{M})^T.$$

We claim that S_T is $(T, B, \frac{1+\varepsilon/2}{1+\varepsilon}, (1+\varepsilon)r)$-spanning. Indeed,

$$P(\tilde{\Omega}_T) \ge \frac{\varepsilon}{2(1+\varepsilon)} = 1 - \frac{1+\frac{\varepsilon}{2}}{1+\varepsilon},$$

for every $\omega \in \tilde{\Omega}_T$ we have $\bar{u}_{[0,T-1]}(\omega) \in S_T$, and the number of t's in $[0; T-1]$ with $x_t(\omega) = \varphi(t, x_0(\omega), \bar{u}(\omega), \bar{w}(\omega)) \in B$ is $\ge T - (1+\varepsilon)rT = (1-(1+\varepsilon)r)T$. Hence,

$$s_B\Big(T, \frac{1+\frac{\varepsilon}{2}}{1+\varepsilon}, (1+\varepsilon)r\Big) \le \#S_T \le (\#\mathcal{M})^T \quad \text{for all } T \ge T_0.$$

Taking logarithms, dividing by T and letting $T \to \infty$ yields the assertion. The case $Q(B) = 1$ is handled by replacing (14.80) with the inequality

$$\lim_{T\to\infty}\frac{1}{T}\sum_{t=0}^{T-1} P(x_t \in B^c) = Q(B^c) = 0 \le r$$

for an arbitrarily chosen $r \in (0, 1)$, and using the same arguments. □

14.8 Concluding Remarks

In this chapter, conditions on information channels leading to stochastic stability of non-linear systems controlled over noisy channels have been studied. Stochastic stability notions considered were asymptotic mean stationarity, ergodicity, and stationarity. Results for linear systems have been recovered as a special case. We presented both information theoretic methods as well as stochastic geometric methods with their complementary results.

14.8.1 Comparison with Invariance Entropy and Deterministic Non-linear Systems Controlled Over Noiseless Channels

It is important to note that for linear systems studied earlier, any local dynamical or control-theoretic property is also a global property. As such, the problems of local stabilization (stabilization to a point), semi-global stabilization (set-invariance), and global stabilization (stochastic stability) can all be handled with similar methods, leading to the aforementioned data-rate theorem in each case. For nonlinear systems, however, the three stability problems are fundamentally different and require distinct approaches. For example, linearization techniques work well for local problems, for semi-global problems only under specific assumptions, and almost never for global problems. In addition, the presence of (possibly unbounded and additive) noise requires an approach fundamentally different from the machinery utilized for local stabilization problems.

In view of the above, a related result in the literature is with regard to stabilization to a point: Under the assumptions that (i) f has the form in (14.5) (without noise) with continuous partial derivatives, (ii) there exists a fixed point (equilibrium) x^* so that $x^* = f(x^*, u^*)$, (iii) a local strong invariability condition is satisfied which relates the size of an invariant set and the size of a control action set in the sense that for any $\epsilon > 0$, there exists $\rho > 0$ so that for all $\epsilon' \in (0, \rho]$, the set $\{x : |x - x^*| \leq \epsilon'\}$ is invariant with the control action set $U = \{u : |u - u^*| \leq \epsilon\}$, and (iv) the pair (A, B) is controllable where A, B are the Jacobians of f with respect to state and control at x^*, u^*; [553] has reported that for convergence to the equilibrium an average rate of $R > \sum_{|\lambda_i|>1} \log_2(|\lambda_i|)$ is sufficient, where λ_i are the eigenvalues of the Jacobian at the equilibrium point.

A further related result in the spirit of this chapter is on a case where there exists an invariant set with a non-empy interior: For continuous-time systems of the form $\frac{dx}{dt} = f(x, u), u \in U$, Colonius and Kawan [184] have established a lower bound on *invariance entropy* as

$$\max(0, \min_{x,u \in Q \times U} \sum_i \frac{\partial f_i}{\partial x_i}(x, u)), \tag{14.81}$$

where Q is a weakly invariant set and f_i is the ith coordinate function of f. More refined bounds are present if further structural properties are imposed: in [198], under a uniform hyperbolicity assumption (see Definition 4.4 in [198]), Theorem

4.8 therein states a similar lower bound by considering the unstable components in an invariant set.

These results can be viewed to be related to Theorem 14.2.3 and Theorem 14.2.5, as well as Theorem 14.5.1, in that the average entropy growth as measured by the eigenvalues of the Jacobian matrix under the invariant probability measure is lower bounded by a minimum over the elements in the support set, or is upper bounded by a maximizing element in the support set. In the stabilization to the point example of [553], the invariant measure is a delta measure on a single point. In the invariant set example leading to (14.81), the set Q can be viewed to be the support set of some invariant measure under the system dynamics if such a measure were to exist. Likewise, [479] and [478] have obtained conditions for noise-free systems controlled over noiseless channels. Due to the absence of noise, one could identify an invariant compact set, and consider a bound on the Lipschitz growth parameter for the system over this invariant set to obtain sufficiency conditions. When the system is (Lebesgue) irreducible, however, due to the effect of noise, local properties are not descriptive and the invariant probability measure reflects the rate conditions and entropy growth in the system. In this case, the local growth integrated under an invariant measure gives a proper bound.

14.8.2 Some Open Directions on Stationary Coding and Control Policies and Information Theory

It would be interesting to show, for a class of systems, that stationary coding and control policies can be used to arrive at stability with a stationary closed loop-process provided that the capacity of the channel satisfies the entropy growth bound and the channel satisfies certain ergodicity conditions. However, except for linear Gaussian systems controlled over Gaussian channels and erasure channels to be studied in Chap. 16, this question has not been answered even for linear systems controlled over general discrete memoryless channels (that is, non-stationary coding schemes have been used for more general discrete memoryless channels). Furthermore, the tightness of the converse results is another direction. A further direction is the causal coding problem for non-ergodic sources: In the information theory literature, through non-causal codes, a class of source coding theorems for non-ergodic sources exist (see e.g. [316]), however, the extensions of these for even control-free non-linear systems under causal coding require further research. Some related results will be presented later in the book in the following three chapters.

14.9 Bibliographic Notes

In the literature, the study of non-linear systems have typically involved noise-free controlled systems controlled over discrete noiseless channels. Many of the studies on control of non-linear systems over communication channels have focused on constructive schemes (and not on converse theorems), primarily for noise-free sources and channels, see. e.g. [56, 479] (see also [147, 478]), and [207]. For noise-free systems, it typically suffices to only consider a sufficiently small *invariant* neighborhood of an equilibrium point to obtain stabilizability conditions.

There have been some studies which have adopted a measure-theoretic entropic view for the control of non-linear dynamical systems over communication channels. Relevant contributions include [528] and [803]: Building on [508] and [840]; [528] develops an entropy analysis for non-linear system dynamics to obtain the relation between the entropy rates of a measurement disturbance, output and the dynamical system, and generalizing a *Bode-type* entropy analysis for non-linear systems. A related entropy analysis for a class of stochastic non-linear systems has been considered in [803]. [730] and [729] have considered fading and erasure channels between the controller and the actuator and have studied ergodicity properties using Lyapunov theoretic arguments under a class of structures imposed on control policies; these contributions do not consider finite-rate information and coding restrictions which may arise due to the presence of a channel.

Finally, we note an important related discussion in view of Bode's integral formula as extended to a class of non-linear systems in [840] under somewhat restrictive conditions, see [840, Thm. 9]. Relevant work includes [237, 246, 508] and [566] for linear systems. For non-linear systems the entropy and mutual information arguments provide the appropriate fundamental bounds instead of a sensitivity integral/transfer function analysis which is commonly used for linear systems as is also advocated in [840]. An earlier contribution utilizing measure theoretic entropy for the study and classification of controlled stochastic systems is [374]. The findings in the chapter provide further generalizations; see Theorem 14.2.2 and Remark 14.2.4. We note also that recently a metric entropy generalization of some of the results in [403] have been developed [183].

In information theory, a well-established result is that for noiseless coding of information stable sources (this includes all finite state stationary and ergodic sources) over a class of information stable noisy channels (which includes the channels we have considered), an asymptotically noise-free recovery is possible if the channel capacity is greater than the source entropy through the use of non-causal codes, see e.g. [741] and [418]. However, for the problems considered in this chapter, (i) the source is non-stationary and may be open-loop unstable, (ii) the encoding is causal, and (iii) the source process space is not finite-alphabet. Nonetheless, we saw that the invariance properties of the (closed-loop) state process does appear in the rate bounds that were presented in the chapter.

For the task of proving more general data rate theorems for set invariance of non-linear (and linear) deterministic systems, two a priori distinct notions have proven particularly fruitful. The first is that of *topological feedback entropy*, a concept modeled after the open cover definition of topological entropy for classical dynamical systems prescribed in [5]. Presented in [553] by Nair, Evans, Mareels, and Moran, the *topological feedback entropy* measures the lowest data-rate required between a coder and controller to ensure the existence of a coding and control policy which guarantees that the system's state process remain inside a compact controlled invariant set.

The second notion is *invariance entropy*, which was introduced in [184]. One can use Bowen's formula for the topological entropy to obtain the invariance entropy in the case of a linear control system [403]. To define *invariance entropy*, let $r(\tau, Q)$ denote the minimum number of control functions required to establish invariance of Q on the time interval $[0, \tau]$ given an arbitrary initial state in the set. The invariance entropy of a compact subset Q of the state space is then defined to be

$$h_{\text{inv}}(Q) := \lim_{\tau \to \infty} \frac{1}{\tau} \log r(\tau, Q).$$

This quantity measures how fast in time the number of open-loop control functions required to render Q invariant grows. It is a measurement of the smallest average rate of information that must be transmitted to a controller to render Q invariant, and is thus intimately linked to minimum data rates.

It turns out that under a strong invariance condition, the notions of *topological feedback entropy* and *invariance entropy* are equivalent, as explained in [243]. We also note the more recent introduction of *metric invariance entropy* in [183]. This quantity provides a measure-theoretic analogue to *invariance entropy* for discrete-time systems, and is based on conditionally invariant measures. Finally, [719] has generalized the notion of invariance entropy to systems with bounded noise/uncertainty under the criterion of set invariance.

Building on work done in the context of deterministic non-linear systems, the authors in [408] generalized the notion of invariance entropy for use in the stability analysis of discrete time stochastic systems controlled over finite-capacity channels. The introduced quantity, called *stabilization entropy* (see Definition 14.3.1, and see also a definition tailored for the ergodicity criterion called Multi-Set Stabilization Entropy (see Definition 14.4.2)), is inspired by both invariance entropy and measure-theoretic entropy of dynamical systems. This is based on a characterization of measure-theoretic entropy due to Katok [401], and a generalization of said notion found in [599] and [408].

For deterministic systems, under uniformly hyperbolicity assumptions of a non-linear system (which amounts to a strict splitting of the tangent space into stable and unstable components) refinements on the rate requirements have been reported in [198] (see also [405]).

Theorem 14.4.3 is from [288]. Theorems 14.2.5, 14.4.1 and 14.4.2 build on [816] [408] and [287]. These papers also provide general overviews of the relevant literature.

Chapter 15
Optimal Real-Time Coding and Control Under Information Constraints: Structural and Existence Results

15.1 Introduction

We have so far addressed the problem of *stabilization* of networked control systems. In this chapter we move beyond stabilization and study *optimization* of such systems, from the points of view of both encoding and control policies.

The chapter considers the optimal causal (also called *zero-delay*) coding/quantization problem for Markov sources or networked control systems. It presents structural results on optimal causal coding of Markov sources in a large class of settings: fully observed and partially observed Markov sources as well as multi-sensor systems, and systems driven by control. For the optimal causal coding of a fully observed or a partially observed Markov source, the structure of optimal causal coders are obtained, which feature a separation structure. It is also shown that real-time decentralized coding of a partially observed i.i.d. source admits a memoryless optimal solution. Such a result does not, in general, extend to decentralized coding of partially observed Markov sources. We also establish in the chapter the existence of optimal control and quantization policies under appropriate technical conditions. Linear systems with quadratic cost will also be considered. The infinite horizon case is also studied, for which both existence, structure and near-optimality results are obtained.

The contents of the chapter are as follows: In Sect. 15.2, we introduce the problem structure while also revisiting the setup of Sect. 12.2.2. We then present, in Sect. 15.3, structural results on optimal encoders, more precisely on optimal real-time coding of Markov sources when there is only one encoder. We study both fully observed and partially observed settings for systems without control, then systems driven by control, and then the case with noisy channels. Section 15.4 considers the existence of optimal quantization policies. Section 15.5 studies the infinite horizon setup, with existence, structural, and approximation results presented. In Sect. 15.6, we move to a decentralized setting, and show through a counterexample the

S. Yüksel, T. Başar, *Stochastic Teams, Games, and Control under Information Constraints*, Systems & Control: Foundations & Applications,
https://doi.org/10.1007/978-3-031-54071-4_15

difficulty one encounters in obtaining structural results for decentralized coding in the absence of a controlled Markov state construction, while providing a separation result when the source is memoryless. An appendix to the chapter includes proofs of the main results.

For background reading on Markov Decision Processes, we refer the reader to Appendix E.

15.2 Policies and Action Spaces for Encoding

We consider a typical causal encoding/quantization setup of the type introduced earlier in Sect. 12.2.2. For simplicity in exposition, but without much loss of conceptual generality, we consider the case of only two encoders and within this context introduce the causality and measurability constraints in quantizer design for decentralized systems.

Consider first a control-free partially observed Markov process, defined on a probability space, $(\Omega, \mathcal{F}, P)$, and generated by the following scalar discrete-time equations for $t \geq 0$:

$$x_{t+1} = f(x_t, w_t), \tag{15.1}$$

$$y_t^i = g^i(x_t, v_t^i), \tag{15.2}$$

for (Borel) measurable functions $f, g^i, i = 1, 2$, with $\{w_t, v_t^i, i = 1, 2\}$ zero-mean noise processes with finite second moments, which are independent across time and space. We further have $x_t \in \mathbb{X}$, and $y_t^i \in \mathbb{Y}^i$, where $\mathbb{X}$, $\mathbb{Y}^i$ are Polish spaces. Let an encoder, Encoder i, be located at one end of a measurement channel characterized by (15.2), this being so for $i = 1, 2$. The encoders transmit their information to a receiver (see Fig. 15.1, as well as Fig. 15.2 as an important particular instance), over

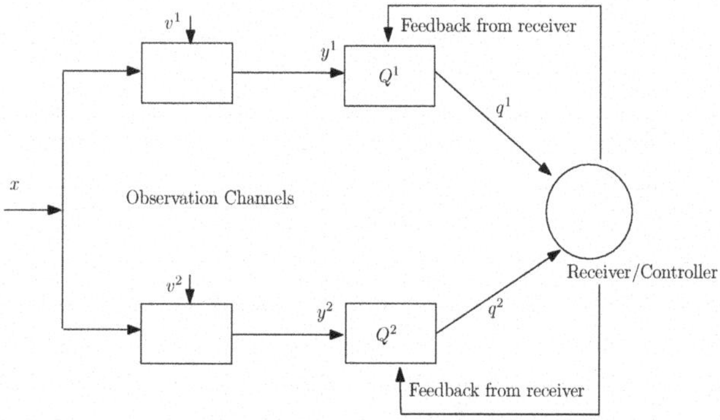

Fig. 15.1 Partially observed source under a decentralized structure

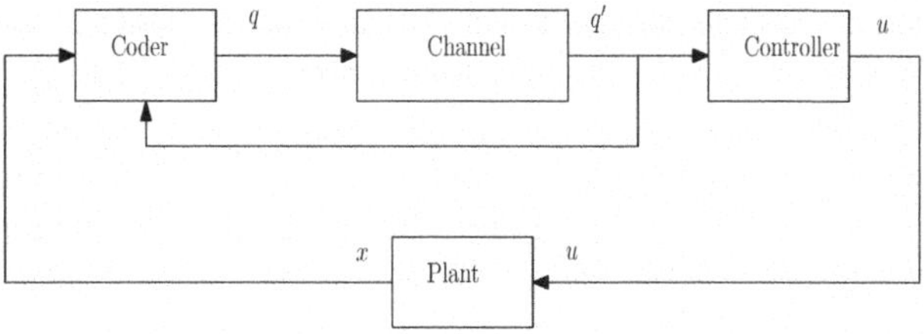

Fig. 15.2 Optimal coding with a single encoding terminal. We will study various scenarios, a general version being a controlled model over a noisy channel

a discrete noiseless channel with finite capacity or over a discrete noisy channel (with feedback), and hence, they have to quantize their input.

We let, as before in Sect. 12.2.2, $\Pi^{comp,i}$ denote a *composite quantization policy* for Encoder i, defined as a sequence of functions $\{Q_t^{comp,i}, t \geq 0\}$ which are causal such that the quantization output at time t, q_t^i, under $\Pi^{comp,i}$ is generated by a function of its local information, that is, a mapping measurable with respect to the sigma-algebra generated by

$$I_t^i = \{y^i_{[0,t]}, q^i_{[0,t-1]}, z^i_{[0,t-1]}\}, \quad t \geq 1,$$

and $I_0^i = \{y_0^i\}$, with image space $\mathcal{M}_t^i$, where $\mathcal{M}_t^i := \{1, 2, \ldots, |\mathcal{M}_t^i|\}$, for $0 \leq t \leq T-1$ and $i = 1, 2$. Here z^i denotes some additional side information available, such as feedback from the receiver.

Let $\mathbb{I}_t^i$ denote the space I_t^i belongs to; hence

$$Q_t^{comp,i} : \mathbb{I}_t^i \to \mathcal{M}_t^i.$$

As discussed in Sect. 12.2.2, equivalently, we can express the policy $\Pi^{comp,i}$ as a composition of a *Quantization Policy* $\underline{\gamma}^i$ and a *Quantizer*: A quantization policy of encoder i, $\underline{\gamma}^i$, is a sequence of functions $\{\gamma_t^i\}$, such that for each $t \geq 0$, γ_t^i is a mapping from the information space $\mathbb{I}_t^i$ to the space of quantizers $\mathbb{Q}_t^i$. A quantizer is subsequently used to generate the quantizer output. Without any loss of generality, a quantizer action will be generated based on the common information at the encoders and the receiver, and the quantizer will map the relevant private information at the encoder to the quantization output. Let the information at the receiver at time t be $I_t^r = \{q^1_{[0,t-1]}, q^2_{[0,t-1]}\}$, for $t \geq 1$. Let the common information, under feedback, at the encoders and the receiver be I_t^c. Thus, we can express any measurable composite quantization policy as:

$$Q_t^{comp,i}(I_t^i) = (\gamma_t^i(I_t^c))(I_t^i \setminus I_t^c), \tag{15.3}$$

mapping the information space to $\mathcal{M}_t^i$.

Viewing each encoder as an agent or a decision maker (DM), we let DMi have policy $\underline{\gamma}^i$ and under this policy generate **quantizers** $\{Q_t^i, t \geq 0\}$, $Q_t^i \in \mathbb{Q}_t^i$. Under action Q_t^i, the encoder generates q_t^i, as the *quantization output* at time t.

The receiver (or the controller), upon receiving the information from the encoders, generates its decision at time t, also causally: An admissible causal receiver policy is a sequence of measurable functions $\underline{\gamma}^0 = \{\gamma_t^0\}$ such that

$$\gamma_t^0 : \prod_{s=0}^{t} \left(\mathcal{M}_s^1 \times \mathcal{M}_s^2 \right) \to \mathbb{U}, \qquad t \geq 0$$

where $\mathbb{U}$ denotes the decision set for the receiver.

With the above formulation, one typical objective functional for the decision makers would be the following

$$\inf_{\mathbf{\Pi}^{comp}} \inf_{\underline{\gamma}^0} E_{\nu_0}^{\mathbf{\Pi}^{comp}, \underline{\gamma}^0} [\sum_{t=0}^{T-1} c(x_t, u_t)], \tag{15.4}$$

with initial condition distribution ν_0. Here $c(\cdot, \cdot)$, is a non-negative, measurable function and $u_t = \gamma_t^0(\mathbf{q}_{[0,t]})$ (with $\mathbf{q} = (q^1, q^2)$) for $t \geq 0$.

Before concluding this section, it may be worth emphasizing the operational nature of causality, as different approaches could be adopted. The encoders at any given time can only use their local information to generate the quantization outputs. The receiver, at any given time, can only use its local information to generate its decision/estimate. These happen with zero delay, that is if there is a common-clock at the encoders and the receiver, the receiver at time t needs to make its decision before the realizations $x_{t+1}, y_{t+1}^1, y_{t+1}^2$ have taken place. This corresponds to the *zero-delay* coding schemes of, for example, [778] and [483].

15.3 Optimal Causal Coding of a Markov Source

In the following, we will restrict the coverage to the single-terminal case, with a generic depiction being that presented in Fig. 15.2. We will first consider the case with discrete noiseless channels (for both fully observed and partially observed setups) without control, then with control, and finally with noisy channels.

15.3.1 Single-Terminal, Fully Observed Case

We first consider the single-encoder, fully observed case: In this setup, (15.1) and (15.2) hold with one encoder, that is,

$$x_{t+1} = f(x_t, w_t), \qquad y_t = x_t, t = 0, 1, \ldots \tag{15.5}$$

For the noiseless channel setup, as noted earlier, a *Coding Policy* Π, a sequence of functions $\{\gamma_t^e, t \geq 0\}$ which are causal such that the channel input at time t, $q_t \in \mathcal{M}$, under Π^{comp}, is generated by a function of its local information, that is,

$$q_t = \gamma_t^e(\mathcal{I}_t^e),$$

where $\mathcal{I}_t^e = \{x_{[0,t]}, q_{[0,t-1]}\}$ and $q_t \in \mathcal{M}$, the channel input alphabet given by $\mathcal{M} := \{1, 2, \ldots, M\}$, for $0 \leq t \leq T-1$.

The receiver/controller, upon receiving the information from the channel, generates its decision at time t, also causally: An admissible causal controller policy is a sequence of functions $\gamma = \{\gamma_t\}$ such that

$$\gamma_t : \mathcal{M}^{t+1} \to \mathbb{U}, \qquad t \geq 0,$$

so that $u_t = \gamma_t(q_{[0,t]})$.

Let $\mathcal{P}(\mathbb{X})$ denote the space of probability measures on $\mathcal{B}(\mathbb{X})$ under the topology of weak convergence and define $\pi_t \in \mathcal{P}(\mathbb{X})$ to be the regular conditional probability measure given by $\pi_t(\cdot) = P(x_t \in \cdot | q_{[0,t-1]})$, that is

$$\pi_t(A) = P(x_t \in A | q_{[0,t-1]}), \quad A \in \mathcal{B}(\mathbb{X}).$$

We first state the following theorem on the structure of optimal causal quantization policies, due to Witsenhausen [778].

Theorem 15.3.1 (Witsenhausen [778]) *For system (15.5) and optimization problem (15.4), any composite quantization policy can be replaced, without any loss in performance, by one which only uses x_t and $q_{[0,t-1]}$ at time $t \geq 1$.*

Proof See Sect. 15.7.1. ◇

The following result is essentially due to Walrand and Varaiya [756], however, the form below is more general since the spaces considered are not necessarily finite, as e.g. $\mathbb{R}^k$-valued sources would also be covered.

Theorem 15.3.2 ([814]) *For system (15.5) and optimization problem (15.4), any composite quantization policy can be replaced, without any loss in performance, by one which only uses the conditional probability measure $\pi_t(\cdot) = P(x_t \in \cdot | q_{[0,t-1]})$, the state x_t, and the time information t, at time $t \geq 1$.*

Proof See Sect. 15.7.2 ◇

Remark 15.3.1 The difference between the structural results above is the following: In the setup of Theorem 15.3.1, the encoder's memory space is not fixed and keeps expanding as the decision horizon in the optimization, $T-1$, increases. In the setup of Theorem 15.3.2, the memory space of an optimal encoder is fixed. In general, the space of probability measures is a very large one; however, it may be the case that different quantization outputs may lead to the same conditional probability measure on the state process, leading to a reduction in the required memory. Furthermore, Theorem 15.3.2 allows one to apply the theory of Markov Decision Processes, an aspect which we will elaborate on further in this chapter, leading to existence, structure, and approximation results. ◇

As we observed in Remark 8.5.2, the set (see (8.7))

$$\Theta := \{\zeta \in P(\mathbb{R}^n \times \mathcal{M}) : \zeta = PQ, Q \in \mathcal{Q}\},$$

(with $\mathcal{Q}$ denoting the set of $|\mathcal{M}|$-cell quantizers) is the Borel measurable set of the extreme points of the set of probability measures on $\mathbb{R}^n \times \mathcal{M}$ with a fixed input marginal P. In view of this observation, and that the class of quantization policies which admit the structure suggested in Theorem 15.3.2 is an important one, we henceforth define

$$\mathbf{\Pi}_W := \Big\{ \Pi^{comp} = \{Q_t^{comp}, t \geq 0\} : \quad \exists \gamma_t^1 : \mathcal{P}(\mathbb{X}) \to \mathcal{Q} \\ Q_t^{comp}(I_t) = (\gamma_t^1(\pi_t))(x_t), \forall I_t \Big\}, \tag{15.6}$$

to represent this class of policies and call this the Walrand-Varaiya class of policies.

Later in the chapter, in Sect. 15.4, we will build on the structural results developed in this section to arrive at existence and approximation results for optimal policies, while also extending the analysis to infinite horizon optimality criteria.

15.3.2 Partially Observed Markov Source

We consider here the setup of (15.1) and (15.2) but with a single encoder. Thus, the system considered is a discrete-time scalar system described by

$$x_{t+1} = f(x_t, w_t), \qquad y_t = g(x_t, v_t), t = 0, 1, \ldots \tag{15.7}$$

where x_t, $\{w_t, v_t\}$ are as introduced earlier. Let the quantizer, as described earlier, map its information to a finite set $\mathcal{M}_t$. At any given time, the receiver generates a quantity u_t as a function of its received information, that is as a function of $\{q_0, q_1, \ldots, q_t\}$. The goal is to obtain a solution to (15.4) subject to constraints on the number of quantizer bins in $\mathcal{M}_t$, and the causality restriction in encoding and decoding.

Now, define $\tilde{\pi}_t \in \mathcal{P}(\mathbb{X})$ to be the regular conditional probability measure (whose existence for every realization of observation variables follows from the fact that both the state and the observation spaces are Polish) given by $P(dx_t|y_{[0,t]})$, that is

$$\tilde{\pi}_t(A) = P(x_t \in A | y_{[0,t]}), \quad A \in \mathcal{B}(\mathbb{X}).$$

Under the topology of weak convergence for $\mathcal{P}(\mathbb{X})$, $\{\tilde{\pi}_t\}$ evolves according to a non-linear filtering equation (see [689]), and is itself a Markov process. Let us also define $\Xi_t \in \mathcal{P}(\mathcal{P}(\mathbb{X}))$ as the regular conditional measure

$$\Xi_t(A) = P(\tilde{\pi}_t \in A | q_{[0,t-1]}), \quad A \in \mathcal{B}(\mathcal{P}(\mathbb{X})).$$

The following are the main results of this subsection.

Theorem 15.3.3 ([814]) *For system (15.7) and optimization problem (15.4) with c bounded, any composite quantization policy can be replaced, without any loss in performance, by one which only uses $\{\tilde{\pi}_t, q_{[0,t-1]}\}$ as a sufficient statistic for $t \geq 1$. This can be expressed as a quantization policy which only uses $q_{[0,t-1]}$ to generate a quantizer, where the quantizer uses $\tilde{\pi}_t$ to generate the quantization output at time $t \geq 1$.*

Theorem 15.3.4 ([814]) *For system (15.7) and optimization problem (15.4) with c bounded, any composite quantization policy can be replaced, without any loss in performance, by one which only uses $\{\Xi_t, \tilde{\pi}_t, t\}$ for $t \geq 1$. This can be expressed as a quantization policy which only uses $\{\Xi_t, t\}$ to generate a quantizer, where the quantizer uses $\tilde{\pi}_t$ to generate the quantization output at time $t \geq 1$.*

A number of remarks are now in order.

Remark 15.3.2 From the proof of Theorem 15.3.4, we will see that (Ξ_t, Q_t) forms a controlled Markov Chain. Defining the actions as the quantizers allows one to define a Markov Decision Problem with well-defined cost functions, and state and action spaces. ◇

Remark 15.3.3 The results above can be viewed as direct extensions of the ones in the previous subsection with perfect state measurements. In fact, once one recognizes the fact that $\{\tilde{\pi}_t\}$ forms a Markov source, and the cost function can be expressed as $\tilde{c}(\tilde{\pi}, u)$, for some function $\tilde{c} : \mathcal{P}(\mathbb{X}) \times \mathbb{U} \to \mathbb{R}$, one could almost directly apply Theorems 15.3.1 and 15.3.2 to recover the structural results above. Note that this can be obtained via an iterated expectations argument:

$$E[c(x_t, u_t)|q_{[0,t]}] = E[E[c(x_t, u_t)|y_{[0,t]}, q_{[0,t]}]|q_{[0,t]}] = E[E[c(x_t, u_t)|y_{[0,t]}]|q_{[0,t]}].$$

The results of Theorems 15.3.3 and 15.3.4 are also generalizable to settings where, (a) the source is Markov of order $m > 0$, (b) a finite delay d is allowed at the decoder, and c) the observation process depends also on past source outputs in a sense described in (15.8) below. For these cases, we consider the following generalization of the source by expanding the state space.

Suppose that the partially observed source is such that either the source is Markov of order m or there is a finite delay $d > 0$ which is allowed at the decoder. Then we can augment the source to obtain $z_t = \{x_{[t-\max(d+1,m)+1,t]}\}$. Note that $\{z_t\}$ is

Markov. We can thus consider the following representation:

$$z_{t+1} = \tilde{f}(z_t, \tilde{w}_t), \qquad y_t = \tilde{g}(z_t, \tilde{v}_t), \tag{15.8}$$

for some $\tilde{f}, \tilde{g}$, and where $z_t = \{x_{[t-\max(d+1,m)+1,t]}\} \in \mathbb{X}^{\max(d+1,m)}$, and $\tilde{w}_t, \tilde{v}_t$ are mutually independent, i.i.d. processes.

Any per-stage cost function of the form $c(x_t, u_t)$ can be written as for some $\tilde{c}$: $\tilde{c}(z_t, u_t)$. For the finite delay case, the cost per-stage can further be specialized as $\tilde{c}(x_{t-d}, u_t)$. For the Markov case with memory, the cost function per-stage writes as $\tilde{c}(x_{[t-m+1,t]}, u_t)$.

Now, by replacing $\mathbb{X}$ with $\mathbb{X}^{\max(d+1,m)}$, let $\tilde{\pi}_t \in \mathcal{P}(\mathbb{X}^{\max(d+1,m)})$ be given by

$$\tilde{\pi}_t(A) = P(z_t \in A | y_{[0,t]}), \quad A \in \mathcal{B}(\mathbb{X}^{\max(d+1,m)})$$

and $\Xi_t \in \mathcal{P}(\mathcal{P}(\mathbb{X}^{\max(d+1,m)}))$ be the regular conditional measure defined by

$$\Xi_t(A) = P(\tilde{\pi}_t \in A | q_{[0,t-1]}), \quad A \in \mathcal{B}(\mathcal{P}(\mathbb{X}^{\max(d+1,m)})).$$

Hence, we have the following result, which is a direct extension of Theorems 15.3.3 and 15.3.4.

Theorem 15.3.5 *Suppose that the partially observed source is such that either the source is Markov of order m or there is a finite delay $d > 0$ which is allowed at the decoder. With $z_t = \{x_{[t-\max(d+1,m)+1,t]}\}$ and y_t generated by (15.8), the following holds.*

(i) Any (causal) composite quantization policy can be replaced, without any loss in performance, by one which only uses $\{\tilde{\pi}_t, q_{[0,t-1]}\}$ as a sufficient statistic for $t \geq 1$. This can be expressed as a quantization policy which only uses $q_{[0,t-1]}$ to generate a quantizer, where the quantizer uses $\tilde{\pi}_t$ to generate the quantization output at time $t \geq 1$.

(ii) Any (causal) composite quantization policy can be replaced, without any loss in performance, by one which only uses $\{\Xi_t, \tilde{\pi}_t, t\}$ for $t \geq 1$. This can be expressed as a quantization policy which only uses $\{\Xi_t, t\}$ to generate a quantizer, where the quantizer uses $\tilde{\pi}_t$ to generate the quantization output at time $t \geq 1$.

For a further case where the decoder's memory is limited or imperfect, the results apply by replacing the full information considered so far at the receiver with the limited one with additional assumptions on the decoder's update of its memory. However, an equivalent result of Theorem 15.3.3 applies also for the limited memory setting. Such memory settings have been considered in [756, 778], and [492].

15.3.3 Structural Results for Systems with Control

Theorem 15.3.2 applies also for Markov sources driven by control. That is, instead of (15.1) and (15.2), consider a system described by the following equations

$$x_{t+1} = f(x_t, u_t, w_t),$$
$$y_t = x_t. \tag{15.9}$$

Suppose that the goal is the minimization of (15.4), with the information restrictions stated in Sect. 15.2.

For this system, we have the following result (which extends the finite state-action space analysis in [755] and [494]).

Theorem 15.3.6 ([812])

(i) For system (15.9) and optimization problem (15.4), any composite quantization policy (with a given control policy) can be replaced, without any loss in performance, by one which only uses x_t and $q_{[0,t-1]}$ at time $t \geq 1$, while keeping the control policy unaltered. This can be expressed as a quantization policy which only uses $q_{[0,t-1]}$ to generate a quantizer, where the quantizer uses x_t to generate the quantization output at time t.

(ii) For system (15.9) and optimization problem (15.4), any composite quantization policy can be replaced, without any loss in performance, by one which only uses the conditional probability measure $\pi_t(\cdot) = P(x_t \in \cdot | q_{[0,t-1]})$, the state x_t, and the time information t, at time t. This can be expressed as a quantization policy which only uses $\{\pi_t, t\}$ to generate a quantizer, where the quantizer uses x_t to generate the quantization output at time t.

Proof See Sect. 15.7.3. ◇

The result also applies to the partially observed case with the conditional probability replacing the state as in Theorem 15.3.4. The proof follows from those of Theorem 15.3.6 and Theorem 15.3.4.

15.3.4 Case with Noisy Channels and Noiseless Feedback

The results and the general program presented heretofore in this chapter apply also to coding over discrete memoryless (noisy) channels (DMCs) with feedback. In this context, consider the setup in Sect. 12.2.2, with one encoder and with $y_t = x_t$ and with the channel being a DMC. The equivalent results of Theorems 15.3.1 and 15.3.2 apply with q'_t terms replacing q_t, if q'_t is the output of a DMC at time t, as we state in the following.

In this context, let again $\mathcal{P}(\mathbb{X})$ denote the set of probability measures on $\mathcal{B}(\mathbb{X})$ under the topology of weak convergence and define $\pi_t \in \mathcal{P}(\mathbb{X})$ to be the regular conditional probability measure given by $\pi_t(\cdot) = P(x_t \in \cdot | q'_{[0,t-1]})$, where q'_t is the channel output when the input is q_t. That is, $\pi_t(A) = P(x_t \in A | q'_{[0,t-1]}), \quad A \in \mathcal{B}(\mathbb{X})$. The goal is the minimization

$$\inf_{\Pi^{comp}} \inf_{\underline{\gamma}^0} E_{\nu_0}^{\Pi^{comp},\underline{\gamma}^0} [\sum_{t=0}^{T-1} c(x_t, u_t)], \tag{15.10}$$

with initial condition distribution ν_0. Here $c(\cdot, \cdot)$, is a non-negative, measurable function and $u_t = \gamma_t^0(q'_{[0,t]})$. In connection with this problem, we have the following results.

Theorem 15.3.7 *Any composite encoding policy can be replaced, without any loss in performance, by one which only uses x_t and $q'_{[0,t-1]}$ at time $t \geq 1$ to generate the channel input q_t.*

Theorem 15.3.8 *Any composite quantization policy can be replaced, without any loss in performance, by one which only uses the conditional probability measure $\pi_t(\cdot) = P(x_t \in \cdot | q'_{[0,t-1]})$, the state x_t, and the time information t, at time $t \geq 1$ to generate the channel input q_t.*

Proofs of these results follow from those of Theorems 15.3.1 and 15.3.2 with almost identical steps with q_t being replaced with q'_t in the information available at the receiver and the encoder.

Likewise, for a partially observed setup, extensions of Theorems 15.3.3 and 15.3.4 also apply to this case.

15.4 Controlled Markov Optimality Analysis: Existence of Optimal Zero-Delay Quantizers and a Topology on Quantizers

We discuss in this section the problem of existence of optimal composite quantization policies, given the structural results for a fully observed setting developed in Sect. 15.3.1. We assume that the source to be quantized is an $\mathbb{R}^n$ valued Markov source. The goal is to minimize the cost

$$J_{\pi_0}(\Pi^{comp}, \underline{\gamma}^0, T) := E_{\pi_0}^{\Pi^{comp},\underline{\gamma}^0} \left[\sum_{t=0}^{T-1} c(x_t, u_t) \right], \tag{15.11}$$

for some $T \geq 1$, where $c : \mathbb{R}^n \times \mathbb{U} \to \mathbb{R}_+$ is a (measurable) stage-wise cost function where $\mathbb{U}$ is an action set.

We make the following assumptions on the source $\{x_t\}$ and the cost function:

Assumption 15.4.1 *The following hold.*

(i) The evolution of the Markov source $\{x_t\}$ in (15.1) and (15.2) is given by

$$x_{t+1} = f(x_t) + w_t, \quad t \geq 0$$
$$y_t = x_t, \tag{15.12}$$

where $\{w_t, t \geq 0\}$ is an i.i.d. Gaussian noise sequence and $f : \mathbb{R}^n \to \mathbb{R}^n$ is measurable and bounded.

(ii) The cost function $c : \mathbb{R}^n \times \mathbb{U} \to \mathbb{R}_+$ is continuous and bounded.

(iii) The initial probability measure π_0 is Gaussian.

(iv) $\mathbb{U}$ is compact (the compactness condition will be relaxed for LQG problems).

In this section, we will assume that the number of bins for the quantizers is constant for every time stage such that $|\mathcal{M}_t| = M$ for all t. As discussed in Sect. 8.5, a quantizer Q with cells $\{B_1, \ldots, B_M\}$ can be characterized as a stochastic kernel Q from $\mathbb{R}^n$ to $\{1, \ldots, M\}$ defined by

$$Q(i|x) = 1_{\{x \in B_i\}}, \quad i = 1, \ldots, M.$$

We endow the quantizers with a topology induced by such a stochastic kernel interpretation as in Sect. 8.5 (see also Sect. F.1). If P is a probability measure on $\mathbb{R}^n$ and Q is a stochastic kernel from $\mathbb{R}^n$ to $\mathcal{M}$, then PQ denotes the resulting joint probability measure on $\mathbb{R}^n \times \mathcal{M}$. That is, a quantizer sequence Q_n converges to Q weakly at P ($Q_n \to Q$ weakly at P) if $PQ^n \to PQ$ weakly. Similarly, Q_n converges to Q in total variation at P ($Q_n \to Q$ at P in total variation at P) if $PQ^n \to PQ$ in total variation.

Suppose we adopt a quantizer policy which is in $\mathbf{\Pi}_W$, that is it admits the form suggested by Theorem 15.3.2. Properties of conditional probability leads to the following expression for $\pi_t(dx_t) = P(dx_t|q_{[0,t-1]})$ for $t \geq 1$:

$$\pi_t(dx_t) = \frac{\int_{x_{t-1}} \pi_{t-1}(dx_{t-1})P(q_{t-1}|\pi_{t-1}, x_{t-1})P(dx_t|x_{t-1})}{\int_{x_{t-1}} \int_{x_t} \pi_{t-1}(dx_{t-1})P(q_{t-1}|\pi_{t-1}, x_{t-1})P(dx_t|x_{t-1})}.$$

Let $\mathcal{P}$ be the set of probability measures on $\mathbb{R}^n$ endowed with the topology of weak convergence. The following is a consequence of Theorem 15.3.4 and Remark 15.3.2.

Theorem 15.4.1 *The sequence of conditional measures and the sequence of quantizers, (π_t, Q_t), form a joint Markov process in $\mathcal{P} \times \mathcal{Q}$.*

Now, under any quantization policy in $\mathbf{\Pi}_W$ and for any $T \geq 1$, by optimizing the receiver policy given a composite quantization policy in (15.11), we can define

$$J_{\pi_0}(\Pi^{comp}, T) = E_{\pi_0}^{\Pi^{comp}}\left[\sum_{t=0}^{T-1} \tilde{c}(\pi_t, Q_t)\right],$$

where, with $B_i = Q_t^{-1}(i), i = 1, \ldots, M$ denoting the cells of Q_t, we have

$$\begin{aligned}
&\tilde{c}(\pi_t, Q_t) \\
&= \sum_{i \in \mathcal{M}} P(q_t = i | q_{[0,t-1]}) \inf_{u \in \mathbb{U}} \left(\int P(dx_t | q_{[0,t-1]}, q_t = i) c(x_t, u) \right) \\
&= \sum_{i \in \mathcal{M}} \inf_{u \in \mathbb{U}} \int_{B_i} \pi_t(dx) c(x, u) \qquad (15.13)
\end{aligned}$$

As in Sect. 8.5, we restrict the set of quantizers considered by only allowing quantizers having convex quantization bins (cells) $B_i, i = 1, \ldots, M$.

Assumption 15.4.2 *The quantizers have convex codecells with at most a given number of cells; that is the quantizers live in $\mathcal{Q}_c(M)$, the collection of k-cell quantizers with convex cells where $1 \leq k \leq M$.*

Let $\mathbf{\Pi}_W^C$ denote the set of all composite quantization policies $\mathbf{\Pi}_W$ (defined in (15.6)) which in addition satisfy the condition that all quantizers $Q_t, t \geq 0$ have convex cells (i.e., $Q_t \in \mathcal{Q}_c$ for all $t \geq 0$).

We have the following result on the existence of optimal quantizers.

Theorem 15.4.2 ([827]) *For any $T \geq 1$ and arbitrary initial condition π_0, under Assumptions 15.4.1 and 15.4.2, there exists a policy in $\mathbf{\Pi}_W^C$ such that*

$$\inf_{\Pi^{comp} \in \mathbf{\Pi}_W^C} \inf_{\gamma^0} J_{\pi_0}(\Pi^{comp}, \gamma^0, T) \qquad (15.14)$$

is achieved. Letting $J_T^T(\cdot) = 0$ and

$$J_0^T(\pi_0) := \min_{\Pi^{comp} \in \mathbf{\Pi}_W^C, \gamma^0} J_{\pi_0}(\Pi^{comp}, \gamma^0, T),$$

the dynamic programming recursion

$$T J_t^T(\pi_t) = \min_{Q \in \mathcal{Q}_c} \left(c(\pi_t, Q_t) + T E[J_{t+1}^T(\pi_{t+1}) | \pi_t, Q_t] \right) \qquad (15.15)$$

holds for all $t = 0, 1, \ldots, T-1$.

Proof See Sect. 15.7.4. ◇

15.5 Infinite Horizon Setup and Optimality of Stationary Codes and Near Optimality of Finite Memory Codes

In this section, we briefly consider infinite horizon problems, for which the structural results of finite-horizon problems do not apply directly as they build on backward induction.

As earlier, the encoder encodes the source samples without delay and transmits the encoded versions to a receiver over a discrete channel with input alphabet $\mathcal{M} = \{1, \ldots, M\}$ and output alphabet $\mathcal{M}' := \{1, \ldots, M'\}$, where M and M' are positive integers.

Henceforth, we assume throughout that an optimal receiver policy is used for a given encoder policy and, with a slight abuse of notation, $\Pi \in \mathbf{\Pi}_A$ will mean the combined encoder and decoder policies $(\Pi, \gamma^*(\Pi))$.

We study the following two performance criteria.

1. *Infinite Horizon Discounted Cost Problem*: In the infinite horizon discounted cost problem, the goal is to minimize the cumulative "discounted" cost

$$J^{\beta}_{\pi_0}(\Pi) := \lim_{T\to\infty} E^{\Pi}_{\pi_0}\left[\sum_{t=0}^{T-1} \beta^t d(x_t, \hat{x}_t)\right] \tag{15.16}$$

for some $\beta \in (0, 1)$.

2. *Infinite Horizon Average Cost Problem*: The more challenging infinite horizon average cost problem has the objective of minimizing the long term average distortion

$$J_{\pi_0}(\Pi) := \limsup_{T\to\infty} E^{\Pi}_{\pi_0}\left[\frac{1}{T}\sum_{t=0}^{T-1} d(x_t, \hat{x}_t)\right]. \tag{15.17}$$

15.5.1 Finite Markov Sources

We first assume that the source $\{X_t\}_{t\geq 0}$ is a discrete time Markov process with finite alphabet $\mathbb{X}$.

Definition 15.5.1 Let $\mathcal{Q}$ denote the set of all quantizers $Q : \mathbb{X} \to \mathcal{M}$. An (admissible) quantization policy $\Pi = \{\eta_t\}$ belongs to $\mathbf{\Pi}_W$ if there exists a sequence of mappings $\{\kappa_t\}$ of the type $\kappa_t : \mathcal{P}(\mathbb{X}) \to \mathcal{Q}$ such that for $Q_t = \kappa_t(\pi_t)$ we have $q_t = Q_t(X_t) = \eta_t(I_t)$. A policy in $\mathbf{\Pi}_W$ is called *stationary* if κ_t does not depend on t. The set of such stationary policies is denoted by $\mathbf{\Pi}_{WS}$.

For any $\Pi \in \mathbf{\Pi}_A$, define

$$J_{\pi_0}(\Pi, T) := E^{\Pi}_{\pi_0}\left[\frac{1}{T}\sum_{t=0}^{T-1} d(x_t, \hat{x}_t)\right].$$

Proposition 15.5.1 *For any $T \geq 1$, there exists a policy Π in $\mathbf{\Pi}_W$ such that*

$$J_{\pi_0}(\Pi, T) = \inf_{\Pi' \in \mathbf{\Pi}_A} J_{\pi_0}(\Pi', T). \tag{15.18}$$

Letting $J_T^T(\cdot) := 0$, $J_0^T(\pi_0) := \min_{\Pi \in \mathbf{\Pi}_W} J_{\pi_0}(\Pi, T)$, the dynamic programming recursion

$$T J_t^T(\pi) = \min_{Q \in \mathcal{Q}} \Big(c(\pi, Q) + T E\big[J_{t+1}^T(\pi_{t+1}) | \pi_t = \pi, Q_t = Q\big]\Big)$$

holds for all $t = T-1, T-2, \ldots, 0$, and $\pi \in \mathcal{P}(\mathbb{X})$.

$$J_{\pi_0}^{\beta} := \inf_{\Pi \in \mathbf{\Pi}_A} J_{\pi_0}^{\beta}(\Pi) \tag{15.19}$$

for given $\beta \in (0, 1)$, where $J_{\pi_0}^{\beta}(\Pi)$ is defined in (15.16).

From the viewpoint of source coding, the discounted cost problem has less significance than the average cost problem. However, the discounted cost approach plays an important role in deriving results for the average cost problem.

Proposition 15.5.2 ([785]) *There exists an optimal (deterministic) quantization policy in $\mathbf{\Pi}_{WS}$ among all policies in $\mathbf{\Pi}_A$ that achieves the infimum in (15.19).*

Proof Observe that

$$\begin{aligned}
&\inf_{\Pi \in \mathbf{\Pi}_A} \lim_{T\to\infty} E^{\Pi}_{\pi_0}\left[\sum_{t=0}^{T-1} \beta^t d(x_t, \hat{x}_t)\right] \\
&\geq \limsup_{T\to\infty} \inf_{\Pi \in \mathbf{\Pi}_A} E^{\Pi}_{\pi_0}\left[\sum_{t=0}^{T-1} \beta^t d(x_t, \hat{x}_t)\right] \\
&= \limsup_{T\to\infty} \min_{\Pi \in \mathbf{\Pi}_W} E^{\Pi}_{\pi_0}\left[\sum_{t=0}^{T-1} \beta^t d(x_t, \hat{x}_t)\right] \\
&= \limsup_{T\to\infty} \min_{\Pi \in \mathbf{\Pi}_W} E^{\Pi}_{\pi_0}\left[\sum_{t=0}^{T-1} \beta^t c(\pi_t, Q_t)\right],
\end{aligned} \tag{15.20}$$

For each T, let Π_T denote the optimal policy in $\mathbf{\Pi}_W$ achieving the minimum in (15.20).

We have seen in Sect. 15.4 that measurable selection criteria apply (see [484]). Thus $\{\Pi_T\}$, can be obtained by using the iteration algorithm

$$J_t(\pi) = \min_{Q \in \mathcal{Q}} \left[c(\pi, Q) + \beta \int_{\mathcal{P}(\mathcal{X})} J_{t-1}(\pi') P(d\pi' | \pi, Q) \right]$$

with $J_0(\pi) \equiv 0$. By the same theorem, the sequence of value functions for the policies $\{\Pi_T\}$, i.e. $\{J_{\pi_0}(\mathbf{\Pi}_W, T)\}$, converges to the value function of some deterministic policy $\Pi \in \mathbf{\Pi}_{WS}$ (i.e., a deterministic stationary Markov policy) which is optimal in the set of policies $\mathbf{\Pi}_W$ for the infinite horizon discounted cost problem. Thus by the chain of inequalities leading to (15.20), Π is optimal among all policies in $\mathbf{\Pi}_A$. ◇

The more challenging average cost case cost criterion given in (15.17) deals with a performance measure that is more commonly studied in source coding problems. The following builds on [785] and also [190] and [820, Theorem 7.3.5].

Revise the following or take out. same with 15.5.2 and 15.5.3.

Theorem 15.5.1 ([785]) *Assume that $\{X_t\}$ is an irreducible and aperiodic Markov chain. Then for any initial distribution π_0,*

$$\inf_{\Pi \in \mathbf{\Pi}_A} J_{\pi_0}(\Pi) = \min_{\Pi \in \mathbf{\Pi}_{WS}} J_{\pi_0}(\Pi).$$

Furthermore, there exist $\Pi^* \in \mathbf{\Pi}_{WS}$ that achieves the minimum above simultaneously for all π_0 and which satisfies for all $T \geq 1$

$$\frac{1}{T} E_{\pi_0}^{\Pi^*} \left[\sum_{t=0}^{T-1} d(x_t, \hat{x}_t) \right] \leq J_{\pi_0} + \frac{K}{T} \tag{15.21}$$

for some positive constant K.

The key step in the proof is a simulation argument in [785] building on *coupling* (see also [134] for a related but different construction) to show that for any two initial distributions μ_0 and ζ_0, the absolute difference of the optimal infinite horizon discounted costs $J^{\beta}_{\mu_0}$ and $J^{\beta}_{\zeta_0}$ is uniformly upper bounded by a constant times the L_1 Wasserstein distance between μ_0 and ζ_0.

Noisy Channel Case We now briefly describe the extension to zero-delay lossy coding over a noisy channel. It is assumed that the source $\{X_t\}_{t \geq 0}$ is a discrete-time Markov process with finite alphabet $\mathcal{X}$. The encoder encodes the source samples without delay and transmits the encoded versions to a receiver over a discrete channel with input alphabet $\mathcal{M} = \{1, \ldots, M\}$ and output alphabet $\mathcal{M}' := \{1, \ldots, M'\}$, where M and M' are positive integers.

In contrast with the setup above, here the channel between the encoder and the decoder is a discrete and memoryless *noisy* channel characterized by the transition probability $T_c(b|a) = \Pr(q' = b | q = a)$, $a \in \mathcal{M}$, $b \in \mathcal{M}'$.

We assume that the encoder has access to the previous channel outputs in the form of *feedback*. In particular, the encoder is specified by a *quantization policy* Π, which is a sequence of functions $\{\eta_t\}_{t\geq 0}$ with $\eta_t : \mathcal{M}^t \times (\mathcal{M}')^t \times \mathbb{X}^{t+1} \to \mathcal{M}$. At time t, the encoder transmits the $\mathcal{M}$-valued message

$$q_t = \eta_t(I_t),$$

where $I_0 = X_0$, $I_t = (q_{[0,t-1]}, q'_{[0,t-1]}, X_{[0,t]})$ for $t \geq 1$, and q'_t is the received (noisy) version of q_t. The collection of all such zero delay policies is called the set of admissible quantization policies and is denoted by $\mathbf{\Pi}_A$. Upon receiving q'_t, the receiver generates the reconstruction, $\hat{X}_t$, also without delay. A zero-delay receiver policy is a sequence of functions $\gamma = \{\gamma_t\}_{t\geq 0}$ of type $\gamma_t : (\mathcal{M}')^{t+1} \to \hat{\mathbb{X}}$, where $\hat{\mathbb{X}}$ is the finite reproduction alphabet. Thus

$$\hat{X}_t = \gamma_t(q'_{[0,t]}) \quad \text{for all } t \geq 0.$$

Note that, due to the presence of feedback, the encoder also has access to $q'_{[0,t]}$ at time $t+1$. The finite and infinite horizon coding problems are defined analogously to the noiseless case.

In view of the structural results Theorems 15.3.7 and 15.3.8 presented above, we can state the following: Under a Walrand-Varaiya type policy (see (15.6)) the filtering equation (15.32) is modified as

$$\pi_{t+1}(x_{t+1}) = \frac{\sum_{x_t,q_t} \pi_t(x_t)T_c(q'_t|q_t)P(q_t|\pi_t, x_t)P(x_{t+1}|x_t)}{\sum_{x_t,q_t}\sum_{x_{t+1}} \pi_t(x_t)T_c(q'_t|q_t)P(q_t|\pi_t, x_t)P(x_{t+1}|x_t)}. \tag{15.22}$$

Thus, as before, given π_t and Q_t, π_{t+1} is conditionally independent of $(\pi_{[0,t-1]}, Q_{[0,t-1]})$ and it follows that $\{\pi_t\}$ can be viewed as $\mathcal{P}(\mathbb{X})$-valued controlled Markov process [342] with $\mathcal{Q}$-valued control $\{Q_t\}$ and average cost up to time $T-1$ given by

$$E^{\Pi,\gamma}_{\pi_0}\left[\frac{1}{T}\sum_{t=0}^{T-1} d(X_t, \hat{X}_t)\right] = E^{\Pi}_{\pi_0}\left[\frac{1}{T}\sum_{t=0}^{T-1} c(\pi_t, Q_t)\right].$$

The set of deterministic Markov coding policies $\mathbf{\Pi}_W$ and deterministic stationary Markov policies $\mathbf{\Pi}_{WS}$ is defined analogously to Definition 15.5.1.

It can be checked that the properties concerning the continuity of the kernel and the existence of invariant measures apply identically to the new controlled Markov state pair (π_t, Q_t). Under the assumption that $\{X_t\}$ is irreducible and aperiodic, the simulation argument for the noise-free channel setup also applies identically by considering the same channel noise realizations for both processes X'_t and X''_t; i.e., in the simulation argument we can compare the performance of the coding schemes by taking the expectations over the channel noise realizations.

Theorem 15.5.2 ([785])

(i) *For the minimization of the finite horizon average distortion (15.11), an optimal solution in* $\mathbf{\Pi}_W$ *exists and a noisy channel analog of Proposition 15.5.1 holds.*

(ii) *For the minimization of the infinite horizon discounted distortion (15.16), an optimal solution exists and such a solution is in* $\mathbf{\Pi}_{WS}$*, i.e., a noisy channel analog of Proposition 15.5.2 holds.*

(iii) *The noisy channel version of Theorem 15.5.1 holds: If* $\{X_t\}$ *is irreducible and aperiodic, there exists a policy in* $\mathbf{\Pi}_{WS}$ *that minimizes the infinite horizon average distortion* (15.17)*. Furthermore, the convergence rate result* (15.21) *holds for this optimal policy.*

(iv) *Under the assumption that* $\{X_t\}$ *is irreducible and aperiodic, if* $X_0 \sim \pi^*$*, where* π^* *is the invariant probability measure, for any* $\epsilon > 0$*, there exists* $K > 0$ *and a finite memory, nonstationary, but periodic quantization policy with period less than* $\frac{K}{\epsilon}$ *that achieves* ϵ*-optimal performance.*

15.5.2 Linear Markov Sources

We now briefly study linear Markov sources and we obtain a counterpart of Theorem 15.5.1.

Assumption 15.5.1 ([484]) *Let* $\{X_t\}$ *be expressed in the stochastic realization form*

$$x_{t+1} = f(x_t, w_t), \quad t = 0, 1, 2, ..., \tag{15.23}$$

where $f : \mathbb{R}^d \times \mathbb{R}^d \to \mathbb{R}^d$ *is a Borel function and* W_t *is an independent and identically distributed (i.i.d.) vector noise sequence which is independent of* X_0*. It is assumed that for each fixed* $x \in \mathbb{R}^d$*, the distribution of* $f(x, W_t)$ *admits the (conditional) density function* $\varphi(\cdot|x)$ *(with respect to the d-dimensional Lebesgue measure) which is positive everywhere. Furthermore,* $\varphi(\cdot|x)$ *is bounded and Lipschitz uniformly in* x*.*

Definition 15.5.4 *Let* $\mathcal{G}$ *denote the set of all probability measures on* $\mathbb{R}^d$ *admitting densities that are bounded by a constant* C *and Lipschitz with constant* C_1*.*

Note that viewed as a class of densities, $\mathcal{G}$ is uniformly bounded. In [484, Lemma 3] it is shown that $\mathcal{G}$ is closed in $\mathcal{P}(\mathbb{R}^d)$ under weak convergence. Note that $\mathcal{G}$ is also closed in $\mathcal{P}_2(\mathbb{R}^d)$, since the Wasserstein convergence is stronger than weak convergence. Define the set of probability measures $\mathcal{Z} := \mathcal{G} \cap \mathcal{P}_2(\mathbb{R}^d)$, as the intersection of $\mathcal{G}$ and $\mathcal{P}_2(\mathbb{R}^d)$.

As in Sect. 15.4, we restrict the set of quantizers by only allowing ones that have convex cells. Likewise, $\mathbf{\Pi}_W^C$ denotes the set of all quantization policies $\Pi = \{\hat{\eta}_t\} \in \mathbf{\Pi}_W$ such that $\hat{\eta}_t : \mathcal{P}(\mathbb{R}^d) \to \mathcal{Q}_c$, i.e., $Q_t = \hat{\eta}_t(\pi_t) \in \mathcal{Q}_c$ for all $t \geq 0$. We also use $\mathbf{\Pi}_{WS}^C$ to denote the set of all quantization policies in $\mathbf{\Pi}_W^C$ that are stationary, i.e., the policy $\{\hat{\eta}_t\}$ does not depend on the time index t. As earlier, building on

Sect. 8.2, we say that a sequence of quantizers Q_n converges to Q at P if the induced joint measure $PQ_n \to PQ$. Here we consider convergence under the order-2 Wasserstein topology. In the setup considered, convergence in quantizers with convex codecells holds equivalently for all admissible input probability measures.

We consider the following system

$$x_{t+1} = Ax_t + w_t, \tag{15.24}$$

where A is a $d \times d$ matrix.

Assumption 15.5.2

(i) Maximum singular value of A is less than 1 (i.e. $\alpha := \max(eig(A^T A)) < 1$*).*
(ii) $\mathcal{U} = \mathbb{R}^d$.
(iii) The cost is given by $c_0(x, u) = \|x - u\|^2$.
(iv) $\sigma^2 := \mathbf{E}[\|W_t\|^2] < \infty$.
(v) The initial distribution ξ *for* X_0 *admits a density such that* $E_\xi[\|X_0\|^2] < \infty$ *or it is a point mass* $\pi_0 = \delta_{x_0}$.

The discounted cost problem for some $\beta \in (0, 1)$ and integer $T \geq 1$ is defined as

$$J_{\pi_0}^{\beta}(\Pi, T) := E_{\pi_0}^{\Pi}\left[\sum_{t=0}^{T-1} \beta^t c(\pi_t, Q_t)\right],$$

and for the infinite horizon case we have

$$J_{\pi_0}^{\beta} = \inf_{\Pi \in \mathbf{\Pi}_W^C} J_{\pi_0}^{\beta}(\Pi) := E_{\pi_0}^{\Pi}\left[\sum_{t=0}^{\infty} \beta^t c(\pi_t, Q_t)\right], \tag{15.25}$$

where $J_{\pi_0}^{\beta}$ is the cost of the optimal solution for this problem. Similarly, the optimal cost for finite horizon defined as

$$J_{\pi_0}^{\beta}(T) := \inf_{\Pi \in \mathbf{\Pi}_W^C} J_{\pi_0}^{\beta}(\Pi, T).$$

By definition, we clearly have

$$J_{\pi_0}^{\beta} \geq \lim_{T \to \infty} J_{\pi_0}^{\beta}(T). \tag{15.26}$$

Lemma 15.5.5 *The value function* $J_{\pi_0}^{\beta}$ *and hence* $J_{\pi_0}^{\beta}(T)$ *is uniformly bounded as*

$$J_{\pi_0}^{\beta}(T) \leq J_{\pi_0}^{\beta} \leq \frac{1}{1-\beta}\left(E_{\pi_0}[\|X_0\|^2] + \frac{1}{1-\alpha}\sigma^2\right).$$

Define the Value Iteration

$$v_n(\pi) = \min_Q (c(\pi, Q) + \beta \boldsymbol{E}[v_{n-1}(\pi')|\pi, Q]), \tag{15.27}$$

for all $n \in \mathbb{N}$ with starting point $v_0 \equiv 0$.

By (15.26), we know that for all $n \geq 0$,

$$J^{\beta}_{\pi_0} \geq \lim_{T \to \infty} J^{\beta}_{\pi_0}(T) \geq v_n(\pi_0).$$

Since $J^{\beta}_{\pi_0}$ is bounded by Lemma 15.5.5, we get that $v_n(\pi_0)$ is bounded and lower semi-continuous. There exists a fixed point in the iteration, denoted by $J^{\beta}_*(\pi)$, which is also lower semi-continuous and bounded, i.e.

$$J^{\beta}_*(\pi) = \min_Q (c(\pi, Q) + \beta \boldsymbol{E}[J^{\beta}_*(\pi_1)|\pi_0 = \pi, Q_0 = Q]).$$

Then, there exists an optimal deterministic quantization policy in $\mathbf{\Pi}^C_{WS}$ among all policies in $\mathbf{\Pi}^C_W$ that achieves the infimum in (15.25), so that

$$J^{\beta}_{\pi_0} = \min_{\Pi \in \mathbf{\Pi}^C_{WS}} J^{\beta}_{\pi_0}(\Pi).$$

We now discuss the average cost problem, building on [299], and also [190] and [820, Theorem 7.3.5].

Theorem 15.5.3 ([299])

(i) Under Assumptions 15.5.1, and 15.5.2, for any initial distribution π_0,

$$\inf_{\Pi \in \mathbf{\Pi}^C_W} J_{\pi_0}(\Pi) = \min_{\Pi \in \mathbf{\Pi}^C_{WS}} J_{\pi_0}(\Pi), \tag{15.28}$$

i.e., the class of stationary quantization policies is optimal. In addition, there exists a deterministic and stationary policy $\Pi^* \in \mathbf{\Pi}^C_{WS}$ *that achieves the minimum above simultaneously for all* π_0*, i.e.,*

$$J_{\pi_0}(\Pi^*) = \inf_{\Pi \in \mathbf{\Pi}^C_W} J_{\pi_0}(\Pi).$$

(ii) Furthermore, there exists $K < \infty$ *so that for all* $T \geq 1$*, if* π_0 *is stationary,*

$$\inf_{\Pi \in \mathbf{\Pi}^C_W} \frac{1}{T} E^{\Pi}_{\pi_0}\left[\sum_{t=0}^{T-1} c_0(x_t, \hat{x}_t)\right] \leq \frac{1}{T} E^{\Pi^*}_{\pi_0}\left[\sum_{t=0}^{T-1} c_0(x_t, \hat{x}_t)\right] \leq J_{\pi_0}(\Pi^*) + \frac{K}{T}.$$

15.6 Multi-Terminal (Decentralized) Setting

15.6.1 Memoryless Sources

Let us first consider a special, but important, case of (15.1) and (15.2) when $\{x_t, t \geq 0\}$ is an i.i.d. sequence. Further, suppose that the observations are generated by

$$y_t^i = g^i(x_t, v_t^i), \tag{15.29}$$

for measurable functions $g^i, i = 1, 2$, with $\{v_t^1, v_t^2\}$ (across time) an i.i.d. noise process. We do not require that v_t^1 and v_t^2 are independent for a given t. We note that the results presented here are also applicable when the process $\{v_t^1, v_t^2\}$ is only independent (across time), but not necessarily identically distributed. One difference with the general setup considered earlier is that we require the observation spaces $\mathbb{Y}^i, i = 1, 2$, to be finite spaces, $\mathbb{X}$ is Polish.

Suppose the goal is again the minimization

$$\inf_{\mathbf{\Pi}^{comp}} \inf_{\underline{\gamma}^0} E_{\nu_0}^{\mathbf{\Pi}^{comp}, \underline{\gamma}^0}[\sum_{t=0}^{T-1} c(x_t, u_t)] \tag{15.30}$$

Toward this end, we introduce the class of non-stationary memoryless team policies, given by:

$$\begin{aligned} \Pi^{NSM} := \Big\{ \mathbf{\Pi}^{comp} : P(\mathbf{q}_t|\mathbf{y}_{[0,t]}) &= P(q_t^1|y_t^1, t)P(q_t^2|y_t^2, t) \\ &= 1_{\{q_t^1 = Q_t^1(y_t^1)\}} 1_{\{q_t^2 = Q_t^2(y_t^2)\}}, \\ Q_t^1 : \mathbb{Y}^1 \to \mathcal{M}_t^1, \quad & Q_t^2 : \mathbb{Y}^2 \to \mathcal{M}_t^2, \quad t \geq 0 \Big\}, \end{aligned} \tag{15.31}$$

where, $\{Q_t^1, Q_t^2\}$ are arbitrary measurable functions.

Theorem 15.6.1 ([814]) *Consider the minimization problem of (15.30). An optimal composite quantization policy over all causal policies exists, and it is an element of* Π^{NSM}.

The result says that an optimal composite quantization policy only uses the product form admitted by a non-stationary memoryless team policy. It ignores the past observations and past quantization outputs without any loss. We note that this result applies also to the case when the source is memoryless, but not necessarily i.i.d..

Remark 15.6.6 If there is an entropy constraint on the quantizer outputs, memory in the encoders might be useful for finite horizon problems as it provides common randomness, which cannot be achieved by time-sharing in a finite-horizon problem.

Neuhoff and Gilbert [562] noted that randomization of two scalar quantizers (operationally achievable through time-sharing) is optimal in causal coding of an i.i.d. source subject to entropy constraints. On the other hand, for the zero-delay setting, when one considers the distortion minimization problem subject to an entropy constraint, [329] observed that the distortion-entropy curve is non-convex (leading to a benefit of common-randomness which can be used to expand the set of achievable rate and distortion pairs) as we elaborated on in Sect. 12.4.1. ◇

15.6.2 Markov Sources and a Counterexample Under Signaling

We now consider general Markov sources and show that a separation result of the type seen in the single-terminal case may not hold when there are multiple terminals.

We have the following (negative) result for the two-encoder setup, where the encoders have access to the feedback from the receiver (Fig. 15.1).

Proposition 15.6.1 ([814]) *Consider the setup in (15.1) and (15.2), and let $\tilde{\pi}_t^i(A) = P(x_t \in A | y^i_{[0,t]})$, $i = 1, 2$ and $A \subset \mathbb{X}$. An optimal composite quantization policy cannot, in general, be replaced by a policy which uses only $\{\mathbf{q}_{[0,t-1]}, \tilde{\pi}_t^i\}$ to generate q_t^i for $i = 1, 2$.*

Proof It suffices to produce an instance where an optimal policy cannot admit the separated structure. Toward this end, let z_1, z_2, z_3 be uniformly distributed, independent, binary numbers; x_0, x_1 be defined by:

$$x_0 = \left[z_1 \; z_2 \; 0 \; 0\right]', \quad x_1 = \left[0 \; 0 \; z_2 \; z_3\right]',$$

such that $x_0(1) = z_1, x_0(2) = z_2, x_0(3) = x_0(4) = 0$. Let the observations be given as follows:

$$y_t^1 = g^1(x_t) = x_t(1) \oplus x_t(3) \oplus x_t(4), \qquad y_t^2 = g^2(x_t) = x_t(1) \oplus x_t(2), \quad t = 0, 1.$$

That is,

$$y_0^1 = \left[z_1\right], \qquad y_0^2 = \left[z_1 \oplus z_2\right]$$

where $\oplus$ is the x-or operation, and

$$y_1^1 = \left[z_2 \oplus z_3\right], \qquad y_1^2 = \left[0\right]$$

Let the cost be:

$$E\left[(x_0(4) - E[x_0(4)|\mathbf{q}_{[0]}])^2 + (x_1(4) - E[x_1(4)|\mathbf{q}_{[0,1]}])^2\right]$$

That is, the cost is $E[(z_3 - E[z_3|\mathbf{q}_{[0,1]}])^2]$, where q_t^i are the information bits sent to the decoder for $t = 0$ and 1.

We further restrict the information rates to satisfy: $|\mathcal{M}_0^1| = |\mathcal{M}_1^1| = |\mathcal{M}_1^2| = 2$, $|\mathcal{M}_0^2| = 1$. That is, the encoder 2 may only send information at time $t = 1$.

Under arbitrary causal composite quantization policies, a cost of zero can be achieved as follows: If the encoder 1 sends the value z_1 to the receiver, and at time 1, encoder 1 transmits $z_2 \oplus z_3$ and encoder 2 transmits z_2 (or $z_1 \oplus z_2$), the receiver can uniquely identify the value of z_3, for every realization of the random variables.

For such a source, an optimal composite policy cannot be written in the separated form, that is, an optimal policy of encoder 2 at time 1 cannot be written as $h_1(\mathbf{q}_0, \tilde{\pi}_1^2)$, for some measurable function h_1. To see this, note the following: The conditional distribution of x_1 at encoder 2 at time 1 is such that the conditional measure on (z_2, z_3) is uniform and independent, that is $P(z_2 = a, z_3 = b|z_1 \oplus z_2) = (1/4)$ for all values of a, b. If a policy of the structure of h_1 is adopted, then it is not possible for encoder 2 to recall its past observation to extract the value of z_2. This is because, $\tilde{\pi}_1^2$ will be a distribution only on z_2 and z_3, which will be uniform and independent, given $z_1 \oplus z_2$. Thus, the information y_0^2 will not be available in the memory and the receiver will have access to at most $z_2 \oplus z_3$ and z_1 and $P(z_2, z_3|z_1 \oplus z_2)$ (the last variable containing no useful information). The optimal estimator will be $E[z_3] = 1/2$, leading to a cost of $1/4$. ◇

Discussion: Further Connections with Team Decision Theory We now interpret the results of this section. With the characterization of information structures for dynamic teams earlier, every lossy coding problem is nonclassical, since a receiver cannot recover the information available at the encoder fully, while its information is clearly affected by the coding policy of the encoder. However, in an encoding problem, the problem itself is the transmission of information. We suggest the following: *Signaling in a coding problem is the policy of an encoder to use the quantizers/encoding functions to transmit a message to other decision makers, or to itself to be used in future stages, through the information sent to the receiver.* We have seen that in decentralized decision making problems, when the information structure is nonclassical, the decision makers might benefit from communicating via their control actions, that is by *signaling*. We also note that, in the information theory literature, signaling has been employed in coding for Multiple Access Channels with feedback, where active information transmission to allow for coordination between encoders (see [24, 148, 185, 187] and [743] for related and further analysis).

The reason for the negative conclusion in Proposition 15.6.1 is that in general for an optimal policy,

$$P(q_t^i|\pi_t^i, \mathbf{q}_{[0,t-1]}, y_{[0,t-1]}^i) \neq P(q_t^i|\pi_t^i, \mathbf{q}_{[0,t-1]}), \tag{15.32}$$

when the encoders have engaged in signaling (in contrast with what we will have in the proof of the separation results). The encoders may benefit from using the received past observation variables explicitly.

As we discussed in detail in Sect. 5.2, separation results for such dynamic team problems can be facilitated by information sharing between the encoders (decision makers), where the shared information is used to establish a sufficient statistic taking values from a fixed state space and which admits a controlled Markov recursion (hence, such a sufficient statistic can serve as a *state* for the decentralized system). For the proof of Theorem 15.3.4, we see that Ξ_t forms such a state. For the proof of Theorem 15.6.1, we see that information sharing is not needed for the encoders to agree on a sufficient statistic, since the source considered is memoryless. Furthermore, for the multi-terminal setting with a Markov source, a careful analysis of the proof of Theorem 15.6.1 reveals that if the encoders agree on $P(dx_t|\mathbf{y}_{[0,t-1]})$ through sharing their beliefs for all $t \geq 1$, then a separation result involving this joint belief can be obtained.

15.7 Appendix: Proofs

15.7.1 Proof of Theorem 15.3.1

At time $t = T - 1$, the per-stage cost function can be written as follows, where γ_t^0 denotes a fixed receiver policy:

$$E[c(x_t, \gamma_t^0(q_{[0,t]}))|q_{[0,t-1]}] = E[F(x_t, q_{[0,t-1]}, q_t)|q_{[0,t-1]}]$$

where, $F(x_t, q_{[0,t-1]}, q_t) = c(x_t, \gamma_t^0(q_{[0,t]}))$.

This is equivalent to, by the smoothing property of conditional expectation, the following:

$$E\left[E[F(x_t, q_{[0,t-1]}, q_t)|x_t, q_{[0,t-1]}]\middle|q_{[0,t-1]}\right]$$

Now, we will apply Witsenhausen's two stage lemma [778], to show that we can obtain a lower bound for the double expectation by picking q_t as a result of a measurable function of $x_t, q_{[0,t-1]}$. Thus, we will find a composite quantization policy which only uses $(x_t, q_{[0,t-1]})$ which performs as well as one which uses the entire memory available at the encoder. To make this precise, let us fix the decision function γ_t^0 at the receiver corresponding to a given composite quantization policy at the encoder Q_t^{comp}, let $t = T - 1$, and define for every $k \in \mathcal{M}_t$:

$$\beta_k := \left\{ x_t, q_{[0,t-1]} : F(x_t, q_{[0,t-1]}, k) \le F(x_t, q_{[0,t-1]}, q'), \forall q' \neq k, q' \in \mathcal{M}_t \right\}.$$

These sets are Borel, by the measurability of F on $\mathbb{X}$. Such a construction covers the domain set consisting of $(x_t, q_{[0,t-1]})$ but with overlaps. It covers the elements in $\mathcal{P}(\mathbb{X}) \times \prod_{t=0}^{T-2} \mathcal{M}_t$, since for every element in this product set, there is a minimizing $k \in \mathcal{M}_t$ ($\mathcal{M}_t$ is finite). To avoid the overlaps, we adopt the following technique which was introduced in Witsenhausen [778]. Let there be an ordering of the elements in $\mathcal{M}_t$ as $1, 2, \ldots, |\mathcal{M}_t|$, and for $k \ge 1$ in this sequence define a function $Q_t^{comp,*}$ as:

$$q_t = Q_t^{comp,*}(x_t, q_{[0,t-1]}) = k, \quad \text{if}(x_t, q_{[0,t-1]}) \in \beta_k - \cup_{i=1}^{k-1} \beta_i,$$

with $\beta_0 = \emptyset$. Thus, for any random variable q_t appropriately defined on the probability space,

$$E\left[E[F(x_t, q_{[0,t-1]}, q_t)|x_t, q_{[0,t-1]}] \Big| q_{[0,t-1]} \right]$$
$$\ge E\left[E[F(x_t, q_{[0,t-1]}, Q_t^{comp,*}(x_t, q_{[0,t-1]}))|x_t, q_{[0,t-1]}] \Big| q_{[0,t-1]} \right] \tag{15.33}$$

Thus, the new composite policy performs at least as well as the original composite coding policy even though it has a restricted structure.

As such, if there is an optimal policy, it can be replaced with one which uses only $\{x_t, q_{[0,t-1]}\}$ without any loss of performance, while keeping the receiver decision function γ_t^0 fixed.

We have thus obtained the structure of the encoder for the last stage. We iteratively proceed to study the other time stages. In particular, since $\{x_t\}$ is Markov, we could proceed as follows (in essence using Witsenhausen's three-stage lemma [778]): For a three-stage cost problem, the cost at time $t = 2$ can be written as, for measurable functions c_2, c_3:

$$E\Big[c_2(x_2, v_2(q_1, q_2), q_1, q_2)$$
$$+ E[c_3(x_3, v_3(q_1, q_2, Q_3^{comp,*}(x_3, q_2, q_1))|x_3, q_2, x_2, q_1, x_1)] \Big| x_2, x_1, q_2, q_1 \Big]$$

Since

$$P(dx_3, q_2, q_1|x_2, x_1, q_2, q_1) = P(dx_3, q_2, q_1|x_2, q_2, q_1),$$

and since under $Q_3^{comp,*}$, q_3 is a function of x_3 and q_1, q_2, the expression above is equal to, for some measurable $F_2(.)$, $F_2(x_2, q_2, q_1)$. By a similar argument as above, a composite quantization policy at time 2 which uses x_2 and q_1 and which performs at least as good as the original policy can be constructed. By similar arguments, an encoder at time t, $1 \leq t \leq T-1$ only uses $(x_t, q_{[0,t-1]})$ can be constructed. The encoder at time $t = 0$ uses x_0, where $x_0 = \nu_0$ is the prior distribution on the initial state.

Now that we have obtained the restricted structure for a composite quantization policy which is without any loss, we can express this as:

$$Q_t^{comp}(x_t, q_{[0,t-1]}) = Q^{q_{[0,t-1]}}(x_t), \quad \forall x_t, q_{[0,t-1]}$$

such that the quantizer action $Q^{q_{[0,t-1]}} \in \mathbb{Q}(\mathbb{X}; \mathcal{M}_t)$ is generated using only $q_{[0,t-1]}$, and the quantizer outcome is generated by evaluating $Q^{q_{[0,t-1]}}(x_t)$ for every x_t. ◇

15.7.2 Proof of Theorem 15.3.2

At time $t = T-1$, the per-stage cost function can be written as:

$$E[c(x_t, \upsilon_t(q_{[0,t]}))|q_{[0,t-1]}] = E[\int_{\mathbb{X}} P(dx_t|q_{[0,t-1]}, q_t)c(x_t, \upsilon_t(q_{[0,t-1]}, q_t))]$$

Thus, at time $t = T-1$, an optimal receiver (which is deterministic without any loss of optimality) will use $P(dx_t|q_{[0,t]})$ as a sufficient statistic for an optimal decision (or any receiver can be replaced with one which uses this sufficient statistic without any loss). Let us fix a receiver policy which only uses the posterior $P(dx_t|q_{[0,t]})$ as its sufficient statistic. Let us further note that:

$$\begin{aligned} P(dx_t|q_{[0,t]}) &= \frac{P(q_t, dx_t|q_{[0,t-1]})}{\int_{x_t} P(q_t, dx_t|q_{[0,t-1]})} \\ &= \frac{P(q_t|x_t, q_{[0,t-1]})P(dx_t|q_{[0,t-1]})}{\int_{x_t} P(q_t|x_t, q_{[0,t-1]})P(dx_t|q_{[0,t-1]})} \end{aligned} \tag{15.34}$$

The term $P(q_t|x_t, q_{[0,t-1]})$ is determined by the quantizer action Q_t (this follows from Theorem 15.3.1). Furthermore, given Q_t, the relation (15.34) is measurable on $\mathcal{P}(\mathbb{X})$ (that is, in $\pi_t(\cdot) = P(x_t \in \cdot|q_{[0,t-1]})$) under weak convergence.

To prove this technical argument, consider the numerator in (15.34) and note that the function $\kappa_B : \mathcal{P}(\mathbb{X}) \to \mathbb{R}$ defined as $\kappa_B(\pi) = \pi(B)$ is measurable under weak convergence topology as a consequence of Theorem C.2.1, for each $B \in \mathcal{B}(\mathbb{X})$. By Theorem C.2.2, this implies that the relation in (15.34) is measurable on $\mathcal{P}(\mathbb{X})$.

Let us denote the quantizer applied, given the past realizations of quantizer outputs as $Q_t^{q_{[0,t-1]}}$. Note that q_t is deterministically determined by (x_t, Q_t) and the optimal receiver function can be expressed as $\gamma_t^0(P(dx_t|q_{[0,t-1]}), q_t)$, given $Q_t^{q_{[0,t-1]}}$. The cost at time $t = T-1$ can be expressed, given the quantizer $Q_t^{q_{[0,t-1]}}$, for some Borel function G as

$$G(P(dx_t|q_{[0,t-1]}), Q_t^{q_{[0,t-1]}})$$
$$= \int_{\mathbb{X}} P(dx_t|q_{[0,t-1]}) \sum_{\mathcal{M}_t} \left(1_{\{q_t = Q_t^{q_{[0,t-1]}}(x_t)\}} \eta^{Q^{q_{[0,t-1]}}} (P(dx_t|q_{[0,t-1]}), q_t)) \right),$$

with $\eta^{Q^{q_{[0,t-1]}}}(P(dx_t|q_{[0,t-1]}), q_t)) = c(x_t, \gamma_t^0(P(dx_t|q_{[0,t-1]}), q_t))$.

Now, one can construct an equivalence class among the past $q_{[0,t-1]}$ sequences which induce the same π_t, and can replace the quantizers in this class with one, which induces a lower cost among the finitely many elements in each class for the final time stage. An optimal quantization output thus may be generated using $\pi_t(\cdot) = P(x_t \in \cdot|q_{[0,t-1]})$ and x_t, by extending Witsenhausen's argument used earlier in the proof of Theorem 15.3.1 for the terminal time stage. Since there are only finitely many past sequences and finitely many π_t, this leads to a Borel measurable selection of x_t for every π_t, leading to a quantizer and a measurable selection in π_t, x_t. Hence, the final stage cost can be expressed as $F_t(\pi_t)$ for some F_t, without any performance loss.

The same argument applies for all time stages: At time $t = T - 2$, the sufficient statistic both for the immediate cost, and the cost-to-go is $P(dx_{t-1}|q_{[0,t-1]})$, and thus for the cost impacting the time stage $t = T - 1$, as a result of the optimality result for Q_{T-1}. To show that the separation result generalizes to all time stages, it suffices to prove that $\{(\pi_t, Q_t)\}$ has a controlled Markov chain form, if the encoders use the structure above.

Now, for $t \geq 1$, for all $B \in \mathcal{B}(\mathcal{P}(\mathbb{X}))$:

$$P\Big(P(dx_t|q_{[0,t-1]}) \in B \Big| P(dx_s|q_{[0,s-1]}), Q_s, s \leq t-1\Big)$$

$$= P\Big(\int_{x_{t-1}} P(dx_t, dx_{t-1}|q_{[0,t-1]}) \in B \Big| P(dx_s|q_{[0,s-1]}), Q_s, s \leq t-1\Big)$$

$$= P\Bigg(\Bigg\{\frac{\int_{x_{t-1}} P(dx_t|x_{t-1})P(q_{t-1}|x_{t-1}, q_{[0,t-2]})P(dx_{t-1}|q_{[0,t-2]})}{\int_{x_{t-1},x_t} P(dx_t|x_{t-1})P(q_{t-1}|x_{t-1}, q_{[0,t-2]})P(dx_{t-1}|q_{[0,t-2]})}\Bigg\} \in B$$
$$\Bigg| P(dx_s|q_{[0,s-1]}), Q_s, s \leq t-1\Bigg)$$

$$= P\Bigg(\Bigg\{\frac{\int_{x_{t-1}} P(dx_t|x_{t-1})P(q_{t-1}|x_{t-1}, q_{[0,t-2]})P(dx_{t-1}|q_{[0,t-2]})}{\int_{x_{t-1},x_t} P(dx_t|x_{t-1})P(q_{t-1}|x_{t-1}, q_{[0,t-2]})P(dx_{t-1}|q_{[0,t-2]})}\Bigg\} \in B$$
$$\Bigg| P(dx_{t-1}|q_{[0,t-2]}), Q_{t-1}\Bigg) \tag{15.35}$$

$$= P\Big(\int_{x_{t-1}} P(dx_t, dx_{t-1}|q_{[0,t-1]}) \in B \Big| P(dx_{t-1}|q_{[0,t-2]}), Q_{t-1}\Big)$$

$$= P\left(P(dx_t|q_{[0,t-1]}) \in B \middle| P(dx_{t-1}|q_{[0,t-2]}), Q_{t-1}\right) \tag{15.36}$$

In the above derivation, (15.35) uses the fact that the term $P(q_{t-1}|x_{t-1}, q_{[0,t-2]})$ is uniquely identified by $P(dx_{t-1}|q_{[0,t-2]})$ and Q_{t-1}, which in turn is uniquely identified by $q_{[0,t-2]}$ and Q_{t-1}. Furthermore, (15.36) defines a regular conditional probability measure since for all $B \in \mathcal{B}(\mathbb{X})$,

$$\begin{aligned}\pi_t(B) &= P(x_t \in B|q_{[0,t-1]}) \\ &= \int_{x_{t-1}} P(x_t \in B, dx_{t-1}|q_{[0,t-1]}) \\ &= \int_{x_{t-1}} P(x_t \in B|x_{t-1})P(dx_{t-1}|q_{[0,t-1]})\end{aligned}$$

is measurable in π_{t-1}, given Q_{t-1} (as a consequence of the measurability of (15.34) in π_t).As a consequence the conditional probability $\pi_t(B)$, $B \in \mathcal{B}(\mathbb{X})$), is a measurable function of π_{t-1}, given Q_{t-1}. By Theorem C.2.2, we conclude that for any measurable function F_t of $P(dx_t|q_{[0,t-1]})$

$$\begin{aligned}&E[F_t(P(dx_t|q_{[0,t-1]}))|P(dx_s|q_{[0,s-1]}), Q_s, s \leq t-1] \\ &\quad = E[F_t(P(dx_t|q_{[0,t-1]}), Q_t)|P(dx_{t-1}|q_{[0,t-2]}), Q_{t-1}],\end{aligned} \tag{15.37}$$

for every given Q_{t-1}. Once again an equivalence relationship between the finitely many past quantizer outputs, based on the equivalence of the conditional measures $P(dx_{t-1}|q_{[0,t-2]})$ they induce, can be constructed and as a consequence the conditional probability measure π_t is measurable in $\{P(dx_{t-1}|q_{[0,t-2]}), Q_{t-1}\}$, given Q_{t-1}. With the controlled Markov structure, we can essentially follow the same argument for earlier time stages. As such, it suffices that the encoder uses only $(P(dx_t|q_{[0,t-1]}), t)$ as its sufficient statistic for all time stages, to generate the optimal quantizer. An optimal quantizer uses x_t to generate the optimal quantization outputs. ◇

15.7.3 Proof of Theorem 15.3.6

We note that the analysis preceding (15.33) applies identically for the case with control, by replacing a receiver policy with a fixed control policy. We can thus obtain the structure of the optimal encoder for the last stage. We iteratively proceed to study the other time stages. The only difference here is that, with control, $\{x_t\}$ is no longer Markov, but $\{x_t, u_t\}$ forms a controlled Markov chain. For a three-stage cost problem, the cost at time $t = 2$ can be written as, for measurable functions c_2, c_3:

$$c_2(x_2, u_2(q_1, q_2)) + E[c_3(x_3, u_3(q_1, q_2, Q_3^{comp}(x_3, q_2, q_1))|q_2, x_2, u_2, q_1, x_1)]$$

Since

$$P(dx_3, q_2, q_1|x_2, u_2, x_1, q_2, q_1) = P(dx_3, q_2, q_1|x_2, u_2, q_2, q_1),$$

and u_2 is a function of q_1, q_2 (with the control policy fixed, as mentioned earlier), and since under $Q_3^{comp,*}$, q_3 is a function of x_3 and q_1, q_2, the expectation above is equal to $F_2(x_2, q_2, q_1)$ for some measurable $F_2(.)$. Thus, an optimal composite quantization policy at time 2 uses x_2 and q_1. The proof follows identically for other time stages. Hence, we have established an analogue of Theorem 15.3.1.

By observing that $P(dx_t|q_{[0,t]})$ is a sufficient statistic for an optimal control policy and the construction of a controlled Markov chain as in (15.35), it follows that the discussion in Theorem 15.3.2 also applies in this case. ◇

15.7.4 Proof of Theorem 15.4.2

We show that the measurable selection hypothesis (see Sect. E.2.4) applies in the set of states which are visited with probability 1. In particular, the elements in the set of reachable probability measures admit densities which furthermore satisfy the equi-continuity condition in view of Lemma 15.7.1 below.

Lemma 15.7.1 *For all $t \geq 1$, $\pi_t(dx)$ is absolutely continuous with respect to the Lebesgue measure, i.e., it has a probability density function, which we will also denote by π_t by an abuse of notation. The density function π_t is uniformly continuous for every t and the sequence $\{\pi_t\}$ is a uniformly bounded and uniformly equicontinuous family.*

Proof Let φ denote the common density of the Gaussian noise variables w_t. Since $x_t = f(x_{t-1}) + w_t$ and w_t is independent of $q_{[0,t-1]}$, it is easy to see that the pdf π_t of the conditional measure $P(dx_t|q_{[0,t-1]})$ is given by

$$\pi_t(z) = \int_{\mathbb{R}^n} \varphi(z - f(x_{t-1}))P(dx_{t-1}|q_{[0,t-1]}), \quad z \in \mathbb{R}^n.$$

The uniform boundedness of $\{\pi_t\}$ is immediate. Since φ is a Gaussian density, there is a $C > 0$ such that $\left|\frac{\partial}{\partial z_j}\varphi(z)\right| \leq C$, $j = 1, \ldots, n$. A standard application of the dominated convergence theorem implies that the partial derivatives of π_t exists and they also satisfy $\left|\frac{\partial}{\partial z_j}\pi_t(z)\right| \leq C$, $j = 1, \ldots, n$. Since C does not depend on t, the sequence of densities $\{\pi_t\}$ is uniformly equicontinuous. ◇

The following lemma plays an important role in the proof.

Lemma 15.7.2 ([827])

(a) *Let $\{\mu_n\}$ be a sequence of density functions on $\mathbb{R}^n$ which are uniformly equicontinuous and uniformly bounded and assume $\mu_n \to \mu$ weakly. Then $\mu_n \to \mu$ in total variation.*
(b) *Let $\{Q_n\}$ be a sequence in $\mathcal{Q}_c$ such that $Q_n \to Q$ weakly at P. If P admits a density, then $Q_n \to Q$ in total variation at P.*
(c) *Let $P_n Q_n \to PQ$ weakly, where $\{Q_n\}$ is a sequence in $\mathcal{Q}_c$. Suppose further that $P_n \to P$ in total variation where P admits a density. Then $P_n Q_n \to PQ$ in total variation.*
(d) *Let $P_n \to P$ in total variation where P admits a density function. Let $\{Q_n\}$ be a sequence in $\mathcal{Q}_c$. Then, there exists some subsequence such that $P_{k_n} Q_{k_n} \to PQ$ for some $Q \in \mathcal{Q}_c$.*

Let $\mathcal{S} \subset \mathcal{P}(\mathbb{R}^n)$ be the set of reachable states for π under any composite coding policy. Note that by Lemma 15.7.1, the set of densities in $\mathcal{S}$ is uniformly bounded and equicontinuous.
Condition D:

(i) The cost function $c(\pi, Q)$ is continuous on $\mathcal{S} \times \mathcal{Q}_c$ in the sense that $P_n Q_n \to PQ$ implies that $c(P_n, Q_n) \to c(P, Q)$.
(ii) $\mathcal{Q}_c$ is compact in total variation at any input π admitting a density.
(iii)

$$\int_{\mathcal{P}(\mathbb{R}^n)} J_{t+1}^T(\pi) P(d\pi_{t+1} | \pi_t, Q_t)$$

is a continuous function on $\mathcal{S} \times \mathcal{Q}_c$, for the value function J_{t+1}^T at time t defined recursively as

$$J_t^T(\pi_t) = \min_{Q_t \in \mathcal{Q}_c} c(\pi_t, Q_t) + E[J_{t+1}^T(\pi_{t+1}) | \pi_t = \pi, Q_t = Q],$$

with $J_T^T = 0$.

We now have the following useful theorem, whose proof follows essentially from dynamic programming equation itself. This is related to (but weaker than) the Measurable Selection Condition 3.3.2 (with a,b,c1) and the subsequent Theorem 3.3.5 in [342]) since here we directly consider the value function.

Theorem 15.7.1 *Under Condition D, there exists an optimal (Borel measurable) policy in $\mathbf{\Pi}_W$ achieving* (15.18).

In view of the preceding theorem, to prove Theorem 15.4.2 it suffices to show that Condition D holds. We note that (ii) in Condition D directly follows from Theorem 8.5.5.

An important lemma is the following.

Lemma 15.7.3 *Let for $\pi_n \in \mathcal{S}$, $Q_n \in \mathcal{Q}_c$*

$$\pi'(m,\pi_n,Q_n)(C) := P(x_{n+1}\in C|\pi_n,Q_n,q_n=m)$$
$$= \frac{1}{\pi_n(B^n_m)}\int_{z\in C}\left\{\int \pi_n(dx)1_{\{x\in B^n_m\}}\varphi(z-f(x))\right\}dz \quad (15.38)$$

As $(\pi_n, Q_n)\to(\pi,Q)$ *in total variation, for every* $m\in\{1,\cdots,M\}$

$$\|\pi'(m,\pi_n,Q_n)-\pi'(m,\pi_n,Q_n)\|_{TV}\to 0,$$

provided that $\pi(B_m)>0$.

Proof We have that

$$\pi'(m,\pi_n,Q_n)(C)-\pi'(m,\pi,Q)(C)$$
$$=\frac{1}{\pi_n(B^n_m)}\int_{z\in C}\int\left\{\pi_n(dx)1_{\{x\in B^n_m\}}\varphi(z-f(x))\right\}dz$$
$$-\frac{1}{\pi(B_m)}\int_{z\in C}\int\left\{\pi(dx)1_{\{x\in B^n_m\}}\varphi(z-f(x))\right\}dz$$
$$\le\frac{\pi_n(B^n_m)\int_{z\in C}\int\left\{\pi_n(dx)1_{\{x\in B^n_m\}}\varphi(z-f(x))\right\}dz}{\pi_n(B^n_m)\pi(B_m)}$$
$$-\frac{\int_{z\in C}\int\{\pi(dx)1_{\{x\in B^n_m\}}\varphi(z-f(x))\}dz}{\pi_n(B^n_m)\pi(B_m)}+\frac{|\pi(B^n_m)-\pi_n(B_m)|}{\pi_n(B^n_m)\pi(B_m)}$$

This term converges uniformly over C (where the convergence of the last terms follows since there exists a uniformly lower bound on the denominator for sufficiently large n). Hence, convergence in total variation follows. ◇

The following lemma is a minor generalization of Theorem 8.3.6.

Lemma 15.7.4 *Let* $P_nQ_n\to PQ$ *in total variation. Then,*

$$\inf_\gamma\int P_n(dx)Q_n(q|x)c(x,\gamma(q))$$
$$\to\inf_\gamma\int P(dx)Q(q|x)c(x,\gamma(q)). \quad (15.39)$$

The next result establishes (i) in Condition D, the continuity of $c(\pi_t,Q_t)$ on $\mathcal{S}\times\mathcal{Q}_c$.

Theorem 15.7.2 *$c(\pi, Q)$ is continuous on $\mathcal{S} \times \mathcal{Q}_c$.*

Proof Let $\{(\pi_n, Q_n)\}$ be a sequence in $\mathcal{S} \times \mathcal{Q}_c$ such that $\pi_n Q_n \to \pi Q$ weakly. It follows from Lemma 15.7.2 (a) that $\pi_n \to \pi$ in total variation, and from Lemma 15.7.2 (c) that $\pi_n Q_n \to \pi Q$ in total variation. Then Lemma 15.7.4 implies that $c(\pi_n, Q_n) \to c(\pi, Q)$. ◇

Next we establish (iii) in Condition D. We wish to prove that $E[J_t^T(\pi_t)|\pi, Q]$ is continuous in π, Q. We apply backwards induction. At $t = T - 1$, let

$$J_{T-1}^T(\pi_{T-1}) = \min_Q c(\pi_{T-1}, Q_{T-1}).$$

By Theorem 15.7.2 and the compactness of the set of quantizers (by Theorem 8.5.6), there exists an optimal quantizer, Q_{T-1}^*. Furthermore, the following holds:

Lemma 15.7.5 *Let $F : \mathcal{S} \times \mathcal{Q}_c \to \mathbb{R}$ be continuous on $\mathcal{S} \times \mathcal{Q}_c$ in the sense that $P_n Q_n \to PQ$ implies that $F(\pi_n, Q_n) \to F(\pi, Q)$. Then, the function $\min_Q F(\pi, Q)$ is continuous in π.*

As a consequence, $J_{T-1}^T(\pi_{T-1})$ is continuous in π_{T-1}. Consider now $t = T - 2$ and we wish to see if there is a solution to the following equality.

$$J_{T-2}^T(\pi_{T-2}) = \min_Q \Big(c(\pi_{T-2}, Q_{T-2}) + E[J_{T-1}^T(\pi_{T-1})|\pi_{T-2}, Q_{T-2}] \Big). \tag{15.40}$$

Note that

$$\begin{aligned} &E[J_{T-1}^T(\pi_{T-1})|\pi_{T-2}, Q_{T-2}] \\ &= \sum_{m=1}^{M} P(\pi'(m, \pi_{T-2}, Q_{T-2})|\pi_{T-2}, Q_{T-2}) J_{T-1}^T(\pi'(m, \pi_{T-2}, Q_{T-2}) \end{aligned}$$

where

$$P(\pi'(m, \pi_{T-2}, Q_{T-2})|\pi_{T-2}, Q_{T-2}) = P(q_{T-2} = m|\pi_{T-2}, Q_{T-2})$$

and

$$\pi'(m, \pi_{T-2}, Q_{T-2})(dz) = \frac{\int \pi_{T-2}(dx_{T-2}) P(q_{t-2}=m|\pi_{T-2}, x_{T-2}) P(dz|x_{T-2})}{\int\int \pi_{T-2}(dx_{T-2}) P(q_{t-2}=m|\pi_{T-2}, x_{T-2}) P(dz|x_{T-2})}.$$

or

$$\pi'(m, \pi, Q)(C) = \frac{1}{\pi(B_m)} \int_{z \in C} \Big\{ \pi(dx) 1_{\{x \in B_m\}} \varphi(z - f(x)) \Big\} dz \tag{15.41}$$

Lemma 15.7.3 shows that as $Q_n \to Q$, $\|\pi'(m,\pi,Q_n) - \pi'(m,\pi,Q)\|_{TV} \to 0$ whenever $\pi'(m,\pi,Q) > 0$. If $\pi'(m,\pi_{T-2},Q_{T-2}) = 0$, by the boundedness of the cost, it follows that

$$P(\pi'(m,\pi_{T-2},Q_{T-2,n})|\pi_{T-2},Q_{T-2,n})J^T_{T-1}(\pi'(m,\pi_{T-2},Q_{T-2,n}) \to 0,$$

since $P(\pi'(m,\pi_{T-2},Q_{T-2,n})|\pi_{T-2},Q_{T-2,n}) \to 0$.

As a consequence, we have that $E[J^T_{T-1}(\pi_{T-1})|\pi_{T-2},Q_{T-2}]$ is continuous in (π_{T-2},Q_{T-2}), since both of the expressions in (15.40) are continuous. Hence,

$$J^T_{T-2}(\pi_{T-2}) = \min_{Q_{T-2}} \left(c(\pi_{T-2},Q_{T-2}) + E[J^T_{T-1}(\pi_{T-1})|\pi_{T-2},Q_{T-2}] \right)$$

exists and by Lemma 15.7.5 is continuous. The recursion applies for all time stages.

This concludes the proof of Theorem 15.4.2. ◇

15.8 Concluding Remarks

This chapter has presented structural results on optimal causal coding of Markov sources in a large class of settings. The structural results are shown to feature a separation structure. For the optimal causal coding of a partially observed Markov source, the structure of the optimal causal coders is obtained and is shown to admit a separation structure. We observed in particular that *separation of estimation (conditional probability computation) and quantization (of this probability)* applies in such a setup. We also observed that optimal real-time decentralized coding of a partially observed i.i.d. source admits separation. Such a separation result does not, in general, extend to decentralized coding of partially observed Markov sources. The chapter has also established the existence of optimal control and quantization policies under some technical conditions.

15.9 Bibliographic Notes

Related papers on real-time coding include the following: [562] established that the optimal optimal causal encoder minimizing the data rate subject to a distortion for an i.i.d. sequence is memoryless. If the source is kth-order Markov, then the optimal causal fixed-rate coder minimizing any measurable distortion uses only the last k source symbols, together with the current state at the receiver's memory [778]. Reference [756] considered the optimal causal coding problem of finite-state Markov sources over noisy channels with feedback. [714], and [493] considered optimal causal coding of Markov sources over noisy channels without feedback. [492] considered the optimal causal coding over a noisy channel with noisy feedback. Reference [485] considered the causal coding of more general sources, stationary sources, under a high-rate assumption. An earlier reference on quantizer design is [195]. Relevant discussions on optimal quantization, randomized decisions, and optimal quantizer design can be found in [283] and [828]. [140] have studied a related problem of coding of a partially observed Markov source. This work also regarded actions as the *quantizer functions.* [556] considered within a multi-terminal setup decentralized coding of correlated sources when the encoders observe conditionally independent messages given a finitely valued random variable, and obtained separation results for the optimal encoders. Their paper also considers noisy channels. Some related studies include optimal control with multiple sensors and sequential decentralized hypothesis testing problems [740] and multi-access communications with feedback [23].

Existence of optimal quantizers for a one-stage cost problem has been investigated in [1, 586], and [828]. For dynamic vector quantizers [140] obtained existence results for an infinite horizon setting. [485] considered causal coding of stationary sources in the limit of low distortion. [516] established the existence of optimal coding and quantizer policies for the LQG setup under the assumption that the controller is memoryless.

Reference [764] considered optimal causal variable-rate coding under side information and [833] considered optimal variable-rate causal coding under distortion constraints.

In this chapter, we also presented structural results for optimal decentralized coding of i.i.d. sources, considered in [814]. There are algorithmic and asymptotic results available in the literature when the encoders satisfy the optimal structure obtained in the chapter; important contributions in this direction include [268, 321] and [766].

For the case with noisy channels, when there is no feedback from the controller or when there is noisy feedback, the analysis requires a Markov chain construction in a larger state space provided that one apriori imposes memory and structural restrictions on the decoders. We refer the reader to Teneketzis [714] and Mahajan and Teneketzis [492, 493] for a class of such settings.

A parallel line of consideration which is of a rate-distortion theoretic nature is the *sequential-rate distortion* proposed in [709], and the *feedforward* setup, which has been investigated in [742] and [231].

This chapter is also related to Witsenhausen's indirect rate distortion problem [779] (see also [219]). Further related papers include [47, 61, 61, 111, 219, 359], and [262].

Related papers considering multi-terminal information theory problems in a team theoretic angle include [185] and [556].

Theorems 15.3.3, 15.3.4, 15.3.5 and 15.6.1 follow from [814] and [807]. Some of these results generalize the approaches in [778] and [756]. Theorem 15.4.2 is due to [827]. Results on infinite horizons build on [299, 484, 785].

Chapter 16
Coding and Control for Linear Gaussian Systems: Separation, Optimality and Linearity

In the previous chapter, we obtained the structure of optimal encoders for general stochastic, controlled and control-free, dynamical systems. Due to the prominence of Linear-Quadratic-Gaussian (LQG) systems (in analogy with the separate treatment of non-linear filtering and Kalman filtering in stochastic control), we dedicate here a stand-alone chapter covering solely LQG systems controlled over information constrained channels. In this context, we have three sets of results: (i) Optimality of separation between optimal quantization and optimal control (which crucially builds on the structural results arrived at in the previous chapter), (ii) Optimality of linear coding and control policies for the Gaussian channel setup, (iii) Information theoretic bounds and relaxations for LQG models.

In particular, we study the following questions: For a linear system, and for a communication channel, e.g. with the setup in Fig. 16.1, what are the optimal quantizer and controller structures? Can we find explicit solutions? Can we find explicit upper and lower bounds on the performance?

In the context of noise-free channels, or noisy channels with feedback, one primary structural result in this chapter will be a (triple-)separation principle between optimal estimation, optimal quantization and finally optimal control. In the context of coding and control over Gaussian channels, we will present information theoretic results on communicating a Gaussian source over a Gaussian channel, and obtain structural results for optimal encoders and controllers. We will obtain conditions under which linear encoding and control policies are optimal, and identify situations where such an optimality does not hold. We have already seen in Chap. 3, in the context of Witsenhausen's counterexample, that even LQG systems with non-classical information structures may fail to admit linear policies as their optimal solutions. Yet, we have also seen that a variation on the Witsenhausen's counterexample, with the cost function modified but still with non-classical information, the Gaussian test channel problem, admits linear optimal policies. Two further important examples will be presented in this chapter, which are similar in spirit to Witsenhausen's counterexample, exhibiting that optimality of linear encoding

S. Yüksel, T. Başar, *Stochastic Teams, Games, and Control under Information Constraints*, Systems & Control: Foundations & Applications,
https://doi.org/10.1007/978-3-031-54071-4_16

policies is a rather particular phenomenon. We will make the point that, while information theoretic methods are to be utilized with caution due to operational limitations to be presented in Sect. 16.3 and those studied in Sect. 12.4.1, they do lead to sharp bounds for a class of Gaussian setups.

We have a separate chapter devoted to such problems because Gaussian source and channel models are widely used in practice and we have already seen in Part I (and will also see later in Part IV) of the book that methods presented in this chapter are versatile in both certifying the optimality of policies as well as for obtaining lower and upper bounds on the performance of team decision and game problems with non-classical information structures. Gaussian systems also give rise to some of the most popular and easily implementable control and filtering algorithms in view of the LQG theory and Kalman Filtering. Gaussian source and channel models are effective at capturing robustness as we will see in Part IV of the book. Quadratic costs are practically important because such costs penalize unstable behavior.

The chapter starts with the noise-free channel model and the joint optimization problem involving encoders and controllers in Sect. 16.1, where separation and existence results will be established. Section 16.2 provides information theoretic ingredients for optimal joint source-channel coding over a Gaussian channel and considers the problem of jointly optimal LQG control and coding schemes for optimization in this case, leading to optimality of linear policies in important special cases. Section 16.3 presents a comprehensive summary of several information theoretic solution methods and bounding techniques.

For some background reading, we refer the reader to Sect. E.4 (in the Appendix of the book) for an overview of LQG control theory and Kalman Filtering.

16.1 Simultaneous Design of Coding and Control Over Discrete Channels: Optimality of Separation of Coding and Control

Consider the LQG setting, where a sensor encodes its noisy information to a controller. Let $x_t \in \mathbb{R}^n$ and the evolution of the system state be described by:

$$x_{t+1} = Ax_t + Bu_t + w_t, \qquad y_t = Cx_t + v_t, \tag{16.1}$$

Here, $\{w_t\}, \{v_t\}$ are mutually independent, zero-mean i.i.d. Gaussian noise sequences, u_t is an $\mathbb{R}^m$-valued control action, $y_t \in \mathbb{R}^p$ is the observation (measurement) variable, and A, B, C are matrices of appropriate dimensions. We assume that x_0 is a zero-mean Gaussian random variable, independent of the noise sequences. As in Fig. 16.1, let there be an encoder which has access to the observation variable y_t, and which transmits its information to a receiver/controller, over a discrete noiseless channel with finite capacity.

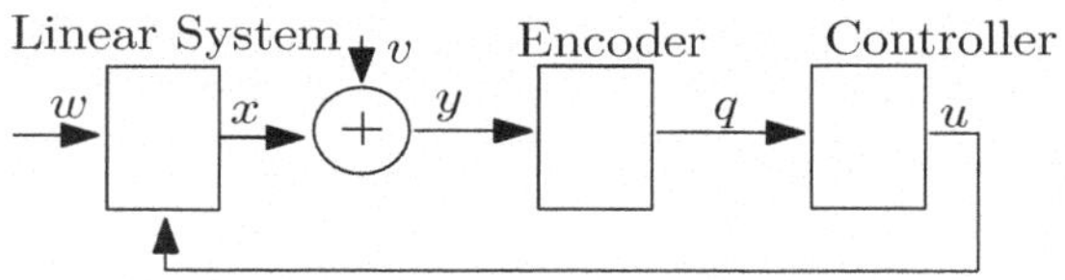

Fig. 16.1 Joint optimal design of coding and control

We will follow here the setup presented in Chap. 15 with regard to *causal* or *admissible* coding and control policies, and use the notation introduced there.

The goal is to obtain the solution to the following minimization problem, over all causal or admissible coding and control policies:

$$\inf_{\Pi^{comp}} \inf_{\gamma} J(\Pi^{comp}, \gamma, T), \tag{16.2}$$

where

$$J(\Pi^{comp}, \gamma, T) := \frac{1}{T} E_{\nu_0}^{\Pi^{comp},\gamma}[\sum_{t=0}^{T-1} x_t' Q x_t + u_t' R u_t].$$

Here, $Q \geq 0$ is a positive semi-definite matrix, $R > 0$ is a positive definite matrix, and ν_0 is the initial Gaussian measure on x_0.

16.1.1 Optimal LQG Coding and Control for the Fully Observed Case: Optimality of Predictive Quantizers Separating Quantization and Control

We study in this subsection a special case of the problem formulated above with full state information available to the encoder, that is in (16.1) we do not have measurement noise $\{v_t\}$, and the matrix C is the identity matrix I. Hence, we write (16.1) for this special case as:

$$\begin{aligned} x_{t+1} &= Ax_t + Bu_t + w_t, \\ y_t &= x_t. \end{aligned} \tag{16.3}$$

As discussed earlier in (8.7), a quantizer Q with cells $\{B_1, \ldots, B_M\}$ can be characterized as a stochastic kernel Q from $\mathbb{R}^n$ to $\{1, \ldots, M\}$ defined by

$$Q(i|x) = 1_{\{x \in B_i\}}, \quad i = 1, \ldots, M$$

and we endow the quantizers with a topology induced by such a stochastic kernel interpretation. If P is a probability measure on $\mathbb{R}^n$ and Q is a stochastic kernel from

$\mathbb{R}^n$ to $\mathcal{M}$, then we use PQ to denote the resulting joint probability measure on $\mathbb{R}^n \times \mathcal{M}$. Consider the set of probability measures

$$\Theta := \{\zeta \in \mathcal{P}(\mathbb{R}^n \times \mathcal{M}) : \zeta = PQ, Q \in \mathcal{Q}\},$$

on $\mathbb{R}^n \times \mathcal{M}$ having fixed input marginal P, equipped with weak topology. This is the Borel measurable set of the extreme points of the set of probability measures on $\mathbb{R}^n \times \mathcal{M}$ with a fixed input marginal P (see Sect. 4.1.1). As before, we have

$$\mathbf{\Pi}_W = \Big\{ \Pi^{comp} = \{Q_t^{comp}, \exists \Upsilon_t : \mathcal{P}(\mathbb{X}) \to \mathcal{Q}, Q_t^{comp}(I_t)$$

$$= (\Upsilon_t(\pi_t))(x_t), \forall I_t, t \geq 0\} \Big\}$$

to represent the class of quantizers that were shown to be optimal in Theorem 15.3.6. Here, the input measure is time varying and is given by π_t.

Separation of Quantization and Control

Consider the LQG problem with the system given in (16.3) and the cost function given in (16.2). By Theorem 15.3.6, an optimal composite quantization policy will be within the class $\mathbf{\Pi}_W$. In the following, we adopt a dynamic programming approach and establish that the optimal controller is linear in its estimate of the state. This fact applies naturally for the terminal time stage control. That this also applies for the previous time stages follows from dynamic programming as we observe in the following.

First consider the terminal time $t = T - 1$. For this time stage, to minimize $E[x_t' Q x_t + u_t' R u_t]$, the optimal control is $u_{T-1} = 0$ a.s. To obtain a solution for $t = T - 2$, we look for a solution to the following quadratic optimization problem, where for notational simplicity we use t for $T - 2$, and we leave out the additive contribution of w_{T-1} through its covariance to this *cost-to-go* function since it is independent of I_{T-2}^c, the information available at time $T - 2$:

$$\min_{\gamma_t} E\left[x_t' Q x_t + u_t' R u_t + E[(Ax_t + Bu_t + w_t)' Q (Ax_t + Bu_t + w_t) | \mathcal{I}_t^c, u_t] \middle| \mathcal{I}_t^c \right],$$

where $\mathcal{I}_t^c = \{q_{[0,t]}, u_{[0,t-1]}\}$. By completing the squares, and using the *Orthogonality Principle*, we obtain that the optimal control is linear in conditional mean of x_{T-2} and is given by

$$u_{T-2} = L_{T-2} E[x_{T-2} | q_{[0,T-2]}],$$

with $L_{T-2} = -(R + B'QB)^{-1}B'QA$. For $t < T - 2$, to obtain the solutions, we will first establish that the estimation errors are uncorrelated. Toward that end, note that

$$\tilde{m}_{t+1} := E[x_{t+1}|\mathcal{I}^c_{t+1}] = E[Ax_t + Bu_t + w_t|\mathcal{I}^c_{t+1}].$$

It then follows that

$$\begin{aligned}\tilde{m}_{t+1} &= E[x_{t+1}|\mathcal{I}^c_{t+1}] = E\left[x_{t+1} - E[x_{t+1}|\mathcal{I}^c_t] + E[x_{t+1}|\mathcal{I}^c_t]\middle|\mathcal{I}^c_{t+1}\right] \\ &= E\left[E[Ax_t + Bu_t + w_t|\mathcal{I}^c_t] + \left(x_{t+1} - E[x_{t+1}|\mathcal{I}^c_t]\right)\middle|\mathcal{I}^c_{t+1}\right] \\ &= A\tilde{m}_t + Bu_t + \left(E[x_{t+1}|\mathcal{I}^c_{t+1}] - E[x_{t+1}|\mathcal{I}^c_t]\right) = A\tilde{m}_t + Bu_t + \bar{w}_t, \end{aligned} \tag{16.4}$$

with

$$\bar{w}_t = E[x_{t+1}|\mathcal{I}^c_{t+1}] - E[x_{t+1}|\mathcal{I}^c_t].$$

Now $\bar{w}_t$ is orthogonal to the control action variable u_t, as control actions are determined by the past quantizer outputs and iterated expectation leads to the result that conditioned on $\mathcal{I}^c_t$, $\bar{w}_t$ is zero mean, and is orthogonal to $\mathcal{I}^c_t$ (in the sense that for any appropriate measurable bounded g, $E[\bar{w}_t g(\mathcal{I}^c_t)] = 0$).

Going backward to earlier time stages, the dynamic programming recursion for linear systems driven by an uncorrelated noise process would normally apply, since the estimate process $\tilde{m}_t$ is driven by an uncorrelated noise (though, not necessarily an independent) process $\bar{w}_t = E[x_{t+1}|\mathcal{I}^c_{t+1}] - E[x_{t+1}|\mathcal{I}^c_t]$. However, this lack of independence may be important, as elaborated on in [554]. Using the completion of the squares method, we can establish that the optimal controller at any time will be linear in its estimate (that is, the conditional mean of the state), provided that the random variable $\bar{w}_t' Q \bar{w}_t$ is not affected by the control policies $\{\gamma_k, k \leq t-1\}$ (that is, the changes in the control actions $\{u_k, k \leq t-1\}$ do not affect $\bar{w}_t' Q \bar{w}_t$) under an optimal coding policy for all time stages t. A sufficient condition for this is that the encoder is a predictive one (see [138], [554] and [709] for related discussions), as is further discussed below.

Definition 16.1.1 ([815, Definition 3.1]) A predictive quantizer policy is one where for each time stage t, the quantization has the form that the quantizer at all time stages subtracts the effect of the past control terms, that is, at time t it has the form $Q_t(x_t - \sum_{k=0}^{t-1} A^{t-k-1}Bu_k)$, and the past control terms are added at the receiver. Hence, the encoder quantizes a control free process, defined by:

$$\bar{x}_{t+1} = A\bar{x}_t + w_t, \tag{16.5}$$

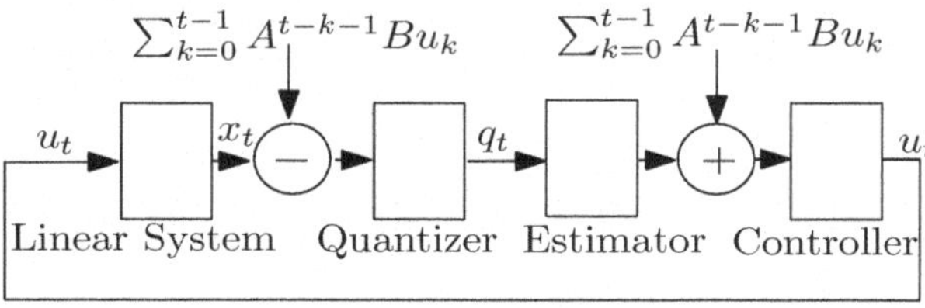

Fig. 16.2 For the LQG problem, a predictive encoder is optimal

the receiver generates the quantized estimate and adds $\sum_{k=0}^{t-1} A^{t-k-1}Bu_k$ to compute the estimate of the state at time t. ◇

A predictive quantizer is depicted in Fig. 16.2. One consequential question is *whether restriction to this class of quantization policies (given in Definition 16.1.1) leads to any performance loss among all quantizers.*. This is resolved in the following lemma on the structure of optimal encoders, the optimality of predictive quantizers, and the associated separation result.

Lemma 16.1.1 ([815, Lemma 3.1] (also [817])) *For problem (16.2), for any quantizer policy in class* $\mathbf{\Pi}_W$ *(which is without any loss as a result of Theorem 15.3.6), there exists a predictive quantizer in the sense of Definition 16.1.1 which attains the same performance given an optimal control policy for problem (16.2). Furthermore, the optimal control is linear in its conditional estimate for all time stages.*

Proof We apply backward induction and dynamic programming. For $t = T-1$, the optimal control is zero, and therefore the quantizer design does not affect the expected cost. We therefore may use a predictive quantizer for $t = T-1$ without any loss. Now, for the time-stage, $t = T-2$, let $f_t(q_{[0,t-1]}) := \sum_{k=0}^{t-1} A^{t-k-1}Bu_k$. If the policy considered is in $\mathbf{\Pi}_W$, the quantization policy is of the form $Q_t\Big(\bar{x}_t + f_t(q_{[0,t-1]}), P(\bar{x}_t + f_t(q_{[0,t-1]}) \in \cdot | q_{[0,t-1]})\Big)$. For this time-stage, the optimal decoder and controller uses a sufficient statistic to generate the optimal control policy, which is $E[x_t|q_{[0,t]}]$. Note that

$$\begin{aligned} E[\bar{x}_t + f_t(q_{[0,t-1]})|q_{[0,t]}] &= E[\bar{x}_t|q_{[0,t]}] + f_t(q_{[0,t-1]}) \\ &= E[\bar{x}_t|q_{[0,t-1]}, q_t] + f_t(q_{[0,t-1]}). \end{aligned}$$

The quantization output q_t represents the bin information for x_t. By shifting each of the finitely many quantizer bins by $f_t(q_{[0,t-1]})$, a new quantizer which quantizes $\bar{x}_t$ (see (16.5)), can generate the same bin information on $\bar{x}_t$ through q_t, that is, can encode the event $1_{\{\bar{x}_t \in B_i\}}$ for some bin B_i almost surely. Hence, there is no information loss due to the elimination of the past control actions. This new quantizer, by adding $f_t(q_{[0,t-1]})$ to the receiver output, generates the same conditional estimate of the state as the original quantizer. Thus, corresponding to a quantizer policy in $\mathbf{\Pi}_W$ at time t, there exists a quantizer of the form $\tilde{Q}_t(\bar{x}_t, P(\bar{x}_t \in \cdot|q_{[0,t-1]}))$ with the following property: The estimation error realization and hence the estimation is the same almost surely. Furthermore, under such a predictive scheme (with $\tilde{Q}_t(\bar{x}_t, P(\bar{x}_t \in \cdot|q_{[0,t-1]}))$ fixed), $\bar{w}_{T-2}$ does not depend on the control actions applied earlier; for a predictive quantizer, the error only depends on the control-free process.

Here, we note that $\bar{w}_{T-2}$ does not (functionally) depend on the control actions, in that if one changes the control policies $\{\gamma_s, 0 \leq s \leq T-3\}$, $\bar{w}_{T-2}$ is not affected. This does not imply that $\bar{w}_{T-2}$ is statistically independent from the past control actions; however, this is not relevant for the analysis as we demonstrate in the following. First, observe that through the law of the iterated expectations, and the orthogonality principle, we have

$$\frac{1}{T}E_{\nu_0}^{\Pi^{comp},\gamma}\left[\sum_{t=0}^{T-1} x_t'Qx_t + u_t'Ru_t\right]$$

$$= \frac{1}{T}E_{\nu_0}^{\Pi^{comp},\gamma}\left[\sum_{t=0}^{T-1}(x_t - \tilde{m}_t)'Q(x_t - \tilde{m}_t) + \tilde{m}_t'Q\tilde{m}_t + u_t'Ru_t\right] \tag{16.6}$$

Now, once we have that the quantizer at time $t = T-2$ is a predictive one, we can write the cost for $t = T-3$ as follows: Through (16.4), with $u_{T-2} = L_{T-2}\tilde{m}_{T-2} = -(R + B'QB)^{-1}B'QA\tilde{m}_{T-2}$, the cost-to-go for a policy γ_{T-3} would write as:

$$E\left[E\left[\tilde{m}_{N-2}'\left(Q + A'QA - A'QB(R + B'QB)^{-1}B'QA\right)\tilde{m}_{N-2}\middle|\mathcal{I}_{N-3}\right]\right.$$

$$+E\left[\bar{w}_{N-2}'Q\bar{w}_{N-2}\middle|\mathcal{I}_{N-3}\right] \tag{16.7}$$

$$+E\left[(x_{N-2} - \tilde{m}_{N-2})'Q(x_{N-2} - \tilde{m}_{N-2})\middle|\mathcal{I}_{N-3}\right] \tag{16.8}$$

$$+E\left[(x_{N-1} - \tilde{m}_{N-1})'Q(x_{N-1} - \tilde{m}_{N-1})\middle|\mathcal{I}_{N-3}\right] \tag{16.9}$$

$$\left.+2E\left[(x_{N-2} - \tilde{m}_{N-2})'A'QBu_{N-2}\middle|\mathcal{I}_{N-3}\right]\right] \tag{16.10}$$

The last term (16.10) is zero since $(x_{T-2} - \tilde{m}_{T-2})$ is orthogonal to u_{T-2}. By the use of the predictive quantizer, the terms (16.8) and (16.9) do not depend on the control policy at time $T-3$. Note that

$$\bar{w}_t = E[x_{t+1}|\mathcal{I}_{t+1}] - E[x_{t+1}|\mathcal{I}_t] = (x_{t+1} - E[x_{t+1}|\mathcal{I}_{t+1}]) - (x_{t+1} - E[x_{t+1}|\mathcal{I}_t])$$
$$= (x_{t+1} - E[x_{t+1}|\mathcal{I}_{t+1}]) - (Ax_t + Bu_t + w_t - E[Ax_t + Bu_t + w_t|\mathcal{I}_t])$$
$$= (x_{t+1} - E[x_{t+1}|\mathcal{I}_{t+1}]) - (Ax_t + w_t - E[Ax_t + w_t|\mathcal{I}_t])$$
$$= (x_{t+1} - E[x_{t+1}|\mathcal{I}_{t+1}]) - (A(x_t - E[Ax_t|\mathcal{I}_t]) + w_t)$$

and this term does not depend on the past applied control policies. Thus, (16.7) is not affected by the control policy at time $T-3$. Hence, these terms can be taken out from the optimization so that the cost-to-go that is relevant for optimization is:

$$E\left[E\left[\tilde{m}'_{T-2}\left(Q + A'QA - A'QB(R + B'QB)^{-1}B'QA\right)\tilde{m}_{T-2}\middle|\mathcal{I}^c_{T-3}\right]\right]$$

To obtain a solution for $t = T-3$, we then look for a solution to:

$$\min_{\gamma_{T-3}} E\left[(x'_{T-3}Qx_{T-3} + u'_{T-3}Ru_{T-3}) + E\left[\tilde{m}'_{T-2}\left(Q + A'QA - A'QB(R + B'QB)^{-1}B'QA\right)\tilde{m}_{T-2} \middle| \mathcal{I}^c_{T-3}, u_{T-3}\right]\middle|\mathcal{I}^c_{T-3}\right]$$

As in (16.6), writing $E[x'_{T-3}Qx_{T-3}]$ as

$$E\left[(x_{T-3} - \tilde{m}_{T-3} + \tilde{m}_{T-3})'Q(x_{T-3} - \tilde{m}_{T-3} + \tilde{m}_{T-3})\right],$$

and noting the orthogonality of $x_{T-3} - \tilde{m}_{T-3}$ and $\tilde{m}_{T-3}$, the estimation cost $E[(x_{T-3} - \tilde{m}_{T-3})'Q(x_{T-3} - \tilde{m}_{T-3})]$ can be left out from the optimization and the cost relevant for the control policy at time $T-3$ is:

$$\min_{\gamma_{T-3}} E\left[E\left[(\tilde{m}'_{T-3}Q\tilde{m}_{T-3} + u'_{T-3}Ru_{T-3}) + E\left[\tilde{m}'_{T-2}\left(Q + A'QA - A'QB(R + B'QB)^{-1}B'QA\right)\tilde{m}_{T-2} \middle| \mathcal{I}^c_{T-3}, u_{T-3}\right]\middle|\mathcal{I}^c_{T-3}\right]\right]$$

Now, using (16.4), by completing the squares, and using the orthogonality principle with u_{T-3} being orthogonal to $\bar{w}_{T-3}$, we obtain that the optimal control is linear and is given by $u_{T-3} = L_{T-3}\tilde{m}_{T-3}$ with $L_{T-3} = -(R + B'K_{T-2}B)^{-1}B'K_{T-2}A$, where K_{T-2} satisfies the following recursion $K_t = A_t' K_{t+1} A_t - P_t + Q$, with $P_t = A_t' K_{t+1} B(R + B' K_{t+1} B)^{-1} B' K_{t+1} A$ and $K_T = P_{T-1} = 0$.

An optimal controller at time $t = T-3$ will then use $\tilde{m}_{T-3}$ as a sufficient statistic (note that the optimal controls for $t = T-1$ and $t = T-2$ have been derived earlier). To design the quantizer at $T-3$, by a similar reasoning as above for $t = T-1$ and $T-2$, a predictive quantizer can be used so that $\bar{w}_k, k \geq T-3$ is independent of the control policies $\{\gamma_s, s < T-3\}$ (and thus does not functionally depend on the control actions) applied earlier. This, inductively, leads to the optimality of linear policies and the optimality of predictive quantizers, for all $t \geq 0$. ◇

Remark 16.1.1 We note that the structure in Definition 16.1.1 separates the estimation from the control process in the sense that the estimation errors do not depend on the control policies. Hence, *there is no dual effect of the control policies in the sense that the estimation error at any given time does not depend on the past applied control policies (or is not affected by the past applied actions).* This subtle separation result indeed gave motivation to a more general separation result in partially observed stochastic control reported in [211].

Remark 16.1.2 For the proof presented, it was essential to show first that the coding policies adopted can be taken to be in class $\mathbf{\Pi}_W$. Indeed, in the absence of such a restriction (which we showed to be without any loss), a counterexample presented in [593, Example 3] utilizing a coding policy which does not belong to $\mathbf{\Pi}_W$ reveals that the aforementioned separation result does not hold.

We have the following (see also [554] which establishes a more restrictive structure than that given in Definition 16.1.1 for a similar result):

Theorem 16.1.1 ([815]) *For the minimization problem (16.2), with the new effective state dynamics in (16.4), an optimal control policy is given by $u_t = L_t E[x_t | q_{[0,t]}]$, where $L_t = -(R + B' K_{t+1} B)^{-1} B' K_{t+1} A$, where K_t satisfies the recursions $K_t = A_t' K_{t+1} A_t - P_t + Q$ with $P_t = A_t' K_{t+1} B(R + B' K_{t+1} B)^{-1} B' K_{t+1} A$ and $K_T = P_{T-1} = 0$.* ◇

With the cost written as (16.6) and with the preceeding analysis, we obtain for $t \geq 0$, the unnormalized value function for any time stage t as

$$\begin{aligned} J_t(\mathcal{I}_t^c) &= E[\tilde{m}_t' K_t \tilde{m}_t | \mathcal{I}_t^c] \\ &+ \sum_{k=t}^{T-1} \left(E[(x_k - E[x_k|\mathcal{I}_k^c])' Q (x_k - E[x_k|\mathcal{I}_k^c])] + E[\bar{w}_k' K_{k+1} \bar{w}_k] \right), \end{aligned}$$

with $J(\Pi^{comp}, \gamma, T) = \frac{1}{T} J_0(\mathcal{I}_0^c)$. To obtain a more explicit expression for the value function J_t, we have the following analysis. Given a positive definite matrix Λ define an inner-product as $\langle z_1, z_2 \rangle_\Lambda = z_1' \Lambda z_2$, and the norm generated by this inner-product as $||z||_\Lambda = \sqrt{z' \Lambda z}$. We now note the following:

$$E\left[||E[x_{t+1}|\mathcal{I}^c_{t+1}] - E[x_{t+1}|\mathcal{I}^c_t]||^2_\Lambda \right]$$

$$= E\left[||(E[x_{t+1}|\mathcal{I}^c_{t+1}] - x_{t+1}) + (x_{t+1} - E[x_{t+1}|\mathcal{I}^c_t])||^2_\Lambda \right]$$

$$= E\left[||(E[x_{t+1}|\mathcal{I}^c_{t+1}] - x_{t+1})||^2_\Lambda] + E[||(x_{t+1} - E[x_{t+1}|\mathcal{I}^c_t])||^2_\Lambda \right]$$

$$+2E[\langle(E[x_{t+1}|\mathcal{I}^c_{t+1}] - x_{t+1}), (x_{t+1} - E[x_{t+1}|\mathcal{I}^c_t])\rangle_\Lambda],$$

and further note that

$$E\left[\langle(E[x_{t+1}|\mathcal{I}^c_{t+1}] - x_{t+1}), (x_{t+1} - E[x_{t+1}|\mathcal{I}^c_t])\rangle_\Lambda \right]$$

$$= E\left[- \langle(E[x_{t+1}|\mathcal{I}^c_{t+1}] - x_{t+1}), (E[x_{t+1}|\mathcal{I}^c_t])\rangle_\Lambda \right.$$

$$\left. +\langle(E[x_{t+1}|\mathcal{I}^c_{t+1}] - x_{t+1}), (x_{t+1})\rangle_\Lambda \right]$$

$$= E[\langle(E[x_{t+1}|\mathcal{I}^c_{t+1}] - x_{t+1}), (x_{t+1})\rangle_\Lambda] = -E[||(E[x_{t+1}|\mathcal{I}^c_{t+1}] - x_{t+1})||^2_\Lambda] \tag{16.11}$$

where (16.11) follows from the orthogonality property of minimum mean-square estimation and that $E[x_{t+1}|\mathcal{I}^c_t]$ is measurable on $\sigma(\mathcal{I}^c_{t+1})$, the sigma-field generated by $\mathcal{I}^c_{t+1}$. Therefore, we have

$$E\left[||(E[x_{t+1}|\mathcal{I}^c_{t+1}] - E[x_{t+1}|\mathcal{I}^c_t])||^2_{K_{t+1}} \right]$$

$$= -E\left[(x_{t+1} - E[x_{t+1}|\mathcal{I}^c_{t+1}])'(K_{t+1})(x_{t+1} - E[x_{t+1}|\mathcal{I}^c_{t+1}]) \right.$$

$$\left. +E[(x_t - E[x_t|\mathcal{I}^c_t])'(A'K_{t+1}A)(x_t - E[x_t|\mathcal{I}^c_t])] + E[w'K_{t+1}w] \right]$$

After some algebra, for $t < T - 1$, the optimal cost can be written as

$$J_t(\mathcal{I}^c_t) = E[\tilde{m}_t' K_t \tilde{m}_t|\mathcal{I}^c_t] + E[(x_t - \tilde{m}_t)'(Q + A'K_{t+1}A)(x_t - \tilde{m}_t)]$$
$$+ \sum_{k=t+1}^{T-1} E[(x_k - \tilde{m}_k)'(Q + A'K_{k+1}A - K_k)(x_k - \tilde{m}_k)] + \sum_{k=t}^{T-1} E[w_k' K_{k+1} w_k]. \tag{16.12}$$

In particular, **it follows that the quantization problem can be separated from the control problem**. For $t = 0$, by writing via the orthogonality principle of conditional expectation,

$$E_{\nu_0}^{\Pi^{comp}}[\tilde{m}_0' K_0 \tilde{m}_0 + (x_0 - \tilde{m}_0)' K_0 (x_0 - \tilde{m}_0)] = E_{\nu_0}^{\Pi^{comp}}[x_0' K_0 x_0]$$

(16.12) simplifies to

$$\begin{aligned} J(\Pi^{comp}, T) &= E_{\nu_0}^{\Pi^{comp}}[x_0' K_0 x_0] \\ &+ \sum_{k=0}^{T-1} E_{\nu_0}^{\Pi^{comp}}[(x_k - \tilde{m}_k)'(Q + A'K_{k+1}A - K_k)(x_k - \tilde{m}_k)] \\ &+ \sum_{k=0}^{T-1} E[w_k' K_{k+1} w_k]. \end{aligned} \tag{16.13}$$

We have thus separated the optimal cost as the summation of two costs due to control and quantization. We address the optimal quantization problem in the following subsection.

Existence of Optimal Quantization Policies

Now that we have separated the costs due to control and quantization, under any such composite policy and $T \in \mathbb{N}$, we can define a cost to be minimized by a composite quantizer policy as:

$$J(\Pi^{comp}, T) = E_{\nu_0}^{\Pi^{comp}}[\frac{1}{T}(x_0' K_0 x_0 + \sum_{t=0}^{T-1} c_t(\pi_t, Q_t))],$$

where

$$c_t(\pi_t, Q_t) = \sum_{i \in \mathcal{M}} \inf_{\gamma_t^0(i)} \int_{\mathbb{R}^n} 1_{\{q_t = i\}} \pi_t(d\bar{x})(\bar{x}_t - \gamma_t^0(i))' P_t (\bar{x}_t - \gamma_t^0(i)),$$

where now $\underline{\gamma}^0 = \{\gamma_t^0, t \geq 0\}$ denotes a receiver policy and $P_t = (Q + A'K_{t+1}A - K_t)$.

We note that, here the process $\bar{x}_t$ is the control-free process given by $\bar{x}_{t+1} = A\bar{x}_t + w_t$.

Therefore, we consider the setting where in (16.3), $u_t = 0$ and the quantizer is designed for this system. We note that as a result of the decoupling from the control actions by the predictive quantization policy (see Definition 16.1.1), the separation results presented in Theorem 15.3.4 directly apply in this context.

In the analysis to follow, we will restrict the quantizers to have convex codecells (see Assumption 15.4.2). In view of the results of Sect. 8.2.1, we present the following. Let $\pi_t(\cdot) = P(x_t \in \cdot | q_{[0,t-1]})$. Recall that the properties of conditional probability lead to the filtering expression in (E.11).

Hence, with $\mathcal{P}(\mathbb{R}^n)$ denoting the set of probability measures on $\mathcal{B}(\mathbb{R}^n)$ under weak convergence, the conditional density process and the quantization process $(\pi_t(x), Q_t)$ form a joint Markov process in $\mathcal{P}(\mathbb{R}^n) \times \mathcal{Q}_c(M)$, as in Theorem 15.4.1.

We have the following result on the existence of optimal quantizers for the finite horizon setting.

Theorem 16.1.2 *For any $T \geq 1$ and arbitrary initial condition π_0, there exists a policy in $\mathbf{\Pi}_W^C$ such that*

$$\inf_{\Pi^{comp} \in \mathbf{\Pi}_W^C} J_{\pi_0}(\Pi^{comp}, T) \tag{16.14}$$

is achieved. Letting $J_T^T(\cdot) = 0$ and

$$J_0^T(\pi_0) := \min_{\Pi^{comp} \in \mathbf{\Pi}_W^C, \gamma^0} J_{\pi_0}(\Pi^{comp}, \gamma^0, T),$$

the dynamic programming recursion

$$T J_t^T(\pi_t) = \min_{Q \in \mathcal{Q}_c(M)} \Big(c(\pi_t, Q_t) + T E[J_{t+1}^T(\pi_{t+1}) | \pi_t, Q_t] \Big) \tag{16.15}$$

holds for all $t = 0, 1, \ldots, T-1$.

Proof See Sect. 16.5.1. ◇

16.1.2 Partially Observed Setup: Separation of Estimation, Quantization and Control

Control-Free Case: Separation of Estimation and Quantization

Consider a control-free LQG setup, where a sensor is connected to an estimator over a discrete noiseless channel. Let $x_t \in \mathbb{R}^n$, $y_t \in \mathbb{R}^m$, and the evolution of the source be given by (16.1) without the control input:

$$\begin{aligned} x_{t+1} &= A x_t + w_t, \\ y_t &= C x_t + v_t, \end{aligned} \tag{16.16}$$

The goal is to obtain a solution to the minimization problem

$$\inf_{\Pi^{comp}} \inf_{\underline{\gamma}^0} E_{\nu_0}^{\Pi^{comp},\underline{\gamma}^0}[\sum_{t=0}^{T-1}(x_t - u_t)'Q(x_t - u_t)], \tag{16.17}$$

with ν_0 denoting a Gaussian distribution for the zero-mean initial state, and $Q > 0$ a positive definite matrix.

The conditional distribution $\widetilde{\pi}_t(\cdot) = P(x_t \in \cdot|y_{[0,t]})$ is Gaussian for all time stages, which is characterized uniquely by its mean and covariance matrix for all time; thus $\widetilde{\pi}_t$ can be uniquely characterized by an element of $\mathbb{R}^{\frac{n^2+3n}{2}}$. Furthermore, the nonlinear filter equation described in (E.11) admits a simpler recursion known as the Kalman Filter (see Sect. B.3). We have the following result (see Fig. 16.3).

Theorem 16.1.3 *For the minimization of the cost in (16.17), any composite quantization policy can be replaced, without any loss in performance, by one which only uses the output of the Kalman Filter and the information available at the receiver.*

Proof The result can be proven by considering a direct approach, rather than as an application of Theorems 15.3.3 and 15.3.4 (which require bounded costs, that, however, can be relaxed for this case), exploiting the specific quadratic nature of the problem. Let, again, $x_t \in \mathbb{R}^n$ and $|\cdot|_Q$ denote the norm generated by an inner product of the form: $\langle x, y\rangle_Q = x^T Q y$ for $x, y \in \mathbb{R}^n$ for positive-definite $Q > 0$. The Projection Theorem for Hilbert Spaces implies that the random variable $x_t - E[x_t|y_{[0,t]}]$ is orthogonal (see Sect. B.3) to the random variables $\{y_{[0,t]}, q_{[0,t]}\}$, where $q_{[0,t]}$ is included due to the Markov chain condition that $P(dx_t|y_{[0,t]}, q_{[0,t]}) = P(dx_t|y_{[0,t]})$. We thus obtain the following identity.

$$\begin{aligned} E[|x_t - E[x_t|q_{[0,t]}]|_Q^2] &= E[|x_t - E[x_t|y_{[0,t]}]|_Q^2] \\ &+ E\left[\left|E[x_t|y_{[0,t]}] - E\left[E[x_t|y_{[0,t]}]\middle|q_{[0,t]}\right]\right|_Q^2\right]. \end{aligned} \tag{16.18}$$

The second term is to be minimized through the choice of the quantizers. Hence, the term $\bar{m}_t := E[x_t|y_{[0,t]}]$, which is computed through a Kalman Filter, is to be quantized (see Fig. 16.3). Recall that by the Kalman Filter (see Sect. E.4), with

$$\Sigma_{0|-1} = E[x_0 x_0']$$

and for $t \geq 0$,

$$\Sigma_{t+1|t} = A\Sigma_{t|t-1}A' + W - (A\Sigma_{t|t-1}C')(C\Sigma_{t|t-1}C' + V)^{-1}(C\Sigma_{t|t-1}A'),$$

the following recursion holds for $t \geq 0$ and with $\bar{m}_{-1} = 0$:

$$\bar{m}_t = A\bar{m}_{t-1} + \Sigma_{t|t-1}C'(C\Sigma_{t|t-1}C' + V)^{-1}(CA(x_{t-1} - \bar{m}_{t-1}) + v_t).$$

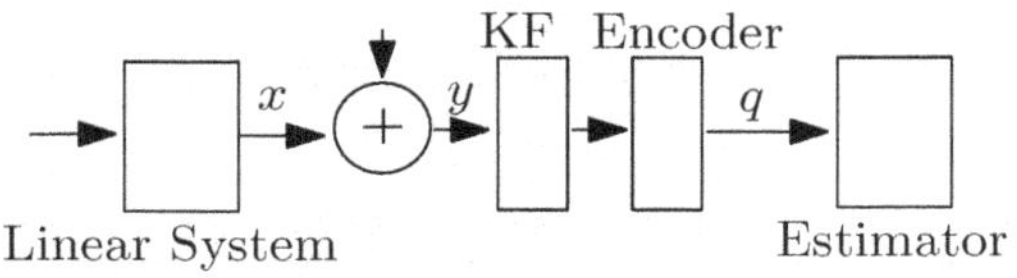

Fig. 16.3 Separation of Estimation and Quantization: when the source is Gaussian, generated by the linear system (16.16), the cost is quadratic, and the observation channel is Gaussian, the separated structure of the encoder above is optimal. That is, first the encoder runs a Kalman filter, and then causally encodes its estimate

Thus, the pair $(\bar{m}_t, \Sigma_{t|t-1})$ is a Markov source, where the evolution of $\Sigma_{t|t-1}$ is deterministic. Even though the cost to be minimized is not bounded, since $\bar{m}_t$ itself is a fully observed process, Theorem 15.3.1 can be used to develop the structural result that any causal encoder can be replaced with one which uses $(\bar{m}_t, \Sigma_{t|t-1})$ and the past quantization outputs. Likewise, the proof of Theorem 15.3.2 shows that, for the fully observed Markov source $(\bar{m}_t, \Sigma_{t|t-1})$, any causal coder can be replaced with one which only uses the conditional probability on $\bar{m}_t$ and the realization $(\bar{m}_t, \Sigma_{t|t-1}, t)$ at time t. ◇

Controlled Case: Separation of Estimation, Quantization and Control

We now return to the original formulation given in (16.1) (and Fig. 16.1) and cost criterion (16.2).

Define $\bar{m}_t := E[x_t|y_{[0,t]}]$, which is computed through a Kalman Filter (see Sect. E.4). With

$$\Sigma_{0|-1} = E[x_0 x_0']$$

and for $t \geq 0$,

$$\begin{aligned}\Sigma_{t+1|t} &= A\Sigma_{t|t-1}A' + W \\ &\quad -(A\Sigma_{t|t-1}C')(C\Sigma_{t|t-1}C' + V)^{-1}(C\Sigma_{t|t-1}A'),\end{aligned}$$

the following recursion holds for $t \geq 0$ and with $\tilde{m}_{-1} = 0$:

$$\begin{aligned}\bar{m}_t &= A\bar{m}_{t-1} + Bu_{t-1} \\ &+\Sigma_{t|t-1}C'(C\Sigma_{t|t-1}C' + V)^{-1}(CA(x_{t-1} - \tilde{m}_{t-1}) + v_t).\end{aligned}$$

Now, note that the cost (16.2) can equivalently be written as

$$J(\Pi^{comp}, \gamma, T) = \frac{1}{T} E_{\nu_0}^{\Pi^{comp},\gamma}[\sum_{t=0}^{T-1} \bar{m}_t' Q \bar{m}_t + u_t' R u_t]$$

$$+\frac{1}{T} E[\sum_{t=0}^{T-1} (x_t - \bar{m}_t)' Q (x_t - \bar{m}_t)], \qquad (16.19)$$

since the quadratic error $(x_t - \bar{m}_t)' Q (x_t - \bar{m}_t)$ is independent of the coding or the control policy.

Thus, the process $(\bar{m}_t, \Sigma_{t+1|t})$ and u_t form a controlled Markov chain and we can invoke Theorem 15.3.6: Any causal quantizer, can be, without any loss replaced, with one in $\mathbf{\Pi}_W$ (where the state is now $(\bar{m}_t, \Sigma_{t+1|t})$ instead of x_t) as a consequence of Theorem 15.3.6. Furthermore, any quantizer in $\mathbf{\Pi}_W$ can be replaced without any loss with a predictive quantizer with the new state $\bar{m}_t$, as a consequence of Lemma 16.1.1 applied to the new state with identical arguments: Observe that the past control actions do not affect the evolution of $\Sigma_{t+1|t}$.

Theorem 16.1.4 *For the minimization problem (16.2), with the new effective state dynamics in (16.4) and with* $\tilde{m}_t := E[\bar{m}_t|\mathcal{I}_t^c]$*, an optimal control policy is given by* $u_t = L_t E[\bar{m}_t|q_{[0,t]}]$*, where* $L_t = -(R + B'K_{t+1}B)^{-1}B'K_{t+1}A$*, where* K_t *satisfies the recursions* $K_t = A_t' K_{t+1} A_t - P_t + Q$ *with* $P_t = A_t' K_{t+1} B(R + B'K_{t+1}B)^{-1} B' K_{t+1} A$ *and* $K_T = P_{T-1} = 0$.

Given that we can use a predictive encoder without any loss, we have the following.

Theorem 16.1.5 *For the minimization of the cost in (16.2), any composite quantization policy can be replaced, without any loss in performance, by an encoder which only uses the output of the Kalman Filter and the information available at the receiver. Furthermore, any causal coder can be replaced with one which only uses the conditional probability on* $\bar{m}_t$*,* $P(d\bar{m}_t|q_{[0,t-1]})$*, and the realization* $(\bar{m}_t, \Sigma_{t|t-1}, t)$ *at time* t *(see Fig. 16.4). An optimal quantization policy exists in* $\mathbf{\Pi}_W^C$. ◇

Thus, the optimality of Kalman Filtering allows the encoder to only use the conditional estimate and the error covariance matrix without any loss of optimality (See Fig. 16.4), and the optimal quantization problem also has an explicit formulation.

16.2 The Gaussian Channel Case: Optimality of Linear Coding and Control Policies

In this section, we consider Gaussian channels and networks of Gaussian channels motivated by LQG problems.

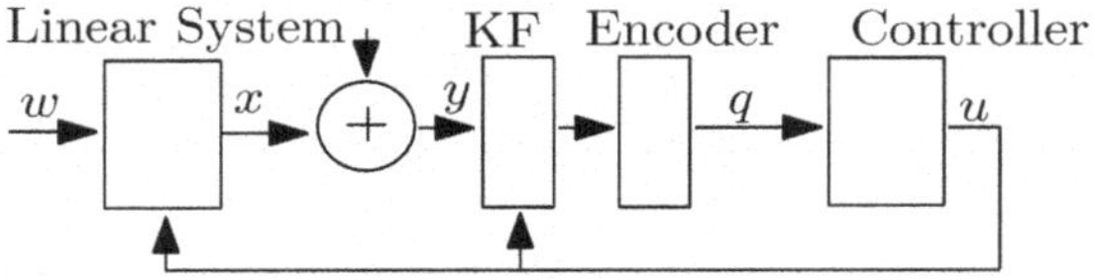

Fig. 16.4 **Separation of Estimation and Quantization**: when the source is Gaussian, generated by the linear system (16.1), the cost is quadratic, and the observation channel is Gaussian, the separated structure of the encoder above (with a **predictive** encoder) is optimal, where KF denotes the Kalman Filter

16.2.1 *Optimality of Linear Coding Policies Over a Gaussian Channel with Matching Between the Source and the Channel*

Converse Bounds on Distortion via Inverse Rate-Distortion Function and Channel Capacity

As seen earlier in Sect. 12.3.3, the *rate-distortion function*, $R(D)$, of a source is the minimum average amount of information that can be transmitted to satisfy a given level of average distortion D. The capacity, $C(P)$, of a channel with cost (power) constraint P, on the other hand, is the maximum amount of information which can be transmitted reliably per channel use given an average cost (power) constraint at the channel input.

These two notions lead to the following classical converse theorem in information theory [188]:

Theorem 16.2.1 *Let $\{x_t\}$ be a real ergodic information source and let a distortion function $\rho : \mathbb{R} \times \mathbb{R} \to \mathbb{R}_+$ be given. This information source can be transmitted over a memoryless channel with capacity $C(P)$ with an average distortion less than or equal to D, that is, there exists a coding policy with* $\lim_{n\to\infty} \frac{1}{n} \sum_{t=0}^{n-1} \rho(x_t, \hat{x}_t) \leq D$, *where $\hat{x}_t$ is the decoder output, only if $R(D) < C(P)$.* ◇

As a consequence of the above result, we obtain a lower bound on the achievable distortion over a channel with capacity $C(P)$ as $D \geq R^{-1}(C(P))$, where R^{-1} denotes the inverse rate-distortion function (known also as the distortion-rate function, see Sect. 12.3.3).

The bound is in fact achievable arbitrarily closely, provided that the encoder and the decoder are allowed to operate with delay, that is block codes are allowed. This is known as the source-channel separation theorem in information theory. However, in the context of delay sensitive real-time systems, the above is typically a very loose bound, as discussed in Sect. 12.4.1. There are very few source and channel pairs which allow the above to be tight. The Gaussian source-channel pair is one such pair; we elaborate on this next.

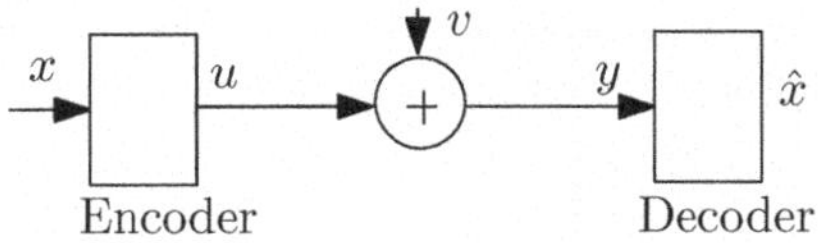

Fig. 16.5 The Gaussian source-channel pair belongs to a special class of pairs where source-channel matching occurs

Optimality and Suboptimality of Linear Policies for Estimation Over a Gaussian Channel

In Sect. 3.3.4, we considered the Generalized Gaussian test channel. One of the findings was that for a scalar Gaussian source transmitted over a scalar Gaussian channel, an optimal encoder is linear, a derivation of which essentially utilized Theorem 16.2.1 above. Let x be a zero-mean real-valued Gaussian random variable transmitted over a scalar Gaussian channel with capacity C, and let $y = u + v$, $u = \gamma_1(x)$, with u being the channel input, v a zero-mean Gaussian channel noise, independent of x and $\hat{x} = \gamma_2(y)$ (see Fig. 16.5).

Let the Gaussian channel have input power constraint P, such that $E[u^2] \leq P$ and the Gaussian channel noise have variance σ_v^2. The capacity then is given by: $C = \frac{1}{2}\log_2(1 + (P/\sigma_v^2))$. In Chap. 3, we saw that the minimum estimation error variance D of a Gaussian source transmitted over a Gaussian channel with capacity C satisfies the following set of inequalities:

$$\begin{aligned} C(P) &= \frac{1}{2}\log_2(1 + (P/\sigma_v^2)) \geq I(u; y) \\ &\geq I(x; \hat{x}) \geq R(D) = \frac{1}{2}\log_2(E[x^2]/D). \end{aligned} \tag{16.20}$$

Furthermore, the above is tight and the minimum attainable distortion is given by $D = E[x^2]/2^{2C}$. This is achieved by a linear scaling of the input. Thus, what is known as the *matching principle* is applicable in this context:

Lemma 16.2.1 ([292]) *The equality $R(D) = C(P)$ holds if and only if*

(a) the distribution P_u of $u = \gamma_1(x)$ achieves the capacity of the channel $P(dy|u)$ with input cost constraint $E[\eta(u)] \leq P$ (for some cost function η),
(b) The conditional distribution $P(d\hat{x}|x)$ with $\hat{x} = \gamma_2(y)$ given x achieves the rate-distortion function of the source x at distortion $D = E[\rho(x, \hat{x})]$,
(c) γ_1 and γ_2 are such that $I(u; y) = I(x; \hat{x})$.

For the Gaussian source and channel case, with $\rho(\cdot, \cdot)$ a quadratic error function, and η a quadratic function, we can take γ_1 and γ_2 to be linear. In this sense, we say that a Gaussian source is matched to a Gaussian channel. We summarize the result for the scalar Gaussian source-channel pair in the following.

Theorem 16.2.2 ([377]) *Consider the minimization of $E[(x - E[x|y])^2]$, subject to a power constraint on the input $E[u^2] \leq P$ over a Gaussian channel considered in Fig. 16.5. The optimal encoder-decoder pair is linear, and the minimum achievable* *cost is given by* $E[x^2]/2^{2C}$. *The encoder applies* $z = \alpha x$, *with* $\alpha^2 = P/E[x^2]$. *The optimal decoder policy is*

$$E[x|y] = \frac{P}{P + \sigma_v^2} \frac{1}{\alpha} y.$$

Multi-Dimensional Source and Channels

We now consider a multi-dimensional source-channel pair. Consider the same setup as above, but now with $u \in \mathbb{R}^m$, $v \in \mathbb{R}^m$ with $u = (u^1, u^2, \cdots, u^m)$. Suppose further that the channel input is constrained to satisfy $E[u'u] \leq P$. The capacity for such a multi-dimensional Gaussian channel is given by what is known as the *water-filling* scheme. Toward that end, let us first define a multi-dimensional (parallel) Gaussian channel as a set of channels

$$y^j = u^j + v^j, \; j = 1, 2, \ldots, m,$$

where $\{v^j\}$ is an independent set of random variables, with each v^j being zero-mean Gaussian with variance N_j. The problem of maximization of the mutual information $I(u; y)$ subject to the power constraint $E[u'u] \leq P$, admits a clean solution, given in the following theorem.

Theorem 16.2.3 ([188]) *The capacity of a multi-dimensional Gaussian channel is given by*

$$\sum_{i=1}^{m} \frac{1}{2} \log_2(1 + \frac{P_i}{N_i}),$$

with

$$P_i = \max(\lambda - N_i, 0),$$

where λ *is chosen such that* $\sum_{i=1}^{m} P_i = P$.

Furthermore, the rate-distortion function for a multi-dimensional source also admits what is known as a *reverse water-filling* solution, as given in the following theorem.

Theorem 16.2.4 ([188]) *Let* x *be an* n*-dimensional, zero-mean Gaussian random vector with independent components, where component* i, x^i, *has a Gaussian distribution with variance* σ_i^2. *Suppose that the distortion criterion is* $\rho(x, \hat{x}) = \sum_{i=1}^{n} (x^i - \hat{x}^i)^2$, *where* $\hat{x}^i$ *denotes the conditional expectation of* x^i. *The rate-distortion function is given by*

$$R(D) = \sum_{i=1}^{n} \frac{1}{2} \log_2(\frac{\sigma_i^2}{D_i}),$$

where

$$D_i = \lambda 1_{\{\lambda < \sigma_i^2\}} + \sigma_i^2 1_{\{\lambda \geq \sigma_i^2\}}$$

and λ *is chosen such that* $\sum_{i=1}^{n} D_i = D$.

Remark 16.2.1 In general, for such multi-dimensional Gaussian source-channel pairs, matching conditions do not hold. Even when the Gaussian source and channel dimensions are the same, matching conditions may not be realized. What is required is that $R(D) = C(P)$ and that the water-filling solutions giving rise to optimal input distributions are such that the rate-distortion achieving test channel is the effective channel from the estimator output to the source (after the encoding and the decoder mappings are performed) itself. To see this aspect, note that linear encoding may well achieve the capacity, but the effective channel from the source to the decoder output also needs to satisfy the rate-distortion constraint, which may not hold. One special case where such a matching holds is the case when the noise and signal power levels are identical in every channel and the distortion criterion is identical for all scalar components [582]. See [703, Section IV.D] for further insight on this problem. ◇

16.2.2 *Optimization of Coding and Control for Linear Systems Controlled Over Noisy Channels*

We now consider a controlled scalar linear system of the form:

$$x_{t+1} = ax_t + bu_t + w_t, \tag{16.21}$$

where $\{w_t\}$ is an i.i.d. sequence of zero-mean Gaussian variables and x_0 is an independent Gaussian random variable.

The encoder transmits its information to a receiver/controller, over an additive Gaussian channel. We assume that the encoder has access to past channel outputs, that is there is noiseless feedback. See Fig. 16.6.

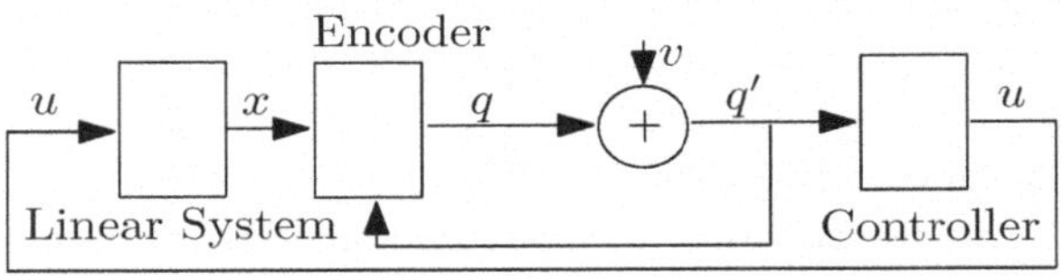

Fig. 16.6 Control over a Gaussian channel

In this context, we mean by a Composite Encoding Policy Π^{comp} a sequence of functions $\{Q_t^{comp}, t \geq 0\}$ which are causal such that the coding output at time t, q_t, under Π^{comp} is generated by a causally measurable function of its local information, that is, a mapping measurable with respect to the sigma-algebra generated by $\mathcal{I}_t^e = \{x_{[0,t]}, q'_{[0,t-1]}\}, \quad t \geq 1$, and $\mathcal{I}_0^e = \{x_0\}$, to $\mathbb{R}$. The setup follows that of Chap. 15 (see Sect. 15.2), except that here the channel is a Gaussian channel. The channel is such that

$$q'_t = q_t + v_t, \tag{16.22}$$

where $\{v_t\}$ is an i.i.d. sequence of zero-mean Gaussian variables with variance σ_v^2. The controller's information at time t is $\mathcal{I}_t^c = \{q'_{[0,t]}\}$. The goal is to minimize

$$E[\sum_{t=0}^{T-1} Qx_t^2 + Ru_t^2], \tag{16.23}$$

with $R > 0, Q > 0$, subject to the constraint that $E[q_t^2] \leq P$, for some power constraint P. We note that if for all time stages, the encoder applies

$$q_t = \alpha_t(x_t - aE[x_{t-1}|\mathcal{I}_{t-1}^c]), \tag{16.24}$$

with $\alpha_t = \frac{P}{E[(x_t - aE[x_{t-1}|\mathcal{I}_{t-1}^c])^2]}$, the estimation error satisfies the recursion

$$E[(x_t - E[x_t|\mathcal{I}_t^c])^2] = \frac{E[a^2(x_{t-1} - E[x_{t-1}|\mathcal{I}_{t-1}^c])^2] + \sigma_w^2}{1 + (P/\sigma_v^2)}.$$

Upon recognizing the capacity expression in the denominator, $C = \frac{1}{2}\log_2(1 + (P/\sigma_v^2))$, we obtain the following: If $C > \log_2(|a|)$, stabilization in the sense of having $\lim_{t\to\infty} E[(x_t - E[x_t|\mathcal{I}_t^c])^2] < \infty$ is possible. This result will be refined in Theorem 16.2.6.

We next consider the optimization problem.

Optimality of Linear Policies

We first note that in the current context there is no dual effect of control, that is, the control actions do not affect the estimation errors for future time stages. The process x_t, conditioned on the past applied control actions is always Gaussian. What is not clear in general, however, is whether conditioning on the past channel outputs leads to a Gaussian measure for the state process under optimal policies. We will consider this in this subsection.

Before we proceed, we present three important lemmas. The first lemma is the following.

Lemma 16.2.2 *Let x be zero-mean Gaussian and $z_{[0,t]}$ be any given collection of random variables for some $t \in \mathbb{Z}_+$. Then,*

$$E[(x - E[x|z_{[0,t]}])^2] \geq E[x^2]2^{-2I(x;z_{[0,t]})}.$$

This inequality is tight if $z_{[0,t]}$ is a Gaussian collection.

Proof By the data-processing inequality (see Lemma 12.3.1) and the fact that conditioning does not increase entropy, it follows that

$$\begin{aligned}
I(x; z_{[0,t]}) &\geq I(x; E[x|z_{[0,t]}]) = h(x) - h(x|E[x|z_{[0,t]}]) \\
&= h(x) - h\left(x - E[x|z_{[0,t]}]\middle|E[x|z_{[0,t]}]\right) \geq h(x) - h(x - E[x|z_{[0,t]}]) \\
&\geq (1/2)\log_2\left(\frac{E[x^2]}{E[(x - E[x|z_{[0,t]}])^2]}\right),
\end{aligned} \tag{16.25}$$

where the last inequality follows from the fact that among all random variables with a given variance, the Gaussian random variable has the largest differential entropy. Tightness for the Gaussian case follows from the following:

$$I(x; z_{[0,t]}) = h(x) - h(x - E[x|z_{[0,t]}]|z_{[0,t]})$$

$$= h(x) - h(x - E[x|z_{[0,t]}]) \tag{16.26}$$

$$= \frac{1}{2}\log_2\left(\frac{E[x^2]}{E[(x - E[x|z_{[0,t]}])^2]}\right) \tag{16.27}$$

Here, (16.26) follows since for a Gaussian collection, $x - E[x|z_{[0,t]}]$ is also Gaussian, and independent of $z_{[0,t]}$. ◇

The second Lemma is the following (see [61, 831]).

Lemma 16.2.3 *Consider the scheme in Fig. 16.7, with $u = \gamma^e(y)$ for a measurable γ^e such that $E[u^2] \leq P$, x is zero-mean Gaussian and v, w are zero-mean Gaussian noise variables with variances σ_v^2, σ_w^2, respectively. Then, $I(x; z)$ is maximized with $u = \gamma^{e,*}(y) = \alpha y$ such that $E[u^2] = P$. Furthermore, under any coding policy γ^e,*

$$E[(x - E[x|z])^2] \geq E[x^2]2^{-2\tilde{C}},$$

where

$$\tilde{C} = \frac{1}{2}\log_2(1 + \frac{\alpha^2 E[x^2]}{\alpha^2\sigma_v^2 + \sigma_w^2}).$$

Under $u = \gamma^{e,*}(y) = \alpha y$, the inequality is an equality.

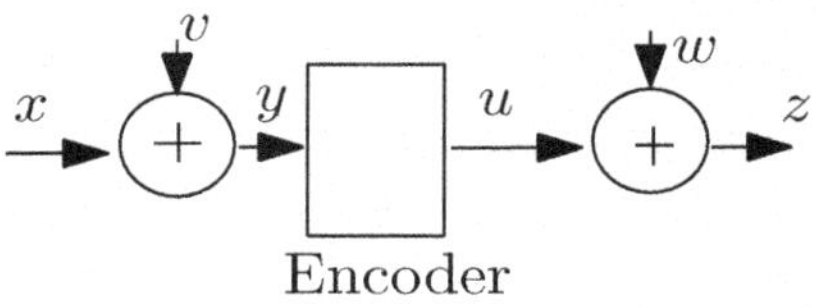

Fig. 16.7 Leak in information through a noisy observation channel

The proof can be established by noting that for any measurable γ^c,

$$E[(x - \gamma^c(z))^2] \geq E[(x - E[x|z])^2]$$
$$= E[(x - E[x|y])^2] + E[(E[x|y] - E[x|z])^2], \quad (16.28)$$

where the equality follows from similar arguments as in (16.18). The term $E[(E[x|y] - E[x|z])^2]$ is minimized by a linear policy in view of (16.20), since $E[x|y]$ is itself a Gaussian random variable. The final important ingredient is the following.

Lemma 16.2.4 *For the linear Gaussian system (16.31) controlled over a Gaussian channel (16.22), the cost (16.23) is equivalent to*

$$E\left[\sum_{t=0}^{T-1} \tilde{R}_t(u_t - \tilde{a}_t x_t)^2\right] + K_0 E[x_0^2] + \sum_{t=0}^{T-1} K_{t+1}\sigma_w^2, \quad (16.29)$$

where for $t = 0, \ldots, T-1$,

$$K_t = Q + a^2 K_{t+1} - \frac{ba^2 K_{t+1}}{K + b^2 K_{t+1}}, \quad \tilde{a}_t = -\frac{baK_{t+1}}{R + b^2 K_{t+1}}, \quad \tilde{R}_t = R + b^2 K_{t+1},$$

with $K_T = 0$.

Hence, the problem reduces to a state estimation problem. Given these lemmas, one can show that optimal coding and control policies are linear, as we state in the following.

Theorem 16.2.5 ([63]) *Consider the linear Gaussian system (16.31) controlled over a Gaussian channel (16.22). For the minimization of (16.23), an optimal encoding policy is of the form:*

$$q_t = \alpha_t(x_t - aE[x_{t-1}|\mathcal{I}_{t-1}^c]), \quad (16.30)$$

with

$$\alpha_t = \frac{P}{E[(x_t - aE[x_{t-1}|\mathcal{I}_{t-1}^c])^2]},$$

and the optimal control policy is given by

$$u_t = \tilde{a}_t E[x_t | \mathcal{I}_t^c],$$

where $\tilde{a}_t$ *is given in (16.29).*

Proof See Sect. 16.5.2. ◇

16.2.3 An Entropy-Power Based Bound for Ergodic Cost Under General Noisy Channels and Tightness for the Gaussian Setup

We now consider again controlled scalar linear system of the form

$$x_{t+1} = ax_t + bu_t + w_t, \tag{16.31}$$

where $\{w_t\}$ is an i.i.d. sequence of zero-mean Gaussian variables with variance σ_w^2, and x_0 is an independent Gaussian random variable.

The encoder transmits its information to a receiver/controller, over a DMC. We assume that the encoder has access to past channel outputs, that is there is noiseless feedback.

In the appendix, we provide a proof for the following result, which is related to Theorem 16.2.5. We point out to the generality of this result in that the bound holds *even if the channel is not Gaussian*, but the Gaussian case will be shown to be tight. The analysis in that case builds on an entropy-power analysis (see Lemma 12.3.2)

Theorem 16.2.6 ([824, Theorem 11.3.2]) *For the scalar linear Gaussian system (16.31) to satisfy*

$$\lim_{t\to\infty} E[x_t^2] \leq d, \tag{16.32}$$

over a memoryless (discrete or continuous) communication channel with noiseless feedback under some policy, the channel capacity must satisfy

$$C \geq \frac{1}{2}\log_2\left(\frac{a^2 d}{d - \sigma_w^2}\right).$$

Furthermore, these inequalities hold with equality for a Gaussian channel and when optimal linear coding and control policies are adopted. Thus, linear policies are optimal.

Remark 16.2.2 As shown in [414, Lemma 3.1], the statement can be revised to replace (16.33) with

$$\limsup_{t\to\infty} \frac{t}{\sum_{k=0}^{t-1} E[x_k^2]} \leq d, \tag{16.33}$$

We also refer the reader to [407, Theorem 4.2] as well as [416, 427] for related results. We also note that one can establish similar results by considering moments other than 2 (see [540] and [491]).

16.2.4 Multi-Dimensional System Case and a Dynamic Program for Optimal Linear Policies

Theorem 16.2.5 does not generalize directly to multi-dimensional sources or channels since source-channel matching needs to take place at every time stage when the channel is used (as discussed in Sect. 16.2.1). This is a restrictive condition in a practical setting. To be more explicit, consider (16.3) To make the ideas more explicit, consider the problem of finding the best linear encoder which minimizes $E[\sum_t \|x_t\|^2]$, with the structure of the encoder being of the form (16.30) above so that

$$q_t = \mathbf{A}_t(x_t - AE[x_{t-1}|\mathcal{I}_{t-1}^c],$$

where $\{\mathbf{A}_t, t \geq 0\}$ are the scaling matrices mapping the estimation errors (or innovations) to the channel input, and are to be optimized.

Under this restriction, the evolution of the estimation error follows a variation of the standard Riccati equation as the estimation error will always be Gaussian leading to a time-varying Kalman filter update equation (see Sect. E.4 in the Appendix). To be more explicit, consider the case where the channel is scalar as in (16.22) but the source is $\mathbb{R}^n$-valued. Now, with $\hat{x}_t = E[x_t|\mathcal{I}_t^c]$ and $\Sigma_t = E[(x_t - \hat{x}_t)(x_t - \hat{x}_t)^T]$, the covariance matrix evolves as (in the Kalman update dynamics)

$$\Sigma_{t+1} = A\Sigma_t A^T - A\Sigma_t \mathbf{A}_t^T(\mathbf{A}_t\Sigma_t\mathbf{A}_t^T + \sigma_v^2)^{-1}\mathbf{A}_t\Sigma_t A^T + W. \tag{16.34}$$

Now, by writing the cost criterion in terms of the covariance matrices, we arrive at the problem of minimization of

$$E[\sum_t \mathrm{Tr}(\Sigma_t)],$$

subject to the evolution of the Kalman update equation (16.34). Under mild conditions on the system, this problem has been studied in[836, Theorem 3] where it is shown that the problem admits a tractable dynamic programming recursion. For generalizations, see [641, Theorem 3.4] (which builds on [836, Theorem 3] where the analysis is restricted to the scalar channel case and which in turn builds on [91],

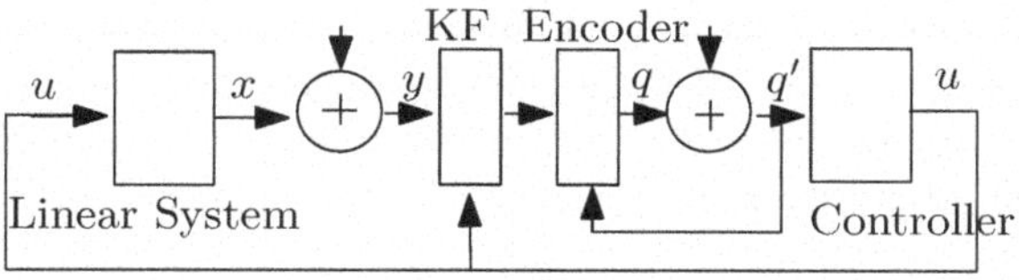

Fig. 16.8 LQG control of a partially observed Linear Gaussian system over a Gaussian channel. Here, the channel encoder has access to the Gaussian channel output at the controller and KF denotes the Kalman Filter. Separation of estimation and coding is optimal as in Theorem 16.2.7

which considers continuous time systems). Such a matrix trace analysis will also be utilized later in the book, in Remark 19.2.5.

Remark 16.2.3 Generalizing the ideas presented for the scalar case and Sect. 16.2.1, via explicit optimality conditions presented in [703] and [688] (see also [427]), sufficient conditions on the system and channel models (at least numerically) on the optimality of linear policies can be obtained.

16.2.5 Structural Results for Optimal Encoders for Partially Observed LQG Problems

Now, we consider the general setup given by (16.1). Building on Theorem 16.17, we have the following result (see Fig. 16.8).

Theorem 16.2.7 *For the minimization of the cost in (16.2) over all coding policies, a coder which uses the Kalman Filter output and the information available at the receiver is as good as any other causal coder.*

A particular application of this result is the encoding problem depicted in Fig. 16.7. Here the encoder uses its estimate without any loss, and as a consequence of (16.28), the optimal encoder is linear in the conditional mean of the state. Thus, one primary message here is that a partially observed Gaussian source is essentially not different from a fully observed Gaussian source.

16.2.6 Revisiting Stabilization: Sufficiency of Shannon Capacity Bounds for Gaussian Channels

Consider an $\mathbb{R}^n$-valued linear Gaussian system of the form

$$x_{t+1} = Ax_t + Bu_t + w_t, \qquad y_t = x_t, \tag{16.35}$$

where we take B to be diagonal. The stabilizability of such a system over a Gaussian channel in the sense of ergodicity or the AMS property follows directly from Chap. 13: By the discussions in Sect. 12.3.1, the capacity of a Gaussian channel can be approximated arbitrarily well by capacities of a sequence of discrete channels obtained through quantization. As a consequence of the results in Sect. 13.5, Shannon capacity being greater than $\sum_{|\lambda_i|>1} \log_2(|\lambda_i|)$ is a sufficient condition for the (AMS) stabilization of the linear system (16.35), where λ_i are the eigenvalues of A. However, not only the AMS property, but also positive Harris recurrence can be established in the special setting of scalar sources and scalar channels through the analysis in the previous section.

In this subsection, we see that this capacity condition is also sufficient for quadratic stability for a large class of settings. In particular, for stabilization across a scalar Gaussian channel with minimum capacity requirements, source-channel matching is not required. This argument is not surprising since the scalar Gaussian channel with noiseless feedback has its error exponent as infinite, see [238] and [651]. In fact, in these papers, Elias [238] and Schalkwijk and Kailath [651] use variations of the linear innovation coding scheme given in (16.24), discussed in the previous section. Thus, as a consequence of Theorem 13.5.4 and arguments in Chap. 13, the infinite exponent implies the existence of coding and control policies which lead to quadratic stability.

However, instead of the constructions in Chap. 13, simpler coding schemes (such as linear time invariant, linear time varying, or memoryless non-linear policies) may also be sufficient in the context of linear Gaussian systems. In the following, we will show that this is the case. Toward this goal, we observe in the following that for a multi-dimensional system controlled over a scalar Gaussian channel, it is sufficient to use linear time-varying coding policies for sequential transmission of scalar components in a Jordan form representation of the system. The proof of the following theorem builds on first stabilizing the lower modes in a Jordan block and then regarding these as noise variables for the upper modes, once one moves up in the Jordan matrix.

Theorem 16.2.8 ([837]) *Consider the n-dimensional Gaussian linear system (16.35) to be controlled over a scalar Gaussian channel with noiseless feedback. If the channel capacity is greater than* $\sum_{|\lambda_i|>1} \log_2(|\lambda_i|)$, *then there exist coding and control schemes such that*

$$\limsup_{t\to\infty} E[|x_t|^2] < \infty.$$

Furthermore, a linear time varying policy is sufficient through sequential linear encoding of scalar components.

The above theorem can be generalized to a class of multi-dimensional sources and multi-dimensional channels through time-varying (periodic) linear policies. What is essential in such a construction is that every scalar mode of a linear system is to be transmitted over at most a single scalar channel at any given time: Consider

m parallel Gaussian channels, each channel with individual rates C_i, $1 \le i \le m$, so that the capacity of the parallel channel is $\sum_{i=1}^{m} C_i$. We have the following.

Theorem 16.2.9 ([838]) *The system (16.35) can be mean square stabilized over m parallel Gaussian channels (having independent noise variables) with feedback, using a linear time varying policy if there exist* $f_{ij} \in \mathbb{Q}$ *such that* $f_{ij} \ge 0$, $\sum_{j=1}^{m} f_{ij} \le 1$, $\sum_{i=1}^{n} f_{ij} \le 1$ *and for all i*

$$\log(|\lambda_i|) < \sum_{j=1}^{m} f_{ij} C_j,$$

where λ_i *are eigenvalues of the system matrix A in (16.35) and* C_i *is the capacity of the i-th channel.*

In Theorem 16.2.9, the rational terms $\{f_{ij}\}$ dictate the period of the *time-sharing* policy for each of the modes.

One question of practical interest is the following: Can one achieve stabilization with time-invariant policies? In general, the answer is clearly negative as the following simple counterexample shows. Consider the system

$$\begin{aligned} x_{t+1} &= Ax_t + Bu_t + w_t, \qquad t \ge 0, \\ y_t &= \gamma^e(x_t) + v_t , \end{aligned} \tag{16.36}$$

where the second equation describes the joint map from the encoder and the channel noise to the output. Consider now, with $|a| > 1$,

$$A = \text{diag}(a, a), \quad \gamma^e(x_t) = Cx_t, \quad C = [\alpha \quad \beta].$$

For this system, independent of the values of α and β, the corresponding noise-free system is unobservable, and therefore the system cannot be stabilized, regardless of the capacity of the channel. For this system a time-varying encoding scheme can be utilized to achieve stability, by transmitting different components of the state at alternating times such that $C_{2t} = [0 \quad \alpha_t]$, $C_{2t+1} = [\beta_t \quad 0]$ for $t \in \mathbb{Z}_+, \alpha_t, \beta_t \in \mathbb{R}$. If the channel is one-dimensional and there is feedback, at every channel use, the channel can carry independent information. Furthermore, the encoder can adjust the signals to be transmitted in a time-varying fashion to ensure that the receiver's estimation error for every (open-loop) unstable mode is stable.

Remark 16.2.4 In [824, Theorem 8.6.7], stabilization of a linear system over a Gaussian forward channel with a noisy Gaussian feedback channel was considered. We had observed that if the forward and the reverse channel capacities satisfy

$$2^{-2C_f} + 2^{-2C_r} - 2^{-2C_f - 2C_r} < 1/a^2,$$

then the steady state variance is finite under linear memoryless policies. In particular, if the reverse channel is noise-free, $C_r = \infty$, the condition reduces to $C_f > \log_2(|a|)$. ◇

16.3 General Information Theoretic Bounds and Methods

As we discussed in detail in Sect. 12.4.1, often information theoretic relaxations are utilized to convexify a non-convex optimal quantization problem, which lead to lower bounds on optimal performance. Operationally, these relaxations are sometimes difficult to convincingly justify, even though some practical setups indeed occur, as we have seen earlier in this chapter. Nonetheless, the performance of such infinite-dimensional relaxations provide a lower bound to the performance of optimal zero-delay/real-time coding and decoding policies. Information theoretic methods can be used at arriving upper bounds on performance as well, as we discuss further below.

16.3.1 Information Theoretic Methods and Design for Upper and Lower Bounds

In the context of optimal zero-delay quantization, a number of methods for obtaining lower and upper bounds are as follows:

(i) Replacing the number of bins constraint with a mutual information constraint, as discussed in Sect. 12.4.1. Notably, $I(x_t; q_t|q_{[0,t-1]}) \leq R$, rather than $\log_2(M) \leq R$ with M being the number of bins. A particular analysis is presented in the proof of Theorem 16.2.5 in Sect. 16.5.2.

(ii) Applying Shannon lower bounding technique (as will be discussed in the proof of Theorem 17.3.4 or Lemma 16.2.2) via the following fact: Among all real random variables V that satisfy a given second moment constraint $E[V^2] \leq P$, the Gaussian maximizes the differential entropy, and the differential entropy in this case is given by $\frac{1}{2}\log(2\pi e P)$. Using this property, a lower bound on conditional entropy $-h(X_t - \hat{X}_t|q_{[0,t]}) \geq -h(X_t - \hat{X}_t)$ can be obtained when the distortion $X_t - \hat{X}_t$ satisfies a second moment constraint.

(iii) Applying the entropy-power inequality lower bounding technique as discussed in Theorem 16.2.6 and its proof. Note that the bound in that theorem holds even if the channel is not Gaussian but is a general memoryless channel.

(iv) [Entropy-Coding with Unbounded Quantization and/or Dithering] Considering upper bounds via entropy-coding and/or dithering methods [839] (see [668], [569] [669] [687]). Dithering is a common-randomness based scheme (see e.g. [622, Section II.B] for a randomized quantization context). Let W be a uniformly distributed random variable with support $[-\frac{\Delta}{2}, \frac{\Delta}{2}]$ and let Q_Δ

be a uniform quantizer with *unbounded* range (i.e., with no over-flow; this unboundedness is to be compensated with entropy coding). Then, let the randomized quantization output for a source X be given by $Q_{\Delta}(X+W)-W$. Then, the expression $N := Q_{\Delta}(X+W)-W-X$ leads to an additive error that is independent of X. In particular, one obtains the following very insightful relation [839]:

$$H(Q_{\Delta}(X+W)|W) = I(X; X+N) = h(Q_{\Delta}(X+W)-W) - h(N)$$

which essentially allows one to view the quantization error as an additive noise model independent of the input signal. Here, it is crucial that W, as common randomness, is known both at the encoder and the decoder and if this is to be implemented in a networked control system, for each time stage, the common randomness needs to be available at the encoder and the decoder. Here N is independent of X and uniformly distributed on $[-\frac{\Delta}{2}, \frac{\Delta}{2}]$. Using ergodicity and invariance properties, [197] has established time-invariant (though still variable-rate) coding schemes using dithering. Avoiding the use of dithering, [426, 427] have presented lattice quantization followed by entropy coding (and thus also with variable rate coding) and established near optimality of such a scheme for high-rates.

(v) [High-rate coding] We note here again that at high rate one can obtain a uniform error approximation without the presence of randomized dithering. Uniform scalar quantizers, in addition to their simplicity, are near-optimal in the limit of high-rate distortion [303] with respect to the mean-square distortion measure. [303] has shown that the uniform quantizer followed by an entropy coder is at most 0.255 bits worse than any optimal quantizer, if one exists, in the context of entropy constrained causal quantization. In the limit of low distortion (high rate), [485] has studied stationary encoding of a stable stationary process and shown that memoryless quantizer followed by a conditional entropy coder is at most 0.25 bits worse than any causal encoder. Near optimality in the context of networked control was also studied in [427] via variable-rate coding of lattice quantized outputs. As reviewed earlier, [414] established near optimality of a two-part fixed rate coding scheme for high rates (to be studied further in Sect. 16.4). We refer the reader also to [407] for high-rate results as well as relations with metric and topological entropy of dynamical systems, going beyond the linear setup (this will be studied in further detail in the following chapter).

(vi) *Sequential rate-distortion theoretic bounds.* A particularly common approach is to minimize the directed information subject to the distortion rate:

$$\limsup_{N\to\infty} \min\{\frac{1}{N} I(X^N \to \hat{X}^N) : \frac{1}{N} E[\sum_{k=0}^{N-1} |X_k - \hat{X}_k|^2] \leq D\}.$$

This leads to an information theoretic lower bound to the optimal estimation error subject to an information rate constraint. There has been a surge of research activity on this problem since [310] where explicit solutions, bounds, as well as convex analytic numerical solutions (including via semi-definite programming) have been presented in [63, 170, 210, 685–687, 702–704, 709].

(vii) *For Noisy Channels: Using channel-source coding separation based methods via the rate-distortion function and Shannon capacity dualities.* As studied in Sect. 12.3.3, the *rate-distortion function*, $R(D)$, of a source is the minimum average amount of *information* that can be transmitted to satisfy a given level of average distortion D. The capacity, $C(P)$, of a channel with cost (power) constraint P is the maximum amount of information which can be transmitted reliably per channel use given an average cost (power) constraint at the channel input. See Theorem 16.2.1 for the utility of an analysis building on the relation $R(D) < C(P)$; and Lemma 16.2.1 for conditions when $R(D) = C(P)$ which can be utilized to establish tightness results, especially for the Gaussian case; as we studied earlier in the current chapter.

When an additive Gaussian channel is present, sequential rate-distortion theoretic ideas presented above in (v) (also via generalizing the scalar Gaussian analysis [63] and the discussion Sect. 16.2.1) lead to explicit optimality conditions studied in [703] and [688] (see also [427]). For the scalar setup, the optimality of linear policies applies as studied in Sect. 16.2.2, building on [63].

We once again caution the reader that the physical realizability of such solutions requires a Shannon theoretic coding-decoding and not a one-shot or finite length coding scheme. A particular instance where such a solution can be *realized optimally* is when a source-channel matching holds, an observation made in the 1960s. As noted above, this was studied earlier in the chapter.

In the following two subsections, we elaborate on the points that while information theoretic methods are useful, often their immediate applications may not lead to sharp bounds, and additionally we argue that linear policies may not be optimal when matching conditions do not apply even for Gaussian source-channel models.

16.3.2 Looseness of Information Theoretic Bounds for Multi-Terminal Models and Counterexamples on Suboptimality of Linear Policies

In the following, we will see that separation results involving optimality of linear policies do not necessarily extend to decentralized settings. We present two counterexamples for Gaussian source-channel pairs. Note that we have already seen that for multi-dimensional channels linear policies may not be optimal.

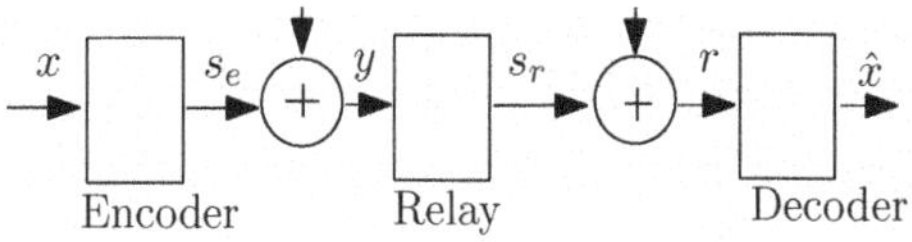

Fig. 16.9 Gaussian relay channel

In particular, the results obtained for scalar sources and channels do not extend to settings where there are two sensors, and to settings when there is a relay in the system. These two are perhaps the most immediate extensions of the scalar setup considered. These, together with Witsenhausen's counter-example considered in Sect. 3.3.2, reveal that optimality of linear policies does not hold for a general class of Gaussian systems, and typically one needs to go beyond the linear structure to find optimal policies.

Gaussian Relay Channels with Two Encoders: Person-by-Person-Optimality of Linear Policies and Lack of Convexity of the Team Problem

Consider the transmission of a Gaussian source over a Gaussian relay channel, as depicted in Fig. 16.9. The relay is an intermediate encoder/decision maker which helps transmit the message from the source to the receiver.

We wish to minimize $E[(x-\hat{x})^2]$ over source encoder and relay encoder policies: We assume that the source x is Gaussian with zero mean and variance σ_x^2. The encoder mapping is γ_e, with $s_e = \gamma_e(x)$ satisfying $\mathbb{E}[s_e^2] \leq P_S$. The transmitted signal s_e is then observed with noise by the relay node as $y = s_e + v_e$, where v_e is a zero-mean independent Gaussian noise of variance N_e. The relay node applies a measurable mapping γ_r on the received signal to produce s_r under the following average relay power constraint: $\mathbb{E}[s_r^2] \leq P_R$. The signal s_r is then transmitted over a Gaussian channel. Accordingly the destination node receives $r = s_r + v_r$, where $\{v_r\}$ is a zero-mean white Gaussian noise with variance N_r. The decoder generates $\hat{x} = g(r)$.

In Lemmas 16.2.2 and 16.2.3, we saw that if the source encoder is restricted to be linear, the optimal relay policy is also linear. On the other hand, if the relay encoder policy is fixed to be linear and memoryless, then the problem becomes equivalent to the transmission of a Gaussian source over a Gaussian channel subject to an average power constraint, for which we already know that linear encoding is optimal. Hence, there exists a person-by-person optimal linear solution to the problem considered in Fig. 16.9 (existence can be seen easily by noting that with encoder, relay and decoder all restricted to linear policies, the best policy in that class can be obtained by minimization of a continuous function on a closed and bounded subset of $\mathbb{R}^3$, and this globally optimal solution over the linear class is person-by-person-optimal over the general class).

We now provide a simple counterexample for the problem depicted in Fig. 16.9 to show that linear policies are not optimal for causal transmission of a Gaussian source over the given relay channel (see [835]). First note that optimal linear policies lead to a cost of $D_L^\star = \sigma_x^2(1 - \frac{P_S P_R}{(P_S+N_e)(P_R+N_r)})$. Consider now the following policies at the source encoder and the relay encoder, respectively:

$$\gamma_e(x) = \begin{cases} a, & \text{for} \quad x > m_1 \\ 0, & \text{for} \quad |x| \leq m_1 \\ -a, & \text{for} \quad x < -m_1 \end{cases}, \qquad \gamma_r(y) = \begin{cases} b, & \text{for} \quad y > m_2 \\ 0, & \text{for} \quad |y| \leq m_2 \\ -b, & \text{for} \quad y < -m_2 \end{cases}$$

where the scalars $a, b, m_1, m_2 \in \mathbb{R}_+$ are to be specified. Under these policies, the signals observed at the relay and the destination are respectively given by

$$y = \begin{cases} a + v_e, & \text{for} \quad x > m_1 \\ v_e, & \text{for} \quad |x| \leq m_1 \\ -a + v_e, & \text{for} \quad x < -m_1 \end{cases}, \qquad r = \begin{cases} b + v_r, & \text{for} \quad y > m_2 \\ v_r, & \text{for} \quad |y| \leq m_2 \\ -b + v_r, & \text{for} \quad y < -m_2 \end{cases}$$

The non-linear policies above have to satisfy the average transmit power constraints. This leads to:

$$a \leq \sqrt{\frac{P_S}{2E\left(\frac{m_1}{\sigma_x}\right)}}, \qquad b \leq \sqrt{\frac{P_R}{2\kappa(m_1, m_2, a, \sigma_x, N_e)}},$$

where

$$\kappa(m_1, m_2, a, \sigma_x, N_e)$$
$$= \left(1 - 2E_q\left(\frac{m_1}{\sigma_x}\right)\right) E_q\left(\frac{m_2}{\sqrt{N_e}}\right) + E_q\left(\frac{m_1}{\sigma_x}\right)\left(E_q\left(\frac{m_2 - a}{\sqrt{N_e}}\right) + E_q\left(\frac{m_2 + a}{\sqrt{N_e}}\right)\right)$$

and $E_q(x) \triangleq \frac{1}{\sqrt{2\pi}} \int_x^\infty e^{-\frac{\tau^2}{2}} d\tau$. Numerical computation [835] leads to the following: Letting $\sigma_x^2 = P_S = P_R = 1$, $N_e = N_d = 4$, $m_1 = 2.45$, and $m_2 = 6.84$, the performance of the given non-linear policy is $D_{NL} = 0.926$, whereas the performance of the best linear policy is $D_L^\star = 0.96$.

A Decentralized Sensing Problem Over Vector Gaussian Channels

As a different setting where optimality of linear policies may not hold, this subsection studies a distributed sensing problem. Consider a two-sensor, single controller, LTI system:

$$\begin{aligned} x_{t+1} &= a x_t + u_t + w_t\,, \\ y_t^i &= x_t + v_t^i\,, \quad i = 1, 2, \end{aligned} \tag{16.37}$$

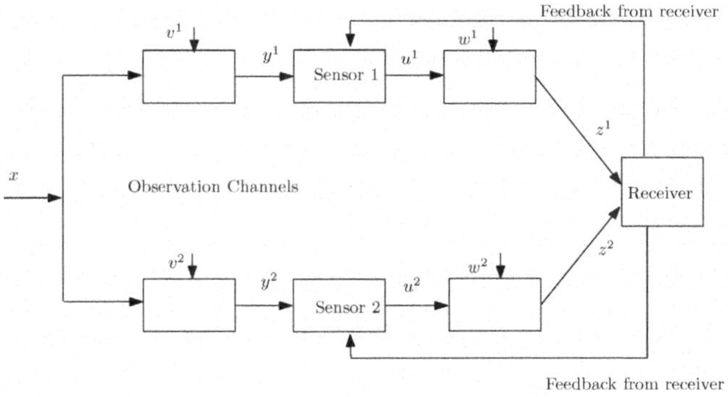

Fig. 16.10 Controller has access to (z_1, z_2)

Here, $x_t \in \mathbb{R}$ is the state of the system with the initial state x_0 a zero-mean Gaussian random variable with variance σ_x^2, $u_t \in \mathbb{R}$ is the control signal, and $y_t^i \in \mathbb{R}$ is the observation available at sensor station i at time t. Here $\{w_t\}$, $\{v_t^1\}$, and $\{v_t^2\}$ are mutually independent, i.i.d. zero-mean Gaussian disturbance processes with variances $\sigma_w^2, \sigma_{v^1}^2$, and $\sigma_{v^2}^2$, respectively.

The sensors transmit their signals at time t, which we denote by $\{u_t^1, u_t^2\}$, over two noisy Gaussian channels where the channel outputs

$$z_t^i = u_t^i + w_t^i, \quad i = 1, 2,$$

are received by the controller. Here w_t^i are zero-mean Gaussian random variables with variances $\{\sigma_{w^i}^2, i = 1, 2\}$. Upon observing the channel outputs, the controller generates its control input u_t.

In the following, we discuss the information structures under which the sensor signals $\{u_t^1, u_t^2\}$ and the control signal u_t are generated. A pictorial description for the sequence of events for a single-stage is presented in Fig. 16.10. Here, the sensor action u_t^i $(i = 1, 2)$ is the output of a mapping γ^i measurable with respect to the sigma-field generated by $\mathcal{I}_t^i = \{y_t^i, \mathbf{z}_t\}$, and is a mapping to $\mathbb{R}$. Here,

$$\mathbf{z}_t = \{\mathbf{y}_{[0,t-1]}, \mathbf{z}_{[0,t-1]}\},$$

is the common past information, where z_t^1, z_t^2 are the information received by the controller. The controller policy γ^0 is measurable with respect to the sigma-algebra generated by $\mathcal{I}_t^c = \{\mathbf{z}_{[0,t-1]}\}$, and is a mapping to $\mathbb{R}$. We also have a power constraint on the sensors: $E[{u_t^i}^2] \leq P^i, \quad i = 1, 2$. The objective is to minimize the following cost function

$$J(\gamma^1, \gamma^2, \gamma^0) = E[\sum_{t=0}^{T} x_t^2 + Q u_t^2], \tag{16.38}$$

for some $Q > 0$, over all admissible sensor and control policies under the information structure presented above.

Theorem 16.3.1 ([832]) *For the minimization of the cost (16.38) under the specified information structure, linear policies are not necessarily optimal. In particular, there exists a non-linear coding scheme which outperforms an optimal linear sensing set of policies for a particular instance of the problem.*

The proof follows by first computing the performance of the best linear sensing policies, and then introduce an alternative sensing scheme which leads to better performance (see [824, Section 11.7.3]).

Looseness of Information Theoretic (Cut-Set) Bounds for Gaussian Networks

We provide now a general discussion on the looseness of the bounds derived from the source-channel separation theorem considered in Theorem 16.2.1. Let us consider the configuration of Fig. 16.7, where v, w are Gaussian noise variables with strictly positive variances. We observed in Lemmas 16.2.2 and 16.2.3 that linear policies are optimal for quadratic error minimization. We note that under such policies

$$I(x; y) - I(x; z) = -h(x|y) + h(x|z) = -h(x|y, z) + h(x|z) = I(x; y|z) > 0$$

Hence, it follows that, the end to end mutual information is less than the mutual information in the first channel, unless the Markov chain condition holds

$$x \leftrightarrow z \leftrightarrow y$$

which implies that there is no information loss with regard to the message at the second channel. Furthermore, it is evident then that

$$I(x; y) < \min(C_1, C_2),$$

where C_1 is the capacity of the additive first channel (where there is no encoding), and C_2 is the capacity of the channel between u and z. As a consequence

$$E[(x - E[x|z])^2] > E[x^2]/2^{2\min(C_1,C_2)}.$$

Likewise, if we consider an encoder in the first channel as in the setup considered in Fig. 16.9, with strictly positive noise variances, we arrive at the following.

Theorem 16.3.2 *The mutual information satisfies:*

$$\max_{P(s_e)} I(s_e; r) \leq \min(\max_{P(s_e)} I(s_e; y), \max_{P(s_r)} I(s_r; r)) < \min(C_3, C_4),$$

where C_3, C_4 denote the Shannon capacities of the additive Gaussian channels between s_e, y and s_r, r, respectively, and $P(s)$ denotes the probability measure on the variable s. Hence, there is a leak in the end-to-end information transmission when the channels are noisy and

$$E[(x - E[x|r])^2] > \frac{E[x^2]}{2^{2\min(C_1,C_2)}}.$$

The above holds despite the fact that, the information theoretic capacity of a block-code is still equal to $C = \min(C_1, C_2)$.

The message here is that such information-theoretic bounds (which are also known as *cut-set* bounds) may only be used as lower bounds on system performance in general. As with the case of optimality of linear policies, tightness of such bounds is also a rare phenomenon and is related to the matching discussion in Lemma 16.2.1.

16.4 An Asymptotically Rate-Optimal Two-Part Stationary Coding Scheme

We now present a coding and control scheme which will be both stabilizing and rate-optimal in the high-rate regime. Our analysis will build on the mathematical results developed earlier in the book and culminate in an efficient two-part coding scheme. The first part, an adaptive coding scheme following the construction in Sect. 13.2.1, will stabilize the system. The second part, with a non-adaptive scheme, will lead to asymptotic order optimality.

To make the analysis efficient, we present the arguments on a simple model. We consider the linear system

$$x_{t+1} = ax_t + bu_t + d_t, \tag{16.39}$$

where $|a| \geq 1$, $b \neq 0$, and d_t is a sequence of i.i.d. Gaussian random variables $d_t \sim \mathcal{N}(0, \sigma^2)$. The state variable x_t is driven by the noise d_t and the aim is to control the state through the action u_t.

The quantizers studied will be of two types. The first one will be what we studied in Fig. 12.1, the uniform quantizer with a single overflow bin: For a given bin size $\Delta > 0$ and even number of bins $M \geq 2$, we define the *modified uniform quantizer* Q_M^Δ as in (13.3): This quantizer uniformly quantizes $x \in \left[-\frac{M}{2}\Delta, \frac{M}{2}\Delta\right]$ into M bins of size Δ and maps all larger x to zero. This requires $M + 1$ output levels.

We will consider two primary applications of this class of quantizers. First, we consider an quantizer, whose bin sizes vary with time, as in Sect. 13.2.1. Fix an even number of bins $K \geq 2$ and let $\{\Delta_t\}_{t=0}^\infty$ be a sequence of strictly positive bin sizes.

In addition, here, we consider a quantizer fixed in time, whose bin size is a function of the number of bins. For a given (even) number of bins $N \geq 2$, we

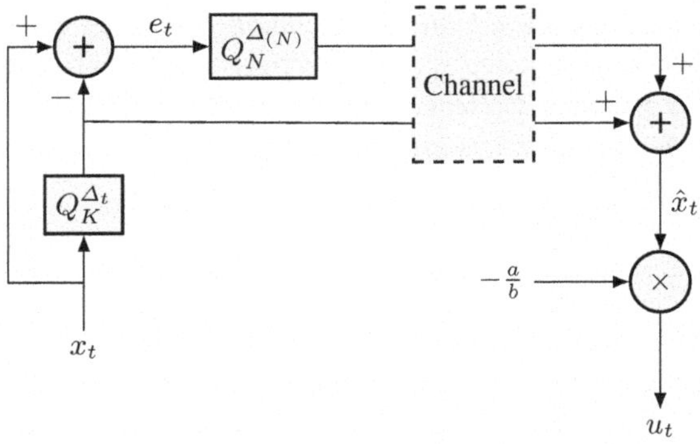

Fig. 16.11 Block diagram of the two-stage coding and control scheme

choose a bin size $\Delta_{(N)}$ and consider the quantizer $Q_N^{\Delta(N)}$. To distinguish this from the adaptive case, we denote $U_N := Q_N^{\Delta(N)}$. We also allow for $N = 0$, for which we write $U_0 \equiv 0$.

We will consider a two-part coding scheme, where the first part is adaptive and the second is fixed. The adaptive part will yield stability, and the fixed part will yield an optimal rate of convergence.

Suppose $\{\Delta_t\}_{t=0}^{\infty}$ is a sequence such that Δ_{t+1} is a function of only Δ_t and the indicator random variable $1_{\left\{|x_t| \leq \frac{K}{2}\Delta_t\right\}}$. Also assume that both the encoder and the decoder (controller) know Δ_0. Then so long as $Q_K^{\Delta_t}(x_t)$ is sent over the channel, it is possible to synchronize knowledge of Δ_t between the quantizer and the controller, since $|x_t| \leq \frac{K}{2}\Delta_t$ if and only if $Q_K^{\Delta_t}(x_t) \neq 0$.

We now describe the proposed communication and control scheme, as depicted in Fig. 16.11. For $\{\Delta_t\}_{t=0}^{\infty}$ as above, we calculate the adaptive quantizer output $Q_K^{\Delta_t}(x_t)$ and the adaptive system error $e_t := x_t - Q_K^{\Delta_t}(x_t)$. Then, using a fixed (i.e., non-adaptive) quantizer U_N with bin size $\Delta_{(N)}$, we calculate the fixed quantizer output $U_N(e_t)$. We then send $Q_K^{\Delta_t}(x_t)$ and $U_N(e_t)$ across the noiseless channel using a total of $C = \log_2(K+1) + \log_2(N+1)$ bits.

The controller then applies the control given by

$$u_t = -\frac{a}{b}\left(Q_K^{\Delta_t}(x_t) + U_N(e_t)\right).$$

This coding and control scheme is illustrated below.

We note that the term $U_N(e_t)$ distinguishes the setup from the analysis in Chap. 13 (see Sect. 13.2.1). This control is chosen to mirror the optimal fully observed control, where $Q_K^{\Delta_t}(x_t) + U_N(e_t)$ is a good estimate of the true state x_t. By (16.39), this results in the state dynamics

$$\begin{aligned} x_{t+1} &= a(x_t - Q_K^{\Delta_t}(x_t) - U_N(e_t)) + d_t \\ &= a(e_t - U_N(e_t)) + d_t. \end{aligned} \tag{16.40}$$

In the case $N = 0$, this reduces to Sect. 13.2.1 with

$$x_{t+1} = a(x_t - Q_K^{\Delta_t}(x_t)) + d_t. \tag{16.41}$$

We briefly motivate the given scheme. For the bin update rules we will describe below, the pair (x_t, Δ_t) forms a Markov chain. The update dynamics (16.41) with no fixed quantization ($N = 0$) ensure that the system is stochastically stable and ergodic, due to results of Sect. 13.2.1.

Through an intricate stochastic stability analysis, we can ultimately prove that the addition of a fixed quantization stage leads to optimal convergence of the second moment with near-optimality in the rate of convergence.

For the adaptive quantizer, we follow the setup as given in (13.7). Then, by Theorem 13.2.3, we have that the state and the quantizer bin process is positive Harris recurrent if $N = 0$. However, it can be shown that this continue to hold even when $N > 0$.

Lemma 16.4.1 *For all $N \geq 0$, $\{(x_t, \Delta_t)\}_{t=0}^{\infty}$ is positive Harris recurrent.*

Now suppose that the system is positive Harris recurrent with fixed parameters $(a, \sigma^2, K, L, \alpha, \delta)$ and given $N \geq 2$ (this is possible so long as $K > |a|$). Let π_N be the resulting invariant measure depending on N, and let $(x_N^*, \Delta_N^*) \sim \pi_N$. Since the adaptive quantization error e_t is a deterministic function of (x_t, Δ_t), this invariant measure will also induce a stationary distribution for e_t, which we denote by $e_N \sim \pi_N^{\text{err}}$.

Lemma 16.4.2 *For all $N \geq 2$, for every initial condition,*

$$\lim_{t\to\infty} E[x_t^2] = E[(x_N^*)^2] = \int \pi_N(dx, \Delta)x^2$$

The following result implies that any rate for the second moment satisfies $r \leq 2$, under *any admissible coding or control policy*.

Lemma 16.4.3 *The following lower bound on $E\left[(x_N^*)^2\right]$ holds (under any admissible coding and control policy):*

$$E\left[(x_N^*)^2\right] - \sigma^2 \geq \frac{a^2\sigma^2}{(K+1)^2(N+1)^2 - a^2}.$$

Proof Sketch This essentially follows from Theorem 16.2.6 ([824, Theorem 11.3.2], via entropy-power inequality) in which the system (16.39) is considered (with $b = 1$) for control across a general memoryless channel of capacity C. Therefore with $C = \log_2(K+1) + \log_2(N+1)$, we obtain

$$\begin{aligned} E\left[(x_N^*)^2\right] - \sigma^2 &\geq \frac{(K+1)^2(N+1)^2\sigma^2}{(K+1)^2(N+1)^2 - a^2} - \sigma^2 \\ &= \frac{a^2\sigma^2}{(K+1)^2(N+1)^2 - a^2} \end{aligned}$$

which completes the proof. ◇

Lemma 16.4.4 ([414]) *Suppose $\{X_N\}_{N=2}^{\infty}$ are random variables satisfying the following uniform moment bound:*

$$E\left[|X_N|^m\right] \leq B_m \quad \textit{for all } m = 1, 2, \ldots \tag{16.42}$$

where $\{B_m\}_{m=1}^{\infty}$ is independent of N. Let $\varepsilon > 0$ and set the bin size for U_N as $\Delta_{(N)} = AN^{-1+\frac{1}{2}\varepsilon}$, for fixed $A > 0$. Then we have

$$E\left[(X_N - U_N(X_N))^2\right] = \mathcal{O}\left(\frac{1}{N^{2-\varepsilon}}\right),$$

i.e., there exists a constant $C > 0$ which depends on ε with

$$E\left[(X_N - U_N(X_N))^2\right] \leq CN^{-2+\varepsilon}, \quad \textit{for all } N \geq 2.$$

To show that we can achieve r arbitrarily close to 2, we will apply Lemma 16.4.4 to the stationary adaptive error e_N, so it will be necessary to show that e_N admits uniformly bounded moments:

Lemma 16.4.5 *The stationary adaptive error $\{e_N\}_{N=2}^{\infty}$ has uniformly bounded moments, i.e.,*

$$E\left[|e_N|^m\right] \leq B_m \quad \textit{for all } m = 1, 2, \ldots$$

for some $\{B_m\}_{m=1}^{\infty}$ independent of N.

Finally, it follows from Lemma 16.4.4 that we can achieve a rate $r \geq 2 - \varepsilon$, for any $\varepsilon > 0$.

Theorem 16.4.1 ([414]) *Choose any $\varepsilon > 0$ and set the bin size for U_N as $\Delta_{(N)} = AN^{-1+\frac{1}{2}\varepsilon}$, for fixed $A > 0$. Then we have*

$$E\left[(x_N^*)^2\right] - \sigma^2 = \mathcal{O}\left(\frac{1}{N^{2-\varepsilon}}\right), \tag{16.43}$$

i.e., there exists $C > 0$ which depends on ε, such that

$$E\left[(x_N^*)^2\right] - \sigma^2 \leq CN^{-2+\varepsilon}, \quad \textit{for all } N \geq 2.$$

Proof Let $(x_N^*, \Delta_N^*) \sim \pi_N$ and let $e_N = x_N^* - Q_K^{\Delta_N^*}(x_N^*)$ so that $e_N \sim \pi_N^{\text{err}}$. Now let x' be the one-step pushforward of x_N^* using e_N, that is

$$x' = a(e_N - U_N(e_N)) + Z, \tag{16.44}$$

where $Z \sim \mathcal{N}(0, \sigma^2)$ is independent of e_N. Since we started under the invariant measure, x' is distributed identically to x_N^*. Furthermore, since $E\left[e_N^2\right]$ is finite by Lemma 16.4.5, it follows from (16.44) that $E\left[(x_N^*)^2\right] = E\left[(x')^2\right]$ is finite. For brevity, let $q_N = e_N - U_N(e_N)$, then by total expectation:

$$
\begin{aligned}
E\left[(x_N^*)^2\right] &= E\left[(x')^2\right] = E\left[E\left[(x')^2 \mid e_N\right]\right] \\
&= E\left[E\left[a^2 q_N^2 + 2aq_N Z + Z^2 \mid e_N\right]\right] \\
&= a^2 E\left[q_N^2\right] + 2aE\left[q_N\right]E\left[Z\right] + E\left[Z^2\right] \\
&= a^2 E\left[(e_N - U_N(e_N))^2\right] + \sigma^2.
\end{aligned}
$$

Therefore, $E\left[(x_N^*)^2\right] - \sigma^2 = a^2 E\left[(e_N - U_N(e_N))^2\right]$. Since e_N has uniformly bounded moments, Lemma 16.4.4 implies that our choice of $\Delta_{(N)}$ causes this MSE to be $\mathcal{O}\left(\frac{1}{N^{2-\varepsilon}}\right)$. The result follows since the optimality gap is linearly related to this MSE. ◇

The analysis can also be generalized to multi-dimensional systems by considering a Jordan block decomposition of the system matrix [414] (see also a related analysis for the first part of the code in Sect. 13.2.2).

16.5 Appendix: Proofs

16.5.1 Proof of Theorem 16.1.2

We will show that *Condition D* used in the proof of Theorem 15.4.2 applies. We need to modify the proof of Theorem 15.4.2 only in view of the unboundedness of the cost, which appears in two places. One is with regard to the continuity of $c(\pi, Q)$ and the other is the weak continuity of the expected value function in the transition kernel. We address both below.

(a) We first establish the continuity of $c(\pi, Q)$. Continuity under total variation can be extended for unbounded functions, provided there is a uniform integrability condition as follows

$$
\lim_{L\to\infty} \sup_{Q_n} \inf_{\gamma} E_\pi^{Q_n}[(x - Q_n(x))'(x - Q_n(x))1_{\{(x-\gamma(Q_n(x)))'(x-\gamma(Q_n(x)))\geq L\}}] = 0.
$$

where by an abuse of notation, the infimization $\inf_\gamma$ is not for the truncated expression

$$
E_\pi^{Q_n}[(x - Q_n(x))'(x - Q_n(x))1_{\{(x-\gamma(Q_n(x)))'(x-\gamma(Q_n(x)))\geq L\}}],
$$

but for the original cost

$$E_{\pi}^{Q_n}[(x - Q_n(x))'(x - Q_n(x))].$$

Now, by the parallellogram law:

$$(x - Q(x))'(x - Q(x)) \leq 2x'x + 2|\sup_x Q(x)|^2$$

As a consequence, for any Q_n, π_n, we have that for some sequence D_n

$$\sup_{\pi_n, Q_n} \inf_{\gamma} \int \pi_n(dx)(x - \gamma(Q_n(x))'(x - \gamma(Q_n(x))1_{\{(x-\gamma(Q_n(x))'(x-\gamma(Q_n(x))\geq L\}}$$

$$\leq \sup_{\pi_n} \inf_{\gamma} \int \pi_n(dx)(2x'x + 2D_n)1_{\{2x'x \geq L - 2D_n\}} \tag{16.45}$$

For every π, Q and every sequence π_n, Q_n converging to π, Q, since the bins converge setwise, so do the minimizing quantizer reconstruction levels in the sense that

$$\int_{B_k^n} \pi_n(dx)x \to \int_{B_k} \pi(dx)x, \quad 1 \leq k \leq M$$

Hence, for some $D < \infty$:

$$\sup_{\pi_n, Q_n} \inf_{\gamma} \int \pi_n(dx)(x - \gamma(Q_n(x))'(x - \gamma(Q_n(x))1_{\{(x-\gamma(Q^(x))'(x-\gamma(Q^(x))\geq L\}}$$

$$\leq \sup_{\pi_n, Q_n} \inf_{\gamma} \int \pi_n(dx)(2x'x + 2D_n)1_{\{2x'x \geq L - 2D_n\}}$$

$$\leq \sup_{\pi_n} \inf_{\gamma} \int \pi_n(dx)(2x'x + 2D)1_{\{2x'x \geq L - 2D\}} \tag{16.46}$$

Hence, one needs to prove that $\{\pi_n\}$ itself is uniformly integrable. If this holds, then, for every ϵ, there exists an L such that for all $(\pi_n, Q_n) \to (\pi, Q)$, it follows that

$$\Big| E_{\pi_n}^{Q}[(x - Q_n(x))'(x - Q_n(x))1_{\{(x-\gamma(Q_n(x)))'(x-\gamma(Q_n(x)))\leq L\}}]$$

$$-E_{\pi_n}^{Q_n}[(x - Q_n(x))'(x - Q_n(x))]\Big| \leq \epsilon/2,$$

and that for sufficiently large n, given L,

$$\left| E_{\pi_n}^{Q_n}[(x - Q_n(x))'(x - Q_n(x))1_{\{(x-\gamma(Q_n(x)))'(x-\gamma(Q_n(x)))\le L\}}] \right. \tag{16.47}$$

$$\left. -E_{\pi}^{Q}[(x - Q(x))'(x - Q(x))1_{\{(x-\gamma(Q(x)))'(x-\gamma(Q(x)))\le L\}}] \right| \le \epsilon/2.$$

Hence, for every $\epsilon > 0$ there exists n_0 such that for all $n \ge n_0$,

$$|E_{\pi_n}^{Q}[(x - Q_n(x))'(x - Q_n(x))] - E_{\pi}^{Q}[(x - Q(x))'(x - Q(x))]| \le \epsilon.$$

Thus, continuity is established under the uniform integrability condition. The following technical lemma addresses the uniform integrability of π_n.

Lemma 16.5.1 ([484]) *Let $\pi_{t,n} \to \pi_t$ be a uniformly integrable sequence. Then, $\pi'(m, \pi_{t,n}, Q_{t,n})$ (defined in (15.38)) is uniformly integrable for $(\pi_{t,n}, Q_{t,n}) \to \pi_t, Q_t$.*

(b) *Continuity of the Value Function in the Quantizer*
We apply backward induction. Let $J_{T-1}^T(\pi_{T-1}) = \min_Q c(\pi_{T-1}, Q_{T-1})$. We have seen in part a) above that for this case the optimal cost function is continuous in π_{T-1}, Q_{T-1}, provided that π_{T-1} varies along a uniformly integrable sequence, and hence by Lemma 15.7.5, the value function J_{T-1}^T is continuous in π_{T-1}. Now, we want to see if

$$J_{T-2}^T(\pi_{T-2}) = \min_Q \left(c(\pi_{T-2}, Q_{T-2}) + E[J_{T-1}^T(\pi_{T-1})|\pi_{T-2}, Q_{T-2}] \right) \tag{16.48}$$

is continuous in (π_{T-2}, Q_{T-2}) (along a uniformly integrable sequence). Lemma 16.5.1 suggests that for $(\pi_{T-2,n}, Q_{T-2,n})$ a uniformly integrable converging sequence, converging to (π_{T-2}, Q_{T-2}), the term $J_{T-1}^T(\pi_{T-1}(m, \pi_{T-2,n}, Q_{T-2,n}))$ also converges for every m, and hence continuity is established. It can be shown that by Lemma 15.7.3, as $Q_n \to Q$, $\|\pi'(m, \pi, Q_n) - \pi'(m, \pi, Q)\|_{TV} \to 0$ for any quantizer Q with M cells of positive measure [484]. Continuity can be established even if the number of cells is less than M, by a bounding argument: if a bin probability decreases to zero, so does the value function (note that an optimal quantizer cannot have fewer than M cells since by splitting a given cell into two leads to the existence of another cell, yet the value function strictly decreases). Thus, in the update equation

$$E[J_{T-1}^T(\pi_{T-1})|\pi_{T-2}, Q_{T-2}]$$
$$= \sum_{m=1}^{M} P(\pi'(m, \pi_{T-2}, Q_{T-2})|\pi_{T-2}, Q_{T-2}) J_{T-1}^T(\pi'(m, \pi_{T-2}, Q_{T-2}),$$

the function $E[J_{T-1}^T(\pi_{T-1})|\pi_{T-2}, Q_{T-2}]$ is continuous in (π_{T-2}, Q_{T-2}). By the continuity of $c(\pi_{T-2}, Q_{T-2})$, we have that $c(\pi_{T-2}, Q_{T-2}) + E[J_{T-1}^T(\pi_{T-1})|\pi_{T-2},$

$Q_{T-2}]$ is continuous in (π_{T-2}, Q_{T-2}), the value function is continuous by Lemma 15.7.5. Thus, (16.48) admits a solution, and

$$J_{T-3}^{T}(\pi_{T-3}) = \min_{Q}\left(c(\pi_{T-3}, Q_{T-3}) + E[J_{T-2}^{T}(\pi_{T-2})|\pi_{T-3}, Q_{T-3}]\right),$$

is continuous in π_{T-3}.

Continuing the same reasoning for the previous time stages, continuity and the existence of an optimal policy follow. ◇

16.5.2 Proof of Theorem 16.2.5

Given Lemma 16.2.4, the goal is to minimize the following expression in (16.29):

$$E\left[\sum_{t=0}^{T-1} \tilde{R}_t(u_t - \tilde{a}_t x_t)^2\right]. \tag{16.49}$$

Note that due to lack of a dual effect, the control actions do not affect the estimation errors. Therefore, the optimal control policies would be of the form $u_t = \tilde{a}_t E[x_t|\mathcal{I}_t^c]$. As a consequence, (16.49) can be regarded as the sum of per-stage estimation errors. Since the control actions do not affect the estimation errors, we can instead consider a control-free process $\bar{x}_t = a\bar{x}_t + w_t$ and let u_t be the estimation of $\tilde{a}_t\bar{x}_t$. Hence, the goal is the minimization of

$$E\left[\sum_{t=0}^{T-1} \tilde{R}_t\left(\tilde{a}_t E[\bar{x}_t|\mathcal{I}_t^c] - \tilde{a}_t\bar{x}_t\right)^2\right].$$

We will first consider a two-stage problem. Note that

$$I(\bar{x}_1; q'_{[0,1]}) = I(\bar{x}_1; q'_1|q'_0) + I(\bar{x}_1; q'_0)$$

Properties of mutual information lead to

$$\begin{aligned} I(\bar{x}_1; q'_1|q'_0) &= h(q'_1|q'_0) - h(q'_1|\bar{x}_1, q'_0) = h(q'_1|q'_0) - h(q'_1|\bar{x}_1, q'_0, Q_1^{comp}(\bar{x}_1, q'_0)) \\ &= h(q'_1|q'_0) - h(q'_1|Q_1^{comp}(\bar{x}_1, q'_0)) \leq h(q'_1) - h(q'_1|Q_1^{comp}(\bar{x}_1, q'_0)) \qquad (16.50) \\ &\leq \frac{1}{2}\log_2(1 + \frac{P}{\sigma_v^2}) =: C, \qquad (16.51) \end{aligned}$$

where C is the Gaussian channel capacity.

Furthermore, $I(\bar{x}_1; q'_0)$ is also maximized by a linear policy at time 0: The relation $\bar{x}_1 = a\bar{x}_0 + w_0$ leads to $E[\bar{x}_0|\bar{x}_1] = \beta_1\bar{x}_1$ for some scalar β_1 so that we can write $z_0 = \beta_1 \bar{x}_1 + \bar{v}_0$ where $\bar{x}_1 \leftrightarrow z_0 \leftrightarrow q'_0$ forms a Markov chain, $\bar{v}_0$ is Gaussian independent of $\bar{x}_1$, and z_0 has the same distribution as $\bar{x}_0$. Therefore, the setup reduces to the setting considered in Lemma 16.2.3 and a linear policy at time 0 also maximizes the mutual information $I(\bar{x}_1; q'_0)$.

As a result, by Lemma 16.2.3, it follows that $I(\bar{x}_1; q'_{[0,1]}) \leq C_1$, where $C_1 = C + \max I(\bar{x}_1; q'_0)$. This upper bound is tight when the optimal encoding policy is of the form

$$q_t = \alpha_t (x_t - aE[x_{t-1}|\mathcal{I}^c_{t-1}]),$$

since Lemma 16.2.3 applies and (16.50) and (16.51) hold with equality since q'_1 is independent of q'_0 and are Gaussian.

By writing

$$I(\bar{x}_2; q'_{[0,2]}) = I(\bar{x}_2; q'_2|q'_{[0,1]}) + I(\bar{x}_2; q'_{[0,1]}) \leq C + \max I(\bar{x}_2; q'_{[0,1]}),$$

the same reasoning applies. Hence, we can obtain an upper bound on $I(\bar{x}_t; q'_{[0,t]}) = I(\bar{x}_t; q'_t|q'_{[0,t-1]}) + I(\bar{x}_t; q'_{[0,t-1]})$, and the recursion is established for all time stages. Hence, the mutual information is maximized for all time stages.

Finally, by Lemma 16.2.2, we observe that the process $\{\bar{x}_t\}$ is zero-mean and

$$E[(\bar{x}_t - E[\bar{x}_t|q'_{[0,t]}])^2] \geq E[\bar{x}_t^2]2^{-2I(\bar{x}_t; q'_{[0,t]})},$$

leading to a tight lower bound on the estimation errors, which is achieved by linear policies. ◇

16.5.3 Proof of Theorem 16.2.6

Let $\sigma^2 = E[w^2]$ and $D_t = E[(x_t + \frac{b}{a}u_t)^2]$, and $d_t = E[x_t^2]$. It follows that $D_t = (d_{t+1} - \sigma^2)/a^2$ for $t \geq 0$. Consider a memoryless channel with inputs q_t and outputs q'_t, and any deterministic control policy. Then,

$$\frac{1}{T}\sum_{t=0}^{T-1} I(x_t; q'_t|q'_{[0,t-1]}) = \frac{1}{T}\sum_{t=0}^{T-1} h(x_t|q'_{[0,t-1]}) - h(x_t|q'_{[0,t]})$$

$$= \frac{1}{T}\sum_{t=0}^{T-1} h(ax_{t-1} + bu_{t-1} + w_{t-1}|q'_{[0,t-1]}) - h(x_t|q'_{[0,t]})$$

$$= \frac{1}{T}\sum_{t=0}^{T-1} h(ax_{t-1} + w_{t-1}|q'_{[0,t-1]}) - h(x_t|q'_{[0,t]})$$

$$
\begin{aligned}
&= \frac{1}{T}\Big(\sum_{t=1}^{T} h(ax_{t-1}+w_{t-1}|q'_{[0,t-1]}) - h(x_{t-1}|q'_{[0,t-1]})\Big)\\
&\qquad +\frac{1}{T}(-h(ax_{T-1}+w_{T-1}|q'_{[0,T-1]})+h(x_0))\\
&\geq \frac{1}{T}\Big(\sum_{t=1}^{T} \frac{1}{2}\log_2(2^{2h(ax_{t-1}|q'_{[0,t-1]})}+2^{2h(w_{t-1})})\\
&\qquad -h(x_{t-1}|q'_{[0,t-1]})\Big)\\
&\qquad +\frac{1}{T}(-h(ax_{T-1}+w_{T-1}|q'_{[0,T-1]})+h(x_0))\\
&= \frac{1}{T}\Big(\sum_{t=1}^{T} \frac{1}{2}\log_2(2^{2h(a(x_{t-1}+(b/a)u_{t-1})|q'_{[0,t-1]})}+2^{2h(w_{t-1})})\\
&\quad -h(x_{t-1}+(b/a)u_{t-1}|q'_{[0,t-1]})\Big)\\
&\qquad +\frac{1}{T}(-h(ax_{T-1}+w_{T-1}|q'_{[0,T-1]})+h(x_0))\\
&\geq \frac{1}{T}\Big(\sum_{t=1}^{T} \frac{1}{2}\log_2(2^{2h(a(x_{t-1}+(b/a)u_{t-1}))}+2^{2h(w_{t-1})}) - h(x_{t-1}+(b/a)u_{t-1})\Big)\\
&\qquad +\frac{1}{T}(-h(ax_{T-1}+w_{T-1}|q'_{[0,T-1]})+h(x_0))\\
&\geq \frac{1}{T}\Big(\sum_{t=1}^{T} \frac{1}{2}\log_2(2\pi e(a^2D_{t-1}+\sigma^2)) - \frac{1}{2}\log_2(2\pi e D_{t-1})\Big)\\
&\qquad +\frac{1}{T}(-h(ax_{T-1}+w_{T-1}|q'_{[0,T-1]})+h(x_0))\\
&= \frac{1}{T}\Big(\sum_{t=1}^{T} \frac{1}{2}\log_2(\frac{(a^2D_{t-1}+\sigma^2)}{D_{t-1}})\Big) + \frac{1}{T}(-h(ax_{T-1}+w_{T-1}|q'_{[0,T-1]})+h(x_0))\\
&= \frac{1}{T}\Big(\sum_{t=1}^{T} \frac{1}{2}\log_2(a^2+\frac{\sigma^2}{D_{t-1}})\Big) + \frac{1}{T}(-h(ax_{T-1}+w_{T-1}|q'_{[0,T-1]})+h(x_0))
\end{aligned}
$$

Here, the first inequality follows from a conditional version of the entropy-power inequality (see Lemma 12.3.2), and the second and the third ones follow from the following argument: $\frac{1}{2}\log_2(a^2e^s+b)-s$, with $b>0$, is a decreasing function of s, hence the entropy term

$$h(x_t|q'_{[0,t]}) = h(x_t + \frac{b}{a}u_t|q'_{[0,t]}) \leq h(x_t + \frac{b}{a}u_t)$$

can be replaced by its maximum, which is achieved by a zero-mean Gaussian random variable with the same second moment: D_t. Hence, for every $T \in \mathbb{N}$, we have that:

$$\frac{1}{T}\sum_{t=0}^{T-1} I(x_t; q'_t|q'_{[0,t-1]})$$

$$\geq \frac{1}{T}\Big(\sum_{t=1}^{T}\frac{1}{2}\log_2(a^2 + \frac{\sigma^2}{D_{t-1}})\Big) + \frac{1}{T}(-h(ax_{T-1} + w_{T-1}|q'_{[0,T-1]}) + h(x_0))$$

$$\geq \frac{1}{2}\log_2\left(a^2 + \frac{\sigma^2}{(\frac{1}{T}\sum_{t=1}^{T} D_{t-1})}\right) + \frac{1}{T}(-h(ax_{T-1} + w_{T-1}|q'_{[0,T-1]}) + h(x_0))$$

where the last line follows from the convexity of $\log(1 + \frac{1}{x})$ in x. Now note that

$$\lim_{T\to\infty}\frac{1}{T}(-h(ax_{T-1} + w_{T-1}|q'_{[0,T-1]}) + h(x_0)) = 0,$$

since

$$\begin{aligned} h(w_{T-1}) &= h(ax_{T-1} + w_{T-1}|x_{T-1}, q'_{[0,T-1]}) \\ &\leq h(ax_{T-1} + w_{T-1}|q'_{[0,T-1]}) \leq h(x_T) < M < \infty, \end{aligned} \quad (16.52)$$

for some $M \in \mathbb{R}_+$ (by the assumption that a limit exists for $E[x_t^2]$, and hence a uniform upper bound exists for the entropy sequence $\{h(x_t)\}$). Now, following the analysis in the proof of Theorem 13.6.2, a memoryless channel is such that its capacity C satisfies

$$\begin{aligned} C &\geq \limsup_{T\to\infty}\frac{1}{T}\sum_{t=0}^{T-1} I(x_t; q'_t|q'_{[0,t-1]}) \\ &\geq \limsup_{T\to\infty}\frac{1}{2}\log_2\left(a^2 + \frac{\sigma^2}{(\frac{1}{T}\sum_{t=0}^{T-1} D_t)}\right) \geq \frac{1}{2}\log_2\left(a^2 + \frac{\sigma^2}{D}\right), \end{aligned} \quad (16.53)$$

where the last inequality follows from the continuity of $\log_2(1 + \frac{1}{x})$ for $x > 0$.

Hence, a lower bound on the capacity is expressed as $C \geq \frac{1}{2}\log(\frac{(a^2 D + \sigma^2)}{D})$ for the second moment to be less than d. This is tight for a Gaussian channel by linear encoding and control policies. ◇

16.6 Concluding Remarks

This chapter has studied optimization of a linear system over noiseless, Gaussian, and general distribution channels, and the conditions for optimality under quadratic cost minimization. We have established optimal coding and control for linear systems under quadratic criteria in the presence of information constraints. We established separation results between optimal quantization and control for general channels. We established conditions for optimality of linear coding and control policies for the special Gaussian setup.

In the context of noise-free channels, or noisy channels with feedback, one primary structural result in this chapter was a (triple-)separation principle between optimal estimation, optimal quantization and finally optimal control. We also established existence and near-optimality results. In the context of coding and control over Gaussian channels, we have presented information theoretic results on communicating a Gaussian source over a Gaussian channel, and obtained structural results for optimal encoders and controllers. We have obtained conditions under which linear encoding and control policies are optimal, and identified situations where such an optimality does not hold. Finally, we have presented a comprehensive analysis of information theoretic relaxations. Section 16.3 presented a general summary of several information theoretic methods and their implications for the general problem considered in this chapter.

In the chapter, we have also presented counterexamples to optimality of linear policies for two basic classes of Gaussian channels. The message is that when one departs from the setting of the transmission of scalar Gaussian sources over scalar Gaussian channels, linearity of optimal policies is a rather rare event unless *matching* occurs.

16.7 Bibliographic Notes

There is a large literature on joint optimal quantization for the LQG problem dating back to early 1960s (see for example [474] and [195]). In this literature, references [65, 138, 264, 516, 554, 709] and [275] have considered the optimal LQG quantization and control design with various results on the optimality or the lack of optimality (with detailed comparisons reported in [554] and [275]) of the separation principle with different assumptions in the setups and various (sometimes inconsistent) conclusions on the structural properties of optimal policies.

The complexity of the problem and the intricacies it entails were, however, not appreciated universally in the literature, which unfortunately led to several inconsistencies. See [815] and [817] for a detailed discussion on some subtleties.

The results in this chapter, building on [815, 817] further refine the existing structural and separation results in [65, 138, 195, 264, 275, 474, 516, 554, 593, 709].

A parallel line of consideration which is of a rate-distortion theoretic nature is the *sequential-rate distortion* proposed in [709], and the *feedforward* setup, which has been investigated in [742] and [231].

It appears that the idea of using information theoretic analysis to study communication constraints in decentralized control was first investigated by Teneketzis in his PhD dissertation [712].

Regarding Gaussian source-channel pairs with noiseless feedback, optimality of linear coding policies for the scalar setting has been known since 1960s, see [238, 377] and [848]. The transmission over scalar Gaussian channels has been considered also [238], [651] and [286], where the error exponents have been shown to be unbounded (and the error probability has been shown to decrease at least doubly exponentially in the block-length). Such a strong result, however, does not hold when there is noisy feedback [423]. The case where there is a mismatch in the source and the channel is a challenging one. Partial results are known regarding optimal zero-delay policies. Regarding the source-channel matching principle, we refer the reader to [110, 194, 292, 707] and [709] for related discussions. For Gaussian source and channel pairs, multi-dimensional source and multi-channel systems have been considered in [468] and [582], and optimal linear encoders have been studied in [84], [82, 468] and [9]. Partially observed settings have been considered in [47, 111, 219], and [262]. Gastpar [291] has considered various settings of multi-access and broadcast channels. A comprehensive discussion and literature review is available in [837]. On the relationship between the minimum mean-square estimation error and mutual information, we further mention the following results (which exhibit that the relation goes beyond the matching principle considered earlier). Duncan [226] has observed that for a general continuous-time stochastic process x_t (not necessarily Gaussian), observed through an additive Gaussian channel in the form $dy_t = x_t dt + dw_t$, with $\{w_t\}$ denoting a standard Brownian process, the estimation error satisfies a relationship of the form $(1/2)\int_0^T E[(x_t - E[x_t|y_{[0,t]}])^2]dt = \int_0^T I(x_{[0,t]}; y_{[0,t]})dt$, with the integration involving a continuous-time generalization of mutual information. Weissman, Kim and Permuter [765] considered an extension of this result involving settings where the process x_t is affected by feedback, and showed that the result holds provided that directed information replaces mutual information. Khina et. al. [416] have considered the setting with erasure channels under feedback.

Along similar lines, Başar and Bansal [91] have considered the continuous-time version of the optimal coding and control problem for quadratic cost functions, establishing an analogous result to Theorem 16.2.5 and the optimality of linear policies for scalar Gaussian source-channel pairs. Functional and topological properties of MMSE decoding have been considered in [786]. References [667, 709] and [169] discussed this for vector sources in a control setting. References [169] and [61] consider the partially observed formulation.

For stabilization of vector sources over scalar and parallel channels, linear time-invariant encoding schemes have been considered in [145, 667] for the case without system/process noise and it was established that for scalar channels such policies are stabilizing for a large class of systems in view of rate requirements provided that the system is observable under a time-invariant policy. Stronger results on linear time-invariant policies have been established for scalar systems in [274]. For noise-free settings, [439] and [441] have considered non-linear policies to establish the sufficiency of capacity bounds for quadratic stabilization.

A counterexample to optimality of linear policies over relay channels was presented in [486] for a sequence of relay channels with the number of relays being greater than 2, [835] generalized this also to the case of a single relay. When there is no relay, linear policies are optimal as the problem is one involving the Gaussian test channel. Sufficient conditions for stabilization of linear systems over relay channels has been considered also in [440]. [684] is a recent study considering linear systems and Gaussian channels in the presence of side information.

We have seen in the chapter that optimal (encoding) policies may not be linear among all possible policies when the source or channels are not scalar. Nonetheless, it may be of practical interest to obtain the best linear policies. A dynamic programming over the space of matrices can be carried out leading to a family of non-linear optimization problems. Such an analysis has been carried out in [91] for continuous-time systems.

We also refer the reader to Sect. 16.3 for related studies on information theoretic methods involving zero-delay coding in the context of linear systems under quadratic distortion criteria. Several of the results in this chapter have appeared in some form in [61, 63, 815, 817, 832, 835, 837, 838] and [836].

Chapter 17
Optimal Coding Under Small Estimation Error and Observability Criteria

Chapter 15 studied the optimal causal coding/quantization problem for Markov sources (control-free or controlled), which were further refined for LQG models in Chap. 16. These chapters presented structural and existence, as well as approximation, results on optimal causal coding of a large class of Markov sources in a variety of settings.

Some applications in estimation and control, such as safety-critical ones, require precision and accuracy, leaving very little room for errors. Accordingly, it is important to be able to further refine structural and performance bounds for such applications. We will undertake such a study in this chapter, which will also let us appreciate the differences between deterministic systems and stochastic ones.

Accordingly, in this chapter, we will first consider the problem of zero-delay coding of a deterministic dynamical system over a discrete noiseless channel under three estimation criteria concerned with the low-distortion regime. For these three criteria, formulated stochastically in terms of a probability distribution for the initial state, we characterize the smallest channel capacities above which the estimation objectives can be achieved. The results establish further connections between topological and metric entropies of deterministic dynamical systems and information theory. We then address these issues for stochastic dynamical systems over communication channels under the same three estimation criteria.

These criteria considered are (i) a strong and (ii) a weak form of almost sure stability of the estimation error, as well as (iii) asymptotic quadratic stability in expectation. For all three objectives, we derive lower bounds on the smallest channel capacity C_0 above which the objective can be achieved with an arbitrarily small error.

For the stochastic setup, we will obtain bounds through a dynamical systems approach by constructing an infinite-dimensional dynamical system and relating the capacity with the topological and the metric entropy of this system. We also consider information-theoretic and probability-theoretic approaches to address the different criteria. Finally, we prove that a memoryless noisy channel in general constitutes no

S. Yüksel, T. Başar, *Stochastic Teams, Games, and Control under Information Constraints*, Systems & Control: Foundations & Applications,
https://doi.org/10.1007/978-3-031-54071-4_17

obstruction to asymptotic almost sure state estimation with arbitrarily small errors, when there is no noise in the system. The results provide new solution methods for the criteria introduced and establish further connections between dynamical systems, networked control, and information theory, especially in the context of nonlinear stochastic systems.

In Sect. 16.3.1 of the previous chapter, we discussed several high-rate coding results which give rise to small distortions. In this chapter, we will present further and more general results, also utilizing tools and concepts from the theory of dynamical systems. As some of the results in this chapter will build on concepts from the theory of dynamical systems and the notions of topological and metric entropies; an overview on relevant concepts from dynamical systems can be found in Appendix H.

In the following section the general model to be studied is presented. The analysis for optimal coding of deterministic dynamical systems is presented in Sect. 17.2. The stochastic counterpart is studied in Sect. 17.3. Comparison of the results between the deterministic and stochastic setups and relations between the various estimation criteria considered are discussed in detail in Sect. 17.4.

17.1 Introduction

Consider stochastic systems described by difference equations of the form

$$x_{t+1} = f(x_t, w_t), \tag{17.1}$$

where x_t is the state at time t and $(w_t)_{t\in\mathbb{Z}_+}$ is an i.i.d. sequence of random variables with common distribution $w_t \sim \nu$, modeling the noise. In general, we assume that

$$f : \mathbb{X} \times W \to \mathbb{X}$$

is a Borel measurable map, where $(\mathbb{X}, d)$ is a complete metric space and W a measurable space, so that for any $w \in W$ the map $f(\cdot, w)$ is a homeomorphism of $\mathbb{X}$. We further assume that x_0 is a random variable on $\mathbb{X}$ with an associated probability measure π_0, independent of $(w_t)_{t\in\mathbb{Z}_+}$. We use the notations

$$f_w(x) = f(x, w), \qquad f^x(w) = f(x, w)$$

so that $f_w : \mathbb{X} \to \mathbb{X}$ and $f^x : W \to \mathbb{X}$.

The system (17.1) is connected over a possibly noisy channel with finite capacity to an estimator, as shown in Fig. 17.1. The estimator has access to the information it has received through the channel. A source coder maps the source symbols (i.e., state values) to corresponding channel inputs. These inputs are transmitted over the channel, which we assume to be discrete with input alphabet $\mathcal{M}$ and output alphabet $\mathcal{M}'$.

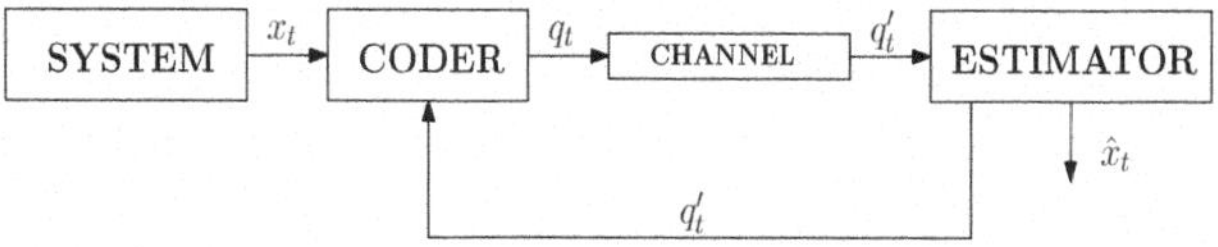

Fig. 17.1 Coding and state estimation over a noisy channel with feedback

As before, we refer by a *coding policy* Π, to a sequence of functions $(\gamma_t^e)_{t \in \mathbb{Z}_+}$ which are causal such that the channel input at time t, $q_t \in \mathcal{M}$, under Π is generated by a function of its local information, i.e.,

$$q_t = \gamma_t^e(I_t^e),$$

where $I_t^e = \{x_{[0,t]}, q'_{[0,t-1]}\}$ and $q_t \in \mathcal{M}$, the channel input alphabet given by $\mathcal{M} = \{1, 2, \ldots, M\}$, for $0 \le t \le T-1$. Here, we use the notation $x_{[0,t-1]} = \{x_s : 0 \le s \le t-1\}$ for $t \ge 1$.

The channel maps q_t to q'_t in a stochastic fashion so that $P(q'_t | q_t, q_{[0,t-1]}, q'_{[0,t-1]})$ is a conditional probability measure on $\mathcal{M}'$ for all $t \in \mathbb{Z}_+$. If this expression is equal to $P(q'_t|q_t)$, the channel is said to be memoryless, i.e., the past variables do not affect the channel output q'_t given the current channel input q_t.

The receiver, upon receiving the information from the channel, generates an estimate $\hat{x}_t$ at time t, also causally: An admissible causal *estimation policy* is a sequence of functions $(\gamma_t^d)_{t \in \mathbb{Z}_+}$ such that $\hat{x}_t = \gamma_t^d(q'_{[0,t]})$ with

$$\gamma_t^d : (\mathcal{M}')^{t+1} \to \mathbb{X}, \quad t \ge 0.$$

In the following, we first consider the special case with deterministic dynamics (in the absence of noise), and then the stochastic setup introduced above.

17.2 Coding for Deterministic Dynamical Systems in the Small Error Regime

17.2.1 The Problem

Let $f : \mathbb{X} \to \mathbb{X}$ be a continuous map with a continuous inverse (i.e., a homeomorphism) on a compact metric space $(\mathbb{X}, d)$ (where the compactness will be relaxed later on). We consider the dynamical system generated by f, i.e.,

$$x_{t+1} = f(x_t), \quad x_0 \in \mathbb{X}, \; t \in \mathbb{Z}.$$

The space $\mathbb{X}$ is endowed with a Borel probability measure π_0 which describes the uncertainty in the initial state.

Suppose that an encoder translates the state values into symbols from a finite coding alphabet $\mathcal{M}$ and sends this information over a noiseless digital channel to an estimator. The input signal sent at time t through the channel is denoted by q_t. At the other end of the channel, the estimator generates an estimate $\hat{x}_t \in \mathbb{X}$ of x_t, using the information it has received through the channel up to time t. We consider arbitrary (but causal / zero-delay) coding and estimation policies, as introduced in Chap. 15. That is, q_t is generated by a (measurable) map

$$q_t = \gamma_t^e(x_0, x_1, \ldots, x_t), \quad \gamma_t^e : \mathbb{X}^{t+1} \to \mathcal{M}, \tag{17.2}$$

and the estimator output at time t is given by a map

$$\hat{x}_t = \gamma_t^d(q_0, q_1, \ldots, q_t), \quad \gamma_t^d : \mathcal{M}^{t+1} \to \mathbb{X}. \tag{17.3}$$

We assume in this subsection that the channel is noiseless with finite input alphabet $\mathcal{M}$. Hence, its capacity is given by

$$C = \log_2 |\mathcal{M}|.$$

Our aim is to characterize the smallest capacity C_0 above which one of the following estimation objectives can be achieved for every $\epsilon > 0$.

(E1) Eventual almost sure estimation: There exists $T = T(\epsilon) \in \mathbb{N}$ such that for some distortion metric d on the state space $\mathbb{X}$,

$$P(d(x_t, \hat{x}_t) \leq \epsilon) = 1 \quad \text{for all } t \geq T.$$

(E2) Asymptotic almost sure estimation:

$$P\Big(\limsup_{t\to\infty} d(x_t, \hat{x}_t) \leq \epsilon\Big) = 1.$$

(E3) Asymptotic estimation in expectation:

$$\limsup_{t\to\infty} E[d(x_t, \hat{x}_t)^p] \leq \epsilon, \quad p > 0.$$

The smallest channel capacity above which one of these objectives can be achieved for a fixed $\epsilon > 0$ will be denoted by C_ϵ. We are particularly interested in

$$C_0 := \lim_{\epsilon \downarrow 0} C_\epsilon.$$

In most of our results we will additionally assume that π_0 is f-invariant, i.e., $\pi_0 = \pi_0 \circ f^{-1}$. In some cases, we will need the stronger assumption that π_0 is ergodic.

Given a pair of encoding and decoding policies, the estimate $\hat{x}_t$ at time t is determined by the initial state x_0, e.g., $\hat{x}_1 = \gamma_1^d(\gamma_0^e(x_0), \gamma_1^e(x_0, f(x_0)))$. To express this, we sometimes write $\hat{x}_t(x_0)$. In terms of the map f and the measure π_0, the estimation criteria described above can be reformulated as follows.

Eventual almost sure estimation:

$$\pi_0(\{x_0 \in \mathbb{X} : d(f^t(x_0), \hat{x}_t(x_0)) \leq \epsilon\}) = 1 \quad \text{for all } t \geq T. \tag{17.4}$$

Asymptotic almost sure estimation:

$$\pi_0(\{x_0 \in \mathbb{X} : \limsup_{t \to \infty} d(f^t(x_0), \hat{x}_t(x_0)) \leq \epsilon\}) = 1. \tag{17.5}$$

Asymptotic estimation in expectation:

$$\limsup_{t \to \infty} \int_{\mathbb{X}} \pi_0(\mathrm{d}x_0) d(f^t(x_0), \hat{x}_t(x_0))^p \leq \epsilon. \tag{17.6}$$

It is easy to see that (17.4) implies both (17.5) and (17.6), and in turn, (17.5) implies (17.6) due to the compactness of $\mathbb{X}$. We will see that this order is monotonically reflected in the conditions required on the information transmission rates for the criteria to be satisfied.

17.2.2 Eventual Almost Sure Estimation

For the estimation objective (E1) we can characterize C_0 as follows.

Theorem 17.2.1 ([406]) *Assume that π_0 is f-invariant and $h_{\text{top}}(f; \operatorname{supp} \pi_0) < \infty$. Then, the smallest channel capacity above which (E1) can be achieved for every $\epsilon > 0$ satisfies*

$$h_{\text{top}}(f; \operatorname{supp} \pi_0) \leq C_0 \leq \log(1 + \lfloor 2^{h_{\text{top}}(f; \operatorname{supp} \pi_0)} \rfloor).$$

The value $+\infty$ for $h_{\text{top}}(f; \operatorname{supp} \pi_0)$, which is excluded in the above theorem, refers to the case when the estimation objective cannot be achieved via a finite-capacity channel.

17.2.3 Asymptotic Almost Sure Estimation

We next address the asymptotic estimation objective (E2). As it turns out, the associated C_0 is always bounded below by the metric entropy $h_{\pi_0}(f)$ when π_0 is f-invariant. Since (E1) implies (E2), the topological entropy on $\operatorname{supp} \pi_0$ still provides an upper bound on C_0. Under an additional condition for the measure-theoretic dynamical system $(\mathbb{X}, f, \pi_0)$, we are able to improve the upper bound by substituting $h_{\pi_0}(f)$ in the place of $h_{\text{top}}(f; \text{supp}\ \pi_0)$.

Theorem 17.2.2 ([406]) *Assume that π_0 is f-invariant. Then the smallest channel capacity above which (E2) can be achieved for every $\epsilon > 0$ satisfies*

$$C_0 \leq \log(1 + \lfloor 2^{h_{\text{top}}(f;\text{supp}\ \pi_0)} \rfloor).$$

Proof This follows from Theorem 17.2.1, since (E1) is stronger than (E2).

We also have the following result.

Theorem 17.2.3 ([406]) *Assume that π_0 is an ergodic measure for f. Furthermore, assume that there exists a finite Borel partition $\mathcal{P}$ of $\mathbb{X}$ of arbitrarily small diameter such that $h_{\pi_0}(f) = h := h_{\pi_0}(f; \mathcal{P})$, and*

$$\mathit{lim\ sup}_{n\to\infty} \frac{1}{n} \log \pi_0\Big(\big\{x : \big| -\frac{1}{n} \log \pi_0(\mathcal{P}^n(x)) - h\big| > \delta\big\}\Big) < 0 \tag{17.7}$$

for all sufficiently small $\delta > 0$, where $\mathcal{P}^n(x)$ denotes the unique element of $\mathcal{P}^n$ containing x. Then the smallest channel capacity C_0 above which (E2) can be achieved for every $\epsilon > 0$ satisfies

$$C_0 \leq \log(1 + \lfloor 2^{h_{\pi_0}(f)} \rfloor).$$

To provide a lower bound on C_0, we have the following theorem.

Theorem 17.2.4 ([406]) *Assume that π_0 is f-invariant. Then, the smallest channel capacity above which (E2) can be achieved for every $\epsilon > 0$ satisfies,*

$$C_0 \geq h_{\pi_0}(f). \tag{17.8}$$

If π_0 is not invariant, but there exists an invariant measure π_* which is absolutely continuous with respect to π_0, then

$$C_0 \geq h_{\pi_*}(f). \tag{17.9}$$

In the above, the cases $h_{\pi_0}(f) = \infty$ and $h_{\pi_*}(f) = \infty$, respectively, are included. They correspond to the situation, when the estimation objective cannot be achieved over a finite-capacity channel.

17.2.4 Asymptotic Estimation in Expectation

Finally, we consider the estimation objective (E3). As we will see, under mild assumptions on the admissible coding and estimation policies, C_0 is characterized by the metric entropy $h_{\pi_0}(f)$.

Theorem 17.2.5 ([406]) *Assume that π_0 is an ergodic measure for f. Then, the smallest channel capacity above which (E3) can be achieved for every $\epsilon > 0$ satisfies,*

$$C_0 \leq \log(1 + \lfloor 2^{h_{\pi_0}(f)} \rfloor).$$

To obtain the analogous lower bound, we need to introduce some notation. Let μ be an ergodic f-invariant Borel probability measure on $\mathbb{X}$. For $x \in \mathbb{X}$, $n \in \mathcal{N}$, $\epsilon > 0$ and $r \in (0, 1)$ define the sets

$$B(x, n, \epsilon, r) := \Big\{y \in \mathbb{X} : \frac{1}{n}\big|\{0 \leq k < n : d(f^k(x), f^k(y)) \leq \epsilon\}\big| > 1 - r\Big\}.$$

A set $F \subset \mathbb{X}$ is called (n, ϵ, r, δ)-spanning for $\delta \in (0, 1)$ if

$$\mu\Big(\bigcup_{x \in F} B(x, n, \epsilon, r)\Big) \geq 1 - \delta.$$

We write $r_{\text{span}}(n, \epsilon, r, \delta)$ for the minimal cardinality of such a set and define

$$h_r(f, \mu, \epsilon, \delta) := \limsup_{n \to \infty} \frac{1}{n} \log r_{\text{span}}(n, \epsilon, r, \delta).$$

According to [599, Thm. A], for every $\rho \in (0, 1)$ we have

$$\lim_{r \downarrow 0} \lim_{\epsilon \downarrow 0} h_r(f, \mu, \epsilon, \rho) = h_\mu(f).$$

Since $h_r(f, \mu, \epsilon, \rho)$ is increasing as $r \downarrow 0$ and as $\epsilon \downarrow 0$, we can write this as

$$h_\mu(f) = \sup_{r>0} \sup_{\epsilon>0} h_r(f, \mu, \epsilon, \rho) = \sup_{(r,\epsilon) \in \mathbb{R}^2_{>0}} h_r(f, \mu, \epsilon, \rho).$$

In particular, this implies

$$h_\mu(f) = \lim_{\epsilon \downarrow 0} h_{\varphi_1(\epsilon)}(f, \mu, \varphi_2(\epsilon), \rho) \tag{17.10}$$

for any functions φ_1, φ_2 with $\lim_{\epsilon \downarrow 0} \varphi_i(\epsilon) = 0, i = 1, 2$.

In the following theorem, we will restrict ourselves to periodic coding and estimation schemes using finite memory.

Theorem 17.2.6 ([406]) *Assume that π_0 is an ergodic measure for f. Additionally, assume that there exists $\tau > 0$ so that the coder map γ_t^e is of the form*

$$q_t = \gamma_t^e(x_{[t-\tau+1,t]}),$$

and $\gamma_{t+\tau}^e \equiv \gamma_t^e$. Further assume that the estimator map is of the form

$$\hat{x}_t = \gamma_t^d(q_{[t-\tau+1,t]}),$$

and also $\gamma_{t+\tau}^d \equiv \gamma_t^d$. Then, the smallest channel capacity above which (E3) can be achieved for every $\epsilon > 0$ satisfies

$$C_0 \geq h_{\pi_0}(f).$$

Remark 17.2.1 In Chap. 15, we studied structural results on optimal zero-delay codes. A notable result (e.g. from Theorem 15.3.1 and its generalizations) was that an optimal encoder only uses the most recent state and some information variable (available at both the encoder and the decoder, such as the past transmitted symbols or a conditional measure on the state given the information at the decoder as in Theorem 15.3.2). For the average cost setup, Theorems 15.5.1 and 15.5.2 established the optimality of stationary and deterministic policies under such a structure. On the other hand, Theorems 15.5.1 and 15.5.2 also showed that for finitely valued Markov sources which are aperiodic, irreducible, and stationary, periodic coding policies (as those considered in Theorem 17.2.6) indeed perform arbitrarily well for the zero-delay setup also (with a rate of convergence of the order $\frac{1}{T}$ with T being the period). Likewise, for a stable linear system under a quadratic cost criterion, Theorem 15.5.3 established the same. For non-causal codes related results exist via asymptotic optimality of block codes [314]: In particular, for stationary and memoryless sources, a rate of convergence for such a result has been shown to be of the type $O(\frac{\log T}{T})$ [581, 841]. See also [428] for a detailed literature review and further finite blocklength performance bounds. Thus, the use of finite memory policies in Theorem 17.2.6 is a practical construction.

17.3 Coding for Stochastic Dynamical Systems in the Small Error Regime

We now consider the more general stochastic setup considered in Sect. 17.1.

For a given $\epsilon > 0$, we denote by C_ϵ the smallest channel capacity above which there exist an encoder and an estimator so that one of the following estimation objectives is achieved:

(E1') Eventual almost sure stability of the estimation error: There exists $T(\epsilon) \geq 0$ so that

$$\sup_{t \geq T(\epsilon)} d(x_t, \hat{x}_t) \leq \epsilon \quad \text{a.s.}$$

(E2') Asymptotic almost sure stability of the estimation error:

$$P\big(\limsup_{t \to \infty} d(x_t, \hat{x}_t) \leq \epsilon\big) = 1.$$

(E3') Asymptotic quadratic stability of the estimation error in expectation:

$$\limsup_{t \to \infty} E[d(x_t, \hat{x}_t)^2] \leq \epsilon.$$

As in the deterministic setup, we are interested in characterizations of C_ϵ, and in particular $C_0 := \lim_{\epsilon \downarrow 0} C_\epsilon$.

17.3.1 A Dynamical Systems Approach

In order to use the concepts of topological and metric entropies defined for deterministic maps, we associate a shift map with the given stochastic system (17.1). More precisely, we consider the space $\mathbb{X}^{\mathbb{Z}_+}$ of all sequences in $\mathbb{X}$, equipped with the product topology. We write $\bar{x} = (x_0, x_1, x_2, \ldots)$ for the elements of $\mathbb{X}^{\mathbb{Z}_+}$ and we fix the product metric

$$D(\bar{x}, \bar{y}) := \sum_{t=0}^{\infty} \frac{1}{2^t} \frac{d(x_t, y_t)}{1 + d(x_t, y_t)}. \tag{17.11}$$

A natural dynamical system on $\mathbb{X}^{\mathbb{Z}_+}$ is the shift map $\theta : \mathbb{X}^{\mathbb{Z}_+} \to \mathbb{X}^{\mathbb{Z}_+}$, $(\theta\bar{x})_t \equiv x_{t+1}$, which is continuous with respect to the product topology. An analogous shift map is defined on $W^{\mathbb{Z}_+}$ and denoted by ϑ.

In the following, we assume that the channel is noiseless. In particular, its capacity is given by $C = \log |\mathcal{M}|$, where $\mathcal{M} = \mathcal{M}'$ is the coding alphabet. We will derive lower bounds on C_0 for the objectives (E1')–(E3').

Observing that the sequence of random variables $(x_t)_{t \in \mathbb{Z}_+}$ forms a Markov chain, when x_0 is fixed, the following lemma shows how a stationary measure of this Markov chain defines an invariant measure for θ.

Lemma 17.3.1 ([407]) *Let π be a stationary measure of the Markov chain $(x_t)_{t\in\mathbb{Z}_+}$. Then, an invariant Borel probability measure Θ for θ is defined on cylinder sets by*

$$\Theta(B_0 \times B_1 \times \cdots \times B_n \times \mathbb{X}^{[n+1,\infty)})$$
$$:= \int_{B_0\times B_1\times\cdots\times B_n} \pi(\mathrm{d}x_0)P(\mathrm{d}x_1|x_0)\ldots P(\mathrm{d}x_n|x_{n-1}),$$

where $B_0, B_1, \ldots, B_n$ are arbitrary Borel sets in $\mathbb{X}$. Here

$$P(x_{n+1} \in B|x_n = x) = P(f(x_n, w) \in B|x_n = x)$$
$$= \nu((f^x)^{-1}(B)).$$

The support of Θ is contained in the closure of the set of all trajectories, i.e.,

$$\operatorname{supp}\Theta \subset \operatorname{cl}\left\{\bar{x} \in \mathbb{X}^{\mathbb{Z}_+} : \exists w_t \in W \text{ with } x_{t+1} \equiv f(x_t, w_t),\ t \in \mathbb{Z}_+\right\}.$$

Theorem 17.3.1 ([407]) *Consider the estimation objective (E1'):*

$$\sup_{t\geq T(\epsilon)} d(x_t, \hat{x}_t) \leq \epsilon \quad a.s. \tag{17.12}$$

Let C_0 denote the smallest channel capacity above which this objective can be achieved for every $\epsilon > 0$ for a fixed initial measure π_0 which is stationary under the Markov chain $(x_t)_{t\in\mathbb{Z}_+}$. Let $\Theta = \Theta(\pi_0)$ be the measure introduced in Lemma 17.3.1. Then the following holds: If supp Θ *is not compact, $C_0 = \infty$, and otherwise,*

$$C_0 \geq h_{\text{top}}(\theta_{|\operatorname{supp}\Theta}).$$

Theorem 17.3.2 ([407]) *Consider the estimation objective (E2'):*

$$P\Big(\limsup_{t\to\infty} d(x_t, \hat{x}_t) \leq \epsilon\Big) = 1. \tag{17.13}$$

Let C_0 denote the smallest channel capacity above which this objective can be achieved for every $\epsilon > 0$ for a fixed initial measure π_0 which is stationary and ergodic under the Markov chain $(x_t)_{t\in\mathbb{Z}_+}$. Let $\Theta = \Theta(\pi_0)$ be the measure introduced in Lemma 17.3.1. Then, if supp Θ *is compact, we have*

$$C_0 \geq h_\Theta(\theta).$$

In the following, we consider (E3'). To obtain a lower bound, we restrict the encoder to have finite memory and be periodic.

Theorem 17.3.3 ([407]) *Consider the estimation objective (E3')*

$$\limsup_{t\to\infty} E[d(x_t, \hat{x}_t)^2] \le \epsilon \tag{17.14}$$

for an initial measure π_0 which is stationary and ergodic under the Markov chain $(x_t)_{t\in\mathbb{Z}_+}$. Additionally, assume that there exists $\tau > 0$ so that the coder map γ_t^e is of the form

$$q_t = \gamma_t^e(x_{[t-\tau+1,t]}) \tag{17.15}$$

and is periodic so that $\gamma_{t+\tau}^e \equiv \gamma_t^e$. Further assume that the estimator map is of the form

$$\hat{x}_t = \gamma_t^d(q_{[t-\tau+1,t]}),$$

and also $\gamma_{t+\tau}^d \equiv \gamma_t^d$. Then, if the support of the measure $\Theta = \Theta(\pi_0)$, introduced in Lemma 17.3.1, is compact, the smallest channel capacity C_0 above which (E3') can be achieved for every $\epsilon > 0$ satisfies,

$$C_0 \ge h_\Theta(\theta).$$

Three remarks are now in order.

Remark 17.3.1 It is worth noting here that for a deterministic dynamical system, the property of being ergodic is typically very restrictive; however, for a stochastic system ergodicity is often simple to satisfy: The presence of noise often leads to strong mixing conditions which directly lead to ergodicity.

Remark 17.3.2 The result of Theorem 17.3.3 leads to an interesting relation between the classical information-theoretic problem of optimally encoding (non-causally) sequences of random variables and metric entropy of an infinite-dimensional dynamical system defined via the shift operator. A close look at the proof of Theorem 17.2.5 reveals that under a stationarity and ergodicity assumption, when the channel is noise-free, the lower bound presented in Theorem 17.3.3 is essentially achievable, provided that the encoder has non-causal access to the source realizations; and in particular, a large enough *look-ahead* is sufficient for an approximately optimal performance. Note though that the decoder is still restricted to be zero-delay.

Remark 17.3.3 A related study is due to [643], which studies non-linear systems of the form

$$x(t+1) = F(x(t), \omega(t)),$$

where $\omega(t)$ is interpreted as an uncertainty input or a disturbance. However, no statistical structure is imposed on ω so that the system can be regarded as deterministic (thus, the formulation is distribution-free). A characterization of the smallest channel capacity C_0 above which the state $x(t)$ can be estimated with arbitrary precision and for every initial state $x(0)$ in a specified compact set via a noiseless channel is given in [643, Thm. 3.1]. A close inspection of this result shows that it characterizes C_0 precisely as the topological entropy of the associated shift operator acting on system trajectories. Moreover, [643, Thm. 3.2] shows that under mild assumptions on the system the entropy of this operator is infinite, and hence observation via a finite-capacity channel is not possible.

17.3.2 An Information-Theoretic and a Probability-Theoretic Approach

We now consider a much broader class of channels, so-called *Class A type channels* (see Definition 13.6.1). We restrict ourselves to systems with state space $\mathbb{X} = \mathbb{R}^N$ and provide lower bounds on the channel capacity under the objectives (E2') and (E3'), using information-theoretic methods. The main results are given in the following two theorems.

Theorem 17.3.4 ([407]) *Consider system* (17.1) *with state space $\mathbb{X} = \mathbb{R}^N$. Suppose that*

$$\limsup_{T\to\infty} \frac{1}{T} \sum_{t=1}^{T-1} h(x_t|x_{t-1}) > -\infty$$

and $h(x_t) < \infty$ for all $t \in \mathbb{Z}_+$. Then, the smallest channel capacity C_ϵ above which (E3') can be achieved over a Class A type channel, satisfies

$$C_\epsilon \geq \left(\limsup_{T\to\infty} \frac{1}{T} \sum_{t=1}^{T-1} h(x_t|x_{t-1}) \right) - \frac{N}{2} \log(2\pi e \epsilon).$$

This, in particular, implies $C_0 = \infty$ for the smallest channel capacity C_0 above which (E3') can be achieved for every $\epsilon > 0$. That is, (E3') cannot be achieved for every $\epsilon > 0$ over a finite-capacity Class A type channel.

Theorem 17.3.5 ([407]) *Suppose that $\mathbb{X} \subset \mathbb{R}^N$ and the system given by $x_{t+1} = f(x_t, w_t)$ satisfies*

$$P(x_{t+1} \in B | x_t = x) \leq K\lambda(B)$$

for all Borel sets $B \subset \mathbb{R}^N$, where λ denotes the Lebesgue measure, $K \in \mathbb{R}_+$, and w_t is an i.i.d. noise process. Then, the smallest channel capacity C_0 above which (E2') *can be achieved for every $\epsilon > 0$ is $C_0 = \infty$. That is, (E2') cannot be achieved for every $\epsilon > 0$ over a finite-capacity Class A type channel.*

A special case for the above is a system of the form

$$x_{t+1} = f(x_t) + w_t,$$

where $w_t \sim \nu$ with the noise measure ν admitting a bounded density function.

17.3.3 An Achievability Result: Coding of Deterministic Dynamical Systems Over Noisy Communication Channels

In this section, we show that for a noise-free system a discrete memoryless noisy communication channel is no obstruction for achieving the objectives (E2') and (E3') with finite capacity. More precisely, we have the following theorem.

Theorem 17.3.6 ([407]) *Consider a nonlinear deterministic system $x_{t+1} = f(x_t)$ given by a continuous map $f : \mathbb{X} \to \mathbb{X}$ on a compact metric space $\mathbb{X}$, estimated via a discrete memoryless channel (DMC). Then, for the asymptotic estimation objectives (E2') and (E3'), the smallest channel capacity C_0 above which these objectives can be achieved for every $\epsilon > 0$ satisfies,*

$$C_0 \leq h_{\text{top}}(f).$$

The proof of the result above crucially depends on the fact that the system is deterministic. The theorem is essentially a possibility result and we note that the proof does not make use of the fact that the encoder has access to the realizations of the channel output. For linear systems, a constructive proof is given in [517, Thm. 6.4.1]. We present this result (without proof) for completeness. It provides a positive answer to the question of whether the estimation objective (E2') can be achieved when no noise is present in the system.

Theorem 17.3.7 *Consider the noiseless linear system*

$$x_{t+1} = Ax_t \tag{17.16}$$

with $A \in \mathbb{R}^{N \times N}$, estimated over a memoryless erasure channel with finite capacity. Then, the smallest channel capacity C_0 above which (E2') can be achieved for every $\epsilon > 0$ satisfies $C_0 \leq \sum_{i=1}^{N} \max\{0, \log\lceil|\lambda_i|\rceil\}$, where $\lambda_1, \ldots, \lambda_N$ are the eigenvalues of A.

For completeness, we also note that [619, Cor. 5.3 and Thm. 4.3] show that for a discrete memoryless channel it suffices that $C > \sum_{|\lambda_i|\geq 1} \log|\lambda_i|$ for the existence of encoder and controller policies leading to almost sure stability.

17.4 Discussion, Comparison and a Review of Related Results in the Literature

17.4.1 Deterministic Model

Hierarchy of Estimation Criteria The results here together with the results in [514] by Matveev and Pogromsky yield a hierarchy of estimation criteria for a deterministic nonlinear dynamical system $x_{t+1} = f(x_t)$ with a continuous map (or a homeomorphism) $f : \mathbb{X} \to \mathbb{X}$ on a compact metric space $(\mathbb{X}, d)$. These estimation criteria, ordered by increasing strength, are the following.

(1) Asymptotic estimation in expectation (criterion (E3)):

$$\limsup_{t\to\infty} \int_{\mathbb{X}} \pi_0(\mathrm{d}x_0) d(f^t(x_0), \hat{x}_t(x_0))^p \leq \epsilon, \quad p > 0$$

with a Borel probability measure π_0 on $\mathbb{X}$.

(2) Asymptotic almost sure estimation (criterion (E2)):

$$\pi_0(\{x_0 \in \mathbb{X} : \limsup_{t\to\infty} d(f^t(x_0), \hat{x}_t(x_0)) \leq \epsilon\}) = 1$$

with a Borel probability measure π_0 on $\mathbb{X}$.

(3) Eventual almost sure estimation (criterion (E1)):

$$\pi_0(\{x_0 \in \mathbb{X} : d(f^t(x_0), \hat{x}_t(x_0)) \leq \epsilon\}) = 1 \quad \forall t \geq T(\epsilon)$$

with a Borel probability measure π_0 on $\mathbb{X}$.

(4) Deterministic estimation (called *observability* in [514]):

$$d(x_t, \hat{x}_t) \leq \epsilon \quad \text{for all } t \geq T(\epsilon).$$

Finally, two even stronger estimation criteria are considered in [514], which do not fit exactly into the above hierarchy, since they are based on an initial error $d(x_0, \hat{x}_0) \leq \delta$, which is known to both the encoder and estimator, and can be much smaller than the desired accuracy ϵ. In terms of this maximal initial error δ, the criteria can be formulated as follows:

(5) There are $\delta_* > 0$ and $G \geq 1$ such that for all $\delta \in (0, \delta_*)$,

$$d(x_t, \hat{x}_t) \leq G\delta \quad \text{for all } t \geq 0.$$

This is called *regular observability* in [514] (see [409] for a detailed analysis and computational aspects on this concept).

(6) There are $\delta_* > 0$, $G \geq 1$ and $g \in (0, 1)$ such that for all $\delta \in (0, \delta_*)$,

$$d(x_t, \hat{x}_t) \leq G\delta g^t \quad \text{for all } t \geq 0.$$

This is called *fine observability* in [514].

Let us denote by $C_0^{(i)}$, $1 \leq i \leq 6$, the smallest channel capacity for the above criteria. Then, modulo specific assumptions such as invertibility of f, ergodicity of π_0 and finiteness of entropies, our results together with those in [514] essentially yield the following chain of inequalities:

$$\begin{aligned} h_{\pi_0}(f) &\overset{(a)}{\leq} C_0^{(1)} \overset{(b)}{\leq} h_{\pi_0}(f) \overset{(c)}{\leq} C_0^{(2)} \overset{(d)}{\leq} h_{\text{top}}(f; \text{supp}\, \pi_0) \\ &\overset{(e)}{\leq} C_0^{(3)} \overset{(f)}{\leq} h_{\text{top}}(f; \text{supp}\, \pi_0) \overset{(g)}{\leq} h_{\text{top}}(f) \overset{(h)}{\leq} C_0^{(4)} \overset{(i)}{\leq} h_{\text{top}}(f) \\ &\overset{(j)}{\leq} C_0^{(5)} \overset{(k)}{=} C_0^{(6)}. \end{aligned}$$

For (h)–(k), see [514, Thm. 8 and Lem. 14], and note that (d) can be replaced by $C_0^{(2)} \leq h_{\pi_0}(f)$ under the assumptions of Theorem 17.2.3. Recently, in [515] the notion of *restoration entropy* was proposed as an intrinsic quantity of a system which characterizes $C_0^{(5)} = C_0^{(6)}$. The results in [514, 515] show that this quantity can be strictly larger than $h_{\text{top}}(f)$.

Except for compactness of $\mathbb{X}$, continuity of f and f-invariance of π_0, only the following inequalities rely on additional assumptions:

(a) Invertibility of f, ergodicity of π_0 and periodic coding and estimation schemes using finite memory (see Theorem 17.2.6).
(b) Ergodicity of π_0 (see Theorem 17.2.5).
(c) Invertibility of f (see Theorem 17.2.4).
(e) Invertibility of f.

We note here that [750] and [333] present analytical or numerical approximation schemes on estimation entropies. Liberzon and Mitra [544] characterizes the critical data rate for exponential state estimation with a given exponent of a continuous-time system on a compact set and introduce *estimation entropy*, which is in agreement with topological entropy when the exponent is 0.

17.4.2 *Stochastic Model*

For the stochastic model setup, we again considered three estimation objectives for stochastic non-linear systems $x_{t+1} = f(x_t, w_t)$ with i.i.d. noise (w_t), assuming that the estimator receives state information via a noisy channel of finite capacity.

(1) For noiseless channels, assuming that the initial measure π_0 is stationary, we proved that C_0 is bounded below by either the topological or the metric entropy of a shift dynamical system on the space of trajectories (Theorems 17.3.1, 17.3.2 and 17.3.3).

(2) For systems on Euclidean space and noisy channels, we provided information-theoretic and probability-theoretic conditions enforcing $C_0 = \infty$. In particular, Theorem 17.3.4 shows that $C_0 = \infty$ for the quadratic stability objective, whenever

$$\limsup_{T\to\infty} \frac{1}{T} \sum_{t=1}^{T-1} h(x_t|x_{t-1}) > -\infty. \tag{17.17}$$

We had a corresponding negative result under (E2) in Theorem 17.3.5 for noisy systems which are sufficiently *irreducible*. Since $h(x_t|x_{t-1})$ is a measure for the uncertainty of x_t given x_{t-1}, the condition (17.17) means that the noise on the long run (in average) influences the state process in a substantial way. Similar to the results in Sect. 17.3.1, this means that the noise makes the space of relevant trajectories too large (or too complicated) to estimate the state with arbitrarily small error over a finite-capacity channel.

(3) Compared with the deterministic setup, the stochastic case has revealed that the rate requirements are not robust with respect to the presence of noise: That is, even an arbitrarily small noise level may lead to drastic effects in the rate requirements. However, the metric or topological entropy bounds are always present and the lower bounds reduce to those established for the deterministic case. We also note that the metric entropy definition for random dynamical systems [467] in the ergodic theory literature is not the answer to the operational questions we have proposed here, unlike the one for the deterministic case which precisely answered the operational question related to (E2).

(4) In Sect. 17.3.3, we assumed that the system is deterministic with a compact state space, but the channel is noisy. We proved that in this case C_0 is bounded from above by the topological entropy of the system for the asymptotic almost sure objective (thus, leading to an achievability result).

Remark 17.4.1 An implication on the achievability for non-causal codes over discrete memoryless channels. In Sect. 17.3.1, we utilized the fact that one can view a stochastic dynamical system as a deterministic one using the shift operator. Building on a similar argument as that in the proof of Theorem 17.3.3, it can be shown that (E2') considered for the stochastic setup implies and is implied by (E2) considered for the deterministic setup with the distortion metric d being the product metric D introduced in (17.11) for the dynamical system θ. Therefore, provided that the encoder has access to future realizations of the state sequence, the proof of Theorem 17.3.6 implies an achievability result: If the encoder has non-causal access to the source realizations, for (E2') we have $C_0 \leq h_{\text{top}}(\theta_{|\text{supp}\ \mu})$, and this can be achieved through the construction in the proof of Theorem 17.3.6 using an encoder which has non-causal access to the future state realizations. Note though that the decoder is still restricted to be zero-delay. We note that in traditional Shannon theory, block codes are allowed to be non-causal.

17.5 Concluding Remarks

We considered the problem of zero-delay coding of a dynamical system over channels under three estimation criteria concerned with the low-distortion regime. For these three criteria, formulated stochastically, we characterized the smallest channel capacities above which the estimation objectives can be achieved. The results establish further connections between topological and metric entropies of dynamical systems, networked control, and information theory.

17.6 Bibliographic Notes

As noted earlier, an overview on dynamical systems theory and the concepts of metric and topological entropy is provided in Appendix H. A review on some related contributions in the literature and their comparison was earlier given in Sect. 17.4.

Additionally, throughout the chapter, several related studies were noted.

The results presented in this chapter build on [406] (deterministic theory) and [407] (stochastic theory).

Part IV
Information and Signaling Games

Chapter 18
Signaling Games: Equilibria Under Criteria or Prior Mismatch Between an Encoder and a Decoder

We consider in this chapter, and the two that follow, the setting where decision makers (DMs) have misaligned criteria or priors, but need to communicate over a communication channel. Such communication may lead to shaping, and sometimes manipulation, of the information received by the DMs, and leads to what is known as *signaling games*.

The present chapter serves as an introduction to signaling games, with also details of some fundamental results, and the next two chapters discuss several variations on signaling games.

Building on a recurrent theme in the book, in this chapter and the next one (Chap. 19) we will view signaling games from the points of view of mismatches in both the objective functions and the priors. A further information game setup, within a slightly different context involving the channel as a third party is considered in Chap. 20.

Regarding the consequences of a mismatch in priors, as we have noted earlier in the book, a team theoretic formulation requires that all DMs in a decentralized stochastic or networked system have the same probabilistic system model, even though they may have different local information. While this also has been the norm for nearly all information and communication theoretic applications, there are also settings where the sender (transmitter) and the receiver in a communication system subscribe to different subjective probabilistic models. This turns the problem (even if the objective functions are fully aligned), once again as we have seen in Chap. 9 and in Sect. 10.4.3, into a game, to be studied also in this chapter.

Section 18.1 presents a general framework for signaling games. Section 18.2 studies quadratic signaling games in detail under a variety of setups and solution concepts. Section 18.3 studies the case with subjective priors and arrives at equilibrium solutions.

S. Yüksel, T. Başar, *Stochastic Teams, Games, and Control under Information Constraints*, Systems & Control: Foundations & Applications,
https://doi.org/10.1007/978-3-031-54071-4_18

18.1 Introduction to Signaling Games: Nash and Stackelberg Solution Concepts, and Some Basic Results

A general *signaling game*, as depicted in Figs. 18.1 and 18.2, can be formulated as follows: There are two DMs, an encoder (sender or transmitter) and a decoder (receiver) where the encoder wants to transmit to the latter an $\mathbb{M}$-valued source random variable M. The encoder maps M into an $\mathbb{X}$-valued random variable X via an encoding policy denoted by $\gamma^e \in \Gamma^e$ where Γ^e denotes the set of all possible encoding strategies. We take Γ^e to be the set of all stochastic kernels from $\mathbb{M}$ to $\mathbb{X}$. The setup may also involve a noisy channel, e.g. an additive Gaussian noise channel between the encoder and the decoder, where the additive noise is independent of the source output. In this noisy channel case, the decoder receives a noise corrupted version of the message sent, denoted by Y, which is a $\mathbb{Y}$-valued random variable. We note that in the case of noise-free channel, Y is the same as X. The decoder generates its optimal decision U, which is also an $\mathbb{M}$-valued random variable, given its observation Y via a decoding policy $\gamma^d \in \Gamma^d$. Here, Γ^d denotes the set of all possible decoding strategies, which is the set of all stochastic kernels from $\mathbb{Y}$ to $\mathbb{M}$. A particular setup, also called *cheap talk* (as communication is *costless*), is where the encoded message is directly observable by the decoder and except for measurability there is no additional (such as power) constraint at the encoder; this setup is depicted in Fig. 18.1.

What makes this setup different from the team theoretic setting is that there may be criteria or model/prior mismatches, or the channel noise may be viewed as (or actually is) chosen by an adversary, which we may call *jammer*.

Accordingly, each agent (i.e., encoder or the decoder or a jammer) chooses a strategy to maximize its own utility which is determined by the joint strategies chosen by all the players (DMs). Such problems are increasingly important in many sub-areas with networked environments with a plethora of applications, including:

(i) Quadratic Signaling Games with Criteria Mismatch (to be studied in the current chapter)
(ii) Privacy Games (where an encoder wishes to preserve privacy, as depicted in Fig. 19.2), to be studied in Chap. 19
(iii) Reputation Games (where a DM wishes to take advantage of the ambiguity on his *type* (e.g. of an undeserved reputation) while acting otherwise with the

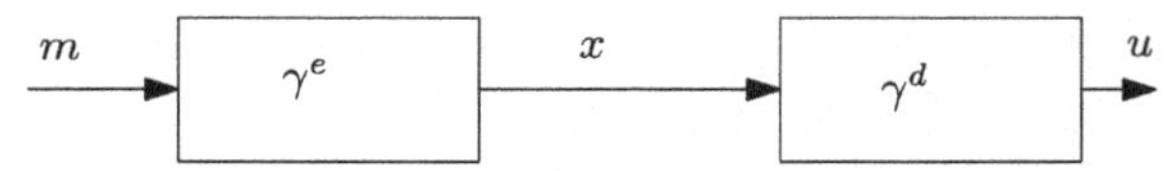

Fig. 18.1 Cheap talk setup

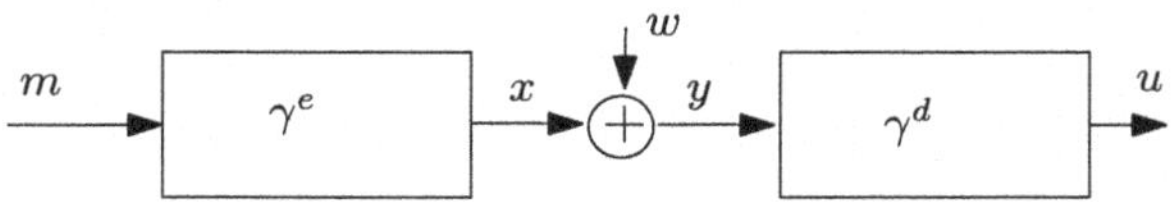

Fig. 18.2 Signaling game setup

actions leading to a signaling game), to be studied in Chap. 19. See Fig. 19.1 for the setup considered.

(iv) Quadratic Signaling Games with Prior Mismatch (to be studied in the current chapter)

(v) Jamming Games (even though the setup here is slightly different in that the channel is subject to adversarial actions while the encoder and decoder are aligned: this will be studied later in Chap. 20)

In the following, we will present the essential ingredients of such problems, while also establishing intrinsic connections with what has been studied earlier in the book in terms of information and communication theoretic analysis as well as various solution concepts. We will consider the following broad sub-areas:

Signaling Games with Criteria or Prior Mismatch
Signaling games and cheap talk constitute a class of Bayesian games where an informed DM transmits information to another DM. Unlike a team, however, the goals of the agents are misaligned. Such a study has been initiated in Crawford and Sobel's seminal work [189], which obtained the striking result that under some technical conditions on the utility functions of the DMs, the cheap talk problem only admits Nash equilibrium policies that are essentially quantization policies (that is, the encoder chooses to hide information or introduces uncertainty into the transmitted message) -quite unlike the team theoretic setup where information is fully revealed-. In contrast to Crawford and Sobel's setup, a Stackelberg formulation (known as the *Bayesian persuasion* problem[386]) investigates signaling scenarios under the Stackelberg equilibrium concept rather than the Nash equilibrium one.

This class also includes **information transmission problems with model or prior mismatch**.

Two further sub-areas are **privacy games** and **reputation games**. Yet another important class of games in this context is **jamming games**. We briefly introduce all three below, as a prelude to their detailed coverage in this and the next two chapters.

Privacy Games In various applications in networked control, users share information with each other or a service provider to improve the quality of service by utilizing information gathered from the users. The users also often desire to maintain a certain level of privacy. The privacy objective arises from the fact that the information they wish to convey to the service provider may be correlated with certain private information they want to protect. This leads to a particular type of criteria mismatch.

Reputation Games In some applications, a *long-lived player* plays a repeated stage game with a sequence of different *short-lived* or *transient* players. The long-lived player has a *reputation*, modeled as an initial prior, and would like to take advantage of his reputation, while acting otherwise. We will see that as the time horizon approaches infinity, an undeserved reputation cannot be sustained under optimal behaviour. The solution concept here is a stage-wise Stackelberg game equilibrium.

Jamming Games The question of how a communication system performs in the worst case scenario is one of vital importance, for if one can guarantee a suitable operation in the least favourable case then the system will operate at least as well in all other (more favorable) situations. In determining the worst case faced by a system, it is useful to personify the noise as an intelligent "jammer" actively working to steer the system toward a deleterious behavior, under of course some constraints on its capabilities. Furthermore, in many scenarios, there may indeed be a malicious agent who intends to suppress communication and control in a decentralized system. These setups motivate the use of game theoretic methods in communications and networked control applications. This topic will be covered in Chap. 20.

Some of the material covered in these chapters is relatively new for the control/information literature, but some variations of the material, especially in the economic theoretic context, has had a long history in the economics community where *information economics* is an established branch of microeconomic theory.

Nash vs. Stackelberg (Bayesian Persuasion) Settings
Now, continuing with the formulation at the beginning of this section, let the encoder be labeled as DM 1 and the decoder as DM 2. Consider first the case where the channel that connects DM 1 and DM 2 is noise-free, as depicted in Fig. 18.1, with the corresponding policy spaces denoted by Γ^e and Γ^d as before. Given $\gamma^e \in \Gamma^e$ and $\gamma^d \in \Gamma^d$, the classical communications theory addresses minimization of

$$J(\gamma^e, \gamma^d) = \int c(m, u)\gamma^e(\mathrm{d}x|m)\gamma^d(\mathrm{d}u|x)P(\mathrm{d}m),$$

where c is some cost function, P is the probability measure on $\mathbb{M}$, and integration is over the product space $\mathbb{X} \times \mathbb{M}$. One frequently adopted cost function, as also studied in the book extensively, is the quadratic one, where $\mathbb{M}$ is a finite-dimensional Euclidean space and $c(m, u) = |m - u|^2$.

Recall that a collection of DMs who have an agreement on the probabilistic description of a system and a common cost function to be minimized, but may not necessarily have access to the same on-line information constitutes a *team*. Hence, the classical communications problem as formulated above can be viewed as a two-person team with the encoder and the decoder as the two DMs.

In many applications, however, the objectives of the encoder and the decoder may not be aligned. For example, DM 1 may aim to minimize

$$J^e(\gamma^e, \gamma^d) = \int c^e(m, u)\gamma^e(\mathrm{d}x|m)\gamma^d(\mathrm{d}u|x)P(\mathrm{d}m),$$

whereas DM 2 may aim to minimize

$$J^d(\gamma^e, \gamma^d) = \int c^d(m, u)\gamma^e(\mathrm{d}x|m)\gamma^d(\mathrm{d}u|x)P(\mathrm{d}m),$$

where c^e and c^d are different functions.

Such a problem is known in the economics literature as *cheap talk* (the transmitted signal, X, does not affect the cost directly, which is why the game is named as such). A more general formulation would be one where the transmitted signal is also an explicit part of the cost functions; in that case, the setup is called a *signaling game*. We will also consider a noisy channel version (as already mentioned at the beginning of this section), with details to be presented later, but note that the setup leads only to minor adjustments in the formulation above at this level of generality, in the sense that the joint map from M to U also entails a noisy channel so that the decoder has access to a noisy output Y of a channel with input X; see Fig. 18.2.

In accordance with the general discussion on game theory presented in Chap. 9, we recall that a pair of policies $(\gamma^{*,e}, \gamma^{*,d})$ constitutes a *Nash equilibrium* if

$$J^e(\gamma^{*,e}, \gamma^{*,d}) \leq J^e(\gamma^e, \gamma^{*,d}) \quad \forall \gamma^e \in \Gamma^e ,$$
$$J^d(\gamma^{*,e}, \gamma^{*,d}) \leq J^d(\gamma^{*,e}, \gamma^d) \quad \forall \gamma^d \in \Gamma^d . \tag{18.1}$$

A setup where there is essentially no information transmission at a Nash equilibrium arises in many applications, as the DMs may choose *not to talk*:

Definition 18.1.1 A non-informative equilibrium (also called a *babbling equilibrium*) is one where the encoder transmits a message that is independent of the source and the decoder takes an action only by considering the prior distribution of the source (since the received message is informationally useless).

Such an equilibrium exists for any signaling game (of the type considered in this section) [189]:

Proposition 18.1.1 *A non-informative (babbling) equilibrium always exists for the cheap talk or the noisy-channel signaling game.*

Moreover, as we will see later in the chapter, there will be many instances where a non-informative equilibrium is the only equilibrium.

So far we have focused on simultaneous game-play between the sender and the receiver, with the underlying solution concept being the Nash equilibrium. It is also possible to consider a sequential game-play where the sender decides on (and announces) a policy by anticipating the response of the receiver, and executes an action in line with that policy, generating a message (signal) received by the receiver over a possibly noisy channel, who then decides on her cost-minimizing action. This sequential (asymmetric) play leads to the *Stackelberg* equilibrium as the solution concept, which was already introduced in Chap. 9. Within the context of signaling games, this decision scenario is commonly referred to as the *Bayesian persuasion* problem in the economics literature [386]). Thus, the sender chooses and announces its strategy and then the receiver acts upon learning the strategy of the sender. Once the sender announces its strategy, it cannot change it, i.e., the sender *commits* to employ this announced strategy.

In mathematical terms, a pair of policies $\gamma^{*,e}, \gamma^{*,d}$ is said to be a *Stackelberg equilibrium* if

$$J^e(\gamma^{*,e}, \gamma^{*,d}(\gamma^{*,e})) \leq J^e(\gamma^e, \gamma^{*,d}(\gamma^e)) \quad \forall \gamma^e \in \Gamma^e,$$

where $\gamma^{*,d}(\gamma^e)$ satisfies

$$J^d(\gamma^e, \gamma^{*,d}(\gamma^e)) \leq J^d(\gamma^e, \gamma^d(\gamma^e)) \quad \forall \gamma^d \in \Gamma^d.$$

We refer the reader to Remark 9.2.4 (see also [385, Footnote 7]) with regard to the potential non-uniqueness of $\gamma^{*,d}(\gamma^e)$ in the definition above.

Remark 18.1.1 (The Kamenica and Gentzkow Formulation and a Relation with Blackwell Ordering) The Stackelberg formulation was studied in the influential paper by Kamenica and Gentzkow [386] (see also [385]), where using the fact that the receiver's belief (or posterior) on the hidden state is a sufficient statistic for $\gamma^{*,d}$ (as we have seen on several occasions in the book, e.g. the proof of Theorem 7.3.2), the sender's policy is formulated as an optimization problem over allowable posteriors so that for each posterior the receiver's criterion is optimized: Let $m \sim P_0$ where m is $\mathbb{M}$ valued and with $x = \gamma^e(m)$, the conditional probability of m given x be $\pi(dm)$, i.e., $\pi(\cdot) = P(m \in \cdot|\gamma^e(m) = y)$. The sender than solves the problem of

$$\min_{\eta \in \mathcal{P}(\mathcal{P}(\mathbb{M}))} \int_{\pi \in \mathcal{P}(\mathbb{M})} \eta(d\pi) c^e(m, \gamma^{*,d}(\pi)),$$

where η is the probability measure on the posteriors induced by the sender policy so that

$$\int_{\pi \in \mathcal{P}(\mathbb{M})} \pi(dm)\eta(d\pi) = P_0(dm)$$

(this formulation has been referred to as *concavification* by the authors). The constraint here entails the smoothing property of conditional expectation (i.e., $E[E[f(m)|\pi]] = E[f(m)]$ for any measure determining collection of functions f). While this is still potentially a very complicated problem as $c^e(m, \gamma^{*,d}(\pi))$ depends on the best-response function $\gamma^{*,d}$ which may depend on π in a poorly structured manner, with this formulation, it can be seen that the sender's policy may be randomized (with also cardinality bounds on the randomization that can be obtained by the Carathéodory Theorem (or what is known as *support lemma*) [236, Appendix C] when $\mathbb{M}$ is finite). We saw, however, in the proof of Theorem 7.3.2 that when the sender and receiver have the same cost function, via an application of Jensen's inequality and the concavity of the value in the posterior, an optimal sender policy is not randomized; however, this Jensen's inequality based argument does not apply in the case with general games since the cost criteria are not aligned.

We now discuss, in the next section, in some detail, signaling games with quadratic cost functions, covering first the scalar (one-dimensional source) case and then the multidimensional case, with different source densities. We discuss both the clean channel setting (cheap talk) and the noisy channel one. Both Nash equilibria and Stackelberg equilibria are covered.

18.2 Quadratic Signaling Games

We assume throughout this section that the cost functions of the sender (encoder) and the receiver (decoder) are quadratic functions of m and u, with the former's also having a bias term, b. Specifically, in the one-dimensional (scalar) case, we have

$$c^e(m, u) = (m - b - u)^2, \quad c^d(m, u) = (m - u)^2 \tag{18.2}$$

Note that if $b = 0$, then we have total alignment, and the problem simply reduces to one of minimum mean-square estimation (MMSE), which in this case is the standard message transmission problem in communications, which is of interest only when the channel is noisy (otherwise the minimization problem is trivial, dictating sending the message as is). Here, of course, our interest is in the case when $b \neq 0$. Later in the section we will generalize the scalar formulation to the multidimensional case. First we have the cheap talk game.

18.2.1 Quadratic Signaling Game with Costless Communication (Cheap Talk)

Nash Equilibria

As noted, let the cost functions be defined as above.

The following builds on [189, Lemma 1], which considers sources supported on $[0, 1]$ that admit densities. This general result, however, does not impose the objective functions to be quadratic and it deals with a general pair of misaligned c^e and c^d.

Theorem 18.2.1 (Crawford and Sobel [189, Theorem 1]) *Let M be a real-valued random variable which admits a density supported on $[0, 1]$. Suppose that $c^e(m, b) \neq c^d(m)$ for all m. Then, there exists at least one equilibrium with a quantization policy at the encoder where there are N quantization bins in $[0, 1]$ with $1 \leq N \leq N_{max}(b)$ and $N_{max}(b) \in \mathbb{N}$ denoting the upper bound on the number of bins. Furthermore, an encoding policy at an equilibrium is equivalent to a quantized encoder policy in terms of the performance at the equilibrium.*

In the following, we consider the case of quadratic cost functions with the presence of bias, as noted earlier. The result below generalizes Theorem 18.2.1 in that it applies to arbitrary scalar valued random variables, not necessarily sources on [0, 1] that admit densities (but with quadratic functions).

Theorem 18.2.2 ([638, Theorem 3.2]) *Let M be a real-valued random variable with an arbitrary, given, probability measure. Let the strategy set of the encoder (DM 1) consist of the set of all measurable (deterministic) functions from $\mathbb{M}$ to $\mathbb{X}$. Then, any Nash equilibrium encoder policy under criteria (18.2) has to be a quantizer (or informationally equivalent to being a quantizer) almost surely. Furthermore, the quantization bins are convex under any equilibrium.*

Proof Suppose that there exists a Nash equilibrium in the game (with possibly uncountably many bins, countably infinitely many bins or finitely many bins). Consider two bins, $\mathcal{B}^\alpha$ and $\mathcal{B}^\beta$. Also let m^α denote any point in $\mathcal{B}^\alpha$; i.e., $m^\alpha \in \mathcal{B}^\alpha$. Similarly, let m^β represent any point in $\mathcal{B}^\beta$; i.e., $m^\beta \in \mathcal{B}^\beta$. The decoder chooses action $u^\alpha = E[M|M \in \mathcal{B}^\alpha]$ when the encoder sends $m^\alpha \in \mathcal{B}^\alpha$ and action $u^\beta = E[M|M \in \mathcal{B}^\beta]$ when the encoder sends $m^\beta \in \mathcal{B}^\beta$ in order to minimize its total expected cost. Without loss of generality, we can assume that $u^\alpha < u^\beta$. Let $F(m,u) \triangleq (m-u-b)^2$. By equilibrium, we must have that $F(m^\alpha,u^\alpha) < F(m^\alpha,u^\beta)$ and $F(m^\beta,u^\beta) < F(m^\beta,u^\alpha)$. Hence, $\exists\, \overline{m}$ that satisfies $F(\overline{m},u^\alpha) = F(\overline{m},u^\beta)$, which reduces to

$$\overline{m} = \frac{u^\alpha + u^\beta}{2} + b \iff (\overline{m} - u^\alpha) = (u^\beta - \overline{m}) + 2b \tag{18.3}$$

Since $F(\overline{m}+\Delta,u^\alpha) > F(\overline{m}+\Delta,u^\beta)$ for any $\Delta > 0$, $\mathcal{B}^\beta$ and $\{m|m < \overline{m}\}$ are disjoint sets. Similarly, $\mathcal{B}^\alpha$ and $\{m|m > \overline{m}\}$ are disjoint sets, too. Thus, from the definitions of u^α and u^β, we have $u^\alpha < \overline{m} < u^\beta$, which implies $\overline{m} - u^\alpha > 0$ and $u^\beta - \overline{m} > 0$. Then, from (18.3),

$$u^\beta - u^\alpha = (u^\beta - \overline{m}) + (\overline{m} - u^\alpha) = 2(u^\beta - \overline{m}) + 2b > 2b$$

and

$$u^\beta - u^\alpha = (u^\beta - \overline{m}) + (\overline{m} - u^\alpha) = 2(\overline{m} - u^\alpha) - 2b > -2b$$

follow. Hence,

$$u^\beta - u^\alpha > 2|b|,$$

which implies that there must be at least $2|b|$ distance between the equilibrium points (decoder's actions, centroids of the bins). Further, from the encoder's point of view, given any two bins $\mathcal{B}^\alpha$ and $\mathcal{B}^\beta$, there exists a point $\overline{m}$ which lies in between these two bins. This assures that each bin must be a single interval; i.e., convex cell except for a possible insignificant set of points with measure zero. Since there is an injective and monotonic relation between the convex cells of the encoder's and decoder's actions, the equilibrium policy must be quantized almost surely. ◇

Thus, at an equilibrium, the encoder will be hiding information, unlike in a team theoretic formulation. Note that, the encoder here does not add ambiguity by constructing a noisy channel, but just compresses information in an equilibrium.

We state the following generalization, without proof.

Theorem 18.2.3 ([641]) *The conclusion of Theorem 18.2.2 holds if the policy space of the encoder is extended to the set of all stochastic kernels from $\mathbb{M}$ to $\mathbb{X}$ for any arbitrary source. That is, even when the encoder is allowed to use private randomization, all equilibria are equivalent to those that are attained by quantized equilibria.*

A few questions that come up in this context and are of high relevance are the following:

(i) For either one-sided or two-sided support for the source distribution, is there an upper bound on the number of bins at the equilibrium as a function of the bias b? As a special case, is it possible to have only a non-informative equilibrium, with only a single bin.
(ii) Is it possible to have an equilibrium with infinitely many bins?
(iii) Is the equilibrium unique for a given number of quantization bins?

Finding answers to these questions is highly relevant in this context because if one can show that there may only exist a finite number of bins for any equilibrium, and if the equilibrium is unique for a given number of bins, then the total number of equilibria would be finite. This, in turn, will allow for a feasible setting where the DMs can coordinate their policies.

Furthermore, when signaling games and cheap talk problems are generalized to dynamic (multi-stage) setups, a crucial property that will allow the generalization is the assumption that the number of bins for each stage equilibrium, conditioned on the past actions, is uniformly bounded.

Attaining the upper bound N on the number of bins at the equilibrium implies that there exists at least one equilibrium with $1, 2, \ldots, N$ bins for the bounded support scenario due to [189, Theorem 1], and thus, the following further questions arise:

(iv) if there exists more than one equilibrium, which one of these should be selected by the encoder and the decoder? Note that this is a question that is relevant for any game with multiple Nash equilibria.
(v) to what extent can one argue that more bins (and, thus, a higher rate of communication) would lead to better performance for both players?

Indeed, it is important to know whether in general a higher number of bins implies more desirable equilibria. If such a monotonic behavior holds for a class of sources, then both players will prefer to have an equilibrium with the highest number of bins. This also allows for a well-posed coding problem as one would like to design a coding scheme that is payoff-dominant for both the encoder and the decoder.

Now consider an equilibrium with N bins, and let the k-th bin be the interval $[m_{k-1}, m_k)$ with $m_0 < m_1 < \ldots < m_N$ and let l_k denote the length of the k-th bin; i.e., $l_k = m_k - m_{k-1}$ for $k = 1, 2, \ldots, N$.[1] By Theorem 18.2.2, at the equilibrium, decoder's best response to such a policy by the encoder is characterized by

$$u_k = E[M|m_{k-1} \le M < m_k] \tag{18.4}$$

for the k-th bin; i.e., the optimal decoder action is the centroid for the corresponding bin. From the encoder's point of view, the best response of the encoder to decoder's action is determined by the condition (see the proof of Theorem 18.2.2):

$$u_{k+1} - m_k = (m_k - u_k) - 2b \Leftrightarrow m_k = \frac{u_k + u_{k+1}}{2} + b \tag{18.5}$$

Due to the definition of Nash equilibirum [97], these best responses in (18.4) and (18.5) must match each other, and only then the equilibrium can be characterized; i.e., for a given number of bins, the positions of the bin edges are chosen by the encoder, and the centroids are determined by the decoder. Alternatively, the problem can be considered as a quantization game in which the bin boundaries are determined by the encoder and the reconstruction values are determined by the decoder.

Note that at the equilibrium of this quantization game, the relation between the encoder cost and the decoder cost can be expressed as follows:

$$\begin{aligned}
J^e(\gamma^{*,e}, \gamma^{*,d}) &= \sum_{i=1}^{N} \Pr(m_{i-1} \le M < m_i) \\
&\quad \times E\Big[(M - E[M|m_{i-1} \le M < m_i] - b)^2 | m_{i-1} \le M < m_i\Big] \\
&= \sum_{i=1}^{N} \Pr(m_{i-1} \le M < m_i)\Big(b^2 \\
&\quad + E\Big[(M - E[M|m_{i-1} \le M < m_i])^2 | m_{i-1} \le M < m_i\Big]\Big) \\
&= J^d(\gamma^{*,e}, \gamma^{*,d}) + b^2.
\end{aligned}$$

Since the difference between the encoder cost and the decoder cost is always b^2, regardless of the number of bins, the equilibrium preferences (i.e., which

[1] Note that $m_0 = m_L$ and $m_N = +\infty$, or $m_0 = -\infty$ and $m_N = m_U$ for sources with semi-infinite support with m_L and m_U denoting lower or upper boundary, whereas $m_0 = -\infty$ and $m_N = +\infty$ for sources with two-sided unbounded support.

equilibrium to select) for the encoder and the decoder are aligned when the costs are quadratic.

Now we address the question of whether for a given finite (or infinite) N, there exists an equilibrium; i.e., whether it is possible to find optimal encoder actions (the boundaries of the bins) $m_0, m_1, \ldots, m_N$ and decoder actions (the centroids of the bins) $u_1, u_2, \ldots, u_N$ which satisfy (18.4) and (18.5) simultaneously. Two possible, different, methods toward that end are:

(i) Lloyd's Method I (via iterative updates): After the initial selection of $m_0, m_1, \ldots, m_N$, determine $u_1, u_2, \ldots, u_N$ by (18.4), and after updating $u_1, u_2, \ldots, u_N$, find the corresponding $m_0, m_1, \ldots, m_N$ by (18.5). Then, continue this iteration. For this approach, the convergence of this Lloyd-Max iteration is the key issue.
(ii) Fixed-point approach: By combining (18.4) and (18.5), we obtain

$$m_k = \frac{E[M|m_{k-1} \leq M < m_k] + E[M|m_k \leq M < m_k + 1]}{2} + b$$

for $k = 1, 2, \ldots, N$. Then, the problem reduces to determining whether there exists a fixed vector $m_0, m_1, \ldots, m_N$ satisfying these equations.

We will address these two approaches shortly in this section with some special types of distributions for the source. First we introduce the notion of a strengthened equilibrium. An equilibrium is **more informative** (than another one) if it results in smaller costs for both the encoder and the decoder, and clearly the players would be looking for the most informative equilibrium. Let $J^{d,N}$ denote the decoder cost at an equilibrium with N bins. Then, the question is whether $J^{d,N} > J^{d,N+1}$ for all finite N, or even, whether $J^{d,N} > J^{d,\infty}$ if an equilibrium with infinitely many bins exists.

Nash Equilibria for Sources with Log-Concave Distributions

We first have the following definitions.

Definition 18.2.1 A probability density function f on $\mathbb{R}$ is said to be (strictly) log-concave if $\log f$ is a (strictly) concave function on the support of f where the support is an interval.

In the context of log-concave sources, some of the conclusions in this chapter are closely related to some foundational results by Kieffer [420] where optimal quantization in the classical sense (i.e., in a team theoretic setup with $b = 0$) is considered.

Theorem 18.2.4 (Kieffer [420]) *If the source is log-concave and the common cost function is convex, strictly increasing and continuously differentiable, then there* *exists a unique locally optimal (i.e., satisfying the optimality conditions in* (18.4)–(18.5) *with* $b = 0$*) quantizer.*

Kieffer [420] also shows that under the conditions of Theorem 18.2.4, Lloyd-Max iterations (i.e., Lloyd's Method I) in (18.4) and (18.5) with $b = 0$ converge to the optimal quantizer.

We note that distributions such as Gaussian, exponential, Laplace, Rayleigh and uniform, which are commonly encountered in information and communication theoretic applications, are log-concave. The reader is referred to [52] for a list of common distributions which are log-concave or log-convex.

In the following, for a probability density function f we will say that f has a two-sided unbounded support if its support is $\mathbb{R}$, f has a semi-unbounded support if its support is either an interval in the form $(-\infty, m_U]$ or an interval in the form $[m_L, \infty)$ for some $m_U, m_L \in \mathbb{R}$, and that f has a bounded support if $f(x) = 0$ for all $x \notin [m_L, m_U]$ for some $-\infty < m_L < m_U < \infty$.

We now have the following theorem on existence and uniqueness of Nash equilibrium when the source has a log-concave density with two-sided unbounded support. Recall that an equilibrium with $N = 1$ bins corresponds to the case where the encoder and decoder do not communicate (as mentioned, this is a non-informative (babbling) equilibrium).

Theorem 18.2.5 ([411]) *Consider a source that has a log-concave density with two-sided unbounded support. Then, for any fixed* $N \geq 1$*, there exists a unique equilibrium (among those with* N *bins).*

Theorem 18.2.5 reveals that there does not exist an upper bound on the number of bins for an equilibrium considering log-concave sources with two-sided unbounded support. Namely, there exist countably infinite number of distinct equilibria for log-concave sources with two-sided unbounded support.

Remark 18.2.1 While we do not present here in the book, it is useful to note that while in Theorem 18.2.5 the source distribution is assumed to be strictly log-concave, the result also holds for distributions which are only log-concave. In that case, for the result to hold, it is required that the density is not log affine everywhere. This is indeed true for a log-concave distribution with two-sided unbounded support.

Remark 18.2.2 For a source with log-concave density with two-sided unbounded support (such as a Gaussian), there also exist equilibria with infinitely many bins. The proof for existence of infinitely many bins essentially relies on Tychonoff's fixed-point theorem [411].

The existence and uniqueness result of Theorem 18.2.5 can be generalized to sources with semi-unbounded support as well as sources with bounded support, and the following is a generalization to sources with semi-unbounded support. In contrast, for sources with semi-unbounded support, there may be an upper bound on the number of bins depending on the sign of b. Throughout, we use the following notation for the smallest upper bound (whenever it exists) on the number of bins in equilibrium for a given problem setup.

Theorem 18.2.6 ([411])

(i) Consider a source that has a log-concave density with a support on $[m_L, \infty)$ *for some* $m_L \in \mathbb{R}$*. Then, for* $b > 0$ *and for any fixed* $N \geq 1$*, there exists a unique Nash equilibrium with N bins.*
(ii) Consider a source that has a log-concave density with a support on $[m_L, \infty)$ *for some* $m_L \in \mathbb{R}$*. Then, for a given* $b < 0$*, there exists a finite upper bound on the number of bins,* $N_{\max}(b)$*, and for any fixed N satisfying* $1 \leq N \leq N_{\max}(b)$*, there exists a unique Nash equilibrium with N bins.*
(iii) Consider a source that has a log-concave density with a support on $(-\infty, m_U]$ *for some* $m_U \in \mathbb{R}$*. Then, for* $b < 0$ *and for any* $N \geq 1$*, there exists a unique Nash equilibrium with N bins.*
(iv) Consider a source that has a log-concave density with a support on $(-\infty, m_U]$ *for some* $m_U \in \mathbb{R}$*. Then, for a given* $b > 0$*, there exists a finite upper bound on the number of bins,* $N_{\max}(b)$*, and for any N satisfying* $1 \leq N \leq N_{\max}(b)$*, there exists a unique Nash equilibrium with N bins.*

In the case of semi-unbounded support, the result of Theorem 18.2.6 holds even for densities which are log affine everywhere on its support (recall the strict definition of log-concavity in Definition 18.2.1). The exponential distribution is an example of such a distribution. Furthermore, for a source with a log-concave density and semi-unbounded support, in the case that there does not exist an upper bound on the number of bins in equilibrium, there also exist equilibria with infinitely many bins.

The following theorem specifies a necessary and sufficient condition for the existence of an informative equilibrium.

Theorem 18.2.7 ([411])

(i) For a source that has a log-concave density with a support on $[m_L, \infty)$ *for some* $m_L \in \mathbb{R}$ *and has a mean* $E[M] = \mu$*, the only Nash equilibrium is non-informative if* $2b \leq -(\mu - m_L)$*. Otherwise, there exists a Nash equilibrium with two bins.*
(ii) For a source that has a log-concave density with a support on $(-\infty, m_U]$ *for some* $m_U \in \mathbb{R}$ *and has a mean* $E[M] = \mu$*, the only Nash equilibrium is non-informative if* $2b \geq (m_U - \mu)$*. Otherwise, there exists a Nash equilibrium with two bins.*

As we had stated earlier in this section, for sources with a density supported on $[0, 1]$ Crawford and Sobel [189] proved that for a given b there exists an upper bound on the number of bins, $N_{\max}(b)$, and that for any $1 \leq N \leq N_{\max}(b)$ there exists an equilibrium with N bins (see Theorem 18.2.1). In addition, an additional monotonicity assumption with regard to the behavior of bin edges was made in [189], which ensures the uniqueness of an equilibrium for a given number of bins [189, Lemma 3]. In fact, when sources supported on a bounded interval are considered, it can be shown that the monotonicity assumption is satisfied for sources with a log-concave density. We then have the following.

Theorem 18.2.8 *For a log-concave density with a support on* $[m_L, m_U]$ *with* $m_L < m_U$ *and* $m_L, m_U \in \mathbb{R}$, *an equilibrium with N bins is unique for* $1 \le N \le N_{\max}(b)$.

Properties of Equilibrium Costs in Number of Bins

So far, the number of bins at equilibrium has been investigated for sources with two-sided unbounded support, semi-unbounded support, and bounded support. Next, we examine the dependence of the encoder's and the decoder's expected costs at Nash equilibrium on the number of bins. In the next theorem, we state that the expected costs at an equilibrium with N bins are smaller for both players than at a Nash equilibrium with K bins if $K < N$. This result has consequences, with regard to equilibrium selection [338], as well as on the value of information in games, a subject which was studied in detail earlier in Chap. 11.

Theorem 18.2.9 ([411]) *For sources with a log-concave density, an equilibrium with more bins leads to smaller expected costs for both players. In other words, denoting expected costs of the encoder and the decoder by* $J^{e,N}$ *and* $J^{d,N}$ *(respectively,* $J^{e,K}$ *and* $J^{d,K}$*) at the unique equilibrium with N (respectively, K) bins, we have* $J^{e,N} < J^{e,K}$ *and* $J^{d,N} < J^{d,K}$ *if* $K < N$.

When $b = 0$ (i.e., the team setup), the problem reduces to the classical optimal quantization problem under quadratic cost structure and thus the result of Theorem 18.2.9 is in fact valid also when $b = 0$. In other words, the classical optimal quantization setup with the quadratic cost criterion is a special case of Theorem 18.2.9.

Remark 18.2.3 Without the assumption of log-concavity, such a monotonic behavior of the expected costs at equilibrium with respect to number of bins may not hold in general. See [411] for a counter-example.

Convergence to Equilibria via Best-Response Dynamics (as a Variation of Lloyd's Method)

It would be interesting to investigate whether the best response iterations, via Lloyd's method as mentioned earlier, converge to the unique equilibrium. We will see that when modified Lloyd-Max iterations (i.e., by using (18.4) and (18.5)) are performed starting from an initial set of bin edges, the unique equilibrium (for the given number of bins) is reached for strictly log-concave sources with two-sided unbounded support. These iterations in the classical team theoretic setting is referred to as Lloyd's Method I [487] where bin edges and centroids are updated as described below.

A modified Lloyd-Max iteration (*modified* due to the presence of $b > 0$) can be introduced as follows: Let $m_1 < \cdots < m_{N-1}$ be an initial set of bin edges and let $\boldsymbol{m} := [m_1, \ldots, m_{N-1}]^T$. Given these bin edges, the decoder first determines its best response by computing the centroids via (18.4), for $k = 1, \ldots, N$, where $m_0 = m_L$ and $m_N = m_U$ remain the same during these iterations. We denote this operation by $\mathrm{BR}_d(\boldsymbol{m}) = \boldsymbol{u}$ where $\boldsymbol{u} \triangleq [u_1, \ldots, u_N]^T$. Then, the encoder computes the nearest neighbors for the resulting centroids $\boldsymbol{u}$ via 18.5) for $k = 1, \ldots, N - 1$.

We denote this operation by $\mathrm{BR}_e(\boldsymbol{u}) = \boldsymbol{m}$. With these best response characterizations, a modified Lloyd-Max iteration is defined as

$$T(\boldsymbol{m}) \triangleq \mathrm{BR}_e(\mathrm{BR}_d(\boldsymbol{m})). \tag{18.6}$$

The operation in (18.6) is applied iteratively to update an initial set of bin edges. In the following theorem, we state that by repeating these iterations, it is guaranteed that a set of bin edges that leads to an equilibrium with N bins is obtained when the source is strictly log-concave.

Theorem 18.2.10 ([411]) *For a strictly log-concave source with two-sided unbounded support, under criteria given in (18.2), best response iterations (i.e., modified Lloyd-Max iterations) through* (18.6) *starting with (an arbitrary, but fixed) N bins converge to the unique equilibrium with N bins.*

Remark 18.2.4 We note that [420] proves convergence in a team theoretic setup even for log-concave sources which are not necessarily strictly log-concave. To prove this result, [420] uses the fact that the common cost is strictly decreasing as long as the corresponding set of bin edges is not a fixed point. Since there is no common cost in our signaling games setup, such an approach is not feasible.

In the case of sources with semi-unbounded support, convergence to the unique equilibrium is guaranteed depending on the sign of b.

Theorem 18.2.11 *The following hold.*

(i) Consider a source that has a strictly log-concave density with a support on $[m_L, \infty)$ *for some* $m_L \in \mathbb{R}$. *If* $b > 0$, *then the best response iterations through* (18.6) *starting with any N bins converge to the unique equilibrium with N bins.*

(ii) Consider a source that has a strictly log-concave density with a support on $(-\infty, m_U]$ *for some* $m_U \in \mathbb{R}$. *If* $b < 0$, *then the best response iterations through* (18.6) *starting with any N bins converge to the unique equilibrium with N bins.*

For sources with semi-unbounded support, if the sign of b does not satisfy the conditions of Theorem 18.2.11, convergence is not guaranteed; see the following remark.

Remark 18.2.5 For the two scenarios not covered by Theorem 18.2.11 (i.e., $[m_L, \infty)$ along with $b < 0$, and $(-\infty, m_U]$ along with $b > 0$), even if there exists an equilibrium with N bins for a given bias b, an initial set of bin edges forming a partition with N bins may not converge to the unique equilibrium with N bins. For instance, for a source with a support on $[m_L, \infty)$, if the initial values for the first two bin edges (i.e., m_1 and m_2 with $m_L < m_1 < m_2$) are close to m_L and $b < 0$, then after the first iteration, the resulting first bin edge may be less than the value of m_L. As a result, the sender reduces the number of bin edges during the best response iterations. This implies that even though iterations start with N

bins, they may converge to an equilibrium with K bins with $K < N$. In this case, it is guaranteed that best response iterations converge to an informative equilibrium with at least two bins (assuming that such an equilibrium exists). A similar result also holds for sources with bounded support so that for such sources, best response iterations starting with N bins may not converge to the unique equilibrium with N bins regardless of the value or the sign of b. Nevertheless, best response iterations always converge to an informative equilibrium with at least two bins (assuming one exists).

One can obtain more explicit results for exponential and Gaussian distributions, as captured by the results reported below. We refer the reader to [411] for proofs and a detailed analysis.

Theorem 18.2.12 *Consider the signaling game with criteria given in (18.2).*

(i) When the source has an exponential distribution with parameter λ, for $b > 0$ and any number of bins N, there exists a unique equilibrium. Furthermore, for $b > 0$, there also exists a unique equilibrium with (countably) infinitely many bins.
(ii) For $b < 0$, any Nash equilibrium has at most $\lfloor -\frac{1}{2b\lambda} + 1 \rfloor$ bins.
(iii) For Nash equilibria with K and N bins where $N > K$, the equilibrium with N bins leads to smaller expected costs at equilibrium for both players than the case with K bins, regardless of the value of b. For $b > 0$, the Nash equilibrium with (countably) infinitely many bins yields the smallest expected costs at equilibrium for both players.

When the source has Gaussian distribution, similar results hold. Since the Gaussian pdf is log-concave with unbounded two-sided support, we state the following result which in part is a corollary of Theorem 18.2.5 and Theorem 18.2.9.

Corollary 18.2.1 *Consider the signaling game with criteria given in (18.2). Let the source M have a Gaussian distribution such that $M \sim \mathcal{N}(\mu, \gamma^2)$.*

(i) Given any finite N, there exists a unique Nash equilibrium with N bins regardless of the value of b.
(ii) A Nash equilibrium corresponding to a higher number of bins is more informative, i.e., leads to lower expected costs for both players.
(iii) There exist Nash equilibria with (countably) infinitely many bins. At a Nash equilibrium with infinitely many bins, as the bin edges get very large in absolute value (i.e., $m_i \to \infty$ for $b > 0$ and $m_i \to -\infty$ for $b < 0$ as $i \to \infty$), the bin-lengths converge to $2|b|$.

Another interesting scenario is when the source has a uniform distribution. The following theorem covers this case, where we take the support set to be the unit interval, without any loss of generality.

Theorem 18.2.13 ([189]) *Consider the signaling game with criteria given in (18.2). When the source M is uniformly distributed on the interval $[0, 1]$, the upper bound on the number of bins at a Nash equilibrium is given by*

$N_{max}(b) = \lceil -\frac{1}{2} + \frac{1}{2}\left(1 + \frac{2}{|b|}\right)^{1/2} \rceil$, and there exists a unique equilibrium with any number of bins $N = 1, 2, \ldots, N_{max}(b)$. In particular, when $|b| \geq 1/4$, the only equilibrium is non-informative with a single bin. For $|b| < 1/4$, on the other hand, the most informative Nash equilibrium is reached with $N = N_{max}(b)$.

18.2.2 Multi-Dimensional Cheap Talk

When the source is multi-dimensional, the signaling game possesses substantial structural richness, with also a rich class of Nash equilibria. We show in this subsection that this richness is due to the subtle geometric structure the problem features, with consequences in both system design, presence of linear equilibria, and an information theoretic problem formulation for high-dimensional sources.

We first provide a set of geometric conditions that need to be met for a Nash equilibrium to exist with a multi-dimensional source. Then, we consider independent and identically distributed sources and characterize necessary and sufficient conditions under which an informative linear equilibrium exists. These conditions involve the bias vector, which is the source of misaligned costs. Depending on certain conditions related to the bias vector, the existence of linear equilibria requires that source have a Gaussian or a general symmetric density. Moreover, in the case of Gaussian sources, the results have a rate-distortion theoretic implication in that the achievable rates and distortions in the considered game theoretic setup can be obtained from its team-theoretic counterpart.

Consider the following multi-dimensional cheap talk problem. The encoder knows the value of an n-dimensional random vector $\boldsymbol{M} = [M_1, \ldots, M_n]^T$ where $M_1, \ldots, M_n$ are each $\mathbb{M}$-valued random variables. The encoder conveys a message $Z = [Z_1, \ldots, Z_n]^T$ via an encoding policy $\gamma^e(\cdot)$, i.e., $Z = \gamma^e(\boldsymbol{M})$, where Z_i is a $\mathcal{Z}_i$-valued random variable for $i = 1, \ldots, n$. The decoder directly observes Z and takes an action $\boldsymbol{U} = [U_1, \ldots, U_n]^T$ using a policy γ^d, that is $\mathbf{U} = \gamma^d(Z)$, where U_i's are $\mathbb{M}$-valued random variables. The aim of the encoder is to minimize $J^e(\gamma^e, \gamma^d) = E[c^e(\boldsymbol{M}, \boldsymbol{U})]$, where

$$c^e(\boldsymbol{m}, \boldsymbol{u}) = \sum_{i=1}^{n} (m_i - u_i - b_i)^2 = \|\boldsymbol{m} - \boldsymbol{u} - \boldsymbol{b}\|^2. \tag{18.7}$$

In (18.7), $\boldsymbol{b}$ denotes a deterministic bias vector which is common knowledge between the players and quantifies the degree of misalignment between the objective functions of the encoder and the decoder. The decoder wishes to estimate the source random vector as accurately as possible, and thus, its objective function does not include a bias vector. In particular, the aim of the decoder is to minimize $J^d(\gamma^e, \gamma^d) = E[c^d(\boldsymbol{M}, \boldsymbol{U})]$ where

$$c^d(\boldsymbol{m}, \boldsymbol{u}) = \sum_{i=1}^{n} (m_i - u_i)^2 = \|\boldsymbol{m} - \boldsymbol{u}\|^2. \tag{18.8}$$

We make the following assumption on the source distribution.

Assumption 18.2.1 *For any component M_i of the source random vector $\boldsymbol{M}$, every non-empty open subset of its support set has a positive measure.*

Lemma 18.2.1 ([412]) *Consider an n-dimensional cheap talk problem with the source random vector $\boldsymbol{M} = [M_1, \ldots, M_n]^T$ where each element of $\boldsymbol{M}$ can have different distributions and can be dependent or independent. Let $\mathcal{B}^\alpha$ and $\mathcal{B}^\beta$ be two bins at equilibrium and let $\boldsymbol{u}^\alpha = E[\boldsymbol{M}|\boldsymbol{M} \in \mathcal{B}^\alpha]$ and $\boldsymbol{u}^\beta = E[\boldsymbol{M}|\boldsymbol{M} \in \mathcal{B}^\beta]$ denote their centroids which are the decoder actions taken when the encoder reveals $\boldsymbol{M} \in \mathcal{B}^\alpha$ and $\boldsymbol{M} \in \mathcal{B}^\beta$, respectively.*

(i) These decoder actions must satisfy the following necessary condition at a Nash equilibrium:

$$2\,|(\boldsymbol{u}^\beta - \boldsymbol{u}^\alpha)^T \boldsymbol{b}| \le \|\boldsymbol{u}^\beta - \boldsymbol{u}^\alpha\|^2. \tag{18.9}$$

(ii) At a Nash equilibrium, the encoder decomposes the complete observation space into two regions via a hyperplane orthogonal to $(\boldsymbol{u}^\alpha - \boldsymbol{u}^\beta)$ and intersecting the line connecting $\boldsymbol{u}^\alpha$ and $\boldsymbol{u}^\beta$, and $\mathcal{B}^\alpha$ and $\mathcal{B}^\beta$ are subsets of these respective regions. In particular, $\mathcal{B}^\alpha$ must be a subset of $\boldsymbol{m}$ values satisfying $h(\boldsymbol{m}) \ge 0$ whereas $\mathcal{B}^\beta$ must be a subset $\boldsymbol{m}$ values satisfying $h(\boldsymbol{m}) \le 0$ where

$$h(\boldsymbol{m}) \triangleq \left(\boldsymbol{m} - \left(\frac{\boldsymbol{u}^\beta + \boldsymbol{u}^\alpha}{2} + \boldsymbol{b}\right)\right)^T (\boldsymbol{u}^\beta - \boldsymbol{u}^\alpha), \tag{18.10}$$

and $h(\boldsymbol{m}) = 0$ defines the hyperplane on which the encoder is indifferent between either decoder actions, i.e., these $\boldsymbol{m}$ values can belong to both $\mathcal{B}^\alpha$ and $\mathcal{B}^\beta$.

(iii) At a Nash equilibrium where the encoder uses quantization policies, the quantization bins are necessarily convex.

Remark 18.2.6 In the special case of one-dimensional cheap talk as captured earlier by Theorem 18.2.2, the condition in (18.9) reduces to $|u^\alpha - u^\beta| \ge 2|b|$. In fact, the only way this inequality holds with equality is when one of the decoder actions is at the boundary of the support. In this case, the bin for the corresponding decoder action, say u^β, becomes a singleton consisting only of u^β, i.e., $\mathcal{B}^\beta = \{u^\beta\}$. In this case, the scenario without the bin $\mathcal{B}^\beta = \{u^\beta\}$ is still an equilibrium which is equivalent to the equilibrium with $\mathcal{B}^\beta = \{u^\beta\}$. Therefore, for the one-dimensional cheap talk, a necessary condition for an equilibrium can essentially be stated as $|u^\alpha - u^\beta| > 2|b|$ by disregarding the decoder action at the boundary of the support.

Theorem 18.2.14 ([412]) *Consider a two-dimensional cheap talk problem with sources M_1 and M_2 which are i.i.d., with the corresponding bias terms b_1 and b_2.*

(i) For $b_1 = 0$ and/or $b_2 = 0$, there exists an informative equilibrium with a linear encoder where the encoder completely reveals the source(s) corresponding to zero bias term(s).
(ii) For $|b_1| \neq |b_2|$, there exists an informative equilibrium with a linear encoder if and only if the source distribution is Gaussian.
(iii) For $b_1 = b_2$, there exists an informative equilibrium with a linear encoder if and only if the source distribution is symmetric around its mean, i.e., denoting the density of M_1 by $f(\cdot)$, we have that $f(\mu + x) = f(\mu - x)$ for almost all x where $E[M_1] = \mu$.
(iv) For $b_1 = -b_2$, there exists an informative equilibrium with a linear encoder regardless of the source distribution.

The following theorem provides more explicit conditions for the case when the source distributions are Gaussian; building also on Corollary 18.2.1.

Theorem 18.2.15 ([412]) *If the source distribution is n-dimensional ($n \in \mathbb{N}, n > 1$) i.i.d. Gaussian, then there exists an equilibrium with a linear encoding policy where the encoder reveals all or a subset of $(n-1)$ sub-vectors completely, and applies a signaling game policy for the remaining sub-vectors with any number of bins.*

18.2.3 Quadratic Cheap Talk: Stackelberg Equilibria (Bayesian Persuasion)

Theorem 18.2.16 *For the quadratic multi-dimensional cheap talk, the Stackelberg equilibrium is unique and corresponds to a fully revealing (fully informative) encoder policy.*

Proof Under the solution concept of Stackelberg equilibrium, the encoder knows that the decoder will use $\gamma^d(x) = u = E[m|x]$ as an optimal decoder policy to minimize its expected cost. Then the goal of the encoder is to minimize the following:

$$\begin{aligned} \min_{x=\gamma^e(m)} E[\|m - u - b\|^2] &= \min_{x=\gamma^e(m)} E[\|m - E[m|x] - b\|^2] \\ &= \min_{x=\gamma^e(m)} E[\|m - E[m|x]\|^2] + \|b\|^2 \\ &= \min_{x=\gamma^e(m)} E[\|m - u\|^2] + \|b\|^2 . \end{aligned}$$

Here, the second equality follows from the law of the iterated expectations. Since the goal of the decoder is to minimize $\min_{u=\gamma^d(x)} E[\|m - u\|^2]$, the goals of the encoder and the decoder become essentially the same, which effectively reduces the game setup to a *team* setup. In the team setting, the equilibrium is fully informative; i.e. the encoder reveals all of its information. ◇

18.2.4 Quadratic Signaling Game: Noisy Channel Case

We now move to study the noisy channel setting depicted in Fig. 18.2, with a soft power constraint for the encoder. We assume throughout this discussion that all distributions are Gaussian and that the setting is scalar. Specifically:

The encoder (DM1) encodes a zero-mean Gaussian random variable M with variance σ_m^2, and sends the resulting real-valued random variable X over a channel. During the transmission, the zero mean Gaussian noise with variance σ_w^2 is added to X; hence, the decoder (DM2) receives $Y = X + W$. The policy spaces of the DMs are similarly defined as the noiseless case. The cost functions of the encoder and the decoder are given as follows: DM 1 aims to minimize

$$J^e(\gamma^e, \gamma^d) = \int c^e(m, x, u)\gamma^e(\mathrm{d}x|m)\gamma^d(\mathrm{d}u|y)P(\mathrm{d}y|x)P(\mathrm{d}m),$$

whereas DM 2 aims to minimize

$$J^d(\gamma^e, \gamma^d) = \int c^d(m, u)\gamma^e(\mathrm{d}x|m)\gamma^d(\mathrm{d}u|y)P(\mathrm{d}y|x)P(\mathrm{d}m),$$

over their respective policies, where $P(\mathrm{d}y|x) = P(W \in \mathrm{d}y - x)$ with $W \sim \mathcal{N}(0, \sigma_w^2)$. The cost functions are given as $c^e\,(m, x, u) = (m - u - b)^2 + \lambda x^2$ and $c^d\,(m, u) = (m - u)^2$, where $\lambda > 0$, and b is a constant as before. Note that a power constraint with an associated positive multiplier is appended to the cost function of the encoder, which corresponds to a power limitation for the transmitter.

We first discuss Nash equilibria of this quadratic-Gaussian signalling game with a noisy channel (Theorem 18.2.17) and then the Stackelberg equilibrium (Theorem 18.2.18). Toward that end, let us first introduce the notation $\alpha := \sigma_m^2/\sigma_w^2$, which is the signal-to-noise ratio. Then, we have:

Theorem 18.2.17

(i) Let $0 < \lambda < \alpha$. For any $b \in \mathbb{R}$, there exists an essentially unique informative linear Nash equilibrium, independent of b, with sender's policy $f^e(M) = \pm[\frac{1}{\alpha}(\sqrt{\alpha/\lambda} - 1]^{\frac{1}{2}}$, and receiver's policy $f^d(Y) = E[M|Y]$, where $Y = f^e(M) + W$.

(ii) *If $\lambda \geq \alpha$, there does not exist an informative (affine or non-linear or even randomized) Nash equilibrium. The only Nash equilibrium is the non-informative one.*

One important observation we can make is that the solution above is independent of the bias parameter b, in both regions of λ. This implies that the two Nash equilibria are the same for the case when there is no bias, which is the standard information transmission problem, known also as the Gaussian Text Channel (GTC), which comes in two versions, hard or soft power constraints on the encoder (see Sect. 3.3.4), and in this case we have the latter. We note that the signaling game with $b = 0$ is actually *strategically equivalent* to a team problem, where the two DMs have essentially identical cost functions, which can be seen by adding the term λx^2 to J^d, which does not affect the optimum response of DM2 since this additional term does not involve the decision variable of DM2. This team problem is the soft-constrained GTC, the globally team-optimal solution to which over all policies (linear and nonlinear) is precisely the one given in the theorem.

We now move on to the Stackelberg equilibrium, since the signaling game is strategically equivalent to a team problem, and in a team problem one can follow a sequential process to obtain the team-optimal solution (in this case first DM2 optimizes, and then DM1), the Stackelberg equilibrium policies are also team optimal.

Theorem 18.2.18 *The only equilibrium in the Stackelberg setup of the signaling game is the linear equilibrium.*

For the proof, one can show that an information theoretic lower bound (following Sects. 16.2 and 3.3.4) is achieved under a linear policy.

One could also consider dynamic and multi-dimensional versions of the problems.

18.3 Signaling Games with Subjective Priors Between an Encoder and a Decoder

In this section, our primary focus is on the variation where the priors are different but the cost function may (or may not) be identical. In the following, to present the primary message of the section as clearly as possible, we will consider the scalar signaling game setting, with a noisy channel as depicted in Fig. 18.2. In full generality, the encoder's goal is to minimize

$$J^e(\gamma^e, \gamma^d) = E_e\left[c^e(m, x, u)\right],$$

whereas the decoder's goal is to minimize

$$J^d(\gamma^e, \gamma^d) = E_d\left[c^d(m, x, u)\right],$$

where $c^e(m, u)$ and $c^d(m, u)$ denote the cost functions of the encoder and the decoder, respectively, when the action u is taken for the corresponding message m, γ^e and γ^d are the policies of the encoder and decoder, respectively, belonging to appropriate policy spaces Γ^e and Γ^d, and $E_e[\cdot]$ and $E_d[\cdot]$ denote that the expectation is taken with respect to the probability measures from the perspectives of the encoder and the decoder, respectively.

Remark 18.3.1 Note that, under the Stackelberg solution concept, the encoder must know the decoder's subjective prior so that it, as a leader, can anticipate the decoder's optimal response; but the decoder does not have to know the encoder's prior. For the case of Nash equilibrium, however, the agents do not need to know their subjective priors; they know only their policies as they (simultaneously) announce to each other, see [87].

18.3.1 The Commitment Setup with Quadratic Costs: Stackelberg Equilibria

In this section, we analyze the Stackelberg equilibria when there is either a soft power constraint or a hard power constraint on the encoder, and the objective functions are quadratic. In the case of soft power constraint, the encoder has the following objective function:

$$J^e(\gamma^e, \gamma^d) = E_e\left[c^e(m, x, u)\right], \tag{18.11}$$

and the decoder has the objective function

$$J^d(\gamma^e, \gamma^d) = E_d\left[c^d(m, u)\right], \tag{18.12}$$

where $c^e(m, x, u) = (m - u)^2 + \lambda x^2$ and $c^d(m, x) = (m - u)^2$, and λ is a positive weight parameter on the power at the input to the channel, with the second term in c^e characterizing the soft power constraint included in the sender's objective. Note that it is possible to add this additional λx^2 term to the cost function of the decoder, without affecting the analysis as this term is irrelevant to the optimization problem faced by the decoder. In other words, misalignment between the encoder and the decoder arises solely from subjective probabilistic beliefs of the players and not from the considered costs. For the case where the encoder has a hard power constraint instead of a soft power constraint, the goal of the encoder is again to minimize

$$J^e(\gamma^e, \gamma^d) = E_e\left[c^e(m,u)\right] \text{s.t. } E_e\left[\left(\gamma^e(m)\right)^2\right] \le \overline{P}, \tag{18.13}$$

whereas the decoder aims to minimize

$$J^d(\gamma^e, \gamma^d) = E_d\left[c^d(m,u)\right], \tag{18.14}$$

where

$$c^e(m,u) = (m-u)^2 = c^d(m,u). \tag{18.15}$$

Remark 18.3.2 If both players share the same probabilistic belief on the source distribution, then the problem reduces to the classical team theoretic setup with a power-constrained encoder. Although obtaining optimal coding/decoding policies for general source distributions is in general difficult, for the special case of scalar Gaussian source, known as the Gaussian Test Channel (GTC), the optimal solution involves linear policies at the encoder and decoder, see Theorem 16.2.2 in Chap. 16.

Robustness to Perturbations Around a Nominal Team

We note that if the priors are the same, then we have a team problem, and what has been formulated above then becomes a standard communication problem with message transmission over a noise-additive channel under hard or soft power constraints on the encoder. In this subsection, we investigate robustness of the (Stackelberg) equilibria around a nominal team setup first for general subjective source priors, and then for the Gaussian priors case. What we discuss in this subsection is what happens if there is small independent deviation in the perceptions of the two agents (sender and the receiver) in which case we no longer have a team problem, and the resulting game is viewed under the Stackelberg equilibrium with the sender (encoder) as leader. We will particularly analyze whether the expected cost function of the sender depends continuously on such a variation in the priors.

Toward that goal, let $\mathbb{X} = \mathbb{R}$ and let $\mathcal{P}(\mathbb{X})$ denote the family of all probability measures on $(\mathbb{X}, \mathcal{B}(\mathbb{X}))$ where $\mathcal{B}(\mathbb{X})$ denotes the Borel σ-field on $\mathbb{X}$. In the following, we will use some results from Appendix C on the convergence of sequences of probability measures.

The main result of this subsection is the following theorem on continuity and robustness properties of the encoder's expected cost around the point of identical priors. We focus here on the expected cost of the encoder since in the Stackelberg game the encoder optimizes its expected cost, and thus the encoder's objective function determines whether the equilibrium is robust or not.

Theorem 18.3.1 ([413]) *Consider the quadratic scalar signaling game with subjective priors, as formulated in this section. Suppose that the priors of the source is $\varphi_e(\cdot)$ and $\varphi_d(\cdot)$ from the perspectives of the sender and the receiver, respectively,*

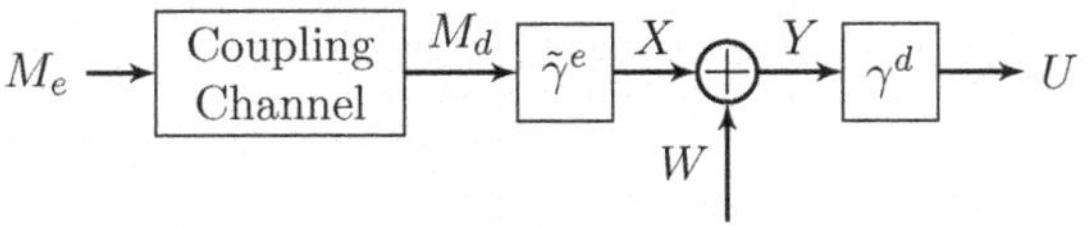

Fig. 18.3 Suboptimal scheme used in proving upper semi continuity result where the coupling channel is such that the marginal distribution of its output is $\varphi_d(\cdot)$

where these prior distributions are arbitrary elements of $\mathcal{P}(\mathbb{X})$. Suppose further that the variance of the source under $\varphi_d(\cdot)$ is finite. Then, we have the following.

(i) The expected Stackelberg equilibrium cost for the encoder under both the soft power constraint and the hard power constraint are upper semi continuous (under the Wasserstein metric) as subjective prior of the encoder approaches the subjective prior of the decoder, i.e., $\varphi_e(\cdot) \to \varphi_d(\cdot)$.

(ii) If the prior distributions $\varphi_e(\cdot)$ and $\varphi_d(\cdot)$ are Gaussian, the expected Stackelberg equilibrium cost for the encoder under both the soft power constraint and the hard power constraint are lower semi continuous as subjective prior of the encoder approaches the subjective prior of the decoder, i.e., $\varphi_e(\cdot) \to \varphi_d(\cdot)$.

(iii) If the prior distributions $\varphi_e(\cdot)$ and $\varphi_d(\cdot)$ are Gaussian, the expected Stackelberg equilibrium cost for the encoder under both the power constraint and the hard power constraint are continuous with respect to perturbations around a corresponding nominal team (under both the Wasserstein metric and the weak convergence topology).

(iv) The expected Stackelberg equilibrium cost for the encoder under the soft power constraint and the hard power constraint are lower semi continuous as subjective prior of the decoder approaches the subjective prior of the encoder, i.e., $\varphi_d(\cdot) \to \varphi_e(\cdot)$.

Proof In the following, we focus on the case with soft power constraint and the case with hard power constraint can be established in a similar manner.

(i) Let M_e and M_d denote random variables distributed according to the priors $\varphi_e(\cdot)$ and $\varphi_d(\cdot)$, respectively. The Stackelberg equilibrium cost for the encoder is given by

$$\inf_{\gamma^e(\cdot)} E[(m_e - u)^2] + \lambda E[\gamma^e(m_e)^2] \tag{18.16}$$

where u is chosen as the best response of the receiver under its own prior, given by $u = E_d[m|\gamma^e(m) + w]$. Now consider a suboptimal scheme in which the sender maps its observation into an auxiliary variable via a function $f(\cdot)$ which is then mapped into the transmitted message through $\tilde{\gamma}^e(\cdot)$ under constraint that the marginal distribution of this auxiliary random variable is fixed to $\varphi_d(\cdot)$. This scheme is depicted in Fig. 18.3. By introducing such an auxiliary random variable which has a fixed marginal distribution, the proposed scheme is in general a suboptimal one. Here, the encoder is given by $\gamma^e(m^e) = \tilde{\gamma}^e(f(m_e))$ where $f(m_e)$ has a fixed marginal distribution $\varphi_d(\cdot)$. Now note that

$$\begin{aligned}
&\inf_{\gamma^e(\cdot)} E_e[(m_e - u)^2] + \lambda E_e[\gamma^e(m_e)^2] \\
&\leq \inf_{\tilde{\gamma}^e(\cdot), f(\cdot)} E_e[(m_e - u)^2] + \lambda E_e[\tilde{\gamma}^e(m_d)^2] \\
&= \inf_{\tilde{\gamma}^e(\cdot), f(\cdot)} E_e[(m_e - m_d + m_d - u)^2] + \lambda E_e[\tilde{\gamma}^e(m_d)^2] \\
&= \inf_{\tilde{\gamma}^e(\cdot), f(\cdot)} E_e[(m_e - m_d)^2] + E_e[(m_d - u)^2] \\
&\qquad + 2E_e[(m_e - m_d)(m_d - u)] + \lambda E_e[\tilde{\gamma}^e(m_d)^2] \\
&\leq \inf_{\tilde{\gamma}^e(\cdot), f(\cdot)} E_e[(m_e - m_d)^2] + E_e[(m_d - u)^2] \\
&\qquad + 2\sqrt{E_e[(m_e - m_d)^2]}\sqrt{E_e[(m_d - u)^2]} \\
&\qquad + \lambda E_e[\tilde{\gamma}^e(m_d)^2] \qquad (18.17)
\end{aligned}$$

where the first inequality is due to suboptimality of the scheme with $f(\cdot)$ and $\tilde{\gamma}^e(\cdot)$, and the second inequality follows from the Cauchy-Schwarz inequality. The final optimization problem in (18.17) can be first solved with respect to $f(\cdot)$ which leads to

$$\begin{aligned}
&\min_{\tilde{\gamma}^e(\cdot), f(\cdot)} E_e[(m_e - m_d)^2] + E_e[(m_d - u)^2] + \lambda E_e[\tilde{\gamma}^e(m_d)^2] \\
&\qquad + 2\sqrt{E_e[(m_e - m_d)^2]}\sqrt{E_e[(m_d - u)^2]} \\
&= \min_{\tilde{\gamma}^e(\cdot)} W_2(\varphi_e, \varphi_d)^2 + E_e[(m_d - u)^2] + \lambda E_e[\tilde{\gamma}^e(m_d)^2] \\
&\qquad + 2\, W_2(\varphi_e, \varphi_d)\sqrt{E_e[(m_d - u)^2]} \qquad (18.18)
\end{aligned}$$

where $W_2(\varphi_e, \varphi_d)$ is the quadratic Wasserstein distance between the distributions $\varphi_e(\cdot)$ and $\varphi_d(\cdot)$. Note that the optimal value for the solution to (18.18) is finite when $W_2(\varphi_e, \varphi_d)$ is finite. To see this, note that if $\tilde{\gamma}^e(m_d) = 0$ for all m_d, then the optimal receiver action becomes $u = \mu_d$ where μ_d denotes the mean of the source message from the decoder's perspective. Then, the objective function in (18.18) takes the value of $(W_2(\varphi_e, \varphi_d) + E_e[(m_d - \mu_d)^2]^{1/2})^2$, which is finite as the variance of the source is finite from the perspective of the decoder. Thus, an optimal solution to (18.18) must yield a finite objective value. From this observation, it follows that as $\varphi_e \to \varphi_d$, the optimal solution of (18.18) leads to a finite value for the term $E_e[(m_d - u)^2]$ since $W_2(\varphi_e, \varphi_d) \to 0$ in this case. As a result, we get

$$\limsup_{\varphi_e \to \varphi_d} \inf_{\gamma^e(\cdot)} E_e[(m_e - u)^2] + \lambda E_e[\gamma^e(m_e)^2]$$

$$\leq \limsup_{\varphi_e \to \varphi_d} \inf_{\tilde{\gamma}^e(\cdot)} W_2(\varphi_e, \varphi_d)^2 + E_e[(m_d - u)^2]$$

$$+ 2\, W_2(\varphi_e, \varphi_d)\sqrt{E_e[(m_d - u)^2]}$$

$$+ \lambda E_e[\tilde{\gamma}^e(m_d)^2]$$

$$= E_e[(m_d - E_d[m_d|\gamma^{e,*}(m_d) + w])^2] + \lambda E_e[\gamma^{*,e}(m_d)^2]$$

where the inequality is due to (18.17) and (18.18), $\gamma^{*,e}(\cdot)$ denotes the optimal encoding policy for the nominal team setup with common prior distribution $\varphi_d(\cdot)$, and the equality follows from the facts that $E[(m_d - u)^2]$ is finite and $W_2(\varphi_e, \varphi_d) \to 0$ as $\varphi_e \to \varphi_d$.

Since the last term corresponds to the expected cost for the nominal team setup with prior $\varphi_d(\cdot)$, this analysis shows that the expected Stackelberg equilibrium cost for the encoder is upper semi continuous around the nominal team.

(ii) We note that if the receiver employs the encoder's prior rather than its own prior, the objective function of the sender is improved. Therefore, the following inequality holds:

$$\inf_{\gamma^e(\cdot)} E_e[(m - E_d[m|\gamma^e(m) + w])^2] + \lambda E_e[x^2]$$

$$\geq \inf_{\gamma^e(\cdot)} E_e[(m - E_e[m|\gamma^e(m) + w])^2] + \lambda E_e[x^2] \tag{18.19}$$

For the latter optimization problem, we have seen earlier (see Theorem 16.2.2 in Chap. 16) that the optimal encoding and decoding policies are affine. By using this observation, we get

$$\liminf_{\varphi_e \to \varphi_d} \inf_{\gamma^e(\cdot)} E_e[(m - E_d[m|\gamma^e(m) + w])^2] + \lambda E_e[x^2]\mathbb{N} \tag{18.20}$$

$$\geq \liminf_{\varphi_e \to \varphi_d} \inf_{\gamma^e(\cdot)} E_e[(m - E_e[m|\gamma^e(m) + w])^2] + \lambda E_e[x^2]\mathbb{N} \tag{18.21}$$

$$= \liminf_{\varphi_e \to \varphi_d} E_e[(m - \beta_1^{*,d}(\alpha_1^{*,e} m + \alpha_2^{*,e} + w) - \beta_2^{*,d})^2]\mathbb{N} \tag{18.22}$$

$$+ \lambda E_e[(\alpha_1^{*,e} m + \alpha_2^{*,e})^2]\mathbb{N} \tag{18.23}$$

$$= E_d[(m - \beta_1^{*,d}(\alpha_1^{*,d} m + \alpha_2^{*,d} + w) - \beta_2^{*,d})^2]\mathbb{N} \tag{18.24}$$

$$+ \lambda E_d[(\alpha_1^{*,d} m + \alpha_2^{*,d})^2] \tag{18.25}$$

where $\alpha_1^{*,e}$, $\alpha_2^{*,e}$, $\beta_1^{*,e}$ and $\beta_2^{*,e}$ (resp. $\alpha_1^{*,d}$, $\alpha_2^{*,d}$, $\beta_1^{*,d}$ and $\beta_2^{*,d}$) are the optimal coefficients at the affine encoder and decoder for the problem of transmitting a Gaussian source M_e (resp. M_d) over an independent additive Gaussian channel under quadratic criterion with soft power constraint. This establishes the lower semi continuity of the Stackelberg equilibrium cost around the point of identical priors.

(iii) The result follows from (i) and (ii).

(iv) The lower semi continuity result in this case can be established via the following argument: The inequality in (18.19) holds for general subjective priors and if the decoder employs the encoder's prior rather than its own prior while computing its estimate, the expected cost of the encoder will decrease. From this inequality, we get

$$\liminf_{\varphi^d \to \varphi^e} \inf_{\gamma^e(\cdot)} E_e[(m - E_d[m|\gamma^e(m) + w])^2] + \lambda E_e[x^2]$$

$$\geq \liminf_{\varphi^d \to \varphi^e} \inf_{\gamma^e(\cdot)} E_e[(m - E_e[m|\gamma^e(m) + w])^2] + \lambda E_e[x^2]$$

$$= \inf_{\gamma^e(\cdot)} E_e[(m - E_e[m|\gamma^e(m) + w])^2] + \lambda E_e[x^2],$$

where the equality follows from the fact that the optimization problem does not depend on $\varphi_d(\cdot)$. This proves the lower semi-continuity of the encoder's expected cost.

◇

Theorem 18.3.1 reveals an interesting duality property of encoder's cost in the sense that if $\varphi_e \to \varphi_d$, then upper semi continuity holds whereas if $\varphi_d \to \varphi_e$, then lower semi continuity holds. On the other hand, for slightly mismatched Gaussian priors, both upper semi continuity and lower semi continuity hold, which proves the robustness of the Stackelberg equilibria around the point of identical priors.

Remark 18.3.3 We note that the property of upper semi continuity also holds when the source is multidimensional [413]. On the other hand, when the source is multidimensional Gaussian from the perspective of both players, then the analysis regarding lower semi continuity in Theorem 18.3.1 does not apply since the optimal encoding policy may not be linear for the nominal team setup with a multidimensional Gaussian source (as we have seen earlier in Chap. 16). Nonetheless, if there is an affine policy restriction for the encoder, a similar analysis to that in (18.25) can be carried out to obtain lower semi continuity result for a setup with multidimensional Gaussian priors.

Affine Policies Are not Optimal for Quadratic Gaussian Signaling with Gaussian Subjective Priors

We have seen earlier that the optimal encoder and decoder policies for the Gaussian Test Channel are affine. The question then comes up whether in the search of the Stackelberg solution of the game arrived at with subjective priors, still Gaussian, one can restrict the search to affine policies without any loss of generality. The following theorem says that this is not necessarily so, and in the quadratic signaling game with subjective Gaussian priors nonlinear encoder policies may lead to better performance for the encoder (leader) than affine policies.

Theorem 18.3.2 ([413]) *Consider the quadratic signaling game with subjective priors with soft or hard power constraint. Then, it is not necessarily true that there exists a Stackelberg equilibrium solution which leads to affine encoder-decoder policies.*

Proof It suffices to provide an example where for a given set of parameters a quantization policy leads to a lower expected cost for the encoder than the case of Stackelberg equilibrium with policies restricted to be affine. The example we provide is for the soft-constrained case. A similar one can be constructed when the constraint is hard. Let the subjective priors be zero-mean Gaussian random variables with variances $\sigma_e^2 = 6.25$, $\sigma_d^2 = 0.25$ for the encoder and the decoder, respectively, and let the channel noise also be zero-mean Gaussian, with variance $\sigma_W^2 = 0.25$. Finally, let the soft power constraint weighting be $\lambda = 1.5$. The aim of the encoder is to solve the optimization problem (18.11) after plugging in the best response of the receiver, which is given by $u = E_d[m|\gamma^e(m) + w]$. Under the affine policy restriction at the encoder, the best response of the decoder also becomes affine. Numerically solving the optimization problem at the sender under the affine policy restriction leads to the optimal encoding policy $\gamma^{*,e}(m) = 0.30\,m$ and the corresponding cost of the encoder is given by $J_{\text{affine}}^{*,e} = 6.12$. Now consider an encoding policy in the form of a quantization policy specified by $\gamma^e(m) = \sqrt{P}\,\text{sgn}(m)$ with $P = 0.5$. The best response of the decoder to such a two-bin quantization policy at the sender can be computed to be $\gamma^d(y) = \sqrt{2/\pi}\sigma_d \tanh(\sqrt{P}y/\sigma_W^2)$. The expected cost of the encoder with these encoding and decoding policies becomes

$$J_{\text{nonlinear}}^e = E_e\left[\left(m - \sqrt{\frac{2}{\pi}}\sigma_d \tanh\left(\frac{P\text{sgn}(m) + \sqrt{P}w}{\sigma_W^2}\right)\right)^2\right] + \lambda P.$$

By computing the expectation via numerical integration, we get $J_{\text{nonlinear}}^e \approx 5.90$. This shows that even such a crude quantization policy outperforms the best affine policy. ◇

Theorem 18.3.2 thus states that when the agents have different perceptions of the prior probability of the message, then the optimality of affine policies (which holds under identical priors) may break down. Similar to the case with soft power constraints, with hard power constraint, it is possible to construct a quantization policy at the encoder which outperforms the best affine policy at the sender.

Gaussian Signaling with Affine Policies

For the quadratic-Gaussian signaling game, with different subjective priors for the agents, and with a soft power constraint on the encoder, we now study the Stackelberg equilibrium when policies are restricted to be affine. We then address the same under a hard constraint on the encoder. Note that for both problems, we already know from Theorem 18.3.2 that, depending on the parameters of the game, there may exist nonlinear policies providing better performance for the encoder than the best affine policy.

In the following, we take the perception of the sender (encoder) on the source to be Gaussian with mean μ_e and variance σ_e^2, and that of the receiver (decoder) to be Gaussian with mean μ_d and variance σ_d^2. The channel noise is zero-mean Gaussian with variance σ_W^2.

Theorem 18.3.3 ([413]) *In the quadratic signaling game with subjective priors as above and with soft power constraint, a unique informative affine Stackelberg equilibrium exists if*

$$\lambda\sigma_e^2\sigma_W^2 < \sigma_d^2(2\sigma_e^2 + 2(\mu_e - \mu_d)^2 - \sigma_d^2). \tag{18.26}$$

When (18.26) *does not hold, there exists an informative affine Stackelberg equilibrium if the following conditions simultaneously hold:*

$$3(\sigma_e^2 + (\mu_d - \mu_e)^2) < 2\sigma_d^2, \tag{18.27}$$

$$4\lambda\sigma_e^2\sigma_W^2(\sigma_d^2 - \sigma_e^2 - (\mu_e - \mu_d)^2) \le \sigma_d^2(\sigma_e^2 + (\mu_e - \mu_d)^2)^2. \tag{18.28}$$

Otherwise, the affine Stackelberg equilibrium is non-informative.

Remark 18.3.4 The affine Stackelberg equilibrium can be non-informative or informative depending on the system parameters. This is because the conditions in (18.26)–(18.28) may or may not hold. For instance, the following are examples of informative and non-informative scenarios:

Let $\sigma_e^2 = 1$, $\sigma_d^2 = 4$, $\sigma_W^2 = 0.25$, $\lambda = 2$ and $\mu_e = \mu_d$. In this case, (18.27) holds whereas (18.26) and (18.28) do not hold, which lead to a non-informative equilibrium.

Let $\sigma_e^2 = 1$, $\sigma_d^2 = 4$, $\sigma_W^2 = 0.25$, $\lambda = 1$ and $\mu_e = \mu_d$. As (18.27) and (18.28) are satisfied in this case, the affine Stackelberg equilibrium is informative.

We now study the counterpart of the previous one when there is instead (of soft power constraint) a hard power constraint on the encoder.

Theorem 18.3.4 ([413]) *In the quadratic signaling game with subjective priors and hard power constraint $\overline{P}$, there exists a unique informative Stackelberg* *equilibrium within the class of affine policies if*[2]

$$\frac{\sigma_e^2}{\sigma_e^2 + (\mu_e - \mu_d)^2} - \frac{2\sigma_e^2}{\sigma_d^2} \leq \frac{\overline{P}}{\sigma_W^2}. \tag{18.29}$$

Otherwise, the affine equilibrium is non-informative. If (18.29) *holds, then the cost of the encoder and decoder at the informative affine Stackelberg equilibrium are*

$$J_s^{*,e} = \frac{\overline{P}\sigma_d^4\sigma_e^2\sigma_W^2 + \sigma_W^4\sigma_e^6 + \sigma_W^4\sigma_e^4(\mu_e - \mu_d)^2}{(\overline{P}\sigma_d^2 + \sigma_e^2\sigma_W^2)^2}, \tag{18.30}$$

$$J_s^{*,d} = \frac{\sigma_e^2\sigma_d^2\sigma_W^2}{\overline{P}\sigma_d^2 + \sigma_e^2\sigma_W^2}. \tag{18.31}$$

On the other hand, if (18.29) *does not hold, the affine Stackelberg equilibrium is non-informative with costs* $J_s^{*,e} = \sigma_e^2 + (\mu_e - \mu_d)^2$ *and* $J_s^{*,d} = \sigma_d^2$.

When $\mu_e = \mu_d = \mu_M$ and $\sigma_e = \sigma_d = \sigma_M$, i.e., the team theoretic setup, the left hand side of (18.29) becomes negative, leading to a fully informative scenario, as expected. Unlike the team theoretic solution (of the GTC) which is always informative, the affine Stackelberg equilibrium solution is either informative or non-informative depending on the system parameters. If both players share a common prior with $\sigma_e = \sigma_d = \sigma_M$ and $\mu_e = \mu_d = \mu_M$ leading to a team theoretic setup, then (18.30) and (18.31) reduces to $J_t^{*,e} = J_t^{*,d} = \frac{\sigma_M^2\sigma_W^2}{\overline{P}+\sigma_W^2}$.

Perturbations Around the Nominal Team (GTC) and Robustness Under Affine Policy Restriction From Theorem 18.3.1, we know that the expected Stackelberg equilibrium cost of the sender is continuous with respect to perturbations around the corresponding nominal team problem with equal priors for the special case of Gaussian distributions where there is no affine policy restriction. It is also possible to analyze the continuity of Stackelberg equilibria under affine policy restriction. Note that the team theoretic solution is always informative whereas the Stackelberg solution may be informative or non-informative depending on whether (18.29) holds. For the team theoretic setup, the left hand side of (18.29) is equal to -1 which leads to an informative scenario. The equilibrium cost values in (18.30) and (18.31) are continuously differentiable, which directly leads to a robustness result as long as the condition in (18.29) holds. Note that a necessary condition for a non-informative equilibrium is $\sigma_d^2 \geq 2\sigma_e^2$ as otherwise the right hand side of (18.29) is always negative. Therefore, when a slight perturbation is applied, the equilibrium is still informative. This shows that the affine Stackelberg equilibrium is robust with respect to perturbations around the nominal team setup.

[2] In the case of equality in (18.29), it is possible to have a non-informative equilibrium as well and the informative and non-informative equilibria induce the same cost for the encoder in this case.

18.3.2 The Non-cooperative/No-commitment Setup (Nash Equilibria)

We now discuss Nash equilibria when the encoder have subjective probabilistic beliefs on the source distribution where these subjective prior distributions are Gaussian, i.e., the source is $M \sim \varphi_e(m) = \mathcal{N}(\mu_e, \sigma_e^2)$ and $M \sim \varphi_d(m) = \mathcal{N}(\mu_d, \sigma_d^2)$ from encoder's and decoder's perspective, respectively.

(a) Gaussian Signaling Under a Soft Power Constraint The following theorem provides a condition under which there exists a Nash equilibrium solution which is affine for both the encoder and the decoder, which is informative. When the condition is not satisfied, then there is a Nash equilibrium with a constant encoder policy, which then is non-informative.

Theorem 18.3.5 *If* $\lambda \geq \frac{\sigma_d^2}{\sigma_W^2}$, *the unique affine equilibrium is non-informative; otherwise, there exists a Nash equilibrium solution which is affine and informative. The affine Nash equilibrium is unique within the class of affine policies.*

Remark 18.3.5 Note that none of the parameters that specify policies at the equilibrium depend on the subjective priors from the perspective of the encoder since the encoder minimizes its cost for every realization m of source M without considering its distribution. We also see that the policies for the equilibrium characterized in Theorem 18.3.5 are the same policies as in the team theoretic setup where both the encoder and decoder take $\varphi_d(m)$, i.e., the subjective prior of the decoder in the game theoretic setup, as the source distribution. This is in contrast with the Stackelberg setup where the equilibrium solution does not reduce to a team theoretic solution when either of the player's prior is taken as the common prior in general.

Comparison with the Team Case We now investigate effects of the subjectivity in priors on the equilibrium cost. Due to Remark 18.3.5, only the objective function of the encoder deviates from its value with the team theoretic solution.

Theorem 18.3.6 *In the quadratic signaling games with subjective priors and soft power constraint, when* $\lambda < \frac{\sigma_d^2}{\sigma_W^2}$, *the encoder expected cost* $J_s^{*,e}$ *and the decoder expected cost* $J_s^{*,d}$ *at the Nash equilibrium are*

$$J_s^{*,e} = \sqrt{\lambda \sigma_d^2 \sigma_W^2} \left(\frac{\sigma_e^2 + \sigma_d^2 + (\mu_e - \mu_d)^2}{\sigma_d^2} \right) - \lambda \sigma_W^2 , \tag{18.32}$$

$$J_s^{*,d} = \sqrt{\lambda \sigma_d^2 \sigma_W^2} . \tag{18.33}$$

Otherwise; i.e., if $\lambda \geq \frac{\sigma_d^2}{\sigma_W^2}$, *the expected costs are* $J_s^{*,e} = \sigma_e^2 + (\mu_e - \mu_d)^2$ *and* $J_s^{*,d} = \sigma_d^2$.

If the priors were equal as $\mu_e = \mu_d = \mu_M$ and $\sigma_e = \sigma_d = \sigma_M$; i.e., the team case with a soft power constraint, then the expected costs would be given by

$$J_t^{*,e} = 2\sqrt{\lambda\sigma_M^2\sigma_W^2} - \lambda\sigma_W^2, \quad J_t^{*,d} = \sqrt{\lambda\sigma_M^2\sigma_W^2}$$

for $\lambda < \frac{\sigma_M^2}{\sigma_W^2}$ (i.e., at the informative equilibrium), and

$$J_t^{*,e} = \sigma_M^2, \quad J_t^{*,d} = \sigma_M^2$$

for $\lambda \geq \frac{\sigma_M^2}{\sigma_W^2}$ (i.e., at the non-informative equilibrium).

(b) Gaussian Signaling Under a Hard Power Constraint Unlike the soft power constrained case, there always exists an informative Nash equilibrium for the hard power constrained case as stated in the following theorem. Nevertheless, this informative Nash equilibrium is not the same as the solution of the corresponding team problem which has the decoder's subjective prior taken as the common prior.

Theorem 18.3.7 *There always exists an informative affine Nash equilibrium in the hard power constrained scalar quadratic signaling game (in contrast to the soft power constrained scalar quadratic signaling game).*

We note that although both the Nash equilibrium solution and the team theoretic solution are always informative, the resulting policies are not the same in general. In other words, the subjectivity in the priors affects the equilibrium solution.

18.3.3 Signaling Under Subjective Priors Without Channel Noise: Cheap Talk Revisited

In this section, we consider the setting with no channel noise, as depicted in Fig. 18.1, but with subjective priors for the encoder and the decoder, who however have otherwise identical costs. Note that in this case the decoder observes the encoded message directly, and there is no power constraint at the encoder. We consider encoders and decoders with subjective priors, and to reflect mainly the effects of the subjectivity in the priors we take $c^e(m, u) = c^d(m, u) = \|m - u\|^2$.

Let $\mathbf{M} \sim f_e$ and $\mathbf{M} \sim f_d$ denote the distributions of the source from the perspectives of the encoder and the decoder, respectively. We assume mutual absolute continuity of f_e and f_d; that is, for any Borel set B, $f_e(B) = 0 \implies f_d(B) = 0$ and $f_d(B) = 0 \implies f_e(B) = 0$. Then, we have the following result.

Theorem 18.3.8 *Consider the cheap talk game introduced above, with subjective priors for the encoder and the decoder. Then, if the priors are mutually absolutely continuous, there exists a fully informative Nash equilibrium and a fully informative Stackelberg equilibrium.*

We thus emphasize the following interesting result: Under the mutually absolutely continuous priors assumption, the subjectivity in priors does not make a difference as both the team setting and the game setting result in fully informative equilibria.

18.4 Concluding Remarks

In this chapter, we have presented a rather general formulation of signaling games and have investigated several specific but diverse applications in this framework. These applications are closely related to various team theoretic information constrained decision and control problems studied earlier in the book, and have thus served to also highlight the conceptual and analytical subtleties, and intricacies, that arise in the study of information in a game theoretic framework.

We first studied quadratic signaling games. Structural results on equilibria were established for both Nash and Stackelberg solution concepts. It was shown that all Nash equilibria have to be quantized (that is, the encoder hides part of the information at any equilibrium) when there is bias mismatch between the objectives of the encoder and the decoder. For general log-concave sources, we have shown that given any finite N, there exists a unique equilibrium with N bins for sources with two-sided unbounded support. Similarly, we have proven that for sources with semi-unbounded support, when the support extends to $+\infty$ and the bias term is positive, or the support extends to $-\infty$ and the bias term is negative, for any given finite N, in each case there exists a unique equilibrium with N bins. On the other hand, conversely, for sources with semi-unbounded support, there exists a finite upper bound on the number of bins at Nash equilibrium. We have further shown that, as the number of bins increases, the expected costs of the encoder and the decoder at Nash equilibrium improve, i.e., a more informative equilibrium is obtained. Moreover, we have proven that best response iterations converge to the unique Nash equilibrium for sources with strictly log-concave density and two-sided unbounded support, generalizing the classical convergence results in optimal quantization via Lloyd's Method I. These results were also generalized to multi-dimensional sources. The Stackelberg setup was studied in detail as well, where the relations with the team theoretic setup are directly evident.

We have also investigated Nash and Stackelberg equilibria for quadratic signaling games with additive channel noise and subjective/inconsistent priors. We have established qualitative (e.g., full revelation, linearity, informativeness, non-informativeness, and robustness around the corresponding nominal team) and quantitative properties (on linearity or explicit computation) of Nash and Stackelberg equilibria.

18.5 Bibliographic Notes

Crawford and Sobel [189] have made foundational contributions to the study of cheap talk with misaligned objectives where the cost functions c^e and c^d satisfy certain monotonicity and differentiability properties but there is a constant bias term in one of the cost functions. Their primary result is that the number of bins in an equilibrium is upper bounded by a function which is inversely related with the bias. For sources with a support on $[0, 1]$ Crawford and Sobel also establish that an equilibrium with more bins induces smaller expected costs from the perspectives of both the encoder and the decoder, under a monotonicity assumption. For the quadratic cost criteria considered in this chapter, if the density of the source is assumed to be differentiable, then this sufficient condition only holds for sources with a uniform distribution. Theorem 18.2.9 has considered sources with a log-concave density, and stated such an informativeness result for sources with two-sided unbounded support, and that this result also generalizes to sources with semi-unbounded support as well as sources with bounded support.

On the topic of signaling games, and the special subclass of cheap talk games, there have been a number of contributions in the economics literature in addition to the seminal work by Crawford and Sobel, including [20, 103, 432, 546, 675] and [655], several of which have studied the multi-dimensional setup which bring in additional mathematical and conceptual richness. The Stackelberg formulation was studied in the influential paper by Kamenica and Gentzkow [386] (see also [385], for a comprehensive review).

In the context of the Bayesian persuasion problem, an important related work [700] considers a multidimensional signaling scenario under a Stackelberg game setup where the sender employs a general quadratic cost structure. An upper bound on the performance of the sender is obtained via formulating a semidefinite program. For jointly Gaussian sources, a linear policy that achieves this upper bound is characterized which shows the optimality of linear policies for such a Stackelberg game.

The strategic information transmission (SIT) problem has attracted attention also in the communication and control theory literature [13, 249, 551, 638, 639, 641, 646], [455, 464–466, 644, 645, 737]. In some of these studies, such as [249], the misalignment in the objectives is due to a bias term which is modeled as a random variable; the authors consider the Stackelberg equilibrium concept and focus on affine policies; we should note that unlike the setup considered in the chapter where the bias was constant, the random bias setup leads to more interesting Stackelberg solutions. A robust formulation of the quadratic Bayesian persuasion problem, where the receiver is a perturbed Bayesian, is studied in [512].

Several studies have considered the strategic information transmission (SIT) problem in which the sender takes privacy of certain information into account by employing a suitable privacy measure, under either the Nash or the Stackelberg criterion [10, 11, 247, 248]. In these studies, a common theme has been to model private and non-private parameters as jointly Gaussian random variables. In [248], a communication scenario between a sender and a receiver has been investigated using the Stackelberg equilibrium concept in which an additional side information is assumed to be available at the receiver. The estimation errors are measured using quadratic costs, and a family of Stackelberg equilibria is characterized under an *a priori* affine policy assumption. The work in [10] also investigates a Stackelberg game where the utility measure for the nonprivate parameter is quadratic and the privacy measure is entropy based. Both noiseless and noisy communication scenarios are considered and essentially unique linear encoding and decoding policies that form a Stackelberg equilibrium are characterized. In [247], a Nash game is studied where the privacy measure is based on mutual information, and the utility measure for the nonprivate parameter is quadratic. In that paper, apart from the previously described Gaussian scenario, another scenario in which private and nonprivate data are treated as discrete random variables is considered.

The material in this chapter builds primarily on [410, 411, 638, 641] and [412]. The case with subjective priors primarily builds on [413]. A related study is binary signaling problems in which the decision makers (the transmitter and the receiver) have subjective priors and/or misaligned objective functions; [640] formulates this as a signaling game problem and establishes equilibrium conditions under both Nash and Stackelberg equilibria.

Chapter 19
Two Classes of Signaling Games: Reputation and Information Privacy Games

In the previous chapter, we introduced and analyzed Nash and Stackelberg equilibria of signaling games where the agents have misaligned objectives or subjective priors where the foci was on quadratic cost criteria. In this chapter we will study further, and operationally and practically important, settings involving reputation management and privacy in communications involving agents connected over information channels. In such applications the objective functions of the agents are misaligned. The chapter has two main sections, devoted to reputation and privacy issues, both within the framework of signaling games.

Section 19.1 deals with a specific class of games with imperfect information, called reputation games, where a long-lived strategic agent interacts with a collection of agents, and controls her own reputation within a signaling game framework. The long-lived player plays a repeated stage game with a sequence of different *short-lived* or *transient* players. The long-lived player has a *reputation*, modeled as an initial prior, and would like to take advantage of his reputation, even while acting against it. We will see that as the time horizon approaches infinity, an undeserved reputation cannot be sustained under optimal behaviour.

Section 19.2 studies information privacy games as signaling games between an encoder and a decoder, capturing the following communication scenario between a sender and a receiver. There is a pair of messages at the sender and the perspective of the sender is such that one message needs to be protected and the other message needs to be conveyed. As opposed to the intention of the sender, the receiver desires to accurately estimate both messages with the aim of acquiring as much information as possible. Under this setting, we investigate the interactions between the sender and the receiver, whose objectives are different from each other due to the privacy concerns of the former. In this context, we will also discuss the information bottleneck problem [718], which provides a general framework in privacy: In the information bottleneck problem, the aim is to compress an observed random variable while trying to preserve information related to a correlated random variable.

S. Yüksel, T. Başar, *Stochastic Teams, Games, and Control under Information Constraints*, Systems & Control: Foundations & Applications,
https://doi.org/10.1007/978-3-031-54071-4_19

We will also have an occasion to revisit yet another foundational contribution from Witsenhausen and Wyner [781] in this context; see Figs. 19.2 and 19.3.

19.1 Reputation Games as Signaling Games

In *reputation games* a typical setup is as follows: A *long-lived player* plays a multi-stage game with a sequence of different *short-lived* or *transient* players. The long-lived player has a *reputation*, modeled as an initial prior as seen by the short-lived players, and would like to take advantage of his reputation. See Fig. 19.1.

19.1.1 The Model

A long-lived player (Player 1) plays a repeated stage game with a sequence of different short-lived players (each of whom is referred to as Player 2).

The stage game is a *sequential-move* game: Player 1 moves first; when action a^1 is chosen by Player 1 in the stage game; a **public** signal $s^2 \in \mathbb{S}^2$ is observed by Player 2, which is drawn according to the probability distribution $\rho^2(.|a^1) \in \mathcal{P}(\mathbb{S}^2)$. Player 2, observing this public signal (and all preceding public signals), moves second and generates a^2. At the end of the stage game, Player 1 observes a **private** signal $s^1 \in \mathbb{S}^1$ which depends on actions of both players in the stage game and is drawn according to the probability distribution $\rho^1(.|(a^1, a^2))$.

Each stage game is considered a Stackelberg game with imperfect monitoring where Player 1 is the leader and Player 2 is the follower. Action sets of Player 1 and Player 2 in the stage game are assumed to be finite and denoted by $\mathbb{A}^1$ and $\mathbb{A}^2$, respectively. We also assume that the set of Player 1's all possible private signals, denoted by $\mathbb{S}^1$, and the set of each Player 2's possible public signals, denoted by $\mathbb{S}^2$, are finite.

There is incomplete information regarding the type of the long-lived Player 1. Player 1 can either be a strategic type (or normal type), denoted by ω^n, or one of finitely many simple commitment types. Each of these commitment types is committed to simply playing the same action $\hat{\omega} \in \mathcal{P}(\mathbb{A}^1)$ at every stage of the

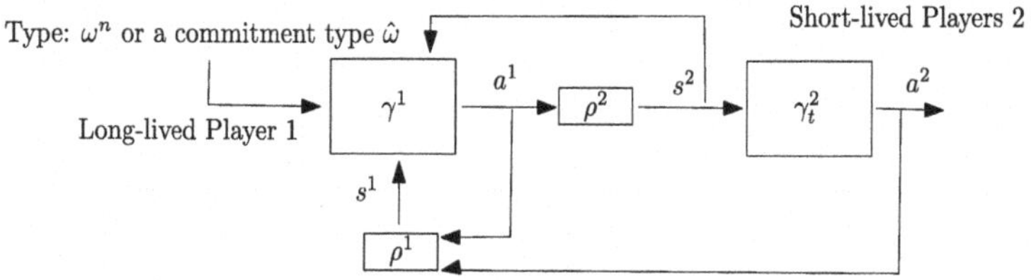

Fig. 19.1 Flow of information in the reputation game

repeated game—independent of the history of the play.[1] The set of all possible commitment types of Player 1 is given by $\hat{\Omega}$. Therefore, the set of all possible types of Player 1 would be $\Omega = \{\omega^n\} \cup \hat{\Omega}$. The type of Player 1 is determined once and for all at the beginning of the game according to a *full-support* prior $\mu_0 \in \mathcal{P}(\Omega)$ (which is common knowledge). However, only Player 1 is informed of his type, i.e., Player 1's type is his private information.

We note that there is a *nested information structure* in the repeated game in the following sense: The signals observed by Player 2s are public, and hence available to all subsequent players, whereas Player 1's signals are his private information. Therefore, the information of Player 2 at time $t-1$ is a subset of the information of Player 1 at time t. Formally, a generic history for Player 2 at time $t-1$ and a generic history for Player 1 at time t are given as follows:

$$h^2_{t-1} = (s^2_0, s^2_1, \cdots, s^2_{t-1}) \in H^2_{t-1} \tag{19.1}$$

$$h^1_t = (a^1_0, s^1_0, s^2_0, \cdots, a^1_{t-1}, s^1_{t-1}, s^2_{t-1}) \in H^1_t \tag{19.2}$$

where $H^2_{t-1} := (\mathbb{S}^2)^t$ and $H^1_t := (\mathbb{A}^1 \times \mathbb{S}^1 \times \mathbb{S}^2)^t$.

That is, each Player 2 observes, before he acts, a finite sequence of public signals which are correlated with Player 1's action in each of his interaction with the preceding Player 2s. On the other hand, Player 1 observes not only these public signals, but also a sequence of private signals for each particular interaction that happened in the past, and his actions in the previous periods—but not necessarily the actions of the preceding Player 2s.[2]

The stage game payoff function of the strategic (or normal) type long-lived Player 1 is given by r^1, and each short-lived Player 2's payoff function is given by r^2, where $r^i : \mathbb{A}^1 \times \mathbb{A}^2 \to \mathbb{R}$. The set of all possible histories for Player 2 of stage t is $H^2_t = H^2_{t-1} \times \mathbb{S}^2$ where $H^2_{t-1} = (\mathbb{S}^2)^t$. On the other hand, the set of all possible histories observable by the long-lived Player 1 prior to stage t is $H^1_t = (\mathbb{A}^1 \times \mathbb{S}^1 \times \mathbb{S}^2)^t$. It is assumed that $H^1_0 := \emptyset$ and $H^2_0 := \emptyset$, which is the standard convention. Let $\mathcal{H}^1 = \bigcup_{t \geq 0} H^1_t$ be the set of all possible histories of the long-lived Player 1.

A (behavioral) strategy for Player 1 is a map:

$$\gamma^1 : \Omega \times \mathcal{H}^1 \to \mathcal{P}(\mathbb{A}^1).$$

that satisfies $\gamma^1(\hat{\omega}, h^1_{t-1}) = \hat{\omega}$ for any $\hat{\omega} \in \hat{\Omega}$ and for every $h^1_{t-1} \in H^1_{t-1}$, since commitment types are required to play the corresponding (fixed) action of the stage game independent of the history. The set of all strategies for Player 1 is denoted by Σ^1, i.e., Σ^1 is the set of all functions from $\Omega \times \mathcal{H}^1$ to $\mathcal{P}(\mathbb{A}^1)$.

[1] $\mathcal{P}(\mathbb{A}^i)$ denotes the set of all probability measures on $\mathbb{A}^i$ for both $i = 1, 2$. That is, the commitment types can be committed to playing mixed stage-game actions as well. We also note here that simple commitment types assumption is a standard one in reputation games.

[2] Note that Player 1 gets to observe the realizations of his earlier possibly mixed actions.

A strategy for Player 2 of stage t is a map:

$$\gamma_t^2 : H_{t-1}^2 \times \mathbb{S}^2 \rightarrow \mathcal{P}(\mathbb{A}^2).$$

We let Σ_t^2 be the set of all such strategies and let $\Sigma^2 = \Pi_{t\geq 0}\Sigma_t^2$ denote the set of all sequences of all such strategies. A history (or path) h_t of length t is an element of $\Omega \times (\mathbb{A}^1 \times \mathbb{A}^2 \times \mathbb{S}^1 \times \mathbb{S}^2)^t$ describing Player 1's type, actions, and signals realized up to stage t. A strategy profile $\sigma = (\gamma^1, \gamma^2) \in \Sigma^1 \times \Sigma^2$ induces a unique probability distribution P_σ over the set of all paths of play $H^\infty = \Omega \times (\mathbb{A}^1 \times \mathbb{A}^2 \times \mathbb{S}^1 \times \mathbb{S}^2)^{\mathbb{Z}_+}$ endowed with the product σ-algebra. We let $a_t = (a_t^1, a_t^2)$ represent the action profile realized at stage t and let $s_t = (s_t^1, s_t^2)$ denote the signal profile realized at stage t. Given $\omega \in \Omega$, $P_{\omega,\sigma}(.) = P_\sigma(.|\omega)$ represents the probability distribution over all paths of play conditional on Player 1 being of type ω. Player 1's discount factor is assumed to be $\delta \in (0, 1)$ and hence, the expected discounted payoff to the strategic (normal type) long-lived Player 1 is given by

$$J^1(\sigma) = (1-\delta)E^{P_{\omega^n,\sigma}}[\sum_{t\geq 0} \delta^t r^1(a_t)].$$

In all the results below, except Lemma 19.1.1, we will assume that Player 2s are Bayesian rational. Hence, we will restrict attention to Perfect Bayesian Equilibrium: In any such equilibrium, the strategic Player 1 maximizes his expected discounted payoff given that the short-lived players play a best response according to their updated beliefs (this will be appropriately modified when we consider the undiscounted setup). Each Player 2, playing the stage game only once, will be best-responding to his expectation according to his beliefs which are updated according to Bayes' Rule.

A strategy of Player 2s, γ^2, is a best response to γ^1 if, for all t,

$$E^{P_\sigma}[r^2(a_t^1, a_t^2)|s_{[0,t]}^2] \geq E^{P_\sigma}[r^2(a_t^1, a^2)|s_{[0,t]}^2]$$
$$\text{for all } a^2 \in A^2 \quad (P_\sigma - a.s.)$$

where $s_{[0,t]}^2 = (s_0^2, s_1^2, \cdots, s_t^2)$ denotes the information available to Player 2 at time t.

19.1.2 *Optimal Strategies and Equilibrium Behavior*

Our first set of results will be regarding the optimal strategies of the strategic long-lived Player 1.

Briefly, since each Player 2 plays the stage game only once, we show that when the information of Player 2 is nested in that of Player 1, under a plausible assumption to be noted, the strategic long-lived Player 1 can, without any loss in payoff performance, formulate his strategy as a controlled Markovian system optimization, and thus through dynamic programming. The discounted nature of the optimization problem then leads to the existence of a stationary solution. This implies that for any Perfect Bayesian Equilibrium, there exists a payoff-equivalent stationary Markov Perfect Equilibrium. Hence, we conclude that the Perfect Bayesian Equilibrium payoff set and the Markov Perfect Equilibrium payoff set of the strategic long-lived Player 1 coincide with each other.

Below, we provide three results on optimal strategies following steps parallel to those introduced in Chap. 15, though with some subtle differences due to the game theoretic context here. These structural results on optimal strategies will be instrumental for the Markov chain construction to follow as well as Lemmas 19.1.1 and 19.1.2.

Optimal Strategies: Finite Horizon

We first consider the finitely repeated game setup where the stage game is to be repeated $T \in \mathbb{N}$ times. In such a case, the strategic long-lived Player 1 is to maximize $J^1(\sigma)$ given by

$$J^1(\sigma) = E^{P_{\omega^n,\sigma}}(1-\delta)\sum_{t=0}^{T-1}\delta^t r^1(a_t).$$

Our first result, Lemma 19.1.1, shows that, given any fixed sequence of strategies of the short-lived Player 2s, any optimal strategy of the strategic long-lived Player 1 can be replaced, without any loss in payoff performance, by another optimal strategy which only depends on the (public) information of Player 2s. More specifically, we show that for any private strategy of the long-lived Player 1 against an arbitrary sequence of strategies of Player 2s, there exists a public strategy of the long-lived Player 1 against the very same sequence of strategies of Player 2s which gives the strategic long-lived player a no worse (or *at least as good*) payoff.[3]

Before we state Lemma 19.1.1, we note here that the signal s_t^2 that will be available to short-lived Player 2s after round t only depends on the action of the long-lived Player 1 at round t and that the following holds for all $t \geq 1$:

$$P_\sigma(s_t^2|a_t^1; a_{t'}^1, a_{t'}^2, t' \leq t-1) = P_\sigma(s_t^2|a_t^1). \tag{19.3}$$

The relationship (19.3) plays an important role in the following:

[3] A public strategy is a strategy that uses only public information that is available to all the players. On the other hand, a strategy that is based on private information of a player is referred to as a private strategy. In particular, any strategy of Player 1 that is based on s_t^1 for some t is a private strategy.

Lemma 19.1.1 *In the finitely repeated setup, given any sequence of strategies of short-lived Player 2s, for any (private) strategy of the strategic long-lived Player 1, there exists a (public) strategy that only conditions on* $\{s_0^2, s_1^2, \cdots, s_{t-1}^2\}$ *which yields the strategic long-lived Player 1 an at least as good payoff against the given sequence of strategies of Player 2s.*

A brief word of caution is in order. Structural results of the type presented in Lemma 19.1.1, while crucially useful in team theory and zero-delay source coding (as utilized in Chap. 15 e.g. in Theorems 15.3.1, 15.3.2, 15.3.3 and 15.3.4), do not always apply to generic games unless one further restricts the setup. In particular, a generic (Nash) equilibrium may be lost if one modifies the strategy structure of one of the players, while keeping the other one fixed (in team problems, the parties can agree to have a better performing team policy even if it is not a strict equilibrium). However, we adopt here the Perfect Bayesian Equilibrium concept. That is, Player 2s respond in a Bayesian fashion to Player 1 who in turn is aware of Player 2s' commitment to such a policy.

This subtle difference is crucial also in signaling games; the features that distinguish Nash equilibria from Stackelberg equilibria in signaling games have been discussed in detail earlier in the chapter.

Lemma 19.1.1 implies that any private information of Player 1 is statistically irrelevant for optimal strategies: for any private strategy of the long-lived Player 1, there exists a public strategy which performs at least as good as the original one against a given sequence of strategies of Player 2s. That is, in the finitely repeated setup, the long-lived Player 1 can make his strategy depend only on the public information and his type without any loss in payoff performance. We note here once again that Lemma 19.1.1 above holds for any sequence of strategies of Player 2s, even non-Bayesian ones.

On the other hand, when Player 2s are Bayesian rational, as is the norm in repeated games, we have a more refined structural result which we state below as Lemma 19.1.2. As mentioned before, in a Perfect Bayesian Equilibrium the short-lived Player 2 at time t, playing the stage game only once, seeks to maximize $\sum_{a^1} P_\sigma(a_t^1 = a^1 | s_{[0,t]}^2) r^2(a^1, a^2)$. However, it may be that his best response set, i.e., the maximizing action set

$$\arg\max \Big(\sum_{a^1} P_\sigma(a_t^1 = a^1 | s_{[0,t]}^2) r^2(a^1, a^2)\Big) \tag{19.4}$$

may not be unique.

To avoid such set-valued correspondence dynamics, we make the following assumption, which requires that the best response of each Player 2 is essentially unique: Note that any strategy for Player 2 of time t who chooses (19.4) in a measurable fashion does not have to be continuous in the conditional probability $\kappa(\cdot) = P_\sigma(a_t^1 = \cdot | s_{[0,t]}^2)$, since such a strategy partitions (or quantizes) the set of probability measures on $\mathbb{A}^1$. The set of κ which borders these partitions is a subset of the set of probability measures $\mathcal{B}_e = \cup_{k,m\in\mathbb{A}^2}\mathcal{B}^{k,m}$, where for any pair $k, m \in \mathbb{A}^2$, the belief set $\mathcal{B}^{k,m}$ is defined as

$$\mathcal{B}^{k,m} = \left\{\kappa \in \mathcal{P}(\mathbb{A}^1) : \sum_{a^1\in\mathbb{A}^1} \kappa(a^1)r^2(a^1,k) = \sum_{a^1\in\mathbb{A}^1} \kappa(a^1)r^2(a^1,m)\right\}. \tag{19.5}$$

These are the sets of probability measures where Player 2 is indifferent between multiple actions.

Assumption 19.1.1 *Either of the following two properties holds:*

(i) The prior measure and the probability space is so that $P_\sigma\left(P_\sigma(a_t^1 = \cdot|s^2_{[0,t]}) \in \mathcal{B}_e\right) = 0$ *for all* $t \geq 0$*. In particular, Player 2s have a unique best response so that the set of discontinuity,* $\mathcal{B}_e$*, is never visited (with probability* 1*).*
(ii) Whenever Player 2s are indifferent between multiple actions, they choose the action that is better for Player 1.

Remark 19.1.1 We make two points regarding Assumption 19.1.1.

(i) In the classical reputations literature, a standard result is that under mild conditions Bayesian rational short-lived players can be *surprised* at most finitely many times, e.g., [277, Theorem 4.1], [679, Lemma 2.4], implying that the jumps in the corresponding belief dynamics of Player 2s will be bounded away from zero in a transient phase until the optimal responses of Player 2s converge to a fixed action. In such cases, the payoff structure can be designed so that the set of discontinuities, $\mathcal{B}_e$, is visited with 0 probability, and hence Assumption 19.1.1(i) holds.
(ii) Assumption 19.1.1(ii) is a standard assumption in the contract theory literature. In a principal-agent model, whenever an agent is indifferent between two actions, he chooses the action that is better for the principal, e.g., when an incentive compatibility condition binds so that the agent is indifferent between exerting a high level and low level efforts, then the agent chooses to exert the high level of effort. Assumption 19.1.1(ii) trivially holds also when the stage game payoff functions are identical for both players (as in *team* setups) or are aligned (as in a *potential game*).

In the following, and the remainder of the section, we will build on Lemma 19.1.1 and Theorem 15.3.2, adapted to the game theoretic setup considered here.

Lemma 19.1.2 *In the finitely repeated setup, under Assumption 19.1.1, given any arbitrary sequence of strategies of Bayesian rational short-lived Player 2s, for any (private) strategy of the strategic long-lived Player 1, there exists a (public) strategy that only conditions on* $P_\sigma(\omega|s^2_{[0,t-1]}) \in \mathcal{P}(\Omega)$ *and* t *which yields the strategic long-lived Player 1 a better payoff against the given sequence of strategies of Player 2s.*

Controlled Markov Chain Construction

Similar to the analysis earlier in Theorem 15.3.2, under Assumption 19.1.1, given any sequence of strategies of Bayesian rational Player 2s, the optimization analysis characterizes the equilibrium behavior of the strategic long-lived player in an associated Markov Perfect Equilibrium. The state space, the action set, the transition kernel, and the per-stage reward function of the controlled Markov chain mentioned above are given as follows:

The state space is $\mathcal{P}(\Omega)$; $\mu_t \in \mathcal{P}(\Omega)$ is often called the *belief*-state. We endow this space with the weak convergence topology, and we note that since Ω is finite, the set of probability measures on Ω is a compact space.

The action set is the set of all maps $\Gamma^1 := \{\gamma^1 : \Omega \to \mathbb{A}^1\}$. We note that since the commitment type policies are given a priori, one could also regard the action set to be the set $\mathbb{A}^1$ itself.[4]

The transition kernel is given by $P : \mathcal{P}(\Omega) \times \Gamma^1 \to \mathcal{B}(\mathcal{P}(\Omega))$ so that for all $B \in \mathcal{B}(\mathcal{P}(\Omega))$:

$$
\begin{aligned}
&P\left(P_\sigma(\omega|s^2_{[0,t-1]}) \in B \,\middle|\, P_\sigma(\omega|s^2_{[0,t'-1]}), \gamma^1_{t'}, t' \le t-1 \right) \\
&= P\left(\left\{ \frac{\sum_{a^1_{t-1}} P_\sigma(s^2_{t-1}|a^1_{t-1}) P_\sigma(a^1_{t-1}|\omega, s^2_{[0,t-2]}) P_\sigma(\omega|s^2_{[0,t-2]})}{\sum_{a^1_{t-1},\omega} P_\sigma(s^2_{t-1}|a^1_{t-1}) P_\sigma(a^1_{t-1}|\omega, s^2_{[0,t-2]}) P_\sigma(\omega|s^2_{[0,t-2]})} \right\} \in B \right. \\
&\qquad\qquad \left| P_\sigma(\omega|s^2_{[0,t'-1]}), \gamma^1_{t'}, t' \le t-1 \right) \\
&= P\left(\left\{ \frac{\sum_{a^1_{t-1}} P_\sigma(s^2_{t-1}|a^1_{t-1}) P_\sigma(a^1_{t-1}|\omega, s^2_{[0,t-2]}) P_\sigma(\omega|s^2_{[0,t-2]})}{\sum_{a^1_{t-1},\omega} P_\sigma(s^2_{t-1}|a^1_{t-1}) P_\sigma(a^1_{t-1}|\omega, s^2_{[0,t-2]}) P_\sigma(\omega|s^2_{[0,t-2]})} \right\} \in B \right. \\
&\qquad\qquad \left| P_\sigma(\omega|s^2_{[0,t-2]}), \gamma^1_{t-1} \right)
\end{aligned}
\tag{19.6}
$$

In the above formulation, we have made use of the fact that the term $P_\sigma(a^1_{t-1}|\omega, s^2_{[0,t-2]})$ is uniquely identified by $P_\sigma(\omega|s^2_{[0,t-2]})$ and γ^1_{t-1}. Here, γ^1_{t-1} is the *control action*.

The per-stage reward function, given γ^2_t, is $U(\mu_t, \gamma^1) : \mathcal{P}(\Omega) \times \Gamma^1 \to \mathbb{R}$ which is defined as follows:

[4] We note that randomized strategies may also be considered by adding a randomization variable.

$$U(\mu_t, \gamma^1) := \sum_{\omega} P_\sigma(\omega|s^2_{[0,t-1]}) \sum_{\mathbb{A}^1} \left(1_{\{a^1_t = \gamma^1(\omega)\}} r^1(a^1_t, \gamma^2_t(P_\sigma(a^1_t|s^2_{[0,t-1]}), s^2_t)) \right) \tag{19.7}$$

where $\mu_t = P_\sigma(\omega|s^2_{[0,t-1]})$. Here, γ^2_t is a given measurable function of the posterior $P_\sigma(a^1_t|s^2_{[0,t]})$. We note again that for each Bayesian rational short-lived Player 2, we have

$$\gamma^2_t(P_\sigma(a^1_t|s^2_{[0,t-1]}), s^2_t)) \in \arg\max \left(\sum_{a^1} P_\sigma(a^1_t|s^2_{[0,t]}) r^2(a^1, a^2) \right).$$

Lemma 19.1.2 implies that in the finitely repeated game setup, under Assumption 19.1.1, when Player 2s are Bayesian rational, the long-lived strategic Player 1 can make his strategy depend only on Player 2s' posterior belief and time without any loss in payoff performance.

Consider now any Perfect Bayesian Equilibrium where the strategic long-lived Player 1 plays a private strategy, since the strategic long-lived Player 1 cannot have a profitable deviation, the public strategy identified in Lemma 19.1.2 must also give him the same payoff against the given sequence of strategies of Player 2s. Hence, in the finitely repeated game setup, under Assumption 19.1.1, any *Perfect Bayesian Equilibrium* payoff of the normal type Player 1 is also a *Perfect Public Equilibrium* payoff.[5] Therefore, given our Markov chain construction:

Theorem 19.1.1 ([201]) *In the finitely repeated game setup, under Assumption 19.1.1, the set of Perfect Bayesian Equilibrium payoffs of the strategic long-lived Player 1 is equal to the set of Markov Perfect Equilibrium payoffs.*

Proof The Markov Perfect Equilibrium payoff set is a subset of Perfect Bayesian Equilibrium payoff set. Hence, it will suffice to show that for each Perfect Bayesian Equilibrium there exists a properly defined Markov Perfect Equilibrium which is payoff equivalent for the strategic long-lived Player 1. This follows from Lemma 19.1.2 and our Markov chain construction. ◇

Lemmas 19.1.1 and 19.1.2 above have a coding theoretic flavor and are not much different from the optimal zero-delay coding results given in Sect. 15.3. However, these results from Chap. 15 are different from those above due to the fact that the utility functions do not depend explicitly on the type of Player 1, but depend explicitly on the actions a^1_t and that these actions are not available to Player 2 unlike the setup in Chap. 15.

Next, we consider the infinitely repeated setting.

[5] A Perfect Public Equilibrium is a Perfect Bayesian Equilibrium where each player uses a public strategy, i.e., a strategy that only depends on the information which is available to both players.

Infinite Horizon Game and Equilibrium Strategies

We proceed with another lemma (Lemma 19.1.3) which is the extension of Lemma 19.1.2 to the infinitely repeated game setting. Lemma 19.1.3 will be the key result that gives us a similar controlled Markov chain construction for the infinitely repeated game, and hence a payoff-equivalent stationary Markov Perfect Equilibrium for each Perfect Bayesian Equilibrium.

Lemma 19.1.3 ([201]) *In the infinitely repeated game, under Assumption 19.1.1, given any arbitrary sequence of strategies of Bayesian rational short-lived Player 2s, for any (private) strategy of the strategic long-lived Player 1, there exists a (public) strategy that only depends on* $P_\sigma(\omega|s^2_{[0,t-1]}) \in \mathcal{P}(\Omega)$ *and* t *which yields the strategic long-lived Player 1 a better payoff against the given sequence of strategies of Player 2s.*

Furthermore, the strategic long-lived Player 1's optimal stationary strategy against this given sequence of strategies of Player 2s can be characterized by solving an infinite horizon discounted dynamic programming problem.

In view of the lemma above, in the infinitely repeated game setup as well, under Assumption 19.1.1, any private strategy of the normal type Player 1 can be replaced, without any loss in payoff performance, with a public strategy which only depends on $P_\sigma(\omega|s^2_{[0,t-1]})$ and t. Hence, for any *Perfect Bayesian Equilibrium*, there exists a *Perfect Public Equilibrium* which is payoff-equivalent for the strategic long-lived Player 1 in the infinitely repeated game as well.

Furthermore, since there is a stationary optimal public strategy for the strategic long-lived Player 1 against any given sequence of strategies of Bayesian rational Player 2s, any payoff the strategic long-lived Player 1 obtains in a *Perfect Bayesian Equilibrium*, he can also obtain in a *Markov Perfect Equilibrium* (see Chap. 10).

Theorem 19.1.2 ([201]) *In the infinitely repeated game, under Assumption 19.1.1, the set of Perfect Bayesian Equilibrium payoffs of the strategic long-lived Player 1 is equal to the set of Markov Perfect Equilibrium payoffs.*

Proof The proof follows from Lemma 19.1.3 and the controlled Markov chain construction. ◇

Note that $\{\mu_t(\bar{\omega}) = E[1_{\omega=\bar{\omega}}|s^2_{[0,t]}]\}$, for every fixed $\bar{\omega}$, is a bounded martingale sequence adapted to the information at Player 2, and as a result as $t \to \infty$, by the submartingale convergence theorem [133] there exists (a random) $\bar{\mu}$ such that $\mu_t \to \bar{\mu}$ almost surely. Let $\bar{\mu}$ be an *invariant posterior*, that is, a (sample-path) limit of the μ_t process. This leads to the following fixed point equation:

$$V^1(\omega, \bar{\mu}) = \max_{a^1=\gamma_t^1(\mu,\omega)} \left(E[r^1(a_t^1, \gamma^2(\mu)) + \delta E[V^1[(\omega, \bar{\mu})]\right)$$

Therefore,

$$V^1(\omega, \bar{\mu}) = \frac{1}{1-\delta} \max_{\gamma_t^1} E[r^1(a_t^1, a_t^2(\bar{\mu}))], \tag{19.8}$$

and since the solution is asymptotically stationary, the optimal strategy of the strategic long-lived Player 1 when $\mu_0 = \bar{\mu}$ has to be a Stackelberg solution for a Bayesian game with prior $\bar{\mu}$; thus, *a Perfect Bayesian Equilibrium strategy for the strategic long-lived Player 1 has to be mimicking the stage game of Stackelberg type forever.* This insight will be useful in the following subsection with further refinements.

19.1.3 Undiscounted Time-Average Payoff Case and an Upper Payoff Bound for the Arbitrarily Patient Long-lived Player

We next analyze the setting where the strategic long-lived Player 1 maximizes his expected *undiscounted* payoff. Not only we identify an optimal strategy for the strategic long-lived Player 1 in this setting, but also we establish an *upper payoff bound* for the arbitrarily patient strategic long-lived Player 1 in the *discounted* payoff case—through an Abelian inequality [342].

The only difference from our original setup is that the strategic long-lived Player 1 now wants to maximize

$$\liminf_{N\to\infty} \frac{1}{N} E^{\gamma^1,\gamma^2\mu_0}[\sum_{t=0}^{N-1} r^1(a_t^1, a_t^2)].$$

Therefore, in any Perfect Bayesian Equilibrium, same as before, the short-lived (Bayesian rational) Player 2s will continue to be best-replying to their updated beliefs. On the other hand, the strategic long-lived Player 1 will be playing a strategy which maximizes his average payoff given that each Player 2 will be best replying to its updated beliefs.

The main problem in analyzing the undiscounted setup is that most of the structural coding/signaling results that we have for finite horizon or infinite horizon discounted optimal control problems do not generalize to the undiscounted case, since the construction of controlled Markov chains is based on backward induction arguments leading to structural results that are applicable only for finite horizon problems.

Let us re-visit the discounted setup: Let $\bar{\mu}$ be an *invariant posterior*, that is, a (sample-path) limit of the μ_t process which exists by the earlier discussion with regard to the submartingale convergence theorem. Equation (19.8) is applicable for every $\delta \in (0, 1)$ so that

$$(1-\delta)V^1(\omega,\bar{\mu}) = \max_{\gamma_t^1} E[r^1(a_t^1, a_t^2(\bar{\mu}))], \tag{19.9}$$

and the optimal strategy of the strategic long-lived Player 1 when $\mu_0 = \bar{\mu}$ is a Stackelberg solution for a Bayesian game with prior $\bar{\mu}$; thus, *a Perfect Bayesian Equilibrium strategy for the strategic long-lived Player 1 has to be mimicking the stage game of Stackelberg type forever*. In the following, we will identify conditions when the limit $\bar{\mu}$ will turn out to be a dirac delta distribution at the normal type, that is $\bar{\mu} = \delta_w$ (basically, as in the complete information case). Furthermore, the above discussion implies the following: By a direct application of the Abelian inequality, we have that when $\mu_0 = \bar{\mu}$,

$$\sup_{\gamma^1,\gamma^2} \liminf_{N\to\infty} \frac{1}{N} E^{\gamma^1,\gamma^2}\left[\sum_{m=0}^{N-1} r^1(a_m^1, a_m^2)\right]$$

$$\le \limsup_{\delta\to 1} \sup_{\gamma^1,\gamma^2} E^{\gamma^1,\gamma^2}(1-\delta)\left[\sum_{m=0}^{\infty} \delta^m r^1(a_m^1, a_m^2)\right]$$

$$= \max_{\gamma_t^1} E[r^1(a_t^1, a_t^2(\bar{\mu}))], \tag{19.10}$$

where the last equality follows from (19.9). In the following, we will elaborate further on these observations and arrive at more refined results. We need the following identifiability assumption.

Assumption 19.1.2 *Uniformly over all stationary and optimal (for sufficiently large discount parameters δ) strategies $\tilde{\sigma}^1, \tilde{\sigma}^2$,*

$$\lim_{\delta\to 1} \sup_{\tilde{\sigma}^1,\tilde{\sigma}^2} \Bigg| E^{\tilde{\sigma}^1,\tilde{\sigma}^2}(1-\delta)\left[\sum_{t=0}^{\infty} \delta^t r^1(a_t^1, a_t^2)\right]$$

$$- \limsup_{N\to\infty} \frac{1}{N} E^{\tilde{\sigma}^1,\tilde{\sigma}^2}\left[\sum_{t=0}^{N-1} r^1(a_t^1, a_t^2)\right]\Bigg| = 0 \tag{19.11}$$

A sufficient condition for Assumption 19.1.2 is the following:

Assumption 19.1.3 *Whenever the strategic long-lived Player 1 adopts a stationary strategy, for any initial commitment prior, there exists a stopping time τ such that for $t \ge \tau$, Player 2s' posterior beliefs become so that his (i.e., each Player 2's) best response does not change (that is, his best-response to his beliefs leads to a constant action). Furthermore, $E[\tau] < \infty$, uniformly over any stationary strategy γ^1.*

Proposition 19.1.1 below shows that Assumption 19.1.2 is indeed implied by one of the most standard identifiability assumptions in the repeated games literature:

Assumption 19.1.4 *Consider a matrix A whose rows consist of the vectors:*

$$\left[P_\sigma(s_t^2 = k|a_t^1 = 1)\ P_\sigma(s_t^2 = k|a_t^1 = 2) \cdots P_\sigma(s_t^2 = k|a_t^1 = |\mathbb{A}^1|)\right]$$

where $k \in \{1, 2, \cdots, |\mathbb{S}^2|\}$. *We have that* $rank(A) = |\mathbb{A}^1|$

Proposition 19.1.1 *Under Assumption 19.1.4,*

$$\|P_\sigma(a_t^1 \in \cdot|h_t^2) - P_\sigma(a_t^1 \in \cdot|h_t^2, \omega)\|_{TV} \to 0.$$

for every σ. *Furthermore, under Assumption 19.1.4, Assumption 19.1.2 holds.*

The sufficient condition stated in Proposition 19.1.1 is a standard identifiability assumption, sometimes referred to as the full-rank monitoring assumption in the reputations literature; see for example [192, Assumption 2]. Under Assumption 19.1.2, we establish that mimicking a Stackelberg commitment type forever is an optimal strategy for the strategic long-lived Player 1 in the undiscounted setup.

Theorem 19.1.3 ([201]) *In the undiscounted setup, under Assumption 19.1.4, an optimal strategy for the strategic long-lived Player 1 in the infinitely repeated game is the stationary strategy mimicking the Stackelberg commitment type forever.*

Remark 19.1.2

(i) We note that we cannot directly use the arguments in Chap. 15 with regard to the optimality of Markovian strategies (those given in Lemma 19.1.2) for average-cost/average-payoff problems since a crucial argument in that chapter is to establish a nearly optimal coding scheme using the fact that more information cannot hurt both the encoder and the decoder; in the case here, we have a game and the value of (or the lack of) information can be positive or negative in the absence of any further analysis (as studied in detail earlier in the book).

(ii) Under the conditions noted, it follows that Player 1 cannot *abuse* his reputation in the undiscounted setup: An optimal policy is an *honest* stage-wise Stackelberg policy. Abusing (through *exploiting*) the reputation is inherently a discounted optimality phenomenon.

As an implication of Theorem 19.1.3, we next state the aforementioned upper bound for Perfect Bayesian Equilibrium payoffs of the arbitrarily patient strategic long-lived Player 1 in the discounted setup as Theorem 19.1.4.

Theorem 19.1.4 ([201]) *Under the assumptions of Theorem 19.1.3,*

$$\limsup_{\delta \to 1} V_\delta^1(\omega, \mu^0) \le \max_{\alpha_1 \in \mathcal{P}(A_1), \alpha_2 \in BR(\alpha_1)} u_1(\alpha_1, \alpha_2).$$

That is, an upper bound for the value of the reputation for an arbitrarily patient strategic long-lived Player 1 in any Perfect Bayesian Equilibrium of the discounted setup is his stage game Stackelberg equilibrium payoff.

Theorem 19.1.4 provides an upper bound on the value of reputation for the strategic long-lived Player 1 in the discounted setup. That is, in the discounted setup, an arbitrarily patient strategic long-lived Player 1 cannot do any better than his best Stackelberg payoff under reputational concerns as well. This upper bound coincides with those provided in [277] and [311].

19.2 Information Privacy Games

The final class of signaling games we consider in this chapter is information privacy games.

We consider an information transmission scenario in which a sender encodes a pair of correlated random vectors X and Y into another random vector Z using an encoding function denoted by $z = \gamma^e(x, y)$. A receiver wants to decode both random variables (X, Y) based on its observation Z. We denote the sizes of the realizations of the random vectors X and Y by n_X and n_Y, respectively. In this communication scenario, the sender wishes to convey information contained in Y whereas it views X as a private information that needs to be hidden from the receiver. The aim of the receiver is to accurately estimate both of the parameters given its observation Z. Let the decoders employed for estimating X and Y at the receiver be denoted by $\gamma^{d_X}(z)$ and $\gamma^{d_Y}(z)$, respectively.

Figure 19.2 depicts the considered information transmission scenario.

Specifically, we model X and Y as jointly Gaussian random vectors with zero-mean and with $\begin{bmatrix} X \\ Y \end{bmatrix}$ having a positive-definite covariance matrix $\Sigma \triangleq \begin{bmatrix} \Sigma_X & \Sigma_{XY} \\ \Sigma_{YX} & \Sigma_Y \end{bmatrix}$ where $\Sigma_{XY} \neq \mathbf{0}$. All this is common knowledge to both agents. Since X and Y are correlated, transmitting Y directly discloses information related to the private vector X. In other words, the objectives of hiding X and conveying Y are conflicting. These conflicting objectives at the transmitter are captured by the following objective function:

$$J^e(\gamma^e, \gamma^{d_X}, \gamma^{d_Y}) = E[\|Y - \gamma^{d_Y}(Z)\|^2] - \delta\, E[\|X - \gamma^{d_X}(Z)\|^2] \tag{19.12}$$

which is to be minimized, where δ is a positive design parameter that determines the level of desired privacy in terms of hiding X. On the other hand, the receiver wishes to deduce the true values of both X and Y. Consistent with this, we have the following objective function on the side of the receiver:

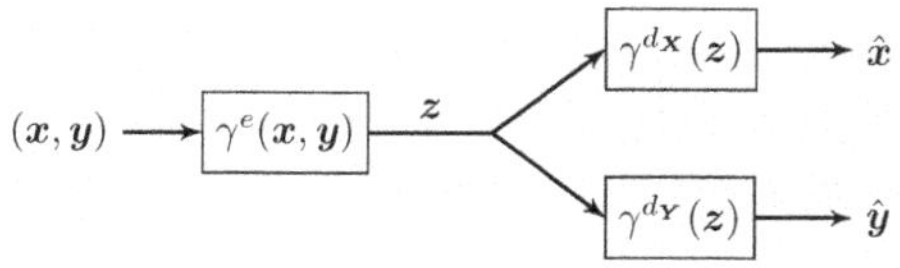

Fig. 19.2 Information flow for a privacy signaling game

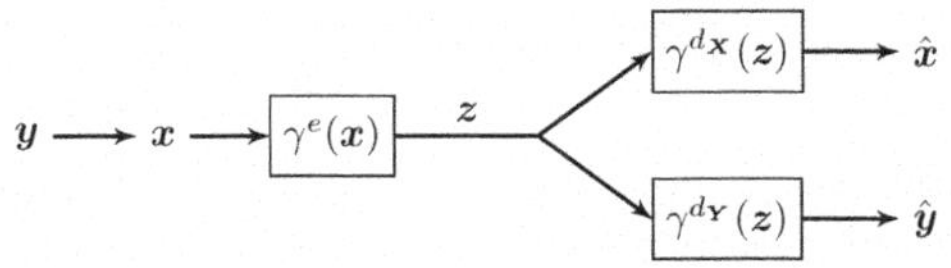

Fig. 19.3 Information flow for information bottleneck

$$J^d(\gamma^e, \gamma^{d_X}, \gamma^{d_Y}) = E[\|Y - \gamma^{d_Y}(Z)\|^2] + E[\|X - \gamma^{d_X}(Z)\|^2] \tag{19.13}$$

In (19.13), the mean squared errors corresponding to vectors X and Y are incorporated into the objective function with equal weights. We now investigate the Nash and the Stackelberg equilibria for this information transmission problem.

Let us recall, using the notation introduced above, that a set of measurable policies $\gamma^{e,*}$, $\gamma^{d_X,*}$ and $\gamma^{d_Y,*}$ constitutes a Nash equilibrium if

$$J^e(\gamma^{e,*}, \gamma^{d_X,*}, \gamma^{d_Y,*}) \leq J^e(\gamma^e, \gamma^{d_X,*}, \gamma^{d_Y,*}) \tag{19.14}$$

for all $\gamma^e \in \Gamma^e$, and

$$J^d(\gamma^{e,*}, \gamma^{d_X,*}, \gamma^{d_Y,*}) \leq J^d(\gamma^{e,*}, \gamma^{d_X}, \gamma^{d_Y}) \tag{19.15}$$

for all $\gamma^{d_X} \in \Gamma^{d_X}$ and $\gamma^{d_Y} \in \Gamma^{d_Y}$.

On the other hand, a set of policies $\gamma^{e,*}$, $\gamma^{d_X,*}$ and $\gamma^{d_Y,*}$ will constitute a Stackelberg equilibrium if

$$\begin{aligned} J^e(\gamma^{e,*}, \gamma^{d_X,*}(\gamma^{e,*}), \gamma^{d_Y,*}(\gamma^{e,*})) \\ \leq J^e(\gamma^e, \gamma^{d_X,*}(\gamma^e), \gamma^{d_Y,*}(\gamma^e)) \end{aligned} \tag{19.16}$$

for all $\gamma^e \in \Gamma^e$, where $\gamma^{d_X,*}(\gamma^e)$ and $\gamma^{d_Y,*}(\gamma^e)$ satisfy

$$\begin{aligned} J^d(\gamma^e, \gamma^{d_X,*}(\gamma^e), \gamma^{d_Y,*}(\gamma^e)) \\ \leq J^d(\gamma^e, \gamma^{d_X}(\gamma^e), \gamma^{d_Y}(\gamma^e)) \end{aligned} \tag{19.17}$$

for all $\gamma^{d_X} \in \Gamma^{d_X}$ and $\gamma^{d_Y} \in \Gamma^{d_Y}$.

We will also consider an instance of the problem above as the *MMSE Gaussian Information Bottleneck Problem*. The difference with the setting above is that in this case the sender has access only to X, which is the message it intends to hide while revealing as much information on Y as possible through transmission of an encoded version of X, which is correlated with Y (see Fig. 19.3). Details are provided in Sect. 19.2.3.

19.2.1 Nash Equilibria

In this section, we characterize Nash equilibria of the privacy-signaling game. Furthermore, we show that special cases of these equilibria lead to payoff dominant Nash equilibria (see Definition 9.2.4). These payoff dominant Nash equilibria are the most desirable equilibria for both players [338] (among all coding/decoding policies, including those that are nonlinear); also known as a payoff dominant Nash equilibrium.

Theorem 19.2.1 ([410]) *Consider the privacy-signaling game formulated above. Then:*

(i) There exist informative linear Nash equilibria with an encoding policy

$$z = \left[\alpha_1 \boldsymbol{q}_1 \cdots \alpha_{n_Y} \boldsymbol{q}_{n_Y}\right]^T \Sigma^{-1/2} \begin{bmatrix} x \\ y \end{bmatrix} \tag{19.18}$$

for scalars $\{\alpha_i\}_{i=1}^{n_Y}$ with at least one of these scalars being nonzero[6] *where $\{\boldsymbol{q}_i\}_{i=1}^{n}$ are normalized eigenvectors of $W = \Sigma^{1/2}\text{diag}(-\delta I, I)\Sigma^{1/2}$ with $n \triangleq n_X + n_Y$ and these eigenvectors are arranged in such a way that the corresponding eigenvalues $\{\lambda_i\}_{i=1}^{n}$ satisfy $\lambda_i > 0$ for $i = 1, \ldots, n_Y$ and $\lambda_i < 0$ for $i = n_Y + 1, \ldots, n$. The corresponding decoding policy is given by*

$$\begin{bmatrix} \gamma^{d_X}(z) \\ \gamma^{d_Y}(z) \end{bmatrix} = \Sigma^{1/2} \left[\beta_1 \boldsymbol{q}_1 \cdots \beta_{n_Y} \boldsymbol{q}_{n_Y}\right] z \tag{19.19}$$

where $\beta_i = 1/\alpha_i$ if $\alpha_i \neq 0$ and $\beta_i = 0$ otherwise for $i = 1, \ldots, n_Y$.These Nash equilibria exist for any set of scalars $\{\alpha_i\}_{i=1}^{n_Y}$. When the indices with $\alpha_i \neq 0$ are the same for two sets of scalars, they lead to the same performance values, i.e., the resulting equilibria with these sets of scalars are informationally equivalent.

(ii) These informative equilibria are payoff dominant Nash equilibria (among all) if $\alpha_i \neq 0$ for all $i = 1, \ldots, n_Y$.

(iii) In addition to informative equilibria, there also exist non-informative Nash equilibria with the transmitted message being independent of the parameters, e.g.,$\gamma^e(x, y) = C$ for some constant C, and with the decoding policies $\gamma^{d_X}(z) = \mathbf{0}$ and $\gamma^{d_Y}(z) = \mathbf{0}$.

Proof We apply a linear transformation of variables, as in [700], that leads to an orthonormal coordinate system and then show that in this transformed coordinate system the sender wishes to convey some of the coordinates and to hide the remaining coordinates. The advantage of this transformation is that as these

[6] The case when $\alpha_i = 0$ for all $i = 1, \ldots, n_Y$ leads to a non-informative Nash equilibrium.

coordinates are orthogonal to each other, conveying one coordinate does not leak information related to other coordinates. Such a transformation does not lead to a change in equilibrium behaviour (as this is policy-independent by the discussion in Sect. 10.3.4).

Since $W = \Sigma^{1/2}\text{diag}(-\delta I, I)\Sigma^{1/2}$ is symmetric, we can decompose it as $W = Q\Lambda Q^T$ for orthonormal Q and diagonal Λ. We denote the columns of Q by $\{\boldsymbol{q}_i\}_{i=1}^n$ and the diagonal elements of Λ by $\{\lambda_i\}_{i=1}^n$. Here, W and $\text{diag}(-\delta I, I)$ have the same number of positive and negative eigenvalues [358, p. 282]. Therefore, W has n_Y positive and n_X negative eigenvalues. We sort these eigenvalues $\{\lambda_i\}_{i=1}^n$ in such a way that the first n_Y of them are positive and the remaining ones are negative. For notational convenience, we define

$$S \triangleq \begin{bmatrix} X \\ Y \end{bmatrix}, \quad \gamma^{d_S}(z) \triangleq \begin{bmatrix} \gamma^{d_X}(z) \\ \gamma^{d_Y}(z) \end{bmatrix}. \tag{19.20}$$

Now, we make the following transformation of variables:

$$T \triangleq Q^T \Sigma^{-1/2} S, \tag{19.21}$$

where $\Sigma^{-1/2}$ is well defined since Σ is assumed to be positive definite. We note that

$$\begin{aligned} E[TT^T] &= E[Q^T \Sigma^{-1/2} S S^T \Sigma^{-1/2} Q] \\ &= Q^T \Sigma^{-1/2} E[S S^T] \Sigma^{-1/2} Q \\ &= Q^T \Sigma^{-1/2} \Sigma \Sigma^{-1/2} Q = I. \end{aligned}$$

Thus, the components of T are independent and identically distributed zero-mean Gaussian random variables each with a unit variance.

Next, we construct an equivalent problem under this linear transformation of variables. The encoder consists of two consecutive mappings, one of which is fixed as above and the other one can arbitrarily be chosen by the sender. In other words, there is a linear mapping from (x, y) to $\boldsymbol{t}$ fixed as (19.21) and then an encoding function $z = \tilde{\gamma}^e(\boldsymbol{t})$. At the receiver side, we also consider a similar setting. The observation at the receiver is mapped into $\tilde{\boldsymbol{t}}$ via $\gamma^{d_T}(z)$, which can arbitrarily be selected by the receiver. Then, these auxiliary variables are mapped into estimates of x and y as follows:

$$\gamma^{d_S}(z) = \Sigma^{1/2} Q \gamma^{d_T}(z). \tag{19.22}$$

Figure 19.4 provides a depiction for the equivalent formulation. The aim is to characterize $\tilde{\gamma}^e(\boldsymbol{t})$ as well as $\gamma^{d_T}(z)$ at a Nash equilibrium. Since the proposed transformation is invertible, the problem in the transform domain is equivalent to the original problem.

Next, we express the objective function of each player in terms of the parameters in the transformed coordinate system. The objective function of the sender can be written as

$$\begin{aligned} J^e(\tilde{\gamma}^e, \gamma^{d_T}) &= E[(S - \gamma^{d_S}(Z))^T \mathrm{diag}(-\delta I, I)(S - \gamma^{d_S}(Z))] \\ &= E[(T - \gamma^{d_T}(Z))^T Q^T \Sigma^{1/2} \mathrm{diag}(-\delta I, I) \Sigma^{1/2} Q(T - \gamma^{d_T}(Z))] \\ &= E[(T - \gamma^{d_T}(Z))^T \Lambda (T - \gamma^{d_T}(Z))] \\ &= \sum_{i=1}^{n} \lambda_i E[(T_i - \gamma^{d_{T_i}}(Z))^2], \end{aligned} \tag{19.23}$$

where the second equality is obtained from (19.12) and (19.20), the third equality is based on (19.21) and (19.22), and $\{\lambda_i\}_{i=1}^n$ are the eigenvalues of W which satisfy $\lambda_i > 0$ for $i = 1, \ldots, n_Y$ and $\lambda_i < 0$ for $i = n_Y + 1, \ldots, n$. Similarly, if we express the objective function of the receiver in terms of the parameters in the new coordinate system, we get

$$\begin{aligned} J^d(\tilde{\gamma}^e, \gamma^{d_T}) &= E[(S - \gamma^{d_S}(Z))^T (S - \gamma^{d_S}(Z))] \\ &= E[(T - \gamma^{d_T}(Z))^T K (T - \gamma^{d_T}(Z))] \end{aligned} \tag{19.24}$$

where $K \triangleq Q^T \Sigma Q$ is a positive definite matrix since Σ is positive definite. In the equivalent problem, the optimal $\gamma^{d_T}(z)$ for a given encoding policy $\tilde{\gamma}^e(x, y)$ is the minimum mean squared error estimator of the corresponding random variable, as stated below. ◇

Lemma 19.2.1 ([410]) *For a fixed encoding function $\tilde{\gamma}^e(x, y)$, the optimal $\gamma^{d_T}(z)$ that minimizes* (19.24) *is given by $E[T|Z = z]$.*

In the equivalent formulation, the objective function of the sender is expressed as a weighted sum of mean squared error terms corresponding to independent random variables where there are both positive and negative weights. We partition the transformed random vector as

$$T \triangleq \begin{bmatrix} U \\ V \end{bmatrix} \tag{19.25}$$

where $\boldsymbol{U}$ is $\mathbb{R}^{n_Y}$-valued and $\boldsymbol{V}$ is $\mathbb{R}^{n_X}$-valued such that the positive and negative coefficients in (19.23) correspond to $\boldsymbol{U}$ and $\boldsymbol{V}$, respectively. The receiver still

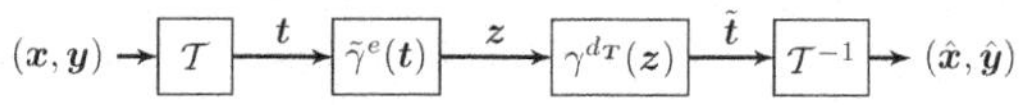

Fig. 19.4 Equivalent formulation where $\mathcal{T}$ denotes the linear transformation specified in (19.21)

employs the minimum mean squared error estimators for the corresponding random variables for a given encoding policy. Next, we use the equivalent formulation to characterize the Nash equilibria. Toward that goal, we first present the following lemma.

Lemma 19.2.2 *At a Nash equilibrium, the sender does not reveal any information related to the linear combinations (of the private and nonprivate parameters)* $\boldsymbol{V}$ *specified in* (19.25).

Proof Consider a set of policies where the sender employs an encoding policy $\tilde{\gamma}^e(\boldsymbol{t}) = f(\boldsymbol{t})$ which conveys information related to T_j for some j with $\lambda_j < 0$. In response to this encoding policy, it is optimal for the receiver to employ the minimum mean squared error estimators of each parameter, as shown in Lemma 19.2.1. Denote these estimators for estimating T_i by $g_i(z)$ for $i = 1, \ldots, n$. Since we assume that the encoding policy $f(\boldsymbol{t})$ conveys information related to T_j, the mean squared error for estimating T_j with the corresponding optimal estimator $g_j(z)$ is lower than $\sigma_{T_j}^2$, i.e., $E[(T_j - g_j(Z))^2] < \sigma_{T_j}^2$. In response to the decoding policies of $\{g_i(\cdot)\}_{i=1}^n$, the sender can switch to the following policy to improve its objective value. Instead of sending $z = f(\boldsymbol{t})$, the sender can transmit $z = f(\bar{\boldsymbol{t}})$ while keeping the encoding function $f(\cdot)$ the same where $\bar{\boldsymbol{t}} \triangleq [t_1, \ldots, t_{j-1}, w, t_{j+1}, \ldots, t_n]^T$ and w is a realization of a random variable that follows the same distribution as T_j and is independent of $\boldsymbol{T}$. In that case, the performance for estimating T_i for $i \neq j$ remains the same since receiving $f(\bar{\boldsymbol{t}})$ or $f(\boldsymbol{t})$ are equivalent for the decoding policy $g_i(z)$. However, the performance for estimating T_j degrades as shown in the following:

$$\begin{aligned}
E[(T_j - g_j(Z))^2] &= E[T_j^2] - 2E[T_j g_j(Z)] + E[g_j(Z)^2] \\
&= E[T_j^2] - 2E[T_j]E[g_j(Z)] + E[g_j(Z)^2] \\
&= E[T_j^2] + E[g_j(Z)^2] \geq \sigma_{T_j}^2,
\end{aligned}$$

where the second equality is due to the fact that T_j and Z are independent in case $f(\bar{\boldsymbol{T}})$ is transmitted. Since the parameter T_j is chosen such that $\lambda_j < 0$ in (19.23), the sender gains by employing $z = f(\bar{\boldsymbol{t}})$ instead of $z = f(\boldsymbol{t})$. As a result, any encoding policy which yields $E[(T_j - \gamma^{d_{T_j}}(Z))^2] < \sigma_{T_j}^2$ for an index j with $\lambda_j < 0$ cannot be a Nash equilibrium since in that case the sender can change its strategy to improve its objective value.

Note that formally in games, when a unilateral change by a decision maker occurs, the perturbed policies may cease to be an equilibrium. However, a subtle aspect of the proof is that the revised sender policy does not alter the policy of the decoder, and therefore the perturbation does not take us off an equilibrium. ◇

From Lemma 19.2.2, it follows that the sender is restricted to send information related to $\boldsymbol{U}$, which is specified in (19.25), at a Nash equilibrium. This implies that the objective function of the sender at a Nash equilibrium reduces to

$$J^e(\tilde{\gamma}^e, \gamma^{d_T}) = \sum_{i=1}^{n_Y} \lambda_i E[(T_i - \gamma^{d_{T_i}}(Z))^2] \tag{19.26}$$

where $\{1, \ldots, n_Y\}$ are the only indices with $\lambda_i > 0$ in $\{\lambda_i\}_{i=1}^{n}$. Since the receiver wishes to extract all the random variables in this transformed coordinate system, including the ones with these indices, the receiver also shares the objective of minimizing these mean squared error terms. As a result, conveying all or a subset of the random variables in $\boldsymbol{U}$ yields a Nash equilibrium and this leads to the linear policies stated in (19.18). In addition, conveying all of these random variables yields the minimum attainable performance at a Nash equilibrium for both players. Thus, if $\alpha_i \neq 0$ for all $i = 1, \ldots, n_Y$ in (19.18), the corresponding linear Nash equilibria are payoff dominant. Namely, there does not exist any other Nash equilibrium which Pareto dominates these characterized equilibria. This completes the proof of Theorem 19.2.1, parts (i) and (ii). Part (iii) follows as earlier due to it being a babbling equilibrium, similar to the setting in Theorem 18.1.1. ◇

Theorem 19.2.1 characterizes linear Nash equilibria in which there is communication between the transmitter and the receiver. Hence, the privacy-signaling game always admits informative linear Nash equilibria for all values of system parameters. More importantly, special cases of these linear equilibria lead to payoff dominant Nash equilibria.

Next, we specialize to the case of scalar sources with the aim to provide an explicit characterization of the payoff dominant Nash equilibria in this case. We take the 2×2 covariance matrix of (X, Y) to be given by σ_X^2 and σ_Y^2, respectively and a nonzero correlation ρ, i.e., $\Sigma = \begin{bmatrix} \sigma_X^2 & \rho \\ \rho & \sigma_Y^2 \end{bmatrix}$, where $\rho \neq 0$. We state the following as a corollary of Theorem 19.2.1.

Corollary 19.2.1 *For scalar sources X and Y, there exist informative linear Nash equilibria with an encoding policy $\gamma^e(x, y) = Ax + By$, where*

$$\frac{B}{A} = -\frac{\delta\sigma_x^2 + \sigma_y^2}{2\delta\rho} - \frac{\sqrt{(\delta\sigma_x^2 + \sigma_y^2)^2 - 4\delta\rho^2}}{2\delta\rho} \tag{19.27}$$

and decoding policies

$$\gamma^{d_X}(z) = \left(\frac{A\sigma_X^2 + B\rho}{A^2\sigma_X^2 + B^2\sigma_Y^2 + 2AB\rho} \right) z, \tag{19.28}$$

$$\gamma^{d_Y}(z) = \left(\frac{A\rho + B\sigma_Y^2}{A^2\sigma_X^2 + B^2\sigma_Y^2 + 2AB\rho} \right) z. \tag{19.29}$$

These equilibria are payoff dominant Nash equilibria. In addition, these constitute the complete class of linear Nash equilibrium policies.

Proof In order to characterize the Nash equilibria, we need to find the eigenvalues and eigenvectors of $W = \Sigma^{1/2}\text{diag}(-\delta, 1)\Sigma^{1/2}$. We note that W has the same eigenvalues as $D \triangleq \Sigma\,\text{diag}(-\delta, 1)$ and these can be computed as

$$\lambda_1 = \frac{-\delta\sigma_X^2 + \sigma_Y^2}{2} + \frac{\sqrt{(\delta\sigma_X^2 + \sigma_Y^2)^2 - 4\delta\rho^2}}{2}, \tag{19.30}$$

$$\lambda_2 = \frac{-\delta\sigma_X^2 + \sigma_Y^2}{2} - \frac{\sqrt{(\delta\sigma_X^2 + \sigma_Y^2)^2 - 4\delta\rho^2}}{2}. \tag{19.31}$$

where $\lambda_1 > 0$ and $\lambda_2 < 0$. Thus, the sender is restricted to transmit $u = \boldsymbol{q}_1^T \Sigma^{-1/2} \begin{bmatrix} x \\ y \end{bmatrix}$ where $\boldsymbol{q}_1$ is the normalized eigenvector of W associated with the eigenvalue λ_1 computed in (19.30). We note that $\boldsymbol{q}_1^T \Sigma^{-1/2}$ is a left eigenvector of D associated with its eigenvalue λ_1. By explicating this left eigenvector, an encoding policy which satisfies (19.27) is obtained. Then, the conditional expectation formula for jointly Gaussian distributions can be employed to obtain (19.28) and (19.29) [587, p. 155]. As a result, these characterized policies lead to payoff dominant Nash equilibria. Moreover, the only possible linear equilibria are attained by transmitting a scaled version of u. ◇

Remark 19.2.1 We note that the ratio of B and A converges to $-\sigma_x^2/\rho$ as $\delta \to \infty$, which can be verified from (19.27). This shows that in the high privacy regime, the encoder makes the revealed information Z and the private parameter X essentially uncorrelated.

19.2.2 Stackelberg Equilibria

In this subsection, we characterize the Stackelberg equilibria of the quadratic privacy-signaling game. The main result is that the payoff dominant Nash equilibria presented in the previous section are also Stackelberg equilibria. It is important to note that even if the solution turns out to be linear, the set of possible encoding strategies, i.e., Γ^e, is not restricted here to be linear. If the sender performs an optimization of its objective function by anticipating the best response of the receiver, these linear policies become the optimal solution among the general set of measurable policies.

Theorem 19.2.2 ([410])

(i) The encoding policies in (19.18) *with* $\alpha_i \neq 0$ *for all* $i = 1, \ldots, n_Y$ *and the decoding policies in* (19.19) *constitute a Stackelberg equilibrium, i.e., a payoff dominant Nash equilibrium and a Stackelberg equilibrium coincide (and these constitute the complete class of (essentially unique) Stackelberg equilibria).*

(ii) *In contrast to the Nash equilibrium setting where there exist both informative and non-informative equilibria, Stackelberg equilibria are necessarily informative, where the sender uses the private and nonprivate random variables in constructing its message.*

(iii) *When the nonprivate random variable is not scalar valued, i.e., $n_Y > 1$, there exist informative Nash equilibria which do not coincide with the Stackelberg equilibria. Such Nash equilibria are attained by an encoding policy in* (19.18) *and a decoding policy in* (19.19) *where the scalars $\{\alpha_i\}_{i=1}^{n_Y}$ can take any value with at least one zero term and one nonzero term.*

Proof We apply a transformation of variables as in (19.21). In this transformed coordinate system, the objective functions of the sender and the receiver are expressed as in (19.23) and (19.24), respectively. If we partition the random vector T in this coordinate system as in (19.25) according to the sign of coefficients in (19.23), we can show that the sender can only convey information related to $\boldsymbol{U}$. The following lemma presents this result in parallel with that in Lemma 19.2.2 with the exception that in this case the sender announces its policy first, anticipating the response of the receiver. ◇

Lemma 19.2.3 *At a Stackelberg equilibrium, the sender does not reveal any information related to the linear combinations (of the private and nonprivate random variables) $\boldsymbol{V}$ specified in* (19.25).

Proof We show that any encoding policy which yields $E[(T_j - \gamma^{d_{T_j}}(Z))^2] < \sigma_{T_j}^2$ for an index j with $\lambda_j < 0$ cannot be a Stackelberg equilibrium via an analysis similar to that employed in Lemma 19.2.2. Towards that goal, we compare the performances of two scenarios from the perspective of the sender. Recall that in a Stackelberg equilibrium the sender chooses a policy and announces this policy to the receiver and the receiver acts with the knowledge of sender's policy. Denote the encoding policy by $\tilde{\gamma}^e(\boldsymbol{t}) = f(\boldsymbol{t})$ in the first scenario. The receiver takes an optimal response to this announced encoding policy. Assume that $E[(T_j - \gamma^{d_{T_j}}(Z))^2] < \sigma_{T_j}^2$ with the corresponding set of policies. In the second scenario, suppose that the encoder chooses the same policy as before with the exception that the sender replaces the realization $T_j = t_j$ by an independent noise following the same distribution as T_j. Namely, the sender uses $f(\bar{\boldsymbol{t}})$ where $\bar{\boldsymbol{t}} = [t_1, \ldots, t_{j-1}, w, t_{j+1}, \ldots, t_n]^T$ and w is a realization of a random variable that follows the same distribution as T_j and is independent of T. As the sender announces its strategy, the optimal response of the receiver for the parameter T_j becomes $\gamma^{d_{T_j}}(z) = E[T_j|Z = z] = E[T_j] = 0$ due to the independence of T_j and Z in this scenario. Therefore, we get $E[(T_j - \gamma^{d_{T_j}}(Z))^2] = \sigma_{T_j}^2$ in this case. Notice that the mean squared error performance in estimating T_i for $i \neq j$ is the same for both scenarios. As a result, the second scenario yields better performance for the sender. Thus, transmitting information related to T_j with $\lambda_j < 0$ in (19.23) is not desirable for the sender. ◇

As a result of Lemma 19.2.3, the sender is restricted to transmit $\boldsymbol{U}$. Since the linear encoding policies in (19.18) with $\alpha_i \neq 0$ for all $i = 1, \ldots, n_Y$ reveals $\boldsymbol{U}$ completely, these encoding policies yield the minimum attainable performance for the sender among all encoding policies. Therefore, the policies in the statement of the theorem lead to a Stackelberg equilibrium.

We note that the random variable $\boldsymbol{U}$, which is desired to be transmitted at a Stackelberg equilibrium, has a size of $n_Y \geq 1$. Therefore, there exists an informative Stackelberg equilibrium where the sender reveals $\boldsymbol{U}$ completely. On the other hand, it is always possible to construct a non-informative Nash equilibrium as stated in Theorem 19.2.1.

When $n_Y > 1$, the random variable $\boldsymbol{U}$ is not scalar valued. Then, there exist informative Nash equilibria with an encoding policy in (19.18) where the scalars $\{\alpha_i\}_{i=1}^{n_Y}$ take any value with at least one zero term and one nonzero term. In these informative Nash equilibria, the sender conveys only a subset of the random variables in $\boldsymbol{U}$. Since at these informative Nash equilibria, the performance of the sender is strictly worse than that at the payoff dominant Nash equilibria, these policies do not lead to a Stackelberg equilibrium.

◇

Remark 19.2.2 In fact, employing any invertible function $h(\boldsymbol{u})$ at the sender also yields a Stackelberg equilibrium. Since the receiver knows the commitment of the sender, it can simply employ $h^{-1}(\boldsymbol{u})$ to perfectly recover $\boldsymbol{u}$.

Remark 19.2.3 We note that the Stackelberg equilibria can be obtained from the analysis presented in [700, Theorem 2] which characterizes linear policies that form a Stackelberg equilibrium for a general quadratic multidimensional signaling setup for jointly Gaussian sources with some generalizations (in the cost setup considered) but also some restrictions, such as the *a priori* restriction of the decoder to an affine class in the conditional expectation.

19.2.3 The MMSE Gaussian Information Bottleneck Problem

We now move on to the information bottleneck problem [718] as depicted in Fig. 19.3. As we have mentioned earlier, we can view this problem as an instance of our formulation under the Stackelberg equilibrium concept in the following sense: in contrast to the privacy game setting, here only the random variable X is observed by the sender, who wants to use it to send to the receiver as much information as possible on another correlated random variable Y, while leaking as little information about X as possible. One way of formulating an optimization problem that captures this scenario is to use mutual information between two random variables as a metric (which is the typical formulation in the information bottleneck problem). Here, we provide an estimation theoretic perspective on the information bottleneck problem by using quadratic distortion criteria as in the privacy game setting. Although the problem can be cast as a constrained optimization problem of maximizing the released (useful) information related to the unobserved random variable under a compression constraint, the Lagrangian dual of this constrained problem is commonly considered (see Remark 19.2.5 for a constrained version in this setting). Accordingly the goal in this approach is to find the optimal solution to the following optimization problem:

$$\min_{Z=\gamma^e(X)} I(X;Z) - \beta I(Y;Z), \tag{19.32}$$

where $I(\cdot;\cdot)$ denotes the mutual information between the two random variables, and $\beta > 0$ is a tradeoff parameter. Here, the goal is to compress an observation X while at the same time to maximize the released information related to Y.

The information bottleneck problem as formulated in (19.32) can in fact be viewed as a Stackelberg game between a sender and a receiver. In this game, the sender wants to compress the observed random variable so as to convey information on the unobserved random variable. The use of mutual information as a performance metric effectively means that the receiver uses all the available information related to both random variables, i.e., it always employs its best response as in a Stackelberg equilibrium. Thus, the receiver is concerned with extracting information related to both random variables, which is also the case in our framework.

The second approach, which we adopt here, employs (instead of the mutual information) mean squared error terms as our metric. Accordingly, the objective functions of the sender and the receiver are as defined in (19.12) and (19.13), respectively. Since the receiver is concerned with estimating both random variables, it employs the minimum mean squared error estimators of each random variable. Hence, under the solution concept of the Stackelberg equilibrium, the objective function of the sender can equivalently be written as (in view of the unique response of the receiver):

$$J^e(\gamma^e) = -\delta\, E[\|X - E[X|Z]\|^2] + E[\|Y - E[Y|Z]\|^2]. \tag{19.33}$$

Theorem 19.2.3 ([410]) *Consider the MMSE Gaussian information bottleneck problem formulated in this subsection, and under the Stackelberg solution concept. Then:*

(i) If $\Sigma_{XY}\Sigma_{YX} - \delta\Sigma_X^2$ is not negative definite, the MMSE Gaussian information bottleneck solution is attained by an encoding policy

$$\gamma^e(x) = \left[\alpha_1 \boldsymbol{q}_1 \ldots \alpha_k \boldsymbol{q}_k\right]^T \Sigma_X^{-1/2} x \tag{19.34}$$

for nonzero scalars $\{\alpha_i\}_{i=1}^k$ where k denotes the number of nonnegative eigenvalues of $W = \Sigma_X^{-1/2}\Sigma_{XY}\Sigma_{YX}\Sigma_X^{-1/2} - \delta\Sigma_X$ and $\{\boldsymbol{q}_i\}_{i=1}^k$ are the normalized eigenvectors of W associated with its nonnegative eigenvalues. The corresponding decoding policy is given by

$$\begin{bmatrix} \gamma^{d_X}(z) \\ \gamma^{d_Y}(z) \end{bmatrix} = \text{diag}(\Sigma_X^{1/2}, \Sigma_{YX}\Sigma_X^{-1/2}) \times \begin{bmatrix} \beta_1 \boldsymbol{q}_1 \cdots \beta_k \boldsymbol{q}_k \end{bmatrix} z \tag{19.35}$$

where $\beta_i = 1/\alpha_i$ *for* $i = 1, \ldots, k$.

(ii) If $\Sigma_{XY}\Sigma_{YX} - \delta\Sigma_X^2$ *is positive definite, the MMSE Gaussian information bottleneck solution is attained by a fully informative encoding policy, where the sender reveals the random variable X completely.*

(iii) If $\Sigma_{XY}\Sigma_{YX} - \delta\Sigma_X^2$ *is negative definite, the MMSE Gaussian information bottleneck solution is non-informative, i.e., the sender does not reveal any information related to its observation.*

Before we present the proof of the theorem, we go back to the mutual information based formulation of the information bottleneck problem given in (19.32). We provide the solution to that in the theorem below, to be contrasted with the results given in Theorem 19.2.3.

Theorem 19.2.4 ([172, Theorem 3.1][410]) *For the Gaussian information bottleneck problem with mutual information used as the metric, that is (19.32), the optimal solution is given by*

$$z = A(\beta)x + \boldsymbol{n} \tag{19.36}$$

where $\boldsymbol{n}$ *is a realization of a zero-mean Gaussian random vector with identity covariance matrix, and*

$$A(\beta) = \begin{cases} [\mathbf{0} \ \cdots \ \mathbf{0}]^T & \text{if } 0 \le \beta < \beta_1^c \\ [\alpha_1 \boldsymbol{p}_1 \ \mathbf{0} \ \cdots \ \mathbf{0}]^T & \text{if } \beta_1^c \le \beta < \beta_2^c \\ [\alpha_1 \boldsymbol{p}_1 \ \alpha_2 \boldsymbol{p}_2 \ \mathbf{0} \ \cdots \ \mathbf{0}]^T & \text{if } \beta_2^c \le \beta < \beta_3^c \\ \vdots & \end{cases}$$

where $\{\boldsymbol{p}_i^T\}_{i=1}^{n_X}$ *are left eigenvectors of*

$$\Sigma_{X|Y}\Sigma_X^{-1} \triangleq (\Sigma_X - \Sigma_{XY}\Sigma_Y^{-1}\Sigma_{YX})\Sigma_X^{-1}$$

with the corresponding eigenvalues $\{\lambda_i\}_{i=1}^{n_X}$, *which are sorted in ascending order,* $\beta_i^c = 1/(1 - \lambda_i)$, *and* $\alpha_i = \sqrt{\frac{\beta(1-\lambda_i)-1}{\lambda_i(\boldsymbol{p}_i^T\Sigma_X\boldsymbol{p}_i)}}$ *for* $i = 1, \ldots, n_X$, *where* n_X *is the dimension of X..*

Remark 19.2.4 Contrasting the solutions of the two formulations (mutual information vs. quadratic error), we note that for the information bottleneck problem under (19.32), the optimal solution given in Theorem 19.2.4 is jointly Gaussian with X which is also the case for the quadratic estimation theoretic solution in

Theorem 19.2.3. However, in the information theoretic formulation, the solution involves applying an independent Gaussian perturbation to a linear function of X as given in (19.36) whereas in the estimation theoretic solution in Theorem 19.2.3, the encoder reveals a linear function of X without applying any random perturbation.

We now present the proof of Theorem 19.2.3.

Proof By the standard property of the conditional expectation the error, we have that $(Y - E[Y|X])$ is orthogonal to the measurement X. Since the sender observes only the random variable X and determines its message based on X, it follows that $Y \to X \to Z$ is a Markov chain in that order. This Markov property implies that $(Y - E[Y|X])$ is orthogonal to Z as well. By using these orthogonality properties, we can express the second term in (19.33) as

$$\begin{aligned} E[\|Y - E[Y|Z]\|^2] &= E[\|Y - E[Y|X] + E[Y|X] - E[Y|Z]\|^2] \\ &= E[\|Y - E[Y|X]\|^2] + E[\|E[Y|X] - E[Y|Z]\|^2] \\ &= E[\|Y - E[Y|X]\|^2] + E[\|E[Y|X] - E[E[Y|X]|Z]\|^2] \end{aligned}$$

where the second equality follows from the fact that $(Y - E[Y|X])$ is orthogonal to X and Z and the third equality is due to iterated expectations. Observing that

$$E[Y|X = x] = \Sigma_{YX}\Sigma_X^{-1}x,$$

the objective function can be written as

$$\begin{aligned} J^e(\gamma^e) &= E[\|Y - E[Y|X]\|^2] - \delta\, E[\|X - E[X|Z]\|^2] \\ &+ E[(X - E[X|Z])^T(\Sigma_X^{-1}\Sigma_{XY}\Sigma_{YX}\Sigma_X^{-1})(X - E[X|Z])], \end{aligned}$$

where the first term is independent of the encoder policy. Thus, we obtain an optimization problem of the form

$$\min_{Z=\gamma^e(X)} E\Big[(X - E[X|Z])^T M(X - E[X|Z])\Big], \tag{19.37}$$

where

$$M \triangleq (\Sigma_X^{-1}\Sigma_{XY}\Sigma_{YX}\Sigma_X^{-1} - \delta I).$$

Now, the optimization problem in (19.37) can be viewed as a quadratic multidimensional signaling game problem considered earlier and the solution can be obtained by using the analysis in Sect. 19.2.2. In particular, we can rewrite the problem in (19.37) as a Stackelberg game between a sender and a receiver with objective functions

$$J^e(\gamma^e, \gamma^{d_X}) = E[(X - \gamma^{d_X}(Z))^T M(X - \gamma^{d_X}(Z))], \tag{19.38}$$

$$J^d(\gamma^e, \gamma^{d_X}) = E[(X - \gamma^{d_X}(Z))^T (X - \gamma^{d_X}(Z))]. \tag{19.39}$$

Notice that for a given encoding policy, the best response of the receiver under (19.39) is given by the minimum mean squared estimator of X given Z, which is consistent with (19.37).

Next, we apply a transformation of variables to express the objective function of the sender in a simplified form. Towards that goal, let $W = \Sigma_X^{-1/2}\Sigma_{XY}\Sigma_{YX}\Sigma_X^{-1/2} - \delta\Sigma_X = Q\Lambda Q^T$ for orthonormal Q and diagonal Λ. Now, consider the following invertible transformation of variables

$$T \triangleq Q^T \Sigma_X^{-1/2} X. \tag{19.40}$$

Under this transformation of variables, we introduce an equivalent problem in a similar manner to Sect. 19.2.1. The observation X is mapped into T via a fixed transformation (19.40). The encoder chooses an arbitrary policy $\tilde{\gamma}^e(\cdot)$ which maps the transformed random vector T into the message Z. The receiver applies an arbitrary policy $\gamma^{d_T}(\cdot)$ to its observation Z and then the estimate of X is obtained via a fixed relation

$$\gamma^{d_X}(z) = \Sigma_X^{1/2} Q \gamma^{d_T}(z). \tag{19.41}$$

In this transformed coordinate system, for a given encoding policy the receiver still employs the minimum mean squared estimator of the parameter T, which can be established via a similar analysis to that in Lemma 19.2.1. The objective function of the sender in this transformed coordinate system can be written as

$$J^e(\tilde{\gamma}^e, \gamma^{d_T}) = \sum_{i=1}^{n_X} \lambda_i E[(T_i - \gamma^{d_{T_i}}(Z))^2], \tag{19.42}$$

where $\{\lambda_i\}_{i=1}^{n_X}$ are eigenvalues of W.

If all of these eigenvalues $\{\lambda_i\}_{i=1}^{n_X}$ are positive, which is equivalent to $\Sigma_{XY}\Sigma_{YX} - \delta\Sigma_X^2$ being positive definite, then the minimum can be attained by revealing T, which corresponds to the fully informative scenario. In case all of these eigenvalues are negative, i.e., $\Sigma_{XY}\Sigma_{YX} - \delta\Sigma_X^2$ is negative definite, then revealing information related to any component of T is not desirable for the sender, and thus, this scenario leads to a non-informative Stackelberg equilibrium. In the remaining case, i.e. when $\Sigma_{XY}\Sigma_{YX} - \delta\Sigma_X^2$ is neither positive definite nor negative definite, we partition the transformed vector according to the sign of the coefficients $\{\lambda_i\}_{i=1}^{n_X}$ in (19.42) as follows:

$$T \triangleq \begin{bmatrix} U \\ V \end{bmatrix} \tag{19.43}$$

where $\boldsymbol{U} \in \mathbb{R}^k$ and $\boldsymbol{V} \in \mathbb{R}^{n_X-k}$ correspond to nonnegative and negative coefficients in (19.42), respectively, with k denoting the number of nonnegative eigenvalues of W. Next, we can apply Lemma 19.2.3 for this particular Stackelberg game setup to establish that the sender cannot convey information related to $\boldsymbol{V}$ and is restricted to send information related to $\boldsymbol{U}$. As the encoding policy in (19.34) reveals $\boldsymbol{U}$ completely, this encoding policy achieves the minimum attainable expected cost for the sender among any set of policies. Thus, the pair of policies (19.34) and (19.35) yield a Stackelberg equilibrium, which gives the solution to the MMSE Gaussian information bottleneck problem. ◇

In the special case of scalar sources, Theorem 19.2.3 simplifies considerably, since now M is a scalar, which is either positive (leading to full revelation) or negative (leading to no revelation), or zero (in which case what the encoder does is irrelevant). Thus, we have, as the corollary to Theorem 19.2.3, the following.

Corollary 19.2.2 *The Stackelberg equilibrium of the MMSE information bottleneck problem with a scalar source is given by one of the following three cases:*

(i) If $(\rho^2/\sigma_X^4) > \delta$, then the sender completely reveals X.
(ii) If $(\rho^2/\sigma_X^4) < \delta$, then the sender does not reveal any information related to X.
(iii) If $(\rho^2/\sigma_X^4) = \delta$, then both informative and non-informative scenarios lead to a Stackelberg equilibrium.

Remark 19.2.5 (Constrained MMSE Information Bottleneck Problem) It is also possible to apply the ideas used in the proof of Theorem 19.2.3 to a constrained version of the MMSE Gaussian information bottleneck problem where the aim is to minimize the mean squared error for estimating Y under the constraint that the mean squared error for estimating X is above a certain threshold α. In this case, the underlying optimization problem is defined as follows

$$\min_{Z=\gamma^e(X)} \operatorname{Tr}\Big(\Upsilon\, E[(X - E[X|Z])(X - E[X|Z])^T]\Big)$$

$$\text{subject to } \operatorname{Tr}\Big(E[(X - E[X|Z])(X - E[X|Z])^T]\Big) \geq \alpha \tag{19.44}$$

where $\Upsilon \triangleq \Sigma_X^{-1}\Sigma_{XY}\Sigma_{YX}\Sigma_X^{-1}$, which is always positive semidefinite. Since any positive semidefinite

$$\Phi \triangleq E[(X - E[X|Z])(X - E[X|Z])^T$$

is attainable via a linear encoding policy with a Gaussian perturbation, the problem equivalently reduces to

$$\min_{\Phi \succeq 0} \operatorname{Tr}(\Upsilon\, \Phi)$$

$$\text{subject to } \operatorname{Tr}(\Phi) \geq \alpha. \tag{19.45}$$

Let the minimum eigenvalue of $\varUpsilon$ be denoted by $\lambda_{\min}$. Note that

$$\begin{aligned}\mathrm{Tr}(\varUpsilon\varPhi) &= \mathrm{Tr}(\varUpsilon\varPhi - \lambda_{\min}\varPhi + \lambda_{\min}\varPhi)\\ &= \mathrm{Tr}((\varUpsilon - \lambda_{\min}I)\varPhi) + \lambda_{\min}\mathrm{Tr}(\varPhi)\\ &\geq \alpha\lambda_{\min}\end{aligned}$$

where the inequality uses the constraint along with the observation that $(\varUpsilon - \lambda_{\min}I)$ and $\varPhi$ are positive semidefinite. As a result, by using a linear encoding policy possibly with a Gaussian perturbation, one can attain $\mathrm{Tr}(\varPhi) = \alpha$ where the solution satisfies the orthogonality condition under the trace inner-product defining a Hilbert space on square matrices:

$$\mathrm{Tr}((\varUpsilon - \lambda_{\min}I)\varPhi) = 0.$$

Since such an encoding policy achieves the characterized lower bound, it becomes the optimal solution to the constrained MMSE Gaussian information bottleneck problem. That the constrained problem with inequality is equivalent to a problem with an equality constraint applies more broadly to information bottleneck problems, see e.g. [781].

Remark 19.2.6 We emphasize that the solution presented in Theorem 19.2.3 is obtained without making an *a priori* linear policy restriction. These policies are optimal among any set of policies for the optimization problem constructed at the sender by anticipating the best response of the receiver.

Remark 19.2.7 In the information bottleneck problem, the sender uses partial information since only parameter X is available at the sender whereas in the privacy-signaling game formulation the sender has access to both random variables. Due to this further restriction that only partial information is available at the sender, the information bottleneck analysis provides a lower bound on the performance of the original Stackelberg game.

Remark 19.2.8 It should be emphasized that the information bottleneck problem involving mutual information corresponds to a Stackelberg equilibrium concept since employing mutual information effectively means that the receiver uses all the available information, i.e., it employs its best response. On the other hand, the Nash solution concept would require an explicit dependence of the functions (considered in the optimization) on the (possibly arbitrary) receiver policy.

19.3 Concluding Remarks

In this chapter, we studied two important classes of signaling games.

We first studied reputation games; studying the reputations problem of an informed long-lived player who controls his reputation against a sequence of uninformed short-lived players by employing tools from stochastic control theory. Results on the structure of equilibrium behavior in finite-horizon, infinite-horizon, and undiscounted settings, as well as continuity results in terms of the prior probabilities, and upper and lower bounds on the value of reputations were presented. In particular, we showed that a control-theoretic formulation can be utilized to characterize the equilibrium behavior. Under some technical assumptions, we have identified that a nested information structure implies the equivalence of the set of Markov Perfect Equilibrium payoffs and the set of Perfect Bayesian Equilibrium payoffs.

We then studied information privacy games. A communication setting between a sender with privacy concerns and a receiver was investigated in a game theoretic framework. The private and nonprivate parameters were modelled as jointly Gaussian random vectors. We showed that a payoff dominant Nash equilibrium is attained by linear policies and that these linear policies at the payoff dominant Nash equilibria lead to Stackelberg equilibria as well. We also provided an estimation-theoretic perspective on the information bottleneck problem under the Stackelberg equilibrium concept. We have shown that the Stackelberg equilibria are attained by a set of linear policies.

19.4 Bibliographic Notes

On reputation and signaling games, Spence's paper on job market signaling [680] is one of the most influential studies in the field. On reputation games with informational ambiguties, Kreps, Milgrom, Roberts, and Wilson [431] introduced the adverse selection approach to study reputations in finitely repeated games. Fudenberg and Levine [276, 277] extended this approach to infinitely repeated games and showed that a patient long-lived player facing infinitely many short-lived players can guarantee himself a payoff close to his Stackelberg payoff when there is a small probability that he is a commitment type who always plays the stage game Stackelberg action. When compared to the *folk theorem* [280, 282], their results imply an intuitive outcome: the equilibria with relatively high payoffs are more likely to arise due to reputation effects. Even though the results of Fudenberg and Levine [276, 277] hold for both perfect and imperfect public monitoring, Cripps, Mailath, and Samuelson [192] have shown that reputation effects are not sustainable in the long-run when there is imperfect public monitoring. In other words, under imperfect public monitoring it is impossible to maintain a permanent reputation for playing a strategy that does not play an equilibrium of the complete information game. There has been further literature which studies the possibility/impossibility of maintaining permanent reputations.

Sorin [679] unified and improved some of the results in the reputations literature by using tools from Bayesian learning and merging results due to Kalai and Lehrer

[380, 381]. Gossner [311] utilized relative entropy (that is, information divergence or Kullback-Leibler divergence) to obtain bounds on the value of reputations; these bounds coincide in the limit (as the strategic long-lived player becomes arbitrarily patient) with the bounds provided by Fudenberg and Levine [276, 277]. Another relevant result on reputation games appears in [278], which shows that sequential equilibrium payoffs and perfect public equilibrium payoffs coincide (See the Appendix B of [278]) in a similar infinitely repeated game setup.

The information bottleneck problem [718] has connections with the privacy funnel problem [498]. In the information bottleneck method [718], the aim is to compress an observed random variable while trying to preserve information related to another correlated random variable which is not observed. It is important to note that the information bottleneck technique is closely related to an earlier seminal work [781] which considers a similar constrained optimization problem by employing conditional entropy to asses the performance. The information bottleneck problem specializing to Gaussian sources has been investigated in [172].

The notion of differential privacy ensures that private information provided by an individual to a database is not compromised by a third party, e.g., data analyst, who retrieves information from this database. In this context, an interesting result from [462] is the application of the Laplacian or Gaussian perturbations to guarantee differential privacy. A comprehensive treatment of differential privacy on such problems as filtering and estimation is given in [461]. Another line of work is the privacy funnel problem [155] where it is desired to convey as much information as possible related to an observed random variable while trying to leak as low information as possible related to an unobserved private random variable. Another related work on the tradeoff between utility and privacy is [40].

A further problem related to reputation games is with regard to information cascades [458] where prior short-term agents may impact, or determine, the behaviour of following short-term agents leading to loss of the value of their private information in an equilibrium.

The material in this chapter builds primarily on [410, 411] and [201].

Chapter 20
Zero-Sum Information Games: Jamming and Witsenhausen's Counterexample as a Game

We have seen earlier in Chaps. 10 and 11 that zero-sum games share many unique properties compared with general non-zero sum games with regard to the value of a game, comparison of information structures and regularity in information structures. Accordingly, it is not surprising that in the context of signaling games also, zero-sum games entail several attributes leading to explicit solutions.

In this chapter, we consider zero-sum information games, and in two different settings. The first one is still a two-agent decision-making problem as in the two previous chapters, but this time their (that is, the sender's and receiver's) objectives are (totally) conflicting. The second one, on the other hand, brings in a third adversarial agent into the earlier setting with fully cooperating sender and receiver, where the third agent (jammer) injects an intruding signal into the channel, with an objective totally opposing that of the sender and the receiver (thus leading again to a zero-sum game).

An important property of zero-sum games that will be utilized throughout the chapter to obtain strong (such as uniqueness and linearity) results is the ordered-interchangeability property of multiple saddle points.

Section 20.1 studies a variation of Witsenhausen's counterexample, formulated as a zero-sum signaling game. Section 20.2 investigates information transmission (from encoder to decoder) in the presence of a jammer, treated as a zero-sum stochastic game.

20.1 Witsenhausen's Counterexample as a Zero-Sum Signaling Game

Before we move on to Witsenhausen's counterexample, for completeness, let us present the following analysis on zero-sum signaling games.

S. Yüksel, T. Başar, *Stochastic Teams, Games, and Control under Information Constraints*, Systems & Control: Foundations & Applications,
https://doi.org/10.1007/978-3-031-54071-4_20

20.1.1 Zero-Sum Signaling Games

Within the setup of Sect. 18.1, zero-sum games result only in non-informative equilibria (Definition 18.1.1). In this case, there is no meaningful information transmission between the encoder and the decoder. Let us adopt the setting of Sect. 18.2, with no bias b, and with c^e and c^d totally conflicting, that is

$$c^e(m, u) = (m - u)^2 = -c^d(m, u)$$

We have the following.

Theorem 20.1.1 *If $c^e = -c^d$ as above, that is, if the setup is a zero-sum quadratic game, then an equilibrium is achieved only when γ^e is non-informative (e.g., a kernel with actions statistically independent of the source) and γ^d uses only the prior information (since the received information is non-informative).*

The argument here is that any kernel which would lead to a message correlated with the source would be used to reduce the uncertainty as seen by the decoder. Therefore, the encoder chooses not to convey any *information*, but he still may *talk* (i.e., babble).

20.1.2 Witsenhausen's Counterexample as a Zero-Sum Signaling Game and Optimality of Linear Policies

In this section, we revisit the counterexample of Witsenhausen in transformed variables as studied extensively in Sect. 3.3, but now with the encoder and the decoder having completely opposing goals. Recall the flow depicted in Fig. 20.1, presented here again (and the representation in state space form in Fig. 3.3).

Staying with the notation of Sect. 3.3, rather than the one adopted in the two previous chapters, we have a zero-mean Gaussian random variable, x, with variance σ_x^2, to be transformed to another real valued random variable, $u_0 = \gamma_0(x)$, which is transmitted over a channel, $y = u_0 + w$, with independent additive zero-mean Gaussian noise w with variance σ_w^2, the output, y, of which is to be further transformed into another random variable, $u_1 = \gamma_1(y)$, again taking values on the real line. The objective is to choose the transformations γ_0 and γ_1 in such a way that a given cost function, $Q_G(x, u_0, u_1)$, quadratic in x, u_0, and u_1 as given below, is minimized with respect to policy γ_1, and maximized with respect to policy γ_0,

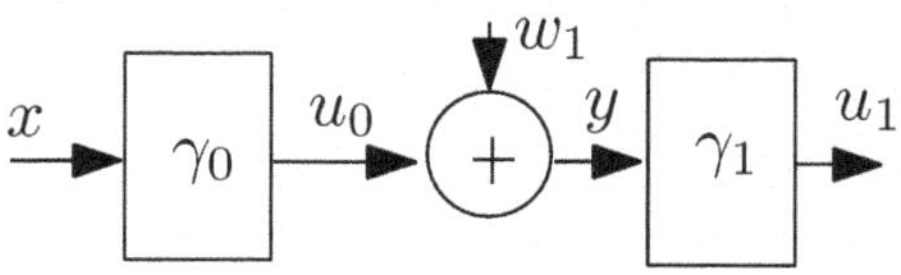

Fig. 20.1 Flow of information in Witsenhausen's counterexample

where both policies are general Borel-measurable mappings. More precisely, we seek a saddle-point pair $\gamma^* := (\gamma_0^*, \gamma_1^*)$, if one exists, such that with

$$J(\gamma_0, \gamma_1) = E^{\gamma_0,\gamma_1}[Q_G(x, u_0, u_1)],$$

we have

$$J(\gamma_0, \gamma_1^*) \leq J(\gamma_0^*, \gamma_1^*) \leq J(\gamma_0^*, \gamma_1),$$

for all admissible γ_0 and γ_1, and with the cost function Q_G given by

$$Q_G(x, u_0, u_1) = -k(u_0 - x)^2 + (u_1 - u_0)^2, \tag{20.1}$$

where $k > 0$ is a given parameter.

Theorem 20.1.2 ([89]) *For the zero-sum game formulated above with $k > 0$, there exists a unique saddle point. This unique saddle point entails linear policies for both players.*

Proof Take $\gamma_0(x) = \alpha x$, for some α; then the uniquely minimizing policy is the conditional mean of αx given y, which is $\gamma_1(y) = \frac{\alpha^2\sigma_x^2}{\alpha^2\sigma_x^2+\sigma_w^2}y$. Now, if γ_1 is taken to be linear, say $\gamma_1(y) = \lambda y$, then provided that $k > (\lambda - 1)^2$ (which leads to strict concavity of the cost in u_0), the γ_0 which maximizes J is also linear and unique (as it is a quadratic maximization problem of a strictly concave function, with conditioning on x):

$$\gamma_0(x) = \frac{-k}{k - (\lambda - 1)^2}x$$

We then look for a fixed point of the two maps (from α to λ, and from λ to α), which are:

$$\lambda = \frac{\alpha^2\sigma_x^2}{\alpha^2\sigma_x^2 + \sigma_w^2} \in [0, 1)$$

$$\alpha = \frac{-k}{k - (\lambda - 1)^2},$$

with the additional condition that $k > (\lambda - 1)^2$. We find it convenient to write this condition equivalently as (since $\lambda < 1$): $\lambda > 1 - \sqrt{k}$, which places a positive lower bound on λ and also not allowing it to be 0, if $k \leq 1$. Now, substitution of the expression for α above into the expression for λ leads to the following 5th order polynomial equation in λ:

$$f(\lambda) := \frac{\sigma_w^2}{\sigma_x^2}\lambda(k - (\lambda - 1)^2)^2 - k^2(1 - \lambda) = 0$$

Let us first consider the case of $k > 1$. Since then $0 \leq \lambda < 1$, it will be sufficient to study the behavior of the function f over $[0, 1)$. We have that $f(0) < 0$, $f(1) > 0$ and f is strictly increasing on $[0, 1)$. Thus, there is a unique solution in $(0, 1)$ if $k > 1$. For $k \leq 1$, using a similar reasoning, one can find a unique solution for $f(\lambda) = 0$ in the interval $(1 - \sqrt{k}, 1)$, since $f(1 - \sqrt{k}) < 0$, $f(1) > 0$ and f is strictly increasing in the region of interest. Thus, we conclude that [89]

$$\gamma_0^*(x) = \frac{-k}{k - (\lambda^* - 1)^2}x, \qquad \gamma_1^*(y) = \lambda^* y$$

with $f(\lambda^*) = 0$, $\lambda^* \in (\max(0, 1 - \sqrt{k}), 1)$, constitutes a saddle-point solution. Since each of these policies is a unique optimizing policy against the other in the general class of Borel-measurable maps, by the ordered interchangeability property of multiple saddle-point policies (Theorem 9.2.1), what we have here is the unique saddle-point pair (that is, there cannot be any other linear or nonlinear pair of policies in saddle-point equilibrium). ◇

20.2 Jamming Games

In this section we consider a class of jamming games: The question of how a communication system performs in the worst case scenario is one of vital importance, for if one can guarantee a suitable operation in the least favorable case then the system will operate at least as well in all other scenarios. In determining the worst case faced by a system, it is useful to personify the noise as an intelligent "jammer" actively working to steer the system towards a deleterious behavior. Furthermore, in many scenarios, there may indeed exist a malicious agent who intends to suppress communication and control in a decentralized system. These setups motivate the use of game theoretic methods in communications and networked control applications.

As a prototype of such scenarios, we consider in this section the case of a Gaussian source-channel system with an intelligent jammer, who injects a disturbing signal into the communication channels, so as to adversely affect the performance of the communication system.

Consider two terminals (decision makers) T_1 and T_2 attempting to concurrently exchange Gaussian independent source signals u_1 and u_2, respectively, where $u_i \sim \mathcal{N}(0, 1)$ has zero mean and unit variance for $i = 1, 2$, across a two-way additive Gaussian noise channel as depicted in Fig. 20.2.

More specifically, each terminal T_i observes signal u_i and uses transmitter policy $\gamma_i : \mathbb{R} \to \mathbb{R}$ to generate signal x_i subject to the power constraint

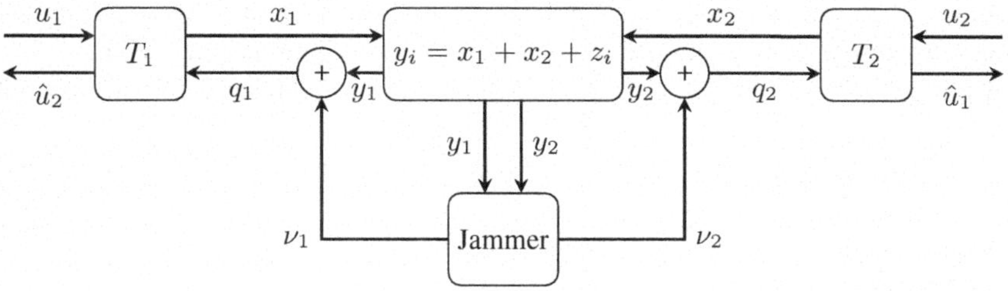

Fig. 20.2 System diagram of a two-way source-channel system with an intelligent jammer

$$E[(\gamma_i(u_i))^2] \le c_i, \quad i = 1, 2. \tag{20.2}$$

The two-way channel inputs are x_1, x_2 and its outputs are

$$y_i = x_1 + x_2 + z_i, \quad i = 1, 2 \tag{20.3}$$

where z_1 and z_2 are Gaussian random variables (which are independent of (u_1, u_2)) with zero mean and covariance matrix

$$\Sigma = \begin{pmatrix} E[z_1^2] & E[z_1 z_2] \\ E[z_2 z_1] & E[z_2^2] \end{pmatrix} = \begin{pmatrix} \zeta_1 & \zeta_{1,2} \\ \zeta_{1,2} & \zeta_2 \end{pmatrix} \tag{20.4}$$

where $\zeta_{1,2}$ takes values in $[-(\zeta_1\zeta_2)^{\frac{1}{2}}, (\zeta_1\zeta_2)^{\frac{1}{2}}]$.

We further assume the presence of a third party, the jammer. The latter taps the channel in both directions and captures signals y_1 and y_2; in return, it sends an adversarial signal v_i to each terminal T_i using the jamming policy $v_i = \beta_i(y_1, y_2)$, $i = 1, 2$, where each β_i is in general a random mapping. Let $\mathcal{M}$ denote the set of the pairs (μ_1, μ_2), where μ_i is the probability measure associated with jamming signal v_i under the power constraint

$$E[v_i^2] \le k_i, \quad i = 1, 2. \tag{20.5}$$

Terminal T_i then receives signal $q_i = y_i + v_i$, which it uses together with side information x_i (i.e., its own signal sent to Terminal T_j) to reconstruct u_j via $\hat{u}_j$ under decoding policy $\delta_i : \mathbb{R} \times \mathbb{R} :\to \mathbb{R}$, $i \neq j$, $i, j = 1, 2$.

The overall mean-square error (MSE) of the system is given by

$$R(\gamma_1, \gamma_2, \delta_1, \delta_2, \mu_1, \mu_2) = \frac{1}{2}\sum_{i=1}^{2}\left(\int_{-\infty}^{\infty} E[(\hat{u}_i - u_i)^2 | v_i] d\mu_i(v_i)\right). \tag{20.6}$$

Let Γ_t^i, Γ_r^i be the set of admissible (as specified above) transmitter and receiver policies for terminal T_i, $i = 1, 2$. Naturally, the objective of terminals T_1 and T_2 is to choose their encoding/decoding policies so that the system MSE is minimized, while the jammer aims at designing its policies in order to maximize MSE.

A policy tuple $(\gamma_1^*, \gamma_2^*, \delta_1^*, \delta_2^*, \mu_1^*, \mu_2^*)$ constitutes a saddle-point equilibrium if

$$R(\gamma_1^*, \gamma_2^*, \delta_1^*, \delta_2^*, \mu_1, \mu_2) \leq R(\gamma_1^*, \gamma_2^*, \delta_1^*, \delta_2^*, \mu_1^*, \mu_2^*) \leq R(\gamma_1, \gamma_2, \delta_1, \delta_2, \mu_1^*, \mu_2^*). \tag{20.7}$$

$\forall \gamma_i \in \Gamma_t^i, \delta_i \in \Gamma_r^i, i = 1, 2, (\mu_1, \mu_2) \in \mathcal{M}$.

An Equivalent Formulation as an Optimal Decentralized Stochastic Linear Quadratic Control Problem As in Witsenhausen's counterexample, there is an equivalent linear systems formulation: We highlight here the connection to decentralized control problems by presenting an explicit control formulation of the problem. We use subscripts to index time steps and superscripts to index the decisions makers. We will use s and θ to indicate states; t and v to indicate control actions; σ and ϵ to indicate control policies.

Dynamics

$$s_0^i \sim \mathcal{N}(0, 1) \qquad\qquad i, j \in \{1, 2\};\ j \neq i$$

$$s_1^i = s_0^j - t_1^i$$

$$\theta_0^i = t_0^1 + t_0^2 + z^i \qquad\qquad (z^1, z^2) \sim \mathcal{N}(0, \Sigma)$$

Transmitter and Receiver Control

$$t_0^i = \sigma_0^i(s_0^i) \qquad\qquad t_1^i = \sigma_1^i(\theta_0^i + v_0^i, u_0^i)$$

Jammer Control

$$v_0^i = \epsilon^i(y_0^1, y_0^2, \mathcal{R}^i)$$

Here r defines a stochastic kernel used by the jammer in constructing its policy since the jamming policy can in general be a random mapping.

Constraints

$$E[(t_0^i)^2] \leq c_i \qquad\qquad E[(v_0^i)^2] \leq k_i$$

Cost Function

$$E[(s_1^1)^2 + (s_1^2)^2]$$

Where the expectation is defined given a set of control policies.

Naturally, the optimal solutions to the two formulations are identical in the sense that an equilibrium solution to the communication theoretic formulation is equivalent to that of the control theoretic formulation.

20.2.1 Equilibrium Solutions

We will separately consider a special case of this problem which can be viewed as a *degenerate case*: the case when the same noise variable affects both channel directions; i.e., $z_1 = z_2$. This yields that $y_1 = y_2$ and the jammer sees two identical signals. As this case changes the course of our analysis, we will treat it separately. We first consider the *non-degenerate* case, where the elements of the channel noise covariance matrix Σ are not all identical. Note that this includes the case when the noise variables are fully correlated, as long as they have different variances. We next specify regions of interest in terms of the system parameters. Setting $C = c_1 + c_2$, consider the case with

$$k_i \geq C + \frac{\zeta_1\zeta_2 - \zeta_{1,2}^2}{\zeta_1 + \zeta_2 - 2\zeta_{1,2}}, \qquad i = 1, 2.$$

Consider an admissible policy for the jammer to use for terminal T_i is given by

$$\beta_i(y_1, y_2) = a_{i,i} y_i + a_{i,j} y_j \tag{20.8}$$

where $i \neq j$,

$$a_{i,i} = -\frac{\zeta_j - \zeta_{1,2}}{\zeta_1 + \zeta_2 - 2\zeta_{1,2}} \qquad a_{i,j} = -\frac{\zeta_i - \zeta_{1,2}}{\zeta_1 + \zeta_2 - 2\zeta_{1,2}}$$

Under this policy, $a_{i,i} + a_{i,j} = -1$ hence the signal received at terminal T_i is purely noise: it contains no trace of either x_i or x_j signals. Therefore, the MSE has a maximal unity value regardless of the transmitter/receiver policies which is considered as an uninteresting case (in terms of finding a saddle-point equilibrium). Thus, for each $i = 1, 2$, we introduce two disjoint regions

$$R_1^i = \left\{ k_i \geq C + \frac{\zeta_1\zeta_2 - \zeta_{1,2}^2}{\zeta_1 + \zeta_2 - 2\zeta_{1,2}} \right\}, \tag{20.9}$$

$$R_2^i = \left\{ k_i < C + \frac{\zeta_1\zeta_2 - \zeta_{1,2}^2}{\zeta_1 + \zeta_2 - 2\zeta_{1,2}} \right\}. \tag{20.10}$$

Equilibrium Policies in the Non-Degenerate Case

If either k_i were to belong to the R_1^i region, then the jammer can just send v_i according to the policy described in (20.8) and the transmitter and receiver policies would be irrelevant. We therefore consider a system where $(k_1, k_2) \in R_2^1 \times R_2^2$. Define

$$\omega = C(\zeta_1 + \zeta_2 - 2\zeta_{1,2}) + \zeta_1\zeta_2 - \zeta_{1,2}^2. \tag{20.11}$$

We then have the following theorem.

Theorem 20.2.1 ([524]) *Consider the two-way Gaussian channel communication problem with an intelligent jammer as formulated in this section. Let $(k_1, k_2) \in R_2^1 \times R_2^2$. Then,*

(i) There exist four saddle-point solutions $(\gamma_1^*, \gamma_2^*, \delta_1^*, \delta_2^*, \mu_1^*, \mu_2^*)$ depending on whether the transmitter uses $\sqrt{c_i}$ or $-\sqrt{c_i}$. Assuming both transmitters use the positive amplification, the equilibrium policies for the system are given by, for $i, j = 1, 2, j \neq i$,

$$\gamma_i^*(u_i) = \sqrt{c_i}u_i \tag{20.12}$$

$$\delta_i^*(q_i, u_i) = \alpha_i(q_i - (1 + a_{i,i}^* + a_{i,j}^*)\sqrt{c_i}u_i) \tag{20.13}$$

$$\beta_i^*(y_1, y_2) = a_{i,i}^* y_i + a_{i,j}^* y_j + \eta_i \tag{20.14}$$

where $\eta_i \sim \mathcal{N}(0, b_i^*)$ is a Gaussian random variable with zero mean and variance b_i^* that is independent of the system signals,

$$\alpha_i = \frac{\sqrt{c_j}(1 + a_{i,i}^* + a_{i,j}^*)}{G}$$

$$G = (1+a_{i,i}^*+a_{i,j}^*)^2 c_j + (1+a_{i,i}^*)^2\zeta_i + (a_{i,j}^*)^2\zeta_j + (1 + a_{i,i}^*)(a_{i,j}^*)\zeta_{1,2} + b_i^* \tag{20.15}$$

and the coefficients $a_{i,i}^*, a_{i,j}^*, b_i^*$ are detailed in Table 20.1 depending on the relationship between ζ_i and $\zeta_{1,2}$.

(ii) Furthermore, the saddle-point equilibria are essentially unique up to the changes of the signs of the encoding/decoding coefficients.

A proof is provided in the appendix to the chapter.

Equilibrium Policies in the Degenerate Case

We next consider the degenerate case of having identical noise signals in both channel directions ($z_1 = z_2$). In this case, y_1 and y_2 are the same signal and hence

Table 20.1 The coefficients that determine the Jammer SP policy for different relationships between the noise variances

	Jammer coefficients ($i, j = 1, 2, i \neq j$)
$\zeta_i > \zeta_{1,2}$	$a^*_{i,i} = \frac{-k_i\omega^{\frac{1}{2}} + (C+\zeta_{1,2})(k_i(C+\zeta_i-k_i))^{\frac{1}{2}}}{(C+\zeta_i)\omega^{\frac{1}{2}}}$ $a^*_{i,j} = -\left(\frac{k_i(C+\zeta_i-k_i)}{\omega}\right)^{\frac{1}{2}}$ $b^*_i = 0$
$\zeta_i < \zeta_{1,2}$	$a^*_{i,i} = \frac{-k_i\omega^{\frac{1}{2}} - (C+\zeta_{1,2})(k_i(C+\zeta_i-k_i))^{\frac{1}{2}}}{(C+\zeta_i)\omega^{\frac{1}{2}}}$ $a^*_{i,j} = +\left(\frac{k_i(C+\zeta_i-k_i)}{\omega}\right)^{\frac{1}{2}}$ $b^*_i = 0$
$\zeta_i = \zeta_{1,2}$	$a^*_{i,i} = -\left(\frac{k_i}{C+\zeta_i}\right)$ $a^*_{i,j} = 0$ $b^*_i = k_i\left(1 - \frac{k_i}{C+\zeta_i}\right)$

the jammer essentially has access to a single signal. We address this scenario by considering the problem where the jamming signal is given by $\nu_i = \beta_i(y_i)$, where again β_i is in general a random mapping. However, we must redefine the regions of interest since the game may become uninteresting in the sense that the signal is fully cancelled at the receiver when $\beta_i(y_i) = -y_i$; the lowest power constraint which leads to such a policy is $k_i = C + \zeta_i$. Therefore, the appropriate regions are

$$\tilde{R}^i_1 = \{k_i \geq C + \zeta_i\} \tag{20.16}$$

$$\tilde{R}^i_2 = \{k_i < C + \zeta_i\}. \tag{20.17}$$

We have the following result.

Theorem 20.2.2 ([524]) *Consider the two-way Gaussian channel communication problem with an intelligent jammer as formulated in this section. Fix* $(k_1, k_2) \in \tilde{R}^1_2 \times \tilde{R}^2_2$. *Then there exist four saddle-point solutions* $(\gamma^*_1, \gamma^*_2, \delta^*_1, \delta^*_2, \mu^*_1, \mu^*_2)$ *depending on whether the transmitters use* $\pm\sqrt{c_i}$. *If the transmitters use positive amplification, the equilibrium policies are, for* $i, j = 1, 2, j \neq i$,

$$\gamma^*_i(u_i) = \sqrt{c_i}u_i \tag{20.18}$$

$$\beta^*_i(y_i) = a^*_i y_i + \eta_i \tag{20.19}$$

$$\delta^*_i(q_i, u_i) = \frac{\sqrt{c_j}}{c_j + \zeta_i - a^*_i c_i}\left(q_i - (1 + a^*_i)(\sqrt{c_i}u_i)\right) \tag{20.20}$$

where $a^*_i = -\frac{k_i}{C+\zeta_i}$ *and* $\eta_i \sim \mathcal{N}\left(0, \left(1 - \frac{k_i}{C+\zeta_i}\right)k_i\right)$.

The proof follows a similar approach as the one for the non-degenerate case [524].

Remark 20.1 This theorem can be considered as providing the solution to the problem of two separate jammers, each one attempting to jam the channel individually without sharing any information. However, we can also apply this result to the degenerate case. Given access to only one signal, the separate jammer's optimal policy consists of sending $\beta^*(y) = -\left(\frac{k_i}{C+\zeta_i}\right) y + \eta_i$. However, since our single jammer has access to the same signal ($y_1 = y_2$), any policy of the form $\beta_i(y_1, y_2) = a_{i,1}y_1 + a_{i,2}y_2 + \eta_i$ which satisfies $a_{i,1} + a_{i,2} = -\left(\frac{k_i}{C+\zeta_i}\right)$ will indeed be an equilibrium policy. We can then see that there is actually an infinite number of equilibrium jamming policies which all produce the same output signal ν_i.

20.2.2 The Case with a One-Way Channel [86] and a Discussion

Let us consider the special case with the parameter values $c_2 = 0$, $\zeta_1 \to \infty$, $\zeta_{1,2} = 0$, corresponding to shutting down the channel in the direction $T_2 \to T_1$, which reduces the system to a one-way system going from T_1 to T_2, which has been treated in a broader context in [86] (see also [99]).

In this case, the boundary between the R_1^2 and R_2^2 regions now becomes

$$C + \left.\frac{\zeta_1\zeta_2 - \zeta_{1,2}^2}{\zeta_1 + \zeta_2 - 2\zeta_{1,2}}\right|_{\zeta_1\to\infty,\zeta_{1,2}=0,c_2=0} = c_1 + \zeta_2$$

which is identical to the definition of the $R2$ region in [86] (where equivalence can be shown via some notational equivalence adjustments).

In general, when faced with a Gaussian system, be it one-way or two-way, there are certain traits that appear in the jamming policies: There is an R_1-type region where the jammer has too much power and can fully cancel the transmitted signal before it reaches the receiver, making the game trivial. When the jammer cannot fully cancel out the signal, its equilibrium policies are either linear, or affine by combining a linear policy with Gaussian noise. For a two-way channel, whether the jamming policy is linear or affine is determined by the covariance matrix of the noise variables, Σ. If $\zeta_i \neq \zeta_{1,2}$, then a linear policy is used. If $\zeta_i = \zeta_{1,2}$ (in either the degenerate or non-degenerate case), then an affine policy is used. The reasoning for this involves an in depth discussion of the best response function used in the proof of Theorem 20.2.1. In essence, if the function admits a fixed point, the policy is linear, and if it does not have a fixed point when an affine policy is used.

We also note the generality given to the jammer's information and policy structure. We have allowed the jammer to have full access to the channel and full knowledge of the system dynamics. Although we impose a power constraint on the jammer, we do not impose any further assumptions on the policy structure itself.

Simple linear/affine jamming policies arise naturally as the optimal choice. We note that further types of information structures for the jammer have been studied in [64, 86, 99].

20.3 Appendix: Proofs

20.3.1 Proof of Theorem 20.2.1

We proceed in two steps by separately validating the right-hand side (RHS) and the LHS inequalities of (20.7) under the stated policies in (20.12)–(20.14). We then examine the uniqueness of the equilibrium.

(i) The RHS Inequality

Assume that for each i, $j \neq i$, $i, j = 1, 2$, the jammer is using an affine policy of the form (20.14), not necessarily with the assumed equilibrium values $(a_{i,i}^*, a_{i,j}^*, b_i^*)$ but any general values $(a_{i,i}, a_{i,j}, b_i)$. We can view the two-way system as two separate one-way channels with each channel going from transmitter $\gamma_j(\cdot)$ to receiver $\delta_i(\cdot)$ as displayed in Fig. 20.3.

In the figure, $\tilde{\gamma}_j(u_j) = (1 + a_{i,i} + a_{i,j})\gamma_j(u_j)$, as the jammer is amplifying the signal x_j by sending back a scaled version of y_i, y_j to the channel. Then the power constraint on the transmitter is $E[\tilde{\gamma}_i(u_i)^2] \leq c_i(1 + a_{i,i} + a_{i,j})^2$. Note that the receiver has side information about the signal x_i since terminal T_i is composed of both the transmitter γ_i and receiver δ_i (see Fig. 20.2). This setup has a classical solution building on the data-processing inequality of information theory: the transmitter will amplify the signal as much as possible and the receiver will use the minimal MSE estimator $\delta_i^*(q_i, u_i) = E[u_j|q_i, u_i]$ (see Chap. 16 or [238], [63]). Thus $\tilde{\gamma}_i^*(u_i) = \sqrt{c_i}(1 + a_{i,i} + a_{i,j})u_i$, which yields that $\gamma_i^*(u_i) = \sqrt{c_i}u_i$. Note that $\gamma_i^*(u_i) = -\sqrt{c_i}u_i$ is also a valid solution; hence there are four possible equilibrium policies as stated in the theorem. We determine the best receiver policy when both transmitters use positive amplification (the policies for the other equilibria can be found by simply changing the signs of $\sqrt{c_i}$ in the transmitter and receiver functions). Writing $q_i = \tilde{q}_i + (1 + a_{i,i} + a_{i,j})\sqrt{c_i}u_i$ and noting that u_i is independent of $\tilde{q}_i$ and u_j, it can be shown that the receiver

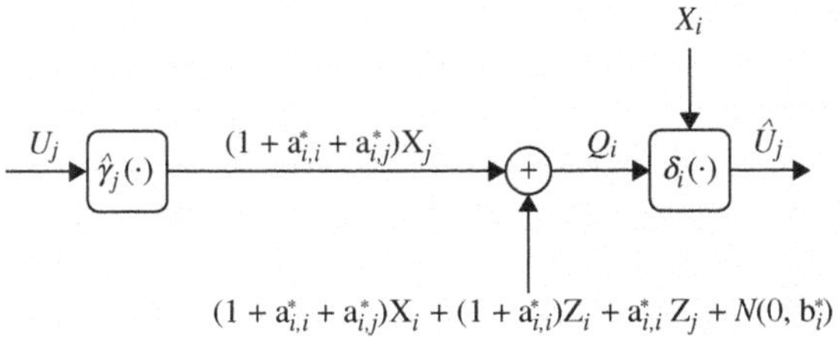

Fig. 20.3 Equivalent one-way Gaussian test system in the direction $T_j \to T_i$ for the non-degenerate case

policy is given by

$$\delta_i^*(q_i, u_i) = \frac{E[u_j \tilde{q}_i]}{E[\tilde{q}_i^2]} \tilde{q}_i = \alpha_i \tilde{q}_i,$$

where α_i is given in (20.15). This verifies the RHS inequality of (20.7).

(ii) The LHS Inequality

Assuming the policies in (20.12) and (20.13) at each terminal, again for some general $(a_{i,i}, a_{i,j}, b_i)$ and not the assumed equilibrium values, the maximization problem faced by the jammer is given by

$$\max_{(\mu_1,\mu_2)\in\mathcal{M}} \frac{1}{2}\left(\sum_{i=1}^{2} \int_{-\infty}^{\infty} E\left[\left(\alpha_i \tilde{q}_i - u_j\right)^2 \Big| \nu_i\right] d\mu_i(\nu_i)\right). \tag{20.21}$$

Setting

$$\tilde{q}_i = w_i + \nu_i \tag{20.22}$$

$$w_i = x_j + z_i - (a_{i,i} + a_{i,j})x_i \tag{20.23}$$

we expand the ith MSE term in the sum above as follows

$$\begin{aligned}
&\int_{-\infty}^{\infty} E\left[\left(\alpha_i \tilde{q}_i - u_j\right)^2 \Big| \nu_i\right] d\mu_i(\nu_i) \\
&= E_{y_1,y_2}\left[\int_{-\infty}^{\infty} E\left[\left(\alpha_i \tilde{q}_i - u_j\right)^2 \Big| \nu_i, y_1, y_2\right] d\mu_i(\nu_i | y_1, y_2)\right] \\
&= \alpha_i^2(\zeta_i + (a_{i,i} + a_{i,j})^2 c_i) + (\alpha_i \sqrt{c}_j - 1)^2 + \\
&E_{y_1,y_2}\left[\int_{-\infty}^{\infty} \alpha_i^2 \nu_i^2 + 2\nu_i E[\alpha_i^2 w_i - \alpha_i u_j | y_1, y_2] d\mu_i(\nu_i | y_1, y_2)\right].
\end{aligned}$$

It can be shown that $E[\alpha_i^2 w_i - \alpha_i u_j | y_1, y_2] = \pi_i(y_1, y_2)$ where

$$\pi_i(y_1, y_2) = \frac{\alpha_i}{\omega}\left(\rho_{i,i} y_i + \rho_{i,j} y_j\right),$$

and

$$\begin{aligned}
\rho_{i,i} &= \alpha_i[c_j(\zeta_i + \zeta_j - 2\zeta_{1,2}) + c_i(\zeta_i - \zeta_{1,2} - (\zeta_j - \zeta_{1,2})(a_{i,i} + a_{i,j})) \\
&\quad + \zeta_1\zeta_2 - \zeta_{1,2}^2] - (\zeta_j - \zeta_{1,2})\sqrt{c}_j,
\end{aligned}$$

$\rho_{i,j} = -\alpha_i(\zeta_i - \zeta_{1,2})c_i(1 + a_{i,i} + a_{i,j}) - (\zeta_i - \zeta_{1,2})\sqrt{c}_j)$. Ignoring the terms which are independent of ν_i, we can focus on determining

$$J = \max_{\mu_1,\mu_2} \sum_{i=1}^{2} E[\alpha_i^2 \nu_i^2] + 2E[\nu_i \pi_i(y_1, y_2)]. \tag{20.24}$$

Applying the Cauchy-Schwarz inequality and the power constraint (20.5) we have

$$J \le \sum_{i=1}^{2} \alpha_i^2 k_i + 2(k_i)^{\frac{1}{2}} E[\pi_i(y_1, y_2)^2]^{\frac{1}{2}}. \tag{20.25}$$

At this point, we break our analysis into two parts.

Case 1 $\zeta_i \neq \zeta_{1,2}$

We have that $E[\pi_i(y_1, y_2)^2]^{\frac{1}{2}} \neq 0$, and thus we can achieve the upper bound with a unique linear jammer policy

$$\begin{aligned} \nu_i = \beta_i^*(y_1, y_2) &= \frac{(k_i)^{\frac{1}{2}}}{E[\pi_i(y_1, y_2)^2]^{\frac{1}{2}}} \pi_i(y_1, y_2) \\ &= \frac{(k_i)^{\frac{1}{2}} \left(\rho_{i,i} y_i + \rho_{i,j} y_j\right)}{\left(C(\rho_{i,i} + \rho_{i,j})^2 + \rho_{i,i}^2 \zeta_i + \rho_{i,j}^2 \zeta_j + 2\rho_{i,i}\rho_{i,j}\zeta_{1,2}\right)^{\frac{1}{2}}}. \end{aligned} \tag{20.26}$$

Therefore, the best response jamming policy to transmission and receiving policies (20.12) and (20.13) is linear, yet these policies were constructed when faced with an affine (linear if $b_i = 0$) jamming policy. Setting $\lambda_i = \left(\frac{k_i}{E[\pi_i(y_1,y_2)^2]}\right)^{\frac{1}{2}}$, for each linear jamming policy with coefficients $(a_{i,i}, a_{i,j})$, we use (20.26) to define the best response mapping

$$T \begin{pmatrix} a_{i,i} \\ a_{i,j} \end{pmatrix} \mapsto \begin{pmatrix} (\lambda_i)\rho_{i,i} \\ (\lambda_i)\rho_{i,j} \end{pmatrix}. \tag{20.27}$$

The equilibrium policy then has the pair of coefficients $(a_{i,i}^*, a_{i,j}^*)$ as the fixed point of this mapping and satisfying (20.5) for a given k_i. However, directly finding this fixed point is computationally difficult. We instead consider an equivalent set of conditions which can be proved via a geometric approach.

Proposition 20.A.1 *A set of necessary and sufficient conditions for equilibria is*

$$a_{i,i}\rho_{i,j} = a_{i,j}\rho_{i,i} \tag{20.28}$$

$$E[(a_{i,i} y_i + a_{i,j} y_j)^2] = k_i \tag{20.29}$$

$$\frac{a_{i,i}}{|a_{i,i}|} = \frac{\rho_{i,i}}{|\rho_{i,i}|}, \qquad \frac{a_{i,j}}{|a_{i,j}|} = \frac{\rho_{i,j}}{|\rho_{i,j}|} \tag{20.30}$$

Proof Suppose that the coefficients $a_{i,i}$ and $a_{i,j}$ are equilibrium policies. Then by being a fixed point of T, we have that $a_{i,i} = (\lambda_i)\rho_{i,i}$ and $a_{i,j} = (\lambda_i)\rho_{i,j}$, thus $a_{i,i}\rho_{i,j} = \lambda_i\rho_{i,i}\rho_{i,j} = a_{i,j}\rho_{i,i}$ and hence condition (20.28) is necessary. Also $E[(a_{i,i}y_i + a_{i,j}y_j)^2] = \lambda_i^2 E[\pi_i(y_1, y_2)^2] = k_i$ and thus condition (20.29) is also necessary. Furthermore, we have

$$\frac{a_{i,i}}{|a_{i,i}|} = \left(\frac{\lambda_i}{|\lambda_i|}\right)\frac{\rho_{i,i}}{|\rho_{i,i}|} = \frac{\rho_{i,i}}{|\rho_{i,i}|}$$

since $\lambda_i > 0$; thus condition (20.30) is necessary as well. This concludes one direction of the proof.

We next show the reverse direction, by proving that any pair of coefficients $(a_{i,i}, a_{i,j})$ satisfying conditions (20.28)–(20.30), forms a fixed point of T; therefore $a_{i,i}$ and $a_{i,j}$ are equilibrium policies. By conditions (20.30), we have that $(a_{i,i}, a_{i,j})$ and $T(a_{i,i}, a_{i,j})$ must exist in the same quadrant of the Cartesian plane. Combining this fact with (20.28) yields that they also both have the same polar angle. Finally, condition (20.29) means that they both live on the same ellipse. Therefore, points $(a_{i,i}, a_{i,j})$ and $T(a_{i,i}, a_{i,j})$ live on the same ellipse, in the same quadrant, and have the same polar angle; thus they form an identical point and hence the conditions are sufficient. ◇

The ellipse described by

$$(C + \zeta_i)(a_{i,i} + 0.5)^2 + 2a_{i,j}(a_{i,i} + 0.5)(C + \zeta_{1,2}) + (C + \zeta_2)a_{i,j}^2 = \frac{C + \zeta_i}{4} \tag{20.31}$$

satisfies condition (20.28). The ellipse defined by

$$a_{i,i}^2(C + \zeta_i) + 2a_{i,i}a_{i,j}(C + \zeta_{1,2}) + a_{i,j}^2(C + \zeta_j) = k_i$$

satisfies condition (20.29). Intersecting these two ellipses gives us two possible solutions, one in each half plane. We can then use the remaining conditions (20.30) to find the only sufficient and necessary solution, which yields the first two rows of Table 20.1.

Case 2 $\zeta_i = \zeta_{1,2}$

If $\zeta_i = \zeta_{1,2}$, then $\rho_{i,j} = 0$ for any affine jamming policy. Thus $E[\pi_i(y_1, y_2)^2] \neq 0$ is not guaranteed. Revisiting (20.24), if there were a choice of $a_{i,i}, a_{i,j}, b_i$ which forced $\rho_{i,i} = 0$, then $J = \max_{\mu_1,\mu_2} \sum_{i=1}^2 \alpha_i^2 E[\nu_i^2] \le \alpha_i k_i^2$, which is maximized by a policy with variance k_i. Therefore, the equilibrium policy when $\zeta_i = \zeta_{1,2}$ must force $\rho_{i,i} = 0$ and achieve variance k_i, which is the policy given in the third row of Table 20.1.

(iii) Uniqueness of the Equilibrium Policies.
Let the presented equilibrium be referred to as $\Lambda^* = (\gamma_1^*, \gamma_2^*, \beta_1^*, \beta_2^*, \delta_1^*, \delta_2^*)$ and denote any other solution by $\Lambda = (\gamma_1, \gamma_2, \beta_1, \beta_2, \delta_1, \delta_2)$. If Λ is also a saddle-point equilibrium, then the *ordered interchangeability* property of multiple saddle points in zero-sum games (Theorem 9.2.1) would require that

$$(\gamma_1^*, \gamma_2^*, \beta_1, \beta_2, \delta_1^*, \delta_2^*) \tag{20.32}$$

$$(\gamma_1, \gamma_2, \beta_1^*, \beta_2^*, \delta_1, \delta_2) \tag{20.33}$$

are also both saddle-point equilibria (note that the fact that there are multiple encoders as a team does not violate the *rectangularity* condition of the policies, that is the policy sets do not depend on the realized policies of the opposing player, the jammer).
Considering the first candidate solution (20.32), when faced with γ_i^*, δ_i^*, an optimal jamming solution must achieve the maximum of (20.24). When $\zeta_i \neq \zeta_{1,2}$, the upper bound of the Cauchy-Schwarz inequality can only be achieved when ν_i is linearly dependent with $\pi_i(y_1, y_2)$ and has variance k_i, with two possibilities: $\pm \left(\frac{k_i}{E[\pi_i(y_1,y_2)^2]}\right)^{\frac{1}{2}} \pi_i(y_1, y_2)$.
However, the negative option would result in a negative second term in (20.24), lowering the overall result. Therefore β_i^* is the only policy to achieve the upper bound in (20.24).
When $\zeta_i = \zeta_{1,2}$, there is only one policy which can force $\rho_{i,i} = 0$ and achieve variance k_i. Any other policy has $E[(\pi_i(y_1, y_2)^2] \neq 0$, forcing the equilibrium policy to be a scaled version of $\pi_i(y_1, y_2)$ which achieves variance k_i. Since $\rho_{i,j} = 0$, there are two possibilities:

$$\beta_i(y_1, y_2) = \pm \left(\frac{k_i}{C + \zeta_i}\right)^{\frac{1}{2}} y_i.$$

The best response mapping (20.27) simplifies to

$$T \begin{pmatrix} a_{i,i} \\ a_{i,j} \end{pmatrix} = \begin{pmatrix} -\left(\frac{k_i}{C+\zeta_i}\right)^{\frac{1}{2}} \frac{a_{i,i}}{|a_{i,i}|} \\ 0 \end{pmatrix}. \tag{20.34}$$

We see that the best response flips the sign of $a_{i,i}$; therefore each of these possible policies is the best response of the other and neither is a fixed point of T. For the second candidate solution (20.33), the transmitter can choose $\pm\sqrt{c_i}$ and the receiver can follow and both of these policies are optimal with respect to the jamming policy. Also, the jamming policy only depends on c_i and c_j, and it is irrelevant to the jammer if the transmitter used positive or negative amplification. Therefore, there exist four saddle-point equilibria which have the same jamming policy, and have transmission and receiving policies which are identical up to the absolute value of $\sqrt{c_i}$. Hence the equilibrium is essentially unique. ◇

20.4 Concluding Remarks

In this chapter, we have considered zero-sum information games in two different settings. The first one was a two-agent decision-making problem with totally conflicting cost criteria, as a variation of Witsenhausen's counterexample. The second one involved jamming.

The identification of optimal linear/affine/Gaussian policies for decentralized systems involving Gaussian variables under quadratic criteria (such as in linear quadratic control (LQG)) is a recurring problem in stochastic networked control and estimation theory. These certainly include the classical problem of communicating a scalar Gaussian source over a Gaussian channel as studied earlier in the book as well as various optimality results in stochastic team theory and optimal control. The chapter presented further settings where linearity preserves optimality and equilibrium behavior, in the context of zero-sum games. In particular, we showed that for a two-way Gaussian scalar networked system with an intelligent jammer, an essentially unique saddle-point equilibrium exists and the equilibrium policies are affine/Gaussian and that the zero-sum formulation of Witsenhausen's counterexample admits the unique saddle-point solution which entails linear policies for both players.

An important property of the saddle-point equilibrium utilized throughout the chapter to obtain strong (such as uniqueness and linearity) results has been the ordered-interchangeability property of multiple saddle points.

20.5 Bibliographic Notes

In the context of dynamic quadratic zero-sum game problems, further papers on min-max optimization include [95, 96, 294, 608] and [89]; [89] also provides an overview on LQG problems under non-classical information including Witsenhausen's counterexample.

One of the earliest works to formulate jamming problems as zero-sum games is is [86], where the transmission of a Gaussian source over a one-way additive Gaussian noise channel is investigated in the presence of an intelligent jammer. [86] establishes complete solutions to a zero-sum game, proving the saddle-point property of linear/affine/Gaussian policies within a general framework. Further relevant studies include [64, 99], and a series of relatively recent papers on jamming [660] and [12]. While building on [86], the chapter has considered a somewhat broader context involving multi-terminal two-way channels introduced by Shannon in [663]. For a setup similar to [86], but with the game being played only between an encoder and a jammer (with the decoder being an impartial Bayesian decision maker), it was shown in [787] that the worst additive channel noise is Gaussian and the optimal encoder is linear. This result may be viewed as a Stackelberg extension of the Nash setup given in [86].

Finally, we note that [375] and [372] consider aspects related to the jamming problem, in a channel coding context. Notably, [375] studies the zero-sum game between a finite blocklength encoder and decoder team, and a jammer; and non-existence of saddle point equilibria in the absence of common randomness for finite length codes is also established together with asymptotic (in the code length) results on existence. Recall from Sect. 11.6 the importance of common randomness in establishing conditions on convexity for the existence of equilibria; we also refer the reader to [22] again, as well as on randomized encoders with common randomness between an encoder and decoder [8, 622, 623]. A relevant application in power systems was studied in [757].

Theorem 20.1.2 is due to [89]. The analysis on jamming builds on [86, 99], and mostly [524].

Part V
Reinforcement Learning Under Full, Partial, Decentralized, and Networked Information

Chapter 21
Reinforcement Learning for Fully Observable Models: Convergence to Near Optimality under Weak Feller Continuity

21.1 Introduction

In this chapter and the following three ones, we will present learning theoretic methods applicable under a variety of information structures.

Let us re-visit a discussion from Sect. 3.1. Consider a system of the form

$$x_{t+1} = f(x_t, u_t, w_t), \quad t \in \mathbb{Z}_+,$$

with actions to be generated using some control policy $\gamma = \{\gamma_t\}$ with $u_t = \gamma_t(I_t)$, $t \in \mathbb{Z}_+$, where I_t is the information available at t, and w_t is i.i.d. noise. Recall that if $I_t = \{x_0, \cdots, x_t; u_0, \cdots, u_{t-1}\}$, we have a *fully observed* system; see Sect. E.2 for a detailed discussion on such models. If the agent, or controller, has instead access to only noisy measurements $y_t = g(x_t, v_t)$, with v_t being another i.i.d. noise, and $I_t = \{y_0, \cdots, y_t; u_0, \cdots, u_{t-1}\}$, we then have a *partially observed* system; see Sect. E.3 for a detailed discussion on such models and regularity properties. If there exist multiple agents each with local information, we have a *decentralized stochastic control* system or a *stochastic team*. If the agents have conflicting information or priors, as we have seen in the book, the setup is one of a *stochastic game*.

We have seen in the book that information structures have significant implications on the mathematical study of optimal stochastic control and more general stochastic game problems.

However, in some such stochastic decision problems, the decision maker(s) may not know the true model (dynamics, information structures or probabilistic description of the present variables) or even the cost functions, and may wish to use past data to obtain an asymptotically optimal solution (that is, via *learning* from past data). In some problems, this may be used as a numerical method to obtain approximately optimal solutions. Yet, in other problems, one may not even know

S. Yüksel, T. Başar, *Stochastic Teams, Games, and Control under Information Constraints*, Systems & Control: Foundations & Applications,
https://doi.org/10.1007/978-3-031-54071-4_21

whether the underlying decision problem is a Markov Decision Process (MDP) or a Partially Observed MDP (POMDP), or a multi-agent system where other agents are present or not. There are many practical scenarios where one indeed works with data and does not know the possibly very complex structure under which the data is generated and tries to respond to the environment.

A major theme of the discussion to follow in this chapter, and the following ones, is that the framework of reinforcement learning is appropriate for a large class of such models, with continuous space and action spaces and under a variety of performance criteria, provided that certain technical continuity conditions apply.

In the following, we will first study fully observed MDPs with both finite and continuous spaces, then POMDPs with continuous spaces, and finally decentralized stochastic control problems and multi-agent systems under a variety of information structures: (i) We first show, in this chapter, that under weak continuity conditions and a technical ergodicity condition, Q-learning can be applied to fully observable MDPs for near optimal performance. (ii) For POMDPs, in Chap. 22, we show that under a uniform filter stability, finite memory policies can be used to arrive at near optimality, and under only asymptotic filter stability leading to a mild unique ergodicity condition and weak Feller continuity of the non-linear filter dynamics, quantization can be used to arrive at near optimality. (iii) For decentralized stochastic control problems and multi-agent models, in Chap. 23, under a variety of information structures with strictly local information or partial global information, we show that synchronous or asynchronous quantized Q-learning can be used to arrive at near optimal performance, or equilibria or subjective equilibria. (iv) Finally, in Chap. 24, we will consider an application involving networked systems and optimal zero-delay coding.

The results will unify and build on several recent studies [18, 391, 392, 398] as well as [36, 794, 796].

In this chapter, in Sect. 21.2 we introduce and review the standard Q-learning algorithm and its convergence for finite models. Toward studying problems with general (standard Borel) state and action spaces, in Sect. 21.3, we establish near optimality of finite approximations for the case with continuous spaces, and using these approximation results, in Sect. 21.4 we introduce a quantized Q-learning algorithm and establish its convergence to near optimality. In particular, we show that the quantized Q-learning algorithm converges to the value/policy corresponding to a finite model MDP, whose near optimality is established in Sect. 21.4.

21.2 Classical Q-Learning for Finite Models and Full State Information

An important class of non-Bayesian learning methods are known as stochastic approximation algorithms: such approximation methods are used extensively in many application areas. A typical stochastic approximation algorithm has the form

$$x_{t+1} = x_t + \alpha_t (F(x_t) - x_t + w_t) \tag{21.1}$$

where w_t is a zero-mean noise variable, x_t is a stochastic process and w_t is some driving noise. The goal is to arrive at a point x^* which satisfies $x^* = F(x^*)$, where F may correspond to an optimality condition.

21.2.1 Classical Q-Learning for Finite Fully Observed Models

Q-learning [57, 115, 696, 698, 724, 760] is a stochastic approximation algorithm used for fully observed finite space MDPs that does not require the knowledge of the transition kernel, or even the cost (or reward) function for its implementation. In this algorithm, the incurred per-stage cost variable is observed through simulation of a single sample path[1]. When the state and action spaces are finite, under mild conditions regarding infinitely often hits for all state-action pairs, this algorithm is known to converge to the optimal cost.

Consider a fully observed Markov Decision Problem with finite state and action sets with a discounted cost criterion for some discount parameter $\beta \in (0, 1)$, in accordance with the setting of Definition E.2.1 in Sect. E.2. We take the state space $\mathbb{X}$ and the action space $\mathbb{U}$ to be finite in this section. Let $Q : \mathbb{X} \times \mathbb{U} \to \mathbb{R}$ denote the *Q-factor*, defined as a matrix. Let us assume that the decision maker applies an arbitrary admissible policy γ and updates its Q-factors as follows for $t \geq 0$,

$$Q_{t+1}(x, u) = Q_t(x, u) + \alpha_t(x, u) \Big(c(x, u) + \beta \min_v Q_t(X_{t+1}, v) - Q_t(x, u) \Big) \tag{21.2}$$

where the initial condition Q_0 is given, $\alpha_t(x, u)$ is the step-size for (x, u) at time t, u_t is chosen arbitrarily -e.g., according to some random exploration policy γ- as long as some technical conditions noted below hold, and the random state X_{t+1} evolves according to $P(X_{t+1} \in \cdot \,|X_t = x, U_t = u)$. It is assumed that, for all (x, u), $t \geq 0$, the following hold

Assumption 21.2.1 *For all* (x, u), $t \geq 0$,

(i) $\alpha_t(x, u) \in [0, 1]$
(ii) $\alpha_t(x, u) = 0$ *unless* $(x, u) = (x_t, u_t)$
(iii) $\alpha_t(x, u)$ *is a (deterministic) function of* $(x_0, u_0), \ldots, (x_t, u_t)$.
(iv) $\sum_{t \geq 0} \alpha_t(x, u) \to \infty$, *almost surely*

[1] In the version we consider here, the algorithm is called asynchronous Q-learning as each state-action is updated sequentially along a given sample path, whereas in a version where parallel updates are possible, the algorithm is called the synchronous Q-learning algorithm, for which a batch of data is available a priori

(v) $\sum_{t\geq 0}\alpha_t^2(x,u) \leq C$*, almost surely, for some (deterministic) constant* $C < \infty$.

A common way to pick α coefficients in the algorithm is to take for every (x, u) pair:

$$\alpha_t(x,u) = \frac{1}{1+\sum_{k=0}^{t} 1_{\{X_k=x,U_k=u\}}}$$

The selection of the control actions for each state can be arbitrary, as long as the assumptions above are guaranteed to hold.

Let F be an operator acting on the Q factors defined by:

$$F(Q)(x,u) = c(x,u) + \beta \sum_{x'} \mathcal{T}(x'|x,u) \min_v Q(x',v), \tag{21.3}$$

where $\mathcal{T}(x'|x,u) = P(x_1 = x'|x_0 = x, u_0 = u)$ is the transition kernel. Consider the following fixed point equation.

$$Q^*(x,u) = F(Q^*)(x,u) = c(x,u) + \beta \sum_{x'} \mathcal{T}(x'|x,u) \min_v Q^*(x',v) \tag{21.4}$$

whose existence and uniqueness follow from contraction arguments (see Theorem E.2.6), by using the norm $\|Q\|_\infty = \max_{(x,u)} |Q(x,u)|$.

Now, note that we can write (21.2) as

$$\begin{aligned} Q_{t+1}(x,u) = Q_t(x,u) + \alpha_t(x,u)\Big(& F(Q_t)(x,u) - Q_t(x,u) \\ & + \Big(c(x,u) + \beta \min_v Q_t(x_{t+1},v) - F(Q_t)(x,u)\Big)\Big) \end{aligned} \tag{21.5}$$

which is in the same form as (21.1) since

$$w_t := \Big(c(x_t,u_t) + \beta \min_v Q_t(x_{t+1},v) - F(Q)(x_t,u_t)\Big),$$

conditioned on the filtration generated by $\{x_t, u_t\}$ up to time t, is a zero-mean random variable. Let us write (21.5) as

$$\begin{aligned} Q_{t+1}(x,u) = (1-\alpha_t(x,u))Q_t(x,u) + \alpha_t(x,u)\Big(& F(Q_t)(x,u) \\ & + \Big(c(x,u) + \beta \min_v Q_t(x_{t+1},v) - F(Q_t)(x,u)\Big)\Big) \end{aligned}$$

and by (21.4)

$$Q_{t+1}(x,u) - Q^*(x,u) = (1-\alpha_t(x,u))(Q_t(x,u) - Q^*(x,u))$$
$$+\alpha_t(x,u)\Big(F(Q_t)(x,u) - F(Q^*)(x,u)$$
$$+\Big(c(x,u) + \beta \min_v Q_t(x_{t+1}, v) - F(Q_t)(x,u)\Big)\Big) \tag{21.6}$$

Since for every realization x_{t+1} of X_{t+1}, $|\min_v Q_t(x_{t+1}, v) - \min_{v'} Q_t^*(x_{t+1}, v')| \leq \max_v |Q_t(x_{t+1}, v) - Q_t^*(x_{t+1}, v)|$, we have that

$$|F(Q_t)(x,u) - F(Q^*)(x,u)| \leq \beta \|Q_t - Q^*\|_\infty := \beta \max_{x,u} |Q_t(x,u) - Q^*(x,u)| \tag{21.7}$$

Theorem 21.2.1 *Consider the dynamics (21.2).*

(i) Under Assumption 21.2.1, the algorithm (21.2) converges almost surely to Q^*.
(ii) A stationary policy f^* *which satisfies* $\min_u Q^*(x,u) = Q^*(x, f^*(x))$ *is an optimal policy.*

Proof Sketch

(i) From (21.6), the process Q_t satisfies the following form, with $S_t = Q_t - Q_t^*$:

$$S_{t+1}(x,u) = (1-\alpha_t(x))S_t(x,u) + \alpha_t(x,u)\Big((F(Q_t)(x,u) - F(Q^*)(x,u)) + w_t\Big),$$

where $\{\alpha_t\}$ satisfies Assumption 21.2.1. We will consider the following two parallel dynamics, as in [368, Theorem 1]:

$$S^a_{t+1}(x,u) = (1-\alpha_t(x,u))S^a_t(x,u) + \alpha_t(x,u)w_t,$$
$$S^b_{t+1}(x,u) = (1-\alpha_t(x,u))S^b_t(x,u) + \alpha_t(x,u)\Big(F(Q_t)(x,u) - F(Q^*)(x,u)\Big), \tag{21.8}$$

We clearly have $S_t(x,u) = S^a_{t+1}(x,u) + S^b_{t+1}(x,u)$. We will study each of these two additive terms separately.

The first step is to show that $S^a_{t+1}(x,u) \to 0$ almost surely. We will show this further below. Once this is attained, we can focus on $S^b_{t+1}(x,u) + S^a_{t+1}(x,u)$, using the fact that, by (21.7)

$$\|(F(Q_t)(\cdot,\cdot) - F(Q^*)(\cdot,\cdot))\|_\infty \leq \beta\|S_t\|_\infty \leq \beta\|S^a_{t+1}\|_\infty + \beta\|S^b_{t+1}\|_\infty$$

Almost surely, for every $\epsilon > 0$, there exists N such that for $t \geq N$, $\|S^a_{t+1}\|_\infty \leq \epsilon$ (where we suppress the sample path dependence). In the following, we assume that $t \geq N$. Now, for some M large enough, let $\hat{\beta} := \beta(M+1)/M < 1$ and for $\|S^b_t\|_\infty > M\epsilon$ note that

$$\beta\|S^b_t(x,u) + \epsilon\| \leq \hat{\beta}\|S^b_t\|_\infty$$

and

$$\begin{aligned} |S^b_{t+1}(x,u)| &\leq (1-\alpha_t(x,u))|S^b_t(x,u)| \\ &\quad + \left|\alpha_t(x,u)\Big(F(Q_t)(x,u) - F(Q^*)(x,u)\Big)\right| \\ &\leq (1-\alpha_t(x,u))|S^b_t(x,u)| + \alpha_t(x,u)(\beta\|S^a_{t+1}\| + \beta\|S^b_{t+1}\|) \qquad (21.9) \\ &\leq (1-\alpha_t(x,u))|S^b_t(x,u)| + \alpha_t(x,u)\hat{\beta}\|S^b_t\|_\infty \qquad (21.10) \\ &< \|S^b_t\|_\infty \qquad (21.11) \end{aligned}$$

Hence $\max_{x,u}(S^b_t(x,u))$ monotonically decreases for $\|S^b_t\|_\infty > M\epsilon$ leading to two possibilities: it either gets below $M\epsilon$ or it never gets below $M\epsilon$ in which case by the monotone non-decreasing property it will converge to some number, say M_1 with $M_1 \geq M\epsilon$. Now, if the former is the case: once $\|S^b_t\|_\infty \leq M\epsilon$ we can show via (21.9) and $\beta(M+1)/M < 1$ that it will remain there thereafter. We now show that the latter, that is with the limit being $M_1 \geq M\epsilon$, is not possible. The relation

$$|S^b_{t+1}(x,u)| \leq (1-\alpha_t(x,u))|S^b_t(x,u)| + \alpha_t(x,u)\hat{\beta}\|S^b_t\|_\infty$$

implies that (via an argument similar to what is known as Grönwall's lemma, as can be inductively shown) the solution is bounded from above by the solution to the equation

$$|S^b_{t+1}(x,u)| = (1-\alpha_t(x,u))|S^b_t(x,u)| + \alpha_t(x,u)\hat{\beta}\|S^b_t\|_\infty$$

which can be shown to converge to zero. This follows from the reasoning that, for any fixed D, the iterate

$$|S^b_{t+1}(x,u)| = (1-\alpha_t(x,u))|S^b_t(x,u)| + \alpha_t(x,u)\hat{\beta}D$$

converges to $\hat{\beta}D$; this follows since the effects of the initial condition diminish by the summability of α_t and hence there exists only one limit solution, which by inspection will be equal to $\hat{\beta}D$ (the uniqueness of the solution can be shown by subtracting $\hat{\beta}D$ from the iterates, whose limit would be zero for any initialization). Therefore, if there is an upper bound on the iterates D_0, the bounds for future iterations eventually get smaller and smaller than $\hat{\beta} D_0 + \delta =: D_1$ for any arbitrarily small $\delta > 0$, and as time progresses by an inductive reasoning, eventually the iterates would have to converge to zero (see [368, Proof of Lemma 3] or [724, p. 196]).

Thus, for any $\epsilon > 0$, for large enough t, we have that $\|S_t^b\|_\infty \leq M\epsilon$. Since $\epsilon > 0$ was arbitrary, the convergence result follows.

The convergence of S_t^a follows from a martingale argument (see also [394] for a more direct argument when the learning rates lead to averaging dynamics).

(ii) Now, consider

$$Q^*(x, u) = F(Q^*)(x, u) = c(x, u) + \beta \sum_{x'} P(x'|x, u) \min_v Q^*(x', v)$$

Note that the minimum of u, for each x, is essentially the solution to the Discounted Cost Optimality Equation (E.6). Hence, the stationary policy $\{f^*\}$ is optimal. ◇

In the following two sections, we consider continuous state and action spaces. To facilitate stochastic reinforcement learning, we first consider near optimality of finite models, and then we will consider learning theory for continuous space models.

21.3 Near Optimal Planning for Continuous Space MDPs: Near Optimality of Finite Models under Weak Feller Continuity

For cases where the spaces are not finite or countable, numerical methods require approximation procedures. In the following, let $\mathbb{X}$, $\mathbb{U}$ be standard Borel spaces with $\mathbb{U}(x) = \mathbb{U}$ compact. We will arrive at rigorously justified approximation algorithms and resulting convergence results, both via analytical as well as learning theoretic methods (to be studied in the following section).

21.3.1 Finite Action Approximation to MDPs

Definition 21.3.1 A measurable function $q : \mathbb{X} \to \mathbb{U}$ is called a *quantizer* from $\mathbb{X}$ to $\mathbb{U}$ if the range of q, i.e., $q(\mathbb{X}) = \{q(x) \in \mathbb{U} : x \in \mathbb{X}\}$, is finite.

The elements of $q(\mathbb{X})$ (the possible values of q) are called the *levels* of q.

Finite Action Approximate MDP: Quantization of the Action Space

Let $d_{\mathbb{U}}$ denote the metric on $\mathbb{U}$. Since the action space $\mathbb{U}$ is compact and thus totally bounded, one can find a sequence of finite sets $\Lambda_n = \{a_{n,1}, \ldots, a_{n,k_n}\} \subset \mathbb{U}$ such that for all n,

$$\min_{i \in \{1,\ldots,k_n\}} d_{\mathbb{U}}(a, a_{n,i}) < 1/n \text{ for all } a \in \mathbb{U}.$$

In other words, Λ_n is a $1/n$-net in $\mathbb{U}$. In the rest of Sect. 21.3.1, we assume that the sequence $\{\Lambda_n\}_{n\geq 1}$ is fixed. To ease the notation in the sequel, let us define the mapping

$$\Upsilon_n(f)(x) := \arg\min_{a \in \Lambda_n} d_{\mathbb{U}}(f(x), a), \tag{21.12}$$

where ties are broken so that $\Upsilon_n(f)(x)$ is measurable.

Our main objective in this section is to find conditions on the components of the MDP under which there exists a sequence of finite subsets $\{\Lambda_n\}_{n\geq 1}$ of $\mathbb{U}$ for which the following holds:

(P) If for each n, MDP$_n$ is defined as the Markov decision process having the components $\{\mathbb{X}, \Lambda_n, p, c\}$, then we would like to find conditions under which the value function of MDP$_n$ converges to the value function of the original MDP as $n \to \infty$.

Near Optimality of Quantized Policies Under Weak Continuity

Consider **(P)** for MDPs with weakly continuous transition probability. We will show that the value function of MDP$_n$ converges to the value function of the original MDP, which is equivalent to **(P)**.

Assumption 21.3.1 *The following hold:*

(i) The one stage cost function c is bounded and continuous.
(ii) The stochastic kernel $\mathcal{T}(\cdot\,|x, a)$ is weakly continuous in $(x, a) \in \mathbb{X} \times \mathbb{U}$.
(iii) $\mathbb{U}$ is compact.
(iv) $\mathbb{X}$ is compact.

For any real-valued measurable function u on $\mathbb{X}$, let $\mathbb{T}$ be given by

$$(\mathbb{T}u)(x) := \min_{a \in \mathbb{U}} (c(x, a) + \beta \int_{\mathbb{X}} u(y)\mathcal{T}(dy|x, a)). \tag{21.13}$$

Here, $\mathbb{T}$ is the Bellman optimality operator for the MDP considered in (E.6). Analogously, let us define the Bellman optimality operator $\mathbb{T}_n$ of MDP$_n$ as

$$\mathbb{T}_n u(x) := \min_{a \in \Lambda_n} (c(x, a) + \beta \int_{\mathbb{X}} u(y)\mathcal{T}(dy|x, a)). \tag{21.14}$$

We have seen that both $\mathbb{T}$ and $\mathbb{T}_n$ are contraction operators. Furthermore, value functions of MDP and MDP$_n$ are fixed points of these operators; that is, $\mathbb{T}J^* = J^*$ and $\mathbb{T}_n J_n^* = J_n^*$. Let us define $v^0 = v_n^0 \equiv 0$, and $v^{t+1} = \mathbb{T}v^t$ and $v_n^{t+1} = \mathbb{T}_n v_n^t$ for

$t \geq 1$; that is, $\{v^t\}_{t\geq 1}$ and $\{v_n^t\}_{t\geq 1}$ are successive approximations to the discounted value functions of the MDP and MDP$_n$, respectively (via value iteration). The following can be shown, inductively for each $t = 0, 1, 2, \cdots$:

Lemma 21.3.1 *Under Assumption 21.3.1, for any $t \geq 1$, we have*

$$\lim_{n\to\infty} \sup_{x\in\mathbb{X}} |v_n^t(x) - v^t(x)| = 0. \tag{21.15}$$

The following theorem states that the discounted value function of MDP$_n$ converges to the discounted value function of the original MDP. The proof builds on Lemma 21.3.1 and taking into account that $\{v^t\}_{t\geq 1}$ and $\{v_n^t\}_{t\geq 1}$ are successive approximations to the value functions J^* and J_n^*, respectively.

Theorem 21.3.1 (Discounted Cost [624]) *Under Assumption 21.3.1, we have*

$$\lim_{n\to\infty} \sup_{x\in\mathbb{X}} |J_n^*(x) - J^*(x)| = 0. \tag{21.16}$$

The proof follows from a successive approximation argument applied iteratively for value iteration updates.

We state an approximation result analogous to Theorem 21.3.1 for the average cost criterion. To do this, some new assumptions are needed on the components of the original MDP in addition to Assumption 21.3.1: For the average cost criterion, we need to impose relatively stronger ergodicity conditions on the controlled Markov chain.

Assumption 21.3.2 *Suppose Assumption 21.3.1 holds. Moreover, suppose that there exist a non-degenerate finite measure λ on $\mathbb{X}$ such that*

(v) $\mathcal{T}(D|x, a) \geq \lambda(D)$ for all $(x, a) \in \mathbb{X} \times \mathbb{U}$ and $D \in \mathcal{B}(\mathbb{X})$.

Note that this assumption is essentially a geometric ergodicity condition.

Suppose that Assumption 21.3.2 holds. This implies that there is a solution to the average cost optimality equation (ACOE) and the stationary policy which minimizes this ACOE is an optimal policy.

Theorem 21.3.2 (Average Cost [624]) *We have, under Assumption 21.3.2, the value functions (that is, the optimal expected average cost) satisfy*

$$\lim_{n\to\infty} |V_n^* - V^*| = 0,$$

where V^* and V_n^* $(n \geq 1)$ (the value functions of the true model and the approximate model sequence, respectively) do not depend on x.

Remark 21.3.1 As we have observed earlier, when one considers partially observed MDPs (POMDPs), any POMDP can be reduced to a (completely observable) MDP whose states are the posterior state distributions or beliefs of the observer. One can show that setwise continuity of the reduced MDP is not possible even under very strict conditions, whereas weak continuity can be verified under reasonable conditions on the transition kernel and the continuity of the measurement channel. Thus the results in this section are applicable to POMDPs as well, as we will see further in Chap. 22.

One can obtain similar results for the case where the kernel $\mathcal{T}$ is setwise continuous in $u \in \mathbb{U}$ for every $x \in \mathbb{X}$. One can also obtain rates of convergence results [626, 628].

21.3.2 Finite State Approximation to MDPs

In this section we study the finite-state approximation problem for MDPs, by reducing them to finite state MDPs obtained through quantization of the state space on a finite grid. Here two questions could be posed:

(i) **Q1** Under what conditions on the components of the MDP do the true cost functions of the policies obtained from finite models converge to the optimal value function as the number of grid points goes to infinity?
(ii) **Q2** Can we obtain bounds on the performance loss due to discretization in terms of the number of grid points if we strengthen the conditions sufficient in **(Q1)**?

We will not discuss **Q2** here, but address **Q1**. For **Q2**, the reader is referred to [628]. We just note that, under further explicit regularity conditions, one can indeed arrive at rates of convergence to optimality. The approach to solve problem **(Q1)** can be summarized as follows: First, we obtain approximation results for the compact-state case. We find conditions under which a compact representation leads to near optimality for non-compact state MDPs: solve the approximate MDP, and apply the optimal solution for the approximate MDP to the original MDP. We then obtain the convergence of the finite-state models to non-compact models. Consider **(Q1)** for MDPs with compact state space.

Finite State Approximate MDP: Quantization of the State Space
Let $d_{\mathbb{X}}$ denote the metric on $\mathbb{X}$. For the case with compact $\mathbb{X}$, for each $m \geq 1$, there exists a finite subset $\{z_{m,i}\}_{i=1}^{k_m}$ of $\mathbb{X}$ such that

$$\min_{i \in \{1,\ldots,k_m\}} d_{\mathbb{X}}(z, z_{m,i}) < 1/m \text{ for all } z \in \mathbb{X}.$$

Let $\mathbb{X}_m := \{x_{m,1}, \ldots, x_{m,k_m}\}$ and define Q_m mapping any $z \in \mathbb{X}$ to the nearest element of $\mathbb{X}_m$, i.e.,

$$Q_m(z) := \arg\min_{z_{m,i} \in \mathbb{X}_m} d_{\mathbb{X}}(z, z_{m,i}). \tag{21.17}$$

For each m, a partition $\{\mathbb{S}_{m,i}\}_{i=1}^{k_m}$ of the state space $\mathbb{X}$ is induced by Q_m by setting

$$\mathbb{S}_{m,i} = \{z \in \mathbb{X} : Q_m(z) = z_{m,i}\}.$$

Let ψ be a probability measure on $\mathbb{X}$, satisfying

$$\psi(\mathbb{S}_{m,i}) > 0 \text{ for all } i, m,$$

and define probability measures $\psi_{m,i}$ on $\mathbb{S}_{m,i}$ by restricting ψ to $\mathbb{S}_{m,i}$:

$$\psi_{m,i}(\cdot) := \psi(\cdot)/\psi(\mathbb{S}_{m,i}).$$

Using $\{\psi_{m,i}\}$, we define a sequence of finite-state MDPs, denoted as f-MDP$_m$, to approximate the compact-state MDP.

For each m, f-MDP$_m$ is defined as: $\big(\mathbb{X}_m, \mathbb{U}, \{\mathbb{U}(z) : z \in \mathbb{X}_m\}, \mathcal{T}_m, d_m\big)$, where $\mathbb{U}(z) = \mathbb{U}$ for all $z \in \mathbb{X}_m$, and the one-stage cost function $d_m : \mathbb{X}_m \times \mathbb{U} \to [0, \infty)$ and the transition probability $\mathcal{T}_m$ on $\mathbb{X}_m$ given $\mathbb{X}_m \times \mathbb{U}$ are given by

$$d_m(z_{m,i}, a) := \int_{\mathbb{S}_{m,i}} d(z, a)\psi_{m,i}(dz)$$

$$\mathcal{T}_m(\cdot|z_{m,i}, a) := \int_{\mathbb{S}_{m,i}} Q_m * \mathcal{T}(\cdot|z, a)\psi_{m,i}(dz),$$

where $Q_m * \mathcal{T}(\cdot|z, a) \in \mathcal{P}(\mathbb{X}_m)$ is the pushforward of the measure $\mathcal{T}(\cdot|z, a)$ with respect to Q_m; that is,

$$Q_m * \mathcal{T}(z_{m,j}|z, a) = \mathcal{T}\big(\{y \in \mathbb{X} : Q_m(y) = z_{m,j}\}|z, a\big),$$

for all $z_{m,j} \in \mathbb{X}_m$.

Upon constructing the finite state MDP, we can obtain an optimal solution and apply it to the original MDP. We state the following results.

Theorem 21.3.3 ([628] Discounted Cost) *Suppose that Assumption 21.3.1 holds. Then, for any $\beta \in (0, 1)$ the discounted cost of the deterministic stationary policy $\hat{f}_m$, obtained by extending the discounted optimal policy f_m^* of f-MDP$_m$ to $\mathbb{X}$ (i.e., $\hat{f}_m = f_m^* \circ Q_m$), converges to the discounted value function J^* of the compact-state MDP:*

$$\lim_{m\to\infty} \|J_\beta(\hat{f}_m, \cdot) - J_\beta^*\| = 0. \tag{21.18}$$

One challenge to be addressed in a proof of Theorem 21.3.3 is that in the quantized models (as an intermediate step in the proof) we do not have the weak continuity condition for each of the quantized kernels $\mathcal{T}_m$. The issue is that the value function in the dynamic programming update iterations is not continuous (and would only be piece-wise continuous), and accordingly the map $\int \mathcal{T}_n(dx_1|x,u)v(x)$, with v being a value function, is not necessarily continuous in the action variables, which violates the measurable selection conditions noted in Sect. E.2.4. Nonetheless, the machinery of universally measurable policies can be utilized and the existence of optimal policies for the approximate kernels does not arise as a problem (see [626, pp. 6–7]) for the proof of the theorem. Alternatively, one can first quantize the action set and work on the approximate (finite-action) MDP, whose near optimality was established earlier.

Theorem 21.3.4 ([628] Average Cost) *Suppose that Assumption 21.3.2 holds. Then,*

$$\lim_{m\to\infty} \|J(\hat{f}_m, \cdot) - J^*\| = 0. \tag{21.19}$$

Remark 21.3.2 We note that [628] imposes total variation continuity, but building on [390, Theorem 16] and adapting the arguments in the proof of [397, Theorem 4.2], the total variation condition continuity condition on the kernel can be relaxed to weak Feller continuity, leading to the above [393].

Non-compact Space Case The results above apply also to the non-compact setup, though under restrictive uniform (in control actions) Lyapunov drift/stability conditions. Let $\mathbb{X}$ be a σ-compact separable metric space. Then, there exists a nested sequence of compact sets $\{K_n\}$ such that $K_n \subset \text{int}\, K_{n+1}$ and $\mathbb{X} = \bigcup_{n=1}^{\infty} K_n$.

One can define a sequence of compact-state MDPs, denoted as c-MDP$_n$, to approximate the original model and then follow the construction above. Similar approximation results can be shown to hold under technical drift conditions for both discounted and average cost problems [628].

The near optimality results on finite model approximations for continuous space/action MDPs presented above facilitate reinforcement learning results to be presented in the next section.

21.4 Reinforcement Learning for Continuous Space Models under weak Feller Continuity: Quantized Q-Learning

In the following, by quantizing the state space and viewing the quantization output as a measurement (and quantizer as a measurement kernel), and thus as a POMDP; we will arrive at an approximate finite MDP. In particular, the proofs of the results to follow in this section essentially follow the proofs of the results to be presented in the next chapter; however, since this chapter studies fully observed models, the results are presented prior to the POMDP setup.

First note that under Assumption 21.3.1, Theorem 21.3.1 leads to near-optimality of finite actions.

Let $\rho \equiv Q_m$ be a quantizer as in (21.17) for the finite set $\mathbb{X}_m$. Then, one runs the following: Using ρ, write for any $(x, u) \in \mathbb{X} \times \mathbb{U}$

$$Q_{k+1}(\rho(x), u) = (1 - \alpha_k(\rho(x), u))Q_k(\rho(x), u) + \alpha_k(\rho(x), u)\left(c(x, u) + \beta \min_v Q_k(\rho(X_1), v)\right) \tag{21.20}$$

that is for any true value of the state, we use its representative state from the finite set $\mathbb{X}_m$: *Thus, one runs Q-learning as if the quantized state is the actual state. The resulting policy is applied for the original model.*

For exploration, we again use polices that choose the control actions randomly and independent of everything else with positive probability for every action: the invariant measure of the state process x_t (under the exploration policy) should put positive measure on these bins and satisfy an ergodicity condition to be presented in the following:

Assumption 21.4.1

(i) *With $y = \rho(x)$, we let $\alpha_t(y, u) = 0$ unless $(Y_t, U_t) = (y, u)$. Otherwise, let*

$$\alpha_t(y, u) = \frac{1}{1 + \sum_{k=0}^{t} 1_{\{Y_k = I, U_k = u\}}}.$$

(ii) *Under the exploration policy γ^*, the state process is either positive Harris recurrent (in which case the initial state can be arbitrary) or uniquely ergodic with invariant measure π_{γ^*} in which case $x_0 \sim \mu_0$ where $\mu_0 \ll \pi_{\gamma^*}$. Furthermore, the invariant measure π_{γ^*} places a positive measure on each quantization bin.*

(iii) *During the exploration phase, every observation-action pair (y, u) is visited infinitely often.*

Recall that any MDP with weakly continuous transition probability can be approximated by MDPs with finite action spaces. Thus, to streamline the presentation, we will either assume that the action set is finite, or that it has been approximated with arbitrarily small approximation error by a finite action set through the construction in Theorem 21.3.1.

Let $\mathbb{Y}$ be a finite set, which will play a role for the approximations of $\mathbb{X}$. Let the quantization map $q : \mathbb{X} \to \mathbb{Y}$ be such that for any $x \in \mathbb{X}$, $q(x) = y_i$ for some $y_i \in \mathbb{Y}$, where y_i's are the representative states for the collection of disjoint sets $\{B_i\}_{i=1}^M$ so that $\bigcup_i B_i = \mathbb{X}$ and $B_i \bigcap B_j = \emptyset$ for any $i \neq j$.

In view of the above, consider Algorithm 1.

Algorithm 1: Learning Algorithm: Quantized Q-Learning

Set Parameters

Input: Q_0 (initial Q-function) $q : \mathbb{X} \to \mathbb{Y}$ (quantizer), γ^* (exploration policy), L (number of data points), $\{N(y, u) = 0\}_{(y,u)\in\mathbb{Y}\times\mathbb{U}}$ (number of visits to state-action pairs).

Initialize Start with Q_0

Iterate If (X_t, U_t) is the current state-action pair $\Longrightarrow$ generate the cost $c(X_t, U_t)$ and the next state $X_{t+1} \sim \mathcal{T}(\cdot\,|X_t, U_t)$,

Iterate set

$$N(q(X_t), U_t) = N(q(X_t), U_t) + 1.$$

Iterate For $t = 0, \dots, L-1$,

Update Q-function Q_t for the inputs $(q(X_t), U_t)$ as follows:

$$\begin{aligned} Q_{t+1}(q(X_t), U_t) &= (1 - \alpha_t(q(X_t), U_t))\, Q_t(q(X_t), U_t) \\ &\quad + \alpha_t(q(X_t), U_t) \left(c(X_t, U_t) + \beta \min_{v\in\mathbb{U}} Q_t(q(X_{t+1}), v) \right), \end{aligned} \tag{21.21}$$

where

$$\alpha_t(q(X_t), U_t) = \frac{1}{1 + N(q(X_t), U_t)}.$$

☞ Generate $U_{t+1} \sim \gamma^*$.

End

Return Q_L

The following result is essentially a corollary of Theorem 22.3.4 to be presented in the next chapter.

Theorem 21.4.1 *Under Assumption 21.4.1, for every pair $(y_i, u) \in \mathbb{Y} \times \mathbb{U}$, the quantized Q-learning algorithm, Algorithm 1, given above converges to*

$$Q^*(y_i, u) = C^*(y_i, u) + \beta \sum_{y_j\in\mathbb{Y}} P^*(y_j|y_i, u) \min_{v\in\mathbb{U}} Q^*(y_j, v).$$

Here, P^* and C^* are defined by

$$C^*(y_i, u) = \int_{B_i} c(x, u)\, \hat{\pi}^*_{y_i}(dx)$$

$$P^*(y_j | y_i, u) = \int_{B_i} \mathcal{T}(B_j | x, u)\, \hat{\pi}^*_{y_i}(dx), \tag{21.22}$$

where

$$\hat{\pi}^*_{y_i}(A) := \frac{\pi_{\gamma^*}(A)}{\pi_{\gamma^*}(B_i)}, \quad \forall A \subset B_i, \quad \forall i \in \{1, \ldots, M\}, \tag{21.23}$$

and π_{γ^*} *is the invariant measure of the state process under the exploration policy* γ^*.

Assumption 21.4.2

(a) $\mathbb{X}$ *is compact.*
(b) There exists a constant $\alpha_c > 0$ *such that* $|c(x, u) - c(x', u)| \leq \alpha_c \|x - x'\|$ *for all* $x, x' \in \mathbb{X}$ *and for all* $u \in \mathbb{U}$.
(c) There exists a constant $\alpha_T > 0$ *such that* $W_1(\mathcal{T}(\cdot|x, u), \mathcal{T}(\cdot|x', u)) \leq \alpha_T \|x - x'\|$ *for all* $x, x' \in \mathbb{X}$ *and for all* $u \in \mathbb{U}$.

Theorem 21.4.2 ([398]) *Let Assumption 21.4.1 hold.*

(i) Under Assumption 21.4.2, for the policy constructed from the limit Q values of the quantized Q-learning algorithm, Algorithm 1, say $\hat{\gamma}$, *we have*

$$\sup_{x_0 \in \mathbb{X}} \left| J_\beta(x_0, \hat{\gamma}) - J^*_\beta(x_0) \right| \leq \frac{2\alpha_c}{(1-\beta)^2(1-\beta\alpha_T)} \bar{L}.$$

where

$$\bar{L}_{\mathbb{X}} := \max_{i=1,\ldots,M} \sup_{x, x' \in B_i} \|x - x'\|.$$

(ii) For asymptotic convergence (without a rate of convergence) to optimality as the quantization rate goes to ∞, *only Assumption 21.3.1 (and thus weak Feller property of* $\mathcal{T}$*) is sufficient for the quantized Q-learning algorithm, Algorithm 1, to be near optimal provided that* $\bar{L}_{\mathbb{X}} \to 0$.

Further error bounds under different set of assumptions, such as non-compact state space $\mathbb{X}$ with non-uniform quantization and models with total variation continuous transition kernels are available [398].

21.5 Concluding Remarks

In this chapter, we introduced the original finite state-action Q-learning algorithm, and its proof of convergence. We then introduced a generalization of this algorithm for continuous spaces. This generalization, whose convergence to near optimality holds under weak Feller continuity, will have significant implications for the analyses to follow in the following chapters. The proof also utilized near optimality results for finite model approximations to MDPs with continuous state and action spaces.

21.6 Bibliographic Notes

Classical Q-learning has been introduced and analyzed by a number of influential studies such as [57, 115, 696, 698, 724, 760]. This is a stochastic approximation algorithm used for fully observed finite space MDPs that does not require the knowledge of the transition kernel, or even the cost (or reward) function for its implementation. An ODE approach to Q-learning presents a rather direct proof of convergence [137] [745]. Two comprehensive resources on reinforcement learning are [697] and [535].

The average cost counterpart of the analysis in this chapter has recently been studied in [393], with a detailed literature review on Q-learning under the average cost control criterion present in the paper.

The chapter primarily builds on the results of [398]. On approximations, we build on [624, 626, 628, 629].

Chapter 22
Reinforcement Learning for Partially Observable Models

22.1 Introduction

Partially Observed Markov Decision Problems (POMDPs) offer a practically rich and relevant, and mathematically challenging class of models. Even in the most basic setup of finite state-action models, the analysis and computation of optimal solutions is complicated. The existence of optimal policies has in general been established via converting, or reducing, the original partially observed stochastic control problem to a fully observed Markov Decision Problem (MDP) with probability measure valued (belief) states, leading to a belief-MDP. However, computing an optimal policy for this fully observed model, and thus for the original POMDP, using classical methods (such as dynamic programming, policy iteration, linear programming) is not simple even if the original system has finite state and action spaces, since the state space of the fully observed (reduced) model is always uncountable.

In this chapter, we will present learning theoretic methods for the partially observable setting, building on [391, 392]. For POMDPs, we show that under a uniform controlled filter stability, finite memory policies can be used to arrive at near optimality, and under a mild asymptotic filter stability and weak Feller continuity of the non-linear filter dynamics quantization can be used to arrive at near optimality. The latter is presented in Sect. 22.2 and the former in Sect. 22.3.

The analysis here highlights a parallel between non-linear partially observed stochastic control and linear partially observed stochastic control: In the theory of linear systems, it is known that one can apply either an observer based controller or, under observability, a finite-memory controller using a measurement-action window (of length at most equal to the dimension of the state space). The following two

S. Yüksel, T. Başar, *Stochastic Teams, Games, and Control under Information Constraints*, Systems & Control: Foundations & Applications,
https://doi.org/10.1007/978-3-031-54071-4_22

methods present similar constructions: for the first approach, we will consider a separation (though with a belief-MDP separation) approach and for the second approach, we will consider a perhaps more practical finite window approach.

22.2 Reinforcement Learning for the Partially Observed Setup I: Near Optimality of Quantized Policies Under Weak Feller Property of Non-linear Filters

As we have shown in Appendix E.3, any POMDP can be reduced to a completely observable Markov process [834], [602], whose states are the posterior state distributions or *beliefs* of the observer; that is, the state at time t is

$$\pi_t(\cdot) := P\{X_t \in \cdot \,|y_0, \ldots, y_t, u_0, \ldots, u_{t-1}\} \in \mathcal{P}(\mathbb{X}).$$

We call this conditional probability measure process the *filter* process. The filter process has state space $\mathcal{P}(\mathbb{X})$ and action space $\mathbb{U}$. Here, $\mathcal{P}(\mathbb{X})$ is equipped with the Borel σ-algebra generated by the topology of weak convergence. As discussed in Appendix C, under this topology $\mathcal{P}(\mathbb{X})$ is a standard Borel space [574]. The transition probability of the filter process is given in (E.12).

Accordingly, we have a fully observed belief-MDP. Now, by combining the approximation results in Sect. 21.3 and reinforcement learning theoretic results in Sect. 21.4, together with the weak Feller continuity results presented in Sect. E.3, we can conclude that the numerical methods can also be applied to POMDPs under the conditions reported in Theorems E.3.2 and E.3.2 [257] [395]. This has explicitly been demonstrated in [629], where also methods for quantizing probability measures have been studied in [629, Section 5]. Notably, one can first quantize the state space and then approximate the probability measures with those with quantized support under the W_1 metric.

Therefore, due to the weak Feller property of controlled non-linear filters, we can apply quantized Q-learning algorithm, given in Algorithm 1, introduced in Sect. 21.4, to also belief-based models to arrive at near optimal control policies. For this setup, unique ergodicity of the filter dynamics during exploration follow from results in the literature, such as, [214, Theorem 2] and [733, Prop 2.1], which holds when the randomized control used in the Quantized Q-learning Algorithm 1 is memoryless, the filter is asymptotically stable and the hidden state is uniquely ergodic. Nonetheless, since positive Harris recurrence cannot typically be assumed for the filter process, the initial state may not be arbitrary (see [394] for details on the initialization during the implementation of the algorithm, which affects both the trained sets, which are those visited infinitely often, and the computation of policies for the set of states reachable from the initialization). A related analysis will be carried out in Chap. 24.

In the following, an alternative approach is presented.

22.3 Reinforcement Learning for the Partially Observed Setup II: Near-Optimality of Finite Window Policies

22.3.1 *Near-Optimality of Finite Window Policies Under Filter Stability*

We will show in what follows that under a uniform controlled filter stability, finite window policies are near optimal [392]. Consider the following discounted-cost stochastic control problem:

$$J_\beta(\mu, \mathcal{T}, \gamma) = E_\mu^{\mathcal{T},\gamma}\left[\sum_{t=0}^{\infty} \beta^t c(x_t, u_t)\right], \qquad J_\beta^*(\mu, \mathcal{T}) = \inf_{\gamma \in \Gamma} J_\beta(\mu, \mathcal{T}, \gamma).$$

Instead of using all available history, we construct an approximate model using the finite window information variables

$$\begin{aligned} I_t^N &= \{y_{[t-N,t]}, u_{[t-N,t-1]}\}, \text{ if } t \geq N, \\ I_t^N &= \{y_{[0,t]}, u_{[0,t-1]}\}, \text{ if } 0 < t < N, \\ I_0 &= \{y_0\}, \end{aligned} \tag{22.1}$$

that is, we observe the information variables through a window whose length is N. We denote the optimal value function of the approximate model by J_β^N and the approximate policy by γ^N.

22.3.2 *An Alternative Finite Window Belief-MDP Reduction and Its Approximation*

In this section we construct an alternative fully observed MDP reduction with the condition that the controller has observed at least N information variables, using the predictor from N stages earlier and the most recent N information variables (that is, measurements and actions). This new construction allows us to highlight the most recent information variables and *compress* the information coming from the past history via the predictor as a probability measure valued variable. In what follows, we will sometimes consider the case with $N = 1$ for some of the proofs to make the presentation and analysis easier to follow. The general case follows from similar arguments.

Inspired from filter stability, consider the following: For any time step $t \geq N$ and for a fixed observation realization sequence $y_{[0,t]}$ and control action sequence $u_{[0,t-1]}$, the state process can be viewed as

$$P^{\mu}(x_t \in \cdot | y_{[0,t]}, u_{[0,t-1]}) = P^{\pi_{t-N_-}}(X_t \in \cdot | y_{[t-N,t]}, u_{[t-N,t-1]})$$

where

$$\pi_{t-N_-}(\cdot) = P^{\mu}(x_{t-N} \in \cdot | y_{[0,t-N-1]}, u_{[0,t-N-1]}).$$

That is, we can view the state as the Bayesian update of π_{t-N_-}, the predictor at time $t-N$, using the observations $y_{t-N}, \ldots, y_t$. Notice that with this representation only the most recent N observation realizations are used for the update and the past information of the observations is embedded in π_{t-N_-}. Consider the following set (state space):

$$\mathcal{Z} = \left\{ (\pi, y_{[0,N]}, u_{[0,N-1]}); \pi \in \mathcal{P}(\mathbb{X}), y_{[0,N]} \in \mathbb{Y}^{N+1}, u_{[0,N-1]} \in \mathbb{U}^N \right\}$$

We place the product metric on this new space: weak convergence on the belief and usual metric on the measurements and actions.

The approach is summarized in Fig. 22.1. Now, for a fixed $\hat{\pi} \in \mathcal{P}(\mathbb{X})$, consider the quantized state space:

$$\mathcal{Z}^N = \left\{ (\hat{\pi}, y_{[0,N]}, u_{[0,N-1]}); y_{[0,N]} \in \mathbb{Y}^{N+1}, u_{[0,N-1]} \in \mathbb{U}^N \right\}$$

The idea is to quantize $\mathcal{Z}$ as follows: collapse all π to a fixed state $\hat{\pi}$, define an approximate finite MDP and establish performance bounds utilizing filter stability.

In the following, we will assume that $\mathbb{X}$ is $\mathbb{R}^n$ for some n and that $\mathbb{U}, \mathbb{Y}$ are finite sets. The actual state space and the finite approximation are:

$$\mathcal{Z} = \left\{ (\pi, y_{[0,N]}, u_{[0,N-1]}); \pi \in \mathcal{P}(\mathbb{X}), y_{[0,N]} \in \mathbb{Y}^{N+1}, u_{[0,N-1]} \in \mathbb{U}^N \right\}$$

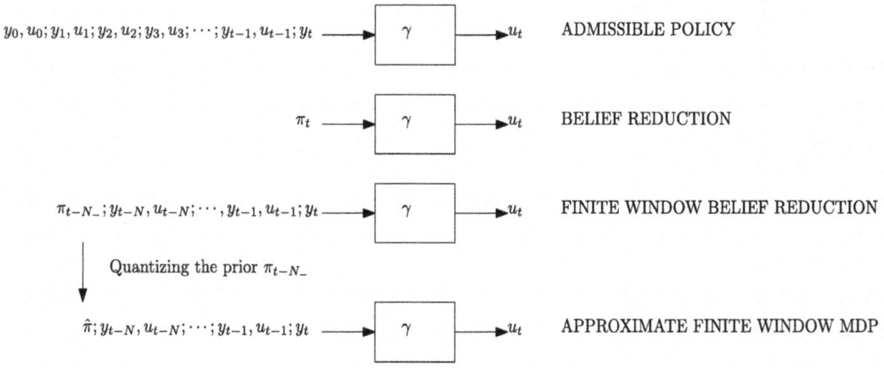

Fig. 22.1 Construction of the finite-window approximate MDP from the finite-window belief-MDP

$$\mathcal{Z}^N = \left\{(\hat{\pi}, y_{[0,N]}, u_{[0,N-1]}); y_{[0,N]} \in \mathbb{Y}^{N+1}, u_{[0,N-1]} \in \mathbb{U}^N\right\}$$

Define the map $F : \mathcal{Z} \to \mathcal{Z}^N$, such that for $(\pi, y_{[0,N]}, u_{[0,N-1]}) \in \mathcal{Z}$

$$F(\pi, y_{[0,N]}, u_{[0,N-1]}) = (\hat{\pi}, y_{[0,N]}, u_{[0,N-1]}).$$

Using the map F and the finite set $\mathcal{Z}^N$, one can define a finite belief MDP, and construct a policy for this finite model, by extending it, we can use the policy, say $\tilde{\varphi}^N$ for the original model.

The cost function for the approximate model is

$$\begin{aligned}\hat{c}(\hat{z}_t^N, u_t) &= \hat{c}(\hat{\pi}, I_t^N, u_t) := \tilde{c}(\varphi(\hat{\pi}, I_t^N), u_t)\\ &= \int_{\mathbb{X}} c(x_t, u_t) P^{\hat{\pi}}(dx_t | y_t, \ldots, y_{t-N}, u_{t-1}, \ldots, u_{t-N}).\end{aligned}$$

We define the controlled transition model for the approximate model by

$$\hat{\eta}^N(\hat{z}_{t+1}^N | \hat{z}_t^N, u_t) = \hat{\eta}^N(\hat{\pi}, I_{t+1}^N | \hat{\pi}, I_t^N, u_t) := \hat{\eta}\Big(\mathcal{P}(\mathbb{X}), I_{t+1}^N | \hat{\pi}, I_t^N, u_t\Big). \tag{22.2}$$

We will write $\mathcal{Z}_{\hat{\pi}}^N$ to make the dependence on $\hat{\pi}$ and N more explicit.

For simplicity, if we assume $N = 1$, then the transitions can be rewritten for some $I_{t+1}^N = (\hat{y}_{t+1}, \hat{y}_t, \hat{u}_t)$ and $I_t^N = (y_t, y_{t-1}, u_{t-1})$

$$\begin{aligned}&\hat{\eta}^N(\hat{\pi}, \hat{y}_{t+1}, \hat{y}_t, \hat{u}_t | \hat{\pi}, y_t, y_{t-1}, u_{t-1}, u_t)\\ &= \hat{\eta}(\mathcal{P}(\mathbb{X}), \hat{y}_{t+1}, \hat{y}_t, \hat{u}_t | \hat{\pi}, y_t, y_{t-1}, u_{t-1}, u_t)\\ &= 1_{\{y_t = \hat{y}_t, u_t = \hat{u}_t\}} P^{\hat{\pi}}(\hat{y}_{t+1} | y_t, y_{t-1}, u_t, u_{t-1}).\end{aligned} \tag{22.3}$$

Denoting the optimal value function for the approximate model by J_β^N, we can write the following fixed-point equation

$$J_\beta^N(\hat{z}^N) = \min_{u \in \mathbb{U}} \left(\hat{c}(\hat{z}^N, u) + \beta \sum_{\hat{z}_1^N \in \hat{\mathcal{Z}}_{\hat{\pi}}^N} J_\beta^N(\hat{z}_1^N) \hat{\eta}^N(\hat{z}_1^N | \hat{z}^N, u) \right). \tag{22.4}$$

E.g., for $N = 1$, we rewrite the fixed point equation for some $\hat{z}_0^N = (\hat{\pi}, y_1, y_0, u_0)$ as

$$J_\beta^N(\hat{\pi}, y_1, y_0, u_0) = \min_{u_1 \in \mathbb{U}} \Bigg(\hat{c}(\hat{\pi}, y_1, y_0, u_0, u_1) \\ + \beta \sum_{y_2 \in \mathbb{Y}} J_\beta^N(\hat{\pi}, y_2, y_1, u_1) P^{\hat{\pi}}(y_2 | y_1, y_0, u_1, u_0) \Bigg). \quad (22.5)$$

We can now investigate the following approximation error terms:

$$|\tilde{J}_\beta^N(\hat{z}) - J_\beta^*(\hat{z})|, \; J_\beta(\hat{z}, \tilde{\varphi}^N) - J_\beta^*(\hat{z}).$$

The first one is the difference between the optimal value function of the original model and that for the approximate model. The second term is the performance loss due to the policy calculated for the approximate model being applied to the true model.

Building on [389, 390], we can show that the loss is related to the term:

$$L_t := \sup_{\hat{\gamma} \in \hat{\Gamma}} E_{\pi_0^-}^{\hat{\gamma}} \Big[\| P^{\pi_t^-}(X_{t+N} \in \cdot | Y_{[t,t+N]}, U_{[t,t+N-1]}) \\ - P^{\hat{\pi}}(X_{t+N} \in \cdot | Y_{[t,t+N]}, U_{[t,t+N-1]}) \|_{TV} \Big] \quad (22.6)$$

Notice that this term is directly related to filter stability studied in Sect. 8.7.3.

Theorem 22.3.1 ([391, 392]) *[Continuity of Value Functions] For $\hat{z}_0 = (\pi_0^-, I_0^N)$, if a policy $\hat{\gamma}$ acts on the first N steps of the process which produces I_0^N, we then have*

$$E_{\pi_0^-}^{\hat{\gamma}} \Big[\Big| \tilde{J}_\beta^N(\hat{z}_0) - J_\beta^*(\hat{z}_0) \Big| \, | I_0^N \Big] \le \frac{\|c\|_\infty}{(1-\beta)} \sum_{t=0}^{\infty} \beta^t L_t$$

Theorem 22.3.2 ([391, 392]) *[Near Optimality of Approximate Finite Window Model Solution applied to Actual Model] For $\hat{z}_0 = (\pi_0^-, I_0^N)$, with a policy $\hat{\gamma}$ acting on the first N steps,*

$$E_{\pi_0^-}^{\hat{\gamma}} \Big[\Big| J_\beta(\hat{z}_0, \tilde{\varphi}^N) - J_\beta^*(\hat{z}_0) \Big| \, | I_0^N \Big] \le \frac{2\|c\|_\infty}{(1-\beta)} \sum_{t=0}^{\infty} \beta^t L_t.$$

As one example, we now show that the term L_t can be bounded by the filter stability result [521, Theorem 3.3]. Recall that this states that

$$E^{\mu,\gamma}\left[\|\pi_n^{\mu,\gamma} - \pi_n^{\nu,\gamma}\|_{TV} \right] \le 2\alpha^n.$$

which holds uniformly for all $\mu \ll \nu$ where $\alpha := (1 - \tilde{\delta}(\mathcal{T}))(2 - \delta(Q))$, and $\delta(\cdot)$ denotes the Dobrushin coefficient of its argument (stochastic kernel). Since $\tilde{\delta}(\mathcal{T})$ is a uniform Dobrushin coefficient over all control actions, the above bound is valid under any control policy. Thus we have that $L_t \leq 2\alpha^N$.

Theorem 22.3.3 ([391, 392]) *Assume that the following two conditions hold:*

(i) $\alpha := (1-\tilde{\delta}(\mathcal{T}))(2-\delta(Q)) < 1$ *(this is the exponential filter stability condition),*
(ii) The transition kernel $\mathcal{T}$ *is dominated, i.e. there exists a dominating measure* $\hat{\pi} \in \mathcal{P}(\mathbb{X})$ *such that for every* $x \in \mathbb{X}$ *and* $u \in \mathbb{U}$, $\mathcal{T}(\cdot|x, u) \ll \hat{\pi}(\cdot)$.

Then, by choosing the dominating measure $\hat{\pi}$ *for the approximate model,*

$$E_{\pi_0^-}^{\hat{\gamma}}\left[\left|J_\beta(\hat{z}_0, \tilde{\varphi}^N) - J_\beta^*(\hat{z}_0)\right| |I_0^N\right] \leq \frac{4\|c\|_\infty}{(1-\beta)^2}\alpha^N. \tag{22.7}$$

Implementing the above is still tedious, though not impossible. Can reinforcement learning be feasible? Can we view the finite history as *state*? We address these two questions in the following.

22.3.3 Q-Learning Convergence of Finite Window Policies

Recall that under mild conditions, for finite state and action MDPs, the Q-learning algorithm given in (21.2) converges to a fixed point which leads to the Discounted Cost Optimality Equation (DCOE).

Learning in POMDPs is challenging, mainly due to the non-Markovian behavior of the observation process. For POMDPs, an attempt may be to study the iterations given by

$$\begin{aligned} Q_{k+1}(y_k, u_k) &= (1 - \alpha_k(y_k, u_k))Q_k(y_k, u_k) \\ &\quad + \alpha_k(y_k, u_k)\left(C_k(y_k, u_k) + \beta \min_v Q_k(Y_{k+1}, v)\right) \end{aligned}$$

However, the observation process y_t is not a controlled Markov process and the cost that is realized is $c(x_k, u_k)$, which is not a function of y_k and u_k only. A two-part question then is the following:

(i) Would the Q-learning iterates for such a setup indeed converge?,
(ii) And, if they do converge, where do they converge to?

The answer to the first part of the question is positive under mild conditions [672] and [699]; and the answer to the second part of the question is that under filter stability conditions, the convergence is to near optimality with an explicit error bound between the performance loss and the memory window size.

Assume that we start keeping track of the last $N+1$ observations and the last N control action variables after at least $N+1$ time steps. That is, at time t, we keep track of the information variables

$$I_t^N = \begin{cases} \{y_t, y_{t-1}, \ldots, y_{t-N}, u_{t-1}, \ldots, u_{t-N}\} & \text{if } N > 0 \\ y_t \quad \text{if } N = 0. \end{cases}$$

We will construct the Q-value iteration using these information variables. In what follows, we will drop the N dependence on I_t^N and sometimes we will use $N = 1$ for simplicity of the notation. For these new approximate states, we follow the usual Q learning algorithm such that for any $I \in \mathbb{Y}^{N+1} \times \mathbb{U}^N$ and $u \in \mathbb{U}$

$$Q_{t+1}(I, u) = (1 - \alpha_t(I, u))Q_t(I, u) + \alpha_t(I, u)\left(C_t(I, u) + \beta \min_v Q_t(I_1^t, v)\right), \tag{22.8}$$

where $I_1^t = \{Y_{t+1}, y_t, \ldots, y_{t-N+1}, u_t, \ldots, u_{t-N+1}\}$, we note the t dependence to emphasize that the distribution of Y_{t+1} and hence I_1^t are different for every t.

To choose the control actions, we use polices that choose the control actions randomly and independent of everything else such that at time t

$$u_t = u_i, \text{ w.p } \sigma_i$$

for any $u_i \in \mathbb{U}$ with $\sigma_i > 0$ for all i.

We note that for the convergence of the learning algorithm, it is sufficient for the hidden state process to converge to its invariant distribution under the exploration policy. Hence, any policy that leads the hidden state process to its invariant measure and visits every action with positive probability can be used for the exploration. For example, the control action can also be chosen to be a function of the most recent measurement and randomized (as long as all actions have positive probability of being selected for every measurement realization); this would again lead to a uniquely ergodic hidden state process under our assumptions. The algorithm is summarized in Algorithm 2.

The algorithm differs from the usual Q-value iteration in view of the following:

(i) The distribution of I_1^t, which is the consecutive N-window information variable when we hit the (I, u), is generally different for every t and the pair (I, u) is not a controlled Markov process.

In other words, the controlled transitions are time dependent, that is, if we assume $N = 1$ then for some $I = (y_t, y_{t-1}, u_{t-1})$ and $u = u_t$:

$$\begin{aligned} &Pr(I_1^t = (y'_{t+1}, y'_t, u'_t) | z = (y_t, y_{t-1}, u_{t-1}), u_t) \\ &= \mathbb{1}_{\{y_t = y'_t, u_t = u'_t\}} Pr(y_{t+1} | y_t, y_{t-1}, u_t, u_{t-1}) \end{aligned}$$

Algorithm 2: Learning Algorithm: Finite Window Q-Learning

Set Parameters

Input: Q_0 (initial Q-function), γ^* (exploration policy), N (memory window length for I^N), L (number of data points), $\{M(I,u)=0\}_{I,u)\in(\mathbb{Y}^{N+1}\times\mathbb{U}^N)}\times\mathbb{U}$ (number of visits to finite memory, state pairs)

Initialize Start with Q_0

Iterate If (I_t, U_t) is the current memory state-action pair $\Longrightarrow$ generate the cost $c(X_t, U_t)$ and the next state $Y_{t+1}\sim\mathcal{T}(\cdot|X_t,U_t)$, and update I_{t+1} given I_t, Y_t, U_t.

Iterate set

$$N(I_t, U_t) = N(I_t, U_t) + 1.$$

Iterate For $t = 0, \ldots, L-1$,

Update Q-function Q_t for the inputs (I_t, U_t) as follows:

$$Q_{t+1}(I_t, U_t) = (1-\alpha_t(I_t, U_t))\, Q_t(I_t, U_t) + \alpha_t(I_t, U_t)\left(c(X_t, U_t) + \beta \min_{v\in\mathbb{U}} Q_t(I_{t+1}, v)\right),$$

where

$$\alpha_t(I_t, U_t) = \frac{1}{1+N(I_t, U_t)}.$$

Generate $U_{t+1}\sim\gamma^*$.

End

Return Q_L

is not stationary and might change at every time step t, since $Pr(y_{t+1}|y_t, y_{t-1}, u_t, u_{t-1})$ depends on the marginal distribution of x_{t-1} (x_{t-N} in the general case).

(ii) Here, we only observe the cost realizations of the underlying state process $\{x_t\}_t$ and the control actions. For example, if we assume that $N=1$ then the cost we observe is $c(x_t, u_t)$. However, $c(x_t, u_t)$ depends on (I, u) pair randomly and in a time dependent way so that for some $I = (y_t, y_{t-1}, u_{t-1})$ and $u = u_t$:

$$C_t(I,u) = c(x_t, u_t) \in B, \quad \text{w.p. } Pr(X_t \in \{x : c(x, u_t)\in B\}|y_t, y_{t-1}, u_{t-1})$$

where $Pr(dx_t|y_t, y_{t-1}, u_{t-1})$ can be seen as some *pseudo-belief* on the underlying state variable given $I = (y_t, y_{t-1}, u_{t-1})$, the most recent $N=1$ information variables. In other words, $Pr(dx_t|y_t, y_{t-1}, u_{t-1})$ is the Bayesian update of π_{t-1}, the marginal distribution of the true state x_{t-1} at the time step $t-1$, using $I = (y_t, y_{t-1}, u_{t-1})$ and thus, it is time dependent. ◇

We will observe that, if one assumes that the hidden state process, $\{x_t\}_t$ admits a unique invariant probability measure π^* under a stationary exploration policy γ, then, with probability one, the average of approximate state transitions gets closer to

$$P^*(I_{t+1}|I_t, u_t) := \hat{\eta}^N((\pi^*, I_{t+1})|(\pi^*, I_t), u_t) \tag{22.9}$$

with $\hat{\eta}^N$ is defined as in (22.2) and (22.3). In particular, if we assume $N = 1$, then we write

$$\begin{aligned} &P^*(I_{t+1} = (y'_{t+1}, y'_t, u'_t)|I_t = (y_t, y_{t-1}, u_{t-1}), u_t) \\ &= \mathbb{1}_{\{y'_t = y_t, u'_t = u_t\}} P^{\pi^*}(y_{t+1}|y_t, y_{t-1}, u_t, u_{t-1}) \end{aligned} \tag{22.10}$$

where $P^{\pi^*}(y_{t+1}|y_t, y_{t-1}, u_t, u_{t-1})$ denotes the distribution of y_{t+1} when the marginal distribution on x_{t-1} is given by the invariant measure π^*.

We also have that the sample path averages of the random cost realizations get close to,

$$C^*(I, u) = \hat{c}(\pi^*, I, u) = \int_{\mathbb{X}} c(x, u) P^{\pi^*}(dx|I)$$

where, $P^*(x|I)$ is the Bayesian update of π^*, using I. If we assume $N = 1$, we can write for some $I = (y_1, y_0, u_0)$ and $u = u_1$

$$C^*(y_1, y_0, u_0, u_1) = \hat{c}(\pi^*, (y_1, y_0, u_0), u_1) = \int_{\mathbb{X}} c(x_1, u_1) P^{\pi^*}(dx_1|y_1, y_0, u_0). \tag{22.11}$$

Now consider the following fixed point equation

$$Q^*(I, u) = C^*(I, u) + \beta \sum_{I'} P^*(I'|I, u) \min_{v} Q^*(I', v) \tag{22.12}$$

where P^* is defined in (22.9) and C^* is defined in (22.11).

The existence of a such fixed point follows from usual contraction arguments. The same fixed equation can also be written as, for $N = 1$, and for $I = (y_1, y_0, u_0)$ and $u = u_1$

$$\begin{aligned} Q^*\left((y_1, y_0, u_0), u_1\right) &= C^*\left((y_1, y_0, u_0), u_1\right) \\ &\quad + \beta \sum_{y_2 \in \mathbb{Y}} P^{\pi^*}(y_2|y_1, y_0, u_1, u_0) \min_{v \in \mathbb{U}} Q^*\left((y_2, y_1, u_1), v\right). \end{aligned} \tag{22.13}$$

We note that the stationary distribution π^* does not have to be calculated by the decision maker. The Q value iterations given in (22.8) only use the finite-memory variables I, and π^* is not used in the iterations. The performance loss will depend on the stationary distribution π^* that is learned via the exploration policy, however, we will establish further upper bounds that are uniform over such π^* which decrease exponentially with the window size N.

That is, one runs Q-learning algorithm by pretending that the finite window is the *state*. We first need to specify some conditions that would be needed.

Assumption 22.3.1

(i) $\alpha_k(I,u) = \frac{1}{k}$ *if* $I_k = I, u_k = u$.
(ii) Under the stationary {memoryless or finite memory exploration} policy, say γ, *the true state process,* $\{X_t\}_t$, *admits a unique invariant probability measure* π^*_γ *and the initial* I_0 *is distributed so that the ergodic theorem holds with probability* 1.
(iii) During the exploration phase, every (I,u) *pair is visited infinitely often.*

Theorem 22.3.4 ([391]) *Under the previous assumption, the algorithm given by*

$$Q_{k+1}(I,u) = (1-\alpha_k(I,u))Q_k(I,u) + \alpha_k(I,u)\left(C_k(I,u) + \beta \min_v Q_k(I_1,v)\right),$$

converges almost surely to Q^* *which satisfies*

$$Q^*(I,u) = C^*(I,u) + \beta \sum_{I'} P^*(I'|I,u) \min_v Q^*(I',v)$$

which are the Q values for the approximate belief MDP.

Corollary 22.3.1 ([391]) *Under the filter stability conditions, finite window Q learning is nearly optimal:*

(i) Under the exponential filter stability condition, for any policy γ^N *that satisfies* $Q^*(I,\gamma^N(I)) = \min_u Q^*(I,u)$

$$E\left[J_\beta(\pi_N^-,\mathcal{T},\gamma^N) - J_\beta^*(\pi_N^-,\mathcal{T})|I_0^N\right] \le \frac{4\|c\|_\infty}{(1-\beta)^2}\alpha^N.$$

(ii) If we only have asymptotic filter stability (uniform in policies) in total variation, as $N \to \infty$,

$$E\left[J_\beta(\pi_N^-,\mathcal{T},\gamma^N) - J_\beta^*(\pi_N^-,\mathcal{T})|I_0^N\right] \to 0.$$

Remark 22.3.1 We caution the reader that our result assumes that the cost starts running after time N: that is the effective cost is:

$$E[\sum_{k=N}^{\infty} \beta^{k-N} c(x_k, u_k)]. \tag{22.14}$$

Of course, this criterion is also applicable if the system starts running prior to time $-N$ and the costs become in effect after time 0. If this criterion is not applicable, and the first N stages are also crucial, (i) if β is large enough, we can conclude that the first N stages are not as critical for the analysis as their contributions will be minor in comparison with the future stages for the criterion. (ii) On the other hand, if β is not large and if the cost starts running at time 0, then, we can first run the Q-learning algorithm above to find the best N-window policies which optimizes (22.14). The remaining question would be to optimize:

$$E[\sum_{k=0}^{N-1} c(x_k, u_k) + V(I_k)] \tag{22.15}$$

as a finite-horizon optimal control problem with a terminal cost and the terminal cost V can be estimated by (22.14) via Theorem 22.3.1 and Theorem 22.3.4. The question then becomes how to select the first N actions, leading to a problem with a finite search complexity for a finite horizon problem, without knowing the system dynamics. For this, one can run a MCMC algorithm in parallel simulations to find the optimal policy for the first N time stages. Since the resulting policy minimizing (22.15) will be at least as good as the first N-window policy under the optimal (belief-MDP) policy (which is not designed to optimize (22.15) but the original cost, the bounds presented in Theorem 22.3.2 will be applicable even when the cost criterion includes the first N time stages.

22.4 Concluding Remarks

In this chapter, we studied reinforcement learning for partially observable models. We presented two alternative approaches, one via quantization of belief spaces, and the latter via finite memory Q-learning; with the latter being applicable under filter stability conditions.

22.5 Bibliographic Notes

The problem of approximate optimality for POMDPs is significantly more challenging compared to the fully observed counterpart. Even for the discounted setup, most of the studies in the literature are algorithmic and computational contributions. These include [583, 589, 674, 747].

Building on [626], [629] introduces a rigorous approximation analysis (and explicit methods for quantization of probability measures) after establishing weak continuity conditions on the transition kernel defining the (belief-MDP) via the non-linear filter [256, 395], and shows that finite model approximations obtained through quantization are asymptotically optimal and the control policies obtained from the finite model can be applied to the actual system with asymptotically vanishing error as the number of quantization bins increases. Another rigorous set of studies is [846] and [847] where the authors provide an explicit quantization method for the set of probability measures containing the belief states.

We refer the reader to [337] and the book [433] for further structural results as well as algorithmic and computational methods for approximating POMDPs.

Q-learning for partially observed MDPs have been studied in [289, 529, 673, 699]. A natural approach (see [391, 672]) to attempt to learn POMDPs would be to ignore the partial observability and pretend the noisy observations reflect the true state perfectly. This was studied in [672] and [391], where [391] rigorously constructed the approximate belief MDP that the limit equation satisfies which gave an operational and practical conclusion regarding the analysis of the algorithm. [392] established near optimality of finite window policies using a different approach by considering the belief-MDP directly and quantizing the belief space with a nearest neighbor map (under a metric on probability measures which induces the weak convergence topology) that uses finite window information variables.

Reference [769] is a related work that studies approximation techniques for POMDPs using finite memory with finite state, action, measurements. In [802], the authors study near optimality of finite window policies for average cost problems where the state, action and observation spaces are finite; under the condition that the liminf and limsup of the average cost are equal and independent of the initial state, the paper establishes the near-optimality of (non-stationary) finite memory policies. Here, a concavity argument building on [252] (which becomes consequential by the equality assumption) and the finiteness of the state space is crucial. In [307], a finite-memory based approximate planing method is studied for POMDPs.

As we noted, another approach is to quantize the belief states and apply *quantized Q-learning* [398] via weak Feller properties of the belief MDP [257] [395]. Further related studies include [502, 843].

The results of this chapter primarily build on [391, 392]. We refer the reader also to [672] for the convergence of measurement based Q-learning updates.

Chapter 23
Multi-Agent Reinforcement Learning with General Information Structures: Convergence to Equilibria

23.1 Introduction

So far we have studied in Chaps. 21 and 22 learning in single-agent models with continuous state and action spaces, in fully observed as well as partially observed settings. In this chapter, we move on to multi-agent systems and discuss learning theoretic methods for decentralized information structure models, for both stochastic teams and games.

For decentralized stochastic decision problems, under a variety of information structures with strictly local information or partial global information, we show that synchronous or asynchronous quantized Q-learning can be used to arrive at near optimal performance or equilibrium. The underlying equilibria may refer to a rich collection of stationary behavior, as well as to an actual or a subjective equilibrium. In fact, an argument we will make in the chapter is that, the latter notion (that is, subjective equilibrium) is perhaps the more appropriate notion of equilibrium for a class of problems in which there is a complex environment consisting of agents with different on-line or prior information, and agents respond to the environment under some modelling assumptions. Nonetheless, the analysis we present will cover a rich class of setups, including stochastic dynamic teams, weakly acyclic stochastic games, symmetric stochastic dynamic games, mean-field games as well as a large class of 2-player stochastic dynamic games.

As we have seen in the previous two chapters, reinforcement learning (RL) algorithms use past experience and feedback information to improve performance in a control task [693]. Multi-agent reinforcement learning (MARL) is the study of emergent behavior in complex, strategic and multi-agent environments, and constitutes one of the important frontiers in modern artificial intelligence research, and possesses inherent challenges.

S. Yüksel, T. Başar, *Stochastic Teams, Games, and Control under Information Constraints*, Systems & Control: Foundations & Applications,
https://doi.org/10.1007/978-3-031-54071-4_23

The first such challenge is due to decentralization of information, as we have seen extensively in the book. The second challenge inherent to MARL comes from the non-stationarity of the environment from the point of view of any individual agent. As an agent learns how to improve its performance, it will alter its behaviour, and this can have a destabilizing effect on the learning processes of the remaining agents. Notably, this issue arises when one tries to apply single-agent RL algorithms—which typically rely on state-action value estimates or gradient estimates that are made using historical data—in multi-agent settings. A number of studies, including [701] and [179], have reported non-convergent play when single-agent algorithms using local information are employed, without modification, in multi-agent settings.

We will consider stochastic games with discounted costs. In this setting, our overarching goal is to provide MARL algorithms that are suitable for independent learners in a complex system, require little coordination among agents, and come with provable guarantees. To inform the development of such algorithms, this chapter identifies structural properties of games that can be leveraged in algorithm design. We then illustrate the usefulness of the identified structure by providing decentralized or independent learning algorithms and proving that, under mild conditions, such algorithms lead to approximate equilibrium policies.

There are several competing paradigms for studying multi-agent learning: As noted in Sect. 10.6, the Bayesian learning paradigm studies agents who maintain beliefs on the policies used by other agents. Bayesian learners select policies to best-respond to their beliefs, while beliefs are updated as new information arrives during the course of play. In addition to elegant results on the convergence of play to (objective) Nash equilibrium under certain conditions, a notion of subjective equilibrium was proposed by Kalai and Lehrer [382] in the Bayesian learning framework. Notions of subjective equilibrium are relevant to MARL at large, as they allow for players to be uncertain of the model of the game being played, rather than allowing only for uncertainty in the policies of the other agents or unobservable hidden variables whose distributions are known. Several other notions of subjective equilibrium have been proposed in the literature; we refer the reader to Sect. 10.6 and [38] for an overview. In the chapter, we present a further angle, where agents take part in Q-learning, and equilibrium behaviour of such decentralized Q-learning updates is studied, which will be regarded as *subjective Q-learning equilibrium*: Independent or decentralized learners constitute a class of MARL algorithms that are characterized by obliviousness to the strategic environment: independent learners ignore the presence of other players, effectively treating other players as part of the environment rather than as the non-stationary learning agents that they actually are. By, subjectively, running a single-agent reinforcement learning algorithm using only local information, independent learners are relieved of the burden of excessive information, which may lead to scalable algorithms for large-scale decentralized systems.

One issue with designing MARL algorithms that use global information about the local states and actions of all players is that such algorithms do not scale with the number of players. In the case of many agents, however, the structure of mean-field models may facilitate mathematical analysis, as we will see in the chapter. This is not surprising, given the analysis earlier in the book for stochastic teams and games with many and infinitely many agents.

In the following section, Sect. 23.2, we present a general learning paradigm, via a Markov chain approach on a policy update dynamics, and to be adapted to two general classes of games. In Sect. 23.3, this is applied to stochastic team problems [36] via best-response or better-response dynamics with inertia. In Sect. 23.4, we introduce a paradigm of satisficing paths [794, 796] and its application to two classes of stochastic dynamic games: Two-player games and symmetric games. In Sects. 23.5 and 23.6 we study mean-field models, as an important instance of symmetric games, and a general learning theory for such models, including learning to a new and practically consequential class of subjective equilibria arrived via subjective ϵ-satisficing paths under a variety of information structures.

23.2 A General Paradigm for Multi-Agent Learning: Two Time Scales and a Markov Chain over Play Path Graphs

The learning setup involves specifying the information that DMs have access to. To make the discussion more concrete, let us consider the following model.

We assume that each DMi knows its own set $\mathbb{U}^i$ of decisions and its own discount factor β^i. In addition, before choosing its decision u_t^i at any time t, each DMi may have more specific information. In particular, if the agents have access to the global state information, we specify the following information structure (this will be relaxed later on):

(i) its own past decisions $u_0^i, \ldots, u_{t-1}^i$, and
(ii) past and current state realizations $x_0, \ldots, x_t$, and
(iii) its own past cost realizations

$$c^i(x_0, u_0^i, u_0^{-i}), \ldots, c^i(x_{t-1}, u_{t-1}^i, u_{t-1}^{-i}).$$

Each DMi has access to no other information such as the state transition probabilities or any information regarding the other DMs (not even the existence of the other DMs). In effect, the problem of decision making from the perspective of each DMi appears to be a stationary Markov decision problem. It is reasonable that each DMi with this view of its environment would use the standard Q-learning algorithm (see Sect. 21.2) to learn its optimal Q-factors and its optimal decisions. This would lead to the following Q-learning dynamics for each DMi:

$$Q_{t+1}^i(x, u^i) = Q_t^i(x, u^i), \quad \text{for all } (x, u^i) \neq (x_t, u_t^i)$$
$$Q_{t+1}^i(x_t, u_t^i) = Q_t^i(x_t, u_t^i) + \alpha_t^i(c^i(x_t, u_t^i, u_t^{-i}) + \beta^i \min_{v^i \in \mathbb{U}^i} Q_t^i(x_{t+1}, v^i) - Q_t^i(x_t, u_t^i))$$

where $\alpha_t^i \in [0, 1]$ denotes DMi's step size at time t.

If only one DM, say DMi, were to use Q-learning and the other DMs used constant policies π^{-i}, then DMi would asymptotically learn its corresponding optimal Q-factors, i.e.,

$$P[Q_t^i \to Q_{\pi^{-i}}^i] = 1$$

provided that all state-control pairs x, u^i are visited infinitely often and the step sizes are reduced at a proper rate. This follows from the convergence of Q-learning in a stationary environment as we studied earlier in Sect. 21.2.

However, when all DMs use Q-learning and select their decisions as described above, the environment is non-stationary for all DMs, and there is no reason to expect convergence in that case. In fact, one can construct examples where DMs using Q-learning are caught up in persistent oscillations; see Section 4 in [471] for the non-convergence of Q-learning in Shapley's game. This discussion reveals that the standard Q-learning may not be convergent even in repeated coordination games. The main obstacle to convergence of Q-learning in games is due to the presence of multiple active learners leading to a non-stationary environment for all learners. To overcome this obstacle, also building on inspiration from prior work [298, 504], on repeated games, [36] modifies the Q-learning for stochastic games as follows: In their variation of Q-learning, DMs are allowed to use constant policies for extended periods of time called *exploration phases*.

As illustrated in Fig. 23.1, the k-th exploration phase runs through times $t = t_k, \ldots, t_{k+1} - 1$, where

$$t_{k+1} = t_k + T_k \qquad (\text{with } t_0 = 0)$$

for some integer $T_k \in [1, \infty)$ denoting the length of the k-th exploration phase. During the k-th exploration phase, DMs use some constant policies $\pi_k^1, \ldots, \pi_k^N$ as their baseline policies with occasional experimentation.

The main idea is to create a stationary environment over each exploration phase so that DMs can almost accurately learn their optimal Q-factors corresponding to

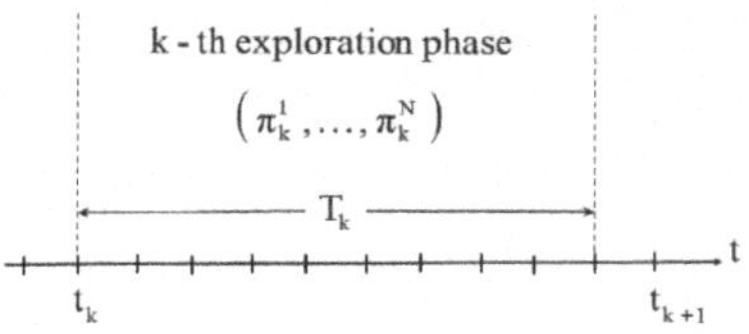

Fig. 23.1 An illustration of the k-th exploration phase

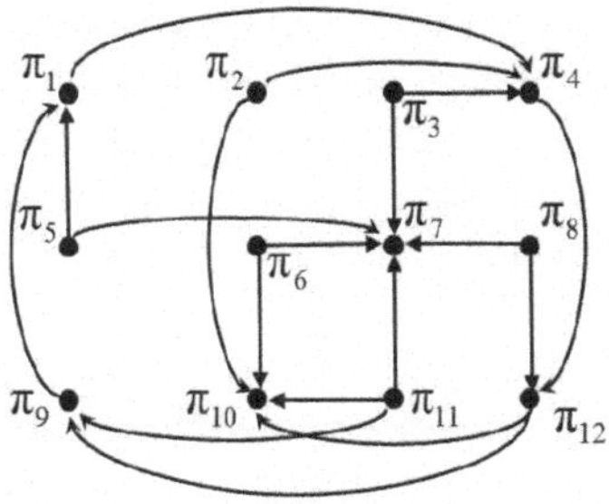

Fig. 23.2 The response graph of a stochastic game

the constant policies used during each exploration phase and update their policies; see Fig. 23.2.

This machinery will be adopted under two types of policy updates: (1) *Best response dynamics with inertia* for weakly acyclic games [36] considered for the case where each agent has access to the global state but only local action (requiring typically deterministic policies), and (2) a different response dynamics referred to as *satisficing paths* dynamics [794, 796] which assumes that the agents have access to a variety of information states and the policies may be randomized. These will be considered in the following two sections, respectively. Generalizations to asynchronous updates, where the DMs do not have to have identical policy update time instances ($\tau_k, k \geq 0$) will also be briefly discussed [795].

23.3 Learning with Local Information for Stochastic Teams and Weakly Acyclic Games with Deterministic Baseline Policies

23.3.1 Stochastic Teams and Weakly Acyclic Games with Local Action and Global State Information

Consider a finite discounted stochastic game $\mathcal{G}$:

(i) A finite set $\mathcal{N} = \{1, \ldots, N\}$ of DMs
(ii) a finite set $\mathbb{X}$ of states
(iii) a finite set $\mathbb{U}^i$ of control decisions for each DMi
(iv) a cost function c^i for each DMi determining DMi's cost $c^i(x, u^1, \ldots, u^N)$ at each state $x \in \mathbb{X}$ and for each joint decision $(u^1, \ldots, u^N) \in \mathbb{U}^1 \times \cdots \mathbb{U}^N$
(v) a discount factor $\beta^i \in (0, 1)$ for each DMi
(vi) a random initial state $x_0 \in \mathbb{X}$
(vii) a transition kernel for the probability $P[x'|x, u^1, \ldots, u^N]$ of each state transition from $x \in \mathbb{X}$ to $x' \in \mathbb{X}$ for each joint decision $(u^1, \ldots, u^N) \in \mathbb{U}^1 \times \cdots \mathbb{U}^N$.

Such a stochastic game induces a discrete-time controlled Markov process where the state at time t is denoted by $x_t \in \mathbb{X}$ starting with the initial state x_0. At any time $t \geq 0$, each DMi makes a control decision $u_t^i \in \mathbb{U}^i$ (possibly randomly) based on the available information. The state x_t and the joint decisions $(u_t^1, \ldots, u_t^N)$ together determine each DMi's cost $c^i(x_t, u_t^1, \ldots, u_t^N)$ at time t as well as the probability distribution $P[\,\cdot \mid x_t, u_t^1, \ldots, u_t^N]$ with which the next state x_{t+1} is selected.

We will focus on stationary policies of the form where a DM's decision at time t is determined solely based on the state x_t. We will denote the set of such policies by Γ_S^i for each DMi. In this section, we will primarily be interested in deterministic (stationary) policies, denoted by Π^i for each DMi, where each policy $\pi^i \in \Pi^i$ is identified by a mapping from $\mathbb{X}$ to $\mathbb{U}^i$. This will be generalized in the following sections.

The objective of each DMi is to find a policy $\pi^i \in \Gamma_S^i$ that minimizes its expected discounted cost

$$J_x^i(\pi^1, \ldots \pi^N) = E_x[\sum_{t \geq 0} (\beta^i)^t c^i \left(x_t, u_t^1, \ldots, u_t^N\right)] \tag{23.1}$$

for all $x \in \mathbb{X}$, where E_x denotes the conditional expectation given $x_0 = x$. Since DMs have possibly different cost functions and each DM's cost may depend on the control decisions of the other DMs, we adopt the notion of equilibrium to represent those policies that are *person-by-person optimal*. For ease of notation, we denote the policies of all DMs other than DMi by π^{-i}. We also define $\mathbf{\Pi}^{-i} := \times_{j \neq i} \Pi^j$ and $\mathbf{\Gamma}_S^{-i} := \times_{j \neq i} \Gamma_S^j$ as well as $\mathbf{\Pi} := \times_j \Pi^j$ and $\Delta := \times_j \Gamma_S^j$. Using this notation, we write a joint policy $(\pi^1, \ldots \pi^N)$ as (π^i, π^{-i}) and $J_x^i(\pi^1, \ldots \pi^N)$ as $J_x^i(\pi^i, \pi^{-i})$. Recall (see Chap. 10) that a joint policy $(\pi^{*1}, \ldots, \pi^{*N}) \in \Delta$ constitutes an (Markov perfect) equilibrium if, for all i, x,

$$J_x^i(\pi^{*i}, \pi^{*-i}) = \min_{\pi^i \in \Gamma_S^i} J_x^i(\pi^i, \pi^{*-i}).$$

It is known that any finite discounted stochastic game possesses an equilibrium policy as defined above. Although the minimum above can always be achieved by a deterministic policy in Π^i (since each DMi's problem is a stationary Markov decision problem when the policies of the other DMs are fixed at π^{*-i}), a deterministic equilibrium policy may not exist in general. However, many interesting classes of games do possess equilibrium in deterministic policies. In particular, large classes of games arising from applications where all DMs benefit from cooperation possess equilibrium in deterministic policies. The primary examples of such games of cooperation are team problems where all DMs have the same cost function, as we have extensively studied. In such team problems, deterministic policies minimizing the common cost function are clearly equilibrium policies although non-optimal deterministic equilibrium policies may also exist. A more general set of games of cooperation are those in which some function, called the potential function, decreases whenever a single DM decreases its own cost by unilaterally switching from one deterministic policy to another one. In this class of games, the deterministic policies minimizing the potential function are equilibrium policies. As such, we are primarily interested in the set of deterministic equilibrium policies denoted by $\mathbf{\Pi}_{\text{eq}}$, where $\mathbf{\Pi}_{\text{eq}} \subset \mathbf{\Pi}$.

We next formally introduce the set of games considered in this section.

23.3.2 Weakly Acyclic Games

Let $\Pi^i_{\pi^{-i}}$ denote DMi's set of (deterministic) best replies to any $\pi^{-i} \in \mathbf{\Gamma}_S^{-i}$, i.e.,

$$\Pi^i_{\pi^{-i}} := \Big\{\hat{\pi}^i \in \Pi^i : J_x(\hat{\pi}^i, \pi^{-i}) = \min_{\pi^i \in \Gamma^i_S} J_x(\pi^i, \pi^{-i}), \quad \text{for all } x\Big\}.$$

DMi's best replies to any $\pi^{-i} \in \mathbf{\Gamma}^{-i}$ can be characterized by its optimal Q-factors $Q^i_{\pi^{-i}}$ satisfying the fixed-point equation

$$\begin{aligned} Q^i_{\pi^{-i}}(x, u^i) =& E_{\pi^{-i}(x)}\Big[c^i(x, u^i, u^{-i}) \\ &+ \beta^i \sum_{x' \in \mathbb{X}} P[x'|x, u^i, u^{-i}] \min_{v^i \in \mathbb{U}^i} Q^i_{\pi^{-i}}(x', v^i)\Big] \end{aligned} \tag{23.2}$$

for all x, u^i, where $E_{\pi^{-i}(x)}$ denotes the expectation with respect to the joint distribution of u^{-i} given by $\pi^{-i}(x) = \pi^1(x) \times \cdots \times \pi^{i-1}(x) \times \pi^{i+1}(x) \times \cdots \times \pi^N(x)$. The optimal Q-factor $Q^i_{\pi^{-i}}(x, u^i)$ represents DMi's expected discounted cost-to-go from the initial state x assuming that DMi initially chooses u^i and uses an optimal policy thereafter while the other DMs use π^{-i}. One can then write $\Pi^i_{\pi^{-i}}$ as

$$\Pi^i_{\pi^{-i}} = \Big\{\hat{\pi}^i \in \Pi^i : Q^i_{\pi^{-i}}(x, \hat{\pi}^i(x)) = \min_{v^i \in \mathbb{U}^i} Q^i_{\pi^{-i}}(x, v^i), \quad \text{for all } x\Big\}.$$

The set of (deterministic) joint best replies is denoted by $\mathbf{\Pi}_\pi := \Pi^1_{\pi^{-1}} \times \cdots \times \Pi^N_{\pi^{-N}}$. Any best reply $\hat{\pi}^i \in \Pi^i_{\pi^{-i}}$ of DMi is called a *strict best reply* with respect to (π^i, π^{-i}) if

$$J^i_x(\hat{\pi}^i, \pi^{-i}) < J^i_x(\pi^i, \pi^{-i}), \quad \text{for some } x.$$

Such a strict best reply $\hat{\pi}^i$ achieves DMi's minimum cost given π^{-i} for all initial states and results in a strict improvement over π^i for at least one initial state.

Definition 23.3.1 We call a (possibly finite) sequence of deterministic joint policies $\pi_0, \pi_1, \ldots$ a *strict best reply path* if, for each k, π_k and π_{k+1} differ in exactly one DM position, say DMi, and π^i_{k+1} is a strict best reply with respect to π_k.

Definition 23.3.2 A discounted stochastic game is called *weakly acyclic* under strict best replies if there is a strict best reply path starting from each deterministic joint policy and ending at a deterministic equilibrium policy.

Figure 23.2 shows the strict best reply graph of a game where the nodes represent the deterministic joint policies and the directed edges represent the single-DM strict best replies. Each deterministic equilibrium policy is represented by a sink, i.e., a node with no outgoing edges, in such a graph. Note that the game illustrated in Fig. 23.2 is weakly acyclic under strict best replies since there is a path from every node to a sink (π_7 or π_{10}). Note also that a weakly acyclic game may have cycles in its strict best reply graph, for example, $\pi_1 \to \pi_4 \to \pi_{12} \to \pi_9 \to \pi_1$ in Fig. 23.2.

Weakly acyclic games constitute a large class of games: in the case of single-stage games, all potential games as well as dominance solvable games are examples of weakly acyclic games; see [244]. We note that the concept of weak acyclicity introduced here is with respect to the stationary Markov policies for stochastic games, and constitutes a generalization of weak acyclicity introduced in [801] for single-stage games. The primary examples of weakly acyclic games in the case of stochastic games are the team problems with finite state and control sets where DMs have identical cost functions and discount factors.

Consider a policy adjustment process in which only one DM updates its policy at each step by switching to one of its strict best replies. Such a process would terminate at an equilibrium policy if the game has no cycles in its strict best reply graph and the process continues until no DM has strict best replies. A weakly acyclic game may contain cycles in its strict best reply graph but there must be some edges leaving each cycle because otherwise there would not be a path from each node to a sink. Therefore, as long as each updating DM considers each of its strict best replies with positive probability, the adjustment process would terminate at an equilibrium policy in a weakly acyclic game with probability (w.p.) one. This adjustment process would require a criterion to determine the updating DM at each step and the DMs would have to a priori agree to this criterion. An equilibrium policy can be reached through a similar adjustment process without a pre-game agreement on the selection of the updating DM, if all DMs update their policies at each step but with some inertia (so that agents may choose also not to alter their policies so that the strict best reply play path as defined in Definition 23.3.1 is realizable with non-zero probability).

We now pursue the learning paradigm reviewed earlier in which DMs would be able to learn their near best replies with minimal information and adjust their policies (approximately) along the strict best reply paths as in the best reply process with inertia. Let us now introduce the variation of Q-learning introduced in [36] more precisely.

Each DM using Algorithm 3 approximately learns its optimal Q-factors during each exploration phase with limited observations. Accordingly, each DM updates its (baseline) policy to one of its near best replies with inertia based on its learnt Q-factors. Hence, Algorithm 3 can be regarded as an approximation to a best reply process (with inertia).

We require the following assumptions.

Algorithm 3: Learning algorithm for stochastic teams or weakly acyclic games with local action information

---

Set parameters
$\mathbb{Q}^i$: some compact subset of the Euclidian space $\mathbb{R}^{|\mathbb{X}\times\mathbb{U}^i|}$ where $|\mathbb{X}\times\mathbb{U}^i|$ is the number of pairs (x, u^i)
$\{T_k\}_{k\geq 0}$: sequence of integers in $[1,\infty)$; $\rho^i \in (0,1)$: experimentation probability; $\lambda^i \in (0,1)$: inertia
$\delta^i \in (0,\infty)$: tolerance level for sub-optimality
$\{\alpha^i_n\}_{n\geq 0}$: sequence of step sizes where $\alpha^i_n \in [0,1]$, $\sum_n \alpha^i_n = \infty$, $\sum_n \left(\alpha^i_n\right)^2 < \infty$
(e.g., $\alpha^i_n = 1/n^r$ where $r \in (1/2, 1]$)
Initialize $\pi^i_0 \in \Pi^i$ (arbitrary), $Q^i_0 \in \mathbb{Q}^i$ (arbitrary) Receive x_0
Iterate $k \geq 0$
(k-th exploration phase)
Iterate $t = t_k, \ldots, t_{k+1}-1$
$u^i_t = \begin{cases} \pi^i_k(x_t), & \text{w.p. } 1-\rho^i \\ \text{any } u^i \in \mathbb{U}^i, & \text{w.p. } \rho^i/|\mathbb{U}^i| \end{cases}$
Receive $c^i(x_t, u^i_t, u^{-i}_t)$
Receive x_{t+1} (selected according to $P[\,\cdot \mid x_t, u^i_t, u^{-i}_t])$
n^i_t = the number of visits to (x_t, u^i_t) in the k-th
exploration phase up to t
$Q^i_{t+1}(x_t, u^i_t) = (1-\alpha^i_{n^i_t})Q^i_t(x_t, u^i_t)$
$+\alpha^i_{n^i_t}\left[c^i(x_t, u^i_t, u^{-i}_t) + \beta^i \min_{v^i} Q^i_t(x_{t+1}, v^i)\right]$
$Q^i_{t+1}(x, u^i) = Q^i_t(x, u^i)$, for all $(x, u^i) \neq (x_t, u^i_t)$
End
$\Pi^i_{k+1} = \big\{\hat{\pi}^i \in \Pi^i : Q^i_{t_{k+1}}(x, \hat{\pi}^i(x))$
$\leq \min_{v^i} Q^i_{t_{k+1}}(x, v^i) + \delta^i, \text{ for all } x\big\}$
If $\pi^i_k \in \Pi^i_{k+1}$
$\pi^i_{k+1} = \pi^i_k$
Else
$\pi^i_{k+1} = \begin{cases} \pi^i_k, & \text{w.p. } \lambda^i \\ \text{any } \pi^i \in \Pi^i_{k+1}, & \text{w.p. } (1-\lambda^i)/|\Pi^i_{k+1}| \end{cases}$
End
Reset $Q^i_{t_{k+1}}$ to any $Q^i \in \mathbb{Q}^i$ (e.g., project $Q^i_{t_{k+1}}$ onto $\mathbb{Q}^i$)
End

Assumption 23.3.1 *For all (x', x), there exists a finite integer $H \geq 0$ and joint decisions $\tilde{u}_0, \ldots, \tilde{u}_H$ such that*

$$P[x_{H+1} = x' \mid (x_0, u_0, \ldots, u_H) = (x, \tilde{u}_0, \ldots, \tilde{u}_H)] > 0.$$

Assumption 23.3.2 *For all i, $0 < \delta^i < \bar{\delta}$ and $0 < \rho^i < \bar{\rho}$, where $\bar{\delta}$ and $\bar{\rho}$ depend only on the parameters of the game at hand.*

Assumption 23.3.2 requires that the tolerance levels for sub-optimality used in the computation of near best replies as well as the experimentation probabilities be nonzero but sufficiently small.

Theorem 23.3.1 ([36]) *Consider a discounted stochastic game that is weakly acyclic under strict best replies. Suppose that each DMi updates its policies by Algorithm 3. Let Assumptions 23.3.1 and 23.3.2 hold.*

(i) For any $\epsilon > 0$, there exist $\tilde{T} < \infty$, $\tilde{k} < \infty$ such that if $\min_\ell T_\ell \geq \tilde{T}$, then

$$P(\pi_k \in \mathbf{\Pi}_{\text{eq}}) \geq 1 - \epsilon, \qquad \text{for all } k \geq \tilde{k}.$$

(ii) If $T_k \to \infty$, then

$$P(\pi_k \in \mathbf{\Pi}_{\text{eq}}) \to 1.$$

(iii) There exist finite integers $\{\tilde{T}_k\}_{k\geq 0}$ such that if $T_k \geq \tilde{T}_k$, for all k, then

$$P(\pi_k \to \pi^*, \text{ for some } \pi^* \in \mathbf{\Pi}_{\text{eq}}) = 1.$$

We now discuss the main idea behind this result. Since all DMs use constant policies throughout any particular exploration phase, each DM indeed faces a stationary Markov decision problem in each exploration phase. Therefore, if the length of each exploration phase is long enough and the experimentation probabilities $\rho^1, \ldots, \rho^N$ are small enough (but non-zero), each DMi can learn its corresponding optimal Q-factors in each exploration phase with arbitrary accuracy with arbitrarily high probability. This allows each DMi to accurately compute its near best replies to the other DMs' policies π_k^{-i} at the end of the k-th exploration phase. Intuitively, allowing each DMi to update its policy π_k^i to its near best replies (to π_k^{-i}) at the end of the k-th exploration phase with some inertia $\lambda^i \in (0, 1)$ results in a policy adjustment process that approximates the best reply process with inertia.

Remark 23.3.1 One may also wish to find explicit lower-bounds on T_k to achieve almost sure convergence based on the convergence rates of the standard Q-learning with a single DM; we refer the reader to [242] for bounds on the convergence rates for standard Q-learning.

It is also possible to generalize such a result by considering better reply paths, instead of *strict best reply paths* [36].

The Case with Continuous Spaces

The above can be extended to the continuous state space setting. The algorithm incorporates quantization as in Sect. 21.4. Like the finite space variant, the algorithm operates on two timescales. On the finer timescale, each agent learns an estimate of the optimal value function for their current environment, composed of the underlying stochastic game and the policies of the other agents. On the coarser timescale, agents switch policies to an estimated best-reply based on their learned value function.

In the extension to the continuous space setting, each agent has a state quantizer, q^i which maps the state space $\mathbb{X}$ to a finite set of representative states $\mathbb{Y}^i$. The agents' quantizers need not be the same. Hence, while all agents observe the same global state, the Q-learning algorithm they each run will in general see different quantizations of the state. Each agent's quantization induces a finite quantized *baseline* policy set, $\hat{\Pi}^i_q := \{\hat{\pi} : \mathbb{Y}^i \to \mathbb{U}^i\}$. When choosing actions, agents follow a perturbation of their baseline policies: with probability ρ^i, player i chooses its action uniformly at random. We denote the perturbed policy set by $\hat{\Pi}^i_{q,\rho} = \{(1-\rho^i)\hat{\pi} + \rho^i \text{Uniform}(\mathbb{U}^i) : \hat{\pi} \in \hat{\Pi}^i_q\}$.

Each environment that an agent can see corresponds to some *perturbed* quantized joint policy in the set $\hat{\boldsymbol{\Pi}}_{q,\rho} := \times_{i\in\mathcal{N}} \hat{\Pi}^i_{q,\rho}$. The *baseline* quantized policy set of agent i, $\hat{\Pi}^i_q$, is the set of policies that the agent considers when determining a response to the environment it sees.

During each exploration phase, the agents use some perturbed quantized policies in their policy space. This induces a stationary MDP environment from the perspective of each agent. Each agent runs the quantized Q-learning algorithm described in Sect. 21.4 on the MDP it sees. Hence, for each 'environment' $\hat{\boldsymbol{\pi}}^{-i}_\rho \in \hat{\boldsymbol{\Pi}}^{-i}_{q,\rho}$, $\hat{Q}^{*i}_{\boldsymbol{\pi}^{-i}_\rho}$ induces a best-reply for agent i inside $\hat{\Pi}^i_q$, which is estimated using the learned Q^i_t. At the end of each exploration phase, each agent transitions to an estimated best-reply.

Assumption 23.3.3

(i) $\mathcal{T}$ *is weakly continuous in* $(x, \boldsymbol{u})$

(ii) $c^i : \mathbb{X} \times \boldsymbol{U} \to \mathbb{R}_+$ *is continuous and bounded for every* i.

(iii) Under every stationary joint policy $\boldsymbol{\gamma} \in \hat{\boldsymbol{\Pi}}_{q,\rho}$, *the state process is positive Harris recurrent with a unique invariant probability measure,* $\mu^*_{\boldsymbol{\gamma}}$

(iv) For each stationary joint policy $\boldsymbol{\gamma} \in \hat{\boldsymbol{\Pi}}_{q,\rho}$, *let* $\mathbb{P}^{\boldsymbol{\gamma}}$ *denote the probability measure associated with* $\boldsymbol{\gamma}$. *For every agent* $i \in \mathcal{N}$, *every observation-action pair* $(y^i, u^i) \in \mathbb{Y}^i \times \mathbb{U}^i$, *we suppose that* (y^i, u^i) *is visited infinitely often* $\mathbb{P}^{\boldsymbol{\gamma}}$-*almost surely.*

Algorithm 4: Continuous-space decentralized Q-learning algorithm (for agent i)

Set Parameters
q^i: agent i's quantizer of the state space $\mathbb{X}$; $\mathbb{Y}^i := q^i(\mathbb{X})$
$\mathbb{Q}^i_q$: some compact subset of $\mathbb{R}^{|\mathbb{Y}^i|\times|\mathbb{U}^i|}$; $\{T_k\}_{k\geq 0}$: sequence of exploration phase lengths in $\mathbb{Z}_{\geq 1}$
$\rho^i \in (0, 1)$: experimentation probability; $\delta^i \in (0, \infty)$: tolerance for suboptimality
Initialize $\pi^i_0 \in \hat{\Pi}^i_q$, $Q^i_0 \in \mathbb{Q}^i_q$ (arbitrary); **Receive** x_0
Iterate $k \geq 0$
(kth exploration phase)
Iterate $t = 1, \ldots, T_k$
Quantize state: $y^i_t = q^i(x_t)$
Choose action: $u^i_t = \begin{cases} \pi^i_k(y_t) & \text{w.p. } 1-\rho^i \\ \text{any } u^i \in \mathbb{U}^i & \text{w.p. } \rho^i \end{cases}$
Receive $c^i(x_t, \boldsymbol{u}_t)$
Receive $x_{t+1} \sim \mathcal{T}\left(\cdot \,\middle|\, x_t, u^i_t, \boldsymbol{u}^{-i}_t\right)$,
Quantize: $y^i_{t+1} = q^i(x_{t+1})$
$\alpha^i_t(y^i_t, u^i_t) = \left(1 + \sum_{s=t_k}^{t} \mathbb{1}\left\{(y^i_s, u^i_s) = (y^i_t, u^i_t)\right\}\right)^{-1}$
[ ]

$$
\begin{aligned}
Q^i_{t+1}(y^i_t, u^i_t) &= \left(1 - \alpha^i_t(y^i_t, u^i_t)\right) Q^i_t(y^i_t, u^i_t) \\
&\quad + \alpha^i_t(y^i_t, u^i_t)\left(c^i(x_t, \boldsymbol{u}_t) + \beta \min_{v^i \in \mathbb{U}^i} Q^i_t(y^i_{t+1}, v^i)\right) \\
Q^i_{t+1}(y^i, u^i) &= Q^i_t(y^i, u^i), \quad \forall (y^i, u^i) \neq (y^i_t, u^i_t)
\end{aligned}
$$

End
[ ]

$$
\widehat{\mathrm{BR}}^i_\delta(Q^i_t) = \left\{\hat{\gamma} \in \hat{\Pi}^i_q : Q^i_t(y, \hat{\gamma}(y)) \leq \min_{u \in \mathbb{U}^i} Q^i_t(y, u) + \delta^i, \ \forall y \in \mathbb{Y}^i_q\right\}
$$

If $\pi^i_k \in \widehat{\mathrm{BR}}^i_\delta(Q^i_t)$
$\pi^i_{k+1} = \pi^i_k$
Else
Choose $\pi^i_{k+1} \in \widehat{\mathrm{BR}}^i_\delta(Q^i_t)$
End
Reset Q^i_t to any $Q^i \in \mathbb{Q}^i_q$ (e.g.: $Q^i_t = 0$)
End

Theorem 23.3.2 ([18]) *Suppose all players use Algorithm 4 to select their actions. For any $\epsilon > 0$, there exists $\tilde{T}$ such that $T_k \geq \tilde{T}$ implies*

$$\mathbb{P}(\| Q^i_{T_k} - \hat{Q}^{*i}_{\pi^{-i}_{k,\rho}} \|_\infty < \epsilon) \geq 1 - \epsilon, \quad \forall k \geq 0, \tag{23.3}$$

where $\boldsymbol{\pi}_k$ is the baseline joint policy during the kth exploration phase and $\boldsymbol{\pi}_{k,\rho}$ is the perturbation of $\boldsymbol{\pi}_k$ that is used for action selection.

Furthermore, for any $\eta > 0$, there exists a fine-enough quantization q^i such that the greedy policy with respect to $\hat{Q}^{*i}_{\pi^{-i}_\rho}$ is η-optimal in the MDP environment where the other agents act according to π^{-i}_ρ.

Remark 23.3.2 Two extensions are as follows:

(i) Convergence to global optimality for stochastic teams with increasing memory: By slightly generalizing the algorithm above, to allow for the presence of past memory, we can also ensure high probability convergence to globally optimal policies [793]. We should note that best response dynamics are bound to get stuck at possibly undesirable equilibria, unless some memory is added.

(ii) Asynchronous Learning: One can also consider asynchronous learning where the analysis is applicable [795], where the DMs do not have to have identical policy update time instances. The main argument is the persistence of learning via constant step-sizes; that is, the learning coefficients α^i_n in Algorithm 3 do not depend on n. In this case, we can guarantee the asymptotic presence of equilibrium behavior with arbitrarily high probability.

In the following, we consider a more general class of stochastic games via a further application of the *two time scales and a Markov chain over play path graphs* paradigm.

23.4 Satisficing Paths and Independent Multi-Agent Reinforcement Learning with Possibly Randomized Policies

In this section, we introduce a different policy revision process, termed as *satisficing* [796].

Let us as in the previous section denote a finite stochastic game with discounted cost by

$$\mathcal{G} = (\mathcal{N}, \mathbb{X}, \{\mathbb{U}^i, c^i, \beta^i : i \in \mathcal{N}\}, P, \nu_0), \tag{23.4}$$

where $\mathcal{N}$ is a finite set of $N \in \mathbb{N}$ players/agents, $\mathbb{X}$ is a finite set of states; for agent $i \in \mathcal{N}$, $\mathbb{U}^i$ is a finite set of actions, and we write $\mathbf{U} := \times_{i \in \mathcal{N}} \mathbb{U}^i$. An element $\mathbf{u} \in \mathbf{U}$ is called a *joint action*. For agent i, $c^i : \mathbb{X} \times \mathbf{U} \to \mathbb{R}$ is a stage cost function, and $\beta^i \in [0, 1)$ is a discount factor. A random initial state $x_0 \in \mathbb{X}$ is given by $x_0 \sim \nu_0$ where $\nu_0 \in \mathcal{P}(\mathbb{X})$. State transitions are governed by the transition kernel $P \in \mathcal{P}(\mathbb{X}|\mathbb{X} \times \mathbf{U})$.

Consider a finite stochastic game as above. Each Agent $i \in \mathcal{N}$ uses a *policy* to select its sequence of actions $\{u_t^i\}_{t\geq 0}$, using only information that is locally available at the time of each decision. We use y_t^i to denote the observation variable for player i at time $t \geq 0$, and we let h_t^i denote player i's information variable at time t, according to which player i selects u_t^i.

Assumption 23.4.1 *For each $i \in \mathcal{N}$, player i's observation variables $\{y_t^i\}_{t\geq 0}$ and information variables $\{h_t^i\}_{t\geq 0}$ are given by*

$$y_0^i = x_0 \text{ and } y_{t+1}^i = (u_t^i, c^i(x_t, \mathbf{u}_t), x_{t+1}) \text{ for } t \geq 0;$$
$$h_0^i = y_0^i \text{ and } h_{t+1}^i = (h_t^i, y_{t+1}^i) \text{ for } t \geq 0;$$

In the following, let Γ_S^i denote a stationary policy for Agent i.

Definition 23.4.1 Let $i \in \mathcal{N}$, $\epsilon \geq 0$, and let $\Pi^i \subseteq \Gamma_S^i$. For $\boldsymbol{\pi}^{-i} \in \boldsymbol{\Gamma}_S^{-i}$, a policy $\pi^{*i} \in \Pi^i$ is called an ϵ-*best-response* to $\boldsymbol{\pi}^{-i}$ *over* Π^i if

$$J^i(\pi^{*i}, \boldsymbol{\pi}^{-i}, x) \leq \inf_{\pi^i \in \Pi^i} J^i(\pi^i, \boldsymbol{\pi}^{-i}, x) + \epsilon, \quad \forall x \in \mathbb{X}. \tag{23.5}$$

Definition 23.4.2 For fixed $i \in \mathcal{N}$, $\epsilon \geq 0$, $\Pi^i \subseteq \Gamma_S^i$, and $\boldsymbol{\pi}^{-i} \in \boldsymbol{\Gamma}_S^{-i}$, we let $\mathrm{BR}_\epsilon^i(\boldsymbol{\pi}^{-i}, \Pi^i)$ denote player i's (possibly empty) set of ϵ-best-responses to $\boldsymbol{\pi}^{-i}$ over Π^i.

Definition 23.4.3 Let $\epsilon \geq 0$. A joint policy $\boldsymbol{\pi}^* \in \boldsymbol{\Gamma}_S$ constitutes an ϵ-equilibrium if $\pi^{*i} \in \mathrm{BR}_\epsilon^i(\boldsymbol{\pi}^{*-i}, \Gamma_S^i)$ for every Agent $i \in \mathcal{N}$.

When $\epsilon = 0$, a 0-best-response is simply called a best-response and a 0-equilibrium is called an equilibrium. When the set Π^i over which i is optimizing is clear from context (typically, $\Pi^i = \Gamma_S^i$), we may omit "over Π^i" and simply write $\mathrm{BR}_\epsilon^i(\boldsymbol{\pi}^{-i})$.

Definition 23.4.4 Let $\epsilon \geq 0$ and let T^i be a policy update rule for Agent $i \in \mathcal{N}$. The policy update rule T^i is said to be ϵ-*satisficing* if, for any $\boldsymbol{\pi} = (\pi^i, \boldsymbol{\pi}^{-i}) \in \boldsymbol{\Gamma}_S$, we have that $\pi^i \in \mathrm{BR}_\epsilon^i(\boldsymbol{\pi}^{-i})$ implies $T^i(\boldsymbol{\pi}) = \pi^i$.

A policy revision process is called ϵ-satisficing if it is associated with a collection of policy update rules $\mathbf{T} = \{T^i\}_{i\in\mathcal{N}}$ such that T^i is ϵ-satisficing for each player $i \in \mathcal{N}$.

Definition 23.4.5 Let $\epsilon \geq 0$. A (possibly finite) sequence $(\boldsymbol{\pi})_{k\geq 0}$ of stationary joint policies is called an ϵ-*satisficing path* if, for every $k \geq 0$ and $i \in \mathcal{N}$, $\pi_k^i \in \mathrm{BR}_\epsilon^i(\boldsymbol{\pi}_k^{-i})$ implies $\pi_{k+1}^i = \pi_k^i$.

Definition 23.4.6 Let $\epsilon \geq 0$ and let $\mathbf{\Pi} \subseteq \mathbf{\Gamma}_S$ be a subset of stationary joint policies. A game $\mathcal{G}$ is said to have the ϵ-satisficing paths property within $\mathbf{\Pi}$ if for every $\boldsymbol{\pi} \in \mathbf{\Pi}$, there exists an ϵ-satisficing path $(\boldsymbol{\pi}_t)_{t\geq 0}$ and an integer $K = K(\boldsymbol{\pi})$, such that (i) $\boldsymbol{\pi}_0 = \boldsymbol{\pi}$, (ii) $\boldsymbol{\pi}_t \in \mathbf{\Pi}$ for every $t \geq 0$, and (iii) $\boldsymbol{\pi}_K \in \mathbf{\Gamma}_S^{\epsilon-eq}$ (which is the set of ϵ-equilibrium stationary policies).

In some applications, the environment being modelled exhibits symmetry in the agents, as we have seen earlier in the book. For such symmetric games: (1) each agent has the same set of actions; (2) the state dynamics depend only on the profile of actions taken by all players, without special dependence on the identities of the agents. That is, permuting the agents' actions in a joint action leaves the conditional probabilities for the next state unchanged; and (3) such a permutation results in a corresponding permutation of costs incurred. If $\mathbb{U}^i = \mathbb{U}^j$ for all $i, j \in \mathcal{N}$, given a permutation $\sigma : \mathcal{N} \to \mathcal{N}$ and joint action $\mathbf{a} = (a^i)_{i\in\mathcal{N}}$, we define $\sigma(\mathbf{a}) \in \mathbf{U}$ to be the joint action in which i's component is given by $a^{\sigma(i)}$. That is, player i's action in $\sigma(\mathbf{a})$ is given by player $\sigma(i)$'s action in $\mathbf{a}$: $\sigma(\mathbf{a})^i = a^{\sigma(i)}$.

Definition 23.4.7 (Symmetric Game) A stochastic game $\mathcal{G}$ given by (23.4) is called *symmetric* if the following holds:

(i) For any players $i, j \in \mathcal{N}$, we have $\beta^i = \beta^j$ and $\mathbb{U}^i = \mathbb{U}^j$;
(ii) For any $i \in \mathcal{N}$, permutation σ on $\mathcal{N}$, and $(\mathbf{s}, \mathbf{a}) \in \mathbb{X} \times \mathbf{U}$, we have $c^i(x, \sigma(\boldsymbol{a})) = c^{\sigma(i)}(x, \boldsymbol{a})$ and $P(\cdot|x, \boldsymbol{a}) = P(\cdot|x, \sigma(\boldsymbol{a}))$.

Lemma 23.4.1 *Let $\mathcal{G}$ be a symmetric game and let $\boldsymbol{\pi} \in \mathbf{\Gamma}_S$ be a stationary joint policy. For $i, j \in \mathcal{N}$, if $\pi^i = \pi^j$, then* $\mathbf{y}J^i(\pi^i, \boldsymbol{\pi}^{-i}, x) = J^j(\pi^j, \boldsymbol{\pi}^{-j}, x)$, *for any $x \in \mathbb{X}$.*

Corollary 23.4.1 *Let $\mathcal{G}$ be a symmetric game and let $\boldsymbol{\pi} \in \mathbf{\Gamma}_S$ be a joint policy. For $i, j \in \mathcal{N}$, if $\pi^i = \pi^j$, then $\pi^i \in \mathrm{BR}^i_\epsilon(\boldsymbol{\pi}^{-i}, \Gamma^i_S) \iff \pi^j \in \mathrm{BR}^j_\epsilon(\boldsymbol{\pi}^{-j}, \Gamma^j_S)$.*

We have two classes of games which exhibit the ϵ-satisficing paths property, symmetric games and games with two players:

Theorem 23.4.1 ([796]) *Let $\mathcal{G}$ be a symmetric stochastic game given by* (23.4). *Then $\mathcal{G}$ has the ϵ-satisficing paths property in $\mathbf{\Gamma}_S$ for all $\epsilon \geq 0$.*

Theorem 23.4.2 ([796]) *Let $\mathcal{G}$ be a symmetric game given by* (23.4) *and let $\epsilon \geq 0$. Let $\mathbf{\Pi} \subseteq \mathbf{\Gamma}_S$ be a subset of stationary joint policies satisfying $\Pi^i = \Pi^j$ for all $i, j \in \mathcal{N}$ and suppose $\mathbf{\Pi} \cap \mathbf{\Gamma}_S^{\epsilon-eq} \neq \varnothing$. Then, $\mathcal{G}$ has the ϵ-satisficing paths property within $\mathbf{\Pi}$.*

Theorem 23.4.3 ([796]) *Let $\mathcal{G}$ be a stochastic game given by* (23.4) *with $|\mathcal{N}| = 2$. Then, $\mathcal{G}$ has the ϵ-satisficing paths property within $\mathbf{\Gamma}_S$ for any $\epsilon \geq 0$.*

One can design independent learners that exploit the ϵ-satisficing paths property. Toward that end, we first consider the simplified setting where learning is black-boxed and players update their policies using an ϵ-satisficing rule that incorporates random search when not ϵ-best-responding. A complete independent learning algorithm suitable for online learning in symmetric games is then presented in the next section and analyzed as a two-timescale process.

Quantization of Stationary Policies The set of stationary policies is uncountable and, we can discretize this set using a mapping $q^i : \Gamma_S^i \to \Gamma_S^i$. In the following, let d^i be an appropriate metric on the spaces of policies Γ_S^i for which the expected cost is continuous on Γ_S^i under this metric (see (23.7) as well as Appendix F).

Definition 23.4.8 Let $\xi > 0$, $i \in \mathcal{N}$, and $\tilde{\Pi}^i \subseteq \Gamma_S^i$. A mapping $q^i : \tilde{\Pi}^i \to \Gamma_S^i$ is called a ξ-*quantizer* (on $\tilde{\Pi}^i$) if

(i) the set $q^i(\Gamma_S^i) = \{q^i(\pi^i) : \pi^i \in \Gamma_S^i\}$ is finite, and
(ii) For all $\pi^i \in \Gamma_S^i$, we have that $d^i(\pi^i, q^i(\pi^i)) < \xi$.

Definition 23.4.9 Let $i \in \mathcal{N}$, $\xi > 0$, and let $\tilde{\Pi}^i \subseteq \Gamma_S^i$. A set of policies $\Pi^i \subseteq \Gamma_S^i$ is called a ξ-*quantization* of $\tilde{\Pi}^i$ if $\Pi^i = q^i(\tilde{\Pi}^i)$ for some ξ-quantizer q^i.

For any $\epsilon > 0$, there exists $\xi > 0$ s.t. if $\mathbf{\Pi} \subset \mathbf{\Gamma}_S$ is a ξ-quantization of $\mathbf{\Gamma}_S$, then $\mathbf{\Pi} \cap \mathbf{\Gamma}_S^{\epsilon-eq} \neq \varnothing$. Thus, if the quantization is fine enough, it contains an ϵ-equilibrium.

Let $\mathcal{G}$ be a symmetric stochastic game. If $\mathbf{\Pi}$ is a sufficiently fine quantization of $\mathbf{\Gamma}_S$ and $\Pi^j = \Pi^i$ for all i, j, then $\mathcal{G}$ has the ϵ-satisficing paths property within $\mathbf{\Pi}$.

That is, quantization and satisficing are compatible. This will be useful for algorithm design, when the finiteness of the policy set can be exploited. In the following, we will consider a particular setup, though the analysis is applicable to a variety of problems.

23.5 Subjective Reinforcement Learning for Mean-Field Problems with General Information Structures

23.5.1 N-Player Weakly Coupled (Mean-Field) Games

We consider models with a possibly large but finite number of symmetric, weakly coupled agents (see, e.g. [627]).

For $N \in \mathbb{N}$, a partially observed N-player mean-field game (MFG) is described by the following list:

$$\mathcal{G} = (\mathcal{N}, X_{\text{loc}}, \mathbb{Y}, \mathbb{U}, \{\varphi^i\}_{i\in\mathcal{N}}, c, \beta, P_{\text{loc}}, \nu_0). \tag{23.6}$$

The list defining $\mathcal{G}$ is made up of the following components:

(i) $\mathcal{N}$ is a set of N players/agents;
(ii) X_{loc} is a finite set of states, and we let $\mathbf{X} := \times_{i\in\mathcal{N}} X_{\text{loc}}$. We refer to an element $s \in X_{\text{loc}}$ as a local state, and we refer to an element $(s^i)_{i\in\mathcal{N}} = \mathbf{s} \in \mathbf{X}$ as a

global state, with the ith component of $\mathbf{s}$ denoting player i's local state in global state $\mathbf{s}$. For each $\mathbf{s} \in \mathbf{X}$, we define an empirical measure $\mu(\cdot|\mathbf{s}) \in \mathcal{P}(X_{\text{loc}})$ as follows:

$$\mu(B|\mathbf{s}) = \frac{1}{N} \sum_{i \in \mathcal{N}} \delta_{s^i}(B), \quad \forall B \subseteq X_{\text{loc}},$$

(iii) and we denote the set of all empirical measures by $\text{Emp}_N := \{\mu(\cdot|\mathbf{s}) : \mathbf{s} \in \mathbf{X}\}$. An element $\mu \in \text{Emp}_N$ is called a *mean-field state*;
(iv) For each $i \in \mathcal{N}$, $\varphi^i : \mathbf{X} \to \mathbb{Y}$ is an observation function, where $\mathbb{Y}$ is a finite set of observation symbols. We refer to the pair $(\mathbb{Y}, \{\varphi^i\}_{i \in \mathcal{N}})$ as the observation channel;
(v) $\mathbb{U}$ is a finite set of actions, and we let $\mathbf{U} := \times_{i \in \mathcal{N}} \mathbb{U}$. An element $a \in \mathbb{U}$ is called an (individual) action, and an element $(a^i)_{i \in \mathcal{N}} = \mathbf{a} \in \mathbf{U}$ is called a *joint* action;
(vi) $c : X_{\text{loc}} \times \mathcal{P}(X_{\text{loc}}) \times \mathbb{U} \to \mathbb{R}$ is a stage cost function; $\beta \in (0, 1)$ is a discount factor;
(vii) $P_{\text{loc}} \in \mathcal{P}(X_{\text{loc}} | X_{\text{loc}} \times \mathcal{P}(X_{\text{loc}}) \times \mathbb{U})$ is a transition kernel governing local state transitions for each player;
(viii) $\nu_0 \in \mathcal{P}(\mathbf{X})$ is an initial probability distribution for the global state variable.

Play of the MFG $\mathcal{G}$ is described as follows: at time $t \in \mathbb{Z}_{\geq 0}$, player i's local state is denoted x^i_t, while the global state variable is denoted by $\mathbf{x}_t$ and the mean-field state is denoted by $\mu_t := \mu(\cdot|\mathbf{x}_t)$. Player i observes its local observation variable $y^i_t := \varphi^i(\mathbf{x}_t)$ and uses its locally observable history variable, defined below, to select an action $u^i_t \in \mathbb{U}$. The joint action random variable at time t is denoted $\mathbf{u}_t$. Player i then incurs a cost $c^i_t := c(x^i_t, \mu_t, u^i_t)$, and player i's local state variable evolves according to $x^i_{t+1} \sim P_{\text{loc}}(\cdot|x^i_t, \mu_t, u^i_t)$. This process is then repeated at time $t + 1$, and so on.

We now formalize the high-level description above in a manner similar to the formalization of POMDPs, with distinctions between the overall system history and each player's locally observable histories. For any $t \in \mathbb{Z}_{\geq 0}$, we define the sets

$$\mathbf{H}_t := (\mathbf{X} \times \mathbf{U})^t \times \mathbf{X}, \quad \text{and} \quad H_t := \mathcal{P}(\mathbf{X}) \times (\mathbb{Y} \times \mathbb{U} \times \mathbb{R})^t \times \mathbb{Y}.$$

For any $t \geq 0$, the set $\mathbf{H}_t$ represents the set of overall system histories of length t, while the set H_t is the set of histories of length t that an individual player in the game $\mathcal{G}$ may observe. Elements of $\mathbf{H}_t$ are called *system histories of length* t, and we use $\mathbf{h}_t = (\mathbf{x}_0, \mathbf{u}_0, \cdots, \mathbf{u}_{t-1}, \mathbf{x}_t)$, a random quantity taking values in $\mathbf{H}_t$, to denote the tth *system history variable*. Similarly, elements of H_t are called *observable histories of length* t, and for player $i \in \mathcal{N}$, we use $h^i_t = (\nu_0, y^i_0, u^i_0, c^i_0, \cdots, c^i_{t-1}, y^i_t)$, a random quantity taking values in H_t, to denote player i's tth *locally observable history variable*.

A policy for player $i \in \mathcal{N}$ is defined as a sequence $\pi^i = (\pi^i_t)_{t \geq 0}$ such that $\pi^i_t \in \mathcal{P}(\mathbb{U}|H_t)$ for every $t \geq 0$. We let Γ^i denote the set of all policies for player i.

Let $i \in \mathcal{N}$. A policy $\pi^i \in \Gamma^i$ is called *stationary* if there exists a transition kernel $f^i \in \mathcal{P}(\mathbb{U}|\mathbb{Y})$ such that for any $t \geq 0$ and any $\tilde{h}_t = (\nu, \tilde{y}_0, \ldots, \tilde{y}_t) \in H_t$, we have $\pi^i_t(\cdot|\tilde{h}_t) = f^i(\cdot|\tilde{y}_t)$. We let Γ^i_S denote the set of stationary policies for player i.

We let $\mathbf{\Gamma} := \times_{i\in\mathcal{N}} \mathbf{\Gamma}^i$ denote the set of *joint policies*. To isolate player i's component in a particular joint policy $\boldsymbol{\pi} \in \mathbf{\Gamma}$, we write $\boldsymbol{\pi} = (\pi^i, \boldsymbol{\pi}^{-i})$, where $-i$ is used in the agent index to represent all agents other than i. Similarly, we write the joint policy set as $\mathbf{\Gamma} = \Gamma^i \times \mathbf{\Gamma}^{-i}$, a joint action may be written as $\mathbf{a} = (a^i, \mathbf{a}^{-i}) \in \mathbf{U}$, and so on.

For each player $i \in \mathcal{N}$, we identify the set Γ^i_S with the set $\mathcal{P}(\mathbb{U}|\mathbb{Y})$ of transition kernels on $\mathbb{U}$ given $\mathbb{Y}$. When convenient, a stationary policy $\pi^i \in \Gamma^i_S$ is treated as if it were an element of $\mathcal{P}(\mathbb{U}|\mathbb{Y})$, and reference to the locally observable history variable is omitted. For each $i \in \mathcal{N}$, we introduce the metric d^i on Γ^i_S, defined by

$$d^i(\pi^i, \tilde{\pi}^i) := \max\{|\pi^i(a^i|y) - \tilde{\pi}^i(a^i|y)| : y \in \mathbb{Y}, a^i \in \mathbb{U}\}, \quad \forall \pi^i, \tilde{\pi}^i \in \Gamma^i_S. \tag{23.7}$$

We metrize the set of stationary joint policies $\mathbf{\Gamma}$ with a metric $\mathbf{d}$, defined as

$$\mathbf{d}(\boldsymbol{\pi}, \tilde{\boldsymbol{\pi}}) := \max_{i\in\mathcal{N}} d^i(\pi^i, \tilde{\pi}^i), \quad \forall \boldsymbol{\pi}, \tilde{\boldsymbol{\pi}} \in \mathbf{\Gamma}_S.$$

A metric $\mathbf{d}^{-i}$ for the set $\mathbf{\Gamma}^{-i}_S$ is defined analogously to $\mathbf{d}$. We have that the sets $\{\Gamma^i_S\}_{i\in\mathcal{N}}$, $\{\mathbf{\Gamma}^{-i}_S\}_{i\in\mathcal{N}}$ and $\mathbf{\Gamma}_S$ are all compact in the topologies induced by the corresponding metrics.

For any joint policy $\boldsymbol{\pi} = (\pi^i)_{i\in\mathcal{N}}$ and initial distribution $\nu \in \mathcal{P}(\mathbf{X})$, there exists a unique probability measure $\Pr^{\boldsymbol{\pi}}_\nu$ on trajectories in $(\mathbf{X} \times \mathbf{U})^\infty$.

We let $E^{\boldsymbol{\pi}}_\nu$ denote the expectation associated with $\Pr^{\boldsymbol{\pi}}_\nu$ and we use it to define player i's (state) value function:

$$J^i_{\boldsymbol{\pi}}(\nu) := E^{\boldsymbol{\pi}}_\nu[\sum_{t=0}^{\infty} \beta^t c(x^i_t, \mu_t, u^i_t)].$$

Definition 23.5.1 (Best-Response) Let $\epsilon \geq 0$, $\nu \in \mathcal{P}(\mathbf{X})$, $i \in \mathcal{N}$, and $\boldsymbol{\pi}^{-i} \in \mathbf{\Gamma}^{-i}$. A policy $\pi^{*i} \in \Gamma^i$ is called an *ϵ-best-response to $\boldsymbol{\pi}^{-i}$ with respect to ν* if

$$J^i_{(\pi^{*i}, \boldsymbol{\pi}^{-i})}(\nu) \leq \inf_{\tilde{\pi}^i \in \Gamma^i} J^i_{(\tilde{\pi}^i, \boldsymbol{\pi}^{-i})}(\nu) + \epsilon.$$

For $i \in \mathcal{N}$, $\boldsymbol{\pi}^{-i} \in \mathbf{\Gamma}^{-i}$, $\epsilon \geq 0$, and $\nu \in \mathcal{P}(\mathbf{X})$, we let $\mathrm{BR}^i_\epsilon(\boldsymbol{\pi}^{-i}, \nu) \subseteq \Gamma^i$ denote player i's set of ϵ-best-responses to $\boldsymbol{\pi}^{-i}$ with respect to ν. If, additionally, $\pi^i \in \mathrm{BR}^i_\epsilon(\boldsymbol{\pi}^{-i}, \nu)$ for all $\nu \in \mathcal{P}(\mathbf{X})$, then π^i is called a *uniform ϵ-best-response to $\boldsymbol{\pi}^{-i}$*. The set of uniform ϵ-best-responses to a policy $\boldsymbol{\pi}^{-i}$ is denoted $\mathrm{BR}^i_\epsilon(\boldsymbol{\pi}^{-i})$.

Definition 23.5.2 (Equilibrium) Let $\epsilon \geq 0$, $\nu \in \mathcal{P}(\mathbf{X})$, and $\boldsymbol{\pi}^* \in \boldsymbol{\Gamma}$. The joint policy $\boldsymbol{\pi}^*$ is called an *ϵt-equilibrium with respect to* ν if π^{*i} is an ϵ-best-response to $\boldsymbol{\pi}^{*-i}$ with respect to ν for every player $i \in \mathcal{N}$. Additionally, if $\boldsymbol{\pi}^* \in \boldsymbol{\Gamma}$ is an ϵ-equilibrium with respect to every $\nu \in \mathcal{P}(\mathbf{X})$, then $\boldsymbol{\pi}^*$ is called a *perfect* ϵ-*equilibrium.*

For $\epsilon \geq 0$ and $\nu \in \mathcal{P}(\mathbf{X})$, we let $\boldsymbol{\Gamma}^{\epsilon-eq}(\nu) \subset \boldsymbol{\Gamma}$ denote the set of ϵ-equilibrium policies with respect to ν, and we let $\boldsymbol{\Gamma}^{\epsilon-eq}$ denote the set of perfect ϵ-equilibrium policies. Furthermore, we let $\boldsymbol{\Gamma}_S^{\epsilon-eq}(\nu) := \boldsymbol{\Gamma}^{\epsilon-eq}(\nu) \cap \boldsymbol{\Gamma}_S$ for each $\nu \in \mathcal{P}(\mathbf{X})$ and we let $\boldsymbol{\Gamma}_S^{\epsilon-eq} := \boldsymbol{\Gamma}^{\epsilon-eq} \cap \boldsymbol{\Gamma}_S$. In the next section, we will describe conditions under which

$$\boldsymbol{\Gamma}_S^{\epsilon-eq} \neq \varnothing, \tag{23.8}$$

and we will state criteria for verifying whether a particular stationary policy $\pi^i \in \Gamma_S^i$ is an ϵ-best-response to the stationary joint policy $\boldsymbol{\pi}^{-i} \in \boldsymbol{\Gamma}_S^{-i}$.

Definition 23.5.3 (Subjective Function Family) [794] Let $\mathcal{G}$ be the N-player MFG in (23.6). Let $\mathcal{V} = \{V_{\boldsymbol{\pi}}^i : \mathbb{Y} \to \mathbb{R} | i \in \mathcal{N}, \boldsymbol{\pi} \in \boldsymbol{\Gamma}_S\}$ and let $\mathcal{W} = \{W_{\boldsymbol{\pi}}^i : \mathbb{Y} \times \mathbb{U} \to \mathbb{R} | i \in \mathcal{N}, \boldsymbol{\pi} \in \boldsymbol{\Gamma}_S\}$ be two families of functions. Then, the pair $(\mathcal{V}, \mathcal{W})$ is called a *subjective function family for* $\mathcal{G}$.

Definition 23.5.4 (Subjective Best-Responding) [794] Let $\epsilon \geq 0$, $\boldsymbol{\pi}^{-i} \in \boldsymbol{\Gamma}_S^{-i}$, and let $(\mathcal{V}, \mathcal{W})$ be a subjective function family for $\mathcal{G}$. A policy $\pi^{*i} \in \Gamma_S^i$ is called a $(\mathcal{V}, \mathcal{W})$-subjective ϵ-best-response to $\boldsymbol{\pi}^{-i}$ if we have

$$V_{(\pi^{*i}, \boldsymbol{\pi}^{-i})}^i(y) \leq \min_{a^i \in \mathbb{U}} W_{(\pi^{*i}, \boldsymbol{\pi}^{-i})}^i(y, a^i) + \epsilon, \quad \forall y \in \mathbb{Y}.$$

For our learning algorithm to be presented later, W^i will represent the subjective Q functions and V^i will represent the perceived value functions.

For fixed $i \in \mathcal{N}$, stationary $\boldsymbol{\pi}^{-i} \in \boldsymbol{\Gamma}_S^{-i}$, and subjective function family $(\mathcal{V}, \mathcal{W})$, we let

$$\text{Subj-BR}_\epsilon^i(\boldsymbol{\pi}^{-i}, \mathcal{V}, \mathcal{W}) \subseteq \Gamma_S^i$$

denote player i's (possibly empty) set of $(\mathcal{V}, \mathcal{W})$-subjective ϵ-best-responses to $\boldsymbol{\pi}^{-i}$.

Definition 23.5.5 (Subjective Equilibrium) [794] Let $\epsilon \geq 0$ and let $(\mathcal{V}, \mathcal{W})$ be a subjective function family for $\mathcal{G}$. A joint policy $\boldsymbol{\pi}^* \in \boldsymbol{\Gamma}_S$ is called a $(\mathcal{V}, \mathcal{W})$-subjective ϵ-equilibrium for $\mathcal{G}$ if, for every $i \in \mathcal{N}$, π^{*i} is a $(\mathcal{V}, \mathcal{W})$-subjective ϵ-best-response to $\boldsymbol{\pi}^{*-i}$.

For any subjective function family $(\mathcal{V}, \mathcal{W})$, we let $\text{Subj}_\epsilon(\mathcal{V}, \mathcal{W}) \subseteq \boldsymbol{\Gamma}_S$ denote the (possibly empty) set of $(\mathcal{V}, \mathcal{W})$-subjective ϵ-equilibrium policies for $\mathcal{G}$.

23.5.2 On the Observation Channel

We have left the particular observation channel $(\mathbb{Y}, \{\varphi^i\}_{i \in \mathcal{N}})$ unspecified. We consider four alternatives for the observation channel. The particular choice used in practice will depend on the application area: in some instances, there will be a natural restriction of information leading to a particular observation channel. In other instances, information may be plentiful in principle but agents may voluntarily compress a larger/more informative observation variable for the purposes of function approximation.

Assumption 23.5.1 (Global State Observability) $\mathbb{Y} = \mathbf{X}$ *and* $\varphi^i(\mathbf{s}) = \mathbf{s}$ *for each global state* $\mathbf{s} \in \mathbf{X}$ *and player* $i \in \mathcal{N}$.

Assumption 23.5.2 (Mean-Field State Observability) $\mathbb{Y} = X_{\text{loc}} \times \text{Emp}_N$ *and* $\varphi^i(\mathbf{s}) = (s^i, \mu(\cdot|\mathbf{s}))$ *for each global state* $\mathbf{s} \in \mathbf{X}$ *and player* $i \in \mathcal{N}$.

Assumption 23.5.3 (Compressed State Observability) *For* $k \in \mathbb{N}$, *let* $[k] := \{1, 2, \ldots, k\}$ *and let* $f : \mathcal{P}(X_{\text{loc}}) \to [k]$. *Then,* $\mathbb{Y} = X_{\text{loc}} \times [k]$ *and for each* $i \in \mathcal{N}$, $\mathbf{s} \in \mathbf{X}$, *we have* $\varphi^i(\mathbf{s}) = (s^i, f(\mu(\cdot|\mathbf{s}))$.

Assumption 23.5.4 (Local State Observability) $\mathbb{Y} = X_{\text{loc}}$ *and for each* $i \in \mathcal{N}$, $\mathbf{s} \in \mathbf{X}$, *we have* $\varphi^i(\mathbf{s}) = s^i$.

The mean-field state observability assumption of Assumption 23.5.2 is the standard observation channel considered in works on mean-field games, see e.g. [627] and the references therein.

Remark 23.5.1 By taking $k = 1$, we see that local state observability is in fact a special case of compressed state observability, where the compressed information about the mean-field state is uninformative. We include Assumption 23.5.4 separately to highlight the importance of this set-up, even though mathematically all results under Assumption 23.5.4 will automatically follow from those involving Assumption 23.5.3.

23.5.3 Stationary Equilibrium Policies: Existence Results

We now present some results relating partially observed N-player mean-field games to MDPs and POMDPs. We then leverage these connections to present results on the existence of stationary equilibrium policies.

Lemma 23.5.1 *Let* $\mathcal{G}$ *be a partially observed* N*-player mean-field game. Fix* $i \in \mathcal{N}$ *and let* $\boldsymbol{\pi}^{-i} \in \boldsymbol{\Gamma}_S^{-i}$ *be a stationary policy for the remaining players. Then, player* i *faces a partially observed Markov decision problem* $M_{\boldsymbol{\pi}^{-i}}$ *with partially observed state process* $\{\mathbf{x}_t\}_{t \geq 0}$.

Lemma 23.5.1 gives conditions under which a player faces a POMDP. Under certain additional conditions, such as those presented in Corollary 23.5.1 and in Lemma 23.5.6, one can show that player $i \in \mathcal{N}$ faces a fully observed MDP. When player i faces an MDP in its observation variable, the classical theory of MDPs and reinforcement learning can be brought to bear on player i's optimization problem, leading to results on the existence of certain equilibrium policies and characterization of one's best-response set.

23.5.4 Existence of Stationary Equilibrium Under Global State Observability

Corollary 23.5.1 ([794]) *Let $\mathcal{G}$ be a partially observed N-player MFG in which Assumption 23.5.1 holds. Fix player $i \in \mathcal{N}$ and let $\boldsymbol{\pi}^{-i} \in \boldsymbol{\Gamma}_S^{-i}$ be a stationary policy for the remaining players. Then, player i faces a (fully observed) Markov decision problem $M_{\boldsymbol{\pi}^{-i}}$ with controlled state process $\{y_t^i\}_{t\geq 0}$, where $y_t^i = \varphi^i(\mathbf{x}_t) = \mathbf{x}_t$ for every $t \geq 0$.*

Under Assumption 23.5.1, Lemma 23.5.1 immediately yields Corollary 23.5.1, which says that if $\boldsymbol{\pi}^{-i}$ is stationary, then player i faces a multi-agent environment that is equivalent to a single-agent MDP $M_{\boldsymbol{\pi}^{-i}}$. As such, we can consider player i's Q-function for this environment, which we denote by $Q^{*i}_{\boldsymbol{\pi}^{-i}} : \mathbf{X} \times \mathbb{U} \to \mathbb{R}$.

$$Q^{*i}_{\boldsymbol{\pi}^{-i}}\left(\mathbf{s}, a^i\right) := E^{\boldsymbol{\pi}^*}_{\nu_0}[\sum_{t=0}^{\infty} \beta^t c(x_0^i, \mu(\cdot|\mathbf{x}_t), u_t^i)|\mathbf{x}_0 = \mathbf{s}, u_0^i = a^i],$$

for each $(\mathbf{s}, a^i) \in \mathbf{X} \times \mathbb{U}$, where $\boldsymbol{\pi}^* = (\pi^{*i}, \boldsymbol{\pi}^{-i})$ and $\pi^{*i} \in \Gamma_S^i \cap \mathrm{BR}_0^i(\boldsymbol{\pi}^{-i})$.

The value $Q^{*i}_{\boldsymbol{\pi}^{-i}}(\mathbf{s}, a^i)$ represents the *optimal cost-to-go* to player i when play begins at global state $\mathbf{s} \in \mathbf{X}$, player i takes action $a^i \in \mathbb{U}$ at time 0 and follows the policy π^{*i} thereafter, while the remaining players play according to the stationary policy $\boldsymbol{\pi}^{-i}$ at all times.

Player i can verify whether a given stationary policy $\pi^i \in \Gamma_S^i$ is in $\mathrm{BR}_\epsilon^i(\boldsymbol{\pi}^{-i})$ by verifying whether $\max_{\mathbf{s}\in\mathbf{X}}(J^i_{(\pi^i, \boldsymbol{\pi}^{-i})}(\mathbf{s}) - \min_{a^i\in\mathbb{U}} Q^{*i}_{\boldsymbol{\pi}^{-i}}(\mathbf{s}, a^i)) \leq \epsilon$. We use the following lemmas, whose proofs are mechanical and may be found in [796].

Lemma 23.5.2 *Let $\mathcal{G}$ be a partially observed N-player MFG satisfying Assumption 23.5.1. Let $i \in \mathcal{N}$ and $(\mathbf{s}, a^i) \in \mathbf{X} \times \mathbb{U}$. The mapping $\boldsymbol{\pi}^{-i} \mapsto Q^{*i}_{\boldsymbol{\pi}^{-i}}(\mathbf{s}, a^i)$ is continuous on $\boldsymbol{\Gamma}_S^{-i}$.*

Lemma 23.5.3 *Let $\mathcal{G}$ be a partially observed N-player MFG satisfying Assumption 23.5.1. Let $i \in \mathcal{N}$ and $\mathbf{s} \in \mathbf{X}$. Then, the mapping $\boldsymbol{\pi}^{-i} \mapsto \min_{a^i\in\mathbb{U}} Q^{*i}_{\boldsymbol{\pi}^{-i}}(\mathbf{s}, a^i)$ is continuous on $\boldsymbol{\Gamma}_S^{-i}$.*

Lemma 23.5.4 *Let $\mathcal{G}$ be a partially observed N-player MFG satisfying Assumption 23.5.1. For any player $i \in \mathcal{N}$ and $\boldsymbol{\pi}^{-i} \in \boldsymbol{\Gamma}_S^{-i}$, the mapping*

$$\pi^i \mapsto \max_{\mathbf{s} \in \mathbf{X}} \left(J^i_{(\pi^i, \boldsymbol{\pi}^{-i})}(\mathbf{s}) - \min_{a^i \in \mathbb{U}} Q^{*i}_{\boldsymbol{\pi}^{-i}}(\mathbf{s}, a^i) \right)$$

is continuous on Γ^i_S.

Lemma 23.5.5 ([794]) *Let $\mathcal{G}$ be a partially observed N-player MFG satisfying Assumption 23.5.1. Then, there exists a stationary policy that is a perfect equilibrium. That is, (23.8) holds with $\epsilon = 0$.*

A partially observed N-player mean-field game with global state observability (Assumption 23.5.1) is a special case of the finite N-player stochastic games studied by Fink et al. [263], and thus Lemma 23.5.5 follows from [263, Theorem 2]: Let us first refer the reader to Sect. 10.5 on existence conditions for equilibria. Accordingly, the proof of Lemma 23.5.5 involves invoking Kakutani-Fan-Glicksberg's fixed point theorem (Theorem 10.5.2) on a product best-response mapping from $\boldsymbol{\Gamma}_S$ to its power set, where the ith component maps $\boldsymbol{\pi} \mapsto \text{BR}^i_0(\boldsymbol{\pi}^{-i}) \cap \Gamma^i_S$ for each $\boldsymbol{\pi} \in \boldsymbol{\Gamma}_S$. By Corollary 23.5.1, one sees that each component mapping maps to non-empty, convex, and compact sets. The upper semicontinuity of the component mappings can be established via Lemmas 23.5.2–23.5.4, above.

From the proof sketch for Lemma 23.5.5, one can see the crucial role played by the MDP structure facing a given player when the remaining players follow a stationary policy. When Assumption 23.5.1 does not hold and the observation channel compresses the global state information, in general player i will not face an MDP with controlled state process $\{y^i_t\}_{t \geq 0}$, and as a result replicating this line of proof is not possible for general observation channels $(\mathbb{Y}, \{\varphi^i\}_{i \in \mathcal{N}})$. However, as we discuss below, it is possible to employ the same proof technique in the special case of mean-field state observability (Assumption 23.5.2), although additional care must be given to account for the loss of global state observability.

23.5.5 Existence of Stationary Equilibrium Under Mean-Field State Observability

For the definition below, let $\mathcal{G}$ be a partially observed N-player mean-field game with mean-field state observability (i.e. let Assumption 23.5.2 hold).

Definition 23.5.6 (Mean-Field Symmetric Policies) Let $i, j \in \mathcal{N}$ and let $\pi^i \in \Gamma^i_S, \pi^j \in \Gamma^j_S$ be stationary policies. We say that the policies π^i and π^j are *mean-field symmetric* if both are identified with the same transition kernel in $\mathcal{P}(\mathbb{U}|\mathbb{Y})$. For any subset of players $I \subset \mathcal{N}$, a collection of policies $\{\pi^i\}_{i \in I}$ is called mean-field symmetric if, for every $i, j \in I$, we have that π^i and π^j are mean-field symmetric.

Lemma 23.5.6 *Let $\mathcal{G}$ be a partially observed N-player MFG, let $i \in \mathcal{N}$, and let Assumption 23.5.2 hold. If $\boldsymbol{\pi}^{-i} \in \boldsymbol{\Gamma}_S^{-i}$ is mean-field symmetric, then i faces a fully observed MDP $M_{\boldsymbol{\pi}^{-i}}$ with controlled state process $\{y_t^i\}_{t\geq 0}$, where $y_t^i = \varphi^i(\mathbf{x}_t) = (x_t^i, \mu(\cdot|\mathbf{x}_t))$ for all $t \geq 0$.*

Lemma 23.5.6 is proved in [794]. From this proof, one observes two things: first, the condition that $\boldsymbol{\pi}^{-i}$ is mean-field symmetric cannot be relaxed in general; second, if $\boldsymbol{\pi}^{-i}$ is mean-field symmetric and $\nu, \nu' \in \mathcal{P}(\mathbf{X})$ are arbitrary initial distributions, then for any policy $\pi^i \in \Gamma^i$, putting $\boldsymbol{\pi} = (\pi^i, \boldsymbol{\pi}^{-i})$, we have

$$\Pr_{\nu}^{\boldsymbol{\pi}}(\cdot|y_0^i = y, u_0^i = a^i) = \Pr_{\nu'}^{\boldsymbol{\pi}}(\cdot|y_0^i = y, u_0^i = a^i), \quad \forall (y, a^i) \in \mathbb{Y} \times \mathbb{U}. \tag{23.9}$$

In light of Lemma 23.5.6, we define the Q-function for player i when playing $\mathcal{G}$ against a mean-field symmetric policy $\boldsymbol{\pi}^{-i} \in \boldsymbol{\Gamma}_S^{-i}$ as

$$Q^{*i}_{\boldsymbol{\pi}^{-i}}(y, a^i) := E_{\nu}^{(\pi^{*i}, \boldsymbol{\pi}^{-i})}\left[\sum_{t=0}^{\infty} \beta^t c(x_t^i, \mu(\cdot|\mathbf{x}_t), u_t^i) \middle| y_0^i = y, u_0^i = a^i\right], \quad \forall a^i \in \mathbb{U},$$

for every $y \in \varphi^i(\mathbf{X}) = \{\varphi^i(\mathbf{s}) : \mathbf{s} \in \mathbf{X}\}$, where $\pi^{*i} \in \mathrm{BR}_0^i(\boldsymbol{\pi}^{-i}) \cap \Gamma_S^i$ is a best-response to $\boldsymbol{\pi}^{-i}$ and $\nu \in \mathcal{P}(\mathbf{X})$ is arbitrary. (That ν can be arbitrarily chosen follows from the preceding discussion culminating in (23.9).) For elements $y \in \mathbf{X} \setminus \varphi^i(\mathbf{X})$, we may define $Q^{*i}_{\boldsymbol{\pi}^{-i}}(y, a^i)$ arbitrarily, say $Q^{*i}_{\boldsymbol{\pi}^{-i}}(y, \cdot) \equiv 0$.

For any player $i \in \mathcal{N}$, we let $\boldsymbol{\Gamma}_{S,\mathrm{sym}}^{-i} \subset \boldsymbol{\Gamma}_S^{-i}$ denote the set of mean-field symmetric joint policies for the remaining players, and we let $\boldsymbol{\Gamma}_{S,\mathrm{sym}} \subset \boldsymbol{\Gamma}_S$ denote the set of stationary joint policies that are mean-field symmetric. We note that the sets Γ_S^i and $\boldsymbol{\Gamma}_{S,\mathrm{sym}}^{-i}$ are in bijection, and we define $\mathrm{sym}^i : \Gamma_S^i \to \boldsymbol{\Gamma}_{S,\mathrm{sym}}^{-i}$ by $\mathrm{sym}^i(\pi^i) = (\pi^i)_{j \in \mathcal{N}\setminus\{i\}}$, for all $\pi^i \in \Gamma_S^i$.

We metrize $\boldsymbol{\Gamma}_{S,\mathrm{sym}}^{-i}$ using the metric d^i on Γ_S^i:

$$\mathbf{d}_{\mathrm{sym}}^{-i}\left(\mathrm{sym}^i(\pi^i), \mathrm{sym}^i(\tilde{\pi}^i)\right) := d^i(\pi^i, \tilde{\pi}^i), \quad \forall \pi^i, \tilde{\pi}^i \in \Gamma_S^i.$$

We note that the metric $\mathbf{d}_{\mathrm{sym}}^{-i}$ is equivalent to the metric $\mathbf{d}^{-i}$ restricted to $\boldsymbol{\Gamma}_{S,\mathrm{sym}}^{-i}$. In the lemmas below—whose proofs resemble those of Lemmas 23.5.2–23.5.4—continuity of functions with domain $\boldsymbol{\Gamma}_{S,\mathrm{sym}}^{-i}$ is with respect to the topology induced by $\mathbf{d}_{\mathrm{sym}}^{-i}$.

Lemma 23.5.7 *Let $\mathcal{G}$ be a partially observed N-player MFG satisfying Assumption 23.5.2. Fix $i \in \mathcal{N}$, $(y, a^i) \in \mathbb{Y} \times \mathbb{U}$. The mapping $\boldsymbol{\pi}^{-i} \mapsto Q^{*i}_{\boldsymbol{\pi}^{-i}}(y, a^i)$ is continuous on $\boldsymbol{\Gamma}_{S,\mathrm{sym}}^{-i}$.*

Lemma 23.5.8 *Let $\mathcal{G}$ be a partially observed N-player MFG satisfying Assumption 23.5.2. Fix $i \in \mathcal{N}$, $y \in \mathbb{Y}$. The mapping $\boldsymbol{\pi}^{-i} \mapsto \min_{a^i \in \mathbb{U}} Q^{*i}_{\boldsymbol{\pi}^{-i}}(y, a^i)$ is continuous on $\boldsymbol{\Gamma}_{S,\mathrm{sym}}^{-i}$.*

Lemma 23.5.9 *Let $\mathcal{G}$ be a partially observed N-player MFG satisfying Assumption 23.5.2. Fix $i \in \mathcal{N}$ and $\boldsymbol{\pi}^{-i} \in \boldsymbol{\Gamma}_{S,\mathrm{sym}}^{-i}$. Then, the mapping*

$$\pi^i \mapsto \max_{s \in \mathbf{X}} \left(J^i_{(\pi^i, \boldsymbol{\pi}^{-i})}(\mathbf{s}) - \min_{a^i \in \mathbb{U}} Q^{*i}_{\boldsymbol{\pi}^{-i}}(\varphi^i(\mathbf{s}), a^i) \right)$$

is continuous on Γ^i_S.

The above will lead to the following theorem, which states that under mean-field state observability a stationary perfect equilibrium exists for the partially observed N-player mean-field game $\mathcal{G}$.

Theorem 23.5.1 ([794]) *Let $\mathcal{G}$ be a partially observed N-player MFG satisfying Assumption 23.5.2. For any $\epsilon \geq 0$, there exists a perfect ϵ-equilibrium policy in $\boldsymbol{\Gamma}_S$, that is (23.8) holds.*

Proof Fix player $i \in \mathcal{N}$. We define a point-to-set mapping $B^i : \Gamma^i_S \to 2^{\Gamma^i_S}$ with $B^i(\pi^i) = \mathrm{BR}^i_\epsilon(\mathrm{sym}^i(\pi^i)) \cap \Gamma^i_S$ for every $\pi^i \in \Gamma^i_S$. By Lemma 23.5.6, we have that $B^i(\pi^i) \subseteq \Gamma^i_S$ is non-empty, convex, and compact for each $\pi^i \in \Gamma^i_S$.

If $(\mathrm{sym}^i(\tilde{\pi}^i_k))_{k \geq 0}$ is a sequence of mean-field symmetric joint policies in $\boldsymbol{\Gamma}^{-i}_{S,\mathrm{sym}}$ converging to some $\mathrm{sym}^i(\tilde{\pi}^i_\infty) \in \boldsymbol{\Gamma}^{-i}_{S,\mathrm{sym}}$ and $(\pi^i_k)_{k \geq 0}$ is a sequence in Γ^i_S such that (1) $\lim_{k\to\infty} \pi^i_k = \pi^i_\infty \in \Gamma^i_S$ and (2) for every $k \geq 0$, we have that $\pi^i_k \in \mathrm{BR}^i_\epsilon(\mathrm{sym}^i(\tilde{\pi}^i_k))$, then one can use Lemmas 23.5.7–23.5.9 to conclude that $\pi^i_\infty \in \mathrm{BR}^i_\epsilon(\mathrm{sym}^i(\tilde{\pi}^i_\infty)) \cap \Gamma^i_S = B^i(\tilde{\pi}^i_\infty)$. This implies that the point-to-set mapping B^i is upper hemicontinuous.

We invoke Kakutani's fixed point theorem on B^i to obtain a fixed point $\pi^{*i} \in B^i(\pi^{*i}) = \mathrm{BR}^i_\epsilon(\mathrm{sym}^i(\pi^{*i})) \cap \Gamma^i_S$. It follows by symmetry that $\left(\pi^{*i}, \mathrm{sym}^i(\pi^{*i})\right) \in \Gamma^i_S \times \boldsymbol{\Gamma}^{-i}_{S,\mathrm{sym}} \subset \boldsymbol{\Gamma}_S$ is a perfect ϵ-equilibrium for $\mathcal{G}$. ◇

Remark (Compressed State Observability and Stationary Equilibrium) Under compressed state observability (Assumption 23.5.3), it is generally not true that a given player $i \in \mathcal{N}$ faces a fully observed MDP with state process $\{y^i_t\}_{t \geq 0}$ when the remaining players follow a stationary policy. In fact, partial analogs of Corollary 23.5.1 and Lemma 23.5.6 do not generally hold even when restricting the remaining players to use a policy in a small subset of the stationary policies. Consequently, it cannot be guaranteed that any stationary policy for player $i \in \mathcal{N}$ is an ϵ-best-response to a stationary joint policy for the remaining players. Moreover, the proof technique of Theorem 23.5.1 cannot be used, as the required regularity properties of the best-response correspondence may fail to hold. Accordingly, we treat this set-up differently in our results below. For a result on existence of objective ϵ-equilibrium under Assumption 23.5.4 as well as further discussion, we refer the reader to [794].

23.5.6 Convergence of Subjective Single-Agent Learning Under Stationary Policies

In this section, we study the convergence of learning iterates obtained when an independent learning agent, player $i \in \mathcal{N}$ in our mean-field game $\mathcal{G}$, naively runs single-agent reinforcement learning algorithms that treat its observation variable $\{y_t^i\}_{t\geq 0}$ as though it were the state variable of a MDP. This learning process is formalized in Algorithm 5, below, where we have fixed the policies of all players to be stationary. By fixing the policies of the remaining players $-i$ to be stationary, this section focuses on the effect that decentralized information has on independent learners, leaving aside the well-known challenge of non-stationarity.

We begin with an assumption on the transition kernel P_{loc}. For intuitive simplicity, we state this assumption in terms of the underlying state process.

Assumption 23.5.5 *Under any soft stationary policy $\boldsymbol{\pi} \in \boldsymbol{\Gamma}_S$ and initial distribution $\nu \in \mathcal{P}(\mathbf{X})$, the global state process $\{\mathbf{x}_t\}_{t\geq 0}$ is an irreducible, aperiodic Markov chain on $\mathbf{X}$.*

Theorem 23.5.2 ([794]) *Let $\mathcal{G}$ be a partially observed N-player mean-field game for which Assumption 23.5.5 holds, and let $\boldsymbol{\pi} \in \boldsymbol{\Gamma}_S$ be soft. Suppose that Agent $i \in \mathcal{N}$ uses Algorithm 5. We have the following:*

(i) There exist deterministic functions $\tilde{V}_{\boldsymbol{\pi}}^{*i} : \mathbb{Y} \to \mathbb{R}$ and $\tilde{W}_{\boldsymbol{\pi}}^{*i} : \mathbb{Y} \times \mathbb{U} \to \mathbb{R}$ such that, for any $\nu \in \mathcal{P}(\mathbf{X})$,

$$\lim_{t\to\infty} \bar{J}_t^i = \tilde{V}_{\boldsymbol{\pi}}^{*i} \quad \textit{and} \quad \lim_{t\to\infty} \bar{Q}_t^i = \tilde{W}_{\boldsymbol{\pi}}^{*i} \quad \mathrm{Pr}_\nu^{\boldsymbol{\pi}}\textit{-almost surely.}$$

(ii) If Assumption 23.5.1 holds, then $\tilde{V}_{\boldsymbol{\pi}}^{*i}(\mathbf{s}) = J_{\boldsymbol{\pi}}^i(\mathbf{s})$ for all $\mathbf{s} \in \mathbf{X}$ and $\tilde{W}_{\boldsymbol{\pi}}^{*i} = Q_{\boldsymbol{\pi}^{-i}}^{*i}$.

(iii) If Assumption 23.5.2 holds and $\boldsymbol{\pi}^{-i}$ is mean-field symmetric, then $\tilde{V}_{\boldsymbol{\pi}}^{*i}(\varphi^i(\mathbf{s})) = J_{\boldsymbol{\pi}}^i(\mathbf{s})$ for all $\mathbf{s} \in \mathbf{X}$ and $\tilde{W}_{\boldsymbol{\pi}}^{*i} = Q_{\boldsymbol{\pi}^{-i}}^{*i}$.

Proof By Lemma 23.5.1, player i faces a POMDP with observation sequence $\{y_t^i\}_{t\geq 0}$ and underlying state process $\{\mathbf{x}_t\}_{t\geq 0}$. By Assumption 23.5.5 and the assumed softness of $\boldsymbol{\pi}$, we have that all pairs $(\mathbf{s}, a^i) \in \mathbf{X} \times \mathbb{U}$ are visited infinitely often $\mathrm{Pr}_\nu^{\boldsymbol{\pi}}$-almost surely for any $\nu \in \mathcal{P}(\mathbf{X})$. We can therefore invoke Theorem 22.3.4 (see [391, Theorem 4.1]) to establish the almost sure convergence of Q-factor iterates $\{\bar{Q}_t^i\}_{t\geq 0}$. The same analysis can be used to establish the convergence of the value function iterates $\{\bar{J}_t^i\}_{t\geq 0}$. This proves the first part.

By Corollary 23.5.1, under Assumption 23.5.1, player i faces an MDP with state process $\{\mathbf{x}_t\}_{t\geq 0}$. Again by the softness of $\boldsymbol{\pi}$ and Assumption 23.5.5, each $(\mathbf{s}, a^i) \in \mathbf{X}\times\mathbb{U}$ is visited infinitely often almost surely. The convergence of Q-function iterates then follow as we studied earlier in Chap. 21. The same analysis can be used to prove that $\lim_{t\to\infty} \bar{J}_t^i(\mathbf{s}) = J_{\boldsymbol{\pi}}^i(\mathbf{s})$ almost surely for each $\mathbf{s} \in \mathbf{X}$, proving the second part.

Algorithm 5: Naive (Subjective Q-Learning) Learning in a partially observed N-player MFG

Initialize Soft $\pi \in \mathbf{\Gamma}_S$, $\bar{Q}_0^i = 0 \in \mathbb{R}^{\mathbb{Y}\times\mathbb{U}}$ and $\bar{J}_0^i = 0 \in \mathbb{R}^{\mathbb{Y}}$

for $t \geq 0$ (tth stage)

  Simultaneously, each player $j \in \mathcal{N}$ selects $u_t^j \sim \pi^j(\cdot|y_t^j)$

  Player i observes y_{t+1}^i and cost $c_t^i := c(x_t^i, \mu(\cdot|\mathbf{x}_t), u_t^i)$

  $n_t^i := \sum_{k=0}^t \mathbf{1}\{(y_k^i, u_k^i) = (y_t^i, u_t^i)\}$

  $m_t^i := \sum_{k=0}^t \mathbf{1}\{y_k^i = y_t^i\}$

  Q-factor update:

$$\bar{Q}_{t+1}^i(y_t^i, u_t^i) = \left(1 - \frac{1}{n_t^i}\right) \bar{Q}_t^i(y_t^i, u_t^i) + \frac{1}{n_t^i}\left(c_t^i + \beta \min_{a^i \in \mathbb{U}} \bar{Q}_t^i(y_{t+1}^i, a^i)\right),$$

  and $\bar{Q}_{t+1}^i(y, a) = \bar{Q}_t^i(y, a)$ for all $(y, a) \neq (y_t^i, u_t^i)$.

  Value function update:

$$\bar{J}_{t+1}^i(y_t^i) = \left(1 - \frac{1}{m_t^i}\right) \bar{J}_t^i(y_t^i) + \frac{1}{m_t^i}\left(c_t^i + \beta \bar{J}_t^i(y_{t+1}^i)\right),$$

  and $\bar{J}_{t+1}^i(y) = \bar{J}_t^i(y)$ for all $y \neq y_t^i$.

The proof of the third part parallels that of the second part, replacing Corollary 23.5.1 by Lemma 23.5.6 and replacing $\mathbf{x}_t$ by $y_t^i = (x_t^i, \mu(\cdot|\mathbf{x}_t))$ as needed. ◇

Remark 23.5.2 As discussed in Sect. 22.3.3 building on [391], $\tilde{W}_{\pi}^{*i}$, restricted to $\{\varphi^i(\mathbf{s}) : \mathbf{s} \in \mathbf{X}\} \times \mathbb{U}$, is in fact the Q-function for *some* fully observed Markov decision problem with state space $\varphi^i(\mathbf{X}) := \{\varphi^i(\mathbf{s}) : \mathbf{s} \in \mathbf{X}\}$. The same argument used there can be used to establish that $\tilde{V}_{\pi}^{*i}$, restricted to $\varphi^i(\mathbf{X})$, is the state value function for the *same* MDP. A complete specification of this MDP—with explicit characterization of its transition probabilities and its stage cost function—is possible.

(i) The MDP with state space $\varphi^i(\mathbf{X})$ described above is an instance of *an approximate belief MDP with memory length 0*, with the memory length studied in Sect. 22.3.3 being 0.

(ii) Under Assumption 23.5.5, each soft policy $\tilde{\pi} \in \mathbf{\Gamma}_S$ gives rise to a unique invariant measure $\nu_{\tilde{\pi}} \in \mathcal{P}(\mathbf{X})$, and we have that, for any $\nu \in \mathcal{P}(\mathbf{X})$, $\Pr_{\nu}^{\tilde{\pi}}(\mathbf{x}_t \in \cdot) \to \nu_{\tilde{\pi}}(\cdot)$ in total variation as $t \to \infty$.

(iii) The MDP described above, with state space $\varphi^i(\mathbf{X})$, depends on the joint policy $\boldsymbol{\pi}$ through the invariant measure $\nu_{\boldsymbol{\pi}}$. In particular, this MDP depends on the entire joint policy $(\pi^i, \boldsymbol{\pi}^{-i})$, not merely on $\boldsymbol{\pi}^{-i}$. Crucially, this means that the limiting quantities depend, in general, on the policy π^i used by player i. This is in contrast to the MDP settings (including the specific cases in Parts (2) and (3) of Theorem 23.5.2), where the limiting values of Q-learning and value function learning are the same for different (soft/sufficiently exploratory) policies.

(iv) In general, player i does not face the particular MDP on $\varphi^i(\mathbf{X}) \subset \mathbb{Y}$ described in item (iii). The limiting quantities $\tilde{V}_{\pi}^{*i}$ and $\tilde{W}_{\pi}^{*i}$ do not, in general, have any inherent relevance to player i's objective function in the game $\mathcal{G}$. These quantities should instead be interpreted as the subjective beliefs of player i, which were arrived at through a naive independent learning process.

We conclude this section by introducing notation for the subjective function family corresponding to each agent's subjective beliefs obtained through the learning process described in Algorithm 5.

Definition 23.5.7 (Naively Learned Subjective Function Family) Let $(\mathcal{V}^*, \mathcal{W}^*)$ be a subjective function family for $\mathcal{G}$ defined as follows: for any $i \in \mathcal{N}$, $\boldsymbol{\pi} \in \boldsymbol{\Gamma}_S$, let

$$V_{\pi}^{*i}(y) := \begin{cases} \tilde{V}_{\pi}^{*i}(y), & \text{if } \boldsymbol{\pi} \text{ is soft,} \\ \infty & \text{otherwise.} \end{cases} \quad \forall y \in \mathbb{Y}.$$

$$W_{\pi}^{*i}(y,a) := \begin{cases} \tilde{W}_{\pi}^{*i}(y,a), & \text{if } \boldsymbol{\pi} \text{ is soft,} \\ 0 & \text{otherwise.} \end{cases} \quad \forall (y,a) \in \mathbb{Y} \times \mathbb{U}.$$

The pair $(\mathcal{V}^*, \mathcal{W}^*)$ is called the *naively learned subjective function family* for $\mathcal{G}$.

We have chosen to define the subjective functions in this way to avoid introducing $(\mathcal{V}^*, \mathcal{W}^*)$-subjective ϵ-equilibrium that may not be the outcome of the naive learning process. We next discuss conditions under which the existence of such subjective equilibrium can be guaranteed and challenges that arise in guaranteeing existence under decentralized information.

23.5.7 Existence of Subjective Equilibrium Under Global or Mean-Field State Observability

Lemma 23.5.10 ([794]) *Let $\mathcal{G}$ be a partially observed N-player mean-field game satisfying Assumptions 23.5.1 and 23.5.5. Let $\epsilon > 0$. Then, there exists $(\mathcal{V}^*, \mathcal{W}^*)$-subjective ϵ-equilibrium policies. That is,* $\text{Subj}_{\epsilon}(\mathcal{V}^*, \mathcal{W}^*) \neq \emptyset$.

Proof Recall that the set $\boldsymbol{\Gamma}_S$ is compact under the topology induced by the metric $\mathbf{d}$. Combining this compactness with Lemma 23.5.3, it follows that there exists $\xi > 0$ such that if two joint policies $\boldsymbol{\pi}, \boldsymbol{\pi}' \in \boldsymbol{\Gamma}_S$ satisfy $\mathbf{d}(\boldsymbol{\pi}, \boldsymbol{\pi}') < \xi$, then

$$\left| J_{\pi}^i(\mathbf{s}) - J_{\pi'}^i(\mathbf{s}) \right| < \frac{\epsilon}{2} \quad \text{and} \quad \left| \min_{a^i \in \mathbb{U}} Q_{\pi^{-i}}^{*i}(\mathbf{s}, a^i) - \min_{a^i \in \mathbb{U}} Q_{\pi'^{-i}}^{*i}(\mathbf{s}, a^i) \right| < \frac{\epsilon}{2}, \tag{23.10}$$

for any $i \in \mathcal{N}$, $\mathbf{s} \in \mathbf{X}$. We fix $\boldsymbol{\pi}^*$ to be a perfect equilibrium policy, which exists by Lemma 23.5.5. Let $\boldsymbol{\pi}_{\text{soft}} \in \boldsymbol{\Gamma}_S$ be a soft joint policy satisfying $\mathbf{d}(\boldsymbol{\pi}^*, \boldsymbol{\pi}_{\text{soft}}) < \xi$.

By Theorem 23.5.2 and the construction of $(\mathcal{V}^*, \mathcal{W}^*)$, since $\boldsymbol{\pi}_{\text{soft}}$ is soft, we have that

$$W^{*i}_{\boldsymbol{\pi}_{\text{soft}}} = Q^{*i}_{\boldsymbol{\pi}^{-i}_{\text{soft}}} \quad \text{and} \quad V^{*i}_{\boldsymbol{\pi}_{\text{soft}}}(\mathbf{s}) = J^i_{\boldsymbol{\pi}_{\text{soft}}}(\mathbf{s}), \quad \forall i \in \mathcal{N}, \mathbf{s} \in \mathbf{X}.$$

Combining this with (23.10), it follows that, for any $\mathbf{s} \in \mathbf{X}$ and $i \in \mathcal{N}$, we have $V^{*i}_{\boldsymbol{\pi}_{\text{soft}}}(\mathbf{s}) \leq \min_{a^i \in \mathbb{U}} W^{*i}_{\boldsymbol{\pi}_{\text{soft}}}(\mathbf{s}, a^i) + \epsilon$, which shows that $\boldsymbol{\pi}_{\text{soft}} \in \text{Subj}_\epsilon(\mathcal{V}^*, \mathcal{W}^*)$. $\diamond$

(We note that the proof of Lemma 23.5.10 demonstrates that when players have global state information, this notion of subjective equilibrium essentially coincides with an (objective) equilibrium.)

Lemma 23.5.11 ([794]) *Let $\mathcal{G}$ be a partially observed N-player mean-field game satisfying Assumptions 23.5.2 and 23.5.5. Let $\epsilon > 0$. Then, there exists $(\mathcal{V}^*, \mathcal{W}^*)$-subjective ϵ-equilibrium policies. That is,* $\text{Subj}_\epsilon(\mathcal{V}^*, \mathcal{W}^*) \neq \varnothing$.

Replacing $\boldsymbol{\Gamma}_S$ by $\boldsymbol{\Gamma}_{S,\text{sym}}$ as required, the proof of Lemma 23.5.11 parallels that of Lemma 23.5.10 and is omitted. Note that Lemma 23.5.11 does not rule out the existence of *subjective-but-not-objective* ϵ-equilibrium policies, i.e. policies in $\text{Subj}_\epsilon(\mathcal{V}^*, \mathcal{W}^*) \setminus \boldsymbol{\Gamma}^{\epsilon-eq}$. If such joint policies exist, one observes that they are not mean-field symmetric.

23.5.8 Equilibria Under Compressed Mean-Field State and Local State Information

We note that the proof methods used above to guarantee the existence of equilibria do not apply to partially observed N-player mean-field games with the observation channels of Assumptions 23.5.3 or 23.5.4. In the convergence results and analysis to follow, we will be assuming the existence of subjective equilibria when one of Assumptions 23.5.3 or 23.5.4 holds.

23.6 Subjective Satisficing, Policy Update Dynamics and Subjective Q-Learning

With the goal of better informing algorithm design, this section studies the structure and limitations of a class of discrete-time dynamical systems on $\boldsymbol{\Gamma}_S$, the set of stationary joint policies in a partially observed N-player mean-field game. We are interested in a class of dynamical systems that arise when, at every time step, each agent updates its policy in response to the prevailing joint policy. The defining characteristic of the class of dynamical systems we consider is that players do not switch policies when they are subjectively ϵ-best-responding to the prevailing joint policy, but are otherwise free to change their policy arbitrarily when they are not subjectively ϵ-best-responding.

For the following definitions, let $\mathcal{G}$ be a partially observed N-player game, let $i \in \mathcal{N}$, let $\epsilon \geq 0$, and let $(\mathcal{V}, \mathcal{W})$ be a subjective function family for $\mathcal{G}$.

Definition 23.6.1 (Policy Update Rules) A mapping $T^i : \mathbf{\Gamma}_S \to \Gamma_S^i$ is called a policy update rule (for player i). A collection $\mathbf{T} = \{T^j\}_{j\in\mathcal{N}}$ of policy update rules for each player is called a *joint policy update rule.*

Definition 23.6.2 If a policy update rule T^i satisfies

$$\pi^i \in \text{Subj-BR}_\epsilon^i(\boldsymbol{\pi}^{-i}, \mathcal{V}, \mathcal{W}) \Rightarrow T^i(\boldsymbol{\pi}) = \pi^i,$$

for every $\boldsymbol{\pi} \in \mathbf{\Gamma}_S$, then T^i is said to be $(\mathcal{V}, \mathcal{W})$-subjective ϵ-satisficing. A joint policy update rule $\mathbf{T} = \{T^i\}_{i\in\mathcal{N}}$ is called $(\mathcal{V}, \mathcal{W})$-subjective ϵ-satisficing if each player's component policy update rule is $(\mathcal{V}, \mathcal{W})$-subjective ϵ-satisficing.

Definition 23.6.3 (Subjective Satisficing Paths) A sequence of policies $(\boldsymbol{\pi}_k)_{k\geq 0}$ in $\mathbf{\Gamma}_S$ is called a $(\mathcal{V}, \mathcal{W})$*-subjective* ϵ*-satisficing path* if, for every $i \in \mathcal{N}$ and $k \geq 0$, we have

$$\pi_k^i \in \text{Subj-BR}_\epsilon^i(\boldsymbol{\pi}_k^{-i}, \mathcal{V}, \mathcal{W}) \Rightarrow \pi_{k+1}^i = \pi_k^i.$$

Given a joint policy update rule $\mathbf{T}$ and a joint policy $\boldsymbol{\pi} \in \mathbf{\Gamma}_S$, we let $\mathbf{T}(\boldsymbol{\pi}) \in \mathbf{\Gamma}_S$ be the joint policy given by $\mathbf{T}(\boldsymbol{\pi}) = \{T^i(\boldsymbol{\pi})\}_{i\in\mathcal{N}}$; that is, player i's component of $\mathbf{T}(\boldsymbol{\pi})$ is given by $T^i(\boldsymbol{\pi})$. We define $\mathbf{T}^0(\boldsymbol{\pi}) := \boldsymbol{\pi}$ for any joint policy update rule $\mathbf{T}$ and joint policy $\boldsymbol{\pi} \in \mathbf{\Gamma}_S$. We further define $\mathbf{T}^{k+1}(\boldsymbol{\pi}) = \mathbf{T}\left(\mathbf{T}^k(\boldsymbol{\pi})\right)$ for each $k \geq 0$. This allows us to denote the orbits of $\mathbf{T}$ from a given initial policy $\boldsymbol{\pi} \in \mathbf{\Gamma}_S$ as $\{\mathbf{T}^k(\boldsymbol{\pi})\}_{k\geq 0}$.

Intuitively, when an agent follows a $(\mathcal{V}, \mathcal{W})$-subjective ϵ-satisficing rule, it only considers switching its policy when it (subjectively) deems the policy to be performing poorly. No further restrictions are placed on how an agent is allowed to switch (or not switch) its policy when it is subjectively unsatisfied. We note that given a $(\mathcal{V}, \mathcal{W})$-subjective ϵ-satisficing rule $\mathbf{T}$ and any initial policy $\boldsymbol{\pi} \in \mathbf{\Gamma}_S$, the resulting orbit $\{\mathbf{T}^k(\boldsymbol{\pi})\}_{k\geq 0}$ is a $(\mathcal{V}, \mathcal{W})$-subjective ϵ-satisficing path. Moreover, we observe that $(\mathcal{V}, \mathcal{W})$-subjective ϵ-equilibrium policies are invariant under any $(\mathcal{V}, \mathcal{W})$-subjective ϵ-satisficing joint policy update: if $\boldsymbol{\pi}^* \in \text{Subj}_\epsilon(\mathcal{V}, \mathcal{W})$ and $\mathbf{T}$ is $(\mathcal{V}, \mathcal{W})$-subjective ϵ-satisficing, then $\mathbf{T}^k(\boldsymbol{\pi}^*) = \boldsymbol{\pi}^*$ for all $k \geq 0$. In addition to stability, it is natural to ask whether a given joint policy update rule $\mathbf{T}$ drives its orbits to $\text{Subj}_\epsilon(\mathcal{V}, \mathcal{W})$. This motivates the following definition.

Definition 23.6.4 (Subjective Satisficing Paths Property) Let $\mathbf{\Pi} \subseteq \mathbf{\Gamma}_S$ be a subset of joint policies. The game $\mathcal{G}$ is said to have the $(\mathcal{V}, \mathcal{W})$*-subjective* ϵ*-satisficing paths property within* $\mathbf{\Pi}$ if the following holds: for every $\boldsymbol{\pi} \in \mathbf{\Pi}$, there exists a $(\mathcal{V}, \mathcal{W})$-subjective ϵ-satisficing path $(\boldsymbol{\pi}_k)_{k\geq 0}$ such that (i) $\boldsymbol{\pi}_0 = \boldsymbol{\pi}$, (ii) $\boldsymbol{\pi}_k \in \boldsymbol{\Pi}$ for all $k \geq 0$, and (iii) for some $K < \infty$, $\boldsymbol{\pi}_K \in \text{Subj}_\epsilon(\mathcal{V}, \mathcal{W})$.

When the game $\mathcal{G}$ does not have the $(\mathcal{V}, \mathcal{W})$-subjective ϵ-satisficing paths property in a policy subset $\boldsymbol{\Pi} \subset \Gamma_S$, there exist initial policies for which *no* $(\mathcal{V}, \mathcal{W})$-subjective ϵ-satisficing update rule can drive the play to subjective equilibrium in finitely many steps. In such instances, it would be inappropriate to employ MARL algorithms that rely on the "win-stay, lose-shift" paradigm to guide exploration of the policy set $\boldsymbol{\Pi}$. There are many examples of MARL algorithms that rely on this policy updating paradigm; see e.g. [156, 171, 175, 793].

23.6.1 *Naively Learned Subjective Functions and ϵ-Satisficing*

We now shift our attention to our subjective function family of interest, $(\mathcal{V}^*, \mathcal{W}^*)$, which arises out of the naive independent learning process of Sect. 23.5.6. We will argue that any partially observed N-player mean-field game will admit $(\mathcal{V}^*, \mathcal{W}^*)$-subjective ϵ-satisficing paths to equilibrium under global state observability and mean-field state observability. For compressed state observability, we show that if subjective equilibria exist then subjective ϵ-satisficing paths exist.

Subjective Satisficing Paths Under Global State Observability

Definition 23.6.5 Let $\mathcal{G}$ be a partially observed N-player mean-field game for which Assumption 23.5.1 is satisfied, and let $i \in \mathcal{N}$. A stationary policy $\pi^i \in \Gamma_S^i$ is said to be *of the mean-field type* if there exists $f^i \in \mathcal{P}(\mathbb{U}|X_{\text{loc}} \times \text{Emp}_N)$ such that $\pi^i(\cdot|\mathbf{s}) = f^i\left(\cdot\middle|s^i, \mu(\cdot|\mathbf{s})\right)$ for every global state $\mathbf{s} \in \mathbf{X}$.

We identify each stationary policy of the mean-field type with its associated transition kernel in $\mathcal{P}(\mathbb{U}|X_{\text{loc}} \times \text{Emp}_N)$.

Definition 23.6.6 Let $\mathcal{G}$ be a partially observed N-player mean-field game for which Assumption 23.5.1 is satisfied. For $i, j \in \mathcal{N}$, let $\pi^i \in \Gamma_S^i$ and $\pi^j \in \Gamma_S^j$ both be of the mean-field type. We say that π^i and π^j are *mean-field symmetric* if they are identified with the same transition kernel in $\mathcal{P}(\mathbb{U}|X_{\text{loc}} \times \text{Emp}_N)$.

This definition of mean-field symmetry is effectively the same as the definition given earlier. Here, we redefine it to emphasize that this definition pertains to stationary policies for a different observation channel, namely that of global state observability rather than that of mean-field state observability, for which the earlier definition was tailored.

Lemma 23.6.1 *Let $\mathcal{G}$ be a partially observed N-player mean-field game for which Assumptions 23.5.1 and 23.5.5 are satisfied, and let $\epsilon \geq 0$. Let $\boldsymbol{\pi} \in \boldsymbol{\Gamma}_S$ be a stationary joint policy for which π^p is of the mean-field type for each player $p \in \mathcal{N}$.*

Suppose that, for some $i, j \in \mathcal{N}$, we have that π^i and π^j are mean-field symmetric. Then we have

$$\pi^i \in \text{Subj-BR}^i_\epsilon(\boldsymbol{\pi}^{-i}, \mathcal{V}^*, \mathcal{W}^*) \iff \pi^j \in \text{Subj-BR}^j_\epsilon(\boldsymbol{\pi}^{-j}, \mathcal{V}^*, \mathcal{W}^*).$$

A proof of Lemma 23.6.1 can be found in [794]. Lemma 23.6.1 enables the following important result, our first on the existence of $(\mathcal{V}^*, \mathcal{W}^*)$-subjective ϵ-satisficing paths.

Theorem 23.6.1 *Let $\mathcal{G}$ be a partially observed N-player mean-field game for which Assumptions 23.5.1 and 23.5.5 hold. Let $\epsilon > 0$, and let $(\mathcal{V}^*, \mathcal{W}^*)$ be the naively learned subjective function family for $\mathcal{G}$. Suppose $\boldsymbol{\Pi} \subset \boldsymbol{\Gamma}_S$ is a subset of stationary joint policies satisfying the following properties: (i) Every $\boldsymbol{\pi} \in \boldsymbol{\Pi}$ is of the mean-field type; (ii) $\boldsymbol{\Pi} \cap \text{Subj}_\epsilon(\mathcal{V}^*, \mathcal{W}^*) \neq \varnothing$; (iii) the set $\boldsymbol{\Pi}$ is symmetric, i.e. $\Pi^i = \Pi^j$ for each $i, j \in \mathcal{N}$.*

Then, $\mathcal{G}$ has the $(\mathcal{V}^*, \mathcal{W}^*)$-subjective ϵ-satisficing paths property within $\boldsymbol{\Pi}$.

The proof of Theorem 23.6.1 parallels that of [796, Theorem 8], which was in the context of symmetric, fully observed stochastic games.

Corollary 23.6.1 *Let $\mathcal{G}$ be a partially observed N-player mean-field game satisfying Assumptions 23.5.1 and 23.5.5. Let $\epsilon > 0$ and let $\boldsymbol{\Pi} := \{\boldsymbol{\pi} \in \boldsymbol{\Gamma}_S : \pi^i \text{ is of the mean-field type for all } i \in \mathcal{N}\}$. Then, $\mathcal{G}$ has the $(\mathcal{V}^*, \mathcal{W}^*)$-subjective ϵ-satisficing paths property within $\boldsymbol{\Pi}$.*

Proof Using the proof of Theorem 23.5.1, we have that $\boldsymbol{\Pi} \cap \text{Subj}_\epsilon(\mathcal{V}^*, \mathcal{W}^*) \neq \varnothing$. The set $\boldsymbol{\Pi}$ thus satisfies the conditions of Theorem 23.6.1, and the result follows. ◇

Subjective Satisficing Paths Under Mean-Field State Observability

We now provide a result analogous to Theorem 23.6.1 under mean-field state observability. To facilitate this result, we now state an analog to Lemma 23.6.1.

Lemma 23.6.2 *Let $\mathcal{G}$ be a partially observed N-player mean-field game for which Assumptions 23.5.2 and 23.5.5 hold, and let $\epsilon \geq 0$. Let $\boldsymbol{\pi} \in \boldsymbol{\Gamma}_S$. For $i, j \in \mathcal{N}$, suppose π^i and π^j are mean-field symmetric. Then, we have*

$$\pi^i \in \text{Subj-BR}^i_\epsilon(\boldsymbol{\pi}^{-i}, \mathcal{V}^*, \mathcal{W}^*) \iff \pi^j \in \text{Subj-BR}^j_\epsilon(\boldsymbol{\pi}^{-j}, \mathcal{V}^*, \mathcal{W}^*).$$

A proof of Lemma 23.6.2 can be found in [794].

Theorem 23.6.2 ([794]) *Let $\mathcal{G}$ be a partially observed N-player mean-field game for which Assumptions 23.5.2 and 23.5.5 hold. Let $\epsilon > 0$, and let $(\mathcal{V}^*, \mathcal{W}^*)$ be the naively learned subjective function family for $\mathcal{G}$. Suppose $\boldsymbol{\Pi} \subset \boldsymbol{\Gamma}_S$ is a subset of stationary joint policies satisfying the following properties: (i) $\boldsymbol{\Pi} \cap \text{Subj}_\epsilon(\mathcal{V}^*, \mathcal{W}^*) \neq \varnothing$; (ii) the set $\boldsymbol{\Pi}$ is symmetric, i.e. $\Pi^i = \Pi^j$ for each $i, j \in \mathcal{N}$. Then, $\mathcal{G}$ has the $(\mathcal{V}^*, \mathcal{W}^*)$-subjective ϵ-satisficing paths property within $\boldsymbol{\Pi}$.*

The argument used in the proof of Theorem 23.6.1 can be suitably modified (e.g. replacing Lemma 23.6.1 by Lemma 23.6.2) to prove Theorem 23.6.2, and so the proof is omitted.

Corollary 23.6.2 *Let $\mathcal{G}$ be a partially observed N-player mean-field game for which Assumptions 23.5.2 and 23.5.5 hold. Let $\epsilon > 0$, and let $(\mathcal{V}^*, \mathcal{W}^*)$ be the naively learned subjective function family for $\mathcal{G}$. Then, $\mathcal{G}$ has the $(\mathcal{V}^*, \mathcal{W}^*)$-subjective ϵ-satisficing paths property within $\mathbf{\Gamma}_S$.*

Subjective Satisficing Paths Under Compressed State Observability

In light of our previous discussion on the possible non-existence of (naively learned) subjective ϵ-equilibrium in games with compressed state observability, we now provide a qualified result analogous to Theorems 23.6.1 and 23.6.2 for the compressed state observation channel. In contrast to the earlier results, here we must assume the existence of a subjective equilibrium.

Definition 23.6.7 Let $\mathcal{G}$ be a partially observed N-player mean-field game and let $i \in \mathcal{N}$. A policy $\pi^i \in \Gamma_S^i$ is said to be *of the local type* if there exists a transition kernel $g^i \in \mathcal{P}(\mathbb{U}|X_{\text{loc}})$ such that

$$\pi^i(\cdot|\varphi^i(\mathbf{s})) = g^i(\cdot|s^i), \quad \forall \mathbf{s} \in \mathbf{X}.$$

Each policy of the local type is identified with the corresponding transition kernel in $\mathcal{P}(\mathbb{U}|X_{\text{loc}})$. For players $i, j \in \mathcal{N}$, if the policies $\pi^i \in \Gamma_S^i, \pi^j \in \Gamma_S^j$ are both of the local type and are identified with the same transition kernel, we say π^i and π^j are *locally symmetric*.

Lemma 23.6.3 *Let $\mathcal{G}$ be a partially observed N-player mean-field game for which Assumptions 23.5.3 and 23.5.5 hold. Let $\boldsymbol{\pi} \in \mathbf{\Gamma}_S$. For $i, j \in \mathcal{N}$, suppose π^i and π^j are locally symmetric. Then, we have $\pi^i \in \text{Subj-BR}_\epsilon^i(\boldsymbol{\pi}^{-i}, \mathcal{V}^*, \mathcal{W}^*)$ if and only if $\pi^j \in \text{Subj-BR}_\epsilon^j(\boldsymbol{\pi}^{-j}, \mathcal{V}^*, \mathcal{W}^*)$.*

The proof of Lemma 23.6.3 mirrors the proof of Lemma 23.6.2 and is therefore omitted.

Theorem 23.6.3 *Let $\mathcal{G}$ be a partially observed N-player mean-field game for which Assumptions 23.5.3 and 23.5.5 hold. Let $\epsilon > 0$, and let $(\mathcal{V}^*, \mathcal{W}^*)$ be the naively learned subjective function family for $\mathcal{G}$. Assume that $\text{Subj}_\epsilon(\mathcal{V}^*, \mathcal{W}^*) \neq \varnothing$. Suppose that $\mathbf{\Pi} \subset \mathbf{\Gamma}_S$ is a subset of stationary joint policies satisfying the following properties: (i) $\mathbf{\Pi} \cap \text{Subj}_\epsilon(\mathcal{V}^*, \mathcal{W}^*) \neq \varnothing$; (ii) the set $\mathbf{\Pi}$ is symmetric, i.e. $\Pi^i = \Pi^j$ for each $i, j \in \mathcal{N}$. Then, $\mathcal{G}$ has the $(\mathcal{V}^*, \mathcal{W}^*)$-subjective ϵ-satisficing paths property within $\mathbf{\Pi}$.*

The argument used in the proof of Theorem 23.6.1 can be suitably modified (e.g. replacing Lemma 23.6.1 by Lemma 23.6.3) to prove Theorem 23.6.3.

23.6.2 Quantization of the Policy Space

For algorithm design purposes, it will be advantageous to restrict policy selection to a finite subset of stationary policies. We argue that if the restricted set of policies is obtained via a sufficiently fine quantization of the original set, then the performance loss for an agent optimizing over this subset will be negligible. Moreover, we argue that if the restricted subset of policies is suitably fine and symmetric, then the structural properties of the previous subsection will be inherited: in particular, subjective near-equilibrium policies will exist in the restricted set, and furthermore the subjective satisficing paths property will hold within this finite subset of policies.

For the following definitions, let $\mathcal{G}$ be a partially observed N-player mean-field game, let $i \in \mathcal{N}$. Recall that d^i is a metric on the set Γ_S^i, suitably defined to match the assumed observation channel.

Definition 23.6.8 Let $\xi > 0$ and $\tilde{\Pi}^i \subseteq \Gamma_S^i$. A mapping $q^i : \tilde{\Pi}^i \to \tilde{\Pi}^i$ is called a ξ-quantizer (on $\tilde{\Pi}^i$) if (1) $q^i(\tilde{\Pi}^i) := \{q^i(\pi^i) : \pi^i \in \tilde{\Pi}^i\}$ is a finite set and (2) $d^i(\pi^i, q^i(\pi^i)) < \xi$ for all $\pi^i \in \tilde{\Pi}^i$.

Definition 23.6.9 Let $\xi > 0$ and let $\tilde{\Pi}^i \subseteq \Gamma_S^i$. A set of policies $\Pi^i \subseteq \tilde{\Pi}^i$ is called a ξ-quantization of $\tilde{\Pi}^i$ if $\Pi^i = q^i(\tilde{\Pi}^i)$, where q^i is some ξ-quantizer on $\tilde{\Pi}^i$.

A set $\Pi^i \subseteq \Gamma_S^i$ is called a quantization of Γ_S^i if it is a ξ-quantization of Γ_S^i for some $\xi > 0$. A quantization Π^i is called *soft* if each policy $\pi^i \in \Pi^i$ is soft. The expression "fine quantization" will be used to reflect that a policy subset is a ξ-quantization for suitably small ξ. We extend the definitions above and each of these terminological conventions to also refer to quantizers and quantizations of sets of joint policies. For instance, $\mathbf{\Pi} = \times_{i \in \mathcal{N}} \Pi^i \subset \mathbf{\Gamma}_S$ is a ξ-quantization of $\mathbf{\Gamma}_S$ if each Π^i is a ξ-quantization of Γ_S^i, and so on.

Definition 23.6.10 Let $\mathbf{\Pi} \subset \mathbf{\Gamma}_S$ be a quantization of $\mathbf{\Gamma}_S$. We say that $\mathbf{\Pi}$ is *symmetric* if $\Pi^i = \Pi^j$ for each $i, j \in \mathcal{N}$.

The following results are straightforward.

Lemma 23.6.4 *Let $\mathcal{G}$ be a partially observed N-player mean-field game satisfying Assumptions 23.5.1 and 23.5.5. Let $\epsilon > 0$. There exists $\xi = \xi(\epsilon) > 0$ such that if $\mathbf{\Pi} \subset \mathbf{\Gamma}_S$ is any soft ξ-quantization of $\mathbf{\Gamma}_S$, then we have*

(i) $\mathbf{\Gamma}^{\epsilon-eq} \cap \mathbf{\Pi} \neq \varnothing$ *and* $\text{Subj}_\epsilon(\mathcal{V}^*, \mathcal{W}^*) \cap \mathbf{\Pi} \neq \varnothing$.

Moreover, if $\tilde{\mathbf{\Pi}} \subset \mathbf{\Gamma}_S$ is the set of joint stationary policies of the mean-field type, there exists $\xi = \xi(\epsilon) > 0$ such that if $\mathbf{\Pi}$ is a soft, symmetric ξ-quantization of $\tilde{\mathbf{\Pi}}$, then we have

(ii) $\text{Subj}_\epsilon(\mathcal{V}^*, \mathcal{W}^*) \cap \mathbf{\Pi} \neq \varnothing$;
(iii) *$\mathcal{G}$ has the $(\mathcal{V}^*, \mathcal{W}^*)$-subjective ϵ-satisficing paths property within $\mathbf{\Pi}$.*

Lemma 23.6.5 *Let $\mathcal{G}$ be a partially observed N-player mean-field game satisfying Assumptions 23.5.2 and 23.5.5. Let $\epsilon > 0$. There exists $\xi = \xi(\epsilon) > 0$ such that if $\mathbf{\Pi} \subset \mathbf{\Gamma}_S$ is any soft, symmetric ξ-quantization of $\mathbf{\Gamma}_S$, then we have*

(i) $\mathbf{\Gamma}^{\epsilon-eq} \cap \mathbf{\Pi} \neq \varnothing$ *and* $\mathrm{Subj}_\epsilon(\mathcal{V}^*, \mathcal{W}^*) \cap \mathbf{\Pi} \neq \varnothing$;
(ii) *$\mathcal{G}$ has the $(\mathcal{V}^*, \mathcal{W}^*)$-subjective ϵ-satisficing paths property within $\mathbf{\Pi}$.*

Lemmas 23.6.4 and 23.6.5 guarantee that the game $\mathcal{G}$ has the subjective satisficing paths property within finely quantized subsets of policies. This has two desirable consequences for algorithm design purposes. First, players can restrict their policy search from an uncountable set (all stationary policies) to a finite subset of policies with only a small loss in performance.

The second desirable consequence is the following: since the game $\mathcal{G}$ has the $(\mathcal{V}^*, \mathcal{W}^*)$-subjective ϵ-satisficing paths property within $\mathbf{\Pi}$, the play can be driven to (subjective) ϵ-equilibrium by changing only the policies of those players that are "ϵ-unsatisfied". We thus obtain a stopping condition, whereby player i can settle on a policy whenever it is subjectively ϵ-best-responding. In contrast, the stability of ϵ-equilibrium may be threatened when an algorithm requires a player to change policies while already ϵ-best-responding. We cite [518, 533] and [519] as recent works showing non-convergence of policies to equilibrium when each agent runs a policy gradient algorithm.

Taken together, these points remove the need for *coordinated* search of the joint policy space $\mathbf{\Pi}$: since $(\mathcal{V}^*, \mathcal{W}^*)$-subjective ϵ-satisficing paths to equilibrium exist within $\mathbf{\Pi}$ and $\mathbf{\Pi}$ is finite, the play can be driven to an ϵ-equilibrium even by random policy updating by those players that are not subjectively ϵ-best-responding. Moreover, this structure also removes the need for specialized policy updating rules (such as inertial best-responding, as used in [37] or gradient ascent, as studied in [204]).

For each $i \in \mathcal{N}$, suppose $\Pi^i \subset \Gamma^i_S$ is a finite quantization of either $\{\pi^i \in \Gamma^i_S : \pi^i \text{ is of the mean-field type}\}$ (under global state observability, Assumption 23.5.1) or Γ^i_S (under mean-field observability, Assumption 23.5.2). Letting $\mathbf{\Pi} = \times_{i\in\mathcal{N}} \Pi^i$, suppose that $\Pi^i = \Pi^j$ for each $i, j \in \mathcal{N}$ and furthermore $\mathrm{Subj}_\epsilon(\mathcal{V}^*, \mathcal{W}^*) \neq \varnothing$. Consider the time homogenous Markov chain $\{\boldsymbol{\pi}_k\}_{k\geq 0}$ in the set $\mathbf{\Pi}$ generated according to Algorithm 6. Since the game has the $(\mathcal{V}^*, \mathcal{W}^*)$-subjective ϵ-satisficing paths property within $\mathbf{\Pi}$ and players randomly explore their policy sets when not ϵ-best-responding, this process converges almost surely to some $(\mathcal{V}^*, \mathcal{W}^*)$-subjective ϵ-equilibrium in $\mathbf{\Pi}$. This is formalized in Lemmas 23.6.6 and 23.6.7, below.

Lemma 23.6.6 *Let $\mathcal{G}$ be a partially observed N-player mean-field game satisfying Assumptions 23.5.1 and 23.5.5, and let $\epsilon > 0$. Let $\mathbf{\Pi} \subset \mathbf{\Gamma}_S$ be a quantization of the set $\{\boldsymbol{\pi} \in \mathbf{\Gamma}_S : \pi^i \text{ is of the mean-field type } \forall i \in \mathcal{N}\}$, and suppose $\mathbf{\Pi}$ satisfies (1) $\mathbf{\Pi} \cap \mathrm{Subj}_\epsilon(\mathcal{V}^*, \mathcal{W}^*) \neq \varnothing$; (2) $\Pi^i = \Pi^j$ for all $i, j \in \mathcal{N}$; and (3) every policy $\boldsymbol{\pi} \in \mathbf{\Pi}$ is soft.*

Suppose that each Agent $i \in \mathcal{N}$ updates its policy sequence $\{\pi^i_k\}_{k\geq 0}$ according to Algorithm 6 and that, for each $k \geq 0$, the policy updates for $\boldsymbol{\pi}_{k+1}$ are conditionally independent across agents given $\boldsymbol{\pi}_k$. Then, $\lim_{k\to\infty} \Pr(\boldsymbol{\pi}_k \in \mathbf{\Pi} \cap \mathbf{\Gamma}^{\epsilon-eq}) = 1$.

Algorithm 6: ϵ-satisficing Policy Revision (for player $i \in \mathcal{N}$)

1 **Set Parameters**
2 $\quad e^i \in (0, 1)$: experimentation probability when not subjectively ϵ-best-responding
3 $\quad \Pi^i \subset \Gamma_S^i$: a fine quantization of Γ_S^i.
4 **Initialize** $\pi_0^i \in \Pi^i$: initial policy
5 **for** $k \geq 0$ (kth policy update)
6 $\quad$ **Receive** $V_{\pi_k}^{*i}$ and $W_{\pi_k}^{*i}$ (from oracle)
7 $\quad$ **if** $V_{\pi_k}^{*i}(y) \leq \min_{a^i} W_{\pi_k}^{*i}(y, a^i) + \epsilon$ for all $y \in \mathbb{Y}$ **then**
8 $\quad\quad \pi_{k+1}^i = \pi_k^i$
9 $\quad$ **else**
10 $\quad\quad \pi_{k+1}^i \sim (1 - e^i)\delta_{\pi_k^i} + e^i \mathrm{Unif}(\Pi^i)$
11 $\quad$ Go to $k + 1$

Lemma 23.6.7 *Let $\mathcal{G}$ be a partially observed N-player mean-field game satisfying Assumptions 23.5.2 and 23.5.5, and let $\epsilon > 0$. Let $\mathbf{\Pi} \subset \mathbf{\Gamma}_S$ be a quantization of $\mathbf{\Gamma}_S$ and suppose $\mathbf{\Pi}$ satisfies (1) $\mathbf{\Pi} \cap \mathrm{Subj}_\epsilon(\mathcal{V}^*, \mathcal{W}^*) \neq \varnothing$; (2) $\Pi^i = \Pi^j$ for all $i, j \in \mathcal{N}$; and (3) every policy $\boldsymbol{\pi} \in \mathbf{\Pi}$ is soft.*

Suppose that each agent $i \in \mathcal{N}$ updates its policy sequence $\{\pi_k^i\}_{k \geq 0}$ according to Algorithm 6 and that, for each $k \geq 0$, the policy updates for $\boldsymbol{\pi}_{k+1}$ are conditionally independent across agents given $\boldsymbol{\pi}_k$. Then, $\lim_{k \to \infty} \Pr(\boldsymbol{\pi}_k \in \mathbf{\Pi} \cap \mathrm{Subj}_\epsilon(\mathcal{V}^*, \mathcal{W}^*)) = 1$.

The main difference between Lemmas 23.6.6 and 23.6.7 is the following: in the former, convergence to an objective equilibrium is guaranteed, while in the latter there is only convergence to a subjective equilibrium. The proofs are essentially the same, and we therefore only include a proof of Lemma 23.6.7.

Proof We have that $\{\boldsymbol{\pi}_k\}_{k \geq 0}$ is a time homogenous Markov chain on $\mathbf{\Pi}$. For any subjective ϵ-equilibrium $\boldsymbol{\pi}^* \in \mathbf{\Pi} \cap \mathrm{Subj}_\epsilon(\mathcal{V}^*, \mathcal{W}^*)$, we have that the singleton $\{\boldsymbol{\pi}^*\}$ is an absorbing set for this Markov chain, by Lines 6-7 of Algorithm 6. (The condition in Line 6 is equivalent to verifying whether $\pi_k^i \in \text{Subj-BR}_\epsilon^i(\boldsymbol{\pi}_k^{-i}, \mathcal{V}^*, \mathcal{W}^*)$.)

By Theorem 23.6.2, the game $\mathcal{G}$ has the $(\mathcal{V}^*, \mathcal{W}^*)$-subjective ϵ-satisficing paths within $\mathbf{\Pi}$. For any $\boldsymbol{\pi} \in \mathbf{\Pi}$, let $L_{\boldsymbol{\pi}} < \infty$ denote the length of a shortest $(\mathcal{V}^*, \mathcal{W}^*)$-subjective ϵ-satisficing path within $\mathbf{\Pi}$ that starts at $\boldsymbol{\pi}$ and terminates at a policy in $\mathbf{\Pi} \cap \mathrm{Subj}_\epsilon(\mathcal{V}^*, \mathcal{W}^*)$. Call such a policy $\boldsymbol{\pi}^*$ and note it depends on $\boldsymbol{\pi}$. We also define $p_{\boldsymbol{\pi}} > 0$ as the probability the Markov chain follows this path when starting at $\boldsymbol{\pi}$, i.e.

$$p_{\boldsymbol{\pi}} := \Pr(\boldsymbol{\pi}_{L_{\boldsymbol{\pi}}} = \boldsymbol{\pi}^* | \boldsymbol{\pi}_0 = \boldsymbol{\pi}) > 0, \quad \forall \boldsymbol{\pi} \in \mathbf{\Pi}.$$

Define $L := \max\{L_{\boldsymbol{\pi}} : \boldsymbol{\pi} \in \mathbf{\Pi}\}$ and $\hat{p} := \min\{p_{\boldsymbol{\pi}} : \boldsymbol{\pi} \in \mathbf{\Pi}\} > 0$. For any $m \geq 0$, we have

$$\Pr\left(\bigcap_{j=1}^{m}\{\boldsymbol{\pi}_{jL} \notin \boldsymbol{\Pi} \cap \mathrm{Subj}_{\epsilon}(\mathcal{V}^*, \mathcal{W}^*)\}\right) \leq (1-\hat{p})^m.$$

Taking $m \to \infty$ leads to the result. $\diamond$

We note that, in Line 9 of Algorithm 6, the choice to update the policy according to $\pi^i_{k+1} \sim (1-e^i)\delta_{\pi^i_k} + e^i \mathrm{Unif}(\Pi^i)$ was somewhat arbitrary. In particular, the choice to uniformly mix over Π^i with probability $e^i > 0$ was made to ensure the paths to equilibrium exist in the Markov chain of Lemma 23.6.7. The choice to remain with one's old policy with probability $1-e^i$ was arbitrarily picked for ease of exposition, and can be replaced by any suitable distribution over Π^i according to the taste of the system designer. Some choices may include gradient descent projected back onto Π^i or selecting a best-response to $\boldsymbol{\pi}^{-i}_k$ within Π^i. Such changes may result in a significant speed-up of convergence to equilibrium when they are well-suited to the underlying game, but the guarantee holds in any case due to the uniform randomization.

In its present form, Algorithm 6 cannot be implemented in an online MARL setting, as we have assumed oracle access to the quantities $V^{*i}_{\boldsymbol{\pi}_k}$ and $W^{*i}_{\boldsymbol{\pi}_k}$. In a decentralized, independent learning setting, these quantities may be estimated through the use of (intentionally) naive single-agent learning algorithms, as described in Sect. 23.5.6. The algorithm presented in the next subsection will effectively mimic the mechanics of Algorithm 6, replacing knowledge that $\pi^i_k \in \text{Subj-BR}^i_{\epsilon}(\boldsymbol{\pi}^{-i}_k, \mathcal{V}^*, \mathcal{W}^*)$ by a learned estimate. That is, the learning algorithm of the next subsection (Algorithm 7) can be viewed as a noise-perturbed, learning-based approximation to Algorithm 6.

23.6.3 Learning Algorithm and Convergence to Subjective Q-Learning Equilibria

In this section, we present Algorithm 7 [794], a decentralized learning algorithm suitable for playing partially observed N-player mean-field games. At a high level, Algorithm 7 is an analog to Algorithm 6: whereas an oracle provides the subjective function information in Algorithm 6, here the subjective functions are learned using system feedback via a Q-learning algorithm under an assumed information/model. The learned subjective functions are then used to estimate whether the agent is (subjectively) ϵ-best-responding.

Learning with Global State

We begin by presenting convergence results for Algorithm 7 under global state observability, the richest of the information structures that we consider. Under

Algorithm 7: Independent learning

Set Parameters

$\Pi^i \subset \Gamma_S^i$: a fine quantization of Γ_S^i

$\{T_k\}_{k\geq 0}$: a sequence in $\mathbb{N}$ of learning phase lengths

set $t_0 = 0$ and $t_{k+1} = t_k + T_k$ for all $k \geq 0$.

$e^i \in (0, 1)$: random policy updating probability

$d^i \in (0, \infty)$: tolerance level for sub-optimality

Initialize $\pi_0^i \in \Pi^i$ (arbitrary), $\widehat{Q}_0^i = 0 \in \mathbb{R}^{\mathbb{Y}\times\mathbb{U}}$, $\widehat{J}_0^i = 0 \in \mathbb{R}^{\mathbb{Y}}$

for $k \geq 0$ (kth exploration phase)

  for $t = t_k, t_k + 1, \ldots, t_{k+1} - 1$

    Observe $y_t^i = \varphi^i(\mathbf{x}_t)$

    Select $u_t^i \sim \pi_k^i(\cdot|y_t^i)$

    Observe $c_t^i := c(x_t^i, \mu(\cdot|\mathbf{x}_t), u_t^i)$ and y_{t+1}^i

    Set $n_t^i = \sum_{\tau=t_k}^{t} \mathbf{1}\{(y_\tau^i, u_\tau^i) = (y_t^i, u_t^i)\}$

    Set $m_t^i = \sum_{\tau=t_k}^{t} \mathbf{1}\{y_\tau^i = y_t^i\}$

    $\widehat{Q}_{t+1}^i(y_t^i, u_t^i) = \left(1 - \frac{1}{n_t^i}\right)\widehat{Q}_t^i(y_t^i, u_t^i) + \frac{1}{n_t^i}\left[c_t^i + \beta \min_{a^i} \widehat{Q}_t^i(y_{t+1}^i, a^i)\right]$

    $\widehat{J}_{t+1}^i(y_t^i) = \left(1 - \frac{1}{m_t^i}\right)\widehat{J}_t^i(y_t^i) + \frac{1}{m_t^i}\left[c_t^i + \beta \widehat{J}_t^i(y_{t+1}^i)\right]$

  if $\widehat{J}_{t_{k+1}}^i(y) \leq \min_{a^i} \widehat{Q}_{t_{k+1}}^i(y, a^i) + \epsilon + d^i \ \forall y \in \mathbb{Y}$, **then**

    $\pi_{k+1}^i = \pi_k^i$

  else

    $\pi_{k+1}^i \sim (1 - e^i)\delta_{\pi_k^i} + e^i \mathrm{Unif}(\Pi^i)$)

  Reset $\widehat{J}_{t_{k+1}}^i = 0 \in \mathbb{R}^{\mathbb{Y}}$ and $\widehat{Q}_{t_{k+1}}^i = 0 \in \mathbb{R}^{\mathbb{Y}\times\mathbb{U}}$

Assumption 23.5.1, strong convergence guarantees can be made. These are presented below in Theorem 23.6.4. In order to state this result, we now fix $\epsilon > 0$ and make the following assumptions on the various parameters of Algorithm 7.

Assumption 23.6.1 *Fix $\epsilon > 0$. For each $i \in \mathcal{N}$ define $\Gamma_{S,\mathrm{MF}}^i$ as*

$$\Gamma_{S,\mathrm{MF}}^i := \{\pi^i \in \Gamma_S^i : \pi^i \text{ is of the mean-field type.}\}.$$

Assume that $\mathbf{\Pi} \subset \prod_{i\in\mathcal{N}} \Gamma_{S,\mathrm{MF}}^i$ is a fine quantization of $\prod_{i\in\mathcal{N}} \Gamma_{S,\mathrm{MF}}^i$ satisfying: (1) $\Pi^i = \Pi^j$ for all $i, j \in \mathcal{N}$; (2) $\mathbf{\Pi} \cap \mathbf{\Gamma}^{\epsilon-eq} \neq \varnothing$; (3) For any $\boldsymbol{\pi} \in \mathbf{\Pi}$, $\boldsymbol{\pi}$ is soft.

Next we present a restriction on the parameters $\{d^i\}_{i\in\mathcal{N}}$. For each player $i \in \mathcal{N}$, the tolerance parameter d^i is taken to be positive, to account for noise in the learned estimates, but small, so that poorly performing policies are not mistaken for ϵ-best-responses [796].

Theorem 23.6.4 ([794]) *Let $\mathcal{G}$ be a partially observed N-player mean-field game satisfying Assumptions 23.5.1 and 23.5.5, and let $\epsilon > 0$. Suppose the policy set $\mathbf{\Pi}$ and the tolerance parameters $\{d^i\}_{i\in\mathcal{N}}$ satisfy mild assumptions and Assump-*

tions 23.6.1, and suppose that all agents follow Algorithm 7. For any $\xi > 0$, *there exists* $\tilde{T} = \tilde{T}(\xi, \epsilon, \mathbf{\Pi}, \{d^i\}_{i\in\mathcal{N}})$ *such that if* $T_k \geq \tilde{T}$ *for all k, then*

$$\Pr\left(\boldsymbol{\pi}_k \in \mathbf{\Pi} \cap \mathbf{\Gamma}^{\epsilon-eq}\right) \geq 1 - \xi,$$

for all sufficiently large k.

For the proof, one shows that if the exploration phases are sufficiently long, then the learning iterates approximate the subjective functions. Then, the policy process $\{\boldsymbol{\pi}_k\}_{k\geq 0}$ obtained from Algorithm 7 can be shown to approximate the policy process of Algorithm 6, and the convergence to near equilibrium is shown using the theory of satisficing paths.

Remark 23.6.1 The policy update $\pi^i_{k+1} \sim (1 - e^i)\delta_{\pi^i_k} + e^i \text{Unif}(\Pi^i)$ was chosen for expositional clarity. One can replace $\delta_{\pi^i_k}$ by any distribution over Π^i without losing the convergence guarantees of Theorem 23.6.4. More generally, $\delta_{\pi^i_k}$ can be replaced by any transition kernel in $\mathcal{P}(\Pi^i | \Pi^i \times \mathbb{R}^{\mathbb{Y}} \times \mathbb{R}^{\mathbb{Y}\times\mathbb{U}})$, where the distribution over Π^i depends on π^i_k, $\widehat{J}^i_{t_{k+1}}$, and $\widehat{Q}^i_{t_{k+1}}$. Furthermore, each agent $i \in \mathcal{N}$ may use a different transition kernel for this update, allowing for heterogeneity in the learning dynamics. This remark parallels a comment made in Sect. 23.6.2.

Learning with Mean-Field State Information

We now present a result analogous to Theorem 23.6.4 for the case of mean-field state observability. Here, we do not guarantee convergence to objective equilibrium but rather subjective equilibrium. Recall that we use $(\mathcal{V}^*, \mathcal{W}^*)$ to denote the naively learned subjective function family for the game $\mathcal{G}$.

Assumption 23.6.2 *Fix* $\epsilon > 0$. *Assume that* $\mathbf{\Pi}$ *is a fine quantization of* $\mathbf{\Gamma}_S$ *satisfying: (1)* $\Pi^i = \Pi^j$ *for all* $i, j \in \mathcal{N}$; *(2)* $\mathbf{\Pi} \cap \text{Subj}_\epsilon(\mathcal{V}^*, \mathcal{W}^*) \neq \varnothing$; *(3) For any* $\boldsymbol{\pi} \in \mathbf{\Pi}$, $\boldsymbol{\pi}$ *is soft.*

Theorem 23.6.5 *Let* $\mathcal{G}$ *be a partially observed N-player mean-field game satisfying Assumptions 23.5.2 and 23.5.5, and let* $\epsilon > 0$. *Suppose the policy set* $\mathbf{\Pi}$ *and the tolerance parameters* $\{d^i\}_{i\in\mathcal{N}}$ *satisfy mild assumptions, and suppose that all agents follow Algorithm 7. For any* $\xi > 0$, *there exists* $\tilde{T} = \tilde{T}(\xi, \epsilon, \mathbf{\Pi}, \{d^i\}_{i\in\mathcal{N}})$ *such that if* $T_k \geq \tilde{T}$ *for all k, then*

$$\Pr\left(\boldsymbol{\pi}_k \in \mathbf{\Pi} \cap \text{Subj}_\epsilon(\mathcal{V}^*, \mathcal{W}^*)\right) \geq 1 - \xi,$$

for all sufficiently large k.

Unlike the setting with global state observability, however, here it is possible that play settles at a subjective equilibrium that is not an objective equilibrium. That is, it is possible that $\left(\mathbf{\Pi} \cap \text{Subj}_\epsilon(\mathcal{V}^*, \mathcal{W}^*)\right) \setminus \mathbf{\Gamma}^{\epsilon-eq} \neq \varnothing$, and that Algorithm 7 drives the play to such a policy.

Learning with Compressed State or Local State Observability

We conclude this section with a discussion of convergence guarantees for Algorithm 7 under compressed state observability (Assumption 23.5.3). As we have discussed previously, there is no guarantee that the set $\mathbf{\Gamma}_S$ contains ϵ-equilibrium policies—either in the objective sense or in the subjective sense using the naively learned subjective function family $(\mathcal{V}^*, \mathcal{W}^*)$. As a result, the convergence guarantees in this setting are constrained, and must be made with a potentially restrictive assumption that subjective ϵ-equilibrium exist within the set of policies $\mathbf{\Pi}$.

Assumption 23.6.3 *Fix $\epsilon > 0$. The set of $(\mathcal{V}^*, \mathcal{W}^*)$-subjective ϵ-equilibrium is non-empty; i.e.,* $\mathrm{Subj}_\epsilon(\mathcal{V}^*, \mathcal{W}^*) \neq \varnothing$.

Assumption 23.6.4 *Fix $\epsilon > 0$. $\mathbf{\Pi}$ is a fine quantization of $\mathbf{\Gamma}_S$ satisfying: (1) $\Pi^i = \Pi^j$ for all $i, j \in \mathcal{N}$; (2) $\mathbf{\Pi} \cap \mathrm{Subj}_\epsilon(\mathcal{V}^*, \mathcal{W}^*) \neq \varnothing$; and, (3) for any $\boldsymbol{\pi} \in \mathbf{\Pi}$, $\boldsymbol{\pi}$ is soft.*

Theorem 23.6.6 *Let $\epsilon > 0$ and let $\mathcal{G}$ be a partially observed N-player mean-field game satisfying Assumptions 23.5.3, 23.5.5, and 23.6.3. Suppose the policy set $\mathbf{\Pi}$ and the tolerance parameters $\{d^i\}_{i\in\mathcal{N}}$ satisfy Assumptions 23.6.4 and mild assumptions, and suppose that all agents follow Algorithm 7. For any $\xi > 0$, there exists $\tilde{T} = \tilde{T}(\xi, \epsilon, \mathbf{\Pi}, \{d^i\}_{i\in\mathcal{N}})$ such that if $T_k \geq \tilde{T}$ for all k, then*

$$\Pr\left(\boldsymbol{\pi}_k \in \mathbf{\Pi} \cap \mathrm{Subj}_\epsilon(\mathcal{V}^*, \mathcal{W}^*)\right) \geq 1 - \xi,$$

for all sufficiently large k.

Theorem 23.6.6 also applies under local state observability, provided that, notably, Assumption 23.6.3. Identification of the precise conditions under which such subjective equilibria exist is an open problem.

Learning Under Non-Markovian Environments

The analysis in this section, and Theorem 22.3.4 presented earlier, imply that one may need to be content with convergence to some subjective equilibrium, when agents are in a complex environment and they only have local information. In particular, the Q-learning algorithms for each exploration phase may converge to a limit, and if subjective equilibria exist, then the algorithm is guaranteed to converge in a precise probabilistic sense under the conditions (on existence of paths) presented in this chapter. This argument holds true even if the environment does not possess a hidden Markovian state, by generalizing the proof of Theorem 22.3.4 [391] to asymptotically ergodic environments, as in [394]. This practically consequential and mathematically interesting problem invites further exploration.

23.7 Concluding Remarks

We studied multi-agent learning under decentralized information structure models, for both stochastic teams and games.

With strictly local information or partial global information, we showed that synchronous or asynchronous quantized Q-learning can be used to arrive at near optimal performance or equilibrium via a two time scales approach. We studied a rich class of setups, including stochastic dynamic teams, weakly acyclic stochastic games, symmetric stochastic dynamic games, mean-field games as well as a large class of 2-player stochastic dynamic games. The underlying equilibria may refer to a rich collection of stationary behavior, as well as to an actual or a subjective equilibrium. An argument we made in the chapter is that subjective equilibrium is an appropriate notion of equilibrium for a class of problems in which there is a complex environment consisting of agents with different on-line or prior information, and agents respond to the environment under some modelling assumptions.

23.8 Bibliographic Notes

Standard Q-learning has been applied to several specific multi agent applications [656, 701]. The long-term behavior for stateless repeated games, for which Q-learning reduces to averaging dynamics, is analyzed in [471], and is strongly connected to the long-term behavior of the well-known Stochastic Fictitious Play (SFP) dynamics [279] in the case of two DMs; see Lemma 4.1 in [471]. In two-DM SFP, each DMi tracks the empirical frequencies of the past decisions of its opponents DM^{-i} and chooses a nearly optimal response (with some experimentation) based on the incorrect assumption that DM^{-i} will choose its decisions according to the empirical frequencies of its past decisions

$$q_t^{-i}(u^{-i}) = \frac{1}{t}\sum_{k=0}^{t-1} 1_{\{u_t^{-i}=u^{-i}\}}, \quad \text{for all } u^{-i}$$

Using the connection between Q-learning dynamics and SFP dynamics, the convergence of Q-learning is established in zero-sum games as well as in partnership games with two DMs; see Proposition 4.2 in [471].

Several other papers extend the Fictitious Play dynamics [150, 605] to stochastic games [179, 654, 752]. Beginning with Brown's fictitious play algorithm [149, 604], there is a large literature on fictitious play and its variants, with most works in this line considering a different information structure than the decentralized one studied here. The bulk of work on fictitious play focuses on settings with perfect monitoring of the actions of other players. This tradition includes the recent works of [472] and [648], which study fictitious play-type algorithms with perfect monitoring in stochastic games. Additionally, multiple recent studies have considered fictitious play-type algorithms for various decentralized information structures, such as [235, 694], and [647].

Due in part to the challenges posed by non-stationarity and decentralized information, most contributions to the literature on independent learners have focused either on the stateless case of repeated games and produced formal results, such as the works of [171, 272, 298, 471, 503, 505], or otherwise studied the multi-state setting and presented only empirical results. In [203], the authors have studied the convergence of single-agent policy gradient algorithms employed in episodic two-player zero-sum games. A complementary study was conducted in [647]. Non-zero-sum games were considered recently in [501]. Further relevant studies for a variety of games and state spaces include [501, 659, 842, 844, 845].

Early contributions to learning in mean-field games include the works of [424] and [792], which studied learning in specific classes of MFGs. Another relatively early contribution to learning in a mean-field environment is that of [791], which studies a model inspired by the mean-field theory of physics; the model used there is closely related to (but different from) typical models of mean-field games. We also note [239, 324, 692, 728, 789] and [21]. These studies generally approach learning in MFGs by analyzing the single-agent problem for a representative agent in the limiting case as $N \to \infty$, and equilibrium is defined using a best-responding condition as well as a consistency condition.

Perhaps a common theme uniting the works cited above is that they are *centralized* methods for finding equilibrium in a mean-field game modelling a decentralized system. The use of a simulator for obtaining data allows these algorithms to sample data as if it were generated by a population of agents using symmetric policies at each step of the algorithm. The analysis presented in some of the results in this chapter, building on [794], has been to understand the patterns of behaviour that emerge when agents use reasonable (if naive and/or limited) learning algorithms in a shared environment.

Establishing existence of subjective equilibria as presented is a generally open problem, the resolution of which is closely related to other solution concepts in both N-player mean-field games and the $N \to \infty$ limit models. Under various names, several solution concepts have been proposed for mean-field games, and a number of approximation results relate these solutions concepts to one another. Some of these approximation results relate solution concepts for models with finitely many players to solution concepts for models with an infinite number of players, while others relate alternative solution concepts within a single model. For a sampling of works in this line, we cite [4, 627, 762, 763] and [38] and the references therein.

This chapter has built on [18, 36, 793, 794, 796].

Chapter 24
Reinforcement Learning for Networked Control and Optimal Quantization

Consistent with the theme of the book, we will cover in this last chapter an application of reinforcement learning to networked control and optimal zero-delay coding; where an MDP formulation was established in Chap. 15.

A theme of the chapter is that all of the problems in which one can formulate the problem via dynamic programming with a weak Feller continuity condition on the Markov transition kernels, under additional mild conditions on the exploration policies, can be investigated via reinforcement learning.

The case study will be zero-delay coding and optimal quantization. As we studied earlier in the book, in Shannon's classical information-theoretic lossy coding problem, one is allowed to encode long sequences of source symbols at once in order to achieve a lower distortion, which is optimal in the limit of unbounded block lengths. Such a block-coding approach is undesirable in many delay-sensitive applications, such as networked control, sensor networks and live-streaming, among others. Accordingly, we are interested in a variant of Shannon's lossy coding problem, where one wishes to send an information source causally at a fixed rate with no delay over a channel with feedback, while minimizing the average distortion at the receiver.

This problem has been studied using stochastic control techniques earlier in the book, leading to existence, structural, and general approximation results. However, these techniques do not provide closed-form solutions for either optimal performance or code designs, and they lead to algorithmic implementations that are computationally difficult. To address this problem, we undertake here a reinforcement learning approach. After revisiting and developing some supporting technical results on regularity and stability properties of the associated processes, we rigorously justify convergence of a quantized Q-learning algorithm to a near-optimal policy for this problem.

S. Yüksel, T. Başar, *Stochastic Teams, Games, and Control under Information Constraints*, Systems & Control: Foundations & Applications,
https://doi.org/10.1007/978-3-031-54071-4_24

24.1 Introduction

Consider the setup in Sects. 15.3.2 and 15.3.4 with a noisy channel with feedback; see Fig. 12.3 with a single encoder/decoder and $y_t = x_t$ or Fig. 15.2 without control u_t. In the following, we consider a source $\{x_t\}_{t\geq 0}$ that is a time-homogeneous, discrete-time Markov process taking values in a finite set $\mathbb{X}$. We assume that this process is irreducible and aperiodic, and has its transition kernel described by $P(x_{t+1}|x_t)$. The distribution of x_0, which we denote by π_0, is available at the encoder and decoder. The channel is a discrete memoryless channel (DMC) with input and output alphabets $\mathcal{M}$ and $\mathcal{M}'$, respectively, and a channel transition matrix given by $T(q_t'|q_t)$ (where $q_t \in \mathcal{M}$ and $q_t' \in \mathcal{M}'$). Finally, we denote the reconstruction sequence as $\{\hat{x}_t\}_{t\geq 0}$, taking values in a finite set $\hat{\mathbb{X}}$.

As in Chap. 15, we denote the *encoder policy* by the sequence $\gamma^e = \{\gamma_t^e\}_{t\geq 0}$ and the *decoder policy* by $\gamma^d = \{\gamma_t^d\}_{t\geq 0}$. At time t, we let the encoder have access to all past channel inputs, all past channel outputs (in the form of channel feedback), and all past and present source symbols in order to generate the channel input. That is, $\gamma_t^e : \mathcal{M}^t \times (\mathcal{M}')^t \times \mathbb{X}^{t+1} \to \mathcal{M}$, and $q_t = \gamma_t^e(q_{[0,t-1]}, q'_{[0,t-1]}, x_{[0,t]})$. We call the set of all such encoder policies the set of *admissible encoder policies*, and denote it by Γ^e. Similarly, we allow the decoder to have access to all past and present channel outputs in order to generate the reconstruction symbol, so that $\gamma_t^d : (\mathcal{M}')^{t+1} \to \hat{\mathbb{X}}$ and $\hat{x}_t = \gamma_t^d(q'_{[0,t]})$, and we denote these admissible decoder policies by Γ^d. We will denote an admissible joint encoding-decoding policy by $\gamma^{ed} \in \Gamma := \Gamma^e \times \Gamma^d$.

For the lossy coding problem, the goal is to minimize the average distortion. In the infinite-horizon case, this is given by

$$J(\pi_0, \gamma) := \limsup_{T\to\infty} \mathbf{E}_{\pi_0}^{\gamma^{ed}} \left[\frac{1}{T} \sum_{t=0}^{T-1} d(x_t, \hat{x}_t) \right],$$

where $d : \mathbb{X} \times \hat{\mathbb{X}} \to [0, \infty)$ is a distortion measure and $\mathbf{E}_{\pi_0}^{\gamma^{ed}}$ is the expectation with initial distribution $X_0 \sim \pi_0$ under policy γ^{ed}. We denote the optimal average cost by $J^*(\pi_0) := \inf_{\gamma^{ed}\in\Gamma} J(\pi_0, \gamma^{ed})$.

Note that for fixed $q_{[0,t-1]}$, $q'_{[0,t-1]}$ and $x_{[0,t-1]}$, the map $\gamma_t^e(q_{[0,t-1]}, q'_{[0,t-1]}, x_{[0,t-1]}, \cdot)$ is a *quantizer* (i.e., a map from $\mathbb{X}$ to $\mathcal{M}$), which we denote by Q_t. Thus we can view an encoder policy at time t as selecting a quantizer Q_t based on $(q_{[0,t-1]}, q'_{[0,t-1]}, X_{[0,t-1]})$, and then quantizing x_t as $q_t = Q_t(x_t)$. Also, since the source alphabet is finite, there clearly exists an optimal decoding policy for every encoding policy. Thus in the following we denote the encoding policy by $\gamma := \gamma^e$, and assume that a corresponding optimal decoding policy is used. We can then restrict our search to finding optimal encoding policies.

Let $\mathcal{P}(\mathbb{X})$ be the space of all probability measures on $\mathbb{X}$, and define $\pi_t \in \mathcal{P}(\mathbb{X})$ as

$$\pi_t(A) := P(X_t \in A|q'_{[0,t-1]}). \tag{24.1}$$

Recall (15.6). We say that an encoder policy $\gamma = \{\gamma_t\}_{t\geq 0}$ is of the *Walrand-Varaiya type* if, at time t, the policy uses only π_t and x_t to generate q_t. That is, γ selects a quantizer $Q_t = \gamma_t(\pi_t)$ and q_t is generated as $q_t = Q_t(x_t)$. Such a policy is called *stationary* if it does not depend on t.

For technical reasons, we also introduce the discounted cost problem, where the goal is to minimize, for some $\beta \in (0, 1)$,

$$J_\beta(\pi_0, \gamma) := \lim_{T\to\infty} \mathbf{E}_{\pi_0}^{\gamma}\left[\sum_{t=0}^{T-1} \beta^t d(X_t, \hat{X}_t)\right].$$

We recall Theorem 15.5.2 for the existence and structure of optimal solutions for the discounted cost criterion. In particular, an optimal policy exists and this policy is stationary and of the Walrand-Varaiya type. We denote the optimal discounted cost by $J_\beta^*(\pi_0) := \inf_{\gamma\in\Gamma} J_\beta(\pi_0, \gamma)$.

Despite the existence results presented in Chap. 15, actually finding an optimal policy (either in closed form or algorithmically) is difficult. Under certain setups for the source and channel, analytical solutions exist. For example, [756] showed memoryless encoding (that is, $q_t = x_t$) is optimal when $\mathbb{X} = \mathcal{M}$ and the channel is symmetric. Furthermore, if the source is independent and identically distributed (iid), one might attempt a Lloyd-Max style algorithm (in the noisy-channel case, this is often called channel-optimized quantization). See for example [250] for discussions regarding these types of quantizers. However, even in the i.i.d. case, such algorithms generally rely on necessary (and not sufficient) conditions for optimality and thus may only obtain local rather than global optima, as we have seen in Chap. 18. These comparisons will be made explicit using simulations.

Thus, for a general source and channel, finding an optimal encoding policy is a generally open problem. It is natural then to pursue a reinforcement-learning approach to find a solution.

24.2 Zero-Delay Source-Channel Coding as an MDP

Recall the setup from Sects. 15.4 and 15.5.1. We have the following.

Proposition 24.2.1 ([785]) *Under a Walrand-Varaiya type policy, the update equation for π_t is given by*

$$\pi_{t+1}(x_{t+1}) = \frac{1}{\sum_{q_t} T(q_t'|q_t)\pi_t(Q_t^{-1}(q_t))} \sum_{q_t} \sum_{x_t\in Q_t^{-1}(q_t)} P(x_{t+1}|x_t)T(q_t'|q_t)\pi_t(x_t), \tag{24.2}$$

where $Q_t^{-1}(q_t) = \{x \in \mathbb{X} : Q_t(x) = q_t\}$.

By the analysis in Chap. 15, under a Walrand-Varaiya type policy, π_{t+1} is conditionally independent of $(\pi_{[0,t-1]}, Q_{[0,t-1]})$ given π_t and Q_t. Then, we have that $\{\pi_t, Q_t\}_{t\geq 0}$ is a controlled Markov chain, and we denote the resulting transition kernel by $P(d\pi_{t+1}|\pi_t, Q_t)$. We also define a cost function for this process in terms of π_t and Q_t:

Lemma 24.2.1 *For a given encoder Q_t, if an optimal decoder is used, per-stage cost can be expressed as*

$$c(\pi_t, Q_t) := \sum_{q_t'} \min_{\hat{x}\in\hat{\mathbb{X}}} \Big(\sum_{q_t} \sum_{x_t\in Q_t^{-1}(q_t)} \pi_t(x_t) T(q_t'|q_t) d(x_t, \hat{x})\Big). \tag{24.3}$$

Proof Note that, for a fixed encoder Q_t, the optimal decoder will be one which chooses $\hat{x}_t$ according to

$$\arg\min_{\hat{x}\in\hat{\mathbb{X}}} \mathbf{E}\Big[d(X_t, \hat{x})|q'_{[0,t]}\Big].$$

Now, consider the posterior distribution

$$P(x_t|q'_{[0,t]}) = \frac{P(x_t, q_t'|q'_{[0,t-1]})}{P(q_t'|q'_{[0,t-1]})} = \frac{\sum_{q_t} P(x_t, q_t', q_t|q'_{[0,t-1]})}{P(q_t'|q'_{[0,t-1]})}$$

$$= \frac{\sum_{q_t} P(x_t|q'_{[0,t-1]}) P(q_t'|q_t, x_t, q'_{[0.t-1]}) P(q_t|x_t, q'_{[0,t-1]})}{P(q_t'|q'_{[0,t-1]})}$$

$$= \frac{\sum_{q_t} \pi_t(x_t) T(q_t'|q_t) \mathbf{1}_{\{Q_t(x_t)=q_t\}}}{P(q_t'|q'_{[0,t-1]})},$$

where the fourth equality is due to the fact that $Q_t(x_t) = q_t$ is deterministic and that given q_t, the probability of q_t' is described entirely by T. Thus, the distortion for each q_t' is given by

$$\min_{\hat{x}\in\hat{\mathbb{X}}} \sum_{x_t} \frac{\sum_{q_t} \pi_t(x_t) T(q_t'|q_t) \mathbf{1}_{\{Q_t(x_t)=q_t\}}}{P(q_t'|q'_{[0,t-1]})} d(x_t, \hat{x})$$

$$= \min_{\hat{x}\in\hat{\mathbb{X}}} \frac{\sum_{q_t} \sum_{x_t\in Q_t^{-1}(q_t)} \pi_t(x_t) T(q_t'|q_t)}{P(q_t'|q'_{[0,t-1]})} d(x_t, \hat{x}).$$

Taking the expectation (conditioned on $q'_{[0,t-1]}$) over all q_t' gives

$$\sum_{q'_t} P(q'_t|q'_{[0,t-1]}) \min_{\hat{x}\in\hat{\mathbb{X}}} \frac{\sum_{q_t}\sum_{x_t\in Q_t^{-1}(q_t)} \pi_t(x_t)T(q'_t|q_t)}{P(q'_t|q'_{[0,t-1]})} d(x_t,\hat{x})$$

$$= \sum_{q'_t} \min_{\hat{x}\in\hat{\mathbb{X}}} \sum_{q_t} \sum_{x_t\in Q_t^{-1}(q_t)} \pi_t(x_t)T(q'_t|q_t)d(x_t,\hat{x}).$$

◇

By Lemma 24.2.1, since we assume an optimal decoder for a given encoder, we have

$$\mathbf{E}_{\pi_0}^{\tilde{\gamma}}\left[\frac{1}{T}\sum_{t=0}^{T-1} c(\pi_t, Q_t)\right] = \mathbf{E}_{\pi_0}^{\gamma}\left[\frac{1}{T}\sum_{t=0}^{T-1} d(X_t,\hat{X}_t)\right].$$

Finally, letting $\tilde{\mathcal{Q}}$ be the set of all quantizers, we can write the zero-delay lossy source-channel coding problem as an MDP:

$$(\mathsf{Z}, \mathsf{U}, P, c) := (\mathcal{P}(\mathbb{X}), \tilde{\mathcal{Q}}, P(\cdot|\pi, Q), c).$$

That is, the state space is the space of probability measures on $\mathbb{X}$, the action space is the set of maps from $\mathbb{X}$ to $\mathcal{M}$, the transition kernel is determined by (24.2), the cost function is determined by (24.3), and an admissible MDP policy $\tilde{\gamma}$ is given by an admissible encoding policy γ.

24.2.1 Topology on Quantizers

Recall the Young topology on quantizers presented in Sects. 8.2 and 15.4. Let $Q : \mathbb{X} \to \mathcal{M}$ be a quantizer, and denote the ith *bin* of Q by $B_i = Q^{-1}(i)$, $i \in \mathcal{M}$. Following [484, 828], with a slight abuse of notation, a quantizer Q can alternatively be represented as a stochastic kernel from $\mathbb{X}$ to $\mathcal{M}$ such that $Q(i|x) = 1_{\{x\in B_i\}}$, $i \in \mathcal{M}$. Then for $P \in \mathcal{P}(\mathbb{X})$, we denote by PQ the joint probability measure $PQ(x, y) = P(x)Q(y|x)$, $(x, y) \in \mathbb{X}\times\mathcal{M}$. If we introduce the equivalence relation $Q \equiv Q'$ iff $PQ = PQ'$, then we can endow the equivalence classes with the weak convergence topology. Under this topology, [484] showed several properties of the above MDP, including the following, which was essentially studied in the context of Theorem 15.7.1:

Lemma 24.2.2 *[484, Lemma 11]. The transition kernel $P(d\pi_{t+1}|\pi_t, Q_t)$ is weakly continuous in (π_t, Q_t). That is, $\int_{\mathcal{P}(\mathbb{X})\times\tilde{\mathcal{Q}}} f(\pi')P(d\pi'|\pi, Q)$ is continuous on $\mathcal{P}(\mathbb{X}) \times \tilde{\mathcal{Q}}$ for all continuous bounded f on $\mathcal{P}(\mathbb{X})$.*

24.2.2 Predictor and Filter Stability and Unique Ergodicity Under an Exploration Policy

For the analysis toward applying the Quantized Q-learning Algorithm 1, we will need some supporting results. We introduce the *filter*, which is obtained by further conditioning π_t on q'_t:

$$\overline{\pi}_t(A) := P(x_t \in A | q'_{[0,t]}).$$

The filter admits a recursion equation similar to (24.2). Note that these recursions are dependent on the initialization of π_0, which is called the *prior*. We denote the predictor (respectively, filter) process resulting from the prior $\pi_0 = \nu$ as $\{\pi_t^\nu\}_{t\geq 0}$ (respectively, $\{\overline{\pi}_t^\nu\}_{t\geq 0}$).

Definition 24.2.1 *We say that the predictor (respectively, filter) process is stable in total variation in expectation if for any $\mu, \nu \in \mathcal{P}(\mathbb{X})$ such that μ is absolutely continuous with respect to ν, we have*

$$\lim_{n\to\infty} E^\mu[\|\pi_t^\mu - \pi_t^\nu\|_{TV}] = 0.$$

We will use the following three lemmas to deduce predictor stability.

Lemma 24.2.3 ([523]) *The filter is stable in total variation in expectation if and only if the predictor is stable in total variation in expectation.*

Lemma 24.2.4 *[732, Corollary 5.5] Let $\{x_t\}_{t\geq 0}$ be a discrete-time Markov chain and $\{Y_t\}_{t\geq 0}$ be a stochastic process such that the Y_t are conditionally independent given $\{x_t\}_{t\geq 0}$ so that for each $t \geq 0$*

$$P(Y_t \in A | X_{[0,t]}) = \int_A g(X_t, y)\psi(dy),$$

where $g(x, y)$ is a probability density with respect to the σ-finite measure ψ. If g is strictly positive, and $\{X_t\}_{t\geq 0}$ is positive Harris recurrent and aperiodic, then the filter $\overline{\pi}_t(A) := P(X_t \in A | y_{[0,t]})$ is stable in total variation in expectation.

Lemma 24.2.5 *Under a uniform exploration policy, the filter process $\{\overline{\pi}_t\}_{t\geq 0}$ is stable in total variation in expectation.*

Proof We have

$$\begin{aligned} P(q'_t|x_t) &= \sum_{q_t}\sum_{Q_t} P(q_t|x_t, Q_t)T(q'_t|q_t)P(Q_t|x_t) \\ &= \sum_{q_t}\sum_{Q_t} 1_{\{Q_t(x_t)=q_t\}}T(q'_t|q_t)P(Q_t|x_t) \end{aligned}$$

$$= \sum_{q_t} T(q_t'|q_t) \sum_{\{Q_t : Q_t(x_t) = q_t\}} P(Q_t|x_t).$$

where the second equality follows from the fact that $q_t = Q_t(x_t)$ is deterministic. Now, under a uniform exploration policy, the quantizer Q_t is chosen independently and randomly, so $P(Q_t|x_t) = P(Q_t) = |\tilde{\mathcal{Q}}|^{-1}$. Thus we have

$$P(q_t'|x_t) = \sum_{q_t} T(q_t'|q_t) \sum_{\{Q_t : Q_t(x_t) = q_t\}} |\tilde{\mathcal{Q}}|^{-1}.$$

Since we are considering the set of all possible quantizers, for any $(x_t, q_t) \in \mathbb{X} \times \mathcal{M}$, the set $\{Q_t : Q_t(x_t) = q_t\}$ is nonempty. Thus the inner sum is positive for all (x_t, q_t). Furthermore, for at least one $q_t \in \mathcal{M}$, we have $T(q_t'|q_t) > 0$ (otherwise, q_t' is not a valid channel output). Thus $P(q_t'|x_t) > 0$ everywhere. Then we use Lemma 24.2.4 with $Y_t = q_t'$ to conclude that the filter is stable in total variation in expectation. ◇

The above lead to the following.

Corollary 24.2.1 *Under a uniform exploration policy, the predictor process* $\{\pi_t\}_{t \geq 0}$ *is stable in total variation in expectation.*

The following result is inspired by [689, Theorem 3], while using additional Markov properties of $\{\pi_t\}_{t \geq 0}$. See [191] for a proof.

Theorem 24.2.1 *Under a uniformly distributed exploration policy, the process* $\{\pi_t\}_{t \geq 0}$ *admits a unique invariant measure.*

By Theorem 24.2.1, if x_0 is distributed according to its invariant probability measure the convergence of time averages will hold (in view of filter stability), and we will have that the Quantized Q-learning Algorithm 1 and its convergence results in Sect. 21.4 will be applicable.

Remark 24.1 (Controlled Models) Note that the results above also hold when the source $\{x\}_{t \geq 0}$ is controlled; see Sect. 15.3.3. That is, the controlled chooses an action $u_t = \gamma_t^d(q'_{[0,t]}, u_{[0,t-1]})$ and the source evolves according to $P(x_t|x_{t-1}, u_{t-1})$. Accordingly, one can apply variations in the setup where a controlled Markov source is encoded over a channel and controlled simultaneously.

24.3 Algorithms

We now present the ingredients to implement quantized Q-learning for the zero-delay lossy source-channel coding problem.

Algorithm 8: Reznik's algorithm

$n \geq 1, \pi_t = (p_1, \ldots, p_m)$
For $i = 1$ **to** m
$k'_i = \lfloor np_i + \frac{1}{2} \rfloor$
$n' = \sum_i k'_i$
If $n = n'$
Return $(\frac{k'_1}{n}, \ldots, \frac{k'_m}{n})$
For $i = 1$ **to** m
$\delta_i = k'_i - np_i$
Sort δ_i s.t. $\delta_{i_1} \leq \ldots \leq \delta_{i_m}$
$\Delta = n' - n$
If $\Delta > 0$
$k_{i_j} = \begin{cases} k'_{i_j} & j = 1, \ldots, m - \Delta \\ k'_{i_j} - 1 & j = m - \Delta + 1, \ldots, m \end{cases}$
Else
$k_{i_j} = \begin{cases} k'_{i_j} + 1 & j = 1, \ldots, |\Delta| \\ k'_{i_j} & j = |\Delta| + 1, \ldots, m \end{cases}$
Return $(\frac{k_1}{n}, \ldots, \frac{k_m}{n})$

24.3.1 *Quantizing* π_t

Since the state space $\mathbb{X}$ is finite, say with $|\mathbb{X}| = m$, then $\mathcal{P}(\mathbb{X})$ is a simplex in $\mathbb{R}^m$. Let $n \in \mathbb{N}$ and consider the set $\mathcal{P}_n(\mathbb{X}) := \{\pi' \in \mathcal{P}(\mathbb{X}) : \pi' = [\frac{k_1}{n}, \ldots, \frac{k_m}{n}], k_i = 0, \ldots, n\}$. For a given π_t, we want to find the nearest $\hat{\pi}_t \in \mathcal{P}_n(\mathbb{X})$. We use an algorithm developed by Reznik [601, Algorithm 1] for this purpose.

Via [601, Proposition 2], the maximum radius of the quantizer for $\mathcal{P}(\mathbb{X})$ is given by

$$d_\infty = \frac{1}{n}\left(1 - \frac{1}{m}\right).$$

Also note that the number of levels when using Reznik's algorithm is given by $N = \binom{n+m-1}{m-1}$ [601].

24.3.2 *Quantized* Q*-Learning for Source-Channel Coding*

We can now state the adaptation of the Quantized Q-learning Algorithm 1 for the zero-delay coding problem:

Theorem 24.3.1 *In Algorithm 9, the* Q*-factors* $\{\mathsf{Q}_t\}_{t\geq 0}$ *converge almost surely to a limit* Q^*. *Furthermore, if the algorithm is used with large enough* n, *the encoding policy* γ^* *given by*

Algorithm 9: Quantized Q-learning for source-channel coding

Set Parameters
source alphabet $\mathbb{X}$, transition kernel $P(x_{t+1}|x_t)$, initial distribution π_0, channel kernel $T(q_t'|q_t)$, quantization parameter n, quantizer set $\widetilde{\mathcal{Q}}$, uniform exploration policy γ
Initialize Q-factor $\mathsf{Q}_0 : \mathcal{P}_n(\mathbb{X}) \times \widetilde{\mathcal{Q}} \to \mathbb{R}$
Sample x_0 according to invariant measure π_0
Quantize π_0 using Reznik's algorithm with parameter n, call this $\hat{\pi}_0$
Select quantizer Q_0 according to γ
$q_0 = Q_0(x_0)$
Generate q_0' according to $T(q_0'|q_0)$
For $t \geq 0$
Compute $c(\pi_t, Q_t)$ ▷ see (24.3)
Generate x_{t+1} according to $P(x_{t+1}|x_t)$
Compute π_{t+1} ▷ see (24.2)
Quantize π_{t+1} using Reznik's algorithm with parameter n, call this $\hat{\pi}_{t+1}$
Update Q-factor Q_t: ▷ see (21.21)
Select quantizer Q_{t+1} according to γ
$q_{t+1} = Q_{t+1}(x_{t+1})$
Generate q_{t+1}' according to $T(q_{t+1}'|q_{t+1})$
End

$$\gamma^*(\pi) = \arg\min_{Q \in \widetilde{\mathcal{Q}}} \mathsf{Q}^*(\hat{\pi}, Q) \quad \textit{for all } \pi \in B_i,$$

is near-optimal for the discounted cost problem when $x_0 \sim \pi_0$ *with* π_0 *invariant and where* B_i *are the bins under Reznik's algorithm* $\left(\textit{thus, } i = 1, \ldots, \binom{n+m-1}{m-1}\right)$.

24.3.3 Connection to the Average Cost Problem

Note that we have studied the discounted cost problem, even though the average cost problem is typically the objective in source-channel coding. Here we make explicit some connections to the average cost problem.

Theorem 24.3.2 *[820, Theorem 7.3.3] Consider an MDP* $(\mathsf{X}, \mathsf{U}, P, c)$, *and suppose that the following hold:*

(i) c *is continuous, nonnegative, and bounded.*
(ii) X *is compact.*
(iii) U *is bounded.*
(iv) $P(\cdot|x, u)$ *is weakly continuous in* (x, u).
(v) The family of functions $\{h_\beta : \beta \in (0, 1)\}$, *where*

$$h_\beta(x) := J_\beta^*(x) - J_\beta^*(x_0)$$

for some fixed $x_0 \in \mathsf{X}$, *is uniformly bounded and equicontinuous.*

Then there exist a constant $g^* \geq 0$ *and a measurable function* $f^* : \mathsf{X} \to \mathsf{U}$ *such that the policy* $\gamma^* = \{f^*\}_{t\geq 0}$ *is optimal for the average cost problem and* g^* *is the optimal value function, i.e.,*

$$g^* = J^*(x) = J(x, \gamma^*).$$

Furthermore, $g^* = \lim_{n\to\infty}(1-\beta_n)J^*_{\beta_n}(x)$ *for any sequence* $\{\beta_n\}_{n\geq 0}$ *such that* $\lim_{n\to\infty}\beta_n = 1$.

Corollary 24.3.1 *For the zero-delay source-channel coding problem, for any sequence* $\{\beta_n\}_{n\geq 0}$ *with* $\lim_{n\to\infty}\beta_n = 1$ *we have* $J^*(\pi_0) = \lim_{n\to\infty}(1-\beta_n)J^*_{\beta_n}(\pi_0)$.

Proof We verify that the conditions of Theorem 24.3.2 hold for the zero-delay coding problem. Indeed (i) holds from the definition of the cost equation (24.3), (ii) and (iii) hold since the source $\mathbb{X}$ and the set of quantizers $\widetilde{\mathcal{Q}}$ are finite, (iv) holds from Lemma 24.2.2, and (v) holds from [785, Eqn 26, Lemma 1]. Then the result follows by applying Theorem 24.3.2. ◇

Note that the above result only concerns the value functions, and does not necessarily give that the *policy* attained by the discounted cost problem is near-optimal for the average cost problem. Nonetheless, the obtained policy can be slightly modified to arrive at a near optimal policy for the average cost criterion as well [191].

24.3.4 Simulations

For the following simulations, we will use the squared-error distortion measure $d(x, \hat{x}) = (x - \hat{x})^2$. Each data point below is generated by running a finite-horizon ($T = 10^5$) discounted cost problem with low discounting ($\beta = 0.9999$).

Optimal Quantizers for Symmetric Channels Here, we consider the case where $\mathbb{X} = \hat{\mathbb{X}} = \mathcal{M} = \{1, \ldots, 4\}$ (that is, we consider 4-level scalar quantizers). Also, $\mathcal{M}' = \{1, \ldots, 5\}$ (the event $q'_t = 5$ can be interpreted as an "erasure" by the channel). The channel is a 4-ary symmetric erasure channel with erasure probability 0.1 and error probability 0.05. The Markov source transition matrix is given by

$$P = \begin{pmatrix} 0 & 1 & 0 & 0 \\ \frac{1}{2} & 0 & \frac{1}{2} & 0 \\ \frac{2}{3} & 0 & 0 & \frac{1}{3} \\ \frac{1}{4} & \frac{1}{2} & \frac{1}{4} & 0 \end{pmatrix}.$$

The case where $\mathbb{X} = \mathcal{M}$ and the channel is symmetric was studied by Walrand and Varaiya [756]. It was shown that encoding is "useless" in this case, i.e., the policy that chooses $Q_t(x_t) = x_t$ is optimal. We see that the algorithm approaches this optimal cost as n, the number of bins, increases (Fig. 24.1).

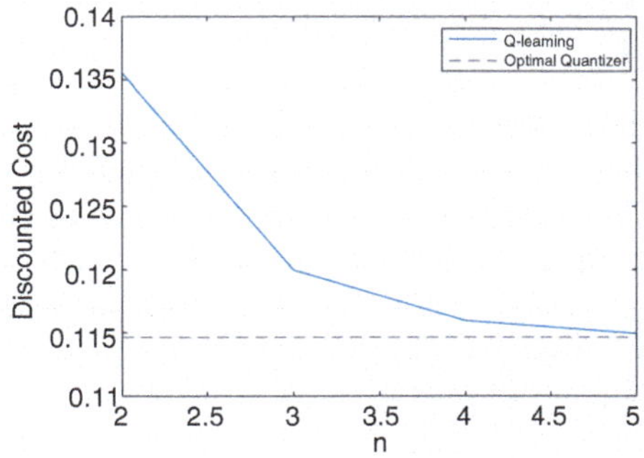

Fig. 24.1 Comparison with an optimal quantizer

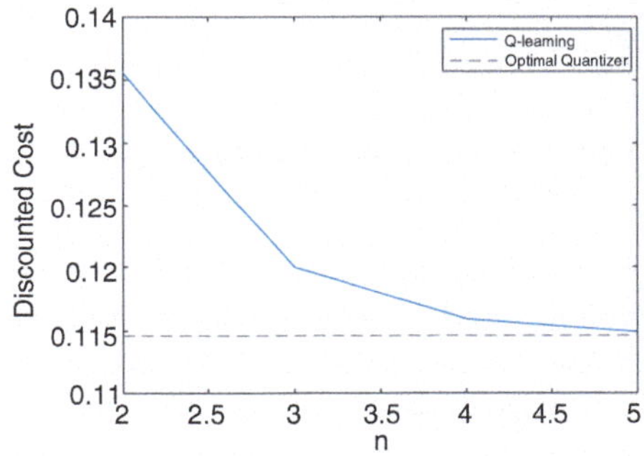

Fig. 24.2 Comparison with COSQ in i.i.d. case

Comparison with a Channel-Optimized Quantizer *A note on optimality:* In the following, we consider an i.i.d. source and compare the algorithm to a channel-optimized scalar quantizer (COSQ). While Lloyd-Max type algorithms (see Theorem 18.2.4) may only converge to local optimality, the algorithm presented here for large enough n is guaranteed to give (near) *global* optimality with probability 1. In the following, we use the source alphabet $\mathbb{X} = \hat{\mathbb{X}} = \{1, \ldots, 4\}$ and the channel input and output are given by $\mathcal{M} = \mathcal{M}' = \{1, 2\}$ (that is, we consider 2-level scalar quantizers). Finally, the channel is given by a binary symmetric channel with error probability 0.1. The source distribution is given by $P = (0.2, \ 0.05, \ 0.4, \ 0.35)$. Simulation results are shown in Fig. 24.2. Further simulations are available in [191].

24.4 Concluding Remarks

We considered an application of reinforcement learning to networked control and optimal zero-delay coding. The proof combined a number of preceding results on quantized Q-learning, weak Feller continuity, as well as unique ergodicity of the predictor process under an exploration policy. We have noted that the algorithm is quite general and is applicable for near global optimality, whereas the popular Lloyd-Max algorithm is only guaranteed to converge to local optimality in the absence of strong regularity conditions.

24.4.1 Controlled Models

The approach presented in the chapter can be applied to the setup depicted in Fig. 14.1 where there is a controlled system which is being controlled by only information received over a noisy/noiseless channel. This will lead to a jointly optimal coding and control problem whose computational analysis is not easy and, accordingly, reinforcement learning would be an appropriate solution approach toward near optimal design. The weak Feller conditions presented in this chapter apply identical to such a model. It would be an interesting problem to fully investigate such general models building on the paradigm presented in this, and the previous few chapters.

24.5 Bibliographic Notes

An extensive literature review on zero-delay coding and networked control was provided earlier in the book, in particular in Chap. 15. The results in the chapter primarily build on [190, 191] which provide further analysis.

Appendix A
Topological Notions and Optimization

This appendix provides some background material on those aspects of real analysis and optimization that are frequently used in the text; it also serves to introduce the reader to some of the notation and terminology used in the book. For a more detailed exposition on the topics covered here, two standard references are [489] and [196].

A.1 Sets

A set S is a collection of elements. If s is a member (element) of S, we write $s \in S$; if s does not belong to S, we write $s \notin S$. If S contains a finite number of elements, it is called a *finite set*; otherwise it is called an *infinite set*. If the number of elements of an infinite set is countable (i.e., if there is a one-to-one correspondence between its elements and positive integers), we say that it is a *denumerable* (*countable*) set, otherwise it is a *nondenumerable* (*uncountable*) set. An example for the latter is the set of all real numbers, $\mathbb{R}$

A set $\mathbb{X}$ with some specific structure attached to it is called a *space*, and it is called a linear (*vector*) *space* if this specific structure is of algebraic nature with certain well-known properties which we assume the reader is familiar with. If S is a vector space, a subset of S which is also a vector space is called a *subspace*. An example of a vector space is the *n-dimensional Euclidean space* (denoted as $\mathbb{R}^n$) each element of which is determined by n real numbers. An $x \in \mathbb{R}^n$ can either be written as a *row vector* $x = (x_1, \dots, x_n)$ where $x_1, \dots, x_n$ are real numbers and denote the components of x, or as a *column vector* which is the "transpose" of $(x_1, \dots, x_n)$ [written as $x = (x_1, \dots, x_n)'$]. We shall adopt the latter convention in this text, unless indicated otherwise.

Linear Independence

Given a finite set of vectors $\{s_1, \dots, s_n\}$, in a vector space S, we say that this set of vectors is *linearly independent* if the equation $\sum_{i=1}^{n} \alpha_i s_i = 0$ implies $\alpha_i = 0 \; \forall i = 1, \dots, n$. Furthermore, if every element of S can be written as a linear combination of these vectors, we say that this set of vectors *generates* S. Now, if S is generated by such a linearly independent finite set (say, X), it is said to be *finite dimensional* with its unique "dimension" being equal to the number of elements of X; otherwise, S is *infinite dimensional*.

S. Yüksel, T. Başar, *Stochastic Teams, Games, and Control under Information Constraints*, Systems & Control: Foundations & Applications,
https://doi.org/10.1007/978-3-031-54071-4

A.2 Linear, Metric, Hilbert, Banach and Standard Borel Spaces

Normed Linear (Vector) Spaces

A *normed linear vector space* is a linear (vector) space S which has some additional structure of topological nature. This structure is induced on S by a real-valued function which maps each element $u \in S$ into a real number $\|u\|$ called the *norm* of u. The norm satisfies the following three axioms, where $\underline{o}$ is the *zero* element:

$$
\begin{aligned}
&(1)\ \|u\| \geq 0\ \ \forall u \in S; \quad \|u\| = 0 \text{ if, and only if, } u = \underline{o}.\\
&(2)\ \|u + v\| \leq \|u\| + \|v\|, \quad \text{for each pair } u, v \in S.\\
&(3)\ \|\alpha u\| = |\alpha| \cdot \|u\|\ \ \forall \alpha \in \mathbb{R} \text{ and for each } u \in S.
\end{aligned}
$$

Convergent Sequences, Cauchy Sequence

An infinite sequence of vectors $\{s_1, s_2, \ldots\}$ in a normed vector space S is said to *converge* to a vector s if, given an $\epsilon > 0$, there exists an integer N such that $\|s - s_i\| < \epsilon$ for all $i \geq N$. In this case, we write $s_i \rightarrow s$, or $\lim_{i\rightarrow\infty} s_i \rightarrow s$, and call s the *limit point* of an infinite sequence $\{s_i\}$. More generally, a point s is said to be a *limit point* of the infinite sequence $\{s_i\}$ if the latter has an infinite subsequence $\{s_i\}$ that converges to s.

An infinite sequence $\{s_i\}$ in a normed vector space is said to be a *Cauchy sequence* if, given an $\epsilon > 0$, there exists an integer N such that $\|s_n - s_m\| < \epsilon$ for all $n, m \geq N$. A normed vector space S is said to be *complete*, or a Banach space, if every Cauchy sequence in S is convergent to an element of S.

Open, Closed and Compact Sets

Let S be a normed vector space. Given an $s \in S$ and an $\epsilon > 0$, the set $N_\epsilon(s) = \{x \in S : \|x - s\| < \epsilon\}$ is said to be an ϵ- *neighborhood* of s. A subset X of S is open if, for every $x \in X$, there exists an $\epsilon > 0$ such that $N_\epsilon(x) \subset X$. A subset X of S is *closed* if its complement in S is open; equivalently, X is closed if every convergent sequence in X has its limit point in X. Given a set $X \subset S$, the largest subset of X which is open is called the *interior* of X.

A subset X of a normed vector space S is said to be *(sequentially) compact* if every infinite sequence in X has a convergent subsequence whose limit point is in X. If X is finite dimensional, compactness is equivalent to being closed and bounded.

Denseness and Separability

Let S be a normed linear space, and X be a subset of S. X is said to be *dense* in S if given any $s \in S$ and $\epsilon > 0$, there exists $x \in X$ such that $\|x - s\| \leq \epsilon$. S is said to be *separable* if there exists a countable set $X \subset S$ which is dense in S.

Metric Spaces

A metric defined on a set X is a function $d : X \times X \rightarrow \mathbb{R}_+$ such that:

$$
\begin{aligned}
&(1)\ d(x, y) \geq 0, \quad \forall x, y \in X, \quad d(x, y) = 0 \text{ if, and only if, } x = y.\\
&(2)\ d(x, y) = d(y, x), \quad \forall x, y \in X.\\
&(3)\ d(x, y) \leq d(x, z) + d(z, y), \quad \forall x, y, z \in X.
\end{aligned}
$$

A metric space (X, d) is a set equipped with a metric d. A normed linear space is also a metric space, with metric $d(x, y) = \|x - y\|$. A metric space (X, d) is complete, if every Cauchy sequence in the space has a limit in X (where the definition of a Cauchy sequence is the same as that for a normed linear space with $\|\cdot\|$ replaced by $d(\cdot, \cdot)$). A complete, separable metric space is also called a Polish metric space. More generally, a space which admits a metric under which it is complete and separable is called a Polish space; while the metric is often not specified for such a space, we will assume that a metric is given.

A Borel subset of a Polish space is called a *Standard Borel* space. Such spaces are related, bijectively, to either finite sets, or countable sets, or $\mathbb{R}$. Most of the spaces we deal with in this book are standard Borel spaces.

Inner-Product Spaces

Given a linear vector space S, an *inner product* on S is a bilinear map $\langle\cdot,\cdot\rangle : S \times S \to \mathbb{R}$ which is positive definite and symmetric, i.e.,

(1) $\langle u, u\rangle \geq 0 \; \forall u \in S$; $\langle u, u\rangle = 0$ if, and only if, $u = \underline{o}$, the *zero* element.
(2) $\langle u, v\rangle = \langle v, u\rangle, \quad \forall u, v \in S$
(3) $\langle \alpha u + \beta v, \gamma w\rangle = \alpha\gamma \langle u, w\rangle + \beta\gamma\langle v, w\rangle \quad \forall \alpha, \beta, \gamma \in \mathbb{R}; u, v, w \in S$.

An important property of the inner product is that it satisfies the Cauchy-Schwartz inequality:

$$|\langle u, v\rangle|^2 \leq \langle u, u\rangle \langle v, v\rangle.$$

A vector space S, equipped with an inner product $\langle\cdot,\cdot\rangle$ is called an *inner-product space*. The inner product induces a natural norm, $\|u\| := \sqrt{\langle u, u\rangle}$, so that an inner-product space is also a normed vector space. Furthermore, if this inner-product space is also complete (that is, a Banach space) it is called a real *Hilbert space*, which we denote by $\mathbf{H}$.

An important notion in a Hilbert space $\mathbf{H}$ is orthogonality. We say that $u \in \mathbf{H}$ and $v \in \mathbf{H}$ are *orthogonal* if $\langle u, v\rangle = 0$, in which case we write $u \perp v$. If S is a subspace of $\mathbf{H}$, then the *orthogonal complement* $S^\perp$ is defined by

$$S^\perp = \{u \in \mathbf{H} : \langle u, v\rangle = 0 \quad \forall v \in S\}.$$

An important property of $S^\perp$ is that it is a *closed linear subspace* of $\mathbf{H}$, and that given any $x \in \mathbf{H}$ there is a unique element in $S^\perp$ (say $x_{S^\perp}$) and a unique element in the closure of S ($\bar{S}$) (say $x_{\bar{S}}$) such that

$$x = x_{\bar{S}} + x_{S^\perp}.$$

Equivalently, $\mathbf{H}$ can be written as the direct sum of $\bar{S}$ and $S^\perp$:

$$\mathbf{H} = \bar{S} \oplus S^\perp.$$

In the above, $x_{S^\perp}$ is called the *orthogonal projection* of x *on* $S^\perp$, and we have the norm identity

$$\|x\|^2 = \|x_{\bar{S}}\|^2 + \|x_{S^\perp}\|^2$$

which is also known as the *Pythagorean Theorem.*

A fundamental result in functional analysis, which we will have occasion to use in the text, is the celebrated *Orthogonal Projection Theorem*:

The Orthogonal Projection Theorem Let $\mathbf{H}$ be a Hilbert space and S a closed subspace of it. Then, given $x \in \mathbf{H}$, there exists a unique $s_0 \in S$, such that

$$\|x - s_0\| = \min_{s \in S} \|x - s\|.$$

Furthermore, a necessary and sufficient condition for $s_0 \in S$ to be the minimizing vector is that $(x - s_0) \perp S$.

Transformations and Continuity

A mapping f of a vector space S into a vector space T is called a *transformation* or a *function*, and is written symbolically $f : S \to T$ or $y = f(x)$, for $x \in S$, $y \in T$. f is said to be a *functional* if $T = \mathbb{R}$.

Let $f : S \to T$ where S and T are normed linear spaces. f is said to be *continuous at a point* $x_0 \in S$ if, for every $\epsilon > 0$, there exists a $\delta > 0$ such that $f(x) \in N_\epsilon(f(x_0))$ for every $x \in N_\delta(x_0)$, where the notation $N_\delta(z) := \{x : \|x - z\| \leq \delta\}$ denotes the ball with radius δ centered at z. If f is continuous at every point of S it is said to be *continuous everywhere* or, simply, *continuous*. f is said to be *uniformly continuous* if δ depends only on ϵ and not on x_0. Furthermore, f is said to be *linear* (or a *linear operator*) if $f(\alpha u + \beta v) = \alpha f(u) + \beta f(v)$, for all $u, v \in S$; $\alpha, \beta \in \mathbb{R}$.

Let S and T be as above, and F be a family of functions from S to T. The family F is said to be *equicontinuous at a point* $x_0 \in S$ if, for every $\epsilon > 0$, there exists a $\delta > 0$ such that $f(x) \in N_\epsilon(f(x_0))$ for all $f \in F$ and for every $x \in N_\delta(x_0)$. The family F is said to be *equicontinuous* if it is equicontinuous at each $x \in S$. The family F is *uniformly equicontinuous* if δ depends only on ϵ and not on x_0.

Let S and T be Banach spaces, and $A : S \to T$ a linear operator. The quantity

$$\ll A \gg = \sup_{\|x\| \leq 1} \|Ax\|$$

is called the (*operator*) *norm* of A, and A is said to be *bounded* if there exists a scalar $k > 0$ such that $\ll A \gg \leq k$. A is said to be a *contraction operator* if this bound is less than 1, i.e., $\ll A \gg < 1$.

If $S = T$, another related quantity is the *spectral radius* of a bounded linear operator $A : S \to S$, defined by

$$\rho(A) = \lim_{k\to\infty} \sup\, [\ll A^k \gg]^{1/k}.$$

An important relationship between the norm and the spectral radius is the inequality

$$\rho(A) \leq \ll A \gg,$$

for $A : S \to S$.

Now again let S and T not necessarily be the same, but assume that each is a Hilbert space. Then, the adjoint of a linear bounded operator $A : S \to T$, denoted by A^*, is defined by the equality

$$\langle Au, v\rangle = \langle u, A^* v\rangle, \quad \forall u \in S,\ \forall v \in T.$$

Note that here the two inner products are not the same, unless S and T are identical. Now let $S = T =: \mathbf{H}$, a common Hilbert space, and introduce the notation $\mathcal{L}(\mathbf{H})$ to denote the space of all linear bounded operators from $\mathbf{H}$ into $\mathbf{H}$. Then we have the following useful terminology and properties.

1. $A \in \mathcal{L}(\mathbf{H})$ is said to be *self-adjoint* if $A^* = A$.
2. For a self-adjoint operator A,

$$\ll A \gg = [\rho(AA^*)]^{\frac{1}{2}} = [\rho(A^*A)]^{\frac{1}{2}}.$$

3. A self-adjoint operator A is *positive* (equivalently, *nonnegative*) if

$$\langle u, Au\rangle \geq 0, \quad \forall u \in \mathbf{H}.$$

A is *strongly positive* if there exists a scalar $\alpha > 0$ such that

$$\langle u, Au\rangle \geq \alpha\langle u, u\rangle, \quad \forall u \in \mathbf{H}.$$

4. The norm of a self-adjoint operator $A \in \mathcal{L}(\mathbf{H})$ is also given by

$$\ll A \gg = \sup_{\|x\| \leq 1} |\langle x, Ax \rangle|.$$

5. Every strongly positive operator $A \in \mathcal{L}(\mathbf{H})$ has a unique strongly positive square root $A^{\frac{1}{2}} \in \mathcal{L}(\mathbf{H})$, so that $A^{\frac{1}{2}} A^{\frac{1}{2}} = A$.
6. Given an $A \in \mathcal{L}(\mathbf{H})$, its inverse A^{-1} is defined by

$$A^{-1}A = AA^{-1} = I$$

where $I \in \mathcal{L}(H)$ is the *identity* operator (i.e., $Ix = x, \forall x \in \mathbf{H}$). A^{-1} also belongs to $\mathcal{L}(\mathbf{H})$.
7. Let S be a closed linear subspace of $\mathbf{H}$. $P \in \mathcal{L}(H)$ is called a *projection operator* (into S) if $Px \in S$ for every $x \in \mathbf{H}$. Here S is called the *range space* of P. Two important properties of P are that (1) $P^2 = I$ (i.e., P is *idempotent*), (2) $\ll P \gg = 1$. (Projection operators are also defined on more general Banach spaces).
8. Given two operators $A, B \in \mathcal{L}(\mathbf{H})$, the following inequalities hold

$$\begin{aligned} \ll AB \gg &\leq \ll A \gg \ \ll B \gg \\ \rho(AB) &\leq \rho(A)\rho(B). \end{aligned}$$

(The first inequality is valid also on Banach spaces, S and T, where $A : M \to T$, $B : S \to M$, and M is another Banach space. Here the operator norm is the one induced by the norm defined on T.)

If a linear operator has finite dimensional domain and range spaces, it is called a matrix, whose properties are further discussed in the following section.

A.3 Matrices

An $(m \times n)$ matrix A is a rectangular array of numbers, called *elements* or *entries*, arranged in m rows and n columns. The element in the ith row and jth column of A is denoted by a subscript ij, such as a_{ij} or $[A]_{ij}$, in which case we write $A = \{a_{ij}\}$.

A matrix is said to be *square* if it has the same number of rows and columns; an $(n \times n)$ square matrix A is said to be an *identity matrix* if $a_{ii} = 1, i = 1, \ldots, n$, and $a_{ij} = 0, i \neq j$, $i, j = 1, \ldots, n$. An $(n \times n)$ identity matrix will be denoted by I_n or, simply, by I whenever its dimension is clear from the context.

The *transpose* of an $(m \times n)$ matrix A is the $(n \times m)$ matrix A' with elements $a'_{ij} = a_{ji}$. A square matrix A is symmetric if $A = A'$; it is nonsingular if there is an $(n \times n)$ matrix called the inverse of A, denoted by A^{-1}, such that $A^{-1}A = I = AA^{-1}$.

Eigenvalues and Quadratic Forms

Corresponding to a square matrix A, a scalar (possibly complex valued) λ and a nonzero vector (also possibly complex valued) x satisfying the equation $Ax = \lambda x$ are said to be, respectively, an *eigenvalue* and an *eigenvector* of A.

A square symmetric matrix A has all its eigenvalues real. If they are all positive (respectively, nonnegative), then A is said to be *positive definite* (respectively, *nonnegative definite* or *positive semidefinite*). An equivalent definition is as follows. A symmetric $(n \times n)$ matrix A is said to be positive definite (respectively, nonnegative definite) if $x'Ax > 0$ (respectively, $x'Ax \geq 0$) for all nonzero $x \in \mathbb{R}^n$. The matrix A is said to be *negative definite* (respectively, *nonpositive definite*) if the matrix $(-A)$ is positive (respectively, nonnegative) definite. We symbolically write $A > 0$ (respectively, $A \geq 0$) to denote that A is positive (respectively, nonnegative) definite.

Norm and Spectral Radius

The *norm* of an $(m \times n)$ matrix A, consistent with the operator norm introduced in Sect. A.2, is given by

$$\ll A \gg = \max\{\lambda : \det(A'A - \lambda I) = 0\} ;$$

in other words, $\ll A \gg$ is the maximum eigenvalue of the symmetric matrix $A'A$ (or equivalently that of AA').

If A is a square matrix, its *spectral radius* is equal to

$$\rho(A) = \max\{|\lambda| : \det(A - \lambda I) = 0\},$$

that is, it is the magnitude of that eigenvalue of A with maximum absolute value.

A.4 Convex Sets and Functionals

A subset C of a vector space S is *convex* if for every $u, v \in C$ and every $\alpha \in [0, 1]$, we have $\alpha u + (1 - \alpha)v \in C$. A functional $f : C \to \mathbb{R}$ defined over a convex subset C of a vector space S is *convex* if, for every $u, v \in C$ and every scalar $\alpha \in [0, 1]$, we have $f(\alpha u + (1 - \alpha)v) \leq \alpha f(u) + (1 - \alpha) f(v)$. If this is a strict inequality for every $\alpha \in (0, 1)$, then f is *strictly convex*. The functional f is *concave* if $(-f)$ is convex, and *strictly concave* if $(-f)$ is strictly convex.

A functional $f : \mathbb{R}^n \to \mathbb{R}$ is *continuously differentiable* if, with $x \in \mathbb{R}^n$, the partial derivatives of f with respect to the components of x exist and are continuous, in which case we write

$$\nabla f(x) = [\partial f(x)/\partial x_1, \ldots, \partial f(x)/\partial x_n].$$

$\nabla f(x)$ is the *gradient* of f at x and is a row vector. We shall also use the notation $f_x(x)$ or $df(x)/dx$ to denote the same quantity. If we partition x into two vectors y and z of dimensions n_1 and $n - n_1$, respectively, and are interested only in the partial derivatives of f with respect to the components of y, then we use the notation $\nabla_y f(y, z)$ or $\partial f(y, z)/\partial z$ to denote this partial gradient.

Let $g : \mathbb{R}^n \to \mathbb{R}^n$ be a vector-valued function whose components are continuously differentiable with respect to the components of $x \in \mathbb{R}^n$. Then, we say that $g(x)$ is differentiable, with the derivative $dg(x)/dx$ being an $(m \times n)$ matrix whose ijth element is $\partial g_i(x)/\partial x_j$. (Here g_i denotes the ith component of g.) The gradient $\nabla f(x)$ being a vector, its derivative (which is the second derivative of $f : \mathbb{R}^n \to \mathbb{R}$) will thus be an $(n \times n)$ matrix, assuming that $f(x)$ is twice continuously differentiable in terms of components of x. This matrix, denoted by $\nabla^2 f(x)$, is symmetric, and is called the *Hessian matrix* of f at x.

Properties of Convex Functionals

1. A functional $f : \mathbb{R}^n \to \mathbb{R}$ which is twice continuously differentiable on $\mathbb{R}^n$ is *convex* (respectively, *strictly convex*) if, and only if, its Hessian matrix $\nabla^2 f(x)$ is *nonnegative definite* (respectively, *positive definite*) for all $x \in \mathbb{R}^n$.
2. Let C be a convex subset of a vector space S, and $f : C \to \mathbb{R}$ be a *convex* (respectively, *strictly convex*) functional. Let u and v be two vectors in C, and let $[a, b]$ be a subinterval of $\mathbb{R}$ such that for all $\epsilon \in [a, b]$, $u + \epsilon v \in C$. Then, the function $h : \mathbb{R} \to \mathbb{R}$, defined by $h(\epsilon) := f(u + \epsilon v)$, is a *convex* (respectively, *strictly convex*) functional on $[a, b]$.
3. If $h : \mathbb{R} \to \mathbb{R}$ is convex on an interval $[a, b]$, then for c_1, c_2, c_3 such that $a \leq c_1 < c_2 < c_3 \leq b$,

$$\frac{h(c_2) - h(c_1)}{c_2 - c_1} \leq \frac{h(c_3) - h(c_1)}{c_3 - c_1}$$

If h is strictly convex, then the inequality above is strict.

4. Let h, and (c_1, c_2) be as above, and let $\dot{h}_+(c_1)$ denote the right-hand derivative of h at the point c_1. Then,

$$\dot{h}_+(c_2) \leq \frac{h(c_2) - h(c_1)}{c_2 - c_1}$$

with the inequality being strict if h is strictly convex. Of course, if h is differentiable at c_1, then the derivative of h satisfies this bound.

A.5 Optimization of Functionals

Given a functional $f : S \to \mathbb{R}$, where S is a vector space, and a subset $X \subseteq S$, by the optimization problem

$$\text{minimize } f(x) \text{ subject to } x \in X$$

we mean the problem of finding an element $x^* \in X$ (called a *minimizing element* or an *optimal solution*) such that

$$f(x^*) \leq f(x) \quad \forall x \in X.$$

This is sometimes also referred to as a *globally minimizing solution* in order to differentiate it from the other alternative—a *locally minimizing solution*. An element $x^\circ \in X$ is called a locally minimizing solution if there exists an $\epsilon > 0$ such that

$$f(x^\circ) \leq f(x) \quad \forall x \in N_\epsilon(x^\circ) \cap X,$$

i.e., we compare $f(x^\circ)$ with values of $f(x)$ in that part of a certain ϵ-neighborhood of x°, which lies in X.

For a given optimization problem, it is not necessary that an optimal solution exists; an optimal solution will exist if the set of real numbers $\{f(x) : x \in X]$ is bounded below and there exists an $x^* \in X$ such that $\inf\{f(x) : x \in X\} = f(x^*)$, in which case we write

$$f(x^*) = \inf_{x \in X} f(x) = \min_{x \in X} f(x).$$

If such an x^* cannot be found, even though $\inf\{f(x) : x \in X\}$ is finite, we simply say that an optimal solution does not exist, but we declare the quantity

$$\inf\{f(x) : x \in X\} \quad \text{or} \quad \inf_{x \in X} f(x)$$

as the *optimal value* of the optimization problem. In this case, given any $\epsilon > 0$, one can find $x_\epsilon \in X$ with the property that $f(x_\epsilon) < \inf_{x \in X} f(x) + \epsilon$, that is one can get arbitrarily close to $\inf_{x \in X} f(x)$ by picking an element out of X. If $\{f(x) : x \in X\}$ is not bounded below, i.e., $\inf_{x \in X} f(x) = -\infty$, then neither an optimal solution nor an optimal value exists.

An optimization problem which involves maximization instead of minimization may be converted into a minimization problem by simply replacing f by $-f$. Any optimal solution of this minimization problem is also an optimal solution for the initial maximization problem, and the optimal value of the former. When a *maximizing element* $x^* \in X$ exists, then $\sup_{x \in X} f(x) = \max_{x \in X} f(x) = f(x^*)$.

Existence of Optimal Solutions and Weierstrass Theorem

In the minimization problem formulated above, an optimal solution exists if f is continuous (or lower semi-continuous) and X is compact—a result known as the *Weierstrass theorem*. For the special case when X is finite dimensional, we should recall that compactness is equivalent to being closed and bounded.

Necessary and Sufficient Conditions for Optimality

Let $S = \mathbb{R}^n$, and $f : \mathbb{R}^n \to \mathbb{R}$ be a differentiable function. If X is an open set, a first order necessary condition for an optimal solution to satisfy is

$$\nabla f(x^*) = 0.$$

If, in addition, f is twice continuously differentiable on $\mathbb{R}^n$, a second order necessary condition is

$$\nabla^2 f(x^*) \geq 0.$$

The pair of conditions $\{\nabla f(x^*) = 0, \nabla^2 f(x^*) > 0\}$ is sufficient for $x^* \in X$ to be a locally minimizing solution. These conditions are also sufficient for global optimality if, in addition, X is a convex set and f is a convex functional on X.

These results, by and large, hold also for the case when S is infinite dimensional, but then one has to replace the gradient vector and the Hessian matrix by first and second Gateaux (or Frèchet) derivatives, and positive definiteness requirement is replaced by "strong positiveness" of an operator. See [489] for these extensions.

A.6 Contraction Mappings and Fixed-Point Theorems

Let S be a normed linear vector space, with norm $\|\cdot\|$. A mapping f of S into itself, is called a *contraction of* S (or equivalently, a *contraction mapping*) if there exists a scalar $p \in [0, 1)$ such that

$$\|f(u) - f(v)\| \leq p\|u - v\|, \quad \forall u, v \in S.$$

It is possible to extend this definition to *metric* spaces where the "distance" function $\|u - v\|$ is replaced by a more general metric $d(u, v)$, but this more general definition is not given here. Consider an equation of the form

$$f(u) = u$$

defined on S. Such an equation is known as a *fixed-point equation*, and any solution to it (i.e., a $u^\circ \in S$ such that $f(u^\circ) = u^\circ$) is called a *fixed point* of f.

Theorem A.6.1 (Banach) *If S is a Banach space under the norm $\|\cdot\|$, and $f : S \to S$ is a continuous mapping which is also a contraction, then f has a unique fixed point.*

It is possible for f to have a fixed point without being a contraction mapping. A sufficient condition for such an existence is provided in the following theorem.

Theorem A.6.2 (Schauder) *Let S be a Banach space, X a nonempty convex set in S, and Y a compact subset of X. Let $f : X \to Y$ be a continuous map. Then there is a fixed point of f (not necessarily unique).*

When specialized to finite-dimensional spaces, this result is known as *Brouwer's fixed-point theorem*, which of course predates the more general one:

Theorem A.6.3 (Brouwer) *Let $S \equiv \mathbb{R}^n$, and X be a closed and bounded subset of S. Let $f : X \to X$ be continuous. Then, f has a fixed point.*

One way of computing the fixed point of a mapping $f : S \to S$, is to use the iteration

$$u_{(k+1)} = f(u_{(k)}), \quad k = 0, 1, \ldots$$

which is called "successive approximation". If S is a Banach space, and f is contraction mapping, then the iteration above converges to the unique fixed point of f, for all starting points $u_{(0)} \in S$. If f is an affine mapping, then this result can be strengthened, as given in the following theorem.

Theorem A.6.4 (Successive Approximation) *Let S be a Banach space and A a linear bounded operator mapping S into itself. Consider the equation*

$$u = Au + b \tag{$*$}$$

defined on S, where $b \in S$ is given. Furthermore, consider the "successive approximation"

$$u_{(k+1)} = Au_{(k)} + b, \quad k = 0, 1, \ldots \tag{$**$}$$

to the solution of ($*$). Then, the sequence $\{u_{(k)}\}$ generated by ($**$) converges to a unique element of S, for any starting point $u_{(0)} \in S$, if, and only if, the spectral radius of A is less than 1 (i.e., $\rho(A) < 1$). The limit of $\{u_{(k)}\}$ is the unique fixed point of ($*$).

Appendix B
Probability Theory and Stochastic Processes

This appendix presents some notions of probability theory and stochastic processes which are used in the text. For a more complete exposition the reader should consult with the standard texts on probability theory, such as [39, 260, 573] and [488], and texts on stochastic processes, such as [784].

B.1 Probability

B.1.1 Measurable Spaces

Let $\mathbb{X}$ be a collection of points (elements).

Definition B.1.1 Let $\mathcal{F}$ be a collection of subsets of $\mathbb{X}$ with the following properties.

- $\mathbb{X} \in \mathcal{F}$
- If $A \in \mathcal{F}$, then $\mathbb{X} \setminus A \in \mathcal{F}$
- If $A_k \in \mathcal{F}, k = 1, 2, 3, \ldots,$ then their countable union is also in $\mathcal{F}$, *i.e.*,

$$\bigcup_{k=1}^{\infty} A_k \in \mathcal{F}$$

Then, $\mathcal{F}$ is said to be a σ-field.

By De Morgan's laws, $\mathcal{F}$ has to be closed under countable intersections as well. Note for example, that the full power-set of any set is a σ-field. If the countable union above is replaced with finite union, $\mathcal{F}$ is called a field.

With the above, the pair $(\mathbb{X}, \mathcal{F})$ is called a *measurable space*.

If elements of $\mathbb{X}$ correspond to outcomes of a random experiment, then $\mathbb{X}$ is called a *sample space* and its subsets are called *events*. In this case one generally uses the notation Ω instead of $\mathbb{X}$.

Let $\mathcal{A}$ be a collection of sets, and $\mathcal{J}$ be the smallest σ-field containing the sets in $\mathcal{A}$; in this case, we write $\mathcal{J} = \sigma(\mathcal{A})$. An important class of σ-fields is the Borel σ-field on a metric (or more

S. Yüksel, T. Başar, *Stochastic Teams, Games, and Control under Information Constraints*, Systems & Control: Foundations & Applications,
https://doi.org/10.1007/978-3-031-54071-4

generally topological) space, generated by open sets. We will denote the Borel σ-field on such a space $\mathbb{X}$ by $\mathcal{B}(\mathbb{X})$. A measurable space $(\mathbb{X}, \mathcal{B}(\mathbb{X}))$ is called *standard Borel*, if $\mathbb{X}$ is a Polish space.

Definition B.1.2 If $(\mathbb{X}, \mathcal{B}(\mathbb{X}))$ and $(\mathbb{Y}, \mathcal{B}(\mathbb{Y}))$ are two measurable spaces, we say a mapping $h : \mathbb{X} \to \mathbb{Y}$ is measurable if

$$h^{-1}(B) = \{x : h(x) \in B\} \in \mathcal{B}(\mathbb{X}), \quad \forall B \in \mathcal{B}(\mathbb{Y})$$

A positive measure μ on $(\mathbb{X}, \mathcal{B}(\mathbb{X}))$ is a function from $\mathcal{B}(\mathbb{X})$ to $[0, \infty]$ which is *countably additive*, that is for $A_k, A_j \in \mathcal{B}(\mathbb{X})$ and $A_k \cap A_j = \emptyset \quad \forall k, j \in \mathbb{N}$:

$$\mu\left(\bigcup_{k=1}^{\infty} A_k\right) = \sum_{k=1}^{\infty} \mu(A_k).$$

Definition B.1.3 μ is a probability measure if it is a positive measure and $\mu(\mathbb{X}) = 1$.

Definition B.1.4 A measure μ is finite if $\mu(\mathbb{X}) < \infty$, and σ−finite, if there exists a collection of subsets $\{A_k\}$ such that $\mathbb{X} = \bigcup_{k=1}^{\infty} A_k$ with $\mu(A_k) < \infty$ for all k.

Theorem B.1.1 (Carathéodory's Extension Theorem) *Let $\mathcal{M}$ be a collection of subsets, which is a field, and suppose that there exists a countably additive (pre-)measure P on $\mathcal{M}$. Then, there exists a unique measure P' on the σ-field generated by $\mathcal{M}$, $\sigma(\mathcal{M})$, which is consistent with P on $\mathcal{M}$.*

Theorem B.1.2 (Kolmogorov's Extension Theorem) *Let $\mathbb{X}$ be a complete, separable metric space and let μ_n be a probability measure on $\mathbb{X}^n$, the n product of $\mathbb{X}$, for each $n = 1, 2, \ldots$, such that*

$$\mu_n(A_1 \times A_2 \times \cdots \times A_n) = \mu_{n+1}(A_1 \times A_2 \times \cdots \times A_n \times \mathbb{X}),$$

for every n and every sequence of Borel sets A_k. Then, there exists a unique probability measure μ on $(\mathbb{X}^{\infty}, \mathcal{B}(\mathbb{X}^{\infty}))$ which is consistent with each of the μ_n's.

The above result is useful because to determine whether two measures are equal it suffices to check if they are equal on the collection of sets which generate the σ-field, and not necessarily on the entire σ-field. For example, if the σ-field on a product space $\mathbb{X}^{\mathbb{Z}}$ is generated by a sequence of finite-dimensional distributions, one can define a measure on the product space which is consistent with the finite-dimensional distributions. In particular, the Borel σ-field defined on a product space is generated by the open sets in the product topology, which in turn allows one to uniquely define a probability measure on the infinite dimensional space by its measures on cylinder sets of the form:

$$\{x = (x_0, x_1, \cdots) \in \mathbb{X}^{\mathbb{Z}_+} : x_m \in A_m \in \mathcal{B}(\mathbb{X}), m \in I \subset \mathbb{Z}_+\}.$$

One could construct the Lebesgue measure on the Borel σ-field on $\mathbb{R}$ by restricting it to intervals only, since a field can be constructed by finitely many disjoint unions of intervals. The Lebesgue measure λ is defined on the Borel σ-field such that for $A = (a, b)$, $\lambda(A) = b - a$.

B.1.2 Integration

Let h be a non-negative measurable function from $(\mathbb{X}, \mathcal{B}(\mathbb{X}))$ to $(\mathbb{R}, \mathcal{B}(\mathbb{R}))$. The Lebesgue integral of h with respect to a measure μ can be defined in three steps:

First, for $A \in \mathcal{B}(\mathbb{X})$, define $1_{\{x \in A\}}$ (or $1_{(x \in A)}$, or $1_A(x)$) as an indicator function for event $x \in A$, that is, the function that takes the value 1 if $x \in A$, and 0 otherwise. In this case, $\int_{\mathbb{X}} 1_{\{x \in A\}} \lambda(dx) =: \lambda(A)$.

Next, define simple functions such that for $A_1, A_2, \ldots, A_n$, all in $\mathcal{B}(\mathbb{X})$, and positive numbers $b_1, b_2, \ldots, b_n$, $h_n(x) = \sum_{k=1}^{n} b_k 1_{\{x \in A_k\}}$. Then, $\int_{\mathbb{X}} h_n(x)\lambda(dx) =: \sum_{k=1}^{n} b_k \lambda(A_k)$.

Finally, observe that for any given non-negative measurable function h there exists a sequence of simple functions h_n such that $h_n(x) \uparrow h(x)$ monotonically. Then, the Lebesgue integral is defined as:

$$\int h(x)\mu(dx) := \lim_{n \to \infty} \int h_n(x)\mu(dx) = \lim_{n \to \infty} \sum_{k=1}^{n} b_k^n \lambda(A_k^n).$$

There are four important convergence theorems related to integration.

Theorem B.1.3 (Monotone Convergence Theorem) *If μ is a σ-finite positive measure on $(\mathbb{X}, \mathcal{B}(\mathbb{X}))$ and $\{f_n, n \in \mathbb{Z}_+\}$ is a sequence of measurable functions from $\mathbb{X}$ to $\mathbb{R}$ which pointwise, monotonically, converges to f, that is, $0 \le f_n(x) \le f_{n+1}(x)$ for all n, and*

$$\lim_{n \to \infty} f_n(x) = f(x),$$

for μ-almost every x, then

$$\lim_{n \to \infty} \int_{\mathbb{X}} f_n(x)\mu(dx) = \int_{\mathbb{X}} f(x)\mu(dx)$$

Theorem B.1.4 (Extended Monotone Convergence Theorem) *Let h be a measurable and integrable function from $\mathbb{X}$ to $\mathbb{R}$, μ be a σ-finite positive measure on $(\mathbb{X}, \mathcal{B}(\mathbb{X}))$ and $\{f_n, n \in \mathbb{Z}_+\}$ be a monotonically non-decreasing sequence of measurable functions from $\mathbb{X}$ to $\mathbb{R}$ which pointwise converges to f such that $h(x) \le f_n(x) \le f_{n+1}(x)$ for all n, and*

$$\lim_{n \to \infty} f_n(x) = f(x),$$

for μ-almost every x, and $\int \mu(dx)h(x) > -\infty$. Then,

$$\lim_{n \to \infty} \int_{\mathbb{X}} f_n(x)\mu(dx) = \int_{\mathbb{X}} f(x)\mu(dx)$$

Theorem B.1.5 (Fatou's Lemma (also Fatou-Lebesgue Theorem)) *If μ is a σ-finite positive measure on $(\mathbb{X}, \mathcal{B}(\mathbb{X}))$ and $\{f_n, n \in \mathbb{Z}_+\}$ is a sequence of non-negative measurable functions from $\mathbb{X}$ to $\mathbb{R}$, then*

$$\int_{\mathbb{X}} \mathit{lim\ inf}_{n \to \infty} f_n(x)\mu(dx) \le \mathit{lim\ inf}_{n \to \infty} \int_{\mathbb{X}} f_n(x)\mu(dx)$$

Theorem B.1.6 (Dominated Convergence Theorem) *If (i) μ is a σ-finite positive measure on $(\mathbb{X}, \mathcal{B}(\mathbb{X}))$, (ii) $g(x) \ge 0$ is a Borel measurable function such that*

$$\int_{\mathbb{X}} g(x)\mu(dx) < \infty,$$

and (iii) $\{f_n, n \in \mathbb{Z}_+\}$ is a sequence of measurable functions from $\mathbb{X}$ to $\mathbb{R}$ which satisfy $|f_n(x)| \le g(x)$ for μ-almost every x, and $\lim_{n \to \infty} f_n(x) = f(x)$, then

$$\lim_{n \to \infty} \int_{\mathbb{X}} f_n(x)\mu(dx) = \int_{\mathbb{X}} f(x)\mu(dx)$$

B.1.3 Probability Spaces and Random Variables

Let $(\Omega, \mathcal{F})$ be a measurable space, and $\mathbf{P}$ be a probability measure on this space. Then, the triple $(\Omega, \mathcal{F}, \mathbf{P})$ is known as a *probability space.*

If $\Omega = \mathbb{R}^n$, then its subsets of interest are the n-dimensional rectangles, and the smallest σ-algebra generated by these rectangles is called the n-dimensional *Borel σ-algebra* and is denoted by $\mathcal{B}(\mathbb{R}^n)$. Elements of $\mathcal{B}(\mathbb{R}^n)$ are *Borel sets*, and the pair $(\mathbb{R}^n, \mathcal{B}(\mathbb{R}^n))$ is a *Borel (measurable) space*. A probability measure defined on this space is known as a *Borel probability measure*.

If Ω is a finite set (say, $\Omega = \{\omega_1, \omega_2, \ldots, \omega_n\}$), we can assign probability weights on individual elements of Ω, instead of on subsets of Ω, in which case we write p_i to denote the probability of the single event ω_i. We call the n-tuple $\{p_1, p_2, \ldots, p_n\}$ a *probability mass function (pmf)* on Ω. Clearly, we have the restriction $0 \leq p_i \leq 1$ $\forall i = 1, \ldots, n$, and $\sum_{i=1}^{n} p_i = 1$. The same convention applies when Ω is countable (i.e., $\Omega = \{\omega_1, \omega_2, \ldots, \omega_i, \ldots\}$), in which case we simply replace n by ∞.

Let $(\Omega, \mathcal{F})$ and $(E, \mathcal{E})$ be two measurable spaces, and $X : (\Omega, \mathcal{F}) \to (E, \mathcal{E})$ be a measurable map. We call X an *E-valued random variable*, which is in fact a function.[1] If $\mathbf{P}$ is a probability measure on $(\Omega, \mathcal{F})$, then the image under X constitutes a probability measure on $(E, \mathcal{E})$, called the law of X and denoted by $\mathbf{P}_X$. If $E = \mathbb{R}^N$, we call X an N-dimensional random vector. In this case we can restrict, $\mathbf{P}_X$, without any loss of generality, to semi-infinite open rectangles, $\{\xi \in \mathbb{R}^N : \xi_i < a_i, i = 1, \ldots, N\}$, in $\mathcal{B}(\mathbb{R}^N)$, resulting in a function $P_X(a_1, a_2, \ldots, a_N)$ of N variables, $\{a_1, a_2, \ldots, a_N\}$, which are real numbers. This function is known as the *cumulative (probability) distribution function (cdf)* of the random vector X. The relationship with the probability measure $\mathbf{P}$ on the original probability space is:

$$P_X(a_1, a_2, \ldots, a_N) = \mathbf{P}(\{\omega \in \Omega : X_1(\omega) < a_1, X_2(\omega) < a_2, \ldots, X_N(\omega) < a_N\})$$

where X_i is the i'th component of X.

The σ-field generated by the events $\{\{\omega : X(\omega) \in A\}, A \in \mathcal{E}\}$ is called the σ-field generated by X and is denoted by $\sigma(X)$. This is the smallest sub-σ-field of $\mathcal{F}$ on which X is measurable.

Probability Density Function

Let $\mathbf{P}$ be a Borel probability measure on $(\mathbb{R}^N, \mathcal{B}(\mathbb{R}^N))$ such that any element of $\mathcal{B}(\mathbb{R}^N)$ which is of Lebesgue measure zero is also of $\mathbf{P}$-measure zero; then we say that $\mathbf{P}$ is *absolutely continuous* with respect to the Lebesgue measure. Now, if $X : (\Omega, \mathcal{F}, \mathbf{P}) \to (\mathbb{R}^N, \mathcal{B}(\mathbb{R}^N), \mathbf{P}_X)$ is random vector and if $\mathbf{P}_X$ is absolutely continuous with respect to the Lebesgue measure, there exists a nonnegative Borel function $p_X(\cdot)$ such that, for every $A \in \mathcal{B}(\mathbb{R}^N)$,

$$\mathbf{P}_X(A) = \int_A p_X(\xi) d\xi.$$

Such a function $p_X(\cdot)$ is called the *probability density function (pdf)* of the random vector X, and X is said to be a *continuous random vector*. In terms of the cumulative distribution function P_X, the preceding relationship can be written as

$$\mathbf{P}_X(a_1, \ldots, a_N) = \int_{-\infty}^{a_1} \cdots \int_{-\infty}^{a_N} p_X(\xi_1, \ldots, \xi_N) d\xi_1 \ldots d\xi_N$$

for every N scalars $a_1, \ldots, a_N$.

Let $X : (\Omega, \mathcal{F}, \mathbf{P}) \to (\mathbb{R}^N, \mathcal{B}(\mathbb{R}^N), \mathbf{P}_X)$ be a random vector and $f : (\mathbb{R}^N, \mathcal{B}(\mathbb{R}^N)) \to \mathbb{R}^M, \mathcal{B}(\mathbb{R}^M))$ be a nonnegative Borel function. Then, f can also be considered as a random

[1] It is common to use upper case letters for a random variable, and lower case letters for its realization, that is $X(\omega) = x$, for $\omega \in \Omega$. In the book, to conserve notation, we will occasionally use lower case letters for the random variable itself as well, when there is no ambiguity from context.

vector from $(\Omega, \mathcal{F})$ into $(\mathbb{R}^M, \mathcal{B}(\mathbb{R}^M))$, and its *average value* (*expected value*) is defined either by $\int_\Omega f(X(\omega))\mathbf{P}(d\omega)$ or by $\int_{\mathbb{R}^n} f(\xi)\mathbf{P}_X(d\xi)$ depending on which interpretation one adopts. Both of these integrals are well defined and are uniquely equal in value. If f changes signs, then we take $f = f^+ - f^-$ where both f^+ and f^- are nonnegative, and write the expected value of f as

$$E_X f(X) \equiv E[f(X)] = \int_{\mathbb{R}^N} f^+(\xi)\mathbf{P}_X(d\xi) - \int_{\mathbb{R}^N} f^-(\xi)\mathbf{P}_X(d\xi) =: \int_{\mathbb{R}^N} f(\xi)\mathbf{P}_X(d\xi),$$

provided that at least one of the pair $E[f^+(X)]$ and $E[f^-(X)]$ is finite. Since, by definition, $\mathbf{P}_X(d\xi) = \mathbf{P}_X(\xi + d\xi) - \mathbf{P}_X(\xi) =: d\mathbf{P}_X(\xi)$, this integral can further be written as

$$E[f(X)] = \int_{\mathbb{R}^N} f(\xi)d\mathbf{P}_X(\xi)$$

which is a Lebesgue-Stieltjes integral. For the special case when $f(x) = x$, we have

$$E[X] := \int_{\mathbb{R}^N} \xi\mathbf{P}_X(d\xi) =: \bar{X} \equiv \mu_X$$

which is known as the *mean* (*expected*) *value* of X. The *covariance* of the N-dimensional random vector X is defined as

$$E[(X - \bar{X})(X - \bar{X})'] = \int_{\mathbb{R}^N} (\xi - \bar{X})(\xi - \bar{X})'\mathbf{P}_X(d\xi) =: \mathrm{cov}(X)$$

which is a nonnegative definite matrix of dimensions $N \times N$.

Let $\mathcal{L}$ be the class of all N-dimensional random vectors $X : (\Omega, \mathcal{F}, \mathbf{P}) \to (\mathbb{R}^N, \mathcal{B}(\mathbb{R}^N), \mathbf{P}_X)$, with bounded second moments (i.e., $E[X'X] := \sum_{i=1}^N E[(X_i)^2] < \infty$). Then $\mathcal{L}$ is a *Hilbert space* (see Appendix A for a definition) under the *inner product* $\langle X, Y\rangle := E[X'Y]$, for $X, Y \in \mathcal{L}$. A consequence of this is the important (and useful) *Cauchy-Schwarz inequality*

$$| E[X'Y] |^2 \leq E[X'X]E[Y'Y]$$

Sequences of Random Variables and Convergence

A *random sequence* $\{X_i\}_{i=1}^\infty$ is a denumerable family of random variables

$$X_i : (\Omega, \mathcal{F}, \mathbf{P}) \to (\mathbb{R}, \mathcal{B}, P_{X_i}), i = 1, 2, \ldots$$

Given a random sequence $\{X_i\}$, we say that it *converges in probability* to $X : (\Omega, \mathcal{F}, \mathbf{P}) \to (\mathbb{R}, \mathcal{B}, \mathbf{P}_X)$ (and write $X_i \to_{\mathrm{P}} X$) if, for every $\epsilon > 0$, Prob$\{| X_i - X | \geq \epsilon\} \to 0$ as $i \to \infty$. We say that $\{X_i\}$ *converges almost surely* (*a.s.*) to X (and write $X_i \to_{\mathrm{as}} X$), if $X_i(\omega) \to X(\omega)$ pointwise, except possibly on a subset of Ω which receives *zero* probability measure (under $\mathbf{P}$).

B.2 Conditional Expectation and Estimation

Definition B.2.1 Given a measurable space $(\Omega, \mathcal{F})$, let $\mathcal{G} \subset \mathcal{F}$ be a (sub-) sigma field. A function $g : (\Omega, \mathcal{F}) \to (\mathbb{R}, \mathcal{B}(\mathbb{R}))$ is said to be *measurable with respect to* $\mathcal{G}$ if

$$\text{for every } B \in \mathcal{B}(\mathbb{R}),\ \{\omega \in \Omega : g(\omega) \in B\} \in \mathcal{G}$$

where $\mathcal{B}$ is the Borel field. In this case we say that the function g is $\mathcal{G}$*-measurable*.

Fact Suppose that X and Y are two real random variables on $(\Omega, \mathcal{F}, \mathbf{P})$, where X is $\sigma(Y)$-measurable. Then, there is a Borel function $g : \mathbb{R} \to \mathbb{R}$ such that $X = g(Y)$. ◇

Definition B.2.2 Let X be an integrable random variable on a probability space $(\Omega, \mathcal{F}, \mathbf{P})$ such that $E[|X|] < \infty$, and Y be another random variable (on the same probability space) with generated sigma field $\sigma(Y)$. *Conditional expectation of X given $\sigma(Y)$* is any $\sigma(Y)$-measurable random variable Z satisfying

$$\int_A Z(\omega)\mathbf{P}(d\omega) = \int_A X(\omega)\mathbf{P}(d\omega)$$

for every $A \in \sigma(Y)$. Such a Z is *essentially unique*, and is written as $E[X \mid \sigma(Y)]$, and sometimes as $E[X \mid Y]$.

Note In the definition above, we can replace $\sigma(Y)$ by any sigma field $\mathcal{G} \subset \mathcal{F}$, in which case we call $E[X \mid \mathcal{G}]$ the "conditional expectation of X given $\mathcal{G}$". ◇

Some Properties

1. If Y_1 and Y_2 are two random variables, defined on the same probability space, then

$$\sigma(Y_1, Y_2) \supset \sigma(Y_1) \text{ and } \sigma(Y_1, Y_2) \supset \sigma(Y_2).$$

Hence, by the law of the iterated expectation

$$E\{E[X \mid \sigma(Y_1, Y_2)] \mid \sigma(Y_i)\} = E[X \mid \sigma(Y_i)], \text{ a.s., } i = 1, 2. \tag{B.1}$$

2. Let $\mathcal{A} = \{\Omega, \theta\}$, which is the trivial sigma field. Then,

$$E[X \mid \mathcal{A}] = E[X].$$

3. Let $\mathcal{G} \subset \mathcal{F}$ be a (sub-) sigma field, and W be a $\mathcal{G}$-measurable random variable. Then,

$$E[WX \mid \mathcal{G}] = WE[X \mid \mathcal{G}] \text{ a.s..}$$

4. Let X and Y be two integrable random variables defined on a common probability space $(\Omega, \mathcal{F}, \mathbf{P})$, and $\mathcal{G} \subset \mathcal{F}$ be a (sub-) sigma field. Then, for any two scalars α and β,

$$E\{[\alpha X + \beta Y] \mid \mathcal{G}\} = \alpha E[X \mid \mathcal{G}] + \beta E[Y \mid \mathbf{G}].$$

5. *Orthogonality principle*. Let X and Z be two random variables defined on the common probability space $(\Omega, \mathcal{F}, \mathbf{P})$, where Z is $\mathcal{G}$-measurable, with $\mathcal{G} \subset \mathcal{F}$ (a sub-sigma field on $\mathcal{F}$). Then the random variables $Y := X - E[X \mid \mathcal{G}]$ and Z are uncorrelated, i.e., $E[YZ] = E[Y]E[Z] = 0$, and this is 0 since $E[Y] = 0$. ◇

The last property has important implications in *minimum mean-square estimation*: Given a random variable X as above, and a sub-sigma field $\mathcal{G}$ (of $\mathcal{F}$), we wish to find a $\mathcal{G}$-measurable random variable W that yields the minimum value to $E[(X - W)^2]$ among all such $\mathcal{G}$-measurable random variables.

The unique solution to this "estimation" problem is $W = E[X \mid \mathcal{G}]$, as given below as a *fact*:

Fact Let X be a random variable on $(\Omega, \mathcal{F}, \mathbf{P})$, and $\mathcal{G} \subset \mathcal{F}$ be a given sub-sigma field. Then,

$$E[(X - E[X \mid \mathcal{G}])^2] \leq E[(X - Z)^2], \text{ for all } \mathcal{G}\text{-measurable random variables } Z,$$

and equality holds if, and only if, $Z = E[X \mid \mathcal{G}]$. Here $Y := X - E[X \mid \mathcal{G}]$ is the "error" in the estimation of X using $\mathcal{G}$, and $E[Y^2]$ is the minimum mean-square error (*mmse*), which can also be written as $E[Y^2] = E[X^2] - E[(E[X \mid \mathcal{G}])^2]$. ◇

B.3 Stochastic Processes

A stochastic process $X := \{x_t, t \in \mathbf{T}\}$ is a parameterized family of random variables defined on the same probability space, with the parameter t being the "time variable". If the *time index* set $\mathbf{T}$ is finite or countably infinite, then the stochastic process is called a *discrete-time process*, whereas if it is uncountably infinite, such as the closed interval $[0, T]$, then the process is called a *continuous-time process*. In addition to being a well-defined random variable for every $t \in T$, a stochastic process should also have the property that for every finite subset S of T, the finite family of random variables $\{x_t, t \in S\}$ is a well-defined random vector.

A stochastic process $X := \{x_t, t \in \mathbf{T}\}$ is said to be a *second-order process*, if $E[|x_t|^2] < \infty$ for all $t \in \mathbf{T}$. For a second-order stochastic process, three quantities of interest (and of importance) are:

Mean function: $\mu_X(t) := E[x_t]$
Autocorrelation function: $R_X(t, s) := E[x_t x_s]$
Autocovariance function: $C_X(t, s) := E[(x_t - \mu_X(t))(x_s - \mu_X(s))] \equiv R_X(t, s) - \mu_X(t)\mu_X(s)$

Estimation

Let $X := \{x_t, t \in \mathbf{T}\}$ and $Z := \{z_s, s \in \mathbf{T}\}$ be two stochastic processes, where $\mathbf{T}$ is a discrete or a continuous time interval. Suppose we are interested in estimating the value of X at some point, say $t \in \mathbf{T}$, based on some observed values of Z. Denote this estimate by $\hat{x}_t = \widehat{g}_t(y_t)$, where y_t is the collection of observed values of Z by time t, and the function $\widehat{g}_t(\cdot)$ is known as an *estimator*. There are several possibilities for y_t:

(i) $\mathbf{T} = [0, T]$ and $y_t = \{z_{t_1}, \ldots, z_{t_n}\}$, where $z_{t_i}, t_i \in [0, T]$ is some sampled value of Z. If $t_n \leq t$, we say that $\widehat{g}$ is a *causal estimator*.
(ii) $y_t = \{z_s, s \leq t\} \Rightarrow \hat{x}_t = \widehat{g}_t(z_s, s \leq t)$. Here all past values of the measurement are used in the construction of the estimate $\hat{x}_t$. Here $\widehat{g}$ is also a causal estimator.
(iii) $y_t = \{z_s, s \in \mathbf{T}\} \Rightarrow \hat{x}_t = \widehat{g}_t(z_s, s \in \mathbf{T})$. This is a *noncausal* estimator, which is also called a *smoother*.

We say that an estimator $\widehat{g}$ is linear (affine) if it is given by a linear (affine) map, e.g., for a causal estimator,

$$\widehat{g}_t(y_t) = k(t) + \int_0^t h(t, s) z_s ds, \qquad T = [0, \infty),$$

or

$$\widehat{g}_n(y_n) = k_n + \sum_{i=1}^{n} h(n, i) z_i, \qquad \text{for } n \in T = \{0, 1, \ldots\}$$

where we have let $t = n$ for the discrete-time process. In all cases, we call $\widehat{g}_t$ a *mms estimator* if it minimizes the quantity $E\{[x_t - g(y_t)]^2\}$ over the allowable class of g's. For each fixed y_t, the lowest possible value is attained by

$$\hat{x}_t = \widehat{g}_t(y_t) = E[x_t \mid \sigma(y_t)] \equiv E[x_t \mid y_t]$$

provided that the conditional mean is well defined, where $\sigma(y_t)$ is the smallest sigma field generated by y_t. Here again the *Orthogonality Principle* applies:

$$E\{(x_t - \hat{x}_t)g_t(y_t)\} = 0 \text{ for all functions } g_t.$$

Suppose that now we restrict our attention to affine estimators, where $g_t(\cdot)$ is *affine* in its arguments, as described above. Denote the class of such random variables (estimators) under any one of the information schemes above by H_y and let $\hat{x}_t$ denote the affine estimate for x_t based on Z for a given information scheme, under the mean-squared error criterion, i.e.,

$$E\{(x_t - \hat{x}_t)^2\} \leq E\{(x_t - w)^2\}, \text{ for all } w \in H_y.$$

Then, it again follows from the *Orthogonality Principle* that $\hat{x}_t \in H_y$ solves the estimation problem if, and only if,

$$E\{[x_t - \hat{x}_t]w\} = 0 \text{ for all } w \in H_y.$$

Appendix C
On Spaces of Probability Measures

In this appendix we present various topologies, and when applicable, several metrics on the sets of probability measures.

C.1 Convergence of Sequences of Probability Measures

Let $\mathbb{X}$ be a Polish space and let $\mathcal{P}(\mathbb{X})$ denote the family of all probability measures on $(\mathbb{X}, \mathcal{B}(\mathbb{X}))$. Let $\{\mu_n,\ n \in \mathbb{N}\}$ be a sequence in $\mathcal{P}(\mathbb{X})$.

The sequence $\{\mu_n\}$ is said to converge to $\mu \in \mathcal{P}(\mathbb{X})$ *weakly* if

$$\int_{\mathbb{X}} c(x)\mu_n(dx) \to \int_{\mathbb{X}} c(x)\mu(dx) \tag{C.1}$$

for every continuous and bounded $c : \mathbb{X} \to \mathbb{R}$.[1]

On the other hand, $\{\mu_n\}$ is said to converge to $\mu \in \mathcal{P}(\mathbb{X})$ *setwise* if

$$\int_{\mathbb{X}} c(x)\mu_n(dx) \to \int_{\mathbb{X}} c(x)\mu(dx)$$

for every measurable and bounded $c : \mathbb{X} \to \mathbb{R}$. Setwise convergence can also be defined through pointwise convergence on Borel subsets of $\mathbb{X}$ (see, e.g., [344]), that is

$$\mu_n(A) \to \mu(A), \quad \text{for all } A \in \mathcal{B}(\mathbb{X})$$

[1] It is important to emphasize that what is typically studied in probability as weak convergence is not the exact weak convergence notion used in functional analysis: The topological dual space of the set of probability measures does not consist of only expectations of continuous and bounded functions. However, the dual space of the space of continuous and bounded functions with the supremum norm does admit a representation in terms of expectations [489]; hence, the weak convergence here is in actuality the weak* convergence in analysis and distribution theory.

S. Yüksel, T. Başar, *Stochastic Teams, Games, and Control under Information Constraints*, Systems & Control: Foundations & Applications,
https://doi.org/10.1007/978-3-031-54071-4

since the space of simple functions are dense in the space of bounded and measurable functions under the supremum norm.

For two probability measures $\mu, \nu \in \mathcal{P}(\mathbb{X})$, the *total variation* metric is given by

$$\|\mu - \nu\|_{TV} := 2 \sup_{B \in \mathcal{B}(\mathbb{X})} |\mu(B) - \nu(B)|$$
$$= \sup_{f: \|f\|_\infty \leq 1} \left| \int f(x)\mu(dx) - \int f(x)\nu(dx) \right|, \tag{C.2}$$

where the supremum is over all measurable real f such that $\|f\|_\infty = \sup_{x \in \mathbb{X}} |f(x)| \leq 1$. A sequence $\{\mu_n\}$ is said to converge to $\mu \in \mathcal{P}(\mathbb{X})$ in total variation if $\|\mu_n - \mu\|_{TV} \to 0$.

Setwise convergence is equivalent to pointwise convergence on Borel sets whereas total variation requires uniform convergence on Borel sets. Thus these three convergence notions are in increasing order of strength: convergence in total variation implies setwise convergence, which in turn implies weak convergence.

On the other hand, total variation is a stringent notion for convergence. For example a sequence of discrete probability measures never converges in total variation to a probability measure which admits a density function with respect to the Lebesgue measure and such a space is not separable. Setwise convergence also induces a topology on the space of probability measures and channels which is not easy to work with since the space under this convergence is not metrizable [300, p. 59].

However, the space of probability measures on a complete, separable, metric (Polish) space endowed with the topology of weak convergence is itself a complete, separable, metric space [116].

There are various ways to metrize weak convergence. One immediate metric builds on the following reasoning: One can construct (since the space of continuous functions on a compact set is separable under the supremum norm) a countable collection of continuous functions $\{c_k, k \in \mathbb{N}\}$ such that it suffices to only consider these functions in (C.1) to establish weak-convergence. We can thus use these *weak-convergence determining* functions (see e.g. [241, Theorem 3.4.5]) to define a countable collection of semi-norms $d_k(\mu, \nu) := |\int c_k(x)\mu(dx) - \int c_k(x)\nu(dx|$, and from these we can construct a locally convex space which is metrizable. Thus, we have the following metric which metrizes the weak topology:

$$\rho(\mu, \nu) = \sum_{m=1}^{\infty} 2^{-(m+1)} \left| \int_{\mathbb{S}} f_m(x)\mu(dx) - \int_{\mathbb{S}} f_m(x)\nu(dx) \right|, \tag{C.3}$$

where $\{f_m\}_{m \geq 1}$ is an appropriate sequence of continuous and bounded functions such that $\|f_m\|_\infty \leq 1$ for all $m \geq 1$ (see [574, Theorem 6.6, p. 47]).

The Prohorov metric [116] also can be used to metrize this convergence topology.

As a more practical metric, the Wasserstein metric can also be used (for compact $\mathbb{X}$) to metrize the weak convergence space topology.

Definition C.1.1 (Wasserstein Metric) The *Wasserstein metric* of order p, $1 \leq p < \infty$, for two distributions $\mu, \nu \in \mathcal{P}(\mathbb{X})$ with finite pth moments (thus defined only on such a subset of $\mathcal{P}(\mathbb{X})$) is defined as

$$W_p(\mu, \nu) = \inf_{\eta \in \mathcal{H}(\mu, \nu)} \left(\int_{\mathbb{X} \times \mathbb{X}} \eta(dx, dy) \|x - y\|^p \right)^{\frac{1}{p}},$$

where $\mathcal{H}(\mu, \nu)$ denotes the set of probability measures on $\mathbb{X} \times \mathbb{X}$ with first marginal μ and second marginal ν, and $\| \cdots \|$ is a norm.

For compact $\mathbb{X}$, the Wasserstein distance of order p metrizes the weak topology on the set of probability measures on $\mathbb{X}$ (see [746, Theorem 6.9]). For non-compact $\mathbb{X}$, weak convergence combined with convergence of moments up to order p (that is of $\int \mu_n(dx)\|x\|^p \to \int \mu(dx)\|x\|^p$) is equivalent to convergence in W_p. Finally, the bounded-Lipschitz metric ρ_{BL} [746, p.109] can also be used to metrize weak convergence:

$$\rho_{BL}(\mu, \nu) = \sup_{\|f\|_{BL} \le 1} \left| \int_{\mathbb{X}} f(e)\mu(de) - \int_{\mathbb{X}} f(e)\nu(de) \right|, \tag{C.4}$$

where

$$\|f\|_{BL} := \|f\|_\infty + \sup_{e \neq e'} \frac{f(e) - f(e')}{d_{\mathbb{X}}(e, e')},$$

and $d_{\mathbb{X}}$ is the metric on $\mathbb{X}$.

We note that W_1 can equivalently be written as [746, Remark 6.5]:

$$W_1(\mu, \nu) := \sup_{\|f\|_{Lip} \le 1} \left| \int_{\mathbb{X}} f(e)\mu(de) - \int_{\mathbb{X}} f(e)\nu(de) \right|,$$

where

$$\|f\|_{Lip} := \sup_{e \neq e'} \frac{f(e) - f(e')}{d_{\mathbb{X}}(e, e')}.$$

Comparing this with (C.4), it follows that

$$\rho_{BL} \le W_1. \tag{C.5}$$

Another important distance measure (though not a metric) that is commonly used is relative entropy:

Definition C.1.2 Consider two probability measures P and Q. The relative entropy between P and Q is defined as $D(P\|Q) = \int \log \frac{dP}{dQ} dP = \int \frac{dP}{dQ} \log \frac{dP}{dQ} dQ$ where $P \ll Q$ and $\frac{dP}{dQ}$ denotes the Radon-Nikodym derivative of P with respect to Q.

Total variation is related to relative entropy via Pinsker's inequality [193]: $\|P - Q\|_{TV} \le \sqrt{\frac{2}{\log(e)} D(P\|Q)}$. This also shows that convergence in relative entropy implies that under total variation.

Weak convergence is very important in applications of stochastic control and probability in general. Prohorov's theorem [223] provides a way to characterize compactness properties under weak convergence. Furthermore, such a convergence notion has very important measurability properties.

C.2 Some Measurability Results on Spaces of Probability Measures

Weak convergence topology leads to important measurability properties, as we state in the following two theorems. The first one appears in [16] (see Theorem 15.13 in [16] or p. 215 in [128]).

Theorem C.2.1 *Let $\mathbb{S}$ be a Polish space and M be the set of all measurable and bounded functions $f : \mathbb{S} \to \mathbb{R}$. Then, for any $f \in M$, the integral*

$$\int \pi(dx) f(x)$$

defines a measurable function on $\mathcal{P}(\mathbb{S})$ under the topology of weak convergence.

This is a useful result since it allows us to define measurable functions in integral forms on the space of probability measures when we work with the topology of weak convergence. The second useful result (below) follows from Theorem C.2.1, [222, Theorem 2.1] and [114, Proposition 7.25].

Theorem C.2.2 *Let $\mathbb{S}$ be a Polish space. A function $F : \mathcal{P}(\mathbb{S}) \to \mathcal{P}(\mathbb{S})$ is measurable on $\mathcal{B}(\mathcal{P}(\mathbb{S}))$ (under weak convergence), if for all $B \in \mathcal{B}(\mathbb{S})$ $(F(\cdot))(B) : \mathcal{P}(\mathbb{S}) \to \mathbb{R}$ is measurable under weak convergence on $\mathcal{P}(\mathbb{S})$, that is for every $B \in \mathcal{B}(\mathbb{S})$, $(F(\pi))(B)$ is a measurable function when viewed as a function from $\mathcal{P}(\mathbb{S})$ to $\mathbb{R}$.*

C.3 A Generalized Dominated Convergence Theorem

Under weak and setwise convergences, we can arrive at generalized forms of the dominated convergence theorem. In particular, from [454, Theorem 3.5] and [658, Theorem 3.5], we have the following:

Theorem C.3.1 *The following generalized convergence results hold:*

(i) Suppose that $\{\mu_n\}_n \subset \mathcal{P}(\mathbb{X})$ converges weakly to some μ. For a bounded real valued sequence of functions $\{f_n\}_n$ such that $\|f_n\|_\infty < C$ for all $n > 0$ with $C < \infty$, if $\lim_{n\to\infty} f_n(x_n) = f(x)$ for all $x_n \to x$, i.e. f_n continuously converges to f, then

$$\lim_{n\to\infty} \int_{\mathbb{X}} f_n(x)\mu_n(dx) = \int_{\mathbb{X}} f(x)\mu(dx).$$

(ii) Suppose that $\{\mu_n\}_n \subset \mathcal{P}(\mathbb{X})$ converges setwise to some μ. For a bounded real valued sequence of functions $\{f_n\}_n$ such that $\|f_n\|_\infty < C$ for all $n > 0$ with $C < \infty$, if $\lim_{n\to\infty} f_n(x) = f(x)$ for all x, i.e. f_n pointwise converges to f, then

$$\lim_{n\to\infty} \int_{\mathbb{X}} f_n(x)\mu_n(dx) = \int_{\mathbb{X}} f(x)\mu(dx).$$

C.4 The w-s Topology

Let $\mathbb{X}$ and $\mathbb{Y}$ be Polish spaces, as before.

Definition C.4.1 The w-s topology on the set of probability measures $\mathcal{P}(\mathbb{X} \times \mathbb{Y})$ is the coarsest topology under which $\int f(x, y)\mu(dx, dy) : \mathcal{P}(\mathbb{X} \times \mathbb{Y}) \to \mathbb{R}$ is continuous for every measurable and bounded $f(x, y)$ which is continuous in y for every x (but unlike the weak topology, f does not need to be continuous in x).

Theorem C.4.1 ([650, Theorem 3.10] [58, Theorem 2.5]) *Let $\mu_n \in \mathcal{P}(\mathbb{X} \times \mathbb{Y})$. If $\mu_n \to \mu$ weakly where the marginals $\mu_n(dx \times \mathbb{Y}) \to \mu(dx \times \mathbb{Y})$ setwise, then the convergence $\mu_n \to \mu$ is also in the w-s sense.*

C.5 Lusin's Theorem

Lusin's theorem is a very consequential result in mathematical analysis.

Theorem C.5.1 ([223, Theorem 7.5.2]) *Let $(\mathbb{X}, T)$ be any topological space and μ a finite, closed regular Borel measure on $\mathbb{X}$. Let $(\mathbb{S}, d)$ be a separable metric space and let f be a Borel-measurable function from $\mathbb{X}$ into $\mathbb{S}$. Then, for any $\epsilon > 0$ there is a closed set $F \subset \mathbb{X}$ such that $\mu(\mathbb{X} \setminus F) < \epsilon$ and the restriction of f to F is continuous.*

We also recall Tietze's extension theorem, which is often used in conjunction with Lusin's theorem to construct a continuous extension of the continuous function defined on F in Theorem C.5.1 to $\mathbb{X}$.

Theorem C.5.2 (Tietze's Extension Theorem [225, Theorem 4.1]) *Let $\mathbb{X}$ be an arbitrary metric space, A a closed subset of $\mathbb{X}$, L a locally convex linear space, and $f : A \to L$ a continuous map. Then there exists a continuous function $f_C : \mathbb{X} \to L$ such that $f_C(a) = f(a) \forall a \in A$. Furthermore, the image of f_C satisfies $f_C(\mathbb{X}) \subset$ [convex hull of $f(A)$].*

Appendix D
Markov Chains, Martingales and Ergodic Processes

This appendix provides some preliminary background on Markov chains, martingales, and ergodic processes. More comprehensive treatment of Markov chains can be found in [537] and [564], martingales in any standard book on stochastic processes, such as [784], and ergodic processes from an information theoretic angle in [665], [315], and [319].

D.1 Markov Chains

A sequence of random variables, ordered temporally, is a Markov Chain if the conditional probability of any element in the sequence given observed values of past elements in the sequence depends only on the most recent past element. That is, if $x_{[0,N]} = \{x_0, x_1, \ldots, x_N\}$ denotes the sequence defined over a probability space $(\Omega, \mathcal{F}, \mathbf{P})$, with each element taking values in a measurable space $(\mathbb{X}, \mathcal{B}(\mathbb{X}))$, then for each $B \in \mathcal{B}(\mathbb{X})$, and each $t \in [0, N-1]$,

$$\mathbf{P}(\{\omega \in \Omega : x_{t+1}(\omega) \in B\} \mid x_t, x_{t-1}, \ldots, x_0) = \mathbf{P}(\{\omega \in \Omega : x_{t+1}(\omega) \in B\} \mid x_t)$$

If P is the probability measure induced by $x_{[0,N]}$ on $(\mathbb{X}^{N+1}, \mathcal{B}(\mathbb{X}^{N+1}))$, then the above can equivalently be written as

$$P(x_{t+1} \in D \mid x_t, x_{t-1}, \ldots, x_0) = P(x_{t+1} \in D \mid x_t)$$

for all $D \in \mathcal{B}(\mathbb{X})$, and each $t \in [0, N-1]$.

Thus a Markov chain is completely determined by the transition probability and the probability of the initial state, $P(dx_0)$, denoting $P(x_0 \in \cdot)$. Hence, the probability of the event $\{x_{t+1} \in D\}$ for any t can be computed recursively by starting at $t = 0$, with $P(x_1 \in D) = \int_{\mathbb{X}} P(x_1 \in D|x_0)P(dx_0)$, and iterating in a similar manner for $t = 1, 2, \ldots$, that is for an arbitrary t, $P(x_{t+1} \in D) = \int_{\mathbb{X}} P(x_{t+1} \in D|x_t)P(dx_t)$.

In the remainder of this section, we discuss a number of properties of Markov chains.

Let $\{x_t, t \geq 0\}$ be a Markov chain taking values in a complete, separable, metric state space $(\mathbb{X}, \mathcal{B}(\mathbb{X}))$, and defined on a probability space $(\Omega, \mathcal{F}, \mathbf{P})$. For each $D \in \mathcal{B}(\mathbb{X})$, let $P(x, D) := P(x_{t+1} \in D|x_t = x)$ denote the transition probability from x to D, that is the probability of the event $\{x_{t+1} \in D\}$ given that $x_t = x$, which as we have seen completely determines the evolution

S. Yüksel, T. Başar, *Stochastic Teams, Games, and Control under Information Constraints*, Systems & Control: Foundations & Applications,
https://doi.org/10.1007/978-3-031-54071-4

of the Markov chain. We now have a number of definitions and properties. We first note that given the one-step transition kernel, $P(x, A)$, the n-step transitions can be obtained via composition, $P(x_{t+n} \in A \mid x_t = x) = P^n(x, A)$, for any $n \geq 1$. The transition law acts on measurable functions $f: \mathbb{X} \to \mathbb{R}$ and measures μ on $\mathcal{B}(\mathbb{X})$ via,

$$Pf(x) := \int_{\mathbb{X}} P(x, dy) f(y), \quad x \in \mathbb{X}, \qquad \mu P(A) := \int_{\mathbb{X}} \mu(dx) P(x, A), \quad A \in \mathcal{B}(\mathbb{X}).$$

A probability measure π on $\mathcal{B}(\mathbb{X})$ is called invariant if $\pi P = \pi$. That is,

$$\int \pi(dx) P(x, A) = \pi(A), \qquad A \in \mathcal{B}(\mathbb{X}).$$

For any initial probability measure ν on $\mathcal{B}(\mathbb{X})$ we can construct a stochastic process with transition law P, and satisfying $x_0 \sim \nu$. We let P_ν denote the resulting probability measure on the sample space, with the usual convention for $\nu = \delta_x$ when the initial state is $x \in \mathbb{X}$. When $\nu = \pi$, then the resulting process is stationary.

There is at most one stationary solution under the following irreducibility assumption. For a set $A \in \mathcal{B}(\mathbb{X})$, let

$$\tau_A := \min(t \geq 1 : x_t \in A) \tag{D.1}$$

Definition D.1.1 Let φ denote a sigma-finite measure on $\mathcal{B}(\mathbb{X})$.

(i) The Markov chain is called *φ-irreducible* if for any $x \in \mathbb{X}$, and any $B \in \mathcal{B}(\mathbb{X})$ satisfying $\varphi(B) > 0$, we have

$$P_x\{\tau_B < \infty\} > 0 .$$

(ii) A φ-irreducible Markov chain is *aperiodic* if for any $x \in \mathbb{X}$, and any $B \in \mathcal{B}(\mathbb{X})$ satisfying $\varphi(B) > 0$, there exists $n_0 = n_0(x, B)$ such that

$$P^n(x, B) > 0 \qquad \text{for all } n \geq n_0.$$

(iii) A φ-irreducible Markov chain is *Harris recurrent* if $P_x(\tau_B < \infty) = 1$ for any $x \in \mathbb{X}$, and any $B \in \mathcal{B}(\mathbb{X})$ satisfying $\varphi(B) > 0$. It is *positive Harris recurrent* if in addition there is an invariant probability measure π.

Definition D.1.2 A set α is called an atom if there exists a probability measure ν such that

$$P(x, A) = \nu(A), \quad \forall x \in \alpha, A \in \mathcal{B}(\mathbb{X}).$$

If the chain is μ-irreducible and $\mu(\alpha) > 0$, then α is called an *accessible atom.*

In case there is an atom α, we have the following:

Theorem D.1.1 *For a Markov chain for which $E_\alpha[\tau_\alpha] < \infty$, the following is an invariant probability measure:*

$$\pi(A) = E_\alpha[\frac{\sum_{k=0}^{\tau_\alpha - 1} 1_{x_k \in A}}{E[\tau_\alpha]} | x_0 = \alpha], \quad \forall A \in \mathcal{B}(\mathbb{X}), \mu(A) > 0$$

In case there is no atom, one can construct atoms in an extended space through Nummelin's splitting technique [564] provided one of the following sets exist:

Definition D.1.3 A set $A \subset \mathbb{X}$ is μ-small on $(\mathbb{X}, \mathcal{B}(\mathbb{X}))$ if for some n, and some positive measure μ

$$P^n(x, B) \geq \mu(B), \quad \forall x \in A, \text{ and } B \in \mathcal{B}(\mathbb{X}).$$

Definition D.1.4 A set $A \subset \mathbb{X}$ is μ-petite on $(\mathbb{X}, \mathcal{B}(\mathbb{X}))$ if for some distribution $\mathcal{T}$ on $\mathbb{N}$ (set of natural numbers), and some positive measure μ,

$$\sum_{n=0}^{\infty} P^n(x, B)\mathcal{T}(n) \geq \mu(B), \quad \forall x \in A, \text{ and } B \in \mathcal{B}(\mathbb{X}).$$

Note By Theorem 5.5.7 of [537], under aperiodicity and irreducibility, every petite set is small. ◇

The following is a key result.

Theorem D.1.2 (Thm. 4.1 of Meyn-Tweedie [536]) *Suppose that* $\mathbf{X}$ *is a* φ*-irreducible Markov chain, and suppose that there is a set* $A \in \mathcal{B}(\mathbb{X})$ *satisfying the following:*

(i) *A is* μ*-petite for some* μ.
(ii) *The chain is recurrent in the sense that* $P_x(\tau_A < \infty) = 1$ *for any* $x \in \mathbb{X}$.
(iii) *A is finite mean recurrent:* $\sup_{x \in A} E_x[\tau_A] < \infty$.

Then $\mathbf{X}$ *is positive Harris recurrent and there exists a unique invariant probability measure.*

An invariant measure satisfies the following:

Theorem D.1.3 *For a* μ*-irreducible Markov chain with the unique invariant probability measure* π*, the following is satisfied by the invariant probability measure:*

$$\pi(A) = \int_C \pi(dx) E_x[\sum_{k=0}^{\tau_C - 1} 1_{\{x_k \in A\}}], \quad A \in \mathcal{B}(\mathbb{X}),$$

where C satisfies $\sup_{x \in C} E_x[\tau_C] < \infty$. *Furthermore, for measurable* f,

$$\pi(f) = \int_C \pi(dx) E_x[\sum_{k=0}^{\tau_C - 1} f(x_k)].$$

The existence of an invariant distribution is important primarily because of the following:

Theorem D.1.4 (Birkhoff's Sample Path Ergodic Theorem for Markov Chains) *Consider a positive Harris recurrent Markov chain* $\{x_t\}$ *taking values in* $\mathbb{X}$*, with invariant distribution* $\pi(.)$. *Let* $f : \mathbb{X} \to \mathbb{R}$ *be such that* $\int f(x)\pi(dx) < \infty$. *Then, the following holds almost surely:*

$$\lim_{T \to \infty} \frac{1}{T} \sum_{t=0}^{T-1} f(x_t) = \int f(x)\pi(dx).$$

On Petite and Small Sets

Establishing petiteness may be difficult to directly verify. In the following, we present two conditions that may be used to establish the petiteness properties.

A kernel A from $(\mathbb{X}, \mathcal{B}(\mathbb{X}))$ to itself is called *sub-stochastic* if for every $x \in \mathbb{X}$, $A(x, \cdot)$ is non-negative and $A(x, \mathbb{X}) \leq 1$ and for every $B \in \mathbb{X}$, $A(\cdot, B)$ is a measurable function on $(\mathbb{X}, \mathcal{B}(\mathbb{X}))$.

For a Markov chain with transition kernel P on a Borel space $\mathbb{X}$, and for $\mathcal{K}$ a probability measure on $\mathbb{N}$, if there exists for every $B \in \mathcal{B}(\mathbb{X})$, a lower semi-continuous function $A(\cdot, B)$ such that

$$\sum_{n=0}^{\infty} P^n(x, B)\mathcal{K}(n) \geq A(x, B),$$

for a sub-stochastic kernel $A(\cdot, \cdot)$, the chain is called a $T-$chain. By Meyn and Tweedie [537], Theorem 6.0.1, every compact set in an irreducible $T-$chain is petite.

Theorem D.1.5 ([537]) *For a T-chain which is irreducible, every compact set is petite.*

For a countable state space, under irreducibility, every finite set S is petite.

To establish the petite set property, Tweedie [726] considers the following test which only depends on the one-stage transition kernel of a Markov chain: If a set S is such that, the following *uniform countable additivity* condition

$$\lim_{n\to\infty} \sup_{x\in S} P(x, B_n) = 0, \tag{D.2}$$

is satisfied for $B_n \downarrow \emptyset$, and if the Markov chain is irreducible, then S is petite. This condition may be easier to directly verify in a large class of applications.

D.2 Discrete-Time Martingales

Let $(\Omega, \mathcal{F}, \mathbf{P})$ be a probability space. An increasing family $\{\mathcal{F}_n\}$ of sub σ-fields of $\mathcal{F}$ is called a *filtration*. A sequence of $\mathbb{X}$-valued random variables x_n defined on $(\Omega, \mathcal{F}, \mathbf{P})$ is said to be adapted to $\mathcal{F}_n$ if x_n is $\mathcal{F}_n$-measurable, that is $x_n^{-1}(B) = \{\omega \in \Omega : x_n(\omega) \in B\} \in \mathcal{F}_n$ for all $B \in \mathcal{B}(\mathbb{X})$. This holds for example if $\mathcal{F}_n = \sigma(x_m, m \leq n), n \geq 0$. Given a filtration $\mathcal{F}_n$ and a sequence of real random variables adapted to it, $(x_n, \mathcal{F}_n)$ is said to be a Martingale if

$$E[|x_n|] < \infty$$

and

$$E[x_{n+1}|\mathcal{F}_n] = x_n$$

We will occasionally take the sigma-fields to be $\mathcal{F}_n = \sigma(x_1, x_2, \ldots, x_n)$. Let $n > m \in \mathbb{Z}_+$. If $\{x_n\}$ is a Martingale sequence, $E[x_n|\mathcal{F}_m] = x_m$. If we have that $E[x_n|\mathcal{F}_m] \geq x_m$, $\{x_n\}$ is called a submartingale. If $E[x_n|\mathcal{F}_m] \leq x_m$, then $\{x_n\}$ is called a supermartingale.

A useful concept related to filtration is that of a stopping time. A stopping time is a random time, whose occurrence is measurable with respect to the filtration in the sense that for each $n \in \mathbb{N}$, $\{T \leq n\} \subset \mathcal{F}_n$.

The following is known as Doob's optional sampling theorem [220].

Theorem D.2.1 *Suppose $(x_n, \mathcal{F}_n)$ is a supermartingale, and ρ, τ are bounded stopping times (say bounded by $n \in \mathbb{Z}_+$) with $\rho \leq \tau$ almost surely. Then, $E[x_\tau|\mathcal{F}_\rho] \leq x_\rho$.*

Using Doob's upcrossing lemma [220], together with the optional sampling theorem above, one obtains the following:

Theorem D.2.2 (Submartingale Convergence Theorem) *Suppose x_n is a submartingale and $\sup_{n\geq 0} E[|x_n|] < \infty$. Then $x := \lim_{n\to\infty} x_n$ exists (almost surely) and $E[|x|] < \infty$.*

D.3 Stochastic Stability of Dynamical Systems and Random Processes

D.3.1 Stationary, Ergodic, and Asymptotically Mean Stationary Processes

In this subsection, we review ergodic theory, in the context of information theory (that is with the transformations being specific to the shift operation). A comprehensive discussion is available in Shields [665] and Gray [315], [319].

Let $\mathbb{X}$ be a complete, separable, metric space. Let $\mathcal{B}(\mathbb{X})$ denote the Borel sigma-field of subsets of $\mathbb{X}$. Let $\Sigma = \mathbb{X}^\infty$ denote the sequence space of all one-sided or two-sided infinite sequences drawn from $\mathbb{X}$. Thus, for a two-sided sequence space if $x \in \Sigma$ then $x = \{\ldots, x_{-1}, x_0, x_1, \ldots\}$ with $x_i \in \mathbb{X}$. Let $X_n : \Sigma \to \mathbb{X}$ denote the coordinate function such that $X_n(x) = x_n$. Let T denote the shift operation on Σ, that is $X_n(Tx) = x_{n+1}$. That is, for a one-sided sequence space $T(x_0, x_1, x_2, \ldots) = (x_1, x_2, x_3, \ldots)$.

Let $\mathcal{B}(\Sigma)$ denote the smallest sigma-field containing all cylinder sets of the form $\{x : x_i \in B_i, m \leq i \leq n\}$ where $B_i \in \mathcal{B}(\mathbb{X})$, for all integers m, n. Observe that $\cap_{n\geq 0} T^{-n}\mathcal{B}(\Sigma)$ is the tail σ-field $\cap_{n\geq 0}\sigma(x_n, x_{n+1}, \cdots)$, since $T^{-n}(A) = \{x : T^n x \in A\}$.

Let μ be a stationary measure on $(\Sigma, \mathcal{B}(\Sigma))$ in the sense that $\mu(T^{-1}B) = \mu(B)$ for all $B \in \mathcal{B}(\Sigma)$. The sequence of random variables $\{x_n\}$ defined on the probability space $(\Sigma, \mathcal{B}(\Sigma), \mu)$ is a stationary process.

Definition D.3.1 Let μ be the measure on a process. This random process is ergodic if $A = T^{-1}A$ implies that $\mu(A) \in \{0, 1\}$.

That is, the events that are unchanged with a shift operation are trivial events. Mixing is a sufficient condition for ergodicity. Thus, a source is ergodic if $\lim_{n\to\infty} P(A \cap T^{-n}B) = P(A)P(B)$, since the process forgets its initial condition. For the special case of Markov sources, we have the following: A positive Harris recurrent Markov chain is ergodic, since such a process is mixing and stationary.

Definition D.3.2 A random process with measure μ is N-stationary, (cyclo-stationary or periodically stationary with period N) if $\mu(T^{-N}B) = \mu(B)$ for all $B \in \mathcal{B}(\Sigma)$, or equivalently for any $n \in \mathbb{N}$ samples $t_1, t_2, \ldots, t_n$:

$$\mu(x_{t_1} \in A_1, \ldots, x_{t_n} \in A_n) = \mu(x_{t_1+N} \in A_1, \ldots, x_{t_n+N} \in A_n)$$

Definition D.3.3 A random process is N-ergodic if $A = T^{-N}A$ implies that $\mu(A) \in \{0, 1\}$.

Definition D.3.4 A set $A \in \mathcal{B}(\mathbb{X})$ is coordinate-recurrent if for some $m \in \mathbb{Z}_+$

$$\sum_{m=0}^{\infty} 1_{\{X_m(x)\in A\}} = \infty, \qquad a.s.$$

Definition D.3.5 A process on a probability space $(\Omega, \mathcal{F}, \mathbf{P})$ with process measure P is asymptotically mean stationary (AMS) if there exists a probability measure $\bar{P}$ such that

$$\lim_{N\to\infty} \frac{1}{N} \sum_{k=0}^{N-1} P(T^{-k}F) = \bar{P}(F)$$

for all events $F \in \mathcal{B}(\Sigma)$. Here $\bar{P}$ is called the stationary mean of P, and is a stationary measure.

$\bar{P}$ is stationary since, by definition $\bar{P}(F) = \bar{P}(T^{-1}F)$. A cyclo-stationary process is AMS; see for example [143, 319] or [315] (Theorem 7.3.1). Asymptotic mean stationarity is an important property: (1) The Shannon-McMillan-Breiman Theorem (The Entropy Ergodic Theorem) applies to finite alphabet AMS sources [319] (see an extension for a more general class [72]). In this case, the ergodic decomposition of the AMS process leads to almost sure convergence of the conditional entropies. (2) Birkhoff's ergodic theorem applies to bounded measurable functions f, if and only if the process is AMS [319]. Let

$$F = \{x : \lim_{N\to\infty} \frac{1}{N} \sum_{i=0}^{N-1} f(T^i x) \quad \text{exists.}\}$$

It follows that for an AMS process, $m(F) = 1$, with m being the stationary mean of the process. Birkhoff's Almost-Sure Ergodic Theorem states the following: If a dynamical system is AMS with stationary mean m, then all bounded measurable functions f have the ergodic property, and with probability 1,

$$\lim_{N\to\infty} \frac{1}{N} \sum_{i=0}^{N-1} f(T^i x) = E_{m_x}[f], \quad x \in F,$$

where E_{m_x} denotes the expectation under measure m_x and m_x is the resulting ergodic measure with initial state x in the ergodic decomposition of the asymptotic mean (see Theorem 1.8.2 in [314]): $m(A) = \int m_x(A) m(dx)$.

Definition D.3.6 (AMS Ergodic) Consider a stochastic process with process measure P, which is AMS with asymptotic mean $\bar{P}$. If $\bar{P}$ is ergodic, we call the process *AMS ergodic*.

For any fixed Borel B, let

$$F_B = \{x : \lim \frac{1}{N} \sum_k 1_{\{T^k \in B\}} = \bar{P}(B)\}$$

It then follows that for an *AMS Ergodic* process

$$P(F_B) = 1$$

Definition D.3.7 A random process x is second-moment stable if the following holds:

$$\lim_{N\to\infty} \frac{1}{N} \sum_{m=0}^{N-1} E[(X_m(x))^2] < \infty$$

Definition D.3.8 A random process x is quadratically stable (almost surely) if the following limit exists and is finite almost surely:

$$\lim_{N\to\infty} \frac{1}{N} \sum_{m=0}^{N-1} X_m(x)^2 < \infty$$

Appendix E
Markov Decision Theory and Optimality of Markov Policies

This appendix provides some preliminary background on optimization of controlled Markov chains. A comprehensive treatment can be found in [112, 153, 342, 445, 534] and [33].

E.1 Controlled Markov Models

Consider the following model:

$$x_{t+1} = f(x_t, u_t, w_t), \tag{E.1}$$

where x_t is a $\mathbb{X}$-valued state variable, u_t is $\mathbb{U}$-valued control action variable, w_t is a $\mathbb{W}$-valued an i.i.d noise process, and f is a measurable function. We assume that $\mathbb{X}$, $\mathbb{U}$, $\mathbb{W}$ are subsets of Polish spaces. The model above in (E.1) contains (see [132]) the class of all stochastic processes which satisfy the following for all Borel sets $B \in \mathcal{B}(\mathbb{X})$, $t \geq 0$, and all realizations $x_{[0,t-1]}, u_{[0,t-1]}$:

$$P(x_{t+1} \in B | x_{[0,t]} = a_{[0,t]}, u_{[0,t]} = b_{[0,t]}) = \mathcal{T}(x_{t+1} \in B | a_t, b_t), \tag{E.2}$$

where $\mathcal{T}(\cdot|x, u)$ is a *stochastic kernel* from $\mathbb{X} \times \mathbb{U}$ to $\mathbb{X}$.

A stochastic process which satisfies (E.2) is called a controlled Markov chain.

E.2 Fully Observed Markov Control Problem Model

Definition E.2.1 A Fully Observed Markov Control Problem is a five tuple $(\mathbb{X}, \mathbb{U}, \mathbb{K}, \mathcal{T}, c)$, where

- $\mathbb{X}$ is the state space, a subset of a Polish (that is, a complete, separable, metric) space.
- $\mathbb{U}$ is the action space, a subset of a Polish space.
- $\mathbb{K} = \{(x, u) : u \in \mathbb{U}(x), x \in \mathbb{X}\}$ is the set of state and control pairs that are feasible so that $\mathbb{U}(x)$ denotes the set of feasible actions at state x.

S. Yüksel, T. Başar, *Stochastic Teams, Games, and Control under Information Constraints*, Systems & Control: Foundations & Applications,
https://doi.org/10.1007/978-3-031-54071-4

- $\mathcal{T}$ is a state transition kernel, that is $\mathcal{T}(A|x_t, u_t) = P(x_{t+1} \in A|x_t, u_t)$.
- $c : \mathbb{K} \to \mathbb{R}$ is the cost function.

The objective function to be minimized for a finite-stage setup is given by:

$$J(x_0, \gamma) := E_{x_0}^{\gamma}[\sum_{t=0}^{T-1} c(x_t, u_t) + c_T(x_T)], \tag{E.3}$$

where γ is a control policy (synonymously, *strategy* or *control law*), $E_{x_0}^{\gamma}[\cdot]$ denotes the expectation under policy γ and initial state x_0, and c_T is a terminal state cost. Given a class of admissible policies (to be defined below), the goal is to find an admissible policy γ^* such that

$$J(x_0, \gamma^*) \leq J(x_0, \gamma),$$

for all admissible policies γ and all initial states x_0. Such a γ^* is an *optimal policy*. Here γ can also be called a *strategy*, or a *control law*.

E.2.1 Classes of Control Policies

Admissible Control Policies

Let $H_0 := \mathbb{X}, \quad H_t = H_{t-1} \times \mathbb{X} \times \mathbb{U}$ for $t = 1, 2, \ldots$. We let h_t denote an element of H_t, where $h_t = \{x_{[0,t]}, u_{[0,t-1]}\}$. A *deterministic admissible control policy* γ is a sequence of functions $\{\gamma_t\}$ from $H_t \to \mathbb{U}$; in this case $u_t = \gamma_t(h_t)$. A randomized control policy is a sequence $\gamma = \{\gamma_t, t \geq 0\}$ such that $\gamma : H_t \to \mathcal{P}(\mathbb{U})$ (with $\mathcal{P}(\mathbb{U})$ being the space of probability measures on $\mathbb{U}$) such that

$$\gamma_t(u_t \in U(x_t)|h_t) = 1, \quad \forall h_t \in H_t.$$

Markov Control Policies

A policy is *randomized Markov* if

$$P_{x_0}^{\gamma}(u_t \in C|h_t) = \gamma_t(u_t \in C|x_t), \quad C \in \mathcal{B}(\mathbb{U}).$$

Hence, the control action only depends on the state and the time, and not the past history. If the control strategy is deterministic, that is if

$$\gamma_t(u_t = f_t(x_t)|x_t) = 1$$

for some function f_t, the control policy is said to be *deterministic Markov*.

Stationary Control Policies

A policy is *randomized stationary* if

$$P_{x_0}^{\gamma}(u_t \in C|h_t) = \gamma(u_t \in C|x_t), \quad C \in \mathcal{B}(\mathbb{U}).$$

Hence, the control action only depends on the state, and not the past history or on time. If the control strategy is deterministic, that is if $\gamma(u_t = f(x_t)|x_t) = 1$ for some function f, the control policy is said to be *deterministic stationary*.

Let Γ_A, Γ_M, and Γ_S denote the sets of all admissible, Markov and stationary policies respectively.

E.2.2 *Optimality of Markov Policies and Elimination of Irrelevant Information*

The following is a useful result on the structure of optimal control policies.

Theorem E.2.1 (Blackwell's Irrelevant Information Theorem [124, 127]) *Let $\mathbb{X}$, $\mathbb{Y}$, and $\mathbb{U}$ be standard Borel, and P be a probability measure on $\mathcal{B}(\mathbb{X} \times \mathbb{Y})$, and let $c : \mathbb{X} \times \mathbb{U} \to \mathbb{R}$ be a Borel measurable cost function. Then, for any Borel measurable function $\gamma : \mathbb{X} \times \mathbb{Y} \to \mathbb{U}$, there exists another Borel measurable function $\gamma^* : \mathbb{X} \to \mathbb{U}$ such that*

$$\int_{\mathbb{X}} c(x, \gamma^*(x))P(dx) \leq \int_{\mathbb{X}\times\mathbb{Y}} c(x, \gamma(x, y))P(dx, dy)$$

Furthermore, policies based only on x almost surely are optimal.

E.2.3 *Markov Decision Processes (MDPs) and Optimality of Markov Policies*

The following result is a well-known theorem in the control theory literature (for example see [342] and [691]). A proof can be obtained through the dynamic programming algorithm and applying Theorem E.2.1 at every time stage.

Theorem E.2.2 *Given an MDP as above, consider the minimization of (E.3) over all admissible control policies. There is no loss in restricting policies to be Markov, that is a policy which only uses the current state x_t and the time information t.*

E.2.4 *Dynamic Programming and Measurable Selection Criteria*

The following dynamic programming arguments hold when there exists a minimizing control policy (sequence of selectors for each time stage) for (E.3).

Dynamic Programming

Let $\{J_t, t = T, T-1, \ldots, 0\}$, with $J_t : \mathbb{X} \to \mathbb{R}$, be a sequence of functions generated by the recursion

$$J_t(x) = \inf_{u \in \mathbb{U}_t(x)} (c(x, u) + \int_{\mathbb{X}} J_{t+1}(y)P(x_{t+1} \in dy|x_t = x, u_t = u)), \tag{E.4}$$

for all $x \in \mathbb{X}$, for $t \in \{0, 1, 2 \ldots, T-1\}$, with

$$J_T(x_T) = c_T(x_T).$$

Let the infimum in (E.4) be achieved for all t. Then, there exists a sequence of measurable functions $\{f_t\}$ such that

$$J_t(x_t) = c(x_t, f_t(x_t)) + \int_{\mathbb{X}} J_{t+1}(y) P(dy|x_t, f_t(x_t))),$$

and $\{f_t\}$ is optimal for (E.3). ◇

Such a sequence of optimal functions exists under either one of the following conditions.

Condition 1 The stage-wise cost function to be minimized, $c(x_t, u_t)$, is continuous on both $\mathbb{U}$ and $\mathbb{X}$, $\mathbb{U}_t(x) = \mathbb{U}$ is compact, and $\int_{\mathbb{X}} Q(dy|x, u)v(y)$ is a continuous function on $\mathbb{X} \times \mathbb{U}$ for every continuous and bounded v on $\mathbb{X}$. ◇

Condition 2 For every $x \in \mathbb{X}$ the stage-wise cost function to be minimized, $c(x_t, u_t)$, is continuous on $\mathbb{U}$; $\mathbb{U}_t(x) = \mathbb{U}$ is compact; and $\int_{\mathbb{X}} Q(dy|x, u)v(y)$ is a continuous function on $\mathbb{U}$ for every bounded, measurable function v on $\mathbb{X}$. ◇

Theorem E.2.3 *Under Condition 1 or Condition 2, the Measurable Selection Hypothesis applies, and there exists an optimal control policy (which is a Markov policy) obtained through the Dynamic Programming recursions in (E.4); that is, there exists a minimizing control policy given by $f_t : \mathbb{X} \to \mathbb{A}_t(x_t)$. Furthermore, under Condition 1, the function $J_t(x)$ is continuous, if $c_N(x_N)$ is continuous.*

We can replace the compactness condition with an *inf-compactness* condition, and modify Condition 1 as below:

Condition 3 For every $x \in \mathbb{X}$, $c(x, u)$ is continuous on $\mathbb{X} \times \mathbb{U}$; is non-negative; $\{u : c(x, u) \leq \alpha\}$ is compact for all $\alpha > 0$ and all $x \in \mathbb{X}$, and $\int_{\mathbb{X}} Q(dy|x, u)v(y)$ is a continuous function on $\mathbb{X} \times \mathbb{U}$ for every continuous and bounded v. ◇

The results of Lemma 5.4.1 also apply when $\mathbb{U}(x)$ depends on x so that it is compact for each x and $\{(x, u) : u \in \mathbb{U}(x), x \in \mathbb{X}\}$ is a Borel subset of $\mathbb{X} \times \mathbb{U}$.

Lemma E.2.1 ([347, Theorem 2], [442, 649]) *Let $\mathbb{X}$ and $\mathbb{U}$ be standard Borel spaces and $G = (x, \psi(x))$ where $\psi(x) \subset \mathbb{U}$ is such that, $\psi(x)$ is compact for each $x \in \mathbb{X}$ and G is a Borel measurable set in $\mathbb{X} \times \mathbb{U}$. Let $c(x, u)$ be a continuous function on $\psi(x)$ for every x. Then,*

(i) There exists a Borel measurable function $f : \mathbb{X} \to \mathbb{U}$ such that

$$c(x, f(x)) = \min_{u \in \psi(x)} c(x, u)$$

(ii) If continuity is also to be attained for the value function $c(x, f(x))$ it suffices if $\mathbb{U}(x)$ is compact and $\mathbb{U}(x)$ is an upper semi-continuous set-valued function (the implication being that: for any $x^n \to x$ and $u'^n \in \mathbb{U}(x^n)$, there exists a subsequence u'^{n_k} which converges to some u' with the property that $u' \in \mathbb{U}(x)$) and c is continuous.

We could relax the continuity condition and replace it with lower semi-continuity. A function is lower semi-continuous at x_0 if $\liminf_{x \to x_0} f(x) \geq f(x_0)$. We state the following theorem; see [342, Theorem 3.3.5].

Theorem E.2.4 *The following hold:*

(a) Suppose that (i) $\mathbb{U}(x)$ is compact for every x and $\{(x, u) : u \in \mathbb{U}(x)\}$ is a Borel subset of $\mathbb{X} \times \mathbb{U}$, (ii) c is lower semi-continuous on $\mathbb{U}(x)$ for every $x \in \mathbb{X}$, and (iii) $\int v(x_{t+1}) P(dx_{t+1}|x_t =$

$x, u_t = u)$ is lower semi-continuous on $\mathbb{U}(x)$ for every $x \in \mathbb{X}$ and every measurable and bounded v on $\mathbb{X}$. Then, the measurable selection hypothesis applies.

(b) *If (i) c is lower semi-continuous on $\{(x,u) : u \in \mathbb{U}(x), x \in \mathbb{X}\}$, (ii) for every lower semi-continuous function v on $\mathbb{X}$, $\int v(x_{t+1})P(dx_{t+1}|x_t = x, u_t = u)$ is lower semi-continuous on $\{(x,u) : u \in \mathbb{U}(x), x \in \mathbb{X}\}$, and (iii) $\mathbb{U}(x)$ is compact for every $x \in \mathbb{X}$ and $\mathbb{U}(x)$ is an upper semi-continuous set-valued function; then the value function v is lower semi-continuous.*

For further related relaxations, see [342, Appendix D]).

Universally Measurable Value Functions and Dynamic Program Recursions Not Requiring Continuity/Compactness Conditions Additionally, as we saw in Sect. 5.4.2, studying the class of *universally measurable* and *semi-analytic* functions allows one to even further relax the conditions required for carrying out dynamic programming recursions for their well-posedness properties in the absence of continuity/compactness conditions: It suffices for a function to be universally measurable (and not necessarily Borel measurable) for its integration with respect to some probability measure to be well defined: a function f is μ-measurable if there exists a Borel measurable function g which agrees with f μ-a.e. A function that is μ-measurable for every probability measure μ is called universally measurable. A set B is universally measurable if the function $1_{\{x \in B\}}$ is universally measurable.

A measurable image of a Borel set is called an *analytic* set [230, Appendix 2]. We note that this is equivalent to the seemingly more restrictive condition of being a *continuous* image of some Borel set. The image of a Borel set under a measurable function, and hence an analytic set, is universally measurable.

The integral of a universally measurable function is well defined and is equal to the integral of a Borel measurable function which is μ-almost equal to that function (with μ being the measure for integration). While applying dynamic programming, we often seek to establish the existence of measurable functions through the operation:

$$J_t(x_t) = \inf_{u \in \mathbb{U}(x_t)} \left(c(x_t, u) + \int J_{t+1}(x_{t+1}) Q(dx_{t+1}|x_t, u) \right)$$

However, here we need a stronger condition than universal measurability for the recursions to be well defined: a function f is called lower semi-analytic if $\{x : f(x) < c\}$ is analytic for each scalar $c \in \mathbb{R}$. Lower semi-analytic functions are universally measurable [230]. In Sect. 5.4.2, we discuss some related results (see [114, Theorems 7.47 and 7.48]), which present general dynamic programming equations leading also to ϵ-optimal policies [666].

E.2.5 Discounted Cost Criterion

For a given $T \in \mathbb{Z}_+$, the expected discounted cost function is given as:

$$J_\beta^T(x_0, \gamma) = E_{x_0}^\gamma[\sum_{t=0}^{T-1} \beta^t c(x_t, u_t)], \tag{E.5}$$

for some $\beta \in (0,1)$. If there exists a policy γ^* which minimizes this cost function, the policy is said to be optimal. We often consider an infinite horizon problem by taking the limit (when c is non-negative)

$$J_\beta(x_0, \gamma) = \lim_{T \to \infty} E_{x_0}^\gamma[\sum_{t=0}^{T-1} \beta^t c(x_t, u_t)],$$

and invoking the monotone convergence theorem:

$$J_\beta(x_0, \gamma) = E^{\gamma}_{x_0}[\sum_{t=0}^{\infty} \beta^t c(x_t, u_t)].$$

We seek to find

$$J_\beta(x_0) = \inf_{\gamma \in \Gamma_A} J_\beta(x_0, \gamma).$$

Define $\mathbb{T}$ as follows:

$$(\mathbb{T}(v))(x) := \min_u \left(c(x, u) + \beta \int_{\mathbb{X}} v(y) \mathcal{T}(dy|x, u) \right),$$

and define the *Discounted Cost Optimality Equation (DCOE)* as follows:

$$v(x) = (\mathbb{T}(v))(x), \qquad x \in \mathbb{X} \tag{E.6}$$

Lemma E.2.2 (Verification Theorem [342])

(i) If v is a measurable $\mathbb{R}_+$-valued function under Condition 2 (or continuous and bounded function under Condition 1) with $v \geq \mathbb{T}v$, then $v(x) \geq J_\beta(x)$.
(ii) If $v \leq \mathbb{T}v$ and

$$\lim_{n\to\infty} \beta^n E^{\gamma}_x [v(x_n)] = 0, \tag{E.7}$$

for every policy and initial condition, then $v(x) \leq J_\beta(x)$. As a result, a fixed point to (E.6) leads to an optimal policy under (E.7).

Theorem E.2.5 (Value Iteration Algorithm: General Cost Setup) *Suppose that the cost function is non-negative. Consider the successive iteration*

$$v_n(x) = \min_u \{c(x, u) + \beta \int_{\mathbb{X}} v_{n-1}(y) \mathcal{T}(dy|x, u)\}, \forall x, n \geq 1 \tag{E.8}$$

with $v_0(x) = 0$ for all $x \in \mathbb{X}$. Then, v_n is a monotonically non-decreasing sequence. If this sequence converges pointwise to a function v where

$$v(x) = c(x, f(x)) + \beta \int v(y) \mathcal{T}(dy|x, f(x))$$

is such that with $\gamma = \{f, f, \cdots\}$, (E.7) holds, then γ is optimal and v is the value function.

Theorem E.2.6 (Value Iteration Algorithm—Bounded Cost Setup) *Suppose the cost function is bounded, non-negative, and one of the measurable selection conditions (Condition 1 or Condition 2) applies. Then, there exists a unique solution to the discounted cost problem, which solves the fixed point equation.*

$$v(x) = \min_u \{c(x, u) + \beta \int_{\mathbb{X}} v(y) \mathcal{T}(dy|x, u)\}, \qquad x \in \mathbb{X}$$

Furthermore, the optimal cost (value function) is obtained by a successive iteration of policies (known as the Value Iteration Algorithm):

$$v_n(x) = \min_u\{c(x,u) + \beta \int_{\mathbb{X}} v_{n-1}(y)\mathcal{T}(dy|x,u)\}, \qquad \forall x, n \in \mathbb{N} \tag{E.9}$$

For any $v_0 \in L_\infty(\mathbb{X})$, the sequence converges to a unique fixed point. If $v_0(x) = 0, x \in \mathbb{X}$, then $v_n(x) \uparrow v(x)$ for all $x \in \mathbb{X}$ (that is, v_n monotonically converges to v). If Condition 1 holds, then v is also continuous.

E.3 Partially Observable MDPs (POMDPs)

Consider the following dynamics.

$$x_{t+1} = f(x_t, u_t, w_t), \quad y_t = g(x_t, v_t)$$

Here, as before, x_t is the state, $u_t \in \mathbb{U}$ is the control, $(w_t, v_t) \in \mathbb{W} \times \mathbb{V}$ are zero-mean, i.i.d noise processes, and w_t is independent of v_t. In addition to the previous fully observed model, y_t denotes an observation variable taking values in $\mathbb{Y}$, taken here to be a subset of $\mathbb{R}^n$. Then, the joint distribution of the state, control, and observation processes is determined by the following system dynamics:

$$\Pr\big((x_0, y_0) \in B\big) = \int_B Q(dy_0|x_0)P_0(dx_0), \; B \in \mathcal{B}(\mathbb{X} \times \mathbb{Y}),$$

where P_0 is the prior distribution of the initial state x_0 and Q_0 is the observation channel, and for $t \in \mathbb{N}$,

$$\Pr\Big((x_t, y_t) \in B \,\Big|\, (x, y, u)_{[0,t-1]} = (x, y, u)_{[0,t-1]}\Big)$$
$$= \int_B Q(dy_t|x_t)\mathcal{T}(dx_t|x_{t-1}, u_{t-1}), \; B \in \mathcal{B}(\mathbb{X} \times \mathbb{Y}),$$

where $\mathcal{T}(\cdot|x,u)$ is a stochastic kernel from $\mathbb{X} \times \mathbb{U}$ to $\mathbb{X}$. This completes the probabilistic setup of the partially observed model.

An *admissible policy* γ is a sequence of control functions $\{\gamma_t, t \in \mathbb{Z}_{\geq 0}\}$ such that γ_t is measurable with respect to the σ-algebra generated by the information variables

$$I_t = \{y_{[0,t]}, u_{[0,t-1]}\}, \quad t \in \mathbb{N}, \qquad I_0 = \{y_0\},$$

where

$$u_t = \gamma_t(I_t), \quad t \in \mathbb{Z}_+ \tag{E.10}$$

are the $\mathbb{U}$-valued control action variables. We define Γ_A to be the set of all such admissible policies. Suppose that we are faced with an optimal control problem. Let a one-stage cost function $c : \mathbb{X} \times \mathbb{U} \to [0,\infty)$, which is a Borel measurable function from $\mathbb{X} \times \mathbb{U}$ to $[0,\infty)$, be given. Then, we denote by $J(\gamma)$ the cost corresponding to the policy $\gamma \in \Gamma_A$, which can be, for instance, finite horizon, discounted cost or average cost criteria. With these definitions, the goal of the control problem is to find an optimal policy γ^* that minimizes J.

As we will discuss in the following, any such problem can be reduced to a completely observable Markov process [602, 834], whose states are the posterior state distributions or 'beliefs' of the observer; that is, the state at time t is

$$\pi_t(\cdot) := P\{X_t \in \cdot | y_0, \ldots, y_t, u_0, \ldots, u_{t-1}\} \in \mathcal{P}(\mathbb{X}).$$

We call this conditional probability measure process the *filter* process. The filter process has state space $\mathcal{P}(\mathbb{X})$ and action space $\mathbb{U}$. Here, $\mathcal{P}(\mathbb{X})$ is equipped with the Borel σ-algebra generated by the topology of weak convergence. As discussed in Appendix C, under this topology, $\mathcal{P}(\mathbb{X})$ is a standard Borel space [574]. The transition probability of the filter process can be constructed as follows.

Let $Q(B|x) := P(y_t \in B|x_t = x)$ for any $B \in \mathcal{B}(\mathbb{Y})$, and assume that Q is *dominated* in the sense that there exists a dominating reference measure λ such that $\forall x \in \mathbb{X}$, $Q(dy|x_n = x) \ll \lambda$. Then, defining the Radon-Nikodym derivative

$$g(x, y) = \frac{dG(y_n \in \cdot | x_n = x)}{d\lambda}(y)$$

as the likelihood function (serving as a conditional probability density function), we can write the filter π_{n+1} recursively in terms of π_n and y_{n+1}, u_n explicitly as a Bayesian update:

$$\begin{aligned}\pi_{n+1}(dx_{n+1}) &= F(\pi_n, y_{n+1}, u_n)(dx_{n+1}) \\ &= \frac{\int_{\mathbb{X}} g(x_{n+1}, y_{n+1})\mathcal{T}(dx_{n+1}|x_n, u_n)\pi_n(dx_n)}{\int_{\mathbb{X}}\int_{\mathbb{X}} g(x_{n+1}, y_{n+1})\mathcal{T}(dx_{n+1}|x_n, u_n)\pi_n(dx_n)}\end{aligned} \tag{E.11}$$

The conditional measure process becomes a controlled Markov chain in $\mathcal{P}(\mathbb{X})$ (equipped with the weak convergence topology).

Theorem E.3.1 *The process $\{\pi_t, u_t\}$ is a controlled Markov chain. That is, under any admissible control policy, given the action at time $t \geq 0$ and π_t, π_{t+1} is conditionally independent of $\{\pi_s, u_s, s \leq t-1\}$.*

Then, the transition probability η of the filter process can be constructed as follows. If we define the measurable function $F(\pi, u, y) := F(\cdot|y, u, \pi) = \Pr\{x_{t+1} \in \cdot | \pi_t = \pi, u_t = u, y_{t+1} = y\}$ from $\mathcal{P}(\mathbb{X}) \times \mathbb{U} \times \mathbb{Y}$ to $\mathcal{P}(\mathbb{X})$ and use the stochastic kernel $P(\cdot|\pi, u) = \Pr\{y_{t+1} \in \cdot | \pi_t = \pi, u_t = u\}$ from $\mathcal{P}(\mathbb{X}) \times \mathbb{U}$ to $\mathbb{Y}$, we can write η as

$$\eta(\cdot|\pi, u) = \int_{\mathbb{Y}} 1_{\{F(\pi,u,y)\in \cdot\}} P(dy|\pi, u). \tag{E.12}$$

As before, let the objective function to be minimized be $\sum_{t=0}^{T-1} E_{x_0}^{\gamma}[c(x_t, u_t)]$. Via iterated expectations, by writing this as: $\sum_{t=0}^{T-1} E_{x_0}^{\gamma}[E_{x_0}^{\gamma}[c(x_t, u_t)|I_t]]$, we transform the system into a fully observed Markov model as follows. Define the one-stage cost function $\tilde{c} : \mathcal{P}(\mathbb{X}) \times \mathbb{U} \to [0, \infty)$ of the filter process as given by

$$\tilde{c}(\pi, u) := \int_{\mathbb{X}} c(x, u)\pi(dx),$$

which is a Borel measurable function. Hence, the filter process defines a completely observable Markov process with the components $(\mathcal{P}(\mathbb{X}), \mathbb{U}, \tilde{c}, \eta)$.

Now, building all these and Theorem E.2.2 together, it follows that an optimal control policy of the original POMDP will use the belief π_t as a sufficient statistic for optimal policies (see [602, 834]). More precisely, the filter process is equivalent to the original POMDP in the sense that for any optimal policy for the filter process, one can construct a policy for the original POMDP which is optimal.

As such, one can obtain the optimal solution by using the filtering equation as a sufficient statistic, as Markov policies (policies that use the Markov state as their sufficient statistics) are optimal for control of Markov chains, under sufficiency conditions for the existence of optimal selectors.

We first present a supporting result.

Assumption E.3.1 *The following hold.*

(i) The transition probability $P(x_1 \in \cdot|x_0 = x, u_0 = u) =: \mathcal{T}(\cdot|x, u)$ *is weakly continuous in* (x, u)*, i.e., for any* $(x_n, u_n) \to (x, u)$*,* $\mathcal{T}(\cdot|x_n, u_n) \to \mathcal{T}(\cdot|x, u)$ *weakly.*
(ii) Assumption 8.7.1 holds; that is, the measurement kernel Q *is continuous in total variation in the sense that as* $x_n \to x$*,* $\|Q(dy|x_n) - Q(dy|x)\|_{TV} \to 0$.

Assumption E.3.2 *The following hold.*

(i) The transition probability $\mathcal{T}(\cdot|x, u)$ *is continuous in total variation in* (x, u)*, i.e., for any* $(x_n, u_n) \to (x, u)$*,* $\mathcal{T}(\cdot|x_n, u_n) \to \mathcal{T}(\cdot|x, u)$ *in total variation.*
(ii) The observation channel $Q(\cdot|x)$ *does not depend on the control variable.*

Theorem E.3.2 *The following hold.*

(i) [257] Under Assumption E.3.1, the transition probability $\eta(\cdot|z, u)$ *of the filter process is weakly continuous in* (z, u).
(ii) [395] Under Assumption E.3.2, the transition probability $\eta(\cdot|z, u)$ *of the filter process is weakly continuous in* (z, u).

See [209, 255, 258, 392] for further results on the above weak Feller property. We also refer the reader to [152] for a control-free context. If the cost function c is continuous and bounded, an application of the dominated convergence theorem implies that $\tilde{c}(\pi, u)$ is also continuous and bounded. If the action set is compact, then under the weak continuity condition noted above on the non-linear filter, we have that the measurable selection conditions apply, and accordingly an optimal control policy exists for J_β (the average cost problem is slightly more nuanced, but under additional mild conditions an optimal policy exists in this case also). Furthermore, by Theorem E.2.3, the optimal cost $J_\beta^*(\mu)$ is continuous in the prior under weak convergence. The above also facilitates rigorous approximation and learning theoretic results (see [391, 392, 626, 628] for a further analysis).

E.4 Kalman Filter and the Linear Quadratic Gaussian Optimal Control Problem

As we have seen above, for controlling partially observed controlled Markov sources, one can enlarge the state space and define a probability measure valued state which is fully observed. For a Gaussian source, however, the above can be done in a computationally very efficient manner, as a Gaussian measure is uniquely characterized by only its mean and covariance matrix. Accordingly, consider now the partially observed linear system:

$$x_{t+1} = Ax_t + Bu_t + w_t,$$

$$y_t = Cx_t + v_t,$$

where $x \in \mathbb{R}^n$, $u \in \mathbb{R}^m$ and $w \in \mathbb{R}^n$, $y \in \mathbb{R}^p$, $v \in \mathbb{R}^p$. Let $\{w_t, v_t\}$ be i.i.d. zero-mean random Gaussian vectors with given covariance matrices $E[w_t w_t'] = W$ and $E[v_t v_t'] = V$ for all $t \geq 0$. Further, let x_0 be zero-mean Gaussian. Let us now define

$$m_t = E[x_t | y_{[0,t-1]}, u_{[0,t-1]}],$$

and

$$\Sigma_{t|t-1} = E[(x_t - E[x_t | y_{[0,t-1]}, u_{[0,t-1]}])(x_t - E[x_t | y_{[0,t-1]}, u_{[0,t-1]}])' \\ | y_{[0,t-1]}, u_{[0,t-1]}].$$

Then, $\{m_t, \Sigma_{t|t-1}\}$ are generated by the following recursions, which constitute the Kalman Filter equations:

$$m_{t+1} = Am_t + Bu_t + A\Sigma_{t|t-1}C'(C\Sigma_{t|t-1}C' + V)^{-1}(y_t - Cm_t)$$

$$m_0 = E[x_0]$$

$$\Sigma_{t+1|t} = A\Sigma_{t|t-1}A' + W - (A\Sigma_{t|t-1}C')(C\Sigma_{t|t-1}C' + V)^{-1}(C\Sigma_{t|t-1}A')$$

$$\Sigma_{0|-1} = E[x_0 x_0']$$

Let us now introduce

$$I_t = \{y_{[0,t]}, u_{[0,t-1]}\},$$

and define $\bar{m}_t = E[x_t | I_t]$. The following readily holds

$$\bar{m}_t = m_t + \Sigma_{t|t-1}C'(C\Sigma_{t|t-1}C' + V)^{-1}(y_t - Cm_t),$$

or

$$\bar{m}_t = A\bar{m}_{t-1} + Bu_t \\ + \Sigma_{t|t-1}C'(C\Sigma_{t|t-1}C' + V)^{-1}(CA(x_{t-1} - \bar{m}_{t-1}) + v_t). \tag{E.13}$$

We note that the zero-mean variable $x_t - \bar{m}_t$ is orthogonal to $I_t = \{y_{[0,t]}, u_{[0,t-1]}\}$, in the sense that the error is independent of the information available at the controller, and since the information available is Gaussian, independence and orthogonality are equivalent. In view of this, the variable $\Sigma_{t|t-1}C'(C\Sigma_{t|t-1}C' + V)^{-1}(CA(x_{t-1} - \bar{m}_{t-1}) + v_t)$ is an additive Gaussian noise acting additively on the new state variable $\bar{m}_t$.

LQG Optimal Control Problem

Now, given the Gaussian POMDP setup considered, suppose that the goal is to obtain a solution to

$$\inf_{\gamma} J(\gamma, \mu_0),$$

where

$$J(\gamma, \mu_0) = E^{\gamma}_{\mu_0}[\sum_{t=0}^{T-1} x_t' Q x_t + u_t' R u_t + x_T' Q_T x_T],$$

with $R > 0$, and $Q, Q_T \geq 0$ (that is, these matrices are positive definite and positive semi-definite), and μ_0 is the prior measure on the initial state, which is taken to be Gaussian.

The quadratic optimization problem can be reformulated as a function of $\bar{m}_t$, u_t and the error $(x_t - \bar{m}_t)$, which makes the optimal control problem equivalent to optimal control of fully observed state $\bar{m}_t$, with an additive time-varying independent Gaussian noise process. This, thus, features the separation of estimation and control, and a more special version of it known as the certainty equivalence principle. The absence of *dual effect* plays a key role in this analysis, in taking $E[(x_t - \bar{m})'Q(x_t - \bar{m})]$ out of the conditioning on I_t. Thus, with J_t being the *value* function at time t, one obtains

$$J_t(I_t) = E[x_t'|I_t]K_t E[x_t|I_t] + \sum_{k=t}^{T-1}\Big(E[(x_k - E[x_k|I_k])'Q(x_k - E[x_k|I_k])] + E[\tilde{w}_t' K_{t+1}\tilde{w}]\Big), \tag{E.14}$$

where $\tilde{w}_t = \Sigma_{t|t-1}C'(C\Sigma_{t|t-1}C' + V)^{-1}(y_t - CE[x_t|I_{t-1}])$.

Solving a fully observed LQG optimal control problem, the optimal minimizing control is linear and has the form [445]:

$$u_t = -(BK_{t+1}B + R)^{-1}B'K_{t+1}AE[x_t|I_t]$$

where K_t is generated by the Discrete-Time Riccati Equation

$$K_t = Q + A'K_{t+1}A - A'K_{t+1}B((BK_{t+1}B + R))^{-1}B'K_{t+1}A,$$

with terminal condition $K_T = Q_T$.

Appendix F
Relaxed Control Topologies

In deterministic as well as stochastic control theory, relaxed or randomized control policies allow for versatility in mathematical analysis, leading to continuity, compactness, convexity and approximation properties, in a variety of system models, cost criteria, and information structures.

Within the relaxed/randomized control framework, with $\mathbb{X}$ a state space, $\mathbb{U}$ a control space and with an $\mathbb{X}$-valued random variable $X \sim \mu$, instead of considering the set of deterministic admissible policies:

$$\Gamma = \Big\{\gamma : \gamma \text{ is a measurable function from } \mathbb{X} \text{ to } \mathbb{U}\Big\}, \tag{F.1}$$

one considers

$$\Gamma_R = \Big\{\gamma : \gamma \text{ is a measurable function from } \mathbb{X} \text{ to } \mathcal{P}(\mathbb{U})\Big\}, \tag{F.2}$$

where $\mathcal{P}(\mathbb{U})$ is endowed with the Borel σ-algebra generated by the weak convergence topology.

On Γ_R, two commonly studied topologies are the following.

F.1 Young Topology on Control Policies

A prominent approach since Young's seminal paper [800] has been via the study of topologies on *Young measures* defined by randomized/relaxed controls, where one views policies to be identified with probability measures defined on a product space with a fixed marginal at an input/state space (typically taken to be the Lebesgue measure in optimal deterministic control) [526, 800], [162, Section 2.1], [759, p. 254], [511], [59, Theorem 2.2]. Thus, under the Young topology, one associates with Γ_R in (F.2) the probability measure induced on the product space $\mathbb{X} \times \mathbb{U}$ with a fixed marginal μ on $\mathbb{X}$.

The generalization to stochastic control problems by considering more general input measures has been commonplace, with applications also to partially observed stochastic control and decentralized stochastic control. See Sects. 5.5.1, 5.4.1, 4.2, 8.2, and 15.4 for explicit applications considered in this book.

S. Yüksel, T. Başar, *Stochastic Teams, Games, and Control under Information Constraints*, Systems & Control: Foundations & Applications,
https://doi.org/10.1007/978-3-031-54071-4

To appreciate the Young topology on control policies, we first present a relevant representation result (see Borkar [132]). Let $\mathbb{X}$, $\mathbb{M}$ be Borel spaces. Let $\mathcal{P}(\mathbb{X})$ denote the set of probability measures on $\mathbb{X}$. Consider the set of probability measures

$$\Theta := \Big\{ \zeta \in \mathcal{P}(\mathbb{X} \times \mathbb{M}) : \tag{F.3}$$

$$\zeta(dx, dm) = P(dx)\, Q^f(dm|x),\ Q^f(\cdot|x) = 1_{\{f(x)\in\cdot\}},\ f : \mathbb{X} \to \mathbb{M} \Big\}$$

on $\mathbb{X} \times \mathbb{M}$ with fixed input marginal P on $\mathbb{X}$ and with the stochastic kernel from $\mathbb{X}$ to $\mathbb{M}$ realized by any measurable function $f : \mathbb{X} \to \mathbb{M}$. We equip this set with the weak convergence topology. This set is the (Borel measurable) set of the extreme points of the set of probability measures on $\mathbb{X} \times \mathbb{M}$ with a fixed marginal P on $\mathbb{X}$. For compact $\mathbb{M}$, the Borel measurability of Θ follows [580] since the set of probability measures on $\mathbb{X} \times \mathbb{M}$ with a fixed marginal P on $\mathbb{X}$ is a *convex and compact* set in a complete separable metric space, and therefore, the set of its extreme points is Borel measurable; measurability for the non-compact case follows from [132, Lemma 2.3]. Furthermore, given a fixed marginal P on $\mathbb{X}$, any stochastic kernel Q from $\mathbb{X}$ to $\mathbb{M}$ can almost surely be identified by a probability measure $\Xi \in \mathcal{P}(\Theta)$ such that

$$Q(\cdot|x) = \int_{\Theta} \Xi(dQ^f)\, Q^f(\cdot|x). \tag{F.4}$$

In particular, a randomized policy can thus be viewed as a mixture of deterministic policies.

Definition F.1.1 (Convergence of Policies in Γ_S Under Young Topology at Reference (Input) Measure μ) Let μ be a σ-finite measure. A sequence of stationary policies $\gamma_n \to \gamma \in \Gamma_S$ at input μ if the joint measure $(\mu\gamma_n) \to (\mu\gamma)$ weakly at input P, i.e., for every bounded $g : \mathbb{X} \times \mathbb{U} \to \mathbb{R}$, which is continuous in u, with

$$\int \mu(dx) \sup_{u\in\mathbb{U}} |g(x,u)| < \infty,$$

$$\int \mu(dx)\Big(\gamma_n(du|x) g(x,u)\Big) \to \int \mu(dx)\Big(\gamma(du|x) g(x,u)\Big) \tag{F.5}$$

With the above, we observe that the Young topology allows for a convex and compact formulation, when the input measure is finite.

We also note that in the above, the reference measure does not need to be a probability measure (and we view weak convergence to be one on signed measures that defines a locally convex space with (F.5) defining the semi-norms).

We finally note that since the marginal of the joint measure $(\mu\gamma_n)$ on $\mathbb{X}$ is fixed, the convergence in (F.5) is also in the setwise-weak sense (with $g(x,u)$ bounded but only continuous in u for every fixed $x \in \mathbb{X}$, see Sect. C.4), following Lemma C.4.1.

F.2 Borkar (Weak*) Topology on Control Policies

In the stochastic setup, another topology is the one introduced by Borkar on relaxed controls [131] (see also [34, Section 2.4], and [120] which [131] notes to be building on), formulated as a weak* topology on randomized policies viewed as maps from states/measurements to the space of signed measures with bounded variation $\mathcal{M}(\mathbb{U})$ of which probability measures $\mathcal{P}(\mathbb{U})$ is a subset. We also refer the reader to [217, 251] for further references on such a weak* formulation on relaxed controls, in particular when instead of countably additive signed measures, finitely additive such measures are also considered.

Under the Borkar topology one studies Γ_R in (F.2), with a weak* topology formulation, as a bounded subset of the set of maps from $\mathbb{X}$ to the space of signed measures with finite variation viewed as the topological dual of continuous functions vanishing at infinity, leading to a compact metric space by the Banach-Alaoglu theorem [269, Theorem 5.18] (and thus, as the unit ball of $L_\infty(\mathbb{X}, \mathcal{M}(\mathbb{U})) = (L_1(\mathbb{X}, C_0(\mathbb{U})))^*$ is compact under the weak* topology, this leads to a compact metric topology on relaxed control policies). We note that the presentations in [131, Section 3] and [34, Section 2.4] are slightly different, though the induced topologies are identical. An equivalent representation of this topology is given in [34, Lemma 2.4.1] (see also [131, Lemma 3.1]).

See [32, 131, 590] for a detailed analysis on some implications in stochastic control theory in continuous time (such as continuity of expected cost in control policies [131], approximation results [590] under various cost criteria, and continuity of invariant measures of diffusions in control policies [32]).

Definition F.2.1 ([131] [34, Lemma 2.4.1] Convergence of Policies in Γ_S Under Borkar Topology) With $\mathbb{X} = \mathbb{R}^d$, a sequence of stationary policies $\gamma_n \to \gamma \in \Gamma_S$ in the Borkar topology if for every continuous and bounded $g : \mathbb{X} \times \mathbb{U} \to \mathbb{R}$ and every $f \in L_1(\mathbb{X}) \cap L_2(\mathbb{X})$

$$\int f(x) \int \gamma_n(du|x) g(x, u) dx \to \int f(x) \int \gamma_n(du|x)\gamma(du|x) g(x, u) dx \tag{F.6}$$

Building on Lemma C.4.1, the functions g in Definition F.2.1 may be relaxed to be continuous only in u for every $x \in \mathbb{X}$.

While $\mathbb{X}$ was taken to be $\mathbb{R}^n$ in [131], Saldi [620] generalized this to setups where $\mathbb{X}$ is a general standard Borel space with a fixed input (probability) measure. The generalization by Saldi [620] is the following, where the input space $\mathbb{X}$ is arbitrary standard Borel, though with a fixed input measure μ: Let $C_0(\mathbb{U})$ be the Banach space of all continuous real functions on $\mathbb{U}$ vanishing at infinity, endowed with the norm $\|g\|_\infty = \sup_{u \in \mathbb{U}} |g(u)|$. Saldi [620] formulated this topology via noting that with $L_1(\mu, C_0(\mathbb{U}))$ denoting the set of all Bochner-integrable functions from $\mathbb{X}$ to $C_0(\mathbb{U})$ endowed with the norm

$$\|f\|_1 := \int_{\mathbb{X}} \|f(x)\|_\infty \, \mu(dx),$$

using the fact that $C_0(\mathbb{U})^* = \mathcal{M}(\mathbb{U})$, and that the topological dual of $\left(L_1(\mu, C_0(\mathbb{U}), \|\cdot\|_1\right)$ can be identified with $\left(L_\infty(\mu, \mathcal{M}(\mathbb{U})), \|\cdot\|_\infty\right)$ [165, Theorem 1.5.5, p. 27] (see also [217, 251] for further context on such duality results, in particular when instead of countably additive signed measures, finitely additive such measures are considered); that is,

$$L_1\big(\mu, C_0(\mathbb{U})\big)^* = L_\infty\big(\mu, \mathcal{M}(\mathbb{U})\big).$$

This topology was used for optimal decentralized stochastic control for an existence analysis, as we studied in Sect. 5.5.3.

F.3 Some Properties of Young and Borkar Topologies

Lemma F.3.1 ([819]) *Let $\eta \ll \kappa$, where κ is a σ-finite and η is a finite measure. Then, $\gamma_n \to \gamma$ under Young topology at input κ implies $\gamma_n \to \gamma$ (under Young topology) at input η.*

Theorem F.3.1 ([819]) *Let $\mathbb{X} = \mathbb{R}^n$ and λ be the Lebesgue measure. Consider convergence in Young topology at some input probability measure ψ.*

(i) If $\psi \ll \lambda$ with $h(x) = \frac{d\psi}{d\lambda}(x)$ is positive everywhere, then convergence in Young topology at input measure ψ implies convergence in Borkar topology.
(ii) If $\psi \ll \lambda$, then convergence in Borkar topology implies convergence in Young topology at input ψ.

In [819], several results on the significance of these topologies on existence of optimal policies and approximations (on near optimality of continuous policies or quantized policies in both measurement and action) have been presented.

Appendix G
Stochastic Stability of Markov Chains and Random-Time State-Dependent Stochastic Drift

One essential aspect of a networked control system is the presence of randomness and uncertainty in measurement information or action transmissions. The evolution of the dynamics is generally event driven and this does not necessarily admit a time-homogeneous dynamics based on a fixed clock or a deterministic sequence. To understand such random event-based updates, it is necessary to develop the requisite mathematical tools to study such processes.

We will present in this appendix an important mathematical program to study such dynamics; this is referred to as the *random-time state-dependent drift criteria for stabilization*, building on [537] and [829].

G.1 Stochastic Stability and Drift Criteria

Let $\mathbf{X} = \{x_t, t \geq 0\}$ be a Markov chain with state space $\mathbb{X}$. Assume that the state space is a complete, separable, metric space, whose Borel σ-field is denoted by $\mathcal{B}(\mathbb{X})$. Let the transition probability be denoted by P, so that for any $x \in \mathbb{X}$, $A \in \mathcal{B}(\mathbb{X})$, the probability of moving from x to A in one step is given by $P(x_{t+1} \in A \mid x_t = x) = P(x, A)$. The n-step transitions are obtained via composition in the usual way, $P(x_{t+n} \in A \mid x_t = x) = P^n(x, A)$, for any $n \geq 1$. The transition law acts on measurable functions $f : \mathbb{X} \to \mathbb{R}$ and measures μ on $\mathcal{B}(\mathbb{X})$ via,

$$Pf(x) := \int_{\mathbb{X}} P(x, dy) f(y), \quad x \in \mathbb{X}, \quad \mu P(A) := \int_{\mathbb{X}} \mu(dx) P(x, A), \quad A \in \mathcal{B}(\mathbb{X}).$$

A probability measure π on $\mathcal{B}(\mathbb{X})$ is called invariant if $\pi P = \pi$. That is,

$$\int \pi(dx) P(x, A) = \pi(A), \qquad A \in \mathcal{B}(\mathbb{X}).$$

The existence of an invariant probability measure is very important since such a measure represents the asymptotic behavior of the Markov chain. In particular, $\lim_{n \to \infty} \frac{1}{n} \sum_{k=0}^{n-1} P^k(x, A)$ converges to $\pi(A)$ for some invariant measure π, which may depend on x. If the Markov chain is positive Harris recurrent (see Sect. D.1), there exists a unique invariant probability measure π and furthermore, for every x, $\lim_{n \to \infty} P^n(x, A) = \pi(A)$.

S. Yüksel, T. Başar, *Stochastic Teams, Games, and Control under Information Constraints*, Systems & Control: Foundations & Applications,
https://doi.org/10.1007/978-3-031-54071-4

One sufficient, and very general, characterization for the existence of an invariant probability measure is through the use of Lyapunov functions and the drift criteria, which we discuss in the following subsections.

For any initial probability measure ν on $\mathcal{B}(\mathbb{X})$ we can construct a stochastic process with transition law P, and satisfying $x_0 \sim \nu$. We let P_ν denote the resulting probability measure on the sample space, with the usual convention $\nu = \delta_x$ when the initial state is $x \in \mathbb{X}$. When $\nu = \pi$, then the resulting process is stationary.

G.1.1 One-Stage Foster-Lyapunov Drift Criteria

An increasing family $\{\mathcal{F}_n\}$ of sub-σ−fields of $\mathcal{F}$ is called a filtration (see Sect. D.2). In the following, let $\mathcal{F}_t$ be such a filtration generated by the state sequence, that is $\mathcal{F}_t = \sigma(x_m, m \leq t), t \geq 0$.

The following results are known as Foster-Lyapunov drift criteria [537].

Theorem G.1.1 ([537]) *Suppose that* **X** *is a φ-irreducible Markov chain. Suppose moreover that there are functions $V : \mathbb{X} \to (0, \infty)$, $\epsilon > 0$, a petite set C, and a constant $b \in \mathbb{R}$, such that the following holds:*

$$E[V(x_{t+1}) \mid \mathcal{F}_t] \leq V(x_t) - \epsilon + b 1_{\{x_t \in C\}} \tag{G.1}$$

Then **X** *is positive Harris recurrent.*

Theorem G.1.2 ([537]) *Suppose that* **X** *is a φ-irreducible Markov chain. Suppose moreover that there are functions $V : \mathbb{X} \to (0, \infty)$, $f : \mathbb{X} \to [1, \infty)$, a petite set C, and a constant $b \in \mathbb{R}$, such that the following holds:*

$$E[V(x_{t+1}) \mid \mathcal{F}_t] \leq V(x_t) - f(x_t) + b 1_{\{x_t \in C\}} \tag{G.2}$$

Then **X** *is positive Harris recurrent, and moreover $\pi(f) := E_\pi(f(x)) < \infty$, with π being the invariant distribution.*

G.1.2 State-Dependent Drift Criteria

The following establishes a set of sufficient conditions for positive Harris recurrence under deterministic but state-dependent drift.

Theorem G.1.3 ([537]) *Suppose that* **X** *is a φ-irreducible Markov chain. Let $V : \mathbb{X} \to \mathbb{R}_+$, C be a petite set on which V is bounded, $b \in \mathbb{R}_+$. Consider the following inequality for some $\epsilon \geq 0$, and some function $n(\cdot) : \mathbb{X} \to \mathbb{Z}_+$:*

$$E[V(x_{t+n(x_t)}) \mid \mathcal{F}_t] \leq (1-\epsilon)^{(n(x_t))} V(x_t) - n(x_t) + b 1_{\{x_t \in C\}},$$

$\forall x \in \mathbb{X}$.

If the inequality holds with $\epsilon = 0$, then a finite invariant measure exists for the Markov chain. If $\epsilon > 0$, then the chain is exponentially ergodic, that is there exists $r > 1$ and a positive function $M(x)$ such that $lim_{n \to \infty} M(x) r^n ||P^n(x, \cdot) - \pi||_{TV} \to 0$. ◇

We now partially generalize the above in the following to random-time state-dependent stochastic drift criteria.

G.1.3 Random-Time State-Dependent Stochastic Drift Criteria

Throughout this sub-section, the sequence of stopping times $\{\tau_i : i \in \mathbb{N}_+\}$ is assumed to be non-decreasing, with $\tau_0 = 0$, measurable on the filtration generated by the state process. Additional assumptions are made in the results that follow.

Theorem G.1.4 ([829]) *Suppose that* $\mathbf{X}$ *is a* φ*-irreducible and aperiodic Markov chain. Suppose moreover that there are functions* $V : \mathbb{X} \to (0, \infty)$, $\delta : \mathbb{X} \to [1, \infty)$, $f : \mathbb{X} \to [1, \infty)$, *a petite set* C *on which* V *is bounded, and a constant* $b \in \mathbb{R}$, *such that the following hold:*

$$
\begin{aligned}
&E[V(x_{\tau_{z+1}}) \mid \mathcal{F}_{\tau_z}] \le V(x_{\tau_z}) - \delta(x_{\tau_z}) + b 1_{\{x_{\tau_z} \in C\}} \\
&E\Big[\sum_{k=\tau_z}^{\tau_{z+1}-1} f(x_k) \mid \mathcal{F}_{\tau_z}\Big] \le \delta(x_{\tau_z}), \qquad z \ge 0.
\end{aligned}
\tag{G.3}
$$

Then the following hold:

(i) $\mathbf{X}$ *is positive Harris recurrent, with unique invariant distribution* π
(ii) $\pi(f) := \int f(x)\pi(dx) < \infty$
(iii) *For any function* g *that is bounded by* f, *in the sense that* $\sup_x |g(x)|/f(x) < \infty$, *we have convergence of moments in the mean, and the Law of Large Numbers holds:*

$$
\lim_{t\to\infty} E_x[g(x_t)] = \pi(g) \le b
$$

$$
\lim_{N\to\infty} \frac{1}{N}\sum_{t=0}^{N-1} g(x_t) = \pi(g) \le b \qquad a.s., \ x \in \mathbb{X}
$$

Remark G.1 We note that, for (ii) in Theorem G.1.4, the condition that $f : \mathbb{X} \to [1, \infty)$, $\delta : \mathbb{X} \to [1, \infty)$, can be relaxed to $f : \mathbb{X} \to [0, \infty)$, $\delta : \mathbb{X} \to [0, \infty)$, provided that one can establish (i), that is the positive Harris recurrence of the Markov chain $\mathbf{X}$ first. ◇

Remark G.2 The assumption of $\tau_0 = 0$ is important for establishing recurrence. If $\tau_0 \neq 0$, additional technical, but mild, conditions would have to be added to verify recurrence. ◇

By taking $f(x) = 1$ for all $x \in \mathbb{X}$, we obtain the following corollary to Theorem G.1.4.

Corollary G.1.1 *Suppose that* $\mathbf{X}$ *is a* φ*-irreducible Markov chain with filtration* $\mathcal{F}_t$. *Suppose moreover that there is a function* $V : \mathbb{X} \to (0, \infty)$, *a petite set* C, *and a constant* $b \in \mathbb{R}$, *such that the following hold:*

$$
\begin{aligned}
&E[V(x_{\tau_{z+1}}) \mid \mathcal{F}_{\tau_z}] \le V(x_{\tau_z}) - 1 + b 1_{\{x_{\tau_z} \in C\}} \\
&\sup_{z\ge 0} E[\tau_{z+1} - \tau_z \mid \mathcal{F}_{\tau_z}] < \infty.
\end{aligned}
\tag{G.4}
$$

Then $\mathbf{X}$ *is positive Harris recurrent.*

To relax the φ-irreducibility assumption, we can impose instead the following continuity assumption: A Markov chain is (weak) Feller if the function Pf is continuous on $\mathbb{X}$, for every continuous and bounded function $f: \mathbb{X} \to \mathbb{R}$. The following can be useful in a number of applications, including the study of adaptive quantizers in networked control:

Theorem G.1.5 ([457, 537]) *Suppose that the chain* $\mathbf{X}$ *is Feller, living in a locally compact space, and that there is a compact set A satisfying*

$$\sup_{x \in A} E_x[\tau_A] < \infty,$$

where $\tau_A = \inf(k > 0 : x_k \in A)$. *Then, the Markov chain admits an invariant probability measure* π *which is positive on A.* ◇

The following is a consequence of Theorems G.1.4 and G.1.5.

Theorem G.1.6 *If* (G.4) *holds for a measurable set C and a function* $V : \mathbb{X} \to (0, \infty)$, *then C satisfies:*

$$\sup_{x \in C} E[\tau_C] < \infty.$$

In view of the above, we have the following. Note that here the Markov chain need not be irreducible.

Theorem G.1.7 ([829]) *Suppose that* $\mathbf{X}$ *is a Feller Markov chain living in a locally compact space, not necessarily* φ*-irreducible. Then, if* (G.3) *holds with C compact, there exists at least one invariant probability measure. Moreover, there is a constant* b_f *so that under any invariant probability measure* π,

$$E_\pi[f(x_t)] = \int_{\mathbb{X}} \pi(dx) f(x) \le b_f. \tag{G.5}$$

Finally, we state a result on transience. This will be useful in establishing that bounded range quantizers lead to transience (see Theorems 13.2.1 and 14.2.4).

Theorem G.1.8 ([537, Proposition 8.4.1]) *Let* $V : \mathbb{X} \to \mathbb{R}_+$. *If there exists a set A such that*

$$E[V(x_{t+1}) | x_t = x] \le V(x), \tag{G.6}$$

for all $x \notin A$, *and* $\exists \bar{x} \notin A$ *such that* $V(\bar{x}) < \inf_{z \in A} V(z)$, *then* $\{x_t\}$ *is not recurrent, in the sense that* $P_{\bar{x}}(\tau_A < \infty) < 1$.

The treatment in this appendix is a condensed discussion from [537], [824, Chapter 6], and [344]. The results on random-time drift criteria in Sect. G.1.3 and the proofs are based on [805, 829]. Theorem G.1.8 is available in [537] as well as [334].

Appendix H
Topological and Metric Entropy in Dynamical Systems

This appendix provides some background material on topological and metric entropy for dynamical systems. Metric entropy was first introduced by Kolmogorov and Sinai in the 1950s as a measure-theoretic invariant of dynamical systems that preserve a probability measure on their state space. In information-theoretic terms, metric entropy measures the expected average (over time) growth of uncertainty about the initial state of a system as time tends to infinity. In the 1960s, an analogous notion of topological entropy was introduced by Adler et al. for dynamical systems in the topological category. While it has no direct information-theoretic interpretation, topological entropy turns out to be the supremal metric entropy, when all probability measures preserved by the system are considered at once. For further details about these concepts and their operational meaning, we refer the reader to the excellent survey [402] and the monograph [221].

In the following, all logarithms are taken to the base 2. We define $\log\infty := \infty$. By $|A|$ we denote the cardinality of a set A. If (X, d) is a metric space and $\emptyset \neq A \subset X$, we write $\mathrm{diam}A := \sup\{d(x, y) : x, y \in A\}$ for the diameter of A. If $(x_t)_{t\in\mathbb{Z}_+}$ is a sequence, we write $x_{[0,t]} = (x_0, x_1, \ldots, x_t)$ for any $t \geq 0$.

Let $f : X \to X$ be a uniformly continuous map on a metric space (X, d). We define the iterates of f recursively by

$$f^0 := \mathrm{id}_X, \quad f^{n+1} := f \circ f^n, \quad \forall n \in \mathcal{N}.$$

If f is invertible, we additionally put $f^{-n} := (f^{-1})^n$ for all $n \in \mathcal{N}$.

For $n \in \mathcal{N}$ and $\epsilon > 0$, a subset $F \subset X$ is said to (n, ϵ)-span another subset $K \subset X$ provided that for each $x \in K$ there is $y \in F$ with $d(f^i(x), f^i(y)) \leq \epsilon$ for $0 \leq i < n$. Alternatively, we can describe this in terms of the metric

$$d_{n,f}(x, y) := \max_{0\leq i<n} d(f^i(x), f^i(y)),$$

which is topologically equivalent to d. Any $d_{n,f}$-ball of radius $\epsilon > 0$ is called a *Bowen-ball of order n and radius* ϵ. In these terms, a set F (n, ϵ)-spans K if the closed Bowen-balls of order n and radius ϵ centered at the points in F form a cover of K. If K is compact, we write $r_{\mathrm{span}}(n, \epsilon; K)$ to denote the minimal cardinality of a set which (n, ϵ)-spans K, observing that this number is finite. The *topological entropy of* f *on* K is then defined by

$$h_{\mathrm{top}}(f; K) := \lim_{\epsilon\downarrow 0} \limsup_{n\to\infty} \frac{1}{n} \log r_{\mathrm{span}}(n, \epsilon; K).$$

S. Yüksel, T. Başar, *Stochastic Teams, Games, and Control under Information Constraints*, Systems & Control: Foundations & Applications,
https://doi.org/10.1007/978-3-031-54071-4

By the monotonicity properties of $r_{\text{span}}(n, \epsilon; K)$, the limit for $\epsilon \downarrow 0$ can be replaced by the supremum over $\epsilon > 0$. An alternative definition can be given in terms of (n, ϵ)-separated sets. A subset $E \subset X$ is called *(n, ϵ)-separated* if for any $x, y \in E$ with $x \neq y$ one has $d_{n,f}(x, y) > \epsilon$. Writing $r_{\text{sep}}(n, \epsilon; K)$ for the maximal cardinality of an (n, ϵ)-separated subset of a compact set K, one finds that $r_{\text{span}}(n, \epsilon; K) \leq r_{\text{sep}}(n, \epsilon; K) \leq r_{\text{span}}(n, \epsilon/2; K)$, and hence

$$h_{\text{top}}(f; K) = \lim_{\epsilon \downarrow 0} \limsup_{n \to \infty} \frac{1}{n} \log r_{\text{sep}}(n, \epsilon; K).$$

If X is compact, we also write $h_{\text{top}}(f) = h_{\text{top}}(f; X)$ and call this number the *topological entropy of f*.

Now assume that X is compact and μ is an f-invariant Borel probability measure on X, i.e., $\mu \circ f^{-1} = \mu$. Let $\mathcal{P}$ be a finite Borel partition of X and put

$$h_\mu(f; \mathcal{P}) := \lim_{n \to \infty} \frac{1}{n} H_\mu(\mathcal{P}^n), \quad \mathcal{P}^n := \bigvee_{i=0}^{n-1} f^{-i}\mathcal{P}, \tag{H.1}$$

where $H_\mu(\mathcal{P}^n) = -\sum_{P \in \mathcal{P}^n} \mu(P) \log \mu(P)$ denotes the entropy of the partition $\mathcal{P}^n$ and $\vee$ is the standard join operation for partitions, i.e., $\bigvee_{i=0}^{n-1} f^{-i}\mathcal{P}$ is the partition of X whose members are the sets

$$P_{i_0} \cap f^{-1}(P_{i_1}) \cap \ldots \cap f^{-n+1}(P_{i_{n-1}}), \quad P_{i_j} \in \mathcal{P}.$$

The existence of the limit in (H.1) follows from subadditivity. The metric entropy of f with respect to (w.r.t.) μ is defined by

$$h_\mu(f) := \sup_{\mathcal{P}} h_\mu(f; \mathcal{P}),$$

the supremum taken over all finite Borel partitions $\mathcal{P}$. If μ is ergodic (w.r.t. f), i.e.,

$$f^{-1}(A) = A \quad \Rightarrow \quad \mu(A) \in \{0, 1\},$$

there is an alternative characterization of $h_\mu(f)$ in terms of (n, ϵ)-spanning sets, due to Katok [401]. For a fixed $\delta \in (0, 1)$ put

$$r_{\text{span}}(n, \epsilon, \delta) := \min \left\{ |F| : F \ (n, \epsilon)\text{-spans a set of } \mu\text{-measure} \geq 1 - \delta \right\}.$$

Then the metric entropy of f satisfies

$$h_\mu(f) = \lim_{\epsilon \downarrow 0} \limsup_{n \to \infty} \frac{1}{n} \log r_{\text{span}}(n, \epsilon, \delta),$$

and this limit is independent of δ; see [401, Thm. 1.1].

Note that both $h_\mu(f)$ are $h_{\text{top}}(f)$ are nonnegative quantities, which can assume the value $+\infty$. They are related by the variational principle, i.e.,

$$h_{\text{top}}(f) = \sup_\mu h_\mu(f),$$

where the supremum is taken over all f-invariant Borel probability measures (see [543] for a short proof). In the following, we present some classical results of ergodic theory. We start with the Shannon-McMillan-Breiman Theorem.

Theorem H.0.1 *Let $(X, \mathcal{F}, \mu, f)$ be an ergodic measure-preserving dynamical system, i.e., $(X, \mathcal{F}, \mu)$ is a probability space and μ is ergodic w.r.t. the map $f : X \to X$. Let $\mathcal{P}$ be a finite measurable partition of X. Then, for almost every $x \in X$, as $n \to \infty$,*

$$-\frac{1}{n} \log \mu(\mathcal{P}^n(x)) \to h_\mu(f; \mathcal{P}),$$

where $\mathcal{P}^n(x)$ is the element of $\mathcal{P}^n$ containing x.

If $\mathcal{P} = \{P_1, \ldots, P_r\}$, for $n \in \mathcal{N}$ and $x \in X$, let $i^n(x) = (i_0^n(x), \ldots, i_{n-1}^n(x))$ be the element of $\{1, \ldots, r\}^n$ with $f^j(x) \in P_{i_j^n(x)}$ for $0 \leq j < n$. We call $i^n(x)$ the *length-n itinerary of x w.r.t. $\mathcal{P}$*. Putting $h := h_\mu(f; \mathcal{P})$, we also define for each $\epsilon > 0$ and $n \in \mathcal{N}$ the set

$$\Sigma_{n,\epsilon} := \left\{a \in \{1, \ldots, r\}^n : 2^{-n(h+\epsilon)} \leq \mu(\{x : i^n(x) = a\}) \leq 2^{-n(h-\epsilon)}\right\}.$$

An immediate corollary to the Shannon-McMillan-Breiman Theorem now follows:

Corollary H.0.1 *Under the assumptions of Theorem H.0.1, it holds that*

$$\mu(\{x \in X : \exists m \in \mathcal{N} \text{ with } i^n(x) \in \Sigma_{n,\epsilon},\ \forall n \geq m\}) = 1 \tag{H.2}$$

and

$$|\Sigma_{n,\epsilon}| \leq 2^{n(h+\epsilon)}. \tag{H.3}$$

Proof The identity (H.2) follows from the observation that $\mu(\mathcal{P}^n(x)) = \mu(\{y \in X : i^n(y) = i^n(x)\})$. From the estimates

$$1 \geq \mu(\{x : i^n(x) \in \Sigma_{n,\epsilon}\}) = \sum_{a \in \Sigma_{n,\epsilon}} \mu(\{x : i^n(x) = a\}) \geq |\Sigma_{n,\epsilon}| \cdot 2^{-n(h+\epsilon)}$$

the inequality (H.3) immediately follows. □

In the ergodic theory of smooth dynamical systems, a fundamental result is Oseledec's Theorem, also known as the Multiplicative Ergodic Theorem:

Theorem H.0.2 *Let $(\Omega, \mathcal{F}, \mu)$ be a probability space and $\theta : \Omega \to \Omega$ a measurable map, preserving μ. Let $T : \Omega \to \mathbb{R}^{d\times d}$ be a measurable map such that* $\max\{0, \log \|T(\cdot)\|\} \in L^1(\Omega, \mu)$ *and write*

$$T_x^n := T(\theta^{n-1}(x)) \cdots T(\theta(x))T(x).$$

Then, there is $\tilde{\Omega} \subset \Omega$ with $\mu(\tilde{\Omega}) = 1$ so that for all $x \in \tilde{\Omega}$ the following holds:

$$\lim_{n\to\infty} \left[(T_x^n)^*(T_x^n)\right]^{1/(2n)} =: \Lambda_x$$

exists and, moreover, if $\exp \lambda_x^{(1)} < \cdots < \exp \lambda_x^{s(x)}$ denote the eigenvalues of Λ_x and $U_x^{(1)}, \ldots, U_x^{s(x)}$ the associated eigenspaces, then

$$\lim_{n\to\infty} \frac{1}{n} \log \|T_x^n v\| = \lambda_x^{(r)} \text{ if } v \in V_x^{(r)} \setminus V_x^{(r-1)},$$

where $V_x^{(r)} = U_x^{(1)} + \cdots + U_x^{(r)}$ and $r = 1, \ldots, s(x)$.

The numbers $\lambda_x^{(i)}$ are called (μ-)*Lyapunov exponents*. We apply the theorem to the situation when θ is a diffeomorphism on a compact Riemannian manifold and $T(x)$ is the derivative of θ at x. In this case, the Lyapunov exponents measure the rates at which nearby orbits diverge (or converge). Despite the fact that this is a purely geometric concept, the Lyapunov exponents are closely related to the metric entropy. Fundamental results here are the Margulis-Ruelle inequality [614] and Pesin's entropy formula [577], which we capture in the following theorem.

Theorem H.0.3 *Let M be a compact Riemannian manifold and $f : M \to M$ be a C^1-diffeomorphism preserving a Borel probability measure μ. Then*

$$h_\mu(f) \leq \int \mu(dx) \sum_i \max\{0, \dim U_x^{(i)} \cdot \lambda_x^{(i)}\}$$

with summation over all Lyapunov exponents at x. If f is a C^2-diffeomorphism and μ is absolutely continuous w.r.t. Riemannian volume, then equality holds. If μ is ergodic, then the Lyapunov exponents are constant almost everywhere, and hence the integration can be omitted.

References

1. E.F. Abaya, G.L. Wise, Convergence of vector quantizers with applications to optimal quantization. SIAM J. Appl. Math. **44**, 183–189 (1984)
2. E.F. Abaya, G.L. Wise, Some remarks on the existence of optimal quantizers. Stat. Probab. Lett. **2**, 349–351 (1984)
3. D. Acemoglu, A. Makhdoumi, A. Malekian, A. Ozdaglar, Informational Braess' paradox: The effect of information on traffic congestion. Oper. Res. **66**(4), 893–917 (2018)
4. S. Adlakha, R. Johari, G.Y. Weintraub, Equilibria of dynamic games with many players: Existence, approximation, and market structure. J. Econ. Theory **156**, 269–316 (2015)
5. R.L. Adler, A.G. Konheim, M.H. McAndrew, Topological entropy. Trans. Am. Math. Soc. **114**(2), 309–319 (1965)
6. M. Aicardi, F. Davoli, R. Minciardi, Decentralized optimal control of Markov chains with a common past information set. IEEE Trans. Autom. Control **32**, 1028–1031 (1987)
7. A. Ajorlou, A. Jadbabaie, Local optimality of almost piecewise linear quantizers for Witsenhausen's problem. IEEE Trans. Autom. Control **66**, 3152 (2020)
8. E. Akyol, K. Rose, On constrained randomized quantization, in *Proc. Data Compress. Conf.*, Snowbird, Utah, USA, Apr. (2012), pp. 72–81
9. E. Akyol, K. Rose, On linear transforms in zero-delay Gaussian source channel coding, in *Proceedings of the IEEE International Symposium on Information Theory*, Boston, MA (2012), pp. 1548–1552
10. E. Akyol, C. Langbort, T. Başar, Privacy constrained information processing, in *IEEE Conference on Decision and Control (CDC)*, December (2015), pp. 4511–4516
11. E. Akyol, C. Langbort, T. Başar, Strategic compression and transmission of information, in *IEEE Information Theory Workshop - Fall*, October (2015), pp. 219–223
12. E. Akyol, K. Rose, T. Başar, Optimal zero-delay jamming over an additive noise channel. IEEE Trans. Inf. Theory **61**(8), 4331–4344 (2015)
13. E. Akyol, C. Langbort, T. Başar, Information-theoretic approach to strategic communication as a hierarchical game. Proc. IEEE **105**(2), 205–218 (2017)
14. D.J. Aldous, *Weak Convergence and the General Theory of Processes* Editeur inconnu (1981)
15. D.J. Aldous, I.A. Ibragimov, J. Jacod, *Ecole d'Ete de Probabilites de Saint-Flour XIII, 1983*, vol. 1117 (Springer, 1985)
16. C.D. Aliprantis, K.C. Border, *Infinite Dimensional Analysis*, 3rd edn. (Springer, Berlin, 2006)
17. C. Alós-Ferrer, K. Ritzberger, Trees and extensive forms. J. Econ. Theory **143**(1), 216–250 (2008)

S. Yüksel, T. Başar, *Stochastic Teams, Games, and Control under Information Constraints*, Systems & Control: Foundations & Applications,
https://doi.org/10.1007/978-3-031-54071-4

18. A. Altabaa, B. Yongacoglu, S. Yüksel, Decentralized multi-agent reinforcement learning for continuous-space stochastic games. Preprint. arXiv:2303.13539 (2023 American Control Conference) (2023)
19. L. Ambrosio, N. Gigli, G. Savaré, *Gradient Flows in Metric Spaces and in Space of Probability Measures* (Birkhäuser, Basel, 2008)
20. A. Ambrus, S. Takahashi, Multi-sender cheap talk with restricted state space. Theoret. Econ. **3**(1), 1–27 (2008)
21. B. Anahtarci, C.D. Kariksiz, N. Saldi, Learning in discrete-time average-cost mean-field games, in *IEEE Conference on Decision and Control (CDC)* (IEEE, 2021), pp. 3048–3053
22. V. Anantharam, V.S. Borkar, Common randomness and distributed control: A counterexample. Syst. Control Lett. **56**(7), 568–572 (2007)
23. A. Anastasopoulos, A sequential transmission scheme for the multiple access channel with noiseless feedback, in *Proceedings of the Annual Allerton Conference on Communications, Control and Computing*, Monticello, IL (2009)
24. A. Anastasopoulos, S. Pradhan, Decentralized sequential active hypothesis testing and the MAC feedback capacity, in *IEEE International Symposium on Information Theory (ISIT)* (IEEE, 2020), pp. 2085–2090
25. M. Andersland, Decoupling non-sequential stochastic control problems. Syst. Control Lett. **16**(1), 65–69 (1991)
26. M. Andersland, D. Teneketzis, Information structures, causality, and non-sequential stochastic control, I: design-independent properties. SIAM J. Control Optim. **30**, 1447–1475 (1992)
27. M. Andersland, D. Teneketzis, Information structures, causality, and non-sequential stochastic control, II: design-dependent properties. SIAM J. Control Optim. **32**, 1726–1751 (1994)
28. G.E. Andrews, *Number Theory* (Dover, New York, 1995)
29. B.R. Andrievskii, A.S. Matveev, A.L. Fradkov, Control and estimation under information constraints: Toward a unified theory of control, computation and communications. Autom. Remote Control **71**(4), 572–633 (2010)
30. M. Aoki, On decentralized linear stochastic control problems with quadratic cost. IEEE Trans. Autom. Control **18**, 243–250 (1973)
31. J. Arabneydi, A. Mahajan, Team-optimal solution of finite number of mean-field coupled LQG subsystems, in *IEEE 54th Annual Conference on Decision and Control (CDC)* (2015), pp. 5308–5313
32. A. Arapostathis, V.S. Borkar, Uniform recurrence properties of controlled diffusions and applications to optimal control. SIAM J. Control Optim. **48**(7), 4181–4223, (2010)
33. A. Arapostathis, V.S. Borkar, E. Fernandez-Gaucherand, M.K. Ghosh, S.I. Marcus, Discrete-time controlled Markov processes with average cost criterion: A survey. SIAM J. Control Optim. **31**, 282–344 (1993)
34. A. Arapostathis, V.S. Borkar, M.K. Ghosh, *Ergodic Control of Diffusion Processes*, vol. 143 (Cambridge University Press, 2012)
35. K.J. Arrow, The informational structure of the firm. Am. Econ. Rev. **75**, 303–307 (1985)
36. G. Arslan, S. Yüksel, Decentralized Q-learning for stochastic teams and games. IEEE Trans. Autom. Control **62**, 1545–1558 (2017)
37. G. Arslan, S. Yüksel, Decentralized Q-learning for stochastic teams and games. IEEE Trans. Autom. Control **62**(4), 1545–1558 (2017)
38. G. Arslan, S. Yüksel, Subjective equilibria under beliefs of exogenous uncertainty for dynamic games. SIAM J. Control Optim. **61**(3), 1038–1062 (2023)
39. R.B. Ash, *Real Analysis and Probability* (Academic Press, New York, 1972)
40. S. Asoodeh, M. Diaz, F. Alajaji, T. Linder, Estimation efficiency under privacy constraints. IEEE Trans. Inf. Theory **65**(3), 1512–1534 (2019)
41. A. Aspect, Proposed experiment to test the nonseparability of quantum mechanics. Phys. Rev. D **14**, 1944–1951 (1976)
42. K.K. Åström, R.M. Murray, *Feedback Systems: An Introduction for Scientists and Engineers* (Princeton University Press, 2010)

43. M. Athans, Survey of decentralized control methods, in *3rd NBER/FRB Workshop on Stochastic Control*, Washington D.C. (1974)
44. R.J. Aumann, Mixed and behavior strategies in infinite extensive games. Technical report, Princeton University NJ, 1961
45. R.J. Aumann, Agreeing to disagree. Ann. Stat. **4**, 1236–1239 (1976)
46. R.J. Aumann, Correlated equilibrium as an expression of Bayesian rationality. Econometrica J. Econom. Soc., 1–18 (1987)
47. E. Ayanoğlu, R.M. Gray, The design of joint source and channel trellis waveform coders. IEEE Trans. Inf. Theory **33**(6), 855–865 (1987)
48. J. Backhoff-Veraguas, D. Bartl, M. Beiglböck, M. Eder, Adapted Wasserstein distances and stability in mathematical finance. Finance Stoch. **24**(3), 601–632 (2020)
49. J. Backhoff-Veraguas, D. Bartl, M. Beiglböck, M. Eder, All adapted topologies are equal. Probab. Theory Related Fields **178**(3), 1125–1172 (2020)
50. A. Bagchi, T. Başar, Team decision theory for linear continuous-time systems. IEEE Trans. Autom. Control **27**, 1154–1161 (1980)
51. M. Baglietto, T. Parisini, R. Zoppoli, Numerical solutions to the Witsenhausen counterexample by approximating networks. IEEE Trans. Autom. Control **46**(9), 1471–1477 (2001)
52. M. Bagnoli, T. Bergstrom, Log-concave probability and its applications. Econ. Theory **26**(2), 445–469 (2005)
53. R.R. Bahadur, Sufficiency and statistical decision functions. Ann. Math. Stat. **25**, 423–462 (1954)
54. R.R. Bahadur, Statistics and subfields. Ann. Math. Stat. **26**, 490–497 (1955)
55. J. Baillieul, Feedback designs for controlling device arrays with communication channel bandwidth constraints, in *Proceedings of the 4th ARO Workshop on Smart Structures, State College, PA*, August (1999)
56. J. Baillieul, Data-rate requirements for nonlinear feedback control, in *Proc. 6th IFAC Symp. Nonlinear Control Syst., Stuttgart, Germany*, pp. 1277–1282 (2004)
57. W.L. Baker, Learning via stochastic approximation in function space, PhD Dissertation, Harvard University, Cambridge, MA, 1997
58. E.J. Balder, On WS-convergence of product measures. Math. Oper. Res. **26**(3), 494–518 (2001)
59. E.J. Balder, Generalized equilibrium results for games with incomplete information. Math. Oper. Res. **13**(2), 265–276 (1988)
60. B. Bamieh, P. Voulgaris, A convex characterization of distributed control problems in spatially invariant systems with communication constraints. Syst. Control Lett. **54**, 575–583 (2005)
61. R. Bansal, T. Başar, Solutions to a class of linear-quadratic-Gaussian (LQG) stochastic team problems with nonclassical information. Syst. Control Lett. **9**, 125–130 (1987)
62. R. Bansal, T. Başar, Stochastic team problems with nonclassical information revisited: When is an affine law optimal? IEEE Trans. Autom. Control **32**, 554–559 (1987)
63. R. Bansal, T. Başar, Simultaneous design of measurement and control strategies in stochastic systems with feedback. Automatica **45**, 679–694 (1989)
64. R. Bansal, T. Başar, Communication games with partially soft power constraints. J. Optim. Theory Appl. **61**(3), 329–346 (1989)
65. L. Bao, M. Skoglund, K.H. Johansson, Iterative encoder-controller design for feedback control over noisy channels. IEEE Trans. Autom. Control **57**, 265–278 (2011)
66. Y. Bar-Shalom, E. Tse, Dual effect, certainty equivalence, and separation in stochastic control. IEEE Trans. Autom. Control **19**(5), 494–500 (1974)
67. J.S. Baras, Noncommutative probability models in quantum communication and multi-agent stochastic control. Richerche Autom. **55**, 217–264 (1979)
68. J.S. Baras, Multi-agent stochastic control: Models inspired from quantum physics, in *International Conference, Physics and Control (PhysCon)*, Saint Petersburg, Russia (2003), pp. 747–758
69. M. Barbie, A. Gupta, The topology of information on the space of probability measures over Polish spaces. J. Math. Econ. **52**, 98–111 (2014)

70. M. Bardi, M. Fischer, On non-uniqueness and uniqueness of solutions in finite-horizon mean field games. ESAIM Control Optim. Calc. Variations **25**, 44 (2019)
71. J. Barrett, N. Linden, S. Massar, S. Pironio, S. Popescu, D. Roberts, Nonlocal correlations as an information-theoretic resource. Phys. Rev. A **71**, 022101 (2005)
72. A.R. Barron, The strong ergodic theorem for densities: Generalized Shannon-McMillan-Breiman theorem. Ann. Probab. **13**, 1292–1303 (1985)
73. A.R. Barron, Information theory and martingales, in *IEEE International Symposium on Information Theory, Recent Results Session*, Budapest, Hungary (1991)
74. K. Barty, J.-P. Chancelier, G. Cohen, M. De Lara, T. Guilbaud, P. Carpentier, Dual effect free stochastic controls. Ann. Oper. Res. **142**(1), 41–62 (2006)
75. A. Barvinok, *A Course in Convexity*, volume 54 of Graduate Studies in Mathematics (American Mathematical Society, Providence, RI, 2002)
76. T. Başar, A counterexample in linear-quadratic games: Existence of nonlinear Nash solutions. J. Optim. Theory Appl. **14**(4), 425–430 (1974)
77. T. Başar, On the uniqueness of the Nash solution in linear-quadratic differential games. Int. J. Game Theory **5**, 65–90 (1976)
78. T. Başar, Some thoughts on saddle-point conditions and information structures in zero-sum differential games. J. Optimization Theory Appl. **18**(1), 165–170 (1976)
79. T. Başar, Two general properties of the saddle-point solutions of dynamic games. IEEE Trans. Autom. Control **22**(1), 124–126 (1977)
80. T. Başar, Decentralized multicriteria optimization of linear stochastic systems. IEEE Trans. Autom. Control **23**, 233–243 (1978)
81. T. Başar, Two-criteria LQG decision problems with one-step delay observation sharing pattern. Inf. Control **38**, 21–50 (1978)
82. T.Ü. Başar, Performance bounds and optimal linear coding for multichannel communication systems, PhD Dissertation, Bogazici University, 1978
83. T. Başar, Stochastic stagewise Stackelberg strategies for linear quadratic systems, in *Stochastic Control Theory and Stochastic Differential Systems*, ed. by M. Kohlmann, W. Vogel (Springer, 1979), pp. 264–276
84. T. Başar, A trace minimization problem with applications in joint estimation and control under nonclassical information. J. Optim. Theory Appl. **31**, 343–359 (1980)
85. T. Başar, Hierarchical decision making under uncertainty, in *Dynamic Optimization and Mathematical Economics* (Springer, 1980), pp. 205–221
86. T. Başar, The Gaussian test channel with an intelligent jammer. IEEE Trans. Inf. Theory **29**(1), 152–157 (1983)
87. T. Başar, An equilibrium theory for multi-person decision making with multiple probabilistic models. IEEE Trans. Autom. Control **30**, 118–132 (1985)
88. T. Başar, Time consistency and robustness of equilibria in non-cooperative dynamic games, in *Contributions to Economic Analysis*, vol. 181 (Elsevier, 1989), pp. 9–54
89. T. Başar, Variations on the theme of the Witsenhausen counterexample, in *Proceedings of the IEEE Conference on Decision and Control*, Cancun, Mexico, December (2008), pp. 1614–1619
90. T. Başar, Stochastic differential games and intricacy of information structures, in *Dynamic Games in Economics*, ed. by J. Haunschmied, V.M. Veliov, S. Wrzaczek, volume 16 of *Dynamic Modeling and Econometrics in Economics and Finance* (Springer, Berlin Heidelberg, 2014), pp. 23–49
91. T. Başar, R. Bansal, Optimum design of measurement channels and control policies for linear-quadratic stochastic systems. Eur. J. Oper. Res. **73**, 226–236 (1994)
92. T. Başar, J. Cruz, Concepts and methods in multiperson coordination and control, in *Optimization and Control of Dynamic Operational Research Models*, ed. by S.G. Tzafestas, Chapter 11, North Holland (1982), pp. 351–387
93. T. Başar, A. Haurie, Feedback equilibria in differential games with structural and modal uncertainties, in *Advances in Large Scale Systems*, ed. by J.B. Cruz, Jr. (JAE Press, 1979), pp. 163–201

94. T. Başar, Y.-C. Ho, Informational properties of the Nash solutions of two stochastic nonzero-sum games. J. Econ. Theory **7**(4), 370–387 (1974)
95. T. Başar, M. Mintz, Minimax estimation under generalized quadratic loss, in *IEEE Conference Decision Contr., Miami, FL*, December (1971), pp. 456–461
96. T. Başar, M. Mintz, On the existence of linear saddle-point strategies for a two-person zero-sum stochastic game, in *IEEE Conference Decision Contr., New Orleans, Louisiana*, December (1972), pp. 188–192
97. T. Başar, G.J. Olsder, *Dynamic Noncooperative Game Theory*. SIAM Classics in Applied Mathematics (SIAM, Philadelphia, PA, 1999)
98. T. Başar, H. Selbuz, Closed-loop Stackelberg strategies with applications in the optimal control of multilevel systems. IEEE Trans. Autom. Control **24**(2), 166–179 (1979)
99. T. Başar, Y.-W. Wu, A complete characterization of minimax and maximin encoder-decoder policies for communication channels with incomplete statistical description. IEEE Trans. Inf. Theory **31**(4), 482–489 (1985)
100. T. Başar, G. Zaccour, *Handbook of Dynamic Game Theory: Volume I (Theory of Dynamic Games)* (Springer, 2018)
101. T. Başar, G. Zaccour, *Handbook of Dynamic Game Theory: Volume II (Applications of Dynamic Games)* (Springer, 2018)
102. B. Bassan, O. Gossner, M. Scarsini, S. Zamir, Positive value of information in games. Int. J. Game Theory **32**, 17–31 (2003)
103. M. Battaglini, Multiple referrals and multidimensional cheap talk. Econometrica **70**(4), 1379–1401 (2002)
104. E. Bayraktar, Yan Y. Dolinsky, J. Guo, Continuity of utility maximization under weak convergence. Math. Financ. Econ. **14**, 1–33 (2020)
105. M. Beiglböck, D. Lacker, Denseness of adapted processes among causal couplings. arXiv, pp. arXiv–1805 (2018)
106. J.S. Bell, On the Einstein Podolsky Rosen paradox. Phys. Phys. Fiz. **1**, 195–200 (1964)
107. V.E. Beneš, Existence of optimal stochastic control laws. SIAM J. Control **9**(3), 446–472 (1971)
108. A. Bensoussan, S. Chen, S.P. Sethi, The maximum principle for global solutions of stochastic Stackelberg differential games. SIAM J. Control Optim. **53**(4), 1956–1981 (2015)
109. T. Berger, Information rates of Wiener processes. IEEE Trans. Inf. Theory **16**, 134–139 (1970)
110. T. Berger, *Rate Distortion Theory* (Prentice- Hall, Englewood Cliffs, NJ, 1971)
111. T. Berger, Multiterminal source coding, in *Lecture Notes Presented at the CISM Summer School*, Udine, Italy, July (1977)
112. D. Bertsekas, *Nonlinear Programming* (Athena Scientific, Belmont, MA, 1999)
113. D. Bertsekas, *Dynamic Programming and Optimal Control Vol. 1* (Athena Scientific, 2000)
114. D.P. Bertsekas, S. Shreve, *Stochastic Optimal Control: The Discrete Time Case* (Academic Press, New York, 1978)
115. D.P. Bertsekas, J.N. Tsitsiklis, *Neuro-Dynamic Programming* (Athena Scientific, 1996)
116. P. Billingsley, *Convergence of Probability Measures* (Wiley, New York, 1968)
117. P. Billingsley, *Probability and Measure*, 2nd edn. (Wiley, New York, 1986)
118. P. Billingsley, *Convergence of Probability Measures*, 2nd edn. (Wiley, New York, 1999)
119. J.-M. Bismut, An example of interaction between information and control: The transparency of a game. IEEE Trans. Autom. Control **18**(5), 518–522 (1972)
120. J.-M. Bismut, *Théorie Probabiliste du Contrôle des Diffusions*, vol. 181 (1973)
121. J.-M. Bismut, Partially observed diffusions and their control. SIAM J. Control Optim. **20**(2), 302–309 (1982)
122. D. Blackwell, The comparison of experiments, in *Proceedings of the Second Berkeley Symposium on Mathematical Statistics and Probability* (1951), pp. 93–102
123. D. Blackwell, Equivalent comparison of experiments. Ann. Math. Stat. **24**, 265–272 (1953)
124. D. Blackwell, Memoryless strategies in finite-stage dynamic programming. Ann. Math. Stat. **35**, 863–865 (1964)

125. D. Blackwell, The stochastic processes of Borel gambling and dynamic programming. Ann. Stat. **4**, 370–374 (1976)
126. D. Blackwell, L. Dubins, Merging of opinions with increasing information. Ann. Math. Stat. **33**, 882–887 (1962)
127. D. Blackwell, C. Ryll-Nadrzewski, Non-existence of everywhere proper conditional distributions. Ann. Math. Stat. **34**, 223–225 (1963)
128. V.I. Bogachev, *Measure Theory* (Springer, Berlin, 2007)
129. C. Boll, Comparison of experiments in the infinite case, PhD Dissertation, Stanford University, 1955
130. V.S. Borkar, The probabilistic structure of controlled diffusion processes. Acta Appl. Math. **11**(1), 19–48 (1988)
131. V.S. Borkar, A topology for Markov controls. Appl. Math. Optim. **20**(1), 55–62 (1989)
132. V.S. Borkar, White-noise representations in stochastic realization theory. SIAM J. Control Optim. **31**, 1093–1102 (1993)
133. V.S. Borkar, *Probability Theory: An Advanced Course* (Springer, New York, 1995)
134. V.S. Borkar, Average cost dynamic programming equations for controlled Markov chains with partial observations. SIAM J. Control Optim. **39**(3), 673–681 (2000)
135. V.S. Borkar, Convex analytic methods in Markov decision processes, in *Handbook of Markov Decision Processes*, ed. by E.A. Feinberg, A. Shwartz (Kluwer Academic Publisher, 2002)
136. V.S. Borkar, Dynamic programming for ergodic control of Markov chains under partial observations: A correction. SIAM J. Control Optim. **45**(6), 2299–2304 (2007)
137. V.S. Borkar, S.P. Meyn, The ODE method for convergence of stochastic approximation and reinforcement learning. SIAM J. Control Optim. **38**, 447–469 (2000)
138. V.S. Borkar, S.K. Mitter, LQG control with communication constraints, in *Kailath Festschrift* (Kluwer Academic Publishers, Boston, 1997)
139. V.S. Borkar, P. Varaiya, Asymptotic agreement in distributed estimation. IEEE Trans. Autom. Control **27**, 650–655 (1982)
140. V.S. Borkar, S.K. Mitter, S. Tatikonda, Optimal sequential vector quantization of Markov sources. SIAM J. Control Optim. **40**, 135–148 (2001)
141. V.S. Borkar, S.K. Mitter, A. Sahai, S. Tatikonda, Sequential source coding: An optimization viewpoint, in *Proceedings of the IEEE Conference on Decision and Control*, December (2005), pp. 1035–1042
142. E. Boylan, Equiconvergence of martingales. Ann. Math. Stat. **42**(2), 552–559 (1971)
143. R.A. Boyles, W.A. Gardner, Cycloergodic properties of discrete-parameter non-stationary stochastic processes. IEEE Trans. Inf. Theory **29**, 105–114 (1983)
144. A. Brandenburger, E. Dekel, Common knowledge with probability 1. J. Math. Econ. **16**, 237–245 (1987)
145. J.H. Braslavsky, R.H. Middleton, J.S. Freudenberg, Feedback stabilization over signal-to-noise ratio constrained channels. IEEE Trans. Autom. Control **52**(8), 1391–1403 (2007)
146. R.W. Brockett, Asymptotic stability and feedback stabilization. Differ. Geom. Control Theory **27**(1), 181–191 (1983)
147. R. Brockett, D. Liberzon, Quantized feedback stabilization of linear systems. IEEE Trans. Autom. Control **45**, 1279–1289 (2000)
148. S. Bross, A. Lapidoth, An improved achievable region for the discrete memoryless two-user multipleaccess channel with noiseless feedback. IEEE Trans. Inf. Theory **51**, 811–833 (2005)
149. G.W. Brown, Iterative solution of games by fictitious play. Activity Anal. Prod. Alloc. **13**(1), 374–376 (1951)
150. G.W. Brown, Iterative solutions of games by fictitious play, in *Activity Analysis of Production and Allocation*, ed. by T.C. Koopmans (Wiley, New York, 1951), pp. 374–376
151. N. Brunner, D. Cavalcanti, S. Pironio, V. Scarani, S. Wehner, Bell nonlocality. Rev. Mod. Phys. **86**(2), 419 (2014)
152. A. Budhiraja, On invariant measures of discrete time filters in the correlated signal-noise case. Ann. Appl. Probab. **12**(3), 1096–1113 (2002)
153. P.E. Caines, *Linear Stochastic Systems*. John Wiley & Sons, New York, NY (1988)

154. P. Caines, M. Huang, and R. Malhamé, Mean field games. *Handbook of Dynamic Game Theory*, pages 345–372 (2017)
155. F.P. Calmon, A. Makhdoumi, M. Médard, Fundamental limits of perfect privacy, in *IEEE International Symposium on Information Theory (ISIT)*, June (2015), pp. 1796–1800
156. O. Candogan, A. Ozdaglar, P.A. Parrilo, Near-potential games: Geometry and dynamics. ACM Trans. Econ. Comput. (TEAC) **1**(2), 1–32 (2013)
157. P. Cardaliaguet, Notes on mean field games, Technical report (2010)
158. R. Carmona, F. Delarue, *Probabilistic Theory of Mean Field Games with Applications I-II* (Springer, 2018)
159. R. Carmona, F. Delarue, D. Lacker, Mean field games with common noise. Ann. Probab. **44**(6), 3740–3803 (2016)
160. P. Cartier, J.M.G. Fell, P.-A. Meyer, Comparaison, des mesures portées par un ensemble convexe compact. Bull. Soc. Math. France **92**, 435–445 (1964)
161. G. Casalino, F. Davoli, R. Minciardi, P.P. Puliafito, R. Zoppoli, Partially nested information structures with a common past. IEEE Trans. Autom. Control **29**, 846–850 (1984)
162. C. Castaing, P.R. De Fitte, M. Valadier, *Young Measures on Topological Spaces: With Applications in Control Theory and Probability Theory*, vol. 571 (Springer Science & Business Media, 2004)
163. D.A. Castanon, N. Sandell, Equilibrium solutions in sequential stochastic games. IEEE Trans. Autom. Control **22**(4), 624–627 (1977)
164. D.A. Castanon, D. Teneketzis, Further results on the asymptotic agreement problem. IEEE Trans. Autom. Control **33**, 515–523 (1988)
165. P. Cembranos, J. Mendoza, *Banach Spaces of Vector-Valued Functions* (Springer, 1997)
166. C.D. Charalambous, Decentralized optimality conditions of stochastic differential decision problems via Girsanov's measure transformation. Math. Control Signals Syst. **28**(3), 1–55 (2016)
167. C.D. Charalambous, N.U. Ahmed, Centralized versus decentralized optimization of distributed stochastic differential decision systems with different information structures-part I: A general theory. IEEE Trans. Autom. Control **62**(3), 1194–1209 (2017)
168. C.D. Charalambous, A. Farhadi, D. Denic, F. Rezaei, Robust control over uncertain communication channels, in *Proceedings of the IEEE Med*, June (2005), pp. 737–742
169. C.D. Charalambous, A. Farhadi, S.Z. Denic, Control of continuous-time linear Gaussian systems over additive Gaussian wireless fading channels: A separation principle. IEEE Trans. Autom. Control **53**, 1013–1019 (2008)
170. C.D. Charalambous, P.A Stavrou, N.U. Ahmed, Nonanticipative rate distortion function and relations to filtering theory. IEEE Trans. Autom. Control **59**(4), 937–952 (2014)
171. G.C. Chasparis, A. Arapostathis, J.S. Shamma, Aspiration learning in coordination games. SIAM J. Control Optim. **51**(1), 465–490 (2013)
172. G. Chechik, A. Globerson, N. Tishby, Y. Weiss, Information bottleneck for Gaussian variables. J. Mach. Learn. Res. **6**(Jan), 165–188 (2005)
173. C.T. Chen, *Linear Systems Theory and Design* (Oxford University Press, Oxford, 1999)
174. H. Chernoff, Rational selection of decision functions. Econometrica **22**(4), 422–443 (1954)
175. S. Chien, A. Sinclair, Convergence to approximate Nash equilibria in congestion games. Games Econ. Behav. **71**(2), 315–327 (2011)
176. P. Chigansky, R. Liptser, R. van Handel, Intrinsic methods in filter stability, in *Handbook of Nonlinear Filtering* (2009)
177. C.Y. Chong, M. Athans, On the periodic coordination of linear stochastic systems. Automatica **12**, 321–335 (1976)
178. C. Choudhuri, U. Mitra, On Witsenhausen's counterexample: the asymptotic vector case, in *2012 IEEE Information Theory Workshop* (IEEE, 2012), pp. 162–166
179. C. Claus, C. Boutilier, The dynamics of reinforcement learning in cooperative multiagent systems, in *Proceedings of the Tenth Innovative Applications of Artificial Intelligence Conference, Madison, Wisconsin*, July (1998), pp. 746–752

180. J.F. Clauser, M.A. Horne, A. Shimony, R.A. Holt, Proposed experiment to test local hidden-variable theories. Phys. Rev. Lett. **23**(15), 880 (1969)
181. R. Cleve, P. Hoyer, B. Toner, J. Watrous, Consequences and limits of nonlocal strategies, in *Proceedings. 19th IEEE Annual Conference on Computational Complexity (2004)*, June (2004), pp. 236–249
182. M. Colombino, R.S. Smith, T.H. Summers, Mutually quadratically invariant information structures in two-team stochastic dynamic games. IEEE Trans. Autom. Control **63**(7), 2256–2263 (2017)
183. F. Colonius, Metric invariance entropy and conditionally invariant measures. Ergodic Theory Dynam. Syst. **38**(3), 921–939 (2018)
184. F. Colonius, C. Kawan, Invariance entropy for control systems. SIAM J. Control Optim. **48**(3), 1701–1721 (2009)
185. G. Como, S. Yüksel, On the capacity of memoryless finite state multiple access channels with asymmetric state information at the encoders. IEEE Trans. Inf. Theory **56**, 1267–1273 (2011)
186. J.P. Corfmat, A.S. Morse, Decentralized control of linear multivariable systems. Automatica **11**, 479–497 (1976)
187. T.M. Cover, C.S.K. Leung, An achievable rate region for the multiple-access channel with feedback. IEEE Trans. Inf. Theory **27**, 292–298 (1981)
188. T.M. Cover, J.A. Thomas, *Elements of Information Theory* (Wiley, New York, 1991)
189. V.P. Crawford, J. Sobel, Strategic information transmission. Econometrica **50**, 1431–1451 (1982)
190. L. Cregg, F. Alajaji, S. Yüksel, Reinforcement learning for zero-delay coding over a noisy channel with feedback, in *Proc. IEEE Conference on Decision and Control* (IEEE, 2023)
191. L. Cregg, T. Linder, S. Yüksel, Reinforcement learning for near-optimal design of zero-delay codes for Markov sources. arXiv (2023)
192. M.W. Cripps, G.J. Mailath, L. Samuelson, Imperfect monitoring and impermanent reputations. Econometrica **72**(2), 407–432 (2004)
193. I. Csiszár, Information-type measures of difference of probability distributions and indirect observation. Stud. Sci. Math. Hungar. **2**, 229–318 (1967)
194. I. Csiszár, J. Korner, *Information Theory: Coding Theorems for Discrete Memoryless Channels* (Akademiai Kiado, Budapest, 1981)
195. R.E. Curry, *Estimation and Control with Quantized Measurements* (MIT Press, Cambridge, MA, 1969)
196. R.F. Curtain, A.J. Pritchard, *Functional Analysis in Modern Applied Mathematics* (Academic Press, London, 1977)
197. T. Cuvelier, T. Tanaka, R.W. Heath Jr., Time-invariant prefix coding for LQG control. Preprint. arXiv:2204.00588 (2022)
198. A. Da Silva, C. Kawan, Invariance entropy of hyperbolic control sets. Discrete Contin. Dyn. Syst. **36**(1), 97–136 (2016)
199. A. Da Silva, C. Kawan, Robustness of critical bit rates for practical stabilization of networked control systems. Automatica **93**, 397–406 (2018)
200. R. Dabora, A. Goldsmith, On the capacity of indecomposable finite-state channels with feedback, in *Proceedings of the Allerton Conf Commun Control Comput*, September (2008), pp. 1045–1052
201. N.A. Dalkıran, S. Yüksel, Stochastic control approach to reputation games. IEEE Trans. Autom. Control **65**(11), 4710–4725 (2020)
202. P. Dasgupta, E. Maskin, The existence of equilibrium in discontinuous economic games, I: Theory. Rev. Econ. Stud. **53**(1), 1–26 (1986)
203. C. Daskalakis, D.J. Foster, N. Golowich, Independent policy gradient methods for competitive reinforcement learning. Adv. Neural Inf. Process. Syst. **33**, 5527–5540 (2020)
204. C. Daskalakis, D.J. Foster, N. Golowich, Independent policy gradient methods for competitive reinforcement learning. Preprint. arXiv:2101.04233 (2021)
205. B. De Meyer, E. Lehrer, D. Rosenberg, Two remarks on Blackwell's theorem. J. Appl. Probab. **45**, 580–586 (2008)

206. B. De Meyer, E. Lehrer, D. Rosenberg, Evaluating information in zero-sum games with incomplete information on both sides. Math. Oper. Res. **34**, 851–863 (2010)
207. C. De Persis, A. Isidori, Stabilizability by state feedback implies stabilizability by encoded state feedback. Syst. Control Lett. **53**(3), 249–258 (2004)
208. P.R. de Waal, J.H. van Schuppen, A class of team problems with discrete action spaces: Optimality conditions based on multimodularity. SIAM J. Control Optim. **38**, 875–892 (2000)
209. Y.E. Demirci, S. Yüksel, Geometric ergodicity, unique ergodicity and wasserstein continuity of non-linear filters with compact state space. Preprint. arXiv:2307.15764 (2023)
210. M.S. Derpich, J. Østergaard, Improved upper bounds to the causal quadratic rate-distortion function for Gaussian stationary sources. IEEE Trans. Inf. Theory **58**(5), 3131–3152 (2012)
211. M.S. Derpich, S. Yüksel, Dual effect, certainty equivalence, and separation revisited: A counterexample and a relaxed characterization for optimality. IEEE Trans. Autom. Control **68**(2), 1259–1266 (2023)
212. L. Devroye, L. Györfi, *Non-parametric Density Estimation: The L_1 View* (John Wiley, New York, 1985)
213. L. Devroye, L. Györfi, No empirical measure can converge in the total variation sense for all distributions. Ann. Stat. **18**, 1496–1499 (1990)
214. G. Di Masi, L. Stettner, Ergodicity of hidden Markov models. Math. Control Signals Syst. **17**(4), 269–296 (2005)
215. P. Diaconis, D. Freedman, Finite exchangeable sequences. Ann. Probab. **8**(4), 745–764 (1980)
216. J. Diestel, J.J. Uhl, *Vector Measures* (American Mathematical Society, 1977)
217. J. Dieudonné, Sur le théorème de lebesgue-nikodym (v). Canad. J. Math. **3**, 129–139 (1951)
218. R.L. Dobrushin, An asymptotic bound for the probability error of information transmission through a channel without memory using the feedback. Probl. Kibemetiki, **8**, 161–168 (1962)
219. R.L. Dobrushin, B.S. Tsybakov, Information transmission with additional noise. IRE Trans. Inf. Theory **18**, 293–304 (1962)
220. J.L. Doob, *Measure Theory* (Springer-Verlag, New York, 1994)
221. T. Downarowicz, *Entropy in Dynamical Systems*, vol. 18 (Cambridge University Press, 2011)
222. L. Dubins, D. Freedman, Measurable sets of measures. Pacific J. Math. **14**, 1211–1222 (1964)
223. R.M. Dudley, *Real Analysis and Probability*, 2nd edn. (Cambridge University Press, Cambridge, 2002)
224. R.M. Dudley, E. Gine, J. Zinn, Uniform and universal Glivenko-Cantelli classes. J. Theoret. Probab. **4**, 485–510 (1991)
225. J. Dugundji, An extension of Tietze's theorem. Pacific J. Math. **1**(3), 353–367 (1951)
226. T.E. Duncan, On the calculation of mutual information. SIAM J. Appl. Math. **19**, 215–220 (1970)
227. N. Dunford, J. Schwartz, *Linear Operators I* (Interscience, New York, 1958)
228. P.G. Dupuis, M.R. James, I. Petersen, Robust properties of risk-sensitive control. Math. Control Signals Syst. **13**(4), 318–332 (2000)
229. R. Durrett, *Probability: Theory and Examples*, vol. 3 (Cambridge University Press, 2010)
230. E.B. Dynkin, A.A. Yushkevich, *Controlled Markov Processes*, vol. 235 (Springer, 1979)
231. H. Ebeid, T.P. Coleman, Source coding with feedforward using the posterior matching scheme, in *Proceedings of the IEEE Int. Symp. on Information Theory (ISIT)*, Austin, Texas, USA (2010)
232. A. Einstein, B. Podolsky, N. Rosen, Can quantum-mechanical description of physical reality be considered complete? Phys. Rev. **47**, 777–780 (1935)
233. E. Einy, O. Haimanko, D. Moreno, B. Shitovitz, Uniform continuity of the value of zero-sum games with differential information. Math. Oper. Res. **33**(3), 552–560 (2008)
234. E. Einy, O. Haimanko, B. Tumendemberel, Continuity of the value and optimal strategies when common priors change. Int. J. Game Theory **41**, 829–849 (2012)
235. C. Eksin, A. Ribeiro, Distributed fictitious play for multiagent systems in uncertain environments. IEEE Trans. Autom. Control **63**(4), 1177–1184 (2017)
236. A. El Gamal, Y.H. Kim, *Network Information Theory* (Cambridge University Press, Cambridge, UK, 2012)

237. N. Elia, When Bode meets Shannon: control-oriented feedback communication schemes. IEEE Trans. Autom. Control **49**(9), 1477–1488 (2004)
238. P. Elias, Channel capacity without coding, Quarterly progress report, MIT Research Laboratory of Electronics (also in Lectures on Communication System Theory, Eds. E. Baghdady, New York: McGraw Hill, 1961) (1956), pp. 90–93
239. R. Elie, J. Perolat, M. Laurière, M. Geist, O. Pietquin, On the convergence of model free learning in mean field games, in *Proceedings of the AAAI Conference on Artificial Intelligence*, vol. 34, pp. 7143–7150 (2020)
240. R.J. Elliott, The optimal control of a stochastic system. SIAM J. Control Optim. **15**(5), 756–778 (1977)
241. S.N. Ethier, T.G. Kurtz, *Markov Processes: Characterization and Convergence*, vol. 282 (John Wiley & Sons, 2009)
242. E. Even-Dar, Y. Mansour, Learning rates for q-learning. J. Mach. Learn. Res. **5**, 1–25 (2004)
243. G.N. Nair, F. Colonius, C. Kawan, A note on topological feedback entropy and invariance entropy. Syst. Control Lett. **62**(5), 377–381 (2013)
244. A. Fabrikant, A.D Jaggard, M. Schapira, On the structure of weakly acyclic games, in *Algorithmic Game Theory* (Springer, 2010), pp. 126–137
245. K. Fan, Minimax theorems. Proc. Natl. Acad. Sci. **39**(1), 42–47 (1953)
246. S. Fang, J. Chen, H. Ishii, *Towards Integrating Control and Information Theories* (Springer, 2017)
247. F. Farokhi, G. Nair, Privacy-constrained communication. IFAC-PapersOnLine **49**(22), 43–48 (2016)
248. F. Farokhi, H. Sandberg, I. Shames, M. Cantoni, Quadratic Gaussian privacy games, in *IEEE Conference on Decision and Control (CDC)*, pp. 4505–4510 (2015)
249. F. Farokhi, A.M.H. Teixeira, C. Langbort, Estimation with strategic sensors. IEEE Trans. Autom. Control **62**(2), 724–739 (2017)
250. N. Farvardin, A study of vector quantization for noisy channels. IEEE Trans. Inf. Theory **36**(4), 799–809 (1990)
251. H.O. Fattorini, Existence theory and the maximum principle for relaxed infinite-dimensional optimal control problems. SIAM J. Control Optim. **32**(2), 311–331 (1994)
252. E.A. Feinberg, Controlled Markov processes with arbitrary numerical criteria. Theory Probab. Appl. **27**, 486–503 (1982)
253. E.A. Feinberg, Non-randomized Markov and semi-Markov strategies in dynamic programming. Theory Probab. Appl. **27**, 116–126 (1982)
254. E.A. Feinberg, *On Measurability and Representation of Strategic Measures in Markov Decision Processes*, ed. by T.S. Ferguson, L.S. Shapley, J.B. MacQueen. Institute of Mathematical Statistics Lecture Notes, pp. 29–43 (1996)
255. E.A. Feinberg, P.O. Kasyanov, Equivalent conditions for weak continuity of nonlinear filters. Syst. Control Lett. **173**, 105458 (2023)
256. E.A. Feinberg, P.O. Kasyanov, N.V. Zadioanchuk, Average cost Markov decision processes with weakly continuous transition probabilities. Math. Oper. Res. **37**(4), 591–607 (2012)
257. E.A. Feinberg, P.O. Kasyanov, M.Z. Zgurovsky, Partially observable total-cost Markov decision process with weakly continuous transition probabilities. Math. Oper. Res. **41**(2), 656–681 (2016)
258. E.A. Feinberg, P.O. Kasyanov, M.Z. Zgurovsky, Markov decision processes with incomplete information and semiuniform feller transition probabilities. SIAM J. Control Optim. **60**(4), 2488–2513 (2022)
259. A.A. Feldbaum, Dual control theory. Autom. Remote Control **21**, 1033–1039 (1961)
260. W. Feller, *An Introduction to Probability Theory and Its Applications* (John Wiley and Sons, New York, 1971)
261. C. Fershtman, Fixed rules and decision rules: Time consistency and subgame perfection. Econ. Lett. **30**(3), 191–194 (1989)
262. T. Fine, Optimum mean-square quantization of a noisy input. IEEE Trans. Inf. Theory **11**, 293–294 (1965)

263. A.M. Fink et al., Equilibrium in a stochastic n-person game. J. Sci. Hiroshima Univ. Ser. AI (Math.) **28**(1), 89–93 (1964)
264. T.R. Fischer, Optimal quantized control. IEEE Trans. Autom. Control **27**, 996–998 (1982)
265. M. Fischer, On the connection between symmetric N-player games and mean field games. Ann. Appl. Probab. **27**(2), 757–810 (2017)
266. W.H. Fleming, *Functions of Several Variables* (Addison-Wesley, Reading, MA, 1965)
267. W.H. Fleming, E. Pardoux, Optimal control for partially observed diffusions. SIAM J. Control Optim. **20**(2), 261–285 (1982)
268. T.J. Flynn, R.M. Gray, Encoding of correlated observations. IEEE Trans. Inf. Theory **33**, 773–787 (1987)
269. G.B. Folland, *Real Analysis: Modern Techniques and Their Applications* (John Wiley and Sons, 1999)
270. F. Forges, An approach to communication equilibria. Econometrica J. Econom. Soc. **54**(6), 1375–1385 (1986)
271. D.P. Foster, H.P. Young, On the impossibility of predicting the behavior of rational agents. Proc. Nat. Acad. Sci. **22**, 12848–12853 (2001)
272. D.P. Foster, H.P. Young, Regret testing: learning to play Nash equilibrium without knowing you have an opponent. Theoret. Econ. **1**, 341–367 (2006)
273. M. Franceschetti, P. Minero, Elements of information theory for networked control systems, in *Information and Control in Networks* (Springer, New York, 2014), pp. 3–37
274. J.S. Freudenberg, R.H. Middleton, V. Solo, Stabilization and disturbance attenuation over a Gaussian communication channel. IEEE Trans. Autom. Control **55**, 795–799 (2010)
275. M. Fu, Lack of separation principle for quantized linear quadratic Gaussian control. IEEE Trans. Autom. Control **57**, 2385–2390 (2012)
276. D. Fudenberg, D.K. Levine, Reputation and equilibrium selection in games with a patient player. Econometrica **57**(4), 759–778 (1989)
277. D. Fudenberg, D.K. Levine, Maintaining a reputation when strategies are imperfectly observed. Rev. Econ. Stud. **59**(3), 561–579 (1992)
278. D. Fudenberg, D.K. Levine, Efficiency and observability with long-run and short-run players. J. Econ. Theory **62**(1), 103–135 (1994)
279. D. Fudenberg, D.K. Levine, *The Theory of Learning in Games* (MIT Press, Cambridge, MA, 1998)
280. D. Fudenberg, E. Maskin, The folk theorem in repeated games with discounting or with incomplete information. Econometrica **54**(3), 533–554 (1986)
281. D. Fudenberg, J. Tirole, *Game Theory* (MIT Press, 1991)
282. D. Fudenberg, D.K. Levine, E. Maskin, The folk theorem with imperfect public information. Econometrica **62**(5), 997–1039 (1994)
283. G. Gabor, G. Szekeres, Z. Györfi, On the Gaarder-Slepian tracking system conjecture. IEEE Trans. Inf. Theory **37**, 1165–1168 (1991)
284. R.G. Gallager, A simple derivation of the coding theorem and some applications. IEEE Trans. Inf. Theory **11**, 3–18 (1965)
285. R.G. Gallager, *Information Theory and Reliable Communication* (John Wiley and Sons, New York, 1968)
286. R.G. Gallager, B. Nakiboğlu, Variations on a theme by Schalkwijk and Kailath. IEEE Trans. Inf. Theory **12**, 6–17 (2010)
287. N. Garcia, C. Kawan, S. Yüksel, Ergodicity conditions for controlled stochastic non-linear systems under information constraints. SIAM J. Control Optim. **59**(1), 534–560 (2021)
288. N. Garcia, C. Kawan, S. Yüksel, Ergodicity of controlled stochastic nonlinear systems under information constraints: Refined bounds via coordinate splitting, in *Proc. IEEE Conference on Decision and Control* (2021)
289. C. Gaskett, A. Zelinsky, D. Wettergreen, Q-learning in continuous state and action spaces, in *Australasian Joint Conference on Artificial Intelligence* (Springer, 1999), pp. 417–428
290. M. Gastpar, Cut-set arguments for source-channel networks, in *Proceedings of the IEEE Int Symp Info Theory*, vol. 34 (2004)

291. M. Gastpar, Uncoded transmission is exactly optimal for a simple Gaussian sensor network. IEEE Trans. Inf. Theory **54**, 5247–5251 (2008)
292. M. Gastpar, B. Rimoldi, M. Vetterli, To code, or not to code: Lossy source-channel communication revisited. IEEE Trans. Inf. Theory **49**, 1147–1158 (2003)
293. A. Gattami, Team decision problems with convex quadratic constraints. IEEE Trans. Control Netw. Syst. **4**(3), 555–563 (2016)
294. A. Gattami, B.M. Bernhardsson, A. Rantzer, Robust team decision theory. IEEE Trans. Autom. Control **57**, 794–798 (2012)
295. J. Geanakoplos, Common knowledge, in *Handbook of Game Theory*, ed. by R.J. Aumann, S. Hart, Chapter 40 (1994), pp. 1437– 1496
296. J. Geanakoplos, H.M. Polemarchakis, We can't disagree forever. J. Econ. Theory **28**, 192–200 (1982)
297. F. Gensbittel, M. Pęski, J. Renault, Value-based distance between the information structures. Theoret. Econ. **17**(3), 1225–1267 (2022)
298. F. Germano, G. Lugosi, Global Nash convergence of foster and young's regret testing. Games Econ. Behav. **60**(1), 135–154 (2007)
299. M. Ghomi, T. Linder, S. Yüksel, Zero-delay lossy coding of linear vector Markov sources: Optimality of stationary codes and near optimality of finite memory codes. IEEE Trans. Inf. Theory **68**(5), 3474–3488 (2021)
300. J.K. Ghosh, R.V. Ramamoorthi, *Bayesian Nonparametrics* (Springer, New York, 2003)
301. I.I. Gihman, A.V. Skorohod, *Controlled Stochastic Processes* (Springer Science & Business Media, 2012)
302. I.V. Girsanov,. On transforming a certain class of stochastic processes by absolutely continuous substitution of measures. Theory Probab. Appl. **5**(3), 285–301 (1960)
303. H. Gish, J.N. Pierce, Asymptotically efficient quantization. IEEE Trans. Inf. Theory **14**, 676–683 (1968)
304. G. Gnecco, M. Sanguinetti, Suboptimal solutions to network team optimization problems, in *INOC 2009*, Pisa (2009)
305. G. Gnecco, M. Sanguinetti, New insights into Witsenhausen's counterexample. Optim. Lett. **6**, 1425–1446 (2012)
306. G. Gnecco, M. Sanguinetti, M. Gaggero, Suboptimal solutions to team optimization problems with stochastic information structure. SIAM J. Control Optim. **22**(1), 212–243 (2012)
307. N. Golowich, A. Moitra, D. Rohatgi, Planning in observable pomdps in quasipolynomial time. Preprint. arXiv:2201.04735 (2022)
308. Z. Gong, M. Aldeen, Stabilization of decentralized control systems. J. Math. Syst. Estimation Control **7**, 1–16 (1997)
309. D.J. Goodman, A. Gersho, Theory of an adaptive quantizer. IEEE Trans. Commun. **22**, 1037–1045 (1974)
310. A.K. Gorbunov, M.S. Pinsker, Nonanticipatory and prognostic epsilon entropies and message generation rates. Probl. Peredachi Inf. **9**(3), 12–21 (1973)
311. O. Gossner, Simple bounds on the value of a reputation. Econometrica **79**(5), 1627–1641 (2011)
312. O. Gossner, J.F. Mertens, The value of information in zero-sum games. Preprint (2001)
313. R.M. Gray, Information rates of autoregressive processes. IEEE Trans. Inf. Theory **16**, 412–421 (1970)
314. R.M. Gray, *Entropy and Information Theory* (Springer, New York, 1990)
315. R.M. Gray, *Probability, Random Processes, and Ergodic Properties* (Springer, 2009)
316. R.M. Gray, L.D. Davisson, Source coding theorems without the ergodic assumption. IEEE Trans. Inf. Theory **20**(4), 502–516 (1974)
317. R.M. Gray, L.D. Davisson, Quantizer mismatch. IEEE Trans. Commun. **23**, 439–443 (1975)
318. R.M. Gray, T. Hashimoto, A note on rate-distortion functions for nonstationary Gaussian autoregressive processes. IEEE Trans. Inf. Theory **54**, 1319–1322 (2008)
319. R.M. Gray, J.C. Kieffer, Asymptotically mean stationary measures. Ann. Probab. **8**, 962–973 (1980)

320. R.M. Gray, D.L. Neuhoff, Quantization. IEEE Trans. Inf. Theory **44**, 2325–2383 (1998)
321. R.M. Gray, S. Boyd, T. Lookabaugh, Low rate distributed quantization of noisy observations, in *Proceedings of the Allerton Conference Communication, Control and Computing* (1985), pp. 354–358
322. E.J. Green, Noncooperative price taking in large dynamic markets. J. Econ. Theory **22**(2), 155–182 (1980)
323. P. Grover, S.Y. Park, A. Sahai, Approximately optimal solutions to the finite-dimensional Witsenhausen counterexample. IEEE Trans. Autom. Control **58**(9), 2189–2204 (2013)
324. X. Guo, A. Hu, R. Xu, J. Zhang, Learning mean-field games. Adv. Neural Inf. Process. Syst. **32** (2019)
325. A. Gupta, Existence of team-optimal solutions in static teams with common information: A topology of information approach. SIAM J. Control Optim. **58**(2), 998–1021 (2020)
326. A. Gupta, A. Nayyar, C. Langbort, T. Başar, Common information based Markov perfect equilibria for linear-Gaussian games with asymmetric information. SIAM J. Control Optim. **52**(5), 3228–3260 (2014)
327. A. Gupta, S. Yüksel, T. Başar, C. Langbort, On the existence of optimal policies for a class of static and sequential dynamic teams. SIAM J. Control Optim. **53**, 1681–1712 (2015)
328. L. Györfi, M. Kohler, Nonparametric estimation of conditional distributions. IEEE Trans. Inf. Theory **53**, 1872–1879 (2007)
329. A. György, T. Linder, Optimal entropy-constrained scalar quantization of a uniform source. IEEE Trans. Inf. Theory **46**, 2704–2711 (2000)
330. A. György, T. Linder, Codecell convexity in optimal entropy-constrained vector quantization. IEEE Trans. Inf. Theory **49**, 1821–1828 (2003)
331. A. György, T. Linder, P.A. Chou, B.J. Betts, Do optimal entropy constrained quantizers have a finite or infinite number of codewords? IEEE Trans. Inf. Theory **49**, 3031–3037 (2003)
332. A. György, T. Linder, G. Lugosi, Tracking the best quantizer. IEEE Trans. Inf. Theory **54**, 1604–1625 (2008)
333. S. Hafstein, C. Kawan, Numerical approximation of the data-rate limit for state estimation under communication constraints. J. Math. Anal. Appl. **473**(2), 1280–1304 (2019)
334. M. Hairer, *Convergence of Markov Processes*. Lecture Notes (University of Warwick, 2010)
335. B. Hajek, *Random Processes for Engineers* (Cambridge University Press, 2015)
336. B. Hajek, *An Introduction to Game Theory*. Lecture Notes (Department of Electrical and Computer Engineering University of Illinois at Urbana-Champaign, 2018)
337. E.A. Hansen, Solving pomdps by searching in policy space. Preprint. arXiv:1301.7380 (2013)
338. J. Harsanyi, R. Selten, *A General Theory of Equilibrium Selection in Games* (The MIT Press, Cambridge, MA, 1988)
339. T. Hashimoto, S. Arimoto, On the rate-distortion function for the nonstationary Gaussian autoregressive process. IEEE Trans. Inf. Theory **26**, 478–480 (1980)
340. M.F. Hellwig, Sequential decisions under uncertainty and the maximum theorem. J. Math. Econ. **25**(4), 443–464 (1996)
341. R. Hermann, A. Krener, Nonlinear controllability and observability. IEEE Trans. Autom. Control **22**(5), 728–740 (1977)
342. O. Hernández-Lerma, J.B. Lasserre, *Discrete-Time Markov Control Processes: Basic Optimality Criteria* (Springer, 1996)
343. O. Hernández-Lerma, J.B. Lasserre, *Further Topics on Discrete-time Markov Control Processes* (Springer, 1999)
344. O. Hernández-Lerma, J.B. Lasserre, *Markov Chains and Invariant Probabilities* (Birkhäuser, Basel, 2003)
345. J.P. Hespanha, *Noncooperative Game Theory: An Introduction for Engineers and Computer Scientists* (Princeton University Press, 2017)
346. B. Heymann, M. De Lara, J.-P. Chancelier, Kuhn's equivalence theorem for games in product form. Games Econ. Behav. **135**, 220–240 (2022)
347. C.J. Himmelberg, T. Parthasarathy, F.S. Van Vleck, Optimal plans for dynamic programming problems. Math. Oper. Res. **1**(4), 390–394 (1976)

348. J. Hirshleifer, The private and social value of information and the reward to inventive activity. Am. Econ. Rev., 561–574 (1971)
349. Y.C. Ho, Team decision theory and information structures. Proc. IEEE **68**, 644–654 (1980)
350. Y.C. Ho, K.C. Chu, Team decision theory and information structures in optimal control problems - part I. IEEE Trans. Autom. Control **17**, 15–22 (1972)
351. Y.C. Ho, K.C. Chu, On the equivalence of information structures in static and dynamic teams. IEEE Trans. Autom. Control **18**(2), 187–188 (1973)
352. I. Hogeboom-Burr, Comparison and continuity properties of equilibrium values in information structures for stochastic games. MSc Thesis, Department of Mathematics and Statistics, Queen's University (2022)
353. I. Hogeboom-Burr, S. Yüksel, Comparison of information structures for zero-sum games and a partial converse to Blackwell ordering in standard Borel spaces. SIAM J. Control Optim. **59**(3), 1781–1803 (2021)
354. I. Hogeboom-Burr, S. Yüksel, Continuity properties of value functions in information structures for zero-sum and general games and stochastic teams. SIAM J. Control Optim. **61**, 398 (2023)
355. I. Hogeboom-Burr, S. Yüksel, Sequential stochastic control (single or multi-agent) problems nearly admit change of measures with independent measurements. Appl. Math. Optim. **87**, 51 (2023)
356. I. Hogeboom-Burr, S. Yüksel, Zero-sum games involving teams against teams: Existence of equilibria, and comparison and regularity in information. Syst. Control Lett. **172**, 105454 (2023)
357. A.S. Holevo, *Probabilistic and Statistical Aspects of Quantum Theory* (Edizioni della Normale, 2011)
358. R.A. Horn, C.R. Johnson, *Matrix Analysis* (Cambridge University Press, New York, 2013)
359. M. Huang, S. Dey, Dynamic quantizer design for hidden Markov state estimation via multiple sensors with fusion center feedback. IEEE Trans. Signal Process. **54**, 2887–2896 (2006)
360. M. Huang, S.L. Nguyen, Linear-quadratic mean field teams with a major agent, in *IEEE 55th Conference on Decision and Control (CDC)* (2016), pp. 6958–6963
361. M. Huang, P.E. Caines, R.P. Malhamé, Large population stochastic dynamic games: closed-loop Mckean-Vlasov systems and the Nash certainty equivalence principle. Commun. Inf. Syst. **6**, 221–251 (2006)
362. M. Huang, P.E. Caines, R.P. Malhamé, Large-population cost-coupled LQG problems with nonuniform agents: Individual-mass behavior and decentralized ϵ-Nash equilibria. IEEE Trans. Autom. Control **52**, 1560–1571 (2007)
363. M. Huang, P.E. Caines, R.P. Malhamé, Social optima in mean field LQG control: centralized and decentralized strategies. IEEE Trans. Autom. Control **57**(7), 1736–1751 (2012)
364. J. Huang, G. Wang, W. Wang, A general linear quadratic stochastic control and information value. J. Math. Anal. Appl. **516**, 126486 (2022)
365. P.J. Huber, Robust estimation of a location parameter. Ann. Math. Stat. **35**, 73–101 (1964)
366. J. Hunter, B. Nachtergaele, *Applied Analysis* (World Scientific, Singapore, 2005)
367. O.C. Imer, S. Yüksel, T. Başar, Optimal control of LTI systems over unreliable communication links. Automatica **42**(9), 1429–1440 (2006)
368. T. Jaakkola, M.I. Jordan, S.P. Singh, On the convergence of stochastic iterative dynamic programming algorithms. Neural Comput. **6**(6), 1185–1201 (1994)
369. S. Jacka, M. Vidmar, On the informational structure in optimal dynamic stochastic control. Preprint. arXiv:1503.02375 (2015)
370. J. Jackson, D. Lacker, Approximately optimal distributed stochastic controls beyond the mean field setting. Preprint. arXiv:2301.02901 (2023)
371. S. Jafarpour, A.D Lewis, Locally convex topologies and control theory. Math. Control Signals Syst. **28**(4), 29 (2016)
372. S. Jaggi, M. Langberg, Two-way interference channels with jammers, in *2017 IEEE International Symposium on Information Theory (ISIT)* (IEEE, 2017), pp. 491–495

373. A. Johnston, S. Yüksel, Stochastic stabilization of partially observed and multi-sensor systems driven by unbounded noise under fixed-rate information constraints. IEEE Trans. Autom. Control **59**, 792–798 (2014)
374. E.A. Jonckheere, A.A. Hammad, B.F. Wu, Chaotic disturbance rejection a Kolmogorov-Sinai entropy approach, in *Proceedings of the IEEE Conference Decision and Control*, San Antonio, Texas (1993)
375. S.T. Jose, A.A. Kulkarni, Shannon meets von Neumann: A minimax theorem for channel coding in the presence of a jammer. IEEE Trans. Inf. Theory **66**(5), 2842–2859 (2020)
376. B. Jovanovic, R.W. Rosenthal, Anonymous sequential games. J. Math. Econ. **17**(1), 77–87 (1988)
377. T. Goblick Jr., Theoretical limitations on the transmission of data from analog sources. IEEE Trans. Inf. Theory **11**, 558–567 (1965)
378. D. Kadnikov, Game theory with information. games in intrinsic Witsenhausen form. PhD thesis, Université Paris-Est, 2020
379. A. Kajii, S. Morris, Payoff continuity in incomplete information games. J. Econ. Theory **82**(1), 267–276 (1998)
380. E. Kalai, E. Lehrer, Rational learning leads to Nash equilibrium. Econometrica **61**(5), 1019–1045 (1993)
381. E. Kalai, E. Lehrer, Weak and strong merging of opinions. J. Math. Econ. **23**, 73–100 (1994)
382. E. Kalai, E. Lehrer, Subjective games and equilibria. Games Econ. Behav. **8**(1), 123–163 (1995)
383. O. Kallenberg, Canonical representations and convergence criteria for processes with interchangeable increments. Z. Wahrsch. Verw. Geb. **27**(1), 23–36 (1973)
384. O. Kallenberg, *Probabilistic Symmetries and Invariance Principles* (Springer Science & Business Media, 2006)
385. E. Kamenica, Bayesian persuasion and information design. Annu. Rev. Econ. **11**, 249–272 (2019)
386. E. Kamenica, M. Gentzkow, Bayesian persuasion. Am. Econ. Rev. **101**(6), 2590–2615 (2011)
387. M.I. Kamien, Y. Tauman, S. Zamir, On the value of information in a strategic conflict. Games Econ. Behav. **2**(2), 129–153 (1990)
388. A.D. Kara, S. Yüksel, Robustness to incorrect priors in partially observed stochastic control. SIAM J. Control Optim. **57**(3), 1929–1964 (2019)
389. A.D. Kara, S. Yüksel, Robustness to incorrect system models in stochastic control. SIAM J. Control Optim. **58**(2), 1144–1182 (2020)
390. A.D. Kara, S. Yüksel, Robustness to approximations and model learning in MDPs and POMDPs, in *Modern Trends in Controlled Stochastic Processes: Theory and Applications, Volume III*, ed. by A.B. Piunovskiy, Y. Zhang (Luniver Press, 2021)
391. A.D. Kara, S. Yüksel, Convergence of finite memory Q-learning for POMDPs and near optimality of learned policies under filter stability. Math. Oper. Res. **48**, 2066–2093 (2023). arXiv:2103.12158
392. A.D. Kara, S. Yüksel, Near optimality of finite memory feedback policies in partially observed Markov decision processes. J. Mach. Learn. Res. **23**(11), 1–46 (2022)
393. A.D. Kara, S. Yüksel, Q-learning for continuous state and action MDPs under average cost criteria. Preprint. arXiv:2308.07591 (2023)
394. A.D. Kara, S. Yüksel, Q-learning for stochastic control under general information structures and non-Markovian environments. Queen's University Technical Report, August (2023)
395. A.D. Kara, N. Saldi, S. Yüksel, Weak Feller property of non-linear filters. Syst. Control Lett. **134**, 104–512 (2019)
396. A.D. Kara, E. Bayraktar, S. Yüksel, Near optimality of finite memory policies for POMPDs with continuous spaces, in *Proc. IEEE Conference on Decision and Control* (IEEE, 2022)
397. A.D. Kara, M. Raginsky, S. Yüksel, Robustness to incorrect models and data-driven learning in average-cost optimal stochastic control. Automatica **139**, 110179 (2022)

398. A.D. Kara, N. Saldi, S. Yüksel, Q-learning for MDPs with general spaces: Convergence and near optimality via quantization under weak continuity. J. Mach. Learn. Res. **24**, 1–34 (arXiv:2111.06781) (2023)
399. J. Karlsson, A. Gattami, T.J. Oechtering, M. Skoglund, Iterative source-channel coding approach to Witsenhausen's counterexample, in *ACC 2011*, San Francisco, July (2011)
400. L. Karp, I.H. Lee, Time-consistent policies. J. Econ. Theory **112**(2), 353–364 (2003)
401. A. Katok, Lyapunov exponents, entropy and periodic orbits for diffeomorphisms. Publ. Math. Inst. Hautes Études Sci. **51**(1), 137–173 (1980)
402. A. Katok, Fifty years of entropy in dynamics: 1958–2007. J. Mod. Dyn. **1**(4), 545–596 (2007)
403. C. Kawan, *Invariance Entropy for Deterministic Control Systems* (Springer, New York, 2013)
404. C. Kawan, J.-Ch. Delvenne, Network entropy and data rates required for networked control. IEEE Trans. Control Netw. Syst. **3**(1), 57–66 (2015)
405. C. Kawan, S. Yüksel, Invariance and stationarity in non-linear networked control: Deterministic and stochastic formulations, in *Proc. IEEE Conference on Decision and Control* (IEEE, 2016), pp. 7353–7358
406. C. Kawan, S. Yüksel, On optimal coding of non-linear dynamical systems. IEEE Trans. Inf. Theory **64**(10), 6816–6829 (2018)
407. C. Kawan, S. Yüksel, Metric and topological entropy bounds for optimal coding of stochastic dynamical systems. IEEE Trans. Autom. Control **65**(6), 2466–2479 (2020)
408. C. Kawan, S. Yüksel, Invariance properties of nonlinear stochastic dynamical systems under information constraints. IEEE Trans. Autom. Control **66**, 4514–4529 (2021)
409. C. Kawan, A.S. Matveev, A.Y. Pogromsky, Remote state estimation problem: Towards the data-rate limit along the avenue of the second Lyapunov method. Automatica **125**, 109467 (2021)
410. E. Kazıklı, S. Gezici, S. Yüksel, Quadratic privacy-signaling games and the MMSE gaussian information bottleneck problem. Preprint. arXiv:2005.05743 (2020)
411. E. Kazıklı, S. Sarıtaş, S. Gezici, T. Linder, S. Yüksel, Signaling games for arbitrary distributions: Number of bins and properties of equilibria. Preprint. arXiv:2012.08265 (2020)
412. E. Kazıklı, S. Gezici, S. Yüksel, Signaling games in higher dimensions: Geometric properties of equilibrium solutions. Preprint. arXiv:2108.05240 (2021)
413. E. Kazıklı, S. Sarıtaş, S. Gezici, S. Yüksel, Optimal signaling with mismatch in priors of an encoder and decoder. Preprint. arXiv:2101.00799 (2021)
414. J. Keeler, T. Linder, S. Yüksel, An asymptotically optimal two-part fixed-rate coding scheme for networked control with unbounded noise. IEEE Trans. Inf. Theory **69**(10), 6765–6786 (arXiv:2202.02841) (2023)
415. J.H.B. Kemperman, Strong converses for a general memoryless channel with feedback, in *Proceedings 6th Prague Conference on Information Theory, Statistical Decision Functions, and Random Processes* (1971), pp. 375–409
416. A. Khina, V. Kostina, A. Khisti, B. Hassibi, Tracking and control of Gauss-Markov processes over packet-drop channels with acknowledgments. IEEE Trans. Control Netw. Syst. **6**(2), 549–560 (2018)
417. M.J. Khojasteh, P. Tallapragada, J. Cortés, M. Franceschetti, The value of timing information in event-triggered control. IEEE Trans. Autom. Control **65**(3), 925–940 (2019)
418. J.C. Kieffer, Zero-error stationary coding over stationary channels. Z. Wahrsch. Verw. Geb. **56**(1), 113–126 (1981)
419. J.C. Kieffer, Stochastic stability for feedback quantization schemes. IEEE Trans. Inf. Theory **28**, 248–254 (1982)
420. J. Kieffer, Uniqueness of locally optimal quantizer for log-concave density and convex error weighting function. IEEE Trans. Inf. Theory **29**(1), 42–47 (1983)
421. J.C. Kieffer, J.G. Dunham, On a type of stochastic stability for a class of encoding schemes. IEEE Trans. Inf. Theory **29**, 793–797 (1983)
422. J.-W. Kim, P.G. Mehta, S.P. Meyn, What is the Lagrangian for nonlinear filtering?, in *Proc. IEEE Conference on Decision and Control* (IEEE, 2019), pp. 1607–1614

423. Y.H. Kim, A. Lapidoth, T. Weissman, Error exponents for the Gaussian channel with active noisy feedback. IEEE Trans. Inf. Theory **57**, 1223–1236 (2011)
424. A.C. Kizilkale, P.E. Caines, Mean field stochastic adaptive control. IEEE Trans. Autom. Control **58**(4), 905–920 (2012)
425. H. Kobayashi, H. Hanafusa, T. Yoshikawa, Controllability under decentralized information structure. IEEE Trans. Autom. Control **23**, 182–188 (1978)
426. V. Kostina, Data compression with low distortion and finite blocklength. IEEE Trans. Inf. Theory **63**(7), 4268–4285, (2017)
427. V. Kostina, B. Hassibi, Rate-cost tradeoffs in control. IEEE Trans. Autom. Control **64**(11), 4525–4540 (2019)
428. V. Kostina, S. Verdú, Fixed-length lossy compression in the finite blocklength regime. IEEE Trans. Inf. Theory **58**(6), 3309–3338 (2012)
429. J.C. Krainak, J.L. Speyer, S.I. Marcus, Static team problems – part I: Sufficient conditions and the exponential cost criterion. IEEE Trans. Autom. Control **27**, 839–848 (1982)
430. J.C. Krainak, J.L. Speyer, S.I. Marcus, Static team problems – part II: Affine control laws, projections, algorithms, and the LEGT problem. IEEE Trans. Autom. Control **27**, 848–859 (1982)
431. D.M. Kreps, P. Milgrom, J. Roberts, R. Wilson, Rational cooperation in the finitely repeated prisoners' dilemma. J. Econ. Theory **27**(2), 245–252 (1982)
432. V. Krishna, J. Morgan, A model of expertise. Q. J. Econ. **116**(2), 747–775 (2001)
433. V. Krishnamurthy, *Partially Observed Markov Decision Processes: From Filtering to Controlled Sensing* (Cambridge University Press, 2016)
434. N.V. Krylov, On the rate of convergence of finite-difference approximations for Bellman's equations with variable coefficients. Probab. Theory Related Fields **117**(1), 1–16 (2000)
435. H. Kuhn, Extensive games and the problem of information, in *Contributions to the Theory of Games*, ed. by H. Kuhn, A. Tucker (1953), pp. 193–216
436. A.A. Kulkarni, Local and networked mean-square estimation with high dimensional log-concave noise. IEEE Trans. Inf. Theory **64**(4), 2759–2773 (2018)
437. A.A. Kulkarni, Near optimality of linear strategies for static teams with "big" non-gaussian noise. IEEE Trans. Autom. Control **65**(2), 534–545 (2019)
438. A.A. Kulkarni, T.P. Coleman, An optimizer's approach to stochastic control problems with nonclassical information structures. IEEE Trans. Autom. Control **60**(4), 937–949 (2014)
439. U. Kumar, V. Gupta, J. Laneman, Stability across a Gaussian product channel: Necessary and sufficient conditions. IEEE Trans. Autom. Control **59**(9), 2530–2535 (2014)
440. U. Kumar, V. Gupta, J.N. Laneman, Sufficient conditions for stabilizability over Gaussian relay and cascade channels, in *Proceedings of the IEEE Conf. Decision and Control*, Atlanta, GA, USA (2010)
441. U. Kumar, V. Gupta, J. Laneman, On stability across a gaussian product channel, in *IEEE Conference on Decision and Control*, Florida, USA (2011), pp. 3142–3147
442. K. Kuratowski, C. Ryll-Nardzewski, A general theorem on selectors. Bull. Acad. Polon. Sci. Ser. Sci. Math. Astron. Phys. **13**(1), 397–403 (1965)
443. B. Kurtaran, A concise derivation of the LQG one-step-delay sharing problem solution. IEEE Trans. Autom. Control **20**, 808–810 (1975)
444. B. Kurtaran, Corrections and extensions to decentralized control with delayed sharing information pattern. IEEE Trans. Autom. Control **24**, 656–657 (1979)
445. H.J. Kushner, *Introduction to Stochastic Control Theory* (Holt, Rinehart and Winston, New York, 1972)
446. H.J. Kushner, Numerical methods for stochastic control problems in continuous time. SIAM J. Control Optim. **28**(5), 999–1048 (1990)
447. D. Lacker, Mean field games via controlled martingale problems: existence of Markovian equilibria. Stoch. Process. Appl. **125**(7), 2856–2894 (2015)
448. D. Lacker, A general characterization of the mean field limit for stochastic differential games. Probab. Theory Related Fields **165**(3–4), 581–648 (2016)

449. D. Lacker, Limit theory for controlled McKean–Vlasov dynamics. SIAM J. Control Optim. **55**(3), 1641–1672 (2017)
450. D. Lacker, On the convergence of closed-loop nash equilibria to the mean field game limit. Preprint. arXiv:1808.02745 (2018)
451. D. Lacker, Probabilistic compactification methods for stochastic optimal control and mean field games (2018)
452. A. Lamperski, L. Lessard, Optimal decentralized state-feedback control with sparsity and delays. Automatica **58**, 143–151 (2015)
453. S. Lang, *Real and Functional Analysis* (Springer, 1993)
454. H.J. Langen, Convergence of dynamic programming models. Math. Oper. Res. **6**(4), 493–512 (1981)
455. B. Larrousse, O. Beaude, S. Lasaulce, Crawford-Sobel meet Lloyd-Max on the grid, in *IEEE International Conference on Acoustics, Speech and Signal Processing (ICASSP)*, May (2014), pp. 6127–6131
456. J.M. Lasry, P.L. Lions, Mean field games. Jpn. J. Math. **2**, 229–260 (2007)
457. J.B. Lasserre, Invariant probabilities for Markov chains on a metric space. Stat. Probab. Lett. **34**, 259–265 (1997)
458. T.N. Le, V.G. Subramanian, R.A. Berry, Information cascades with noise. IEEE Trans. Signal Inf. Process. Netw. **3**(2), 239–251 (2017)
459. L. Le Cam, Sufficiency and approximate sufficiency. Ann. Math. Stat. **35**, 1419–1455 (1964)
460. L. Le Cam, Comparison of experiments - A short review, in *Statistics, Probability and Game Theory Papers in Honor of David Blackwell*, ed. by T. Ferguson, L. Shapley. IMS Lecture Notes Monograph Ser. (1996)
461. J. Le Ny, *Differential Privacy for Dynamic Data* (Springer, 2020)
462. J. Le Ny, G.J. Pappas, Differentially private filtering. IEEE Trans. Autom. Control **59**(2), 341–354 (2014)
463. M. Le Treust, T.J. Oechtering, Optimal control designs for vector-valued Witsenhausen counterexample setups, in *56th Annual Allerton Conference on Communication, Control, and Computing (Allerton)* (IEEE, 2018), pp. 532–537
464. M. Le Treust, T. Tomala, Information design for strategic coordination of autonomous devices with non-aligned utilities, in *54th Annual Allerton Conference on Communication, Control, and Computing*, Monticello, IL (2016)
465. M. Le Treust, T. Tomala, Information-theoretic limits of strategic communication. Preprint. arXiv:1807.05147 (2018)
466. M. Le Treust, T. Tomala, Persuasion with limited communication capacity. J. Econ. Theory **184**, 104940 (2019)
467. F. Ledrappier, L.-S.Young, Entropy formula for random transformations. Probab. Theory Related Fields **80**(2), 217–240 (1988)
468. K.H. Lee, D.P. Petersen, Optimal linear coding for vector channels. IEEE Trans. Commun. **24**, 1283–1290 (1976)
469. J.T. Lee, E. Lau, Y.C. Ho, The Witsenhausen counterexample: a hierarchical search approach for nonconvex optimization problems. IEEE Trans. Autom. Control **46**(3), 382–397 (2001)
470. E. Lehrer, D. Rosenberg, E. Shmaya, Signaling and mediation in games with common interests. Games Econ. Behav. **68**, 670–682 (2010)
471. D. Leslie, E. Collins, Individual Q-learning in normal form games. SIAM J. Control Optim. **44**, 495–514 (2005)
472. D. Leslie, S. Perkins, Z. Xu, Best-response dynamics in zero-sum stochastic games. J. Econ. Theory **189**, 105095 (2020)
473. L. Lessard, S. Lall, Optimal control of two-player systems with output feedback. IEEE Trans. Autom. Control **60**(8), 2129–2144 (2015)
474. J.B. Lewis, J.T. Tou, Optimum sampled-data systems with quantized control signals. IEEE Trans. Appl. Ind. **82**, 229–233 (1965)

475. K. Leyton-Brown, Y. Shoham, *Essentials of Game Theory: A Concise Multidisciplinary Introduction*. Synthesis Lectures on Artificial Intelligence and Machine Learning, vol. 2, no. 1 (2008), pp. 1–88
476. S. Li, T. Başar, Asymptotic agreement and convergence of asynchronous stochastic algorithms. IEEE Trans. Autom. Control **32**, 612–618 (1987)
477. N. Li, R. Marden, J.S. Shamma, Learning approach to the Witsenhausen counterexample from a view of potential games, in *Proceedings of the IEEE Conference on Decision and Control*, Shanghai, Dec. (2009), pp. 157–162
478. D. Liberzon, Stabilization by quantized state or output feedback: A hybrid control approach, in *Proc. IFAC 15th Triennial World Congress* (2002)
479. D. Liberzon, J.P. Hespanha, Stabilization of nonlinear systems with limited information feedback. IEEE Trans. Autom. Control **50**(6), 910–915 (2005)
480. B. Light, G.Y. Weintraub, Mean field equilibrium: uniqueness, existence, and comparative statics. Exist. Comp. Stat. **70**(1), 585–605 (2022)
481. T. Linder, On the training distortion of vector quantizers. IEEE Trans. Inf. Theory **46**, 1617–1623 (2000)
482. T. Linder, A note on the existence of optimal quantizers. *Private communication* (2012)
483. T. Linder, G. Lugosi, A zero-delay sequential scheme for lossy coding of individual sequences. IEEE Trans. Inf. Theory **47**, 2533–2538 September (2001)
484. T. Linder, S. Yüksel, On optimal zero-delay quantization of vector Markov sources. IEEE Trans. Inf. Theory **60**, 2975–5991 (2014)
485. T. Linder, R. Zamir, Causal coding of stationary sources and individual sequences with high resolution. IEEE Trans. Inf. Theory **52**, 662–680 (2006)
486. G.M. Lipsa, N.C. Martins, Optimal memoryless control in Gaussian noise: A simple counterexample. Automatica **47**, 552–558 (2011)
487. S. Lloyd, Least squares quantization in PCM. IEEE Trans. Inf. Theory **28**(2), 129–137 (1982)
488. M. Loève, *Probability Theory* (Van Nostrand, New York, 1963)
489. D.G. Luenberger, *Optimization by Vector Space Methods* (John Wiley & Sons, New York, NY, 1969)
490. D.G. Luenberger, *Linear and Nonlinear Programming* (Addison-Wesley, Reading, MA, 1984)
491. E. Lutwak, D. Yang, G. Zhang, Crame/spl acute/r-Rao and moment-entropy inequalities for Renyi entropy and generalized Fisher information. IEEE Trans. Inf. Theory **51**(2), 473–478 (2005)
492. A. Mahajan, D. Teneketzis, On the design of globally optimal communication strategies for real-time noisy communication with noisy feedback. IEEE J. Sel. Areas Commun. **26**, 580–595 (2008)
493. A. Mahajan, D. Teneketzis, Optimal design of sequential real-time communication systems. IEEE Trans. Inf. Theory **55**, 5317–5338 (2009)
494. A. Mahajan, D. Teneketzis, Optimal performance of networked control systems with non-classical information structures. SIAM J. Control Optim. **48**, 1377–1404 (2009)
495. A. Mahajan, S. Yüksel, Measure and cost dependent properties of information structures, in *Proceedings of the 2010 American Control Conference* (IEEE, 2010), pp. 6397–6402
496. A. Mahajan, N.C. Martins, M. Rotkowitz, S. Yüksel, Information structures in optimal decentralized control, in *IEEE Conference on Decision and Control*, Hawaii, USA (2012)
497. A. Mahajan, N.C. Martins, S. Yüksel, Static LQG teams with countably infinite players, in *Proceedings of the IEEE Conference on Decision and Control* (IEEE, 2013), pp. 6765–6770
498. A. Makhdoumi, S. Salamatian, N. Fawaz, M. Medard, From the information bottleneck to the privacy funnel, in *IEEE Information Theory Workshop (ITW)* (2014), pp. 501–505
499. A.A. Malikopoulos, On team decision problems with nonclassical information structures. IEEE Trans. Autom. Control **68**(7), 3915–3930 (2023)
500. J.W. Mamer, K.E. Schilling, A zero-sum game with incomplete information and compact action spaces. Math. Oper. Res. **11**(4), 627–631 (1986)

501. W. Mao, T. Başar, Provably efficient reinforcement learning in decentralized general-sum Markov games. Dyn. Games Appl. **13**(1), 165–186 (2023)
502. W. Mao, K. Zhang, E. Miehling, T. Başar, Information state embedding in partially observable cooperative multi-agent reinforcement learning, in *2020 59th IEEE Conference on Decision and Control (CDC)* (IEEE, 2020), pp. 6124–6131
503. J.R. Marden, J.S. Shamma, Revisiting log-linear learning: Asynchrony, completeness and payoff-based implementation. Games Econ. Behav. **75**(2), 788–808 (2012)
504. J.R. Marden, H.P. Young, G. Arslan, J.S. Shamma, Payoff based dynamics for multi-player weakly acyclic games. SIAM J. Control Optim. **48**(1), 373–396 (2009)
505. J.R. Marden, H.P. Young, L.Y. Pao, Achieving Pareto optimality through distributed learning. SIAM J. Control Optim. **52**(5), 2753–2770 (2014)
506. J. Marschak, Elements for a theory of teams. Manag. Sci. **1**, 127–137 (1955)
507. J. Marshak, R. Radner, *Economic Theory of Teams* (Yale University Press, New Haven, CT, 1972)
508. N.C. Martins, M.A. Dahleh, Feedback control in the presence of noisy channels: "Bode-like fundamental limitations of performance. IEEE Trans. Autom. Control **53**, 1604–1615 (2008)
509. A. Mas-Colell, On a theorem of Schmeidler. J. Math. Econ. **13**(3), 201–206 (1984)
510. Ll. Masanes, A. Acin, N. Gisin, General properties of nonsignaling theories. Phys. Rev. A **73**, 012112 (2006)
511. E. Mascolo, L. Migliaccio, Relaxation methods in control theory. Appl. Math. Optim. **20**(1), 97–103 (1989)
512. O. Massicot, C. Langbort, Almost-Bayesian quadratic persuasion. Preprint. arXiv:2212.13619 (2022)
513. A.S. Matveev, State estimation via limited capacity noisy communication channels. Math. Control Signals Syst. **20**, 1–35 (2008)
514. A.S. Matveev, A.Y. Pogromsky, Observation of nonlinear systems via finite capacity channels: Constructive data rate limits. Automatica **70**, 217–229 (2016)
515. A.S. Matveev, A.Y. Pogromsky, Observation of nonlinear systems via finite capacity channels, part II: Restoration entropy and its estimates. Automatica **103**, 189–199 (2019)
516. A.S. Matveev, A.V. Savkin, The problem of LQG optimal control via a limited capacity communication channel. Syst. Control Lett. **53**, 51–64 (2004)
517. A.S. Matveev, A.V. Savkin, *Estimation and Control over Communication Networks* (Birkhäuser, Boston, 2008)
518. E. Mazumdar, L.J. Ratliff, M.I. Jordan, S.S. Sastry, Policy-gradient algorithms have no guarantees of convergence in linear quadratic games. Preprint. arXiv:1907.03712 (2019)
519. E. Mazumdar, L.J. Ratliff, S.S. Sastry, On gradient-based learning in continuous games. SIAM J. Math. Data Sci. **2**(1), 103–131 (2020)
520. C. McDonald, S. Yüksel, Observability and filter stability for partially observed Markov processes, in *Proc. IEEE Conference on Decision and Control* (IEEE, 2019), pp. 1623–1628
521. C. McDonald, S. Yüksel, Exponential filter stability via Dobrushin's coefficient. Electron. Commun. Probab. **25**, 1–13 (2020)
522. C. McDonald, S. Yüksel, Robustness to incorrect priors and controlled filter stability in partially observed stochastic control. SIAM J. Control Optim. **60**(2), 842–870 (2022)
523. C. McDonald, S. Yüksel, Stochastic observability and filter stability under several criteria. IEEE Trans. Autom. Control **69**(5), 2931–2946 (2024)
524. C. McDonald, F. Alajaji, S. Yüksel, Two-way Gaussian networks with a jammer and decentralized control. IEEE Trans. Control Netw. Syst. **7**(1), 446–457 (2019)
525. W.M. McEneaney, S.H. Han, Optimization formulation and monotonic solution method for the Witsenhausen problem. Automatica **55**, 55–65 (2015)
526. E.J. McShane, Relaxed controls and variational problems. SIAM J. Control **5**(3), 438–485 (1967)
527. M. Mehmetoglu, E. Akyol, K. Rose, A deterministic annealing approach to Witsenhausen's counterexample, in *ISIT 2014*, Honolulu, July (2014)

528. P.G. Mehta, U. Vaidya, A. Banaszuk, Markov chains, entropy, and fundamental limitations in nonlinear stabilization. IEEE Trans. Autom. Control **53**(3), 784–791 (2008)
529. F.C. Melo, S.P. Meyn, I.M. Ribeiro, An analysis of reinforcement learning with function approximation, in *Proceedings of the 25th International Conference on Machine Learning* (2008), pp. 664–671
530. N. Merhav, On optimum parameter modulation-estimation from a large deviations perspective. IEEE Trans. Inf. Theory **58**, 172–182 (2012)
531. N. Merhav, Exponential error bounds on parameter modulation-estimation for discrete memoryless channels. IEEE Trans. Inf. Theory **60**(2), 832–841 (2013)
532. J.-F. Mertens, S. Sorin, S. Zamir, *Repeated Games*, vol. 55 (Cambridge University Press, 2015)
533. P. Mertikopoulos, C. Papadimitriou, G. Piliouras, Cycles in adversarial regularized learning, in *Proceedings of the Twenty-Ninth Annual ACM-SIAM Symposium on Discrete Algorithms* (SIAM, 2018), pp. 2703–2717
534. S.P. Meyn, *Control Techniques for Complex Networks* (Cambridge University Press, 2007)
535. S. Meyn, *Control Systems and Reinforcement Learning* (Cambridge University Press, 2022)
536. S.P. Meyn, R. Tweedie, Stability of Markovian processes I: Criteria for discrete-time chains. Adv. Appl. Probab. **24**, 542–574 (1992)
537. S.P. Meyn, R. Tweedie, *Markov Chains and Stochastic Stability* (Springer, London, 1993)
538. R.H. Middleton, Trade-offs in linear control system design. Automatica **27**(2), 281–292 (1991)
539. P.R. Milgrom, R.J. Weber, Distributional strategies for games with incomplete information. Math. Oper. Res. **10**(4), 619–632 (1985)
540. P. Minero, M. Franceschetti, Anytime capacity of a class of Markov channels. IEEE Trans. Autom. Control **62**(3), 1356–1367 (2016)
541. P. Minero, M. Franceschetti, S. Dey, G.N. Nair, Data rate theorem for stabilization over time-varying feedback channels. IEEE Trans. Autom. Control **54**(2), 243–255 (2009)
542. L. Miretti, E. Björnson, D. Gesbert, Team MMSE precoding with applications to cell-free massive MIMO. IEEE Trans. Wirel. Commun. **21**(8), 6242–6255 (2022)
543. M. Misiurewicz, Topological entropy and metric entropy. Enseign. Math **29**, 61–66 (1981)
544. S. Mitra, D. Liberzon, Entropy and minimal bit rates for state estimation and model detection. IEEE Trans. Autom. Control **63**(10), 3330–3344 (2017)
545. S. Mitter, A. Sahai, Information and control: Witsenhausen revisited, in *Learning, Control, and Hybrid Systems* (1999), pp. 281–293
546. S. Miura, Strategic communication games: theory and applications, PhD thesis, Washington University in St. Louis, 2012
547. E. Mossel, O. Tamuz, Opinion exchange dynamics. Probab. Surv. **14**, 155–204 (2017)
548. H. Moulin, Advanced Game Theory. Online Notes (2009)
549. N. Muraki, M. Ohya, Note on continuity of information rate. Illinois J. Math. **36**, 529–550 (1992)
550. J.H. Nachbar, Prediction, optimization, and learning in games. Econometrica **65**, 275–309 (1997)
551. V.S.S. Nadendla, C. Langbort, T. Başar, Effects of subjective biases on strategic information transmission. IEEE Trans. Commun. **66**(12), 6040–6049 (2018)
552. G.N. Nair, R.J. Evans, Stabilizability of stochastic linear systems with finite feedback data rates. SIAM J. Control Optim. **43**, 413–436 (2004)
553. G.N. Nair, R.J. Evans, I.M.Y. Mareels, W. Moran, Topological feedback entropy and nonlinear stabilization. IEEE Trans. Autom. Control **49**(9), 1585–1597 (2004)
554. G.N. Nair, F. Fagnani, S. Zampieri, J.R. Evans, Feedback control under data constraints: an overview. Proc. IEEE **95**, 108–137 (2007)
555. J. Nash, Non-cooperative games. Ann. Math. **54**(2), 286–295 (1951)
556. A. Nayyar, D. Teneketzis, On the structure of real-time encoders and decoders in a multi-terminal communication system. IEEE Trans. Inf. Theory **57**, 6196–6214 (2011)

557. A. Nayyar, D. Teneketzis, Common knowledge and sequential team problems. IEEE Trans. Autom. Control **64**(12), 5108–5115 (2019)
558. A. Nayyar, A. Mahajan, D. Teneketzis, Optimal control strategies in delayed sharing information structures. IEEE Trans. Autom. Control **56**, 1606–1620 (2011)
559. A. Nayyar, A. Gupta, C. Langbort, T. Başar, Common information based Markov perfect equilibria for stochastic games with asymmetric information: Finite games. IEEE Trans. Autom. Control **59**(3), 555–570 (2013)
560. A. Nayyar, A. Mahajan, D. Teneketzis, The common-information approach to decentralized stochastic control, in *Information and Control in Networks*, ed. by G. Como, B. Bernhardsson, A. Rantzer (Springer, 2013)
561. A. Nayyar, A. Mahajan, D. Teneketzis, Decentralized stochastic control with partial history sharing: A common information approach. IEEE Trans. Autom. Control **58**, 1644–1658 (2013)
562. D.L. Neuhoff, R.K. Gilbert, Causal source codes. IEEE Trans. Inf. Theory **28**, 701–713 (1982)
563. L. Nielsen, Common knowledge, communication and convergence of beliefs. Math. Soc. Sci. **8**, 1–14 (1984)
564. E. Nummelin, *General Irreducible Markov Chains and Non-negative Operators* (Cambridge University Press, 1984)
565. T. Oechtering, M. Le Treust, Coordination coding with causal decoder for vector-valued Witsenhausen counterexample setups, in *2019 IEEE Information Theory Workshop (ITW)* (2019)
566. K. Okano, S. Hara, H. Ishii, Characterization of a complementary sensitivity property in feedback control: An information theoretic approach. Automatica **45**(2), 504–509 (2009)
567. J. Ooi, S. Verbout, J. Ludwig, G. Wornell, A separation theorem for periodic sharing information patterns in decentralized control. IEEE Trans. Autom. Control **42**, 1546–1550 (1997)
568. M. Osborne, A. Rubinstein, *A Course in Game Theory* (MIT Press, 1994)
569. J. Østergaard, Stabilizing error correction codes for controlling LTI systems over erasure channels, in *Proc. IEEE Conference on Decision and Control* (2021)
570. G. Owen, *Game Theory* (Emerald Group Publishing, 2013)
571. A. Özcelikkale, S. Yüksel, H.M. Ozaktas, Unitary precoding and basis dependency of MMSE performance for Gaussian erasure channel. IEEE Trans. Inf. Theory **60**(11), 7186–7203 (2014)
572. G. Papavassilopoulos, J. Cruz, Nonclassical control problems and Stackelberg games. IEEE Trans. Autom. Control **24**(2), 155–166 (1979)
573. A. Papoulis, *Probability, Random Variables and Stochastic Processes* (McGraw-Hill, New York, 1965)
574. K.R. Parthasarathy, *Probability Measures on Metric Spaces* (AMS Bookstore, 1967)
575. J.W. Patchell, O.L.R. Jacobs, Separability, neutrality and certainty equivalence. Int. J. Control **13**(2), 337–342 (1971)
576. H.H. Permuter, T. Weissman, A.J. Goldsmith, Finite state channels with time-invariant deterministic feedback. IEEE Trans. Inf. Theory **55**(2), 644–662 (2009)
577. Ja.B. Pesin, Characteristic Lyapunov exponents and smooth ergodic theory. Uspekhi Mat. Nauk **32**(4), 55–112 (1977)
578. M. Pęski, Comparison of information structures in zero-sum games. Games Econ. Behav. **62**, 732–735 (2008)
579. M. Pęski, Comparison of information structures in zero-sum games. Games Econ. Behav. **62**(2), 732–735 (2008)
580. R.P. Phelps, *Lectures on Choquet's Theorem* (Van Nostrand, New York, 1966)
581. R. Pilc, Coding theorems for discrete source-channel pairs, PhD thesis, Massachusetts Institute of Technology, 1967
582. R. Pilc, The optimum linear modulator for a Gaussian source used with a Gaussian channel. IEEE Trans. Autom. Control **48**, 3075–3089 (1969)

583. J. Pineau, G. Gordon, S. Thrun, Anytime point-based approximations for large pomdps. J. Artif. Intell. Res. **27**, 335–380 (2006)
584. A.B. Piunovskiy, Multicriteria control problems for stochastic sequences, in *European Control Conference* (IEEE, 1997), pp. 531–534
585. A.B. Piunovskiy, Controlled random sequences: methods of convex analysis and problems with functional constraints. Russ. Math. Surv. **53**(6), 1233–1293 (1998)
586. D. Pollard, Quantization and the method of k-means. IEEE Trans. Inf. Theory **28**, 199–205 (1982)
587. V. Poor, *An Introduction to Signal Detection and Estimation* (Springer, 1994)
588. S. Popescu, D. Rohrlich, Quantum nonlocality as an axiom. Found. Phys. **24**(3), 379–385 (1994)
589. J.M. Porta, N. Vlassis, M.T.J. Spaan, P. Poupart, Point-based value iteration for continuous pomdps. J. Mach. Learn. Res. **7**(Nov), 2329–2367 (2006)
590. S. Pradhan, S. Yüksel, Continuity of cost in Borkar control topology and implications on discrete space and time approximations for controlled diffusions under several criteria. Preprint. arXiv:2209.14982 (2022)
591. S. Pradhan, S. Yüksel, Controlled diffusions under full, partial and decentralized information: Existence of optimal policies and discrete-time approximations. arXiv preprint arXiv:2311.03254 (2023)
592. Q. Qi, L. Xie, H. Zhang, Optimal control for stochastic systems with multiple controllers of different information structures. IEEE Trans. Autom. Control **66**(9), 4160–4175 (2020)
593. M. Rabi, C. Ramesh, K.H. Johansson, Separated design of encoder and controller for networked linear quadratic optimal control. SIAM J. Control Optim. **54**(2), 662–689 (2016)
594. R. Radner, Team decision problems. Ann. Math. Stat. **33**, 857–881 (1962)
595. M. Raginsky, Shannon meets Blackwell and Le Cam: Channels, codes, and statistical experiments, in *Proceedings of the IEEE International Symposium on Information Theory* (St. Petersburg, Russia, 2011), pp. 1220–1224
596. F. Ramponi, A. Milias-Argeitis D. Chatterjee, P. Hokayem, and J. Lygeros. Attaining mean square boundedness of a marginally stable stochastic linear system with a bounded control input. IEEE Trans. Autom. Control **55**, 2414–2418 (2010)
597. G. Ranade V. Kostina, Y. Peres, M. Sellke, Exact minimum number of bits to stabilize a linear system, in *2018 IEEE Conference on Decision and Control (CDC)* (IEEE, 2018), pp. 453–458
598. A. Rantzer, Linear quadratic team theory revisited, in *Proceedings of the American Control Conference* (2006), pp. 1637–1642
599. Y.L. Ren, L.F. He, J.F. Lü, G.P. Zheng, Topological r-entropy and measure-theoretic r-entropy of a continuous map. Sci. China Math. **54**(6), 1197–1205 (2011)
600. F. Rezaei, C.D. Charalambous, N.U. Ahmed, Optimization of stochastic uncertain systems with variational norm constraints, in *Proceedings of the IEEE Conference on Decision and Control*, December (2007), pp. 2159–2163
601. Y.A. Reznik, An algorithm for quantization of discrete probability distributions, in *Data Compression Conference* (2011), pp. 333–342
602. D. Rhenius, Incomplete information in Markovian decision models. Ann. Stat. **2**, 1327–1334 (1974)
603. R. Rishel, Necessary and sufficient dynamic programming conditions for continuous time stochastic optimal control. SIAM J. Control **8**(4), 559–571 (1970)
604. J. Robinson, An iterative method of solving a game. Ann. Math. **54**, 296–301 (1951)
605. C. Robinson, *Dynamical Systems: Stability, Symbolic Dynamics, and Chaos* (CRC Press, 1998)
606. R.T. Rockafellar, *Convex Analysis*, Number 28 (Princeton University Press, 1970)
607. C.G. Rothschild, Payoff continuity in incomplete information games: A comment. J. Econ. Theory **120**(2), 270–274 (2005)

608. M. Rotkowitz, Linear controllers are uniformly optimal for the Witsenhausen counterexample, in *Proceedings of the IEEE Conference on Decision and Control*, San Diego, CA, December (2006), pp. 553–558
609. M. Rotkowitz, On information structures, convexity, and linear optimality, in *IEEE 47th Annual Conference on Decision and Control (CDC)* (2008), pp. 1642–1647
610. M. Rotkowitz, S. Lall, A characterization of convex problems in decentralized control. IEEE Trans. Autom. Control **51**, 274–286 (2006)
611. A. Rubinstein, The electronic mail game: Strategic behavior under "almost common knowledge". Am. Econ. Rev. **79**(3), 385–391 (1989)
612. W. Rudin, *Real and Complex Analysis*, 3rd edn. (McGraw-Hill, New York, 1987)
613. W. Rudin, *Functional Analysis* (McGraw-Hill, 1991)
614. D. Ruelle, An inequality for the entropy of differentiable maps. Bol. Soc. Bras. Mat. Bull. Braz. Math. Soc. **9**(1), 83–87 (1978)
615. E.P. Ryan, On Brockett's condition for smooth stabilizability and its necessity in a context of nonsmooth feedback. SIAM J. Control Optim. **32**(6), 1597–1604 (1994)
616. O. Sabag, V. Kostina, B. Hassibi, Stabilizing dynamical systems with fixed-rate feedback using constrained quantizers, in *IEEE International Symposium on Information Theory* (IEEE, 2020), pp. 2855–2860
617. A. Sahai, Anytime information theory. Ph.D. dissertation, Massachusetts Institute of Technology, Cambridge, MA (2001)
618. A. Sahai, Coding unstable scalar Markov processes into two streams, in *Proceedings of the IEEE International Symposium on Information Theory* (2004), p. 462
619. A. Sahai, S. Mitter, The necessity and sufficiency of anytime capacity for stabilization of a linear system over a noisy communication link part I: Scalar systems. IEEE Trans. Inf. Theory **52**(8), 3369–3395 (2006)
620. N. Saldi, A topology for team policies and existence of optimal team policies in stochastic team theory. IEEE Trans. Autom. Control **65**(1), 310–317 (2020)
621. N. Saldi, S. Yüksel, Geometry of information structures, strategic measures and associated control topologies. Probab. Surv. **19**, 450–532 (2022)
622. N. Saldi, T. Linder, S. Yüksel, Randomized quantization and source coding with constrained output distribution. IEEE Trans. Inf. Theory **61**(1), 91–106 (2014)
623. N. Saldi, T. Linder, S. Yüksel, Output constrained lossy source coding with limited common randomness. IEEE Trans. Inf. Theory **61**(9), 4984–4998 (2015)
624. N. Saldi, S. Yüksel, T. Linder, Near optimality of quantized policies in stochastic control under weak continuity conditions. J. Math. Anal. Appl. **435**(1), 321–337 (2016)
625. N. Saldi, S. Yüksel, T. Linder, Finite model approximations and asymptotic optimality of quantized policies in decentralized stochastic control. IEEE Trans. Autom. Control **62**, 2360–2373 (2017)
626. N. Saldi, S. Yüksel, T. Linder, On the asymptotic optimality of finite approximations to Markov decision processes with Borel spaces. Math. Oper. Res. **42**(4), 945–978 (2017)
627. N. Saldi, T. Başar, M. Raginsky, Markov-Nash equilibria in mean-field games with discounted cost. SIAM J. Control Optim. **56**(6), 4256–4287 (2018)
628. N. Saldi, T. Linder, S. Yüksel, *Finite Approximations in Discrete-Time Stochastic Control: Quantized Models and Asymptotic Optimality* (Springer, Cham, 2018)
629. N. Saldi, S. Yüksel, T. Linder, Finite model approximations for partially observed Markov decision processes with discounted cost. IEEE Trans. Autom. Control **65**, 130–142 (2020)
630. N. Sandell, On open-loop and closed-loop Nash strategies. IEEE Trans. Autom. Control **19**(4), 435–436 (1974)
631. N. Sandell, M. Athans, Solution of some nonclassical LQG stochastic decision problems. IEEE Trans. Autom. Control **19**, 108–116 (1974)
632. S. Sanjari, S. Yüksel, Optimal solutions to infinite-player stochastic teams and mean-field teams. IEEE Trans. Autom. Control **66**(3), 1071–1086 (2020)
633. S. Sanjari, S. Yüksel, Optimal policies for convex symmetric stochastic dynamic teams and their mean-field limit. SIAM J. Control Optim. **59**(2), 777–804 (2021)

634. S. Sanjari, N. Saldi, S. Yüksel, Optimality of independently randomized symmetric policies for exchangeable stochastic teams with infinitely many decision makers. Preprint. arXiv:2008.11570 (2020)
635. S. Sanjari, N. Saldi, S. Yüksel, Nash equilibria for exchangeable team against team games, their mean field limit, and role of common randomness. SIAM J. Control Optim. (to appear), arXiv (2024)
636. S. Sanjari, N. Saldi, S. Yüksel, Optimality of independently randomized symmetric policies for exchangeable stochastic teams with infinitely many decision makers. Math. Oper. Res. **48**(3), 1254–1285 (2023). arXiv:2008.11570
637. S. Sanjari, T. Başar, S. Yüksel, Isomorphism properties of optimality and equilibrium solutions under equivalent information structure transformations: Stochastic dynamic games and teams. SIAM J. Control Optim. (arXiv:2104.05787) (2023)
638. S. Sarıtaş, S. Yüksel, S. Gezici, Quadratic multi-dimensional signaling games and affine equilibria. IEEE Trans. Autom. Control **62**(2), 605–619 (2017)
639. S. Sarıtaş, P. Furrer, S. Gezici, T. Linder, S. Yüksel, On the number of bins in equilibria for signaling games, in *IEEE International Symposium on Information Theory (ISIT)* (2019), pp. 972–976
640. S. Sarıtaş, S. Gezici, S. Yüksel, Hypothesis testing under subjective priors and costs as a signaling game. IEEE Trans. Signal Process. **67**(19), 5169–5183 (2019)
641. S. Sarıtaş, S. Yüksel, S. Gezici, Dynamic signaling games with quadratic criteria under Nash and Stackelberg equilibria. Automatica **115**, 108883 (2020)
642. L.J. Savage, *The Foundations of Statistics* (John Wiley and Sons, New York, 1954)
643. A.V. Savkin, Analysis and synthesis of networked control systems: Topological entropy, observability, robustness and optimal control. Automatica **42**(1), 51–62 (2006)
644. M.Ö. Sayın, E. Akyol, T. Başar, On the structure of equilibrium strategies in dynamic Gaussian signaling games, in *IEEE Multi-Conference on Systems and Control* (2016)
645. M.Ö. Sayın, E. Akyol, T. Başar, Strategic control of a tracking system, in *IEEE Conference on Decision and Control (CDC)*, Dec. (2016), pp. 6147–6153
646. M.O. Sayin, E. Akyol, T. Başar, Hierarchical multistage Gaussian signaling games in noncooperative communication and control systems. Automatica **107**, 9–20 (2019)
647. M.O. Sayin, K. Zhang, D.S. Leslie, T. Basar, A. Ozdaglar, Decentralized Q-learning in zero-sum Markov games. Preprint. arXiv:2106.02748 (2021)
648. M.O. Sayin, F. Parise, A. Ozdaglar, Fictitious play in zero-sum stochastic games. SIAM J. Control Optim. **60**(4), 2095–2114 (2022)
649. M. Schäl, Conditions for optimality in dynamic programming and for the limit of n-stage optimal policies to be optimal. Z. Wahrsch. **32**, 179–296 (1975)
650. M. Schäl, On dynamic programming: compactness of the space of policies. Stoch. Process. Appl. **3**(4), 345–364 (1975)
651. J. Schalkwijk, T. Kailath, A coding scheme for additive noise channels with feedbackl part I: No bandwidth constraint. IEEE Trans. Inf. Theory **12**, 172–182 (1966)
652. L. Schenato, B. Sinopoli, M. Franceschetti, K. Poolla, S.S. Sastry, Foundations of control and estimation over lossy networks. Proc. IEEE **95**(1), 163–187 (2007)
653. D. Schmeidler, Equilibrium points of nonatomic games. J. Stat. Phys. **7**(4), 295–300 (1973)
654. G. Schoenmakers, J. Flesch, F. Thuijsman, Fictitious play in stochastic games. Math. Methods Oper. Res. **66**(2), 315–325 (2007)
655. S. Sémirat, Strategic information transmission despite conflict. Int. J. Game Theory **48**(3), 921–956 (2019)
656. S. Sen, M. Sekaran, J. Hale, Learning to coordinate without sharing information, in *Proceedings of the 12th National Conference on Artificial Intelligence* (1994), pp. 426–431
657. N. Şen, F. Alajaji, S. Yüksel, Feedback capacity of a class of symmetric finite-state Markov channels. IEEE Trans. Inf. Theory **56**, 4110–4122 (2011)
658. R. Serfozo, Convergence of Lebesgue integrals with varying measures. Sankhyā Indian J. Stat. Ser. A, **44**(3), 380–402 (1982)

659. X. Sha, J. Zhang, K. You, K. Zhang, T. Başar, Fully asynchronous policy evaluation in distributed reinforcement learning over networks. Automatica **136**, 110092 (2022)
660. S. Shafiee, S. Ulukus, Mutual information games in multiuser channels with correlated jamming. IEEE Trans. Inf. Theory **55**(10), 4598–4607 (2009)
661. P. Shah, P. Parrilo, H2 optimal decentralized control over posets: A state-space solution for state-feedback. IEEE Trans. Autom. Control **58**(12), 3084–3096 (2013)
662. C.E. Shannon, The zero error capacity of a noisy channel. IEEE Trans. Inf. Theory **2**, 8–19 (1956)
663. C.E. Shannon, Two-way communication channels, in *Proceedings of the Fourth Berkeley Symposium on Mathematical Statistics and Probability, Volume 1: Contributions to the Theory of Statistics* (The Regents of the University of California, 1961)
664. C.E. Shannon, W. Weaver, *The Mathematical Theory of Communication* (University of Illinois Press, Urbana, IL, 1949)
665. P.C. Shields, *The Ergodic Theory of Discrete Sample Paths* (AMS, Providence RI, 1996)
666. S. Shreve, D.P. Bertsekas, Universally measurable policies in dynamic programming. Math. Oper. Res. **4**(1), 15–30 (1979)
667. Z. Shu, R.H. Middleton, Stabilization over power-constrained parallel Gaussian channels. IEEE Trans. Autom. Control **56**, 1718–1724 (2011)
668. E.I. Silva, M.S. Derpich, J. Østergaard, A framework for control system design subject to average data-rate constraints. IEEE Trans. Autom. Control **56**, 1886–1899 (2011)
669. E.I. Silva, M.S. Derpich, J. Østergaard, M.A. Encina, A characterization of the minimal average data rate that guarantees a given closed-loop performance level. IEEE Trans. Autom. Control **61**(8), 2171–2186 (2015)
670. H.A. Simon, Dynamic programming under uncertainty with a quadratic criterion function. Econometrica J. Econom. Soc. **24**(1), 74–81 (1956)
671. R. Simpson, S. Yüksel, Non-sequential decentralized stochastic control revisited: Causality and static reducibility. Preprint. arXiv:2211.03949 (2022)
672. S.P. Singh, T. Jaakkola, M.I. Jordan, Learning without state-estimation in partially observable Markovian decision processes, in *Machine Learning Proceedings 1994* (Elsevier, 1994), pp. 284–292
673. S.P. Singh, T. Jaakkola, M.I. Jordan, Reinforcement learning with soft state aggregation, in *Advances in Neural Information Processing Systems* (1995), pp. 361–368
674. T. Smith, R. Simmons, Point-based pomdp algorithms: Improved analysis and implementation. Preprint. arXiv:1207.1412 (2012)
675. J. Sobel, Complexity versus conflict in communication, in *46th Annual Conference on Information Sciences and Systems (CISS)*, Mar. (2012), pp. 1–6
676. Y. Socratous, F. Rezaei, C.D. Charalambous, Nonlinear estimation for a class of systems. IEEE Trans. Inf. Theory **55**, 1930–1938 (2009)
677. E.D. Sontag, A concept of local observability. Syst. Control Lett. **5**(1), 41–47 (1984)
678. E.D. Sontag, *Mathematical Control Theory: Deterministic Finite Dimensional Systems*, vol. 6 (Springer Science & Business Media, 2013)
679. S. Sorin, Merging, reputation, and repeated games with incomplete information. Games Econ. Behav. **29**(1), 274–308 (1999)
680. M. Spence, Job market signaling, in *Uncertainty in Economics* (Elsevier, 1978), pp. 281–306
681. R. Srikant, T. Başar, Asymptotic solutions to weakly coupled stochastic teams with nonclassical information. IEEE Trans. Autom. Control **37**, 163–173 (1992)
682. W.E. Stark, R.J. McEliece, On the capacity of channels with block memory. IEEE Trans. Inf. Theory **3**, 322–324 (1988)
683. H. Stark, J.W. Woods, *Probability, Statistics, and Random Processes for Engineers* (Pearson, Upper Saddle River, NJ, 2012)
684. P.A. Stavrou, M. Skoglund, LQG control and linear policies for noisy communication links with synchronized side information at the decoder. Automatica **123**, 109306 (2021)
685. P.A. Stavrou, M. Skoglund, Asymptotic reverse waterfilling algorithm of NRDF for certain classes of vector Gauss-Markov processes. IEEE Trans. Autom. Control **67**(6), 3196–3203 (2022)

686. P.A. Stavrou, T. Charalambous, C.D. Charalambous, S. Loyka, Optimal estimation via nonanticipative rate distortion function and applications to time-varying Gauss–Markov processes. SIAM J. Control Optim. **56**(5), 3731–3765 (2018)
687. P.A. Stavrou, J. Østergaard, C.D. Charalambous, Zero-delay rate distortion via filtering for vector-valued Gaussian sources. IEEE J. Sel. Top. Signal Process. **12**(5), 841–856 (2018)
688. P.A. Stavrou, T. Tanaka, S. Tatikonda, The time-invariant multidimensional gaussian sequential rate-distortion problem revisited. IEEE Trans. Autom. Control **65**(5), 2245–2249 (2019)
689. L. Stettner, Ergodic control of partially observed Markov control processes with equivalent transition probabilities. Appl. Math. **22**, 25–38 (1993)
690. V. Strassen, The existence of probability measures with given marginals. Ann. Math. Stat. **36**(2), 423–439 (1965)
691. C. Striebel, Sufficient statistics in the optimum control of stochastic systems. J. Math. Anal. Appl. **12**, 576–592 (1965)
692. J. Subramanian, A. Mahajan, Reinforcement learning in stationary mean-field games, in *Proceedings of the 18th International Conference on Autonomous Agents and MultiAgent Systems* (2019), pp. 251–259
693. R.S. Sutton, A.G. Barto, *Reinforcement Learning* (MIT Press, Cambridge, MA, 1998)
694. B. Swenson, C. Eksin, S. Kar, A. Ribeiro, Distributed inertial best-response dynamics. IEEE Trans. Autom. Control **63**(12), 4294–4300 (2018)
695. J. Swigart, S. Lall, Optimal controller synthesis for decentralized systems over graphs via spectral factorization. IEEE Trans. Autom. Control **59**(9), 2311–2323 (2014)
696. C. Szepesvari, *Algorithms for Reinforcement Learning* (Morgan and Claypool, 2010)
697. C. Szepesvári, *Algorithms for Reinforcement Learning*, vol. 4 (2010), pp. 1–103
698. C. Szepesvári, M.L. Littman, A unified analysis of value-function-based reinforcement-learning algorithms. Neural Comput. **11**(8), 2017–2060 (1999)
699. C. Szepesvàri, W.D. Smart, Interpolation-based q-learning. ICML '04: Proceedings of the Twenty-first International Conference on Machine Learning (2004)
700. W. Tamura, Bayesian persuasion with quadratic preferences. Available at SSRN 1987877 (2018)
701. M. Tan, Multi-agent reinforcement learning: Independent vs. cooperative agents, in *Proceedings of the Tenth International Conference on Machine Learning, Amherst, MA* (1993), pp. 330–337
702. T. Tanaka, Semidefinite representation of sequential rate-distortion function for stationary Gauss-Markov processes, in *2015 IEEE Conference on Control Applications (CCA)* (IEEE, 2015), pp. 1217–1222
703. T. Tanaka, K.K. Kim, P.A. Parrilo, S.K. Mitter, Semidefinite programming approach to Gaussian sequential rate-distortion trade-offs. IEEE Trans. Autom. Control **62**(4), 1896–1910 (2016)
704. T. Tanaka, P.M. Esfahani, S.K. Mitter, LQG control with minimum directed information: Semidefinite programming approach. IEEE Trans. Autom. Control **63**(1), 37–52 (2018)
705. D. Tang, H. Tavafoghi, V.G. Subramanian, A. Nayyar, D. Teneketzis, Dynamic games among teams with delayed intra-team information sharing. Dyn. Games Appl. **13**, 1–59 (2022)
706. S. Tatikonda, Control under Communication Constraints. PhD dissertation, Massachusetts Institute of Technology, Cambridge, MA, 2000
707. S. Tatikonda, S. Mitter, Control under communication constraints. IEEE Trans. Autom. Control **49**(7), 1056–1068 (2004)
708. S. Tatikonda, S. Mitter, The capacity of channels with feedback. IEEE Trans. Inf. Theory **55**(1), 323–349 (2009)
709. S. Tatikonda, A. Sahai, S. Mitter, Stochastic linear control over a communication channels. IEEE Trans. Autom. Control **49**, 1549–1561 (2004)
710. H. Tavafoghi, Y. Ouyang, D. Teneketzis, A unified approach to dynamic decision problems with asymmetric information: Non-strategic agents. IEEE Trans. Autom. Control **67**(3), 1105–1119 (2021)

711. H. Tembine, Q. Zhu, T. Başar, Risk-sensitive mean-field games. IEEE Trans. Autom. Control **59**(4), 835–850 (2013)
712. D. Teneketzis, Communication in decentralized control, PhD Dissertation, M.I.T., 1979
713. D. Teneketzis, On information structures and nonsequential stochastic control. CWI Q. **9**, 241–260 (1996)
714. D. Teneketzis, On the structure of optimal real-time encoders and decoders in noisy communication. IEEE Trans. Inf. Theory **52**, 4017–4035 September (2006)
715. D. Teneketzis, P. Varaiya, Consensus in distributed estimation, in *Advances in Statistical Signal Processing*, ed. by H.V. Poor (JAI Press, 1988), pp. 361–386
716. H. Theil, A note on certainty equivalence in dynamic planning. Econometrica J. Econom. Soc. **25**(2), 346–349 (1957)
717. H. Theil, Econometric models and welfare maximization, in *Henri Theil's Contributions to Economics and Econometrics* (Springer, 1992), pp. 1055–1075
718. N. Tishby, F.C. Pereira, W. Bialek, The information bottleneck method, in *Annual Allerton Conference on Communication, Control, and Computing (Allerton)* (1999), pp. 368–377
719. M.S. Tomar, M. Rungger, M. Zamani, Invariance feedback entropy of uncertain control systems. IEEE Trans. Autom. Control **66**, 5680 (2020)
720. F. Topsœ, Uniformity in convergence of measures. Z. Wahrscheinlichkeitsth **39**, 1–30 (1977)
721. E. Torgersen, *Comparison of statistical experiments*, vol. 36 (Cambridge University Press, 1991)
722. J. Tsitsiklis. Decentralized detection by a large number of sensors. Math. Control Signals Syst. **1**(2), 167–182 (1988)
723. J.N. Tsitsiklis, Extremal properties of likelihood-ratio quantizers. IEEE Trans. Commun. **41**, 550–558 (1993)
724. J.N. Tsitsiklis, Asynchronous stochastic approximation and Q-learning. Mach. Learn. **16**, 185–202 (1994)
725. J.N. Tsitsiklis, M. Athans, Convergence and asymptotic agreement in distributed decision problems. IEEE Trans. Autom. Control **29**(1), 42–50 (1984)
726. R.L. Tweedie, Drift conditions and invariant measures for Markov chains. Stoch. Process. Appl. **92**, 345–354 (2001)
727. V.A. Ugrinovskii, I.R. Petersen, Minimax LQG control of stochastic partially observed uncertain systems. SIAM J. Control Optim. **40**, 1189–1226 (2001)
728. M.A. uz Zaman, K. Zhang, E. Miehling, T. Başar, Reinforcement learning in non-stationary discrete-time linear-quadratic mean-field games, in *IEEE Conference on Decision and Control (CDC)* (IEEE, 2020), pp. 2278–2284
729. U. Vaidya, N. Elia, Stabilization of nonlinear systems over packet-drop links: Scalar case. Syst. Control Lett. **61**(9), 959–966 (2012)
730. U. Vaidya, N. Elia, Limitations for nonlinear stabilization over uncertain channel. Preprint, arXiv:1605.04432 (2014)
731. R. van Handel, Observability and nonlinear filtering. Probab. Theory Related Fields **145**(1–2), 35–74 (2009)
732. R. van Handel, The stability of conditional Markov processes and Markov chains in random environments. Ann. Probab. **37**, 1876–1925 (2009)
733. R. van Handel, Uniform time average consistency of Monte Carlo particle filters. Stoch. Process. Appl. **119**(11), 3835–3861 (2009)
734. T. van Zandt, Decentralized information processing in the theory of organizations, in *Contemporary Economic Issues, Vol. 4: Economic Design and Behavior*, ed. by M. Sertel (MacMillan Press, London, 1999), pp. 125–160
735. V.N. Vapnik, *The Nature of Statistical Learning Theory*, 2nd edn. (Springer, New York, 2000)
736. P. Varaiya, J. Walrand, On delayed-sharing patterns. IEEE Trans. Autom. Control **23**, 443–445 (1978)
737. D. Vasal, A. Anastasopoulos, Signaling equilibria for dynamic LQG games with asymmetric information, in *55th IEEE Conference on Decision and Control (CDC)*, Las Vegas, NV, Dec. (2016), pp. 6901–6908

738. D. Vasal, A. Sinha, A. Anastasopoulos, A systematic process for evaluating structured perfect Bayesian equilibria in dynamic games with asymmetric information. IEEE Trans. Autom. Control **64**(1), 81–96 (2018)
739. L. Vasudevan, J. Hespanha, A. Ortega, Towards the control of linear systems with minimum bit-rate, in *Proc. 15th Int. Symp. on Mathematical Theory of Networks and Systems (MTNS)* (2002)
740. V.V. Veeravalli, T. Başar, H.V. Poor, Decentralized sequential detection with a fusion center performing the sequential test. IEEE Trans. Inf. Theory **39**, 433–442 (1993)
741. S. Vembu, S. Verdú, Y. Steinberg, The source-channel separation theorem revisited. IEEE Trans. Inf. Theory **41**, 44–54 (1995)
742. R. Venkataramanan, S.S. Pradhan, Source coding with feedforward: Rate-distortion theorems and error exponents for a general source. IEEE Trans. Inf. Theory **53**, 2154–2179 (2007)
743. R. Venkataramanan, S.S. Pradhan, A new achievable rate region for the discrete memoryless multiple-access channel with feedback, in *Proceedings of the IEEE Int. Symp. on Information Theory (ISIT)*, Seoul, South Korea (2009)
744. S. Verdú, T.S. Han, A general formula for channel capacity. IEEE Trans. Inf. Theory **40**, 1147–1157 (1994)
745. M. Vidyasagar, Convergence of stochastic approximation via martingale and converse Lyapunov methods. Math. Control Signals Syst. **35**, 1–24 (2023)
746. C. Villani, *Optimal Transport: Old and New* (Springer, 2008)
747. N. Vlassis, M.T.J. Spaan, Perseus: Randomized point-based value iteration for pomdps. J. Artif. Intell. Res. **24**, 195–220 (2005)
748. J. von Neumann, Zur theorie der gesellschaftsspiele. Math. Ann. **100**, 295–320 (1928)
749. J. von Neumann, A certain zero-sum two-person game equivalent to the optimal assignment problem. Contrib. Theory Games **2**, 5–12 (1953)
750. Q. Voortman, A.Y. Pogromsky, A.S. Matveev, H. Nijmeijer, Data-rate constrained observers of nonlinear systems. Entropy **21**(3), 282 (2019)
751. P.G. Voulgaris, A convex characterization of classes of problems in control with specific interaction and communication structures, in *Proceedings of the American Control Conference*, June (2001), pp. 3128–3133
752. O.J. Vrieze, S.H. Tijs, Fictitious play applied to sequence of games and discounted stochastic games. Int. J. Game Theory **11**, 71–85 (1980)
753. A.B. Wagner, V. Anantharam, An improved outer bound for multiterminal source coding. IEEE Trans. Inf. Theory **54**(5), 1919–1937 (2008)
754. A. Wald, *Statistical Decision Functions* (John Wiley and Sons, New York, 1950)
755. J.C. Walrand, P. Varaiya, Causal coding and control of Markov chains. Syst. Control Lett. **3**, 189–192 (1983)
756. J.C. Walrand, P. Varaiya, Optimal causal coding-decoding problems. IEEE Trans. Inf. Theory **19**, 814–820 (1983)
757. Y. Wang, J.P. Hespanha, Distributed estimation of power system oscillation modes under attacks on GPS clocks. IEEE Trans. Instrum. Meas. **67**(7), 1626–1637 (2018)
758. B.C. Wang, J.F. Zhang, Social optima in mean field linear-quadratic-gaussian models with markov jump parameters. SIAM J. Control Optim. **55**(1), 429–456 (2017)
759. J. Warga, *Optimal Control of Differential and Functional Equations* (Academic Press, 2014)
760. C.J.C.H. Watkins, P. Dayan, Q-learning. Mach. Learn. **8**, 279–292 (1992)
761. J. Watrous, *The Theory of Quantum Information* (Cambridge University Press, 2018)
762. G.Y. Weintraub, C.L. Benkard, B. van Roy, Oblivious equilibrium: A mean field approximation for large-scale dynamic games, in *Advances in Neural Information Processing Systems*, 18 (2005)
763. G.Y. Weintraub, C.L. Benkard, B. van Roy, Markov perfect industry dynamics with many firms. Econometrica **76**(6), 1375–1411 (2008)
764. T. Weissman, N. Merhav, On causal source codes with side information. IEEE Trans. Inf. Theory **51**, 4003–4013 (2005)

765. T. Weissman, Y.H. Kim, H.H. Permuter, Directed information, causal estimation, and communication in continuous time. IEEE Trans. Inf. Theory **59**(3), 1271–1287 (2012)
766. N. Wernersson, J. Karlsson, M. Skoglund, Distributed quantization over noisy channels. IEEE Trans. Commun. **57**, 1693–1700 (2009)
767. D.B. West, *Introduction to Graph Theory*, vol. 2 (Prentice Hall, Upper Saddle River, NJ, 1996)
768. R.L. Wheeden, A. Zygmund, *Measure and Integral* (Marcel Dekker, New York, 1977)
769. C.C. White III, W.T. Scherer, Finite-memory suboptimal design for partially observed markov decision processes. Oper. Res. **42**(3), 439–455 (1994)
770. J.L. Willems, Time-varying feedback for the stabilization of fixed modes in decentralized control systems. Automatica **25**, 127–131 (1989)
771. H.S. Witsenhausen, A counterexample in stochastic optimal control. SIAM J. Control **6**, 131–147 (1968)
772. H.S. Witsenhausen, On information structures, feedback and causality. SIAM J. Control **9**, 149–160 (1971)
773. H.S. Witsenhausen, On the relations between the values of a game and its information structure. Inf. Control **19**(3), 204–215 (1971)
774. H.S. Witsenhausen, Separation of estimation and control for discrete time systems. Proc. IEEE **59**, 1557–1566 (1971)
775. H.S. Witsenhausen, A standard form for sequential stochastic control. Math. Syst. Theory **7**, 5–11 (1973)
776. H.S. Witsenhausen, Alternatives to the tree model for extensive games, in *The Theory and Application of Differential Games* (Springer, 1975), pp. 77–84
777. H.S. Witsenhausen, The intrinsic model for discrete stochastic control: Some open problems, in *Lecture Notes in Econ. and Math. Syst.*, vol. 107 (Springer, 1975), pp. 322–335
778. H.S. Witsenhausen, On the structure of real-time source coders. Bell Syst. Tech. J. **58**, 1437–1451 (1979)
779. H.S. Witsenhausen, Indirect rate-distortion problems. IEEE Trans. Inf. Theory **26**, 518–521 (1980)
780. H.S. Witsenhausen, Equivalent stochastic control problems. Math. Control Signals Syst. **1**, 3–11 (1988)
781. H. Witsenhausen, A. Wyner, A conditional entropy bound for a pair of discrete random variables. IEEE Trans. Inf. Theory **21**(5), 493–501 (1975)
782. W.S. Wong, R.W. Brockett, Systems with finite communication bandwidth constraints - part II: Stabilization with limited information feedback. IEEE Trans. Autom. Control **42**, 1294–1299 (1997)
783. P.W. Wong, R.M. Gray, Sigma-delta modulation with i.i.d. Gaussian inputs. IEEE Trans. Inf. Theory **36**, 784–778 (1990)
784. E. Wong, B.E. Hajek, *Stochastic Processes in Engineering Systems* (Springer, New York, 1985)
785. R.G. Wood, T. Linder, S. Yüksel, Optimal zero delay coding of Markov sources: Stationary and finite memory codes. IEEE Trans. Inf. Theory **63**, 5968–5980 (2017)
786. Y. Wu, S. Verdú, Witsenhausen's counterexample: a view from optimal transport theory, in *Proceedings of the IEEE Conference on Decision and Control, Florida, USA*, December (2011), pp. 5732–5737
787. Y. Wu, S. Verdú, Functional properties of minimum mean-square error and mutual information. IEEE Trans. Inf. Theory **58**(3), 1289–1301 (2012)
788. A.D. Wyner, Another look at the coding theorem of information theory: A tutorial. Proc. IEEE **58**, 894–913 (1970)
789. Q. Xie, Z. Yang, Z. Wang, A. Minca, Learning while playing in mean-field games: Convergence and optimality, in *International Conference on Machine Learning* (2021), pp. 11436–11447. PMLR
790. J. Xu, H. Zhang, Open-loop decentralized LQ control problem with multiplicative noise. IEEE Trans. Control Netw. Syst. **9**, 1887 (2022)

791. Y. Yang, R. Luo, M. Li, M. Zhou, W. Zhang, J. Wang, Mean field multi-agent reinforcement learning, in *International Conference on Machine Learning* (2018), pp. 5571–5580. PMLR
792. H. Yin, P.G. Mehta, S.P. Meyn, U.V. Shanbhag, Learning in mean-field games. IEEE Trans. Autom. Control **59**(3), 629–644 (2013)
793. B. Yongacoglu, G. Arslan, S. Yüksel, Decentralized learning for optimality in stochastic dynamic teams and games with local control and global state information. IEEE Trans. Automatic Control **67**, 5230–5245 (2022)
794. B. Yongacoglu, G. Arslan, S. Yüksel, Independent learning in mean-field games: Satisficing paths and convergence to subjective equilibria. Preprint. arXiv:2209.05703 (2022)
795. B. Yongacoglu, G. Arslan, S. Yüksel, Decentralized multi-agent reinforcement learning for continuous-space stochastic games. arXiv:2308.03239 (2023)
796. B. Yongacoglu, G. Arslan, S. Yüksel, Satisficing paths and independent multi-agent reinforcement learning in stochastic games. SIAM J. Math. Data Sci. (arXiv:2110.04638) (2023)
797. T. Yoshikawa, Dynamic programming approach to decentralized control problems. IEEE Trans. Autom. Control **20**, 796–797 (1975)
798. T. Yoshikawa, Decomposition of dynamic team decision problems. IEEE Trans. Autom. Control **23**, 627–632 (1978)
799. K. You, L. Xie, Minimum data rate for mean square stabilization of discrete LTI systems over lossy channels. IEEE Trans. Autom. Control **55**(10), 2373–2378 (2010)
800. L.C. Young, Generalized curves and the existence of an attained absolute minimum in the calculus of variations. C. R. Soc. Sci. Lett. Varsovie **30**, 212–234 (1937)
801. H.P. Young, *Strategic Learning and Its Limits* (Oxford University Press, New York, 2004)
802. H. Yu, D.P. Bertsekas, On near optimality of the set of finite-state controllers for average cost pomdp. Math. Oper. Res. **33**(1), 1–11 (2008)
803. S. Yu, P.G. Mehta, Bode-like fundamental performance limitations in control of nonlinear systems. IEEE Trans. Autom. Control **55**(6), 1390–1405 (2010)
804. X. Yu, Y. Zhang, Z. Zhou, Teamwise mean field competitions. Appl. Math. Optim. **84**, 1–40 (2021)
805. S. Yüksel, A random time stochastic drift result and application to stochastic stabilization over noisy channels, in *Proceedings of the Annual Allerton Conference on Communications, Control and Computing*, Monticello, IL (2009)
806. S. Yüksel, Stochastic nestedness and the belief sharing information pattern. IEEE Trans. Autom. Control **54**, 2773–2786 (2009)
807. S. Yüksel, On optimal causal coding of partially observed Markov sources under classical and non-classical information structures, in *Proceedings of the IEEE International Symposium on Information Theory*, Austin, TX (2010), pp. 81–85
808. S. Yüksel, Stochastic stabilization of noisy linear systems with fixed-rate limited feedback. IEEE Trans. Autom. Control **55**, 2847–2853 (2010)
809. S. Yüksel, Stochastic stability and ergodicity of linear Gaussian systems controlled over discrete channels, in *Proceedings of the IEEE Conference on Decision and Control*, Florida, USA (2011), pp. 4680–4685
810. S. Yüksel, A tutorial on quantizer design for networked control systems: Stabilization and optimization. Appl. Comput. Math. **11**, 365–403 (2011)
811. S. Yüksel, Characterization of information channels for asymptotic mean stationarity and stochastic stability of non-stationary/unstable linear systems. IEEE Trans. Inf. Theory **58**, 6332–6354 (2012)
812. S. Yüksel, Jointly optimal LQG quantization and control policies for multi-dimensional linear Gaussian sources, in *Proceedings of the Annual Allerton Conference on Communications, Control and Computing*, Monticello, IL (2012)
813. S. Yüksel, Design of information channels for optimization and stabilization in networked control, in *Information and Control in Networks*, ed. by G. Como, B. Bernhardsson, A. Rantzer (Springer, 2013)

814. S. Yüksel, On optimal causal coding of partially observed Markov sources in single and multi-terminal settings. IEEE Trans. Inf. Theory **59**, 424–437 (2013)
815. S. Yüksel, Jointly optimal LQG quantization and control policies for multi-dimensional linear Gaussian sources. IEEE Trans. Autom. Control **59**, 1612–1617 (2014)
816. S. Yüksel, Stationary and ergodic properties of stochastic non-linear systems controlled over communication channels. SIAM J. Control Optim. **54**, 2844–2871 (2016)
817. S. Yüksel, A note on the separation of optimal quantization and control policies in networked control. SIAM J. Control Optim. **57**(1), 773–782 (2019)
818. S. Yüksel, A universal dynamic program and refined existence results for decentralized stochastic control. SIAM J. Control Optim. **58**(5), 2711–2739 (2020)
819. S. Yüksel, On Borkar and Young relaxed control topologies and continuous dependence of invariant measures on control policy. SIAM J. Control Optim. Preprint. arXiv:2304.07685 (2024, to appear)
820. S. Yüksel, *Optimization and Control of Stochastic Systems*. Queen's University, Lecture Notes. Available Online. https://mast.queensu.ca/ yuksel/LectureNotesOnStochasticOptCon trol.pdf
821. S. Yüksel, T. Başar, Coding and control over discrete noisy forward and feedback channels, in *Proceedings of the IEEE Conference on Decision and Control*, Seville, Spain (2005), pp. 2517–2522
822. S. Yüksel, T. Başar, Minimum rate coding for LTI systems over noiseless channels. IEEE Trans. Autom. Control **51**(12), 1878–1887 (2006)
823. S. Yüksel, T. Başar, Control over noisy forward and reverse channels. IEEE Trans. Autom. Control **56**, 1014–1029 (2011)
824. S. Yüksel, T. Başar, *Stochastic Networked Control Systems: Stabilization and Optimization under Information Constraints* (Springer, New York, 2013)
825. S. Yüksel, T. Linder, On optimization and convergence of observation channels and quantizers in stochastic control, in *Proceedings of the Annual Allerton Conference on Communications, Control and Computing*, Monticello, IL (2010), pp. 286–293
826. S. Yüksel, T. Linder, Optimization and convergence of observation channels in stochastic control, in *American Control Conference*, San Francisco, CA (2011), pp. 637–642
827. S. Yüksel, T. Linder, On optimal zero-delay quantization of vector Markov sources, in *Proceedings of the IEEE Conference on Decision and Control, Hawaii* (2012)
828. S. Yüksel, T. Linder, Optimization and convergence of observation channels in stochastic control. SIAM J. Control Optim. **50**, 864–887 (2012)
829. S. Yüksel, S.P. Meyn, Random-time, state-dependent stochastic drift for Markov chains and application to stochastic stabilization over erasure channels. IEEE Trans. Autom. Control **58**, 47–59 (2013)
830. S. Yüksel, N. Saldi, Convex analysis in decentralized stochastic control, strategic measures and optimal solutions. SIAM J. Control Optim. **55**, 1–28 (2017)
831. S. Yüksel, S. Tatikonda, Distributed sensing and control over communication channels and a counterexample, in *Proceedings of the Annual Allerton Conference on Communications, Control and Computing*, Monticello, IL, September (2007)
832. S. Yüksel, S. Tatikonda, A counterexample in distributed optimal sensing and control. IEEE Trans. Autom. Control **54**, 841–844 (2009)
833. S. Yüksel, T. Başar, S.P. Meyn, Optimal causal quantization of Markov sources with distortion constraints, in *Proceedings of the Information Theory and Applications (ITA) Workshop, UCSD* (2008)
834. A.A. Yushkevich, Reduction of a controlled Markov model with incomplete data to a problem with complete information in the case of Borel state and control spaces. Theory Prob. Appl. **21**, 153–158 (1976)
835. S.A. Zaidi, S. Yüksel, T. Oechtering, M. Skoglund, On optimal policies for control and estimation over a Gaussian relay channel, in *Proceedings of the IEEE Conference on Decision and Control*, December (2011)

836. A.A. Zaidi, T.J. Oechtering, S. Yüksel, M. Skoglund, Stabilization and control over Gaussian networks, in *Information and Control in Networks*, ed. by G. Como, B. Bernhardsson, A. Rantzer (Springer, 2013)
837. S.A. Zaidi, T. Oechtering, S. Yüksel, M. Skoglund, Stabilization over Gaussian networks. IEEE Trans. Autom. Control **59**, 2369–2384 (2014)
838. S.A. Zaidi, S. Yüksel, T. Oechtering, M. Skoglund, On the tightness of linear policies for stabilization of linear systems over Gaussian networks. Syst. Control Lett. **88**, 32–38 (2016)
839. R. Zamir, M. Feder, On universal quantization by randomized uniform/lattice quantizers. IEEE Trans. Inf. Theory **38**(2), 428–436 (1992)
840. H. Zang, P.A. Iglesias, Nonlinear extension of Bode's integral based on an information-theoretic interpretation. Syst. Control Lett. **50**(1), 11–19 (2003)
841. Z. Zhang, E.-H. Yang, V.K. Wei, The redundancy of source coding with a fidelity criterion. 1. Known statistics. IEEE Trans. Inf. Theory **43**(1), 71–91 (1997)
842. K. Zhang, Z. Yang, T. Başar, Networked multi-agent reinforcement learning in continuous spaces, in *IEEE Conference on Decision and Control* (IEEE, 2018), pp. 2771–2776
843. K. Zhang, E. Miehling, T. Başar, Online planning for decentralized stochastic control with partial history sharing, in *2019 American Control Conference (ACC)* (IEEE, 2019), pp. 3544–3550
844. K. Zhang, S. Kakade, T. Başar, L. Yang, Model-based multi-agent RL in zero-sum Markov games with near-optimal sample complexity, in *Advances in Neural Information Processing Systems*, vol. 33 (2020), pp. 1166–1178
845. K. Zhang, Z. Yang, T. Başar, Multi-agent reinforcement learning: A selective overview of theories and algorithms, in *Handbook of Reinforcement Learning and Control* (2021), pp. 321–384
846. E. Zhou, M.C. Fu, S.I. Marcus, A density projection approach to dimension reduction for continuous-state POMDPs, in *Decision and Control (2008) CDC 2008. 47th IEEE Conference on*, pp. 5576–5581 (2008)
847. E. Zhou, M.C. Fu, S.I. Marcus, Solving continuous-state POMDPs via density projection. IEEE Trans. Autom. Control **55**(5), 1101–1116 (2010)
848. J. Ziv, The behavior of analog communication systems. IEEE Trans. Inf. Theory **16**, 587–594 (1970)

Index

S. Yüksel, T. Başar, *Stochastic Teams, Games, and Control under Information Constraints*, Systems & Control: Foundations & Applications,
https://doi.org/10.1007/978-3-031-54071-4

SPRINGER NATURE

GPSR Compliance

The European Union's (EU) General Product Safety Regulation (GPSR) is a set of rules that requires consumer products to be safe and our obligations to ensure this.

If you have any concerns about our products, you can contact us on ProductSafety@springernature.com

In case Publisher is established outside the EU, the EU authorized representative is:

Springer Nature Customer Service Center GmbH
Europaplatz 3
69115 Heidelberg, Germany

The manufacturer's authorised representative in the EU is Springer Nature Customer Service Centre GmbH, Europaplatz 3, 69115 Heidelberg, Germany. If you have any concerns regarding our products, please contact ProductSafety@springernature.com

Printed and bound by CPI Group (UK) Ltd, Croydon, CR0 4YY
01/10/2026
02225874-0002